Environmental Encyclopedia

Third Edition

Environmental Encyclopedia

Third Edition

Volume 2
N-Z

Historical Chronology
U.S. Environmental Legislation
Organizations
General Index

Marci Bortman, Peter Brimblecombe,
Mary Ann Cunningham, William P. Cunningham,
and Bill Freedman, *Editors*

Detroit • New York • San Diego • San Francisco • Cleveland • New Haven, Conn. • Waterville, Maine • London • Munich

Environmental Encyclopedia 3

Marci Bortman, Peter Brimblecombe, Bill Freedman, Mary Ann Cunningham, William P. Cunningham

Project Coordinator
Jacqueline L. Longe

Editorial
Deirdre S. Blanchfield, Madeline Harris, Chris Jeryan, Kate Kretschmann, Mark Springer, Ryan Thomason

Editorial Systems Support
Andrea Lopeman

Permissions
Shalice Shah-Caldwell

Imaging and Multimedia
Robert Duncan, Mary Grimes, Lezlie Light, Dan Newell, David Oblender, Christine O'Bryan, Kelly A. Quin

Product Design
Michelle DiMercurio, Tracey Rowens, Jennifer Wahi

Manufacturing
Evi Seoud, Rita Wimberley

For more information, contact
The Gale Group, Inc.
27500 Drake Road
Farmington Hills, MI 48331-3535
Or visit our Internet site at
http://www.gale.com

ISBN 0-7876-5486-8 (set), ISBN 0-7876-5487-6 (Vol. 1),
ISBN 0-7876-5488-4 (Vol. 2), ISSN 1072-5083

Printed in the United States of America
10 9 8 7 6 5 4

CONTENTS

C

D

E

G

O

P

ADVISORY BOARD

A number of recognized experts in the library and environmental communities provided invaluable assistance in the formulation of this encyclopedia. Our panel of advisors helped us shape this publication into its final form, and we would like to express our sincere appreciation to them:

Dean Abrahamson: Hubert H. Humphrey Institute of Public Affairs, University of Minnesota, Minneapolis, Minnesota

Maria Jankowska: Library, University of Idaho, Moscow, Idaho

Terry Link: Library, Michigan State University, East Lansing, Michigan

Holmes Rolston: Department of Philosophy, Colorado State University, Fort Collins, Colorado

Frederick W. Stoss: Science and Engineering Library, State University of New York—Buffalo, Buffalo, New York

Hubert J. Thompson: Conrad Sulzer Regional Library, Chicago, Illinois

CONTRIBUTORS

Margaret Alic, Ph.D.: Freelance Writer, Eastsound, Washington

William G. Ambrose Jr., Ph.D.: Department of Biology, East Carolina University, Greenville, North Carolina

James L. Anderson, Ph.D.: Soil Science Department, University of Minnesota, St. Paul, Minnesota

Monica Anderson: Freelance Writer, Hoffman Estates, Illinois

Bill Asenjo M.S., CRC: Science Writer, Iowa City, Iowa

Terence Ball, Ph.D.: Department of Political Science, University of Minnesota, Minneapolis, Minnesota

Brian R. Barthel, Ph.D.: Department of Health, Leisure and Sports, The University of West Florida, Pensacola, Florida

Stuart Batterman, Ph.D.: School of Public Health, University of Michigan, Ann Arbor, Michigan

Eugene C. Beckham, Ph.D.: Department of Mathematics and Science, Northwood Institute, Midland, Michigan

Milovan S. Beljin, Ph.D.: Department of Civil Engineering, University of Cincinnati, Cincinnati, Ohio

Heather Bienvenue: Freelance Writer, Fremont, California

Lawrence J. Biskowski, Ph.D.: Department of Political Science, University of Georgia, Athens, Georgia

E. K. Black: University of Alberta, Edmonton, Alberta, Canada

Paul R. Bloom, Ph.D.: Soil Science Department, University of Minnesota, St. Paul, Minnesota

Gregory D. Boardman, Ph.D.: Department of Civil Engineering, Virginia Polytechnic Institute and State University, Blacksburg, Virginia

Marci L. Bortman, Ph.D.: The Nature Conservancy, Huntington, New York

Pat Bounds: Freelance Writer,

Peter Brimblecombe, Ph.D.: School of Environmental Sciences, University of East Anglia, Norwich, United Kingdom

Kenneth N. Brooks, Ph.D.: College of Natural Resources, University of Minnesota, St. Paul, Minnesota

Peggy Browning: Freelance Writer,

Marie Bundy: Freelance Writer, Port Republic, Maryland

Ted T. Cable, Ph.D.: Department of Horticulture, Forestry and Recreation Resources, Kansas State University, Manhattan, Kansas

John Cairns Jr., Ph.D.: University Center for Environmental and Hazardous Materials Studies, Virginia Polytechnic Institute and State University, Blacksburg, Virginia

Liane Clorfene Casten: Freelance Journalist, Evanston, Illinois

Ann S. Causey: Prescott College, Prescott, Arizona

Ann N. Clarke: Eckenfelder Inc., Nashville, Tennessee

David Clarke: Freelance Journalist, Bethesda, Maryland

Sally Cole-Misch: Freelance Writer, Bloomfield Hills, Michigan

Edward J. Cooney: Patterson Associates, Inc., Chicago, Illinois

Terence H. Cooper, Ph.D.: Soil Science Department, University of Minnesota, St. Paul, Minnesota

Gloria Cooksey, C.N.E.: Freelance Writer, Sacramento, California

Mark Crawford: Freelance Writer, Toronto, Ontario, Canada

Neil Cumberlidge, Ph.D.: Department of Biology, Northern Michigan University, Marquette, Michigan

John Cunningham: Freelance Writer, St. Paul, Minnesota

Mary Ann Cunningham, Ph.D.: Department of Geology and Geography, Vassar College, Poughkeepsie, New York

William P. Cunningham, Ph.D.: Department of Genetics and Cell Biology, University of Minnesota, St. Paul, Minnesota

Richard K. Dagger, Ph.D.: Department of Political Science, Arizona State University, Tempe, Arizona

Tish Davidson, A.M.: Freelance Writer, Fremont, California

Stephanie Dionne: Freelance Journalist, Ann Arbor, Michigan

Frank M. D'Itri, Ph.D.: Institute of Water Research, Michigan State University, East Lansing, Michigan

Teresa C. Donkin: Freelance Writer, Minneapolis, Minnesota

David A. Duffus, Ph.D.: Department of Geography, University of Victoria, Victoria, British Columbia, Canada

Douglas Dupler, M.A.: Freelance Writer, Boulder, Colorado

Cathy M. Falk: Freelance Writer, Portland, Oregon

L. Fleming Fallon Jr., M.D., Dr.P.H.: Associate Professor, Public Health, Bowling Green State University , Bowling Green, Ohio

George M. Fell: Freelance Writer, Inver Grove Heights, Minnesota

Gordon R. Finch, Ph.D.: Department of Civil Engineering, University of Alberta, Edmonton, Alberta, Canada

Paula Anne Ford-Martin, M.A.: Wordcrafts, Warwick, Rhode Island

Janie Franz: Freelance Writer, Grand Forks, North Dakota

Bill Freedman, Ph.D.: School for Resource and Environmental Studies, Dalhousie University, Halifax, Nova Scotia, Canada

Rebecca J. Frey, Ph.D.: Writer, Editor, and Editorial Consultant, New Haven, Connecticut

Cynthia Fridgen, Ph.D.: Department of Resource Development, Michigan State University, East Lansing, Michigan

Andrea Gacki: Freelance Writer, Bay City, Michigan

Brian Geraghty: Ford Motor Company, Dearborn, Michigan

Robert B. Giorgis, Jr.: Air Resources Board, Sacramento, California

Debra Glidden: Freelance American Indian Investigative Journalist, Syracuse, New York

Eville Gorham, Ph.D.: Department of Ecology, Evolution and Behavior, University of Minnesota, St. Paul, Minnesota

Darrin Gunkel: Freelance Writer, Seattle, Washington

Malcolm T. Hepworth, Ph.D.: Department of Civil and Mineral Engineering, University of Minnesota, Minneapolis, Minnesota

Katherine Hauswirth: Freelance Writer, Roanoke, Virginia

Richard A. Jeryan: Ford Motor Company, Dearborn, Michigan

Barbara J. Kanninen, Ph.D.: Hubert H. Humphrey Institute of Public Affairs, University of Minnesota, Minneapolis, Minnesota

Christopher McGrory Klyza, Ph.D.: Department of Political Science, Middlebury College, Middlebury, Vermont

John Korstad, Ph.D.: Department of Natural Science, Oral Roberts University, Tulsa, Oklahoma

Monique LaBerge, Ph.D.: Research Associate, Department of Biochemistry and Biophysics, University of Pennsylvania, Philadelphia, Pennsylvania

Royce Lambert, Ph.D.: Soil Science Department, California Polytechnic State University, San Luis Obispo, California

William E. Larson, Ph.D.: Soil Science Department, University of Minnesota, St. Paul, Minnesota

Ellen E. Link: Freelance Writer, Laingsburg, Michigan

Sarah Lloyd: Freelance Writer, Cambria, Wisconsin

James P. Lodge Jr.: Consultant in Atmospheric Chemistry, Boulder, Colorado

William S. Lynn, Ph.D.: Department of Geography, University of Minnesota, Minneapolis, Minnesota

Alair MacLean: Environmental Editor, OMB Watch, Washington, DC

Alfred A. Marcus, Ph.D.: Carlson School of Management, University of Minnesota, Minneapolis, Minnesota

Gregory McCann: Freelance Writer, Freeland, Michigan

Cathryn McCue: Freelance Journalist, Roanoke, Virginia

Mary McNulty: Freelance Writer, Illinois

Jennifer L. McGrath: Freelance Writer, South Bend, Indiana

Robert G. McKinnell, Ph.D.: Department of Genetics and Cell Biology, University of Minnesota, St. Paul, Minnesota

Nathan H. Meleen, Ph.D.: Engineering and Physics Department, Oral Roberts University, Tulsa, Oklahoma

Liz Meszaros: Freelance Writer, Lakewood, Ohio

Muthena Naseri: Moorpark College, Moorpark, California

B. R. Niederlehner, Ph.D.: University Center for Environmental and Hazardous Materials Studies, Virginia Polytechnic Institute and State University, Blacksburg, Virginia

David E. Newton: Instructional Horizons, Inc., San Francisco, California

Robert D. Norris: Eckenfelder Inc., Nashville, Tennessee

Teresa G. Norris, R.N.: Medical Writer, Ute Park, New Mexico

Karen Oberhauser, Ph.D.: University of Minnesota, St. Paul, Minnesota

Stephanie Ocko: Freelance Journalist, Brookline, Massachusetts

Kristin Palm: Freelance Writer, Royal Oak, Michigan
James W. Patterson: Patterson Associates, Inc., Chicago, Illinois
Paul Phifer, Ph.D.: Freelance Writer, Portland, Oregon
Jeffrey L. Pintenich: Eckenfelder Inc., Nashville, Tennessee
Douglas C. Pratt, Ph.D.: University of Minnesota: Department of Plant Biology, Scandia, Minnesota
Jeremy Pratt: Institute for Human Ecology, Santa Rosa, California
Klaus Puettman: University of Minnesota, St. Paul, Minnesota
Stephen J. Randtke: Department of Civil Engineering, University of Kansas, Lawrence, Kansas
Lewis G. Regenstein: Author and Environmental Writer, Atlanta, Georgia
Linda Rehkopf: Freelance Writer, Marietta, Georgia
Paul E. Renaud, Ph.D.: Department of Biology, East Carolina University, Greenville, North Carolina
Marike Rijsberman: Freelance Writer, Chicago, Illinois
L. Carol Ritchie: Environmental Journalist, Arlington, Virginia
Linda M. Ross: Freelance Writer, Ferndale, Michigan
Joan Schonbeck: Medical Writer, Nursing, Massachusetts Department of Mental Health, Marlborough, Massachusetts
Mark W. Seeley: Department of Soil Science, University of Minnesota, St. Paul, Minnesota
Kim Sharp, M.Ln.: Freelance Writer, Richmond, Texas
James H. Shaw, Ph.D.: Department of Zoology, Oklahoma State University, Stillwater, Oklahoma
Laurel Sheppard: Freelance Writer, Columbus, Ohio
Judith Sims, M.S.: Utah Water Research Laboratory, Utah State University, Logan, Utah
Genevieve Slomski, Ph.D.: Freelance Writer, New Britain, Connecticut
Douglas Smith: Freelance Writer, Dorchester, Massachusetts
Lawrence H. Smith, Ph.D.: Department of Agronomy and Plant Genetics, University of Minnesota, St. Paul, Minnesota
Jane E. Spear: Freelance Writer, Canton, Ohio
Carol Steinfeld: Freelance Writer, Concord, Massachusetts
Paulette L. Stenzel, Ph.D.: Eli Broad College of Business, Michigan State University, East Lansing, Michigan
Les Stone: Freelance Writer, Ann Arbor, Michigan
Max Strieb: Freelance Writer, Huntington, New York
Amy Strumolo: Freelance Writer, Beverly Hills, Michigan
Edward Sucoff, Ph.D.: Department of Forestry Resources, University of Minnesota, St. Paul, Minnesota
Deborah L. Swackhammer, Ph.D.: School of Public Health, University of MinnesotaMinneapolis, Minnesota
Liz Swain: Freelance Writer, San Diego, California
Ronald D. Taskey, Ph.D.: Soil Science Department, California Polytechnic State University, San Luis Obispo, California
Mary Jane Tenerelli, M.S.: Freelance Writer, East Northport, New York
Usha Vedagiri: IT Corporation, Edison, New Jersey
Donald A. Villeneuve, Ph.D.: Ventura College, Ventura, California
Nikola Vrtis: Freelance Writer, Kentwood, Michigan
Eugene R. Wahl: Freelance Writer, Coon Rapids, Minnesota
Terry Watkins: Indianapolis, Indiana
Ken R. Wells: Freelance Writer, Laguna Hills, California
Roderick T. White Jr.: Freelance Writer, Atlanta, Georgia
T. Anderson White, Ph.D.: University of Minnesota, St. Paul, Minnesota
Kevin Wolf: Freelance Writer, Minneapolis, Minnesota
Angela Woodward: Freelance Writer, Madison, Wisconsin
Gerald L. Young, Ph.D.: Program in Environmental Science and Regional Planning, Washington State University, Pullman, Washington

HOW TO USE THIS BOOK

- The third edition of *Environmental Encyclopedia* has been designed with ready reference in mind.
- Straight **alphabetical arrangement** of topics allows users to locate information quickly.
- **Bold-faced terms** within entries direct the reader to related articles.
- **Contact information** is given for each organization profiled in the book.
- **Cross-references** at the end of entries alert readers to related entries not specifically mentioned in the body of the text.
- The **Resources** sections direct readers to additional sources of information on a topic.
- **Three appendices** provide the reader with a chronology of environmental events, a summary of environmental legislation, and a succinct alphabetical list of environmental organizations.
- A comprehensive **general index** guides readers to all topics mentioned in the text.

INTRODUCTION

Welcome to the third edition of the Gale *Environmental Encyclopedia*! Those of us involved in writing and production of this book hope you will find the material here interesting and useful. As you might imagine, choosing what to include and what to exclude from this collection has been challenging. Almost everything has some environmental significance, so our task has been to select a limited number of topics we think are of greatest importance in understanding our environment and our relation to it. Undoubtedly, we have neglected some topics that interest you and included some you may consider irrelevant, but we hope that overall you will find this new edition helpful and worthwhile.

The word environment is derived from the French *environ*, which means to "encircle" or "surround." Thus, our environment can be defined as the physical, chemical, and biological world that envelops us, as well as the complex of social and cultural conditions affecting an individual or community. This broad definition includes both the natural world and the "built" or technological environment, as well as the cultural and social contexts that shape human lives. You will see that we have used this comprehensive meaning in choosing the articles and definitions contained in this volume.

Among some central concerns of environmental science are:

- how did the natural world on which we depend come to be as it is, and how does it work?
- what have we done and what are we now doing to our environment—both for good and ill?
- what can we do to ensure a sustainable future for ourselves, future generations, and the other species of organisms on which—although we may not be aware of it—our lives depend?

The articles in this volume attempt to answer those questions from a variety of different perspectives.

Historically, environmentalism is rooted in natural history, a search for beauty and meaning in nature. Modern environmental science expands this concern, drawing on almost every area of human knowledge including social sciences, humanities, and the physical sciences. Its strongest roots, however, are in ecology, the study of interrelationships among and between organisms and their physical or nonliving environment. A particular strength of the ecological approach is that it studies systems holistically; that is, it looks at interconnections that make the whole greater than the mere sum of its parts. You will find many of those interconnections reflected in this book. Although the entries are presented individually so that you can find topics easily, you will notice that many refer to other topics that, in turn, can lead you on through the book if you have time to follow their trail. This series of linkages reflects the multilevel associations in environmental issues.

As our world becomes increasingly interrelated economically, socially, and technologically, we find evermore evidence that our global environment is also highly interconnected. In 2002, the world population reached about 6.2 billion people, more than triple what it had been a century earlier. Although the rate of population growth is slowing—having dropped from 2.0% per year in 1970 to 1.2% in 2002—we are still adding about 200,000 people per day, or about 75 million per year. Demographers predict that the world population will reach 8 or 9 billion before stabilizing sometime around the middle of this century. Whether natural resources can support so many humans is a question of great concern.

In preparation for the third global summit in South Africa, the United Nations released several reports in 2002 outlining the current state of our environment. Perhaps the greatest environmental concern as we move into the twenty-first century is the growing evidence that human activities are causing global climate change. Burning of fossil fuels in power plants, vehicles, factories, and homes release carbon dioxide into the atmosphere. Burning forests and crop residues, increasing cultivation of paddy rice, raising billions of ruminant animals, and other human activities also add to the rapidly growing atmospheric concentrations of heat trapping gases in the atmosphere. Global temperatures have begun

to rise, having increased by about 1°F (0.6°C) in the second half of the twentieth century. Meteorologists predict that over the next 50 years, the average world temperature is likely to increase somewhere between 2.7–11°F (1.5–6.1°C). That may not seem like a very large change, but the difference between current average temperatures and the last ice age, when glaciers covered much of North America, was only about 10°F (5°C).

Abundant evidence is already available that our climate is changing. The twentieth century was the warmest in the last 1,000 years; the 1990s were the warmest decade, and 2002 was the single warmest year of the past millennium. Glaciers are disappearing on every continent. More than half the world's population depends on rivers fed by alpine glaciers for their drinking water. Loss of those glaciers could exacerbate water supply problems in areas where water is already scarce. The United Nations estimates that 1.1 billion people—one-sixth of the world population—now lack access to clean water. In 25 years, about two-thirds of all humans will live in water-stressed countries where supplies are inadequate to meet demand.

Spring is now occurring about a week earlier and fall is coming about a week later over much of the northern hemisphere. This helps some species, but is changing migration patterns and home territories for others. In 2002, early melting of ice floes in Canada's Gulf of St. Lawrence apparently drowned nearly all of the 200,000 to 300,000 harp seal pups normally born there. Lack of sea ice is also preventing polar bears from hunting seals. Environment Canada reports that polar bears around Hudson's Bay are losing weight and decreasing in number because of poor hunting conditions. In 2002, a chunk of ice about the size of Rhode Island broke off the Larsen B ice shelf on the Antarctic Peninsula. As glacial ice melts, ocean levels are rising, threatening coastal ecosystems and cities around the world.

After global climate change, perhaps the next greatest environmental concern for most biologists is the worldwide loss of biological diversity. Taxonomists warn that one-fourth of the world's species could face extinction in the next 30 years. Habitat destruction, pollution, introduction of exotic species, and excessive harvesting of commercially important species all contribute to species losses. Millions of species—most of which have never even been named by science, let alone examined for potential usefulness in medicine, agriculture, science, or industry—may disappear in the next century as a result of our actions. We know little about the biological roles of these organisms in the ecosystems and their loss could result in an ecological tragedy.

Ecological economists have tried to put a price on the goods and services provided by natural ecosystems. Although many ecological processes aren't traded in the market place, we depend on the natural world to do many things for us like purifying water, cleansing air, and detoxifying our wastes. How much would it cost if we had to do all this ourselves? The estimated annual value of all ecological goods and services provided by nature are calculated to be worth at least $33 trillion, or about twice the annual GNPs of all national economies in the world. The most valuable ecosystems in terms of biological processes are wetlands and coastal estuaries because of their high level of biodiversity and their central role in many biogeochemical cycles.

Already there are signs that we are exhausting our supplies of fertile soil, clean water, energy, and biodiversity that are essential for life. Furthermore, pollutants released into the air and water, along with increasing amounts of toxic and hazardous wastes created by our industrial society, threaten to damage the ecological life support systems on which all organisms—including humans—depend. Even without additional population growth, we may need to drastically rethink our patterns of production and disposal of materials if we are to maintain a habitable environment for ourselves and our descendants.

An important lesson to be learned from many environmental crises is that solving one problem often creates another. Chlorofluorocarbons, for instance, were once lauded as a wonderful discovery because they replaced toxic or explosive chemicals then in use as refrigerants and solvents. No one anticipated that CFCs might damage stratospheric ozone that protects us from dangerous ultraviolet radiation. Similarly, the building of tall smokestacks on power plants and smelters lessened local air pollution, but spread acid rain over broad areas of the countryside. Because of our lack of scientific understanding of complex systems, we are continually subjected to surprises. How to plan for "unknown unknowns" is an increasing challenge as our world becomes more tightly interconnected and our ability to adjust to mistakes decreases.

Not all is discouraging, however, in the field of environmental science. Although many problems beset us, there are also encouraging signs of progress. Some dramatic successes have occurred in wildlife restoration and habitat protection programs, for instance. The United Nations reports that protected areas have increased five-fold over the past 30 years to nearly 5 million square miles. World forest losses have slowed, especially in Asia, where deforestation rates slowed from 8% in the 1980s to less than 1% in the 1990s. Forested areas have actually increased in many developed countries, providing wildlife habitat, removal of excess carbon dioxide, and sustainable yields of forest products.

In spite of dire warnings in the 1960s that growing human populations would soon overshoot the earth's carrying capacity and result in massive famines, food supplies have more than kept up with population growth. There is

more than enough food to provide a healthy diet for everyone now living, although inequitable distribution leaves about 800 million with an inadequate diet. Improved health care, sanitation, and nutrition have extended life expectancies around the world from 40 years, on average, a century ago, to 65 years now. Public health campaigns have eradicated smallpox and nearly eliminated polio. Other terrible diseases have emerged, however, most notably acquired immunodeficiency syndrome (AIDS), which is now the fourth most common cause of death worldwide. Forty million people are now infected with HIV—70% percent of them in sub-Saharan Africa—and health experts warn that unsanitary blood donation practices and spreading drug use in Asia may result in tens of millions more AIDS deaths in the next few decades.

In developed countries, air and pollution have decreased significantly over the past 30 years. In 2002, the Environmental Protection Agency declared that Denver—which once was infamous as one of the most polluted cities in the United States—is the first major city to meet all the agency's standards for eliminating air pollution. At about the same time, the EPA announced that 91% of all monitored river miles in the United States met the water quality goals set in the 1985 clean water act. Pollution-sensitive species like mayflies have returned to the upper Mississippi River, and in Britian, salmon are being caught in the Thames River after being absent for more than two centuries.

Conditions aren't as good, however, in many other countries. In most of Latin America, Africa, and Asia, less than two % of municipal sewage is given even primary treatment before being dumped into rivers, lakes, or the ocean. In South Asia, a 2-mile (3-km) thick layer of smog covers the entire Indian sub-continent for much of the year. This cloud blocks sunlight and appears to be changing the climate, bringing drought to Pakistan and Central Asia, and shifting monsoon winds that caused disastrous floods in 2002 in Nepal, Bangladesh, and eastern India that forced 25 million people from their homes and killed at least 1,000 people. Nobel laureate Paul Crutzen estimates that two million deaths each year in India alone can be attributed to air pollution effects.

After several decades of struggle, a world-wide ban on the "dirty dozen" most dangerous persistent organic pollutants (POPs) was ratified in 2000. Elimination of compounds such as DDT, Aldrin, Dieldrin, Mirex, Toxaphene, polychlorinated biphenyls, and dioxins has allowed recovery of several wildlife species including bald eagles, perigrine falcons, and brown pelicans. Still, other toxic synthetic chemicals such as polybrominated diphenyl ethers, chromated copper arsenate, perflurooctane sulfonate, and atrazine are now being found accumulating in food chains far from anyplace where they have been used.

Solutions for many of our pollution problems can be found in either improved technology, more personal responsibility, or better environmental management. The question is often whether we have the political will to enforce pollution control programs and whether we are willing to sacrifice short-term convenience and affluence for long-term ecological stability. We in the richer countries of the world have become accustomed to a highly consumptive lifestyle. Ecologists estimate that humans either use directly, destroy, coopt, or alter almost 40% of terrestrial plant productivity, with unknown consequences for the biosphere. Whether we will be willing to leave some resources for other species and future generations is a central question of environmental policy.

One way to extend resources is to increase efficiency and recycling of the items we use. Automobiles have already been designed, for example, that get more than 100 mi/gal (42 km/l) of diesel fuel and are completely recyclable when they reach the end of their designed life. Although recycling rates in the United States have increased in recent years, we could probably double our current rate with very little sacrifice in economics or convenience. Renewable energy sources such as solar or wind power are making encouraging progress. Wind already is cheaper than any other power source except coal in many localities. Solar energy is making it possible for many of the two billion people in the world who don't have access to electricity to enjoy some of the benefits of modern technology. Worldwide, the amount of installed wind energy capacity more than doubled between 1998 and 2002. Germany is on course to obtain 20% of its energy from renewables by 2010. Together, wind, solar, biomass and other forms of renewable energy have the potential to provide thousands of times as much energy as all humans use now. There is no reason for us to continue to depend on fossil fuels for the majority of our energy supply.

One of the widely advocated ways to reduce poverty and make resources available to all is sustainable development. A commonly used definition of this term is given in *Our Common Future*, the report of the World Commission on Environment and Development (generally called the Brundtland Commission after the prime minister of Norway, who chaired it), described sustainable development as: "meeting the needs of the present without compromising the ability of future generations to meet their own needs." This implies improving health, education, and equality of opportunity, as well as ensuring political and civil rights through jobs and programs based on sustaining the ecological base, living on renewable resources rather than nonrenewable ones, and living within the carrying capacity of supporting ecological systems.

Several important ethical considerations are embedded in environmental questions. One of these is intergenerational

justice: what responsibilities do we have to leave resources and a habitable planet for future generations? Is our profligate use of fossil fuels, for example, justified by the fact that we have technology to extract fossil fuels and enjoy their benefits? Will human lives in the future be impoverished by the fact that we have used up most of the easily available oil, gas, and coal? Author and social critic Wendell Berry suggests that our consumption of these resources constitutes a theft of the birthright and livelihood of posterity. Philosopher John Rawls advocates a "just savings principle" in which members of each generation may consume no more than their fair share of scarce resources.

How many generations are we obliged to plan for and what is our "fair share?" It is possible that our use of resources now—inefficient and wasteful as it may be—represents an investment that will benefit future generations. The first computers, for instance, were huge clumsy instruments that filled rooms full of expensive vacuum tubes and consumed inordinate amounts of electricity. Critics complained that it was a waste of time and resources to build these enormous machines to do a few simple calculations. And yet if this technology had been suppressed in its infancy, the world would be much poorer today. Now nanotechnology promises to make machines and tools in infinitesimal sizes that use minuscule amounts of materials and energy to carry out valuable functions. The question remains whether future generations will be glad that we embarked on the current scientific and technological revolution or whether they will wish that we had maintained a simple agrarian, Arcadian way of life.

Another ethical consideration inherent in many environmental issues is whether we have obligations or responsibilities to other species or to Earth as a whole. An anthropocentric (human-centered) view holds that humans have rightful dominion over the earth and that our interests and well-being take precedence over all other considerations. Many environmentalists criticize this perspective, considering it arrogant and destructive. Biocentric (life-centered) philosophies argue that all living organisms have inherent values and rights by virtue of mere existence, whether or not they are of any use to us. In this view, we have a responsibility to leave space and resources to enable other species to survive and to live as naturally as possible. This duty extends to making reparations or special efforts to encourage the recovery of endangered species that are threatened with extinction due to human activities.Some environmentalists claim that we should adopt an ecocentric (ecologically centered) outlook that respects and values nonliving entities such as rocks, rivers, mountains—even whole ecosystems—as well as other living organisms. In this view, we have no right to break up a rock, dam a free-flowing river, or reshape a landscape simply because it benefits us. More importantly, we should conserve and maintain the major ecological processes that sustain life and make our world habitable.

Others argue that our existing institutions and understandings, while they may need improvement and reform, have provided us with many advantages and amenities. Our lives are considerably better in many ways than those of our ancient ancestors, whose lives were, in the words of British philosopher Thomas Hobbes: "nasty, brutish, and short." Although science and technology have introduced many problems, they also have provided answers and possible alternatives as well.

It may be that we are at a major turning point in human history. Current generations are in a unique position to address the environmental issues described in this encyclopedia. For the first time, we now have the resources, motivation, and knowledge to protect our environment and to build a sustainable future for ourselves and our children. Until recently, we didn't have these opportunities, or there was not enough clear evidence to inspire people to change their behavior and invest in environmental protection; now the need is obvious to nearly everyone. Unfortunately, this also may be the last opportunity to act before our problems become irreversible.

We hope that an interest in preserving and protecting our common environment is one reason that you are reading this encyclopedia and that you will find information here to help you in that quest.

[*William P. Cunningham, Managing Editor*]

N

NAAQS

see **National Ambient Air Quality Standard**

Ralph Nader (1934 –)

American consumer advocate

Born in Connecticut, Ralph Nader is the son of Lebanese immigrants who emphasized citizenship and democracy and stressed the importance of justice rather than power. Nader earned his bachelor's degree in government and economics from Princeton University in 1955 and his law degree from Harvard University in 1958, having served as editor of the prestigious *Harvard Law Record.* Nader reads some 15 publications daily and speaks several languages, including Arabic, Chinese, and Russian.

Nader published his first article, "American Cars: Designed for Death," as editor of the *Harvard Law Record.* He later made his reputation with *Unsafe at Any Speed* (1965), a condemnation of the American **automobile** industry's record on safety. The book was initially commissioned by then-assistant secretary of labor Daniel Patrick Moynihan in 1963 as a report to congress, and was then brought out as a trade book by a small publisher. In the book Nader condemned U.S. automakers for valuing style over safety in developing their products and specifically targeted General Motors and its Chevrolet Corvair. General Motors hired a private investigator to dig dirt on Nader, resulting in a scandal that propelled the book to best-seller status. The controversy that *Unsafe at Any Speed* generated led to passage of the Traffic and Motor Vehicle Safety Act of 1966, which gave the government the right to enact and regulate safety standards for all automobiles sold in the United States.

With the money and fame generated by his book, Nader went on to champion a wide variety of causes. He played a key role in the establishment of the **Occupational Safety and Health Administration** (OSHA) as well as the Consumer Product Safety Commission. One of his most important victories was the establishment of the **Environmental Protection Agency** (EPA) in 1970, during the Nixon administration. He also worked to ensure passage of the Freedom of Information Act (1974). Nader got many young people to work on consumer and environmental issues through the Public Information Research Groups found on college campuses across the country. To assist him in his far-reaching investigative efforts, Nader created a watchdog team, known as Nader's Raiders. This group of lawyers was the core of what became the Center for Study of Responsive Law (CSRL), which has been Nader's headquarters since 1968. A network of public interest organizations branched off from CSRL, including Public Citizen, Health Research Group, Critical Mass Energy Project, and Congress Watch. Other groups have been established by Nader's associates, and though not run by Nader himself, they follow the same ideals as CSRL and work toward similar goals. Among these groups are: the Clean Water Action Project, the Center for Auto Safety, and the **Center for Science in the Public Interest**. A 1971 Harris poll placed Nader as the sixth most popular figure in the nation, and he is still recognized as the most important consumer advocate in the country.

Nader was at the height of his influence during the 1970s, but through the 1980s and early 1990s he was less in the public eye. The Reagan years saw a loosening of the regulations for which he had fought, and Nader himself suffered many personal setbacks, including the death of his brother and his own neurological illness. In 1987, Nader vigorously campaigned for Proposition 103 in California, which would roll back automobile insurance rates. The bill passed and exit polls showed that Nader's efforts had made the difference. Though he had been asked to run for president by the Beatles' John Lennon and by writer Gore Vidal in the early 1970s, Nader had always claimed that he needed to stay outside government in order to reform it. This changed in 1992, when Nader asked voters to write him in on the presidential ballot. He voiced his disgust at the two major parties, the Republicans and the Democrats, and asked

Ralph Nader. (AP/Wide World Photos. Reproduced by permission.)

to be considered as a "none-of-the-above" option. In 1996 he made a wider effort as the candidate for the Green Party. Nader was on the ballot in 21 states and polled a minuscule.68 percent of the vote (Reform Party candidate Ross Perot by contrast earned over 8 percent of the total national vote). Nader ran an informal campaign, mainly touring college campuses, and he did not appear to have much of an effect on the election. In 2000 Nader ran a much more concerted campaign, again as the Green Party candidate. He famously declared the leading candidates, George W. Bush and Al Gore to be "tweedledum and tweedledee," meaning they were virtually indistinguishable because of their support for corporate interests. Late in the campaign, it appeared that Gore and Bush were tied, and the Democratic party began to excoriate Nader. Though it appeared that Nader might siphon crucial votes away from Gore and thus let Bush win, Nader refused to alter his position or throw his support to Gore. **Environmentalism** became a key issue in the three-way race. Though Gore had a strong record on the **environment** and Bush what might be called an abyssmal one, Nader continued to criticize Gore for not having done enough. Finally, Gore won the popular vote with 48.39%, Bush came in with 47.88%, and Nader with 2.72%. But the election went to Bush after the Supreme Court halted the recount of contested Florida ballots, and that state's electoral college votes put Bush in the White House. Nader had won some 97,000 votes in Florida, and arguably if these votes had all gone to Gore, the outcome of the presidential race would have been different.

After the election, Nader was anathemized in some circles, and blamed for Gore's loss. Nader claimed to have no regrets except that he had not gotten at least 5% of the vote, which would have qualified the Green Party for federal matching funds for an election campaign in 2004. Nader published a memoir about the campaign, called *Crashing the Party: How to Tell the Truth and Still Run for President.* He refused to speculate about whether he would run for president again.

[*Kimberley Peterson*]

RESOURCES

PERIODICALS

Alterman, Eric. "Left in Shambles." *Nation* 271, no. 17 (November 27, 2000): 10.

Colapinto, John. "Ralph Nader Is Not Sorry." *Rolling Stone*, September 13, 2001, 64–71.

Heilbrunn, Jacob. "Leftover." *Commentary* 113, no. 3 (March 2002): 74.

Kennedy, Robert F. "Nader's Threat to the Environment." *New York Times*, August 10, 2000, A21.

McLaughlin, Abraham. "Nader's Voters: Steadfast...Or Switchable?" *Christian Science Monitor* (October 30, 2000): 2.

Dr. Arne Naess (1912 –)

Norwegian philosopher and naturalist

Arne Naess is a noted mountaineer and philosopher and the founder of the **deep ecology** movement. He was born to Ragnar and Christine Naess in Oslo, Norway, on January 27, 1912, the youngest of five children. Naess was an introspective child, and he displayed an early interest in logic and philosophy. After undergraduate work at the Sorbonne in Paris, he did graduate work at the University of Vienna, the University of California, Berkeley, and the University of Oslo. While in Vienna, his interests in logic, language, and methodology drew him to the Vienna Circle of logical empiricism. He was awarded the Ph.D. in philosophy from the University of Oslo in 1938.

A year later, in 1939, Naess was made full professor of philosophy at the same university. He promptly reorganized Norwegian higher education, making the history of ideas a prerequisite for all academic specializations and encouraging greater conceptual sophistication and tolerance. From the beginning, Naess was interested in empirical semantics, that is, how ordinary persons use words to communicate. In *"Truth" as Conceived by Those Who Are Not Professional Philosophers* (1939), he was one of the first to use statistical methods

and questionnaires to survey philosophical beliefs. Shortly after his appointment at the university, the Germans occupied Norway. Naess resisted any changes in academic routine, insisting that education be separate from politics. The increasing brutality of the Quisling regime, however, impelled him to join the **resistance** movement. While in the resistance he helped avert the shipment of thousands of university students to concentration camps. Immediately after the war, he confronted Nazi atrocities by mediating conversations between the families of torture victims and their pro-Nazi Norwegian victimizers.

In the post-War period, Naess's academic interests and accomplishments were many and varied. He continued his work on language and communication in *Interpretation and Preciseness* (1953) and *Communication and Argument* (1966), concluding that communication is not based on a precise and shared language. Rather we understand words, sentences, and intentions by interpreting their meaning. Language is thus a double-edged sword. Communication is often difficult and requires successive interpretations. Even so, the ambiguity of language allows for a tremendous flexibility in verbal meaning and content. Because of his work in communication and his resistance to the Nazis, Naess was selected by the United Nations Educational, Scientific and Cultural Organization [UNESCO] in 1949 to explore the meanings of democracy. This project resulted in *Democracy, Ideology and Objectivity* (1956). In 1958, he founded the journal *Inquiry*, serving as its editor until 1976. The magazine explores the relations of philosophy, science, and society, especially as they reflect normative assumptions and implications. He also published on diverse topics, including Gandhian nonviolence, the philosophies of science, and the Dutch philosopher Baruch Spinoza (1632–1677). These works include *Gandhi and the Nuclear Age* (1965), *Skepticism* (1968), *Four Modern Philosophers: Carnap, Wittgenstein, Heidegger, Sartre* (1968), *The Pluralist and Possibilist Aspect of the Scientific Enterprise* (1972), and *Freedom, Emotion and Self-Subsistence: The Structure of a Central Part of Spinoza's "Ethic"* (1972).

Naess began examining humanity's relationship with **nature** during the early 1970s. The genesis of this interest is best understood in the context of Norway's **environment**, culture, and politics. Nature, not humanity, dominates the landscape of Norway. The nation has the lowest population density in Europe, and over 90% of the land is undeveloped. As a consequence, the interior of Norway is relatively wild and diverse, a mixture of mountains, glaciers, fjords, forests, **tundra**, and small human settlements. Moreover, Norwegian culture deeply values nature; environmental themes are common in Norwegian literature, and the majority of Norwegians share a passion for outdoor activities and **recreation**. This passion is known as *friluftsliv*, meaning "open air life." Friluftsliv is widely touted as one means of reconnecting with the natural world. Norwegian environmental politics has been wracked by a **succession** of ecological and resource conflicts. These conflicts involve **predator control**, recreation areas, national parks, industrial **pollution**, **dams** and hydroelectric power, nuclear energy, North Sea oil, and economic development. During the 1960s Naess became deeply involved in environmental activism. Indeed, his participation in protests lead to his arrest for nonviolent civil disobedience. He even wrote a manual to help environmental and community activists participate in nonviolent resistance. Growing up, Naess was deeply moved by his experiences in the wild places of Norway. He became an avid mountaineer, leading several ascents of the Tirich Mir (25,300 ft [7700 m]) in the Hindu Kush range. In 1937, he built a small cabin near the final assent to the summit of Hallingskarvet, a mountain approximately 111 mi (180 km) northeast of Oslo. He named it *Tvergastein*, meaning "across the stones."

Since his retirement, Naess has published widely on environmental topics. His main contributions are in ecophilosophy, **environmental policy**, and **conservation biology**. The insight and controversy surrounding these writings have propelled him to the forefront of **environmental ethics** and politics.

Naess regards philosophy as "wisdom in action." He notes that many policy decisions are "made in a state of philosophical stupor" wherein narrow and short-term goals are all that is considered or recommended. Lucid thinking and clear communication help widen and lengthen the options available at any point in time. Naess describes this work as a labor in "ecophilosophy," that is, an inquiry where philosophy is used to study the natural world and humanity's relationship to it. Recalling the ambiguity of language and communication, he distinguishes ecophilosophy from ecosophy—a personal philosophy whose conceptions guide one's conduct toward nature and human beings. While important elements of our ecosophies may be shared, we each proceed from assumptions, norms, and hypotheses that vary in substance and/or interpretation. Of central importance to Naess's exploration of ecophilosophy are norms and beliefs about what one should or ought to do based on what is prudent or ethical. Norms play a leading role in any ecophilosophy. While science may explain nature and **human ecology**, it is norms that justify and motivate our actions in the natural world. Along with the concept of norms, Naess stresses the importance of depth. By depth, Naess means reflecting deeply on our concepts, emotions and experiences of nature, as well as digging to the cultural, personal, and social roots of our environmental problems. Thinking deeply means taking a broad and incisive look at our values, lifestyles, and community life. In so doing, we discover if our way of life is consistent with our most deeply felt norms.

An ecophilosophy which deeply investigates and clarifies its norms is called a "deep ecological philosophy." A social movement that incorporates this process, shares significant norms, and seeks deep personal and social change is termed "deep **ecology** movement." Naess coined the term "deep ecology" in 1973, intending to highlight the importance of norms and social change in environmental decision-making. He also coined the term "shallow ecology" to describe what he considered short-term technological solutions to environmental concerns. Naess's own ecophilosophy is called "Ecosophy T." The "T" symbolizes Tvergastein. **Ecosophy** T stresses a number of themes, including the **intrinsic value** of nature, the importance of cultural and natural diversity, and the norm of self-realization for persons, cultures, and non-human life-forms. Naess offers his ecosophy as a tentative template, encouraging others to construct their own ecosophies.

Deep ecology is therefore a rubric representing many philosophies and practices, each differing in significant ways. Naess encourages this diversity, recognizing that many mutually acceptable interpretations of deep ecology are possible. On the whole, therefore, Naess is philosophically and environmentally non-dogmatic. He avoids rigid dichotomies pitting individual versus social accounts, liberal versus radical solutions, or **wilderness** versus justice concerns. To paraphrase Naess, "the frontier is long and there are many places to stand." Some may focus their thoughts and efforts on nature, others on society, still others on culture. According to Naess, these are legitimate and necessary foci and should be encouraged as separate endeavors and joint undertakings.

[*William S. Lynn*]

Resources

Books

Naess, A. "The Shallow and the Deep, Long-Range Ecology Movements." *Inquiry* 16 (1973): 95–100.

———. "Intrinsic Value: Will the Defenders of Nature Please Rise?" In *Conservation Biology: The Science of Scarcity and Diversity*, edited by M. E. Soule. Sunderland, MA: Sinauer Associates, 1986.

———. *Ecology, Community and Lifestyle: Outline of an Ecosophy.* Cambridge, UK: Cambridge University Press, 1989.

Reed, P., and D. Rothenberg, eds. *Wisdom in the Open Air: The Norwegian Roots of Deep Ecology.* Minneapolis: University of Minnesota Press, 1993.

Rothenberg, D. *Is It Painful to Think? Conversations with Arne Naess.* Minneapolis: University of Minnesota Press, 1993.

NAFTA

see **North American Free Trade Agreement**

Nagasaki, Japan

Nagasaki is a harbor city located at the southwestern tip of Japan. It is Japan's oldest open port. Traders from the western world started arriving in the mid-sixteenth century. It became a major shipbuilding center by the twentieth century. It was the second city to be devastated by an atomic bomb which was dropped on August 9, 1945, just 3 days after the bomb that destroyed Hiroshima. The bomb that was dropped at Nagasaki resulted in 35,000-40,000 men, women, and children killed with an equal number injured. Like Hiroshima, most of the city was destroyed. The Nagasaki survivors of the atomic bomb have an increased risk for radiation-induced **cancer** and are part of a continuing study by the Atomic Bomb Casualty Commission which is now known as the Radiation Effects Research Foundation (RERF). The city was rebuilt after the end of World War II and is now a tourist attraction.

[*Robert G. McKinnell*]

National Academy of Sciences

The National Academy of Sciences (NAS) is a private, non-profit, self-governing organization that is responsibility for advising the U.S. federal government, upon request and without fee, on questions of science and technology that affect policy decisions. NAS was created in 1863 by a congressional charter approved by President Abraham Lincoln. Under this same charter, the institution was expanded to include sister organizations: in 1916 the **National Research Council** was established, in 1964 the National Academy of Engineering, and in 1970 the Institute of Medicine. Collectively these organizations are called the National Academies.

NAS publishes a scholarly journal, *Proceedings of the National Academy of Sciences*, organizes symposia, and calls meetings on issues of national importance and urgency. Most of its study projects are conducted by the National Research Council rather than by committees within NAS. However, NAS sponsors two committees, the the Committee on International Security and Arms Control and the Committee on Human Rights.

NAS is an honorary society that elects new members each year, in recognition of their distinguished and continuing achievements in original research. New members are nominated and voted on by existing members. Election to membership is one of the highest honors that a scientist can receive. As of early 1999, NAS had 1,798 Regular Members, 87 Members Emeriti, and 310 Foreign Associates. More

The aftermath from the atom bomb that was dropped on Nagasaki during World War II. (UPI/Corbis-Bettmann. Reproduced by permission.)

than 170 members have won Nobel Prizes. No formal duties are required of members, but they are invited to participate in the governance and advisory activities of NAS and the National Research Council. NAS is governed by a Council, comprised of twelve councilors and five officers elected from the Academy membership. Committee members are not paid for their work, but are reimbursed for travel and subsistence costs. Foreign Associates are elected using the same standards that apply to Regular Members, but do not vote in the election of new members or in other deliberations of the NAS.

NAS does not receive federal appropriations directly, but is funded through contracts and grants with appropriations made available to federal agencies. Most of the work done by NAS is at the request of federal agencies. NAS is responsive to requests from both the executive and legislative branches of government for guidance on scientific issues.

NAS also cooperates with foreign scientific organizations. Officers of NAS meet with officers of the Royal Society of England every two or three years. NAS represents U.S. scientists as an institutional member of the International Council of Scientific Unions.

The work and service of NAS and its sister organizations have resulted in significant and lasting improvement in the health, education, and welfare of U.S. citizens.

[*Judith L. Sims*]

Resources

Books

Hilgartner, Stephen. *Science on Stage: Expert Advice as Public Drama (Writing Science).* Palo Alto, CA: Stanford University Press, 2000.

National Academy of Sciences. *The National Academy of Sciences: The First Hundred Years, 1863–1963.* Washington, DC: National Academy Press, 1978.

Organizations

National Academy of Sciences, 2001 Wisconsin Avenue, N.W., Washington, DC United States of America 20007, Email: <http://www.nas.edu>

National Air Toxics Information Clearinghouse

The National Air Toxics Information Clearinghouse (NATIC) is a program administered by the **Environmental Protection Agency** (EPA). This information network was developed by the EPA to provide state and federal **air pollution control** experts with **air pollution** data. The computerized data base is headquartered in North Carolina, but air **pollution control** officers nationwide are able to access the data.

NATIC is a "one-stop shop" for emissions data for state enforcement officers, according to the EPA. Included on the network is information on air toxics programs, **ambient air** quality standards, **pollution** research, non-health related impact of air pollution, permits, emissions inventories, and ambient **air quality** monitoring programs.

National Ambient Air Quality Standard

The key to national **air pollution control** policy since the passage of the **Clean Air Act** amendments of 1970 are the National Ambient Air Quality Standards (NAAQSs). Under this law, the **Environmental Protection Agency** (EPA) had to establish NAAQSs for six of the most common air pollutants, sometimes referred to as criteria pollutants, by April 1971. Included are **carbon monoxide**, **lead** (added in 1977), **nitrogen oxides**, **ozone**, **particulate** matter, and **sulfur dioxide**. **Hydrocarbons** originally appeared on the list of pollutants, but were removed in 1978 since they were adequately regulated through the ozone standard. The provisions of the law allow the EPA to identify additional substances as pollutants and add them to the list.

For each of these pollutants, primary and **secondary standards** are set. The **primary standards** are designed to protect human health. Secondary standards are to protect crops, forests, and buildings if the primary standards are not capable of doing so; a secondary standard presently exists only for sulfur dioxide. These standards apply uniformly throughout the country, in each of 247 Air Quality Control Regions. All parts of the country were required to meet the NAAQSs by 1975, but this deadline was extended, in some cases, to the year 2010. The states monitor **air pollution**, enforce the standards, and can implement stricter standards than the NAAQSs if they desire.

The primary standards must be established at levels that would "provide an adequate margin of safety ... to protect the public ... from any known or anticipated adverse effects associated with such air pollutant[s] in the ambient air." This phrase was based on the belief that there is a threshold effect of **pollution**: pollution levels below the threshold are safe, levels above are unsafe. Although such an approach to setting the standards reflected scientific knowledge at the time, more recent research suggests that such a threshold probably does not exist. That is, pollution at any level is unsafe. The NAAQSs are also to be established without consideration of how much it will cost to achieve them; they are to be based on the **Best Available Control Technology** (BAT). The secondary standards are to "protect the public welfare from any known or anticipated adverse effects."

The NAAQSs are established based on the EPA's "criteria documents," which summarize the effect on human health caused by each pollutant, based on current scientific knowledge. The standards are usually expressed in parts of pollutant per million parts of air and vary in the duration of time a pollutant can be allowed into the **environment**, so that only a limited amount of contaminant may be emitted per hour, week, or year, for example. The 1977 Clean Air Act amendments require the EPA to submit criteria documents to the Clean Air Scientific Advisory Committee and the EPA's Science Advisory Board for review. Several revisions of the criteria documents are usually required. Although standards should be based on scientific evidence, politics often become involved as environmentalists and public health advocates battle industrial powers in setting standards.

The six criteria pollutants come from a variety of sources and have a variety of health effects. **Carbon** monoxide is a gas produced by the incomplete **combustion** of **fossil fuels**. It can lead to damage of the cardiovascular, nervous, and pulmonary systems, and can also cause problems with short-term attention span and sensory abilities.

Lead, a heavy metal, has been traced to many health effects, mainly brain damage leading to learning disabilities and retardation in children. Most of the lead found in the air came from **gasoline** fumes until a 1973 court case, *Natural Resources Defense Council v. EPA*, prompted its inclusion as a **criteria pollutant**. Lead levels in gasoline are now monitored.

Nitrogen oxide is formed primarily by fossil fuel combustion. It not only contributes to **acid rain** and the formation of ground-level ozone, but it has been linked to respiratory illness.

Ground-level ozone is produced by a combination of hydrocarbons and nitrogen oxides in the presence of sunlight, and heat. It is the prime component of **photochemical smog**, which can cause respiratory problems in humans, reduce crop yields, and cause forest damage. In 1979, the first revision of an original NAAQS slightly relaxed the photochemical oxidant standard and renamed it the ozone standard. Experts are currently debating whether this standard is low enough, and in 1991 the American Lung Associa-

tion sued the EPA for failure to review the ozone NAAQS in light of new evidence.

Particulate matter is composed of small pieces of solid and liquid matter, including soot, dust, and organic matter. It reduces **visibility** and can cause eye and throat irritation, respiratory ailments, and **cancer**. Through 1987, total suspended particulates were the basis of the NAAQS. In 1987, the standard was revised and based on particulate matter with an aerodynamic diameter of 10 micrometers or less (PM-10), which was identified as the main health risk.

Sulfur dioxide is a gas produced primarily by coal-burning utilities and other fossil fuel combustion. It is the chief cause of **acid** rain and can also cause respiratory problems.

To achieve the NAAQSs for these six pollutants, the Clean Air Act incorporated three strategies. First, the federal government would establish New Source Performance Standards (NSPSs) for stationary sources such as factories and **power plants**; they would also establish **emission standards** for mobile sources. Finally, the states would develop State Implementation Plans (SIPs) to address existing sources of air pollution. If the federal government determined that an SIP was not adequate to assure that the state would meet the NAAQSs, it could impose federal controls to meet them. According to the EPA, the SIPs must be designed to bring sub-standard air quality regions up to the NAAQSs, or to make sure any area already meeting the requirements continued to do so. The SIPs should prevent increased air pollution in areas of noncompliance, either by preventing significant expansions of existing industries or the opening of new plants. In order to allow economic development and growth in such non-attainment areas while still working to reduce air pollution, the 1977 amendments to the Clean Air Act required that new sources of pollution in non-attainment areas must control emissions to the **lowest achievable emission rate** (LAER) for that type of source and pollution and demonstrate that the new pollution would be offset by new **emission** reductions from existing sources in the area, reductions that went beyond existing permits and compliance plans. So, new sources were allowed, but only if the additional pollution were offset by reductions at existing sources.

Between 1978 and 1987, data collected nationally at 84 to 1,726 sites (varying for each pollutant) indicate that annual average concentrations of total suspended particulates fell by 21%, sulfur dioxide levels fell by 35%, carbon monoxide levels fell by 32%, lead levels fell by 88%, nitrogen dioxide levels fell by 12%, and ozone levels fell by 9%. With the exception of ozone, the average concentrations are below the NAAQSs. It is unclear how these reductions have improved human health conditions, but illnesses due to chronic lead exposure are down. Ozone probably causes the most health problems in affected areas, since the NAAQS for it is most often exceeded. Due to other complex factors, though, it is unclear how much of a problem it is.

Though the Clean Air Act allows one violation per year for each of the one-hour, eight-hour, and twenty-four hour NAAQSs for each pollutant, most urban areas, called non-attainment areas, violate standards much more often. For example, in Los Angeles the ozone standard was exceeded an average of 123 times annually in 1983, 1984, and 1985. As of 1989, 341 counties had ozone non-attainment, 317 had particulate non-attainment, 123 had carbon monoxide non-attainment, sixty-seven had sulfur dioxide non-attainment, and four had nitrogen dioxide non-attainment. A total of 529 counties had not met standards on at least one criteria pollutant.

The 1990 amendments to the Clean Air Act dealt with the problem of such non-attainment areas. Six categories of ozone non-attainment were established, ranging from marginal to extreme, and two categories for both carbon monoxide and particulate matter were established. Deadlines to achieve the NAAQSs were extended from three to twenty years. Increased restrictions were required in non-attainment areas; existing controls were tightened and smaller sources were made subject to regulation. Also, annual reduction goals were mandated. Areas considered to have made inadequate progress toward reaching attainment are subjected to stringent regulations on new plants and limited use of federal highway funds. *See also* Air pollution control; Air pollution index; Air quality criteria; Heavy metals and heavy metal poisoning; Nonpoint source; Point source

[*Christopher McGrory Klyza*]

Resources
Books

Bryner, G. C. *Blue Skies, Green Politics: The Clean Air Act of 1990.* Washington, DC: CQ Press, 1993.

Melnick, R. S. *Regulation and the Courts: The Case of the Clean Air Act.* Washington, DC: Brookings Institution, 1983.

Portney, P. R. "Air Pollution Policy." In *Public Policies for Environmental Protection*, edited by P. R. Portney. Washington, DC: Resources for the Future, 1990.

National Audubon Society

The National Audubon Society (NAS) is one of the largest and oldest **conservation** organizations in the world. Founded in New York City in February 1886, its original purpose was to protect American birds from destruction for the millinery trade. Many **species** of birds were being killed and sold as adornments to women's hat and bonnets, as well as other clothing. The first preservation battle taken on by

the NAS was the snowy egret—a white, wading marshland bird—whose long plumes were in high demand. The group was instrumental in securing passage of the New York Bird Law in 1886, an act for the preservation of the state's avifauna.

The NAS was named after John James Audubon, the nineteenth-century artist and naturalist. Audubon was not a conservationist; he often killed dozens of birds to get a single individual that was right for his paintings. But he had published his life-sized renderings of all the birds in North America in 1850, and his name was the one most closely associated with birds when the society was founded.

The protection of **wildlife** and habitats continues to be the primary focus of the NAS. Since their founding, much of their conservation work has been accomplished through education. Their *Audubon* magazine is perhaps their most important educational tool. They also publish *American Birds*, which records many of the field observations reported to the society from around the country.

One of the most important observations published in *American Birds* is the data recorded on Christmas Bird Counts, an annual winter bird census maintained by NAS since 1900. During a three-week period from December to early January, teams of observers count and record every different species of bird observed within a 24-hour period in a designated circle, 15 mi (24 km) in diameter. The count circles remain constant from year to year which allows for comparison of data collected. The first circle was Central Park in New York City; the number of circles have grown from one to over 1,600 in recent years. Thousands of people participate in this annual event.

Additional educational and conservation programs have proven successful for NAS over the years, including a nature club for children and a series of **ecology** camps for the field study of different ecosystems in North America. The society has also produced nature films and nationally televised programs on environmental issues facing the world today. NAS also addresses these issues through lobbying on a national level in Washington, D.C., and on regional and local levels by members of the several hundred local Audubon Societies, clubs, and state coalitions nationwide. With over half a million environmentally-conscious members, the National Audubon Society continues to make a difference in the battle to conserve the natural world and the wildlife in it.

[*Eugene C. Beckham*]

Organizations

National Audubon Society, 700 Broadway, New York, NY USA 10003 (212) 979-3000, Fax: (212) 979-3188, <http://www.audubon.org>

National Biological Service

see **Biological Resources Division; Biological Resources Division**

National Coalition Against the Misuse of Pesticides

see **Beyond Pesticides**

National Emission Standards for Hazardous Air Pollutants

Emission standards for hazardous air pollutants are set forth in the Section 112 of the **Clean Air Act** of 1970 and the **National Ambient Air Quality Standard**. Section 112 directed the **Environmental Protection Agency** (EPA) to issue a list of hazardous pollutants that were to be regulated. However, the provisions of the 1970 Clean Air Act that were designed to regulate hazardous pollutants proved to be ineffective. Although more than 300 toxic substances are known to be emitted into the air, between 1970 and 1990 **emission** standards were established for only eight substances: **asbestos**, **benzene**, beryllium, coke oven emissions, inorganic **arsenic**, **mercury**, **radionuclides**, and **vinyl chloride**. Although the EPA had epidemiologic and laboratory data that linked more than sixty other airborne **toxins** with **cancer**, **birth defects**, or neurological disease, they remained unregulated. The regulatory process specified by Section 112 was expensive and conservative. It required the EPA to gather sufficient evidence to demonstrate that an airborne toxin was hazardous before it could act. Frequently, there was a lack of scientific information on the health effects of these toxins. There was also a strong potential for legal challenges by industry.

Records dating from 1987 indicated that between 1987 and 1989, over 2 billion pounds of toxic **chemicals** were released into the air each year. Furthermore, more toxic chemicals were released into the air than into any other medium (water or **soil**) during these three years. Environmentalists and their supporters in Congress argued that the 1970 program had not been successful. The issue of airborne toxins was a major **air pollution** problem that was not being adequately addressed by Section 112. For this reason, an improved program to regulate airborne toxins became a central component in the efforts to amend the Clean Air Act.

One of the major components in the 1990 Clean Air Act included provisions to regulate hazardous air pollutants. Title III of the law required the thousands of sources of airborne toxins in the country to use the best available technologies to control hazardous air pollutants (HAPs). The act listed 189 chemicals that are considered HAPs. This

effectively removed the power to designate hazardous air pollutants from the EPA. However, the legislation did provide a mechanism to designate other HAPs for regulation in the future. Since 2000, all major stationary sources, regardless of age, location, or size, have been required to meet emission standards for every HAP established by the EPA. Major sources of HAPs (those emitting at least 10 tons of one pollutant or 25 tons of a combination of pollutants, per year) are required to apply "maximum achievable control technology" (MACT) and are required to receive emission permits. An incentive-based provision of the 1990 Clean Air Act ensured that sources that voluntarily and permanently achieved HAP reductions of 90% (from 1987 levels) could postpone instillation of MACT for up to six years.

Under the revised law, the EPA was directed to focus on those HAPs and sources that are deemed the most dangerous. Further, the EPA was directed to devise strategies to control all sources of HAPs in urban areas. By the year 2000, the EPA was mandated to identify the 30 HAPs that posed the greatest threat to human health and to identify the sources generating 90% of these emissions. The EPA was also to devise a strategy to reduce HAP-related cancer risk by 75%. It ws estimated that these additional provisions would reduce hazardous air pollutants from industrial sources by 75–90%. The EPA has undertaken programs to comply with these directives. As of 2002, the EPA has not completed any of the mandates.

[*L. Fleming Fallon Jr., M.D., Dr.P.H.*]

Resources

Books

Alley, E. R., Cleland, W.L. and Stevens, L.B. *Air Quality Control Handbook.* New York: McGraw-Hill, 1998.

Davis, Wayne T. *Air Pollution Engineering Manual.* 2nd ed. New York: John Wiley and Sons, 2000.

Liu, D. H., and B. G. Liptbak. *Air Pollution* Boca Raton: Lewis Publishers, 1999.

Wirth, John D. *Smelter Smoke in North America: The Politics of Transborder Pollution.* Lawrence, KS: University Press of Kansas, 1999.

Other

American Meteorological Society. *Clean Air Act.* [July 2002]. <http://www.ametsoc.org/AMS/sloan/cleanair/cleanairlegisl.html>.

National Environmental Policy Act (1969)

The National Environmental Policy Act (NEPA) of 1969 was signed into law on January 1, 1970, launching a decade marked by passage of key environmental legislation and increased awareness of environmental problems. The act established the first comprehensive national policies and goals for the protection, maintenance, and use of the **environment**. The act also established the **Council on Environmental Quality** (CEQ) to oversee NEPA and advise the president on environmental issues.

Title I of NEPA declares that the federal government will use all practicable means and measures to create and maintain conditions under which people and **nature** can exist in productive harmony, while fulfilling the social and economic requirements of the American people. Included in this declaration are goals to attain the widest range of beneficial uses of the environment without undesirable consequences and to preserve culturally and aesthetically important features of the landscape. The declaration also commits each generation of Americans to stewardship of the environment for **future generations**.

To achieve the national environmental goals, the act directs all federal agencies to evaluate the impacts of major federal actions upon the environment. Before taking an action, each federal agency must prepare a statement describing (1) the environmental impact of the proposed action, (2) adverse environmental effects that cannot be avoided, (3) alternatives to the proposal, (4) short-term versus long-term impacts, and (5) any irreversible effects on resources that would result if the action were implemented. The act requires any federal, state, or local agency with jurisdiction over the impacted environment to take part in the decision-making process. The general public also is given opportunities to take part in the NEPA process through hearings and meetings, or by submitting written comments to agencies involved in a project.

Title II of NEPA created the CEQ, as part of the Executive Office of the President, to oversee implementation of NEPA and assist the president in preparing an annual environmental quality report. The CEQ existed until February 1993, when President Clinton abolished the council and established the White House Office on Environmental Policy, with broader powers to coordinate national environmental policy. The CEQ issued regulations in 1978 which implement NEPA and are binding on all federal agencies. The regulations (40 CFR Parts 1500-1508) cover procedures and administration of NEPA, including preparation of environmental assessments and environmental impact statements. Many federal agencies have established their own NEPA regulations, following CEQ regulations, but customized for the particular activities of the agencies. In addition, 11 states have passed state environmental policy acts, sometimes called "little NEPAs."

Federal agencies must incorporate the NEPA review process early in project planning. A complete environmental analysis can be very complex, involving potential effects on physical, chemical, biological and social factors of the proposed project and its alternatives. Various systematic meth-

ods have been developed to deal with the complexity of environmental analysis. There are three levels at which an action may be evaluated, depending on how large an impact it will have on the environment.

The first level is categorical exclusion which allows an undertaking to be exempt from detailed evaluation if it meets previously determined criteria designated as having no significant environmental impact. Some federal agencies have lists of actions which have been thus categorically excluded from evaluation under NEPA.

When an action cannot be excluded under the first level of analysis, the agency involved may prepare an environmental assessment to determine if the action will have a significant environment effect. An environmental assessment is a brief statement of the impacts of the action and alternatives. If the assessment determines there will not be significant environmental consequences, the agency issues a finding of no significant impact (FONSI). The finding may describe measures that will be taken to reduce potential impacts. An agency can skip the second level if it anticipates in advance that there will be significant impacts.

An action moves to the third level of analysis, an **environmental impact statement** (EIS), if the environmental assessment determines there will be significant environmental impacts. An EIS is a detailed evaluation of the action and alternatives, and it is used to make decisions on how to proceed with the action. An agency preparing an EIS must release a draft statement for comment and review by other agencies, local governments, and the general public. A final statement is released with modifications based on the results of the public review of the draft statement. When more than one agency is involved in an action, a lead agency is designated to coordinate the environmental analysis. An agency also may be called upon to cooperate in an environmental analysis if it has expertise in an area of concern. The CEQ regulations describe a process for settling disagreements which arise between agencies involved in an environmental analysis.

In addition to having to prepare their own environmental assessments and environmental impact statements, the **Environmental Protection Agency** (EPA) is involved in all NEPA review processes of other federal agencies, as mandated by Section 309 of the **Clean Air Act**. Section 309 was added to the Clean Air Act in 1970, after NEPA was passed and the EPA formed, with the purpose of ensuring independent reviews of all federal actions impacting the environment. As a result, the EPA reviews and comments on all federal environmental impact statements in draft and final form, on proposed environmental regulations and legislation, and on other proposed federal projects the EPA considers to have significant environmental impacts. The EPA's procedures for carrying out the Section 309 requirements are contained in the publication "Policies and Procedures for the Review of Federal Actions Impacting the Environment" (revised 1984). The EPA is also responsible for many of the administrative aspects of the EIS filing process. The NEPA review process includes an evaluation of a project's compliance with other environmental laws such as the Clean Air Act. Federal agencies often integrate NEPA reviews with review requirements of other environmental laws to expedite decision-making and reduce costs and effort.

NEPA requires that previously unquantified environmental amenities be given consideration in decision-making along with more technical considerations. That means that the environmental analysis can be a subjective process, guided by the values of the particular players in any given project. The act does not define what constitutes a major action or what is considered a significant effect on the environment. Some agencies have developed their own guidelines for what types of actions are considered major. Examples of major actions include construction projects such as highway expansion and creek channelization. However, major actions do not always involve construction. For example, legislative changes which may affect the environment may come under NEPA review. When an action will be controversial or clearly violate a pre-set environmental standard, it also may be categorized as a major action with significant impacts. Actions which will have less measurable effects, such as disrupting scenic beauty, must be categorized more subjectively. The courts are frequently used to settle disputes about whether actions require compliance with NEPA.

Although NEPA is targeted to federal agencies, its implementation has resulted in closer scrutiny of major environmental actions other than those sponsored by the government. Environmental analyses also are required for private developments which need federal **pollution control** permits such as water discharges, air emissions, waste disposal, and **wetlands** filling. *See also* Environmental impact assessment

[*Teresa C. Donkin*]

Resources

Books

Fogelman, V. M. *Guide to the National Environmental Policy Act: Interpretations, Applications and Compliance.* Westport, CT: Greenwood, 1990.

Periodicals

Caldwell, L. K. "20 Years With NEPA Indicates the Need." *Environment* 31 (December 1989): 6–15.

Parenteau, P. A. "NEPA at Twenty: Great Disappointment or Whopping Success?" *Audubon* 92 (March 1990): 104–07.

Other

Facts about the National Environmental Policy Act. Washington, DC: U.S. Environmental Protection Agency, September 1989.

National Environmental Satellite, Data and Information Service

see **National Oceanic and Atmospheric Administration (NOAA)**

National Estuary Program

The National Estuary Program (NEP) was established in 1987 when amendments to the **Clean Water Act** provided that the significant estuaries of the United States must be identified, and protected. According to the **National Oceanic and Atmospheric Administration**, (NOAA) in its publication *Where the Rivers Meet the Sea,* "An estuary occurs where saltwater from the sea meets and mixes with fresh water from the land." Estuaries are places where fresh and salt water mix. They are known as bays, harbors, sounds, or lagoons, as well. As much as 80% of the fish that are caught for food or for **recreation** depend on estuaries for all or part of their lives, according to the *National Association of Estuary Programs.* Because of their vital role in the impact of marine life, estuaries are often referred to as "cradles of the sea." The water bodies of the United States that play close to hearts of Americans and foreign visitors alike are estuaries—San Francisco Bay, **Chesapeake Bay**, Puget Sound, and Long Island Sound.

The extensive **ecosystem** that comprises all estuaries creates a similar balance of characteristics, as well as inherent problems. A continual force of tides and winds mix salt and fresh water constantly. Excessive **nutrient pollution** and loss of natural habitats—many due to human manipulation—can upset that mix and create an inhospitable **environment** for the living organisms that reside there. The estuary is not alone in its environmental home. The surrounding **wetlands**, rivers, and streams, as well as the land that courses through it and around it are all an integral piece of a whole that affects both humans and **wildlife** in their struggles for survival. The NOAA points out also that, "Estuaries are among the most biologically productive systems on earth. More than two-thirds of the fish and shellfish commercially harvested in coastal waters spend part or all their lives in estuaries."

As of 2002, the EPA had identified the key issues in managing estuaries, particularly those in the NEP, as well as the challenges they face in their survival. Common to each of the 28 areas named in the NEP are:

- overenrichment of nutrients—nitrogen and **phosphorus** are vital to a healthy aquatic system; in excess, they cause or add to fish disease, red or brown tide, algae blooms, and low dissolved oxygen
- pathogen contamination—viruses, bacteria, and **parasites** that indicate a health hazard to swimmers, surfers, divers, and seafood consumers
- toxic chemicals—metals, **polycyclic aromatic hydrocarbons** (PAHs) **polychlorinated biphenyls** (PCBs), **heavy metals**, and pesticides, all particularly threatening to humans who would consume any fish or seafood from the water body
- alteration of fresh water inflow
- **habitat** loss and degradation—the ecological balance of any estuary depends on the health of a habitat for survival
- decline in the fish and wildlife population.
- introduction of invasive species—either by intention or accident, such an addition can upset the balance of the system and create unexpected impacts economically, socially, and ecologically

As a voluntary, community based **watershed** program, the NEP has not only set as its goal to improve the quality of water in the estuaries. The program has sought to improve the entire ecological system—its chemical, physical, and biological properties, in addition to its economic, recreational, and aesthetic values. As of 2002, there were 102 estuaries in the United States. Only 28 of those have been designated as "nationally significant" and are therefore the focus of the entire effort of restoration and preservation. These 28 estuaries include 42% of the continental United States shoreline, with over half of the population living in the nation's coastal counties.

Since June 1995, the 28 nationally significant estuaries are Albemarle-Pimlico, NC; Barartaria-Terrebone, LA; Barnegat Bay, NJ; Buzzards Bay, MA; Casco Bay, ME; Charlotte Harbor, FL; Corpus Christi, TX; Delaware Estuary Program, DE, PA, NJ; Delaware Inland Bays, DE; Galveston Bay, TX; Indian River **Lagoon**, FL; Long Island Sound, NY and CT; Lower Columbia River, OR and WA; Maryland Coastal Bay, MD; Massachusetts Bay, MA; Mobile Bay, AL; Morro Bay, CA; Narragansett Bay, RI; New Hampshire Estuaries, NH; New York–New Jersey Harbor, NY and NJ; Peconic Bay, NY; Puget Sound, WA; San Francisco Bay, CA; San Juan Bay, Puerto Rico; Santa Monica Bay, CA; Sarasota Bay, FL; Tampa Bay, FL; and, Tillamook Bay, OR. The Chesapeake Bay in Maryland by itself is not specifically included in the NEP but is related to it. The Chesapeake Bay Program is a separate, federally-mandated program.

Economic factors as well as environmental concerns play a significant role in the maintenance of estuaries, primarily in estimating their worth. Each year, these particular estuaries account for over $7 billion in revenue from commercial and recreational fishing and related marine indus-

tries; $16 billion is generated by the attendant tourism and recreation activities.

The program as it has been established is a descendant of the early environmental programs of the 1970s and 1980s that sought to restore the **Great Lakes** and the Chesapeake Bay—all of which had become polluted to such an extent that concerns of whether the problems could be reversed were tantamount to the discussions and burgeoning activism that sought to fix them. When Cleveland's **Cuyahoga River** set fire in 1970 the laughter that was provoked publicly was only to hide a city's, and a nation's horror, that such an event had occurred.

Chesapeake Bay was about to meet a similarly dark fate with out of control pollution, unbridled industrial and private development, and a fishing industry that was dying. In 1983 the governors of Maryland, Virginia and Pennsylvania, the Mayor of Washington, D.C., and the EPA administrator signed the Chesapeake Bay Agreement committing their states and the District of Columbia to prepare plans for protecting and improving the **water quality** and living resources of the bay.

The implementation of the Chesapeake Bay Program, along with the 1965 **Water Resources** and Planning Act, the 1972 Federal **Water Pollution** Control Act, and the 1977 Clean Water Act laid the groundwork for the Comprehensive Conservation Management Plans (CCMP) for the estuaries that had been identified.

The process for developing the CCMP for the national estuaries involves four steps:

- building a management and decision-making framework
- scientifically characterizing existing resources and identifying priority problems
- developing both conventional and innovative solutions to identified problems
- implementing the management recommendations, with support from public and private sectors

Each local community has a particular stake in maintaining the estuary in its midst. The NEP is designed so that each community takes responsibility for managing it. Representatives from local, state, and federal governmental agencies join together to serve as managers of the estuaries. Local citizens, business leaders, educators, and researchers serve on a volunteer basis, too, and all groups work together to create their individual CCMPs. On the national level, the **Environmental Protection Agency** (EPA) administers the NEP—but the programs are carried out by the previously mentioned people. Nominations for determining significant estuaries are also handled by the EPA submitted to them by the governors of the states where the estuary is located. It is after an estuary has been named that the committees continue with their participation.

The Association of National Estuary Programs (ANEP) is a trade association that supports the 28 NEP estuaries, and provides for a joint effort in addressing the issues facing citizens and lawmakers. The ANEP has been successful in getting key legislation passed as in the instance of Senate Bill 835, signed by President Clinton; and responsible for committing federal dollars to restore a million acres of estuary habitat, and ensuring the continued survival of the National Estuary Program. *The Estuaries and Clean Waters Act of 2000* "authorizes a total of $275 million" until 2005 for matching funds for local restoration projects and also designates an additional $35 million per year until 2004 in support of the NEP. That bill also served to clear up the issue that NEP funds could be used to carry out existing, locally-crafted CCMPs for the 28 NEPs throughout the United States.

In celebration of the NEPs tenth anniversary in 1987, the EPA prepared a "Ten-year perspective" it has posted through it web site. It notes that, "A number of key lessons have been learned over the past 10 years. The NEP has demonstrated that community-based resource management achieves results. Although it takes time to see environmental changes such as improvements in water quality, progress is being made. In order to demonstrate improvements in an estuary, we have seen the importance of NEPs setting measurable environmental goals and indicators." Of the many local programs, EPA noted that, "the NEPs have created innovative management approaches to solve these [of the estuaries] common problems. They have employed alternative on-site **wastewater** treatment technologies to control **nitrogen**; established marina pump-out facilities; provided education and training for owners, installers, and pumpers of septic systems to reduce pathogens; promoted beneficial uses of dredged material to restore and create wetland habitat; installed fish passages to increase spawning; and, helped citizen volunteers remove invasive plant **species** from public areas."

There are a total of 130 estuaries in the United States—the question remains whether or not all of these should also become a part of the NEP. What improvements have been accomplished through the designated 28, however, can serve as focal points for improvements for those not named. **Restore America's Estuaries (RES)** is an organization dedicated to preserving the coastal integrity through the improvement of all estuaries, including those other then the 28 in the NEP. RES has prepared its document entitled, *A National Strategy to Restore Coastal and Estuarine Habitat* whose stated purpose it is to "provide a framework for restoring function to coastal and estuarine habitat." As with the ANEP, the work of RES includes professionals and citizens—scientists, community leaders, nongovernmental organizations, and governmental representatives from various agencies and lev-

els. **Communities Actively Restoring Estuaries (C.A.R.E.)** as of 2002, is scheduled to work in a three-year, $12 million in partnership with the NOAA's **Community-Based Restoration Program** to activate "on-the-ground" habitat restoration in 11 major estuaries throughout the United States. The program funds salt marsh restoration, oyster reef restoration, the installation of fishways, shoreline restoration, and other projects.

[*Jane E. Spear*]

Resources

Periodicals

Texas Water Resources. *Precursors of the National Estuary Program.* 1995 [cited June 2002]. <http://www.twri.tamu.edu>.

Other

Associaation of National Estuary Programs. 2002 [cited July 2002]. <http://www.anep-usa.org>.

Environmental Protection Agency. *The National Estuary Program.* [cited July 2002]. <http://www.epa.gov>.

Estuarine Research Federation. [cited July 2002]. <http://www.erf.org>.

Restore America's Estuaries. [cited July 2002]. <http://www.estuaries.org>.

Organizations

Association of National Estuary Programs, 4505 Carrico Drive, Annandale, VA USA 22003 (703) 333-6150, Fax: (703) 658-5353, , <http://www.anep-usa.org>

Estuarine Research Foundation, P.O. Box 510, Port Republic, MD USA 20676 (410) 586-0997, Fax: (410) 586-9226, , <http://www.erf.org>

Restore America's Estuaries, 3801 N. Fairfax Drive, Suite 53, Arlington, VA USA 22203 (703) 524-0248, Fax:)703) 524-0287, , <http://www.estuaries.org>

U.S. Environmental Protection Agency, 1200 Pennsylvania Avenue, N.W., Washington, D.C. USA 20460 (202) 260-2090, , <http://www.epa.gov>

National forest

A national forest is forest land owned and administered by a national government. Mandates designating national forest ownership, administration and the distribution of benefits vary greatly around the world. Some nations (e.g. Canada) retain little or no national forest, delegating **public land** ownership to regional provinces or communities. Other nations (e.g. Albania and other formerly Communist States) retain all public forests as national forests. In many former colonies, national forest administration is patterned after that of the colonizing nation, and lands now comprising national forests were appropriated from **indigenous peoples**. In all cases, the term national forest refers to a type of *state* (i.e. government) property and must not be confused with forests that are owned as *common* property (i.e. private forest owned by a group) or *private* property. The precepts of modern national forests originated in early eighteenth century France and Germany where feudal lords set aside forests to preserve **hunting** grounds. National forest administration stems from this tradition of managing the forests for the direct benefit of central government authorities.

As in several nations in Southeast Asia, national forests in Indonesia comprise the vast majority of national territory, and colonial-style legal and organizational structures continue to dominate management. Forests are centrally administered by the State Forest Corporation with provinces given some authority over harvesting and marketing. Shared government and private party forestry successions are common. Indonesia derives a substantial portion of national income from timber production and has been charged with excluding the concerns of indigenous peoples from management decision.

In 1957 the government of Nepal nationalized all forested land, the majority of which was previously managed as community common property. This act resulted in the breakdown of community management systems and accelerated **deforestation** and forest degradation. The Forest Act of 1992 seeks to reverse these trends and divides national forests into five classes: community forests (managed by community user groups); leasehold forests (leased to private entities for timber production); government managed forests (retained by government for national purposes); protection forests (managed by government for environmental protection); and religious forests.

In the United States the term national forest refers to specific federal land units that are administered by the **U.S. Department of Agriculture Forest Service** (USDAFS). The USDAFS currently administers 156 national forests entailing 223 million acres and covering about one-tenth of the nation's land area. Establishment of these forests beginning in 1903 was largely a response to growing public fears of timber **famine**, wildfires, **flooding**, depleted **wildlife** populations, and damage to the beauty of the national landscape. National **forest management** is governed by federal law and edicts from the executive branch of government. The USDAFS is currently mandated to: 1) manage forests for multiple uses (wood, water, **recreation**, wildlife, and **wilderness** in perpetuity); 2) protect the habitats of **endangered species**; 3) impartially serve the public interest in choosing between management alternatives. National forests have been the focus of great conflict since the 1960s when the environmental movement began challenging USDAFS actions. Recent legislation, and numerous federal court decisions, have resulted in greater integration of public opinion into the USDAFS decision-making process.

[*T. Anderson White*]

RESOURCES
BOOKS

Peluso, N. L. *Rich Forests, Poor People: Resource Control and Resistance in Java.* Berkeley: University of California Press, 1992.

Poffenberger, M., ed. *Keepers of the Forest: Land Management Alternatives in Southeast Asia.* West Hartford, CT: Kumarian Press, 1990.

PERIODICALS

Bromley, D. W. "Property Relations and Economic Development: The Other Land Reform." *World Development* 17 (1989): 867–77.

Gericke, K. L., J. Sullivan, and J. D. Wellman. "Public Participation in National Forest Planning: Perspectives, Procedures and Costs." *Journal of Forestry* 90 (1992): 35–38.

Pardo, R. "Back to the Future: Nepal's New Forestry Legislation." *Journal of Forestry* 91 (1993): 22–26.

National Forest Management Act (1976)

The National Forest Management Act (NFMA), passed in 1976, is the law that established the guidelines for the management of national forests. The act replaced the Organic Act of 1897, which had supplied such guidelines for the previous seventy-nine years. The impetus for this new bill was a court case related to **clear-cutting** on the Monongahela National Forest in West Virginia. Such clear-cutting had begun on this forest in the 1960s, and many environmental and **recreation** groups opposed it. When the **Forest Service** continued this management technique, the **Izaak Walton League** filed suit against the agency, claiming that clear-cutting violated the Organic Act on three counts: that only mature timber was to be harvested, that all timber cut had to be marked and designated, and that all timber cut had to be removed from the forest. The Court ruled in favor of the Izaak Walton League; clear-cutting did violate the Organic Act on each of these counts (*Izaak Walton League v. Butz*, (1973)). The judge issued an injunction against any cutting in West Virginia that violated the Organic Act. The Forest Service appealed, but lost, and the injunction was spread throughout the Fourth Circuit of the Court of Appeals (*Izaak Walton League v. Butz* (1975)). Injunctions based on similar cases were soon issued in Alaska and Texas. With timber harvesting on the national forests in chaos, the Forest Service and timber industry turned to Congress for new management authority.

Congressional action came quickly due to these court decisions. The debate centered around bills introduced by Senator Hubert Humphrey of Minnesota, more favorable to industry, and by Senator Jennings Randolph of West Virginia, supported by environmentalists. The final act was based primarily on the Humphrey bill. The act basically gave legislative approval to how the Forest Service had been managing the national forests. It granted the agency a great deal of administrative discretion to manage the forests based on the general philosophy of multiple use and sustained yield.

Five of the most contentious issues regarding national **forest management** were discussed in the law, but the fundamental issues were left unresolved. Clear-cutting is recognized as a legitimate management technique, and the Forest Service was given qualified discretion to determine when and how this approach would be used. **Species** diversity is to be considered in forest management, but the specifics are left to agency discretion. The Forest Service is directed to identify marginal lands and, based on agency discretion, to move away from timber management on these lands. Trees should be cut at a mature age when possible, but rotation age can be lessened if the agency deems it necessary. Finally, timber is to be harvested based on nondeclining even flow, which is a very conservative approach to sustained yield. Forest Service management decisions on these decisions have continued to stir debate. The NFMA also substantially amended the interdisciplinary national forest planning provisions of the Resources Planning Act of 1974. *See also* Biodiversity; Deforestation; Forest decline; Environmental law; Sustainable development

[*Christopher McGrory Klyza*]

RESOURCES
BOOKS

Clary, D. A. *Timber and the Forest Service.* Lawrence, KS: University Press of Kansas, 1986.

Dana, S. T., and S. K. Fairfax. *Forest and Range Policy.* 2nd ed. New York: McGraw-Hill, 1980.

National Institute for the Environment

The National Institute for the Environment (NIE) was a proposed federal agency conceived in 1989 by a group of scientists, environmentalists, and policy analysts who were concerned that the nation's decision makers did not have adequate scientific information to make suitable environmental policies. They were also concerned that the existing federal environmental research and development activities were spread over many uncoordinated agencies, each with different missions and interests therefore preventing an integrated effort to resolve environmental problems. Led by Dr. Stephen Hubbell of Princeton University and Dr. Henry Howe of the University of Illinois, Chicago, the group developed a proposal to establish the NIE. This initiative to create the NIE was promoted by the National Council on Science and the Environment (NCSE), formerly called the Committee for the National Institute for the Environment (CNIE).

Formed in the late 1980s, the NCSE is a national nonprofit organization comprising 15 board members and

over 9,000 scientists, educators, business leaders, state and local government officials, environmental advocates, and other interested individuals from around the country whose goal it was to work with the United States Congress to establish the National Institute for the Environment (NIE).

The goal of the NIE was to have an independent, nonregulatory, federal institute similar to the National Institutes of Health that focuses on environmental problems. The NIE was to promote better communication between scientists and policy makers to help balance social, economic, and environmental goals. The NIE would also have concentrate activities in four integrated areas: identification and assessment of current knowledge of environmental problems of national interest through a Center for Environmental Assessment; the funding of peer-reviewed, interdisciplinary research to increase scientific understanding of these environmental problems through three research divisions; development of an electronic, on-line library (the National Library for the Environment) to disseminate environmental information; and support of education and training of scientists and other professionals interested in working on environmental issues. The NIE was to have been led by a President-appointed Board of Governors representing federal and nonfederal scientists, state and local government, environmental organizations, citizen groups, academia, and industry.

Within the NIE, a Center for Environmental Assessment was to be established to perform evaluations on the current state of knowledge of particular environmental issues. Assessments would include a review of existing scientific information on an identified environmental problem, an assessment of policy decisions made based on existing information, and the identification of additional research needed to fill information gaps to improve the scientific basis for policy development.

The centerpiece of the NIE was the research it would sponsor. Experts in the natural sciences, engineering, economics, and the social sciences were to work cooperatively on complex environmental topics that could potentially require long-term investigations. Research funded by NIE was to be administered by three directors: a Director of **Environmental Resources**, a Director of Environmental Systems, and a Director of Environmental Sustainability. The Director of Environmental Resources would develop research programs to inventory **natural resources**, monitor and predict change to environmental systems, and develop tools for improved environmental assessment. The Director of Environmental Systems would focus on creating a research program to improve understanding of the functioning of environmental systems and human impact on these systems. For example, possible research programs could include an evaluation of the ecological and social effects of global **climate** change and research to distinguish between natural **ecosystem** variation and environmental change due to human activities. The director of Environmental Sustainability would concentrate on issues related to maintaining **environmental health** while utilizing resources in such a way to ensure their availability and viability in the future.

A key component of the NIE would be to provide information to decision makers, researchers, environmental managers, educators and other professionals, and the public at large. An electronic, on-line National Library for the Environment would be developed. The Committee for the National Institute for the Environment is in the process of establishing a prototype that can be accessed electronically. The prototype library will include research reports from the Congressional Research Service, an environmental encyclopedia, and peer-reviewed and nonpeer-reviewed research articles.

The NIE would support higher education training (i.e., education above the high school level) through research grants, fellowships, teacher training, and grants to develop environmental programs at colleges and universities. The NIE would also maintain a small in-house Center for Integrative Studies on the Environment that would invite visiting scholars to work with a core of scientific staff to evaluate emerging environmental issues on which NIE and other federal agencies need to focus.

The NIE proposal was endorsed by over 150 colleges and universities, scientific, professional, environmental, and business organizations, and state and local governments. Legislation to create the NIE was first introduced in 1989. The legislation was reintroduced in the 103rd Congress and again in 1995 during the 104th Congress by Representative Jim Saxton (R-NJ). The bill was co-sponsored by numerous Democratic and Republican members of Congress. The bill was referred to the House Committee on Science where in 1997 it awaited consideration by the Subcommittee on **Energy and the Environment**.

According to an NCSE update issued in January 2000, in 1999 they supsended the creation of the NIE and the "National Science Board approved an interim report recommending that the National Science Foundation implement most of the activities initially proposed for a National Institute for the Environment. In October 1999, [NCSE] announced its support for the full and effective implementation of this report and suspended its call for the creation of a National Institute for the Environment to work in support of the National Science Foundation initiative."

[*Marci L. Bortman Ph.D.*]

Resources

Periodicals

Benedick, R. "NIE: Its Time Has Come," *Chemistry and Industry* (1996): 64.

National Institute for Urban Wildlife

Since its founding in 1973 as the Urban **Wildlife** Research Center, this organization has promoted the preservation of wildlife in urban settings. In 1983, it became the National Institute for Urban Wildlife and continues to provide support to individuals and organizations involved in maintaining a place for wildlife in the expanding American cities and suburbs. The institute conducts research exploring the relationship between humans and wildlife in urban areas, publicizes methods of urban **wildlife management**, and raises public awareness of the value of wildlife in city settings. Its activities are divided into four programs: research, urban **conservation** education, technical services, and urban wildlife sanctuaries.

The research program provides specific information on the interplay between urban dwellers and wildlife. Developers, engineers, government agencies, industry, planners, students, and the general public use the research conducted by the institute. Some of the studies published by the institute have included *Planning for Wildlife in Cities and Suburbs*; *Urban Wetlands for Stormwater Control and Wildlife Enhancement*; *Planning for Urban Fishing and Waterfront Recreation*; *Highway-Wildlife Relationships: A State-of-the-Art Report*; and *An Annotated Bibliography on Planning and Management for Urban-Suburban Wildlife*.

Through its educational arm, the institute publishes numerous documents on the major issues of wildlife to professional environmentalists as well as to the general public. Among these publications is the quarterly *Urban Wildlife News*, the official publication of the organization. In addition to the newsletter, the institute publishes other resources such as the *Urban Wildlife Manager's Notebook*, the *Wildlife Habitat Conservation Teacher's PAC* series, and *A Guide to Urban Wildlife Management*.

Technical services are provided by the institute to urban planners, developers, land managers, state and federal non-game programs, and homeowners. Among the information and services offered are environmental assessments and impact statements, open space planning and management, recreational planning, experimental design, urban **wetlands** enhancement, data analysis, literature research, and expert testimony.

The urban wildlife sanctuaries program is designed to create a network of certified sanctuaries on public and private land across the United States. Landowners who dedicate their land to wildlife preservation and are certified by the institute receive support from the institute's wildlife biologists.

The institute's 1,000 members include organizations and individuals who are concerned with the preservation of wildlife in urban settings. Although there is no official volunteer program offered through the institute, guidance is available to members working in grassroots organizations.

[*Linda M. Ross*]

Resources

Organizations

National Institute for Urban Wildlife, 10921 Trotting Ridge Way, Columbia, MD USA 21044

National Institute of Environmental Health Sciences (Research Triangle Park, North Carolina)

The National Institute of Environmental Health Sciences (NIEHS) is one of 24 components of the National Institutes of Health (NIH), which is part of the **U.S. Department of Health and Human Services** (DHHS). The mission of NIEHS is to conduct research on environment-related diseases. The focus of research is on understanding how environmental factors, individual susceptibility, and age interrelate to cause human illness and on the development of methods to reduce these illnesses. NIEHS achieves its mission through biomedical research programs, prevention and intervention activities, and communication strategies that include training, education, technology transfer, and community outreach.

NIEHS was established as the Division of Environmental Health Sciences within NIH in 1966. In 1967, the Research Triangle Foundation in North Carolina presented the U.S. Surgeon General with 509 acres (206 ha) in the Research Triangle Park to serve as a site for NIEHS. In 1969, the Division of Environmental Health Sciences was raised to Institute status.

Research is conducted through both on-site resources and an extramural science program. The Division of Extramural Research and Training (DERT) supports a network of university-based environmental health science centers and also provides research and training grants and contracts for research and development. Through DERT, NIEHS supported the research of Dr. Mario J. Molina of the Massachusetts Institute of Technology (MIT), who was a co-recipient of the 1995 Nobel Prize in Chemistry for work showing the loss of the earth's protective **ozone** shield.

The purpose of the Division of Intramural Research (DIR) is to provide research to address the environmental components of many different diseases. Dr. Martin Rodbell, an NIEHS scientist in DIR, was a co-recipient of the Nobel Prize in Medicine for discoveries about the communication system that regulates cellular activity. DIR is organized into programs for Environmental Biology, Environmental Diseases and Medicine, and Environmental Toxicology. The

program for Environmental Biology includes the work of three laboratories. The Laboratory of Molecular Genetics studies the basic mechanisms of the mutational process, fundamental mechanisms of genomic **stability**, and the impact of environmental agents on the genetic apparatus. The Laboratory of Signal Induction studies the effects of environmental agents on physiological processes, and the Laboratory of Structural Biology studies environmentally associated diseases resulting from perturbations in biological.

The second DIR program, for Environmental Diseases and Medicine, includes three branches: **Epidemiology**, Biostatistics, and Comparative Medicine. The Epidemiology Branch studies the impacts of environmental toxicants on human health and reproduction using sensitive health endpoints, susceptible sub-groups, and highly exposed populations; four laboratories support this research. The Laboratory of Reproductive and Developmental Toxicology develops an understanding of the basic mechanisms underlying normal and abnormal development and reproduction. The Laboratory of Pulmonary Pathobiology studies the respiratory tract system biology at the cellular, biochemical, and molecular level in order to develop an understanding of pathogenic mechanisms involved in the onset of diseases of the airways. The Laboratory of Molecular Carcinogenesis studies the mechanisms of environmental carcinogenesis by identifying the target genes in the process and by defining how **chemicals** act on these genes to influence **cancer** development. The Laboratory of Environmental Carcinogenesis and Mutagenesis focuses on chemical, physical, and environmental causes of cancer and uses transgenic and computer models to distinguish carcinogens and the mechanisms by which they act.

The second branch of the program for Environmental Diseases and Medicine, Biostatistics, conducts research in biomathematics and population genetics and in design and analysis of laboratory animal toxicology and carcinogenicity studies to develop methodology for epidemiological and clinical studies, and to provide statistical and computational support to NIEHS scientists. The third branch, Comparative Medicine, provides services and collaborative support to NIEHS scientists in the areas of animal facilities management, animal procurement, health surveillance and disease diagnosis, clinical veterinary services, rodent breeding, technical and surgical assistance, and quality assurance support.

The third DIR program, that for Environmental Toxicology, uses the research conducted at three laboratories. Laboratory of Computational Biology and **Risk Analysis** conducts and coordinates research into the development and use of mechanistic data and models for characterizing and quantifying human health risks associated with exposure to environmental agents. It maintains a research program in computer-based mathematical **modeling**, ranging from modeling cellular and molecular levels to whole animals and focusing on describing and evaluating chemical structures, biological response mechanisms, and their perturbations by potentially hazardous environmental agents. It also develops and uses cellular and molecular markers to investigate the link between environmental exposures and toxicity. It provides approaches for the evaluation of human susceptibility factors that affect toxic environmental effects.

Two other laboratories aid DIR's Environmental Toxicology program. The Laboratory of Toxicology conducts studies to characterize chemical toxic effects, including immune, reproductive, genetic, respiratory, and nervous system toxicities. It studies interactions of chemicals and metabolites with subcellular macromolecules, and it develops methods for characterizing toxicity of chemicals and other agents. The Laboratory of Pharmacology and Chemistry studies the exposure and disposition of environmental chemicals; it studies the **enzyme** systems involved in the **metabolism** of environmental chemicals and drugs; and, it studies the mechanisms responsible for the toxic effects of xenobiotics and their metabolites, including photochemical and free radical mechanisms. The lab utilizes alternative model systems from comparative and marine biology to study the pharmacology and toxicology of chemicals and drugs, and it provides chemical support for NIEHS scientists, including the assessment of chemical purity, stability, and biotransformation.

NIEHS sponsors research on the effects of environmental impacts in several areas. Some research topics include birth and developmental defects and sterility; women's health issues, including breast cancer susceptibility and osteoporosis; and Alzheimer's and other neurologic disorders. NIEHS sponsors research on hazards to the poor due to likely exposure to **lead** paint, hazardous chemicals at work, air and **water pollution**, and **hazardous waste** sites in their communities; some researchers focus on **agricultural pollution**, including natural materials (e.g., grain dust) and **agricultural chemicals**, while other researchers study signal error, i.e., whether environmental chemicals can mimic hormonal growth factors and contribute to the development of cancer or reproductive disorders. Still other research investigates animal alternatives, to reduce the number of animals used in research, to refine the design of experiments to obtain more information at lower cost, and to replace animals with microbial and tissue cultures. Some research identifies biomarkers to measure the uptake and exposure to environmental **toxins**.

NIEHS is the headquarters for the National Toxicology Program (NTP), an interagency program within DHHS. The NTP was established in 1978 to coordinate toxicology research and testing activities within the Department of Health and Human Services, to provide information about potentially toxic chemicals to regulatory and research

agencies, and to strengthen the scientific basis of toxicology. The NTP coordinates toxicology activities of NIEHS, the National Institute for Occupational Safety and Health of the Centers for Disease Control (NIOSH/CDC), and the National Center for Toxicological Research of the **Food and Drug Administration** (NCTR/FDA). The director of NIEHS is also the director of the NTP. Primary research support within NIEHS for NTP is in the Environmental Toxicology Program.

The Superfund Basic Research Program (SBRP), in coordination with the U.S. **Environmental Protection Agency** (U.S. EPA), is also administered by NIEHS. The Superfund Amendments and Reauthorization Act (SARA) of 1986 established a university-based program of basic research within NIEHS. The SBRP receives funding from the U.S. EPA through an interagency agreement using Superfund Trust monies. The funding is used to study human health effects of hazardous substances in the **environment**, especially those found at uncontrolled, leaking waste disposal sites. The primary objectives of SBRP are to find methods, through basic research, to reduce the amount and toxicity of hazardous substances and to prevent adverse human health effects. The SARA legislation specifically mandates that the basic research program administered by NIEHS focus on: methods and technologies to detect hazardous substances in the environment; development of advanced techniques for the detection, assessment, and evaluation of the effects on human health of hazardous substances; basic biological, chemical, and physical methods to reduce the amount and toxicity of hazardous substances.

In support of these mandates, NIEHS supports projects in the areas of engineering, ecological, and hydrogeological research in conjunction with biomedically related components, thus encouraging collaborative efforts among researchers. Specific research areas included in the SBRP include health effects, exposure/risk assessment, **ecology**, fate and transport, **remediation**, **bioremediation**, analytical chemistry, biomarkers, epidemiology, metals, and waste site characterization.

NIEHS was also given responsibility for initiating a training grants program under SARA. The major objective of the NIEHS Worker Education and Training Program, initiated in 1987, is to prevent work-related harm by assisting the training of workers to know how to protect themselves and their communities from exposure to hazardous materials during hazardous waste operations, hazardous materials **transportation**, environmental restoration of **nuclear weapons** facilities, or chemical emergency response. Through this program, non-profit organizations with a demonstrated record of providing occupational safety and health education develop safety and health curriculum for workers involved in handling hazardous waste or in responding to emergency releases of hazardous materials. Information concerning this program is disseminated through the National Clearinghouse for Worker Safety and Health Training.

The SBRP, through an Interagency Agreement, provides additional training by support of the NIOSH Hazardous Substance Continuing Education Program (HST), initiated in 1988 for hazardous substance professionals, and the NIOSH Hazardous Substance Academic Training Program (HSAT), a graduate academic program initiated in 1993 allowing occupational safety and health professionals to specialize in the study of hazardous substances.

In coordination with the Department of Energy, NIEHS is overseeing the implementation of the Electric and Magnetic Fields (EMF) Research and Public Information Dissemination (RAPID) program, established by the 1992 **Energy Policy** Act. The EMF RAPID program is a five-year federally coordinated effort to conduct research on potential effects on biological systems of exposure to 60 Hz electric and magnetic fields produced by the generation, transmission, and use of electric energy and to evaluate developing technologies for characterizing and mitigating these fields. The program also involves collection, compilation, publication, and dissemination of information to the public concerning possible human effects of electric and magnetic fields; types and extent of human exposure to electrical and magnetic fields in various occupational and residential settings; technologies to measure and characterize electric and magnetic fields; and methods to assess and manage exposure to electric and magnetic fields.

The NIEHS Clearinghouse is an information service staffed with scientists who respond to questions concerning environmental health issues. Information on research is also provided through an NIEHS-sponsored journal, *Environmental Health Perspectives.*

[*Judith Sims*]

Resources

Organizations

National Institute of Environmental Health Sciences, P.O. Box 12233, Research Triangle Park, NC USA 27709 (919) 541-3345, <http://www.niehs.nih.gov>.

National Institute for Occupational Safety and Health

The National Institute for Occupational Safety and Health (NIOSH) is a research institute created by the **Occupational Safety and Health Act** of 1970, and since 1973 it has been a division of the Centers for Disease Control (CDC). The purpose of NIOSH is to gather data documenting incidences of occupational disease, exposure, and injury in the United

States. After gathering and evaluating data, the agency develops "Criteria Documents" for specific hazards; in some cases the **Occupational Safety and Health Administration** (OSHA) has used these documents as the basis for specific legal standards to be followed by industry. NIOSH has developed databases which are available to other federal agencies, as well as state governments, academic researchers, industry, and private citizens. The organization also conducts seminars for those in the field of occupational safety and health, as well as for industry, labor, and other government agencies. NIOSH prepares various publications for sale to the public, and it provides a telephone hotline in its Cincinnati, Ohio office to answer inquiries.

In April of 1996, NIOSH and the Occupational Safety and Health Administration (OSH) commemorated their twenty-fifth anniversary in an event that was jointly sponsored by those two agencies and by the Smithsonian Institute. CDC Director David Satcher stated, "Thanks in large measure to NIOSH's efforts, the Nation has made dramatic advancements in recognizing that safe and healthful workplaces are an integral part of good public health, and that the tools we use to curb infectious diseases also work against occupational diseases—knowledge, timely intervention, and prevention."

Resources

Organizations

National Institute for Occupational Safety and Health, 4676 Columbia Pkwy., Cincinnati, OH USA 45226 Fax: 513-533-8573, Toll Free: (800) 35-NIOSH, Email: eidtechinfo@cdc.gov, <http://www.cdc.gov/niosh>

National lakeshore

National lakeshores are part of a system of United States coastlines administered by the **National Park Service** and preserved for their scenic, recreational, and **habitat** resources. The national lakeshore system is an extension of the national seashores system established in the 1930s to preserve the nation's dwindling patches of publicly-owned coastline on the Atlantic, Pacific, and Gulf coasts. Since before 1930 the movement to preserve both seashores and lakeshores has been a conservationist response to the rapid privatization of coastlines by industrial interests and private home owners. In 1992 the United States had four designated National Lakeshores: Indiana **Dunes** on the southern tip of Lake Michigan, Sleeping Bear Dunes on Lake Michigan's eastern shore, and the Apostle Islands and Pictured Rocks, both on Lake Superior's southern shore.

Attention focused on disappearing **Great Lakes** shorelines, sometimes called the United States' "fourth coastline," as midwestern development pressures increased after World War II. During the 1950s lakeshore industrial sites became especially valuable with the impending opening of the **St. Lawrence Seaway**. The seaway, giving landlocked lake ports access to Atlantic trade from Europe and Asia, promised to boost midwestern industry considerably. Facing this threat to remaining wild lands, the **National Park** Service conducted a survey in 1957-58, attempting to identify and catalog the Great Lakes' remaining natural shoreline. The survey produced a list of 66 sites qualified for preservation as natural, scenic, or recreational areas. Of these, five sites were submitted to Congress in the spring of 1959.

The Indiana Dunes site was a spearhead for the movement to designate national lakeshores. Of all the proposed preserves, this one was immediately threatened in the 1950s and 1960s by northern Indiana's expanding steel industries. Residents of neighboring Gary were eager for jobs and industrial development, but conservationists and politicians of nearby Chicago argued that most of the lake was already developed and lobbied intensely for preservation. The Indiana Dunes provided a rare patch of undeveloped acreage that residents of nearby cities valued for **recreation**. Equally important, the dunes and their intradunal ponds, **grasslands**, and mixed deciduous forests provided habitat for animals and migratory birds, most of whose former range already held the industrial complexes of Gary and Chicago. In addition, the dunes harbored patches of relict boreal habitat left over from the last **ice age**. After years of debate, the Indiana Dunes and Great Sleeping Bear Dunes National Lakeshores were established in 1966, with the remaining two lakeshores designated four years later.

The other three national lakeshores are less threatened by industrial development than Indiana Dunes, but they preserve important scenic and historic resources. Sleeping Bear Dunes contains some of Michigan's sandy pine forests as well as **arid** land forbs, grasses, and sedges that are rare in the rest of the Midwest. Two prized aspects of this national lakeshore are its spectacular bluffs and active dunes, some standing hundreds of feet high along the edge of Lake Michigan.

Wisconsin's Apostle Islands, a chain of 22 glacier-scarred, rocky islands, bear evidence of perhaps 12,000 years of human habitation and activity. However most of the historic relics date from the nineteenth century, when loggers, miners, and sailors left their mark. In this area the coniferous boreal forest of Canada meets the deciduous Midwestern forests, producing an unusual mixture of sugar maple, hemlock, white cedar, and black spruce forests. Nearly 20 **species** of orchids find refuge in these islands. Pictured Rocks National Lakeshore preserves extensive historic navigation relics, including sunken ships, along with its scenic and recreational resources. *See also* Coniferous forest; Conservation; Ecosystem; Glaciation; National Parks and Conservation Association; Privatization movement; Wilderness

[*Mary Ann Cunningham Ph.D.*]

Grand Sable dunes in Rocks National Lakeshore, Michigan. (Photograph by Rod Planck. Tom Stack & Associates. Reproduced by permission.)

RESOURCES

BOOKS

Cockrell, R. *A Signature of Time and Eternity: The Administrative History of Indiana Dunes National Lakeshore, Indiana.* Omaha, NE: U.S. National Park Service, Midwest Regional Office, 1988.

PERIODICALS

Bowles, M. L., et al. "Endangered Plant Inventory and Monitoring Strategies at Indiana Dunes National Lakeshore." *Natural Areas Journal* 6 (1986): 18–26.

Herbert, R. D., D. A. Wilcox, and N. B. Pavlovic. "Vegetation Patterns in and among Pannes (Calcerious Intradunal Ponds) at the Indiana Dunes National Lakeshore, Indiana." *American Midland Naturalist* 116 (1986): 276–81.

National Marine Fisheries Service

see **National Oceanic and Atmospheric Administration (NOAA)**

National Mining and Minerals Act (1970)

The Mining and Minerals Policy Act of 1970 was the first of a series of efforts by the United States Congress to address the seeming lack of a coordinated and comprehensive federal minerals policy. The act directed the Secretary of the Interior to follow a policy that encouraged the private mining sector in four ways: to develop a financially viable and stable domestic mining sector, to develop domestic mineral sources in an orderly manner, to conduct research to further "wise and efficient use" of these minerals, and to develop methods of mineral extraction and processing that would be as environmentally benign as possible. Given the broad and vague nature of these directives, it is difficult to determine what, if any, effect the law has had. For example, the U.S. Department of the Interior's report on the bill found that it did not provide the department any new authority. There was one clear directive included in the law: the Secretary of the Interior was to report to Congress annually on the mining

industry and to make any legislative recommendations to help the industry at that time.

Two additional general mining policy laws were passed within fifteen years of the Mining and Minerals Policy Act. The National Materials and Minerals Policy, Research and Development Act was passed in 1980. Like the 1970 Act, this law was an effort to develop a coordinated national minerals policy. But, like its predecessor, this law also had little effect due to its lack of specifics. The 1984 National Critical Materials Act underscored the concern among some in Congress that the United States had become vulnerable, due to foreign dependence, in the supply of such strategic defense and high technology minerals as cobalt and platinum group metals. This law was also filled with vague generalities, though it did create a three-person National Critical Materials Council and charged this Council with overseeing minerals policy. As with the prior two acts, the 1984 law has seemingly had little effect.

The growing concern for mineral policy in the 1970s and 1980s can be traced to three sources: administrative law and capacity, environmental concerns, and national security concerns. The federal government had little control over mining policy. On federal lands, mining policy had been essentially privatized by the 1872 Mining Law. Individuals or private firms could stake claims on public lands and remove the minerals; the government received no royalties and issued no permits. The mining policy regime was also fragmented. Three agencies had major roles in mining policy: the **Bureau of Land Management**, the Bureau of Mines, and the United States **Geological Survey**. Throughout the 1970s and 1980s, concern for the **environment** grew in the country, and a significant concern of environmentalists was mining operations, which often generated large amounts of **pollution** and had significant effects on the land. Hence, environmentalists successfully sought to restrict mining in certain areas and regulate the mining that continued. Partially in response to the reduced access and increased regulation, the mining industry and its supporters began to focus on the United States's foreign dependence on **strategic minerals**. They argued that more lands should be open to mining, and regulations should be relaxed. Despite the passage of these three laws, though, no significant change in federal mining policy has occurred. *See also* Environmental law; Environmental policy; Natural resources

[*Christopher McGrory Klyza*]

Resources

Books

Leshy, J. D. *The Mining Law: A Study in Perpetual Motion.* Washington, DC: Resources for the Future, 1987.

National Ocean Service

see **National Oceanic and Atmospheric Administration (NOAA)**

National Oceanic and Atmospheric Administration

The National Oceanic and Atmospheric Administration (NOAA) is a multi-faceted agency of the United States government that concerns itself with a variety of challenges, from making five-day weather forecasts to protecting **sea turtles**. Under the U.S. Department of Commerce, NOAA is unusual because it functions not only as a service agency, providing weather reports, for example, but also as a regulatory and research agency. Many of its regulatory functions appear to overlap with those of the **Environmental Protection Agency** (EPA), and much of its research parallels that of National Aeronautics and Space Administration (NASA). NOAA's National Ocean Service, for example, assessed the damage to the Alaskan coast and **wildlife** from the disastrous oil spill of the ***Exxon Valdez***; NOAA also monitors the rate and speed of the earth's rotation.

Created in 1807 by President Thomas Jefferson as the National Ocean Service, its first duty was to survey the coasts to set up shipping lanes for trade routes in the United States. Later, its function as an air and sea chart-making agency, when known as the U.S. Coast and Geodetic Survey, was in high demand, especially during World Wars I and II.

In 1970, when the National Oceanic and Atmospheric Administration was instituted, it settled into five major service branches and a NOAA Corps. The Corps, called "the seventh and smallest uniformed service," consists of about 400 men and women trained to perform diverse duties, such as fly into the eye of a **hurricane** and make descents in ocean submersibles to do deep ocean research.

NOAA's five branches are as follows:

- National Weather Service (NWS): For a long time, NOAA was the U.S. Weather Bureau. NWS has provided daily weather forecasts for several decades by gathering data from manned weather forecast offices around the country. To perfect its capability, to increase **accuracy**, and to lengthen predictions for severe weather, NOAA is implementing a long-term program to modernize the National Weather Service by installing 1,000 automatic sensors in all the states. Already the Hydrometeorological Information Center can issue spring flood warnings from river forecast centers well in advance of their occurrence. Nexrad ("Next generation radar"), a new Doppler radar that can pick up severe weather 125 mi (201 km) away, predicts tornadoes with a 19-minute lead time, giving residents more time to find shelter.

- The Office of Oceanic and Atmospheric Research collects data from polar-orbiting and geostationary satellites. These not only provide televised weather pictures, but also monitor elements of **climate** change (such as **greenhouse gases** in the **stratosphere**). At NOAA's Environmental Research Labs, studies of the **ozone** layer are additionally being conducted jointly with NASA; NOAA and Japan have also joined forces to study the thermal vents 20,000 ft (6096 m) down in the Marianna Trench under the Pacific Ocean. NOAA funds university research through the National Sea Grant College Program and the National Undersea Research Program.
- The National Environmental Satellite, Data and Information Service (NESDIS), one of the world's largest banks of information on earth geophysics, solar activity, geomagnetic variations, and paleoclimates, stores data collected from NOAA's satellites and environmental research labs.
- The National Ocean Service still makes coastal charts for trade ships as well as for weekend sailors. But now it also assesses ocean and coastal **pollution**, protects **wetlands**, and is charged with creating and maintaining sanctuaries for various sea creatures including **whales** and living coral reefs.
- The National Marine Fisheries Service (NMFS) manages fisheries by overseeing coastal fish-breeding habitats, restoring **endangered species**, collecting abandoned drift nets, and trying to save turtles, **dolphins**, and whales that swallow sunken plastic balloons or get caught in plastic six-pack circles. The NMFS instituted the use of Turtle Excluder Devices (TEDs) on shrimp nets to keep turtles from being strangled.

NOAA also operates a rescue satellite, known as COSPAS-SARSAT (Satellite-Aided Search and Rescue). Stranded fishermen and sailors, whose boats are equipped with an Emergency Position Indicating Radio Beacon (EPIRB), send a signal to the satellite that relays it to a ground station which responds by sending a rescue vessel within hours of the initial signal.

Because NOAA is unusual in its functions as a regulatory, research, and service agency, it is often an easy target for budget-cutting. Underfunded during most of the 1980s and always dependent on Congressional funding, NOAA is currently struggling to streamline its duties in the face of rapid climate change and increased **marine pollution**.

[*Stephanie Ocko*]

Resources

Periodicals

Kerr, R. A. "NOAA Revived for the Green Decade." *Science* 248 (June 8, 1990): 1177–79.

Other

National Implementation Plan for the Modernization and Restructuring of the National Weather Service. Washington, DC: U. S. Department of Commerce, 1992.

Toward a New National Weather Service. Second Report. Washington, DC: National Research Council, 1992.

Organizations

National Oceanic and Atmospheric Administration, 14th Street & Constitution Avenue, NW, Room 6013, Washington, D.C. USA 20230 (202) 482-6090, Fax: (202) 482-3154,

National park

National parks are areas that have been legally set apart by national governments because they have cultural or **natural resources** which are deemed significant for the particular country. National parks are typically large areas that are mostly undisturbed by human occupation or exploitation. They are characterized by spectacular scenery, abundant **wildlife**, unique geologic features, or interesting cultural or historic sites.

National parks are managed to eliminate or minimize human disturbances, while allowing human visitation for recreational, educational, cultural, or inspirational purposes. Activities consistent with typical national park management include hiking, camping, picnicking, wildlife observation, and photography. Fishing is usually allowed, but **hunting** is often prohibited. In the United States, national parks are distinguished from national forests and other federal lands because timber harvesting, cattle grazing, and mining are, with a few exceptions, not permitted in national parks, whereas they are permitted on most other federal lands.

The United States was the first country to establish national parks. The Yosemite Grant of 1864 was the first act that formally set aside land by the federal government for "public use, resort and recreation." Twenty square miles (52 sq km) of land in the Yosemite valley and four square miles (10 sq km) of giant sequoia were put under the care of the State of California to be held "inalienable for all time." In 1872, President Ulysses S. Grant signed into law the establishment of **Yellowstone National Park**. Yellowstone differed from Yosemite in that it was to be managed and controlled by the federal government, not the state, and therefore has received the honor of being considered the first national park. Years later, Yosemite was turned over to the federal government for federal management also. Since 1916, national parks in the United States have been administered by the **National Park Service** an agency in the **U.S. Department of the Interior**.

The concept of national parks has caught on all over the world and continues to spread. Between the years 1972 and 1982 the number of national parks in the world increased

by 47% and the area encompassed in the parks increased 82%. Today over 1000 national parks can be found worldwide in more than 120 countries.

Many national parks in both developed and developing countries are facing threats, however. The most commonly reported threats are illegal removal of wildlife, destruction of vegetation, and increased **erosion**. Often there is a lack of personnel to deal with these threats. Management problems also arise because demand for use of park resources is increasing. Many of these uses are conflicting, and virtually all would have significant impacts on the resources that characterize the parks.

Although parks consist of natural resources, they are conceived, established, maintained, and often threatened by humans. It is necessary for a society to derive benefits from the parks to maintain public support for them. In light of the need for public support, merely putting a fence around the park and keeping people out is likely to fail in the long term. Paradoxically then, some development and use is necessary for **conservation** of the resources. Deciding on the appropriate amounts and kind of uses compatible with the resources is the key to successful park management.

During the summer of 1962 the first World Conference on National Parks was held in Seattle, Washington. This historic conference and subsequent ones have given people of many nations a forum to discuss threats facing their parks and strategies for meeting the demand for conflicting uses. Only through such international dialogue and continued diligence will these treasures we call national parks be saved for **future generations**.

[*Ted T. Cable*]

RESOURCES
BOOKS

Machlis, G. E., and D. L. Tichnell. *The State of the World's Parks*. Boulder, CO: Westview Press, 1985.

Runte, A. *National Parks and American Experience*. 2nd ed. Lincoln: University of Nebraska Press, 1987.

National Park Service

The **National Park** Service, an agency in the **U.S. Department of the Interior**, was established by the National Park Service Act of 1916 making it the first such agency in the world. Its mission as stated in the Act is: "to conserve the scenery and the natural and historic objects and the wild life therein and to provide for the enjoyment of the same in such a manner and by such means as will leave them unimpaired for the enjoyment of future generations."

The mission has not changed over the years but the ambiguous wording of the act has sparked much debate over what the primary use of a national park should be. Some people feel the parks should be commercially developed in order to best "provide for the enjoyment" of the resources, whereas others think the highest priority should be preserving the resources in an "unimpaired" state for **future generations**. These conflicting views have existed ever since the agency was established, and the debate will likely continue. The dual mandate of providing enjoyment from the resources and leaving them unchanged has been called the "preservation/use paradox." Striking the balance between use and preservation is still at the forefront of current national park policy issues.

National parks existed in the United States prior to the establishment of the National Park Service. The Yosemite Grant of 1864 was the first act that formally set aside land by the federal government for "public use, resort and recreation." Twenty square miles (52 km^2) of land in the Yosemite valley and 4 mi^2 (10 km^2) of giant sequoia were put under the care of the State of California by President Abraham Lincoln. This land was to be held "inalienable for all time." In 1872, President Ulysses S. Grant signed into law the establishment of **Yellowstone National Park**. Yellowstone differed from Yosemite in that it was to be managed and controlled by the federal government, not the state, and therefore has received the honor of being considered the first national park. Years later, Yosemite was also turned over to the federal government for management.

Originally no money had been set aside for the protection and care of the national parks. The parks suffered from illegal timber harvesting, grazing, **poaching**, and vandalism. The U.S. Cavalry was sent in to protect the lands and did so until 1916 when the National Park Service was formed.

Horace Albright was named Assistant for Parks in 1913 by the Secretary of Interior Franklin Lane. Albright, in turn, recruited self-made millionaire and **nature** lover Stephen Mather to help him establish an agency to manage the parks. All the parks thus far had been created through separate acts of Congress with no unified guidelines to manage and control them. Bills to form an agency to oversee the parks had been introduced to Congress, but none had yet passed. Through the intensive lobbying efforts of Albright, Mather, and others, the National Park Service was formed in 1916. Upon its creation the National Park Service took over the management of 14 national parks and 21 national monuments. Mather was the named the agency's first Director and Albright became his assistant. Later Albright would become the second Director of the National Park Service.

Albright and Mather believed that if the park system was to successfully defend itself from attacks by individuals with utilitarian philosophies (i.e., those believing timber cutting, grazing, and mining should be allowed on federal lands) it would have to have a strong base of public support. They

Grand Canyon National Park in Arizona. (Photograph by Robert J. Huffman. Fieldmark Publications. Reproduced by permission.)

emphasized a tourism effort to "See America First" and embarked on a policy of development and expansion to include tennis courts, **golf courses**, and swimming pools. As the **automobile** became popular they encouraged visitation by car and tried to make the parks easily accessible. Through their efforts park use soared.

Although public support was strong, lands still had to be deemed worthless for agriculture or economic development to be set aside as parks by Congress. Therefore, parks typically were high altitude with steep rocky terrain. Also, because they were still being justified and preserved as tourist destinations, only those areas of spectacular scenery were included within the park boundaries. This has resulted in many management problems, as today park managers attempt to look after park resources that are a part of ecosystems which extend beyond the park boundaries.

The number of visitors to the parks increased so greatly that by the 1920s the limited facilities could no longer keep up with the growing demand. Overcrowding and **pollution** were becoming problems. Conservationists complained that tourism was given priority over preservation and that the parks were being degraded. Lodges and other amenities continued to be added to the parks to handle the increased visitation.

Until the establishment of **Everglades** National Park in 1934, all of the national parks had been designated from land that was already under federal government control. The Everglades was the first park taken from annexed private land. It was also the first park that was not preserved for its spectacular scenery. Rather it was primarily established to preserve its fragile **ecosystem**, especially its colorful and conspicuous **wildlife**. Shortly thereafter, the Great Smoky Mountain and Shenandoah National Parks were established. These were developed from private lands that were purchased using donations from private citizens, including John D. Rockefeller.

With the growing environmental awareness of the 1970s increased public scrutiny was focused on the balance between providing an enjoyable experience for visitors while preserving the park resources. The National Park Service looked for ways to accommodate the increasing numbers of visitors while minimizing the impacts to the **environment**. For example, to cut down on pollution and congestion, shuttle buses now take visitors into the Yosemite, Grand

Canyon, and Denali national parks. The National Park Service continues to seek creative ways to simultaneously achieve their dual missions of use and preservation.

The National Park Service administers 386 sites covering 80 million acres (32.4 million ha) in 49 states. Fifty of these sites are national parks, sometimes referred to as the "crown jewels" of the United States. The largest of the parks is Wrangell-St. Elias covering 13.2 million acres (5.3 million ha) in Alaska. In 1989 use at National Park Service facilities was measured at 114 million visitor days. The most visited national park is Great Smoky Mountain National Park.

In addition to the national parks, the National Park Service also administers 79 National Monuments, 69 National Historic Sites, and 29 National Historic Parks. They manage sites in 22 different categories including wild and scenic rivers, seashores, lakeshores, scenic trails, parkways, preserves, and even the White House. Each of the sites are placed into one of these categories based on such attributes as size, level of development, and significance. They are managed under policies appropriate for that type of site. For example, a national preserve is typically set aside to protect a particular resource. **Hunting**, fishing, mining, or extraction of fuels can be allowed on the preserve so long as it does not threaten the specific resource being preserved.

Each year millions of people from around the world come to the National Parks to recreate, to learn, and to be inspired in these wondrous places. Truly, the National Park Service is the caretaker of America's "crown jewels."

[*Ted T. Cable*]

Resources
Books

Albright, Horace M., and Marian Albright Schenck. *Creating the National Park Service: The Missing Years.* Norman: University of Oklahoma Press, 1999.

Runte, A. *National Parks and American Experience.* 2nd ed. Lincoln: University of Nebraska Press, 1987.

Other

National Parks Service. [June 2002]. <http://www.cr.nps.gov>.

National Parks and Conservation Association

The National Parks Conservation Association (NPCA) was founded in 1919 for the defense, promotion, and improvement of the United States' **national park** system and the education of the general public regarding national parks. Throughout its history, the NPCA has worked toward a three-fold goal: to save the parks from exploitation, **pollution**, and degradation of their **natural resources**; to facilitate easy access to the national parks for all Americans; and to preserve the great variety of **wildlife** found within the boundaries of the parks.

With a membership of 280,000—including both individuals and organizations—the NPCA boasts considerable political influence and is particularly effective in organizing ground swell support. Members are kept informed of political, environmental, and other issues through *National Parks*, the official NPCA publication. The magazine also contains articles of interest on state parks, international public lands, and matters of general environmental importance.

For more than 70 years, the NPCA has helped increase the holdings of the National Park System by lobbying for the expansion of existing grounds and for the establishment of new parks. Through the National Park Trust Fund, the NPCA has acquired and donated land to the National Park System. It has labored to protect national parks from overdevelopment and encroaching **urban sprawl** and has participated in studies to analyze and minimize the impact of visitor traffic.

Other support programs sponsored by the NPCA include fund raisers such as the March for Parks program, independent studies of resource management in national parks, and efforts to protect and revitalize populations of wild animals within the parks. Working in conjunction with other conservation groups, the NPCA has dealt with such wide ranging environmental issues as **air pollution**, **acid rain**, **water pollution**, **endangered species**, and damage to coastal areas.

Through its bimonthly magazine, *National Parks*, the NPCA helps coordinate grassroots activities such as the Park Watchers Program, a network of concerned local parties who keep track of potential threats to the Park System, and the Contact System which alerts members to make their views known to public officials on issues concerning the **environment**. Services to its members include special tours of national parks, general information about environmental issues and their impact on the Park System, and periodic legislative updates and action alerts.

A second bimonthly publication, *Exchanges*, is directed to those who participate in NPCA volunteer programs. These volunteers serve in a variety of capacities, including public relations, trip coordination, event planning, and fund-raising.

The NPCA also has significant political influence, supporting such proposals as the American Heritage Trust Act, which would "protect our historic and natural heritage" with funding provided by royalties from offshore **oil drilling**. Internships are offered by the NPCA to college students and graduates who want to work in the legislative process.

[*Linda M. Ross*]

RESOURCES

ORGANIZATIONS

National Parks and Conservation Association, 1300 19th Street, NW, Suite 300, Washington, D.C. USA 20036 Fax: (202) 659-0650, Toll Free: (800) 628-7275, Email: npca@npca.org, <http://www.eparks.org>

National pollutant discharge elimination system

The National Pollutant **Discharge** Elimination System (NPDES) seeks to end the discharge of all types of industrial, municipal, and agricultural waste from point sources into American waters. These point sources include industrial facilities and publicly owned treatment works (POTWs), as well as **runoff** from urban areas.

The NPDES was established by the Federal **Water Pollution** Control Act Amendments of 1972 and was expanded under the **Water Quality** Act of 1987. Most NPDES programs are administered by the states, under the direction of the Office of **Wastewater** Management of the U. S. **Environmental Protection Agency** (EPA).

Nationwide more than 400,000 facilities are required to obtain NPDES permits. NPDES permits usually license a facility to discharge a given amount of a pollutant under specific conditions, although permits also are granted for other types of waste disposal. Permits are required only for facilities that discharge directly into bodies of water. The NPDES National Pretreatment Program regulates indirect discharges, such as discharges to POTWs. Although the NPDES encompasses all of the nation's surface waters, including **wetlands** and seasonal streams, discharges into ground water that is not connected to surface water are not regulated unless required by the state. Individual permits are issued to a single facility for a maximum of five years. General permits license multiple facilities that have similar types of discharge within a specific geographical area. Most agricultural facilities, with the exception of concentrated-animal-feeding facilities, are exempt from NPDES regulations.

NPDES permits require that discharges be treated using the best technology available. If these technology-based standards do not protect the water, water-quality-based standards must be met instead. Permits may contain both technology-based limits, such as a nationwide limit on total suspended solids in a discharge, and water-quality-based limits, such as ammonia limitations based on aquatic toxicity. Permits issued by the EPA must meet state water-quality standards and local discharge limits. Permits include monitoring and reporting requirements and draft permits are open for public comment.

The NPDES initially focused on controlling the discharge of conventional pollutants such as suspended solids, oil and grease, and disease-causing organisms. By 1989 facilities were required to meet "Best **Conventional Pollutant** Control Technology" (BCT) limits, as well as "Best Available Technology Economically Achievable" (BAT) limits for the discharge of toxic and non-conventional pollutants. Toxic and priority pollutants include 126 metals, organic substances, and pollutant classes. Non-conventional pollutants include **nitrogen**, **phosphorus**, **chlorine**, and ammonia, as well as parameters such as whole **effluent** toxicity. BAT standards require facilities to utilize the best control and treatment methods that are achievable. Unlike earlier NPDES standards, the benefits do not have to be weighed against the costs of control and treatment.

NPDES permits typically require POTWs to provide physical separation and settling of wastes, biological treatment, and disinfection. The NPDES also regulates **sludge** from POTWs, as well as municipal separate **storm sewer** systems and combined sewer overflow systems. The latter discharge untreated effluent during storms. Polluted storm water discharges include runoff from land, streets, parking lots, and building roofs. Industrial storm-water discharges must meet both water-quality-based standards and the storm-water equivalents of BCT and BAT standards. By March 2003 NPDES storm water permits will be required for construction sites of at least one acre, as well as some smaller sites.

The NPDES has significantly reduced the discharge of pollutants into American waterways and has brought about major improvements in the quality of the nation's **water resources**. In 1972 only one-third of these waters were safe for swimming and fishing compared with two-thirds in 2001. Wetlands loss has been reduced from 460,000 acres annually to 70–90,000 acres per year. **Soil erosion** from agricultural runoff has been reduced by one billion tons annually. Nitrogen and phosphorous levels in waterways have decreased. Modern wastewater treatment facilities now serve 173 million people, compared to only 85 million in 1972. As of 2001, more than 50 categories of industry and hundreds of thousands of businesses, as well as more than 16,000 municipal **sewage treatment** systems, were in compliance with the NPDES standards.

However the remaining sources of water **pollution** are more difficult to regulate. The EPA estimates that over 20,000 river sections, lakes, and estuaries, including about 300,000 miles of shoreline, are polluted. The vast majority of the American population lives within 10 miles of these waters. The millions of smaller sources of pollution are varied and less amenable to available technological solutions. These include hundreds of thousands of storm-water sources and about 450,000 animal-feeding operations. Therefore the NPDES is focusing more on the pollution problems of individual watersheds, developing total maximum daily loads,

or pollution budgets, for up to 40,000 bodies of water and rivers segments.

[*Margaret Alic Ph.D.*]

Resources
Periodicals

Treadway, Elizabeth, Andrew Reese, and Douglas Noel. "EPA Finalizes Phase II Regs." *The American City and County* 115, no. 4 (March 2000): 44–9.

Other

Office of Wastewater Management. *Water Permitting 101.* U.S. Environmental Protection Agency. [cited May 25, 2002]. <www.epa.gov/npdes/pubs/101pape.htm>.

Office of Water. *Wastewater Primer.* U.S. Environmental Protection Agency. [cited May 26, 2002]. <www.epa.gov/npdes/pubs/primer.pdf>.

Water Permits Division, Office of Water. *Protecting the Nation's Waters Through Effective NPDES Permits: A Strategic Plan, FY 2001 and Beyond.* U. S. Environmental Protection Agency. June 2001 [cited May 25, 2002]. <www.epa.gov/npdes/pubs/strategicplan.pdf>.

Organizations

Friends of the Earth, 1025 Vermont Avenue, NW, Washington, DC, USA 20005, (202) 783-7400, Fax: (202) 783-0444 , Toll Free: (877) 843-8687, Email: foe@foe.org, <www.foe.org>

Office of Wastewater Management, U.S. Environmental Protection Agency, Mail Code 4201, 401 M Street, SW, Washington, DC, USA 20460, (202) 260-5922, Fax: (202) 260-6257, <www.epa.gov/npdes>

National Priorities List

The National Priorities List, or NPL, is a list compiled by the **Environmental Protection Agency** (EPA) as a part of the Superfund program under the Comprehensive Environmental Response, Compensation and Liability Act (CERCLA). CERCLA was enacted by Congress on December 11, 1980. This law levied a tax on the chemical and **petroleum** industries and gave the Federal government broad authority to respond directly to actual or threatened releases of hazardous substances. $1.6 billion was collected in the five years between 1980 and 1985. The tax money was placed in a trust fund for cleaning up abandoned or uncontrolled **hazardous waste** sites. CERCLA was amended by the Superfund Amendments and Reauthorization Act (SARA) on October 17, 1986.

CERCLA requires the EPA to identify hazardous waste sites on an annual basis throughout the United States. The law allows for two types of response: short-term removal of hazardous materials in emergency situations; and long-term remedial response actions. Long-term response actions can be conducted only at sites placed on the National Priorities List (NPL) by the EPA.

After in-depth testing and studies, the EPA places sites on the NPL in order of severity of contamination. Contamination is measured by the **Hazard Ranking System** (HRS), which yields a numerical score calculated on the basis of the site's risk factors. Risk factors are grouped into three categories: the site's potential for releasing hazardous substances into the **environment**; the amount and toxicity of the waste materials; and the number of people or sensitive environments affected by the hazardous materials. The HRS evaluates four possible pathways of contamination: ground water; surface water; **soil** exposure; and air **migration**.

HRS scores range between 0 and 100. Sites that score above 28.5 qualify for listing on the NPL. A site may also be listed if the **Agency for Toxic Substances and Disease Registry** (ATSDR) issues a health advisory for the site. In addition, each state or territory of the United States is allowed to place one top-priority site on the NPL without regard to its HRS score. The proposal is then published in the *Federal Register*. At this point, the public has an opportunity to comment in writing on the site's addition to the National Priorities List. As of May 29, 2002, 812 sites on the NPL were classified as having construction completed; 1221 sites remained on the NPL, with 74 new sites proposed for addition to the list.

The EPA administers the Superfund program in cooperation with individual states and Native American tribal governments. The Office of Emergency and Remedial Response (OERR) oversees management of the program. *See also* Hazardous waste site remediation; Hazardous waste siting; Superfund Amendments and Reauthorization Act

[*Rebecca J. Frey Ph.D.*]

Resources
Books

Probst, Katherine N., and David M. Konisky. *Superfunds Future: What Will It Cost?* Baltimore: Johns Hopkins University Press, 2001.

Other

United States Environmental Protection Agency. *Fact Sheets.* [cited July 2002]. <http://www.epa.gov/superfund/action/index.htm>.

Organizations

EPA primary contact: Carolyn Offutt, 1200 Pennsylvania Avenue, NW, Mail Code 5202G, Washington, DC USA 20460 (703) 603-8797, Email: offutt.carolyn@epa.gov, <http://www.epa.gov/superfund/contacts/index.htm>

National Recycling Coalition

Founded in 1978, the National **Recycling** Coalition (NRC) is a non-profit organization comprised of concerned individuals and environmental, labor, and business organizations who wish to promote the recovery and **reuse** of materials and energy. Believing that recycling is vital to the nation's

well-being, NRC encourages collection, the development of more processing venues, and the purchase and promotion of recycled materials.

Originally formed by a small group of recycling professionals to convince others that recycling provides many benefits to America's economy, NRC has since grown to 4,500 members. They participate in several different projects, including the Peer Match Program—in which NRC gives technical advice and direct assistance in recycling—and technical councils such as the Rural Recycling Council and the Minority Council. The most prominent of these councils is the Recycling Advisory Council (RAC)

The Coalition established the RAC in 1989 to build consensus on recycling issues. Partially funded by the U.S. **Environmental Protection Agency** (EPA), the mission of the RAC is to "examine the current status of recycling in the United States with the aim of recommending consensus public policies and private initiatives to increase recycling, consistent with the protection of public health and the environment." The NRC Board of Directors selects members of the RAC from among high-ranking environmental and public interest groups, recycling professionals, and business and industry. The RAC examines information on a specific recycling topic and then issues recommendations; both public and private organizations recognize the standards the RAC sets. Some topics about which the Council has made decisions include the definition of recycling; solid **waste management** costs; and policies and initiatives to promote the recycling of paper and **plastics**.

One of the most productive campaigns of the NRC has been the "Buy Recycled Business Alliance." Twenty-five major American businesses in cooperation with NRC have agreed to increase the number of their products and packages made with recycled materials. In conjunction with the Alliance, NRC launched the "Buy Recycled" campaign. Aimed at consumers, it promotes a market for recycled products by encouraging the patronage of companies that sell such products. NRC maintains that first, one must request and buy recycled products, then one must ask governments and businesses to buy recycled material, and third, one should help develop a lasting system for procuring recyclable material.

NRC sponsors an annual conference, the National Recycling Congress and Exposition. Regarded as the nation's most significant recycling event, this membership meeting exhibits the latest innovations in recycling and allows networking within the recycling industry. NRC also publishes a quarterly newsletter and the *National Policy on Recycling and Policy Resolutions*, thereby keeping members and the concerned public abreast of recycling matters.

[*Andrea Gacki*]

RESOURCES

ORGANIZATIONS

National Recycling Coalition, 1325 G Street NW, Suite 1025, Washington, D.C. USA 20005-3104 (202) 347-0450, Fax: (202) 347-0449, Email: info@nrc-recycle.org, <http://www.nrc-recycle.org>

National Research Council

The National Research Council (NRC) was organized by the **National Academy of Sciences** (NAS) in 1916 to connect the broad community of science and technology with the goals of the NAS, i.e., advancing the state of knowledge and advising the federal government. The NRC has become the principal agency of both the NAS and the National Academy of Engineering (NAE) in providing services to the government, the public, and the scientific and engineering communities. The NRC is administered by both Academies and the Institute of Medicine.

During the early years of World War I, there was concern about the lack of American preparedness if the United States should enter the war. The NAS had been active during the Civil War and for some years after the end of the war, but had become inactive during the latter part of the nineteenth century and into the first decade of the twentieth century. In 1915, George Ellery Hale, an astrophysicist and member of the NAS since 1902, advocated to the NAS president the offering of the services of the NAS to the U.S. Government in case of entry in the war. In April of 1916, Hale introduced a resolution to the NAS, which passed unanimously, that stated:

"That the President of the Academy be requested to inform the President of the United States that in the event of a break in diplomatic relations with any other country the academy desires to place itself at the disposal of the Government for any service within its scope."

President Woodrow Wilson agreed to the proposal, and a Committee for the Organization of the Scientific Resources of the Country for National Service was appointed. Members of the committee included some of the leading scientists of the time: physicist Robert A. Millikan, biologist Edwin G. Conklin, medical laboratory director Simon Flexner, and physical chemist Arthur A. Noyes. In June of 1916 the committee called for:

"A National Research Council, the purpose of which shall be to bring into cooperation government, educational, industrial, and other research organizations with the object of encouraging the investigation of natural phenomena, and increased use of scientific research in the development of American industries, the employment of scientific methods in strengthening the national defense, and such other applications of science as will promote the national security and welfare."

The plan was approved by the NAS and by President Wilson. The NRC was organized in September, 1916, with Hale as the first chairman. Members represented the government, various branches of the military, universities, and private research laboratories. Members were grouped into Divisions, which were organized according to disciplines or functions. Smaller units within the Divisions worked on specific projects and studies.

Although organized to meet a specific emergency, the NRC, due to its valuable wartime service, was made a permanent organization by an executive order of President Wilson in 1918.

Presently, the activities of the NRC are supervised by Major Program Units. The Major Program Units may conduct projects on the their own, but more frequently supervise projects organized under numerous subsidiary standing boards and committees. These boards and committees, which provide independent advice to the government, are comprising experts who serve without compensation. They provide draft reports on issues of concern and subject the draft reports to rigorous review before release to ensure quality and integrity. The composition and balance of the study committees are carefully monitored to avoid potential conflict of interest and bias. Major Program Units include: Policy Division; Center for Sciences, Mathematics, and Engineering Education, Commission on Behavioral and Social Sciences and Education; Commission on Engineering and Technical Systems; Commission on Geosciences, **Environment**, and Resources; Commission on Life Sciences; Commission on Physical Sciences, Mathematics, and Applications; Office of International Affairs, Office of Scientific and Engineering Personnel; Board on Agriculture; and **Transportation** Research Board.

The NRC also provides library services to the staffs and committee members of the NAS, NAE, and the Institute of Medicine as well as to the NRC.

[*Judith Sims*]

Resources

Organizations

The National Academies, National Research Council, 2101 Constitution Avenue, NW, Washington, D.C. USA 20418 (202) 334-2000, <http://www.nas.edu/nrc>

National seashore

The **National Park Service**, under the **U.S. Department of the Interior**, manages ten tracts of coastal land known as national seashores. Over 435 miles (700 km) of Atlantic, Gulf, and Pacific coastline, including over 592,800 acres (240,000 hectares) of beaches, **dunes**, sea cliffs, maritime forests, fresh ponds, marshes, and estuaries comprise the National Seashore System.

Protection of the sensitive natural habitats is only one of the objectives of the National Seashores System that the Park Service has established. These areas are also lightly developed for recreational purposes, including roads, administrative buildings, and some commercial businesses. In fact, until recently it was stipulated that public access must be provided to these areas. A third objective is to combat coastal **erosion**, as beaches and dunes are important as buffers to coastal storms.

The need to preserve coastal areas in their natural states was recognized as long ago as 1934, when the Park Service surveyed the Gulf and Atlantic coasts and identified 12 areas deserving of federal protection. The first of these to be authorized was Cape Hatteras National Seashore, a narrow strip of **barrier island** on the North Carolina outer banks. Acquiring the land, however, remained a problem until after World War II, when the Mellon Foundation matched state contributions and purchased the first of what is now over 100 mi (160 km) of beaches, dunes, marsh, and maritime forest.

In 1961, Cape Cod National Seashore in Massachusetts was the second area to be so designated. Protecting beach and dune areas of biological and geologically significance, this site was acquired with legislation that set the standard for future Park Service acquisitions. The "Cape Cod Formula" is the model for current regulation and purchase of private improved lands by the Park Service.

Five more sites were authorized between 1962 and 1966, including Fire Island on Long Island, New York, and Point Reyes National Seashore, the only national seashore on the west coast. Assateague Island, on the eastern shore of Maryland and Virginia, was deemed too developed to become a protected area but a nor'easter storm in March 1962 destroyed or seriously damaged nearly all of the development. By 1965, about 31 mi (50 km) of shoreline were purchased and became part of the National Seashore System.

In the 1970s, the final three national seashores were authorized. Gulf Islands is a non-continuous collection of estuarine, barrier island, and marsh habitats in Mississippi and Florida. It also includes an historic Spanish fort. Cumberland Island, Georgia, is the most "natural" of the ten national sea shores and is completely undeveloped with the only access being a public ferry from the mainland. The other National Seashores are Padre Island, in Texas; Cape Lookout, in North Carolina; and the newest national seashore, Florida's Cape Canaveral.

The National Historic Preservation Act (passed in 1966), the **Coastal Zone Management Act** (1972), and the National Seashore Act (1976) have been written to ensure that not all natural coastal areas fall to development. Public

recognition and subsequent congressional action have saved these few areas that are important as fish and shellfish spawning and nursery areas, bird and sea turtle nesting grounds, and refuges for threatened vegetation and **wildlife**. *See also* Wetlands

[*William G. Ambrose Jr.*]

Resources

Books

Kaufman, W., and O. Pilkey Jr. *The Beaches Are Moving: The Drowning of America's Shoreline.* Durham, NC: Duke University Press, 1983.

Mackintosh, B. *The National Historic Preservation Act and the National Park Service: A History.* Washington, DC: U.S. Department of the Interior, 1986.

Sutton, A., and M. Sutton. *Wilderness Areas of North America.* New York: Funk and Wagnalls, 1974.

National Weather Service

see **National Oceanic and Atmospheric Administration (NOAA)**

National Wildlife Federation

With over four million members or supporters, the National **Wildlife** Federation (NWF) is the country's largest and one of its most influential **conservation** organizations. The NWF has worked since 1936 to "promote wise use of the nation's wildlife and natural resources," and their primary goal is "to educate citizens about the need for sustainable use and proper management of our natural resources."

NWF sponsors National Wildlife Week, held during Earth Action Month, and it distributes over 620,000 education kits to more than 20 million students. NWF also operates the Institute for Wildlife Research, which concentrates on such creatures as bears, wild cats, and birds of prey. Their Backyard Wildlife **Habitat** Program distributes information and encourages homeowners to set up their own refuges for wild animals by providing food, water, and shelter, and by planting or preserving trees and shrubs and building backyard ponds. The program has certified more than 6,000 backyard habitats nationwide, and the group maintains a model backyard habitat at Laurel Ridge Conservation Center in Vienna, Virginia.

The National Wildlife Federation also sponsors a variety of nature-oriented meetings and programs. These include workshops and training sessions for grassroots leaders, children's wildlife camps, Earth Tomorrow pilot launched in Detroit, Michigan, to encourage conservation among high school students, and NatureQuest, a leadership training program for young people. Other educational outreach programs designed to develop next generation conservationists include NatureLink for children 9-17 and Campus Ecology for universities. The organization also offers conservation internships, allowing young people to work with their professional staff; outdoor vacations for adults and families; symposiums and lectures on urban wildlife and gardening at the Laurel Ridge Conservation Center; and the National Conservation Achievement Awards, recognizing outstanding work in the field. Other NWF activities include producing large-format films, television programs, a biweekly column called "The Backyard Naturalist," and publishing books on wildlife for adults and children.

NWF's magazines have a combined circulation of 1.8 million, and include the bimonthly publications *National Wildlife* and *International Wildlife* for adults, *Ranger Rick* for elementary school children, *Your Big Backyard* for preschoolers and *Wild Animal Baby* for toddlers. NWF also publishes *The Conservation Directory*, a comprehensive listing of regional, national, and international conservation groups, agencies, and officials, and *EnviroAction*, an environmental news digest covering the latest topics and legislative/regulatory updates.

Each year, NWF issues its Environmental Quality Index, assessing the status of air, water, wildlife, and other **natural resources**, as well as the quality of life in America. It also coordinates a nationwide **bald eagle** count each year and distributes millions of its famous wildlife stamps. NWF has been involved in a wide range of political campaigns. They have challenged government **strip mining** and timber cutting programs, worked to upgrade the **Environmental Protection Agency** (EPA) to cabinet-level status, helped strengthen enforcement of the **Safe Drinking Water Act**, and created the Corporate Conservation Council to encourage dialogue with industry on conservation issues.

Under the leadership of Mark Van Putten, the National Wildlife Federation differs from many other conservation groups in that the organization, and especially its 46 state affiliates, emphasizes "educating and empowering people to make a difference." This grassroots approach has led the NWF in such conservation efforts as returning gray wolves to **Yellowstone National Park** and Idaho, rallying support for the restoration of Florida's Everglades, imploring Congress to allow a portion of off-shore oil and gas leasing revenues to be used to fund environmental programs, publishing the "Higher Ground" report that illustrated how voluntary buyouts of flood-prone property would save tax dollars and restore natural habitat, and continuing to address global climate change and its effects on the world's animal and human populations.

[*Jacqueline L. Longe*]

RESOURCES
ORGANIZATIONS

National Wildlife Federation, 111000 Wildlife Center Dr., Reston, VA 20190. (703) 438-6000, (800) 822-9919, <info@nwf.org>, <http://www.nwf.org>.

National wildlife refuge

The United States began establishing **wildlife** refuges under **Theodore Roosevelt**. In 1903 he declared Pelican Island in Florida a refuge for the **brown pelican**, protecting a **species** that was close to **extinction**. In 1906 Congress closed all refuges to **hunting**, and in 1908, it established the National **Bison** Range refuge in Montana to protect that **endangered species**.

The refuge system continued to expand later in the century, with a primary emphasis on migratory waterfowl. In 1929 Congress passed the Migratory Bird Convention Act, followed by the Migratory Bird Hunting Stamp Act of 1934, which provided funding for waterfowl reserves. **Wetlands** now make up roughly 75 percent of the national wildlife refuges and serve as linked management units along the major waterfowl flyways.

The second focus in the expansion of the refuge system has been the need to protect endangered and **rare species**. Several acts have been passed for this purpose, going as far back as the Migratory Bird Treaty Act of 1918, but most recently the **Endangered Species Act** of 1973. For example, the **whooping crane** is protected by a series of national wildlife refuges in Texas, Oklahoma, and Kansas which are linked to **wildlife management** areas in Kansas and Nebraska, and these are linked to a national wildlife area in Saskatchewan and their primary breeding grounds in a Canadian **national park**.

A third recent focus has been on the addition of high-latitude **wilderness** in Alaska. In 1980 Congress passed the Alaska National Interest Conservation Lands Act which added roughly 54 million acres (22 million ha) of Alaskan land to the national **wildlife refuge** system.

The refuge system now consists of slightly over 400 refuges, which encompass 99 million acres (40 million ha) of land in parcels ranging from 2.5 acres (1 ha) to 17 million acres (7 million ha) in size. It is administered by the U.S. **Fish and Wildlife Service**, which was consolidated under the **U.S. Department of the Interior** in 1940. As it is officially written, the mandate is to provide, preserve, restore, and manage a national network of land and water for the benefit of society. The wording of the mandate has resulted in an open policy approach, and NWRs have been used for a variety of activities, some of which seem to many incompatible with the ideal of a refuge. For example, a consequence of having financial support from the sale of duck stamps has been the opening of refuges to waterfowl hunting. Other activities, such as **fertilizer runoff** and other **agricultural pollution** have seriously impacted the **ecological integrity** of several reserves in the western United States, including **Kesterson National Wildlife Refuge** in California's San Joaquin Valley.

Perhaps the most significant test of the government's ability to maintain the ecological integrity of the refuge system came over the fight to open the **Arctic National Wildlife Refuge** for oil exploration. This is the second largest reserve in the system, and perhaps the most spectacular in terms of wilderness values. Although the proposal had the support of two successive Presidents, Congress rejected it. Critics argued that it was inconsistent to maintain that oil production would benefit the country when it was earmarked for export. And the disaster caused by the ***Exxon Valdez*** **made oil companies' claims of environmental sensitivity seem to lack credibility.**

The general challenges now facing the refuge system arise both from pressure to develop in order to accommodate increased visitor demand and the vague criteria defined in the National Wildlife Refuge Administration Act of 1966 for multiple use. *See also* Forest Service; Migration; National Park Service; Oil drilling; Pinchot, Gifford; Wilderness Study Area; Wildlife management

[*David A. Duffus*]

RESOURCES
BOOKS

Reed, N. P., and D. Drabelle. *The United States Fish and Wildlife Service.* Boulder, CO: Westview Press, 1984.

PERIODICALS

Doherty, J. "Refuges on the Rocks." *Audubon* 85 (July 1983): 4, 6, 74–117.

"The Wildlife Refuges (National Wildlife Refuge System)." *Wilderness* 47 (Fall 1983): 2–38.

Native landscaping

Landscaping with native plants (also called indigenous plants) can help improve the **environment** in several ways. North America is home to thousands of indigenous plants that over many thousands of years have evolved and adapted to specific North American habitats such as prairies, deciduous forests, deserts, or coastal plains. However, many plants that are common today were brought to North America from different biogeographic areas by European and Asian immigrants. Some of the most common landscaping plants used in North America today, such as zinnias (*Zinnia elegans*) or Kentucky bluegrass (*Poa pratensis*), are not native to areas where they now thrive.

Plants brought from their native **habitat** to another **ecosystem** are called non-native or alien plants. For example, early colonists brought with them seeds from their homes in Europe to plant for food or use for landscaping. Likewise, some native North American plants, such as **tobacco** *Nicotiana* sp.), were taken back to Europe. Many of these plants thrived in their new homes.

While new plants were being imported, increased colonization altered the character of the land in North America. Forests were felled and prairies plowed under to convert large swatches of land to farming. Some of the newly introduced plants grew better under these altered conditions than did native North American plants. Some **species** of non-native plants begin to take over the habitat of native plants, forcing the native plants out. Sometimes this was intentional, as when a non-native plant was grown for food or planted for landscaping. Other times, it was accidental, as when a non-native plant used in landscaping escaped the garden and began to grow uncontrolled in the wild.

One way to maintain native plants is to set aside large **nature** preserves and protected parks where native plants can grow undisturbed. Native landscaping hopes to protect our plant heritage in another way, by planting native species in home gardens and landscaped areas such as the grounds around corporate headquarters, on college campuses, and in city parks. Native landscaping in many cases saves water, reduces **pesticide** use, and maintains unique systems of indigenous plants and animals.

Landscaping around urban and suburban homes usually employs only several dozen kinds of plants, from turf grass to evergreens for foundation plantings to decorative flowers. This typical landscape supports little variety in insect and animal life. Lawns and evergreen hedges offer little protective cover for small mammals, and streams or ponds that provide water and a home for animals are often diverted or drained to make way for development. Gardeners may plant beds of decorative annual flowers, but these are usually not adequate for feeding bees, butterflies, and other insects that depend on nectar.

Native plants, on the other hand, have evolved in complex communities. Often the land can support a greater variety of native plant species. In some cases these plant communities naturally enrich the **soil** to make it fertile enough to support still more species. These communities shelter a variety of animals and insects that in turn help plant species thrive by pollinating the plants and distributing their seeds. The plants and animals are tied together in ecosystems where all have evolved to live with specific natural conditions (for example, high soil acidity, low rainfall, or high mineral content of the soil).

The number of different species of plants and animals that form an ecological community determines the **biodiversity** of that community. Many urban and suburban landscapes have low biodiversity, because they support only a few different species. Even landscaped lawns and gardens that support a high level of biodiversity are considered of low value by ecologists when the plants are almost entirely alien.

Native landscaping helps to develop biocommunities that improve the **ecological integrity** of the land. The biodiversity of the area returns more closely to its original state. This is considered a desirable goal by ecologists. Often when native plants are grown in their natural habitat, they require fewer resources (water, **fertilizer**) than alien species because they have evolved to suit the natural conditions of the area. For example, the turf grass that covers most American yards evolved in the cool, rainy summers of Northern Europe. Consequently, in many parts of the United States there is inadequate natural rainfall in the summer for this grass to thrive and, people must water their lawns to keep the non-native grass green and healthy. Alien flowers and non-native ornamental trees may also be more susceptible to disease and pests. Native landscaping is a valuable way to cut back on the use of chemical pesticides and fertilizers, as well as water, because native plants in the proper site can support themselves without this supplemental care.

In some cases, plants introduced from other places have become pests, growing abundantly and choking off native growth. These plants are called invasive plants, because they invade or take over the habitat of native plants. One example is **purple loosestrife** (*Lythrum salicaria*), an ornamental plant brought from Europe in the 1800s. It has grown so well in some areas of North America that it is now a significant threat to the biodiversity of wetland habitats. Alien insects that arrive on introduced plants can also threaten native plants with **extinction**. Native landscapers help root out invasive plants and restore the original plant communities.

The trend toward landscaping with native plants emerged in the United States in the 1980s and 1990s. Some communities (for example in the San Francisco Bay area) now have local native plant societies that hold annual sales of native plants to help the gardener interested in native landscaping. The first step towards native landscaping is usually to reduce the amount of lawn. Some gardeners have replaced traditional turf lawn with a meadow or **prairie** planting of native grasses. Some environments, such as **desert**, do not naturally support any kind of grass. Native landscapers in these regions tend to favor drought-tolerant perennials. In dry areas, landscaping with native plants can be an important element in landscaping for **water conservation** (called xeriscaping).

Another move toward native landscaping is to border a turf lawn with plantings of native perennials, shrubs, and trees. Many native plants are quite ornamental, but the native

landscaper may also select plants based on kind of ecological community they support. Native planting might include shrubs that bear fruits eaten by local birds and mammals, early spring wildflowers that provide nectar for insects just emerging from winter dormancy, or nut trees to feed squirrels through the winter. A native landscape might also include flowers that will attract local butterflies. The resulting landscape may prove to easier and cheaper to maintain than a traditional planting, need few or no pesticides or fertilizers, use less water, and provide food and habitat for a diverse animal population.

[*Angela Woodward*]

RESOURCES
BOOKS

Stein, Sara. *Noah's Garden.* New York: Houghton Mifflin, 1993.

Stein, Sara. *Planting Noah's Garden.* New York: Houghton Mifflin, 1997.

Wasowski, Andy, and Sally Wasowski. *The Landscaping Revolution.* Chicago: Contemporary Books, 2000.

PERIODICALS

"Going Native: Biodiversity in Our Own Backyards." *Brooklyn Botanic Garden Handbook* 140 (1994).

"Native Perennials." *Brooklyn Botanic Garden Handbook* 146 (1996).

Orecklin, Michele, and David Schwartz. "Say Goodbye to Grass" *Time* (July 2, 2001): 56.

OTHER

Environmental Protection Agency. *Landscaping with Native Plants.* January 24, 2000 [cited July 2002]. <http://www.epa.gov/glnpo/native plants/index.html>.

Natural gas

Naturally occurring gas that primarily contains **methane**, CH_4. A fossil fuel like **coal** and **petroleum**, it is the hottest and cleanest burning of these fuels and is increasingly touted as a substitute for petroleum. It is, however, an explosion hazard in coal mines and an unwanted byproduct of **anaerobic digestion** in landfills. It is most commonly found with petroleum or at depths which vaporize oil. Use is hampered by **transportation** difficulties and risks, and thus it is limited to pipelines and compression tanks. Huge quantities exist as unconventional sources, including **coal gasification** and conversion to **methanol**. *See also* Liquified natural gas

Natural radiation

see **Background radiation**

Natural resource accounting

see **Environmental accounting**

Natural resources

Natural resources, unlike man-made resources, exist independently of human labor. Natural resources can be viewed as an endowment or a gift to humankind. These resources are, however, not unlimited and must be used with care. Some natural resources are called "fund resources" because they can be exhausted through use, like the burning of **fossil fuels**. Other fund resources such as metals can be dissipated or wasted if they are discarded instead of being reused or recycled. Some natural resources can be used up like fund resources, but they can renew themselves if they're not completely destroyed. Examples of the latter would include the **soil**, forests, and fisheries.

Because of **population growth** and a rising standard of living, the demand for natural resources is steadily increasing. For example, the rising demand for minerals, if continued, will deplete the known and expected reserves within the coming decades.

The world's industrialized nations are consuming **nonrenewable resources** at an accelerating pace, with the United States ranking first on a per capita basis. With only 5% of the global population, Americans consumes 30% of the world's resources. Because of their tremendous demand for goods, Americans have also created more waste than is generated by any other country. The **environment** in the United States has been degraded with an ever-increasing volume and variety of contaminants. In particular, a complex of synthetic **chemicals** with a vast potential for harmful effects on human health has been created. The long-term effects of a low dosage of many of these chemicals in our environment will not be known for decades. The three most important causes for global environmental problems today are population growth, excessive resource consumption, and high levels of **pollution**. All of these threaten the natural resource base.

[*Terence H. Cooper*]

RESOURCES
BOOKS

Craig, J. R. *Resources of the Earth.* Englewood Cliffs, NJ: Prentice-Hall, 1988.

Meadows, D. H., et al. *The Limits to Growth.* New York: Universe Books, 1972.

Simmons, I. G. *Earth, Air, and Water: Resources and Environment in the Late 20th Century.* London: Edward Arnold, 1991.

OTHER

World Resources, 1990–91: A Report. New York: Oxford University Press, 1990.

Natural Resources Defense Council

Founded in 1970, the Natural Resources Defense Council (NRDC) is a major national environmental organization active in most areas of concern to environmentalists and conservationists. It boasts a membership of 500,000 nationwide and had an annual budget in 2001 of nearly $41 million dollars. These resources are employed in the organization's quest to use science and law to create "a world in which human beings live in harmony with our environment." NRDC seeks a sustainable society and believes that economic growth can continue only if it does not destroy the natural resources that fuel it. NRDC's stated goals include access to pure air and water and safe food for every human being. NRDC's philosophy is based increasingly on a belief in the inherent sanctity of the natural **environment**. As its mission statement proclaims, NRDC ultimately "strives to help create a new way of life for humankind, one that can be sustained indefinitely without fouling or depleting the resources that support all life on Earth." The organization is headquartered in New York City, with regional offices in Washington, D.C., San Francisco, and Los Angeles.

Generally regarded as an influential and highly effective mainstream environmental organization, NRDC's successes have often come from its expertise in lobbying governmental officials, helping to draft environmental laws, and working with public utility officials to reduce wasteful and environmentally destructive practices. NRDC has also made considerable use of the judicial system in defense of environmental causes and retains on its staff some of the nation's top environmental lawyers.

Among its many notable achievements, NRDC has been instrumental in setting national standards for building and appliance efficiency. It collaborated with over a dozen states and Canadian provinces on the reform of **electric utilities** and helped the Soviet Union design an electrical planning program in the late 1980s. NRDC has helped protect dozens of national forests from abusive **logging** practices, including forests in the Greater Yellowstone **ecosystem** and old-growth forests in the Sierra Nevada. In addition, NRDC led the fight for the national phase-out of **lead** from **gasoline** and lobbied intensively for passage of the 1990 **Clean Air Act**. It was a sponsor of the Nuclear Non-Proliferation Act and helped prove that an underground **nuclear test ban** treaty could be verified.

The NRDC is active in several areas, including Air and Energy; Global Warming; Clean Water and Oceans; Wildlife and Fish; Parks, Forests, and Wildlands; Toxic **Chemicals** and Health; **Nuclear Weapons** and Waste; Cities and Green Living; and Environmental Legislation. Its ongoing projects are numerous and include the continuing protection of the **Arctic National Wildlife Refuge** and California coastal areas from **oil drilling**. NRDC continues to monitor the policies and practices of the **Environmental Protection Agency** (EPA) and has convinced the EPA to strengthen its rules regarding the reporting of releases of toxic substances into the environment. It has lobbied the government to designate new funds for the protection of **endangered species**; to include environmental standards and controls in the **North American Free Trade Agreement**; to close loopholes in proposed rules for controlling **acid rain**; and to strengthen **automobile pollution** standards, as well as on many other environmental issues. In addition, the organization publishes a quarterly magazine of environmental thought and opinion, entitled *OnEarth* (formerly the *Amicus Journal*).

Internationally, NRDC seeks the eventual elimination of nuclear weapons, the protection of **habitat** and ecosystems in developing countries, and the worldwide phaseout of **chlorofluorocarbons**. It promotes alternative agricultural techniques to reduce the use of pesticides, the strengthening of regulations governing the disposal of waste in landfills, and the testing and treatment of poor children for lead **poisoning**. NRDC is urging states and municipalities to stop the **ocean dumping** of sewage **sludge** and encouraging **mass transit** programs as an alternative to automobiles and new highways.

The NRDC urged President Bill Clinton to create six new National Monuments and protect 4 million new acres from development in the United States. By 2002, the NRDC had initiated a "BioGems Defender" program, which relies upon the e-mails of 250,000 members to pressure lawmakers to protect endangered areas or **species**. With the help of actor Robert Redford, who e-mailed letters on behalf of the BioGems program, the NRDC helped protect the Arctic **National Wildlife Refuge** (ANWR) from oil drilling in 2002. The NRDC also made headlines in 2002 when it successfully sued the George W. Bush Administration, under the Freedom of Information Act, to release documents produced by the National **Energy Policy** Development Group concerning U.S. energy policies.

[*Lawrence J Biskowski and Douglas Dupler*]

Resources

Organizations

Natural Resources Defense Council, 40 W 20th Street, New York, NY 10011 (212) 727-2700, Fax: (212) 727-1773, Email: nrdcinfo@nrdc.org, <www.nrdc.org/>

Natural selection

see **Evolution**

Nature

The word nature stems from the Latin *natura*, whose meaning ranges from "birth" to "the order of things." In English, nature comprises all plants, animals, and ecosystems, as well as the biological and nonbiological materials and processes of our planet. This range of meaning narrows if we consider the use of the word *natural*. Something is natural if it is not artificial, if it pertains to or comes from the natural world. When we speak of bears, mountains, or **evolution** and say, "they are natural," we mean they are neither human nor created by humans. Thus the concept of nature is often restricted to beings, things, or processes which are not human in origin.

Conceptual and empirical inquiries into nature loosely center on eight questions. The first question is a scientific one, which simply asks what constitutes nature: What are its essential properties and how shall we classify them? The second seeks purpose in nature, asking whether the earth is a designed home for humankind or an accident of cosmic history. A third question explores the effect of the natural **environment** on human physiology, psychology and culture, examining whether nature determines who we are or constrains what we can do. A fourth examines the metaphors used to understand nature: Is nature an organism subject to death or a machine of fungible parts? The fifth question studies how nature changes: Is it dynamic, changing as a result of internal processes, or static, changing in response to external human disturbances? A sixth surveys how people have transformed the natural world, and the seventh queries whether human beings and their societies are part of or separate from nature. The final question investigates whether nature has a structure of intrinsic moral values: Does nature have values or a good of its own, and if so, should human beings respect these values or promote this good?

We lack unambiguous answers to these questions, though the questions are important in themselves, for they help us clarify our assumptions about nature. Humans, for example, have emerged from the evolutionary processes of nature. Thus our thoughts and actions are certainly natural, yet our societies and technologies are unique in the natural world, and the scope of our impact on nature is without precedent in other **species**. It seems that while we are part of nature biologically, we are separate from the rest of nature in our social and technological characteristics, and to declare humanity natural or non-natural is unhelpful. The best alternative may be to accept the ambiguity of our relationship to nature, conceiving of humankind as relatively natural and non-natural depending on the time, place, and activity. *See also* Bioregionalism; Deep ecology; Ecosophy; Environmental attitudes/values; Environmental ethics; Environmentalism; Humanism; Speciesism

[*William S. Lynn*]

Resources
Books

Collingwood, R. G. *The Idea of Nature.* Oxford, UK: Oxford University Press, 1945.

Glacken, C. J. *Traces on the Rhodian Shore: Nature and Culture in Western Thought From Ancient Times to the End of the Eighteenth Century.* Berkeley: University of California Press, 1967.

Lyon, T. J., ed. *This Incomparable Lande: A Book of American Nature Writing.* New York: Houghton Mifflin, 1989.

Lyon, T. J., and P. Stine, eds. *On Nature's Terms.* College Station, TX: Texas A & M University Press, 1992.

Nash, R. F. *The Rights of Nature.* Madison: University of Wisconsin, 1989.

Williams, R. "Nature." In *Keywords: A Vocabulary of Culture and Society.* New York: Oxford University Press, 1985.

Nature Conservancy, The

The Nature Conservancy is America's fourth-largest environmental organization, with 1 million members and a budget of over $300 million. It owns or helps manage over 1,400 nature preserves worldwide, covering more than 12 million acres (4.8 million ha) (2 million acres [809,000 ha] of which is owned outright by the organization) in the United States alone. It is the largest private system of nature sanctuaries in the world, helping like-minded partner organizations to preserve more than 80 million acres (3.2 million ha) (7 million acres [2.8 million ha] of which The Nature Conservancy manages directly) in the Asia Pacific, Canada, the Caribbean and Latin America.

The organization is dedicated to protecting nature and its animal inhabitants by preserving their habitats. This is accomplished with a strategic, science-based planning process, called Conservation by Design. Conservation by Design helps The Nature Conservancy identify the highest-priority places (landscapes and seascapes) that, if conserved, promise to ensure **biodiversity** (the variety of life that can be found on Earth—plants, animals, **fungi**, and micro-organisms—as well as to the communities that they form and the habitats in which they live) over the long term. Conservation is achieved by the organization through: buying and managing ecologically important areas; negotiating agreements to manage areas without holding title to them, partnerships or **conservation easements**; offering training to partner organizations and government agencies that manage land; educating people who live in ecologically sensitive areas about the importance of biodiversity and helping them to live in better harmony with the natural **environment**; working with resource-based industries to alter their business practices to

have less environmental impact; and helping government agencies work together and better allocate public resources toward conservation.

The organization was incorporated in 1951 by American Richard Pough, who had been inspired by the British government's nature preserves of the same name. Unlike the English version, Pough's organization was funded privately. The Conservancy's early years were difficult financially, but Pough eventually secured two major endowments that made the society solvent: a $100,000 donation from Mrs. DeWitt Wallace, co-owner of *Readers Digest*, and a $55-million bequest from Minnesota Mining heiress Katharine Ordway. Most of the latter money went for land purchases. Subsequent purchases have acquired millions of acres, although some tracts are later traded or sold to the government, individuals, or other environmental organizations.

One example of The Nature Conservancy's international programs is the *Debt for Nature Swaps* that have been contracted in Central America. These allow The Nature Conservancy, in conjunction with the Department of Defense, to conduct ecological inventories and supervise over 25 million acres (10 million ha). Other projects, such as Parks in Peril, have given The Conservancy charge of over 28 million acres (11.3 million ha) in Latin America and the Caribbean. In the United States, a partnership with **Ducks Unlimited** restored riparian forests and **wetlands** in several states. A three-year program called The Rivers of the Rockies, spent $5 million to preserve 20 sites along Colorado rivers and tributaries. Similar projects are underway in other states as well.

The Nature Conservancy cooperates with federal land and **wildlife** agencies, as well as other environmental associations, and is instrumental in helping to identify land of critical environmental interest. In addition to its local and foreign activities in land preservation, the Nature Conservancy identifies and protects animal life, especially the "rarest of the rare," those **species** most in need of preservation.

In November, 2001, the Nature Conservancy and the United States **Forest Service** signed a Memorandum of Understanding (MOU) to promote cooperation and partnership as the two organizations work to preserve, protect, and monitor forests and grassland ecosystems. The MOU is in effect for five years, during which The Nature Conservancy and the Forest Service will work together to, among other things, protect migratory bird **habitat**, prevent the spread of invasive species, and to carry out prescribed burns in natural areas to promote biodiversity and **ecosystem health**. The two organizations will also jointly establish outreach, educational, and training programs for communities, and will promote partnerships and liaisons with foreign and local governments, tribal interests, landowners, and non-government agencies.

In response to its mission statement, which is "to preserve the plants, animals and natural communities that represent the diversity of life on Earth by protecting the lands and waters they need to survive." The Nature Conservancy instigated its "$1 Billion Campaign," which commits the organization to the investment of one billion dollars to save 200 of "the world's last great places,". As of the middle of 2002, more than $960 million had been raised toward this goal. The organization maintains its Heritage database at its headquarters in Arlington, Virginia, in which **endangered species** are ranked by status both globally and within each state.

[*Marie H. Bundy*]

Organizations

The Nature Conservancy, 1815 N. Lynn Street, Arlington, VA USA 22209, <http://www.tnc.org>

Nature reserve

see **Biosphere reserve**

Dr. Scott Nearing (1883 – 1983)

American conservationist

A prolific and iconoclastic writer, a socialist, and a conservationist, Scott Nearing—along with his wife Helen—is now considered the "great-grandparent" of the back-to-the-land movement. Nearing was born August 6, 1883, in Morris Run, Pennsylvania. He is the author of nearly fifty books and thousands of pamphlets and articles, some of which have become classics of modern **environmentalism**.

Originally a law student at the University of Pennsylvania, Nearing received his B.S. from that university in 1905 and a Ph.D. in economics in 1909. He was the secretary of the Pennsylvania Child Labor Commission from 1905 to 1907. He taught economics at the Wharton School, at Swarthmore College, and at the University of Toledo in Toledo, Ohio. In 1915, he was dismissed from his position at the Wharton School for his politics, particularly his position on child labor. A pacifist, he was arrested during World War I and charged with obstructing recruitment for the armed forces. He was a Socialist candidate for United States Congress in 1919. His socialism, his pacifism, and his politics in general made it increasingly difficult for him to find and maintain teaching positions after 1917, and he supported himself for the rest of his life by writing, lecturing, and farming.

In 1932, Nearing and his wife bought a small, dilapidated farm in Vermont, hoping, in their words, "to live

sanely and simply in a troubled world." They restored the land and removed or rebuilt the buildings there. They farmed organically and without machinery, feeding themselves from their garden, and making and selling maple sugar. They wrote about their experiences in *The Maple Sugar Book* (1950) and *Living the Good Life* (1954). Their small farm became a mecca for people seeking both to simplify their lives and to live in harmony with **nature**, as well as each other.

In his 1972 autobiography, *The Making of a Radical,* Nearing wrote that the attempt "to live simply and inexpensively in an affluent society dedicated to extravagance and waste" was the most difficult project he had ever undertaken, but also the most physically and spiritually rewarding. He defined the good life he was striving towards as life stripped to its essentials—a life devoted to labor and learning, to doing no harm to humans or animals, and to respecting the land. As he wrote in his autobiography, he tried to live by three basic principles: "To learn the truth, to teach the truth, and to help build the truth into the life of the community."

Confronted by the growth of the recreational industry in Vermont during the 1950s, particularly the construction of a ski resort near them, the Nearings sold their property and moved their farm to Harborside, Maine, where Scott Nearing died on August 24, 1983, just over two weeks after his one-hundredth birthday.

Reviled as a radical for most of his life, Nearing's critique of the wastefulness of modern consumer society, his emphasis on smallness, self-reliance, and the restoration of ravaged land, as well as his **vegetarianism**, have struck a responsive chord in many Americans since the late 1960s. He is considered by some as the representative of twentieth century American counterculture, and he has inspired activists throughout the environmental movement, from the editors of *The Whole Earth Catalog* and the founders of the first **Earth Day** to proponents of appropriate technology. He was made an honorary professor emeritus of economics at the Wharton School in 1973.

[*Terence Ball*]

Resources
Books

Nearing, S. *The Making of a Radical: A Political Autobiography.* New York: Harper and Row, 1972.

———, and H. Nearing. *Living the Good Life: How to Live Sanely and Simply in a Troubled World.* Reissued with an introduction by Paul Goodman. New York: Schocken Books, 1970.

———, and H. Nearing. *The Maple Sugar Book.* New York: Schocken Books, 1971.

Nekton

Nekton are aquatic animals that swim or move freely in the water. Their movement is generally not controlled by waves and currents. Nekton include fish, squid, marine mammals, and marine reptiles. They live in the sea, lakes, rivers, ponds, and other bodies of water.

Fish are a major segment of the nektonic animals, with approximately 14,500 kinds of fish living in the ocean. Many nekton live near the ocean surface because food is abundant there. Other nekton live in the deep ocean.

Most nekton are chordates, animals with bones or cartilage. This category of nekton includes **whales**, **sharks**, bony fish, turtles, snakes, eels, **dolphins**, porpoises, and **seals**.

Molluscan nekton like squid and octopus are invertebrates, animals with no bones. Squid, octopus, clams, and oysters are mollusks. However, molluscan nekton have no outer shells.

Molluscan nekton like squid and octopus are invertebrates, animals with no bones. Squid, octopus, clams, and oysters are mollusks. However, molluscan nekton have no outer shells.

Arthopod nekton are invertebrates like shrimp. Many arthropod live on the ocean floor.

Most nektonic mammals live only in water. However, walruses, seals, and sea otters can exist on land for a time. Other nekton mammals include **manatees** and dugongs, whale-like animals that live in the Indian Ocean.

[*Liz Swain*]

Nematicide

see **Pesticide**

Neoplasm

Neoplasm results in the formation of both benign and, more particularly, malignant tumors or cancers. A neoplasm is a mass of new cells, which proliferate without control and serve no useful function. This lack of control is particularly marked in malignant tumors (**cancer**). The difference between a benign tumor and a malignant one is that the former remains at its site of origin while the latter acquires the ability to escape from its original location, migrate to another place, and invade and colonize other tissues and organs. This results in the death of the surrounding cells as the neoplasm grows rapidly.

Neotropical migrants

Birds that migrate each year between the American tropics and higher latitudes, especially in North America, are known as neotropical migrants. So called because they migrate from the tropics of the "new world" (the Western hemisphere), neotropical migrants are a topic of concern because many **species** have been declining in recent years, and because they are vulnerable to **habitat** destruction and other hazards in both winter and summer ranges. One survey of 62 forest-dwelling migrant bird species found that 44 declined significantly from 1978–1987. Migrants' dependence on nesting and feeding habitat in multiple regions, usually in different countries, makes **conservation** difficult, since habitat protection often requires cooperation between multiple countries. In addition, it is often unclear whose fault it is when populations fall. Most of the outcry over neotropical migrants has been raised in the United States and Canada, where summer bird populations have thinned noticeably in recent decades. Conservationists in northern latitudes tend to attribute bird disappearances to the destruction of tropical forests and other winter habitat. In response, governments in Mexico, Central America, and South America argue that loss of suitable summer nesting habitat, as well as summer feeding and shelter requirements, are responsible for the decline of migratory birds.

About 250 species of birds breed in North America and winter in the south. In the southern and western United States 50-60% of breeding birds are migrants, a number that rises to 80% in southern Canada and to 90% in the Canadian sub-arctic. Among the familiar migrants are species of orioles, hummingbirds, sandpipers and other wading birds, herons and bitterns, flycatchers, swallows, and almost all warblers. Half of these winter in Mexico, the Bahamas, and the Greater Antilles islands. Some 30 species winter as far south as the western **Amazon Basin** and the foothills of the Andes Mountains, where forest clearance poses a significant threat. Species most vulnerable to habitat loss may be those that require large areas of continuous habitat, especially the small woodland birds, and aquatic birds whose wetland habitats are being drained, filled, or contaminated in both summer and winter ranges.

In addition to habitat loss, neotropical migrants suffer from agricultural pesticides, which poison both food sources and the birds themselves. The use of pesticides, including such persistent **chemicals** as **dichlorodiphenyl-trichloroethane** (DDT), has risen in Latin America during the same period that **logging** has decimated forest habitat. Furthermore, some ecologists argue that migrants are especially vulnerable in their winter habitat because winter ranges tend to be geographically more restricted than summer ranges. Sometimes an entire population winters in just a few islands, swamps, or bays. Any habitat damage, **pesticide** use, or predation could impact a large part of the population if the birds concentrate in a small area.

Population declines are by no means limited to loss of wintering grounds. An estimated 40% of North America's eastern deciduous forests, the primary breeding grounds of many migrants, have either been cleared or fragmented. **Habitat fragmentation** is the term used to describe breaking up a patch of habitat into small, dispersed units, or dissecting a patch of habitat with roads or suburban developments. This problem is especially severe on the outskirts of urban areas, where suburbs continue to cut into the surrounding countryside. For reasons not fully understood, neotropical migrants appear to be more susceptible to habitat fragmentation than short-distance migrants or species that remain in residence year round. Hazards of fragmentation include nesting failure due to predation (often by raccoons, snakes, crows, or jays), nesting failure due to **competition** with human-adapted species such as starlings and English sparrows, and **mortality** due to domestic house cats. (An Australian study of house cat predation estimated that 500,000 cats in the state of Victoria had killed 13 million small birds and mammals, including 67 different native bird species.) In addition to hazards in their summer and winter ranges, threats along **migration** routes can impact migratory bird populations. Loss of stop-over **wetlands**, forests, and **grasslands** can reduce food sources and protective cover along migration routes. Sometimes birds are even forced to find alternative migration paths.

European-African migrants suffer similar fates and risks to those of American migrants. In addition to wetland **drainage**, habitat fragmentation, and increased pesticide use, Europeans and Africans also continue to hunt their migratory birds, a hazard that probably impacts American migrants but that is little documented here. It is estimated that in Italy alone, some 50 million songbirds are killed each year as epicurean delicacies. **Hunting** is also practiced in Spain and France, as well as in African countries where people truly may be short of food.

One of the principal sources of data on bird population changes is the North American Breeding Bird Survey, an annual survey conducted by volunteers who traverse a total of 3,000 established transects each year and report the number of breeding birds observed. While this record is by no means complete, it provides the best available approximation of general trends. Not all birds are disappearing. Some have even increased slightly, as a consequence of reduced exposure to DDT, a pesticide outlawed in the United States because it poisons birds, and because of habitat restoration efforts. However the BBS has documented significant and troubling declines in dozens of species.

Declines in migrant bird populations alarm many people because birds are often seen as indicators of more general **ecosystem health**. Birds are highly visible and their disappearance is noticeable, but they may indicate the simultaneous declines of insects, amphibians, fish, and other less visible groups using the same habitat areas.

[*Mary Ann Cunningham Ph.D.*]

RESOURCES
BOOKS

Terbogh, J. *Where Have All the Birds Gone?* Princeton, NJ: Princeton University Press, 1989.

PERIODICALS

Friesen, L., P. F. Eagles, and R. J. Mackay. "Effects of Residential Development on Forest-Dwelling Neotropical Migrant Songbirds," *Conservation Biology* 9, no. 6 (1995): 1408–14.

Keast, A. "The Nearctic-Neotropical Bird Migration System." *Israel Journal of Zoology* 4, no. 1 (1995): 455–70.

"Neotropical Migratory Birds." *Conservation Biology* 7, no. 3 (1993): 501–09.

Youth, H. "Flying Into Trouble" *Worldwatch* 7 (1994): 110–17.

Neritic zone

The portion of the marine **ecosystem** that overlies the world's continental shelves. This subdivision of the **pelagic zone** includes some of the ocean's most productive water. This productivity supports a food chain culminating in an abundance of commercially important fish and shellfish and is largely a consequence of abundant sunlight and nutrients. In the shallow water of the neritic zone, the entire water column may receive sufficient sunlight for **photosynthesis**. Nutrients are delivered to the area by terrestrial **runoff** and through resuspension from the bottom by waves and currents. Runoff and dumping of waste also deliver pollutants to the neritic zone, and this area is consequently among the ocean's most polluted.

Neurotoxin

Neurotoxins are a special class of metabolic poisons that attack nerve cells. Disruption of the nervous system as a result of exposure to neurotoxins is usually quick and destructive. Neurotoxins are categorized according to the nature of their impact on the nervous system. Anesthetics (ether, chloroform, halothane), **chlorinated hydrocarbons** (DDT, Dieldrin, Aldrin), and **heavy metals** (**lead**, **mercury**) disrupt the **ion** transport across cell membranes essential for nerve action. Common pesticides, including carbamates such as Sevin, Zeneb and Maneb and the organophosphates such as Malathion and Parathion, inhibit acetylcholinesterase, an **enzyme** that regulates nerve signal transmission between nerve cells and the organs and tissues they innervate.

Environmental exposure to neurotoxins can occur through a variety of mechanisms. These include improper use, improper storage or disposal, occupational use, and accidental spills during distribution or application. Since the identification and ramifications of all neurotoxins are not fully known, there is risk of exposure associated with this lack of knowledge.

Cell damage associated with the introduction of neurotoxins occurs through direct contact with the chemical or a loss of oxygen to the cell. This results in damage to cellular components, especially in those required for the synthesis of protein and other cell components.

The symptoms associated with **pesticide poisoning** include eye and skin irritation, miosis, blurred vision, headache, anorexia, nausea, vomiting, increased sweating, increased salivation, diarrhea, abdominal pain, slight bradycardia, ataxia, muscle weakness and twitching, and generalized weakness of respiratory muscles. Symptoms associated with poisoning of the central nervous system include giddiness, anxiety, insomnia, drowsiness, difficulty concentrating, poor recall, confusion, slurred speech, convulsions, coma with the absence of reflexes, depression of respiratory and circulatory centers, and fall in blood pressure.

The link between environmental neurotoxin exposure and neuromuscular and brain dysfunction has recently been identified. Physiological symptoms of Alzheimer's disease, amyotrophic lateral sclerosis (ALS, or Lou Gehrig's disease), and lathyrism have been identified in populations exposed to substances containing known neurotoxins. For example studies have shown that heroin addicts who used synthetic heroin contaminated with methylphenyltetrahydropyridine developed a condition which manifests symptoms identical to those associated with Parkinson disease. On the island of Guam, the natives who incorporate the seeds of the false sago plant (*Cycas circinalis*) into their diet develop a condition very similar to ALS. The development of this condition has been associated with the specific nonprotein amino **acid**, B methylamino-1-alanine, present in the seeds.

[*Brian R. Barthel*]

RESOURCES
BOOKS

Aldrich, T., and J. Griffith. *Environmental Epidemiology.* New York: Van Nostrand Reinhold, 1993.

PERIODICALS

Griffith, J., R.C. Duncan, and J. Konefal. "Pesticide Poisonings Reported By Florida Citrus Field Workers." *Environmental Science and Health* 6 (1985): 701–27.

OTHER

Agency for Toxic Substances and Disease Registry, Annual Report 1989 and 1990. Atlanta, GA: U.S. Department of Health and Human Services.

Neutron

A subatomic particle with a mass of a proton and no electric charge. It is found in the nuclei of all atoms except hydrogen-1. A free neutron is unstable and decays to form a proton and an electron with a half life of 22 minutes. Because they have no electric charge, neutrons easily penetrate matter, including human tissue. They constitute, therefore, a serious health hazard. When a neutron strikes certain nuclei, such as that of uranium-235, it fissions those nuclei, producing additional neutrons in the process. The **chain reaction** thus initiated is the basis for **nuclear weapons** and **nuclear power**. *See also* Nuclear fission; Radioactivity

Nevada Test Site

The Nevada Test Site (NTS) is one of two locations (the South Pacific being the other) at which the United States has conducted the majority of its **nuclear weapons** tests. The site was chosen for weapons testing in December 1950 by President Harry S. Truman and originally named the Nevada Proving Ground. The first test of a nuclear weapon was carried out at the site in January 1951 when a B-50 bomber dropped a bomb for the first of five tests in "Operation Ranger."

The Nevada Test Site is located 65 miles (105 km) northwest of Las Vegas. It occupies 1,350 square miles (3,497 km^2), an area slightly larger than the state of Rhode Island. Nellis Air Force Base and the Tonopath Test Range surround the site on three sides.

Over 3,500 people are employed by NTS, 1,500 of whom work on the site itself. The site's annual budget is about $450 million, a large percentage goes for weapons testing and a small percentage which is spent on the **radioactive waste** storage facility at nearby **Yucca Mountain**.

The first nuclear tests at the site were conducted over an area known as Frenchman Flat. Between 1951 and 1962, a total of fourteen atmospheric tests were carried out in this area to determine the effect of nuclear explosions on structures and military targets. Ten underground tests were also conducted at Frenchman Flat between 1965 and 1971.

Since 1971, most underground tests at the site have been conducted in the area known as Yucca Flat. These tests are usually carried out in **wells** 10 ft (3 m) in diameter and 600 ft (182 m) to one mile (1.6 km) in depth. On an average, about twelve tests per year are carried out at NTS.

Some individuals have long been concerned about possible environmental effects of the testing carried out a NTS. During the period of atmospheric testing, those effects (**radioactive fallout**, for example) were relatively easy to observe. But the environmental consequences of underground testing have been more difficult to determine.

One such consequence is the production of earthquakes, an event observed in 1968 when a test code-named "Faultless" produced a fault with a vertical displacement of 15 ft (5 m). Such events are rare, however, and of less concern than the release of radioactive materials into **groundwater** and the escape of radioactive gases through venting from the test well.

In 1989, the Office of Technology Assessment (OTA) of the United States Congress carried out a study of the possible environmental effects from underground testing. OTA concluded that the risks to humans from underground testing at NTS are very low indeed. It found that, in the first place, there is essentially no possibility that any release of radioactive material could go undetected. OTA also calculated that the total mount of radiation a person would have received by standing at the NTS boundary for every underground test conducted at the site so far would be about equal to 1/1000 of a single chest x-ray or equivalent to 32 minutes more of exposure to natural **background radiation** in a person's lifetime. *See also* Groundwater pollution; Hazardous waste; Nuclear fission; Radiation exposure; Radioactive decay; Radioactivity

[*David E. Newton*]

RESOURCES
PERIODICALS

"New Bomb Factory to Open Soon at Test Site." *Bulletin of the Atomic Scientists* 46 (April 1990): 56.

"Press Releases Don't Tell All." *Bulletin of the Atomic Scientists* 46 (January-February 1990): 4–5.

Slonit, R. "In the State of Nevada." *Sierra* (September-October 1991): 90–101.

OTHER

Office of Technology Assessment. *The Containment of Underground Nuclear Explosions.* OTA-ISC-414. Washington, D.C.: U.S. Government Printing Office, 1989.

New Madrid, Missouri

Those who think that earthquakes are strictly a California phenomenon might be amazed to learn that the most powerful **earthquake** in recorded American history occurred in the middle of the country near New Madrid (pronounced MAD-rid), Missouri. Between December 16, 1811, and February 7, 1812, about 2,000 tremors shook southeastern

Missouri and adjacent parts of Arkansas, Illinois, and Tennessee. The largest of these earthquakes is thought to have had a magnitude of 8.8 on the Richter scale, making it one of the most massive ever recorded.

Witnesses reported shocks so violent that trees 6 feet (2 meters) thick were snapped like matchsticks. More than 150,00 acres (60,000 ha) of forest were flattened. Fissures several yards wide and many miles long split the earth. Geysers of dry sand or muddy water spouted into the air. A trough 150 miles (242 km) long, 40 miles (64 km) wide, and up to 30 feet (9 m) deep formed along the fault line. The town of New Madrid sank about 12 feet (4 m). The Mississippi River reversed its course and flowed north rather than south past New Madrid for several hours. Many people feared that it was the end of the world.

One of the most bizarre effects of the tremors was **soil** liquefaction. Soil with high water content was converted instantly to liquid mud. Buildings tipped over, hills slid into the valleys, and animals sank as if caught in quicksand. Land surrounding a hamlet called Little **Prairie** suddenly became a soupy swamp. Residents had to wade for miles through hip-deep mud to reach solid ground. The swamp was not drained for nearly a century.

Some villages were flattened by the earthquake, while others were flooded when the river filled in subsided areas. The tremors rang bells in Washington, D.C., and shook residents out of bed in Cincinnati, Ohio. Since the country was sparsely populated in 1812, however, few people were killed.

The situation is much different now, of course. The damage from an earthquake of that magnitude would be calamitous. Much of Memphis, Tennessee, only about 100 miles (160 km) from New Madrid, is built on **landfill** similar to that in the Mission District of San Francisco where so much damage occurred in the earthquake of 1990. St. Louis had only 2,000 residents in 1812; nearly a half million live there now. Scores of smaller cities and towns lie along the fault line and transcontinental highways and pipelines cross the area. Few residents have been aware of earthquake dangers or how to protect themselves. Midwestern buildings generally are not designed to survive tremors.

Anxiety about earthquakes in the Midwest was aroused in 1990 when climatologist Iben Browning predicted a 50-50 chance of an earthquake 7.0 or higher on or around December 3, in or near New Madrid. Browning based his prediction on calculations of planetary motion and gravitational forces. Many geologists were quick to dismiss these techniques, pointing out that seismic and geochemical analyses predict earthquakes much more accurately than the methods he used. Although there were no large earth quakes along the New Madrid fault in 1990, the **probability** of a major tremor there remains high.

While the general time and place of some earthquakes have been predicted with remarkable success, mystery and uncertainty still abound concerning when and where "the next big one" will occur. Will it be in California? Will it be in the Midwest? Or will it be somewhere entirely unexpected? Meanwhile, residents of New Madrid are planning emergency exit routes and stocking up on camping gear and survival supplies.

[*William P. Cunningham Ph.D.*]

Resources

Periodicals

Finkbeiner, A. "California's Revenge: Someday a Major Earthquake Will Ravage the United States–in the East." *Discover* 11 (September 1990): 78–82, 84–5.

Johnson, A. C., and L. R. Kanter. "Earthquakes in Stable Continental Crust." *Scientific American* 262 (March 1990): 68–75.

New Source Performance Standard

The Clean Air Acts of 1963 and 1967 gave to the **Environmental Protection Agency** (EPA) the authority to establish **emission standards** for new and modified stationary sources. These standards are called new source performance standards (NSPS) and are determined by the best **emission** control technology available, the energy needed to use the technology, and its overall cost. An example of an NSPS is the standard set for plants that make Portland cement. Such plants are allowed to release no more than 0.30 pounds of emissions for each ton of raw materials used and to produce an emission with no more than 20 percent opacity.

New York Bight

A bight is a coastal embayment usually formed by a curved shoreline. The New York Bight forms part of the Middle Atlantic Bight, which runs along the east coast of the United States. The dimensions of the New York Bight are roughly square, encompassing an area that extends out from the New York-New Jersey shore to the eastern limit of Long Island and down to the southern tip of New Jersey. The apex of the bight, as it is known, is the northwestern corner, which includes the **Hudson River** estuary, the Passaic and Hackensack River estuaries, Newark Bay, Arthur Kill, Upper Bay, Lower Bay, and Raritan Bay.

The New York Bight contains a valuable and diverse **ecosystem**. The waters of the bight vary from relatively fresh near the shore to **brackish** and salty as one moves eastward, and the range of **salinity**, along with the islands and shore areas present within the area, have created a diver-

sity of environmental conditions and habitats, which include marshes, woods, and beaches, as well as highly developed urban areas. The portion of the bight near the shore lies directly in the path of one of the major transcontinental migratory pathways for birds, the North Atlantic **flyway**.

The New York Bight has a history of extremely intensive use by humans, especially at the apex, and here environmental impacts have been most severe. Beginning with the settlement of New York City in the 1600s, the bight area has supported one of the world's busiest harbors and largest cities. It receives more than two billion gallons per day of domestic sewage and industrial **wastewater**. Millions of gallons of **nonpoint source runoff** also pours into the bight during storms, and regulated **ocean dumping** of dredge spoils also occurs.

Numerous studies have shown that the sediments of the bight, particularly at the apex, have been contaminated. Levels of **heavy metals** such as **lead**, **cadmium** and **copper** in the sediments of the apex are of special concern because they far exceed current guidelines on acceptable concentrations. Similarly, organic pollutants such as **polychlorinated biphenyls** (PCBs) from transformer oil and **polycyclic aromatic hydrocarbons** (PAHs) degraded from **petroleum** compounds are also in the sediments at levels high enough to be of concern. Additionally, the sewage brings enormous quantities of **nitrogen** and **phosphorus** into the bight which promotes excessive growth of algae.

The continual polluting of the bight since the early days of settlement has progressively reduced its capacity as a food source for the surrounding communities. The oyster and shellfishing industry that thrived in the early 1800s began declining in the 1870s, and government advisories currently prohibit shellfishing in the waters of the bight due to the high concentrations of contaminants that have accumulated in shellfish. Fishing is highly regulated throughout the area, and health advisories have been issued for consumption of fish caught in the bight. The bottom-dwelling worms and insect larvae in the sediments of the apex consist almost entirely of **species** that are extremely tolerant of **pollution**; sensitive species are absent and **biodiversity** is low.

There are, however, some reasons to be optimistic. The areas within the bight closest to the open ocean are much cleaner than the highly degraded apex. In these less-impacted areas, the bottom-dwelling communities have higher species diversity and include species that prefer unimpacted conditions. Since the 1970s, enforcement of the **Clean Water Act** has helped greatly in reducing the quantities of untreated wastewater entering the bight. Some fish species that had been almost eliminated from the area have returned, and today striped bass again swim up the Hudson River to spawn. *See also* Algal bloom; Environmental stress; Marine pollution; Pollution control; Sewage treatment; Storm runoff

[*Usha Vedagiri*]

Resources

Books

Meyer, G. *Ecological Stress and the New York Bight.* Crownsville, MD: Estuarine Research Federation, 1982.

New York-New Jersey Harbor Estuary Program. *Toxics Characterization Report.* Washington, DC: U.S. Environmental Protection Agency, 1992.

Periodicals

Payton, B. M. "Ocean Dumping in the New York Bight." *Environment* 27 (1985): 26.

NGO

see **Nongovernmental organization**

Niche

The term niche is used in **ecology** with a variety of distinct meanings. It may refer to a spatial unit or to a function unit. One definition focuses on niche as a role claimed exclusively by a **species** through **competition**. The word is also used to refer to "utilization distribution" or the frequency with which populations use resources. Still, niche is well enough established in ecology that Stephen Jay Gould can label it as "the fundamental concept" in the discipline, "an expression of the location and function of a species in a habitat." Niche is used to address such questions as what determines the species diversity of a **biological community**, how similar organisms coexist in an area, how species divide up the resources of an **environment**, and how species within a community affect each other over time.

Niche has not been applied very satisfactorily in the ecological study of humans. Anthropologists have used it perhaps most successfully in the study of how small pre-industrial tribal groups adapt to local conditions. Sociologists have not been very successful with niche, subdividing the human species by occupations or roles, creating false analogies that do not come very close to the way niche is used in biology. More recently, sociologists have extended niche to help explain organizational behavior, though again distorting it as an ecological concept.

Some attempts were made to build on the vernacular sense of niche as in "he found his niche," a measure of how individual human beings attain multidimensional "fit" with their surroundings. But this usage was again criticized as too much of a distortion of the original meaning of niche in biology. The word and related concepts remain common, however, and are widely understood in vernacular usage to

describe how individual human beings make their way in the world.

The niche concept has not been much employed by environmental scientists, though it might be helpful in attempts to understand the relationships between humans and their environments, for instance. Efforts to formulate niche or a synonym of some sort for use in the study of such relationships will probably continue. The best use of niche might be in its utility as an indicator of the richness and diversity of **habitat**, serving in this way as an indicator of the general health of the environment.

[*Gerald L. Young Ph.D.*]

Resources

Books

Schoener, T. W. "The Ecological Niche." In *Ecological Concepts*, edited by J. M. Cherrett. Oxford, England: Blackwell Scientific Publications, 1989.

Periodicals

Broussard, C. A., and G. L. Young. "A Reorientation of Niche Theory in Human Ecology: Toward a Better Explanation of the Complex Linkages between Individual and Society." *Sociological Perspectives* 29 (April 1986): 259–283.

Colinvaux, P. A. "Towards a Theory of History: Fitness, Niche, and Clutch of *Homo Sapiens*." *Coevolution Quarterly* 41 (Spring 1984): 94–107.

Mark, J., G. M. Chapman, and T. Gibson. "Bioeconomics and the Theory of Niches." *Futures* 17 (December 1985): 632–51.

Nickel

Nickel is a heavy metal, and it can be an important toxic chemical in the **environment**. Natural **pollution** by nickel is associated with soils that have a significant presence of a mineral known as serpentine. Serpentine-laced soils are toxic to nonadapted plants, and although the most significant toxic stressor is a large concentration of nickel, sometimes the presence of cobalt and/or chromium, along with high **pH** and an impoverished supply of nutrients also create a toxic environment. Serpentine sites often have a specialized **flora** dominated by nickel-tolerant **species**, many of which are endemic to such sites. Nickel pollution can occur through human influence as well—most often in the vicinity of nickel smelters or refineries. The best-known example of a nickel-polluted environment occurs around the town of **Sudbury, Ontario**, where smelting has been practiced for a century. *See also* Heavy metals and heavy metal poisoning

Nickel mining

see **Sudbury, Ontario**

NIMBY

see **Not In My Backyard**

NIOSH

see **National Institute of Occupational Safety and Health**

Nitrates and nitrites

Nitrates and nitrites are families of chemical compounds containing atoms of **nitrogen** and oxygen. Occurring naturally, nitrates and nitrites are critical to the continuation of life on the earth, since they are one of the main sources from which plants obtain the element nitrogen. This element is required for the production of amino acids which, in turn, are used in the manufacture of proteins in both plants and animals.

One of the great transformations of agriculture over the past century has been the expanded use of synthetic chemical fertilizers. Ammonium nitrate is one of the most important of these fertilizers. In recent years, this compound has ranked in the top fifteen among synthetic **chemicals** produced in the United States.

The increased use of nitrates as **fertilizer** has led to some serious environmental problems. All nitrates are soluble, so whatever amount is not taken up by plants in a field is washed away into **groundwater** and, eventually, into rivers, streams, ponds, and lakes. In these bodies of water, the nitrates become sources of food for algae and other plant life, resulting in the formation of algal blooms. Such blooms are usually the first step in the eutrophication of a pond or lake. As a result of eutrophication, a pond or lake slowly evolves into a marsh or swamp, then into a bog, and finally into a meadow.

Nitrates and nitrites present a second, quite different kind of environmental issue. These compounds have long been used in the preservation of red meats. They are attractive to industry not only because they protect meat from spoiling, but also because they give meat the bright red color that consumers expect.

The use of nitrates and nitrites in meats has been the subject of controversy, however, for at least twenty years. Some critics argue that the compounds are not really effective as preservatives. They claim that preservation is really effected by the table salt that is usually used along with nitrates and nitrites. Furthermore, some scientists believe that nitrates and nitrites may themselves be carcinogens or may be converted in the body to a class of compounds known as the nitrosamines, compounds that are known to be carcinogens.

In the 1970s, the **Food and Drug Administration** (FDA) responded to these concerns by dramatically cutting back on the quantity of nitrates and nitrites that could be added to foods. By 1981, however, a thorough study of the issue by the **National Academy of Sciences** showed that nitrates and nitrites are only a minor source of nitrosamine compared to smoking, drinking water, cosmetics, and industrial chemicals. Based on this study, the FDA finally decided in January 1983 that nitrates and nitrites are safe to use in foods. *See also* Agricultural revolution; Cancer; Cigarette smoke; Denitrification; Drinking-water supply; Fertilizer runoff; Nitrification; Nitrogen cycle; Nitrogen waste

[*David E. Newton*]

Resources

Books

Canter, L. W. *Nitrates in Ground Water.* Chelsea, MI: Lewis, 1992.

Cassens, R. G. *Nitrate-Cured Meat: A Food Safety Issue in Perspective.* Trumbull, CT: Food and Nutrition Press, 1990.

Selinger, B. *Chemistry in the Marketplace.* 4th ed. Sydney, Australia: Harcourt Brace Jovanovich, 1989.

Periodicals

"Clearest Lake Clouding Up." *Environment* 30 (January-February 1988): 22–3.

Raloff, J. "New Acid Rain Threat Identified." *Science News* 133 (30 April 1988): 276.

Nitrification

A biological process involving the conversion of nitrogen-containing organic compounds into **nitrates and nitrites**. It is accomplished by two groups of chemo-synthetic bacteria that utilize the energy produced. The first step involves the oxidation of ammonia to nitrite, and is accomplished by *Nitrosomas* in the **soil** and *Nitrosoccus* in the marine **environment**. The second step involves the oxidation of **nitrites** into **nitrates**, releasing 18 kcal of energy. It is accomplished by *Nitrobacter* in the soil and *Nitrococcus* in salt water. Nitrification is an integral part of the **nitrogen cycle**, and is usually considered a beneficial process, since it converts organic **nitrogen** compounds into nitrates which can be absorbed by green plants. The reverse process of nitrification, occurring in oxygen-deprived environments, is called **denitrification** and is accomplished by other **species** of bacteria.

Nitrites

see **Nitrates and nitrites**

Nitrogen

Comprising about 78% of the earth's **atmosphere**, nitrogen (N_2) has an atomic number of seven and an atomic weight of 14. It has a much lower solubility in water than in air—there is approximately 200 times more nitrogen in the atmosphere than in the ocean. The main source of gaseous nitrogen is volcanic eruptions; the major nitrogen sinks are synthesis of nitrate in electrical storms and biological **nitrogen fixation**. All organisms need nitrogen. It forms part of the chlorophyll molecule in plants, it forms the nitrogen base in DNA and RNA, and it is an essential part of all amino acids, the building blocks of proteins. Nitrogen is needed in large amounts for **respiration**, growth, and reproduction. **Nitrogen oxides** (NO_x), produced mainly by motor vehicles and internal **combustion** engines, are one of the main contributors to **acid rain**. They react with water molecules in the atmosphere to form nitric **acid**. *See also* Nitrates and nitrites; Nitrogen cycle

Nitrogen cycle

Nitrogen is a macronutrient essential to all living organisms. It is an integral component of amino acids which are the building blocks of proteins; it forms part of the nitrogenous bases common to DNA and RNA; it helps make up ATP, and it is a major component of the chlorophyll molecule in plants. In essence, life as we know it cannot exist without nitrogen.

Although nitrogen is readily abundant as a gas (it comprises 79 percent of atmospheric gases by volume), most organisms cannot use it in this state. It must be converted to a chemically usable form such as ammonia (NH_3) or nitrate (NO_3) for most plants, and amino acids for all animals. The processes involved in the conversion of nitrogen to its various forms comprise the nitrogen cycle. Of all the **nutrient** cycles, this is considered the most complex and least well understood scientifically. The processes that make up the nitrogen cycle include **nitrogen fixation**, ammonification, **nitrification**, and **denitrification**.

Nitrogen fixation refers to the conversion of atmospheric nitrogen gas (N_2) to ammonia (NH_3) or nitrate (NO_3). The latter is formed when lightning or sometimes cosmic radiation causes oxygen and nitrogen to react in the **atmosphere**. Farmers are usually delighted when electrical storms move through their areas because it supplies "free" nitrogen to their crops, thus saving money on **fertilizer**. Ammonia is produced from N_2 by a special group of **microbes** in a process called biological fixation, which accounts for about 90 percent of all fixed N_2 each year worldwide. This process is accomplished by a relatively small number of **species** of bacteria and blue-green algae, or blue-green

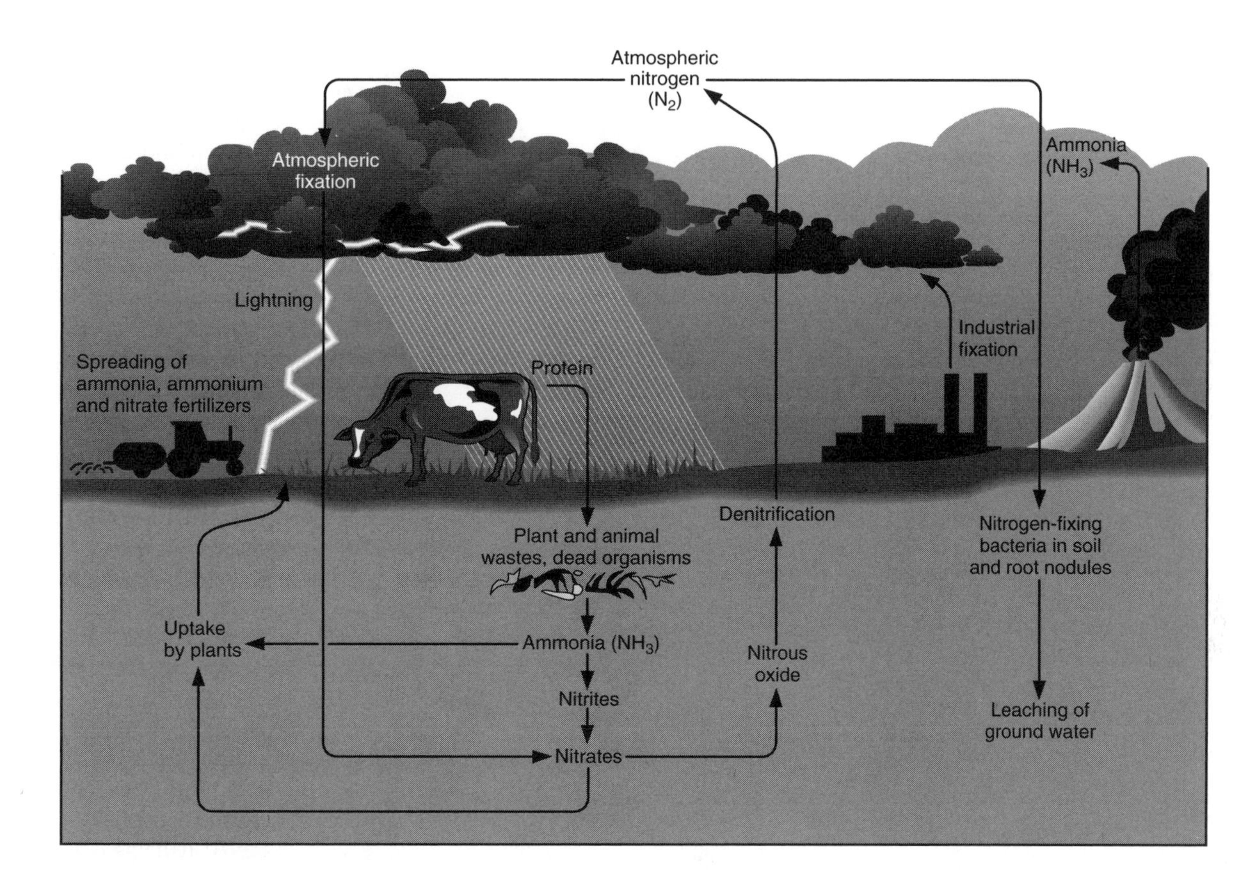

The nitrogen cycle. (Illustration by Hans & Cassidy.)

bacteria. The most well known of these nitrogen-fixing organisms are the bacteria in the genus *Rhizobium* which are associated with the root nodules of legumes. The legumes attract these bacteria by secreting a chemical into the **soil** that stimulates the bacteria to multiply and enter the root hair tips. The resultant swellings contain millions of bacteria and are called root nodules, and here, near the soil's surface, they actively convert atmospheric N_2 to NH_3, which is taken up by the plant. This is an example of a symbiotic relationship, where both organisms benefit. The bacteria benefit from the physical location in which to grow, and they also utilize sugars supplied by the plant photosynthate to reduce the N_2 to NH_3. The legumes, in turn, benefit from the NH_3 produced. The energetic cost for this nutrient is quite high, however, and legumes typically take their nitrogen directly from the soil rather than from their bacteria when the soil is fertilized.

Although nitrogen fixation by legume bacteria is the major source of biological fixation, other species of bacteria are also involved in this process. Some are associated with non-legume plants such as toyon (*Ceanothus*), silverberry (*Elaeagnus*), water fern (*Azolla*), and alder (*Alnus*). With the exception of the water fern, these plants are typically pioneer species growing in low-nitrogen soil. Other nitrogen-fixing bacteria live as free-living species in the soil. These include microbes in the genera *Azotobacter* and *Clostridium*, which live in **aerobic** and **anaerobic** sediments, respectively.

Blue-green algae are the other major group of living organisms which fix atmospheric N_2. They include approximately forty species in such genera as *Aphanizomenon*, *Anabaena*, *Calothrix*, *Gloeotrichia*, and *Nostoc*. They inhabit both soil and freshwater and can tolerate adverse and even extreme conditions. For example, some species grow in hot springs where the water is 212°F (100°C), whereas other species inhabit glaciers where the temperature is 32°F (0°C). The characteristic bluish-green coloration is a telltale sign of their presence. Some blue-green algae are found as pioneer species invading barren soil devoid of nutrients, particularly nitrogen, either as solitary individuals or associated with other organisms such as **lichens**. Flooded rice fields are another prime location for nitrogen-fixing blue-green algae.

Perhaps the most common environments where blue-green algae are found are lakes and ponds, particularly when the body of water is eutrophic—containing high concentrations of nutrients, especially **phosphorus**. Algae can reach bloom proportions during the warm summer months and are often considered a nuisance because they float on the surface, forming dense scum. The resultant odor following **decomposition** is usually pungent, and fish such as catfish often acquire an off-flavor taste from ingesting these algae.

The next major component of the nitrogen cycle is ammonification. It involves the breakdown of organic matter (in this case amino acids) by decomposer organisms to NH_3, yielding energy. It is therefore the reverse reaction of amino **acid** synthesis. Dead plant and animal tissues and waste materials are broken down to amino acids and eventually NH_3 by the saprophagous bacteria and **fungi** in both soil and water.

Nitrification is a biological process where NH_3 is oxidized in two steps, first to NO_3 and next to nitrate (NO2). It is accomplished by two genera of bacteria, *Nitrosomonas* and *Nitrobacter* in the soil, and *Nitrosococcus* and *Nitrococcus* in salt water. Since nitrification is an oxidation reaction, it requires oxygenated environments.

Dentrification is the reverse reaction of nitrification and occurs under anaerobic conditions. It involves the breakdown of **nitrates and nitrites** into gaseous N_2 by **microorganisms** and fungi. Bacteria in the genus *Pseudomonas* (e.g., *P. dentrificans*) reduce NO_3 in the soil.

The cycling of nitrogen in an **ecosystem** is obviously complex. In aquatic ecosystems, nitrogen can enter the food chain through various sources, primarily surface **runoff** into lakes or rivers, mixing of nutrient-rich bottom waters (normally only during spring and fall turnovers in north temperate lakes), and biological fixation of atmospheric nitrogen by blue-green algae. **Phytoplankton** (microscopic algae) then rapidly take up the available nitrogen in the form of NH_3 or NO_3 and assimilate it into their tissues, primarily as amino acids. Some nitrogen is released by leakage through cell membranes. Herbivorous **zooplankton** that ingest these algae convert their amino acids into different amino acids and excrete the rest. Carnivorous or omnivorous zooplankton and fish that eat the herbivores do the same. Excretion (usually as NH_3 and urea) is thus a valuable nutrient **recycling** mechanism in aquatic ecosystems. Different species of phytoplankton actively compete for these nutrients when they are limited. Decomposing bacteria in the lake, particularly in the top layer of the bottom sediments, play an important role in the breakdown of dead organic matter which sinks to the bottom. The cycle is thus complete.

The cycling of nitrogen in marine ecosystems is similar to that in lakes, except that nitrogen lost to the sediments in the deep open water areas is essentially lost. Recycling only occurs in the nearshore regions, usually through a process called upwelling. Another difference is that marine phytoplankton prefer to take up nitrogen in the form of NH_3 rather than NO_3.

In terrestrial ecosystems, NH_3 and NO_3 in the soil is taken up by plants and assimilated into amino acids. As in aquatic habitats, the nitrogen is passed through the **food chain/web** from plants to herbivores to carnivores, which manufacture new amino acids. Upon death, **decomposers** begin the breakdown process, converting the organic nitrogen to inorganic NH_3. Bacteria are the main decomposers of animal matter and fungi are the main group that break down plants. Shelf and bracket fungi, for example, grow rapidly on fallen trees in forests. The action of termites, bark beetles, and other insects that inhabit these trees greatly speed up the process of decomposition.

There are three major differences between nitrogen cycling in aquatic versus terrestrial ecosystems. First, the nitrogen reserves are usually much greater in terrestrial habitats because nutrients contained in the soil remain accessible, whereas nitrogen released in water and not taken up by phytoplankton sinks to the bottom where it can be lost or held for a long time. Secondly, nutrient recycling by herbivores is normally a more significant process in aquatic ecosystems. Thirdly, terrestrial plants prefer to take up nitrogen as NO_3 and aquatic plants prefer NH_3.

Forces in **nature** normally operate in a balance, and gains are offset by losses. So it is with the nitrogen cycle in freshwater and terrestrial ecosystems. Losses of nitrogen by detritrification, runoff, **sedimentation**, and other releases equal gains by fixation and other sources.

Humans, however, have an influence on the nitrogen cycle that can greatly change normal pathways. Fertilizers used in excess on residential lawns and agricultural fields add tremendous amounts of nitrogen (typically as urea or ammonium nitrate) to the target area. Some of the nitrogen is taken up by the vegetation, but most washes away as surface runoff, entering streams, ponds, lakes, and the ocean. This contributes to the accelerated eutrophication of these bodies of water. For example, periodic unexplained blooms of toxic dinoflagellates off the coast of southern Norway have been blamed on excess nutrients, particularly nitrogen, added to the ocean by the fertilizer runoff from agricultural fields in southern Sweden and northern Denmark. These algae have caused massive dieoffs of **salmon** in the **mariculture** pens popular along the coast, resulting in millions of dollars of damage. Similar circumstances have contributed to blooms of other species of dinoflagellates, creating what are known as red tides. When filter-feeding shellfish ingest these algae, they become toxic, both to other fishes and humans. Paralytic shellfish **poisoning** may result within thirty minutes, leading to impairment of normal nerve conduction, difficulty in breathing, and possible death. Saxo-

toxin, the toxin produced by the dinoflagellate *Gonyaulax*, is fifty times more lethal than strychnine and curare.

Other forms of human intrusion into the nitrogen cycle include harvesting crops, **logging**, sewage, animal wastes, and exhaust from automobiles and factories. Harvesting crops and logging remove nitrogen from the system. The other processes are point sources of excess nitrogen. Autos and factories produce nitrous oxides (NO_x) such as nitrogen dioxide (NO_2), a major air pollutant. NO_2 contributes to the formation of **smog**, often irritating eyes and leading to breathing difficulty. It also reacts with water vapor in the atmosphere to form weak nitric acid (HNO_3), one of the major components of **acid rain**. *See also* Agricultural pollution; Air pollution; Aquatic weed control; Marine pollution; Nitrogen waste; Soil fertility; Urban runoff

[*John Korstad*]

Resources
Books

Ehrlich, P. R., A. H. Ehrlich, and J. P. Holdren. *Ecoscience: Population, Resources, Environment.* New York: W. H. Freeman, 1977.

Ricklefs, R. E. *Ecology.* 3rd ed. New York: W. H. Freeman, 1990.

Smith, R. E. *Ecology and Field Biology.* 4th ed. New York: Harper and Row, 1990.

Periodicals

Brill, W. J. "Biological Nitrogen Fixation." *Scientific American* 236 (1977): 68–81.

Delwiche, C. C. "The Nitrogen Cycle." *Scientific American* 223 (1970): 136–46.

Nitrogen fixation

Nitrogen fixation is the biological process by which atmospheric nitrogen gas (N_2) is converted into ammonia (NH_3). Nitrogen is an essential **nutrient** for the growth of all organisms, as it is a component of proteins, nucleic acids, amino acids, and other cellular constituents. Nearly 79% of the earth's **atmosphere** is N_2, but N_2 is unavailable for use by most organisms, because it is nearly inert due to the triple bonds between the two nitrogen atoms. To utilize nitrogen, organisms require that it be fixed (combined) in the form of ammonium (NH_4) or nitrate ions (NO_3).

The bacteria that accomplish nitrogen fixation are either free-living or form symbiotic associations with plants or other organisms such as termites or protozoa. The free-living bacteria include both **aerobic** and **anaerobic** bacteria, including some **species** that are photosynthetic.

The most well-known nitrogen-fixing symbiotic relationships are those of legumes (e.g., peas, beans, clover, soybeans, and alfalfa) and the bacteria *Rhizobium*, which is present in nodules on the roots of the legumes. The plants provide the bacteria with carbohydrates for energy and a stable **environment** for growth, while the bacteria provide the plants with usable nitrogen and other essential nutrients. The bacteria invade the plant and cause formation of the nodule by inducing growth of the plant host cells. The *Rhizobium* are separated from the plant cells by being enclosed in a membrane. The cultivation of legumes as green manure crops to add nitrogen by incorporation into the **soil** is a long-established agricultural process.

Frankia, a type of bacteria belonging to the bacterial group actinomycetes, form nitrogen-fixing root nodules with several woody plants, including alder and sea buckhorn. These trees are pioneer species that invade nutrient-poor soils.

The photosynthetic nitrogen-fixing bacteria cyanobacteria (also known as blue-green algae) live as free-living organisms or as symbionts with **lichens** in pioneer habitats such as **desert** soils. They also form a symbiotic association with the fern *Azolla*, which is also found in lakes, swamps, and streams. In rice paddies, *Azolla* has been grown as a green manure crop; it is plowed into the soil to release nitrogen before planting of the rice crop.

The nitrogen fixed by the nitrogen-fixing bacteria is utilized by other living organisms. After their death, the nitrogen is released into the environment through **decomposition** processes and continues to move through the **nitrogen cycle**.

[*Judith L. Sims*]

Resources
Books

Postgate, John R., and J. R. Postgate. *Nitrogen Fixation.*Cambridge, UK: Cambridge University Press, 1998.

Other

Biology Teaching Organization. *The Microbial World: The Nitrogen Cycle and Nitrogen Fixation.* Edinburgh School of Biology, The University of Edinburgh. [cited June 21, 2002]. <http://helios.bto.ed.ac.uk/bto/microbes/nitrogen.htm>.

Nitrogen oxides

Five oxides of **nitrogen** are known: N_2O, NO, N_2O_3, NO_2, and N_2O_5. Environmental scientists usually refer to only two of these, nitric oxide and nitrogen dioxide, when they use the term nitrogen oxides. The term may also be used, however, for any other combination of the five compounds. Nitric oxide and nitrogen dioxide are produced during the **combustion** of **fossil fuels** in automobiles, jet aircraft, industrial processes, and electrical power production. The gases have a number of deleterious effects on human health, including irritation of the eyes, skin, respiratory tract, and

lungs; chronic and acute **bronchitis**; and heart and lung damage. *See also* Ozone; Smog

Nitrogen scrubbing

see **Raprenox (nitrogen scrubbing)**

Nitrogen waste

Nitrogen waste is a component of sewage that comes primarily from human excreta and **detergents** but also from fertilizers and such industrial processes as steel-making. Nitrogen waste consists primarily of **nitrates and nitrites** as well as compounds of ammonia. Because they tend to clog waterways and encourage algae growth, nitrogen wastes are undesirable. They present a problem for **wastewater** treatment since they are not removed by either primary or secondary treatment steps. Their removal at the tertiary stage can be achieved only by specialized procedures. Another approach is to use wastewater on farmlands. This use removes about 50 percent of nitrogen wastes from water by turning it into nutrients for the **soil**. *See also* Sewage treatment

Nitrous oxide

Nitrous oxide (N_2O), also called di-nitrogen monoxide, is one of several gaseous oxides of **nitrogen**. It is sometimes referred to as laughing gas, because when inhaled it causes a feeling of intoxication, mild hysteria, and sometimes laughter. The gas is colorless and has a faint odor and slightly sweet taste. It can lessen the sensation of pain and is used as an anesthetic in dentistry and in minor surgery of short duration. For more complex surgery, it is combined with other anesthetics to produce a deeper and longer-lasting state of anesthesia. Nitrous oxide also is commonly used as a propellant for pressurized food products, and sometimes is used in race cars to boost the power of high performance engines. It is one of the few gases capable of supporting **combustion**. In this process, it transfers its oxygen to the material being combusted and is converted to molecular nitrogen (N_2).

Nitrous oxide is an important component of the Earth's upper **atmosphere** at heights above 30 mi (45 km). It is one of the **greenhouse gases**, together with **carbon dioxide**, **methane**, and **ozone**, that allow radiation from the Sun to reach the Earth's surface but prevent the infrared or heat component of sunlight from re-irradiating into space. This leads to the so-called **greenhouse effect**, and results in warmer temperatures at the earth's surface. Greenhouse warming is important in creating surface temperatures suitable for life, but recent studies have shown that the levels of some greenhouse gases are increasing at rates that are a cause for concern. Human-made gases, such as **chlorofluorocarbons**, are also contributing to the greenhouse effect. An increase in the Earth's ability to trap infrared radiation can be expected to result in global **climate** change with wide-ranging consequences.

Concern over greenhouse gases has centered on reports of increasing concentrations of atmospheric **carbon** dioxide (CO_2). Two factors thought to be responsible are the widespread use of **fossil fuels** (which add CO_2 to the atmosphere) and the rapid destruction of tropical rain forests which remove CO_2). The levels of other greenhouse gases may also be changing. Studies indicate that atmospheric levels of nitrous oxide may also be increasing as a result of human activity. Nitrous oxide is produced in **nature** by **microorganisms** acting on nitrogen-containing compounds in the **soil**. Increased use of nitrogen **fertilizer** in lawns, gardens, and agricultural fields may stimulate microbial production of nitrous oxide. Manure and municipal **sludge**, applied as soil enhancers and fertilizers, and burning fossil fuels also may contribute to an increase.

Nitrous oxide production in soils is greatly influenced by soil temperature, moisture level, organic content, soil type, **pH**, and oxygen availability. In agricultural soil, rates of fertilizer application, fertilizer type, tillage practices, crop type, and **irrigation** all influence nitrous oxide production. High soil temperature, moisture, and organic content tend to enhance production, whereas tilling of the soil tends to lower it. The relative importance of each of these factors has not been determined, and further study is needed to develop recommendations to limit harmful nitrous oxide emissions from soils.

[*Douglas C. Pratt Ph.D.*]

Resources

Books

Minnesota Pollution Control Agency, Air Quality Division. *Minnesota Greenhouse Gas Inventory 1990.* St. Paul, MN: Minnesota Pollution Control Agency, 1995.

Umarov, M. "Biotic Sources of Nitrous Oxide (N_2O) in the Context of Global Budgets of Nitrous Oxide," *Soils and the Greenhouse Effect,* A. Bowman, ed. Chicester, U.K.: John Wiley and Sons, 1990.

Periodicals

Smith, S.C. "N_2O Laughing Gas: Has the NHRA Been Looking the Other Way after Allegations That Pro Stock Champions Got There with the Help of Nitrous Oxide?" *Car and Driver* 41, no. B (February 1996): 105.

Other

U.S. Department of Energy. *Emissions of Greenhouse Gases in the United States, 1985–1990.* DOE/EIA-0573. Washington, D.C.: GPO, 1993.

U.S. Environmental Protection Agency. *State Workbook: Methodologies for Estimating Greenhouse Gas Emissions.* EPA-230-B-92-002, Washington, D.C.: GPO, 1992.

NOAA

see **National Oceanic and Atmospheric Administration (NOAA)**

NOAEL

see **No-observable-adverse-effect-level**

Noise pollution

Every year since 1973, the U.S. Department of Housing and Urban Development has conducted a survey to find out what city residents dislike about their **environment**. And every year the same factor has been named most objectionable. It is not crime, **pollution**, or congestion; it is noise--something that affects every one of us every day.

We have known for a long time that prolonged exposure to noises, such as loud music or the roar of machinery, can result in hearing loss. Evidence now suggests that noise-related stress also causes a wide range of psychological and physiological problems ranging from irritability to heart disease. An increasing number of people are affected by noise in their environment. By age forty, nearly everyone in America has suffered hearing deterioration in the higher frequencies. An estimated ten percent of Americans (24 million people) suffer serious hearing loss, and the lives of another 80 million people are significantly disrupted by noise.

What is noise? There are many definitions, some technical and some philosophical. What is music to your ears might be noise to someone else. Simply defined, noise pollution is any unwanted sound or any sound that interferes with hearing, causes stress, or disrupts our lives. Sound is measured either in dynes, watts, or decibels. Note that decibels (db) are logarithmic; that is, a 10 db increase represents a doubling of sound energy.

Noises come from many sources. Traffic is generally the most omnipresent noise in the city. Cars, trucks, and buses create a roar that permeates nearly everywhere. Around airports, jets thunder overhead, stopping conversation, rattling dishes, some times even cracking walls. Jackhammers rattle in the streets; sirens pierce the air; motorcycles, lawnmowers, snowblowers, and chain saws create an infernal din; and music from radios, TVs, and loudspeakers fills the air everywhere.

We detect sound by means of a set of sensory cells in the inner ear. These cells have tiny projections (called microvilli and kinocilia) on their surface. As sound waves pass through the fluid-filled chamber within which these cells are suspended, the microvilli rub against a flexible membrane lying on top of them. Bending of fibers inside the microvilli sets off a mechanico-chemical process that results in a nerve signal being sent through the auditory nerve to the brain where the signal is analyzed and interpreted.

The sensitivity and discrimination of our hearing is remark able. Normally, humans can hear sounds from about 16 cycles per second (hz) to 20,000 hz. A young child whose hearing has not yet been damaged by excess noise can hear the whine of a mosquito's wings at the window when less than one quadrillionth of a watt per cm^2 is reaching the eardrum.

The sensory cell's microvilli are flexible and resilient, but only up to a point. They can bend and then spring back up, but they die if they are smashed down too hard or too often. Prolonged exposure to sounds above about 90 decibels can flatten some of the microvilli permanently and their function will be lost. By age thirty, most Americans have lost 5 db of sensitivity and cannot hear anything above 16,000 Hertz (Hz); by age sixty-five, the sensitivity reduction is 40 db for most people, and all sounds above 8,000 Hz are lost. By contrast, in the Sudan, where the environment is very quiet, even seventy-year-olds have no significant hearing loss.

Extremely loud sounds—above 130 db, the level of a loud rock band or music heard through earphones at a high setting—actually can rip out the sensory microvilli, causing aberrant nerve signals that the brain interprets as a high-pitched whine or whistle. Many people experience ringing ears after exposure to very loud noises. Coffee, aspirin, certain antibiotics, and fever also can cause ringing sensations, but they usually are temporary.

A persistent ringing is called tinnitus. It has been estimated that 94 percent of the people in the United States suffer some degree of tinnitus. For most people, the ringing is noticeable only in a very quiet environment, and we rarely are in a place that is quiet enough to hear it. About 35 out of 1,000 people have tinnitus severely enough to interfere with their lives. Sometimes the ringing becomes so loud that it is unendurable, like shrieking brakes on a subway train. Unfortunately, there is not yet a treatment for this distressing disorder. One of the first charges to the **Environmental Protection Agency** (EPA) when it was founded in 1970 was to study noise pollution and to recommend ways to reduce the noise in our environment. Standards have since been promulgated for noise reduction in automobiles, trucks, buses, motorcycles, mopeds, refrigeration units, power lawnmowers, construction equipment, and airplanes. The EPA is considering ordering that warnings be placed on power tools, radios, chain saws, and other household equipment. The **Occupational Safety and Health Administration** (OSHA) also has set standards for noise in the workplace that have considerably reduced noise-related hearing losses.

Noise is still all around us, however. In many cases, the most dangerous noise is that to which we voluntarily subject ourselves. Perhaps if people understood the dangers of noise and the permanence of hearing loss, we would have a quieter environment.

[*William P. Cunningham Ph.D.*]

RESOURCES
BOOKS

Chatwal, G. R., ed. *Environmental Noise Pollution and Its Control.* Columbia: South Asia Books, 1989.

Energy and Environment 1990: Transportation-Induced Noise and Air Pollution. Washington, DC: Transportation Research Board, 1990.

OECD Staff. *Fighting Noise in the Nineteen Nineties.* Washington, DC: Organization for Economic Cooperation and Development, 1991.

PERIODICALS

Bronzaft, A. "Noise Annoys." *E Magazine* 4 (March-April 1993): 16–20.

O'Brien, B. "Quest for Quiet." *Sierra* 77 (July-August 1992): 41–2.

Nonattainment area

Any locality found to be in violation of one or more National Ambient Air Quality Standards set by the **Environmental Protection Agency** (EPA) under the provisions of the **Clean Air Act**. However, a nonattainment area for one standard may be an **attainment area** for a different standard. The seven criteria pollutants for which standards were established in 1970 under the Clean Air Act are **carbon monoxide**, **lead**, **nitrogen** dioxide, **ozone** (a key ingredient in **smog**), **particulate** matter, **sulfur dioxide**, and **hydrocarbons**. Violation of National Ambient Air Quality Standards for one of the seven criteria pollutants can have a variety of consequences for an area, including restrictions on permits for new stationary sources of **pollution** (or significant modifications to existing ones), mandatory institution of vehicle emissions inspection programs, or loss of federal funding (including funding unrelated to pollution problems). *See also* Automobile emissions

Noncriteria pollutant

Pollutants for which specific standards or criteria have not been established. Although some air pollutants are known to be toxic or hazardous, they are released in relatively small quantities or in locations where individual regulation is not required. Others are not yet regulated because data is insufficient to set definite criteria for acceptable ambient levels or control methods. Political and economic interests have also blocked regulatory action. The **Clean Air Act** Amendments of 1990 required the **Environmental Protection Agency** (EPA) to establish **emission standards** for some 189 toxic air pollutants and 250 source categories, thus changing many noncriteria pollutants to criteria ones. *See also* Criteria pollutant

Nondegradable pollutant

A pollutant that is not broken down by natural processes. Some nondegradable pollutants, like the **heavy metals**, create problems because they are toxic and persistent in the **environment**. Others, like synthetic **plastics**, are a problem because of their sheer volume. One way of dealing with nondegradable pollutants is to reduce the quantity released into the environment either by **recycling** them for **reuse** before they are disposed of, or by curtailing their production. A second method is to find ways of making them degradable. Scientists have been able to develop new types of bacteria, for example, that do not exist in **nature**, but that will degrade plastics. *See also* Decomposition; Pollution

Nongame wildlife

Terrestrial and semi-aquatic vertebrates not normally hunted for sport. The majority of wild vertebrates are contained in this group. In the United States **wildlife** agencies are funded largely by **hunting** license fees and by excise taxes on arms and ammunition used for hunting, and they have had to develop other revenue sources for nongame wildlife. The most common method is the state income-tax checkoff, by which citizens may donate portions of their tax returns to nongame wildlife programs. A limited amount of federal aid for such programs has recently been made available to state wildlife agencies through the Nongame Wildlife Act of 1980. *See also* Game animal

Nongovernmental organization

A nongovernmental organization (NGO) is any group outside of government whose purpose is the protection of the **environment**. The term encompasses a broad range of indigenous groups, private charities, advisory committees, and professional organizations; it includes mainstream environmental groups such as the **Sierra Club** and **Defenders of Wildlife**, and more radical groups such as **Greenpeace** and **Earth First!**

In the United States, NGOs have played a pivotal role in the creation of **environmental policy**, directing lobbying efforts and mobilizing the kind of popular support which have made such changes possible. They have been involved in the protection of many **endangered species** and threatened

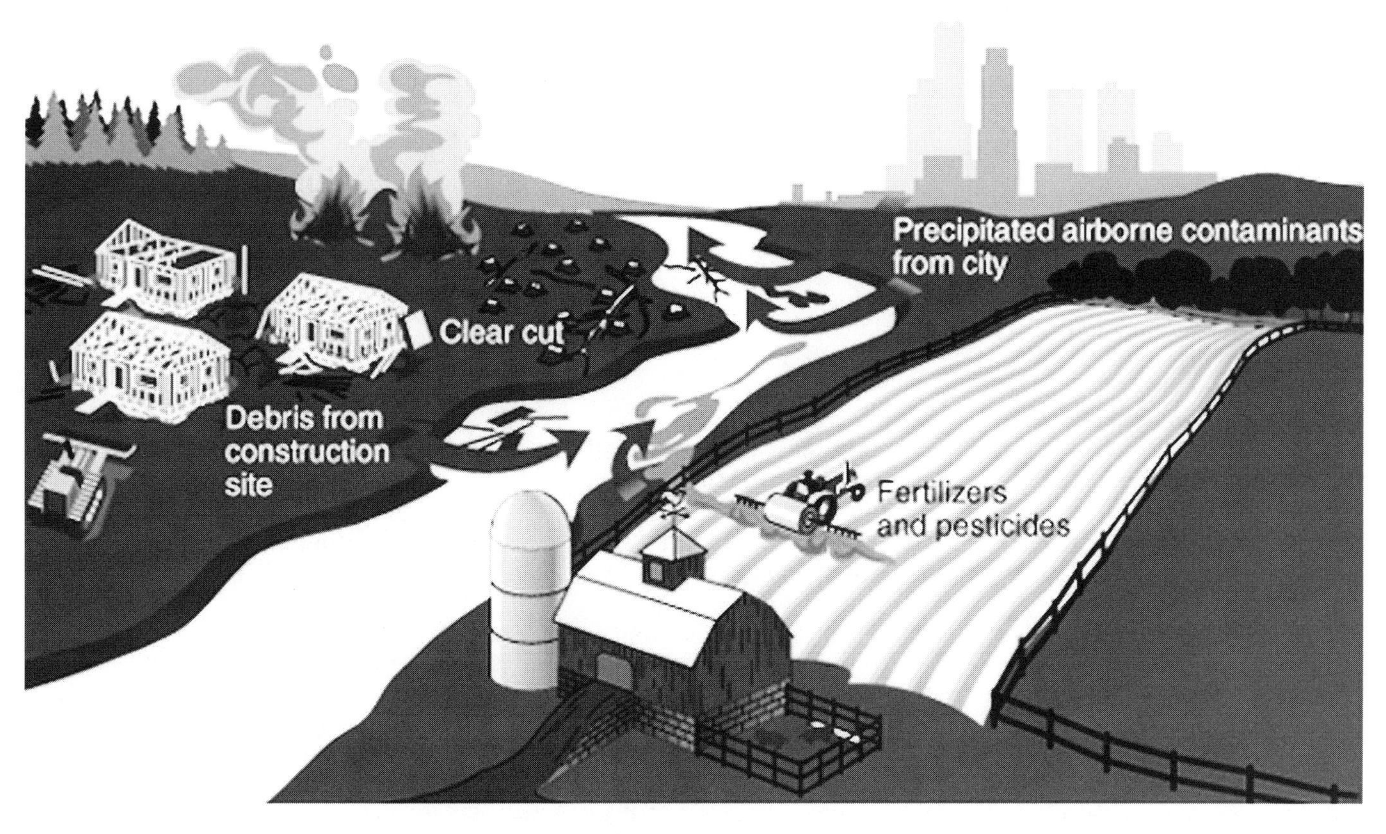

Nonpoint sources of water pollution. (Illustration by Hans & Cassidy.)

habitats, including the **northern spotted owl** and the old-growth forests in the Pacific Northwest. Organizations such as **Earthwatch**, **Earth Island Institute**, and **Sea Shepherd Conservation Society** raised international awareness about the environmental dangers of using drift nets in the **commercial fishing** industry. Their campaign included drift-net monitoring, public education, and direct action, and their efforts led to an international ban on this method. NGOs are extensively involved in the current debate about the future of environmental protection and issues such as **sustainable development** and **zero population growth**.

The number of NGOs worldwide is estimated at over 12,000. They have grown rapidly in number and influence during the last 20 years. In 1972, NGOs had little representation at the **United Nations Conference on the Human Environment** in Stockholm, which was called by industrialized nations primarily to discuss **air pollution**. But these groups had become a much more significant international presence by 1992, and over 9,000 NGOs sent delegates to the Earth Summit in Rio de Janeiro, Brazil. The political pressure NGOs were able to bring to bear had an important, if indirect, effect on the long and complicated preparations for the summit. During the summit itself, NGOs organized a "shadow assembly" or **Global Forum** in a park near Guanabara Bay, where they monitored official negotiations and held conferences of their own. *See also* Animal rights; Bioregionalism; Environmental education; Environmental ethics; Environmental monitoring; Environmentalism; Green politics; Greens; United Nations Earth Summit

[*Douglas Smith*]

Nonpoint source

A diffuse, scattered source of **pollution**. Nonpoint sources have no fixed location where they **discharge** pollutants into the air or water as do chimneys, outfall pipes, or other point sources. Nonpoint sources include **runoff** from agricultural fields, **feedlots**, lawns, **golf courses**, construction sites, streets, and parking lots, as well as emissions from quarrying operations, forest fires, and the evaporation of volatile substances from small businesses such as dry cleaners. Unlike pollutants discharged by point sources, nonpoint pollution is difficult to monitor, regulate, and control. Also, it frequently occurs episodically rather than predictably. Where treatment plants have been installed to control discharge from point sources, nonpoint sources can be responsible for most of the pollution found in bodies of water. As much as 90% of the pollution load in a body of water may come from nonpoint sources. *See also* Water pollution

Nonrenewable resources

Any naturally occurring, finite resources that diminish with use, such as oil and **coal**. In terms of the human timescale, a nonrenewable resource cannot be renewed once it has been consumed. Most nonrenewable resources can only be renewed over geologic time, if at all. All the **fossil fuels** and mineral resources fall into this category. Renewable resources occur naturally and cannot be used up, such as **solar energy** or **wave power**. As resource depletion has become more common, the process of **recycling** has somewhat reduced reliance on virgin nonrenewable resources.

Non-timber forest products

Forest covers 30% of the world's land area. Fifty-six percent of the world's forest is classified as tropical and subtropical and 44% is considered temperate and boreal. Forests are most often valued for their timber resources, however timber is not the only product available from the forest. Forests are also host to an array of non-timber forest products (NTFPs).

NTFPs include berries and fruits, wild mushrooms, honey, gums, spices, nuts, ornamental foliage, mosses and **lichens**, and botanicals used for medicinal, cosmetic, and handicraft purposes. Although there is debate on the matter, some definitions of NTFPs also include fish, wild game, insects, and firewood. NTFPs can also be referred to as; special forest products, non-wood forest products, minor forest products, alternative forest products and secondary forest products. Many NTFPs come from mature, intact forests, illustrating the need to conserve the forests for the vast renewable resources they provide.

People use NTFPs everyday and don't even think about it. Most medicines contain ginseng or other roots that must be harvested in the wild. A very important drug for the treatment of ovarian and other cancers, Taxol, is extracted from the bark of Pacific yew trees. These trees are scarce in number and found predominately in old-growth forests between California and Alaska.

Indigenous people around the world have a long history of using NTFPs in everyday life. They also have a great knowledge and tradition of the medicinal, nutritional, cultural, and spiritual uses of NTFPs. Settler populations moving into areas inhabited by native peoples throughout history have learned of these diverse uses of NTFPs and also developed their own traditions and culture of use.

NTFPs provide market and non-market benefits and commodities for households, communities, and enterprises around the world without the large-scale extraction of timber. In most countries in the world, large-scale industrial **logging** has become the main economic focus in the forest, with timber being at the center of **forest management**. In the last 50 years industrial logging operations have become larger and more mechanized and often controlled by large corporations or central governments. This has left many forest-based communities and forest workers with diminished access to the forest and reduced economic means connected with the forest. Increasing concern for community and **ecosystem health** has brought NTFPs into the spotlight. NTFPs and the development of community-based NTFP enterprises can be seen as important steps towards a diverse, sustainable, multiple-use forest **ecology** and economy.

Although not as large as timber trade values, NTFPs are also a valuable commodity on the global market. According to a 1995 assessment by the Food and Agricultural Organization of the United Nations (FAO), the annual value of international trade in NTFPs for natural honey is $268.2 million, mushrooms and truffles $210.7 million, plants used in pharmacy $689.9 million, nuts $593.1 million, ginseng roots $389.3 million, and spices $175.7 million. The general direction of trade is from developing to developed countries. The **European Union** countries, the United States, and Japan import 60% of the NTFPs traded on the world market. China is the dominant exporter of NTFPs along with India, Indonesia, Malaysia, Thailand and Brazil.

Developed countries import many NTFPs, however, NTFP harvesting and production are also important locally and domestically in the countries of the developed world. For example, researchers estimate that in 1994 the floral and Christmas greens markets alone in Washington, Oregon in the United States and southwest British Columbia, Canada reached a level of US $106.8 million. Mushrooms are also an important product. It is estimated that as many as 36 mushroom **species** are traded commercially in the states of Washington, Oregon and Idaho. In 1992 the wild mushroom market was estimated to be valued at $41.1 million increasing from $21.5 million in 1985.

Because of the nature of the NTFPs and the markets dictating their value, it is difficult to come up with exact figures for the value of different products. The figures above for the value of international trade record only those NTFPs that come onto the world market. Because many people in the world are using NTFPs for their own household

subsistence the figures above only give a partial picture of the total value and use of NTFPs. For instance, the FAO estimates that 80% of the population of the developing world uses NTFPs for health and nutritional needs.

Furthermore, NTFPs have a social and spiritual value for many people making it difficult to assign a dollar value. Economists stumble when trying to assign a value to activities such as, enjoying a couple hours in the forest picking a bucket full of berries to give to a relative or friend. Economists and policy-makers are focusing more attention on these non-market values and attempts are being made to set these values. Quantifying these important non-market values may provide decision makers with tools to better incorporate NTFPs into land-use planning and management.

Knowledge of the economics of NTFPs is limited. And knowledge of the ecology of NTFPs is patchy at best. Of course there are some pockets of extensive knowledge. For example, researchers in Finland forecast the volume of several berry species on a regular basis throughout the harvest season, alerting harvesters and buyers of where the best picking will be, based on weather and on-the-ground reporting.

On the other side of the spectrum, little is known about many of the individual NTFP species and their interaction with other species in the greater forest **ecosystem**. There is a risk that as markets grow for individual NTFPs, they will experience greater harvesting pressures. Lack of scientific knowledge and regulation of harvests may bring over-harvesting in concentrated areas or unexpected uncontrolled market expansion, which could lead to unchecked stress on species with unknown effects on species population and viability. Many NTFP development programs are in fact starting to look to cultivation and agro-forestry alternatives as a way of taking the pressure off of the wild growing species.

As more attention is paid to the opportunities that NTFPs present for multiple-use forestry and locally based economies, it is important that gaps in ecological and economic knowledge be filled. Generally speaking it can be said, that the later years of the twentieth century many regions have witnessed a paradigm shift in forest management from management for timber resources only to **ecosystem management**.

Management of NTFPs presents complex challenges for policy-makers, managers, scientists, enterprises, and communities. There is a need for greater flow of information about the ecology, economy and social issues surrounding NTFP development and management. The list below exemplifies the difficult challenges in developing sustainable management systems for NTFPs. Traditional policy-makers and forest resource managers will be pushed to expand their grasp of the issue and their way of working. NTFP management demands, among other things, **adaptive management** planning, non-linear thinking, and the involvement of multiple stakeholders.

Primary considerations for the sustainable management of NTFPs include

- Understanding the unique biology and ecology of special forest product species
- Anticipating the dynamics of forest communities on a landscape level, delineating present and future areas of high production potential and identifying areas requiring protection
- Developing silvicultural and vegetation management approaches to sustain and enhance production
- Integrating human behavior by monitoring and **modeling** people's responses to management decisions about special forest products
- Conducting necessary inventory, evaluation, and research monitoring

The current non-formal nature of much of the NTFP activity does have its benefits for households and communities. It must be kept in mind when developing NTFP management schemes that regulatory systems must be designed so that they protect NTFP ecology but do not result in reduced access for small enterprises and individual users or user groups. For example, regulatory systems involving difficult bureaucratic processes and fee based permits may make it difficult for small actors with less capital to participate in the commercial and recreational harvest.

In many countries in the developing world, women are the primary actors in NTFP trade. In many countries and cultures women are denied land ownership and decision making power but are able to have access to the forest resources to earn income and provide for the subsistence nutritional and medicinal needs of their families. However, according to the Center for International Forestry Research, in many cases, attempts to formalize the NTFP production and trade have pushed women out of these traditional roles. NTFP development programs designed with women specifically in mind have managed to avoid some of this displacement.

NTFP development offers opportunities for communities and enterprises in the forests of the world. The challenge today is to broaden the view of the forest's economic, ecological and social values to include NTFPs. In adopting this expanded view, policy makers, citizens, scientists, and industry must come together to fill in the gaps of knowledge and to plan the continued and sustainable use of the resources.

[*Sarah E. Lloyd Ph.D.*]

RESOURCES
BOOKS

Emery, Marla R., and Rebecca J. McLain, eds. *Non-Timber Forest Products: Medicinal Herbs, Fungi, Edible Fruits and Nuts, and Other Natural Products from the Forest.* New York: Food Products Press/Haworth, 2001.

Jones, Eric T., Rebecca J. McLain, and James Weigand, eds. *Nontimber Forest Products in the United States.* Lawrence: University Press of Kansas, 2002.

Molina, Randy, Nan Vance, et al. "Special Forest Products: Integrating Social, Economic, and Biological Considerations into Ecosystem Management." In *Creating a Forestry for the 21st Century: the Science of Ecosystem Management.* Edited by Kathryn A. Kohm and Jerry F. Franklin. Washington DC: Island Press, 1996.

Non-Western environmental ethics

Ordinary people are powerfully motivated to do things that can be justified in terms of their religious beliefs. Therefore, distilling **environmental ethics** from the world's living religions is extremely important for global **conservation**. Christianity is a world religion, but so are Islam and Buddhism. Other major religious traditions, such as Hinduism and Confucianism, while more regionally restricted, nevertheless claim millions of devotees. The well-documented effort of Jewish and Christian conservationists to formulate the Judeo-Christian Stewardship Environmental Ethic in biblical terms suggests an important new line of inquiry: How can effective conservation ethics be formulated in terms of other sacred texts? In *Earth's Insights: A Multicultural Survey of Ecological Wisdom*, a comprehensive survey is offered, but to provide even a synopsis of that study would be impossible in this entry. However, a few abstracts of traditional non-Western conservation ethics may be suggestive.

Muslims believe that Islam was founded in the seventh century A.D., by Allah (God) communicating to humanity through the Arabian prophet, Mohammed, who regarded himself to be in the same prophetic tradition as Moses and Jesus. Therefore, since the Hebrew Bible and the New Testament are earlier divine revelations underlying distinctly Muslim belief, the basic Islamic worldview has much in common with the basic Judeo-Christian worldview. In particular, Islam teaches that human beings have a privileged place in **nature**, and, going further in this regard than Judaism and Christianity, that indeed, all other natural beings were created to serve humanity. Hence, there has been a strong tendency among Muslims to take a purely instrumental approach to the human-nature relationship. As to the conservation of **biodiversity**, the Arabian oryx was hunted nearly to **extinction** by oil-rich sheikhs armed with military assault rifles in the cradle of Islam. But callous indifference to the rest of creation in the Islamic world is no longer sanctioned religiously.

Islam does not distinguish between religious and secular law. Hence, new conservation regulations in Islamic states must be grounded in the Koran, Mohammed's book of divine revelations. In the early 1980s, a group of Saudi scholars scoured the Koran for environmentally relevant passages and drafted *The Islamic Principles for the Conservation of the Natural Environment.* While reaffirming "a relationship of utilization, development, and subjugation for man's benefit and the fulfillment of his interests," this landmark document also clearly articulates an Islamic version of stewardship: "he [man] is only a manager of the earth and not a proprietor, a beneficiary not a disposer or ordainer: (Kadr, et al., 1983). The Saudi scholars also emphasize a just distribution of "natural resources," not only among members of the present generation, but among members of **future generations**. And as Norton (1991) has argued, conservation goals are well served when future human beings are accorded a moral status equal to that of those currently living. The Saudi scholars have found passages in the Koran that are vaguely ecological. For example, God "produced therein all kinds of things in due balance" (Kadr, et al., 1983).

Ralph Waldo Emerson and **Henry David Thoreau**, thinkers at the fountainhead of North American conservation philosophy, were influenced by the subtle philosophical doctrines of Hinduism, a major religion in India. Hindu thought also inspired Arne Naess's (1989) contemporary "Deep Ecology" conservation philosophy. Hindus believe that at the core of all phenomena there is only one Reality or Being. God, in other words, is not a supreme Being among other lesser and subordinate beings, as in the Judeo-Christian-Islamic tradition. Rather, all beings are a manifestation of the one essential Being, called *Brahman.* And all plurality, all difference, is illusory or at best only apparent.

Such a view would not seem to be a promising point of departure for the conservation of biological diversity, since the actual existence of diversity, biological or otherwise, seems to be denied. Yet in the Hindu concept of *Brahman*, Naess (1989) finds an analogue to the way ecological relationships unite organisms into a **systemic** whole. However that may be, Hinduism, unambiguously invites human beings to identify with other forms of life, for all life-forms share the same essence. Believing that one's own inner self, *atman*, is identical, as an expression of *Brahman*, with the selves of all other creatures leads to compassion for them. The suffering of one life-form is the suffering of all others; to harm other beings is to harm oneself. As a matter of fact, this way of thinking has inspired and helped motivate one of the most persistent and successful conservation movements in the world, the Chipko movement, which has managed to rescue many of India's Himalayan forests from commercial exploitation (Guha 1989b; Shiva 1989).

Jainism is a religion with a relatively few adherents, but a religion of great influence in India. Jains believe that every living thing is inhabited by an immaterial soul, no less pure and immortal than the human soul. Bad deeds in past lives, however, have crusted these souls over with *karma*-matter. *Ahimsa* (noninjury of all living things) and asceticism (eschewing all forms of physical pleasure) are parallel paths that will eventually free the soul from future rebirth in the material realm. Hence, Jains take great care to avoid harming other forms of life and to resist the fleeting pleasure of material consumption. Extreme practitioners refuse to eat any but leftover food prepared for others, and carefully strain their water to avoid ingesting any waterborne organisms--not for the sake of their own health, but to avoid inadvertently killing other living beings. Less extreme practitioners are strict vegetarians and own few material possessions. The Jains are bidding for global leadership in environmental ethics. Their low-on-the-food-chain and low-level-of-consumption lifestyle is held up as a model of ecological right livelihood (Chappel 1986). And the author of the *Jain Declaration on Nature* claims that the central Jain moral precept of ahimsa "is nothing but environmentalism" (Singhvi, n.d.).

Though now virtually extinct in its native India, Buddhism has flourished for many hundreds of years elsewhere in Asia. Its founder, Siddhartha Gautama, first followed the path of meditation to experience the oneness of Atman-Brahman, and then the path of extreme asceticism in order to free his soul from his body—all to little effect. Then he realized that his frustration, including his spiritual frustration, was the result of desire. Not by obtaining what one desires—which only leads one to desire something more—but by stilling desire itself can one achieve enlightenment and liberation. Further, desire distorts one's perceptions, exaggerating the importance of some things and diminishing the importance of others. When one overcomes desire, one can appreciate each thing for what it is.

When the Buddha realized all this, he was filled with a sense of joy, and he radiated loving-kindness toward the world around him. He shared his enlightenment with others, and formulated a code of moral conduct for his followers. Many Buddhists believe that all living beings are in the same predicament: we are driven by desire to a life of continuous frustration. And all can be liberated if all can attain enlightenment. Thus Buddhists can regard other living beings as companions on the path to Buddhahood and *nirvana*.

Buddhists, no less than Jains and Christians, are assuming a leadership role in the global conservation movement. Perhaps most notably, the Dalai Lama of Tibet is the foremost conservationist among world religious leaders. In 1985, the Buddhist Perception of Nature Project was launched to extract and collate the many environmentally relevant passages from Buddhist scriptures and secondary literature. Thus, the relevance of Buddhism to contemporary conservation concerns could be demonstrated and the level of conservation consciousness and conscience in Buddhist monasteries, schools, colleges, and other institutions could be raised (Davies 1987). Bodhi (1987) provides a succinct summary of Buddhist environmental ethics: "With its philosophic insight into the interconnectedness and thoroughgoing interdependence of all conditioned things, with its thesis that happiness is to be found through the restraint of desire, with its goal of enlightenment through renunciation and contemplation and its ethic on non-injury and boundless loving-kindness for all beings, Buddhism provides all the essential elements for a relationship to the natural world characterized by respect, care, and compassion."

One-fourth of the world's population is Chinese. Fortunately, traditional Chinese thought provides excellent conceptual resources for a conservation ethic. The Chinese word *tao* means way or road. The Taoists believe that there is a *Tao*, a Way, of Nature. That is, natural processes occur not only in an orderly but also in a harmonious fashion. Human beings can discern the *Tao*, the natural well-orchestrated flow of things. And human activities can either be well adapted to the tao, or they can oppose it. In the former case, human goals are accomplished with ease and grace and without disturbing the natural **environment**; but in the latter, they are accomplished, if at all, with difficulty and at the price of considerable disruption of neighboring social and natural systems. Capital-intensive Western technology, such as nuclear **power plants** and industrial agricultural, is very "unTaoist" in esprit and motif.

Modern conservationists find in Taoism an ancient analogue of today's countermovement toward appropriate technology and **sustainable development**. The great Mississippi Valley flood of 1993 is a case in point. The river system was not managed in accordance with the *Tao*. Thus, levees and flood walls only exacerbated the big flood when it finally came. Better to have located cities and towns outside the flood plain and allowed the mighty Mississippi River occasionally to overflow. The rich alluvial soils in the river's floodplains could be farmed in dryer year, but no permanent structures should be located there. That way, the floodwaters could periodically spread over the land, enriching the **soil** and replenishing **wetlands** for **wildlife**, and the human dwellings on higher ground could remain safe and secure. Perhaps the officers of the U.S. **Army Corps of Engineers** should study Taoism. We can hope that their counterparts in China will abandon newfangled Maoism for old-fashioned Taoism before going ahead with their plans to contain, rather than cooperate with, the Yangtze River.

The other ancient Chinese religious worldview is Confucianism. To most people, Asian and Western alike, Confucianism connotes conservativism, adherence to rigid customs and social forms, filial piety, and resignation to feudal inequality. Hence, it seems to hold little promise as an intellectual soil in which to cultivate a conservation ethic. Ames (1992), however, contradicts the received view: "There is a common ground shared by the teachings of classical Confucianism and Taoism...Both express a 'this-worldly' concern for the concrete details of immediate experience rather than.-..grand abstractions and ideals. Both acknowledge the uniqueness, importance, and primacy of particular persons and their contributions to the world, while at the same time expressing the ecological interrelatedness and interdependence of this person with his context."

From a Confucian point of view, a person is not a separate immortal soul temporarily residing in a physical body; a person is, rather, the unique center of a network of relationships. Since his or her identity is constituted by these relationships, the destruction of one's social and environmental context is equivalent to self-destruction. Biocide, in other words, is tantamount to suicide.

In the West, since individuals are not ordinarily conceived to be robustly related to and dependent upon their context--not only for their existence but for their very identity--then it is possible to imagine that they can remain themselves and be "better off" at the expense of both their social and natural environments. But from a Confucian point of view, it is expanded from its classic social to its current environmental connotation, Confucianism offers a very firm foundation upon which to build a contemporary Chinese conservation ethic.

[J. Baird Callicott]

Resources

Books

Ames, R.T. "Taoist Ethics." In *Encyclopedia of Ethics*, edited by L. Becker. New York: Garland Press, 1992.

Bodhi, B. "Foreword." In *Buddhist Perspectives on the Ecocrisis*. Kandy, Sri Lanka: Buddhist Publication Society.

Callicott, J.B. *Earth's Insights: A Multicultural Survey of Ecological Wisdom*. Berkeley, CA: University of California Press, 1994.

Davies, S. *Tree of Life: Buddhism and Protection of Nature*. Hong Kong: Buddhist Perception of Nature Project, 1987.

Kadr, A., et al. *Islamic Principles for the Conservation of the Natural Environment*. Gland, Switzerland: International Union for the Conservation of Nature and Natural Resources, 1983.

Naess, A. *Ecology, Community, and Lifestyle*. Cambridge, MA: Cambridge University Press, 1989.

Norton, B. G. *Toward Unity among Environmentalists*. New York: Oxford University Press, 1991.

Shiva, V. *Staying Alive: Women, Ecology, and Development*. London: Zed Books, 1989.

Other

Chappel, C. "Contemporary Jaina and Hindu Responses to the Ecological Crisis." Paper presented at the 1990 meeting of the College Theological Society, Loyola University, New Orleans, 1990.

No-observable-adverse-effect-level

The NOAEL is the lowest "dose" or exposure level to a non-cancer-causing chemical at which no increase in adverse effects is seen in test animals compared with animals in a control group. First, a NOAEL is determined. For instance, it is established that no adverse effects are seen in test animals after they have been exposed to a set amount, such as an ounce per day for a year, whereas a higher exposure causes adverse effects. Then this level is divided by one or more uncertainty factors to establish an acceptable human exposure level. Regulatory agencies typically use a factor of 10, 100, or 1,000 to account for the uncertainty arising from the fact that the NOAEL is based on an animal study but is being extrapolated to set a "safe" level of exposure for human beings.

North

Often the term "North" refers to the countries located in the northern hemisphere of the globe. Scholars who are concerned with worldwide problems such as global **climate** change, sometimes tend to think of the planet as consisting of two halves, the North and the South. The North, or northern hemisphere, is a region where only about one-fifth of the Earth's population lives, but where four-fifths of its goods and services are consumed. Environmental issues of interest to the North are those related to high technology, high consumption, and high energy use. These are areas are of relatively less concern to those who live in the South. Although the North/South dichotomy may be simplistic, it highlights differences in the way peoples of various nation view global environmental problems.

North American Association for Environmental Education

The North American Association for Environmental Education, founded in 1971, is a nonprofit network for professionals and students working in the field of environmental education. It claims to be the world's largest association of environmental educators, with members in the United States and more than 55 countries. The organization's stated mission is to "go beyond consciousness-raising about these issues. [It] must prepare people to think together about the difficult decisions they have to make concerning environmental stewardship, and to work together to improve and try to solve environmental problems." It advocates a cooperative and nonconfrontational approach to environmental education.

NAAEE conducts an annual conference and produces publications, including a member newsletter, the *Environmental Communicator*, and a *Directory of Environmental Educators*. It also operates a Skills Bank and numerous programs. Several of these programs focus on community-based activities. The VINE (Volunteer-led Investigations of Neighborhood Ecology) Network promotes programs using volunteers to lead children in explorations of the ecology of their urban neighborhoods. The Urban Leadership Collaboratives Program aims to develop collaborative environmental education programs in cities. The Environmental Issues Forums (EIF) Program assists students and adults in working with controversial environmental issues in their communities and nationally.

Other NAAEE programs have a wider focus. The NAAEE Training and Professional Development Institute offers workshops and training events on such topics as fundraising and long-range planning, as well as courses for educators from developing countries in cooperation with United States Assistance for International Development (USAID) and the Smithsonian Institute. The Environmental Education and Training Partnership (EETAP), under a U.S. **Environmental Protection Agency** grant, coordinates a consortium of 18 environmental education organizations that promote training. The NAAEE Policy Institute aims to develop standards for environmental education. Additional NAAEE activities include writing reports to the United States Congress on environmental education issues, reviewing proposals submitted for funding under the National Environmental Education Act of 1990, and testifying to support environmental education legislation and innovative programs. Internationally, the organization has helped to establish environmental education centers in Moscow and Kiev and a program in Thailand, and has assisted with developing educational materials in Sri Lanka.

NAAEE maintains four member sections that represent different contexts of environmental education and conduct individual activities: **Conservation** Education Section, Elementary and Secondary Education Section, Environmental Studies Section, and Nonformal Section (for members working outside of formal school settings).

In 1996, the Conservation Education Association was incorporated into NAAEE.

[*Carol Steinfeld*]

RESOURCES
ORGANIZATIONS

North American Association for Environmental Education, 410 Tarvin Road, Rock Spring, GA USA 30739 (706) 764-2926, Fax: (706) 764-2094, Email: email@naaee.org, <http://www.naaee.org>

North American Free Trade Agreement

The North American Free Trade Agreement (NAFTA) is an international trade agreement among the United States, Canada, and Mexico that became effective on January 1, 1994. Pursuant to NAFTA most tariffs among the three countries are being phased out over a period of fifteen years. In addition, the agreement liberalizes rules regulating investment by United States and Canadian firms in Mexico. NAFTA includes side agreements on **environment**, labor, and dealing with import surges. Controversy surrounding the adoption of NAFTA by the United States in 1993 included issues related to trade, employment, immigration, labor law, and environmental protection. And, several years after its effective date, observers continue to debate the wisdom of the agreement. The economic slowdown in the United States that began in 2001 has intensified public debate over NAFTA's provisions.

While NAFTA's adoption was being debated, environmentalists were divided in response to whether or not we are environmentally better off with NAFTA. Thus, its adoption was opposed by some environmental groups and supported by others. Major groups opposing it included the **Sierra Club**, **Greenpeace**, the United States Public Interest Research Group (U.S. PIRG), Citizen's Action, Public Citizen: the Clean Water Fund, and the **Student Environmental Action Coalition**. NAFTA was supported, however, by a coalition of major environmental organizations including the **National Wildlife Federation**, the **World Wildlife Fund**, the **National Audubon Society**, the **Environmental Defense** Council, the **Natural Resources Defense Council**, **Conservation International**, and **Defenders of Wildlife**, and **the Nature Conservancy**. Such division among environmentalists with respect to NAFTA and its consequences for the environment continues through the present day.

NAFTA's provisions

NAFTA is first and overall a trade agreement. Thus, the majority of the provisions within its over 1,200 pages focus on elimination of barriers to trade. Over a period of fifteen years, beginning on January 1, 1994, tariffs on over 9,000 products traded among the United States, Mexico, and Canada are being phased out. In addition, other "non-tariff" barriers to trade are being eased or eliminated.

Elimination of tariffs

Tariffs on approximately 4,500 products traded among the NAFTA parties were eliminated on January 1, 1994, and by 1999 tariffs will remain on only about 3,000 products. The remainder will be phased out by the year 2009. The period during which tariffs are phased out gives producers

time to adjust gradually to competition from products of other NAFTA countries. Those products for which tariffs are being phased out gradually are labeled "sensitive products" and include many farm commodities. For example, to protect Mexican producers, corn and dry beans are being protected through a "phase-out" period for tariffs. Producers in the United States are being protected with respect to sugar, asparagus, orange juice concentrate, and melons.

The **automobile** industry is another area for which NAFTA includes special provisions. NAFTA sets out formulas requiring a certain minimum percentage of content from North America (the three NAFTA parties) for an automobile to qualify for duty-free treatment. By 2002, automobiles must contain 62.5% North American content to qualify for such treatment.

The fact that Mexico agreed to abolish tariffs in certain sectors of the economy is important to U.S. investors. For example, telecommunications in Mexico are underdeveloped, and that country has a poor quality telephone system. By 1998, Mexico phased out all tariffs on telecommunications service and equipment.

Removal of non-tariff barriers to trade

NAFTA removes certain barriers to trade that are not in the form of tariffs. For example, NAFTA allows financial service providers of a NAFTA country to establish banking, insurance, securities operations, and other types of financial services in another NAFTA country. NAFTA also provides U.S. and Canadian firms with greater access to Mexico's energy markets. Under NAFTA, U.S. and Canadian firms can sell products to PEMEX, Mexico's federally owned **petroleum** company. In addition, Mexico has opened operation of self-generation, **cogeneration**, and independent **power plants** in Mexico to investment by U.S. or Canadian firms.

Transportation has also been facilitated by NAFTA. Since 1995, U.S., Mexican, and Canadian firms have been allowed to establish cross-border routes. In the early summer of 2002, the United States signed an agreement allowing Mexican trucks to travel throughout the United States. The long-term effects of transportation agreements are difficult to evaluate, however. **Statistics** for 2001 indicate that truck crossings into the United States from Canada and Mexico fell by 4.2% from their level in 2000. This decrease represents the first annual decline since the agreement among the three countries took effect in 1994. On the other hand, this decrease may be only temporary, as it reflects the impact of tightened border security measures following the terrorist attacks of September 11, 2001.

NAFTA does not create a common market for movement of labor, but it does provide for temporary entry of business people from one NAFTA country into another. The four different categories for such movement of workers include business visitors in marketing, sales and related activities, traders and investors, specified kinds of intracompany transfers, and specified professionals meeting educational or special knowledge requirements.

Dispute resolution

Administration of NAFTA is handled by a Commission of cabinet-level officers each of whom has been appointed by one of the NAFTA countries. The Commission includes a Secretary, who is the chief administrator.

The process for resolution of disputes under NAFTA is cumbersome and time-consuming. When a dispute arises with respect to a NAFTA country's rights under the agreement, a consultation can be requested. If the consultation does not resolve the dispute, the Commission will seek to resolve the dispute through alternative dispute resolution procedures. If those are unsuccessful, the complaining country can request the establishment of a five member arbitration panel which will recommend solutions, monitor results, and, if necessary, impose sanctions.

NAFTA's side agreements

In areas involving labor law and environmental protection, NAFTA's provisions are limited.

Labor side agreement

The North American Agreement on Labor Cooperation, also known as the "Side Agreement on Labor" or "Labor Side Agreement," was negotiated in response to concerns that NAFTA itself did little to protect workers in Mexico, the United States, and Canada. In the Labor Side Agreement, the three countries articulate their desire to improve labor conditions and encourage compliance with labor laws. Each party agrees to enforce its own labor laws, but new laws protecting workers are not established. The Agreement establishes a multiple-step process for dealing with instances in which a NAFTA country has exhibited a "consistent pattern of ... failure to effectively enforce its own occupational safety and health, child labor, or minimum wage labor standards." If the three NAFTA parties cannot agree to a resolution, then a committee of experts can be convened to assist in resolving the dispute.

Seven years after NAFTA was enacted, a number of observers consider its side agreement on labor a conspicuous failure that protected international investors at the expense of ordinary workers in all three countries. In the United States, NAFTA led to the elimination of 800,000 manufacturing jobs. Contrary to predictions in 1995, the agreement did not result in an increased trade surplus with Mexico, but the reverse. As manufacturing jobs disappeared, workers in the United States were downscaled to lower-paying, less-secure services jobs. Employers' threats to move production to Mexico became a powerful weapon for undercutting workers' bargaining power. In Canada, a similar process has

brought about a decline in stable full-time employment as well as damaging Canada's social safety net.

Mexican workers have not benefited from this redistribution of jobs. While some manufacturing jobs did move to Mexico, they are largely confined to *maquiladoras* just across the border. *Maquiladoras* are factories that assemble electronics equipment and small appliances for export. Workers are typically paid low wages and have few rights. The *maquiladoras* are isolated from the rest of the Mexican economy and contribute very little to its development. In addition, the expiration of the duty-free status of *maquiladora* products in 2001 means that these factories are less competitive in the global market. As a result, many foreign companies began to shut down their plants in Mexico in 2002.

Environmental side agreement

The North American Agreement on Environmental Cooperation, also known as the "Environmental Side Agreement," was adopted along with the Labor Side Agreement and was negotiated in response to concerns that the body of NAFTA did little to protect the environment. In the Environmental Side Agreement, the three parties articulate their desire to promote environmental concerns without harming the economy, and promote open public discussion of environmental issues. As in the Labor Side Agreement with respect to labor laws, each party agrees to enforce its own environmental laws, but no new environmental laws are made. The agreement establishes a **Commission for Environmental Cooperation** (CEC) that includes a Secretariat and a Council, which is charged with implementing the side agreement.

The Environmental Side Agreement allows a **nongovernmental organization** (such as an environmental group) or a private citizen to file a complaint with the Commission asserting that a NAFTA country "has shown a persistent pattern of failure to effectively enforce its environmental laws." Thus, such groups and individuals can serve as watchdogs for the Commission.

The Environmental Side Agreement allows a NAFTA country to request a consultation with a second NAFTA country regarding that second country's persistent pattern of failure to enforce its environmental laws. If a resolution is not reached as a result of the consultation, a party can request that the Council meet. If the Council is unable to resolve the dispute, it can convene an arbitration panel, which will study the dispute and issue a report. After the parties submit comments, the report is made final. The parties are given an opportunity to reach their own action plan, but, if there is no such agreement, the panel can be reconvened to establish a plan. If an imposed fine is not paid, as under the Labor Side Agreement, NAFTA benefits can be suspended.

Environmental consequences of NAFTA

Environmentalists share many concerns with respect to NAFTA, three of which will be introduced here. First, environmentalists are concerned that NAFTA represents a threat to the sovereignty of the United States because U.S. environmental laws may be challenged under NAFTA. For example, it is feared that the U.S. will not be allowed to refuse entry of goods produced in Mexico even if those goods are produced using production methods that do not meet U.S. environmental, health, or food safety standards. This is based on U.S. experience under the General Agreement on Tariffs and Trade (GATT) when a GATT panel declared that a U.S. embargo on Mexican tuna caught using methods violating the U.S. Marine Mammal Protection Act was illegal. The GATT panel held that products must be treated equally regardless of how they are produced.

Ironically, the first major challenge to national sovereignty came from Canadian corporations. A Canadian conglomerate known as Methanex made use of a little-known section of NAFTA known as Chapter 11 to sue the state of California, demanding financial compensation for loss of market share. Methanex is the world's largest producer of the key ingredient in a **gasoline** additive known as MBTE, a substance that California ordered to be phased out after it began to contaminate public water systems in 1995. A phrase in Chapter 11 written to protect investors from having their property seized by foreign governments has been invoked to justify corporate lawsuits against state and federal environmental regulations. Other examples of cross-border lawsuits include another Canadian company's suit against the state of Mississippi, and a United States corporation's suit against the government of Canada for banning another gasoline additive.

Second, environmentalists label Mexico as a "pollution haven" for U.S. businesses seeking less-stringent enforcement of environmental laws than that which occurs in the United States. Concerns about **environmental degradation** in Mexico due to increased industrialization do seem to be valid. Environmentalists, business leaders, and government officials from the three NAFTA countries agree that environmental contamination had reached serious proportions in northern Mexico in the decades preceding the 1994 effective date of NAFTA. Such contamination continues to spread as Mexico becomes more heavily industrialized under NAFTA. A report published by the CEC in 2002 indicates that chemical producers accounted for the largest amount of pollutants.

On paper, Mexico has environmental laws that parallel many, but not all, of the federal environmental laws of the United States. Mexico's 1988 Código Ecológico, known in English as the "General Ecology Law," is a comprehensive statute addressing **pollution**, resource **conservation**, and

environmental enforcement. Often, however, Mexico's environmental laws are not enforced. This lack of enforcement can be attributed to a variety of reasons. For one thing, the statutes are relatively new, dating only from 1988, and regulations have not been made to supplement the laws. Also, Mexico is a country in which people distrust the legal system and where there are widespread violations of law in various areas (not limited to environmental laws.) Noncompliance with law is accepted and expected in many areas in Mexico. Since Mexico is a developing country striving to implement open, democratic systems of government, this casual attitude toward legislation is a significant hurdle to be overcome throughout Mexican society.

Third, Mexico is a debt-ridden country that continues to struggle with periodic financial crises. Thus, even if Mexico's leaders were convinced that there should be vigorous enforcement of environmental laws and regulations, Mexico lacks money for such enforcement. In connection with its lack of capital, Mexico lacks technology and it lacks infrastructure such as **water treatment** plants, sewage systems, and waste disposal facilities. For example, as of mid-1997, in the entire country of Mexico, there was only one facility licensed for disposal of hazardous wastes. Lacking a place to dispose of toxics and knowing the low **probability** of being cited by government officials for violations, most companies and individuals continue to dump hazardous wastes into Mexico's waters and air and onto its land. Mexico's citizens do not want to live with the contamination any more than we do in the United States. They feel, however, that they are less empowered to "do something about it" than we are as U.S. citizens.

Looking ahead

NAFTA represents a significant turning point in international law, because it recognizes explicitly that trade policy and **environmental policy** are inextricably linked. Through its Environmental Side Agreement, it establishes mechanisms for dealing with citizens' concerns and for dealing with that party or country that systematically fails to enforce its own environmental laws. However, NAFTA's environmental provisions provide only a foundation upon which citizens and governments of the U.S., Mexico, and Canada must build if the environment of North America is to be protected from the often detrimental effects of expanded trade. As of 2002, those detrimental effects have been sufficiently obvious to produce a growing chorus of criticisms of NAFTA. Proposals to create a Free Trade Area of the Americas (FTAA), which would represent an extension of NAFTA's provisions to every country in Central America, South America, and the Caribbean except Cuba, have provoked widespread protest from a variety of environmental and social justice organizations.

[*Rebecca J. Frey, Ph.D.*]

Resources
Books

Bello, J. H., A. F. Holmer, and J. J. Norton, eds. *NAFTA: a New Frontier in International Trade and Investment in the Americas.* American Bar Association Section of International Law & Practice, 1994.

Hufbauer, G. C., and J. J. Schott, *NAFTA: An Assessment.* The Institute for International Economics, 1993.

MacArthur, John R. *The Selling of Free Trade: NAFTA, Washington, and the Subversion of American Democracy.* New York: Farrar, Straus and Giroux, 2000.

Moran, R. T., and J. Abbott, *NAFTA: Managing the Cultural Differences.* Gulf Publishing Company, 1994.

Periodicals

Franz, Neil. "Report Tracks NAFTA Region Emissions." *Chemical Week* 164 (June 5, 2002): 39.

Knight, Danielle. "Environment Groups Organise Against NAFTA Rules." *International Third World Press Service (IPS)*, September 8, 2000.

Smith, Geri. "The Decline of the Maquiladora." *Business Week.* April 29, 2002.

Stenzel, P.L. "Can NAFTA's Environmental Provisions Promote Sustainable Development?" 59 *Albany Law Review* 59 (1995): 423–480.

Other

Economic Policy Institute. *NAFTA at Seven.* Washington, DC: Economic Policy Institute, 2001.

Global Exchange. *Documentary Exposing How NAFTA's Chapter 11 Has Become Private Justice for Foreign Companies.* January 7, 2002.

Williams, Edward J. *The Maquiladora Industry and Environmental Degradation in the United States–Mexican Borderlands.* Paper presented at the annual meeting of the Latin American Studies Association, Washington, DC, September 1995.

Organizations

NAFTA Information Center, Texas A&M International University, College of Business Administration, 5201 University Blvd., Laredo, TX USA 78041-1900 (956)326-2550, Fax: (956)326-2544, Email: nafta@tamiu.edu, <www.tamiu.edu/coba/usmtr/>

North American Water and Power Alliance

Numerous schemes were suggested in the 1960s to accomplish large-scale water transfers between major basins in North America, and one of the best known is the North American Water and Power Alliance (NAWAPA). The plan was devised by the Ralph M. Parson Company of Los Angeles "to divert 36 trillion gallons of water (per year) from the Yukon River in Alaska (through the Great Bear and Great Slave Lakes) southward to 33 states, seven Canadian provinces, and northern Mexico."

The proposed NAWAPA system would bring water in immense quantities from western Canada and Alaska through the plains and **desert** states all the way down to the Rio Grande **watershed** and into the Northern Sonora

and Chihuahua provinces of Mexico. The Rocky Mountain Trench, Peace River, Great Slave Lake, Lesser Slave Lake, North Saskatchewan River, Columbia River, Lake Winnipeg, **Hudson River**, James Bay, and numerous tributaries are part of the proposed NAWAPA feeder system designed to channel water from Canada to Mexico.

A second feeder system in the plan would channel large quantities of water into the western portion of Lake Superior. This influx of water would wash the pollutants dumped into the **Great Lakes** out into the Atlantic Ocean. It would also boost the capacity of the area for generating hydroelectricity.

The NAWAPA plan was also designed to develop hydroelectric plants within northern Quebec and Ontario which would produce power that would be diverted to the United States. The **James Bay hydropower project** in Quebec was completed in the early 1970s. It flooded an area 4,250 mi^2 (11,000 km^2), and 90% of its power goes directly into the Northeastern United States and Ohio. The James Bay II Project in Ontario will eventually incorporate over 80 **dams**, divert three major rivers, and flood traditional Cree land. The majority of its hydroelectric output will also go to the United States.

Proponents of inter-basin transfers tend to focus on the impending water shortages in the western and southwestern United States. In the Great Plains, the **Ogallala Aquifer** is rapidly being depleted. The Black Mesa **water table** is almost exhausted, due to the excessive quantities of water used in mining operations, and California has been consistently unable to meet the needs of both its industries and its population. Supporters of NAWAPA have long argued that this plan is the only way the nation can solve these problems. On February 22, 1965, *Newsweek* hailed the NAWAPA plan as "the greatest, the most colossal, stupendous, super-splendificent public works project in history."

NAWAPA was described as "a monstrous concept—a diabolical thesis" by a former chairperson of the **International Joint Commission**. Much of the opposition to the plan in the 1960s was nationalist rather than environmental in character: The plans were viewed as an attempt to appropriate Canadian resources. Today, many people are asking whether it is necessary or even right to hydrologically re-engineer the ecosystems of North America in order to meet the water needs of the United States. Environmentalists point out that entire ecosystems in many western states have already been disrupted by various water projects. They argue that it is time to investigate other methods, such as **conservation**, which would bring water consumption to levels sustainable by the watersheds of the plains and deserts.

[*Debra Glidden*]

Resources

Books

Higgins, J. "Hydro-Quebec and Native People." *Cultural Survival Quarterly* 11 (1987).

Reisner, M. *Cadillac Desert: The American West and Its Disappearing Water.* New York: Viking Press, 1986.

Royal Society of Canada. *Water Resources of Canada.* Ottawa: Royal Society of Canada, 1968.

Welsh, F. *How to Create a Water Crisis.* Boulder: Johnson Publishing, 1985.

Periodicals

Canadian Council of Resource Ministers. *Water Diversion Proposals of North America.* Ottawa: Canadian Council of Resource Ministers, 1968.

Northern spotted owl

The northern spotted owl (*Strix occidentalis caurina*) is one of three subspecies of the spotted owl (*Strix occidentalis*). Adults are brown, irregularly spotted with white or light brown spots. The face is round with dark brown eyes and a dull yellow colored bill. They are 16–19 in (41–48 cm) long and have wing spans of about 42 in (107 cm). The average weight of a male is 1.2 lb (544 g), whereas the average female weighs 1.4 lb (635 g).

This subspecies of the spotted owl is found only in the southwestern portion of British Columbia, western Washington, western Oregon, and the western coastal region of California south to the San Francisco Bay. Occasionally the bird can be found on the eastern slopes of the Cascade Mountains in Washington and Oregon. It is estimated that there are about 3,000–5,000 individuals of this subspecies.

The other two subspecies of spotted owl are the California spotted owl (*S. o. occidentalis*) found in the coastal ranges and western slopes of the Sierra Nevada mountains from Tehama to San Diego counties, and the Mexican spotted owl (*S. o. lucida*) found from northern Arizona, southeastern Utah, southwestern Colorado, south through western Texas to central Mexico.

It is thought that spotted owls mate for life and are monogamous. Breeding does not occur until the birds are two to three years of age. The typical clutch size is two, but sometimes as many as four eggs are laid in March or early April. The incubation period is 28–32 days. The female performs the task of incubating the eggs while the male bird brings food to the newly-hatched young. The owlets leave the nest for the first time when they are around 32–36 days old. Without fully mature wings, the young are not yet able to fly well and must often climb back to the nest using their talons and beak. Juvenile **survivorship** may be only 11%.

Spotted owls hunt by sitting quietly on elevated perches and diving down swiftly on their prey. They forage during the night and spend most of the day roosting. Mammals make up over 90% of the spotted owl's diet. The most

important prey **species** is the northern flying squirrel (*Glaucomys sabrinus*) which makes up about 50% of the owl's diet. Woodrats and hares also are important. In all, 30 species of mammals, 23 species of birds, two reptile species, and even some invertebrates have been found in the diets of spotted owls.

Northern spotted owls live almost exclusively in very old coniferous forests. They are found in virgin stands of Douglas fir, western hemlock, grand fir, red fir, and areas of **redwoods** that are at least 200 years old. They favor areas that have an old-growth overstory with layers of second-growth understory beneath. The overstory is the preferred nesting site and the owls tend to build their nests in trees that have broken tops or cavities, or on stick platforms. In one study, 64% of the nests were in cavities, and the remainder were on stick platforms or other debris on tree limbs. All of the nests in this study were in conifers, all but two of which were living.

Little is known about what features of a stand are critical for spotted owls. The large trees that have nest sites may be important, particularly those producing a multi-layered canopy in which the owls can find a benign **microclimate**. A thick canopy may be critical in sheltering juvenile owls from avian predators, whereas the understory may be important in providing a cool place for the birds to roost during the warm summer months.

Because of this subspecies' dependence on old-growth coniferous forests and because it feeds at the top **trophic level** in the **old-growth forest food chain/web**, it is considered an "indicator species." Indicator species are used by ecologists to measure the health of the **ecosystem**. If the indicator species is endangered, then it is likely that scores of other species in the ecosystem are just as endangered.

The owls are nonmigratory, with dispersal of young being the only regularly observed movement out of established home ranges. The home range size of spotted owls varies from an average of 4,200 acres (1,700 ha) in Washington to about 2,000 acres (800 ha) in California. In 1987, a team of scientists recommended that in order to be reasonably sure of the species' survival that **habitat** for 1,500 pairs be set aside. This would necessitate preserving 4–5 million acres (1.5–2 million ha) of old-growth forests—most of what remains.

Unfortunately for these owls, old-growth forests are a scarce habitat which is commercially valuable for timber. Because of the demand for old-growth timber these birds have been the center of controversy between timber interests and environmentalists. The declining numbers of owls alarm preservationists who want old-growth forests set aside to protect the owls, while the loggers feel it is in the public's best interest to continue to cut the economically valuable old-growth timber. Timber companies claim that 12,000 jobs will be lost along with about $300 million annually if felling is restricted.

Northern spotted owl. (Photograph by John and Karen Hollingsworth. U.S. Fish & Wildlife Service.)

It has been argued that since old-growth forests are being destroyed, these jobs and revenue will be lost eventually anyway. It has also been argued that the U. S. **Forest Service**, which manages most of the remaining old-growth forests, subsidizes the timber industry by building expensive access roads and selling the timber at artificially low prices. Environmentalists suggest that the social costs associated with not cutting old growth could be mitigated by redirecting these monies to retraining programs and income supplements.

In 1990, the northern spotted owl was designated by the U. S. **Fish and Wildlife Service** as a "threatened species." This requires that the owl's habitat be protected from **logging**. Although the decision to list the spotted owl as "threatened" did not affect existing logging contracts, timber companies are trying to avoid compliance with the decision. Specifically, they are trying to persuade the President and Congress to revise the **Endangered Species Act** to allow consideration of economic impacts or to make a specific exception for some or all of the spotted owl's habitat. Currently, under certain circumstances, economic factors can take precedence over biological criteria in deciding whether it is necessary to comply with habitat protection measures.

In these cases a special seven-person interdisciplinary committee can assess the economic impacts of protecting the habitat and circumvent the **Endangered Species** Act if they believe it is warranted. President Bush's Secretary of Interior Manuel Lujan convened the committee to consider allowing logging in spotted owl habitat on some **Bureau of Land Management** lands.

A team of scientists appointed by the federal government to study the situation recommended that the annual harvest on old-growth forests be reduced by 47%. However, former President George Bush rejected this recommendation and instead proposed that harvest be reduced by 21%. This angered both the environmentalists and the timber industry and the two sides became deadlocked. In the meantime, spotted owl policy is being determined by federal judges rather than biologists. For example, it was a court order that forced the U. S. Fish and **Wildlife** Service to identify 11.6 million acres (4.7 million ha) as **critical habitat**. In 1991, a federal judge issued an injunction stopping all new timber sales in areas where the spotted owls live on **national forest** land. The judge also mandated that the Forest Service produce a **conservation** plan and **Environmental Impact Statement** by March 1992. This controversy continued into the presidency of Bill Clinton who convened a Forest Summit in Portland, Oregon on April 2, 1993, to gather information from loggers and environmentalists. Following the summit President Clinton asked his cabinet to devise a balanced solution to the old-growth forest dilemma within 60 days.

Ultimately the fate of the northern spotted owl will be decided in the court rooms and halls of government, where environmentalists and timber interests continue to battle. It is important to realize that the dispute is not merely over one species of owl. The spotted owl is just one of many species dependent on old-growth forests, and may not be in the greatest danger of **extinction**. As an indicator of the prosperity of old-growth ecosystems in the Pacific Northwest, though, its survival means continued health for the entire **biological community**.

[*Ted T. Cable*]

Resources

Books

Hunter Jr., M. L. *Wildlife, Forests, and Forestry.* Englewood Cliffs, NJ: Prentice-Hall, 1990.

"Northern Spotted Owl." *Beacham's Guide to the Endangered Species of North America.* Farmington Hills, MI: Gale Group, 2001.

Periodicals

Casey, C. "The Bird of Contention." *American Forests* 97 (1991): 28–68.

Other

"Northern Spotted Owl." *The Sierra Club.* [cited May 2002]. <http://www.sierraclub.org/lewisandclark/species/owl.asp>.

Not In My Backyard

NIMBY is an acronym for "Not In My Backyard" and is often heard in discussions of **waste management**. While every community needs a site for waste disposal, frequently no one wants it near his or her home. In the early part of this century, in fact, up until the early 1970s, the town dump was a smelly, rodent-infested place that caught fire on occasion. Loose debris from these facilities would also blow onto adjacent property. Citizens were justified in their aversion to landfills because **hazardous waste** and **chemicals** were often dumped into landfills, which contaminated **groundwater** and surface water. After the passage of the **Resource Conservation and Recovery Act** (RCRA) in 1976, many of these facilities were closed and dumps converted into sanitary landfills that required daily cover, fencing, leachate collection systems, and other design elements that made them much better neighbors. Still, many citizens refused to have landfills close to their homes.

In addition to landfills, citizens are also concerned about having other waste management facilities near them. Large open air **composting** facilities are not popular because of the odor they produce. A materials **recycling** facility (MRF) is not always a desirable neighbor because of the noise it generate.

Despite assurances by experts of the improvements and safety of modern waste management facilities, communities continue to be wary of them. In order to overcome the NIMBY attitude, professionals in the waste management field must collaborate with communities so that citizens gain understanding and ownership of **solid waste** management problems. The problem of waste is generated at the community level so the solution must be generated at the community level. With sensible planning and patience there may be less NIMBY-mentality in the future.

[*Cynthia Fridgen*]

Resources

Books

Brion, D. J. *Essential Industry and the NIMBY Phenomenon: A Problem of Distributive Justice.* Westport, CT: Greenwood, 1991.

Piller, C. *The Fail-Safe Society: Community Defiance and the End of American Technological Optimism.* New York: Basic Books, 1991.

Periodicals

Guerra, S. "NIMBY, NIMTOF, and Solid Waste Facility Siting." *Public Management* 73 (October 1991): 11–15.

Shields, P. "Overcoming the NIMBY Syndrome." *American City and County* 105 (May 1990): 54.

NRC

see **Nuclear Regulatory Commission**

Nuclear accidents

see **Chelyabinsk, Russia; Chernobyl Nuclear Power Station; Radioactive pollution; Three Mile Island Nuclear Reactor; Windscale (Sellafield) plutonium reactor**

Nuclear fission

When a **neutron** strikes the nucleus of certain isotopes, the nucleus breaks apart into two roughly equal parts in a process known as nuclear fission. The two parts into which the nucleus splits are called fission products. In addition to fission products, one or more neutrons is also produced. The fission process also results in the release of large amounts of energy.

The release of neutrons during fission makes possible a **chain reaction**. That is, the particle needed to initiate a fission reaction—the neutron—is also produced as a result of the reaction. Each neutron produced in a fission reaction has the potential for initiating one other fission reaction. Since the average number of neutrons released in any one fission reaction is about 2.3, the rate of fission in a block of material increases rapidly.

A chain reaction will occur in a block of fissionable material as long as neutrons (1) do not escape from the block and (2) are not captured by nonfissionable materials in the block. Two steps in making fission commercially possible, then, are (1) obtaining a block of fissionable material large enough to sustain a chain reaction—the critical size—and (2) increasing the ratio of fissionable to nonfissionable material in the block—enriching the material.

Atomic bombs, developed in the 1940s, obtain all of their energy from fission reactions while **hydrogen** bombs use fission reactions to trigger **nuclear fusion**. A long-term environmental problem accompanying the use of these weapons is their release of radioactive fission products during detonation.

The energy available from fission reactions is far greater, pound for pound, than can be obtained from the **combustion** of **fossil fuels**. This fact has made fission reactions highly desirable as a source of energy in weapons and in power production.

Many experts in the post-World War II years argued for a massive investment in nuclear **power plants**. Such plants were touted as safe, reliable, nonpolluting sources of energy. When operating properly, they release none of the pollutants that accompany power generation in fossil fuel plants. By the 1970s, more than a hundred **nuclear power** plants were in operation in the United States.

Then, questions began to arise about the safety of nuclear power plants. These concerns reached a peak when the cooling water system failed at the **Three Mile Island Nuclear Reactor** at Harrisburg, Pennsylvania, in March 1979. That accident resulted in at least a temporary halt in nuclear power plant construction in the United States. No new plants have been authorized since that time. A much more serious accident occurred at Chernobyl, Ukraine, in 1986 when one of four reactors on the site exploded, spreading a cloud of radioactive material over parts of the USSR, Poland, and northern Europe.

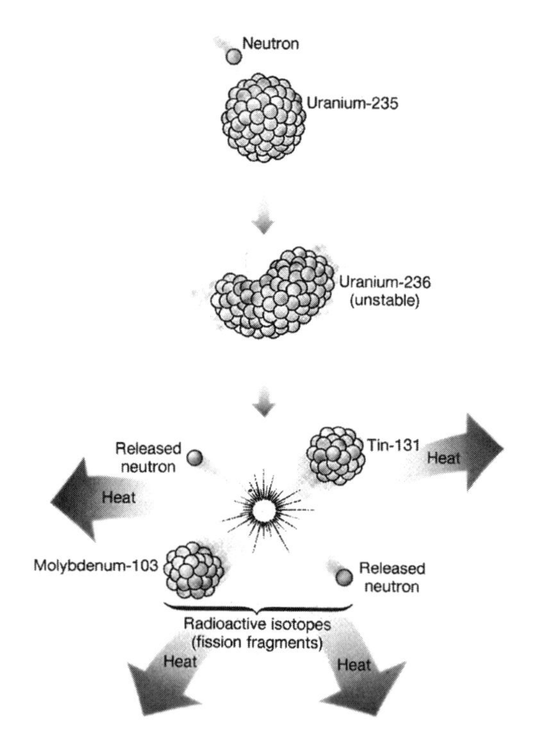

The process of nuclear fission as carried out in the core of a nuclear reactor. A neutron strikes the unstable isotope Uranium-235. This isotope absorbs the neutron and splits or fissions into tin-131 and molybdenum-103. Two or three neutrons are released per fission event and continue the chain reaction. The reaction product has a total mass slightly less than the starting material with the residual mass converted into energy (primarily heat). (McGraw-Hill Inc. Reproduced by permission.)

Perhaps the most serious environmental concern about fission reactions relates to fission products. The longer a fission reaction continues, the more fission products accumulate. These fission products are all radioactive, some with short half lives, other with longer half lives. The former can be stored in isolation for a few years until their **radioactivity** has reduced to a safe level. The latter, however, may remain hazardous for hundreds or thousands of years. As of the early 1990s, no completely satisfactory method for storing

these nuclear wastes had been developed. *See also* Nuclear fusion; Nuclear weapons; Radiation exposure; Radioactive pollution; Radioactive waste; Radioactive waste management; Radioactivity

[*David E. Newton*]

RESOURCES
BOOKS

Fowler, J. M. *Energy-Environment Source Book.* Washington, DC: National Science Teachers Association, 1975.

Inglis, D. R. *Nuclear Energy: Its Physics and Social Challenge.* Reading, MA; Addison-Wesley, 1973.

Joesten, M. D., et al. *World of Chemistry.* Philadelphia: Saunders, 1991.

Nuclear fusion

The process by which stars produce energy has always been of great interest to scientists. Not only would the answer to that puzzle be of value to astronomers, but it might also suggest a method by which energy could be generated for human use on earth.

In 1938, German-American physicist Hans Bethe suggested a method by which **solar energy** might be produced. According to Bethe's hypothesis, four **hydrogen** atoms come together and fuse—join together—to produce a helium atom. In the process, very large amounts of energy are released.

This process is not a simple one, but one that requires a series of changes. In the first step, two hydrogen atoms fuse to form an atom of deuterium, or "heavy" hydrogen. In later steps, hydrogen atoms are regenerated, providing the materials needed to start the process over again. Like **nuclear fission**, then, nuclear fusion is a **chain reaction**.

Since Bethe's original research, scientists have discovered other fusion reactions. One of these was used in the first practical demonstration of fusion on earth, the "hydrogen" bomb. It involved the fusion of two hydrogen isotopes, deuterium and tritium.

Nuclear fusion reactions pose a difficult problem. Fusing isotopes of hydrogen requires that two particles with like electrical charges be forced together. Overcoming the electrical repulsion of these two particles requires the initial input into a fusion reaction of very large amounts of energy. In practice, this means heating the materials to be fused to very high temperatures, a few tens of millions of degrees Celsius. Because of these very high temperatures, fusion reactions are also known as thermonuclear reactions.

Temperatures of a few millions of degrees Celsius are common in the center of stars, so nuclear fusion can easily be imagined there. On earth, the easiest way to obtain such

Tokamak nuclear fusion reactor in Oak Ridge, Tennessee. (Phototake. Reproduced by permission.)

temperatures is to explode a fission (atomic) bomb. That explosion momentarily produces temperatures of a few tens of millions of degrees Celsius. A fusion weapon such as a hydrogen bomb consists, therefore, of nothing other than a fission bomb surrounded by a mass of hydrogen isotopes.

As with nuclear fission, there is a strong motivation to find ways of controlling nuclear fusion reactions so that they can be used for the production of power. This research, however, has been hampered by some extremely difficult technical challenges. Obviously, no ordinary construction material can withstand the temperatures of the hot, gaseous-like material, or **plasma**, involved in a fusion reaction. Efforts have been aimed, therefore, at finding ways of containing the reaction with a magnetic field. The tokamak reactor, originally developed by Russian scientists, appears to be one of the most promising methods of solving this problem.

Research on controlled fusion has been a slow, but continuous, progress. Some researchers are confident that a solution is close at hand. Others doubt the possibility of bringing nuclear fusion under human control. All agree, however, that successful completion of this research could provide humans with perhaps the "final solution" to their energy needs.

[*David E. Newton*]

RESOURCES
BOOKS

Hippenheimer, T. A. *The Man-Made Sun: The Quest for Fusion Power.* Boston: Little Brown, 1984.

Inglis, D. R. *Nuclear Energy: Its Physics and Social Challenge.* Reading, MA; Addison-Wesley, 1973.

PERIODICALS

Lidsky, L. M. "The Trouble With Fusion." *Technology Review* (1984): 52–6.

Rafelski, J., and S. E. Jones. "Cold Nuclear Fusion." *Scientific American* 257 (1987): 84–9.

Nuclear power

When nuclear reactions were first discovered in the 1930s, many scientists doubted they would ever have any practical application. But the successful initiation of the first controlled reaction at the University of Chicago in 1942 quickly changed their views.

In the first controlled nuclear reaction, scientists discovered a source of energy greater than anyone had previously imagined possible. They discovered that the nuclei of **uranium** isotopes could be split, thus releasing tremendous energy. The reaction occurred when the nuclei of certain isotopes of uranium were struck and split by neutrons. This is now known as **nuclear fission**, and the fission reaction results in the formation of three types of products: energy, neutrons, and smaller nuclei about half the size of the original uranium nucleus.

Neutrons are actually produced in a fission reaction, and this fact is critical for energy production. The release of neutrons in a fission reaction means that the particles required to initiate fission are also a product of the reaction. Once initiated in a block of uranium, fission occurs over and over again, in a **chain reaction**. Calculations done during these early discoveries showed that the amount of energy released in each fission reaction is many times greater than that released by the chemical reactions that occur during a conventional chemical explosion.

The possibility to release such high energies with nuclear reactions was used in the development of the atomic bomb. After the dropping of this bomb brought World War II to an end, scientists began researching the harnessing of nuclear energy for other applications, primarily the generation of electricity. In developing the first **nuclear weapons**, scientists only needed to find a way to initiate nuclear fission—there was no need to control it once it had begun. In developing the peacetime application of nuclear power however, the primary challenge was to develop a mechanism for keeping the reaction under control once it had begun so that the energy released could be managed and used. This is the main purpose of nuclear power plants—controlling and converting the energy produced by nuclear reactions.

There are many types of nuclear **power plants**, but all plants have a reactor core and every core consists of three elements. First, the fuel rods; these are long, narrow, cylindrical tubes that hold small pellets of some fissionable material. At present only two such materials are in practical use, uranium-235 and plutonium-239. The uranium used for nuclear fission is known as enriched uranium, because it is actually a mixture of uranium-235 with uranium-238. Uranium-238 is not fissile and the required chain reaction will not occur if the fraction of uranium-235 present is not at least 3%.

The second component of a reactor core is the moderator. Only slow-moving neutrons are capable of initiating nuclear fission, but the neutrons produced as a result of nuclear fission are fast-moving. These neutrons move too fast to initiate other reactions, thus moderators are used to slow them down. Two of the most common moderators are graphite (pure **carbon**) and water.

The third component of a reactor core is the control rods. In operating a nuclear power plant safely and efficiently, it is of the utmost importance to have exactly the right amount of neutrons in the reactor core. If there are too few, the chain reaction comes to an end and energy ceases to be produced. If there are too many, fission occurs too quickly, too much energy is released all at once, and the rate of reaction increases until it can no longer be controlled or contained. Control rods decrease the number of neutrons in the core because they are made of a material that has a strong tendency to absorb neutrons. **Cadmium** and boron are materials that are both commonly used. The rods are mounted on pulleys allowing them to be raised or lowered into the reactor core as need may be. When the rods are fully inserted, most of the neutrons in the core are absorbed and relatively few are available to initiate a chain reaction. As the rods are withdrawn from the core, more and more neutrons are available to initiate fission reactions. The reactions reach a point where the number of neutrons produced in the core is almost exactly equal to the number being used to start fission reactions, and it is then that a controlled chain reaction occurs.

The heat energy produced in a reactor core is used to boil water and make steam, which is then used to operate a turbine and generate electricity. The various types of nuclear power plants differ primarily in the way in which heat from the core is used to do this. The most direct approach is to surround the core with a huge tank of water, some of which can be boiled directly by heat from the core. One problem with boiling-water reactors is that the steam produced can be contaminated with radioactive materials. Special precautions must be taken with these reactors to prevent contaminated

steam from being released into the **environment**. A second type of nuclear reactor makes use of a heat exchanger. Water around the reactor core is heated and pumped to a heat exchange unit, where this water is used to boil water in an external system. The steam produced in this exchange is then used to operate the turbine and generator.

There is also a type of nuclear reactor known as a breeder, or fast-spectrum reactor because it not only produces energy but also generates more fuel in the form of plutonium-239. In conventional reactors, water is used as a coolant as well as a moderator, but in breeder reactors the coolant used is sodium. Neutrons have to be moving quickly to produce **plutonium**, and sodium does not moderate their speed as much as water does. Another design, the CANDU reactor, (acronym for CANada Deuterium Uranium) uses deuterium oxide (heavy water) as moderator and natural uranium as fuel. With the uranium fuel surrounded by heavy water, chain reaction fission takes place, releasing energy in the form of heat. The heat is transferred to a second heavy water system pumped at high pressure through the tubes to steam generators, from which the heat is transferred to ordinary water which boils to become the steam that drives the turbine generator.

Nuclear power plants could never explode with the power of an atomic bomb, because the quantity of uranium-235 required is never present in the reactor core. However, they do pose a number of well-known safety hazards. From the very beginning of the development of nuclear reactors, safety was an important consideration as scientists and engineers tried to anticipate the dangers associated with nuclear reactions and radioactive materials. Thus, control rods were developed to prevent the fission reactions from generating too much heat. The reactor and its cooling system are always enclosed in a containment shell made of thick sheets of steel to prevent the escape of radioactive materials. Nuclear power plants are highly complex facilities, with back-up systems for increased safety which are themselves supported by other back-up systems. But the components of these systems age, and human errors can and do occur; safety measures do not always function the way there were designed.

On December 2, 1957, the first nuclear power plant opened in Shippingport, Pennsylvania, and to many the nuclear age seemed to have begun. Over the next two decades, more than 50 plants were commissioned, with dozens more ordered. But safety problems plagued the industry. An experimental reactor in Idaho Falls, Idaho, had already experienced a partial meltdown as a result of operator error in 1955. In October 1957, just months before the Shippingport plant came on line, a production reactor near Liverpool, England, caught fire, releasing radiation over Great Britain and northern Europe.

The most critical event in the history of nuclear power in the United States was the accident at the **Three Mile Island nuclear reactor** near Harrisburg, Pennsylvania. In March 1979, fission reactions in the reactor core went out of control, generating huge amounts of heat, and a meltdown resulted. Fuel rods and the control rods were melted; the cooling water was turned to steam and the containment structure itself was threatened. No new plants have been ordered in the United States since this accident, and 65 plants on order at that time were cancelled. The explosion at the Chernobyl reactor near Kiev, Ukraine, dealt a second blow to the industry.

Even without these accidents, another problem with nuclear power would remain. This is the problem of spent radioactive wastes. About a third of the 10 million fuel pellets used in any reactor core must be removed each year because they have been so contaminated with fission by-products that they no longer function efficiently. These highly radioactive pellets must be disposed of in a safe fashion, but 50 years after the first controlled reaction, no method has yet been discovered to address this issue. Today, these wastes are most commonly stored on a temporary basis at or near the power plant itself. Many have argued that further development of nuclear power should not even be considered until better methods for **radioactive waste management** have been developed.

The International Nuclear Safety Center (INSC), which operates under the guidance of the Director of International Nuclear Safety and Cooperation in the **U.S. Department of Energy** (DOE), has the mission of improving nuclear power reactor safety worldwide. The INSC is dedicated to developing improved nuclear safety technology and promoting the open exchange of nuclear safety information among nations, sponsoring scientific research activities as collaborations between the U.S. and its international partners. Safety issues are addressed at several levels, including: **risk assessment**, containment, structural integrity of reactors, assessment of their seismic reliability, equipment operability, fire protection, and reactor safeguards.

The security of nuclear facilities has also been a point of growing concern. In 1991, the NRC instituted an Operational Safeguards Response Evaluation (OSRE) program, which evaluated the ability of nuclear facility security personnel to withstand a staged commando-style attack by intruders. Unfortunately, six of the 11 evaluations performed in 2000 and 2001 resulted in the "attackers" being able to penetrate security and simulate damage to reactor equipment.

In the post-September 11 terrorist attack environment, the vulnerability of America's 104 nuclear power facilities are a critical national security concern. All nuclear facilities were placed on high alert immediately following the attacks,

Nuclear power plants are a peacetime application of energy initiated through nuclear fission. (Photograph by Robert J. Huffman. Field Mark Publications. Reproduced by permission.)

and in early 2002 the **Nuclear Regulatory Commission** (NRC) issued interim confidential security orders for all licensees to comply with. In addition, decommissioned nuclear plants and spent fuel storage facilities were also required to implement the security orders. The NRC also conducted a thorough review of its Internet web site, taking the site offline temporarily to analyze content and remove all documents deemed sensitive to national security. In April 2002, the NRC announced the establishment of a dedicated department for plant security, the Office of Nuclear Security and Incident Response.

The future role of nuclear power in energy production throughout the world is uncertain. But the current absence of nuclear power plant development in the United States should not be taken as indicative of future trends, as well as trends in the rest of world. France, for example, obtains more than half of its electricity from nuclear power, despite the safety problems. And many believe that nuclear power should be an important part of energy production in the United States as well. Proponents of nuclear power argue that its dangers have been greatly exaggerated in this country. The risks, they argue, must be compared with the health and environmental hazards of other fuels, particularly **fossil fuels**.

In 1999, the U.S. Department of Energy announced a new initiative called Generation IV. Designed to generate interest and scientific research in nuclear power advances, the program set out the following goals for the next generation (i.e., generation IV) of nuclear power plants: 1) highly economical; 2) enhanced safety; 3) sustainable with minimal waste; and 4) commercially viable by 2025. Ten countries, including the United States, United Kingdom, Canada, South Africa, and Japan, were participating in the Generation IV International Forum (GIF) as of early 2002. One of the most promising prototypes reactor designers have come up with is the gas-cooled pebble-bed reactor, which uses a non-reactive helium coolant and operates more efficiently and at less cost than current water-cooled systems.

For some, hope for the future of nuclear power rests with the development of **nuclear fusion**. A nuclear power plant based on a fusion reaction would amount to a controlled

version of a **hydrogen** bomb, just as conventional nuclear plants are equivalent to a controlled version of an atomic bomb. But the problem of managing the reaction is far more difficult with fusion than it is with fission, and scientists have been working on this issue unsuccessfully for more than 40 years. Some believe that a fusion power plant could become a reality in the next century, but many now doubt that such a plant will ever be feasible.

[*David E. Newton and Paula Anne Ford-Martin*]

Resources

Books

Cheney, Glenn A. *Journey to Chernobyl: Encounters in a Radioactive Zone.* Chicago: Academy Chicago Pub, 1995.

Garwin, Richard L. and Georges Charpak. *Megawatts and Megatons: A Turning Point in the Nuclear Age.* New York: Knopf, 2001.

Hodgson, Peter. *Nuclear Power, Energy, and the Environment.* London: World Scientific Publishing, 1999.

Wilpert, Bernard, et al., eds. *Safety Culture in Nuclear Power Organizations.* London: Taylor & Francis, 2001.

Periodicals

Lake, James et al. "Next-Generation Nuclear Power." *Scientific American* 286, no. 1 (January 2002): 72(8).

Other

Union of Concerned Scientists. *Briefing: Nuclear Reactor Security.* [cited July 2002]. <http://www.ucsusa.org/energy/br_safenplants.html>.

U.S. Department of Energy, Office of Nuclear Energy, Science, and Technology. *Generation IV: The Future of Nuclear Power.* [cited July 2002]. <http://gen-iv.ne.doe.gov>.

Organizations

Sandia National Laboratories: Energy and Critical Infrastructure, PO Box 5800, Albuquerque, NM USA 87185 (505) 284-5200, <http://www.sandia.gov>.

U.S. Nuclear Regulatory Commission, Office of Public Affairs (OPA), Washington, DC USA 20555 (301) 415-8200, Toll Free: (800) 368-5642, Email: opa@nrc.gov, <http://http://www.nrc.gov>.

Nuclear Regulatory Commission

The development of **nuclear weapons** during World War II raised a number of difficult nonmilitary questions for the United States. Most scientists and many politicians realized that the technology used in weapons research had the potential for use in a variety of peacetime applications. In fact, research on techniques for controlling **nuclear fission** for the production of electricity was well under way before the end of the war.

An intense Congressional debate over the regulation commercial **nuclear power** resulted in the creation in 1946 of the **Atomic Energy Commission** (AEC). The AEC had two major functions: to support and promote the development of nuclear power in the United States and to monitor and regulate the applications of nuclear power.

Some critics pointed out early on that these two functions were inherently in conflict. How could the AEC act as a vigorous protector of the public safety, they asked, if it also had to encourage industry growth? The validity of that argument did not become totally obvious for nearly two decades. It was not until the early 1970s that the suppression of information about safety hazards from existing plants by the AEC became public knowledge.

The release of this information prompted Congress and the President to rethink the government's role in nuclear power issues. The result of that process was the **Energy Reorganization Act** of 1973 and Executive Order 11834 of January 15, 1975. These two actions established two new governmental agencies, the Nuclear Regulatory Commission (NRC) and the Energy Research and Development Agency (ERDA). NRC was assigned all of AEC's old regulatory responsibilities, while ERDA assumed its energy development functions.

The mission of the Nuclear Regulatory Commission is to ensure that the civilian uses of nuclear materials and facilities are conducted in a manner consistent with public health and safety, environmental quality, national security, and anti-trust laws. The single most important task of the commission is to regulate the use of nuclear energy in the generation of electric power.

In order to carry out this mission, the commission has a number of specific functions. It is responsible for inspecting and licensing every aspect of nuclear power plant construction and operation, from initial plans through actual construction and operation to disposal of **radioactive waste** materials. The commission also contracts for research on issues involving the commercial use of nuclear power and holds public hearings on any topics involving the use of nuclear power. An important ongoing NRC effort is to establish safety standards for nuclear **radiation exposure**.

A fair amount of criticism is still directed at the Nuclear Regulatory Commission. Critics feel that the NRC has not been an effective watchdog for the public in the area of nuclear safety. For example, the investigation of the 1979 **Three Mile Island Nuclear Reactor** accident found that the commission was either unaware of existing safety problems at the plant or failed to inform the public adequately about these problems.

[*David E. Newton*]

Resources

Organizations

U.S. Nuclear Regulatory Commission, Office of Public Affairs, Washington, D.C. USA 20555 (301) 415-8200, Toll Free: (800) 368-5642, Email: opa@nrc.gov, <http://www.nrc.gov>

Nuclear test ban

The Comprehensive Test Ban Treaty (CTBT), which prohibits all nuclear explosions either for military or "peaceful" purposes, was adopted by the United Nations General Assembly in September 1996. Although disputes remain in some countries over the ratification of the treaty, the adoption of the text by the General Assembly marks the culmination of several decades of intermittent negotiation aimed at the worldwide prohibition of nuclear explosions within the United Nations nuclear nonproliferation framework. The CTBT aims at preventing the development and proliferation of **nuclear weapons** and is viewed by the United Nations as an important step toward eventual nuclear disarmament.

President Dwight D. Eisenhower first suggested a ban on atmospheric testing in 1955, but it was not until August 1963, in the aftermath of the 1962 superpower showdown over the stationing of nuclear-capable missiles in Cuba, that the United States and the U.S.S.R. agreed to a Partial Test Ban Treaty (PTBT). This agreement banned all atmospheric tests of nuclear weapons, but allowed both parties to set off explosions in the more frequently-used underground tests. With both superpowers concerned about issues of sovereignty and espionage, the PTBT limited procedures for verification of compliance to what were called "national technical means," which consisted mainly of satellite, communications, and long-distance seismic monitoring. While the PTBT's overall impact on arms control is generally thought to have been negligible, it did serve to prevent a great deal of **pollution** of the **atmosphere** and the Earth's surface by **plutonium** and other radioactive substances.

In 1974, the PTBT was supplemented by the signing of a Threshold Test Ban Treaty (TTBT) that limited the explosive yield of underground nuclear tests. Initially, the TTBT also relied on national technical means for verification. But in 1990, the original protocol was replaced with a new agreement on verification means that allowed for on-site inspections and in-country seismic monitoring.

Fourteen years after the adoption of the PTBT, the United States, U.S.S.R., and United Kingdom convened trilateral talks aimed at achieving a comprehensive test ban. Some limited progress was made toward resolving several points of disagreement, but in general the parties achieved few tangible results in these negotiations. This lack of success combined with a general toughening of United States policy towards the U.S.S.R. after the 1980 elections to all but end discussion of a test ban treaty.

Against the wishes of both the United States and the United Kingdom, a series of amendments was suggested in January 1991, under the authority of the existing treaties' revision procedures. A conference of the nuclear powers was convened to consider the proposed amendments. These negotiations were inconclusive, but they did signal considerable international interest in a comprehensive ban. Political momentum toward this end was given a boost in 1992, when Russia, France, and the United States declared a nine-month moratorium on all testing. In doing so, the United States reversed official doctrine introduced in the 1980s, which held that testing via explosions was a necessary component of national security. The radical geopolitical changes associated with the disintegration of the U.S.S.R. also played a role in changing the climate of negotiations. As a result, the 1994 Conference on Disarmament created a committee with a mandate to establish a framework for a comprehensive ban. By extending its own moratorium on nuclear tests, the Clinton Administration signaled its willingness to sign the resulting CTBT, giving the final process of negotiation an important boost.

Technological and political developments, however, are already raising new issues for arms control experts and negotiators. Advances in communications and surveillance technology aid the verification of compliance. But advances in nuclear physics and weapons technology, such as the advent of so-called micro-nukes and hydronuclear tests with greatly reduced yields, make the very definition of nuclear testing problematic. Several countries, especially India and Pakistan, have expressed strong misgivings about the CTBT. Although the United States strongly supports the treaty on an official level, substantial and deep-rooted opposition to the CTBT persists, including some notable critics among those charged with oversight of its huge nuclear arsenal. Opponents of the treaty argue that limited testing is necessary for checking the safety and reliability of existing nuclear warheads and for maintaining the credibility of the nuclear deterrent.

[*Lawrence J. Biskowski*]

Nuclear waste

see **Radioactive waste**

Nuclear weapons

There are two types of nuclear weapons, each of which utilizes a different nuclear reaction: **nuclear fission** and **nuclear fusion**. The bomb developed by the Manhattan Project and dropped on Japan during World War II were fission weapons, also known as atomic bombs. **Hydrogen** bombs are fusion weapons, and these were first developed and produced during the early 1950s.

In fission weapons, the explosive energy is derived from nuclear fission, in which large atomic nuclei are split into two roughly equal parts. Every time a nucleus divides,

it releases a large amount of energy. Each fission reaction also produces one or more neutrons, the subatomic particles that are needed to initiate a fission reaction. Thus, once a fission reaction has been started in a few nuclei, it rapidly spreads to other nuclei around it, creating a **chain reaction**. The two nuclei most commonly used for this type of nuclear reaction are uranium-235 and plutonium-239.

The necessary fission reaction will not occur if an atomic bomb carries any single piece of fissionable material that is more than a few kilograms. The bomb can contain more than one piece of this size, but it seldom contains more than three or four. Thus there is a limit to the size of a fission weapon, as well as the energy it can release. Nuclear weapons and the force of their blasts are measured in kilotons, each of which is equivalent to a thousand tons of TNT. Fission weapons are limited to 20 or 30 kilotons.

Fusion weapons derive their explosive power from a reaction that is the opposite of fission. In fusion, two small nuclei combine or fuse, releasing large amounts of energy in the process. The materials needed to initiate another fusion reaction are produced as a byproduct. Fusion, like fission, is a cyclic reaction.

In contrast to fission weapons, a hydrogen bomb can be of almost any size. Such a bomb consists of a fission bomb at the core, surrounded by a mass of hydrogen isotopes used in the fusion reactions. No limit exists to the mass of hydrogen that can be used, and there is thus no theoretical limit to the size of fusion weapons. The practical limit is simply the necessity of transporting it to a target; the bomb cannot be so large that a rocket or an airplane is unable to carry it effectively. Both fission and fusion weapons are often classified as strategic or tactical. Strategic weapons are long-range weapons intended primarily for attack on enemy land. Tactical weapons are designed for use on the battlefield, and their destructive power is adjusted for their shorter range.

Nuclear weapons cause destruction in a number of different ways. They create temperatures upon explosion that are, at least initially, millions of degrees hot. Some of their first effects are heat effects, and materials are often incinerated on contact. The heat from the blast also causes rapid expansion of air, resulting in very high winds that can blow over buildings and other structures. A weapon blast also releases high levels of radiation, such as neutrons, x rays, and gamma rays. Humans and other animals close to the center of the blast suffer illness and death from **radiation exposure**. The set of symptoms associated with such exposure is known as **radiation sickness**. Many individuals who survive radiation sickness eventually develop **cancer** and their offspring frequently suffer genetic damage. Finally, a weapon's blast releases huge amounts of radioactive materials. Some of these materials settle out of the **atmosphere** almost immediately, creating widespread contamination. Others remain in the atmosphere for weeks or months, resulting in long-term **radioactive fallout**.

Because of their destructive power, the nations of the world have been trying for many years to reach agreements on limiting the manufacture and possession of nuclear weapons. Between July 16, 1945, and September 23, 1992, the United States of America conducted (by official count) 1054 nuclear tests, and two nuclear attacks. In 1963, the United States and the former Soviet Union agreed to a Limited Nuclear Test Ban Treaty that banned explosions in the atmosphere, outer space, and underwater. After 1963, both nations continued testing underground. The 1974 Threshold Test Ban Treaty restricted the underground testing of nuclear weapons by the United States and the former Soviet Union to yields no greater than 150 kilotons.

In the 1980s and 1990s, debates over arms control and continued testing of nuclear weapons were affected by a number of international developments. The first was the growing internal weakness and eventual disintegration of the former Soviet Union. After a decade of difficult negotiations, the United States and the Soviet Union signed a bilateral Strategic Arms Reduction Agreement, or START I, on July 31, 1991. Five months later the Soviet Union dissolved. It was succeeded by four states that had nuclear weapons on their territories: Russia, the Ukraine, Belarus, and Kazakhstan. On December 5, 2001, the United States and the Russian Federation succeeded in reducing their number of deployed warheads to the level specified by START I. The Ukraine, Belarus, and Kazakhstan have eliminated or removed from their territory all nuclear weapons left over from the Soviet period. The START II treaty, which specifies further reductions in the number of nuclear weapons possessed by both powers, was ratified by the U.S. Senate on January 26, 1996, and by the Russian Duma on April 14, 2000. On May 24, 2002, President George W. Bush of the United States and President Vladimir Putin of Russia signed the Strategic Offensive Reductions Treaty, which commits both nations to reduce the number of their nuclear warheads to two-thirds of their 2002 levels by December 31, 2012.

The second international development of concern has been the steady increase in the number of states possessing nuclear weapons, and the potential for their use in international conflicts. As of the spring of 2002, the Carnegie Institute's Non-Proliferation Project estimates that 35 nations have some type of nuclear ballistic missile. It is difficult to obtain precise information about the number and location of these weapons because many countries in the so-called "nuclear club" have not openly acknowledged their membership.

The third development since the 1980s is the growing possibility that terrorist groups might acquire the informa-

tion and materials to build nuclear devices. This fear acquired new prominence after the terrorist attacks of September 11, 2001. As early as 1997 Alexander Lebed, a Russian general, claimed that the former Soviet Union had developed nuclear weapons small enough to fit in a suitcase. Although technical information is lacking on the Russian side, the United States' experimental development of small-sized nuclear weapons suggests that it is technically possible to design a bomb small enough to fit in a large suitcase. In addition, it is known that members of al-Qaeda made several attempts to purchase commercial reactor-grade **plutonium** during the 1990s. As of 2002, however, the likelihood of a terrorist group's constructing a small portable nuclear device is not high. The construction of a suitcase bomb using plutonium would require a very high degree of sophistication, and would be extremely hazardous to its creators. On the other hand, there are no substantial technical obstacles for a small terrorist group to manufacture a simple but highly destructive nuclear bomb about the size of a truck. One scientific expert has stated, "There is no doubt that if a group like al-Qaeda were to obtain sufficient fissile material—no more than 12 kg of plutonium, or 50 kg of highly enriched **uranium**, and quite possibly less—a highly destructive [nuclear] bomb could be constructed."

[*Rebecca J. Frey Ph.D.*]

Resources

Books

Dresser, P. D., ed. *Nuclear Power Plants Worldwide.* Detroit, MI: Gale Research, 1993.

Jagger, J. *The Nuclear Lion: What Every Citizen Should Know about Nuclear Power and Nuclear War.* New York: Plenum, 1991.

Weinburg, A. M. *Continuing the Nuclear Dialogue.* La Grange Park, IL: American Nuclear Society, 1985.

Periodicals

Ahearne, J. F. "Nuclear Power after Chernobyl." *Science* (May 8, 1987): 673–79.

Other

National Academy of Sciences, Committee on International Security and Arms Control. *The Future of U.S. Nuclear Weapons Policy.* Washington, DC: National Academy Press, 1997.

Robinson, C. Paul. *A White Paper: Pursuing a New Nuclear Weapons Policy for the Twenty-First Century.* Albuquerque, NM: Sandia National Laboratories, 2002.

Sublette, Carey. "Could Al-Qaeda Go Nuclear?" *EnviroWeb.* May 18, 2002 [cited July 11, 2002]. <http://www.nuketesting.enviroweb.org>.

Organizations

Federation of American Scientists, 1717 K Street, NW, Suite 209, Washington, DC USA 20036 (202) 546-3300, <www.fas.org>.

National Nuclear Security Administration, Sandia National Laboratories, New Mexico, P. O. Box 5800, Albuquerque, NM USA 87185 (925) 294-3000, <www.sandia.gov>.

Nuclear weapons testing

see **Bikini atoll; Nevada Test Site**

Nuclear winter

Nuclear winter is a term given to what would happen to Earth's **environment** following a large-scale nuclear war. In effect, the entire planet would be plunged into a very bitter winter that would last months or years. The most severe consequence would be that Earth's **climate** would become too cold to grow vital food crops, including rice, wheat, and corn. It would likely lead to the starvation of hundreds of millions of people who survived the initial nuclear blasts. In effect, it would likely be the end of civilization.

However, with the end of the Cold War and the breakup of the Soviet Union, the threat of a full-scale nuclear exchange between the United States and Russia has greatly diminished. Today, the term nuclear winter is also used to describe the aftermath of any event that sends huge amounts of **particulate** matter into the **atmosphere**, blocking the sun and drastically cooling Earth's atmosphere. Such events could include the simultaneous eruption of a number of large volcanoes, or the impact upon the planet of a large comet or asteroid.

Although eight countries are known to possess **nuclear weapons** as of 2002, it was the thousands of warheads held by the United States and Russia that most worried scientists during the Cold War. A nuclear exchange between other countries, such as India and Pakistan, would not create a nuclear winter because most of the world's croplands would be spared.

Many of the horrible consequences of a global nuclear conflict have been well known for nearly half a century. Though much research has been conducted in this area, there is still much to be learned. Many scientists believe that the **smoke** produced as a result of a nuclear conflict would reduce the transmission of sunlight to Earth's surface over a significant portion of the planet. According to the theory, this reduction in sunlight would then cause a cooling on land surfaces of anywhere from 18–65°F (10–35°C). Interior parts of a continent would experience greater cooling than would coastal regions. Some parts of the planet that now experience a temperate climate might be plunged into winter-like conditions.

The idea of nuclear winter has been controversial ever since it was first proposed in 1983 by famed astronomer Carl Sagan (1934–1996). The U.S. **National Academy of Sciences**, the U.S. Office of Science and Technology Policy, the World Meteorological Organization, the International Council of Scientific Unions, and the **Scientific Committee on Problems of the Environment**, as well as dozens of

individual scientists, have analyzed the potential for a nuclear winter and the problems that it might engender. These studies have not resolved the many issues surrounding a possible nuclear winter effect, but they have clarified a number of factors involved in that effect.

First, the fundamental logic behind a possible nuclear winter effect has been considerably strengthened. Experts estimate that about 9,000 teragrams (9 trillion grams) of finished lumber exist in the world. A large portion of that is found in urban areas that are likely to be the focus of a nuclear attack. By one estimate, anywhere from 25–75% of the wood in an urban area would be ignited in a nuclear attack. A second resource vulnerable to nuclear attack is the world's **petroleum** reserves. Most experts agree that oil refineries, oil storage centers, and other concentrations of oil deposits would be likely targets in a nuclear attack. These materials would also serve as fuel in a widespread fire storm.

The main product of concern in the **combustion** of wood, petroleum, and other materials (**plastics**, tar, asphalt, vegetation, etc.) would be sooty smoke. This smoke would decrease solar radiation by as much as 50%. In addition, it would not easily be washed out of the atmosphere by rain, snow, and other forms of precipitation. Computer models have also shown that soot in the atmosphere may be more stable than first imagined. Studies also suggest that a nuclear winter effect could produce serious consequences for the **ozone** layer. One effect would be the dislocation of the ozone layer over the Northern Hemisphere towards the Southern Hemisphere. Another effect would involve actual destruction of the ozone layer by **nitrogen** oxide molecules carried aloft by smoke.

Scientists have studied a number of natural phenomena with nuclear winter-like effects. Volcanic eruptions, massive forest fires, natural dust clouds, urban fires, and extensive wild fires all produce massive amounts of smoke similar to what would be expected in a nuclear conflict. For example, massive wildfires in China during May of 1987 were found to reduce daytime temperatures in Alaska by 4–12°F (2–6°C) in ensuing months. Possible climatic effects from the enormous oil well fires during the **Persian Gulf War**, as well as recent volcanic eruptions, are also being studied. One scientist studying this phenomenon has said, however, that "severe environmental anomalies—possible leading to more human casualties globally than the direct effects of nuclear war—would be not just a remote possibility, but a likely outcome."

A number of scientists have challenged Sagan's findings, saying he overestimated the devastation that would result from a large-scale nuclear war. Several scientists suggested the result of such a war would be a "nuclear fall" rather than nuclear winter. While this would also be a disaster, the scientists suggested that it would not mean the end of civilization as Sagan predicted. The scientists explained that this is because Sagan greatly overestimated both the amount of smoke and dust that would be created and its duration in the atmosphere.

[*Ken R. Wells*]

Resources

Books

Sagan, Carl, and Richard Turco. *A Path Where No Man Thought: Nuclear Winter and the End of the Arms Race.*New York: Random House, 1990.

Periodicals

Marshall, Eliot. "Nuclear Winter Debate Heats Up." *Science* (January 16, 1987): 271–274.

Overbye, D. "Prophet of the Cold and Dark." *Discover* (January 1985): 24–32.

Seitz, Russell. "An Incomplete Obituary." *Forbes* (February 10, 1997): 123–124.

Sinberg, Stan. "Springtime for Nuclear Winter." *Omni* (April 1992): 98–99.

Turco, R. P., et al. "Climate and Smoke: An Appraisal of Nuclear Winter." *Science* (January 12, 1990): 166–167.

Organizations

Manchester University Computing Society, Oxford Road, Room G33, Manchester, Great Britain M13 9PL 44(0)161-275-7329, Email: admin@compsoc.man.ac.uk, <http://www.compsoc.man.ac.uk/~samp/nuclearage/effect.html>

Nucleic acid

Nucleic acids are macromolecules composed of polymerized nucleotides. Nucleotides, in turn, are structured of phosphoric **acid**, pentose sugars, and organic bases. Deoxyribonucleic acid (DNA) most commonly exists as a double stranded helix. The genetic information of some viruses, bacteria, and all higher organisms is encoded in DNA, and the physical basis of heredity of these organisms is dependent upon the molecular structure of DNA.

DNA is transcribed into single stranded **ribonucleic acid** (RNA), which is then translated into protein. The conversion of the genetic information of a **species** into the fabric of that organism involves several kinds of RNA, *viz.*, messenger RNA (mRNA), transfer RNAs (tRNA), and ribosomal RNAs (rRNA). The genetic material of some viruses is RNA.

Nutrient

All plants and animals require certain **chemicals** for growth and survival. These chemicals are called biogenic salts or nutrients (from the Latin word *nutrio* meaning to feed, rear, or nourish). They can be categorized as those needed in large amounts called macronutrients, and those needed in

minute amounts called micronutrients or trace elements. Macronutrients include **nitrogen** (an essential building block of chlorophyll and protein), **phosphorus** (used to make DNA and ATP), calcium (a component of cell walls and bones), sulfur (a component of amino acids), and magnesium (a component of bones and chlorophyll). Micronutrients, although needed only in trace amounts, are still essential for survival. Examples include cobalt (used in the synthesis of vitamin B_{12}), iron (essential for **photosynthesis** and blood respiratory pigment), and sodium (used in the maintenance of proper acid-based balance called osmoregulation, nerve transmission, and several other functions). Some micronutrients, such as **copper** and zinc, can be harmful in large amounts. The borderline between necessary and excessive is often narrow and varies among **species**.

O

Oak Ridge, Tennessee

Along with towns such as Los Alamos, New Mexico, and the area around the Savannah River in Georgia, Oak Ridge is central to the history of the development of **nuclear weapons** in the United States. It has also come to represent many of the environmental consequences of nuclear research and weapons production.

Oak Ridge was a small, sleepy town when it was selected as a research site in the 1940s for the development of the atomic bomb. Amidst an **atmosphere** of intense secrecy, the government built the Oak Ridge National Laboratories within a period of months and assembled a force of 75,000 scientists. These physicists, engineers, and others worked under extreme security to design various components of the **hydrogen** bomb. Their research was carried out under the auspices of the Manhattan Project, though the tasks were compartmentalized and few scientists are thought to have been aware of the larger significance of their work.

After the end of World War II, the laboratories at Oak Ridge were used for the research activities and weapons production of the Cold War. Although the area did experience decreases in the number of highly trained personnel, the research center remained an important part of the government's network of national laboratories. During this period, the **U.S. Department of Energy** (DOE) took over the management of the Oak Ridge laboratories and subcontracted administrative operations to such private corporations as Union Carbide and the Martin Marietta Corporation. Peacetime activities also included major research and educational initiatives developed in association with the local university.

Because of the urgency and secrecy under which military research was conducted at Oak Ridge, little attention was paid to the health impacts of radiation or the safe disposal of **hazardous waste**. Starting in 1951, the research facilities were responsible for storing 2.7 million gallons (10.2 million liters) of concentrated acids and radioactive wastes in open ponds. 76,000 rusting barrels and drums containing mixed radioactive wastes remained on the site. Millions of cubic yards of toxic and **radioactive waste** were also buried in the ground with no containment precautions. 2.4 million pounds (1 million kg) of **mercury** and an unknown amount of **uranium** are estimated to have been released into the ambient **environment** through the air, water and **soil** pathways. The DOE has spent $1.5 billion evaluating the level of contamination and planning **remediation** and treatment activities, and that figure is expected to grow exponentially. Cleanup programs are now the focus of much of the research done by the nuclear scientists at Oak Ridge.

Vegetation and **wildlife** (water fleas, **frogs** and deer) around the laboratories have set off high readings of **radioactivity** in Geiger counters. **Radionuclides** such as strontium, tritium and **plutonium** have been traced in surface waters 40 miles (64 km) downstream of the plant. However, few published studies exist on human health effects from hazardous waste disposal in the area. Studies that examined short term **cancer** rates in the male worker population found no correlation between cancer risk and worker **radiation exposure**; these studies also noted that Oak Ridge employees were actually 20% less likely to die from cancer as the rest of the country—a fact which may be due to the quality of their medical care. Yet studies that followed the health patterns of workers over a period of 40 years documented that cancer risk did indeed increase by 5% with each rem of increasing radiation exposure. These studies also found that white male workers at the laboratories had a 63% higher **leukemia** death rate than the national average.

In 1977 large amounts of employee health records were deliberately destroyed at the Oak Ridge National Laboratories, and the DOE has attempted to influence the interpretation and publication of health studies on the effects of radiation exposure. These facts raise troubling questions for many about the level of knowledge at the research center of the effects of low-level radiation. Both the laboratories and the town itself are considered case studies of the environmental problems that can be caused by large federal facilities operating under the protection of national security.

[*Douglas Smith*]

RESOURCES

PERIODICALS

"Low-Level Radiation: Higher Long-Term Risk?" *Science News* 139 (March 23, 1991): 181.

Thompson, D. "Living Happily Near a Nuclear Trash Heap." *Time* 139 (May 11, 1992): 53–54.

Occupational Safety and Health Act (1970)

The Occupational Safety and Health Act (1970) was intended to reduce the incidence of personal injuries, illness, and deaths as a result of employment. It requires employers to provide each of their employees with a workplace that is free from recognized hazards, which may cause death or serious physical harm. The act directed the creation of the **Occupational Safety and Health Administration** (OSHA) within the U. S. Department of Labor. This agency is responsible for developing and promulgating safety and health standards, issuing regulations, conducting inspections, and issuing citations as well as proposing fines for violations. *See also* Right-to-act legislation

Occupational Safety and Health Administration

The Occupational Safety and Health Administration (OSHA) was established pursuant to the **Occupational Safety and Health Act** (1970). Within the Department of Labor, OSHA has the responsibility for occupational safety and health activities pursuant to the Act which covers virtually every employer in the country except all types of mines which are regulated separately under the Mine Safety and Health Act (1977). The Occupational Safety and Health Administration develops and promulgates standards, develops and issues regulations, conducts inspections to insure compliance, and issues citations and proposes penalties. In the case of a disagreement over the results of safety and health inspections performed by OSHA, employers have the right of appeal to the Occupational Safety and Health Review Commission, which works to ensure the timely and fair resolution of these cases.

RESOURCES

ORGANIZATIONS

U.S. Department of Labor, Occupational Safety & Health Administration, 200 Constitution Avenue, Washington, D.C. USA 20210 Toll Free: (800) 321-OSHA, <http://www.osha.gov>

Ocean Conservatory, The

"The will to understand, conserve, and protect ocean life is at the very core of the Ocean Conservatory's mission. To fulfill this mission, the Ocean Conservatory seeks to: protect marine ecosystems, prevent **marine pollution**, protect endangered marine **species**, manage fisheries for **conservation**, and conserve marine biodiversity." Since its founding in 1972, the Ocean Conservatory (formerly the Center for Marine Conservation) has worked toward these goals. The **endangered species** the group helps to protect include, among many others, **whales**, **dolphins**, seabirds, **seals and sea lions**, and **sea turtles**.

Over 110,000 Ocean Conservatory members nationwide volunteer their time in many different ways, including writing to Congress asking for support of marine conservation, and organizing or participating in beach cleanups across the country. In 1988, for example, more than 16,000 people took part in beach cleanups in Florida and Texas. This number grew and in 2000, 850,000 people cleaned over 20,000 mi (32,187 km) of coast. More than 100 countries have helped the United States in this effort to clean the beaches.

Because of human actions and the ever-changing **environment**, the Ocean Conservatory's goals constantly grow and change. Over the years they have challenged formidable opponents, including Exxon. After the ***Exxon Valdez*** oil spill, the Ocean Conservatory participated in the cleanup and forced Exxon to step up the rescue and rehabilitation of sea otters (*Enhydra lutris*) and other mammals and birds injured by the accident.

Another important Ocean Conservatory activity is the Marine **Habitat** Program. Through this program, the Ocean Conservatory has played a pivotal role in establishing six marine sanctuaries in the United States (there are only eight total), as well as in other countries. One of these, the Silver Back Humpback Whale Sanctuary, protects humpback whales (*Megaptera novaeangliae*) near the Dominican Republic. Established in 1986, it was the world's first whale sanctuary.

Research is an important component of all Ocean Conservatory programs. The group's ongoing research and advocacy has resulted in the adoption of stricter regulations on commercial **whaling** by the International Whaling Commission. The Ocean Conservatory has also helped to pass federal and state regulations requiring the use of turtle ex-

cluder devices (TEDs) on shrimp nets to help protect thousands of endangered sea turtles. Turtles have also benefited from the Ocean Conservatory's research in artificial lighting. The group convinced counties and cities throughout Florida to control the use of artificial light on the state's beaches after proving that it prevents turtles from nesting and lures baby turtles away from their natural habitat.

In 1988 the Ocean Conservatory established a database on marine debris and, subsequently, the group created two Marine Debris Information offices, one in Washington, D.C. and one in San Francisco. These offices provide information on marine debris, especially **plastics**, to scientists, policy makers, teachers, students, and the general public. While many of the Ocean Conservatory's projects are exemplary, one has received so much attention that it has been used as a model by other environmental groups, including the **Environmental Protection Agency** (EPA). The California Marine Debris Action Plan, which went into effect in 1994, is a comprehensive strategy for combating marine debris, and a collaborative effort of the Ocean Conservatory and a wide network of public and private organizations. In 2000 President Clinton enacted the Oceans Act which set up an Oceans Commission to review and revise all policies dealing with the protection of the ocean and the coast. The most recent win for the Ocean Conservatory came in 2001, when Tortugus (200 square nautical miles) was established as the largest no-take marine reserve near Key West, Florida.

[*Cathy M. Falk*]

Resources

Organizations

The Ocean Conservancy, 1725 DeSales Street, Suite 600, Washington, D.C. USA 20036 (202) 429-5609, Email: info@oceanconservancy.org, <http://www.oceanconservancy.org>

Ocean dumping

Ocean dumping is internationally defined as "any deliberate disposal at sea of wastes or other matter from vessels, aircraft, platforms, or other man-made structures at sea, and any deliberate disposal at sea of vessels, aircraft, platforms, or other man-made structures at sea." The **discharge** of sewage and other effluents from a pipeline and the discharge of waste incidental to, or derived from the normal operations of, ships are not considered ocean *dumping*. Wastes have been dumped into the ocean for thousands of years. Fish and fish processing wastes, rubbish, industrial wastes, sewage **sludge**, dredged material, **radioactive waste**, pharmaceutical wastes, drilling fluids, munitions, **coal** wastes, cryolite, ocean **incineration** wastes, and wastes from ocean mining have all been dumped at sea. Ocean dumping has historically been more economically attractive, when compared with other land-based **waste management** options.

The 1972 **Convention on the Prevention of Marine Pollution by Dumping of Wastes and Other Matter**, commonly called the London Convention, came into force in 1975 to control ocean dumping activities. The framework of the London Convention includes a *black list* of materials that may not be dumped at sea under any circumstances (including radioactive and industrial wastes), a *grey list* of materials considered less harmful that may be dumped after a special permit is obtained, and criteria that countries must consider before issuing an ocean dumping permit. These criteria require the consideration of the effects dumping activities can have on marine life, amenities, and other uses of the ocean, and encompass factors related to disposal operations, waste characteristics, attributes of the site, and availability of land-based alternatives. Most ocean dumping permits (90–85%) reported by parties to the London Convention are for dredged material. Currently only three parties to the London Convention dump sewage sludge at sea. Other categories of wastes that are permitted for ocean dumping include inert, geological materials, vessels, and fish wastes. The International Maritime Organization is responsible for administrative activities related to the London Convention, and it facilitates cooperation among the countries party to the Convention. As of 2001, 78 countries had ratified the Convention.

In the early 1990s, the parties to the London Convention began a comprehensive review of the treaty. This resulted in the adoption of the 1996 Protocol to the London Convention. The purpose of the Protocol is similar to that of the Convention, but the Protocol is more restrictive: application of a "precautionary approach" is included as a general obligation; a "reverse list" approach is adopted; incineration of wastes at sea is prohibited; and export of wastes for the purpose of dumping or incineration at sea is prohibited. Under the reverse list approach, dumping is prohibited unless explicitly permitted, and only seven categories of materials may be considered for dumping: dredge material; sewage sludge; fish waste; vessels and platforms; inert, inorganic geological material; organic material of natural origin; and bulky items of unharmful materials when produced at locations with no other disposal options. The Protocol will enter into force after 26 states ratify it (15 of which are party to the London Convention). So far, 16 countries have ratified it (13 of which are party to the London Convention).

Ocean dumping has been used as a method for municipal waste disposal in the United States for about 80 years, and even longer for dredged material. The law to regulate ocean dumping in the United States is the Marine Protection, Research, and Sanctuaries Act of 1972. This Act, also known as the Ocean Dumping Act, banned the disposal of

A ship dumping jarosite waste into the ocean off the Australia coast. (Photograph by Hewetson. Greenpeace. Reproduced by permission.)

radiological, chemical, and biological warfare agents, high-level radioactive waste, and medial waste. It requires a permit for the ocean dumping of any other materials. Such a permit can only be issued where it is determined that the dumping will not unreasonably degrade or endanger human health, welfare, or amenities, or the marine environment, ecological systems, or economic potentialities. Furthermore, all permits require notice and opportunity for public comment. EPA was directed to establish criteria for reviewing and evaluating ocean dumping permit applications that consider the effect of, and need for, the dumping. The Ocean Dumping Act has been revised by the United States Congress in the years since its enactment. In 1974, Congress amended this Act to conform with the London Convention. In 1977, the Act was amended to incorporate a ban by 1982 on ocean dumping of wastes that may unreasonably degrade the marine **environment**. As a result of this ban, approximately 150 permittees dumping sewage sludge and industrial waste stopped ocean dumping of these materials. In 1983, the law was amended to make any dumping of low-level radioactive waste require specific approval by Congress. In 1988, the **Ocean Dumping Ban Act** was passed to prohibit ocean dumping of all sewage sludge and industrial waste by 1992. Virtually all material ocean dumped in the United States today is dredged material. Other materials include fish wastes, human remains, and vessels.

Physical properties of wastes, such as density, and its chemical composition affect the dispersal, and settling of the wastes dumped at sea. Many contaminants such as metals found in trace amounts in natural materials are enriched in wastes. After material is dumped at sea, there is usually initial, rapid dispersion of the waste. For example, some wastes are dumped from a moving vessel, with dilution rates of 1,000 to 100,000 from the ship's wake. As the wake subsides, naturally occurring turbulence and currents further disperse wastes eventually to nondetectable background levels in water in a matter of hours to days, depending on the type of wastes and physical oceanic processes. Dilution is greater if the material is released slower and in smaller amounts, with dispersal rates decreasing over time. Wastes, such as dredged material, that are

much denser than the surrounding seawater sink rapidly. Less dense waste sinks more slowly, depending on the type of waste and physical processes of the dumpsite. As sinking waste particles reach seawater of equal density, they begin to spread horizontally, with individual particles slowly settling to the sea bottom. The accumulation of the waste on the sea floor varies with the location and characteristics of the dumpsite. Quiescent waters with little tidal and wave action in enclosed shallow environments have more waste accumulate on the bottom in the general vicinity of the dumpsite. Wastes dumped in more open, well-mixed ocean waters are transported away from the dumpsite by currents and can disperse over a very large area, as much as several hundred square kilometers.

The physical and chemical properties of wastes change after the material has been dumped into the ocean. As wastes mix with seawater, acid-base neutralization, dissolution or precipitation of waste solids, particle **adsorption** and desorption, volatilization at the sea surface, and changes in the oxidation state may occur. For example, when acid-iron waste is dumped at sea, the buffering capacity of seawater rapidly neutralizes the waste. Hydrous iron oxide precipitates are formed as the iron reacts with seawater and changes from a dissolved to a solid form.

The fraction of the waste that settles to the bottom is further changed as it undergoes geochemical and biological processes. Some of the elements of the waste may be mobilized in organic-rich **sediment**; however, if sulfide ions are present, metals in waste may be immobilized by precipitation of metal sulfides. Organisms living on the sea floor may ingest waste particles, or mix waste deeper in the sediment by burrowing activities. **Microorganisms** decompose **organic waste**, potentially **recycling** elements from the waste before it becomes part of the sea floor sediments. Generally sediment-related processes will act on waste particles over a longer time scale than hydrologic processes in the water column.

The effects of dumping on the ocean are difficult to measure and depend on complex interactions of factors including type, quantity, and physical and chemical properties of the wastes; method and rate of dumping; toxicity to the **biotic community**; and numerous site-specific characteristics, such as water depth, currents (turbulence), water column density structure, and sediment type. Many studies have found that the effects of ocean dumping on the water column are usually temporary and that the ocean floor receives the most impact. Impacts from dumping dredged material are usually limited to the dumpsite. Burial of some benthic organisms (organisms living at or near the sea floor) may occur; however, burrowing organisms may be able to move vertically through the deposited material and fishes typically leave the area. Seagrasses, coral reefs, and oyster beds may never recover after dredged material is dumped on them. There also can be topographic changes to the sea floor. *See also* Dredging

[*Marci L. Bortman Ph.D.*]

Resources

Other

ArmyCorps of Engineers Ocean Disposal Database. [cited July 2002]. <http://ered1.wes.army.mil/ODD>.

"EPA's Ocean Dumping and Dredged Material Management." *EPA Web Site.* July 16, 2002 [cited July 2002]. <http://www.epa.gov/owow/oceans/regulatory/dumpdredged/dumpdredged.html>.

London Convention. [cited July 2002]. <http://www.londonconvention.org>.

Ocean dumping act

see **Marine Protection, Research and Sanctuaries Act (1972)**

Ocean Dumping Ban Act (1988)

The Ocean Dumping Ban Act of 1988 (Public Law 100-688) marked an end to almost a century of sewage **sludge** and industrial waste dumping into the ocean. The law was enacted amid negative publicity about beach closures from high levels of pathogens and **floatable debris** washing up along New York and New Jersey beaches and strong public sentiment that ending ocean dumping may improve coastal **water quality**. The Ban Act prohibits sewage sludge and industrial wastes from being dumped at sea after December 31, 1991. This law is an amendment to the Marine Protection, Research, and Sanctuaries Act of 1972 (Public Law 92-532), which regulates the dumping of wastes into ocean waters. These laws do not cover wastes that are discharged from outfall pipes such as from **sewage treatment** plants or industrial facilities or that are generated by vessels.

The Ocean Dumping Ban Act was not the first attempt to prohibit dumping of sewage sludge and industrial wastes at sea. An earlier ban was developed by the U. S. **Environmental Protection Agency** (EPA) and later passed by the U.S. Congress (Public Law 95-153) in 1977, amending the 1972 act. This 1977 law prohibits ocean dumping that "may unreasonably degrade the marine environment" by December 31, 1981. Approximately 150 entities dumping sewage sludge and industrial waste sought alternative disposal options. However, New York City, and eight municipalities in New York and New Jersey filed a lawsuit against the EPA objecting to the order to end their ocean dumping practices. A Federal district court granted judgment in their favor, allowing them to continue ocean dumping under a

court order. The court held that the EPA must balance, on a case-by-case basis, all relevant statutory criteria with the economics of ocean dumping against land-based alternatives. After 1981, the New York and New Jersey entities were the only dumpers of sewage sludge, and there were only two companies dumping industrial waste at sea.

In anticipation of the 1988 Ocean Dumping Ban Act, one of the two industries stopped its dumping activities in 1987. The remaining industry, which was dumping hydrochloric **acid** waste, also ceased its activities before the 1988 Ban Act became law. The entities from New Jersey and New York continued to dump a total of approximately eight million wet metric tonnes of sewage sludge (half from New York City) annually into the ocean.

From 1924 to 1987, sludge dumpers used a site approximately 12 miles (19 km) off the coasts of New Jersey and New York. The EPA, working with the **National Oceanic and Atmospheric Administration** (NOAA), determined that ecological impacts such as shellfish bed closures, elevated levels of metals in sediments, and introduction of human pathogens into the marine **environment** were attributed entirely or in part to sludge dumping at the 12 mile site. As a result the EPA decided to phase out the use of this site by December 31, 1987. The sewage sludge dumpers were required to move their activities farther offshore to the 106-mile (171-km) deep water dumpsite, located at the edge of the continental shelf off southern New Jersey. Industries had used this dumpsite from 1961 to 1987.

The Ocean Dumping Ban Act prohibits all dumping of sewage sludge and industrial waste into the ocean, without exception. The law also prohibits any new dumpers, and required existing dumpers to obtain new permits that included plans to phase-out sewage sludge dumping at sea. The Ban Act also established ocean dumping fees and civil fines for any dumpers that continue their activities after the mandated end date. The fines were included in the law in part because legislators assumed that some sludge dumpers would not be able to meet the December 31, 1991 deadline. The law required fees of $100 per dry ton of sewage sludge or industrial waste in 1989, $150 per dry ton in 1990, and $200 per dry ton in 1991. After the 1991 deadline penalties rose to $600 per ton for any sludge dumped, and increased incrementally in each subsequent year. Those ocean dumpers that continue beyond December 31, 1991 are allowed to use a portion of their penalties for developing and implementing alternative sewage sludge management strategies. While the amount of the penalty increased each year after 1991, the amount that could be devoted to developing land-based disposal alternatives decreased.

As part of the law, the EPA, in cooperation with NOAA, is responsible for implementing an **environmental monitoring** plan at the 12 mile site, the 106 mile site, and surrounding areas potentially influenced by dumping activities to determine the effects of dumping on living marine resources. The Ban Act also includes provisions not directly associated with dumping of sewage sludge or industrial waste at sea. Massachusetts Bay-MA, Barataria Terrebonne Estuary Complex-LA, Indian River Lagoon-FL, and Peconic Bay-NY were named as priority areas for consideration to the **National Estuary Program** by EPA. The law also includes a prohibition on the disposal of **medical waste** at sea by public vessels. Finally, the Ban Act requires vessels transporting **solid waste** over the New York Harbor to the Staten Island **Landfill** to use nets to secure the waste to minimize the amount that may spill overboard.

The New Jersey dumpers ceased ocean dumping by March 1991. The two New York counties stopped ocean dumping by December 1991 and New York City, the last entity to dump sewage sludge into the ocean, stopped zn June 1992. Landfilling is currently used as an alternative to ocean dumping and other sewage sludge management strategies are under consideration including **incineration** (after the sludge is dewatered), **composting**, land application, and pelletization. *See also* Convention on the Prevention of Marine Pollution by Dumping of Waste and Other Matter

[*Marci L. Bortman Ph.D.*]

Resources

Periodicals

Kitsos, T. R., and J. M. Bondareff. "Congress and Waste Disposal at Sea." *Oceanus* 33 (Summer 1990): 23–28.

Millemann, B. "Wretched Refuse Off Our Shores." *Sierra* 74 (January-February 1989): 26–28.

"Ocean Dumping Ban Advances." *Journal of the Water Pollution Control Federation* 60 (August 1988): 1320+.

Weis, J. S. "Ocean Dumping Revisited." *BioScience* 38 (December 1988): 749.

Ocean farming

Although oceans cover about 70% of the Earth's surface, most food is produced on land areas through agriculture and animal husbandry. This is because 90% of the ocean is unproductive. The productive areas are near shore, such as continental shelfs, **upwellings**, coral reefs, mangrove swamps (also known as mangal communities), and estuaries. The primary reason why these near shore areas are more productive than the open ocean is because of the supply of nutrients which are needed for plant (primarily **phytoplankton**) productivity, which fuels the **food chain/web**.

A worldwide plateau in harvesting of natural fish catch was reached in about 1989, so commercial and sport fisheries

in the oceans cannot be increased unless something is changed. One possibility is to enhance production through fertilization of the water. Another, more practical, possibility is through **aquaculture**. As wild harvests decrease and per capita seafood consumption increases, aquaculture should have an important role. For example, the U.S. imports 60% of its seafood, which contributes to its trade imbalance. Aquaculture can help by providing jobs as well as food products for domestic consumption and for export.

Aquaculture done in seawater is known as **mariculture**. Fish and shellfish are grown in improved conditions to produce more and better food in environments with lower predation and disease. There are several forms of aquaculture: intensive, extensive, and open ocean. In extensive aquaculture, relatively little control is exerted by the mariculturist. These typically occur near shore where organisms are grown in several different ways: floating cages or pens; cordoned-off bodies of water which can be fertilized for enhanced production; or racks and other structures (to grow shellfish). For example, oyster farmers in the **Chesapeake Bay** place clean substrates in selected areas of the bottom mud each summer to collect young oyster larvae (called "spat") as they settle down from the **plankton**. These young oysters attach and grow on the substrates, which are periodically transferred to new locations to prevent them from being covered with mud and **silt**. After several years, the oysters are large enough to be marketed. Other culture methods for bivalves such as oysters and clams utilize the stake method, which involves the use of bamboo or other poles driven into the **sediment** and placed approximately 6 ft (2 m) apart. Nylon ropes hanging from floats can also be used to collect larvae. Following growout, the shellfish are collected and sold on the market. Farmers in the Philippines are able to produce nearly 1.5 tons per acre (0.60 tons per ha) of mussels and 1.7 tons per acre (0.70 tons per ha) of oysters annually. These techniques all have relatively low cost and maintenance. Sometimes there are problems with **pollution** from domestic and industrial waste, and periodic blooms of toxic dinoflagellates (a type of phytoplankton), which create red tides. Red tides can be unhealthy and even lethal to humans.

Intensive mariculture systems involve the production of large quantities of marine animals in relatively small areas. They are relatively expensive to build and maintain, so typically only highly-valued **species** are used, such as lobsters, shrimp, halibut, and certain other fish. Intensive systems can sometimes be located in tanks on land, using either pumped salt water from the nearby ocean or recirculated artificial seawater if the system is located farther away from the ocean. Typically, however, nearshore or offshore floating tanks or pens are used. The organisms are raised from eggs to market size in these controlled environments. **Salmon** have been raised in sea pens (netted cages) for over 20 years along the Pacific coast of the United States and Canada and along the Atlantic coasts of France, Scotland, and the Scandinavian countries, especially Norway. This process typically involves the growth of salmon from fertilized eggs through the fry stage in indoor fresh water hatcheries. This is followed by the transfer of parr (the stage after the fry in which the fish develop vertical lines along their sides) to outdoor tanks or cages in salt water. The fish are fed either commercial pellets or "trash" fish such as herring, capelin, menhaden, and anchovies. The feeding process can be quite labor-intensive, and thus translates into one of the highest expenses for the salmon farmer. After one to two years, the salmon go through a process called smoltification, which involves a physiological change that allows them to live in salt water. The smolt are then transferred to salt water cages for another one to two years until they reach a market size of about 4-12 lb (2-5 kg). Rainbow and sea trouts are handled in a similar way, although they are usually larger when moved into sea pens. As with the extensive method, problems can arise through pollution and red tides.

Most mariculture industries use both nearshore and offshore cages for raising the fish. The nearshore cages are necessary for raising young fish, but the offshore cages are better for raising fish to maturity because of enhanced growing conditions (cleaner water and lower **mortality**). Improvements are still needed, such as methods to remove dead fish, make grading (sorting) of fish possible, make harvesting easier and feeding possible (particularly during adverse weather), and make fouling removal easier. It will also be helpful to construct 24-hour living accommodations on some of the larger offshore units.

A third form of mariculture is known as ocean ranching, or "enhancement aquaculture." This can be compared to cattle ranching on land. For example, salmon are raised in hatcheries until they reach the smolt stage, and are then released at a particular point along the shore where they swim away until they reach reproductive maturity. Like other anadromous fish, they return to that same area several years later, where they are recaptured, processed, and sold on the market. Salmon are ideal fish for this because they are "self-herding." However, normal returns range from only 1-20%, depending upon the **water quality** of the return **environment** and the release size of the fish. For example, Coho salmon fry that were around 0.5 oz (14 g) showed a 1-2% return while those twice this size had a 7-8% return rate. Recent research has shown that released salmon can be imprinted to return to salt water sites, which has the added advantages of lower cost (due to not needing more expensive property located along streams that enter the ocean), and better meat quality (which declines when salmon swim from salt to fresh water). The Sea Run, Inc. salmon company of Kennebunkport, Maine, rears Pacific pink and chum salmon

from eggs to fingerlings in a hatchery located near the **discharge** of an oil-fired power plant. The water has a characteristic temperature and smell which attracts the salmon back to the same area each year. They have also experimented with the release of morpholine and phenyethyl alcohol (synthetic organic compounds) to the water to aid in imprinting.

Similar ocean ranching techniques have been used in other countries. For example, in Japan the natural catch of Red Sea bream has declined for the last 24 years (more than a 51% decrease) due to over-exploitation, so these fish have been raised in tanks and cages since 1962. Other species commonly raised in Japan include salmon, flounder, mackerel, bluefin tuna, shrimp, scallops, and abalone. Many of these species have a high consumer demand and thus are highly priced. Abalone can also be grown to a small size of approximately 1 in (3 cm) and used to re-seed areas where the natural population has declined. They can then be harvested as adults about three to five years later. However, recapture rates are typically low (only 0.5–10%).

Ocean farming or ranching is becoming popular worldwide. In Equador, Mazatlan Yellowtail fish are grown to commercial size (2.7 lb; 1.2 kg) in six months with a 90% survival rate. Other species raised in this country include flounder, snook, red drum, and Pacific pompano. In **Australia**, oysters are commonly raised via sea ranching. Scallops and salmon are now raised in Canada. Mussels, oysters, clams, bream, turbot, and salmon are common mariculture species in Spain, where the warm water allows these species to be grown to market size in a shorter time.

Recent work has been done on algal turf mariculture around some Caribbean islands in natural, unfertilized waters. Test plots have shown that they can raise up to 25 million tons (23 million metric tonnes) of dry algae per year without harming the natural **ecosystem**. This product can then be used for direct human ingestion or as food to raise marine invertebrates (such as West Indian spider crabs, conchs, and whelks) as well as some herbivorous fish. This would be particularly helpful because conch are now an **endangered species** through over-harvesting in this area.

Ocean ranching, along with other forms of aquaculture, is an exciting field that should be in greater demand in the future. Our progeny may depend on it more for global food production to feed the increasing human population.

[*John Korstad*]

Resources

Books

Landau, M. *Introduction to Aquaculture.* New York: Wiley, 1992.

Periodicals

Adey, W. H. "Food Production in Low-Nutrient Seas." *BioScience* 37 (1987): 340–8.

Shelty, H. P. C., and G. P. Satyanarayana Rao. "Aquaculture in India." *World Aquaculture* 27 (1996): 20–24.

Other

Reinertsen, H., L.A. Dahle, L. Jørgensen, and K. Tvinnereim. *Fish Farming Technology.* Proceedings of the First International Conference on Fish Farming Technology, Trondheim, Norway, August 9-12, 1993. Brookfield, VT: Balkema Publishers, 1993.

Ocean outfalls

Pipelines extending into coastal and ocean waters that are used by various industries and municipal **wastewater** treatment facilities to **discharge** treated **effluent**. Some may be simple pipes serving as conveyances from land-based facilities; others include diffusers that help to rapidly dilute effluent or risers that ensure effluent is discharged at a certain height above the ocean floor. The conveyances may extend more than three miles (5 km) offshore, beyond coastal waters into open ocean. Offshore oil and gas exploration, development, and production rigs also possess ocean outfalls. Discharges from these pipes must be permitted under the **Clean Water Act national pollutant discharge elimination system**. *See also* Sewage treatment

Ocean pollution

see **Marine pollution; Ocean dumping**

Ocean thermal energy conversion

For many years, scientists have been aware of one enormous **reservoir** of energy on the earth's surface: the oceans. As sunlight falls on the oceans, its energy is absorbed by seawater. The oceans are in one sense, therefore, a huge "storage tank" for **solar energy**. The practical problem is finding a way to extract that energy and make it available for human use.

The mechanism suggested for capturing heat stored in the ocean depends on a thermal gradient always present in seawater. Upper levels of the ocean may be as much as 36°F (20°C) warmer than regions 0.6 mile (1 km) deeper. The technology of ocean thermal energy conversion (OTEC) takes advantage of this temperature gradient.

An OTEC plant would consist of a very large floating platform with pipes at least 100 feet (30 m) in diameter reaching to a depth of up to 0.6 mile (1 km). The "working fluid" in such a plant would be ammonia, propane, or some other liquid with a low boiling point.

Warm surface waters would be pumped into upper levels of the plant, causing the working fluid to evaporate. As the fluid evaporates, it will also exert increased pressure. That

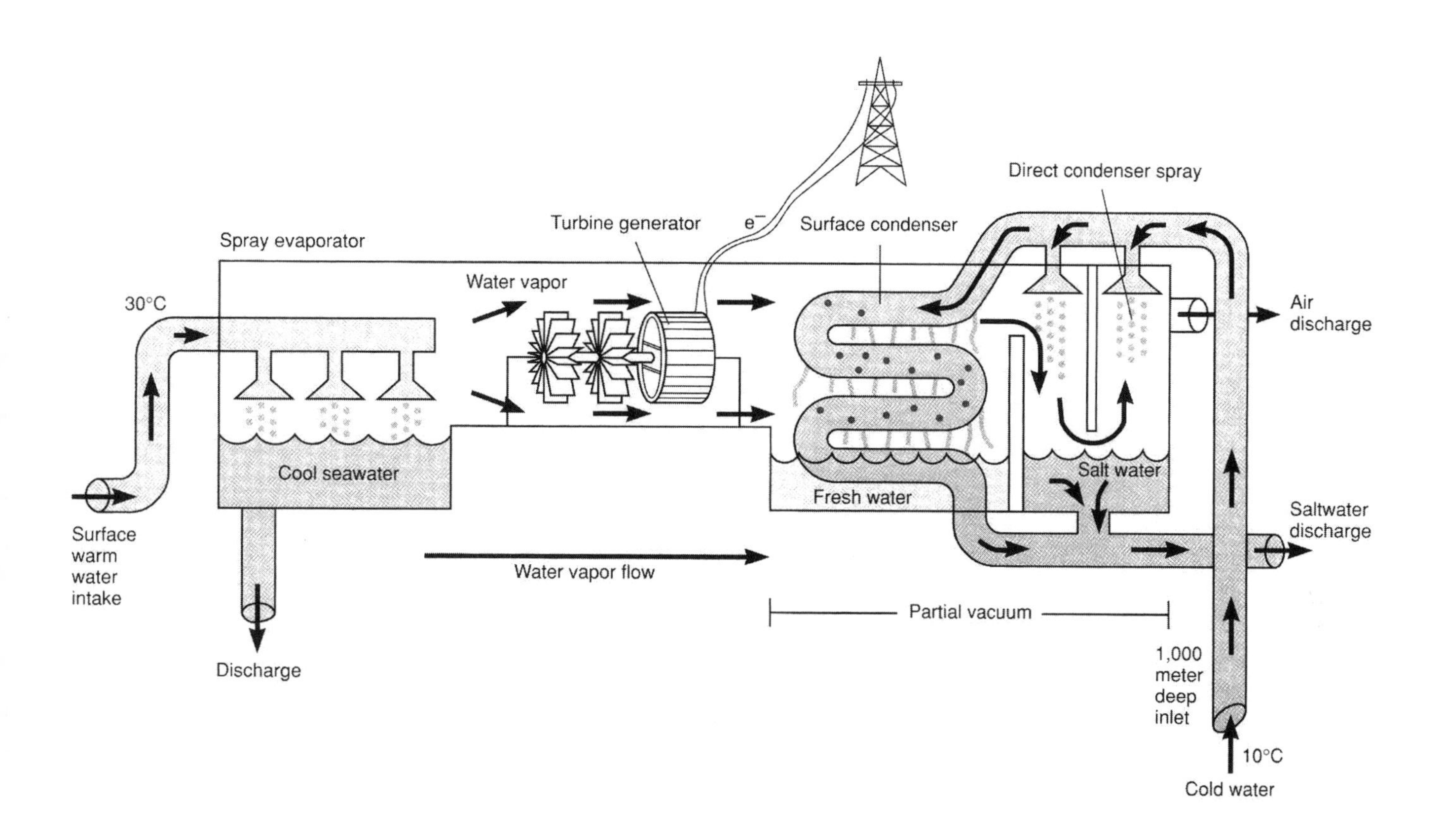

The Open Cycle Ocean Thermal Electric Generator requires a water temperature differential of at least 36°F (20° C) to produce both fresh water and electricity. (McGraw-Hill Inc. Reproduced by permission.)

pressure can be used to drive a turbine that, in turn, generates electricity. The electricity could be carried to shore along large cables or used directly on the OTEC plant to desalinize water, electrolyze water, or produce other chemical changes.

In the second stage of operation, cold water from deeper levels of the ocean would be brought to the surface and used to cool the working fluid. Once liquified, the working fluid would be ready for a second turn of the generating cycle.

OTEC plants are attractive alterative energy sources in regions near the equator, where surface temperatures may reach 77°F (25°C) or more. These parts of the ocean are often adjacent to **less developed countries**, where energy needs are growing.

Wherever they are located, OTEC plants have a number of advantages. For one thing, oceans cover nearly 70 percent of the planet's surface so that the raw material OTEC plants need—seawater—is readily available. The original energy source—sunlight—is also plentiful and free. Such plants are also environmentally attractive since they produce no **pollution** and cause no disruption of land resources. Planners suggest that a by-product of OTEC plants might be nutrients brought up from deeper ocean levels and used to feed "farms" of fish or shellfish.

Unfortunately, many disadvantages exist also. The most important is the enormous cost of building and maintaining the mammoth structures needed for an OTEC plant. Also, the temperature differential available under even the best of conditions means that an OTEC plant will not be more than about 3% efficient.

Currently, the disadvantages of OTEC plants are so great that none has even been built. Research continues in a number of countries, but some experts believe that the low efficiency of OTEC means that this technology will never be able to compete economically with other alternative sources of energy. *See also* Desalinization; Energy efficiency; Power plant; Thermal stratification (water)

[*David E. Newton*]

Resources

Periodicals

Fisher, A. "Energy from the Sea." *Popular Science* (June 1975): 78–83.

Haggin, J. "Ocean Thermal Energy Conversion Experiment Slated for Hawaii." *Chemical & Engineering News* (February 10, 1986): 24–26.

Penney, T. R., and D. Bharathan, "Power from the Sea." *Scientific American* 256 (January 1987): 86–92.

Walters, S. "Power in the Year 2001, Part 2—Thermal Sea Power." *Mechanical Engineering* (October 1971): 21–25.

Whitmore, W. "OTEC: Electricity from the Ocean." *Technology Review* (October 1978): 58–63.

OCRWM

see **Office of Civilian Radioactive Waste Management**

Octane rating

Octane rating is a method for describing antiknock properties of **gasoline**. Knocking is a pinging sound produced by internal **combustion** engines when fuel ignites prematurely during the engine's compression cycle. Because knocking can damage an engine and rob it of power, gasoline formulations have been developed to minimize the problem. Gasolines containing relatively large amounts of straight-chain **hydrocarbons** (such as n-heptane) have an increased tendency to knock, whereas those containing branched-chain forms (such as isooctane) burn more smoothly. In addition to isooctane, other compounds also reduce engine knocking. By using an index called octane number, it is possible to compare the antiknock properties of gasoline mixtures. A higher octane number indicates that a mixture has the equivalent antiknock properties of a gasoline containing a higher percentage of isooctane.

Although gasoline used as automotive fuel is now "deleaded," gasoline used as aviation fuel still contains **tetraethyl lead** as an octane enhancer and antiknock agent.

Svante Odén

Swedish agricultural scientist

One of the great environmental issues of the 1970s and 1980s was the problem of **acid** precipitation. Research studies suggested that rain, snow, and other forms of precipitation in certain parts of the world had become increasingly acidic over the preceding century. The southern parts of Scandinavia and England, the Northeastern United States, and Eastern Canada were four regions in which the phenomenon was particularly noticeable.

Evidence began to accumulate that the increasing level of acidity might be associated with environmental damage, such as the death of trees and aquatic life. Scientists began to ask how extensive this damage might be and what sources of acid precipitation could be identified.

If any single person could be credited with raising international awareness of this problem, it would probably be the Swedish agricultural scientist Svante Odén. Odén was certainly not the first person to recognize the existence, effects, or origins of acid precipitation. That honor belongs to an English chemist, **Robert Angus Smith**. As early as 1852, Smith hypothesized a connection between **air pollution** in Manchester and the high acidity of rains falling in the area. He first used the term **acid rain** in a book he published in 1872.

Smith's research held relatively little interest to most scientists, however. Those who did study acid rain approached their work with little concern about the **environment** and did so, as one of them later said, "with no environmental consciousness," but simply because "it was an interesting situation."

Odén's attitude was quite different. He had been asked by the Swedish government to prepare a report on his hypothesis that acid rain falling on Swedish land and lakes had its origins hundreds or thousands of miles away. In preparing his report, he came to the conclusion that acid rain might be responsible for widespread **fish kills** then being reported by Swedish fishermen. Odén was shocked by this discovery because, as he later said, it was the "first real indication that acid precipitation had an impact on the biosystem."

Odén's method of dealing with his discoveries was unorthodox. In most cases, a researcher sends the report of his or her work to a scientific journal, which has the report reviewed by other scientists in the same field. If the research is judged to have been well done, the report is published.

In this instance, however, Odén sent his report to a Stockholm newspaper, *Dagens Nyheter*, where it was published on October 24, 1967. Odén's decision undoubtedly disturbed some scientists, but it did have the effect of bringing the issue of acid rain to the attention of the general public.

A year later, Odén published a more formal report of his research, "The **Acidification** of Air and Precipitation, and Its Consequences," in the *Ecology Committee Bulletin*. The article was later translated into English. Odén carried his message about acid precipitation to the United States in person in 1971, when he presented a series of 14 lectures on the topic at various institutions across the country. In his presentations, he argued that acid rain was an international phenomenon that, in Europe, originated especially in England and Germany and was spreading over thousands of miles to other parts of the continent, especially Scandinavia. His work also laid the foundation for Sweden's case study for the **United Nations Conference on the Human Environment** "Air **Pollution** Across National Boundaries" presented at Stockholm in 1972. He further suggested that a

number of environmental effects, such as the death of trees and fish and damage to buildings, could be traced to acid precipitation. Odén's passionate commitment to publicizing his findings about acid precipitation was certainly a critical factor in awakening the world's awareness to the potential problems of this environmental danger.

[*David E. Newton*]

Resources
Books

Boyle, R. H., and R. A. Boyle. *Acid Rain.* New York: Nick Lyons Books, 1983.

Luoma, J. *Troubled Skies, Troubled Waters.* New York: Viking Press, 1984.

Park, C. *Acid Rain: Rhetoric and Reality.* London: Methuen, 1987.

Periodicals

Cowling, E. B. "Acid Precipitation in Historical Perspective." *Environmental Science and Technology* 16 (1982): 110A–123A.

Odor control

Refuse-handlers and many industries release unpleasant odors into the air which can travel for miles. Odors inside factories can also make it difficult for people to work, and pollutants can impart a strong odor to water.

Odors can be released from **chemicals**, chemical reactions, fires, or rotting material. The air carries odor-producing gas molecules, and they are detected by breathing or sniffing. The molecules stimulate receptor cells in the nose, which in turn send nerve impulses to the brain where they are processed into information about the odor.

Research is being performed to quantify and characterize odors. Tests such as sniff **chromatography**, **emission** rate measurement, and hedonics are being used in an effort to develop a better definition of what offensive odors are.

Scientists have found that the perception of odors is highly subjective. What one person might not like, another person might not be able to smell at all, and what people define as an offensive odor depends on age and sex, as well as other characteristics.

One method of controlling unpleasant smells is deodorizers. Deodorizers either disguise offensive odors with an agreeable smell or destroy them. Some deodorizers do this by chemically changing odor-producing particles while others merely remove them from the air. Disinfectants, such as formaldehyde, can kill bacteria, **fungi** or molds that create odors. Odors can also be removed through ventilation systems. The air can be "scrubbed" by forcing it through liquid or through **filters** containing such materials as charcoal, methods which trap and remove odor-producing particles from the air. Factories can also employ a process known as "re-odorization," which works on the principle that there are seven basic odor types: camphoraceous, mint, floral, musky, ethereal, putrid, and pungent. The process is based on the theory that different combinations of these odor types produce different smells, and re-odorization releases chemicals into the air to combine with the regular factory odors and generate a more pleasant smell.

Aeration is one method of removing objectionable odors from water. The surface of the water is mixed with air and the oxygen oxidizes various materials that would otherwise turn the water foul. Aeration can be accomplished by running the water over steps or spraying the water through nozzles. Trickling the water over trays of coke also helps eliminate offensive odors, and adding activated charcoal can have the same effect.

Even though science has shown that the perception of odors is subjective, many people are offended by them. It is often considered a quality-of-life issue, and politicians have been strongly influenced by their constituencies. Nuisance regulations have been passed in many states and municipalities, and though they often vary, their attempts to distinguish between acceptable and unacceptable odors are often vague and difficult to apply. Proving that odors interfere with the quality of life is not only subjective but nearly impossible. Refuse-handlers and factories are perhaps the most adversely affected; they often receive heavy fines for odors, although there is no proven method for eliminating them. Many believe a more concrete system of regulations is needed, but science has not been able to provide the basis on which to build one. *See also* Noise pollution; Pollution control

[*Nikola Vrtis*]

Resources
Books

Bowker, P. G. *Odor and Corrosion Control in Sanitary Sewerage Systems and Treatment Plants.* New York: Hemisphere, 1989.

Hesketh, H. E. *Odor Control Including Hazardous-Toxic Odors.* Lancaster, PA: Technomic, 1988.

Kreis, R. D. *Control of Animal Production Odors: The State-of-the-Art.* Ada, OK: U.S. Environmental Protection Agency, 1978.

Periodicals

Hunt, P., and K. Hauck. "Raising a Stink Over Composting Odors." *American City and County* 105 (December 1990): 64–65.

Dr. Eugene P. Odum (1913 –)

American ecologist

Eugene P. Odum has been called "the ecologist's ecologist," meaning that he has been of special significance to his colleagues in the development of the discipline, and in formulat-

ing its content, outlines, and boundaries. He helped lay the foundations for what might be called the "modern" study of **ecology** in the 1940s and 1950s by redirecting the field to studies based on energetics, on the relatively new **ecosystem** concept, and on functional as well as structural analyses of living communities. From early in his career, he has worked to define general principles in ecology. His colleagues and associates have recognized him by electing him president of the **Ecological Society of America**, through election to the **National Academy of Sciences**, and through the awarding of prestigious prizes, including the Tyler Award in 1977 and the Crafoord Prize in 1987.

Odum was born in Lake Sunapee, New Hampshire, to an academic family. His father, Howard W. Odum, was a well-known sociologist, remembered today especially for his work on regionalism. His brother, Howard T. Odum, is a well-known systems ecologist. Eugene Odum received his bachelor's and master's degrees in biology from the University of North Carolina, and his Ph.D. in ecology and ornithology from the University of Illinois in 1939. He spent brief stints as an instructor at Western Reserve University and as a research biologist at the Edmund Niles Huck Preserve in New York. In 1940, he accepted a teaching position at the University of Georgia, spent the next four decades there, and since his retirement in 1984, has been an emeritus professor. During his tenure in Georgia, the university has become a world center for the study of ecology, through various departments but also through the Institute of Ecology, which he helped initiate and of which he was the long-time director.

Odum's best-known work is his "landmark" general text, *Fundamentals of Ecology*, first published in 1953, with a third edition published in 1971. This text dominated the market for two decades and is still a book to consult, including an extensive and still useful reference list. Odum revised the third edition "in light of the increasing importance of the subject in human affairs." That edition is organized into three parts: basic ecological principles and concepts, emphasizing energy and ecosystems; "the **habitat** approach," which discusses freshwater, marine, estuarine, and terrestrial ecology; and applications and technology, which includes chapters on resources, **pollution** and **environmental health**, radiation ecology, remote sensing, and microbial ecology. Since Odum is a well-known radiation ecologist, that chapter especially is still timely.

Odum's early work was on birds, an interest threaded into his publications for many years. His research focused for several years on various aspects of radiation ecology. He turned early to an emphasis on energetics and on ecosystem studies, becoming almost as well known for systems ecology as his younger brother. A lot of his work has been on productivity in estuaries and marshes. And he is known for his work on old fields in the southern United States. That research has, in many instances, been applied to problems in environmental management, and to issues of human impact and reciprocity.

In recent years, a considerable amount of his energy has been devoted to human interactions with and impacts on environmental systems. That last interest has turned Odum into something of a philosopher, resulting in the publication of several papers on values and ethics. One influence of his thinking that will be debated for a long time is his repeated attempts to outline ecology as an integrative discipline, and to locate ecologists as scientists interested in every level of integration, from micro to macro. That work has led him several times, with mixed success, to move away from the classical reductionist approach in science and to try an outline for a **holistic approach** in ecology.

[*Gerald L. Young Ph.D.*]

Resources
Books

Odum, E. P. *Ecology: A Bridge Between Science and Society.* Sunderland, MA: Sinauer Associates, 1996.

———. *Fundamentals of Ecology.* 3rd ed. Philadelphia: W.B. Saunders Company, 1971.

Periodicals

Odum, E. P. "The Emergence of Ecology as a New Integrative Discipline." *Science* 195, no. 4284 (March 25, 1977).

———. "Great Ideas in Ecology for the 1990s." *BioScience* 42, no. 7 (July 1992): 542–545.

Office of Civilian Radioactive Waste Management

Humans have been using nuclear materials for nearly 50 years. Nuclear reactors and **nuclear weapons** account for the largest volume of these materials, while industrial, medical, and research applications account for smaller volumes. One of the largest single problems involved with the use of nuclear materials is the volume of wastes resulting from these applications. By one estimate, 8,000–9,000 metric tons (8,816–9,918 tons) of high-level radioactive wastes alone are produced in the United States every year. It is something of a surprise, therefore, to learn that as late as 1982, the United States had no plan for disposing of the radioactive wastes produced by its commercial, industrial, research, and defense operations.

In that year, the United States Congress passed the Nuclear Waste Policy Act establishing national policy for the disposal of **radioactive waste**. Responsibility for the implementation of this policy was assigned to the **U.S. Department of Energy** through the Office of Civilian **Radio-**

active Waste Management (OCRWM). OCRWM manages federal programs for recommending, constructing, and operating repositories for the disposal of high-level radioactive wastes and spent nuclear fuel. It is also responsible for arranging for the interim storage of spent nuclear fuel and for research, development, and demonstration of techniques for the disposal of **high-level radioactive waste** and spent nuclear fuel.

In addition, OCRWM oversees the Nuclear Waste Fund, also established by the 1982 act. The fund was established to enable the federal government to recover all costs of developing a disposal system and of disposing of high-level waste and spent nuclear fuel. It is paid for by companies that produce **nuclear power**, power consumers, and those involved in the use of nuclear materials for defense purposes.

OCRWM has published a number of short pamphlets dealing with the problem of waste disposal. They cover topics such as "Nuclear Waste Disposal," "What Will a Nuclear Waste Repository Look Like?" "What Is Spent Nuclear Fuel?" "Can Nuclear Waste Be Transported Safely?" and "How Much High-Level Nuclear Waste Is There?"

OCRWM experienced a number of setbacks in the first decade of its existence. No state was willing to allow the construction of a high-level nuclear waste repository within its borders. The technology for immobilizing wastes seemed still too primitive to guarantee that wastes would not escape in to the **environment**.

Eventually, however, OCRWM announced that it had chosen a site under **Yucca Mountain** in southeastern Nevada. The site lies on the boundaries of the **Nevada Test Site** and Nellis Air Force Base. It is near the town of Beatty, 100 mi (161 km) northwest of Las Vegas. The site has been studied since 1977 and should be ready to receive wastes early sometime around 2010. Some residents of Nevada are unhappy with the choice of Yucca Mountain as a nuclear waste repository, however, and continue to fight OCRWM's decision.

[*David E. Newton*]

Resources

Organizations

Office of Civilian Radioactive Waste Management, Toll Free: (800) 225-6972, Email: mowebmaster@rw.doe.gov, <http://www.rw.doe.gov>

Office of Energy Research

see **U.S. Department of Energy**

Office of Management and Budget

The Office of Management and Budget (OMB), established in 1939 within the office of the President, determines both how much money will be spent by the federal government and what kinds of regulations will be adopted to implement all environmental and other legislated programs.

According to the 1946 Administrative Procedures Act, regulations "interpret, implement or prescribe law or policy." Draft regulations must be publicized, and agencies must incorporate public comments into the final regulations. Executive Orders 12291 and 12498, issued in 1981 and 1984, respectively, gave the Office of Information and Regulatory Affairs (OIRA), a division of OMB, the authority to review and approve all regulations and paperwork drafted by the **Environmental Protection Agency**, as well as all other federal agencies.

Resources

Organizations

Office of Management and Budget, 725 17th Street, NW, Washington, D.C. USA 20503 (202) 395-3080, <http://www.whitehouse.gov/omb>

Office of Oceanic and Atmospheric Research

see **National Oceanic and Atmospheric Administration (NOAA)**

Office of Surface Mining

In 1977 the Office of Surface Mining(OSM) was created to police **coal** extraction within the United States and to enforce the federal **Surface Mining Control and Reclamation Act** (SMRCA). Under the auspices of OSM, citizens are empowered to enforce **reclamation** of surface mines, as well as deep mines, that have damaged surface features.

The government, industry, or a combination of both is required to pay all expenses, including legal fees, when citizens bring successful administrative or judicial complaints for noncompliance under the SMRCA. The reclamation of abandoned mines is funded by a tax on coal. To fund the reclamation of new mines OSC requires performance bonds that must cover the cost of reclamation should the operator not complete the process.

The majority of mining states have been granted "primacy" by OSM, which means that the individual states are authorized to handle reclamation enforcement, with OSM stepping in only when states fail to enforce the SMRCA. There are almost 30,000 known violations that remain uncorrected, yet OSM issues less than 25 citations a year. The

majority of citations issued are disposed of by vacating the citation or by making arrangements with the company.

There are 24,000 operations including surface mines, deep mines, refuse piles and prep plants that come under the act, (including those that went into operation after 1977); about 17,000 of these have been reclaimed. Many of these reclaimed sites are merely grass planted over **desert**, incapable of supporting any form of **wildlife**. The reclamation plans developed for many **surface mining** sites designate the post mining **land use** to be pastureland, but due to the remoteness and inaccessibility, many reclaimed sites sit idle devoid of all wildlife.

Environmentalists and other opponents of the mining industry charge that coal industries and OSM personnel are the same group. It is interesting to note that Harry Snyder, the current director of OSM, was a former lobbyist for CSX railroad and is a heavy investor in coal. According to many critics the coal industry has become adept at manipulating the OSM and SMCRA for its own purposes. The OSM is currently trying to push through legislation that would eliminate the SMRCA's attorney fee provisions. The agency is also in the process of throwing out numerous other SMCRA regulations that are burdensome to the coal industry, including one that bans deep mining from national parks and wildlife refuges.

[*Debra Glidden*]

Resources

Periodicals

Brown, F. "The Miner's Watchdog Doesn't Bite." *Sierra* 34 (September/October 1986): 28–31.

"Coal Mining: Profit Reclamation." *The Economist* 319 (April 1991): A24–26.

Sherwood, T. "Strip Search". *Common Cause Magazine* 15 (May/June 1989): 8–10.

Organizations

Office of Surface Mining, 1951 Constitution Ave. NW, Washington, D.C. USA 20240 (202) 208-2719, Email: getinfo@osmre.gov, <http://www.osmre.gov>.

Office of Surface Mining, Reclamation and Enforcement

see **Office of Surface Mining**

Off-road vehicles

Off-road vehicles (ORVs) include motorcycles, dirt bikes, snowmobiles, bicycles, and all-terrain vehicles (ATVs) that can be ridden or driven in areas where there are no paved roads. While the use of off-road vehicles has gained in popularity, conservationists and landowners have prompted some legislatures to restrict their use because of the damage the vehicles do to the **environment**.

During eight years under President Bill Clinton (1946–), environmentalists continued their progress in limiting or banning the use of off-road vehicles in federally protected areas, such as national parks and forests. However, with the election of George W. Bush (1946–) in 2000, environmentalists began to worry their progress might be in jeopardy. As of 2002, the Bush administration had sent mixed signs regarding off-road vehicle use on **public land**.

In October 2001, the U.S. **Environmental Protection Agency** (EPA) announced plans to impose **pollution control** restrictions on off-road vehicle engines. The EPA said off-road vehicles, including snowmobiles, account for 13 percent of hydrocarbon **emission** released into the **atmosphere** each year. The EPA wants off-road vehicle manufacturers to switch from two-cycle to four-cycle engines starting in 2006. Snowmobiles would be required to reduce emissions by 30% in 2006 and 50% in 2010. The Clinton administration had planned to phase-out snowmobiles altogether in **Yellowstone National Park**.

In 2002, the federal **Bureau of Land Management** (BLM) proposed lifting the ban on off-road vehicles in the 50,000-acre Imperial Sand **Dunes Recreation** Area in California. However, that same year, the BLM closed nearly 18,000 acres of federal land in the California's Mojave **Desert** to off-road vehicles. The closing was necessary to protect the **desert tortoise**, a threatened **species**.

A popular sport in desert areas, mountain passes, and riverbeds, owners and drivers of off-road vehicles defend their right to ride wherever they like. They contend that the Clinton administration policy of closing federal land to ORVs was discriminatory, since it favored recreational use of public land only for certain groups of the public, such as hikers, campers, and backpackers. But environmentalists claim that the vehicles scar the land, kill **wildlife**, destroy vegetation, and cause noise, safety and **pollution** problems. The knobby tires of mountain bicycles contribute to **erosion** in delicate desert areas and along the sides of steep mountain trails. Motorized off-road vehicles are more devastating to local **ecology** as landowners across the country are finding.

In Missouri's Black River, off-roaders typically discard beer cans, used baby diapers, and empty motor-oil cans. Usually clear, the Black River in places runs as green as a sewage ditch when algae are stirred up by the commotion of off-road vehicles. Some drivers drain their crankcases into the river. Inevitably, the oil and gas from motorized vehicles seeps into the river **ecosystem**. A bill passed by the Missouri legislature in April 1988 restricted the vehicles to areas of

A rider maneuvers his Honda three-wheeler over sand dunes. (Photograph by Inga Spence. Tom Stack & Associates. Reproduced by permission.)

the river where landowners gave permission. Since much of the area is not posted, however, the law failed to halt a good deal of the use of off-road vehicles in the river.

At the Chincoteague **Wildlife Refuge** in Virginia, an environmental assessment of the effects of off-road vehicles and foot traffic found the effects devastating to some species, particularly the threatened piping plover. The assessment recommended that the area be closed to the vehicles and to all recreation during the nesting season, concluding that the nesting site is subject to damage not only from off-road vehicles, but also from the human intrusion accompanying the use of these vehicles.

In California, about 500,000 acres of public land are open to use by off-road vehicles, and California conservationists have fought since the late 1980s to ban the sport in state parks. Off-roaders, however, waged their own battle. Editorials in magazines for off-road vehicles users urged readers to ignore posted property, sue for the right to use the land, and lobby their state and federal elected officials.

More than 30 states now have enacted legislation that regulates the sale and use of off-road vehicles, especially motorized ones. The legislation was more a reaction to the safety problems inherent in three-wheel vehicles than to the detrimental ecological effects of the vehicles in pristine areas. The Consumer Product Safety Commission listed almost 1,000 deaths related to the use of ATVs from 1982 to 1988, and hundreds of product liability suits have been brought against manufacturers.

[Ken R. Wells]

Resources
Periodicals

Daerr, Elizabeth. "Park Victories Losing Ground." *National Parks* (July–August 2001): 18.

"Environmentalists Decry Bush Administration's Off–Road Vehicle Policy." *Knight–Ridder/Tribune Business News*, June 25, 2001.

Kilian, Michael. "EPA to Limit Pollution From Engines in Boats, Snowmobiles." *Knight–Ridder/Tribune news Service*, October 19, 2001.

Organizations

The Wilderness Society, 1615 M St. NW, Washington, DC USA 20036, 800–843–9453, Email: member@tws.org, http://www.wilderness.org/standbylands/orv

Offshore drilling

see **Oil drilling**

Ogallala Aquifer

The Ogallala **Aquifer** is an extensive underground **reservoir** that supplies water to most of the irrigated agriculture in the central United States. Discovered early in the nineteenth century, this aquifer—also known as the High Plains regional aquifer—became a major economic resource in the 1960s and 1970s when advanced pumping technology made large-scale **irrigation** possible. In 1980 this aquifer supported 170,000 **wells** and provided one third of all irrigation water pumped in the United States. In the last several decades, Ogallala-irrigated agriculture has redefined the landscape of the central United States by fostering economic expansion, **population growth**, and the development of large-scale agribusiness in an **arid** region.

The High Plains regional aquifer underlies eight states, including most of Nebraska and portions of South Dakota, Wyoming, Colorado, Kansas, Oklahoma, Texas, and New Mexico. Covering an area of 175,000 square miles (453,250 sq km), the aquifer runs 800 miles (1,288 km) from north to south, and stretches 200 miles (322 km) at its widest point. Nebraska holds by far the greatest amount of water, with 400–1,200 feet (130–400 m) of saturated thickness, while much of the aquifer's southern extent averages less than 100 feet (30 m) thick. Geologically, this aquifer comprises a number of porous, unconsolidated sand, **silt**, and clay formations that were deposited by wind and water from the Rocky Mountains. The most important of these formations is the

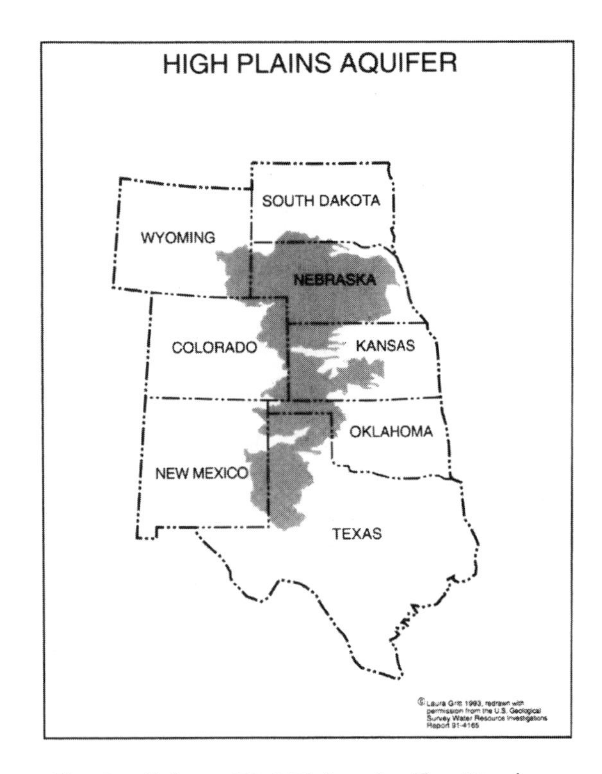

The Ogallala, or High Plains, Aquifer. (Line drawing by Laura Gritt Lawson. U. S. Geological Survey Water Resource Investigations Report.)

Ogallala formation, which makes up seventy-seven percent of the regional aquifer. The aquifer's total drainable water is about 3.25 billion acre feet.

Large volume use of **groundwater** became possible with the invention of powerful pumps and center-pivot irrigation in the late 1950s. The center pivot sprays water from a rotating arm long enough to water a quarter section (160-acre) field. This irrigation technique is responsible for the characteristic circular pattern visible from the air in most agricultural landscapes in the western United States. Although installation is extremely expensive, center pivots and powerful pumps enable farmers to produce great amounts of corn, wheat, sorghum, and cotton where low rainfall had once made most agriculture impossible.

High-volume pumping has also reduced the amount of available groundwater. Most water in the High Plains aquifer is "fossil water," which has been stored underground for thousands or millions of years. New water enters the aquifer extremely slowly. In most of the Ogallala region, pumping exceeds recharge rates, a process known as water mining, and the net volume is steadily decreasing. Region-wide, less than 0.5% of the water pumped each year is replaced by **infiltration** of rainwater. At current rates of pumping, the resource should be 80% depleted by about 2020.

As saturation levels in the aquifer fall, consequences on the surface are clearly visible. Many streams and rivers, dependent on groundwater for base flow, run dry. The state of Kansas alone has lost more than 700 miles (1,126 km) of rivers that once flowed year round. Center pivot irrigation, requiring a well that can pump 750 gallons (2,839 liters) per minute, is beginning to disappear in Texas and New Mexico, where the aquifer can no longer provide this volume. **Groundwater pollution** is becoming more concentrated as more **agricultural chemicals** seep into a shrinking reservoir of Ogallala water. Many high plains towns, formerly rich with pure, clean groundwater, now have tap water that is considered unhealthy for children and pregnant women. As the Ogallala runs dry, the region's demographics are also changing. Many farmers assumed heavy debt burdens in the 1970s when they installed center pivots. Drying wells and falling production have led to farm foreclosures and the depopulation of small, rural towns. Regional populations that swelled with irrigated agriculture are beginning to shrink again. Land ownership, formerly held by independent families and ranchers, is increasingly in the hands of banks, insurance companies, and corporations, which provide little support to local communities.

[*Mary Ann Cunningham Ph.D.*]

Resources
Periodicals

Aucoin, J. *Water in Nebraska—Use, Politics, Policies.* Lincoln: University of Nebraska Press, 1984.

Kromm, D. E., and S. E. White. *Groundwater Exploitation in the High Plains.* Lawrence: University Press of Kansas, 1992.

Oil

see **Petroleum**

Oil drilling

Petroleum occurs naturally in the earth in porous rock, found anywhere from thousands of meters underground all the way to the earth's surface. The porous layer of rock normally lies between two nonporous layers, so oil does not flow out of its **reservoir**. In most cases water and/or **natural gas** occur along with petroleum in the rock. Oil can be extracted from the rock by sinking a pipe into the earth until it penetrates the saturated porous rock. In most cases, oil will then begin to flow of its own accord out of the rock and into the pipe.

The upward flow of oil can be caused by a number of different factors. In some cases, the porous layer of rock is covered by a hollow cap filled with natural gas. Pressure of the gas forces petroleum out of the rock into the pipe. In some cases, gas pressure will be so great as to force oil out of a new well in a fountain-like effect known as a "gusher."

In other cases, the pressure of water also present in the saturated rock pushes oil towards and into the pipe. In still other instances, gases dissolved in petroleum exert pressure on it, forcing it up the pipe.

The ease with which oil flows through a rock layer and into a well depends on a number of factors. In addition to water and gas pressure, viscosity of the oil determines how easily it will move through rock. Even under the most favorable conditions, no more than about 30 percent of the oil in a rock layer can be extracted by any of the natural methods described above.

The term *primary recovery* is used to describe the natural flow of oil by any of the means described above. By adding a pump to the well—*secondary recovery*—an additional quantity of oil can be removed. Even after primary and secondary recovery, however, 40–80% of the oil may still remain in a reservoir. All or most of that oil can be recovered by a variety of techniques described as *tertiary recovery*.

All forms of tertiary recovery involve the injection of some kind of fluid into the oil-bearing stratum. A long pipe is stuck into the ground parallel to the oil-recovery pipe. Into the second pipe is injected a mixture of **carbon dioxide** in water, steam, or some combination of water and **chemicals**. In any one of these cases, the injected material diffuses through the oil-bearing rock, pushing the petroleum out and up into the recovery pipe.

Yet another tertiary recovery approach is to set fire to the oil remaining in one part of the stratum. The heat thus generated reduces the viscosity of the unburned oil remaining in the stratum, allowing it to flow more easily into the recovery pipe. Any form of tertiary recovery is relatively expensive and is not used, therefore, until the price of oil justifies this approach. The same can be said for off-shore drilling.

Organic materials washed into the oceans from rivers often settle on the sloping underwater area known as the continental shelf. In this oxygen-free **environment**, those materials often decay to produce petroleum and natural gas. In recent decades, oil companies have found it profitable to locate and drill for oil in these off-shore reserves.

The technique for drilling from off-shore **wells** is generally the same as that used on land. The major difference is that before drilling may begin a stable platform on the water's surface for the drilling rig and ancillary equipment must be constructed. The drilling platform must be protected from high winds, waves, and serious storms. In addition, special safeguards must be taken to protect pipes from breaking and releasing oil into the environment. *See also* Oil Pollution Act (1990); Oil spills

[*David E. Newton*]

Resources

Books

Freudenburg, W. R. *Oil in Troubled Waters: Perceptions, Politics, and the Battle Over Offshore Drilling.* Albany: State University of New York Press, 1994.

Holing, D. *Coastal Alert: Ecosystems, Energy, and Offshore Oil Drilling.* Covelo, CA: Island Press, 1990.

Periodicals

Carey, J. "Hot Science in Cold Lands." *National Wildlife* 29 (April-May 1991): 4–13.

"Oil and Gas Drilling Threatens Grizzlies." *National Parks* 67 (May-June 1993): 13–14.

Petulla, J. M. *American Environmental History.* 2nd ed. San Francisco: Boyd & Fraser, 1988.

Oil embargo

The year of 1973 marks one of the most important turning points in the history of the twentieth century. Prior to 1973, the world had become accustomed to a plentiful supply of inexpensive **fossil fuels**: **coal**, **petroleum**, and **natural gas**. Developed nations had built economies that depended not just on these fossil fuels, but also on their relatively low cost. Patterns of urban growth in the United States, to take just one example, reflected the fact that the average person could easily afford to drive her **automobile** many miles a day. The average price of a barrel of oil in 1973, for example, was $2.70 and the average cost of **gasoline** at the pump about 35¢ a gallon.

A number of factors combined in 1973, however, to change this picture dramatically. The most obvious of those factors was a decision made by the Arab members of the **Organization of Petroleum Exporting Countries** (OPEC) to cut back on its export of petroleum to many nations of the world. The reason given for this decision was the support given by the United States to Israel during the 18-day war with Syria and Egypt. The action, taken on October 18, 1973, reduced the direct flow of oil from the Arab states to the United States to zero.

OPEC had been established in 1960 by a Venezuelan oil man Juan Perez Alfonso. An extremely conservative businessman who rode a bicycle and read by candlelight to save

Cars line up for gasoline along Route 4 in Fort Lee, New Jersey, in 1974. (AP/Wide World Photos. Reproduced by permission.)

energy, Perez Alfonso saw OPEC as an organization that could encourage **energy conservation** throughout the world.

His objectives were certainly achieved in some places as a result of the 1973 embargo. Americans, along with citizens of other developed nations, found themselves waiting in long lines to buy gasoline, turning their thermostats down to 60°, and learning how to live with less energy in general.

The embargo also had a devastating effect on national economies. In the United States, inflation climbed to more than 10 percent a year, an enormous trade deficit developed, and interest rates climbed to the high teens. Elsewhere, a global recession began.

By the time the embargo ended in March 1974, oil prices had climbed to nearly $12 a barrel, an increase of 330 percent. Gasoline prices had also begun to climb, reaching 57% a gallon by 1975, 86¢ a gallon by 1979, and $1.19 a gallon by 1980.

The embargo caused developed nations to rethink their dependence on fossil fuels. Research on **alternative energy sources** such as wind, tides, geothermal, and **solar energy** suddenly attained a new importance. The United States government responded to the new era of expensive energy by formulating an entirely new **energy policy** expressed in such legislation as the Energy Policy and **Conservation** Act of 1976, the Energy Conservation and Production Act of 1976, the **Energy Reorganization Act** of 1974, and the National Energy Act of 1978.

Interestingly enough, the OPEC embargo had impacts that were relatively little known and discussed at the time. For example, the United States was importing only about 7 percent of its oil from Arab nations in 1973. Yet, almost as soon as the embargo was announced, prices at the gasoline pump began to rise. Such a fast response is, at first, difficult to understand since "old," cheap oil was still on its way from the Middle East and was still being processed, shipped and stored by petroleum companies.

The embargo must also be understood, therefore, as an opportunity for which oil companies had been looking to increase their prices and profits. In the years preceding the embargo, these companies had been feeling increased pressures from domestic environmental groups to cut back on drilling in sensitive areas. They were also losing business to small, independent competitors.

The embargo afforded the companies an opportunity to turn this situation around and once more increase profitability, which they did. In the one year following the embargo, for example, the seven largest oil companies reported profits from 40–85%. Two years after the embargo, one of these companies (Exxon) became the nation's richest corporation, with revenues of $45.1 billion.

Neither have the long term effects of the embargo been what observers in 1973 might have expected. In 1990 the United States imports a larger percentage of its oil from Arab OPEC members than it did in 1973. And the enthusiasm for alternative energy sources of the 1970s has largely

diminished. *See also* Alternative fuels; Energy conservation; Geothermal energy; Wave power; Wind energy

[*David E. Newton*]

RESOURCES
BOOKS

Commoner, Barry. *The Politics of Energy.* New York: Knopf, 1979.

Silber, D., ed. *The Arab Oil Embargo: Ten Years Later.* Washington, DC: Americans for Energy Independence, 1984.

Oil pipeline

see **Trans-Alaska pipeline**

Oil shale

The term oil shale is technically incorrect in that the rock to which it refers, marlstone, is neither oil nor shale. Instead, it is a material that contains an organic substance known as *kerogene*. When heated to a temperature of 900°F (480°C) or more, kerogene decomposes, forming a petroleum-like liquid and a combustible gas. Huge amounts of high-grade oil shale exist in the western United States. By some estimates, these reserves could meet the nation's fuel needs for about a century. Although the technology for tapping these reserves already exists, it is still too expensive to compete with conventional **fossil fuels** or other sources of energy now in use. *See also* Alternative fuels

Oil spills

An oil spill is the common expression used to refer to the release of crude oil or **petroleum** into water or on land. Crude oil released in this way represents an evironmental issue of great concern because spills threaten animals, plant life and other marine resources. Oil can also cause long term environmental and economic damage to the marine **ecosystem** near a spill. According to a 2002 report from the **National Academy of Sciences**, approximately 210 million gal (790 million l) of oil spills into the oceans each year. Sources include the **wells** from which oil is extracted and the ships used to transport it, as well as natural oil **seepage** from geologic formations below the seafloor, as for example in **Coal** Oil Point along the California Coast, where an estimated 2,000–3,000 gal (7,570–11,350 l) of crude oil is released naturally from the ocean floor every day. While accidental tanker oil spills receive the most publicity, they only account for approximately 20% of the crude oil released into the oceans each year by human activity with the remainder largely due to routine oil tanker ship maintenance operations such as **loading**, discharging, and emptying ballast tanks.

Oil and spills are often measured in gal (l) or "barrels of petroleum"; a barrel equals 42 gal (159 l), and a (metric) tonne equals 7.2 barrels. Tankers sometimes transport more than 30,000 barrels of oil (200,000 tonnes).

According to a 2002 study performed by the **National Research Council**, a total of 29 million gal (110 million l) of petroleum are released into North American ocean waters each year as a result of human activities or carelessness. However, only a small fraction of that environmental **pollution** is due to pipeline ruptures or oil tanker spills. Approximately 85% of those spills involve land-based runoffs from cars and trucks, fuel dumping by commercial airplane pilots, and emissions from small boats and crafts.

Oil and its properties

The word oil usually refers to petroleum, a liquid that occurs in **nature** and consists of **hydrocarbons**, a group of organic chemical compounds of **hydrogen** and **carbon** atoms. Petroleum, also known as crude oil, is classified in categories that range from light, volatile oils (Class A) to heavy, sticky oils (Class C), depending on physical properties and charcateristics. Refined oil is crude oil that has been processed for use as **gasoline**, kerosene, lubricating oil, and fuel oils of varying weights.

In spills, the majority of oils spread horizontally to form a smooth, slippery layer on the surface of water. This surface is called a slick. When oil stays on the ocean surface, it cuts off the oxygen supply to the marine life below. The oil also kills birds, animals, and can harm the water supply and the coastline.

"Fate" is the term used by scientists to describe what happens to the various oil components after a spill. Factors that determine fate include the type of oil, the quantity spilled, water temperature, and **climate**. The fates of oil are natural processes. **Weathering** is a series of chemical and physical changes that causes oil to break down and become heavier than water. Evaporation occurs when lighter substances in the oil evaporate and leave the water surface.

About half of a spill may evaporate within several days. Warm water temperatures help this process. Although oil evaporates, the process of emulsification may increase the size of a spill. Emulsions are a mixture of small drops of oil and water. Wave action mixes together a water-in-oil mixture called "chocolate mousse." Mousse may remain in the **environment** for months or years, according to the United States **Environmental Protection Agency** (EPA).

Oxidation reactions may also result from the contact of oil with water and oxygen. Portions of the slick cling together in tar balls. Biodegradation occurs when micro-

organisms like bacteria feed on oil. How oil affects the environment depends on the rate at which the oil spreads. The slick is affected by surface tension, which is the measure of the attraction between the surface molecules of a liquid. Oil with a higher surface tension usually remains in one place. If the surface tension is lower, the oil tends to spread. Wind and water currents also cause oil to spread. Furthermore, higher temperatures can reduce surface tension so oil tends to spread more in warmer water.

Light refined products like gasoline and kerosene spread on the water surface and quickly penetrate porous **soil**. The risk of fire and toxic hazards is high, but these oils evaporate quickly and leave little residue. Heavier refined oil products are less of a fire and toxic hazard risk.

Oil spills and their aftermaths

After a spill, oil can spread very quickly unless contained, for example, by a boom or a boat slip. The lighter the oil, the faster it spreads out. For example, gasoline spreads faster than heavy fuel oil. Faster currents and winds can also cause oil to spread faster and temperature can sometimes make a difference as well because colder oil does not flow as well and spreads more slowly.

Oil spills have occured all over the world. The Cutter Information Corporation tracks oil spills involving at least 10,000 gal (34 tonnes). It reports that spills of that magnitude have occurred in the waters of 112 countries since 1960. Oil spills are also known to happen more often in some parts of the world. Major oil spills from tankers have occurred in the Gulf of Mexico (267 spills); the northeastern United States (140 spills); the **Mediterranean Sea** (127 spills); the Persian Gulf (108 spills); the North Sea (75 spills); Japan (60 spills); the Baltic Sea (52 spills); the United Kingdom and English Channel (49 spills); Malaysia and Singapore (39 spills); the west coast of France and north and west coasts of Spain (33 spills); and Korea (32 spills).

The aftermath of an oil spill always results in environmental damage. During the 1991 Gulf War, oil was spilled onto Kuwaiti land and into the Arabian Gulf when the Iraqi Army began destroying tankers, oil terminals, and oil wells. Approximately 9,000,000 barrels of oil were spilled in the Arabian Gulf, forming a slick measuring some 600 mi^2 (1,600 km^2). When the slick moved close to Saudi Arabia, people pumped water into the area between the beach and the oil. Some 400 mi (640 km) of the western shores of the gulf was oiled, with Saudi Arabian shores the most severely affected, the spill destroying most of its shrimp fields.

In 1989, the ***Exxon Valdez*** ran aground Bligh Reef in **Prince William Sound**, Alaska, spilling more than 11 million gal (41 million l) of crude oil. The spill was the largest in United States history and focused worldwide attention to the damage caused by oil spills. The first day, thousands of animals died. During the month after the spill, more than 7,000 sea otters in Prince William Sound died. Other casualties included more than 100 bald eagles, and 36,000 birds including puffin, auklet and other species. As the oil spread, people tried to rescue wildlife. They had to calm terrified animals. Rescuers fed animals and kept them warm. When the animals were stronger, people started cleaning birds and mammals.

Three methods were used in the effort to clean up the spill, namely burning, mechanical cleanup and the use of chemical dispersants. Burning was conducted during the early stages by placing a fire-resistant boom on tow lines, with two ends of the boom each attached to a different ship. The two ships with the boom between them sailed very slowly throughout the slick until the boom was full of oil. They then towed the boom away from the slick and the oil was set on fire. The procedure did not endanger the main slick nor the *Exxon Valdez*, because a safe distance separated them. Mechanical cleanup was also carried out using booms and skimmers with people cleaning birds and mammals. Chemical dispersants were also used in the cleanup effort but were not very effective because there was not enough wave action to mix the dispersant with the oil in the water.

The aftermath of the *Exxon Valdez* spill included the adoption of the federal Oil Pollution Act of 1990. The law created a spill clean-up fund, set penalties for oil spillers, and directed the federal government to respond quickly to oil spills. In 1993, federal law required a double hull for all tankers carrying oil to the United States.

The United States produces an average of 125 billion gal (473 billion l) of crude oil each year. The country imports an additional 114 billion gal (430 billion l) and 29 million gal (110 million l) of oil enter coastal waters off the United States each year. Nearly 85% of that oil comes from polluted rivers, small boats, vehicles, and street run-off. As a result, the National Academy of Sciences has called on the federal government to work with state environmental agencies to address these issues.

In the United States, various techniques are used to respond to oil spills. Containment and recovery are usually the primary goals of the response team. Equipment includes booms and skimmers that are used to collect the oil and store it. Floating booms are mechanical barriers that extend above and below the surface of the water to stop the spread of oil. They can be used to surround a slick completely and reduce its spread, to protect harbor entrances or biologically sensitive areas, and to divert oil to an area where it can be recovered. Dispersants are **chemicals** used to break up oil and keep it from reaching the land. Furthermore, the spill response team works to keep birds and animals away from the oil spill area. Their equipment includes propane scare cans, floating dummies, and helium balloons.

Workers clean up an oil-fouled California beach. (Corbis-Bettmann. Reproduced by permission.)

Oil spills on a global level

The EPA, the **National Oceanic and Atmospheric Administration**, and the American Petroleum Institute are among the sponsors of the International Oil Spill Conference. Since 1969, the conference has been scheduled every two years. Goals include delineating the overall oil spill problem and exploring ways to prevent and respond to spills. A 2003 conference was planned in Vancouver, British Columbia.

Oil spills are a global problem. In 1999, the tanker *Erika* caused the greatest oil spill in European history. The tanker broke and spilled 3 million gal (11 million l) of oil off the coast of Brittany, France. The following year, a pipeline ruptured and spilled 343,200 gal (1.3 million l) of oil into Guanabara Bay in Brazil. And in 2000, the tanker *Westchester* ran aground south of New Orleans, spilling 567,000 gal (2.1 million l) of oil into the Mississippi River. That was the largest spill in the United States since the *Exxon Valdez*.

Environmental organizations concerned with oil spills include the Sea Shepherd Conservation. Founder Paul Watson's goal in 2002 was to create a coalition of oil producers and conservationists. The coalition would develop airborne teams that would respond to oil spills within 12 hours. Team equipment would include items for cleaning wildlife and pumps for oil.

[*Liz Swain*]

Resources

Books

Burger, Joanna. *Oil Spills.* New Brunswick, NJ: Rutgers University Press, 1997.

Fingas, Mervin F., and Jennifer Charles. *The Basics of Oil Spill Cleanup, Second Edition.* Boca Raton: CRC Press, 2000.

Garcia-Martinez, R., and C. A. Brebbia, eds. *Oil and Hydrocarbon Spills, Modelling, Analysis and Control.* Southampton, UK: Computational Mechanics (WIT Press), 1998.

Hayes, Miles. *Black Tides.* Austin: University of Texas Press, 1999.

Keeble, John, and Natalie Fobes. *Out of the Channel: The Exxon Valdez Oil Spill in Prince William Sound.* Seattle: University of Washington Press, 1999.

Organizations

EPA - Oil Spill Program, UNIDO New York Office, Toll Free: 800-424-9346, Email: oilinfo@epa.gov, <http://www.epa.gov/oilspill/>

International Tanker Owners Pollution Federation Limited, Staple Hall, Stonehouse Court, 87-90 Houndsditch, London, U.K. EC3A 7AX +44(0)20-7621-1255, Toll Free: 800-424-9346, Email: central@itopf.com, <http://www.itopf.com/index2.html>

National Oceanic and Atmospheric Administration., 14th Street and Constitution Avenue NW, Room 6013, Washington, D.C. USA 20230 (202) 482-6090, Fax: (202) 482-3154, Email: answers@noaaa.gov, <http://www.noaa.gov>

Oil Spill Response Ltd (ORSL), 1 Great Cumberland Place, London, U.K. W1H 7AL +44 (0)20-7724-0102, Toll Free: 800-424-9346, Email: orsl@orsl.co.uk, <http://www.oilspillresponse.com/>

Sea Shepherd International., 22774 Pacific Coast Highway, Malibu, CA USA 90165 (310) 456-1141, Fax: (310) 456-2488, Email: seashepherd@seashepherd.org, <http://www.seashepherd.org>

The National Academies, 2101 Constitution Avenue, NW, Washington, DC USA 20418 (202) 334-2000, , <http://www4.nationalacademies.org>

Oil-recycle toilet

see **Toilets**

Old-growth forest

The trees stand tall and thick of girth. The air about them is cool and moist. The **soil** is rich and matted by a thick organic blanket. People marvel at the forest, imbibing its grandeur, coveting its timber. In another area, across the mountains, perhaps, the trees stand stooped and scraggly. The air about them is parched. The soil is coarse, barren, and hardened to the elements. People pass by the forest, ignoring its dignity, rejecting its worth.

These images reflect extremes in old-growth forests. The first is the compelling one: it exemplifies the common perception of a **ecosystem** at the center of a bitter environmental controversy. The second depicts an equally valid old-growth forest, but it is one whose fate few people care to debate.

The controversy over old-growth forests is the result of competition for what has become a scarce natural resource—large, old trees that can be either harvested to produce high value lumber products or preserved as notable relics as a forest proceeds through its stages of ecological **succession**. This competition is a classic environmental struggle, impelled by radically different perceptions of value and conflicting goals of consumptive and nonconsumptive use.

What is an old-growth forest? Before the modern debates, the definition seemed simple. Old-growth was a mature virgin forest; it consisted of giant old trees, many past their prime, which towered over a shady, multilayered understory and a thick, fermenting forest floor. In contrast to second-growth timber, the stand never had been harvested. It was something that existed in the West, having long since been cut in the East.

Old-growth Douglas fir forest, Pacific Northwest. (Photograph by Tom and Pat Lesson. National Audubon Society Collection/Photo Researchers, Inc. Reproduced by permission.)

In the mid to late 1980s several professional and governmental organizations, including the **Society of American Foresters**, the U.S. **Forest Service**, and California's State Board of Forestry, began efforts to define formally "old-growth forest" and the related term "ancient forest" for eco-

logical and regulatory purposes. The task was complicated by the great diversity in forest types, as well as by different views of the purpose and use of the definition. For example, 60 years of age might be considered old for one type, whereas 200 or 1,000 years might be more accurate for other types. Moreover, forest attributes other than age are more important for the wellbeing of certain **species** which are dependent on forests commonly considered old growth, such as the **northern spotted owl** and marbled murrelet. Nonetheless, some common attributes and criteria were developed.

Old-growth forests are now defined as those in a late seral stage of ecological succession, based on their composition, structure, and function. Composition is the representation of plant species—trees, shrubs, forbs, and grasses—that comprise the forest. (Often, in referring to an old-growth stand foresters limit composition to the tree species present). Structure includes the concentration, age, size, and arrangement of living plants, standing dead trees (called "snags"), fallen logs, forest-floor litter, and stream debris. Function refers to the forest's broad ecological roles, such as **habitat** for terrestrial and aquatic organisms, a repository for genetic material, a component in the hydrologic and biogeochemical cycles, and a climatic **buffer**. Each of these factors vary and must be defined and evaluated for each forest type in the various physiographic regions, while accounting for differences in disturbance history, such as wildfires, landslides, hurricanes, and human activities. The problem of specifically defining and determining use of these lands is exceedingly complex, especially for managers of multiple-use public lands who often are squeezed between the opposing pressures of commercial interests, such as the timber industry, and environmental preservation groups. The modern controversy centers primarily around forests in the northwest of United States and Canada—forests consisting of virgin **redwoods**, Douglas firs, and mixed conifers.

As an example of old-growth characteristics, the Douglas-fir forests are characterized by large, old, live trees, many more than 150 feet (46 m) tall, 4 ft (1.2 m) in diameter, and 200 years old. Interspersed among the trees are snags of various sizes—skeletons of trees long dead, now home to birds, small climbing mammals, and insects. Below the giants are one or more layers of understory—subdominant and lower growing trees of the same or perhaps different species, and beneath them are shrubs, either in a thick tangle providing dense cover and blocking passage or separated and allowing easy passage. The trees are not all healthy and vigorous. Some are malformed, with broken tops or multiple trunks, and infected by fungal rots whose conks protrude through the bark. Eventually, these will fall, joining others that fell decades or centuries ago, making a criss-cross pattern of rotting logs on the forest floor. In places, high in the trees, neighboring crowns touch all around, permanently shading the ground; elsewhere, gaps in the canopy allow sunlight to reach the forest floor.

Proponents of harvesting mature trees in old-growth forests assert that the forests cannot be preserved, that they have reached the **carrying capacity** of the site and the stage of decadence and declining productivity that ultimately will result in loss of the forests as well as their high commercial value which supports local lumber-based economies. They feel that society would be better served by converting these aged, slow-growing ecosystems to healthy, productive, managed forests. Management proponents also argue that adequate old-growth forests are permanently protected in designated wildernesses and national and state parks. Moreover, they point out that even though most old-growth forests are on **public land**, many forests are privately owned, and that land owners not only pay taxes on the forests, but they also have made an investment from which they are entitled a reasonable profit. If the forests are to be preserved, land owners and others suffering loss from the preservation should be reimbursed.

Proponents of saving the large old trees and their environments claim that the forests are dynamic, that although the largest, oldest trees will die and rot, they also will be returned to earth to support new growth, foster biological diversity, and preserve genetic linkages. Moreover, protection of the forests will help ensure survival of dependent species, some of which are threatened or endangered. Defenders claim that the trees will not be wasted; they simply will have alternative value. They believe that their cause is one of moral as well as biological imperative. More than 90% of America's old-growth forests have been logged, depriving **future generations** of the scientific, social, and psychic benefits of these forests. As a vestige of North American heritage, the remaining forests, they believe, should be manipulated only insofar as necessary to protect their integrity and minimize threats of natural fire or disease from spreading to surrounding lands. *See also* American Forestry Association; Endangered species; National forest; National Forest Management Act; Restoration ecology

[*Ronald D. Taskey*]

Resources

Books

Arrandale, T. *The Battle for Natural Resources.* Washington, DC: Congressional Quarterly, Inc., 1983.

Kaufmann, M. R., W. H. Moir, and R. L. Bassett. *Old-Growth Forests in the Southwest and Rocky Mountain Regions.* Proceedings of a Workshop. Washington, DC: U.S. Forest Service, Rocky Mountain Forest and Range Experiment Station, 1992.

Other

Spies, T. A., and J. F. Franklin. "The Structure of Natural Young, Mature, and Old-Growth Douglas-Fir Forests in Oregon and Washington." In

Wildlife and Vegetation of Unmanaged Douglas-Fir Forests, edited by L. F. Ruggiero, et al. Washington, DC: U. S. Forest Service, Pacific Northwest Forest and Range Experiment Station, 1991.

Oligotrophic

The term oligotrophic is derived from the Greek term meaning "poorly nourished" and refers to an aquatic system that has low overall levels of primary production, principally because of low concentrations of the nutrients that plants require. The bottom waters of an oligotrophic lake do not become depleted of oxygen in summer, when rates of **primary productivity** in surface waters of the lake are typically at their highest. Oligotrophic bodies of water tend to have a more diverse community of **zooplankton** than waters with high levels of primary production. The term can also be used to describe any organism that only needs a limited supply of nutrients, or an insect that only utilizes a few plants as **habitat**.

[*Marie H. Bundy*]

Frederick Law Olmsted Sr. (1822 – 1903)

American landscape designer

The famous landscape designer of Central Park (on which he collaborated with the architect Calvert Vaux), Frederick Law Olmsted Sr. is widely known for that accomplishment alone. Considering the impact of the park on city life in New York, and on other designs around the world, that would be accomplishment enough. But Olmsted other achievements are so overshadowed by the Central Park project, that he is still not as well known in environmental circles as he should be.

Many people don't know, or are confused by, the fact that there were two Frederick Law Olmsteds: father and son, F. L. O Sr., and F. L. O Jr. Both of them were landscape architects, and F. L. O Jr. followed directly in his father's footsteps, assuming "leadership of the country's largest and most prestigious landscape architecture firm" after his father's death.

Olmsted Sr., was born in "the rural environs" of Hartford, Connecticut, the oldest son of a well-to-do merchant, from an old New England family. He was educated in a series of boarding schools in the rural Connecticut area where he grew up. More unorthodox educational benefits accrued from his apprenticeships as civil engineer and farmer along with his stint as a ship's cabin boy on a trip to China.

Reportedly, Olmsted's firm laid out some 1,000 parks in 200 cities. He also designed university campuses (U.C. Berkeley, Harvard, Amherst, Yale, etc.), cemeteries, hospital grounds (including the grounds for the hospital where he died, unhappy—even while suffering a terminal illness—that his plans for the hospital had not been followed carefully enough). A source of unhappiness in general was the fact that many of his plans were modified in ways he did not agree with (e.g., Stanford University), or even suppressed, as happened with his plan for **Yosemite National Park**.

The crown jewel of Olmsted's creativity is, of course, Central Park in New York City. As Bill Vogt describes it: "the entire area was man-made, literally from the ground up. It sprang from an 843-acre eyesore of stinking quagmire, rubbish heaps, rocky outcroppings, and squatters shacks." An army of workers created a lake, shoveled in enormous quantities of top **soil** to create natural-appearing meadows, and planted whole forests to screen the park from the city. Charles McLaughlin, the editor of Olmsted's papers, claimed that one of the reasons Olmsted was virtually forgotten for so long was that his designs have an "always been there" quality, resulting in landscapes that today "in their maturity...appear so 'natural' that one thinks of them as something not put there by artifice but merely preserved by happenstance." What Olmsted and Vaux created stands today as an oasis and respite from what Olmsted described as the city's "constantly repeated right angles, straight lines, and flat surfaces."

Among Olmsted's lesser known accomplishments were both his role in the Commission to establish Yosemite Park (a task undertaken in 1864, long before **John Muir** first saw Yosemite Valley) and preserve Niagara Falls, a movement which resulted in creation of the Niagara Falls State Reservation in 1888. He also was a prominent participant in the campaign to preserve the Adirondack region in upstate New York.

Even less well known are his writings not directly associated with landscape architecture or ecological design. Olmsted was quite an accomplished travel writer, producing still readable books like a *Saddle-trip on the Southwestern Frontier of Texas*, or his first book titled *Walks and Talks of an American Farmer in England*. Olmsted's travels in the South were as a roving journalist for the *New York [Daily] Times*; he was an acute observer and his travel books on the south were also commentaries on the institution of slavery (that he believed "to be both economically ruinous and morally indefensible.") Books from this period included *The Cotton Kingdom: A Traveler's Observations on Cotton and Slavery in the American Slave States*, *A Journey in the Seaboard Slave States*, and *Slavery and the South, 1852–1857*. He was also a founder of the liberal journal of commentary, *The Nation*. He never dissociated his design work from its social context (believing, in his words, that his works should have "a manifest civilizing effect"), and his biggest disappointments

occurred when he was not allowed to incorporate his ideas on social change into his various designs.

[*Gerald L. Young Ph.D.*]

Resources
Books

Beveridge, C. E., and P. Rocheleau. *Frederick Law Olmsted: Designing the American Landscape.* New York: Rizzoli, 1995.

Fein, A. *Frederick Law Olmsted and the American Environmental Tradition.* New York: George Braziller, 1972.

Hall, L. *Olmsted's America.* Boston: Little, Brown, 1995.

Kalfus, M. *Frederick Law Olmsted: The Passion of a Public Artist.* New York: New York University Press, 1990.

Olmsted, F. L. *The Papers of Frederick Law Olmsted.* Ed. C. C. McLaughlin. Baltimore: The Johns Hopkins University Press, 1977.

White, D. F. "Frederick Law Olmsted, Placemaker." *Two Centuries of American Planning.* Ed. Daniel Schaffer. Baltimore: The Johns Hopkins University Press, 1988.

Onchocerciasis

see **River blindness**

OPEC

see **Organization of Petroleum Exporting Countries**

Open marsh water management

Open Marsh Water Management (OMWM) refers to the practice of controlling the mosquito population in salt marshes by creating an appropriate **habitat** for the natural enemies of the mosquitoes; and by reducing **flooding** in areas that are not wet on an ordinary basis—thus reducing an **environment** that would support mosquitoes but not their predators. Without use of **chemicals** that might be harmful to the **natural resources** surrounding the water body, as well as harmful to **wildlife** and humans, OMWM calls on nature's own ecological balance in order to successfully alter the pest—mosquito. The techniques were eventually employed by individuals and municipalities in order to control mosquitoes even in their own backyards, no matter how far that might be from tidal **wetlands**.

Before this plan was developed, and first used in New Jersey to control the mosquito problem in their tidal wetlands and marshes, the normal practice utilized a network of ditches. Under this method, which started around the time of the Civil War, the shallow ponds and water pools, the natural habitats of the birds and fish that fed on mosquitoes were changed or destroyed through ditch digging to drain the marshes and remove standing water. According to the *Conservationist* in an article published in June 1997, the digging also created a problem because, "Spoil generated by the ditch digging was often disposed of in marsh areas, creating high areas. The high areas were soon invaded by reeds and shrubs—vegetation of lesser value for marsh wildlife. In addition **spoil** mounds restricted the ebb and flow of tides, reducing flushing and creating habitat that actually favored mosquitoes. Many marshes became cut off from tidal flow, causing them to gradually become less saline [salty]. This resulted in changes in the composition of **flora** and **fauna**, enabling such non-native plants as the common reed (Phragmites) to invade the marsh. While somewhat attractive and often collected for dry plant arrangments, this plant has little wildlife value and displaces other more valuable species."

The key to OMWM is creating a habitat for one particular mosquito predator, the mummichogs, or marsh killfish. By restoring the body of water to its natural state, marsh killfish are provided with a habitat for survival, as are other fish and wildlife. The killfish are small but have appetites for mosquito larvae that never subside. Natural habitats are created by the tidal flow that go between being pools of deep water and shallow ponds. Killfish thrive under these conditions—laying their eggs when it is dry, and hatching them when it floods. The mosquito produces a vast number of larvae (all waiting to emerge as mosquitoes) as soon as the marshes are once again full of water. Yet simply producing more killfish does not ensure that they will reach the mosquitoes, and thus control the population. Again, the *Conservationist* has pointed out that, "For a project to be effective over a wide tidal range, the design of the tidal channels must be carefully tailored to the specific conditions of the marsh. Shallow access channels are constructed to allow the fish access to all parts of the marsh surface. However, pool and ditch sections must provide enough water and adequate depth to prevent excessive predation of the fish by wading birds."

Mosquito control in history

The state of New Jersey was the first place where OMWM was used—and the reasons for that go as far back as the early settlers. According to the *New Jersey Mosquito Homepage (NJMH)* Yellow Fever, contracted from mosquitoes, was the known as the "American plague," infecting the Massachusetts colony as early as 1647. In 1793, Philadelphia was hit and the city's population was devastated. When an army surgeon named **Walter Reed** finally traced the **virus** to the mosquito, *Aedes aegypti*, in 1900, the disease was dealt a death blow like the one it once had wagered on its victims. But the mosquito remained more than a simple annoyance that even the window screen introduced in the 1880s could not totally erase. Humans could contract **Eastern Equine Encephalitis** through mosquitoes, a disease of

epidemic proportions even into the middle half of the twentieth century. Horses continue to be infected if they are not inoculated against the disease. Mosquitoes can also transmit heartworm disease to dogs. Even in modern-day America, a variety of the **species** has been found to infect humans with the deadly **West Nile Virus** causing illness and death to wildlife and people from New York to Ohio, and down the east coast, by 2002.

The *NJMH* elaborates further on the history of mosquito control as an early priority in that state, and others. "Considerable public debate was given to the question whether mosquitoes could ever be controlled. Mosquito control operations grew in some towns but not in all towns. Newspaper battles raged when it was painfully noted that mosquitoes ignored municipal and even state borders. Local boards of health funded most of the extermination work. Laws in 1906 required support for local efforts from the state experiment station. Another law in 1912 directed the creation of county mosquito extermination commissions to assure full-time mosquito work With an increase in mosquito control workers and their rapid progress, it became clear than an organization was needed within which these workers could discuss their problems and share their experiences." Following the first convention of county commission in February 1914 in Atlantic City, the permanent organization known as the **New Jersey Mosquito Extermination Association** was formed. In 1935, a regional organization was formed with ten other states, the **Eastern Association of Mosquito Control Workers**, from a meeting at Trenton; it was re-named in 1944. The **American Mosquito Control Association** remains the premier organization concerned with mosquito control.

The common salt marsh mosquitoes include those known as *Aedes sollicitans*; *Aedes contator*,; and, *Aedes taeniorhynchus.* Adult females deposit their eggs on the marsh surface, which dry for 24 hours. The egg-filled depressions fill with water during the monthly high tides when the marsh is flooded, causing the larvae to hatch quickly. In addition to killfish acting as predators for the larvae, they can be controlled through the OMWM technique of utilizing a series of ditches to connect mosquito breeding depressions to more permanent bodies of water. It eliminates standing water, and also provides for predator fish to reach any mosquito larvae that is still there.

One example of a project to turn around the results of the former ditch digging practice that had been used and proven harmful(see History of to the natural environment is a research project operating in 2000, and supervised by the U.S. **Fish and Wildlife Service** (USFWS), Region 5, at the Rachel Carson **Wildlife Refuge** in Maine. The **U.S. Department of the Interior**, U.S. **Geological Survey** information system project of the Patuxent Wildlife Research Center in Laurel, MD provided a history and description of the project. "By the 1930s, most salt marshes throughout the northeastern United States had been ditched for mosquito control purposes. Documented impacts of this extensive network of ditching include **drainage** of marsh pools, lowered **water table** levels, vegetation changes, an associated trophic [feeding] responses. To restore the ecological functions of ditched salt marshes, while maintaining effective mosquito control, the USFWS (Region 5) is in the process of plugging the salt marsh ditches and establishing marsh pools. Ditch plugging is an adaptation of the mosquito control practice known as Open Marsh Water Management. The purpose of this proposed study is to compare and evaluate marshes that have been ditch plugged with unditched and parallet ditched marshes. Physical (tidal **hydrology**, water table level), chemical (**soil salinity**), and ecological (vegetation, **nekton**, marsh invertebrates, waterbirds, mosquito production) factors will be evaluated." Before data could be collected, whatever the results, this study was expected to create an established pattern for long-term salt marsh monitoring.

East coast states, as well as areas around the country, have engaged in OMWM not merely for simple mosquito control; but as a part of the restoration of **estuaries** and other **wildlife management** and restoration projects. OMWM techniques were still being refined as of 2002 as part of a return to an ecologically-balanced system that provides for the survival of natural resources, and for the survival of humans and wildlife.

[*Jane E. Spear*]

Resources

Periodicals

"Tidal Wetlands in New York State Diversity." *Conservationist* 51, no. 6 (June 1997): 4.

Other

"Cape May County Mosquito Extermination Commission." *Cape May County Web Page.* 2001 [June 2002]. <http://www.capemaycountygov.net>.

"Coastal Mosquito Control in Your Neighborhood." *Norfolk County Mosquito Control Project.* May 3, 2001 [June 2002]. <http://www.ultranet.com/~ncmcp/>.

"Connecticut River Estuary and Tidal River Wetlands Complex." *State of Connecticut Department of Environmental Protection.* 2002 [July 2002]. <http://www.dep.state.ct.us>.

James-Pirri, Mary Jane. "Mosquito Beach Salt Marsh OMWM Restoration." *Northeastern Mosquito Control Association.* 2002 [July 2002]. <http://www.nmca.org>.

Johnson, David. "Africana." *Patuxent Wildlife Research Center.* July 25, 2000 [cited June 2002]. <http://www.pwrc.usgs.gov/>.

"Mosquito Breeding Habitats." *Monmouth County, NJ Web Page.* 1999 [June 2002]. <http://www.visitmonmouth.com>.

Patuxent Wildlife Research Center. *Research Project; Ecosystem response of salt marshes to Open Marsh Water management (OMWM): Rachel Carson National Wildlife Refuge (Maine).* "Angolan deepwater fields emerge as world's most exciting oil frontier." 2001. (June 2002). <http://www.internationalspecialreports.com/>

Purdue University/Forestry and Natural Resources. *Did You Know? Healthy Wetlands Devour Mosquitoes?* [cited June 2002]. <http://www.agriculture.-purdue.edu/fnr>.

Rutgers University/Entomology. "History of Mosquito Control in New Jersey (Where it all Began)." *New Jersey Mosquito Control Homepage.* [cited June 2002]. <http://www.-rci.rutgers.edu/~insects/>.

ORGANIZATIONS

U.S. Environmental Protection Agency, 1200 Pennsylvania Avenue, N.W., Washington, DC USA 20460 202-260-2090, <www.epa.gov>

U.S. Fish and Wildlife Service, Washington, D.C. USA , <www.fws.gov>

Open system

The relationship between any system and its surrounding **environment** can be described in one of three ways. In an isolated system, neither matter nor energy is exchanged with its environment. In a closed system, energy, but not matter, is exchanged. In an open system, both matter and energy are exchanged between the system and its surrounding environment. Any **ecosystem** is an example of an open system. Energy can enter the system in the form of sunlight, for example, and leave in the form of heat. Matter can enter the system in many ways. Rain falls upon it and leaves by evaporation or streamflow, or animals migrate into the system and leave in the form of decay products.

Opportunistic organism

Opportunistic organisms commonly refer to animals and plants that tolerate variable environmental conditions and food sources. Some opportunistic **species** can thrive on almost any available **nutrient** source: omnivorous rats, bears, and raccoons are all opportunistic feeders. Many opportunists flourish under varied environmental conditions: the common house sparrow (*Passer domesticus*) can survive both in the warm, humid **climate** of Florida and in the cold, dry conditions of a Midwestern winter. Aquatic opportunists, often aggressive fish species, fast-spreading **plankton**, and water plants, frequently tolerate fluctuations in water **salinity** as well as temperature.

A secondary use of the term "opportunistic" signifies species that can quickly take advantage of favorable conditions when they arise. Such species can postpone reproduction, or even remain dormant, until appropriate temperatures, moisture availability, or food sources make growth and reproduction possible. Some springtime-breeding lizards in Australian deserts, for example, can spend months or years in a juvenile form, but when temperatures are right and a rare rainfall makes food available, no matter what time of year, they quickly mature and produce young while water is still available. More familiar opportunists are viruses and bacteria that reside in the human body. Often such organisms will remain undetected with a healthy host for a long time. But when the host's immune system becomes weak, resident viruses and bacteria seize an opportunity to grow and spread. Thus people suffering from malnutrition, exhaustion, or a prolonged illness are especially vulnerable to common opportunistic diseases such as the common cold or pneumonia.

Adaptable and prolific reproductive strategies usually characterize opportunistic organisms. While some plants can reproduced only when pollinated by a specific, rare insect and many animals can breed only in certain conditions and at a precise time of year, opportunistic species often reproduce at any time of year or under almost any conditions. House mice (*Mus musculus*) are extremely opportunistic breeders: they can produce sizeable litters at any time of year. Opportunistic feeding aids their ability to breed year round; these mice can nourish their young with almost any available vegetable matter, fresh or dry.

The common dandelion (*Taraxacum officinale*) is also an opportunistic breeder. Producing thousands of seeds per plant from early spring through late fall, the dandelion can reproduce despite **competition** from fast-growing grass, under heavy applications of chemical herbicides, and even with the violent weekly disturbance of a lawn mower. Once mature, dandelion seeds disperse rapidly and effectively, riding on the wind or on the fur of passing rodents. The common housefly (*Musca domestica*) is also an opportunistic feeder and reproducer—it can both feed and lay eggs on almost any organic material as long as it is fairly warm and moist.

Because of their adaptability, opportunistic organisms commonly tolerate severe environmental disturbances. Fire, floods, **drought**, and **pollution** disturb or even eliminate plants and animals that require stable conditions and have specialized nutrient sources. Fireweed (*Epilobium angustifolium*), an opportunist that readily takes advantage of bare ground and open sunlight, spreads quickly after land is cleared by fire or by human disturbance. Because they tolerate, or even thrive, in disturbed environments, many opportunists flourish around human settlements, actively expanding their ranges as human activity disrupts the **habitat** of more sensitive animals and plants. Opportunists are especially visible where chemical pollutants contaminate habitat. In such conditions overall species diversity usually declines, but the population of certain opportunistic species may increase as competition from more sensitive or specialized species is eliminated.

Because they are tolerant, prolific, and hardy, many opportunistic organisms, including the house fly, the house mouse, and the dandelion, are considered pests. Where they occur naturally and have natural limits to their spread, however, opportunists play important environmental roles. By

quickly colonizing bare ground, fireweed and opportunistic grasses help prevent **erosion**. Cottonwood trees (*Populus spp.*), highly opportunistic propagators, are among the few trees able to spread into **arid** regions, providing shade and nesting places along stream channels in deserts and dry plains. Some opportunists that are highly tolerant of pollution are now considered indicators of otherwise undetected **chemical spills**. In such hard-to-observe environments as the sea floor, sudden population explosions among certain bottom-dwelling marine mollusks, plankton, and other invertebrates have been used to identify **petrochemical** spills around drilling platforms and shipping lanes. *See also* Adaptation; Contaminated soil; Environmental stress; Flooding; Growth limiting factors; Indicator organism; Parasites; Resilience; Scavenger; Symbiosis

[*Mary Ann Cunningham Ph.D.*]

Resources

Books

Foster, H. D. *Health, Disease and the Environment.* Boca Raton: CRC Press, 1992.

Periodicals

Alexander, S. P. "Oasis Under the Ice." *International Wildlife* 18 (November-December 1988): 32–7.

Arcieri, D. T. "The Undesirable Alien—The House Sparrow." *The Conservationist* 46 (1992): 24–25.

Bradshaw, S. D., H. S. Giron, and F. J. Bradshaw. "Patterns of Breeding in Two Species of Agamid Lizards in the Arid Subtropical Pilbara Region of Western Australia." *General and Comparative Endocrinology* 82 (1991): 407–24.

Moreno, J. M., and W. C. Oechel. "Fire Intensity Effects on Germination of Shrubs and Herbs in Southern California Chaparral." *Ecology* 72 (1991): 1993–2004.

Shafir, A. "Dynamics of a Fish Ectoparasite Population: Opportunistic Parasitism in *Argulus japonicus*." *Crustaceana* 62 (1992): 50–64.

Orangutan

The orangutan (*Pongo pygmaeus*), one of the Old World great apes, has its population restricted to the rain forests of the Indonesian islands of Sumatra and Borneo. The orangutan is the largest living arboreal mammal, and it spends most of the daylight hours moving slowly and deliberately through the forest canopy in search of food. Sixty percent of their diet consists of fruit, and the remainder is composed of young leaves and shoots, tree bark, mineral-rich **soil**, and insects. Orangutans are long-lived, with many individuals reaching between 50 and 60 years of age in the wild. These large, chestnut-colored, long-haired apes are facing possible **extinction** from two different causes: **habitat** destruction and the wild animal trade.

An orangutan. (Photograph by Tim Davis. Photo Researchers Inc. Reproduced by permission.)

The **rain forest ecosystem** on the islands of Sumatra and Borneo is rapidly disappearing. Sumatra loses 370 mi^2 (960 km^2) of forest a year, or about 1.6%, faster than any other Indonesian island. The rest of central Indonesia, of which Borneo comprises a major part, loses about 2,700 square miles (7,000 km^2) per year. Some experts believe this

estimate is too low, and they argue it could be closer to 4,600 mi^2 (12,000 km^2) per year. Another devastating blow was dealt Borneo's rain forests just over a decade ago, when more than 15,400 mi^2 (40,000 km^2) of the island's tropical forest was destroyed by **drought** and fire between 1982 and 1983. The fire was set by farmers who claimed to be unaware of the risks involved in burning off vegetation in a drought stricken area. Even though Indonesia still has over 400,000 mi^2 (1,000,000 km^2) of rain forest habitat remaining, the rate of loss threatens the continued existence of the wild orangutan population, which is now estimated at about 25,000 individuals.

Both the Indonesian government and the **Convention on International Trade in Endangered Species of Wild Fauna and Flora** (CITES) have banned international trade of orangutans, yet their population continues to be threatened by the black market. In order to meet the demand for these apes as pets around the world, poachers kill the mother orangutan to secure the young ones, and the **mortality** rate of these orphans is extremely high, with less than 20% of those smuggled ever arriving alive at their final destination. This high mortality rate is directly due to stress, both emotional and physiological, on the young orangutans. The **transportation** scheme involved in smuggling these animals out of Indonesia to major trade centers throughout the world is intricate and time-consuming, and the way in which they are concealed for shipping is inhumane. These are two more reasons why only one out of five or six orangutan babies will survive the ordeal.

Some hope for the **species** rests in a global effort to manage a captive propagation program in zoos. A potentially self-sustaining captive population of more than 850 orangutans has been established. An elaborate system of networking and recording of all legally held individuals may also aid in the recognition and recapture of smuggled animals. Researchers have also developed methods of determining whether smuggled orangutans are of Bornean or Sumatran origin, thus providing a means of maintaining genetic integrity for those that can be bred in captivity or relocated.

[*Eugene C. Beckham*]

Resources

Books

Galdikas, M. F. Biruté, and Nancy Brigs. *Orangutan Odyssey.* New York: Harry N. Abrams, 1999.

Russon, Anne. *Orangutans: Wizards of the Rain Forest.* Buffalo: Firefly Books, 2000.

Other

Orangutan Sanctuary. [cited May 2002]. <http://www.yorku.ca/arusson>.

Orang-outang. [cited May 2002]. <http://www.orang-outang.com>.

Order of magnitude

A mathematical term used loosely to indicate tenfold differences between values. This concept is crucial to interpreting logarithmic scales such as **pH** or **earthquake** magnitude, where each number differs by a factor of 10, and two numbers differ by a factor of 100 (10 times 10). For example, a pH of 4 is 100 times as acidic as a pH of 6 because they differ by two orders of magnitude. Scientists often generalize with this term; for example "our ability to measure pollutants has improved by several orders of magnitude" means that whereas before we were able to measure **parts per million** (ppm), we can now measure **parts per billion** (ppb).

Oregon silverspot butterfly

The Oregon silverspot butterfly (*Speyeria zerene hippolyta*) is a medium-sized butterfly, predominantly orange and brown with black veins and spots on its hindwings and bright silver spots on its forewings. The length of its forewings is about 1.1 in (2.9 cm). The female is usually slightly larger than the male. This butterfly is listed as threatened by the U. S. **Fish and Wildlife Service** and has been protected under the **Endangered Species Act** since 1980.

Inhabiting a very restricted range, the Oregon silverspot occurs only in salt spray meadows along the Pacific coast in Oregon and Washington. This **habitat** is characterized by heavy rainfall, fog, and mild temperatures. The most critical feature of this habitat, however, is the presence of the western blue violet (*Viola adunca*), the host plant of the butterfly's larva. For two months each spring, larval Oregon silverspots feed on violet leaves before entering the pupa stage of development. This butterfly was historically present at 17 locations along the coasts of Oregon and Washington, but now only five populations in Oregon are known to exist with certainty.

Housing developments and recreational uses of the coast that destroy or degrade butterfly habitat are the major threats to this subspecies's survival. Natural fire patterns in its meadow habitat have been suppressed, allowing non-native vegetation to mix with native plants and changing the habitat's character. An area of Lane County, Oregon with a healthy population of Oregon silverspots has been designated **critical habitat** for this subspecies. Expansion of the population of western blue violets in this area will be encouraged by the control of saplings and other invading plants. Transplantation of western blue violets to other sites with suitable meadow habitat may also be attempted. A recovery plan was put into effect in 1999 and monitored through 2000. It was headed by Lewis and Clark College, the Oregon Zoo in Portland, and the Woodland Park Zoo in Seattle and consisted of rearing the larvae in captivity,

Oregon silverspot butterfly (Speyeria zerene hippolyta). (Photograph by Paul A. Opler. Reproduced by permission.)

then returning the butterflies to the wild. Although the Oregon silverspot butterfly is not in immediate danger of **extinction**, its specific habitat requirements and the vulnerability of that habitat to degradation and destruction, makes intervention necessary to ensure the long-term survival of this subspecies.

[*Christine B. Jeryan*]

Resources

Books

Howe, W. H. *The Butterflies of North America.* Garden City, NY: Doubleday, 1975.

Other

U.S. Fish and Wildlife Service *Revised Recovery Plan Published for the Oregon Silverspot Butterfly.* November 29, 2001 [cited May 2002]. <http://news.fws.gov/NewsReleases/R1/C71501ED-5742-4E26-965408A77BC0875F.html>.

Organic gardening and farming

Agriculture has changed dramatically since the end of World War II. As a result of new technologies, mechanization, increased chemical use, specialization, and government policies, food and fiber productivity has soared. While some of these changes have led to positive effects, there have been significant costs: **topsoil** depletion, **groundwater** contamination, harmful **pesticide residue**, the decline of family farms, and increasing costs of production. To counterbalance these costs, there is a growing movement to grow plants organically.

On a local level, suburban homeowners and city dwellers are finding ways to plant their own food for personal consumption in plots that are not sprayed with **chemicals** or treated with synthetic fertilizers. City dwellers may team up to work a collective, organic garden on a vacant lot. Homeowners have the option to create **mulch** piles—a mixture of leaves and organic materials, vegetable leavings, egg shells, coffee grounds, etc.—that eventually become a rich **soil**, thanks to the breakdown of these elements by tiny organisms. This enriched soil becomes a fertile, clean foundation for a bountiful garden. The inevitable weeds that grow can be picked by hand.

On a larger level, emerging as an answer to some of these farming problems is a movement called "sustainable agriculture." Sustainability rests on the principle that we must meet the needs of the present without compromising the ability of **future generations** to meet their own needs.

Organic farming is part of this movement. Most significantly, organic foods are grown without synthetic fertilizers, pesticides, or herbicides. Organic farming and gardening is the result of the belief that the best food crops come from soil that is nurtured rather than treated. Organic farmers take great care to give the soil nutrients to keep it healthy, just as a person works to keep his or her body healthy with certain nutrients. Thus, organic farmers prefer to provide those essential soil builders in natural ways: using cover crops instead of chemical fertilizers, releasing predator insects rather than spraying pests with pesticides and hand weeding rather than applying herbicides.

Organic production practices involve a variety of farming applications. Specific strategies must take into account **topography**, soil characteristics, **climate**, pests, local availability of inputs, and the individual grower's goals. Despite the site-specific and individual nature of this approach, several general principles can be applied to help growers select appropriate management practices. Growers must:

1. Select **species** and varieties of crops that are well suited to the site and to the conditions of the farm.
2. Diversify the crops, including livestock and cultural practices, to enhance the biological and economic **stability** of the farm.
3. Manage the soil to enhance and protect its quality.

Making the transition to **sustainable agriculture** is a process. For farmers, the transition normally requires a series of small, realistic steps. For example in California, where there has been a water shortage, steps are being taken to develop drought-resistant farming systems, even in normal years. Farmers are encouraged to improve **water conservation** and storage measures, provide incentives for selecting specific crops that are drought-tolerant, reduce the volume of **irrigation** systems, and manage crops to reduce water loss.

In order to stop soil **erosion**, farmers are encouraged to reduce or eliminate tillage, manage irrigation to reduce **runoff**, and keep the soil covered with plants or mulch.

Farmers are also encouraged to diversify, since diversified farms are usually more economically and ecologically resilient. By growing a diversity of crops, farmers spread economic risk and the crops are less susceptible to the infestation of certain predators that feed off one crop. Diversity can buffer a farm biologically. For example, in annual cropping systems, crop rotation can be used to suppress weeds, pathogens, and insect pests.

Cover crops can have a stabilizing effect on the agro-**ecosystem**. Cover crops hold soil and nutrients in place, conserve soil moisture with mowed or standing dead mulches, and increase the water **infiltration** rate and soil water holding capacity. Cover crops in orchards and vineyards can buffer the system against **pest** infestations by increasing beneficial arthropod populations and can therefore reduce the need for chemicals. Using a variety of cover crops is also important in order to protect against the failure of a particular species to grow and to attract and sustain a wide range of beneficial arthropods.

Optimum diversity may be obtained by integrating both crops and livestock in the same farming operation. This was the common practice for centuries until the mid-1900s, when technology, government policy, and economics compelled farms to become more specialized. Mixed crop and livestock operations have several advantages. First, growing row crops only on more level land and pasture or forages on steeper slopes will reduce soil erosion. Second, pasture and forage crops in rotation enhance soil quality and reduce erosion; livestock manure in turn contributes to soil fertility. Third, livestock can buffer the negative impacts of low rainfall periods by consuming crop residue that in "plant only" systems would have been considered failures. Finally, feeding and marketing are more flexible in animal production systems. This can help cushion farmers against trade and price fluctuations and, in conjunction with cropping operations, make more efficient farm labor.

Animal production practices are also sustainable or organic. In the midwestern and northeastern United States, many farmers are integrating crop and animal systems, either on dairy farms or with range cattle, sheep, and hog operations. Many of the principles outlined in the crop production section apply to both groups. The actual management practices will, of course, be quite different.

Animal health is crucial, since unhealthy stock waste feed and require additional labor. A herd health program is critical to sustainable livestock production. Animal nutrition is another major issue. While most feed may come from other enterprises on the ranch, some purchased feed is usually imported. If the animals feed from the outside, this feed should be as free of chemicals as possible.

A major goal of organic farming is a healthy soil. Healthy soil will produce healthy crops and plants that have optimum vigor and less susceptibility to pests. In organic or sustainable systems, the soil is viewed as a fragile and living medium that must be protected and nurtured to ensure its long-term productivity and stability. Fertilizers and other inputs may be needed, but they are minimized as the farmer relies on natural, renewable, and on-farm inputs.

While many·crops have key pests that attack even the healthiest of plants, proper soil, water, and **nutrient** management can help prevent some pest problems brought on by crop stress or nutrient imbalance. Additionally, crop management systems that impair soil quality often result in greater inputs of water, nutrients, pesticides, and/or energy for tillage to maintain yields.

Sustainable approaches are those that are the least toxic and least energy intensive and yet maintain productivity and profitability. Farmers are encouraged to use preventive strategies and other alternatives before using chemical inputs from any source. However, there may be situations where the use of synthetic chemicals would be more "sustainable" than a strictly nonchemical approach or an approach using toxic "organic" chemicals. For example, one grape grower in California switched from tillage to a few applications of a broad spectrum contact **herbicide** in his vine row. This approach may use less energy and may compact the soil less than numerous passes with a cultivator or mower.

Coalitions have been created to address the growing organic movement concerns on a local, regional, and national level. The Organic Food Production Association of North America is the trade and marketing arm of the organic industry in the United States and Canada. The Farm Bill of 1990—Organic Foods Production Act—addressed the growing organic farming movement. Title 21 of the bill is the section that will be dealing with the regulations regarding organic certification. The **U.S. Department of Agriculture**, with the guidance of an advisory committee or National Organic Standards Board, is in the process of establishing the federal regulations that will standardize the rules for the entire organic industry in the United States, from growing and manufacturing to distribution. Presently, the organic movement includes growers, retailers, distributors, traders, urban and individual consumers, processors, and various nonprofit organizations nationwide.

Until the regulations are in place, the standards established by individual states vary, or in some states do not exist at all. The 1990 law requires that organic farmers wait for three years before they are officially certified—to ensure most of the chemicals have been eliminated. With proper documentation, on-site inspectors are then able to certify the farm. Organic groups admit that no organic program can claim absolutely it is residue free. The issue here is a process—of farming and producing product in as chemically free and healthy an **environment** as is possible.

The international community is joining the organic movement. Europe—the European Common Market—Australia, Argentina, and many countries in South America are establishing their own organic standards and have joined together under the organization called International Federation of Organic Agricultural Movements.

In addition to the upcoming United States federal standardization of organic farming, more policies are needed to promote simultaneously **environmental health** and economic profitability.

For example, commodity and price support programs could be restructured to allow farmers to realize the full benefits of the productivity gains made possible through alternative practices. Tax and credit policies could be modified to encourage a diverse and decentralized system of family farms rather than corporate concentration and absentee ownership. Government and land grant university research could be modified to emphasize the development of sustainable alternatives. Congress can become more rigorous in preventing the application of certain pesticides, especially those that have been shown to be carcinogenic. Marketing orders and cosmetic standards (i.e. color and uniformity of product, etc.) could be amended to encourage reduced **pesticide** use.

Consumers play a key role. Through their purchases, they send strong messages to producers, retailers, and others in the system about what they think is important. Food cost and nutritional quality have always influenced consumer choices. The challenge now is to find strategies that broaden consumer perspectives so that environmental quality and resource use are also considered in shopping decisions. Coalitions organized around improving the food system are one specific method of systemizing growing and delivery for the producers, retailers, and consumers.

"Clean" meat, vegetables, fruits, beans, and grains are available, as are products such as organically grown cotton. As consumer demands increase, growers will respond. *See also* Biodiversity; Monoculture

[*Liane Clorfene Casten*]

Resources

Books

Conford, P., ed. *Organic Tradition: An Anthology of Writings on Organic Farming, 1900–1950.* Cincinnati, OH: Seven Hills Book Distributors, 1991.

Dudley, N. *G is for EcoGarden: An A to Z Guide to a More Organically Healthy Garden.* New York: Avon, 1992.

Erickson, J. *Gardening for a Greener Planet: A Chemical-Free Approach.* Blue Ridge Summit, PA: TAB Books, 1992.

National Research Council. *Alternative Agriculture.* Washington, DC: National Academy Press, 1989.

Rodale, R. *Regenerative Farming Systems.* Emmaus, PA: Rodale Press, 1985.

Periodicals

Krueger, S. "Green Acres: Farmers Are Hoping Chemical-Free Crops Will Help Get Them Out of the Red." *Nature Canada* 21 (Spring 1992): 42–48.

Organic waste

Organic wastes contain materials which originated from living organisms. There are many types of organic wastes and they can be found in **municipal solid waste**, industrial **solid waste**, agricultural waste, and wastewaters. Organic wastes are often disposed of with other wastes in landfills or incinerators, but since they are **biodegradable**, some organic wastes are suitable for **composting** and land application.

Organic materials found in municipal solid waste include food, paper, wood, sewage **sludge**, and **yard waste**. Because of recent shortages in **landfill** capacity, the number of municipal composting sites for yard wastes is increasing across the country, as is the number of citizens who compost yard wastes in their backyards. On a more limited basis, some mixed municipal waste composting is also taking place. In these systems, attempts to remove inorganic materials are made prior to composting.

Some of the organic materials in municipal solid waste are separated before disposal for purposes other than composting. For example, paper and cardboard are commonly removed for **recycling**. **Food waste** from restaurants and grocery stores is typically disposed of through **garbage** disposals, therefore, it becomes a component of **wastewater** and sewage sludge.

A large percentage of sewage sludge is landfilled and incinerated, but it is increasingly being applied to land as a **fertilizer**. Sewage sludge may be used as an agricultural fertilizer or as an aid in reclaiming land devastated by **strip mining**, **deforestation**, and over-application of inorganic fertilizers. It may also be applied to land solely as a means of disposal, without the intention of improving the **soil**.

The organic fraction of industrial waste covers a wide spectrum including most of the components of municipal organic waste, as well as countless other materials. A few examples of industrial organic wastes are papermill sludge, meat processing waste, brewery wastes, and textile mill fibers. Since a large variety and volume of industrial organic wastes are generated, there is a lot of potential to recycle and compost these materials. Waste managers are continually experimenting with different "recipes" for composting industrial organic wastes into soil conditioners and soil amendments. Some treated industrial wastewaters and sludges contain large amounts of organic materials and they too can be used as soil fertilizers and amendments.

Production of biogas is another use of organic waste. Biogas is used as an alternative energy source in some **third world** countries. It is produced in digester units by the **anaerobic decomposition** of organic wastes such as manures and crop residues. Beneficial byproducts of biogas production include sludges that can be used to fertilize and improve soil, and the inactivation of pathogens in the waste. In addition, there is ongoing research on using organic wastes in developing countries: (1) in fish farming; (2) to produce algae for human and animal consumption, fertilizer, and other uses; and (3) to produce aquatic macrophytes for animal feed supplements.

[*Teresa C. Donkin*]

Resources

Books

Polprasert, C. *Organic Waste Recycling.* New York: Wiley, 1989.

Periodicals

Logsdon, G. "Composting Industrial Waste Solves Disposal Problems." *Biocycle* 29 (May-June 1988): 48–51.

Organization of Petroleum Exporting Countries

Established in 1960 by Iran, Iraq, Kuwait, and Venezuela, the Organization of **Petroleum** Exporting Countries (OPEC) was created to control the price of oil by controlling the volume of production. Modeled on the Texas Railroad Commission in the United States, the group was also intended to make other decisions about petroleum policy and to provide technical and economic support to its members. Indonesia, Libya, Qatar, Algeria, Nigeria, Saudi Arabia, and the United Arab Emirates have been admitted since 1960, and it is now estimated that the nations in OPEC control nearly three-quarters of the world's oil reserves.

In October 1973, members of OPEC met at their headquarters in Vienna and voted to raise oil prices by 70%. OPEC is dominated by oil-producing countries from the Middle East, and this decision was designed to retaliate against Western support of Israel during its war with Egypt. At a conference in Tehran, Iran, in December of that same year, OPEC countries raised oil prices an additional 130%, and they enacted an embargo on shipments to the United States and the Netherlands. The Iranian revolution in 1979 further restricted the world supply of oil, intensifying the effects of OPEC policies; between 1973 and 1980, the price of a barrel of oil rose from three dollars to 35 dollars.

The steep rise in oil prices had a disruptive effect in the United States, which is the largest oil importer in the world. It had an ever greater economic impact on industrialized countries such as Japan, which have little or no petroleum reserves of their own. But the economies in **less developed countries** (LDC) were the hardest hit; the rising price of oil decreased their purchasing power, increasing their trade deficit as well as their level of debt. There was a rapid transfer of wealth to oil-producing countries during this period, with the annual income of OPEC countries increasing 22.5 billion dollars in 1973 to 275 billion dollars in 1980.

The organization was not, however, able to maintain its influence over the international oil market in the 1980s, and prices dropped to as low as 10 dollars a barrel during this decade. World oil consumption reached a peak in 1979, at a high of 66 million barrels a day, and then dropped sharply in the years that followed. Many Western countries were encouraging **conservation**; they had also invested in other sources of energy, most notably **nuclear power**. Oil exploration had resulted in discoveries in Alaska and the North Sea; Mexico and Soviet Union, oil-exporting countries that were not members of OPEC, had also become an increasingly important part of the international petroleum trade. As their power over the price of oil became more diffused, consensus within OPEC became more difficult, and these internal divisions were made worse by political conflicts within the Middle East, particularly the war between Iran and Iraq.

From an environmental perspective, the most important effect of the oil price shocks of the 1970s may have been how it changed world patterns of energy consumption. Though the increase in prices forced many countries to use **coal** and nuclear power despite the damage they can do to the **environment**, economic pressures also stimulated research into many **alternative energy sources**, such as **solar energy**, **wind energy**, and hydroelectric power. In the last decade of the twentieth century, the sudden decrease in oil prices limited the sense of urgency as well as the funding for many of these research projects, and some environmentalists have suggested that higher oil prices might be better for the environment and the global economy over the long term.

[*Douglas Smith*]

Resources

Organizations

Organization of Petroleum Exporting Countries, Obere Donaustrasse 93, Vienna, Austria A-1020 + 43-1-21112 279, Fax: + 43-1-2149827, <http://www.opec.org>

Organochloride

Usually refers to organochlorine pesticides, although it could also refer to any chlorinated organic compound. The organo-

chlorine pesticides can be classified by their molecular structures. The cyclopentadiene pesticides are aliphatic, cyclic structures made from Diels-Alder reactions of pentachlorocyclopentadiene, and include **chlordane**, nonachlor, heptachlor, heptachlor epoxide, dieldrin, aldrin, endrin, **mirex**, and **kepone**. Other subclasses of organochlorine pesticides are the DDT family and the hexachlorocyclohexane isomers. All of these pesticides have low solubilities and volatilities and are resistant to breakdown processes in the **environment**. Their toxicities and environmental persistence have led to their restriction or suspension for most uses in the United States. *See also* Organophosphate

Organophosphate

Refers to classes of organic compounds that are often used in insecticides. Many of the organophosphate pesticides were developed over the last 30 years to replace organochlorine insecticides. They generally have the highest solubilities and shortest half-lives of the common classes of pesticides, but tend to have higher acute toxicities. Organophosphates act by disrupting the central nervous system, specifically by blocking the action of the **enzyme** acetylcholinesterase, which controls nerve impulse transmission. Organophosphates can impact human health at moderate levels of exposure, causing disorientation, numbness, and tremors. Acute exposures can cause blindness and death. Well-known organophosphate pesticides include parathion, malathion, **diazinon**, and phosdrin. *See also* Cholinesterase inhibitor; Organochloride

David W. Orr (1944 –)

American environmental writer

In recent years, an effort has been undertaken to make every citizen numerate, that is, able to use mathematics at a level that is minimally functional in modern society. David Orr, and writers and thinkers of kindred mind, believe that for citizens to be minimally functional in the complex, interdependent world of the twenty-first century, they must go beyond literacy in language and beyond numeracy and attain also a certain level of ecological literacy.

Orr is best known for his 1992 book titled *Ecological Literacy* and for his brief editorial essays in the journal *Conservation Biology*, written from his position as education editor for that journal. Born in Des Moines, Iowa, and raised in New Wilmington, Pennsylvania, he holds degrees from Westminster College (BA), Michigan State University (MA), and the University of Pennsylvania (Ph.D. in International Relations). He was a professor of political science before becoming Professor and Chair of the Environmental Studies Program at Oberlin College in Ohio. He is well-known for his criticism of the **environmental degradation** brought about by the free market system, his critique of the American educational system, and his activist attempts to change both systems toward more ecologically sustainable behavior.

Though the free market system seems to be in the ascendancy today, Orr believes that capitalism "is failing because it produces too much and shares too little." He especially laments conservatives in the capitalist system who do not understand the history and meaning of what conservative means. He claims that a genuine conservatism would be **conservation** oriented and not intent, for example, on perpetual economic growth that "has become the most radicalizing force for change in the modern world."

Much of this failure in the western system he traces to an ecological crisis that "represents, in large measure, a failure of education." He faults every level of education in general for not teaching students how to learn, while in school and throughout life, and for failing to define knowledge adequately for postmodern life. Too many disciplines do not consider the interrelations between people and the earth on which they live. This failure, Orr claims, can "ultimately...be traced to our schools and to our proudest universities."

Much of Orr's work focuses on changing universities, to eliminate what he considers the nationwide campus emphasis on "fast knowledge," or "ignorant knowledge" that is "cut off from its ecological and social context." He laments the over-professionalization of the professorate, researchers who churn out fast knowledge and create students (and citizens) who are "knowledge technicians instead of broadly thoughtful, liberally educated persons." He relates narrowness and self-interest to "the problem of ecological denial," in which people cannot think in the long term or about the "large and messy questions" associated with environmental issues. Througout his career, Orr has received many awards such as an Honorary Doctorate in Humane Letters from Arkansas College in 1990 and the Benton Box Award from Clemson University in 1995.

Orr believes that campus buildings even teach negative lessons: "the design, construction, and operation of academic buildings can be a liberal education in a microcosm that includes virtually every discipline in the catalog." Thus, he has undertaken what is proving to be a life-long campaign to change campuses to educational laboratories for ecological systems and environmental sustainability.

[*Gerald L. Young Ph.D.*]

Resources

Books

Orr, D. W. *Ecological Literacy: Education and the Transition to a Postmodern World.* Albany: State University of New York Press, 1992.

———. *Earth in Mind: On Education, Environment, and the Human Prospect.* Washington, DC: Island Press, 1994.

Henry Fairfield Osborn (1887 – 1969)

American conservationist and naturalist

›rn in Princeton, New Jersey, Osborn was the son of a renowned paleontologist who was also president of the American Museum of Natural History. Osborn graduated from Princeton in 1909 and attended Cambridge University in England. He then held a variety of jobs, working in freight and railroad yards and serving as a soldier in World War I. He finally took a position at a bank dealing in Wall Street investments.

Throughout his life, Osborn had accompanied his father on paleontological expeditions. As these trips continued into his adulthood, Osborn realized that his true vocation lay not finance but in natural science. In 1935 he accepted a position as secretary of the New York Zoological Society, an organization devoted to the "instruction and **recreation** of the people of New York." He was named president of the Society five years later and remained in office until 1968. As president, he actively pursued the creation of the Marine Aquarium at Coney Island, New Jersey, as well as improvements to the Bronx Zoological Park.

Osborn recognized the crucial need for an organization that would foster the preservation of **endangered species** and their habitats. With this purpose in mind, he founded in 1947 the Conservation Foundation (CF), an adjunct of the New York Zoological Society. CF was absorbed in 1990 by the **World Wildlife Fund** which still works for the conservation of endangered **wildlife** and natural areas.

Osborn used his knowledge of **nature** and his positions within these organizations to influence policy in Washington. He served on the Conservation Advisory Committee of the **U.S. Department of the Interior** as well as on the Planning Committee of the Economic and Social Council of the United Nations.

Like **William Vogt** and later Paul R. Ehrlich, Osborn often expressed concern about **population growth** and available **natural resources**. He fully realized the serious ramifications of environmental and ecological neglect and understood the steps needed to conserve existing resources and develop improved methods of distribution. He published two highly successful books, *Our Plundered Planet* (1948) and *The Limits of the Earth* (1953), each of which outlined the importance of conservationist ideals and the perils of a population explosion. Osborn was an active member of the **Save-the-Redwoods League**, International Council for Bird Preservation, and other related organizations. He died in New York City in 1969 at the age of 82.

[*Kimberley A. Peterson*]

Resources

Books

Osborn, Henry Fairfield. *Cope: Master Naturalist.* Salem, NH: Ayer, 1978.

———. *Major Papers on Early Primates, Compiled From the Publications of the American Museum of Natural History.* New York: AMS Press, 1980.

———. *Naturalist in the Bahamas: October 12, 1861–June 25, 1891.* New York: AMS Press, 1910.

———. *Origin and Evolution of Life: On the Theory of Action, Reaction and Interaction of Energy.* Salem, NH: Ayer, 1980.

OSHA

see **Occupational Safety and Health Administration**

Otter

see **Sea otter**

Our Common Future (Brundtland Report)

In late 1983 **Gro Harlem Brundtland**, the former Prime Minister of Norway, was asked by the Secretary-General of the United Nations to establish and chair the World Commission on **Environment** and Development, a special, independent commission convened to formulate "a global agenda for change."

The Secretary-General's request emerged from growing concern in the General Assembly about a number of issues, including: long-term **sustainable development**; cooperation between developed and developing nations; more effective international management of environmental concerns; the differing international perceptions of long-term environmental issues; and strategies for protecting and enhancing the environment.

The commission worked for three years and produced what is commonly known as "The Brundtland Report." Published in book form in 1987 as *Our Common Future*, the report addresses what it identifies as "common concerns," such as a threatened future, sustainable development, and the role of the international community. The report also examines "common challenges," including **population growth**, food security, **biodiversity**, and energy choices, as well as how to make industry more efficient. Finally, the report lists "common endeavours," such as managing the commons, maintaining peace and security while not sus-

pending development or degrading the environment, and changing institutional and legal structures. A chapter on each one of these concerns, challenges, and endeavors is included in the book.

Two years after publication of the report, Brundtland summarized its findings in a speech to the **National Academy of Sciences** in the United States. The reports's core concepts, she explained, were "that development must be sustainable, and the environment and world economy are totally, permanently intertwined." She went on to assert that these concepts "transcend nationality, culture, ideology, and race." She summarized by repeating the report's urgent warning: "Present trends cannot continue. They must be reversed."

The Brundtland Report issued a multitude of recommendations to help attain sustainable development and to address the problems posed by a global economy that is intertwined with the environment. The report recommends ways to deal with the debt crisis in developing nations, and insists on linking poverty and environmental deterioration: "A world in which poverty is endemic will always be prone to ecological and other catastrophes." The report also argues that security issues should be defined in environmental rather than military terms.

The members of the commission felt that a vast array of institutional changes were necessary if progress was to be made, and the report addresses these issues. It declares that the problems confronting the world are all tied together, "yet most of the institutions facing those challenges tend to be independent, fragmented, working to relatively narrow mandates with closed decision processes." Much of the work done by the commission focused on policy issues such as long-term, multifaceted population policies and ways to create effective incentive systems to encourage production, especially of food crops. Recommendations also include methods for a successful transition from **fossil fuels** to "low-energy" paths based on renewable resources.

The report repeatedly notes the differences between developed and developing nations in terms of energy use, **environmental degradation**, and urban growth, but it also emphasizes that "specific measures must be located in a wider context of effective [international] cooperation, if the problems are to be solved." The report always brings the problems back to people, to the importance of meeting basic human needs and the necessity of decreasing the disparities between developed and developing countries: "All nations will have a role to play in changing trends, and in righting an international economic system that increases rather than decreases inequality, that increases rather than decreases numbers of poor and hungry."

The report is not merely a grim listing of "ever increasing environmental decay, poverty, and hardship in an ever more polluted world among ever decreasing resources." Surprising to some, it is also a documentation of "the possibilities for a new era of economic growth," an on-going era of sustainable, non-destructive growth. *See also* Economic growth and the environment; Environmental economics; Environmental monitoring; Environmental policy; Environmental stress; Green politics; International Geosphere-Biosphere Programme; United Nations Earth Summit; United Nations Environment Programme; World Bank

[*Gerald L. Young Ph.D.*]

Resources
Books

Silver, C. S., and R. S. DeFries. *One Earth One Future: Our Changing Global Environment.* Washington, DC: National Academy Press, 1990.

World Commission on Environment and Development. *Our Common Future.* New York: Oxford University Press, 1987.

Overburden

Refers to the rock and **soil** above a desired economic resource, such as **coal** or an ore body. It is normally associated with **surface mining**, in contrast to underground mining where **tailings** are a more common byproduct. The depth of overburden is a critical economic factor when assessing the feasibility of mining. Unless the **topsoil** is stored for later **reclamation**, overburden removal usually destroys this crucial resource, greatly magnifying the task of natural or cultural revegetation. In North America, overburden removal requires blasting through hard caprock, which leave landscapes resembling fields of glacial debris. *See also* Mine spoil waste

Overfishing

With the tremendous increase in the human population since the industrial revolution, there has been an ever increasing use and, often, exploitation of many of the world's **natural resources**. The demand for fish and shellfish has exemplified this misuse of natural resources. Paralleling the changes in agriculture, the fisheries industry has progressed from a small-scale, subsistence operation to a highly mechanized, ultra-efficient means of securing huge quantities of fish and shellfish to satisfy the burgeoning market demand. These industrialized commercial fisheries have allowed fishermen to easily work the far offshore waters, and more efficient refrigeration has allowed greater travel time, thus allowing for longer excursions.

Overfishing results in the removal of a substantial portion of a species' population so that there are too few

individuals left to reproduce and bring the population back to the level it was the year before. Overfishing has a tremendous negative impact on nontarget **species** as well. More than half, and occasionally as much as 90%, of a catch may be discarded, and this can adversely affect the **environment** by altering predator-prey ratios and adding excess **organic waste** through the dumping of nearly 30 million tons of dead or dying fish and shellfish overboard annually. Overfishing has also negatively impacted diving seabird, sea turtle, and dolphin populations since these animals often get trapped and killed in fishing nets. It is estimated that 44,000 albatrosses are killed each year by Japanese tuna fishermen, and that, over the years, tuna purse seine nets in the Eastern Tropical Pacific have caused the death of millions of **dolphins**.

The offshore waters are not regulated as are the coastal waters of our continents, therefore several populations of commercially important species have been overfished. Often, fishermen simply try to catch as many fish as possible (as fast as possible), because they believe that any fish left in the water will be caught by someone else. The codfish industry of the Atlantic coast of Canada and the United States had been a productive fishery since the time of the early settlers, hundreds of years ago. But with the introduction, in 1954, of a new kind of highly industrialized fishing vessel, the factory trawler, conditions changed for the cod population. In 1968, exceeding the normal catch for the region by over 500,000 tons, 810,000 tons of cod were taken off Labrador and the Grand Banks off Newfoundland. In the decade to follow, the cod population declined to the point that the number of spawning codfish off Newfoundland in 1977 showed a 94% decrease from the number in 1962. Another species to show population decline due to overfishing is the American shad, often referred to as the poor man's **salmon**. Harvested by Native Americans and the early colonists, the shad fishery was supplying one of the most commercially valuable fish to the region surrounding the **Chesapeake Bay** during the 1800's. Overfishing, combined with an increase in **pollution** and the disruption of spawning runs by the damming of rivers, took a fishery with an annual harvest of nearly 18 million pounds at the turn of the century to less than 2 million pounds in the 1970s.

A tragic example of the effect of overfishing is seen at **Georges Bank** in the Atlantic Ocean. Georges Bank, discovered by the Italian explorer Giovanni da Verrazano in the early part of the 16th century, was named after St. George by English settlers in 1605. Home to over 100 species of fish, Georges Bank has been a historically significant site of plentiful fishing for centuries. It is an ovoid embankment that is about 75 miles (120 km) off of the coast of New England. It encompasses an area 149 miles (240 km) in length and 75 miles (120 km) in width, and is over 330 feet (100 meters) above the sea floor of the adjacent Gulf of Maine. A bank, or shoal, is a shallow portion of an ocean that is analogous to a plateau. Georges Bank is part of a large range of shoals that extends from Newfoundland to the edge of the North American continental shelf. Georges Bank itself is larger than the state of New Jersey, and is an extremely important breeding ground for cod, haddo! ck, herring, flounder, lobster, and scallops. Its unique conditions are formed by the mingling of two major Atlantic currents: the cold Labrador current and the warm Gulf stream. The mixing of nutrient-rich and well-oxygenated waters in combination with good light penetration due to shallow depth creates an ideal environment for **plankton** that is a vital food source for young fish.

Known for its abundant fish supply, Georges Bank was heavily fished for centuries. Early fisheries believed the supply to be inexhaustible. The first documented consequence of indiscriminate fishing was in 1850 when, after a concentrated season of overfishing, a dramatic decline in halibut numbers in Georges Bank threatened commerce. The near **extinction** of halibut in this area occurred when ocean fishing was still accomplished using relatively small sailboats and handlines with baited hooks. After World War II, colossal factory-sized fishing ships were created and new large-scale fishing techniques were employed in Georges Bank. Miles of **gill nets** and other efficient methods became standard tools for trolling the bank. Fisheries from the United States, the Soviet Union, Spain, Japan, and Germany collected huge bounties from Georges Bank for decades. Each modern ship could catch as much cod in a single hour as an average boat from the 1600s could collect in an enti! re season. The fishing industry boomed from the 1950s through the 1970s, until it became apparent that even the ocean had limits that could be surpassed.

In 1976, the United States passed the Magnuson Act which established American authority to ban international factory fishing in the Georges Bank area. The intention was to protect and preserve fish stocks for the United States and Canada. Rather than initiate **conservation** strategies for managing the fish populations of Georges Bank, western fisheries free from foreign competition in the area, advanced. The New England Council, created by the Magnuson Act to protect Georges Bank, was effectively controlled by fishing industry interests that did little to address growing concerns about dwindling supplies.

In 1994 the National Marine Fisheries Service estimated a 40% decline in cod stocks in the Georges Bank area that occurred over an alarmingly short four-year period. Survey after survey found that the once abundant bank had been stripped of every prime commercial fish species that once filled its waters. As a result, that same year, fishing was banned in a 9,600 square kilometer area of Georges

Bank. As of 1999, scientists reported that some commercial species were beginning to show signs of a slow increase in numbers, but that cod continued a rapid decline. While the ban on heavy fishing continues, some strictly regulated fishing is permitted in certain areas of Georges Bank. Although the depletion of fish populations in Georges Bank are assumed by some to be temporary, scientists remain cautious about providing assurances of a return to former abundance since recovery from such massive overfishing is not well understood.

Because regions relatively close to land, such as Georges Bank, tend to be more regulated, fishermen have begun to move farther out into unregulated water. Lack of regulations, however, is not the only reason fishermen are going farther offshore. Many populations have been overfished in coastal waters, but fish and shellfish populations have declined in these waters due to near-shore pollution as well. Decreased **water quality** has had an impact on the population numbers and also on the quality of the fish itself, therefore creating a demand for fish and shellfish from distant, presumably cleaner, waters. Being driven farther offshore to make their catches, commercial fishermen may often overfish a population to bring in more revenues to offset their huge debt incurred on their equipment, crew, and ever increasing fuel costs.

A particularly alarming new method of fishing that threatens ocean **fauna** is cyanide fishing. Cyanide fishing is an increasingly popular method of fishing that uses cyanide to temporarily stun fish in exotic areas, like coral reefs. Once stunned, the fish are easily collected, then sold either as aquarium pets or as live fish in restaurants. Live fish sell for much higher prices than frozen fish, and live fish selections in restaurants is a growing trend. One problem with cyanide fishing is that only about half of the fish captured survive for sale, making it a wasteful effort. A second problem is the nonselective **poisoning** of other **wildlife** found in the ecosystems. Coral itself is killed by the cyanide used to flush-out **exotic species** from hiding places for luxury fish markets in Asia. Cyanide fishing constitutes a new threat for overfishing. It is estimated that roughly 65 tons of cyanide are used each year in Philippine coral reefs alone. Cyanide used on one! area spreads and jeopardizes ocean life on a scale that is poorly understood.

Popular trends in cuisine, in addition to live exotic fish selections in restaurants, can effect fishing. During the 1980s, when a New Orleans chef popularized "blackened redfish" and the dish was in demand across the country, the Gulf of Mexico population of "redfish", actually the Red drum, was nearly decimated by overfishing. Because of its unpopularity as a food and game fish, the Red drum was not regulated as are the more important commercial fish and shellfish species of this region. Therefore, when orders for the fish started coming in from across the nation, commercial fishermen began catching them in record numbers. There were cases in which great numbers of dead redfish were washing up on the shores of the Gulf states because fishermen had caught too many to process on their vessels and dumped the excess overboard. In just a couple of years the Red drum population crashed in the Gulf of Mexico, prompting the fish and game officials in all of the Gulf states to halt **commercial fishing** for redfish. (A recovery program, however, is meeting with success.) This problem could have been avoided had the original dish been named "blackened fish" or had the nation's seafood markets and restaurants known that the flavor in "blackened redfish" is in the seasonings and not in the fish itself, thus relying on local fisheries and the redfish of the Gulf of Mexico.

One means of offsetting overfishing by commercial fishing fleets, and reducing fuel use as well, is **aquaculture**, the cultivation and harvesting of fish and shellfish under controlled conditions. With pressures on fish populations from overfishing, increased pollution problems, and a dying industry strapped with higher costs of fuel and equipment, "fish farming" may offer a true alternative to greatly reduced quantities and inferior quality seafood in the future. *See also* Commercial fishing

[*Eugene Beckham*]

Resources

Periodicals

Bricklemyer, E., S. Iudicello, and H. Hartmann. "Discarded Catch in U.S. Commercial Marine Fisheries." *Audubon Wildlife Report.* San Diego: Academic Press, 1990–91.

Kunzig, Robert. 1995. Twilight of the Cod. *Discover*, April 1995,44–49, 52–57, 60.

Lawren, B. "Net Loss." *National Wildlife* 30 (1992): 46–53.

Overgrazing

Overgrazing is one of the most critical environmental problems facing the western United States. **Rangelands** have been mismanaged for over a hundred years, mainly due to cattle grazing. In addition to consuming vegetation, cattle alter the **ecosystem** of rangeland by tramping, urination, defecation, and trashing. Degradation due to heavy livestock grazing continues to occur in many diverse and fragile ecosystems, including **savanna**, **desert**, meadow, and alpine communities.

Riparian lands, highly vegetated, narrow strips of land bordering rivers or other natural watercourses, make up only two percent of rangelands, but have the most diverse populations of vegetation and **wildlife**. Overgrazing has had a devastating effect on these areas. Cattle eat the seedlings of

young trees, which has led to the elimination of some **species**, and this has reduced the species of birds in these areas and disrupted migratory patterns. Lack of new tree growth in riparian areas has also resulted in the drying up of stream beds and the loss of **habitat** for fish and amphibians. It has contributed to the problem of **soil erosion**, **desertification** and the **greenhouse effect**. Other rangeland ecosystems are facing similar disruption.

The **Bureau of Land Management** (BLM) part of the U. S. Department of the Interior is the largest landholder in the United States. They are responsible for 334 million acres of land which fall under multiple-use mandates. The BLM leases much of this **public land** to individuals for grazing purposes, charging ranchers approximately $1.35 per month to graze one cow and a calf. It has been estimated that the fair market value for such forage is $6.65, and the agency has come under attack for leasing grazing rights at extremely low rates. It has also been criticized for allowing abusive land management practices. Critics claim that those who utilize these public lands try to maximize profits by putting excessive numbers of livestock on the range, and they argue that almost half of the range areas in the United States are in dire need of **conservation**.

Environmentalists argue that it is possible to eliminate overgrazing and manage rangelands in a way that both preserves ecosystems and meets the needs of ranchers. They advocate above all the reduction of herds. They also suggest that cattle should not be allowed to roam at will and should be rotated among various pastures, so that all rangeland areas can receive back-to-back spring and summer rest.

Perhaps the main obstacle in conservation of rangelands is the lack of knowledge regarding their diverse ecosystems. Rangelands are regions where natural revegetation tends to be slow. Artificial attempts to introduce and establish plant growth have been frustrated by the fact that development is a long-term process in these environments as well as by other factors. Knowledge of the dynamics of **competition**, reaction, and stabilization of species is minimal. For over a century, rangers have tried to eliminate sagebrush, planting wheatgrass in its place. The solution has been short-lived because sagebrush usually prevails over wheatgrass in the natural **succession** of plants. To further compound the problem, overgrazing has depleted the perennial grasses that compete with sagebrush, and the plant has become even more prolific. A better understanding of the dynamic relationship between plants, animals, **microorganisms**, soil, and the **climate** is necessary to reestablish rangeland areas. *See also* Agricultural pollution; Agricultural Stabilization and Conservation Services; Agroecology; Feedlot runoff; Feedlots; Land use; Sagebrush Rebellion; Taylor Grazing Act (1934)

[*Debra Glidden*]

Resources

Books

Heitschmidt, R. K., ed. *Grazing Management: An Ecological Perspective.* Portland, OR: Timber Press, 1991.

Spedding, C. R. W. *Grassland Ecology.* London: Oxford University Press, 1971.

Periodicals

Senft, R., et al. "Large Herbivore Foraging and Ecological Hierarchies; Landscape Ecology Can Enhance Traditional Foraging Theory." *Bioscience* 37 (December 1987): 789–95.

Strickland, R. "Taking the Bull by the Horns: Conservationists Have Been Wrangling Politely With Land Managers For Years—But Have Failed to Halt Overgrazing Throughout the West." *Sierra* 75 (September-October 1990): 46–48.

Overhunting

Overhunting is any **hunting** activity that has an adverse impact on the total continuing population of a **species**. With the tremendous increase in the human population since the industrial revolution, there has been an ever increasing use and, often, exploitation of many of the world's **natural resources**. The demand for fish and shellfish has exemplified this misuse of natural resources. The amount of hunting pressure that a species can tolerate depends on its productivity, and it may change seasonally and annually because of **drought**, **habitat** alteration, **pollution**, or other **mortality** factors. Hunting which is well-regulated can be sustained, and sportsmen in countries with regulated hunting are quick to point out that they are not responsible for the endangerment or **extinction** of any species.

In unregulated situations, however, overhunting does occur, and it has endangered **wildlife**, even driving some species to extinction. The great auk (*Alca impennis*), a large, flightless, penguin-like bird of the North Atlantic coasts, was so easy to catch that sailors could kill hundreds in a few minutes, and by 1844 they had become extinct. Even the world's most abundant bird, the **passenger pigeon** (*Ectopistes migratorius*), was driven to extinction by overhunting. In the 1800s there were 3 to 5 billion passenger pigeons, about one-fourth of all North American land birds, and enormous flocks of them would darken the sky for days as they flew overhead. One colony in Wisconsin covered 850 square miles (220,149 ha) and included 135 million adult birds. Commercial hunting sent train loads of these birds to markets, including 15 million birds from a single colony in Michigan, and eliminated the species in the early 1900s.

There are several cases where overhunting has nearly exterminated a species. In 1850 there were 60 million American **bison** (*Bison bison*) on the Great Plains of the United States, and within 40 years, overhunting had reduced the wild population to 150 individuals. At the turn of the cen-

tury, snowy egrets (*Leucophoyx thula*) were almost wiped out by hunters who the sold the feathers to be made into fashionable women's hats. Many species of **whales** were also driven to the brink of extinction by whalers.

Many species continue to be overhunted today. Although protected by law, African **rhinoceroses** are endangered by poachers who sell the horns to Yemen and China. In Yemen the horns are used to make dagger handles for wealthy businessmen, and in China they are made into an aphrodisiac and fever-reducing drug which is reportedly useless. Likewise, **elephants** are being slaughtered for their ivory tusks. There were 4.5 million elephants in 1970, and by 1990, only 610,000.

Many large cat species are also threatened by overhunting, because the economic incentive to poach these animals far outweighs the risks of being caught and fined. For example, a Bengal tiger (*Panthera tigris tigris*) fur coat sells for $100,000; an ocelot (*Felis pardalis*) skin sells for $40,000; a **snow leopard** (*Panthera uncia*) skin sells for $14,000; and tiger meat sells for $629 per pound ($286/kg).

Even when appropriate hunting regulations do exist, there are often not enough rangers or **conservation** officers to enforce them. As species become more scarce, the demand increases on the black market, inflating the price, and the economic incentive for **poaching** only becomes greater. Unless appropriate laws are enacted and enforced other species will become extinct from overhunting. *See also* Conservation; Convention on International Trade in Endangered Species of Fauna and Flora; Defenders of Wildlife; Endangered species; Endangered Species Act; Fish and Wildlife Service; IUCN—The World Conservation Union; Wildlife management; Wildlife rehabilitation

[*Ted T. Cable*]

Resources

Books

Ofcansky, T. P. *Paradise Lost: A History of Game Preservation in East Africa.* Morgantown: West Virginia University Press, 1993.

Owens, D., and M. Owens. *Eye of the Elephant: Life and Death in an African Wilderness.* New York: Houghton Mifflin, 1992.

Periodicals

Jackson, P. "They've Shot Miro." *International Wildlife* 22 (November-December 1992): 38–43.

Owl

see **Northern spotted owl**

Oxidation reduction reactions

An oxidation-reduction or redox reaction is transformation involving electron transfer. It consists of a half reaction in which a substance loses an electron or electrons (oxidation) and another half reaction in which a substance gains an electron or electrons (reduction).

The substance that gains electrons is the **oxidizing agent**, while the substance that gives up electrons is the reducing agent. Adding the two half equations algebraically and eliminating the free electrons gives the complete oxidation-reduction equation. This equation, however, is not an assurance that the reaction will proceed spontaneously as written. One way to know the direction of the reaction is to determine the value of the standard free energy change, ΔG^0, and then to calculate the free energy change, ΔG, for the complete redox reaction. If the resulting value of ΔG is negative, the reaction will proceed spontaneously as written. If ΔG is positive, the reaction can proceed spontaneously in the opposite direction. If ΔG is zero, the reaction is in equilibrium.

Simple redox reactions occur with the direct transfer of electrons from the reducing agent to the oxidizing agent. An example is the reaction between **hydrogen** (H_2) and **chlorine** (Cl_2) to form hydrogen chloride. In this reaction, hydrogen, the reducing agent, donates two electrons to chlorine which is the oxidizing agent. Because it gains electrons, the oxidizing agent changes in valence and becomes more negative or less positive. In the given example, the valence of chlorine changes from 0 to -1. The reducing agent, on the other hand, loses electrons and hence becomes more positive. Thus hydrogen, being the reducing agent in the given example, changes in valences from 0 to +1.

Some common applications of redox reactions in **wastewater** treatment are the **detoxification** of cyanide and the precipitation of chromium. Highly toxic cyanide is converted to nontoxic cyanate and finally to **carbon dioxide** and **nitrogen** gas by the action of strong oxidizing agents, such as chlorine gas or sodium hypochlorite. In removing hexavalent chromium, the objective is to reduce it to trivalent chromium, which is less toxic and can be precipitated out in the form of hydroxide. Hexavalent chromium is reduced by sodium bisulfite or ferrous sulfate in an **acid** medium. Lime can then be applied to precipitate out the trivalent chromium as chromium hydroxide, which collects as a chemical **sludge** upon settling.

In biochemical redox reactions, however, the electrons go through a series of transfers before reaching the terminal acceptor, which is oxygen in **aerobic systems**. Examples of these electron carriers are NAD (nicotinamide adenine dinucleotide), NADP (nicotinamide adenine dinucleotide phosphate), and flavoproteins. In **anaerobic systems**, other

materials such as **nitrates**, sulfates, and **carbon** dioxide may become the electron acceptor, as in the process of **denitrification** in which nitrate (NO_3) is reductively degraded to molecular nitrogen (N_2). In either case, organic or inorganic matter, which serves as food for the **microorganisms**, is the substance oxidized. A simplified stoichiometry of the complete **aerobic** biochemical oxidation of glucose, for example, produces six moles of carbon dioxide and six moles of water with the release of 686 kilocalories per mole. *See also* Corrosion and material degradation; Electron acceptor and donor; Industrial waste treatment; Sewage treatment; Waste management

[*James W. Patterson*]

Resources
Books

Eckenfelder Jr., W. W. *Industrial Water Pollution Control.* New York: McGraw-Hill, 1989.

Snoeyink, V. L., and D. Jenkins. *Water Chemistry.* New York: Wiley, 1980.

Oxidizing agent

Any substance that donates oxygen or that gains electrons in a chemical reaction. Perhaps the most common example of an oxidizing agent is the element oxygen itself. When a substance burns, rusts, or decays, that substance combines with oxygen in the air around it. Oxidizing agents can have both beneficial and harmful environmental effects. **Chlorine** gas is used as an oxidizing agent to kill bacteria in the purification of water. **Ozone** damages plant and animal cells by oxidizing them.

Oxygen, dissolved

see **Dissolved oxygen**

Ozonation

The process which takes advantage of the oxidizing properties of **ozone** is known as ozonation. Ozone can be used in a variety of applications including the treatment of drinking water, bottled water, beverages, **wastewater**, industrial wastes, air pollutants, swimming pool water, and cooling tower water. In addition, there are a number of proprietary processes that use ozone, ranging from carpet cleaning to the making of gourmet ice cubes. Ozonation consists of four fundamental tasks: drying and cleaning the oxygen-containing feed-gas; generating ozone in a silent corona **discharge** generator; bringing the ozone into contact with the material being treated; and finally destroying remaining ozone prior to releasing the waste gas to the **atmosphere**. All ozonation processes require ozone generation and contacting, but not all applications require feed-gas preparation and off-gas treatment.

Ozone can be generated by three processes: electrical discharge, **photochemical reaction** with **ultraviolet radiation**, and electrolytic reactions. The latter two methods produce very low concentrations of ozone, limiting their application. Electrical discharge generators are currently the most economical choice for producing large quantities of ozone. Feed-gas must contain oxygen and be free of contaminants such as particles, **hydrocarbons**, and water vapor. For optimum production of ozone, the feed-gas should have a **dew point** of at least -37°F (-40°C) and preferably -73°F (-60°C). Small ozonation systems can use air which is dried before entering the ozone generator. Larger systems may find pure oxygen is economically viable. The temperature of the feed-gas and the generator is one of the most important parameters affecting the production of ozone. High temperatures reduce the concentration of ozone and greatly decrease the life of ozone in the gas phase. The **recycling** of ozone process gases increases the concentration of **nitrogen**, which leads to decreases in ozone production, and, in the presence of water vapor, to the production of corrosive nitric **acid**.

Contacting ozone with the material to be treated is complicated by several factors: ozone is reactive and disappears, so it must be generated at the site where it is used, and the waste ozone in the exhaust gas is toxic. When choosing ozone contacting devices for an aquatic system, two types of reactions must be taken into account. In mass-transfer limited reactions ozone is being consumed faster than it can be transferred to solution. In reaction rate limited reactions ozone is in surplus in solution but the material being oxidized is rate limiting, so that ozone is wasted. Bubble-diffuser systems are commonly used in **water treatment** because they are good compromises for satisfying the need to control both mass-transfer and rate limited reactions. However, in some aquatic applications in-line dissolution and contacting may be the optimum technique for ozonating the water in question.

Exhaust gas or off-gas requires treatment to remove traces of ozone remaining in the gas after contacting. Thermal destruction is one of the most commonly used methods for removing ozone from the waste gas. Other methods for destroying ozone include catalytic destruction, activated **carbon adsorption**, and zeolites. *See also* Industrial waste treatment; Water treatment

[*Gordon R. Finch*]

RESOURCES
BOOKS

Langlais, B., D. A. Reckhow, and D. R. Brink, eds. *Ozone in Water Treatment: Application and Engineering.* Chelsea, MI: Lewis Publishers, 1991.

Rice, R. G., and A. Netzer, eds. *Handbook of Ozone Technology and Applications.* Vol. 1. Ann Arbor, MI: Ann Arbor Science Publishers, 1982.

Rice, R. G., and A. Netzer, eds. *Handbook of Ozone Technology and Applications.* Vol. 2. Boston: Butterworth Publishers, 1984.

Ozone

Ozone is a toxic, colorless gas (but can be blue when in high concentration) with a characteristic acrid odor. A variant of normal oxygen, it had three oxygen atoms per molecule rather than the usual two. Ozone strongly absorbs **ultraviolet radiation** at wavelengths of 220 through 290 nm with peak **absorption** at 260.4 nm. Ozone will also absorb infrared radiation at wavelengths in the range 9-10 μm. Ozone occurs naturally in the ozonosphere (ozone layer), which surrounds the earth, protecting living organisms at the earth's surface from ultraviolet radiation. The ozonosphere is located in the **stratosphere** from 6–31 miles (10–50 km) above the earth's surface, with the highest concentration between 7.5 and 12 miles (12 and 20 km). The concentration of ozone in the ozonosphere is 1 molecule per 100,000 molecules, or if the gas were at standard temperature and pressure, the ozone layer would be 0.12 inch (3 mm) thick. However, the ozone layer absorbs about 90% of incident ultraviolet radiation.

Ozone in the stratosphere results from a chemical equilibrium between oxygen, ozone, and ultraviolet radiation. Ultraviolet radiation is absorbed by oxygen and produces ozone. Simultaneously, ozone absorbs ultraviolet radiation and decomposes to oxygen and other products. **Ozone layer depletion** occurs as a result of complex reactions in the **atmosphere** between organic compounds that react with ozone faster than the ozone is replenished. Compounds of most concern include the byproducts of ultraviolet degradation of **chlorofluorocarbons** (CFCs), **chlorine** and fluorine.

Ozone is also a secondary air pollutant at the surface of the earth as a result of complex chemical reactions between sunshine and primary pollutants, such as **hydrocarbons** and oxides of **nitrogen**. Ozone can also be generated in the presence of oxygen from equipment that gives off intense light, electrical sparks, or creates intense static electricity, such as photocopiers and laser printers. Human olfactory senses are very sensitive to ozone, being able to detect ozone odor at concentrations between 0.02 and 0.05 **parts per million**. Toxic symptoms for humans from exposure to ozone include headaches and drying of the throat and respiratory tracts. Ozone is highly toxic to many plant **species** and destroys or degrades many building materials, such as paint, **rubber**, and some **plastics**. The total losses in the United States each year due to ozone damage to crops, livestock, buildings, natural systems, and human health is estimated to be in the tens of billions of dollars. The threshold limit value (TLV) for **air quality** standards is 0.1 ppm or 0.2 mg O_3 per m^3 of air.

Industrial uses of ozone include chemical manufacturing and air, water, and waste treatment. Industrial quantities of ozone are typically generated from air or pure oxygen by means of silent corona **discharge**. Ozone is used in **water treatment** as a disinfectant to kill pathogenic **microorganisms** or for oxidation of organic and inorganic compounds. Combinations of ozone and **hydrogen** peroxide or ultraviolet radiation in water can generate powerful oxidants useful in breaking down complex synthetic organic compounds. In **wastewater** treatment, ozone can be used to disinfect effluents, or decrease their color and odor. In some industrial applications, ozone can be used to enhance biodegradation of complex organic molecules. Industrial cooling tower treatment with ozone prevents transmission of airborne pathogenic organisms and can reduce odor. *See also* Biodegradable; Ozonation

[*Gordon R. Finch*]

RESOURCES
BOOKS

Horváth, M., L. Bilitzky, and J. Hüttner. *Ozone.* Amsterdam: Elsevier, 1985.

Kaufman, D. G., and C. M. Franz. *Biosphere 2000: Protecting Our Global Environment.* New York: HarperCollins College Publishers, 1993.

Ozone layer depletion

Destroying the ozone shield

Ozone, a form of oxygen consisting of three atoms of oxygen instead of two, is considered an air pollutant when found at ground levels and is a major component of **smog**. It is formed by the reaction of various air pollutants in the presence of sunlight. Ozone is also used commercially as a bleaching agent and to purify municipal water supplies. Since ozone is toxic, the gas is harmful to health when generated near the earth's surface. Because of its high rate of breakdown, such ozone never reaches the upper **atmosphere**.

But the ozone that shields the earth from the sun's radiation is found in the **stratosphere**, a layer of the upper atmosphere found 9–30 mi (15–50 km) above ground. This ozone layer is maintained as follows: the action of ultraviolet light breaks O_2 molecules into atoms of elemental oxygen (O). The elemental oxygen then attaches to other O_2 mole-

cules to form O_3. When it absorbs **ultraviolet radiation** that would otherwise reach the earth, ozone is, in turn, broken down into O_2 + O. The elemental oxygen generated then finds another O_2 molecule to become O_3 once again.

In 1974, chemists F. Sherwood Rowland and Mario J. Molina realized that **chlorine** from chlorofluorocarbon (CFC) molecules was capable of breaking down ozone in the stratosphere. In time, evidence began accumulate that the ozone layer was indeed being broken apart by these industrial **chemicals**, and to a lesser extent by **nitrogen** oxide emissions from jet airplanes as well as **hydrogen** chloride emissions from large volcanic eruptions.

When released into the **environment**, CFCs slowly rise into the upper atmosphere, where they are broken apart by solar radiation. This releases chlorine atoms that act as *catalysts*, breaking up molecules of ozone by stripping away one of their oxygen atoms. The chlorine atoms, unaltered by the reaction, are each capable of destroying ozone molecules repeatedly. Without a sufficient quantity of ozone to block its way, ultraviolet radiation from the sun passes through the upper atmosphere and reaches the surface of the earth.

When damage to the ozone layer first became apparent in 1974, **propellants** in **aerosol** spray cans were a major source of CFC emissions, and CFC aerosols were banned in the United States in 1978. However, CFCs have since remained in widespread use in thermal insulation, as cooling agents in refrigerators and air conditioners, as cleaning solvents, and as foaming agents in **plastics**, resulting in continued and accelerating depletion of stratospheric ozone.

The Antarctic ozone hole

The most dramatic evidence of the destruction of the ozone layer has occurred over **Antarctica**, where a massive "hole" in the ozone layer appears each winter and spring, apparently exacerbated by the area's unique and violent climatological conditions. The destruction of ozone molecules begins during the long, completely dark, and extremely cold Antarctic winter, when swirling winds and ice clouds begin to form in the lower stratosphere. This ice reacts with chlorine compounds in the stratosphere (such as hydrogen chloride and chlorine nitrate) that come from the breakdown of CFCs, creating molecules of chlorine.

When spring returns in August and September, a seasonal vortex—a rotating air mass—causes the ozone to mix with certain chemicals in the presence of sunlight. This helps break down the chlorine molecules into chlorine atoms, which, in turn, react with and break up the molecules of ozone. A single chlorine, **bromine**, or nitrogen molecule can break up literally thousands of ozone molecules.

During December, the ozone-depleted air can move out of the Antarctic area, as happened in 1987, when levels of ozone over southern **Australia** and New Zealand sank by 10% over a three week period, causing as much as a 20% increase in ultraviolet radiation reaching the earth. This may have been responsible for a reported increase in skin cancers and damage to some food crops.

The seasonal hole in the ozone layer over Antarctica has been monitored by scientists at the National Aeronautics and Space Administration's (NASA) Goddard Space Flight Center outside Washington, D.C. NASA's NIMBUS-7 satellite first discovered drastically reduced ozone levels over the Southern Hemisphere in 1985, and measurements are also being conducted with instruments on aircraft and balloons. Some of the data that has been gathered is alarming.

In October 1987, ozone levels within the Antarctic ozone hole were found to be 45% below normal, and similar reductions occurred in October 1989. A 1988 study revealed that since 1969, ozone levels had declined by 2% worldwide, and by as much as 3% or more over highly populated areas of North America, Europe, South America, Australia and New Zealand.

In September 1992, the NIMBUS-7 satellite found that the depleted ozone area over the southern polar region had grown 15% from the previous year, to a size three times bigger than the area of the United States, and was 80% thinner than usual. The ozone hole over Antarctica was measured at approximately 8.9 million mi^2 (23 million km^2), as compared to its usual size of 6.5 million mi^2 (17 km^2). The contiguous 48 states is, by comparison, about 3 million mi^2 (7.8 million km^2), and all of North America covers 9.4 million mi^2 (24.3 million km^2). Researchers attributed the increased thinning not only to industrial chemicals but also to the 1991 volcanic eruptions of **Mount Pinatubo** in the Philippines and Mount Hudson in Chile, which emitted large amounts of **sulfur dioxide** into the atmosphere.

Dangers Of ultraviolet radiation

The major consequence of the thinning of the ozone layer is the penetration of more solar radiation, especially Ultraviolet-B (UV-B) rays, the most dangerous type, which can be extremely damaging to plants, **wildlife**, and human health. Because UV-B can penetrate the ocean's surface, it is potentially harmful to marine life forms and indeed to the entire chain of life in the seas as well.

UV-B can kill and affect the reproduction of fish, larvae, and other plants and animals, especially those found in shallow waters, including **phytoplankton**, which forms the basis of the oceanic **food chain/web**. The National Science Foundation reported in February 1992 that its research ship, on a six week Antarctic cruise, found that the production of phytoplankton decreases at least 6–12% during the period of greatest ozone layer depletion, and that the destructive effects of UV radiation could extend to depths of 90 ft (27 m).

A decrease in phytoplankton would affect all other creatures higher on the food chain and dependent on them,

including **zooplankton**, microscopic ocean creatures that feed on phytoplankton and are also an essential part of the ocean food chain. And marine phytoplankton are the main food source for **krill**, tiny Antarctic shrimp that are the major food source for fish, squid, penguins, **seals**, **whales**, and other creatures in the Southern Hemisphere.

Moreover, phytoplankton are responsible for absorbing, through **photosynthesis**, great amounts of **carbon dioxide** (CO_2) and releasing oxygen. It is not known how a depletion of phytoplankton would affect the planet's supply of life-giving oxygen, but more CO_2 in the atmosphere would exacerbate the critical problem of global warming, the so-called **greenhouse effect**.

There are numerous reports, largely unconfirmed, of animals in the southern polar region being harmed by ultraviolet radiation. Rumors abound in Chile, for example, of pets, livestock, sheep, rabbits, and other wildlife getting cataracts, suffering reproductive irregularities, or even being blinded by solar radiation. Many residents of Chile and Antarctica believe these stories, and wear sunglasses, protective clothing, and sun-blocking lotion in the summer, or even stay indoors much of the day when the sun is out. If the ozone layer's thinning continues to spread, the lifestyles of people across the globe could be similarly disrupted for generations to come.

Particularly frightening have been incidents reported to have taken place in Punta Arenas, Chile's southernmost city, at the tip of Patagonia. After several days of record low levels of ozone were recorded in October 1992, people reported severe burns from short exposure to sunlight. Sheep and cattle became blind, and some starved because they could not find food. Trees wilted and died, and melanoma-type skin cancers seem to have increased dramatically. Similar stories have been reported from other areas of the southern hemisphere. And malignant melanoma, once a rare disorder, is now the fastest rising **cancer** in the world.

Ozone thinning spreads

Indeed, ozone layer depletion is spreading at an alarming rate. In the 1980s, scientists discovered that an ozone hole was also appearing over the Arctic region in the late winter months, and concern was expressed that similar thinning might begin to occur over, and threaten, heavily populated areas of the globe. These fears were confirmed in April 1991, when the **Environmental Protection Agency** (EPA) announced that satellite measurements had recorded an ominous decrease in atmospheric ozone, amounting to an average of 5% over the mid-latitudes (including the United States), almost double the loss previously thought to be occurring.

The data showed that ozone levels measured in the late fall, winter, and early spring over large areas of the United States, Europe, and the mid-latitudes of the Northern and Southern Hemisphere had dropped by 4–6% over the last decade—twice the amount estimated in earlier years. The greatest area of ozone thinning in the United States was found north of a line stretching from Philadelphia to Denver to Reno, Nevada. One of the most alarming aspects of the new findings was that the ozone depletion was continuing into April and May, a time when people spend more time outside, and crops are beginning to sprout, making both more vulnerable to ultraviolet radiation.

The new findings led the EPA to project that over the next 50 years, thinning of the ozone layer could cause Americans to suffer some 12 million cases of skin cancer, 200,000 of which would be fatal. Several years earlier, the agency had calculated that over the next century, there could be an additional 155 million cases of skin cancers and 3.2 million deaths if the ozone layer continued to thin at the then current rate. Another EPA projection made in the 1980s was that the increase in radiation could cause Americans to suffer 40 million cases of skin cancer and 800,000 deaths in the following 88 years, plus some 12 million eye cataracts.

No one can say how accurate such varying projections will turn out to be, but evidence of ozone layer thinning is well-documented. In October 1991, additional data of spreading ozone layer destruction were made public. Dr. Robert Watson, a NASA scientist who co-chairs an 80-member panel of scientists from 80 countries, called the situation "extremely serious," saying that "we now see a significant decrease of ozone both in the Northern and Southern Hemispheres, not only in winter but in spring and summer, the time when people sunbathe, putting them at risk for skin cancer, and the time when we grow crops."

In February 1992, a team of NASA scientists announced that they had found record high levels of ozone-depleting chlorine over the Northern Hemisphere. This could, in turn, lead to an ozone "hole" similar to the one that appears over Antarctica developing over populated areas of the United States, Canada, and England. The areas over which increased levels of **chlorine monoxide** were found extended as far south as New England, France, Britain, and Scandinavia.

Action to protect The ozone layer

As evidence of the critical threats posed by ozone layer depletion has increased, the world community has begun to take steps to address the problem. In 1987, the United States and 22 other nations signed the Montreal Protocol, agreeing, by the year 2000, to cut CFC production in half, and to phase out two ozone-destroying gases, Halon 1301 and Halon 1211. **Halons** are man-made bromine compounds used mainly in fire extinguishers, and can destroy ozone at a rate 10 to 40 times more rapidly than CFCs. Fortunately, these

restrictions appear to already be having an impact. In 1992, it was found that the rate at which these two Halon gases were accumulating in the atmosphere had fallen significantly since 1987. The rate of increase of levels of Halon 1301 was about 8% a year from 1989 to 1992, about half of the average annual rate of growth over previous years. Similarly, Halon 1211 was increasing at only 3% annually, much less than the previous growth of 15% a year.

Since the Montreal Protocol, other international treaties have been signed limiting the production and use of ozone-destroying chemicals. When alarming new evidence on the destruction of stratospheric ozone became available in 1988, the world's industrialized nations convened a series of conferences to plan remedial action. In March 1989, the 12-member European Economic Community (EEC) announced plans to end the use of CFCs by the turn of the century, and the United States agreed to join in the ban. A week later, 123 nations met in London to discuss ways to speed the CFC phase-out. The industrial nations agreed to cut their own domestic CFC production in half, while continuing to allow exports of CFCs, in order to accommodate **third world** nations.

Ironically, the large industrial nations, which have created the CFC problem, are now much more willing to take effective action to ban the compounds than are many developing nations, such as India and China. The latter nations resist restrictions on CFCs on the grounds that the chemicals are necessary for their own economic development.

After the meeting in London, leaders and representatives from 24 countries met in an environmental summit at The Hague, and agreed that the United Nations' authority to protect the world's ozone layer should be strengthened.

In May 1989, members of the EEC and 81 other nations that had signed the 1987 Montreal Protocol decided at a meeting in Helsinki to try to achieve a total phase-out of CFCs by the year 2000, as well as phase-outs as soon as possible of other ozone-damaging chemicals like **carbon** tetrachloride, halons, and methyl chloroform. In London in June 1990, most of the Montreal Protocol's signatory nations formally adopted a deadline of the year 2000 for industrial nations to phase out the major ozone-destroying chemicals, with 2010 being the goal for developing countries.

Finally, in November 1992, 87 nations meeting in Copenhagen decided to strengthen the action agreed to under the Montreal Protocol and move up the phase-out deadline from 2000 to January 1, 1996, for CFCs, and to January 1, 1994, for halons. A timetable was also agreed to for eliminating **hydrochlorofluorocarbons** (HCFCs) by the year 2030. HCFCs are being used as substitutes for CFCs even though they also deplete ozone, albeit on a far lesser scale than CFCs. The conference failed to ban the production of the **pesticide** methyl bromide, which may account for 15% of ozone depletion by the year 2000, but did freeze production at 1991 levels.

Environmentalists were disappointed that stronger action was not taken to protect the ozone layer. But Environmental Protection Agency (EPA) Administrator William K. Reilly, who headed the U.S. delegation, estimated that the reductions agreed to could, by the year 2075, prevent a million cases of cancer and 20,000 deaths.

Although the restrictions apply to developed nations, which produce most of the ozone-damaging chemicals, it was also agreed to consider moving up a phase-out of such compounds by developing nations from 2010 to 1995. A month after the Copenhagen conference, the nations of the European Community agreed to push bans on the use of CFCs and carbon tetrachloride to 1995 and to cut CFC emissions by 85% by the end of 1993.

The private sector has also taken action to reduce CFC production. The world's largest manufacturer of the chemicals, DuPont Chemical Company, announced in 1988 that it was working on a variety of substitutes for CFCs, would phase out production of them by 1996, and would partially replace them with HCFCs. Environmentalists charge that DuPont has been moving too slowly to eliminate production of these chemicals.

There are many ways that individuals can help reduce the release of CFCs into the atmosphere, mainly by avoiding products that contain or are made from CFCs, and by **recycling** CFCs whenever possible. Although CFCs have not generally been used in spray cans in the United States since 1978, they are still used in many consumer and industrial products, such as styrofoam. Other products manufactured using CFCs include solvents and cleaning liquids used on electrical equipment, **polystyrene** foam products, and fire extinguishers that use halons.

Refrigerants in cars and home air conditioning units contain CFCs and must be poured into closed containers to be cleaned or recycled, or they will evaporate into the atmosphere. Using foam insulation to seal homes also releases CFCs. Many alternatives to foam insulation exist, such as cellulose fiber, gypsum, fiberboard, and fiberglass.

Unfortunately, whatever steps are taken in the next few years, the problem of ozone layer depletion will continue even after the release of ozone-destroying chemicals is limited or halted. It takes six to eight years for some of these compounds to reach the upper atmosphere, and once there, they will destroy ozone for another 20–25 years. Thus, even if all emissions of destructive chemicals were stopped, compounds already released would continue to damage the ozone layer for another quarter century.

Understanding ozone depletion

As detailed collection of data about interactions in the stratosphere progresses, the observational support for the ozone depletion theory continues to grow more compelling. Yet atmospheric scientists are beginning to realize that their understanding of the upper atmosphere is still quite crude. While certain key reactions which maintain and destroy ozone are theoretically and observationally supported, scientists will have to comprehend the interaction of dozens, if not hundreds, of reactions between natural and artificial **species** of hydrogen, nitrogen, bromine, chlorine and oxygen before a complete picture of ozone-layer dynamics emerges. The recent eruption of **Mount Pinatubo**, for example, made scientists aware that *heterogenous processes*—those reactions which require cloud surfaces to take place—may play a far greater role in causing ozone depletion than originally believed. Such reactions had previously been observed taking place only at the earth's poles, where stratospheric clouds form during the long winter darkness, but it is now thought that sulfur aerosols ejected by Pinatubo may be serving as a catalyst to speed ozone depletion at nonpolar latitudes.

Ironically, ozone-depleting reactions are best understood around the thinly inhabited polar regions, where stable and isolated conditions over the winter allow scientists to understand stratospheric changes most easily. In contrast, at the temperate latitudes where constantly moving air masses undergo no seasonal isolation, it is difficult to determine whether a fluctuation in a given chemical's density is a result of local reactions or atmospheric turbulence. In 1995, the Nobel Prize for Chemistry was awarded to F. Sherwood Roland, Mario Molina and Paul Crutzen for their work on the formation and destruction of the ozone layer. It is hoped that increasingly detailed measurements using a new generation of equipment (such as NASA's *Perseus* remote-control aircraft) will begin to shed more light on the processes occurring away from the poles. Joe Waters of NASA's Jet Propulsion Laboratory summarizes the urgent task: "We must be able to lay out the catalytic cycles that are destroying ozone at all altitudes all over the globe—from its production region in the tropics to the higher latitudes and the polar regions." *See also* Photochemical reaction

[*Lewis G. Regenstein*]

RESOURCES

BOOKS

Benedick, R. E. *Ozone Diplomacy: New Directions in Safeguarding the Planet.* Washington, DC: World Wildlife Fund, 1991.

Clark, S. L. *Protecting the Ozone Layer: What You Can Do.* New York: Environmental Information Exchange, 1988.

PERIODICALS

Lyman, F. "As the Ozone Thins, the Plot Thickens." *Amicus Journal* 13 (Summer 1991): 20–28, 30.

Monastersky, R. "Ozone Layer Shows Record Thinning." *Science News* 143 (April 24, 1993): 260.

Zurer, P. S. "Ozone Depletion's Recurring Surprises Challenge Atmospheric Scientists." *Chemical and Engineering News* (May 24, 1993): 8–18.

P

PAHs

see **Polycyclic aromatic hydrocarbons**

Paleoecology/paleolimnology

Paleoecology, the scientific study of ancient environments and the interrelationships of their plants and animals, and paleolimnology, the scientific study of evidence of ancient inland waterways, including lakes, ponds, freshwater marshes, and streams, were both established at the end of the nineteenth century. The roots of both disciplines can be traced to early nineteenth-century botanical and chemical studies, and later, geological studies of the consolidated sediments of ancient lake beds, in which their organisms and environments were examined in relationship to both the lakes and the surrounding uplands.

By the early 1920s, limnologists began to collect **sediment** cores from lakes and to interpret stratigraphic data from plant and animal fossils as a record of a lake's history. This provided clues to how lakes had changed over time as a result of either natural events or human activities. Data collected from lakes and **wetlands** provide baselines to compare the impacts of a activities such as land clearance, **drainage**, **water pollution**, and **air pollution**. They can also be used to assess the rate of recovery from the activities once they have ended.

Lake and wetland sediments contain detailed archeological records of how the overlying ecosystems have been affected over time. It is in many ways equivalent to archeological reconstruction of past civilizations by examining their stratified remains. Lake sediments and wetland peat deposits form a selective trap for a variety of plant and animal remains and elements such as **carbon**, **phosphorus**, sulfur, iron, and manganese that are stored at varying concentrations depending on the activities that were occurring at the time the sediment layer was formed. Changes in this profile record not only the history of the lake or peatland but also what happened in the surrounding **watershed**. Initially, limnologists collected core samples from lakes and interpreted them on the basis of plant and animal fossils. The limnologists observed that in some lakes there were thin laminated sediments. These were shown to result from an annual deposition cycle which involves summer deposition of calcium carbonate and deposition of organic matter during the rest of the year. These alternating light and dark layers represent a yearly cycle. Using these bands, or laminae, scientists can count back in time and date individual strata of sediment cores.

Studying plant and animal microfossil remains (such as pollen, diatom shells, and remains of **zooplankton** bodies) in the sediment cores and knowing the environments in which these organisms occur today allow a limnologist to reconstruct historical conditions in a lake as well as in its drainage basin. Such wetland studies were limited, however, until the development of **radioisotope** dating methods in the 1950s and 1960s. Today radioisotopes such as carbon-14 and lead-210 can be used to date the time a sediment was deposited.

Other methods limnologists use include the pollen, which indicates what types of land vegetation were present. They also examine plant and animal fossil remains, including cell fragments and molecules such as plant pigments, all of which provide clues about vegetation and aquatic life. Limnologists also search for organic pollutants and certain trace elements, whose presence indicates the result of human activities.

Analysis of the layers in sediment cores from lakes and wetlands has provided information about regional variations in past climatic conditions and watershed vegetation patterns which indicates the causes of environmental change. Studies on recently deposited lake sediments have provided evidence for the timing and causes of lake **pollution**, including the effects of excess nutrients on lake **ecology** and information about atmospheric transport of various pollutants.

These techniques have also been applied to address basic plant ecology questions regarding plant **succession**, on which there are two major schools of thought. The first

is the community concept which has three basic attributes: vegetation occurs in recognizable and characteristic communities; community change through time occurs because of the vegetation; and the changes occur in a sequence that leads to a mature stable climax **ecosystem**. The community explanation can be contrasted with the continuum concept. Here, the distribution of vegetation is determined by the **environment**. Since each plant **species** adapts differently, no two occupy the exact same location. Additionally, the observed replacement sequence is influenced by the chance occurrence of viable seeds that allows a certain plant to grow on the site. This results in a continuum of overlapping sets of species. In this scenario although ecosystems change, it is not directed toward a particular climax community.

One of the areas where these two concepts directly collide is in the study of wetlands and **peatlands**. The classical community view of succession is that wetlands are a transitional stage in the progression from shallow lake to terrestrial forested climax community. This view requires that lakes gradually fill in as organic material from dying plants accumulates and minerals are carried down from upslope into the water. At first change is slow, but it accelerates when the lake becomes shallow enough to support rooted aquatic plants. When the water becomes even more shallow, it supports the development of a peat mat, allowing trees and shrubs to grow which further modifies the site by not only adding organic matter but also by drying the site through **evapotranspiration**. According to the community version, in the final stages a climax forest occupies the site. This sequence implies that most of the change is caused by plants and not external environmental changes.

Paleoecological analyses of peat beds have provided information refuting the validity of the community concept's explanation. Fossil records including analysis of pollen in northern peat lands provide two generalizations: in some sites the present vegetation has existed for several thousand years and climatic change and **glaciation** had a large impact on plant species and distribution. Generally, bogs expanded during warm, wet periods and contracted during cool, drier periods although the influence of local topographic and drainage conditions often mask climatic shifts.

While this evidence supports the continuum assumptions that other environmental factors overwhelm the vegetation effects, there is evidence within the bogs that says that vegetation can have a large impact on the character of the landscape. In particular, this is evidenced by the patterned landscape of these peatlands as the water flow is changed due to the vegetation. It seems that in the wetlands or bogs themselves there can be changes in vegetation that occur over time as a result of succession. However, these changes do not necessarily point toward a terrestrial climax. In fact, pollen profiles indicate that a bog, not some type of terrestrial forest, is the endpoint. These studies confirmed conclusions from a classical study on the Lake Aggassiz Plain (Minnesota and south-central Canada) that indicates most of the peatlands developed during a moist **climate** about 4,000 years ago when surface water levels rose about 12 ft (4 m). This implies that while there may be changes within the peatlands, they are stable, so the central concepts of succession in this case are not supported.

Examples of the findings and results of some significant types of paleolimnological studies show how they are useful in evaluating changes and as indicators of potential problems due to human activities. The Red Lake peatland in northwestern Minnesota has been used as a study site to document the occurrence of **acid rain** and global warming. Based on past changes in the type of vegetation and changes in **pH**, future impacts on the bog surface can be evaluated. If **acid deposition** were to lower the pH of the bog significantly, different types of mosses would be evident. Likewise, if global warming were to lower the water tables' moss types, characteristics of drier conditions would occur.

Using sediment cores from lakes and observing the changes in **flora**, **fauna**, and chemistry allows the evaluation of impacts of human settlement on lakes in the upper Midwest of the United States. A lake studied in northern Minnesota on the iron range shows these changes. There was a distinct increase in hematite (iron) grains in the sediments as mining began and a decrease as it declined. Settlement was also marked by a rise in the concentration or ragweed pollen which reflects the replacement of the forests with agricultural fields. There were also changes in the type of diatom shells reflecting the **nutrient** enrichment of the lake due to the **discharge** of sewage directly to the lake. By analyzing the pH preferences of individual species of diatoms and algae the past pH conditions in a lake can be determined. Using the results of these studies it has been demonstrated that lakes in the **Adirondack Mountains** were not naturally acidic. Current monitoring suggests that these lakes that were acidified due to **acid** rain have not yet responded reductions in the amount of sulfate deposition.

[*James L. Anderson*]

Resources

Books

Freshwater Ecosystems: Revitalizing Educational Programs in Limnology. Washington, DC: National Academy Press, 1996.

Mitsch, W.J., and J.G. Gosselink. *Wetlands.* New York: Van Nostrand Reinhold, 1993.

PAN

see **Peroxyacetyl nitrate**

Panda

see **Giant panda**

Panther

see **Florida panther**

Panthera tigris

see **Tigers**

Paper mills

see **Pulp and paper mills**

Parasites

Organisms that live in or upon the body of a host organism and are metabolically dependent on the host for completion of their life cycle. Parasites may be plants, animals, viruses, bacteria, or **fungi**. Parasites feed either on their host directly or upon its surplus fluids. Some parasites, known as endoparasites, live inside their host, while ectoparasites live on the outside of their host. Organisms in which parasites reach maturity are called definitive hosts, and hosts harboring parasite stages are called intermediate hosts. Organisms which spread parasite stages between hosts are known as vectors. Full-time, or obligatory, parasites have an absolute dependence on their hosts. Examples of this type are viruses, which can only live and multiply inside living cells, and tapeworms, which can only live and multiply inside other **species**. Part-time, or facultative, parasites, such as wood ticks, have parasitic and free-living stages in their life cycle and are only temporary residents of their hosts.

The effects of parasites on their hosts depend on the health of the host, as well as the severity of the infestation. In diseased, old, or poorly-fed individuals, parasite infestations can be fatal, but parasites do not typically kill their hosts, though they can slow growth and cause weight loss. Some plant parasites do kill their host and then live on its decomposing remains, and certain species of hymenopteran insect are parasitoids, whose larva feeds within the living body of the host, eventually killing it. Some parasites, such as *Sacculina,* castrate their host by infecting its reproductive organs. Cuckoos are brood parasites, and they lay their eggs in the nest of another bird species, evicting the eggs that were there and leaving the young cuckoos to be raised by the parents of the host species.

In **community ecology**, parasites are often grouped together with predators, since both feed directly upon other organisms, either harming them or killing them. Ecologically, small disease-causing organisms (viruses, bacteria, protozoans) are regarded as microparasites, while larger parasites (flatworms, roundworms, lice, fleas, ticks, rusts, mistletoe) are macroparasites. Together with predators and diseases, parasites are one of the natural components of environmental **resistance** which serves to limit **population growth**. Microparasites that are transmitted directly between infected hosts, such as rabies or distemper, target herd animals with a high host density and can significantly reduce population levels. Macroparasites that employ one or more intermediate hosts, such as flukes, tapeworms, or roundworms, have highly effective transmission stages but usually have only a limited effect on the population of the host.

Parasites can play a larger role in altered ecosystems. In the eastern United States parasitic infections have held populations of cottontail rabbits well below the **carrying capacity** of the **habitat**. Parasitic insects have been used to control populations of the olive scale insect, a serious **pest** of olive trees in California. Accidentally introduced parasites have negatively impacted populations of commercial species by altering the balance of the **ecosystem**. For example, a protozoan parasite infecting California oysters was introduced to French oyster beds, wiping out the native European oysters and seriously damaging **commercial fishing** there. *See also* Population biology; Predator-prey interactions

[*Neil Cumberlidge Ph.D.*]

Resources

Books

Bullock, W. L. *People, Parasites, and Pestilence: An Introduction to the Natural History of Infectious Disease.* Minnesota: Burgess Publishing Company, 1982.

Crew, W., and D. R. W. Haddock. *Parasites and Human Disease.* London: Edward Arnold, 1985.

Despommier, D. D., and J. W. Karapelou. *Parasite Life Cycles.* New York: Springer-Verlag, 1987.

Noble, E. R., and G. A. Noble. *Parasitology: The Biology of Animal Parasites.* 6th ed. Philadelphia: Lea & Febiger, 1989.

Schmidt, G. D., and L. S. Roberts. *Foundations of Parasitology.* 4th ed. St. Louis: Times Mirror/Mosby College Pub., 1989.

Smith, R. L. *Elements of Ecology.* 3rd ed. New York: Harper Collins, 1991.

Pareto optimality (Maximum social welfare)

Usually, one thinks of efficiency as not being wasteful or getting the most out of the resources one has available. Economists offer the Pareto optimum—"a situation where no one can be better off without making someone worse off." Derived from the work of the Italian economist and sociologist Vilfredo Pareto, whose late nineteenth-century writings on political economy inspired much thinking about what made an economy efficient, Pareto optimality has come

to mean making at least one person better off without making anyone else worse off. For an economy, it means that the allocation of resources is optimal if no other allocation exists wherein a person is better off and everyone is at least as well off. Economic theory holds that these conditions are met if consumers maximize utility, producers maximize profits, competition prevails, and information is adequate for the making of rational decisions. A free market unconstrained by government involvement, it is assumed, will achieve Pareto optimality by an invisible hand, that is, automatically, provided that production and consumption decisions do not entail substantial environmental disamenities. Imperfect market conditions, which include environmental disamenities on a large scale, challenge the assumptions of Pareto optimality and call for remedial measures such as **pollution** taxes or **emission** rights to restore market efficiency. *See also* Externalities

Parrots and parakeets

These intelligent, brightly colored, affectionate birds are found mainly in warm, tropical regions and have long been popular as pets, in large part because many of them can learn to talk. But capture of wild parrots for the **pet trade**, along with the destruction of forests, have decimated populations of these birds, and many **species** are now threatened with **extinction**.

About half of the approximately 315 species of parrots in the world are native to Central and South America. Members of the parrot order (Psittaciformes) include the macaws of Central and South America, which are the largest parrots, with bright feathers; cockatoos of **Australia** with white feathers and crests on their heads; lorikeets of Australia with orange-red bills and brightly-colored feathers; cockatiels, which are small, long-tailed, crested parrots also from Australia; and parakeets, which are small, natural acrobats, usually with green feathers. Parakeets include lovebirds from Africa and budgies from Australia. Budgies are the most common type of pet parakeet, and they can be trained to say many words. Most parakeets sold as pets are bred for that purpose, thus their trade does not usually threaten wild populations.

Part of the attraction of parrots is their high intelligence, but this can make them unsuitable pets. The birds are often loud and they demand a great deal of attention, and many people who buy parrots give them up because of the frustrations of owning one of these complex birds. In addition, parrots often carry a disease called *psittacosis* which can be transmitted to humans and commercially raised poultry. For this reason, parrots must be examined by officials from the **U.S. Department of Agriculture** (USDA) before entering the United States, and hundreds of thousands of parrots have been destroyed to prevent the spread of psittacosis.

The large scale capture of parrots for the legal and illegal pet trade has been a major factor in the depletion of these beautiful birds. The world trade in wild birds has been estimated at over seven million annually, and the United States is the world's largest consumer of exotic birds. Over 1.4 million wild birds were imported into this country during 1988–1990, and half of these were parrots and other birds supposedly protected under the **Convention on International Trade in Endangered Species of Wild Fauna and Flora** (CITES). The **mortality** rate for birds transported in international trade is massive. For every wild bird that makes it to the pet store, at least five may have died on the way, and USDA **statistics** show that 79,192 birds perished in transit to the United States from 1985 to 1990, and 258,451 died while in quarantine or because they were refused entry due to Newcastle disease. In 1991 one airline shipment alone included 10,000 dead birds. Moreover, the shock of capture and caging the birds before shipment may cause even greater mortality rates.

As a species becomes rare or endangered, it often becomes more valuable and sought-after, and some parrots have been sold for $10,000 or more. In October 1992, shortly before being named Secretary of the Interior, former Arizona governor Bruce Babbitt addressed the annual meeting of the **Humane Society of the United States**. Expressing dismay at "the looming extinction of tropical parrots and macaws in South America," he described their exploitation in the following terms: "These birds are captured for buyers in the United States who will pay up to $30,000 for a hyacinth macaw. You can stand on docks outside Manaus, Brazil, and other towns in the Amazon and see confiscated crates with blue and yellow macaws, their feet taped, their beaks wired, stacked up like cordwood in boxes. They have a fatality rate of 50% by the time they're smuggled into Miami."

Some progress has been made in restricting the international trade in parrots. By the end of 1992 over 100 airlines had agreed to forbid the carrying of wild birds. The Wild Bird Conservation Act, signed into law on October 27, 1992, bans trade immediately for several severely exploited bird species and, within a year of passage, outlaws commerce in all parrots and other birds protected under CITES. The act requires exploiters to prove that a species can withstand removal from the wild. This has greatly reduced the number of birds imported into the United States. Conservation groups, such as the Humane Society of the United States and the **Animal Welfare Institute**, have worked for years to secure passage of this legislation.

Nevertheless, the persistence of smuggling as well as the legal trade in some species of these birds will continue

to threaten wild populations. This is especially true when the trees in which parrots nest are cut down to provide collectors access to chicks in the nest. As of 2002, 16 parrot species, 10 parakeet species, and three macaw species were listed as endangered by the **U.S. Department of the Interior**, and all species of parrots, parakeets, macaws, lories, and cockatoos are listed in the most endangered categories of CITES. IUCN—The World Conservation Union—includes 69 members of the parrot order in its 2000 list of threatened animals.

Nine parrots are listed as extinct by IUCN. One is the beautiful orange, yellow, and green Carolina parakeet (*Conuropsis carolinensis*) once ranged in large numbers from Florida to New York and Illinois, but the demand for its feathers by the millinery trade was so great that the species was hunted into extinction between 1904 and 1920. Conservationists fear that several other species of the parrot family may soon join the Carolina parakeet on the list of extinct birds.

[*Lewis G. Regenstein*]

Resources

Books

Beissinger, S. R., and E. H. Bucher. *New World Parrots in Crisis: Solutions From Conservation Biology.* Washington, DC: Smithsonian Institution Press, 1991.

Periodicals

Beissinger, S. R., and E. H. Bucher. "Can Parrots Be Conserved Through Sustainable Harvesting?" *BioScience* 42 (March 1992): 164–73.

Defreitas, M. "Feathering a Nest in the Windwards. (Saving the Santa Lucian parrot)." *Americas* 43 (March–April 1991): 40–45.

Particulate

An adjective describing anything that consists of, or relates to, particles. The term was formerly used in laboratory slang to stand for "particulate matter," but this use has nearly disappeared since its repudiation by the **Environmental Protection Agency** (EPA). The term particulate matter, which often means the particle content of a given volume of air, is used as a more inclusive variant of "particles."

Typical atmospheric particulate matter may comprise three populations of particles according to size. The smallest of these is found only near sources, since the particles rapidly aggregate to larger sizes. The upper size limit for these particles is near 0.1 micrometer (1 micrometer = 1 μm = 0.001 mm). Number concentrations can be quite high. This population is referred to as the nuclei mode of particles. The next larger size class, which begins at a diameter of about 0.1 μm and extends to about 2.0 μm, is called the accumulation mode, since once in the air it tends to remain for days; number concentrations have become low enough that **agglomeration** is slow, and settling velocities are very small. The sum of the nuclei mode and the accumulation mode is called the fine particles. These are characteristically formed by condensation from the gas phase, or by agglomeration of particles formed from the gas phase. The final particle population in air is called the coarse mode, or simply the coarse particles. These are of any size larger than about 2 μm (about 1/10,000 inch), and are formed by mechanical grinding of larger masses of matter. Coarse particles are usually particles of local **soil** or rocks.

Partnership for Pollution Prevention

The Partnership for **Pollution** Prevention (Ppp) is an international cooperative effort to ensure the environmental sustainability of the Western Hemisphere. Pollution prevention requires the efficient utilization of raw materials and energy, as well as modifications in products and manufacturing processes to minimize environmental impact. The Ppp promotes technical and financial cooperation and information exchange among nations. It also promotes the adoption of strict, internationally compatible environmental laws and regulations and the implementation of international environmental agreements. The Ppp works toward improving technologies for environmental protection and increasing public awareness and participation in environmental issues, particularly among indigenous groups and other affected communities. Finally, the Ppp encourages the inclusion of sustainability objectives in other developmental and national policies.

The Ppp is Action Initiative 23 of the Plan of Action developed at the First Summit of the Americas, held in Miami, Florida, in 1994. It is based on the policies formulated at the 1992 United Nations Conference on Environment and Development and the 1994 Global Conference on the **Sustainable Development** of Small Island Developing States. The initial Ppp priorities were formulated at a meeting in San Juan, Puerto Rico, in November 1995. The meeting also addressed issues of financing, public-private partnerships, new legislation, compliance and enforcement, and public participation. Sponsored by the Organization of American States (OAS), the U.S. **Environmental Protection Agency** (EPA), and the Pan American Health Organization (PAHO), this meeting included representatives from 20 countries, as well as members of international organizations, non-governmental organizations, and development banks.

The initial Ppp priorities are health and environmental problems stemming from **lead** contamination and the misuse of pesticides. The Ppp also addresses problems of waste, air and **water quality**, **marine pollution**, and urbanization.

Under the Ppp, 10 nations have made formal political commitments in the areas of health and the environment. Furthermore, under the Ppp, the environmental policies of the **North American Free Trade Agreement** (NAFTA) are being integrated with its trade and economic policies, and NAFTA is confronting issues of **pesticide** registration and chemical **pollution control**.

An OAS meeting in 1996 that included the Inter-American Development Bank, the **World Bank**, the PAHO, the EPA, and the United States Agency for International Development (USAID) formed a task force for Ppp implementation. In October 1997 the task force merged with a similar task force from the 1996 Santa Cruz (Bolivia) Summit on Sustainable Development. This new task force established a number of working groups to address cleaner production, innovative financing for sustainable development, improved access to drinking water, and the networking of environmental experts. The working group on energy infrastructure has obtained funding for a gas pipeline being constructed between Santa Cruz and São Paolo, Brazil. The working group for sustainable cities and communities initiates projects to strengthen local governments, promote local involvement in planning, and improve access to capital and to low-income housing.

The Ppp working group to phase out leaded **gasoline** has served as a model for other working groups. The EPA, PAHO, and the World Health Organization (WHO) have held lead phase-out training workshops. The World Bank works with individual countries to finance refinery upgrades and conversions. As a result of these efforts, a number of nations have banned the sale of leaded gasoline. Leaded gasoline is expected to be eliminated from the Western Hemisphere by 2007. The World Bank, USAID, and the EPA also have developed a lead-monitoring program for children and adults in Latin America and the Caribbean.

Under the auspices of the Ppp, the Cleaner Production Conference of the Americas was held in São Paulo in 1998. It established the Cleaner Production Roundtable of the Americas, a formal network of cleaner production practitioners who share ideas and promote the benefits of cleaner production for sustainable development. The conference also set up a communications network and identified priority needs for cleaner production and pollution controls.

Under the Ppp, the USAID's Regional Environmental Project for Central America, in collaboration with the EPA and the Central American Commission on Environment and Development, has established regional pollution prevention networks to address **solid waste** and wastewater management and the safe use of pesticides. Wastewater and air-contamination regulations, municipal pollution prevention and environmental action plans, and a training network in **environmental law** have been established in Central American countries. As part of the Ppp, the USAID has carried out projects for pollution prevention, clean industrial technologies, and coastal and marine management throughout Latin America and the Caribbean.

[*Margaret Alic Ph.D.*]

Resources

Books

Bureau of Inter-American Affairs. *Words Into Deeds, Progress Since the Miami Summit: Report on the Implementation of the Decisions Reached at the 1994 Miami Summit of the Americas.* Washington, DC: United States Department of State, 1998.

Other

First Summit of the Americas: Plan of Action. Secretariat for Legal Affairs, Organization of American States. 2001 [cited May 29, 2002]. <www.oas.-org/Juridico/english/PlanI.html>.

Office of the Summit Follow-Up. *Words Into Deeds: Progress Since the Miami Summit.* Organization of American States. 1998 [cited May 28, 2002]. <www.summit-americas.org/WordsintoDeeds-eng.htm>.

Organizations

Free Trade Area of the Americas "FTAA" Administrative Secretariat, Apartado Postal 89-10044, Zona 9, Ciudad de Panamá, Republica de Panamá (507) 270-6900, Fax: (507) 270-6990, , www.ftaa-alca.org/alca_e.asp

National Pollution Prevention Roundtable, 11 Dupont Circle, NW, Suite 201, Washington, DC, USA 20036 (202) 299-9701, Fax: (202) 299-9704, Email: talk@p2.org, www.p2.org

Organization of American States, 17th Street and Constitution Ave., NW., Washington, DC, USA 20006 (202) 458-3000, , <http://www.oas.org>

United States Environmental Protection Agency, 1200 Pennsylvania Avenue, NW, Washington, D.C. USA 20460 Toll Free: (800) 490-9198, Email: public-access@epa.gov, <http://www.epa.gov>

Parts per billion

A means of expressing minute concentrations, often of an element, compound, or particle that contaminates **soil**, water, air, or blood. The expression may be on either a mass or a volume basis, or a combination of the two, such as micrograms of a chemical per kilogram of soil, or micrograms of chemical per liter of water. In the United States, one billion equals $1x10^9$. **Dioxin**, a contaminant associated with the **herbicide 2,4,5-T** and implicated in the formation of **birth defects**, can cause concern at concentrations measured in ppb.

Parts per million

A means of expressing small concentrations, usually of an element, compound, contaminant, or particle in water or a mixed medium such as **soil** and normally on a mass basis. Units that are equivalent to ppm include micrograms per gram, milligrams per kilogram, and milligrams per liter.

One ppm times 10,000 equals one part per hundred or one percent. The **nutrient phosphorus** and the contaminant **lead** found in soil are commonly measured in ppm.

Parts per trillion

A means of expressing extremely minute concentrations of substances in water or air. In the United States, the concentration is the number of units of the substance found in 1×10^{12} units of water or air; an equivalent unit is nanograms per kilogram. Detection of concentrations this low, which was not possible until the late twentieth century, is limited to only a few types of chemical compounds. For example, one gram of sulfur hexafluoride, used as a tracer in studies of ocean mixing, can be detected in a cubic kilometer of sea water, a concentration of 1,000 parts per trillion.

Passenger pigeon

The passenger pigeon (*Ectopistes migratorius*), perhaps the world's most abundant bird **species** at one time, became extinct due directly to human activity. In the mid-1800s passenger pigeons travelled in flocks of astounding numbers. Alexander Wilson, the father of American ornithology, noted a flock he estimated to contain two billion birds. The artist and naturalist John James Audubon once observed a flock over a three-day period and estimated the birds were flying overhead at a rate of 300 million per hour.

The species became extinct within a span of 50 years, several factors having led to its rapid demise. The passenger pigeon was considered an agricultural **pest**, thus providing ample reason to kill large numbers of the birds. It was also in demand as food, largely due to the fact that nesting flocks were easily accessible. Young squabs were easy prey for hunters who knocked them from their nests or forced them out by setting fires below them. Adults were also killed in huge numbers. They were baited with alcohol-soaked grain or with captive pigeons set up as decoys, then trapped and shot. A common practice of the day was to use the live pigeons as targets in shooting galleries. In 1878, near Petoskey, Michigan, a professional market hunter earned $60,000 by killing over three million passenger pigeons near their nesting grounds. Once killed, many of the birds were packed in barrels and shipped to cities where they were sold in markets and restaurants. The demand was particularly high on the East Coast where forest clearing and **hunting** had already eradicated the species from the area.

By the 1880s commercial hunting of passenger pigeons was no longer profitable, because the population had been depleted to only several thousand birds. Michigan provided their last stronghold, but that population became extinct in

Passenger pigeon. (Photograph by John D. Cunningham. Visuals Unlimited. Reproduced by permission.)

1889. The remaining small flocks of birds were so spread out and isolated that their numbers were too low to be maintained. The disruption of the population in the 1860s and 1870s had been so severe that breeding success was permanently reduced. At one time the sheer numbers of passenger pigeons in a flock was enough to discourage potential predators. Once the population was split into small, isolated remnants, however, natural predation also contributed to the species' rapid decline.

The last individual passenger pigeon was a female named Martha, which died in the Cincinnati Zoo in 1914. It is now on display at the National Museum of Natural History in Washington, D.C.

[*Eugene C. Beckham*]

Resources

Books

Schorger, A. W. *Passenger Pigeon: Its Natural History and Extinction.* Norman, OK: University of Oklahoma Press, 1973.

Periodicals

Wilcove, D. "In Memory of Martha and Her Kind." *Audubon* 91 (September 1989): 52–55.

Worsnop, R. L. "Evolving Attitudes." *CQ Researcher* 2 (January 24, 1992): 58–60.

Passive solar design

Looking into the sky on a sunny day, the notion that humans could have an "energy crisis" seems absurd. Each day, the earth receives 1.78×10^{14} kilowatts of energy, more than 10,000 times the amount needed by the whole world this year. All that is required is a way to collect and harness the energy of sunlight.

Humans have explored systems for the capture of **solar energy** for centuries. The Roman architect Vitruvius described a plan in the first century B.C. for building a bathhouse heated by sunlight. He explained that the building should "look toward the winter sunset" because that would make the bathhouse warmer in the late afternoon. More recently, water heaters operated by solar energy were built and widely sold in the early 1990s, especially in California and Florida.

Most historical examples illustrate the principles of passive solar heating, namely constructing a building so that it can take advantage of normal sunlight without the use of elaborate or expensive accessory equipment. A home built on this principle, for example, has as much window space as possible facing toward the south, with few or no windows on other sides of the house. Sunlight enters the south-facing window and is converted to heat, which is then trapped inside the house. To reduce loss of heat produced in this way, the window panes are double- or triple-glazed, that is, consist of two or three panes separated by air pockets. The rest of the house is also as thoroughly insulated as possible. In some cases, a house can be built directly into a south-facing hill so that the earth itself acts as insulation for the north-, east-, and west-facing walls.

Adjustments can also be made to take account of changing sun angles throughout the year. In the winter, when the sun is low in the sky, solar heat is needed most. In the summer, when the sun is high in the sky, heating is less important. An overhang of some kind over the south-facing window can provide the correct amount of sunlight at various times of the year. Changing seasonal temperatures can also be managed by installing insulating screens on the south-facing window. When the screens are open, they allow solar energy to come in. When they are closed (as at night), they keep heat inside the building.

The primary drawback to the use of solar energy is its variability. Often the energy supply and the energy requirements are out of balance for weeks or months depending on season, amount of cloud cover, latitude, etc. Thus, there is frequently the need for storage of energy for later use. Several technologically-simple methods for storing solar energy are commonly used in passive solar design. A large water tank on top of the house, for example, provides one way to store this energy. Sunlight warms the water during the day and the water can then be pumped into the house at night as a heat source. A heat sink such as a dark-colored masonry wall or concrete floor near the south facing windows can also enhance a passive solar system by absorbing heat during the day and slowly radiating it into the room during the night. A Trombe wall may also be installed. This masonry wall, 6-18 inches (15-46 cm) thick with gaps at the top and bottom, faces the sun. The space in front of the wall is enclosed by glass. Air between the glass and the masonry wall is heated by the sun, rises and passes into the room behind the wall through the upper gap. Cooler air is drawn from the room into the space through the lower gap. The masonry wall also acts as a heat sink.

[*David E. Newton*]

Resources

Books

Anderson, B. *Passive Solar Energy: The Homeowner's Guide to Natural Heating and Cooling.* Amherst, MA: Brick House Publishing, 1993.

Balcomb, J. D. *Passive Solar Buildings.* Cambridge: MIT Press, 1992.

Mazria, E. *The Passive Solar Energy Book: A Complete Guide to Passive Solar Home, Greenhouse, and Building Design.* Emmaus, PA: Rodale Press, 1979.

John Arthur Passmore (1914 –)

Australian philosopher and writer

As a philosopher, John Passmore has defended Western civilization against the charge that such societies "can solve their ecological problems only if they abandon the analytical, critical approach which has been their peculiar glory and go in search of a new ethics, a new metaphysics, a new religion." We do indeed face daunting ecological crises, Passmore states, but the best hope for solving them lies in "a more general adherence to a perfectly familiar ethic."

Born in Manly, New South Wales, **Australia**, on September 9, 1914, Passmore earned his B.A. and M.A. degrees from the University of Sydney, where he taught philosophy from 1935 to 1949. He then served on the faculties of Otago University in New Zealand and the Australian National University, holding the chair of philosophy at the latter from 1959 to 1979. The best known of his many publications are *Hume's Intentions* (1952), *A Hundred Years of Philosophy* (1957), *The Perfectibility of Man* (1970), and *Man's Responsibility for Nature* (1974).

Part of Passmore's argument in *Man's Responsibility for Nature* is that people in the West, like people everywhere, cannot simply adopt a new and unfamiliar way of thinking: one must begin where one is at. He also insists that Western civilization encompasses more than one way of thinking. He argues that "central Stoic-Christian traditions are not favourable to the solution of [the West's] ecological problems..." but that these "are not the only Western traditions and their influence is steadily declining." More favorable are traditions such as Jewish thought and, especially, modern science. These are hospitable to two attitudes—a sense of stewardship and a sense of cooperation with nature—that are incompatible with the Stoic-Christian traditions, "which deny that man's relationships with **nature** are governed by any moral principles and assign to nature the very minimum of independent life."

As a champion of stewardship and cooperation with nature, Passmore sharply disagrees with those who see civilization as an enemy of nature or a blight upon it. On the contrary, Passmore sees in civilization "man's great memorials—his science, his philosophy, his technology, his architecture, his countryside...all of them founded upon his attempt to understand and subdue nature." He considers the transformation of the natural **environment** necessary and, when done with care, desirable. For there is "no good ground...for objecting to transforming as such; it can make the world more fruitful, more diversified, and more beautiful."

Passmore's case for a more responsible attitude toward nature rests, finally, on a rejection of mysticism that includes a rejection of religion and the concept of the sacred: "To take our ecological crises seriously...is to recognize, first, man's utter dependence on nature, but secondly, nature's vulnerability to human depredations—the fragility, that is, of both man and nature, for all their notable powers of recuperation. And this means that neither man nor nature is sacred or quasi-divine."

[*Richard K. Dagger*]

RESOURCES
BOOKS

Evory, A., ed. *Contemporary Authors: New Revision Series 6.* Detroit: Gale Research Co., 1982.

Passmore, J. *Hume's Intentions.* 3rd ed. London: Duckworth, 1980.

———. *A Hundred Years of Philosophy.* Rev. ed. New York: Basic Books, 1966.

———. *Man's Responsibility for Nature: Ecological Problems and Western Traditions.* London: Duckworth, 1974.

———. *The Perfectibility of Man.* New York: Scribner Sons, 1970.

Pathogen

A pathogen is an agent that causes disease. Pathology is the scientific study of human disease. One could argue that anything that causes disease is therefore by definition a "pathogen." Sunlight is the environmental agent that (with excessive exposure) induces the potentially fatal skin **cancer** known as melanoma. Ordinarily, however, most people do not consider sunlight to be a pathogen. An unbalanced diet may result in nutritional deficiencies which can lead to diseases such as pellagra (caused by a niacin deficiency) and scurvy (caused by inadequate vitamin C in the diet). Nevertheless, the *failure* to consume a balanced diet is not considered to be a pathogen. Generally, most students of disease refer to biological agents when they use the term pathogen. Such agents, which include viruses, bacteria, **fungi**, protozoa, and worms, cause a tremendous diversity of diseases.

Viruses, while considered biological agents, are not cellular organisms and accordingly, are not living in the usual sense. They are tiny particles consisting of either DNA or RNA as the genetic material and a protein coat, and they are incapable of **metabolism** outside of a living cell. Pathogenic viruses cause diseases of the respiratory system such as colds, laryngitis, croup, and influenza. Skin eruptions such as measles, rubella, chicken pox, and foot-and-mouth disease are viral in origin. The long list of viral diseases include insect-borne Western equine encephalitis and yellow fever. Recently, some human cancers have been thought to be associated with viruses, perhaps because animal cancers such as the Lucké renal carcinoma and mouse mammary carcinoma are known to be caused by viruses. One such human cancer is Burkitt's lymphoma, a childhood malignancy occurring primarily in Africa, which is associated with a herpes **virus**. **AIDS** results from infection with HIV and is a pandemic viral disease.

Bacteria are true cells but their genetic material, DNA, is not packaged in a nucleus as in all higher forms of life. Not all bacteria are pathogens, but many well-known diseases are caused by bacterial infections. Tuberculosis, **cholera**, **plague**, gonorrhea, syphilis, rheumatic fever, typhus, and typhoid fever are some of the very serious diseases caused by pathogenic bacteria.

Probably 100 million cases of **malaria** occur each year primarily in Africa, Asia, and Central and South America. Malaria is caused by four **species** of the protozoan *Plasmodium. Amebiasis* and *giardiasis* are parasitic protozoa infections. Protozoa are single-celled animals. Thrush (*candidiasis*) is a common infection of mucous membranes with a yeast-like fungus. Valley fever (*coccidioidomycosis*) is a fungal infection generally limited to the lungs. Athlete's foot is a common skin infection caused by a fungus.

Worms that infect humans are a significant health problem in those parts of the world where there is inadequate public health protection. Examples of helminth infections are the beef and pork tapeworms. The consumption of inadequately cooked pork can result in trichinosis, which is a roundworm (nematode) infection of human muscle. Trematode flukes cause an extraordinarily important infection in Asia and Africa called **schistosomiasis** with hundreds of millions of individuals infected. Perhaps the most common parasitic helminth infection in the United States is *enterobiasis* known as pinworm or seatworm infection, which is a common condition of children with improper personal hygiene.

Many diseases caused by **microbial pathogens** can be treated with a diversity of antibiotics and other drugs. However, viral pathogens remain intractably difficult to manage with drugs.

[*Robert G. McKinnell*]

Ruth Patrick (1907 –)

American biologist and limnologist

Emphasizing the vital importance of environmental cleanup and **conservation**, Ruth Patrick has spent a lifetime facilitating cooperation among the scientific and political communities to find solutions to biospheric **pollution**. Her aggressive endeavors in freshwater research, including work on diatoms and the biodynamic cycles of rivers, have made much progress toward rectifying damage done by pollutants.

Her enthusiasm and concern for the **biosphere** began at an early age in her hometown of Kansas City, where, as a child, she conducted many expeditions through the countryside surrounding her family home. Collecting specimen materials and identifying them set the course for Patrick's future. She was fascinated by the microscopic world that could be found even in the smallest drop of water.

Patrick continued to foster her interests in science as she grew older. She attended Coker College in Hartsville, South Carolina, and earned a bachelor's degree in botany there in 1929. She then attended the University of Virginia where she earned a doctoral degree in 1934. Throughout her career, Patrick has been awarded honorary degrees from many institutions, including Princeton University and Wake Forest University. In 1975, she received the prestigious John and Alice Tyler Ecology Award (through Pepperdine University); the award amount of $150,000 is the largest amount granted for a scientific award.

Patrick was a key force in the founding of **limnology**, the scientific study of the biological, physical, and chemical conditions of freshwater. By conducting further studies, she also became widely known as an expert on diatoms, the basic food substance of freshwater organisms, and she later developed the diatometer. The diatometer is used to measure the levels of diatoms in freshwater, a measurement which is used in evaluating the pollution levels and general health of freshwater bodies. During the late 1960s and early 1970s, she acted as a consultant in both the private sector and in government, and her vast knowledge and influence enabled her to play an active role in the development of **environmental policy**. She worked with the **U.S. Department of the Interior** and served on numerous committees, including the **Hazardous Waste** Advisory Committee of the **Environmental Protection Agency** (EPA) and the Science Advisory Council of the World **Wildlife** Federation. She has often been called the "Ralph Nader of water pollution," and she has a wide reputation as a dauntless watchdog. Patrick has conducted most of her work through her offices at the Academy of Natural Sciences in Philadelphia, where she has been on staff since the late 1930s. She was the first female board chairperson at the academy.

Patrick also maintains membership in such organizations as the **National Academy of Sciences**, the National Academy of Engineering, the American Academy of Arts and Sciences, the International Limnological Society, the American Society of Limnology and Oceanography, and the American Society of Naturalists. She has been bestowed dozens of awards from schools, institutions, and organizations, and she has written four books, including *Diatoms of the United States* (with C. W. Reimer). In 1996, Patrick was presented with a National Medal of Science by President Clinton for her work in pioneering the field of limnology and inducted into the South Carolina Hall of Science and Technology. In 1999, she was awarded the Governor of South Carolina, and currently hold the Francis Boyer Chair of Limnology and Senior Curator at the National Academy of Sciences.

[*Kimberley A. Peterson*]

Resources

Books

Patrick, Ruth. *Diatoms of the United States, Exclusive of Alaska and Hawaii.* Philadelphia: University of Pennsylvania Press, 1966.

———. *Groundwater Contamination in the United States.* Philadelphia: University of Pennsylvania Press, 1987.

———. *Surface Water Quality: Have the Laws Been Successful?* Princeton, NJ: Princeton University Press, 1992.

Periodicals

Kunreuther, H., and Ruth Patrick. "Managing the Risks of Hazardous Waste." *Environment* 33 (April 1991): 12–21.

PBBs

see **Polybrominated biphenyls**

PCBs

see **Polychlorinated biphenyls**

PCP

see **Pentachlorophenol**

Peat soils

A **soil** that is derived completely from the decomposing remains of plants. Plants that commonly form peat include reeds, sedges, sphagnum moss, and grasses. The plant remains do not decompose but continue to accumulate because the wet and/or cool **environment** in which they occur is not conducive to **aerobic decomposition**. Vegetable crops are often grown on peat soils. Peat can also be harvested for

use in horticulture or as a fuel for heating and cooking. *See also* Peatlands

Peatlands

Expansive areas of **peat soils** are referred to as peatlands. These areas are often located in what were once lakes or oceans. The clay deposits from the former lake provide an impermeable layer so that water accumulates. Plants growing in this wet **environment** will not be able to decompose because of a lack of oxygen. Accumulations of plants will continue to increase the thickness of the peat deposit until a **soil** formed entirely of peat is created. These deposits can be 40 feet (12 m) or more thick. Extensive peatlands occur in Minnesota, Wisconsin, Michigan, New England, Russia, England, and Scandinavian countries.

Pedology

The scientific study of **soil** as a natural body, without emphasizing the practical uses or ecological significance of the soil. Pedology includes soil formation and morphology, its basic physical and chemical properties, its distribution, and the erosional processes affecting it without direct regard to plants or animals. Pedology provides the foundation for the study of soil as a medium of biological activity, called **edaphology**, and for the study of soil necessary in construction and **drainage** projects.

Pelagic zone

The entire water column in marine ecosystems, regardless of depth. Plants and algae in the pelagic zone directly or indirectly support most of the ocean's animal life. Because plant growth is confined to the shallow depths of the pelagic zone reached by light (epipelagic zone, 0-650 ft/0-200 m), much of the biological activity in the pelagic zone is concentrated in near-surface waters. The pelagic zone is further subdivided by depth into the mesopelagic zone (650-3,280 ft/200-1000 m), bathypelagic zone (3,280-13,125 ft/1000-4000 m) and abyssopelagic zone (greater than 13,125 ft/4000 m), each zone inhabited by a distinctive **fauna**. These deeper fauna are nourished largely by organic matter (e.g. dead **plankton**, fecal material) settling from surface waters. *See also* Littoral zone; Neritic zone; Photic zone; Phytoplankton; Zooplankton

Pelecanus occidentalis

see **Brown pelican**

Pelican

see **Brown pelican**

Pentachlorophenol

One of the most widely manufactured **chemicals** in the world, PCP is used extensively as a wood preservative for telephone poles, fences, and indoor or outdoor construction materials. It has also been used for slime control in the pulp and paper manufacturing process, for weed control, termite control, and as a paint preservative. Technical grade PCP contains trace amounts of less-chlorinated phenols and certain chlorinated dioxins and **furans**. Concerns over its toxicity have curtailed the use of PCP in the United States to treat materials that humans or animals will have contact with. Readily absorbed through skin and also volatile, it is acutely toxic from dermal exposure or inhalation.

People for the Ethical Treatment of Animals

Founded in 1980 by Alex Pacheco and Ingrid Newkirk, People for the Ethical Treatment of Animals (PETA) is a nonprofit charitable organization dedicated to protecting and promoting **animal rights**. These rights are not, of course, civil or religious, but rather, the right to live free from human-caused pain or predation. To this end, PETA has mounted campaigns against the use of animals in painful and frequently fatal medical experiments, as well as in the testing of cosmetics and other products. They have also campaigned against the **trapping** of fur-bearing animals, so-called "factory farming," such as confining cattle in crowded **feedlots**, and the cruel use of animals in rodeos, carnivals, and circuses. Through its newsletter, *PETA News*, and its "Factsheets," PETA publicizes these and other abuses and encourages readers to take action. Members write to the offending organizations, organize boycotts of their products or services, and contact their political representatives in support of proposed or pending legislation regarding the treatment of animals.

Drawing many of its ideas and much of its inspiration from philosophers such as **Tom Regan** and **Peter Singer**, PETA is attempting (in Singer's words) to "expand the circle" of creatures considered worthy of respect and protection. Although other forms of discrimination, such as racism and sexism, have been widely discredited (if not yet eliminated), another form of discrimination has barely begun to be recognized. **Speciesism** is the view that one particular **species**, *Homo sapiens*, is superior to all other species and that therefore humans have the unquestionable and unlim-

About 50 members of People for the Ethical Treatment of Animals (PETA) and several dogs protest against a furrier in Atlanta, Georgia. The protestors wore red shirts to symbolize the blood spilled to make fur garments. (Corbis-Bettmann. Reproduced by permission.)

ited right to use or to kill those other species for food, fur, leather, labor, or amusement. Such speciesist beliefs and attitudes are deeply rooted in our culture, Regan and Singer argue, and challenging them is often very difficult.

Animal rights advocates have been accused of putting the interests of animals above or at least on a par with those of human beings. Many critics of PETA quote, out of context, Ingrid Newkirk's statement that "a rat is a pig is a dog is a boy," thereby purporting to prove that PETA and other animal rights organizations equate the rights and interests of human and nonhuman animals. This statement of Newkirk's was actually intended to show that humans, like animals, are sentient beings because they have a central nervous system and are therefore able to feel pleasure and pain. "When it comes to having a central nervous system and the ability to feel pain, hunger, and thirst," said Newkirk, "a rat is a pig is a dog is a boy." More elaborate and extended versions of the argument about sentience can be found in Peter Singer's *Animal Liberation* (1990) and in Tom Regan's *The Case for Animal Rights* (1983). But PETA's primary aim is not to engage in philosophical disputation but to participate in political, economic, and educational campaigns on behalf of creatures who cannot defend themselves against humanly caused pain and predation.

[*Terence Ball*]

Resources

Organizations

PETA, 501 Front St., Norfolk, VA USA 23510 (757) 622-PETA, Fax: (757) 622-0457, Email: info@peta-online.org, <http://www.peta-online.org>

Peptides

A chemical compound consisting of two or more amino acids joined to each other through a bond between the **nitrogen** atom of one amino **acid** to an oxygen atom of its neighbor. A more precise term describes the number of amino acid units involved. A dipeptide or tripeptide consists

Peregrine falcons on ledge, downtown Detroit, Michigan. (Photograph by Karl W. Kenyon. National Audubon Society Collection. Photo Researchers Inc. Reproduced by permission.)

of two or three amino acid units respectively. A few oligopeptides (about ten amino acid units) are of physiological importance. The antibiotics bacitracin, gramicidin S, and tyrocidin A are examples of oligopeptides. The largest polypeptides contain dozens or hundreds of amino acid units and are better known as proteins. The bond between peptide units is especially sensitive to attack by various types of corrosive poisons such as strong acids and bases.

Percina tanasi

see **Snail darter**

Percolation

The movement of water through **soil** and the unsaturated zone into and through the pores of materials in the **zone of saturation**. **Groundwater** is recharged by this movement of water through the unsaturated zone. From a soil science perspective, percolation refers to the **drainage** of initially wetted areas of soil and movement of water beyond the rooting zone of plants toward the **water table**. Sanitarians commonly use percolation, however, in reference to results of the common soil test, known as the Perk test, which evaluates the rate at which soils accept water. The rate water moves into the soil is referred to as percolation. *See also* Aquifer; Aquifer restoration; Drinking water supply; Recharge zone; Vadose zone; Water table draw-down

Peregrine falcon

The peregrine falcon (*Falco peregrinus*), a bird of prey in the family Falconidae, is one of the most wide-ranging birds in the world with populations in both the Eastern and Western Hemispheres. However, with extensive **pesticide** use, particularly DDT, beginning in the 1940s, many populations of these birds were decimated. In the United States by the 1960s, the peregrine falcon was completely extirpated from the eastern half of the country because DDT and related compounds, which are amplified in the **food chain/web**,

caused the birds' eggshells to become thin and fragile. This led to reproductive failures, as eggs were crushed in the nest during incubation. Prior to the DDT-induced losses, there were about 400 breeding pairs of peregrine falcons in the eastern United States. In the early 1970s there were over 300 active nests in the western states, but within a single decade that number dropped to 200. The numbers continued to decline, and in 1978 there were no breeding pairs of peregrine falcons in the eastern United States. By 1984, due to reintroduction efforts, there were 27 nesting pairs, and in 1985, 38 nesting pairs were present in the east with at least 16 pairs fledging young. Also in 1985, 260 young, captive-raised peregrine falcons were released into the wild, 125 in the eastern states, and 135 in the west. By 1986, 43 pairs were nesting, and 25 of those pairs fledged 53 young. By 1991, over 100 breeding pairs were found in the east, and 400 pairs were found in the west. The increase of peregrine falcons has brought the numbers up to 215 breeding pairs in the mid-west regions.

The recovery success of the peregrine falcon is due largely to the efforts of two groups, the Peregrine Fund based at Cornell University in Ithaca, New York, and the **Canadian Wildlife Service** at Camp Wainwright in Alberta. Much of their research centered on captive breeding for release in the wild and finding ways to induce the falcons to nest and raise young in their former range. With great patience, and limited early success, the projects paid off.

The restoration projects also yielded much valuable information as well as innovative approaches to reestablishing peregrine falcon populations. For captive breeding, falcons trapped as nestlings stood a much better chance of reproducing in captivity than when trapped as flying immatures or adults. Since **habitat** destruction and human encroachment limit potential nesting sites—which typically are cliff ledges—researchers found that a potential, and ultimately successful, alternative nest site was the window ledge of tall, city buildings. These locations mimic their natural nest sites, and these "duck hawks," as they were once called, had a readily available prey in their new urban **ecosystem**. Peregrine falcons immediately began killing rock doves for food, which some saw as a service to the cities, since these "pigeons" tended to be regarded as a "nuisance" **species**.

Since it began in the 1970s, this captive breeding and release program has become well established, with over 4,000 captive-bred peregrine falcons released over the past three decades. In August of 1999, the peregrine falcon was removed from the **Endangered Species** list by the U.S. **Fish and Wildlife Service**.

[*Eugene C. Beckham*]

Resources

Books

Ehrlich, P., D. Dobkin, and D. Wheye. *Birds in Jeopardy.* Stanford, CA: Stanford University Press, 1992.

Ratcliffe, D. A. *The Peregrine Falcon.* Vermillion, SD: Buteo Books, 1980.

Other

The Peregrine Fund. [cited May 2002]. <http://peregrinefund.org>.

The Raptor Center. [cited May 2002]. <http://www.raptor.cvm.umn.edu>.

Perfluorooctane sulfonate

Fluoro-organic **chemicals** such as perfluoro-octane sulfonate ($C_8F_{17}SO_3H$)(PFOS), have been used since the 1950s in soil- and stain-resistant coatings for fabrics, carpets, and leather (2.4 million lb [1 millionkg] in 2000), as well as in oil- and grease-resistant coatings for paper products (about 2.7 million lb [1.2 million kg]in 2000). Though PFOS is mostly applied to products in textile mills, leather tanneries, finishers, and carpet manufacturing facilities, the use of Scotchgard fabric protector also allows for the use of PFOS by the consumer. Additional uses of PFOS and related compounds in specialized industrial applications (1.5 millions lb [680,000 kg] in 2000) include fire-fighting foams, mining and oil well surfactants, **acid** mist suppressants for metal plating and electronic etching baths, alkaline cleaners, floor polishes, photographic film, denture cleaners, and shampoos, and ant insecti cide.

In May 2000, 3M Corporation announced that, by 2002, it would phase out a group of perfluorinated chemicals, including PFOS, used in Scotchgard and other products. 3M Corporation is the only manufacturer of PFOS in the United States. All of the fluorochemicals that are being phased out either use PFOS in their manufacture or breakdown to PFOS, which is persistent in the **environment** due to the high energy of the carbon-fluorine bond. 3M Corporation had conducted a major research effort to characterize the environmental presence, environmental and human effects, and environmental fate of PFOS; the results of those research efforts indicated that continued use of PFOS could potentially have severe long-term consequences to human health and the environment.

PFOS, unlike most other persistent organic compounds, does not accumulate in fats, but as a surfactant with both lipophilic and hydrophobic tendencies, binds to blood proteins and accumulates in the liver and gall bladder. In a study published in 2001, PFOS was shown to be present in each of 65 human blood samples from persons who had not been industrially exposed to PFOS. In another 2001 study, 247 tissue samples from marine mammals collected from

Florida, California, and Alaskan coastal waters, as well as the northern Baltic Sea, the Arctic, and Sable Island in Canada were analyzed for PFOS. PFOS was found in the blood and livers of nearly all of the marine mammals. Although animals from sites nearer to developed and industrialized regions had higher levels of PFOS in their tissues, animals from even the most remote locations, including the Arctic Ocean, were found to have contain significant levels of PFOS. Near-shore animals had higher concentrations those that lived off shore. Further, fresh-water **species** had higher concentrations than ocean species. PFOS was also found in samples of 21 species of fish-eating birds, including eagles, ospreys, albatrosses, gulls, herons, loons, ibises, and gannets that had been collected from locations in the United States and the central Pacific Ocean. Birds in more urbanized areas had higher concentrations. Birds from the central Pacific also had measurable PFOS levels, indicating the presence of PFOS in remote marine locations. The lowest concentrations of PFOS were found in an insect-eating gull. Contamination levels observed in these studies were lower than those known to be toxic in **wildlife**; however, additional studies are required to determine the long-term effects of low-level exposure to PFOS.

Toxicological studies have shown that PFOS can result in adverse developmental, reproductive, and **systemic** effects. In a study of two generations of rats and the effects of PFOS on reproduction, PFOS was found to cause postnatal death as well as have detrimental developmental effects. At the highest doses (3.2 mg/kg/day), the offspring in the first generation of rats died, while at one-half of the highest dose (1.6 mg/kg/day), almost one-third of the offspring in the first generation died. At a lower dose (0.4 mg/kg/day), PFOS resulted in lower pup weight compared to controls. Reversible delays in reflex and physical development were also observed in this study.(

In Rhesus monkey studies, no monkeys survived more than three weeks at 10 mg/kg/day or more than seven weeks at doses of 4.5 mg/kg/day. Cynomolgus monkeys also died at a dose of 0.75 mg.kg.day, after first becoming listless and exhibiting loss of appetite. Both species of monkeys showed liver enlargement and reduced cholesterol levels in the blood.

From the results of the toxicological studies and the survey of animals and human blood samples throughout the world, a representative of the U.S. **Environmental Protection Agency** (EPA) stated that PFOS exhibits "persistence, **bioaccumulation**, and toxicity to an extraordinary degree". Research is being conducted to develop safer substitutes for PFOS.

[*Judith L. Sims*]

RESOURCES
PERIODICALS

Giesy, J. P., and K. Kannan.*Global Distribution of Perfluorooctane Sulfonate in Wildlife.*Environmental Science and Technology 35(2001): 1339–1342.

Kannan, Kurunthachalam, et al. "Accumulation of Perfluorooctane Sulfonate in Marine Mammals."*Environmental Science and Technology*35 (2001): 1593–1598.

Kannan, Kurunthachalam, et al. "Perfluorooctane Sulfonate in Fish-eating Water Birds Including Bald Eagles and Albatrosses." *Environmental Science and Technology*35 (2001): 3065–3070.

Renner, Rebecca. "Growing Concern Over Perfluorinated Chemicals." *Environmental Science and Technology* 35 (2001): 154A–160A.

Permaculture

Permaculture is an approach to land management that creates high-yielding, low-energy, self-perpetuating systems by which the functions of animals, plants, humans, and Earth are integrated to maximize their value and create sustainable human habitats. Permaculture brings together disciplines relating to food, shelter, energy, water, **waste management**, economics, and social sciences. It aims to maximize a site's productivity, while maintaining ecosystems and restoring damaged land to a healthy, life-promoting state. Bill Mollison has written five books on this topic: *Introduction to Permaculture, Permaculture: A Designer's Manual*; *Permaculture One*; *Permaculture Two;* and *Permaculture: A Practical Guide for a Sustainable Future.* Several permaculture organizations and model projects exist around the world.

The term was first coined in 1972 by Bill Mollison of Tasmania, **Australia**, by merging the terms, "permanent" and "agriculture." Although originally developed for small subsistence farms, the practice has expanded to apply to gardens and urban settings. Some consider it a lifestyle as much as a design approach.

Permaculture principles focus on designs for small-scale intensive systems that are labor-efficient and use biological resources instead of **fossil fuels**. These designs stress ecological connections and closed energy and material loops. The core of permaculture is integrating working relationships and connections between all things. Each component in a system performs multiple functions, and each function is supported by many elements. Key to efficient permaculture design is observing and replicating natural ecosystems. Designers maximize diversity with polycultures, stress efficient energy planning for houses and settlements, and use and accelerate natural plant **succession**. The philosophy behind permaculture is one of working with rather than against **nature**, looking at systems in all of their functions, and using systems for multiple purposes. It is a method of agriculture that aims to endure without constant human inputs and does not deplete the land.

According to Mollison, permaculture has several distinct characteristics. It makes the most of small landscapes, using intensive practices. Ideally, nothing is wasted, and everything is arranged so that the least amount of effort is exerted and the highest yield from the systems is gained. Systems are designed that use and complement the natural systems that are present. For example, storm water is controlled with planted swales (marshy depressions), not concrete drains. The design harvests the natural flows of energy through the landscape (such as sunlight, rain, and plant and animal behaviors). Diversity is promoted in plant **species**, varieties, yield, **microclimate**, and **habitat**. Permaculture maintains that, in a **monoculture**, a single species cannot make full use of all of the available energy and nutrients. Wild or seldom-selected animal and plant species are used. Each element performs many functions in the system. For example, a fruit tree provides not only a crop, but also wind shelter, a trellis, **soil** conditioning, and shade and roosting for birds. The long-term evolution of the land is recognized, and those changes are incorporated in planning. All activities involving the land—agriculture, animal husbandry, extant **forest management**, animal cropping, and landform engineering—are integrated. Difficult landscapes (rocky, marshy, marginal, or steep) not typically suited to other systems are utilized. Permaculture involves long-term and evolving land-use planning, using diverse **flora** and **fauna** in various ways at different times, recognizing that different species use different nutrients and resources.

Permaculture systems typically feature: passive energy systems and minimal external energy needs; on site **climate** control; planned future developments; on-site provision for food self-sufficiency; safe on-site disposal of wastes; low-maintenance structures and grounds; assured and conserved water supply; and control and direction of fire, cold, excess heat, and wind factors.

Permaculture originated as a strategy for designing systems for "permanent" or perennial agriculture, by creating **agroforestry** systems using tree crops, shrubs, vines and, herbaceous plants in highly productive symbiotic assemblages. The practice was originally oriented to subsistence farms in Tasmania, which typically were small and on poor land. It was then extended to include other landscapes, urban settings, and climates worldwide, and has even been applied to other systems such as houses and factories.

Permaculture as a Life Philosophy

Some consider permaculture a life philosophy. In this context, permaculture emphasizes putting oneself into a symbiotic relationship with Earth and one's community. Permaculture is oriented to place, with reliance on native plants and a close awareness of the **ecosystem**. It revolves around self-reliance, growing food and building attractive energy-efficient structures from local materials. Designing a permaculture landscape begins with assessing a site's native features (such as soil structure, microclimates, cycles of decay, and existing flora and fauna), in order to take advantage of existing resources. This assessment also helps in selecting and adapting technologies to the site. These technologies might include methods of **composting**, gardening, irrigating, or generating electricity. Multifunctional living systems abound in a typical permaculture design.

Examples of Permaculture Practices and Approaches

Animals are raised for their value as producers of food, skins, and manure, and as pollinators, heat sources, gas producers, earth tillers, and **pest** controller. For example, rabbits raised in a rabbit hutch are fed kitchen scraps. Their droppings fall into worm bins (vermiculture) as fodder for worms, and the resulting worm castings are used to fertilize gardens. In the winter, the rabbits are harvested for their meat and fur. As another example, movable chicken hutch can be used to place chickens in gardens where they effectively till and fertilize the soil.

Particular varieties of trees are chosen for their fuel, forage, material (for fences, structures and shelters), heat reflector and windbreak values, as well as for crop diversification. When they are young, the trees may act as hedgerows, then grow to serve as a fuel source. Plants provide many services for other plants. They act as trellises, for other screen and shade them, provide nutrients, cross-fertilize them (as happens with varieties of plums and nuts), help to repel pests, prevent **erosion**, and provide spare parts (grafts) for other plants. Fruit trees and vines are planted strategically around a house to provide shade. In a home's front yard, attractive gardens feature high-yielding food, medicinal and culinary plants where they are easily accessed. Some crops are planted for their self-propagating patterns, such as leeks, onions, potatoes, and garlic.

Water harvesting is an essential function in a permaculture landscape, and is supported by as many components as possible. **Filtration** of water for animals may be provided with shells and water plants. Water pathways are traced, and systems are created for collecting it. Swales may direct rain water and run-off to fruit trees. Run-off water from culverts may be directed to ponds where water plants and fish are raised, and where it can be used for **irrigation** and firefighting. Water can be collected from roofs. Composting **toilets** and septic systems with planted **leaching** beds can be employed.

Insects and crops are used to aerate the soil. Mulching, growing green crops, composting, and strategic planting are employed to build soil health. Permaculture grows forests and shrubs to protect the soil, and uses plows that do not turn the soil. Food crops, such as corn and legumes, are chosen for their low-maintenance qualities and their ability to fix **nitrogen** in the soil.

Permaculture emphasizes reactive homes, sheltered from cold winds with windbreak planting. Such homes are oriented on an east-west axis facing the sun, usually with a greenhouse, and are well-sealed. They should use few resources not found on site. Shelters can be built into the earth, with living turf roofs. Sometimes, living trees and plants are used to create shelters.

Permaculture offers an **environmental design** practice for making better use of resources in a variety of growing settings. What started as a method for cultivating **desert** land has grown into a system that integrates living systems, fostering greater consciousness of ecosystems and helping to ensure economic and ecological sustainability for its practitioners. Contact: Permaculture Resources, 56 Farmersville Rd., Califon, N.J. 07830; phone: 800-832-6285.

[*Carol Steinfeld*]

Resources

Books

Barnes, L. "The Permaculture Connections." In *Southeastern Permaculture Network News.*

Mollison, B. *Permaculture: A Designer's Manual, Permaculture One* and *Permaculture Two.* Australia: Tagari, 1979.

Permafrost

Permafrost is any ground, either of rock or **soil**, which is perennially frozen. Continuous permafrost refers to areas which have a continuous layer of permafrost. Discontinuous permafrost occurs in patches. It is believed that continuous permafrost covers approximately 4% of the earth's surface and can be as deep as 3,281 ft (1,000 m), though normally it is much less. Permafrost tends to occur when the mean annual air temperature is less than the freezing point of water. Permafrost regions are characterized by a seasonal thawing and freezing of a surface layer known as the active layer which is typically 3–10 ft (1–3 m) thick.

Permanent retrievable storage

Permanent retrievable storage is a method for handling highly toxic hazardous wastes on a long-term basis. At one time, it was widely believed that the best way of dealing with such wastes was to seal them in containers and either bury them underground or dump them into the oceans. However, these containers tended to leak, releasing these highly dangerous materials into the **environment**.

The current method is to store such wastes in a quasi-permanent manner in salt domes, rock caverns, or secure buildings. This is done in the expectation that scientists will eventually find effective and efficient methods for converting these wastes into less hazardous states, in which they can then be disposed of by conventional means. One chemical for which permanent retrievable storage has been used so far is the group of compounds known as polychlorinated biphenyl (PCB)s.

Permanent retrievable storage has its disadvantages. Hazardous wastes so stored must be continuously guarded and monitored in order to detect breaks in containers or leakage into the surrounding environment. In comparison with other disposal methods now available for highly toxic materials, however, permanent retrievable storage is still the preferred alternative means of disposal. *See also* Hazardous waste site remediation; Hazardous waste siting; NIMBY; Toxic use reduction legislation

[*David E. Newton*]

Resources

Books

Makhijani, A. *High-Level Dollars, Low-Level Sense: A Critique of Present Policy for the Management of Long-Lived Radioactive Wastes and Discussion of an Alternative Approach.* New York: Apex Press, 1992.

Schumacher, A. *A Guide to Hazardous Materials Management.* New York: Quorum Books, 1988.

Periodicals

Flynn, J., et al. "Time to Rethink Nuclear Waste Storage." *Issues in Science and Technology* 8 (Summer 1992): 42–46.

Kliewer, G. "The 10,000-Year Warning." *The Futurist* 26 (September-October 1992): 17–19.

Permeable

In **soil** science, permeable is a qualitative description for the ease with which water or some other fluid can pass through soil. Permeability in this context is a function not only of the total pore volume but also of the size and distribution of the pores. In geology, particularly with reference to **groundwater**, permeability is the ability of water to move through any water-bearing formation, rock, or unconsolidated material. This condition can be measured in the laboratory by measuring the volume of water that flows through a sample over a certain period of time. *See also* Aquifer; Recharge zone; Soil profile; Vadose zone; Zone of saturation

Peroxyacetyl nitrate (PAN)

The best known member of a class of photochemical oxidizing agents known as the peroxyacyl **nitrates**. The peroxyacyl nitrates are formed when **ozone** reacts with **hydrocarbons**

such as those found in unburned **petroleum**. They are commonly found in **photochemical smog**. The peroxyacyl nitrates attack plants, causing spotting and discoloration of leaves, destruction of flowers, reduction in fruit production and seed formation, and death of the plant. They also cause red, itchy, runny eyes and irritated throats in humans. Cardiac and respiratory conditions, such as **emphysema** and chronic **bronchitis**, may result from long-term exposure to the peroxyacyl nitrates. *See also* Air pollution; Los Angeles Basin

Persian Gulf War

The Persian Gulf War in 1991 had a variety of environmental consequences for the Middle East. The most devastating of these effects were from the **oil spills** and oil fires deliberately committed by the Iraqi army. There were extensive press coverage of these events at the time, and the United States accused the Iraqis of "environmental terrorism." These accusations were seen by some as propaganda effort, and there was almost certainly some political motivation to both how the damage was estimated and how it was characterized during the war. But it is clear now that the note of outrage often struck by the Allies was not out of place. The devastation, though not as extensive as originally supposed, was still substantial.

The Iraqis began discharging oil into the Persian Gulf from the Sea Island Terminal and other supertanker terminals off the coast of Kuwait on January 23, 1991. Allied bombers tried to limit the damage by striking at pipelines carrying oil to these locations, but the flow continued throughout the war. Estimates of the size of the spill have varied widely and the controversy still continues, with a number of diplomatic and political pressures preventing many government agencies from committing themselves to specific figures. But it now seems likely that this was the worst oil spill in history, probably 20 times larger than the ***Exxon Valdez*** spill in **Prince William Sound**, Alaska, and twice the size of the 1979 spill from the blowout of Ixtoc I well in the Gulf of Mexico.

Whatever the actual size of the spill, it occurred in an area that was already one of the most polluted in the world. Oil spills and oil dumping are common in the Persian Gulf; it has been estimated that as many as two million barrels of oil are spilled in these waters every year. Some ecologists believe that the **ecosystem** in the area has a certain amount of **resistance** to the effects of **pollution**. Other scientists have maintained that the high level of **salinity** in the Gulf will prevent the oil from having many long-term effects and that the warm water will increase the speed at which the oil

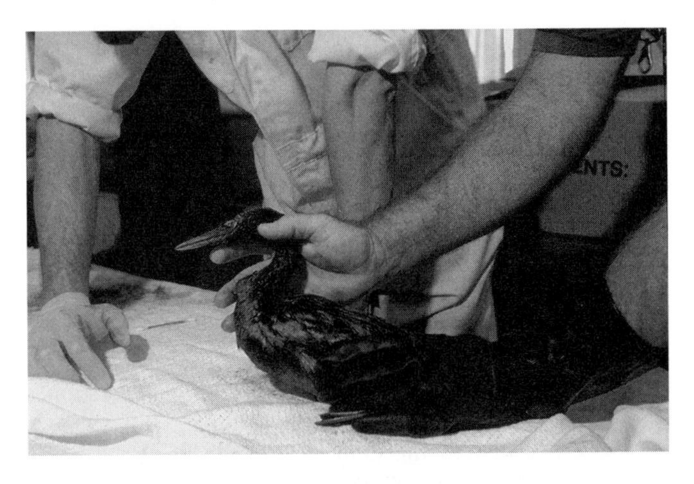

Two men cleaning a bird soaked in oil that was spilled into the Persian Gulf by Iraq during the Persian Gulf War. (Photograph by Wesley Bocxe, National Audubon Society Collection. Photo Researchers Inc. Reproduced by permission.)

degrades. But the Gulf has a slow circulation system and large areas are very shallow; many scientists and environmentalists have predicted that it will be many years before the water can clear itself.

The spill killed thousands of birds within months after it began, and it had an immediate and drastic effect on **commercial fishing** in the region. Oil soaked miles of coastline; coral reefs and **wetlands** were damaged, and the seagrass beds of the Gulf were considered particularly vulnerable. Mangrove swamps, migrant birds, and **endangered species** such as green turtles and the dugong, or sea cow, are still threatened by the effects of the spill. The Saudi Arabian government has protected the water they draw from the Gulf for **desalinization**, but little has been done to limit or alleviate the environmental damage. This has been the result, at least in part, of a shortage of resources during and after the war, as well as obstacles such as floating mines and shallow waters which restricted access for boats carrying cleanup equipment.

At the end of the war, the retreating Iraqi army set over 600 Kuwaiti oil **wells** on fire. When the last burning well was extinguished on November 6, 1991, these fires had been spewing oil **smoke** into the **atmosphere** for months, creating a cloud which spread over the countries around the Gulf and into parts of Asia. It was thought at the time that the cloud of oil smoke would rise high enough to cause global climatic changes. Carl Sagan and other scientists, who had first proposed the possibility of a **nuclear winter** as one of the consequences of a nuclear war, believed that rain patterns in Asia and parts of Europe would be affected by the oil fires, and they predicted failed harvests and widespread starvation as a

result. Though there was some localized cooling in the Middle East, these kinds of global predictions did not occur, but the smoke from the fire has still been an environmental disaster for the region. **Air quality** levels have caused extensive health problems, and **acid rain** and **acid deposition** have damaged millions of acres of forests in Iran.

The oil fires and oil spills were not the only environmental consequences of the Persian Gulf War. The movement of troops and military machinery, especially tanks, damaged the fragile **desert** soils and increased wind **erosion**. Wells sabotaged by the Iraqis released large amounts of oil that was never ignited, and lakes of oil as large as half a mile wide formed in the desert. These lakes continue to pose a hazard for animals and birds, and tests have shown that the oil is seeping deeper into the ground, causing long-term contamination and perhaps, eventually, **leaching** into the Gulf. *See also* Gulf War syndrome

[*Douglas Smith*]

Resources

Periodicals

Peck, L. "The Spoils of War." *The Amicus Journal* 3 (Spring 1991): 6–9.

Zimmer, C. "Ecowar." *Discover* 13 (January 1992): 37–39.

Persistent compound

A persistent compound is slow to degrade in the **environment**, which often results in its accumulation and deleterious effects on human and **environmental health** if the compound is toxic. Toxic metals such as **lead** and **cadmium**, organochlorides such as **polychlorinated biphenyls** (PCBs), and **polycyclic aromatic hydrocarbons** (PAHs) are persistent compounds.

Persistent molecules are termed recalcitrant if they fail to degrade, metabolize, or mineralize at significant rates. Their compounds can be transported through the environment over long periods of time and over long distances, resulting in long-term exposure and possible changes in organisms and ecosystems. However, organisms and ecosystems may adapt to the compounds, and deleterious effects may weaken or even disappear.

Compounds may be persistent for several reasons. A compound can be persistent due to its chemical structure. For example, in a molecule, the number and arrangement of **chlorine** ions or hydroxyl groups can make a compound recalcitrant. It can also persist due to unfavorable environmental conditions such as **pH**, temperature, ionic strength, potential for **oxidation reduction reactions**, unavailability of nutrients, and absence of organisms that can degrade the compound.

The period of persistence can be expressed as the time required for half of the compound to be lost, the **half-life**. It is also often expressed as the time for detectable levels of the compound to disappear entirely. Compounds are classified for environmental persistence in the following categories:(1) Not degradable, compound half-life of several centuries; (2) Strong persistence, compound half-life of several years; (3) Medium persistence, compound half-life of several months; and (4) Low persistence, compound half-life less than several months.

Non-degradable compounds include metals and many radioisotopes, while semi-degradable compounds (of medium to strong persistence) include PAHs and chlorinated compounds. Compounds with low persistence include most organic compounds based on **nitrogen**, sulfur, and **phosphorus**.

In general, the greater the persistence of a compound, the more it will accumulate in the environment and the **food chain/web**. Non-degradable and strongly persistent compounds will accumulate in the environment and/or organisms. For example, because of **bioaccumulation** through the food chain, dieldrin can reduce populations of birds of prey. Compounds with intermediate persistence may or may not accumulate, while non-persistent pollutants generally do not accumulate. However, even compounds with low persistence can have long-term deleterious effects on the environment. **2,4-D** and **2,4,5-T** can cause **defoliation**, which may result in **soil erosion** and long-term effects on the **ecosystem**.

Chemical properties of a compound can be used to assess persistence in the environment. Important properties include: (1) rate of biodegradation (both **aerobic** and **anaerobic**); (2) rate of hydrolysis; (3) rate of oxidation or reduction; and (4) rate of photolysis in air, soil, and water. In addition, the effects of key parameters, such as temperature, concentration, and pH, on the rate constants and the identity and persistence of transformation products should also be investigated. However, measured values for these properties for many persistent compounds are not available, since there are thousands of **chemicals**, and the time and resources required to measure the desired properties for all the chemicals is unrealistic. In addition, the data that are available are often of variable quality.

Persistence of a compound in the environment is dependent on many interacting environmental and compound-specific factors, which makes understanding the causes of persistence of a specific compound complex and in many

cases incomplete. *See also* Chronic effects; Heavy metals and heavy metal poisoning; Toxic substance

[*Judith Sims*]

Resources

Books

Ayres, R. U., F. C. McMichael, and S. R. Rod. "Measuring Toxic Chemicals in the Environment: A Materials Balance Approach." In *Toxic Chemicals, Health, and the Environment*, edited by L. B. Lave and A. C. Upton. Baltimore, MD: Johns Hopkins University Press, 1987.

Govers, H., J. H. F. Hegeman, and H. Aiking. "Long-Term Environmental and Health Effects of PMPs." In *Persistent Pollutants: Economics and Policy*, edited by H. Opschoor and D. Pearce. Dordrecht, Netherlands: Kluwer Academic Publishers, 1991.

Lyman, W. J. "Estimation of Physical Properties." In *Environmental Exposure from Chemicals.* Vol. 1. Edited by W. B. Neely and G. E. Blau. Boca Raton: CRC Press, 1985.

Persistent organic pollutants

Persistent organic pollutants (POPs) are man-made organic compounds that persist in the natural **environment** for long periods of time. Because of their long-lasting presence in air, water, and **soil**, they accumulate in the bodies of fish, animals, and humans over time. Exposure to POPs can create serious health disorders throughout the tiers of the food web. In human beings, POPs can cause **cancer**, autoimmune deficiencies, kidney disorders, **birth defects**, and other reproductive problems.

Because these **chemicals** are derived from manufacturing industries, **pesticide** applications, waste disposal sites, spills, and **combustion** processes, POPs are a global problem. Many POPs are carried long distances through the **atmosphere**. They tend to move from warmer climates to colder ones, which is why even remote regions such as the Arctic contain significant levels of these contaminants. Because of the global creation and transmission of POPs, no country can protect itself against POPs without assistance. An international commitment is essential for eradication of this problem.

In the early 1990s the Organization for Economic Cooperation and Development (OECD), **United Nations Environment Programme** (UNEP), World Health Organization (WHO), and other groups began to assess the impacts of hundreds of chemicals, including POPs. In 1998 36 countries participated in the POPs Protocol, sponsored by the **Convention on Long-range Transboundary Air Pollution**. The purpose was to build an international effort toward controlling POPs in the environment.

In 2001, led by the UNEP, over 100 countries participated in the Stockholm Convention on Persistent Organic Pollutants. The proposal will become a legally binding agreement upon its ratification by 50 or more countries. It is considered by many to be a landmark human-health document of international proportions. Research will address the most effective ways to reduce or eliminate POP production, various import/export issues, disposal procedures, and the development of safe, effective, alternative chemical compounds.

The screening criteria for POP designation are: the potential for long-range atmospheric transport, persistence in the environment, bioaccumlation in the tissues of living organisms, and toxicity. Many chemicals volatilize, which increases the concentrations of these chemicals in the air. Longevity is measured by a chemical's **half-life**, or how much time passes before half of the original amount of a chemical **discharge** or **emission** breaks down naturally and dissipates from the environment. The minimum half-life for a POP in water is two months; for soil or **sediment** it is about six months. Several POPs have half-lives as long as 12 years. Animals accumulate POPs in their fatty tissue (**bioaccumulation**); these POPs are then consumed and reconcentrated by higher-order animals in the food chain (**biomagnification**). In some **species**, biomagnification can result in concentrations up to one million times greater than the background value of the POP. Most humans are exposed through consumption of food products (especially meat, fish, and dairy products) that contain small amounts of these chemicals.

The Stockholm Convention calls for the immediate ban on production and use of 12 POPs (known as the "dirty dozen"): aldrin, dieldrin, endrin, DDT, **chlordane**, heptachlor, **mirex**, **toxaphene**, hexachlorobenzene (HCB), **polychlorinated biphenyls** (PCBs), polychlorinated dioxins, and **furans**. The convention did acknowledge, however, certain health-related use exemptions until effective, environmentally friendly substitutes can be found. For example, DDT can still be applied to control malarial mosquitoes subject to WHO guidelines. Electrical transformers that contain PCBs can also be used until 2025, at which point any old transformers that are still active must be replaced with PCB-free equipment.

A brief description of the 12 banned POPs is shown below:

- Aldrin was a commonly used pesticide for the control of termites, corn rootworm, grasshoppers, and other insects. It has been proven to cause serious health problems in birds, fish, and humans. Aldrin biodegrades in the natural environment to form dieldrin, another POP
- Dieldrin was applied extensively to control insects, especially termites. Its half-life in soil is about five years. Dieldrin is especially toxic to birds and fish.
- Endrin, another insecticide, is used to control insects and certain rodent populations. It can be metabolized in

animals, however, which reduces the risk of bioaccumulation. It has a half-life of about 12 years.

- DDT, also known as dichlorodiphenyl trichloroethane, was applied during World War II to protect soldiers against insect-borne diseases. In the 1960s and 1970s its use on crops resulted in dramatic decreases in bird populations, including the **bald eagle**. DDT is still being manufactured today and is a chemical intermediate in the manufacture of dicofol. Regions that want to continue using DDT for public-health purposes include Africa, China, and India (mosquitoes, however, are becoming resistant to DDT in these areas).
- Chlordane is a broad-spectrum insecticide used in termite control. Its half-life is about one year. Chlordane is easily transported through the air and suspected of causing immune system problems in humans.
- Heptachlor was applied to control **fire ants**, termites, and mosquitoes and has been used in closed industrial electrical junction boxes. Its production has been eliminated in most countries.
- Mirex was used as an insecticide to control fire ants and termites. It is also a fire-retardant component in **rubber**, **plastics**, and electrical goods. It is a very stable POP with a half-life of up to 10 years.
- Toxaphene is an insecticide that was widely used in the United States in the 1970s to protect cereal grains, cotton crops, and vegetables. Toxaphene has a half-life of 12 years. Aquatic life is especially vulnerable to toxaphene toxicity.
- Hexachlorobenzene, also called HCB, was introduced in the 1940s to treat seeds and kill **fungi** that damaged crops. It can also be produced as a by-product of chemical manufacturing. It is suspected of causing reproductive problems in humans.
- Polychlorinated biphenyls, commonly referred to as PCBs, were used extensively in the electrical industry as a heat exchange fluid in transformers and capacitors. They were also used as an additive in plastics and paints. PCBs are no longer produced but are still in use in many existing electrical systems. Thirteen varieties of PCBs create dioxin-like toxicity. Studies have shown PCBs are responsible for immune system suppression, developmental neurotoxicity, and reproductive problems.
- Polychlorinated dibenzo-p-dioxins, usually shortened to the term "dioxins," are chemical by-products of incomplete combustion or chemical manufacturing. Common sources of dioxins are municipal and **medical waste** incinerators, backyard burning of trash, and past emissions from elemental **chlorine** bleach pulp and paper manufacturing. A dioxin's half-life is typically 10–12 years. Related health problems can include birth defects, reproductive problems, immune and **enzyme** disorders, and increased cancer risk.
- Polychlorinated dibenzofurans, usually shortened to "furans" or "PCDFs," are structurally similar to dioxins and by-products of the same processes as for dioxins. The health impacts of dibenzofuran toxicity are considered to be similar to those of dioxins. Over 135 types of PCDFs are known to exist.

The UNEP and other world organizations are continuing to look for safer and more economically viable alternatives to POPs. Manufacturing facilities are being upgraded with cleaner technologies that will reduce or eliminate emissions. Focus is also being placed on regulating the international trade of hazardous substances. Disposal is also an issue; poorer countries don't have the money or proper technological resources to dispose of the growing accumulations of obsolete toxic chemicals. The UNEP is working creatively with countries to secure financing to introduce alternative products, technology, methods enforcement, and critical infrastructure. Research sponsored by UNEP studies the chemical characteristics of current POPs so that chemical companies will be less likely to create new POPs.

The Convention on Long-range Transboundary **Air Pollution** has been extended by eight protocols, including the 1998 protocol on POPs. The third meeting of an ad-hoc group of experts on POPs was held in 2002. These scientists continue to research other chemicals that may qualify in the future as POPs.

Current technical review activities focus on the following chemicals:

- Ugilec, which was used in capacitors, transformers, and as a hydraulic fluid in underground mining. It was originally thought to be a safe replacement for PCBs.
- Hexachlorobutadiene (HCBD) was used to recover chlorine-bearing gas from chlorine plants. It is very toxic to aquatic life. HCBD is on national priority lists for some countries, such as Canada.
- Pentabromodiphenyl ether (PentaBDE) is sold as a brominated flame retardant. Its use has doubled in the last 10–15 years. This compound volatilizes easily and enters the air long after the disposal of treated products.
- Pentachlorobenzene (PeCB) was used as a **fungicide**, flame retardant, and as a component in dielectric fluids. Although it is no longer being produced, there is still an abundance of PeCB in the environment. It is probably released through **landfill** leaching/degassing or incineration.
- Polychlorinated naphthalenes (PCNs) are structurally similar to PCBs. In the 1980s PCNs were a component of products called Halowax, Nibren Wax, and Seekay Wax. Uses included wood preservation, electroplating, dye carriers, and wire insulation. Major points of release are old electrical equipment or incineration.

- Short-chain chlorinated paraffins (SCCP) are still being used today in the rubber, leather, and metal-working industries, especially in China. No suitable substitutes have been developed yet.
- Polychlorinated terphenyls (PCTs) are very similar in structure and chemical behavior to PCBs. Over 8,000 different varieties of PCTs theoretically exist. PCTs have not been commercially produced since the 1980s, but are still part of aging electrical capacitors and transformers.

[*Mark J. Crawford*]

RESOURCES

OTHER

Draft Summaries for Discussion by the Expert Group on POPs, Protocol on POPs. Summary documents. United Nations Economic Commission for Europe, Environmental and Human Settlements Division, June 2002.

The Foundation for Global Action on Persistent Organic Pollutants: A United States Perspective. External review draft. U.S. Environmental Protection Agency, November 2001.

New Protocol on Persistent Organic Pollutants Negotiated. News release. U.S. Environmental Protection Agency, Office of Pesticide Programs, 1998.

Persistent Organic Pollutants: A Global Issue, a Global Response. Brochure. Document EPA 160-F-02-001. U.S. Environmental Protection Agency, Office of Internal Affairs, 2002.

Stockholm Convention on POPs. Pamphlet. United Nations Environment Programme, June 2001

ORGANIZATIONS

United Nations Environment Programme, Chemicals Division, 11-13 chemin des Anemones, 1219 Chatelaine, Geneva, Switzerland (41) (22) 917-8191, Fax: (41) (22) 797-3460

U.S Environmental Protection Agency, 1200 Pennsylvania Avenue NW, Washington, DC USA 20460 (202)260-2090

Pest

A pest is any organism that humans consider destructive or unwanted. Whether or not an organism is considered a pest can vary with time, geographical location, and individual attitude. For example, some people like pigeons, while others regard them as pests. Some pests are merely an inconvenience. In the United States, mosquitoes are thought of as pests because they are an annoyance, not because they are dangerous. The most dangerous pests are those that carry disease or destroy crops. One direct way of controlling pests is by **poisoning** them with toxic **chemicals** (pesticides). A more environmentally sensitive approach is to find natural predators that can be used against them (biological controls).

See also Bacillus thuringiensis; Integrated pest management; Population biology

Pesticide

Pesticides are **chemicals** that are used to kill insects, weeds, and other organisms to protect humans, crops, and livestock. There have been many substantial benefits of the use of pesticides. The most important of these have been: (1) an increased production of food and fibre because of the protection of crop plants from pathogens, **competition** from weeds, **defoliation** by insects, and parasitism by nematodes; (2) the prevention of spoilage of harvested, stored foods; and (3) the prevention of debilitating illnesses and the saving of human lives by the control of certain diseases.

Unfortunately, the considerable benefits of the use of pesticides are partly offset by some serious environmental damages. There have been rare but spectacular incidents of toxicity to humans, as occurred in 1984 at **Bhopal, India**, where more than 2,800 people were killed and more than 20,000 seriously injured by a large **emission** of poisonous methyl isocyanate vapor, a chemical used in the production of an agricultural insecticide.

A more pervasive problem is the widespread environmental contamination by persistent pesticides, including the presence of chemical residues in **wildlife**, in well water, in produce, and even in humans. Ecological damages have included the **poisoning** of wildlife and the disruption of ecological processes such as productivity and **nutrient** cycling. Many of the worst cases of environmental damage were associated with the use of relatively persistent chemicals such as DDT. Most modern pesticide use involves less-persistent chemicals.

Pesticides can be classified according to their intended **pest** target:

- fungicides protect crop plants and animals from fungal pathogens;
- herbicides kill weedy plants, decreasing the competition for desired crop plants;
- insecticides kill insect defoliators and vectors of deadly human diseases such as **malaria**, yellow fever, **plague**, and typhus;
- acaricides kill mites, which are pests in agriculture, and ticks, which can carry encephalitis of humans and domestic animals;
- molluscicides destroy snails and slugs, which can be pests of agriculture or, in waterbodies, the vector of human diseases such as schistosomiasis;
- nematicides kill nematodes, which can be **parasites** of the roots of crop plants;
- rodenticides control rats, mice, gophers, and other rodent pests of human habitation and agriculture;
- avicides kill birds, which can depredate agricultural fields;

- antibiotics treat bacterial infections of humans and domestic animals.

The most important use-categories of pesticides are in human health, agriculture, and forestry:

Human Health

In various parts of the world, **species** of insects and ticks play a critical role as vectors in the transmission of disease-causing pathogens of humans. The most important of these diseases and their vectors are: (1) malaria, caused by the protozoan *Plasmodium* and spread to humans by an *Anopheles*-mosquito vector; (2) yellow fever and related viral diseases such as encephalitis, also spread by mosquitoes; (3) trypanosomiasis or sleeping sickness, caused by the protozoans *Trypanosoma* spp. and spread by the tsetse fly *Glossina* spp.; (4) plague or black death, caused by the bacterium *Pasteurella pestis* and transmitted to people by the flea *Xenopsylla cheops*, a parasite of rats; and (5) typhoid fever, caused by the bacterium *Rickettsia prowazeki* and transmitted to humans by the body louse *Pediculus humanus*.

The incidence of all of these diseases can be reduced by the judicious use of pesticides to control the abundance of their vectors. For example, there are many cases where the local abundance of mosquito vectors has been reduced by the application of insecticide to their aquatic breeding **habitat**, or by the application of a persistent insecticide to walls and ceilings of houses, which serve as a resting place for these insects. The use of insecticides to reduce the abundance of the mosquito vectors of malaria has been especially successful, although in many areas this disease is now re-emerging because of the **evolution** of tolerance by mosquitoes to insecticides.

Agriculture

Modern, technological agriculture employs pesticides for the control of weeds, arthropods, and plant diseases, all of which cause large losses of crops. In agriculture, arthropod pests are regarded as competitors with humans for a common food resource. Sometimes, defoliation can result in a total loss of the economically harvestable agricultural yield, as in the case of acute infestations of locusts. More commonly, defoliation causes a reduction in crop yields. In some cases, insects may cause only trivial damage in terms of the quantity of **biomass** that they consume, but by causing cosmetic damage they can greatly reduce the economic value of the crop. For example, codling moth (*Carpocapsa pomonella*) larvae do not consume much of the apple that they infest, but they cause great esthetic damage by their presence and can render produce unsalable.

In agriculture, weeds are considered to be any plants that interfere with the productivity of crop plants by competing for light, water, and nutrients. To reduce the effects of weeds on agricultural productivity, fields may be sprayed with a **herbicide** that is toxic to the weeds but not to the crop plant. Because there are several herbicides that are toxic to dicotyledonous weeds but not to members of the grass family, fields of maize, wheat, barley, rice, and other grass-crops are often treated with those herbicides to reduce weed populations.

There are also many diseases of agricultural plants that can be controlled by the use of pesticides. Examples of important fungal diseases of crop plants that can be managed with appropriate fungicides include: (1) late blight of potato, (2) apple scab, and (3) *Pythium*-caused seed-rot, damping-off, and root-rot of many agricultural species.

Forestry

In forestry, the most important uses of pesticides are for the control of defoliation by epidemic insects and the reduction of weeds. If left uncontrolled, these pest problems could result in large decreases in the yield of merchantable timber. In the case of some insect infestations, particularly spruce budworm (*Choristoneura fumiferana*) and **gypsy moth** (*Lymantria dispar*), repeated defoliation can cause the death of trees over a large area of forest. Most herbicide use in forestry is for the release of desired conifer species from the effects of competition with angiosperm herbs and shrubs. In most places, the quantity of pesticide used in forestry is much smaller than that used in agriculture.

Pesticides can also be classified according to their similarity of chemical structure. The most important of these are:

- Inorganic pesticides, including compounds of **arsenic**, **copper**, **lead**, and **mercury**. Some prominent inorganic pesticides include Bordeaux mixture, a complex pesticide with several copper-based active ingredients, used as a **fungicide** for fruit and vegetable crops; and various arsenicals used as non-selective herbicides and **soil** sterilants and sometimes as insecticides.
- Organic pesticides, which are a chemically diverse group of chemicals. Some are produced naturally by certain plants, but the great majority of organic pesticides have been synthesized by chemists. Some prominent classes of organic pesticides are:
- Biological pesticides are bacteria, **fungi**, or viruses that are toxic to pests. One of the most widely used biological insecticide is a preparation manufactured from spores of the bacterium ***Bacillus thuringiensis***, or B.t. Because this insecticide has a relatively specific activity against leaf-eating lepidopteran pests and a few other insects such as blackflies and mosquitoes, its non-target effects are small.

The intended ecological effect of a pesticide application is to control a pest species, usually by reducing its abundance to an economically acceptable level. In a few situations, this objective can be attained without important non-target damage. However, whenever a pesticide is broadcast-sprayed over a field or forest, a wide variety of on-site,

non-target organisms are affected. In addition, some of the sprayed pesticide invariably drifts away from the intended site of deposition, and it deposits onto non-target organisms and ecosystems. The ecological importance of any damage caused to non-target, pesticide-sensitive organisms partly depends on their role in maintaining the integrity of their **ecosystem**. From human perspective, however, the importance of a non-target pesticide effect is also influenced by specific economic and esthetic considerations.

Some of the best known examples of ecological damage caused by pesticide use concern effects of DDT and other **chlorinated hydrocarbons** on predatory birds, marine mammals, and other wildlife. These chemicals accumulate to large concentrations in predatory birds, affecting their reproduction and sometimes killing adults. There have been high-profile, local and/or regional collapses of populations of **peregrine falcon** (*Falco peregrinus*), **bald eagle** (*Haliaeetus leucocephalus*), and other raptors, along with **brown pelican** (*Pelecanus occidentalis*), western grebe (*Aechmorphorus occidentalis*), and other waterbirds. It was the detrimental effects on birds and other wildlife, coupled with the discovery of a pervasive presence of various chlorinated **hydrocarbons** in human tissues, that led to the banning of DDT in most industrialized countries in the early 1970s. These same chemicals are, however, still manufactured and used in some tropical countries.

Some of the pesticides that replaced DDT and its relatives also cause damage to wildlife. For example, the commonly used agricultural insecticide carbofuran has killed thousands of waterfowl and other birds that feed in treated fields. Similarly, broadcast-spraying of the insecticides phosphamidon and fenitrothion to kill spruce budworm in infested forests in New Brunswick, Canada, has killed untold numbers of birds of many species.

These and other environmental effects of pesticide use are highly regrettable consequences of the broadcast-spraying of these toxic chemicals in order to cope with pest management problems. So far, similarly effective alternatives to most uses of pesticides have not been discovered, although this is a vigorously active field of research. Researchers are in the process of discovering pest-specific methods of control that cause little non-target damage and in developing methods of **integrated pest management**. So far, however, not all pest problems can be dealt with in these ways, and there will be continued reliance on pesticides to prevent human and domestic-animal diseases and to protect agricultural and forestry crops from weeds, diseases, and depredations caused by economically important pests. *See also* Agent Orange; Agricultural chemicals; Agricultural pollution; Algicide; Chlordane; Cholinesterase inhibitor; Diazinon; Environmental health; Federal Insecticide, Fungicide and Rodenticide Act (1947); Kepone; Methylmercury seed dressings; National Coalition Against the Misuse of Pesticides; Organochloride; Persistent compound; Pesticide Action Network; Pesticide residue; 2,4-D; 2,4,5-T

[*Bill Freedman Ph.D.*]

Resources

Books

Baker, S., and C. Wilkinson, eds. *The Effect of Pesticides on Human Health.* Princeton, NJ: Princeton Scientific Publishing, 1990.

Freedman, B. *Environmental Ecology.* 2nd edition, San Diego: Academic Press, 1995.

Hayes, W. J., and E. R. Laws, eds. *Handbook of Pesticide Toxicology.* San Diego: Academic Press, 1991.

McEwen, F. L., and G. R. Stephenson. *The Use and Significance of Pesticides in the Environment.* New York: Wiley, 1979.

Pesticide Action Network

The **Pesticide** Action Network is an international coalition of more than 300 nongovernmental organizations (NGOs) that works to stop both pesticide misuse and the proliferation of pesticide use worldwide. The Pesticide Action Network North America Regional Center (PANNA) is the North American member.

The Pesticide Action Network serves as a clearinghouse to disseminate information to pesticide action groups and individuals in this country and abroad. It promotes research on and implementation of alternatives to pesticide use in agriculture, such as **integrated pest management** programs. With a library of 6,000 books, reports, articles, slides, and other materials, it sponsors a pesticide issues referral and information service.

Formerly called the Pesticide Education Action Project, which was founded in 1982, PANNA conducts the Dirty Dozen Campaign to "replace the most notorious pesticides with sustainable, ecologically sound alternatives." Various materials published by PANNA include brochures, pamphlets and books on United States pesticide problems and on global pesticide use. It publishes a quarterly newsletter, *Global Pesticide Campaigner*, and a bimonthly publication, *PANNA Outlook*.

The organization also publishes a number of books dealing with pesticides and **pest** control including *Problem Pesticides, Pesticide Problems: A Citizen's Action Guide to the UN Food and Agriculture Organization's (FAO) International Code of Conduct on the Distribution and Use of Pesticides*; *FAO Code: Missing Ingredients*; and *Breaking the Pesticide Habit: Alternatives to 12 Hazardous Pesticides.* While PANNA is not a membership organization for individuals outside of

Pesticides

Type	Examples	Characteristics
Insecticides		
Inorganic chemicals	Mercury, lead, arsenic, copper sulfate	Highly toxic to many organisms, persistent, bioaccumulates
Organochlorines	DDT, methoxychlor, heptachlor, HCH, pentachloraphenol, chlordane, toxaphene, aldrin, endrin, dieldrin, lindane	Mostly neurotoxins, cheap, persistent, fast acting, easy to apply, broad spectrum, bioaccumulates, biomagnifies
Organophosphates	Parathion, malathion, diazinon, dichlorvos, phosdrin, disulfoton, TEPP, DDVP	More soluble, extremely toxic nerve poisons, fast acting, quickly degraded, toxic to many organisms, very dangerous to farm workers
Carbamates and urethanes	Carbaryl (Sevin), aldicarb, carbofuran, methomyl, Temik, mancozeb	Quickly degraded, do not bioaccumulate, toxic to broad spectrum of organisms, fast acting, very toxic to honey bees
Formamidines	Amitraz, chlordimeform (Fundal and Galecron)	Neurotoxins specific for certain stages of insect development, act synergistically with other insecticides
Microbes	*Bacillus thuringensis*	Kills caterpillars
	Bacillus popilliae	Kills beetles
	Viral diseases	Attack a variety of moths and caterpillars
Plant products and synthetic analogs	Nicotine, rotenone, pyrethrum, allethrin, decamethrin, resmethrin, fenvalerate, permethrin, tetramethrin	Natural botanical products and synthetic analogs, fast acting, broad insecticide action, low toxicity to mammals, expensive
Fungicides	Captan, maneb, zeneb, dinocap, folpet, pentachlorphenol, methyl bromide, carbon bisulfide, chlorothalonil (Bravo)	Most prevent fungal spore germination and stop plant diseases; among most widely used pesticides in United States.
Fumigants	Ethylene dibromide, dibromochloro-propane, carbon tetrachloride, carbon disulfide, methyl bromide	Used to kill nematodes, fungi, insects, and other pests in soil, grain, fruits; highly toxic, cause nerve damage, sterility, cancer, birth defects
Herbicides	2,4 D; 2,4,5 T; paraquat, dinoseb, diaquat, atrazine, Silvex, linuron	Block photosynthesis, act as hormones to disrupt plant growth and development, or kill soil micro-organisms essential for plant growth

Different types of pesticides. (McGraw-Hill Inc. Reproduced by permission.)

the member NGOs, it does offer subscriptions to its quarterly newsletter.

[*Linda Rehkopf*]

Resources

Organizations

Pesticide Action Network North America, 49 Powell St., Suite 500, San Francisco , CA USA 94102 (415) 981-1771 , Fax: (415) 981-1991 , Email: panna@panna.org, <http://www.panna.org>

Pesticide residue

Chemical substances—pesticides and herbicides—that are used as weed and insect control on crops and animal feed always leave some kind of residue either on the surface or within the crop itself after treatment. The amounts of residue vary, but they do remain on the crop when ready for harvest and consumption.

While these **chemicals** are generally registered for use on food crops, it does not necessarily mean they are "safe." As part of its program to regulate the use of pesticides, **Environmental Protection Agency** (EPA) sets "tolerances" which are the maximum amount of pesticides that may remain in or on food and animal feed. Tolerances are set at levels that are supposed to ensure that the public (including infants and children) is protected from unreasonable health risks posed by eating foods that have been treated.

Tolerances are initially calculated by measuring the amount of **pesticide** that remains in or on a crop after it is treated with the pesticide at the proposed maximum allowable level. The EPA then calculates the possible risk posed by that proposed tolerance to determine if it is acceptable.

In calculating risk, the EPA estimates exposure based on the "theoretical maximum residue concentration" (TMRC) normally assuming three factors: that all of the crop is treated, that residues on the crop are all at the maximum level, and that all consumers eat a certain fixed percent of the commodity in their diet. This exposure calculation is then compared to an "allowable daily intake" (ADI) and calculated on the basis of the pesticide's inherent toxicity.

The ADI represents the level of exposure in animal tests at which there appears to be no significant toxicological effects. If the residue level for that crop—the TMRC—exceeds the allowable daily intake, then the tolerance will not be granted as the residue level of the pesticide on that crop is presumed unsafe. On the other hand, if the residue level is lower than the ADI, then the tolerance would normally be approved. Exposure rates are based on assumptions regarding the amount of treated commodities that an "average" adult person, weighing 132 lb (60 kg), eating an "average" diet would consume.

To set tolerances that are protective of human health, the needs information about the anticipated amount of pesticide residues found on food, the toxic effects of these residues, and estimates of the types and amounts of foods that make up our diet. The **burden of proof** is on the manufacturer who has a vested interest in getting EPA approval. A pesticide manufacturer begins the tolerance-setting process by proposing a **tolerance level**, which is based on field trials reflecting the maximum residue that may occur as a result of the proposed use of the pesticide. The petitioner must provide food residue and toxicity studies to show that the proposed tolerance would not pose an unreasonable health risk.

The Federal Food, Drug, and Cosmetic Act (FFDCA) requires that the manufacturer of any substances apply to the EPA for a tolerance or maximum residue level to be allowed in or on food. Because of the way pesticide residues are defined and incorporated into the food additive provisions of the FFDCA, there are actually several kinds of tolerances, each subject to different decisional criteria. A given pesticide can thus have many tolerances, both for use on different crops and for any one crop. For example, for each crop for which it is registered, a pesticide may need a raw tolerance, a food additive tolerance, and an animal feed tolerance, each of which will be subject to particular decisional criteria.

The EPA sets tolerances on the basis of data the manufacturer is required to submit on the nature, level, and toxicity of a substance's residue. According to the EPA, one requirement is for the product chemistry data, which is information about the content of the pesticide products, including the concentration of the ingredients and any impurities. The EPA also requires information about how plants and animals metabolize (break down) pesticides to which they are exposed and whether residues of the metabolized pesticides are detectable in food or feed. Any products of pesticide **metabolism**, or "metabolites," that may be significantly toxic are considered along with the pesticide itself in setting tolerances.

The EPA also requires field experiments for pesticides and their metabolites for each crop or crop group for which a tolerance is requested. This includes each type of raw food derived from the crop. Manufacturers must also provide information on residues found in many processed foods (such as raisins), and in animal products if animals are exposed to pesticides directly through their feed.

A pesticide's potential for causing adverse health effects, such as **cancer**, **birth defects**, and other reproductive disorders, and adverse effects on the nervous system or other organs, is identified through a battery of tests. Tests are conducted for both short-term or "acute" toxicity and long-term or "chronic" toxicity. In several series of tests, laboratory animals are exposed to different doses of a pesticide and EPA scientists evaluate the tests to find the highest level of exposure that did not cause any effect. This level is called the "No Observed Effect Level" or NOEL.

Then, according to the EPA, the next step in the process is to estimate the amount of the pesticide to which the public may be exposed through the food supply. The EPA uses the Dietary Risk Evaluation System (DRES) to estimate the amount of the pesticide in the daily diet, which is based upon a national food consumption survey conducted by the **U.S. Department of Agriculture**. The USDA survey provides information about the diets of the overall American population and any number of subgroups, including several different ethnic groups, regional populations, and age groups such as infants and children.

Finally, for non-cancer risks, the EPA compares the estimated amount of pesticide in the daily diet to the Reference Dose. If the DRES analysis indicates that the dose consumed in the diet by the general public or key subgroups exceeds the Reference Dose, then generally the EPA will not approve the tolerance. For potential carcinogens, the EPA ordinarily will not approve the tolerance if the dietary analysis indicates that exposure will cause more than a negligible risk of cancer.

There have been serious problems in carrying out the law. Although the law requires a thorough tolerance review for chemical pesticides prior to registration for health and environmental impacts, the majority of pesticides currently on the market and in use today were registered before modern testing requirements were established. "Thus," according to a 1987 report published by the **National Research Council** Board on Agriculture, "many older pesticides do not have an adequate data base, as judged by current standards, particularly data about potential chronic health effects."

The 1972 amendments to the **Federal Insecticide, Fungicide and Rodenticide Act** (FIFRA) required a review—or registration—of all then-registered products according to contemporary standards. Originally, the EPA was to complete the review of all registered products within three years, but this process broke down completely. In 1978, amendments were added to FIFRA that authorized the EPA to group together individual substances by active ingredients for an initial "generic" review of similar chemicals in order

to identify data gaps and assessment needs. The amendments mandate that after this generic review, the EPA is to establish as "registration standard" for the active ingredient basis. Then, the EPA must wait for additional data in order to establish specific requirements for each substance's specific use on differing crops. (The EPA does not specifically address inert ingredients, although inert ingredients have been found to be just as potentially toxic as the active ingredients.)

The 1988 amendments extended the deadline for re-registration until 1997. To date, however, the EPA only has complete residue data on less than 25% of pesticides used on foods, and less than ten products, as yet, have been registered. Thus, the backlog of substances awaiting re-registration is on the order of some 16,000 to 20,000 compounds, including, according to the General Accounting Office, over 650 active ingredients. As a consequence, most of these untested chemicals temporarily retain their registration status under the former testing requirements.

Once tested in accordance with contemporary standards, many of the older pesticides are likely to be identified as carcinogenic. In fact, the **National Academy of Sciences** estimates that 25–33% of the pesticides to undergo re-registration procedures are likely to be found to be oncogenic. An oncogen causes tumors in laboratory animals and is considered potentially cancerous to humans. In general, such older pesticides, registered prior to 1972, have not been tested adequately for oncogenicity. For a substance to be used on a new crop, however, current data must be submitted in order to obtain a tolerance.

One way pesticide companies are able to completely skirt EPA regulations is to send pesticides that are suspended or banned in the United States to **Third World** countries. At least 25% of American pesticide exports are products that are banned, heavily restricted or have never been registered for use in the United States. Hazardous pesticide exports are a major source of **pollution** in Third World countries where lack of regulation, illiteracy, and repressive working conditions can turn even a "safe" pesticide into a deadly weapon for the field workers. These pesticides are returned to the United States, unexamined, as residues in the imported food American consumers eat.

[*Liane Clorfene Casten*]

Resources

Books

Briggs, S. A. *Basic Guide to Pesticides: Their Characteristics and Hazards.* Washington, DC: Hemisphere, 1992.

Dinham, B. *Pesticide Hazard: Global Health and Environmental Audit.* Highlands, NJ: Humanities Press International, 1993.

Somerville, L. *Pesticide Effects on Terrestrial Wildlife.* London: Taylor and Francis, 1990.

Periodicals

Raloff, J., and D. Pendick. "Pesticides in Produce May Threaten Kids." *Science News* 144 (July 3, 1993): 4–5.

Pet trade

Pets are part of human cultures all around the world. We keep animals to admire their beauty or to enjoy their companionship or devotion. The pet trade is big business throughout the world but with its positive aspects, there are distinct negative ones as well. As with any commodity, economics dictates that the rarer an item is, the more valuable it becomes in the market place. Many **rare species** thus become quite valuable to wealthy collectors, who are willing to pay exorbitant prices to black market dealers for certain species—even though they are protected by international treaties and laws.

Because these illegal pets must be smuggled into the market arenas, concealment almost always requires physical abuse of the animals. This may be in the form of constrictive bindings, overcrowded or confined spaces, exposure to extreme heat, or lack of oxygen due to these restrictions. Such treatment results in the death of many individuals during transit, with **mortality** that often approaches 80–90%. Therefore, collectors and smugglers must over-collect specimens in order to ensure the delivery of a sufficient quantity to realize a profit. Two groups, monkeys and **parrots**, have been the recipients of the bulk of this treatment. Due to their popularity as pets, there is a lucrative business in their illegal trade, and because the rarer **species** bring a higher price, these practices further threaten **extinction** of certain species.

This problem extends beyond the black market trade in birds and mammals. Other animal groups, as well as the legal pet trade, are also involved. Exotic reptiles, snakes and lizards in particular, are a target group. Legitimate pet dealers get involved in this problem when they, often unknowingly, purchase pets that were illegally caught or smuggled.

The **aquarium trade** has also faced illegal collecting. Over-collecting and poor handling of specimens during shipment have been problems in the freshwater segment of this business, but recent years have produced rampant abuses in the marine aquarium trade. Advances in technology over the past twenty years have made marine aquaria more accessible to the general public, and the sheer beauty of the brilliant, often neon-like, colors of many of the **coral reef** fish has fascinated a new breed of aquarists.

The main problems here still include the removal of large numbers of specimens of localized populations of rare or low density species and their illegal trade. Now, new and profound problems exist in the capture of these marine

creatures: the total destruction of their **habitat** and the killing of unwanted, non-commercial species. Although legitimate collecting of marine fish is done with nets, it is estimated that during the 1980s about 80% of specimens for the marine aquarium trade were collected by using poison. Since most of the **target species** use coral reefs as hiding places, collectors use a squeeze bottle filled with sodium cyanide to force the fish out into the open water. The stunned, gasping fish are thus more easily caught, but the section of coral reef sprayed with poison, and the animals that did not escape, are now dead. Lingering effects of the poison and abusive shipping practices lead to mortality rates of 60–80% for these pets.

Public awareness and some governmental regulations have eased some of the abuses of these animals, but the illegal pet trade is still a profitable business. As long as people are willing to purchase rare species, the illegal pet trade will threaten the very species the collectors hold in high esteem. *See also* Endangered Species Act; Overhunting; Parrots and parakeets; Poaching; Wildlife management

[*Eugene C. Beckham*]

Resources

Books

Forshaw, J., and W. Cooper. *Parrots of the World.* Melbourne, Australia: Lansdowne Press, 1973.

Periodicals

Bergman, C. "The Bust!" *Audubon* 93 (1991): 66–77.

Derr, M. "Raiders of the Reef." *Audubon* 94 (1992): 48–56.

McLarney, W. "Still a Dark Side to the Aquarium Trade." *International Wildlife* 18 (1988): 46–51.

Speart, J. "What's Wildlife Worth?" *Wildlife Conservation* 95 (1992): 44–47.

PETA

see **People for the Ethical Treatment of Animals**

Roger Tory Peterson (1908 - 1996)

American ornithologist

A small book, tucked away in innumerable back-packs and car pockets, ever ready to hand for perhaps the majority of birders in the United States, is quite likely to be Roger Tory Peterson's *A Field Guide to the Birds*, first published in 1934, revised and reissued several times, and still a must-have for many bird-watchers, serious or otherwise. In 1941 it was joined by a companion volume, *A Field Guide to Western Birds*. Together, the two guides (the first rejected by four publishers before Houghton Mifflin finally accepted it) have sold on the order of seven million copies by 1997. The critic William Zinser suggested that Peterson's *Field Guide* was "the single most revolutionary development in American birding." He was the first to introduce simplified, comparative drawings and to point out key field marks (distinguishing characteristics) that help identification in the field. Later critics judged Peterson harshly for his last revision of the *Guide* as "not knowing much about contemporary identification skills." But the revolutionary importance of the *Guides* lies in the number of people they have "turned on" to birds and to birding; they still sell, and are still used by birders at every level.

The man who has been called the modern successor to John James Audubon was born in Jamestown, New York, of a Swedish immigrant father and a German immigrant mother. Reportedly a loner, a boy considered strange by other children, Peterson came out of his shell when one of his teachers started a Junior Audubon Club. Birds became a passion for the eleven-year old, and he sought a job as a newspaper delivery boy so that he could buy a camera to photograph birds. Some seven decades later, he was still making pictures of birds. The prescient caption under his photograph in his high school yearbook read "Woods! Birds! Flowers! Here are the makings of a great naturalist."

Peterson studied art rather than ornithology, and he considered himself first a painter, then a writer, and only third a naturalist. Yet such was his reputation and standing among people interested in natural history that the *New York Times* could announce his death as that of "the best-known ornithologist of the twentieth century." Because his work reached so many people, because it helped them become involved in knowing more about the world around them, and because it inspired them to actually get out in the field and get involved, Peterson might even deserve *Sports Afield's* label of "the twentieth century's most influential naturalist." Academically trained or not, he spent his lifetime doing ornithology, i.e., studying birds, and in getting other people to join in that passion. His guides provided clear access to bird identification, even for the most rank amateur. Along with his many awards for art, he was also honored as a Fellow in the American Association for the Advancement of Science, and was bestowed with the Linné gold medal of the Royal Swedish Academy of Sciences.

After art school, he taught at a boys' school in Massachusetts. The success of the first field guide allowed him to return to New York to a position as an educational specialist and art editor for *Audubon Magazine*. He remained associated with the **National Audubon Society** for the rest of his professional life, serving in various capacities as artist, writer, secretary, and two different terms as director.

His passion for **nature** centered on birds, certainly, but he also was enthusiastic for all of the natural world,

worked on many fronts to protect its diversity, and became an advocate for and teacher about managing the **environment** wisely. For example, he and his wife shared a love of flowers, and created and maintained butterfly gardens; one of his best known field guides is for wildflowers. All told, he published almost 50 books, including his edition of *Audubon's Birds of America*, and wrote many articles on birds as well as on other topics in **conservation** and management of gardens and wildlands.

As stated in a *New York Times* editorial, Peterson did indeed become "a great naturalist and more. He was one of the pioneers in teaching twentieth-century Americans to walk more gently on their land."

[*Gerald L. Young Ph.D.*]

Resources
Books

Devlin, J. C., and G. Naismith. *The World of Roger Tory Peterson: an Authorized Biography.* New York: New York Times Books, 1977.

Peterson, R. T., and R. Hoglund, eds. *Roger Tory Peterson; Art and Photography from the World's Foremost Birder.* New York: Rizzoli, 1994.

Petrochemical

Petroleum is probably best known as a source for many important fuels, such as **gasoline**, kerosene, acetylene, and **natural gas**. However, many of the organic compounds that make up the complex mixture known as petroleum have another use. They are raw materials used in the production of a host synthetic products. These **chemicals** are known as petrochemicals. To a large extent, these petrochemicals are **hydrocarbons**. Some of these petrochemicals and the products into which they are made include ethylene (**plastics**, synthetic fibers, and anti-freeze), **benzene** (synthetic **rubber**, latex paints and paper coatings), propylene (drugs and **detergents**), and phenol (adhesives, perfumes, flavorings, and pesticides). *See also* Fossil fuels; Synthetic fuels; Volatile organic compound

Petroleum

Petroleum is a complex mixture of solid, liquid, and gaseous **hydrocarbons** normally found a few miles beneath the earth's surface. It, along with **coal** and **natural gas**, is one of the **fossil fuels**. That name comes from the most common scientific theory that these three materials were formed between 10 and 20 million years ago as a product of the decay of plants and animals. Petroleum is sometimes called crude oil or, simply, oil, although some authorities give slightly different means to these terms.

Most commonly, petroleum and natural gas occur together under dome-shaped layers of rock. When **wells** are drilled into such domes, gas pressure forces petroleum up into the well, producing the "gushers" that are typical of successful new wells.

Petroleum has been known to and used by humans for thousands of years. Collecting the liquid that seeped to the surface, the Egyptians, Babylonians, and Native American Indians used petroleum for waterproofing, embalming, caulking ships, warpaint, and medicine. The first successful oil well in the United States was drilled in 1859 by "Colonel" Edwin Drake.

The number and variety of compounds found in petroleum are large, and its composition varies significantly from location to location. Petroleum from Pennsylvania tends to have a high proportion of aliphatic (open-chain) hydrocarbons, those from the West contain more aromatic (benzene-based) hydrocarbons, and those from the Midwest contain a more even mix of the two.

The most common impurities (non-hydrocarbon compounds) in petroleum are compounds of sulfur and **nitrogen**. The percentage of sulfur in a sample of petroleum determines its status as "sweet" (low-sulfur) or "sour" (high-sulfur) crude oil. The emphasis on sulfur derives from the fact that one of the most undesirable products of the **combustion** of petroleum is **sulfur dioxide**. Oil with low sulfur is, therefore, more desirable than that with high sulfur.

The complex mix of compounds that make up petroleum results in its having relatively few important uses in its natural state. It becomes an important product only when some method is used to separate the native liquid into its component parts. The fractional distillation of petroleum results in the production of portions known as petroleum ether, **gasoline**, kerosene, heating oil, lubricating oil, and a semi-solid residue that includes paraffin, pitch, and tar. Each of these fractions can then be utilized directly or subdivided even further.

Few naturally occurring materials have as many environmental effects as does petroleum. During its removal from the earth, it may escape from wells and contaminate **groundwater**. The combustion of some of its most important components, gasoline, kerosene and fuel oil results in the release of **carbon dioxide**, **carbon monoxide**, **carbon**, sulfur dioxide, and **nitrogen oxides**, all compounds that contribute to **air pollution**, global warming, or other environmental problems.

In recent decades, petroleum has found another major use: in the production of petrochemicals. The compounds found in oil are now important starting materials in the manufacturing of a nearly limitless number of synthetic products. The manufacture, use and disposal of these **petrochemical** products contributes to additional environmental prob-

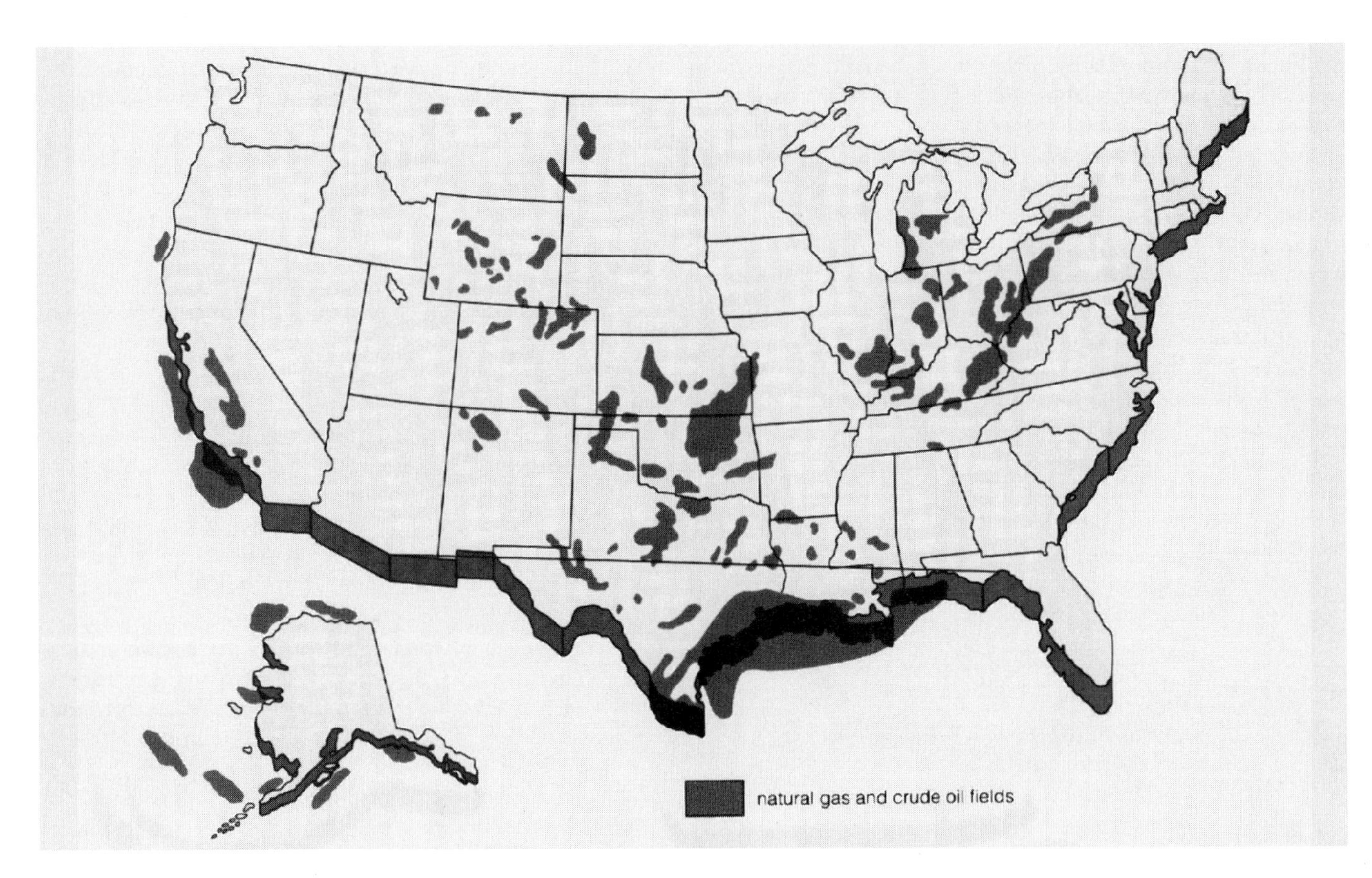

Natural gas and oil deposits in the United States. (Council on Environmental Quality.)

lems such as **hazardous waste** disposal and **solid waste** accumulation. *See also* Alternative fuels; Decomposition; Emission standards; Greenhouse effect; Groundwater pollution; Oil drilling; Oil spills

[*David E. Newton*]

RESOURCES

BOOKS

Fesharaki, F., and R. Reed, eds. *The Petroleum Market in the 1990s.* Boulder, CO: Westview Press, 1989.

Kupchella, C. E. *Environmental Science: Living within the System of Nature.* 3rd ed. Boston: Allyn and Bacon, 1993.

Metzger, N. "Getting More Oil." In *Energy and the Way We Live*, edited by M. Kranzberg and T. A. Hall. San Francisco: Boyd & Fraser, 1980.

Petroleum Exploration: Continuing Need. Washington, DC: American Petroleum Institute, 1986.

Petulla, J. M. *American Environmental History.* 2nd ed. San Francisco: Boyd & Fraser, 1988.

OTHER

Flavin, C. "World Oil: Coping With the Dangers of Success." *Worldwatch Paper #66.* Washington, DC: Worldwatch Institute, 1985.

Pfiesteria

Pfiesteria piscicida (fee-STEER-ee-uh pis-kuh-SEED-uh) is a microscopic, polymorphic organism that belongs to a group of single-celled, free-swimming **phytoplankton** called dinoflagellates. Although many phytoplankton photosynthesize to generate the energy they need to survive, *Pfiesteria* does not—it is more animal-like in that it consumes other organisms such as bacteria and algae. It can transform in and out of the following stages within a matter of minutes: flagellated, amoeboid, and cyst. In the flagellated stage *Pfiesteria* is powered by two tiny flagella that resemble tails. The cyst stage lies dormant in the bottom sediments of estuaries and rivers.

Two **species** are known to release **toxins** that are harmful to aquatic and human life: *Pfiesteria piscicida* and *Pfiesteria shumwayae.*

When it was discovered in 1988, *Pfiesteria* was declared to be a new family (Pfiesteriaceae), genus (*Pfiesteria*), and species (*piscicida*), in the order Dinamoebales. It was named in honor of Dr. Lois Pfiester, an internationally respected

researcher of dinoflagellates. Since its discovery, it has been identified in mud cores from the bottom of **Chesapeake Bay** that are at least 3,000 years old.

Depending on its stage, *Pfiesteria piscicida*, the most studied *Pfiesteria* species, ranges in size from about 5–450 microns in diameter. It is typically found in the bottom sediments and water of **brackish** tidal rivers and estuaries, primarily along the eastern seaboard of the United States.

Pfiesteria is a remarkably hardy organism that can survive over a wide range of temperatures and water salinities. For most of its life cycle *Pfiesteria* is a nontoxic predator, feeding on bacteria, algae, and other small organisms. When schools of fish are present, *Pfiesteria* occasionally responds with a quick release of **biotoxins** into the water. These **chemicals** stun or kill the fish, which allow the *Pfiesteria* to feed on the injured skin, blood, and dying tissue.

These toxic episodes last for several hours before dissipating, sometimes resulting in large-scale **fish kills**, many of which plagued the Atlantic coastline in the late 1980s and 1990s. Fish kills attributed to *Pfiesteria* occurred in estuaries in the Albemarle–Pamlico estuarine system in North Carolina, Chesapeake Bay tributaries in Virginia, the Chicamacomico, Pokomoke, and Manokin Rivers and King's Creek in Maryland, and Indian River in Delaware. *Pfiesteria* has been identified as far north as Long Island, New York, and as far west as the southern tip of Texas. It may also occur in Pacific waters along the California–Oregon–Washington coastline, but has not yet been identified because these waters have not been sampled.

Although *Pfiesteria* is technically not an algae, it's fish kills are usually included under the category of harmful algal blooms (HABs). This is because *Pfiesteria* shares many physiological and toxicological characteristics with the various algae species that cause algal blooms (red or brown tides), some of which are toxic to aquatic life.

Toxic strains of *Pfisteria* have been identified along European, Scandinavian, New Zealand, and **Australia** coastlines. It likely occurs in other international waters that have not been tested for its presence. As the technology of detection improves, it is highly probable the global distribution of *Pfiesteria* will become better defined. However, the United States is the only country in the world that has experienced the serious deleterious effects of its toxins on aquatic life and human health.

When *Pfiesteria* detects fish oils or excretions in the water from a large school of fish, a biological toxic response may be triggered. The organisms change from the cyst stage (excystment) into the predator flagellate or ameoboid stages, which then release potent toxins into the water.

The toxins penetrate the skin of the fish, creating bleeding ulcers. The fish become lethargic or die. The skin becomes so damaged that the fish cannot maintain their internal salt balance. The *Pfiesteria* organisms feed on the dead skin tissue, blood, and other fish fluids leaking into the water. Within several hours *Pfiesteria* transforms back to the cyst stage (encystment) and lays dormant in the bottom sediments. Fish kills usually occur in warmest times of the year and when the **dissolved oxygen** content of the water is low.

There has been a global increase in toxic and harmful algae blooms in the late twentieth century, including *Pfiesteria.* Modern technology increasingly allows for the rapid identification of the species causing the bloom or the fish kill. *Pfiesteria* attacks are becoming a growing problem because they have a negative impact on **commercial fishing**, **recreation**, and human health along the eastern seaboard.

Toxins can be present in the water or aerosolize into the air. Toxic episodes and fish kills close down sections of coastline, which impact the fish and shellfish industries. If fish aren't killed directly by *Pfiesteria,* they are seriously weakened, which makes them susceptible to other predators or bacterial and fungal infections. Humans can suffer serious health problems from coming into contact with, or inhaling, these biotoxins. Effects to humans and aquatic life persist for six to eight weeks after the bloom disappears.

Two kinds of toxins have been discovered to date: a fat-soluble toxin that blisters skin and a water-soluble toxin that affects the nervous system. Injection of *Pfiesteria* toxin has induced significant learning problems in laboratory rats. Symptoms in humans consist of lethargy, sores and rashes, eye sensitivity, headaches, blurred vision, gastrointestinal distress, nausea, memory loss, kidney and liver problems, shortness of breath, and fibromyalgia-like symptoms. Symptoms can recur for up to eight years after exposure. This condition, sometimes referred to as possible estuary-associated syndrome (PEAS) can be developed from exposure to *Pfiesteria* biotoxins, even if there is no evidence of fish kills or fish disease in the water.

Identifying a fish kill as being *Pfiesteria*-related (and taking subsequent actions, such as closing shorelines and issuing fish consumption warnings) can be difficult because *Pfiesteria* has 24 recognized morphological shapes at different growth stages. Scientists use light microscopes, bioassays, scanning electron microscopes, or polymerase **chain reaction** (PCR) methods to establish positive identifications.

An unanswered question remains: Are fish from areas that have experienced *Pfiesteria* outbreaks safe to consume? This may have significant impact to the fishing industry. So far research indicates that no person has become ill from eating fish or shellfish that have been exposed to *Pfiesteria*—the data, however, are limited. Most experts recommend to err on the side of caution: If an area has been closed because of a *Pfiesteria* outbreak, do not catch or consume fish from

that area or swim or waterski in those waters for at least eight weeks after the outbreak.

Data suggest that *Pfiesteria* has a fairly wide global distribution, but only creates health problems under certain conditions, which seem to be best demonstrated in the coastal estuaries of the eastern and southern United States.

Research at North Carolina State University's Center of Applied Aquatic Ecology shows that *Pfiesteria* is stimulated by organic or inorganic **nitrogen** and **phosphorus** in the water—two key components of **fertilizer**. **Carbon** may also be a trigger substance. It has been documented that *Pfiesteria* outbreaks occur in areas that are downstream from septic tanks, sewage-plant discharges, agricultural waste such as excrement from cattle, swine, and poultry, and agricultural or landscaping chemical **runoff**. When these nutrients accumulate in slow-moving or stagnant sections of estuaries and mix with the incoming tidal water, *Pfiesteria* can become more active and toxic. Other stimulators may be airborne pollutants that settle in the water. An abundance of fish is also required. As yet it has not been determined if the **nutrient loading** stimulates the growth of the algae on which *Pfiesteria* feeds, or if directly stimulates *Pfiesteria*.

Federal organizations such as the U.S. **Environmental Protection Agency** (EPA), U.S.Department of Agriculture, **U.S. Department of the Interior**, **National Oceanic and Atmospheric Administration**, Centers for Disease Control and Prevention(CDC), and the **National Institute of Environmental Health Sciences**, state departments of health or **natural resources**, and several prominent university research centers are continuing research into the life cycles of *Pfiesteria* and other *Pfiesteria*-like organisms, their nutritional ecologies, and the trigger mechanisms that result in toxic episodes.

One major areas of study focuses on integrating management approaches with new emerging technologies. This collaborative effort between risk managers, universities, federal agencies, and users—commercial and recreational fishermen, Native peoples, and other water users—will lead toward a systematic approach to managing the risk for exposure to *Pfiesteria* biotoxins. The ultimate goal is to develop an accurate model of prediction—what conditions will result in toxic *Pfiesteria*/algal blooms? Knowing these factors may reduce or eliminate the need to close down sections of coastline and interrupt commercial fishing operations.

Researchers are also attempting to purify and study *Pfiesteria* biotoxins. When the actual toxin can be identified, it will be much easier to determine if fish kill/disease episodes are related to *Pfiesteria* attacks. Once the chemical signature of the toxin is identified, methods can be developed to test humans for exposure. Antitoxins or related drugs may also be developed. It is also possible that, when consumed or injected in prescribed dosages under certain conditions, *Pfiesteria* toxins may be useful medicines that reduce or alleviate other health conditions in humans.

[*Mark J. Crawford*]

Resources

Periodicals

Burkholder, J. M. "Overview and Present Status of the Toxic *Pfiesteria* Complex." *Phycologia.*

Burkholder, J. M., et al. "New 'Phantom' Dinoflagellate Is the Causative Agent of Major Estuarine Fish Kills." *Nature* 358 (1992): 407–410.

Burkholder, J. M. and H. B. Glasgow. "*Pfiesteria piscicida* and Other Toxic *Pfiesteria*-like Dinoflagellates: Behavior, Impacts, and Environmental Controls." *Limnology & Oceanography.* 42 (1997):1052–1075.

Other

Pfiesteria and Related Harmful Blooms: Natural Resources and Human Health Concerns. Pamphlet. U.S. Environmental Protection Agency, 1998.

What You Should Know about Pfiesteria piscicida. Document EPA 842-F-98-011. U.S. Environmental Protection Agency, 2002.

U.S. Department of Public Heath and Human Services.ldquo;*Pfiesteria*: From Biology to Public Health.rdquo; *Environmental Health Perspectives* 109 (2001): Supplement 5.

Organizations

Aquatic Pathobiology Center, University of Maryland, 8075 Greenmead Dr., College Park, MD USA 20742 (301)314-6808, Fax: (301)314-6855

Maryland Sea Grant, University of Maryland, 0112 Skinner Hall, College Park, MD USA 20742 (301)405-6371, Fax: (301)314-9581

North Carolina State University Center for Applied Aquatic Ecology, 620 Hutton Street, Suite 104, Raleigh, NC USA 27608 (919)515-3421, Fax: (919)513-3194

U.S. Environmental Protection Agency, 1200 Pennsylvania Avenue NW, Washington, DC USA 20460 (202)260-2090

Virginia Institute of Marine Science, Route 1208 Greate Road, Gloucester Point, VA USA 23062 (804)684-7000, Fax: (804)685-7097

PFOS

see **Perfluorooctane sulfonate**

pH

A measure of the acidity or alkalinity of a solution based on its **hydrogen ion** (H^+) concentration. The pH of a solution is the negative logarithm (base 10) of its H^+ concentration. Since the scale is logarithmic, there is a tenfold difference in hydrogen ion concentration for each pH unit. The pH scale ranges from 0 to 14 with 7 indicating neutrality ((H^+) = (OH^-)). Values above 7 indicate progressively greater alkalinity, while values below 7 indicate progressively increasing acidity. *See also* Acid and base; Buffer

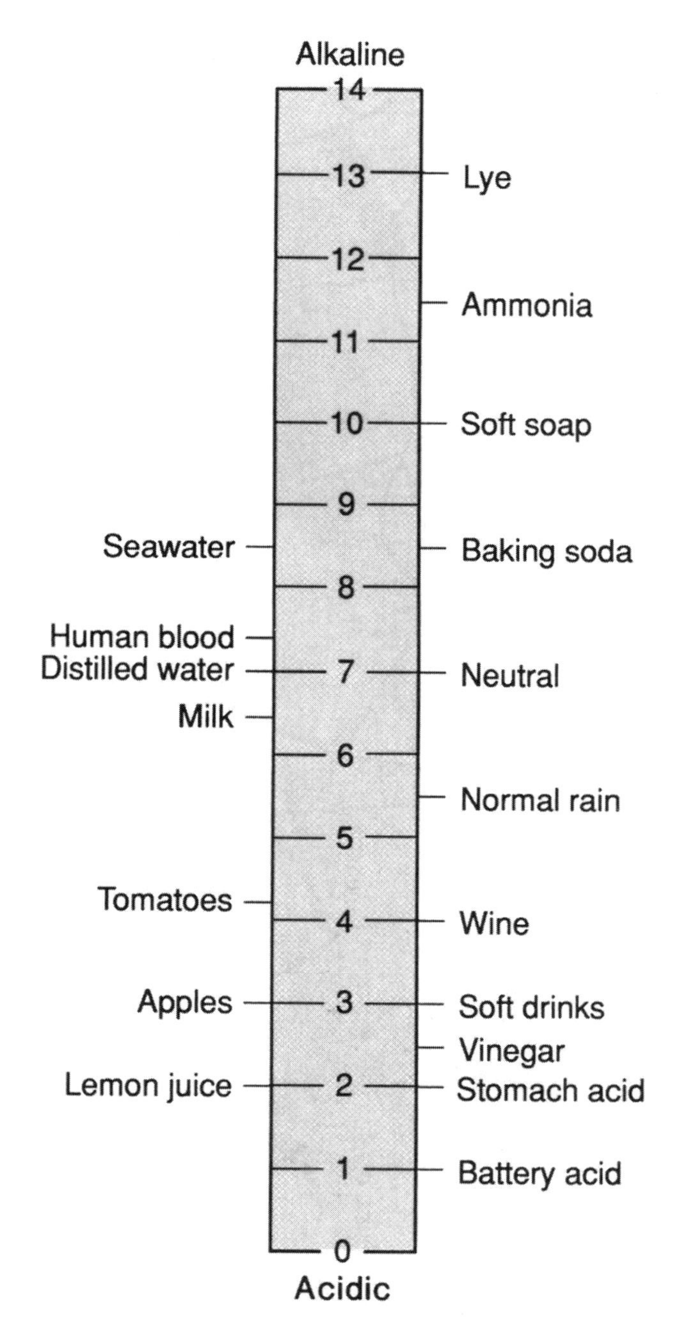

pH scale. (McGraw-Hill Inc. Reproduced by permission.)

Phosphates

Phosphorus is essential to the growth of biological organisms, including both their metabolic and photosynthetic processes. Phosphorus occurs naturally in bodies of water mainly in the form of phosphate (i.e., a compound of phosphorus and oxygen). Addition of phosphates through the activities of humans can accelerate the eutrophication process of **nutrient** enrichment that results in accelerated ecological aging of lakes and streams. Phosphorus, especially in inland waters, is often the nutrient that limits growth of aquatic plants. Thus when it is added to a body of water, it may result in increased plant growth that gradually fills in the lake. Critical levels of phosphorus in water, above which eutrophication is likely to be triggered, are approximately 0.03 mg/l of dissolved phosphorus and 0.1 mg/l of total phosphorus. The **discharge** of raw or treated **wastewater**, agricultural **drainage**, or certain industrial wastes that contain phosphates to a surface water body may result in a highly eutrophic state, in which the growth of photosynthetic aquatic micro- and macroorganisms are stimulated to nuisance levels. Aquatic plants and mats of algal scum may cover the surface of the water. As these algal mats and aquatic plants die, they sink to the bottom, where their **decomposition** by **microorganisms** uses most of the oxygen dissolved in the water. The decrease in oxygen severely inhibits the growth of many aquatic organisms, especially more desirable fish (e.g., recreational catch fish such as trout) and in extreme cases may lead to massive **fish kills**. Excessive input of phosphorus can change clear, oxygen-rich, good-tasting water into cloudy, oxygen-poor, foul smelling, and possibly toxic water. Therefore, control of the amount of phosphates entering surface waters from domestic and industrial waste discharges, natural **runoff**, and **erosion** may be required to prevent eutrophication.

Rocks may contain phosphates; those with calcium phosphate upon heat treatment with sulfuric **acid** serve as a source of fertilizers. Phosphates, in addition to those found in fertilizers, are also present in such consumer products as detergent, baking powders, toothpastes, cured meats, evaporated milk, soft drinks, processed cheeses, pharmaceuticals, and water softeners. Phosphates are classified as orthophosphates (PO_4^{-3}, HPO_4^{-2}, $H_2PO_4^-$, and H_3PO_4); condensed phosphates, or polyphosphates, which are molecules with two or more phosphorus atoms, oxygen atoms and in some cases, **hydrogen** atoms, combined in a complex molecule; and organically-bound phosphates.

Orthophosphates in an aqueous solution can be used for biological **metabolism** without further breakdown. Orthophosphates applied to agricultural or residential cultivated land as fertilizers may be carried into surface waters with **storm runoff** and melting snow.

Polyphosphates can be added to water when it is used for laundering or cleaning, for polyphosphates are present as builders of some commercial cleaning preparations for the public health sector. Massive algal blooms in lakes and floating foam on rivers aroused alarm in the United States in the 1970s. After it was shown that phosphates from **detergents** were a key factor, legislation banning the use

of phosphates in home laundry detergents was passed in many areas. Phosphate legislation usually includes exemptions for products such as hard-surface cleaner and automatic dishwashing detergents used in the public health sector. Polyphosphates may also be added to water supplies during culinary **water treatment** and during treatment of boiler water. Polyphosphates slowly undergo hydrolysis in aqueous solutions and are converted to the orthophosphate forms.

Organic phosphates are formed primarily by biological processes. They enter sewage water through body wastes and food residues. They may also be formed from orthophosphates in biological treatment processes and by receiving water organisms. Like polyphosphates, they are biologically transformed back to orthophosphates.

One means of surface water protection from phosphorus addition in both domestic and industrial wastewaters is the use of **phosphorus removal** processes in wastewater treatment. Phosphates are typically present in raw wastewaters at concentrations near 10 mg/l as P. During wastewater treatment, about 10-30% of the phosphates in raw wastewater is utilized during secondary biological treatment for microbial cell synthesis and energy transport. Additional removal is required to achieve low **effluent** concentration levels from the wastewater treatment process. Effluent limits usually range from 0.1-2 mg/l as P, with many established at 1.0 mg/l. Removal processes for phosphates from wastewaters utilize incorporation into suspended solids and the subsequent removal of those solids. Phosphates can be incorporated into chemical precipitates that are insoluble or of low solubility or into biological solids, (e.g, microorganisms).

Chemical precipitation is accomplished by the addition of metal salts or lime, with polymers often used as flocculant aids. The precipitation of phosphates from wastewater can occur during different phases within the wastewater treatment process. Pre-precipitation, where the **chemicals** are added to raw wastewater in primary **sedimentation** facilities, removes the precipitated phosphates with the primary **sludge**. In co-precipitation, the chemicals are added during secondary treatment to the effluent from the primary sedimentation facilities; to the mixed liquor in the activated-sludge process; or to the effluent from a biological treatment process before secondary sedimentation. They are removed with the waste biological sludge. In post-precipitation, the chemicals are added to the effluent from secondary sedimentation facilities and are removed in separate sedimentation facilities or in effluent **filters**.

The most commonly used metal salts are ferric chloride and **aluminum** sulfate (alum), which combine with phosphate to form the precipitates aluminum phosphate and iron phosphate, with one mole of iron or aluminum precipitating one mole of phosphate. However, because of competing reactions and the effects of alkalinity, **pH**, trace elements, and ligands found in wastewater, appropriate dosages are established on the basis of bench-scale or full-scale testing. Less commonly used metal salts are ferrous sulfate and ferrous chloride, which are available as by-products of steel production operations. Because polyphosphates and organic phosphorus are less easily removed than orthophosphates, adding metal salts after secondary treatment, where organic phosphorus and polyphosphates are transformed into orthophosphates, usually results in the best control.

Lime $(CaOH)_2$ is used less frequently because of the larger amounts of sludge produced as compared with the use of metal salts. Also, operating and maintenance problems are associated with the handling, storage, and feeding of lime. As lime is added to water, it reacts with natural bicarbonate alkalinity to precipitate $CaCO_3$. As the pH of the wastewater increases beyond about 10, excess calcium ions react with phosphates to precipitate hydroxylapatite, $Ca_{10}(PO_4)_6(OH)_2$. The amount of lime required will be independent of the amount of phosphate present and will depend primarily on the alkalinity of the wastewater and the degree of phosphate removal required. The pH of the wastewater must be adjusted by recarbonation with **carbon dioxide** (CO_2) before subsequent treatment or disposal. Lime can be added to primary sedimentation tanks or following secondary treatment.

Biological phosphate removal is accomplished by sequencing and producing appropriate environmental conditions in a reactor system in which microorganisms absorb and store phosphorus. The *Acinetobacter* bacteria are one of the primary microorganisms responsible for phosphorus uptake. The sludge containing the microorganisms with the excess phosphorus can either be wasted or removed and treated in a side stream to release the excess phosphate, which can then be treated with chemical precipitation processes. Wasted sludge, if it has a relatively high phosphorus content (3-5%), may have **fertilizer** value.

In natural treatment systems for wastewater, which include managed land-farms, landscape **irrigation** sites, **groundwater** recharge sites, and constructed **wetlands**, phosphates are removed by **sorption** with **clay minerals** in the **soil** matrix and by chemical precipitation. Sorbed phosphates are held tightly and are generally resistant to **leaching** until the soil sorptive capacity for phosphates becomes saturated. Chemical precipitation with calcium (in soils with neutral or alkaline pH) or with iron or aluminum (at acid pH) occurs at a slower rate than sorption, but is also an important removal mechanism. The degree of phosphorus removal in a natural treatment system depends on the degree of wastewater contact with the soil matrix.

Transport of phosphates into surface waters from nonpoint sources is controlled by practices that minimize soil erosion and runoff water. Phosphate losses from a **water-**

shed can be increased by a number of human activities, including timber harvest, intensive livestock grazing, soil tillage, and soil application of animal manures and phosphate-containing fertilizers. Since most soils, except for very sandy soils or soils with high levels of organic matter, retain phosphates through sorption and precipitation, techniques that prevent soil transport will also prevent transport of phosphates. Control of runoff water is required to prevent dissolved phosphates from entering rivers and lakes.

[*Judith L. Sims*]

RESOURCES
BOOKS

Metcalf & Eddy. *Wastewater Engineering: Treatment, Disposal, and Reuse.* 3rd ed. New York: McGraw-Hill, 1991.

Phosphorus

An important chemical element, which has the atomic number 15 and atomic weight 30.9738. Phosphorus forms the basis of a large number of compounds, by far the most environmentally important of which are **phosphates**. All plants and animals need phosphates for growth and function, and in many natural waters the production of algae and higher plants is limited by the low natural levels of phosphorus. As the amount of available phosphorus in an aquatic **environment** increases, plant and algal growth can increase dramatically leading to eutrophication. In the past, one of the major contributors to phosphorus **pollution** was household **detergents** containing phosphates. These substances have now been banned from these products. Other contributors to phosphorus pollution are **sewage treatment** plants and **runoff** from cattle **feedlots**. (Animal feces contain significant amounts of phosphorus.) **Erosion** of farmland treated with phosphorus fertilizers or animal manure also contributes to eutrophication and **water pollution**. *See also* Cultural eutrophication

Phosphorus removal

Phosphorus (usually in the form of phosphate, $PO^{3-}{}_4$) is a normal part of the **environment**. It occurs in the form of phosphate-containing rocks and as the excretory and decay products of plants and animals. Human contributions to the phosphorus cycle result primarily from the use of phosphorus-containing **detergents** and fertilizers.

The increased load of phosphorus in the environment as a result of human activities has been a matter of concern for more than four decades. The primary issue has been to what extent additional phosphorus has contributed to the eutrophication of lakes, ponds, and other bodies of water. Scientists have long recognized that increasing levels of phosphorus are associated with eutrophication. But the evidence for a direct cause and effect relationship is not entirely clear. Eutrophication is a complex process involving **nitrogen** and **carbon**, as well as phosphorus. The role of each **nutrient** and the interaction among them is still not entirely clear.

In any case, environmental engineers have long explored methods for the removal of phosphorus from **wastewater** in order to reduce possible eutrophication effects. Primary and secondary treatment techniques are relatively inefficient in removing phosphorus with only about ten percent extracted from raw wastewater in each step. Thus, special procedures during the tertiary treatment stage are needed to remove the remaining phosphorus.

Two methods are generally available: biological and chemical. Bacteria formed in the **activated sludge** produced during secondary treatment have an unusually high tendency to adsorb phosphorus. If these bacteria are used in a tertiary treatment stage, therefore, they are very efficient in removing phosphorus from wastewater. The **sludge** produced by this bacterial action is rich in phosphorus and can be separated from the wastewater leaving water with a concentration of phosphorus only about five percent that of its original level.

The more popular method of phosphorus removal is chemical. A compound is selected that will react with phosphate in wastewater, forming an insoluble product that can then be filtered off. The two most common substances used for this process are alum, **aluminum** sulfate [$Al_2(SO_4)_3$] and lime, or calcium hydroxide [$Ca(OH)_2$]. An alum treatment works in two different ways. Some aluminum sulfate reacts directly with phosphate in the waste water to form insoluble aluminum phosphate. At the same time, the aluminum **ion** hydrolyzes in water to form a thick, gelatinous precipitate of aluminum hydroxide that carries phosphate with it as it settles out of solution.

The addition of lime to wastewater results in the formation of another insoluble product, calcium hydroxyapatite, which also settles out of solution.

By determining the concentration of phosphorus in wastewater, these chemical treatments can be used very precisely. Exactly enough alum or lime can be added to precipitate out the phosphate in the water. Such treatments are normally effective in removing about 95 percent of all phosphorus originally present in a sample of wastewater.

[*David E. Newton*]

RESOURCES
BOOKS

Phosphorus Management Strategies Task Force. *Phosphorus Management for the Great Lakes.* Windsor, Ont.: International Joint Commission, 1980.

Retrofitting POTWs for Phosphorus Removal in the Chesapeake Bay Drainage Basin: A Handbook. Cincinnati: U.S. Environmental Protection Agency, 1987.

Symposium on the Economy and Chemistry of Phosphorus. *Phosphorus in the Environment: Its Chemistry and Biochemistry.* New York: Elsevier, 1978.

Photic zone

The surface waters of an ocean or lake that receive sufficient solar radiation to support **photosynthesis**, (i.e. the growth of vascular plants and **phytoplankton**). When sunlight falls on the surface of a lake or ocean, a large part of the light is reflected, scattered, or absorbed. Some light penetrates the water, but its intensity decreases quite rapidly with depth. The photic zone can be up to 328 ft (100 m) deep, and, typically, its limit is reached when the light intensity falls below one percent of the incident radiation.

Photochemical reaction

Photochemical reactions are driven by light or near-visible electromagnetic radiation. In general, incoming units of energy, known as photons, excite effected molecules, raising their energy to a point where they can undergo reactions that would normally be exceedingly difficult. The process is distinguished from thermal reactions, which take place with molecules in their normal energy states. Under sunlit conditions, photochemical processes can generate small amounts of extremely reactive molecules which initiate important chemical reaction sequences.

To initiate a photochemical reaction, two requirements need to be met. First, the photon must have enough energy to initiate the photochemical reaction in the molecule. Second, the compound must be colored, in order to be able to react with visible or near-visible photon radiation.

In **environmental chemistry**, photochemical reactions are of considerable importance to the trace chemistry of the **atmosphere**. The Los Angeles **photochemical smog** is an example of a system of photochemical reactions. Perhaps one of the most important reactions in this system is the photochemically driven conversion of **nitrogen** dioxide, a major component of **automobile** exhaust, to nitric oxide and an atom of oxygen, which usually occurs in pairs in the air. The oxygen atom subsequently attaches to an oxygen molecule to form the secondary pollutant, **ozone**. Photochemically initiated reactions are also responsible for generating the all important hydroxyl radical, a key intermediate in the trace chemistry of the lower atmosphere.

Photochemical reactions are also important in natural waters. Here they may be responsible for enhancing the reaction rate of organic compounds or changing the oxidation state of metallic ions in solution. However, some of these compounds are not colored, and are unable to absorb light. In such cases reactions may be mediated by photosensitizers, compounds capable of being energized by absorbed light which then pass the energy onto other molecules. Seawater appears to contain natural photosensitizers, perhaps in the form of fulvic **acid** or chlorophyll derivatives. Photosensitizers activated by sunlight can react with organic compounds or **dissolved oxygen** to produce the reactive singlet-oxygen, which can react rapidly with many seawater compounds.

Photochemical reactions also occur in solids. Since solids often lack the transparency required for the light to penetrate the surface, reactions are usually limited to the surface, where incoming photons initiate reactions in the top-most molecules. There is also some evidence that, under extremely bright **desert** sunlight, traces of nitrogen and water absorbed on titanium and zinc oxides can be photochemically converted to ammonia. Such processes are no doubt also important on planetary surfaces. *See also* Air pollution; Air quality; Emission; Los Angeles Basin

[*Peter Brimblecombe*]

RESOURCES
BOOKS

Findlayson-Pitts, B. J., and J. N. Pitts. *Atmospheric Chemistry.* New York: Wiley, 1986.

Photochemical smog

A form of **smog** that characterizes polluted atmospheres where high concentrations of **nitrogen oxides** and volatile organic compounds--often from gas-driven automobiles--mixed with sunlight promote a series of photochemical reactions that lead to the formation of **ozone** and a range of oxidized and nitrated organic compounds. The smog turns brownish in color and lowers **visibility** towards the middle of the day as sunlight becomes intense. The smog also causes eye irritation and a range of less distinct short- and long-term health effects. Photochemical smog, though typified by the **atmosphere** of the **Los Angeles Basin**, is increasingly found in cities all over the world where volatile fuels are used.

Photodegradable plastic

Plastics are clearly one of the great chemical inventions of all time. It would probably be impossible to list all the ways

in which plastics are used in everyday life. Suffice to say that, in the early 1990s, more than 60 billion pounds (27 billion kg) of plastics were produced in the United States alone each year.

For their many advantages, plastics also create some serious environmental problems. About one third, 20 billion pounds (9 billion kg), of all plastics now produced are used for short-term purposes, such as shopping bags and wrapping material. These plastics are used once, briefly, and then discarded.

Current estimates are that plastic materials make up about 7 percent by weight and 18 percent by volume of all municipal wastes. Since most plastics do not degrade naturally, they will continue to be a part of the nation's (and the world's) **solid waste** problem for decades.

One solution to the accumulation of plastics is to do a better job of **recycling** them. Technical, economic, and social factors have limited the success of this approach so far, and no more than 1 percent of all discarded plastics are ever recycled. Another solution is to fabricate plastics in such a way that they will degrade. Addition of starch as a filler in plastics, for example, tends to make them more **biodegradable**, that is, degradable by naturally-occurring organisms.

Yet another approach is photodegradable plastics, plastics that will decompose when exposed to sunlight. The process of photodegradation is well understood by chemists. In general, when light strikes a molecule, it may initiate any number of reactions that result in the destruction of that molecule. For example, light may dislocate one or more electrons that make up a **chemical bond** in the molecule. When the bond is broken, the molecule is destroyed.

On a practical level, that means that light can cause certain materials to decompose. Among the plastics, aromatic-based polymers (those that have benzene-like ring structures) are particularly susceptible to photodegradation. Those that lack an aromatic structure are less susceptible to the effects of light. Polyethylene, polypropylene, **polyvinyl chloride**, polymethyl methacrylate, polyamides, and polystyrene--some of the most widely used plastics--fall into that latter category.

In addition to the type of plastic, the kind of light that falls on a material affects the rate of photodegradation. Ultraviolet (UV) light is more effective, in general, in degrading all plastics than are most other forms of light.

The chemist's task, then, is to find a mechanism for converting a substance that is normally non-photodegradable (like **polystyrene**) into a form that is decomposed by light. In principle, that challenge is rather easily met. Certain metallic ions and organic groups are known to absorb visible light strongly. If these ions or groups are included in a plastic material, their **absorption** of light will contribute to the decay of the plastic.

Photodegradable plastics are not a new market item. Webster Industries, of Peabody, Massachusetts, has been making photodegradable plastic bags since 1978. And Mobil Chemical focused some of its advertising in 1990 on Hefty trash bags that were designed to photodegrade.

But critics raise a number of points about photodegradable (as well as biodegradable) plastic materials. For one thing, most plastics are disposed of in sanitary landfills where they are rapidly covered with other materials. The critics question the good of having a photodegradable plastic that is not exposed to sunlight after it is discarded.

Questions have also been raised about the possible byproducts of photodegradation. It is possible that the very additives used to make a plastic photodegradable may have toxic or other harmful environmental effects, critics say.

Finally, photodegradable plastics take so long to decay (typically two months to two years) that they will still pose a hazard to **wildlife** and contribute to solid waste volume. The "false promise" of photodegradability may, in fact, actually encourage the use of more plastics and add to the present problem, rather than helping to solve it. *See also* Decomposition; Green packaging; Green products; Ultraviolet radiation

[*David E. Newton*]

Resources

Books

Denison, R. A., and J. Wirka. *Degradable Plastics: The Wrong Answer to the Right Question.* New York: Environmental Defense Fund, 1989.

Periodicals

Thayer, A. M. "Degradable Plastics Generate Controversy inSolid Waste Issues." *Chemical & Engineering News* (25 June 1990): 7–14.

"The Problem with Plastics." *Horticulture* (August 1989): 18–19.

Donnelly, J. "Degradable Plastics." *Garbage* 2 (May-June 1990): 42–47.

Photoperiod

The light phase in a cycle of alternating periods of light and dark. Changes in light-dark cycles, such as changes in daylength, are the most dependable external time cues (zeitgebers). The physiological response to the length of the day or night, known as photoperiodism, is mediated by an internal or biological clock that synchronizes daily activities with external light-dark cycles. Most **species** studied show a circadian or daily rhythm set to a 24-hour cycle. Organisms also show long-term photoperiodism where their seasonal activities, such as courtship, mating, reproduction, **migration**, flowering, and seed production, are coordinated with others in the population and within the community.

Photosynthesis

Photosynthesis is the process by which green plants capture sunlight and convert its kinetic energy into chemical energy by manufacturing complex sugar molecules or carbohydrates. The plants use **carbon dioxide** from the air and water as the source materials for photosynthesis. Both **carbon** dioxide and water store relatively small amounts of energy. The carbohydrates manufactured are rich in energy. Later, during the process of **respiration**, the plant breaks down these carbohydrates, and the energy that is then released is used to fuel the growth and **metabolism** of the plant. The photosynthetic process also releases oxygen along with the formation of the carbohydrates. The water (H_2O) that is used in the photosynthetic reaction contributes its **hydrogen** to the formation of the carbohydrate. The oxygen that is released comes from the remaining, unused portion of the water molecule. Therefore, water is just as essential a component as carbon dioxide to the photosynthetic process. The entire reaction is carried out with the help of green, light-sensitive pigments within the plant, known as chlorophylls.

Photosynthesis is vital to the earth in two ways. It is the process by which plant life is established, thereby providing food and supporting all the other consumers in the **food chain/web**. The release of oxygen also ensures a livable, breathable **atmosphere** for all other oxygen-dependent life forms. The maintenance of a stable balance between the processes of photosynthesis (which produces oxygen) and respiration (which consumes oxygen) is critical to the **environment**. For example, a large amount of organic matter goes into a lake receiving sewage. This organic matter will be used as a food source by bacteria which break it down by the process of respiration just as humans do. During the respiration process, the oxygen that is dissolved in the water is used up. If there is insufficient plant material in the lake to restore this used-up oxygen by photosynthesis, the total supply of **dissolved oxygen** in the waters of the lake may drop dangerously. Since fish are completely dependent on the dissolved oxygen for their breathing requirements, severe drops in the concentration of dissolved oxygen may kill the fish, leading to sudden, and sometimes massive, **fish kills**.

Another aspect of environmental **pollution** are the effects on the process of photosynthesis itself. A number of contaminants have been shown to affect plant growth and metabolism by inhibiting the plant's ability to photosynthesize. Frequently, as in the case of metals like **copper**, **lead** and **cadmium**, the mechanism of inhibition is due to the contaminant's effect on the chlorophyll pigment. Copper replaces the necessary magnesium in the chlorophyll molecule, and the copper-substituted chlorophyll cannot effectively capture light energy. Therefore, the effectiveness of the photosynthetic process is greatly diminished, which leads to a stunting of plant growth and a depletion of oxygen in the environment. The maintenance of a healthy level of photosynthesis is thus essential to the life of the planet, both on the global and the micro-environmental scales.

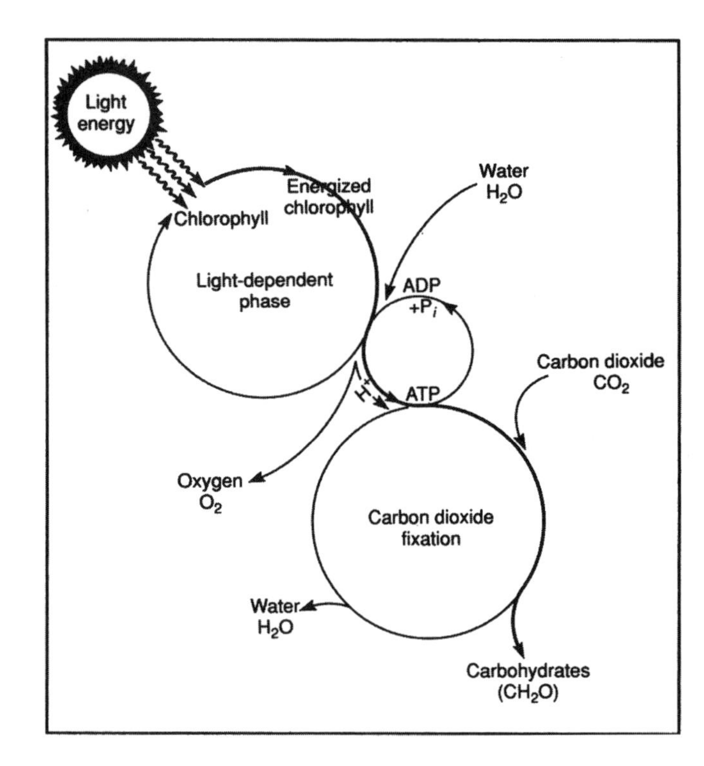

Photosynthesis. (McGraw-Hill Inc. Reproduced by permission.)

[*Usha Vedagiri*]

Resources

Books

Weier, T. E., et al. *Botany: An Introduction to Plant Biology.* New York: Wiley, 1982.

Connell, D. W., and G. J. Miller. *Chemistry and Ecotoxicology of Pollution.* New York: Wiley, 1984.

Photovoltaic cell

Photovoltaics is the direct conversion of sunlight into electrical energy using solar cells. All energy on earth is received from the sun through its electromagnetic spectrum. At any one instant, the sun delivers 1,000 watts (kw) per square meter to the earth's surface. Most of this energy is absorbed as heat by the lithosphere (**soil**), hydrosphere (water), and **atmosphere** but photovoltaic cells (PVC) are capable of converting it into a non-polluting, ecologically sound, and dependable source of electrical power. Although the photo-

voltaic effect was discovered 150 years ago, economically viable applications were not possible until the recent development of efficient semiconductor material and processing methods.

The physics of converting sunlight into electricity is simple. Most photovoltaic cells are a standard negative/positive type with attached leads. The negative terminal **lead** is soldered on the light sensitive side of the cell, and the positive lead is attached to the back side of the cell. When simply exposed to light, each cell produces about the same voltage between the two terminals. But if the cell is exposed to light when a load such as a discharged battery or an electric motor is connected between the two terminals, the voltage difference causes a flow of electrons. This current is caused by the formation of hole-electron pairs by the absorbed light photons, and the amount of current is dependent on the amount of absorbed light, which is dependent in turn on the incident light intensity and the surface area of the absorbing photovoltaic cell.

There are two main types of solar cells: thick-film cells with a thickness greater than 25 microns of crystalline silicon and thin-film cells with a thickness of less than 10 microns. Thin-film cells are made of various materials, including amorphous silicon and **copper** indium diselenide, and by combining several varieties of these in tandem, each with unique absorbing characteristics, the solar flux can be more efficiently utilized. Because these cells use less material, they are less costly to produce and will probably replace thick-film cells.

Both thick-film and thin-film cells are classified by the materials from which they are made—as crystalline (a wafer sliced from a large ingot), amorphous (The condensed gaseous form of a semi-conductor material such as silicon), and polycrystalline. Material combinations known as compound semiconductors have been investigated in the 1990s. These are cell materials whose active layers are comprised of various semiconductor materials, such as gallium arsenide, copper sulfide, and **cadmium** telluride.

To increase voltage, multiple cells are connected in a series by attaching the positive lead of one cell to the negative lead of another. The series most commonly used for both commercial and domestic applications is known as a module and usually consists of 36 cells. One or more modules can be connected directly to a load, such as a battery, a water pump, or an exhaust fan. A typical photovoltaic system consists of the modules, a storage battery, a charge and voltage regulator, and a suitable load. Some compact high performance modules are designed to charge 12-volt batteries or directly power a 12-volt DC motor.

Photovoltaic modules may be installed on a standard ground mount or on a tracker. When installed on a standard ground mount, the modules can be adjusted from 15–65 degrees at 5 degree increments. Ground mounts can support two to eight modules. The tracker utilizes a variable thermal expansion of gas, due to the changing solar exposure, and actually follows the sun at approximately 15 degrees plus latitude valve. To maximize efficiency, several modules are mounted on a tracker supported by a single pole.

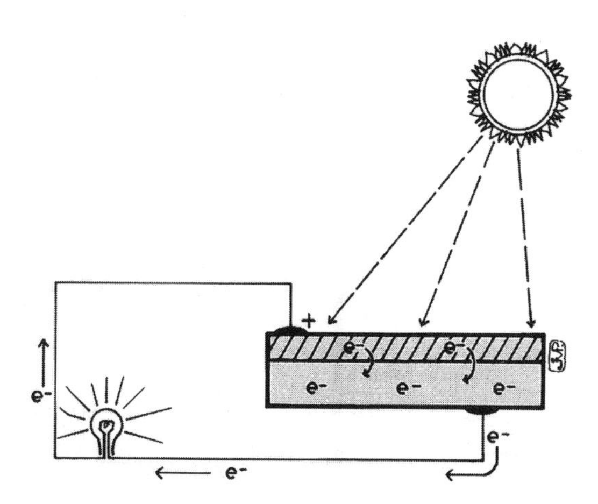

The operation of a photovoltaic cell. (McGraw-Hill Inc. Reproduced by permission.)

The net cost per kilowatt hour is the most important factor in the future of photovoltaic cell production and application. In industrialized countries, the economic viability of this form of **solar energy** is determined by their cost relative to competitive energy sources, particularly **fossil fuels** and **nuclear power**, and the environmental impact of each source. The world market price for solar cells in 1993 was four dollars a watt, based on rated output, and prices are expected to drop 50% by the end of the decade, when most photovoltaic cells will be manufactured with thin-film technology.

Decentralized single dwellings, cattle ranches and tree farms remote from electric power lines, and small villages with limited power demands are one of the three market segments where photovoltaic cells can be utilized competitively. The consumer and leisure market is another, and solar cells are already widely used in boats, motor homes, and camp sites, as well as in calculators and other electronics. The third market is in industrial applications such as offshore buoys, lighthouses, illuminated road signs, and railroad and traffic signals.

The market for photovoltaic cells is increasing at an overall annual rate of about 10%. Current worldwide demand is estimated at about 100 megawatts of electrical power, and predictions for the year 2000 estimate a demand of several times this size. Growth is the fastest in the remote market; rural consumer applications are increasing at an annual rate

of approximately 35%. There are millions of people around the world who are not served by **electric utilities** due to their remote location and the high cost of electrical transmission. These populations generally depend on a 12-volt **automobile** battery powered by a generator for their electrical needs, which include water pumping, lighting, and radio and television reception. The initial cost of a photovoltaic kit offsets the cost of owning and operating a generator within three years. As the life span of a module is usually 15 years, this option is much more economical in many of these situations.

Prior to 1989, the largest manufacturer of photovoltaic cells was ARCO Solar, a division of Atlantic Richfield Oil Company. In 1989, Siemens Solar Industries (SSI), which already had a joint manufacturing enterprise in Munich, Germany, with ARCO, had purchased ARCO Solar. SSI has manufacturing plants in Camarillo, California, and Munich, Germany.

Converting sunlight into electricity with photovoltaic cells is a versatile and simple process. Unlike diesel or **gasoline** generators, all-weather modules have no moving parts to wear out or break down, and they produce electricity without contributing to **air pollution**. Solar cells do not produce any noise and they do not require alternating-current power lines, since photovoltaic electricity is direct current. Maintenance is minimal and requires little technical skill; systems are easy to expand and there are no expensive fuels to purchase on a continuous basis. Photovoltaic cells are a cheap and dependable source of power for a variety of uses. *See also* Alternative energy sources; Alternative fuels; Energy and the environment; Energy policy

[*Muthena Naseri and Douglas Smith*]

Resources

Books

Lasnier, F., and T. Ang. *Photovoltaic Engineering Handbook.* New York: American Institute of Physics, 1990.

Periodicals

Edelson, E. "Solar Cell Update." *Popular Science* 240 (June 1992): 95–99.

Lewis, N. S. "More Efficient Solar Cells." *Nature* 345 (May 24, 1990): 293–94.

Spinks, P. "Plug Into the Sun." *New Scientist* 127 (September 22, 1990): 48–51.

Phthalates

Phthalates are aromatic compounds that are found throughout the **environment** and are used in some **plastics**, solvents, **detergents**, lubricating oils, and cosmetics. There is concern that phthalates can cause **cancer** or reproductive problems in people and animals. In 1998, the U.S. Consumer Product Safety Commission (CPSC) conducted a study of the amount of one type of phthalate, diisononyl phthalate (DINP) that can leach from children's products such as teethers, rattles, and toys. This study set an international standard concentration for DINP in children's products that is 10 times higher than the level found to cause health problems in laboratory animals. The results of a study released in 2000 by the **Centers for Disease Control and Prevention** (CDC) and the National Toxicology Program (NTP) found four types of phthalates in the urine of more than 75% of test subjects, indicating that the phthalates that humans are exposed to in the environment are metabolized. The study was not designed to provide information on the health risks of phthalates or on the degree of exposure of the United States population, and follow-up studies by the CDC are measuring phthalate concentrations in a larger part of the population, including children six years old and older.

[*Marie H. Bundy*]

Phytoplankton

Photosynthetic **plankton**, including microalgae, blue-green bacteria, and some true bacteria. These organisms are produced in surface waters of aquatic ecosystems, to depths where light is able to penetrate. Sunlight is necessary since, like terrestrial plants, phytoplankton use solar radiation to convert **carbon dioxide** and water into organic molecules such as glucose. Phytoplankton form the base of nearly all aquatic food chain/webs, directly or indirectly supplying the energy needed by most aquatic protozoans and animals. Temperature, **nutrient** levels, light intensity, and consumers (grazers) are among the factors that influence phytoplankton community structure.

Phytoremediation

Phytoremediation combines the Greek word *phyton*, "plant," with the Latin word *remediare,* "to remedy," to describe a system whereby certain plants, working together with **soil** organisms, can transform contaminants into harmless and often, valuable, forms. This practice is increasingly used to remediate sites contaminated with **heavy metals** and toxic organic compounds.

Phytoremediation takes advantage of plants' **nutrient** utilization processes to take in water and nutrients through roots, transpire water through leaves, and act as a transformation system to metabolize organic compounds, such as oil and pesticides. Or they may absorb and bioaccumulate toxic trace elements, such as the heavy metals, **lead**, **cadmium**, and selenium. In some cases, plants contain 1,000 times more metal than the soil in which they grow. Heavy metals

are closely related to the elements plants use for growth. In many cases, the plants cannot tell the difference, according to Ilya Raskin, Professor of Plant Sciences in the Center for Agricultural Molecular Biology at Rutgers University.

Phytoremediation is an affordable technology that is most useful when contaminants are within the root zone of the plants (top 3-6 ft [1-2 m]). For sites with contamination spread over a large area, phytoremediation may be the only economically feasible technology. The process is relatively inexpensive because it uses the same equipment and supplies used in agriculture.

Soil **microorganisms** can degrade organic contaminants. This is called **bioremediation** and has been used for many years both as an in situ process and in land farming operations with soil removed from sites.

Ilya Raskin and his interdisciplinary team at Rutgers AgBiotech Center were first to demonstrate the utility of certain varieties of mustard plants in removing such metals as chromium, lead, cadmium, and zinc from **contaminated soil**. Related technology developed by Raskin's group used hydroponic plant cultures to remove toxic metals from aqueous waste streams.

Plants can accelerate bioremediation in surface soils by their ability to stimulate soil microorganisms through the release of nutrients from and the transport of oxygen to their roots. The zone of soil closely associated with the plant root, the rhizosphere, has much higher numbers of metabolically active microorganisms than unplanted soil. The rhizosphere is a zone of increased microbial activity and **biomass** at the root-soil interface that is under the interface of the plant roots. It is this symbiotic relationship between soil **microbes** that is responsible for the accelerated degradation of soil contaminants.

The interaction between plants and microbial communities in the rhizosphere is complex and has evolved to the mutual benefit of both organisms. Plants sustain large microbial populations in the rhizosphere by secreting substances such as carbohydrates and amino acids through root cells and by sloughing root epidermal cells. Also, root cells secrete mucigel, a gelatinous substance that is a lubricant for root penetration through the soil during growth. Using this supply of nutrients, soil microorganisms proliferate to form the plant rhizosphere.

In addition to this rhizosphere effect, plants themselves are able to passively take up a wide range of organic wastes from soil through their roots. One of the more important roles of soil microorganisms is the **decomposition** of organic residues with the release of plant nutrient elements such as **carbon, nitrogen**, potassium, phosphate, and sulfur. A significant amount of the **carbon dioxide** (CO_2) in the **atmosphere** is utilized for organic matter synthesis primarily through **photosynthesis**. This transformation of carbon dioxide and the subsequent sequestering of the carbon as root biomass contributes to balancing the effect of burning **fossil fuels** on global warming or cooling.

Compounds are frequently transformed in the plant tissue into less toxic forms or sequestered and concentrated so they can be removed (harvested) with the plant. For example, mustard greens were used to remove 45% of the excess lead from a yard in Boston, Massachusetts, to ensure the safety of children who play there. The sequestered lead was carefully removed and safely disposed of. Besides mustard greens, pumpkin vines were used to clean up an old Magic Marker factory site in Trenton, New Jersey. Hydroponically grown sunflowers were used to absorb radioactive metals near the Chernobyl nuclear site in the Ukraine and a **uranium** plant in Ohio. The mustard's hyper-accumulation results in much less material for disposal. The **composting** of plant material can be another highly efficient stage in the breakdown of contaminants removed from the soil.

When trees are used, such as poplars, the idea is to move as much water through them as possible, so that they take up as much of the contaminants as possible. Once the heavy metals are absorbed, they are sequestered in the trees' roots. Any organic compounds that are absorbed are metabolized.

Absorption of large amounts of nutrients by plants and only a small amount of plant **toxins** that might be harmful to them, is the key factor. Plants generally absorb large amounts of elements they need for growth and only small amounts of toxic elements that could harm them. Therefore, phytoremediation is a cost-effective alternative to conventional **remediation** methods. Cleaning the top 6 in (15 cm) of contaminated soil with phytoremediation costs an estimated $2,500-$15,000 per hectare (2.5 acres), compared to $7,500-$20,000 per hectare for on-site microbial remediation. If the soil is moved, the costs escalate, but phytoremediation costs are still far below those of traditional remediation methods, such as stripping the contaminants from the soil using physical, chemical, or thermal processes, according to Dr. Scott Cunningham, a scientist at Dupont Central Research for Environmental Biotechnology.

Plants are effective at remediating soils contaminated with organic chemical wastes, such as solvents, petrochemicals, wood preservatives, explosives, and pesticides. The conventional technology for soil clean-up is to remove the soil and isolate it in a **hazardous waste landfill** or incinerate it.

Phytoremediation, says Dr. Ray Hinchman, botanist and plant physiologist at Argonne National Laboratory, is "an *in situ* approach," not reliant on the transport of contaminated material to other sites. Organic contaminants are, in many cases, completely destroyed (converted to CO_2 and H_2O) rather than simply immobilized or stored.

Salt-tolerant plants, called halophytes, have reduced the salt levels in soils by 65% in only two years in one project involving brine-damaged land from run-off from oil and gas production in Oklahoma. After the salt was reduced, the halophytes died and native grasses, which failed to thrive when too much salt entered the soil, naturally returned, replacing the salt-converting plants.

The establishment of vegetation on a site also reduces soil **erosion** by wind and water, which helps to prevent the spread of contaminants and reduces exposure of humans and animals.

Classes of organic compounds that are more rapidly degraded in rhizosphere soil than in unplanted soil include: total **petroleum hydrocarbons**; **polycyclic aromatic hydrocarbons**; chlorinated pesticides (PCP, **2,4-D**); other chlorinated compounds (PCBs, TCE); explosives (TNT, DNT); **organophosphate** insecticides (diazanon and parathion); and surfactants (**detergents**).

Some plants used for phytoremediation are: alfalfa, symbiotic with hydrocarbon-degrading bacteria; arabidopsis, carries a bacterial gene that transforms **mercury** into a gaseous state; bladder campion, accumulates zinc and **copper**; brassica juncea (Indian mustard greens), accumulates selenium, sulfur, lead, chromium, cadmium, **nickel**, zinc, and copper; buxaceae (boxwood), accumulates nickel; compositae family (symbiotic with arthrobacter bacteria), accumulates cesium and strontium; euphorbiaceae (succulent), accumulates nickel; ordinary tomato, accumulates lead, zinc, and copper; poplar, used in the absorption of the **pesticide**, **atrazine**; and thlaspi caerulescens (alpine pennycress), accumulates zinc and cadmium.

[*Carol Steinfeld*]

Resources

Books

Elliott, L. F., and F. J. Stevenson. *Soils for the Management of Wastes and Waste Waters.* Madison, WI: Soil Science Society of America, 1977.

Periodicals

Anderson, G., and W. "Bioremediation, Environmental Science and Technology." *American Chemical Society* 27, no. 13 (1993).

Howe, P. "Plants Doing the Dirty Work in Cleanup of Toxic Waste," *Boston Globe,* March 10, 1997.

Phytotoxicity

Phytotoxicity refers to the damage to plants caused by exposure to some toxic stressor. Phytotoxicity can be evidenced by acute injury such as obviously damaged, necrotic tissues. Chronic phytotoxicity does not present such obvious symptoms and may result in such "hidden injuries" as a decrease in productivity or a change in growth form. Plants can be poisoned by naturally occurring **toxins**, such as **nickel** in serpentine-influenced soils. Humans also release phytotoxic **chemicals** into the **environment**. Such emissions occur when, for example, gaseous and/or metal pollutants around smelters damage vegetation. Deliberate use of phytotoxic chemicals includes the use of herbicides in agriculture or forestry to decrease the abundance of unwanted plants or weeds.

Gifford Pinchot (1865 – 1946)

American conservationist and forester

Pinchot was born in Simsbury, Connecticut, to a prosperous business and industrial family, part of whose wealth came from timber holdings in several states. The Pinchots, like other lumber investors of their day, practiced **clear-cutting** on forests to maximize their profits, shipped the logs to market, and with the returns, repeated the cycle. Young Gifford was pressured by his grandfather to enter the family business, but his father James was beginning to dislike the **deforestation** of his area, and he encouraged his son to pursue forestry.

Pinchot was educated at Exeter and then Yale, graduating in 1889. After graduation, his family supported further education at L'Ecole nationale forestiere in Nancy, France, where he studied silviculture, or forest **ecology**. It was in Nancy that he learned "*le coup d'oeil forestier*—the forester's eye, which sees what it looks at in the woods." In his 1947 book, *Breaking New Ground*, Pinchot noted that it was in France that he began to think of "the forest as a crop," that forests could sustain human use by a "fixed and annual supply of trees ready for the axe."

Returning to the United States in 1892, Pinchot engaged in "the first scientific forestry in America" on the Vanderbilt estate in North Carolina, hoping to prove that "trees could be cut and the forest preserved at one and the same time." Pinchot figured out how to sustain the forest while maintaining its yield, and thus its income. He called his first successful year "a balance on the side of practical forestry."

Pinchot then worked as a consulting forester in New York City, where he attracted increasing attention for his ideas. The National Academy of Science appointed him to the new National Forest Commission, organized by the **U.S. Department of the Interior**. On the Commission's recommendation, President Grover Cleveland added 13 more forest reserves totaling 21 million acres (8.5 million ha). Pinchot, voicing a minority opinion on the Commission,

Gifford Pinchot. (Corbis-Bettmann. Reproduced by permission.)

saw the reserves as property for public use, not as a way to lock up the nation's **natural resources**.

In 1898 Pinchot became head of the U.S. Department of Agriculture's Division of Forestry. The division did not operate the nation's forest reserves, which were then administered by the Department of the Interior's General Land Office. Pinchot lobbied hard to have the reserves transferred to his division, but his efforts were unsuccessful until William McKinley's assassination moved **Theodore Roosevelt** into the White House. Pinchot and Roosevelt together mustered enough support in Congress to approve the transfer to the renamed Bureau of Forestry in 1905, marking the beginning of the national forest system and what became the U.S. **Forest Service**.

Pinchot's long-time collaboration with Roosevelt was one of the benchmarks of the **conservation** movement, which culminated in the creation of what have been called the "midnight forests." These 16 million acres of forest were set aside as reserves late in the night before a Congressional bill to limit the President's power to do so took effect. After Roosevelt left office, Pinchot's association with President William Howard Taft soured quickly, climaxing in a feud with Richard Ballinger, Taft's Secretary of the Interior and ending in 1910, when Pinchot was fired by the President.

Despite leaving before he had accomplished all that he wished, Pinchot left his mark on the Forest Service. Ironically, both are often criticized by modern environmentalists for their utilitarian policies. Yet this early conservationist was one of first to protect national forest lands from the industrial powers who sought to destroy them completely.

Pinchot still generates controversy today. Most writers trace the conflict between conservationists and preservationists to his falling-out with **John Muir** over the Hetch-Hetchy Reservoir in California. Pinchot, with his pragmatic approach, argued that the river valley served the greatest public good as a dependable source of water for the people of San Francisco. Muir maintained that Hetch Hetchy was sacred and too beautiful to flood. He also believed the dam would betray the **national park** ideal, since Hetch Hetchy was in **Yosemite National Park**.

As he grew older, Pinchot did become less utilitarian, creating what he called the "new conservationism," but even his "new" approach was an extension of his long-held conviction that the forests of the nation belonged first to the people, to be used by them wisely and in perpetuity. To Pinchot, conservation meant "the greatest good to the greatest number for the longest time." Pinchot's beliefs, which have been revived in recent years, make sense both ecologically and economically. He claimed that "the central thing for which Conservation stands is to make this country the best possible place to live in, both for us and our descendants."

Pinchot was active on many fronts of the struggle for conservation. The **Society of American Foresters**, a professional association and major force in resource management and conservation today was created in 1900, due in part to his efforts. He served two terms as Governor of Pennsylvania (1923–27; 1931–35), and his efforts to relieve the effects of the Great Depression led to the creation of emergency work camps there, providing a model for the Civilian Conservation Corps started by President Franklin Roosevelt.

Pinchot recognized that people *must* live off the resources of land where they live. In *The Fight for Conservation* (1910), he stated bluntly that "the first principle of conservation is development, the use of the natural resources now existing on the continent for the benefit of the people who live here now." But, he also recognized that use of resources could quickly turn into abuse of resources, especially by exploitation for the few. He believed that destruction of forest resources was detrimental to the **environment** and to the many who depended on it.

Pinchot's ideas and his national forests are still with us today. His concern with producing the greatest totality of **land use** produced the **wilderness** that is preserved but

accessible and forest that is logged and renewed. Pinchot, a giant of the conservation and the environmental movements, leaves his legacy of a vast forest domain that belong to the American people.

[*Gerald L. Young Ph.D.*]

Resources

Books

McGeary, M. N. *Gifford Pinchot: Forester-Politician.* Princeton, NJ: Princeton University Press, 1960.

Nash, R. "Gifford Pinchot." In *From These Beginnings: A Biographical Approach to American History.* Vol. 2. 2nd ed. New York: Harper & Row, 1978.

Norton, B. G. "Moralists and Aggregators: The Case of Muir and Pinchot." In *Toward Unity Among Environmentalists.* New York: Oxford University Press, 1991.

Pinchot, G. *Breaking New Ground.* Covelo, CA: Island Press, 1947.

Periodicals

Miller, C. "The Greening of Gifford Pinchot." *Environmental History Review* 16 (Fall 1992): 1–20.

Watkins, T. H. "Father of the Forests." *American Heritage* 42 (February–March 1991): 86–98.

PIRG

see **U.S. Public Interest Research Group**

Placer mining

Placer deposits are collections of some mineral existing in discrete particles, mixed with sand, gravel, and other forms of eroded rock. Some of the minerals most commonly found in placer deposits are diamond, gold, platinum, magnetite, rutile, monazite, and cassiterite. These deposits are formed by the action of wind, water, and chemical changes on more massive beds of the mineral. Placer deposits can be classified according to the method by which they are produced. Some examples include stream placers, eolian placers (formed by the action of winds), beach placers, and moraine placers (formed by glacial action).

Placer deposits are important sources of some minerals that are more dense than the sand or gravel around them. For example, the gold in a placer deposit tends to settle out and accumulate in a stream as surrounding materials wash away.

With few exceptions, the mining of placer deposits is carried out by traditional surface (compared to underground) mining. Four general methods are in use. The oldest and perhaps most familiar to the general public is the hand method. Placerville, California, commemorates by its name the type of mining that was done during the Gold Rush of 1849. As the worker rocks his or her box or pans back and forth, clay and sand are washed away, and the heavier gold (or other minerals) settles to the bottom. This method is less used today since deposits with which it works best have been largely depleted.

Artificial streams of water are used in a second method of placer mining--hydraulic mining. In this process, intense streams of water are directed with large hoses at a placer deposit. The water washes away the overlying sand and clay, leaving behind the desired mineral.

A number of variations of this procedure exists. In some cases, a natural stream is diverted from its course and directed against a placer deposit. The action of the stream substitutes for that of a hose in the traditional hydraulic method. In another variation, a stream may be dammed rather than diverted. When the dam is broken, the penned-up water is released all at once with a force that can tear apart a placer deposit. This technique is known as "booming" a deposit and although it is no longer used in the United States, the method is used around the world.

An increasingly popular approach to placer mining makes use of familiar **surface mining** machinery used to extract **coal** and other minerals. This machinery is often very large and efficient. It removes **overburden**, extracts mineral-containing earths with streams of water and then separates and processes the extracted material by gravity methods.

A fourth method of mining is used with deposits underwater, as in a lake or along the seashore. In this process, a large machine somewhat similar to a surface mining excavator is placed on a barge. The machine then scoops materials off the lake, stream, or ocean bottom and carries them to a processing area on another part of the barge. This type of **dredging** machine might be thought of as a very large, mechanized version of the old prospector's gold pan and rocker box.

Placer mining poses two kinds of environmental problems. The first is one that is common to all types of surface mining, the disturbance of land. Federal law now requires that land disturbed by surface mining such as placer mining be restored to a condition that approximates its original state.

The extensive use of water during placer mining is a second source of concern. By its nature, placer mining produces huge flows of water that have become contaminated with mud, sand, and other suspended solids. As this water is dumped into natural waterways, these solids may damage aquatic plant and animal life and, as they settle out, create new navigational hazards. In the United States, the Clean Water Acts and other environmental laws carry provisions to reduce the potential harm posed by placer mining.

[*David E. Newton*]

RESOURCES
BOOKS

Sinclair, J. *Quarrying, Opencast and Alluvial Mining.* New York: Elsevier, 1969.

PERIODICALS

"Mining Companies Urged to Provide for Wildlife Habitats." *Engineering and Mining Journal* 194 (April 1993): 16.

Pynn, L. "The Legacy of Klondike Gold: Strange Things Done to the Landscape by the Men and Women Who Moil for Gold." *Canadian Geographic* 112 (May-June 1992): 52–61.

Wuerthner, G. "Hard Rock and Heap Leach." *Wilderness* 55 (Summer 1992): 14–21.

Plague

For centuries past, plague was the scourge of the earth, killing at least an estimated 163 million people during three major worldwide pandemics. Today, plague still exists in the world although the number of cases and deaths is much lower. Plague still infects several thousand people a year and kills several hundred. Most cases are in southern Africa, Asia, and central South America.

Plague is an infectious disease that affects animals and humans. It is caused by the bacterium *Yersinia pestis* and is usually found in rodents and their fleas. When released into the air, the bacterium can survive for up to an hour. People usually contract the disease after being bitten by an infected flea or after handling an infected animal. To reduce the risk of death, antibiotics must be given within 24 hours after the first symptoms occur.

There are three types of plague and they can occur separately or in combination. The types are:

- Bubonic plague, the most common form of the disease. It does not spread from person to person but is transmitted when an infected flea bites a person or when material contaminated with the bacterium enters through a cut in a person's skin. Symptoms include swollen lymph glands, fever, headache, chills, and weakness.
- Pneumonic plague, which infects a person through the lungs. It can be spread in the air from person to person. Symptoms include fever, headache, weakness, shortness of breath, chest pain, cough, and pneumonia.
- Septicemic plague, which occur when plague bacteria multiply in the blood. It can occur by itself or be a complication of pneumonic or bubonic plague. It is not spread from person to person. Symptoms include fever, chills, exhaustion, abdominal pain, shock, and bleeding into the skin and other organs.

To reduce the chance of death, people with plague must receive antibiotics within 24 hours after the first symptoms appear. Effective antibiotics include streptomycin, gentamicin, and tetracycline. Early treatment allows most people with plague to be cured if diagnosed in time. There is no vaccine against plague available.

From 1954 to 1997, there were 80,613 reported plague cases in 38 countries, of which 6,587 people died, according to the World Health Organization (WHO). The highest number of cases, 6,004, occurred in 1967, and the lowest number, 200, occurred in 1981. These numbers likely represent only about a third of the actual number of cases, due to inadequate surveillance and reporting methods, according to WHO.

Plague has been around since the dawn of civilization and probably existed in animals before humans inhabited the earth. The first recorded plague epidemic occurred in 1320 BC and is described by its symptoms in the Bible (I Samuel, V and VI). It is not known how many people were infected. The first known pandemic, or severe worldwide epidemic, occurred in Asia, Africa, and Europe between 542 and 546 AD. It was named Justinian's plague, after the emperor of the Byzantine Empire at the time, and killed an estimated 50 million to 100 million people.

The first pandemic is believed to have started in Egypt, spreading north through the eastern Mediterranean. The plague struck at a critical time. In Constantinople, the capital of the Byzantine empire, Justinian and his generals were in the midst of battles to rejoin Byzantium to the remains of Western Roman Empire. Justinian's dream was to re-establish the former Roman Empire; and he might have succeeded if not for the plague. Procopius, a historian living in Constantinople at the time, vividly described the plague and its effects. His writings also include an accurate description of what would later be called the bubonic plague. He wrote about agitated, feverish disease victims with painfully enlarged lymph nodes (buboes) under their arms and in their groins. According to Procopius, 300,000 people died within the city alone. It is impossible to verify whether that figure is accurate, but it is certain that the plague had a dramatic effect on history. Justinian's plans to re-establish the Roman Empire were very likely derailed because of the loss of manpower. In the west, the remnants of the Western Roman Empire collapsed into the Dark Ages.

Following the initial outbreak in 542, the plague disappeared and reappeared at intervals over the next 200 years. The years 542 to 600 were the most intense plagues years, but local epidemics flared up throughout the Mediterranean region through the mid-eighth century. The population of Europe wasn't able to recover between outbreaks and some historians estimate that the population dropped by half between 542 and 700. After the late-eighth century, plague disappeared from Europe for nearly 600 years.

The second pandemic is the infamous "Black Death" from 1347 to 1350 AD. It killed from 25 million to 50 million

people, half in Asia and Africa and half in Europe (a quarter of the population). Its entry into Europe was through Kaffa, a small Italian trading colony on the shore of the Black Sea in Crimea. Kaffa was besieged by an army and during the siege, plague broke out among the soldiers. According to some sources, the soldiers threw the corpses of those who died over the town walls to spread plague among the men defending the town. Whether that caused plague to break out in the town is unknown, but the defenders were afflicted. They managed to get to their boats and flee to Italy, unknowingly carrying the disease with them.

By late 1347, plague was widespread in the Mediterranean region, and in 1348 it spread throughout Italy, France, and England. The Middle East and the Far East were also severely affected. Medieval physicians were at a loss to explain the disease. Some claimed it was due to person-to-person infection, while others said it arose from a poisonous **atmosphere**. Other explanations put forth by panicked citizens blamed astrological influences, divine punishment, and the Jewish community. Tens of thousands of Jewish citizens were burned in Spain, Germany, Switzerland, and France despite protests by Pope Clement VI, Emperor Charles IV, and medical experts who said that they were innocent. However, the authorities were powerless to stop the spread of plague, and they were not believed. Quarantines of plague victims were ordered, but they were largely ineffective.

The Black Death was the beginning of a number of plague outbreaks that ravaged Europe and Africa in subsequent centuries. In England alone, there were plague outbreaks in 1361, 1369, 1390, 1413, 1434, 1439, and 1464. Additional outbreaks occurred in Scotland in 1401–1403, 1430–1432, and 1455. The epidemics cut England's population of about seven million in half from 1300 to 1380. The last major outbreaks of the Black Death pandemic occurred in London in 1665–1666 and in Marseilles in 1720–1722. Each of these outbreaks resulted in approximately 100,000 deaths.

In a controversial 2001 book, "Biology of Plagues," authors and epidemiologists Susan Scott and Christopher Duncan write they believe the plague of 1347–1351 was not caused by the bubonic plague, but rather by a still unknown disease. They say the quick spread of the disease and the fact that bubonic plague is not transmitted person to person, as the Black Death was, points to another cause. Another possibility, however, is that the ship trade of that time would have allowed for disease-infested rats to move from location to location rapidly and thus, spread the disease quickly through various populations.

The world's third plague pandemic began in 1894 in China and Hong Kong. Called the Modern Pandemic, it quickly spread throughout the world, carried by rats aboard steamships. From 1894 through 1903, plague-infested ships had brought the disease to 77 ports in Asia, Africa, Europe, North America, **Australia**, and South America. In India alone, plague killed about 13 million people. In the year the pandemic started, scientists discovered the cause of the disease, and established that rats and their fleas were the carriers.

According to most sources, the Modern Pandemic continues to the present time. Epidemics linked to the Modern pandemic were reported in San Francisco, New Orleans, and other coastal cities throughout the world and millions died. The Modern pandemic also established in areas that previously had been plague free, including North America, South America, and southern Africa. With the advent of antibiotics and the understanding of how plague spreads, the Modern pandemic has been contained and the plague no longer claims millions of victims. However, health authorities remain vigilant because the plague has not been eradicated.

At the turn of the twentieth century, plague outbreaks twice hit San Francisco. The first outbreak was from 1900 to 1904, and the second from 1907 to 1909. Several hundred infections were confirmed but health officials believe hundreds of cases went unreported. More recently, an outbreak of bubonic plague occurred in the African nation of Malawi in early 2002, infecting at least 10 people

The decrease in the incidence of plague today is due primarily to improved living standards and health services in many underdeveloped nations. In India, plague epidemics have stopped altogether. The reasons for this are not clear, according to WHO scientists. The Government of India, however, has taken a number of stringent measures to prevent spread of the disease. These include mandatory screening at airports of every airline passenger intending to travel abroad for signs and symptoms of plague, **fumigation** of aircraft, including both the passenger cabin and the cargo hold. Additionally, all aircraft are inspected for the presence of rats. At all major sea ports, all out-going vessels are being inspected for the presence of rats and insects. Ships are required to have a certificate, issued by the port health officer, showing that they have been deratted, before they can leave port. All these measures are in accordance with the WHO International Health Regulations.

Plague is often mentioned as a potential biological weapon of terrorists. When released into the air, the bacterium can survive for up to an hour. Of the three types of plague (pneumonic, bubonic, and septicemic), pneumonic is they one most likely to be used by terrorists since large stockpiles were developed by the United States and Soviet Union in the 1950s and 1960s. Much of the Soviet supply is now in the hands of several of its former republics, now independent nations. The U.S. government believes several so-called "rogue" nations, such as Iran, Iraq, and North

Korea, have biological weapons programs. While several other nations presumably have biological weapons programs, these nations pose a more significant threat to the United States and its allies.

[*Ken R. Wells*]

Resources
Books

Kohn, George C., ed. *Encyclopedia of Plague and Pestilence: From Ancient Times to the Present.* New York: Facts on File, Inc., 2001.

Periodicals

Childress, Diana. "Like Black Smoke, the Black Death's Journey." *Calliope* (March 2001): 4.

Ligomeka, Brian. "Bubonic Plague Breaks Out in Lilongwe." *Africa News Service*, May 10, 2002.

Mackenzie, Deborah. "All Fall Down: The Black Death Wasn't Caused by Bubonic Plague, and We Need to Find the Real Culprit Before it Strikes Again." *New Scientist* (November 24, 2001): 35–38.

McCabe, Suzanne. "Dark Death: In 1347, a Terrible Plague Brought Death and Destruction to Europe." *Junior Scholastic* (February 11, 2002): 18–22.

Saul, Nigel. "Britain 1400." *History Today* (July 2000): 38.

"Time Trip. (Black Death)" *Current Events, a Weekly Reader Publication*, January 11, 2002): 2–3.

Truitt, Elly. "The Medical Response to the Black Death." *Calliope* (March 2001): 14.

Organizations

World Health Organization Regional Office, 525 23rd Street NW, Washington, DC USA 20037 (202) 974-3000, Fax: (202) 974-3663, Email: postmaster@paho.org, <http://www.who.int>

Plankton

Organisms that live in the water column, drifting with the currents. Bacteria, **fungi**, algae, protozoans, invertebrates, and some vertebrates are represented, some organisms spending only parts of their lives (e.g. larval stages) as members of the plankton. Plankton is a relative term since many planktonic organisms possess some means by which they may control their horizontal and/or vertical positions. For example, organisms may possess paddle-like flagella for propulsion over short distances, or they may regulate their vertical distributions in the water column by producing oil droplets or gas bubbles. Plankton comprise a major item in aquatic food chain/webs. *See also* Phytoplankton; Zooplankton

Plant pathology

Plant pathology is the study of plant diseases. The most common causes of plant diseases are bacteria, **fungi**, algae, viruses, and roundworms. Plant diseases can drastically affect a country's economy. They may be responsible for the loss of up to ten percent of human crops each year. Plant pathology is also important in environmental studies. Many plants are especially sensitive to pollutants in the air and water. They can be used as "early warning systems" of **pollution** problems that may affect humans as well. For example, **ozone** at a concentration of only 0.005 **parts per million** produces noticeable changes in **tobacco** plants.

Plasma

The term plasma has two major definitions in science. In biology, it refers to the clear, straw-colored liquid portion of blood. In physics, it refers to a state of matter in which atoms are completely ionized. By this definition, a plasma consists entirely of separate positive and negative ions. Plasmas of this kind exist only at high temperatures. The study of plasma is extremely important in research on **nuclear fusion**. One of the most difficult problems in that research is to find ways of physically containing a plasma in which fusion reactions are occurring.

Plastics

The term plastic refers to any material that can be shaped or molded. In this sense, ordinary clay or a soft wax is a plastic material. Perhaps more commonly, plastic has become the term used to describe a class of synthetic materials more accurately known in chemistry as polymers. Some common examples of plastics are the polyethylenes, polystyrenes, vinyl polymers, methyl methacrylates, and polyesters. These synthetic materials may or may not be "plastic" in the pliable sense.

Research on plastic-like materials began in the mid-nineteenth century. At first, this research made use of natural materials. Credit for discovery of the first synthetic plastic is often given to the American inventor, John Wesley Hyatt. In 1869, Hyatt was awarded a patent for the manufacture of a hard, tough material made out a natural cellulose. He called the product "celluloid."

It was not until 1907, however, that an entirely synthetic plastic was invented. In that year, the Belgian-American chemist Leo Baekeland discovered a new compound that was hard, water- and solvent-resistant and electrically non-conductive. He named the product Bakelite. Almost immediately, the new material was put to use in the manufacture of buttons, radio cases, telephone equipment, knife handles, counter tops, cameras, and dozens of other products.

Today, thousands of different plastics are known. Their importance is illustrated by the fact that of the 50 **chemicals** produced in the greatest volume in the United States, 24 are used in the production of polymers. Products

that were unknown until the 1920s are now manufactured by the millions of tons each year in the United States.

Despite the bewildering variety of plastics now available, most can be classified in one of a small number of ways. First, all plastics can be categorized as thermoplastic or thermosetting. A thermoplastic polymer is one that, after being formed, can be re-heated and re-shaped. If you warm the handle of a toothbrush in a flame, for example, you can bend it into another shape. A thermosetting polymer is different, however, since, once formed, it can be re-heated, but not re-shaped.

Polymers can also be classified according to the chemical reaction by which they are formed. Addition polymers are formed when one kind of molecule reacts with a second molecule of the same kind. This type of reaction can occur only when the molecules involved contain a special grouping of atoms containing double or triple bonds.

As an example one molecule of ethylene can react with a second molecule of ethylene. This reaction can continue, with a third ethylene molecule adding to the product. In fact, this reaction can be repeated many times until a very large molecule, polyethylene, results.

The prefix *poly-* means "many" indicating that many molecules of ethylene were used in its production. In the formation of a polymer like polyethylene, "many" can be equal to a few hundred or few thousand ethylene molecules. The basic unit of which the polymer is made (ethylene, in this case) is called the monomer. The process by which monomers combine with each other many times is known as polymerization.

A second type of polymer, the condensation polymer, is formed when two different molecules combine with each other through the loss of some small molecule, most commonly, water. For example, Baekeland's original Bakelite is made by the reaction between a molecule of phenol and a molecule of formaldehyde. Phenol and formaldehyde fragments condense to make a new molecule (C_6H_5CHO) when water is removed. As with the formation of polyethylene, this reaction can repeat hundreds or thousands of times to make a very large molecule.

The names of polymers often reveal how they are made. For example, polyethylene is made by the polymerization of ethylene, polypropylene by the polymerization of propylene, and **polyvinyl chloride** by the polymerization of **vinyl chloride**. The names of other polymers give no hint as to the way they are formed. One would not guess, for example, that nylon is a condensation polymer of adipic **acid** and hexamethylene diamine or that Teflon is an addition polymer of tetrafluorethylene.

Listing all the uses to which plastics have been put is probably impossible. Such a list would include squeeze bottles, electrical insulation, film, indoor-outdoor carpeting, floor tile, garden hoses, pipes for plumbing, trash bags, fabrics, latex paints, adhesives, contact lenses, boat hulls, gaskets, non-stick pan coatings, insulation, dinnerware, and table tops.

Plastics technology has become highly sophisticated in the last few decades. Rarely is a simple polymer used by itself in a product. Instead, all types of additives are available for giving the polymer special properties. Ultraviolet stabilizers, as one example, are added to absorb ultraviolet light that would otherwise attack the polymer itself. Many polymers become stiff and brittle when exposed to ultraviolet light.

Plasticizers are compounds that make a polymer more flexible. Foaming agents are used to convert a polymer to the kind of foam used in insulation or cooler chests. Fillers are materials like clay, alumina, or **carbon** black that add properties such as color, flame **resistance**, hardness, or chemical resistance to a plastic.

Some of the most widely used additives are reinforcing agents. These are fibers made of carbon, boron, glass, or some other material that add strength to a plastic. Reinforced plastics find application in car bodies and boat hulls and in many kinds of sporting equipment such as football helmets, tennis rackets, and bicycle frames.

One of the most interesting new variations in polymer properties is the development of conducting polymers. Until 1970, the word plastic was nearly synonymous with electrical non-conductance. In fact, one important use of plastics has been as electrical insulation. In 1970, however, a Korean university student accidentally produced a form of polyacetylene that conducts electricity.

For a number of reasons, that discovery did not lead to a commercial product until 1975. Then, two researchers at the University of Pennsylvania discovered that adding a small amount of iodine to polyacetylene increases its conductivity a trillion times.

A number of technical problems remain, but the day of plastic batteries is no longer a part of the distant future. Indeed, scientists are now studying dozens of ways in which conduction plastics can be substituted for metals in a variety of applications.

For all their many advantages, plastics have long posed some difficult problems for the **environment**. Perhaps the most serious of those problems, is their **stability**. Since plastics have not occurred in **nature** for very long, **microorganisms** that can degrade them have not yet had an opportunity to evolve. Thus, plastic objects that are discarded may tend to remain in the environment for hundreds or thousands of years.

Sometimes, this problem translates into one of sheer volume. In 1988, for example 19.9 percent by volume of all municipal solid wastes consisted of plastic products. That

translated into 14.4 million tons of waste products, third only to paper products and yard and food wastes. At other times, the stability of plastics is actually life-threatening to various organisms. Stories of aquatic birds who are strangled by plastic beer can holders are no longer news because they occur so often.

The presence of plastics can interfere with some of the methods suggested for dealing with solid wastes. For example, **incineration** has been recommended as a way of getting rid of wastes and producing energy at the same time. But the **combustion** of some types of plastics results in the release of toxic hydrochloric acid, **hydrogen** cyanide, and other hazardous gases.

Scientists are now moving forward in the search for ways of dealing with waste plastic materials. Some success has been achieved, for example in the development of photodegradable plastics, polymers that degrade when exposed to light. One problem with such materials is that they are often buried in landfills and are never exposed to sunlight.

Research also continues to progress on the **recycling** of plastics. A major problem here is that some kinds of polymers are more easily recycled than others, and separating one type from the other is often difficult to accomplish in everyday practice.

The use of plastics is also interconnected with the world's energy problems. In the first place, the manufacture of most plastics is energy intensive. It takes only 24 million **Btu** to make one ton of steel, but 49 million Btu to make one ton of polyvinyl chloride and 106 million Btu to make one ton of low-density polyethylene.

Perhaps even more important is the fact that **petroleum** provides the raw materials from which the great majority of plastics are made. Thus, as our supplies of petroleum dwindle, as they inevitable will, scientists will have to find new ways to produce plastics. While they will probably be able to meet that challenge, the change-over from an industry based on petroleum to one based on other raw materials is likely to be long, expensive, and disruptive.

[*David E. Newton*]

Resources

Books

Denison, R. A., and J. Wirka. *Degradable Plastics: The Wrong Answer to the Right Question.* New York: Environmental Defense Fund, 1989.

Joesten, M. D., et al. *World of Chemistry.* Philadelphia: Saunders, 1991.

Selinger, B. *Chemistry in the Marketplace.* 4th ed. Sydney: Harcourt Brace Jovanovich, 1989.

Williams, A. L., H. D. Embree, and H. J. DeBey. *Introduction to Chemistry.* 3rd ed. Reading, MA: Addison-Wesley, 1981.

Plate tectonics

Anyone who looks carefully at a map of the Atlantic Ocean is likely to be struck by an interesting point. The eastern coastline of South America bears a striking similarity to the western coastline of Africa. Indeed, it looks almost as if the two continents could somehow fit together--provided, of course, one could find a way to slide the land masses across the ocean floor.

The match of coastlines was noticed by scholars almost as soon as good maps were first available in the late fifteenth century. In 1620, for example, the English philosopher **Sir Francis Bacon** commented on this fact and argued that the match was "no mere accidental occurrence."

One obvious explanation for the South America-Africa fit was that the two continents had once been connected and had somehow become separated in the past. A few early scientists tried to use the Biblical story of The Flood to show how that might have happened. But as Biblical explanations for natural phenomena began to lose credibility, this approach was discarded.

That made the concept of moving continents even more difficult to believe. The earth's crust was generally thought to be solid and immoveable. How could one or two whole continents somehow slide through such a material?

Yet, over time, more and more evidence began to accumulate, supporting the notion that South America and Africa might once have been joined to each other. Much of that early evidence came from the research of the German geographer, Alexander von Humboldt. Von Humboldt spent a number of years traveling through South America, Africa, and other parts of the world. In his journeys, he collected plant and animal specimens and studied geological and geographic patterns. He was struck by the many similarities he observed between South America and Africa, similarities that went far beyond an obvious geographic fit of continental coastlines.

For example, he observed that mountain ranges in Brazil that end at the sea appear to match other mountain ranges in Africa that began at the coastline. He noted similar patterns among mountain ranges in Europe and North America.

During the nineteenth century, similarities among **fauna** and **flora** on either side of the Atlantic were observed. Although **species** in eastern South America do differ to some extent from those in western Africa, their similarities are often striking. Before long, similarities across other oceanic gaps began to be noted. Plant and animal fossils found in India, for example, were often remarkably similar to those found in **Australia**.

Attempts to explain the many similarities in various continental properties were consistently stymied by the be-

liefs that the earth's crust was solid and immoveable. One way around this problem was the suggestion that mammoth land bridges existed between continents. These large bridges would have allowed the movement of plants and animals from one continent to another. But no evidence for such bridges could be found, and this idea eventually fell into disrepute.

By the mid-1850s, an important breakthrough in geological thought began to occur. A few geologists started to accept the hypothesis that the earth's crust is not as solid and immovable as it appears. In fact, they said, it may be that the earth's outer layer is actually floating (and, thus, is moveable) on the layer below it, the mantle.

Still, it was not until the early years of the twentieth century that a new theory of "floating continents" was seriously proposed. Then, in a period of less than four years, two distinct theories of this kind were suggested. The first was offered in a December 29, 1908, paper by the American geologist Frank B. Taylor. Taylor outlined a theory that described how the continents had slowly shifted over time with a "mighty creeping movement."

Taylor's paper met largely with indifference. Such was not the fate, however, of the ideas of a German astronomer and meteorologist, Alfred Wegener. While browsing through the University of Marburg library in the fall of 1911, Wegener was introduced to the problem of continental similarities. Almost immediately, he decided to devote his attention to this question and began a study that was to dominate the rest of his professional life and to revolutionize the field of geology.

By January 1912, Wegener had developed a theory to explain continental similarities. Such similarities cannot be explained by sunken land bridges, he said, but are the result of continents having moved slowly across the face of the planet. Three more years of research were needed before Wegener's theory was completed. In 1915, he published *The Origins of Continents and Oceans*, summarizing his ideas about continental similarities.

According to Wegener's theory, the continents were once part of one large land mass, which he called Pangaea. Eventually this land mass broke into two parts, two supercontinents, which he called Gondwanaland and Laurasia. Over millions of years, Gondwanaland broke apart into South America, Africa, India, Australia, and **Antarctica**, he suggested, while Laurasia separated into North America and Eurasia.

The basic problem Wegener faced was to explain how huge land masses like continents can flow. His answer was that the materials of which the earth's crust is made are of two very different types. One, then call "sial," is relatively light, but strong. The other, then called "sima," can be compared to very thick tar. Continents are made of sial, he said, and sea floors of sima. The differences in these materials allows continents to "ride" very slowly across sea floors.

Wegener's theory was met with both rejection and hostility. Fellow scientists not only disagreed with his ideas, but also attacked him personally even for suggesting the ideas. The theory of continental drift did not totally disappear as a result of these attitudes, but it fell into disfavor for more than three decades.

Research dating to the mid-1930s revealed new features of the sea floor which made Wegener's theory more plausible. Scientists found sections of the ocean bottoms through which flows of hot lava were escaping from the mantle, somewhat like underwater volcanoes. These discoveries provided a crucial clue in the development of plate tectonics, the modern theory of continental drift.

According to the theory of plate tectonics, the upper layer of the earth is made of a number of plates, large sections of crust, and the upper mantle. About ten major plates have been identified. The largest plate is the Pacific Plate, underlying the Pacific Ocean. The North and South American, Eurasian, African, Indian, Australian, Nazca, Arabian, Caribbean, and Antarctica are the other major plates.

Scientists believe that plates rest on an especially plastic portion of the mantle known as the "asthenosphere." Hot magma from the asthenosphere seeps upward and escapes through the ocean floor by way of openings known as rifts. As the magma flows out of the rift, it pushes apart the plates adjoining the rift. The edge of the plate opposite the rift is ultimately forced downward, back into the asthenosphere. The region in which plate material moves down into the mantle is a trench.

Plates move at different speeds in different directions at different times. On an average, they travel 0.4–2 in (1-5 cm) per year. To the extent that this theory is correct, a map of the earth's surface ten million years from now will look quite different from the way it does now.

The theory of plate tectonics explains a number of natural phenomena that had puzzled scientists for centuries. Earthquakes, for example, can often be explained as the sudden, rather than gradual, movement of two adjacent plates. One of the world's most famous **earthquake** zones, the San Andreas Fault, lies at the boundary of the Pacific and the North American plates. Volcanoes often accompany the movement of plates and earthquakes. The boundaries of the Pacific Plate, for example, define a region where volcanoes are very common, a region sometimes called the Ring of Fire.

Plate tectonics is now accepted as one of the fundamental theories of geology. Its success depends not only on the discovery of an adequate explanation for continental movement (sea-floor spreading, rifts, and trenches), but also

on the discovery of more and more similarities between continents.

For example, modern geologists have re-examined Humboldt's ideas about the correlation of mountain ranges in South America and Africa. They have found that rock strata, or layers, in Brazil lie almost exactly where they should be expected if strata in Ghana are projected to the west. If the Atlantic Ocean could somehow be removed, the two strata could be made to coincide almost perfectly.

Some interesting data have come from studies of paleomagnetism also. Paleomagnetism refers to the orientation of iron crystals in very old rock. Since the earth's magnetic poles have shifted over time, so has the orientation of iron crystals in rock. Scientists have found that the orientation of iron crystals in one continent correspond very closely to those found in rocks of another continent thousands of miles away.

Fossils continue to be a crucial way of confirming continental drift. For example, scientists have learned that the fossil population of is much less like the fossil population of Africa (300 miles [483 km] to the west) than it is like the fossil population of India (3,000 miles/4,830 km to the northeast). This finding would make almost no sense at all unless one recognized that the theory of continental drift has **Madagascar** breaking off from India millions of years ago, not off Africa.

The flow of magma out of the asthenosphere often results in the formation of ores, regions that are rich in some mineral. For example, the high temperatures characteristic of a rift or a trench may be sufficient to cause the release of metals from their compounds. The flow of magma across rock may also result in a phenomenon known as contact metamorphism, a mechanism that also results in the formation of metals such as **lead** and silver. *See also* Biogeography; Endemic species; Evolution; Geosphere; Topography

[*David E. Newton*]

RESOURCES
BOOKS

McGraw-Hill Encyclopedia of Science & Technology. 7th ed. New York: McGraw-Hill, 1992.

Miller, R., and the Editors of Time-Life Books. *Continents in Collision.* Alexandria, VA: Time-Life Books, 1983.

Moran, J. M., M. D. Morgan, and J. H. Wiersma. *Environmental Science.* Dubuque, IA: W. C. Brown, 1993.

Plow pan

A plow pan is a subsurface **horizon** or **soil** layer having a high **bulk density** and a lower total porosity than the soil directly above or below it as a result of pressure applied by normal tillage operations, such as plows, discs, and other tillage implements. Plow pans may also be called pressure pans, tillage pans, or traffic pans. Plow pans are not cemented by organic matter or **chemicals**. Plow pans are the result of pressure exerted by humans, whereas hard pans occur naturally. *See also* Soil compaction

Plume

A flowing, often somewhat conical, trail of emissions from a continuous **point source**, for example the plume of **smoke** from a chimney. As a plume spreads, its constituents are diluted into the surrounding medium. When plumes disperse in media with high turbulence, they can take on more complex shapes with loops and meanders. This somewhat chaotic nature has lead to probabilistic descriptions of the concentration of materials in plumes; for example, calculations concerning the downstream impact of pollutants released from pipes and chimneys will be couched in terms of average concentrations. Typically plumes are found in air or water, but plumes of trace contaminants may also be found in less-mobile media such as soils.

Plutonium

A synthetically produced transuranium element. It does not exist in measurable amounts in the earth's crust. Glenn Seaborg and his colleagues at the University of California at Berkeley first prepared this element in 1940. Plutonium is by far the most important of all transuranium elements because one of its isotopes, plutonium-239, can be fissioned. Plutonium-239 is the only **isotope** other than uranium-235 that is readily available for use in **nuclear weapons** and nuclear reactors. Unfortunately plutonium is also one of the most toxic substances known to humans, making its commercial use a serious environmental hazard. With a **half-life** of 24,000 years, the isotope presents difficult disposal problems. *See also* Nuclear fission; Nuclear power

Poaching

Poaching is the stealing of game or fish from private property or from a place where shooting, **trapping** or fishing rights are reserved. Until the twentieth century, most poaching was subsistence **hunting** or fishing to augment scanty diets. Today, poaching is usually committed for sport or profit.

The **Fish and Wildlife Service** estimates that illegal trade in U.S. **wildlife** generates $200 million per year, a hefty slice of a $1.5 billion worldwide market. Big game animals are shot as trophies, and animal parts such as bear

A poached rhinoceros with its horn removed. (Photograph by Warren and Jenny Garst. Tom Stack & Associates. Reproduced by permission.)

gallbladders are sold for "medicinal" purposes--usually as tonics to enhance male virility. Poaching also attracts organized crime because wildlife crime sentences, if tendered, tend to be lenient and probation requirements are difficult to enforce.

Poaching has a long history in this country: spotlighting a deer at night or shooting a duck in the family pond for dinner have long been a part of rural life. Today, despite a 94 percent conviction rate for those caught, poachers feed a demand for wildlife on both domestic and global markets. They decapitate walruses for ivory tusks, net thousands of night-roosting robins for Cajun gumbo, and shoot anhingas nesting in the **Everglades** and raptors for decorative feathers. **Wolves** are tracked and shot from airplanes. Sturgeon and paddlefish are caught and killed for their caviar. Poaching poses a serious threat to wildlife because it kills off the biggest and best of a species' **gene pool**. A report by the Fish and Wildlife Service estimates that 3,600 threatened or **endangered species** receive little or no federal protection.

The **extinction** of the Labrador duck in the 1880s and the near extinction of a dozen other waterfowl **species** led to the U.S.-Canadian Migratory Bird Treaty Act of 1918. The enactment of kill limits and the ban on commercial market hunting helped populations of birds. The U.S. Marine Mammal Protection Act of 1972 makes it a crime for non-Native Americans to hunt or sell walruses, sea otters, **seals and sea lions**, and polar bears. And the Lacey Act, passed in 1990, makes it a federal crime to transport illegally taken wildlife across state lines.

The sale of bear parts is legal in some states, creating a convenient outlet for poachers. Since merchandise such as bear meat, galls, and paws is difficult to track, the poachers are fairly safe. Uniform regulations in all 50 states would create a more difficult **environment** for poachers to operate in and might lead to a decrease in poaching, especially of bears.

As economic and political instability increased in many developing countries, so did poaching. A global demand for ivory caused wholesale slaughter of **elephants** and rhinoc-

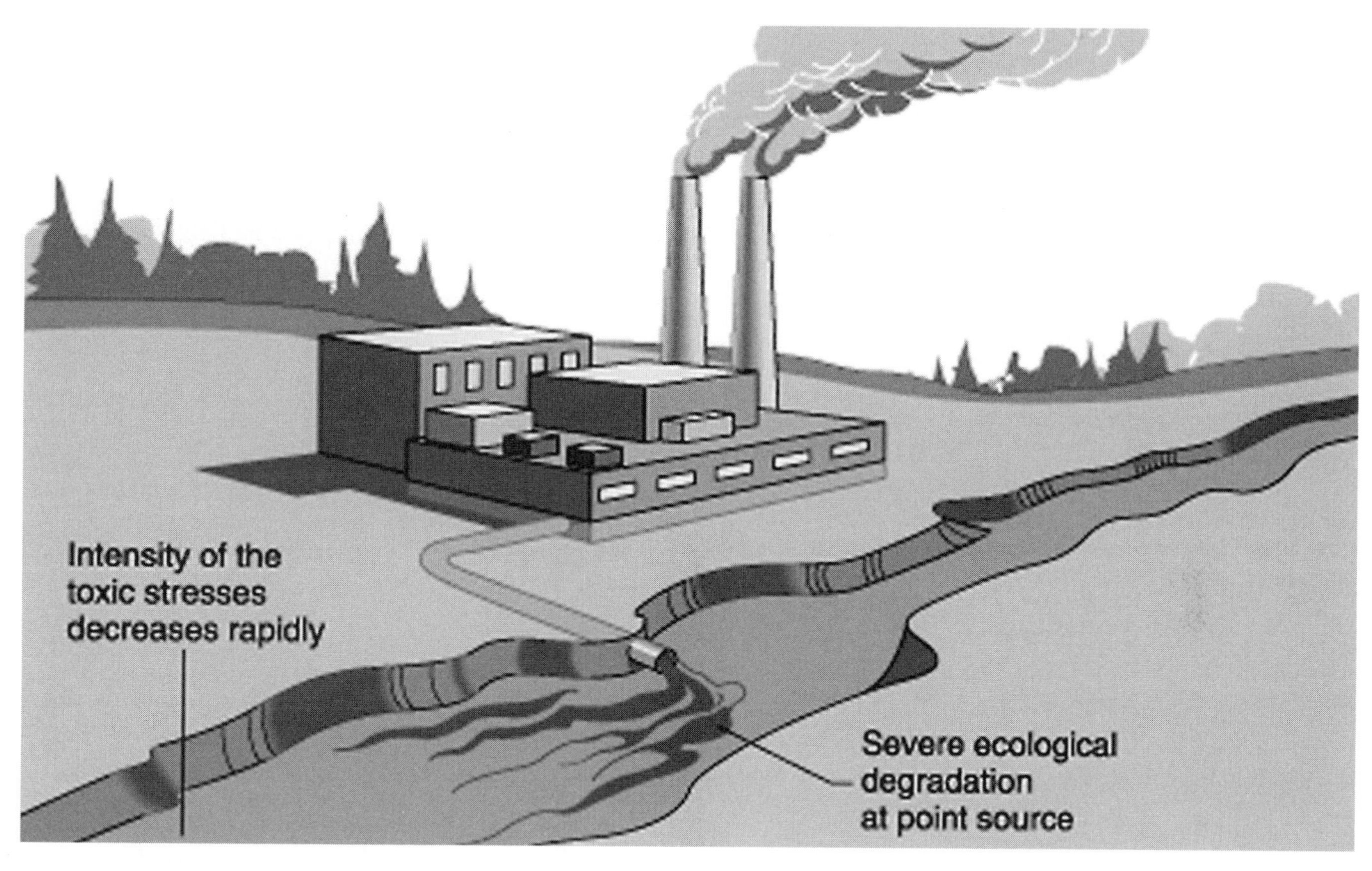

An example of a point source.

eros. As the big male elephants disappeared, poachers turned to females, the primary caretakers of the young. To help stop the carnage of elephant and rhinoceros for their ivory, 105 nations party to the **Convention on International Trade in Endangered Species of Wild Fauna and Flora** (CITES) have agreed to ban the raw ivory trade. The ban has caused ivory prices to plummet to pre-1970s levels, making poaching much less attractive. In Zimbabwe, game wardens now also have orders to shoot on sight poachers who menace the rare black rhino. *See also* Grizzly bear

[*Linda Rehkopf*]

Resources

Books

Chadwick, D. H. *Fate of the Elephant.* San Francisco: Sierra Club Books, 1992.

Periodicals

Hyman, R. "Check the Fine Print, Mate." *International Wildlife* 23 (January-February 1993): 22–5.

Manry, D., and U. Hirsch. "Cliff-hanger in Morocco." *International Wildlife* 23 (March-April 1993): 34–7.

Poten, C. J. "A Shameful Harvest." *National Geographic* 180 (September 1991): 106–132.

Point source

A localized, discrete, and fixed contaminant **emission** source, such as a smokestack or waste **discharge** pipe. Since point sources are usually easily identified, their discharges of **pollution** can be monitored readily. Point sources are distinguished from nonpoint sources and mobile (or vehicular) sources and historically were the first to receive emission controls. Some point sources release exceptionally large quantities of contaminants. For example, a **smelter** in **Sudbury, Ontario**, was for many years the largest single source of **sulfur dioxide** in North America, emitting several million tons of pollutants annually through a smokestack over 1,247 ft (380 m) tall.

Poisoning

Poisoning, either from naturally occurring or manmade **chemicals**, can result from ingestion, inhalation, or skin contact with the toxin. It can also be either acute (a one-time, high amount such as a drug overdose) or chronic (a smaller, amount over a long period of time, such as **lead** poisoning). Poison control centers in many countries can provide information on treatment and prevention of accidental poisoning from household or industrial products. There are more than 13 million known **toxins**, but less than 3,000 cause most incidents of poisoning. Several practices put in place since the 1950's have reduced accidental poisoning by ten-fold. These practices include the accurate labeling of potentially poisonous household compounds, and the use of monitoring devices such as carbon-monoxide detectors. One of the most successful prevention tactics is the use of child-resistant caps on containers of medicine and household products.

[*Marie H. Bundy*]

Pollination

Pollination is the process of moving pollen grains, which contain male sex cells, from the anthers (the pollen-containing part of floral stamens, the male reproductive structure) of flowers to the stigma (the glandular female receptive portion) in the pistil (female reproductive organ). When a pollen grain lands on the female part of the flower, the male sex cell joins with the female sex cells in the flower in a process called fertilization to form a seed from which a new plant can grow. The anthers and stigma can be on the same flower (self-pollination) or on different flowers (cross-pollination), but must be of the same **species**. All higher plants, including flowers, herbs, bushes, grasses, conifers, and broad-leaved trees, use pollination for sexual reproduction.

Pollination can be accomplished by abiotic means such as wind and water. Many pollen grains are small (less than 0.05 mm, or 0.002 in). Thus wind can carry the pollen grains to other members of their species. Many plants, to ensure pollination, grow in dense stands and produce millions of pollen grains. Wind-pollinated plants generally have small, inconspicuous flowers that dangle in the wind (e.g., willow catkins). Grasses have wispy plume-like flowers that catch grains floating in the air. Some water plants such as the hornwort have their pollen transferred by water currents.

Many plants use animals such as insects, birds, and **bats** to transport pollen grains. This process, referred to as biotic pollination, requires a relationship between the pollinator and the flower to be pollinated. Such a relationship is usually established by some kind of direct attractant, such as nectar, sweet-tasting pollen, odor, or visual attraction (e.g., brightly-colored flowers). There may also be an indirect attraction, such as when insects of prey visit flowers to catch other visiting pollinators.

Insects, including bees, beetles, flies, wasps, ants, butterflies, and moths, are common biotic pollinators. As an insect crawls in and out of flowers in its search for nectar or other food source, it receives a dusting of pollen grains from the anther, the male part. When the insect visits another flower, the pollen rubs off on the stigma, the female part. If the pollen is left on the same species of flower, a long tube grows from each pollen grain down the stalk (style) of the stigma and into the ovule at the base, which contains the female egg cells. The male cells from the pollen grains pass along the tubes to the female cells and fertilize them. Plants with trumpet-shaped flowers, such as petunias, have nectar at the bottom, so only insects with long tube-like tongues can act as their pollinators.

The best-known and best-adapted biotic pollinator is the bee. The bee is a relatively large insect with a large demand for food both for itself and for its carefully looked-after brood. It normally gets all of its food from flower blossoms. The bee has an ability to remember plant forms, which aids in its ability to find flowers. Social bees live together in a communal nest and often share foraging and nest activities. The amount of food carried into a hive by a honeybee has been estimated to be 100 times its own requirements. Social bees have developed communication systems that permit them to inform each other about the location and sources of food. These communication systems include odor paths, special buzz tones to alert other individuals, and dances that can indicate direction, distance to, and yield of a source of food. However, most of the world's 40,000 species of bees are solitary. The female bee mates and then constructs a nest underground or within woody stems containing many smooth-walled cells. The cells are filled with a mixture of nectar and pollen, which provides all the food required for larvae to complete their development into adult bees.

Biotic pollination may also be accomplished by such animals as birds and bats. For example, hummingbirds feeding from the hibiscus flower carry pollen on their beak and heads. Bats hover in front of flowers that open at night, licking nectar and covering their faces with pollen. Successful pollination, often mediated by animals but also accomplished abiotically, is extremely important for food production as well as maintenance of biological plant diversity. Pollination of plants is necessary for seed set, fruit yields, and reproduction of most food crops.

Many threats to animal pollinators and pollination processes have been identified. These include use of toxic **chemicals**, decline in pollinator populations, **habitat** loss, and migratory corridor fragmentation. Toxic chemicals can kill

pollinators, and wild pollinators are often more vulnerable to insecticides and herbicides than domestic honeybees. Laws regulate and control the use of many pesticides during periods when pollinators are foraging, and use of toxic chemicals near pollinators' nesting sites should also be controlled. Pesticides that are known to be less toxic to pollinators can be used to reduce stress on pollinator populations. Fewer pollinators will result in fewer plants. When factors such as the use of pesticides and **habitat fragmentation** reduce populations of pollinators, plants will have low reproductive success. Some endangered plant species may even become extinct. Appropriate **pesticide** spraying set-back distances should be based on on-site determinations made by pollination ecologists familiar with the plant and pollinator species involved.

Another threat to pollination processes is the decline in honeybee populations. Since 1990, U.S. beekeepers have lost one-fifth of their domestic managed honeybee colonies for reasons that include two kinds of mite infestations, diseases, spraying of pesticides, and several other factors. The *Varroa* mite is an external parasite that was identified in the United States in 1987 and affects bee colonies in thirty states. This mite lives and feeds on developing bee larvae so that when the bees hatch, they are small and deformed. *Varroa* mites can be controlled by placing medicated plastic strips inside hives to kill the mites. The bees walk on the strips and then carry the medicine on their feet to the larvae growing in the honeycomb. The tracheal mite infects the respiratory system of adult honeybees. These mites were found in the United States in 1984 and are now present in most states. These tracheal mites make bees weak and can kill an entire colony. To control the mites, an antibiotic powder, such as terramycin, is mixed into sugar and oil and is placed inside the bee hive.

Diseases and use of pesticides also take their toll on bee populations. There are several diseases that can kill bees; these include American foul brood, chalk brood, European foul brood, paralysis **virus**, sacbrood disease and kashmir virus. Some bacterial diseases can be treated by stirring antibiotics into feed sugar. Often, if a hive is badly infected, the hive is burned to prevent the infection of other hives. Also, the spraying of pesticides (e.g., by farmers for crop protection or by towns and communities for control of mosquitoes) when bees are foraging can result in their death. Some pesticides kill bees directly while they are in the crops, while others are carried back to the hives with the pollen, where they are stored in the honeycombs. The bees and larvae die when they eat the pollen, which can be at any time of year. Often the pesticide does not kill the entire colony, but makes the colony susceptible to mite infections or freezing in cold weather.

Several other factors have contributed to declining bee populations. **Africanized bees**, a type of highly defensive bee that is also known as the killer bee, became established in the United States in 1990, after years of northward **migration** from South America where they were first released. Beekeepers often are forced to abandon their hives when Africanized bees move into an area. Also, bee populations that have been weakened by other factors are in danger of freezing in winter, due to an insufficient number of bees to provide necessary warmth; or an insufficient supply of food to convert to heat energy. Finally, loss of agricultural subsidies and price supports in the United States adversely affects the economics of managing bee colonies.

Habitat loss and the severing of migratory corridors also constitute threats to animal pollinators and pollination processes. Habitat fragmentation is the division of natural ecosystems into smaller areas due to land conversion for agriculture, forestry, and urbanization. As habitat areas become smaller and widely scattered, they may be insufficient to provide an adequate diversity of host plants and nectar sources that their pollinators require. Habitat fragmentation may also cause reduction in pollinator populations due to loss of nesting habitats. Ecologists need to monitor populations of pollinators and habitat fragmentation trends to determine possible causes of pollinator decline and to develop **land use** plans to protect pollinator populations such as maintenance of habitat set-asides or greenbelts near agricultural fields and timber areas.

The severing of migratory corridors can also disrupt pollination processes. Some pollinating animal species, such as nectarivorous bats, navigate through a variety of nectar-providing plants as they migrate from tropical to **arid** and temperate environments. A bat species may utilize a single type of flower in each local **environment**, but a series of plants are linked into a nectar corridor of successive flowering times along the bat's migration route. For example, a type of long-nosed bat flies a loop of 3,200 mi (5,152 km) to follow the sequential flowering of at least 16 flowering plants species, including tree morning glories, several century plants, and giant columnar cactus. Severing of migratory corridors by habitat and vegetation destruction or by spraying of toxic pesticides may adversely affect the success of the migration. For example, migratory monarch butterflies require critical habitats through their migratory cycle and can be affected if habitat is lost due to activities such as development. Monitoring of plant/pollinator changes in migratory corridors is required to develop appropriate protection strategies.

[*Judith L. Sims*]

Resources

Books

Buchmann, S. L., and G. P. Nabhan. *Forgotten Pollinators.* Washington, DC: Island Press, 1996.

Faegri, K., and L. van der Pijl. *The Principles of Pollination Ecology*, 3rd ed (revised).

Meeuse, R. J. D. *The Story of Pollination.* New York: Ronald Press, 1961.

Pollution

Pollution can be defined as unwanted or detrimental changes in a natural system. Usually, pollution is associated with the presence of toxic **chemicals** in some large quantity, but pollution can also be caused by the presence of excess quantities of heat or by excessive fertilization with nutrients.

Because pollution is judged on the basis of degradative changes, there is a strongly anthropocentric bias to its determination. In other words, humans decide whether pollution is occurring and how bad it is. Of course, this bias favors **species**, communities, and ecological processes that are especially desired or appreciated by humans. In fact, however, some other, *less-desirable* species, communities, and ecological processes may benefit from what we consider pollution.

An important aspect of the notion of pollution is that ecological change must actually be demonstrated. If some potentially polluting substance is present at a concentration or intensity that is less than the threshold required to cause a demonstrable ecological change, then the situation would be referred to as contamination, rather than pollution.

This aspect of pollution can be illustrated by reference to the stable elements, for example, **cadmium**, **copper**, **lead**, **mercury**, **nickel**, selenium, **uranium**, etc. All of these are consistently present in at least trace concentrations in the **environment**. Moreover, all of these elements are potentially toxic. However, they generally affect biota and therefore only cause pollution when they are present at water-soluble concentrations of more than about 0.01 to 1 **parts per million** (ppm).

Some other elements can be present in very large concentrations, for example, **aluminum** and iron, which are important constituents of rock and **soil**. Aluminum constitutes 8-10 percent of the earth's crust and iron 3-4 percent. However, almost all of the aluminum and iron present in minerals are insoluble in water and are therefore not readily assimilated by **biotic community** and cannot cause toxicity. In acidic environments, however, ionic forms of aluminum are solubilized, and these can cause toxicity in concentrations of less than one part per million. Therefore, the bio-availability of a chemical is an important determinant of whether its presence in some concentration will cause pollution.

Most instances of pollution result from the activities of humans. For example, **anthropogenic** pollution can be caused by:

- the **emission** of **sulfur dioxide** and metals from a **smelter**, causing toxicity to vegetation and acidifying surface waters and soil,
- the emission of waste heat from an electricity generating station into a river or lake, causing community change through thermal stress, or
- the **discharge** of nutrient-containing sewage wastes into a water body, causing eutrophication.

Most instances of anthropogenic pollution have natural analogues, that is, cases where pollution is not the result of human activities. For example, pollution can be caused by the emission of sulfur dioxide from volcanoes, by the presence of toxic elements in certain types of soil, by thermal springs or vents, and by other natural phenomena. In many cases, natural pollution can cause an intensity of ecological damage that is as severe as anything caused by anthropogenic pollution. (This does not, of course, in any way justify anthropogenic pollution and its ecological effects.)

An interesting case of natural **air pollution** is the Smoking Hills, located in a remote and pristine **wilderness** in the Canadian Arctic, virtually uninfluenced by humans. However, at a number of places along the 18.63 miles (30 km) of seacoast, bituminous shales in sea cliffs have spontaneously ignited, causing a **fumigation** of the **tundra** with sulfur dioxide and other pollutants. The largest concentrations of sulfur dioxide (more than two parts per million) occur closest to the combustions. Further away from the sea cliffs the concentrations of sulfur dioxide decrease rapidly. The most-important chemical effects of the air pollution are **acidification** of soil and fresh water, which in turn causes a solubilization of toxic metals. Surface soils and pond waters commonly have pHs less than 3, compared with about **pH** 7 at non-fumigated places. The only reports of similarly acidic water are for volcanic lakes in Japan, in which natural pHs as acidic as 1 occur, and pH less than 2 in waters affected by **drainage** from **coal** mines.

At the Smoking Hills, toxicity by sulfur dioxide, acidity, and water-soluble metals has caused great damage to ecological communities. The most-intensively fumigated terrestrial sites have no vegetation, but further away a few pollution-tolerant species are present. About one kilometer away the toxic stresses are low enough that reference tundra is present. There are a few pollution-tolerant algae in the acidic ponds, with a depauperate community of six species occurring in the most-acidic (pH 1.8) pond in the area.

Other cases of natural pollution concern places where certain elements are present in toxic amounts. Surface mineralizations can have toxic metals present in large concentrations, for example copper at 10 percent in peat at a copper-rich spring in New Brunswick, or surface soil with three percent lead plus zinc on Baffin Island. Soils influenced by nickel-rich serpentine minerals have been well-studied by ecologists. The stress-adapted plants of serpentine habitats form distinct communities, and some plants can have nickel concentrations larger than 10 percent in their tissues. Simi-

larly, natural soils with large concentrations of selenium support plants that can hyperaccumulate this element to concentrations greater than one percent. These plants are poisonous to livestock, causing a toxic syndrome known as *blind staggers*.

Of course, there are many well-known cases where pollution is caused by anthropogenic emissions of chemicals. Some examples include:

- Emissions of sulfur dioxide and metals from smelters can cause damage to surrounding terrestrial and aquatic ecosystems. The sulfur dioxide and metals are directly toxic. In addition, the deposition of sulfur dioxide can cause an extreme acidification of soil and water, which causes metals to be more bio-available, resulting in important, secondary toxicity. Because smelters are point sources of emission, the spatial pattern of chemical pollution and ecological damage displays an exponentially decreasing intensity with increasing distance from the source.
- The use of pesticides in agriculture, forestry, and around homes can result in a non-target exposure of birds and other **wildlife** to these chemicals. If the non-target biota are vulnerable to the **pesticide**, then ecological damage will result. For example, during the 1960s urban elm trees in the eastern United States were sprayed with large quantities of the insecticide DDT, in order to kill beetles that were responsible for the transmission of Dutch elm disease, an important **pathogen**. Because of the very large spray rates, many birds were killed, leading to reduced populations in some areas. (This was the "silent spring" that was referred to by Rachel Carson in her famous book by that title.) Birds and other non-target biota have also been killed by modern insecticide-spray programs in agriculture and in forestry.
- The deposition of acidifying substances from the **atmosphere**, mostly as acidic precipitation and the **dry deposition** of sulfur dioxide, can cause an acidification of surface waters. The acidity solubilizes metals, most notably aluminum, making them bio-available. The acidity in combination with the metals causes toxicity to the biota, resulting in large changes in ecological communities and processes. Fish, for example, are highly intolerant of acidic waters.
- Oil spills from tankers and pipelines can cause great ecological damage. When oil spilled at sea washes up onto coastlines, it destroys seaweeds, invertebrates, and fish, and their communities are changed for many years. Seabirds are very intolerant of oil and can die of hypothermia if even a small area of their feathers is coated by petroleum.
- Most of the **lead shot** fired by hunters and skeet shooters miss their target and are dispersed into the environment. Waterfowl and other avian wildlife actively ingest lead shot because it is similar in size and hardness to the grit that they ingest to aid in the mechanical abrasion of hard seeds in their gizzard. However, the lead shot is toxic to these birds, and each year millions of birds are killed by this source in North America.

Humans can also cause pollution by excessively fertilizing natural ecosystems with nutrients. For example, freshwaters can be made eutrophic by fertilization with **phosphorus** in the form of phosphate. The most conspicuous symptoms of eutrophication are changes in species composition of the **phytoplankton** community and, especially, a large increase in algal **biomass**, known as a *bloom*. In shallow waterbodies there may also be a vigorous growth of vascular plants. These primary responses are usually accompanied by secondary changes at higher trophic levels, including arthropods, fish, and waterfowl, in response to greater food availability and other **habitat** changes. However, in the extreme cases of very eutrophic waters, the blooms of algae and other **microorganisms** can be noxious, producing toxic chemicals and causing periods of oxygen depletion that kill fish and other biota. Extremely eutrophic waterbodies are polluted because they often cannot support a fishery, cannot be used for drinking water, and have few recreational opportunities and poor esthetics.

Pollution, therefore, is associated with ecological degradation, caused by environmental stresses originating with natural phenomena or with human activities. The prevention and management of anthropogenic pollution is one of the greatest challenges facing modern society. *See also* Cultural eutrophication

[*Bill Freedman Ph.D.*]

RESOURCES

BOOKS

Ehrlich, P. R., A. H. Ehrlich, and J. P. Holdren. *Ecoscience. Population, Resources, Environment.* San Francisco: W. H. Freeman & Co., 1977.

Freedman, Bill. *Environmental Ecology.* 2nd edition, San Diego: Academic Press, 1995.

Speth, J. G. *Environmental Pollution: A Long-Term Perspective.* Washington, DC: World Resources Institute, 1988.

Pollution control

Pollution control is the process of reducing or eliminating the release of pollutants into the **environment**. It is regulated by various environmental agencies which establish pollutant **discharge** limits for air, water, and land.

Air pollution control strategies can be divided into two categories, the control of **particulate** emissions and the control of gaseous emissions. There are many kinds of equipment which can be used to reduce particulate emissions. Physical separation of the particulates from the air using settling chambers, **cyclone** collectors, impingers, wet

scrubbers, electrostatic precipitators, and **filtration** devices, are all processes that are typically employed.

Settling chambers use gravity separation to reduce particulate emissions. The air stream is directed through a settling chamber, which is relatively long and has a large cross section, causing the velocity of the air stream to be greatly decreased and allowing sufficient time for the settling of solid particles.

A **cyclone collector** is a cylindrical device with a conical bottom which is used to create a tornado-like air stream. A centrifugal force is thus imparted to the particles, causing them to cling to the wall and roll downward, while the cleaner air stream exits through the top of the device.

An impinger is a device which uses the inertia of the air stream to impinge mists and dry particles on a solid surface. Mists are collected on the impinger plate as liquid forms and then drips off, while dry particles tend to build up or reenter the air stream. It is for this reason that liquid sprays are used to wash the impinger surface as well, to improve the collection efficiency.

Wet scrubbers control particulate emissions by wetting the particles in order to enhance their removal from the air stream. Wet scrubbers typically operate against the current by a water spray contacting with the gas flow. The particulate matter becomes entrained in the water droplets, and it is then separated from the gas stream. Wet scrubbers such as packed bed, venturi, or plate scrubbers utilize initial impaction, and cyclone scrubbers use a centrifugal force.

Electrostatic precipitators are devices which use an electrostatic field to induce a charge on dust particles and collect them on grounded electrodes. Electrostatic precipitators are usually operated dry, but wet systems are also used, mainly by providing a water mist to aid in the process of cleaning the particles off the collection plate.

One of the oldest and most efficient methods of particulate control, however, is filtration. The most commonly-used filtration device is known as a **baghouse** and consists of fabric bags through which the air stream is directed. Particles become trapped in the fiber mesh on the fabric bags, as well as the filter cake which is subsequently formed.

Gaseous emissions are controlled by similar devices and typically can be used in conjunction with particulate control options. Such devices include scrubbers, **adsorption** systems, condensers, flares, and incinerators.

Scrubbers utilize the phenomena of adsorption to remove gaseous pollutants from the air stream. There is a wide variety of scrubbers available for use, including spray towers, packed towers, and venturi scrubbers. A wide variety of solutions can be used in this process as absorbing agents. Lime, magnesium oxide, and sodium hydroxide are typically used.

Adsorption can also be used to control gaseous emissions. Activated **carbon** is commonly used as an adsorbent in configurations such as fixed bed and fluidized bed adsorbers.

Condensers operate in a manner so as to condense vapors by either increasing the pressure or decreasing the temperature of the gas stream. Surface condensers are usually of the shell-and-tube type, and contact condensers provide physical contact between the vapors, coolant, and condensate inside the unit.

Flaring and **incineration** take advantage of the combustibility of a gaseous pollutant. In general, excess air is added to these processes to drive the **combustion** reaction to completion, forming **carbon dioxide** and water.

Another means of controlling both particulate and gaseous air pollutant **emission** can be accomplished by modifying the process which generates these pollutants. For example, modifications to process equipment or raw materials can provide effective source reduction. Also, employing fuel cleaning methods such as desulfurization and increasing fuel-burning efficiency can lessen air emissions.

Water pollution control methods can be subdivided into physical, chemical, and biological treatment systems. Most treatment systems use combinations of any of these three technologies. Additionally, **water conservation** is a beneficial means to reduce the volume of **wastewater** generated.

Physical treatment systems are processes which rely on physical forces to aid in the removal of pollutants. Physical processes which find frequent use in water pollution control include screening, filtration, **sedimentation**, and **flotation**. Screening and filtration are similar methods which are used to separate coarse solids from water. Suspended particles are also removed from water with the use of sedimentation processes. Just as in **air pollution** control, sedimentation devices utilize gravity to remove the heavier particles from the water stream. The wide array of sedimentation basins in use slow down the water velocity in the unit to allow time for the particles to drop to the bottom. Likewise, flotation uses differences in particle densities, which in this case are lower than water, to effect removal. Fine gas bubbles are often introduced to assist this process; they attach to the particulate matter, causing them to rise to the top of the unit where they are mechanically removed.

Chemical treatment systems in water pollution control are those processes which utilize chemical reactions to remove water pollutants or to form other, less toxic, compounds. Typical chemical treatment processes are chemical precipitation, adsorption, and disinfection reactions. Chemical precipitation processes utilize the addition of **chemicals** to the water in order to bring about the precipitation of **dissolved solids**. The solid is then removed by a physical process such as sedimentation or filtration. Chemical precip-

itation processes are often used for the removal of **heavy metals** and **phosphorus** from water streams. Adsorption processes are used to separate soluble substances from the water stream. Like air pollution adsorption processes, activated carbon is the most widely used adsorbent. Water may be passed through beds of granulated activated carbon (GAC), or powdered activated carbon (PAC) may be added in order to facilitate the removal of dissolved pollutants. Disinfection processes selectively destroy disease-causing organisms such as bacteria and viruses. Typical disinfection agents include **chlorine**, **ozone**, and **ultraviolet radiation**.

Biological water pollution control methods are those which utilize biological activity to remove pollutants from water streams. These methods are used for the control of **biodegradable** organic chemicals, as well as nutrients such as **nitrogen** and phosphorus. In these systems, **microorganisms** consisting mainly of bacteria convert carbonaceous matter as well as cell tissue into gas. There are two main groups of microorganisms which are used in biological treatment, **aerobic** and **anaerobic** microorganisms. Each requires unique environmental conditions to do its job. Aerobic processes occur in the absence of oxygen. Both processes may be utilized whether the microorganisms exist in a suspension or are attached to a surface. These processes are termed suspended growth and fixed film processes, respectively.

Solid pollution control methods which are typically used include landfilling, **composting**, and incineration. Sanitary landfills are operated by spreading the **solid waste** in compact layers which are separated by a thin layer of **soil**. Aerobic and anaerobic microorganisms help to break down the biodegradable substances in the **landfill** and produce carbon dioxide and **methane** gas which is typically venter to the surface. Landfills also generate a strong wastewater called leachate which must be collected and treated to avoid **groundwater** contamination.

Composting of solid wastes is the microbiological biodegradation of organic matter under either aerobic or anaerobic conditions. This process is most applicable for readily biodegradable solids such as sewage **sludge**, paper, **food waste**, and household **garbage**, including garden waste and organic matter. This process can be carried out in static pile, agitated beds, or a variety of reactors.

In an incineration process, solids are burned in large furnaces thereby reducing the volume of solid wastes which enter landfills, as well as reducing the possibility of groundwater contamination. Incineration residue can also be used for metal **reclamation**. These systems are typically supplemented with air pollution control devices. *See also* Air pollution index; Air quality criteria; Clean Air Act; Clean Water Act; Emission standards; Heavy metals and heavy metal poisoning; Industrial waste treatment; Primary standards; Secondary standards; Sewage treatment; Water quality standards

[*James W. Patterson*]

Resources
Books

Advanced Emission Control for Power Plants. Paris: Organization for Economic Cooperation and Development, 1993.

Handbook of Air Pollution Technology. New York: Wiley, 1984.

Jorgensen, E. P., ed. *The Poisoned Well: New Strategies for Groundwater Protection.* Washington, DC: Island Press, 1989.

Kenworthy, L., and E. Schaeffer. *A Citizen's Guide to Promoting Toxic Waste Reduction.* New York: INFORM, 1990.

Wentz, C. A. *Hazardous Waste Management.* New York: McGraw-Hill, 1989.

Pollution credits

Pollution credits emerged from the federal **Clean Air Act** of 1990 (CAA) as a way for businesses to deal with the regulations that attempt to lower **air pollution**. The **Environmental Protection Agency** (EPA) and individual states have created a "cap and trade" system (in practice since 1995) allowing utilities and manufacturers allowances to emit a certain amount of specific pollutants. This system offers companies some flexibility in their compliance while helping to maintain the standards required for cleaner air.

A company can earn pollution credits by voluntarily reducing polluting emissions below limits dictated by the EPA. Earned credits can then be sold to another company that has trouble keeping its emissions within permissible limits, or saved for future use. According to Dean S. Sommer writing for *The Albany (New York) Business Review* in March 1999, "The 'cap' portion of the system refers to the specific amount of emissions that may be discharged within a state, region or the nation as set forth by the EPA and the CAA. The 'trade' portion refers to a participant's ability to buy or sell the allowances. States, or the EPA, allocate **emission** allowances to affected sources in specified amounts to ensure that the cap is not exceeded by emissions from all participants."

Different "cap and trade" allowances apply to the most common pollutants, **sulfur dioxide** and **nitrogen oxides**, which contribute to **acid rain**, **smog**, and **ozone** depletion. Each allowance, or credit, is a trading unit equivalent to one ton of air emissions. The EPA exacts annual penalties to those companies whose emissions exceed the number of allowances.

According to the EPA, the practice of pollution credits bartering has proven beneficial not only in reducing air pollution, but in providing an incentive for companies to meet

environmental goals. For example, the EPA's national **acid** rain program to reduce sulfur dioxide resulted in a decrease of six million tons of emissions a year compared to 1980 levels. **Nitrogen** oxide levels were reduced by 50% compared to 1990 levels as a result of a program undertaken by numerous northeastern states.

Critics of the trading program, however, have suggested that the system is flawed. They contend that although it was meant to be a way of encouraging industry to find ways of producing less pollution, the system provides a way of merely transferring responsibility—the one polluting is not held responsible for the poor quality of air it has affected.

In 2000, for example, the state of New York identified several midwestern and southeastern states—Delaware, Illinois, Indiana, Ohio, Michigan, Kentucky, Maryland, New Jersey, North Carolina, Pennsylvania, Tennessee, Virginia, West Virginia, and Wisconsin—as the major sources of its acid rain, a serious problem undermining **natural resources** and **wildlife**. Between 1995 and 2000, the state had accumulated more than 700,000 pollution credits—providing the New York businesses selling them with a total of approximately $37.5 million. The result was that pollution within New York had declined only to be blown back over state lines by other states that had purchased the credits. According to a May 2, 2000 report of the *Environment News Service* (ENN) by writer Cat Lazaroff, "Despite the state's efforts to reduce sulfur dioxide emissions, many high altitude lakes in New York's **Adirondack Mountains** remain too acidic to support many native **species**. Some government studies estimate about half of the region's 3,000 lakes and ponds may become too acidic to support life if acid rain is not reduced." The state fought back with legislation authorizing the fining of utility companies that sell their pollution credits to the polluting states.

Attempts to regulate less and provide more incentives have been met with suspicion by program critics, who fear an increase in pollution, and applause by businesses, who view the reduction in standards as a possible financial relief. In any case, pollution credits had become a valuable commodity on the open market by the early twenty-first century, in the United States and throughout the world.

[*Jane E. Spear*]

Resources

Periodicals

"Big Cities a Headache U.N. Summit wants to Address." *Environmental News Network* June 6, 2002 [cited July 7, 2002]. <http://www.enn.com/news/wire-stories/2002/06/06062002/reu_47462.asp>.

Borenstein, Seth. "Bush Vows to Cut Pollution with more Incentives, Fewer Regulations." *Knight Rider Newspapers*, February 14, 2002.

"Environmentalists, EPA Disagree Whether Current Law or Bush Plan would Help Air More." *Environmental News Network*, February 21, 2002 [cited July 7, 2002]. <http://www.enn.com/news/wire-stories/2002/02/02212002/ap_46452.asp>.

"Industry Still Failing on Environment, says U.N. Report." *Environmental News Network*, May 16, 2002 [cited July 7, 2002]. <http://www.enn.com/news/wire-stories/2002/05/05162002/reu_47235.asp>.

Kahn, Joseph. "Whitman Begins to Consider Streamlining Pollution Checks." *New York Times*, July 28, 2001.

Kranish, Michael. "The Politics of Pollution." *Boston Globe Magazine*, February 18, 1998.

Lazaroff, Cat. "New York Fights Acid Rain Through Curb on Pollution Credits." *Environment News Service*, May 2, 2000.

Max, Arthur. "Pollution Trading a Contentious Issue at Climate Talks." *Associated Press*, November 15, 2000.

"Oil Spills in the Sky." December 9, 1997 [cited July 7, 2002]. <http://www.enn.com/features/1997/12/120997/1209fea_20122.asp>.

"Pollution Credits: The Science Isn't In Yet." *Environmental News Network*, February 2, 1999 [cited July 7, 2002]. <http://www.enn.com/news/enn-stories/1999/02/020299/co2findings_1423.asp>.

Sommer, Dean S. "'Retiring' Pollution Credits helps both Business and Environment." *The Business Review*, March 22, 1999.

"States Ready to Trade Air Pollution Credits." *Environmental News Network*, January 3, 2001 [cited July 7, 2002]. <http://www.enn.com/news/enn-stories/2001/01/01032001/pollution_41007.asp>.

"Trading for Clean Air just got Easier." December 5, 2001 [cited July 7, 2002]. <http://www.enn.com/news/enn-stories/2001/12/12052001/s_45776.asp>.

"Trading Credits Does not cut Pollution." *Houma Today*, November 28, 2001.

"U.S. EPA Proposes Pollution Credits to Clean Up Rivers." *Environmental News Network*. May 16, 2002 [cited July 7, 2002]. <http://www.enn.com/news/wire-stories/2002/05/05162002/reu_47230.asp>.

Other

Environmental Protection Agency. "The Plain English Guide to the Clean Air Act." May 13, 2002 [cited July 7, 2002]. <http://www.epa.gov/oar/oaqps/peg_caa/pegcaain.html>.

Organizations

U.S. Environmental Protection Agency, 1200 Pennsylvania Avenue, NW, Washington, DC USA 20460 (202) 260-2090, <http://www.epa.gov>

Pollution Prevention Act (1990)

The **Pollution** Prevention Act of 1990 is a piece of legislation intended to limit the creation of pollution. As part of the Omnibus Budget Reconciliation Act of 1990, the Pollution Prevention Act differed from previous legislation, which had generally treated pollution after it had been created. **Environmental Protection Agency** (EPA) administrator William K. Reilly strongly supported the act, believing that much hazardous or toxic pollution can be more effectively and economically controlled, and the **environment** protected more fully, if the pollution never occurs.

The impetus for this act lies in a 1986 EPA report to Congress entitled "Minimization of Hazardous Wastes." Based on this report, the agency began to take actions designed to reduce pollution. According to a subsequent report by the Office of Technology Assessment (OTA), however,

the EPA needed to do more. The two reports identified four main approaches to prevent pollution: (1) manufacturing changes, (2) equipment changes, (3) product reformulations and substitutions, and (4) improved industrial housekeeping. By using such approaches, companies (and individuals) can both save money by reducing **pollution control** costs and have a cleaner environment because certain pollution will be prevented. The corporation 3M, for instance, has saved over $300 million since 1975 by reducing air, water, and **sludge** pollutants at the source. The OTA estimated that a ten percent reduction in hazardous wastes per year for five years could be achieved through source reduction. The reports also found that pollution prevention is primarily hindered not by a lack of technology but rather by a lack of knowledge and awareness due to institutional hurdles. The Pollution Prevention Act was designed to help overcome these hurdles through information collection, assistance in technology transfer, and financial assistance to state pollution prevention programs.

In addition to developing a general strategy to promote the reduction of pollution at its source, the act crated a new EPA office to administer the program. The major provisions of the law are as follows: a state grant program to aid states in establishing similar pollution prevention programs; EPA technical assistance for business; the development of a Source Reduction Clearinghouse, administered by the EPA and available for public use, to help in the collection and distribution of source reduction information; the establishment of an advisory committee on the issue; the creation of a training program on pollution prevention; and the creation of annual awards to businesses that achieve outstanding source reduction. Businesses using toxic substances were required to meet special standards--to report on how many and the amount of toxic substances released into the environment, how much recycled, how they attempted to prevent the generation of toxic pollutants, and how they learned of source reduction techniques. Lastly, the EPA is to report biennially on the effectiveness of the new program and its components. The Act authorized $16 million each year for three years for the EPA to implement this program.

[*Christopher McGrory Klyza*]

Resources

Books

Environmental Quality: 21st Annual Report. Council on Environmental Quality. Washington, DC: Government Printing Office, 1990.

Periodicals

"Budget Reconciliation Act Provisions." *Congressional Quarterly Almanac* 46 (1990): 141–163.

Other

"Pollution Prevention Act of 1990." Senate Committee on Environment and Public Works, Senate Report 101-526, 1990, 101st Congress, 2d Session.

"Waste Reduction Act." House Committee on Energy and Commerce, House Report 101-555, 1990, 101st Congress, 2d Session.

Nicholas Polunin (1909 – 1997)

English environmentalist

Nicolas Polunin will be long remembered for two unusual achievements. First, he founded and was first editor of not one but two influential journals, *Biological Conservation* (founded in 1967 and edited from that year to 1974) and then *Environmental Conservation* (founded in 1974 and edited from its inception through 1995 when he was in his mid-eighties). Second, the readers of either journal, but particularly the latter, will remember the vivid language Polunin used in numerous editorials and other articles and texts to urge readers to start paying attention to the impact of human activities on other species on earth and on the natural environment.

Polunin credited his very public concern about such issues to a small UNESCO Conference in Finland in 1966, where he "had a sudden realization, amounting to a kind of vision, that if the world's populations went on growing and acting as profligately as they had been doing in recent decades, more and more ecodisasters would be inevitable and ultimately there would be an end to much of what was best in our civilization and conceivably to all life on Earth." He went on to say that "thereupon we decided that it was our solemn duty to do all we could henceforth to warn humanity of these grave dangers." That "quite shocking realization" led to his founding and editing of the international journal on conservation, and after he considered it "sufficiently established," a "more predominately environmental journal." *Biological Conservation* has become a mainstay for scientists investigating the vast range and significance of biological diversity on the planet and on the possibilities for sustainable use by humans of the biosphere. *Environmental Conservation* tries to reach a wider readership, one well outside the community of research biologists and is much more activist in tone and content. Establishment of two such influential journals is a significant accomplishment and they remain Polunin's enduring legacy, one matched by few other ecologists.

On the cover of the first issue of *Biological Conservation*, Polunin listed 30 subject headings as "the content of our subject." These included the history and development of conservation, a focus on "man and environment," disturbance and maintenance of ecosystems and the biosphere, population dynamics, threatened species and nature reserves and

parks, legislation and enforcement, education and training, fresh waters and wetlands, and a list of chemical, physical and biological threats to wildlife and environment. Those 30 topics were listed only on the first six issues, but the editorial description of topics for papers (as noted on an inside cover sheet in most science journals) are almost identical in the latest issues (2002) with those Polunin stated in the first issue. Typically, however, Polunin's topical editorial in that first issue was titled "Some Warnings," suggesting that "our world is beset with many threats," including his concern about the rapid increase in human population, a concern he voiced many times in editorials in both journals. In a still relevant and prescient voice, he called for a "severe limitation of human breeding and cognate desecrating tendencies." When he stepped down as editor, the publishers noted his energy and enthusiasm as reflected in "the great wealth of material published under his guidance."

While the stated purpose in the first journal was, and remains, the widest possible dissemination of original papers on "the preservation of wildlife and all nature," Polunin's editorial directions to authors in the first issue of *Environmental Conservation* differed quite radically, stating that the journal "advocates timely action for the protection and amelioration of the environment of Man and Nature throughout the world...for the lasting future of Earth's fragile biosphere." Polunin's editorial language remained strong, continuing to warn about the plagues human actions were visiting on each other, on other organisms, and on the earth. Words and phrases like 'faustian bargain,' 'ecodisasters,' 'our world menaced' and 'rays of hope' were common in his editorial pleas for sanity.

He published longer statements, such as his "Thoughts on Some Conceivable Ecodisasters" in the Autumn, 1974, issue of *Environmental Conservation*, he organized major international conferences on the environmental future and edited the proceedings, all evidence of his world-wide influence on thinking about human-environment issues and problems. He also never lost hope, claiming in 1991 (Autumn issue of *Environmental Conservation*) that there had been "recent changes for the better in the prospect of an improved future for the global environment" and listing twelve rays of hope, including "a growing realization that there are coming to be far too many human beings on our limited Planet Earth."

Nicholas Polunin was born in England to a Russian father and an English mother. He earned a First Class Honours degree in botany and ecology from Oxford in 1932, a Masters from Yale, and returned to Oxford for a PhD in 1935. One of his mentors was Arthur Tansley. He traveled widely with a number of scientific expeditions while still an undergraduate and published his first book (*Russian Waters*, 1931) before he completed his degree. He was a member of the party that discovered Prince Charles Island, the "last major island to be marked on the world's map." He held academic positions at McGill University, and at universities in Iraq and Nigeria, among others.

Perhaps unknown to general readers more familiar with his visible roles as editor, writer, and environmental activist, Polunin was a productive research biologist, specializing in arctic vegetation and plant ecology and, later, on marine ecosystems. His 1960 text on plant geography is one of the best known in that field and still in print over forty years later.

Myers remembered Polunin as a "scientist in the round," an appropriate epitaph for one whose legacy includes two influential international journals, ever renewed encouragement to scientists to focus more clearly on global issues of conservation and environmental science, a better informed and more aware public around the world, and more realistic hopes for a better quality of life for all humanity and for continued maintenance of the diversity of Earth's natural biota.

[*Gerald L. Young Ph.D.*]

Further Reading

Periodicals

Bataille, A. "Nicholas Polunin C.B.E., M.S., M.A., D.Phil., D.Sc., F.L.S., F.R.G.S. (1909–1997). *Watsonia* 22, no. 2 (August 1998): 203&ndahs;205.

Myers, Norman. "Nicholas Polunin: A Scientist in the Round." *Environmental Conservation* 25, no. 1 (March 1998): 8–10.

Polybrominated biphenyls

A mixture of compounds having from one to ten **bromine** atoms attached to a biphenyl ring, analogous to polychlorinated biphenyl (PCB)s. Manufactured as fire retardants, PBBs were banned after a 1973 Michigan incident when pure product was accidently mixed with cattle feed and distributed throughout the state. PBBs were identified as the cause of weight loss, decreased milk production, and **mortality** in many dairy herds. Approximately 30,000 cattle, 1.5 million chickens, 1,500 sheep, 6,000 hogs, 18,000 pounds (8,100 kg) of cheese, 34,000 pounds (15,300 kg) of dried milk products, 5 million eggs, and 2,700 pounds (1,215 kg) of butter were eventually destroyed at an estimated cost of $1 million. Although human exposures have been well-documented, long term epidemiological studies have not shown widespread health effects.

Polychlorinated biphenyls

Mixture of compounds having from one to ten **chlorine** atoms attached to a biphenyl ring structure. There are 209 possible structures theoretically; the manufacturing process results in approximately 120 different structures. PCBs resist biological and heat degradation and were used in numerous applications, including dielectric fluids in capacitors and transformers, heat transfer fluids, hydraulic fluids, plasticizers, dedusting agents, adhesives, dye carriers in carbonless copy paper, and **pesticide** extenders. The United States manufactured PCBs from 1929 until 1977, when they were banned due to adverse environmental effects and ubiquitous occurrence. They bioaccumulate in organisms and can cause skin disorders, liver dysfunction, reproductive disorders, and tumor formation. They are one of the most abundant organochlorine contaminants found throughout the world. *See also* Organochloride

Polycyclic aromatic hydrocarbons

Polycyclic aromatic **hydrocarbons** are a class of organic compounds having two or more fused **benzene** rings. While usually referring to compounds made of **carbon** and **hydrogen**, PAH also may include fused aromatic compounds containing **nitrogen**, sulfur, or cyclopentene rings. Some of the more common PAH include naphthalene (2 rings), anthracene (3 rings), phenanthrene (3 rings), pyrene (4 rings), chrysene (4 rings), fluoranthene (4 rings), benzo(a)-pyrene (5 rings), benzo(e)pyrene (5 rings), perylene (5 rings), benzo(g,h,i)perylene (6 rings), and coronene (7 rings).

PAH are formed by a variety of human activities including incomplete **combustion** of **fossil fuels**, wood, and **tobacco**; the **incineration** of **garbage**; **coal gasification** and liquefaction processes; smelting operations; and coke, asphalt, and **petroleum** cracking; they are also formed naturally during forest fires and volcanic eruptions. Low molecular-weight PAH (those with four or fewer rings) are generally vapors while heavier molecules condense on submicron, breathable particles. It is estimated that more than 800 tons of PAH are emitted annually in the United States. PAH are found worldwide and are present in elevated concentrations in urban aerosols, and in lake sediments in industrialized countries. They also are found in developing countries due to **coal** and wood heating, open burning, coke production, and vehicle exhaust.

The association of PAH with small particles gives them atmospheric residence times of days to weeks, and allows them to be transported long distances. They are removed from the **atmosphere** by gravitational settling and are washed out during precipitation to the earth's surface, where they accumulate in soils and surface waters. They also directly enter water in discharges, **runoff**, and **oil spills**. They associate with water particulates due to their low water solubility, and eventually accumulate in sediments. They do not bioaccumulate in biota to any appreciable extent, as they are largely metabolized.

Many but not all PAH have carcinogenic and mutagenic activity; the most notorious is benzo(a)pyrene, which has been shown to be a potent **carcinogen**. Coal tar and soot were implicated in the elevated skin **cancer** incidence found in the refining, shale oil, and coal tar industries in the late nineteenth century. Subsequent research led to the isolation and identification of several carcinogens in the early part of this century, including benzo(a)pyrene. More recent research into the carcinogenicity of PAH has revealed that there is significant additional biological activity in urban aerosols and soot beyond that explained by known carcinogens such as benzo(a)pyrene. While benzo(a)pyrene must be activated metabolically, these other components have direct biological activity as demonstrated by the **Ames test**. They are polar compounds, thought to be mixtures of mono- and dinitro-PAH and hydroxy-nitro derivatives. Tobacco smoking exposes more humans to PAH than any other source.

[*Deborah L. Swackhammer*]

Resources

Books

Dias, J. R. *Handbook of Polycyclic Aromatic Hydrocarbons.* New York: Elsevier, 1987.

Harvey, R. G. *Polycyclic Aromatic Hydrocarbons: Chemistry and Carcinogenicity.* New York: Cambridge University Press, 1991.

Polycyclic organic compounds

In organic chemistry, a cyclic compound is one whose molecules consist of three or more atoms joined in a closed ring. A polycyclic compound is one whose molecules contain two or more rings joined to each other. Environmentally, the most important polycyclic organic compounds are the **polycyclic aromatic hydrocarbons**, also known as polynuclear aromatic **hydrocarbons** (PAHs). Some examples of polycyclic organic compounds include naphthalene, anthracene, pyrene, and the benzopyrenes. A number of polycyclic hydrocarbons pose hazards to human health. For example, benzo(a)pyrene from **automobile** exhaust and **tobacco smoke** is known to be a **carcinogen**. *See also* Cigarette smoke

Polynuclear aromatic hydrocarbons

see **Polycyclic aromatic hydrocarbons**

Polystyrene

Polystyrene is a lightweight transparent plastic derived from **petroleum** by-products and **natural gas**. It is widely used in the packaging industry, but most polystyrene is used to make durable goods such as television cabinets, appliances, and furniture. Polystyrene also has excellent insulating properties. As packaging, polystyrene is used both in foam and solid forms. Solid polystyrene is used in yogurt, sour cream, and cottage cheese containers, cutlery, clear clamshells used at salad bars, and video and audio cassette containers. As foam, polystyrene is used in cups, bowls, plates, trays, clamshell containers, meat trays, egg cartons, and packaging for electronics and other delicate items. Polystyrene foam is manufactured by processing solid polystyrene resin pellets with a gaseous expansion agent. In the past, about 30% of polystyrene foam products were made with **chlorofluorocarbons** (CFCs), which were identified as contributors to the deterioration of the **ozone** layer in the Earth's upper **atmosphere**. By 1990, manufacturers had phased out the use of CFCs as a polystyrene expansion agent. Some manufacturers switched to the use of hydrochlorofluorocarbon-22 (HCFC-22), which reduced **ozone layer depletion** potential by about 95% over CFCs. By 1994, as required by federal law, polystyrene foam manufacturers phased out the use of HCFC-22 and by 1997 were using alternative expansion agents, most commonly pentane gas. Pentane does not affect the ozone layer, but because it can contribute to **smog** formation some manufacturers recycle pentane emissions. **Carbon dioxide** (CO_2) is also used by manufacturers as a polystyrene expansion agent. CO_2 is nontoxic, nonflammable, does not contribute to smog, and has no atmospheric ozone depletion potential, but it has been implicated in global warming.

Polystyrene packaging accounts for about 1.2 % (by weight) of the total solid **waste stream** in the United States, and polystyrene makes up from about 2-10% of materials thrown away as litter. However, polystyrene can be recycled and used to make new polystyrene products such as wall insulation, packing filler, and cafeteria trays. Polystyrene has an energy content of 17,000 Btu/pound, about four times that of average **municipal solid waste**, and can be burned with other solid wastes in an incinerator that has appropriate **emission** controls to contain potentially toxic **combustion** products. Another potential environmental concern is the release of toxic metals such as **cadmium** and **lead** from combustion of inks used to tint the polystyrene. The polystyrene manufacturing industry is developing source reduction techniques, including improving resin properties to make stronger products with less material and making lighter-weight foam products.

Polystyrene is manufactured from **styrene**, which is a hazardous chemical. When styrene is heated to 392°F (198°C), it is converted into the polymer polystyrene. Health concerns have been raised about styrene residues remaining in polystyrene products, especially those that are used to contain food for consumption. Styrene is soluble in oil and in **ethanol**, substances commonly found in foods and alcoholic beverages. A U.S. EPA National Human Adipose Tissue Survey in 1986 identified styrene in 100% of the 46 human fat samples collected.

[*Judith Sims*]

Resources
Other

Expansion Agents for Polystyrene Foam. Washington, DC: Polystyrene Packaging Council, 1997.

Polystyrene in the Solid Waste Stream. Washington, DC: Polystyrene Packaging Council, 1997.

Uses and Benefits of Polystyrene. Washington, DC: Polystyrene Packaging Council, 1997.

Polyvinyl chloride

Polyvinyl chloride, also known as PVC, is a plastic produced by the polymerization of **vinyl chloride**. It is used with plasticizers to make packaging films, boots, garden hose, etc. Without plasticizers, PVC is used to make pipe, siding, shingles, window frames, toys, and other items. An attractive aspect of PVC for industry is its ability to withstand **weathering** and its **resistance** to **chemicals** and solvents. However, this attractive aspect is the major environmental concern for PVC and many other **plastics**. The great bulk of such plastic (about 98%) is neither reused nor recycled but occupies ever dwindling **landfill** space.

Pongo pygmaeus

see **Orangutan**

POPs

see **Persistent Organic Pollutants**

Population biology

Population biology is the study of the factors determining the size and distribution of a population, as well as the ways in which populations change over time. The discipline of population biology dates back to the 1960s, when researchers merged aspects of population **ecology** with aspects of population genetics. It employs a traditional empirical approach

which consists of observation of the numbers of individuals in a population and the variation in those numbers over time and space, and the measurement of physical (abiotic) factors and the living (biotic) factors that may affect population numbers.

Given optimum conditions, the populations of most organisms grow at a constant rate of increase, doubling in size at regular intervals, which is known as **exponential growth**. Exponential **population growth** is explosive but it is usually opposed by factors that reduce numbers, such as disease, predation, or harsh climates. The result is **logistic growth**, where the rapidly growing population slows and reaches a stable but dynamic equilibrium at or near the **carrying capacity** of the **environment**. The populations of some **species** such as migratory locusts show boom and bust cycles: in these cases explosive growth produces vast numbers of individuals that eventually overwhelm the carrying capacity and a catastrophic **dieback** follows.

The rate of growth of a population is the net result of the gains and losses from a number of intrinsic factors operating within the population. These include natality, **fecundity**, life span, longevity, **mortality**, and immigration and emigration. Patterns of **survivorship** and age structure created by these interacting factors show how a population is growing and indicate what general role a species plays in the **ecosystem**. Density-dependent biotic factors that decrease natality or increase mortality include the numbers of competitors both interspecific and intraspecific, as well as predators, prey, **parasites**, and other interactive species. Stress and overcrowding are other density-dependent biotic factors that limit population size through excessive intraspecific **competition** for limited resources.

The local distribution pattern of populations of most species is limited by physical factors such as temperature, moisture, light, **pH**, **soil** quality, and **salinity**. Within their areas of distribution, animals occur in varying densities (either scattered thinly or crowded) and in varying dispersal patterns (either spaced evenly or clumped into herds). Population density and dispersion are often studied together, and are important in ecology and management. Density is measured by direct visual count, and by **trapping**, collecting fecal pellets, using pelt records, monitoring vocalization frequencies, and so on. Life tables constructed from these data show precisely how a population is age-structured.

Population genetics recognizes two important attributes of a population--its gene frequencies and its total **gene pool**. Population size exerts some influence on the genetic composition of its members, since the number of sexually interbreeding individuals influences the transfer of genes within a population. It also affects the kinds of genotypes that are available, and the survival and reproductive capacity of individuals with certain genes. The application of the principles of population genetics is vital to the success of programs to improve the breeds of animals and plants for agricultural use, and for the captive breeding of **endangered species**. *See also* Captive propogation and reintroduction; Gene bank; Genetic engineering; Predator-prey interactions; Sustainable development; Wildlife management

[*Neil Cumberlidge Ph.D.*]

Resources

Books

Begon, M., J. L. Harper, and C. R. Townsend. *Ecology: Individuals, Populations, and Communities*. 2nd Edition. Boston: Blackwell Scientific Publications, 1990.

Emmel, T. C. *Population Biology*. New York: Harper and Row, 1976.

Hendrick, P. W. *Population Biology: The Evolution and Ecology of Populations*. Boston: Jones and Bartlett, 1984.

Ricklefs, R. E. *The Economy of Nature*. 3rd ed. New York: W. H. Freeman, 1993.

Smith, R. L. *Elements of Ecology*. New York: HarperCollins, 1991.

Stiling, P. D. *Introductory Ecology*. Englewood Cliffs, NJ: Prentice Hall, 1992.

Population control

see **Family planning; Male contraceptives**

Population Council

The Population Council was established in 1952 to find solutions to the world's population problems. Although based in New York, the Council is an international nonprofit organization with regional offices in Bangkok, Cairo, Dakar, Mexico City, and Nairobi. In conducting research in health, social, and biomedical sciences, the Council has, among other accomplishments, broken ground in contraceptive devices, explored male fertility, and analyzed the social position of women and its concurrent effect on the population.

The Council seeks to use science and technology to provide relief to the population problems of developing countries. Governed by a board of 20 trustees from 16 countries who meet twice a year, the Council has programs in 50 countries throughout Latin America, the Caribbean, Asia, and Africa. It receives funding from world governments, United Nations agencies, foundations, and individuals for its research and programs. The major programs and institutions of the Population Council are the Center for Biomedical Research, the Research Division, and the Programs Division.

The Center for Biomedical Research investigates new ways to regulate human fertility, primarily through development of new methods of contraception. For example the

Center invented the Norplant levonorgestrel implant and a copper-bearing T-shaped intrauterine device (IUD). Both methods are for long-term protection and are reversible. Other methods include a hormone-releasing IUD, Norplant II (with two implants instead of six), and contraceptive vaginal rings. Another idea in progress includes a vaginal contraceptive that provides protection from the human immunodeficiency **virus** (HIV). These methods are designed for women, but the Center for Biomedical Research also concentrates on controlling male fertility through research into male physiology. With particular interest in creating a contraceptive to block sperm production without dampening the male sex drive, the Center is working on a "nonsurgical" or "no-scalpel" vasectomy and male subdermal implants, akin to women's Norplant.

The Population Council does not restrict its research to the biomedical domain, however. Through its Research Division, it looks for solutions to population problems through social research. A small interdisciplinary group of demographers, economists, sociologists, and anthropologists make up the Research Division. Among notable studies is an ongoing project examining the consequences of high fertility in a family. For instance, early research indicates that in Maharashtra, India, the advantages associated with growing up in a small family exist for boys but not for girls. Apparently in many smaller families in India, girls have a proportionately heavier burden of labor. The Research Division also conducts many studies in China, where the state is heavily involved in regulating its population. Through research into **family planning** and fertility, child survival, and women's roles and status—a subject that has taken precedence in the Council's research since 1976—the Research Division provides theories that the Population Council puts to use in controlling worldwide population problems.

The Council applies the research conducted by the Center for Biomedical Research and the Research Division through its Programs Division. The Programs Division collaborates with governments, organizations, and scientific institutions of developing countries to formulate population policy. Many of the programs set forth by this division parallel studies being conducted by the Research Division. For example, the Programs Division conducts family planning and fertility counseling; works to insure reproductive health and child survival; and seeks to improve women's roles and status in developing countries. The contraceptive introduction program is also an important part of the Programs Division.

In addition to its programs and research, the Council publishes two scholarly journals: *Population and Development Review* and *Studies in Family Planning*. Both contain studies of current ideas and theories in population control. The Population Council also sponsors outstanding scholars and scientists to further research into the technology needed to control the world's populations.

[*Andrea Gacki*]

Resources

Organizations

Population Council, 1 Dag Hammarskjold Plaza, New York, NY USA 10017 (212) 339-0515, Fax: (212) 755-6052, Email: pubinfo@popcouncil.org, http://www.popcouncil.org

Population growth

A population is the number of individuals of a given **species**, usually within a specified **habitat** or area. Within the science of **ecology**, the study of population dynamics, or the ways in which the number of individuals in a community expands and contracts, makes up an important subfield known as **population biology**. While ecologists and **wildlife** biologists rely on understanding the nature of population change and the factors that influence population size, one of the main ways population studies have been used is in projecting the growth of the human population. Since the 1960s human population growth has been an issue of fierce theoretical debate. Many scientists and social planners predict that the human population, like animal populations frequently observed in nature, is growing at a perilous rate that will ultimately lead to a catastrophic **die-off** in which billions of people will perish. Other scholars and planners argue that current trends may not lead to disaster and that we may yet find a reasonable and stable population level. Both sides of the debate use ecological principles of population biology in predicting human population dynamics and their effect upon the conditions in which we may live in the future.

The extent of population growth is based on fertility, or the number of offspring that individuals of reproductive age successfully produce. In some animal and plant species (flies, dandelions) every individual produces a great number of offspring. These populations grow quickly, as long as nutrients and space are available. Other species (**elephants**, pandas) produce few offspring per reproductive adult; populations of these species usually grow slowly even when ample food and habitat are available. The maximum number of offspring an organism can produce under ideal conditions is known as the biotic potential of that organism. For instance, if an individual housefly can produce 120 young in her lifetime, and each of those successfully produces 120 young, and so on, the fly's biotic potential in one year (seven generations) is nearly six trillion flies. Clearly biotic potentials are rarely met: reproductive failures and life hazards usually prevent the majority of houseflies, like other species, from achieving their theoretical reproductive potential.

Populations increase through reproduction, but they do not increase infinitely because of environmental limitations, including disease, predation, **competition** for space and nutrients, and **nutrient** shortages. These limits to population growth are collectively known as environmental **resistance**.

An important concept in environmental resistance is the idea of **carrying capacity**. Carrying capacity is the maximum number of individuals a habitat can support. In any finite system, there is a limited availability of food, water, nesting space, and other essentials, which limits that system's carrying capacity. When a population exceeds its environment's carrying capacity, shortages of nutrients or other necessities usually weaken individuals, reduce successful reproduction, and raise death rates from disease, until the population once again falls below its maximum size. A population that grows very quickly and exceeds its environment's carrying capacity is said to overshoot its environment's capacity. A catastrophic **dieback**, when the population plummets to well below its maximum, usually follows an overshoot. In many cases populations undergo repeated overshoot-dieback cycles. Sometimes these cycles gradually decrease in severity until a stable population, in equilibrium with carrying capacity, is reached. In other cases, overshoot-dieback cycles go on continually, as in the well known case of lemmings. Prolific breeding among these small arctic rodents leads to a population explosion every four to six years. In overshoot years, depletion of the vegetation on which they feed causes widespread undernourishment, which results in starvation, weakness, and vulnerability to predators and disease. The lemming population collapses, only to begin rebuilding, gradually approaching overshoot and another dieback. At the same time, related populations fluctuate in response to cycles in the lemming population: populations of owls and foxes surge when lemmings become plentiful and fall when lemmings are few. The grasses and forbs on which lemmings feed likewise prosper and diminish in response to the lemming population.

As a population grows, the number of breeding adults increases so that growth accelerates. Increase at a constant or accelerating rate of change is known as **exponential growth**. If each female lemming can produce four young females, and each of those produces another four, then the population is multiplied by four in each generation. After two generations there are 16 (or 24) lemmings; after three generations there are 64 (34); after four generations there are 256 (44) lemmings, if all survive. When environmental resistance (predation, nutrient limitations, and so on) causes a population to reach a stable level, without significant increases or decreases over time, population equilibrium is achieved. Generally we might consider stable (equilibrium) populations the more desirable situation because repeated diebacks involve extensive suffering and death.

These principles of population biology have strongly informed our understanding of human population changes. Over the centuries the world's human population has tended to expand to the maximum allowed by available food, water, and space. When humans exceeded their environment's carrying capacity, catastrophic diebacks (usually involving disease, **famine**, or war) sometimes resulted. However, in many cases, diebacks have been avoided through emigration, as in European migrations to the Americas, or through technological innovation, including such inventions as agriculture, **irrigation**, and mechanization, each of which effectively expanded environmental carrying capacities. For tens of thousands of years the human population climbed very gradually, until about the year 1000, when it began to grow exponentially. Where we once needed a thousand years (200 to 1200 A.D.) to double our population from 200 million to 400 million, at current rates of growth we would require only 40 years to double our population. Since the eighteenth century, population theorists have increasingly warned that our current pattern of growth is leading us toward a serious overshoot, perhaps one that will permanently damage our **environment** and result in a consequent dieback. The only course to avoid such a catastrophe, argue population theorists, is to stabilize our population somewhere below our environment's carrying capacity.

Popular awareness of population issues and agreement with the principle of population reduction have spread in recent years with the publication of such volumes as *The Population Bomb* by Paul Ehrlich, and *The Limits to Growth* by Donella Meadows and others. The current population debate, however has older roots, especially in the work of Thomas Malthus, an English cleric who in 1798 published *An Essay on the Principle of Population as It Affects the Future Improvement of Society*. Malthus argued that, while unchecked human population growth increases at an exponential rate, food supplies increase only arithmetically (a constant amount being added each year, instead of multiplying by a constant amount). The consequence of such a disparity in growth rates is starvation and death. The remedy is to reduce our reproductive rates, where possible by "moral restraint," but where necessary by force. Malthus' work has remained well-known principally because of its conclusion about social policy: because providing food and shelter to the poor only allows them to increase their rates of "breeding," assistance should be withheld. If the poorer classes should starve as a result, argued Malthus, at least greater rates of starvation at a later date would be avoided. This conclusion continues to be promoted today by neo-Malthusians, who protest the principle of aiding developing countries. Poorer countries, neo-Malthusians point out, have especially dangerous

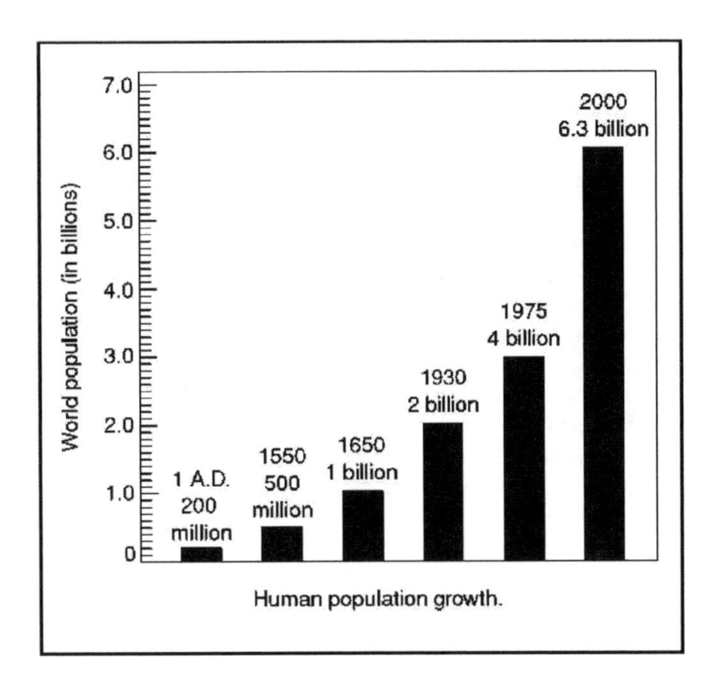

Human population growth.

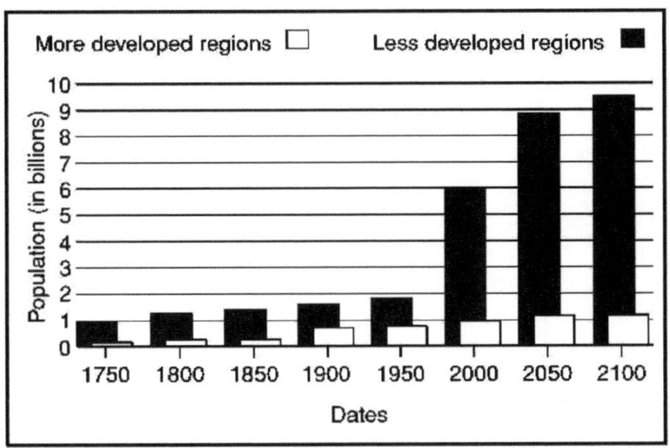

A comparison of population growth in less developed and more developed regions, 1750-2100.

Contrasting the growth (exponential for less developed).

Human population growth. (Illustration by Electronic Illustrators Group.)

growth rates. If wealthy countries provide assistance today because they cannot bear to watch the suffering in poorer nations, they only make the situation worse. By giving aid, donor nations only postpone temporarily the greater suffering that will eventually result from artificially supported growth rates.

Naturally this conclusion is deeply offensive to developing nations and to those who sympathize with the plight of poorer countries. Malthusian conclusions are especially harsh in situations where military and economic repression, often supported by North American and European governments, have caused the poverty. Those who defend aid to developing countries argue that high reproductive rates often result from poverty, rather than vice versa. Where poor nutrition and health care make infant **mortality** rates high, parents choose to have many children, thus increasing the odds that at least some will survive to support them in their old age. Assistance to poor countries, and to poor areas within a single country, argue anti-Malthusians, helps to alleviate the poverty that necessitates high birth rates. When people become confident of their childrens' survival, then birth rates will drop and population equilibrium may be achieved.

There are some ways in which ecological principles of population dynamics do not necessarily fit the human population. Improving health care and increasing life expectancies have caused much of this century's population expansion. Wars, changes in prosperity, development and **transportation** of resources, and social changes sometimes increase or decrease populations locally. Immigration and emigration tend to redistribute populations from one region to another, temporarily alleviating or exacerbating population excesses. Also unlike most animals, we have not historically been subject to a fixed environmental carrying capacity. The **agricultural revolution**, the industrial revolution and other innovations have expanded our environment's capacity to support humans. Some population theorists today, known as technological optimists, insist that, even as the human population continues to grow, our technological inventiveness will help us to feed and shelter the world's population. Even in the past 50 years, when the world's population has jumped from less than three billion to more than six billion, world food production has exceeded population growth. Technological optimists point to such historical evidence as support for their position.

Also unlike most animal populations, humans are, at least in principle, able to voluntarily limit reproductive rates. Most population theorists point to some sort of voluntary restraint as the most humane method of preventing a disastrous population overshoot. **Family planning** programs have been developed around the world in an effort to encourage voluntary population limits. Furthermore, human societies have been observed to go through a **demographic transition** as their economic and social **stability** has improved. A demographic transition is a period when (infant) mortality rates have decreased but people have not yet adjusted their reproductive behavior to improved life expectancies. After 20 to 30 years, people adjust, realizing that more children are surviving and that smaller families now have economic advantages. The birth rate then begins to fall. While families in poor countries often have five to ten children, the average family in developed countries has about two children.

Because of family planning programs and demographic transitions, birth rates around the world have fallen dramatically in just the last 20 years. Where the average number of children per woman was 6.1 in 1970, the average number in 1990 was only 3.4. This rate still exceeds the minimum replacement level (**zero population growth** level) of 2.1 children per woman, but if recent demographic trends continue the world's population will stabilize within the first decade of the twenty-first century.

No one knows exactly what the earth's carrying capacity is. Some warn that we have already surpassed our safe population size. Others argue that the earth can comfortably support far more people than it currently does. Many worry about what would happen to other species if the human population continues to expand. Most people agree that we would benefit from a decrease in population growth rates. How urgent the population question is and how best to curb growth rates are issues that continue to be hotly debated.

[*Mary Ann Cunningham Ph.D.*]

Resources

Books

Cunningham, W. P. *Understanding Our Environment: an Introduction*. Dubuque, IA: Wm. C. Brown, 1993.

Ehrlich, P. R. *The Population Bomb*. New York: Ballantine Books, 1968.

Meadows, D. H., et al. *Limits to Growth*. New York: Universe Books, 1972.

Periodicals

Ehrlich, P. R., and A. H. Ehrlich. "The Population Explosion: Why Isn't Everyone As Scared As We Are?" *Amicus Journal* 12 (Winter 1990): 22–29.

Population Institute

Despite the fact they work with an annual budget of less than $2 million, the Population Institute, an information, education, and communication organization that spreads the word on population control, has been able to reach 172 countries, principally in the developing world. Based on the philosophy that overpopulation affects all social, economic, and political issues, the Institute's goal is to bring world population into balance with available environmental resources. To do this, the organization disseminates information via a World Population News Service which is picked up in developing countries in newspaper articles, editorial pieces, and radio and television broadcasts. In addition to making people aware of the perils of overpopulation, the Institute encourages birth control, without emphasizing any particular method.

Once a year the Institute funds an international journalism award, the Global Media Award in Population Reporting. The winner spends about one week touring a country with a successful population control program, visiting family planning clinics, and talking with government and private officials. Governmental leaders within the country usually present the award. The Institute benefits by keeping in contact not only with leaders and people at the grassroots level, but with local journalists.

Begun in 1969 by a Methodist minister with a strong conviction that overpopulation was a major cause of human misery, the Population Institute was one of the first groups organized specifically around population control. No longer connected to the Methodist church, the Institute is supported by grants from foundations, private individuals, and the sale of its bimonthly newsletter, *Popline*. In 1980 its domestic component split and formed the Center for Population Options, also headquartered in Washington, D.C.

If contraceptive use is a valid indicator, things are looking up for population control. Between 1970 and 2000, contraceptive use rose from 9–64%. If use can be increased to 72% within the next few years, World Bank statistics predict that world population can be stabilized at 8.5–11 billion.

[*Stephanie Ocko*]

Resources

Organizations

Population Institute, 107 Second Street, NE, Washington, D.C. USA 20002 (202) 544-3300, Fax: (202) 544-0068, Email: , <http://www.populationinstitute.org>.

Eliot Furness Porter (1901 – 1990)

American photographer

Born in Winnetka, Illinois, Eliot Porter's fascination with **nature** (and birds in particular) was evident from a very early age. His father, an architect with a keen interest in

Greek, Gothic, and Roman architecture and art, encouraged his son's insightful, artistic talents. Using the camera he received as a gift from his father, the young Porter began photographing primarily landscapes during the family's sojourns on islands off the coast of Maine. There he cultivated an enthusiasm and love for naturalist photography. He noted in the introduction of the classic book, *Birds of America*, "...the most satisfactory outlet for expressing my excitement over birds was the camera, rather than pencil or brush."

Porter began his undergraduate education at Harvard University in 1920. Casting aside his passion for photography, he pursued the more practical major of chemical engineering; he earned his bachelor's degree (cum laude) in 1924. He continued his education at Harvard's Medical School where he earned his doctor of medicine in 1929. Porter used his degrees to teach biochemistry and bacteriology at Harvard and Radcliffe College until 1939. In addition to his teaching, Porter was working through the Harvard Biology Department conducting numerous studies. By 1939, whatever attraction there had been to biochemistry diminished and Porter turned his energies back to photography. Determined to transform bird photography from less-than-professional reportage to art, Porter became involved with the Eastman Kodak company. Porter was awarded Guggenheim Fellowships in 1941 and again in 1946 to finance experiments with Kodachrome photography—a new venture for Kodak. He believed that by using the new color film, the photographer could provide a more sensitive interpretation of the subject.

Finally achieving success with color film, Porter produced a number of bird photography publications, including a photo-essay pairing his photos with specific Thoreau excerpts. The book, *In Wildness Is the Preservation of the World*, was published by the **Sierra Club**. Porter produced subsequent books with the organization (*Baja California—The Geography of Hope* and *The Place No One Knew: Glen Canyon on the Colorado*), and later—until 1971—served on the Sierra Club's Board of Directors.

While Porter's photography took him to locations throughout the world (Greece, Iceland, Turkey, and the Galapagos Islands), his most brilliant work came from his travels in eleven of the United States. *Birds of America—A Personal Selection* was the result of his trip and became one of his best-known books. It combines rich, full-color photographs with anecdotal notes detailing his trials and tribulations in taking them. Porter's other publications include *Galapagos—The Flow of Wilderness, Antarctica*, and *Intimate Landscapes*.

Porter's one-man exhibits have appeared in some of the most influential art institutes and museums in the United States, among them: the George Eastman House, Museum of Modern Art, Metropolitan Museum of Art, and Stieglitz's An American Place. He maintained membership in several organizations including Advocates for the Arts, American Civil Liberties Union, American Ornithologists' Union, and the Audubon Society.

Porter died in November 1990 in New Mexico following a heart attack; he had also suffered from Lou Gehrig's disease.

[*Kimberley A. Peterson*]

Resources

Books

Porter, Eliot. *Eliot Porter's Southwest*. New York: Henry Holt, 1991.

———. *Nature's Chaos*. New York: Viking Penguin, 1990.

Periodicals

"Eliot Porter: For More Than 50 Years, This Master of Color and Light Portrayed and Idealized the American Landscape." *Life* 14 (February 1991): 80–85.

Henry, G. "Eliot Porter." *ARTnews* 89 (Summer 1990): 178.

Portheria dispar

see **Gypsy moth**

Positional goods

A concept coined by English economist Fred Hirsch (1931–1978) to describe goods or activities whose value depends on their exclusivity. For example, fame is considered a positional good since by definition only a few people can be famous and thus enjoy this "privilege." If everyone were famous, no one would be. Similarly, solitude on a mountain peak or in the **wilderness** would qualify as a positional good, since if one shared the peak or the wilderness with many other people, one could not experience solitude. Ironically, a positional good tends to diminish its own value because of the high demand it creates: as more people enjoy positional goods, they no longer become exclusive or valued. For instance, automobiles were once a positional good in America, but as more and more people began owning cars, they were no longer such a status symbol. A number of concerned environmentalists contend that access to wilderness areas should be restricted or limited to prevent the destruction of this positional good.

Posterity

see **Future generations**

Postmodernism and environmental ethics

Environmental ethics are a set of norms that a society accepts as a foundation for mediating and guiding their behavior towards the **environment**. In Western society, we have used our science, as well as other traditions (e.g. capitalist economics, humaneness, aesthetics), to form the foundation of our environmental behavior. We are in the midst of a search for new ethical foundations. **Deep ecology** has offered one possible route, Eastern religions have been another source for advancing our ethics, and yet none of these has proved palatable in our Western social systems.

Western science has guided the environmental interactions of our society which in turn has dominated global development processes, but this pattern has some definite shortcomings. While science has served to develop our understanding of **nature** and enhanced our knowledge of the ecological difficulties that we have created for ourselves, the system of which science is part has been very slow in responding to what many suggest is an ecological crisis. Sorting the science from the politics and the ethics of our society is not an easy task. Yet, science has been the target of criticism in the environmental arena because, while it has the power to expose the depth of our ecological crisis, it has done little in that regard. Critics point to that inaction as a negative structural attribute of how embedded science has become in our society.

An alternative proposed from within the humanist tradition in the social sciences suggests that science can only produce a false construction of nature. From this viewpoint science is tied to traditions that inevitably create an end product related to the position from which the observer views nature. Further it holds that all views of nature are constructed by humans who cannot step outside of ourselves to view the reality we think exists. Thus, no views are more or less closer to reality; they are all merely our creation. This alternative approach falls under the rubric of postmodernism and is elucidated through one of the principal methods in the postmodernist tool kit, deconstruction.

The deconstructionist approach of the postmodernist suggests that, even thought the views of all are equally worthy, an imbalance of power has been created by the dominant classes. The disempowered classes attempt to create a balance, a struggle which proceeds by dissecting everything down to its bare power relationships, so that having been revealed the wrong can be corrected. Deconstruction is a term that does not differ significantly from any other critical analytical method, but it pays particular homage to power and position, supposedly more conscientiously than other methods.

Nature "out there" then becomes a series of constructions made by humans with different positions within a continuum of power relations. Nature is the invention of humans in power and is constantly reinvented by people in those same positions. To the postmodernist, the present invention of nature, held by the current cadre of largely male, largely Caucasian scientists has given us a biased view of nature. The postmodernist wish is to level all points of view, so that life is seen as a struggle of words and nature becomes another, albeit large, "text" (a metatext). A rather curious outcome of this view of life is the idea that things in nature are doing what they do only because we think of them doing it. Furthermore, those who wish to protect nature may then become a force of oppression against those of other social orders who are struggling to gain power. Thus, a thinly disguised structuralist approach emerges. Power creates a framework within which the powerful struggle to remain empowered, and the remainder struggle to gain power. Environmental views supported by science are the handmaiden of the power elite and can only succeed in supporting the status quo.

While there is undoubtedly a cultural context in which nature is understood and in which science operates, this does not necessarily mean that context is equivalent to the content of nature. Even though reality may be seen as having been invented by words and the human shaping of sensory data, nevertheless there is much commonality across cultures regarding what we see in nature. The postmodernist view appears to be remote from the natural world, occupying an anthropocentric stronghold where humans are dominant. Interestingly, while humanists thus refer to a nature made by humans, scientists have since Darwin referred to humans as having been created by nature through natural selection.

The diametrically opposed view points under discussion here are not new; they merely reflect the range of opinions humans have always held about their relationship with nature. However, the postmodernist expression of human opinion about nature outlined above seems to be a position that is distinctly unhelpful, especially at this time in our existence. The heart of the environmental problems we currently experience has been blamed to a large degree on humanity's separation of itself from nature, and any proposal that maintains or widens that gap may exacerbate the problems. The postmodernist critique seems to ignore the power and influence that nature has on us, and instead uses an intricate semantics to describe the power we have over each other.

Most naturalists and ecologists are immediately struck by the apparent naiveté of a position that denies the creative strength in nature and the complex architecture of life provided by natural selection. A postmodern view is that our knowledge of distributions of organisms is a construct of

the way we have studied distributions, and the whole idea of **species** is merely a human distinction. The notion of objective criteria devised by, critiqued by, and used by scientists is rendered unusable. The postmodern view renders the concept that ecological tolerance of the organisms, historical events, and climate create a relationship based on a combination of determinist relationships and stochastic events a mere artifact regardless of how many people agree to the concepts. This of course leads to a conclusion which provides little guidance for environmental ethics.

That does not mean that the entire postmodernist critique leads to the same less than useful conclusions. Rather, the very aspect of a critique that causes us to examine our position may be an important and valuable tool. We have relied heavily on science and found at the center of **ecology** some powerful, yet largely incorrect, metaphors firmly rooted in our conception of nature. The ideas of balance, **succession**, and homogeneity in nature have all fallen by the wayside in the past two decades. Those readjustments have given ecologists pause because they are strong examples of how viewpoint can influence our interpretation of nature, one that was originally thought to have been discovered by an objective science. Are there more misconceptions that we hold about ecology and environment that need to be examined? Clearly the possibility exists, and it serves us well to search for those by any and all reasonable methods. That postmodernism creates the notion that all human perspectives are inherently equal may create an obstacle to clear thinking about proposing a comprehensive ethic towards the environment. At this time we need to synthesize a more positive relationship between nature and humanity to bring all of our cultures into an integrated understanding of nature. However, a revised ethical foundation for humanity's relationship with the environment may benefit from the tradition of critical thinking that various philosophical positions bring. An examination of biases linked to our present position.

[*David A. Duffus*]

Potable water

see **Drinking-water supply**

John Wesley Powell (1834 – 1902)

American philosopher, geologist, anthropologist, and scientific explorer

John Wesley Powell—civil-war veteran, college professor, long-time head of the U.S. **Geological Survey**, member of the **National Academy of Sciences**, president of the American Association for the Advancement of Science, and instrumental in the establishment of the National Geographic Society, the Geological Society of America, the U.S. Geological Survey, the Bureau of Ethnology, and the Bureau of Reclamation—is best-known for two expeditions down the Green and Colorado Rivers.

Born in Mount Morris, New York, of English-born parents, Powell was discouraged from education by his father who believed that the ministry was the only purpose for which one needed to be educated. The younger Powell however, worked at a variety of jobs to support his attendance at numerous schools, but never completed a degree. His hard-won recognition as a scientist is based in large part on self-taught concepts and methods that he applied all his life and in everything he did.

Powell made significant contributions to **conservation** and **environmental science**. He was an early and ardent student of the culture of the fast-disappearing North American Indians tribes, and his work ultimately led to the creation of the Bureau of Ethnology, of which he was the first director. He served as the second Director of the United States Geological Survey (USGS) from 1881 to 1894, and as such he instigated extensive topographic mapping projects and geological studies, stimulated studies of soils, **groundwater**, and rivers, and advocated work on flood control and **irrigation**. His Irrigation Survey led eventually to the creation of first the **Reclamation** Service and then the **Bureau of Reclamation**.

Powell was the author of the *Report on the Lands of the Arid Region of the United States*, published in 1878. Through this document, he became an early advocate of land-use planning. Powell was open to controlled development and suggested that "to a great extent, the redemption of...these lands will require extensive and comprehensive plans..." In his role as director of the USGS, in his writings, his work on various commissions, and in public hearings, Powell advocated that the federal government assume a major role to insure an orderly and environmentally sound settlement of the **arid** lands of the West.

Wallace Stegner argues that Powell's ultimate importance derives from his impact as an agent of change, from setting in motion ideas and agencies that still benefit the country today. He also widens Powell's impact, asserting that Powell's ideas, through his friend and employee W. J. McGee, heavily influenced the whole conservation movement at the beginning of the twentieth century. Powell informed the American public of the grandeur and vulnerability of the arid lands and canyon lands of the West. The legacy of that heritage is the Grand Canyon still largely unspoiled as well as a better understanding of **land use** possibilities on arid and all lands.

[*Gerald L. Young Ph.D.*]

John Wesley Powell. (Corbis-Bettmann. Reproduced by permission.)

RESOURCES

BOOKS

Murphy, D. *John Wesley Powell: Voyage of Discovery.* Las Vegas: KC Publications, 1991.

Stegner, W. *Beyond the Hundredth Meridian: John Wesley Powell and the Second Opening of the West.* New York: Penguin Books, 1992.

PERIODICALS

Robinson, F. G. "A Useable Heroism: Wallace Stegner's *Beyond the Hundredth Meridian.*" *South Dakota Review* 23 (1985): 58–69.

Power lines

see **Electromagnetic field**

Power plants

The term power plant refers to any installation at which electrical energy is generated. Power plants can operate using any one of a number of fuels: oil, **coal**, nuclear material, or geothermal steam, for example. The general principle on which most power plants operate, however, is the same. In such a plant, a fuel such as coal or oil is burned. Heat from the burning fuel is used to boil water, converting it to steam. The hot steam is then used to operate a steam turbine.

A steam turbine is a very large machine whose core is a horizontal shaft of metal. Attached to the horizontal shaft are many fan-shaped blades. As hot steam is directed at the turbine, it strikes the blades and causes the horizontal shaft to rotate on its axis. The rotating shaft is, in turn, attached to the shaft of an electric generator. The pressure of hot steam on turbine blades, is therefore, ultimately converted into the generation of a electric current. One of the first steam turbines designed to be used for the generation of electricity was patented by Charles Parson in 1884. Twenty years later, entrepreneurs had already put such turbines to use for the generation of electricity for street lines, subways, train lines and individual appliances.

Most early power plants were fueled with coal. Coal was the best-know, most abundant fossil fuel at the time. It had several disadvantages, however. It was dirty to mine, transport, and work with. It also did not burn very cleanly. Areas around a coal-fired power plant were characterized by clouds of **smoke** belching from smokestacks and films of ash deposited on homes, cars, grass, and any other exposed surface.

By the 1920s interest in oil-fired power plants began to grow. The number of these plants remained relatively low, however, as long as coal was inexpensive. But, by the 1960s, there was a resurgence of interest in oil-fired plants, largely in response to increased awareness of the many harmful effects of coal **combustion** on the **environment**. One of the most troublesome effects of coal combustion is **acid rain**. Coal contains trace amounts of elements which produce acid-forming oxides when burned. The two most important of these elements are sulfur and **nitrogen**. During the combustion of coal, sulfur is oxidized to **sulfur dioxide** and nitrogen to nitric oxide. Both sulfur dioxide and nitrogen oxide undergo further changes in the **atmosphere**, forming sulfur trioxide and nitrogen dioxide. Finally, these two oxides combine with water in the atmosphere to produce sulfuric and nitric acids.

This series of chemical reactions in the atmosphere may take place over hundreds or thousands of miles. Oxides of sulfur and nitrogen that leave a power plant smokestack are then carried eastward by prevailing winds. It may be many days or weeks before these oxides are carried back to earth—now as acids—in rain, snow, or some other form of precipitation. During the 1960s and 1970s, many authorities became concerned about the possible effects of **acid** precipitation on the environment. They argued that the problem could be solved only at the source—the electric power-generating plant.

One approach to the reduction of pollutants from a power plant is the installation of equipment that will remove undesirable materials from waste gases. An electrostatic precipitator is one such device which removes particulates. It

attaches electrical charges to the particulates in waste gases and then applies an opposing electric field to a plate that attracts the particulates and removes them from the waste gas stream. **Scrubbers, filters**, and **cyclone** collectors also remove harmful pollutants from waste gases.

Another approach to acid rain and other power-plant-generated **pollution** problems is to switch fuel from coal to oil or **natural gas**. Although **petroleum** and natural gas also contain sulfur and nitrogen, the concentration of these elements is much less than it is in many forms of coal. As a result, a number of electrical utilities began to retrofit their generating plants in the 1960s and early 1970s to allow them to burn oil or natural gas rather than coal. This change-over is reflected in the increased use of petroleum by electrical utilities in the United States between 1965 and 1975. Consumption increased more than six-fold in that ten year period, from about 100 million barrels per year to more than 6,700 million barrels per year. Interestingly enough, the use of coal by utilities did not drop off during the same period, but continued to rise from about 200 million short tons to over 300 million short tons annually.

This pattern of **fuel switching** turned out to be short-lived, however. Although oil and natural gas are relatively clean fuels, they also appealed to utilities because of their relatively low cost. The **oil embargo** instituted by the **Organization of Petroleum Exporting Countries** (OPEC) in 1973 altered that part of the equation. Worldwide oil prices jumped from $1.36 per million **Btu** in 1973 to $2.23 per million Btu in 1975, an increase of 64 percent per million Btu in just two years. A decade later, the price of oil had more than doubled. And, the situation was even worse for natural gas. The cost of a million Btu of this resource sky-rocketed from 36.7 cents in 1970 to 69.3 cents in 1975 to $2.23 in the mid-1980s.

Suddenly, the conversion of power plants from coal to oil and natural gas no longer seemed like such a wonderful idea. And the utilities industry began once more to focus on the fossil fuel in most plentiful supply—coal. Between 1975 and 1990, the fraction of power plants powered by oil and natural gas dropped from 12% to 4% for the former and from 24% to less than 10% for the latter. In the same period, the fraction of plants powered by coal rose from 46%–57%.

Power plants using **fossil fuels**, of whatever kind, are not the world's final solution to the generation of electricity. In addition to the environmental problems they cause, there is a more fundamental limitation. The fossil fuels are a nonrenewable resource. The time will come—sooner or later—when coal, oil, and natural gas supplies will be depleted. Thus, research on alternatives to fossil-fuel-fired plants has gone on for many decades.

Thirty years ago, many people believed that **nuclear power** was the best solution to the world's need to produce large amounts of electricity. Indeed, the number of nuclear power plants in the United States grew from 6 in 1965 to 95 in 1985. By 1992, 421 nuclear plants were in commercial operation worldwide, supplying 17% of the world's electricity. But enthusiasm for nuclear power plants began to wane in the 1980s. One reason for this trend was the serious accidents at the **Three Mile Island Nuclear Reactor** (Middletown, Pennsylvania) in 1979 and at the **Chernobyl Nuclear Power Station** (Chernobyl, Ukraine) in 1986. Another reason was the growing concern about disposal of the ever-increasing volume of dangerous radioactive wastes produced by a nuclear power plant. In any case, no new nuclear power plant has been ordered in the United States since the Three Mile Island accident and 65 plant orders have been canceled since that event. Forty-nine nuclear plants are now under construction in other parts of the world, but this number is only a quarter as many as were under construction a decade ago. Indeed, between 1990 and 1991, the total installed nuclear generating capacity worldwide declined for the first time since commercial nuclear power generation began. Given current economic and political conditions, a major revival of the nuclear power industry seems unlikely.

A nuclear power plant operates on much the same principle as does a fossil-fuel power plant. In a nuclear power plant, water is heated not by the combustion of a fuel like coal or oil, but by **nuclear fission** reactions that take place within the reactor core. Some scientists anticipate that another type of nuclear power plant—the fusion reactor—may provide a long-term answer to the world's electrical power needs. In a fusion reactor, atoms of light elements are fused together to make heavier elements, releasing large amounts of energy in the process. **Nuclear fusion** is the process by which stars make their energy and is also the energy source in the **hydrogen** bomb. Research efforts to develop an economic and safe method of controlling fusion power have been underway for nearly 40 years. Although progress has been made, a full-scale commercial fusion power plant currently appears to be many decade in the future.

Other types of power plants also have been under investigation for many years. For example, it is possible to operate a steam turbine with hot gases that come directly from geothermal vents. In locations where geysers or other types of vents exist, geothermal power plants are an economical, safe and dependable alternative to fossil-fueled and nuclear power plants. In the early 1990s, for example Pacific Gas and Electric in northern California obtained about eight percent of its electrical power from geothermal plants.

Finally, power plants that do not use heat at all are possible. One of the oldest forms of power plant is, in fact, the hydroelectric power plant. In a hydroelectric power plant,

the movement of water (as from a river) is directed again turbine blades, causing them to rotate. In 1970, 16 percent of all electric power in the United States came from hydropower although that portion has now fallen to less than eight percent.

Proposals for wind power generated-electricity, **tidal power** plants, and other alterative energy sources as substitutes for fossil fuels and nuclear power have been around for decades. During the 1960s and 1970s, governments aggressively encouraged research on these new technologies, offering both direct grants and tax breaks. Since 1980, however, the United States government has lost interest in alternative methods of power generation, preferring instead to encourage the growth and development of more traditional fuels, such as coal, oil and petroleum. *See also* Geothermal energy; Ocean thermal energy conversion; Solar energy; Tidal power; Wind energy

[*David E. Newton*]

RESOURCES

BOOKS

Brown, L. R., et al. *Vital Signs, 1992.* New York: Norton, 1992.

McGraw-Hill Encyclopedia of Science & Technology. 7th ed. New York: McGraw-Hill, 1992.

PERIODICALS

Balzhiser, R. E., and K. E. Yeager. "Coal-fired Power Plants for the Future." *Scientific American* 257 (September 1987): 100–107.

Swain, H. "Power Plants Threaten Shenandoah's Air." *National Parks* 65 (March-April 1991): 14.

Ppb

see **Parts per billion**

Ppm

see **Parts per million**

PPP

see **Partnership for Pollution Prevention**

Ppt

see **Parts per trillion**

Prairie

An extensive temperate grassland with flat or rolling terrain and moderate to low precipitation. The term is most often applied to North American **grasslands**, which once ex-

A prairie. (Photograph by Robert J. Huffman. Field Mark Publications. Reproduced by permission.)

tended from Alberta to Texas and from Illinois to Colorado, but similar grasslands exist around the world. Perennial bunchgrasses and forbes dominate prairie **flora**. In dry **climate** prairies, a lack of precipitation prohibits tree growth except in isolated patches, such as along stream banks. In wetter prairies, periodic fires are essential in preventing the incursion of trees and preserving open grasslands. Historically, huge herds of large mammals and scores of bird, rodent, and insect **species** composed native prairie **fauna**. During the nineteenth and twentieth centuries most of North America's native prairie **habitat** and its occupants disappeared in the face of agricultural settlement, cattle grazing, and fire suppression. Today the geographic extent of native prairie plants and animals is severely restricted, but pockets of remaining prairie are sometimes preserved for their ecological interest and genetic diversity.

While grasses make up the dominant vegetation type in all prairies, grass sizes and species vary from shortgrass prairies, usually less than 20 inches (50 cm) high, to tallgrass prairies, whose grasses can exceed 7 ft (2 m) in height. Following the gradual east-west decrease in precipitation rates on the Great Plains, eastern tallgrass prairies gradually gave way to midgrass and then shortgrass prairies in western states. On their margins, prairie grasslands extended into a mixed "parkland," or "savanna," with bushes and scattered

trees such as oak and juniper. Characteristic prairie grasses are perennial bunchgrasses, which have deep, extensive root systems and up to a hundred shoots from a single plant. Bunchgrasses typically grow, sprout, and flower early in the summer when rainfall is plentiful in the prairies, and their deep root systems make the long-lived plants resistant to **drought**, hail, grazing, and fire. In addition to the dominant grasses, vast numbers of flowering annual and perennial forbes—herbaceous flowering plants that are not grasses—add color to prairie landscapes.

Prairie plant species are well adapted to live with fire. Under natural conditions fires sweep across prairies every few years. Prairie fires burn intensely and quickly, fed by dry grass and prairie winds, but they burn mainly the dry stalks of previous years' growth. Below the fire's heat, the roots of prairie plants remain unharmed, and new stalks grow quickly, fertilized by ashes and unencumbered by accumulations of old litter. Before European settlement, many prairie fires were ignited by lightening, but some were started by Indians, who had an interest in maintaining good grazing land for **bison**, antelope, and other game animals.

Well into the nineteenth century, prairie habitats supported a tremendous variety of animal species. Vast herds of bison, pronghorn antelope, deer, and elk ranged the grasslands; smaller mammals from mice and **prairie dogs** to badgers and fox lived in or below the prairie grasses. Carnivores including grey **wolves**, coyotes, cougar, grizzlies, hawks, and owls fed on large and small herbivores. Shorebirds and migratory waterfowl thrived on millions of prairie pothole ponds and marshes. Most of these species saw their habitats severely diminished with the expansion of agricultural settlement and cattle grazing range. Prairie soils, mainly mollisols, are rich and black with a thick, fertile organic layer composed of fine roots, **microorganisms**, and **soil**. Where water was available, this soil proved ideal for growing corn, wheat, soybeans, and other annual crops. Where water was scarce, nutritious bunchgrasses provided excellent grazing for cattle, which quickly replaced buffalo, elk, and antelope on the prairies in the middle of the nineteenth century.

Prairie habitat and animal species have also lost ground through fire suppression and wetland **drainage**. Farmers and ranchers have actively suppressed fires, allowing trees, exotic annuals, and other non-prairie species to become established. Widespread **wetlands** drainage has seriously diminished bird populations because of habitat loss. Especially in wetter northern and eastern regions, canals were cut into prairie wetlands, draining them to make arable fields. This practice continues today, even though it destroys scarce breeding grounds for many birds and essential migratory stopovers for others.

In recent decades increased attention has been given to preserving the few remaining patches of native prairie in North America. Public and private organizations now study and protect prairie grasslands and wetlands, but the long-term value of these efforts may be questionable because most remnant patches are small and widely separated. Whether genetic diversity can be maintained, and whether habitat for large or ranging animals can be reestablished under these conditions, is yet to be seen.

[*Mary Ann Cunningham Ph.D.*]

Resources

Books

Cushman, R. C., and S. R. Jones *The Shortgrass Prairie.* Boulder, CO: Pruett Publishing Co., 1988.

Smith, R. L. *Ecology and Field Biology.* 3rd ed. New York: Harper and Row, 1980.

Whittaker, R. H. *Communities and Ecosystems.* 2d ed. New York: Macmillan, 1975.

Prairie dogs

Prairie dogs, members of the genus *Cynomys*, belong to the squirrel family of the order *Rodentia*. There are five **species** of prairie dogs found on North American plains and plateaus from southern Saskatchewan, Canada to northern Mexico: Utah (*C. parridens*), Gunnison's (*C. gunnisoni*), white-tailed (*C. leucurus*), black-tailed (*C. ludovicianus*), and Mexican (*C. mexicanus*).

The explorer Meriwether Lewis, explorer of the American West, described prairie dogs as barking squirrels, which "bark at you as you approach them, their note being much like that of little toy dogs." These barking calls, including barks, chirps, and whistles, are used to communicate greetings, social status, and approaching or retreating danger.

Prairie dogs as social animals live in colonies, or "towns," which consist of extensive and complex underground burrows one to five meters deep. Cone-shaped mounds at the entrance of the burrows are used as look-out points; they also prevent water from entering the burrows. The burrows contain several chambers, including one near the entrance where the prairie dogs can listen for above-ground activity, as well as one or more nesting chambers where young prairie dogs sleep and are cared for. Prairie dogs generally prefer to eat grasses and herbs. They also clip vegetation in the vicinity of their colonies to provide clear views of predators. They spend daylight hours above-ground, grooming each other, grazing on grass, tumbling in play, and defending their family territorial boundaries.

A prairie dog. (U.S. Fish and Wildlife Service. Reproduced by permission.)

More than 200 other **wildlife** species are associated with prairie dog colonies, resulting in the increased biological diversity seen in prairie lands with colonies. Some of these animals are prairie dog predators, such as hawks, eagles, snakes, and coyotes, while others use the colonies as their **habitat**. The reduction and/or spread of certain insects and plants is also dependent on prairie dogs.

However, prairie dog populations have been greatly reduced, in some cases to one to two percent of their historical numbers. This decrease in prairie dog populations was, and continues to be, due to **poisoning**, **hunting**, and extermination campaigns by some ranchers who believe that prairie dogs compete for livestock forage. Long-term **overgrazing** by livestock has resulted in loss of grasses that prairie dogs depend on, as well as in **erosion** and destruction of prairie dog habitat. Additional habitat losses are due to development of lands for human housing and other uses. Prairie dogs are also susceptible to diseases, including the **plague**, which can wipe out entire colonies.

The **IUCN—The World Conservation Union** lists the black-tailed and Utah prairie dogs at low risk and the Mexican prairie dog is already endangered. Environmental and citizen's groups are beginning to act to preserve prairie dog habitats throughout the western grassland areas that serve as their home lands.

[*Judith L. Sims*]

Resources

Books

Graves, Russell A. *The Prairie Dog: Sentinel of the Plains.* Lubbock: Texas Tech University Press, 2001.

Hoogland, John L. *The Black-Tailed Prairie Dog: Social Life of a Burrowing Mammal (Wildlife Behavior and Ecology).* Chicago: University of Chicago Press, 1995.

Long, Kim. *Prairie Dogs: A Wildlife Handbook.* Boulder: Johnson Books, 2002.

Patent, Dorothy Hinshaw, and William Munoz. *Prairie Dogs.* Boston: Houghton Mifflin Company, 1999.

Precision

Precision refers to the ability to repeat the same measurement and arrive at the same value. In **statistics**, precision refers to the difference between repeated measurements of the same quantity, where the closer the measurements are to each other; the greater the precision of the estimate of the quantity. Precise measurements are not necessarily accurate measurements, because the repeated measurements can be very close to the same value and still be far from correct. As an example, if the differences in ten measurements of the same distance are very close to each other, the distance-measuring device is considered to be precise. However, there is no surety that the distance measured is correct unless the device is also accurate.

[*Marie H. Bundy*]

Precycling

Precycling is source reduction and **reuse**. In most **waste management** planning the hierarchy is Reduce, Reuse, and Recycle. Reduction and reuse are the first lines of defense against increasing waste volume. Precycling are those actions that can be taken before **recycling** becomes an option. It is the decision on the part of a consumer to not purchase an unnecessary product or the decision to purchase a reusable as opposed to a disposable item. Precycling is using the china plates in the cupboard for a party instead of purchasing paper or plastic plates. It is also the decision to buy in bulk or to buy refillable containers and then purchasing the product in a bulk container that can refill a dispenser. Precycling is thinking ahead to the ultimate destination of the containers and packaging of products are bought and making the decision to reduce those that might end up in a **landfill**. In addition to individual decisions to reduce waste before it

happens there is the decision on the part of industry to reuse **chemicals** and to buy from parts suppliers that take back packaging and reuse it. Precycling is stopping waste before it happens. *See also* Green packaging; Green products; Solid waste

Predator control

Predator control is a **wildlife management** policy specifically aimed at reducing populations of predatory **species** either to protect livestock or boost populations of game animals. Coyotes, bobcats, grey and red **wolves**, bears, and mountain lions have been the most frequent targets. Historically, efforts were centered in government-run programs to hunt, trap, or poison predators, while bounties offered for particular predatory species encouraged private citizens to do the same. Over time, however, it has become apparent that predator control is disruptive of ecosystems and often inefficient.

Predator control is based on a lack of understanding of the complex interactive mechanisms by which natural environments sustain themselves. Control efforts take the position that prey populations benefit when predators are removed. In the long term, however, removing predators is generally bad for the prey species and their ecosystems. The conservationist **Aldo Leopold**, noticing the devastating effects of **overgrazing** by deer, realized as early as the 1920s that shooting wolves, although it led to an immediate increase in the deer population, ultimately posed a threat to the deer themselves as well as other **fauna** and **flora** in their **environment**. Another example can be found in the undesirable consequences of killing significant numbers of coyotes. Immediate benefits rebound not only to livestock but also to populations of rabbits and other rodents, which are normally controlled by predation. Since rabbits compete with livestock for the same rangeland, an explosion in the rabbit population simply means that the land can sustain fewer livestock animals.

In 1940, the **U.S. Department of the Interior** initiated a large-scale program against livestock predators, particularly aiming at the **coyote** and relying heavily on poisons. The poison most widely used in bait was sodium fluoroacetate, also called "compound 1080," which is highly persistent and wreaks havoc on many animals besides the coyote, including threatened and **endangered species**. Moreover, when a poisoned coyote dies, the carrion enters the **food chain/web**, with lethal effects primarily on scavengers. Eagles, including the rare **bald eagle**, are particularly hard-hit by efforts to poison coyotes.

Rising environmental awareness has caused continuing controversy over predator control. By 1972, the government-appointed Committee on Predator Control concluded that the use of poisons had unacceptably harmful consequences. Government agencies discontinued the use of poisons, and poison was banned on public lands. Nevertheless, private citizens, mostly ranchers and hunters who feel that their interests have been sorely overlooked, have continued using poisons with little regard for the law. In the 1980s the government started taking a less progressive stance, and in 1985 the Reagan administration reinstituted the use of compound 1080 in "livestock collars." While nonpredatory species will not fall victim immediately, the poison still enters the food chain with every coyote killed.

In spite of aggressive methods, the fight against coyotes has mostly been unsuccessful. Even after decades of predator control, total annual losses are given in the hundreds of thousands of animals (mostly sheep and goats) by ranchers who claim they lose as much as 25 percent of their lambs to predators each year. Conservationists and supporters of **wildlife** say these numbers are inflated, but, even if they are accurate, the cost of lost livestock is significantly lower than the cost of the government's predator control programs. **Conservation** groups argue that it would be much cheaper to give up on predator control and reimburse ranchers for livestock lost to predation.

Besides killing predators, ranchers also have a number of other options when it comes to protecting their livestock. Sheep can be protected by letting it graze together with cattle or other animals too big for coyotes to handle. Llamas and donkeys have also been suggested for the purpose. Trained guard dogs are an effective deterrent to coyotes as well.

Although predator control can clearly have devastating consequences, it is by no means a thing of the past. Recently, Alaska began implementing a program to shoot 1000 wolves (out of a total population of 7000 wolves in the state). The program, which is to remove the wolves from a vast, centrally located stretch of **wilderness** over a period of five years, will have government agents tracking and shooting packs of wolves from the air. Private citizens are also allowed to shoot wolves. The object is to boost moose and caribou herds for **hunting** and tourist purposes. The view that the value of animal resources lies primarily in their usefulness to human beings still motivates many government policies and individual behaviors.

[*David A. Duffus and Marijke Rijsberman*]

Resources

Books

Dasmann, R. *Environmental Conservation.* New York: Wiley, 1984.

Gilbert, F. F., and D. G. Dodds. *The Philosophy and Practice of Wildlife Management.* Malabar, FL: R. E. Kreiger, 1987.

Rolston III, H. *Philosophy Gone Wild.* Buffalo, NY: Prometheus Books, 1989.
Wilson, E. O. *Biophilia.* Cambridge: Harvard University Press, 1984.

Predator-prey interactions

The relationship between predators and their prey within an **ecosystem** is often a quite complex array of offenses and defenses. Predation is the consumption of one living organism by another. Predators must have ways of finding, catching, and eating their prey (the offensive strategies), and prey organisms must have ways of avoiding or discouraging this activity (the defensive strategies). Predator-prey interactions are often simply thought of as one animal, a carnivore, catching and eating another animal, another carnivore or an herbivore. The interrelationships of predators and their prey is much more involved than this and encompasses all levels within an ecological **food chain/web**, from plant-herbivore systems to herbivore-carnivore systems to three-way interactions of interdependent plant-herbivore-carnivore systems.

Even though a plant may not be killed outright, removing leaves, stems, roots, bark, or the sap from plants will reduce its fitness and ultimately its ability to survive. The younger, more tender leaves and stems of a plant are typically consumed by herbivores. Because these are sites of active growth, the plants send stores of nutrients from roots and other tissues to these developing regions. When they are eaten the plant is losing a disproportionate amount of nutrients that will be difficult to replenish. Severe **defoliation** results in a tremendous loss of vigor and the loss of the ability to photosynthesize and store energy reserves for the dormant season. Without this energy store available, the plants are more vulnerable to insect or disease attack the following year.

Plants have responded to herbivory in a variety of ways. Some plants have developed spines or thorns, whereas others have developed the ability to produce chemical products which deter herbivory by making the plant unpalatable, barely digestible, or toxic. Many herbivores have found ways to circumvent these defenses, however. Some insects are able to absorb or metabolically detoxify these **chemicals** from the plants. Other insects have learned to cut trenches in the leaves of plants before they eat them, which stops the flow of these chemical substances to the leaves. Other animals have developed digestive modifications, such as bacterial communities that break down cellulose, to enable them to ingest less nutritious, but more abundant, vegetation.

A classic example of predator-prey interactions in carnivores is the relationship of the lynx, snowshoe hare, and woody browse of northern coniferous forests. The hare feeds on buds and twigs of conifer, aspen, alder, and willow trees. Excessive browsing, as the hare population increases, decreases the subsequent year's growth, initiating a decline in the hare population. The lynx population follows that of the snowshoe hare, having greater reproductive success when more prey are available, then declining as hare populations drop. This relationship is cyclical and follows its course over a ten-year period.

The prey of carnivores, both herbivores and other carnivores, have developed a variety of defensive mechanisms to avoid being detected, caught, or eaten. Some employ chemicals, such as odorous secretions, stored toxic substances ingested from plants, or synthesized venom or poisons. Other animals have taken to hiding from predators through cryptic coloration, patterns, shapes, and postures. Mimicry, widespread in the animal kingdom, is a system whereby one animal, a palatable **species**, mimics an inedible species, and is thus avoided by the predator. Additional means of predator avoidance or discouragement involves the employment of armor coats or shells, the use of quills or spines, or living in groups. However, predators can benefit prey populations by removing sick or inferior individuals. Subtle, long-term interactions, in fact, can benefit both predator and prey as they co-evolve together into more highly adapted or finely-tuned forms. *See also* Evolution; Population biology

[*Eugene C. Beckham*]

Resources
Books

Predators and Predation: The Struggle for Life in the Animal World. New York: Facts on File, 1989.

Predator-Prey Relationships: Perspectives and Approaches From the Study of Lower Vertebrates. Chicago: University of Chicago Press, 1986.

Taylor, R. J. *Predation.* New York: Chapman and Hall, 1984.

van Lawick, H. *Among Predators and Prey.* San Francisco: Sierra Club Books, 1986.

Periodicals

Sunquist, F. "The Strange, Dangerous World of Folivory." *International Wildlife* 21 (1991): 4–11.

Prescribed burning

Intentional burning, usually in forests, prairies, and savannas, in order to maintain desired ecological conditions or prevent major wildfires that could result from accumulated dry, woody debris. Prescribed burning, also called controlled burning, has become a widely recognized component of modern **ecosystem management** in recent decades as ecologists and managers have recognized the role of fire in natural ecosystems. In practice, though, prescribed burning is still used only on a small portion of public and private lands. In a forest, prescribed burning is usually aimed at removing accumulated fuel—fallen branches, pine needles, and other

dry debris. In a **prairie** or **savanna environment**, prescribed burns may be used to prevent encroachment of trees, to reduce populations of nonnative grassland **species**, and to enhance the health of native grassland vegetation that is adapted to periodic fires.

The principle of prescribed burning is that fire is a normal component of ecosystems, and that frequent small fires can serve both to maintain species diversity and to prevent large, catastrophic wildfires. Under natural conditions wildfires, usually ignited by lightning, occur frequently in many forest and grassland biomes. Proponents of prescribed burning argue that these natural fires would usually be small and cool, occurring too frequently to allow excessive accumulation of fallen branches, dead grass, and other fuel. Frequent, low-intensity fires should do little damage either to mature trees or to the root systems of most plants. At the same time, by clearing away debris every few years, and by a natural fire regime should provide fresh ground, exposed to sunlight and freshly fertilized by ashes, to encourage new germination and plant growth.

Deliberate burning is also an age-old method of maintaining grazing lands in many parts of the world. Periodic forest burning to maintain pasturage and to clear farm plots has probably been practiced for millennia in Africa and South America. In North America, **indigenous peoples** used fire to maintain open forests and prairies that provide grazing for game animals such as deer and **bison**. In some areas, especially New England and the rural South, early European immigrants to the New World adopted Indian methods of burning. These practices differ from what is usually considered a prescribed burn, however, because they are often carried out without the planning and the careful fire control usually exercised in a controlled or prescribed burn. In most of North America, traditional burning practices have also disappeared as more and more forests have come under professional management for timber production.

The introduction of prescribed burning to professional land management strategies in the United States is often attributed to **Aldo Leopold**, who pointed out in 1924, while he was serving as a forester and firefighter in Arizona, that fire control programs were producing brush-choked forests susceptible to extreme **wildfire**. Leopold argued that decades of fire suppression had actually increased the **probability** of severe fires, and that forests were healthier when subject to regular, natural fires. Since Leopold began arguing for prescribed burning in the 1920s, **ecosystem** managers have widely recognized the ecological benefits of fire. The amount of prescribed burning carried out still remains modest, however. Between 1983–1994, prescribed burns were carried out on only about 0.1% of public lands in the United States each year, as compared to 0.3% burned annually by wildfires in the same interval.

Proponents of prescribed burning sometimes argue fire suppression techniques--bulldozed fire breaks, tree felling, air drops of chemical fire retardants, and other measures--cause more damage than the fires they are meant to control. In addition, these expensive, and dangerous, tactics may ultimately have little effect. Prescribed burn proponents argue that most large natural fires die out as a result of rainfall, cooler weather, or a lack of fuel, not because of the efforts of fire fighters. These opponents of standard military-style fire suppression programs propose that it would be more effective to use fire control budgets for planned burns than for impressive but useless shows of force against large forest fires. However events such as the **Yellowstone National Park** fires of 1988, a series of highly publicized fires that led to loud criticism from the media and politicians that Park officials were negligent in their duties to protect the park, demonstrate that the public is unlikely to settle for inactivity during a forest fire. During fires such as those in Yellowstone, the public is quick to criticize officials for doing nothing even if there is nothing to be done. Most of us would rather see a campaign to fight fires even if wastes money and causes serious damage to the forest.

Benefits of Prescribed Fires

Fire has many ecological benefits. By clearing patches of forest, fire allows tree seeds to germinate. The seedlings of many trees, including Douglas fir (*Pseudotsuga menziesii*) and white pine (*Pinus strobus*), both prized for timber, require the strong sunlight of an open patch in the forest to begin growing. Other species, such as jack pine (*Pinus banksiana*) require intense heat, such as that produced by a fire or extremely hot sun, to release the seeds from their resin-sealed cones. These trees are considered fire-adapted, because they reproduce best, or only, when seedlings can grow in the full sun and fresh ashes of recently burned patches of forest. Many prairie grasses and flowers are also fire adapted, thriving best when periodic fires remove dead stem litter and release stored nutrients to the **soil**.

Fire plays an important role in cleaning out damaged or diseased regions of a forest. Where **parasites**, fungus, or disease have damaged stands of trees, a fire can locally eliminate the **pest** population, reducing the risk of parasite or disease spread, at the same time as it removes standing dead trees and makes room for new seedlings to replace the damaged trees. Land managers in southern Alaska, faced with the worst spruce beetle outbreak ever in the United States, have recently begun to consider using prescribed fire to control beetles.

Undesirable species encroaching on **grasslands** and savannas can also be controlled with fire. Opportunistic trees and brush such as aspen, buckthorn, juniper and sumac, introduced European grasses, as well as nuisance species such as poison oak and poison ivy can all be set back by

periodic burning. At the same time, prescribed burns can rejuvenate fire-adapted plant populations by clearing debris and releasing nutrients to the soil. In a burned area increased sunlight can extend the growing season, an important advantage in cold climates. In addition, periodic patchy burns help maintain **habitat** diversity by producing a mosaic of landscape types and different-aged tree stands. Because many animal species utilize a variety of environments for feeding and nesting, a mosaic landscape maintained by fire can benefit **wildlife** populations.

Methods

A prescribed burn is a fire set under conditions that should allow control of the fire. Usually this means identifying the limits of an area that should be burned and creating or identifying fire breaks that will prevent the uncontrolled spread of the fire once it has been lit. Often a backfire is used. A backfire is line of fire burned on the downwind edge of the area to be burned. The backfire is kept low and forced to burn upwind until it has created a swath of land clear of any combustible fuel. Flank fires, similarly, clear combustible materials from strips along the flanks of the burn area. The main fire, moving forward in a downwind direction, will die out due to a lack of fuel when it reaches the blacked zone created by the backfire. In addition to these precautions, prescribed burning usually requires selecting a day with high humidity and light winds, to keep the fire relatively low.

Ideally areas should be burned periodically in order to mimic the pattern of natural fires. Fire ecologists and managers usually attempt to burn areas with the frequency they estimate would occur in local ecosystems under natural conditions. In some areas natural fire frequency might be as often as every 2-3 years; elsewhere natural fires might normally occur as infrequently as every 10-20 years. Or managers may simply burn any areas in which debris has accumulated to dangerous levels.

Prescribed Natural Fires

Sometimes a prescribed burn policy means establishing a rule to let naturally ignited fires, usually started by lightning, burn themselves out. When rules establish that fires should be allowed to burn, fires that occur are referred to as prescribed natural fires. "Let-burn" policies are widely applied in national parks and forests in the United States. Seventy-five percent of **Yosemite National Park**, for example, is treated as a prescribed natural fire zone. In large part this is because it is more politically palatable to allow **nature** to take its course than it is to deliberately burn a park. In addition, it is less costly to let fires burn naturally than to maintain a burn crew, design burn plans and fire breaks, and carry out and monitor a regular program of prescribed burns. Usually a prescribed natural fire policy sets limits on the conditions and size of fires that will be allowed to burn, and any natural fires that do occur are closely monitored to ensure that they do not become too large or dangerous.

History of Prescribed Burning in the United States

Two factors influenced the development of prescribed burning in the United States. One was the observation that despite decades of fire suppression, the number and intensity of wildfires has actually increased since the advent of major fire suppression programs. One United States **Forest Service** study showed that total acreage burned, after declining from 1917 to 1955, increased steadily from the 1960s to the 1990s. The other factor was a transition in ecological theory toward understanding ecosystems as complex and dynamic, not static structures that should be maintained a single, undisturbed, optimal climax state. The shift in ecological theory developed partly from the observation that ecosystems maintained rigorously by managers were not necessarily healthier, than those that underwent periodic catastrophic events such as floods and fires. In fact, according to this theory, ecosystems may require periodic change and disruption in order to remain healthy.

The Nature Conservancy, which maintains scattered patches of native prairie, savanna, and other ecosystems, was one of the pioneers of prescribed burning. Since it conducted its first prescribed burn on a North Dakota prairie in the 1970s, the Nature Conservancy has expanded its prescribed burning program to a current level of over 40,000 acres (16,400 ha) of native habitat each year. The organization's objectives are controlling invasive plant species and maintaining the vigor of native fire-adapted plant communities.

Controversy over Prescribed Burns

Prescribed burning is controversial for several reasons. First, decades of education about fire suppression, both in the general public and among land management professionals, makes burning seem intuitively wrong to many people. Cultural icons such as Smokey Bear and Bambi, both fictional characters who suffered from forest fires, have symbolized the trauma and destruction of fires. The idea that fire might rejuvenate a forest is often harder to explain than the idea that fire destroys a forest. Second, people living near forests and grasslands fear that a controlled burn may escape and damage their property--an event that occurs rarely but is always a potential risk. Property owners are understandably nervous about the possibility of a fire escaping. On the other hand, prescribed burns have sometimes been used to protect property. A number of homes adjacent to the Wenatchee **National Forest** in Washington State were spared from the Tyee Creek fire of 1994 because previous burns had reduced fuel levels in the surrounding forest.

In addition to objections to fire danger, the **smoke** resulting from a prescribed fire is unpleasant and sometimes hazardous. There can be a real health risk for people with

Prescribed burning of a prairie in South Dakota for land management purposes. (Photograph by Stephen J. Krasemann. National Audubon Society Collection/Photo Researchers, Inc. Reproduced by permission.)

lung ailments or **asthma** if they are exposed to concentrated smoke. In addition, a deliberately set fire often violates **air quality** laws, making fires technically illegal in many populated areas. If the fire occurs near a road, **visibility** is also reduced, so that the road must often be closed while a burn is in progress. Although the smoke produced by a series of prescribed fires may be less than the smoke from a single large wildfire, prescribed fires are sensitive because the individual or group responsible for the fire is also liable for any damage or injury it causes.

Controversy over the use of prescribed burns has become especially serious in areas where human settlement is encroaching on natural areas. Where cities and suburbs spread into forested areas or reach the edges of preserves that are managed with fire, residents of new subdivisions usually object to the potential risks and inconvenience imposed by fire near their property. This problem is especially severe in the **chaparral** country of southern and central California. Chaparral is a fire-adapted ecosystem with highly combustible, dry vegetation that normally burns as every 30 years or so. As suburban development, with its high real estate values, moves into the chaparral, the economic and environmental costs of fire control have risen sharply. In wildlife and habitat preserves near metropolitan areas land managers have sometimes had to abandon successful fire management programs because of suburban sprawl.

[*Mary Ann Cunningham Ph.D.*]

Resources

Books

de V. Booysen, P., and N. Tainton, eds. *Ecological Effects of Fire in South African Ecosystems.* Ecological Studies vol. 48. New York: Springer-Verlag, 1995.

Pyne, S. J. *Fire in America.* Princeton, NJ: Princeton University Press, 1982.

United States Forest Service. *The Use of Fire in Forest Restoration.* USFS General Technical Report INT-GTR-341. Washington, D.C.: General Printing Office, 1996.

Periodicals

Babbit, B. "To Take up the Torch." *American Forests* 101 (July/August 1995):17–18, 59.

McLean, H. "Fighting Fire with Fire." *American Forests* 101 (July/August 1995):13–16, 56–7.

President's Council on Sustainable Development

An environmental advisory group created by President Bill Clinton in June 1993 to develop new approaches for combining environmental protection in the United States with economic development. The council is composed of 25 members drawn from government, industry, and environmental groups. One of the primary mandates of the President's Council on **Sustainable Development** is to develop plans for the United States to fulfill its role in the **biodiversity**, global warming, and other accords negotiated at the 1992 **United Nations Earth Summit**.

Price-Anderson Act (1957)

The Price-Anderson Act, passed in 1957, limits the liability of civilian producers of **nuclear power** in the case of a catastrophic nuclear accident. In the case of such an accident, damages would be recovered from two sources: private insurance covering each plant and a common fund created by contributions from each nuclear power plant. This common fund would cover the difference in damages between the private insurance and the liability limit.

The act, named for its chief sponsors, Senator Clinton Anderson (NM) and Representative Melvin Price (IL), was passed to encourage private investment in nuclear power production. It was part of a general strategy to encourage and stimulate nuclear power production in the private sector. Without such liability limitations, the risk of nuclear power for utilities and manufacturers would be too great. Private insurance companies were not willing to underwrite the risks

due to the uncertainty involved and the potential magnitude of damages.

The law requires the private operators to carry the maximum amount of private insurance available ($65 million in the late 1950s). The upper limit for liability was set at $560 million per accident for personal damages, with the common fund covering the difference (between $65 million and $560 million). Once the liability limit is reached this law protects the operator from further financial liability. In addition to setting liability limits, the law establishes a common fund for victims to draw on if the utility did not have the ability to pay damages up to the liability limit. In return for this limited liability, utilities had to accept sole responsibility for accidents; equipment manufacturers would not be liable for any damages.

The Price-Anderson Act serves as a subsidy to the nuclear power industry. If it were not for this liability limitation, it is unlikely that any private firms would have become involved in the production of nuclear power. As a further incentive to private industry, the federal government set the liability limits at a very low level. This was done despite a 1957 government estimate of the damages due to a catastrophic accident: 3,000 immediate deaths and $7 billion in property damage.

In 1977, the Carolina Environmental Study Group filed a lawsuit, claiming that the Price-Anderson Act was unconstitutional because it was a taking of private property without just compensation, a violation of the Fifth Amendment. If a nuclear accident did occur and compensation was limited, the argument went, a taking of private property had occurred and compensation was limited in advance. A federal district court ruled for the environmental group: the act was unconstitutional. In 1978, however, the Supreme Court overturned this ruling, maintaining the constitutionality of the Price-Anderson Act.

The 1988 amendments to the act increased the liability limit per accident from $720 million to $7.2 billion, with the utilities continuing to contribute to a common fund to help pay for any damages if a major accident did occur. The private insurance coverage was set at $160 million per plant, with the common fund to make up the over $7 billion difference. These liability limits will last through 2003. Despite this tremendous increase in liability limits, the nuclear power industry was pleased to retain any liability limit in the face of significant opposition. The amendments also protect **U.S. Department of Energy** nuclear contractors from damage claims and limit the liability for nuclear waste accidents to $7 billion. *See also* Environmental law; Environmental liability; Nuclear fission; Radioactive waste; Radioactivity; Three Mile Island Nuclear Reactor

[*Christopher McGrory Klyza*]

RESOURCES

BOOKS

Jasper, J. M. *Nuclear Politics: Energy and the State in the United States, Sweden, and France.* Princeton: Princeton University Press, 1990.

Kruschke, E. R., and B. M. Jackson. *Nuclear Energy Policy.* New York: ABC-Clio, 1990.

Mazuzan, G. T., and J. S. Walker. *Controlling the Atom: The Beginnings of Nuclear Regulation 1946–1962.* Berkeley: University of California Press, 1985.

Primary pollutant

Primary pollutants cause **pollution** by their direct release into the **environment**. The substance released may already be present in some quantities, but it is considered a primary pollutant if the additional release brings the total quantity of the substance to pollution levels. For example, **carbon dioxide** is already naturally present in the **atmosphere**, but it becomes toxic when additional releases cause it to rise above its natural concentrations. The rise in levels of **carbon** dioxide is one contributing factor to the **greenhouse effect**.

The direct release into the atmosphere of a chemical that is not normally present in the air is also classified as a primary pollutant. **Hydrogen** fluoride, for example, is released from some coal-fired furnaces, but the compound is not usually present in unpolluted air.

A primary pollutant can be generated by many sources: **pesticide** dust sprayed intensively in agricultural areas, emissions from car and industrial exhausts, or dust kicked about from mining operations, to name a few. Some primary pollutants are composed of **particulate** matter that is not easily dispersed. **Smoke**, soot, dust, and liquid droplets released into the air either by the burning of fuel or other industrial or agricultural processes, are considered primary pollutants.

Primary pollutants also originate from natural sources. Volcanic ash, as well as grit and dust from volcanic explosions belong to this category. Salt dust blown inland by strong oceanic winds or gaseous pollution that originates from bogs, marshes, and other decomposing matter can also be classified as primary pollutants.

When two or more primary pollutants react in the atmosphere and cause additional atmospheric pollution, the result is called secondary pollution. **Nitrogen oxides**, for instance, can react with volatile organic compounds to from **smog**, a secondary pollutant. *See also* Air pollution; Air pollution index; Air quality criteria; National Ambient Air Quality Standard

[*Linda Rehkopf*]

RESOURCES
BOOKS

Ney, R. E. *Where Did That Chemical Go? A Practical Guide to Chemical Fate and Transport in the Environment.* New York: Van Nostrand Reinhold, 1990.

Shineldecker, C. L. *Handbook of Environmental Contaminants: A Guide for Site Assessment.* Boca Raton: Lewis, 1992.

Primary productivity (gross and net)

Primary producers (or autotrophs) are organisms that synthesize their own biochemical constituents using simple inorganic compounds and an external energy source to drive the process. The amount of energy fixed by autotrophs is known as primary production, and the rate of fixation is primary productivity. Both primary productivity and primary production may be measured in units of **carbon dioxide** (CO_2) fixed, energy fixed (as calories or joules), or **biomass** produced. Productivity in terrestrial ecosystems is often expressed in such units as kilograms of dry biomass (or its energy equivalent) per hectare per year (e.g., kg/ha.yr or kJ/ha.yr), while aquatic productivity is often measured on a volumetric basis (e.g., $kg/m^3.yr$). Many studies have been made of the primary productivity of various kinds of ecosystems.

Most primary producers are photoautotrophs that use sunlight as the external source of energy to drive **photosynthesis**. Photoautotrophs capture solar radiation using photosynthetic pigments, such as chlorophyll. Green plants are the most abundant photosynthetic organisms, along with algae and some bacteria. A much smaller number of autotrophs are chemoautotrophs, which capture some of the energy content of certain inorganic **chemicals** to drive their **chemosynthesis**. For example, *Thiobacillus thiooxidans* is a bacterium that oxidizes sulfide minerals to sulfate, utilizing some of the energy liberated during this reaction to drive its chemosynthesis.

The total fixation of solar (or chemical) energy by primary producers within an **ecosystem** is known as gross primary production (or GPP). Some of this production is utilized by autotrophs in support of their own **respiration** (R). Respiration involves physiological functions needed to maintain organisms in a healthy condition. Its complex reactions involve the metabolic oxidation of biochemicals, which requires a supply of oxygen and releases **carbon** dioxide and water as waste products. Net primary production (NPP) refers to the fraction of gross primary production that remains after primary producers have utilized some of their GPP for their own respiration. In other words, NPP = GPP - R.

For example, studies of an oak-pine forest in New York found that the total fixation of **solar energy** by vegetation (i.e., the gross primary productivity) was about 48.1-thousand kilojoules per hectare per year. This fixation rate was equivalent to less than 0.01% of the input of solar radiation to the forest. Because the plants utilized 27.2-thousand kJ/ha.yr in support of their respiration, the net primary productivity was 20.9-thousand kJ/ha.yr, occurring as the accumulating biomass of trees.

The NPP of primary producers supports the productivity of all other organisms, known as heterotrophs, in ecosystems. Heterotrophs can only utilize living or dead biomass as food, and they rely on other organisms to supply this fixed energy. Animal heterotrophs that feed on plants are known as herbivores (or primary consumers). Animals that kill and eat other animals are known as carnivores (or secondary consumers). Animals that feed on both plant and animal biomass are known as omnivores. Many heterotrophs feed on dead organic matter, and are called **decomposers** or **detritivores**.

[*Bill Freedman Ph.D.*]

RESOURCES
BOOKS

Begon, M., C. R. Townsend, and J. L. Harper. *Ecology: Individuals, Populations and Communities.* London: Blackwell Science Inc., 1998.

Lieth, H. and R. H. Whittaker, eds. *Primary Productivity of the Biosphere.* New York: Springer-Verlag.

Odum, E. P. *Basic Ecology.* New York: Saunders College Publishing.

Primary sewage treatment

see **Sewage treatment**

Primary standards

Primary standards are numerical limits of allowable air and water pollutants designed to protect human health but not necessarily other parts of the **environment**. Primary standards differ from **secondary standards**, which are those that protect against all adverse effects on the environment, such as those on animals and vegetation. While the term primary standards usually relates to **air pollution** or **air quality**, there are federal primary standards for the drinking water supply.

The **Clean Air Act** is the major piece of legislation that protects and enhances the nation's air quality. As part of the act, Congress required the **Environmental Protection Agency** (EPA) to establish National Ambient Air Quality Standards (NAAQS) for air pollutants, which were characterized by wide dispersal and **emission** from many sources.

The NAAQS, or primary air quality standards, define the level of air quality to be achieved and maintained nation-

wide for six criteria pollutants: **sulfur dioxide**, **carbon monoxide**, **nitrogen oxides**, **ozone**, **particulate** matter, and **lead**. In general, these standards are not permitted to be exceeded more than once a year. In addition, the EPA is required to identify local hazardous air pollutants which could increase **mortality** or result in serious illnesses such as **cancer**, and establish national standards for them. Although hundreds of potentially hazardous air pollutants exist, only eight had been listed as of 1990: **mercury**, beryllium, **asbestos**, **vinyl chloride**, **benzene**, radioactive substances, coke oven emissions, and inorganic **arsenic**.

The Clean Air Act set a deadline of December 1987 for cities in the United States to meet federal primary standards. Some cities did meet the standards, increasing **automobile** emission inspections and instituting tighter regulations for incinerators, but some sixty cities could not comply. There has been a measurable increase in **nitrogen** oxide emissions in recent years, and more than 100 areas nationwide have exceeded standards at least part of the time for ozone and **carbon** monoxide levels. The EPA has estimated that 100 million people live in areas exceeding these standards. The most heavily-polluted areas of the **Los Angeles Basin**, Houston, and the New York corridor may take years to meet the air quality standards. Suggestions to hasten compliance have included plans to shut down industrial plants, ration gas, and restrict automobile use.

The 1990 Clean Air Act Amendments include provisions to tighten **pollution control** requirements in cities that have not attained the National Ambient Air Quality Standards. These provisions include, among others, requirements for stringent automobile **emission standards**. The new act also requires a 50 percent reduction in sulfur dioxide emissions, and it sets primary and secondary standards for dozens of **chemicals** that were not mentioned in the original act or amendments.

Primary standards are not consistent from country to country. For example, standards are defined for different average times, at varying numerical limits. In thirteen countries the sulfur dioxide standard ranges from 0.30 to 0.75 mg/m^3 for a 30-minute average, and from 0.05 to 0.38 mg/m^3 for a 24-hour average. Since **pollution** does not stay in once place, and since **acid rain** is often a source of controversy between bordering countries, the lack of international standards will continue to contribute to air pollution problems.

In the United States, concern about **water pollution** has resulted in higher standards for water cleanliness. These standards have been adopted by state and federal agencies, but the involvement of so many different agencies on different levels has contributed to a lack of agreement on some water standards.

The primary law to protect the integrity of the nation's water is the **Clean Water Act** (1972). The act requires the EPA to set **water quality** criteria based on the most recent scientific information. Under the Clean Water Act Reauthorization (1987), the EPA is required to publish revised water quality criteria, stress new programs to combat water pollution from toxics, and restrict waivers from national standards that had been easily obtained by discharges.

Criteria are not rules; they are a compilation of data about pollutants that can be used to formulate standards. Congress intended for the EPA to set the criteria and for the states to set the **water quality standards**. By 1990 the EPA had established criteria for 126 priority pollutants. Of these, 109 dealt with primary standards and thirty-four with secondary standards. But although these criteria have been published, few states have actually established standards for toxic pollutants or have incorporated these standards into the regulation of toxic **discharge**.

Under the Safe Drinking Water Reauthorization Act (1986), the EPA is not only required to set standards for contaminants in drinking water, but it also must monitor public drinking water for unregulated contaminants and set deadlines for the issuance of new standards. By 1990, the EPA had set primary standards for drinking water that covered the inorganic chemicals arsenic, barium, **cadmium**, chromium, lead, mercury, nitrate, selenium, silver, and fluoride. Primary drinking water standards are measured in terms of maximum contaminant levels (MCLs), which are the maximum permissible level of a contaminant in water at the tap; they are health related and legally enforceable.

Organic chemicals regulated by primary standards for drinking water include the pesticides Endrin, Lindane, Methoxychlor, and **Toxaphene**; the herbicides **2,4-D** and **2,4,5-T** Silvex; and the organics benzene, carbon tetrachloride, p-dichlorobenzene, 1,2-dichlorobenzene, 1,2-dichloroethane, 1,1-dichloroethylene, 1,1,1-trichloroethane, trichloroethylene, vinyl chloride, and an organic compound referred to as total **trihalomethanes**. Microbiological standards for drinking water have been set for coliform bacterium such as ***Escherichia coli***, which is present in sewage and which can cause gastroenteric infections, dysentery, hepatitis, and other diseases. Standards have also been set for turbidity, the murkiness of treated water that can interfere with disinfection processes. Radionuclides regulated by primary standards include beta particles and photon activity, gross alpha particles, and Radium-226 and -228.

As is the case with primary standards for air quality, an international comparison of primary standards for drinking water reveals a collection of confusing numbers. While the United States and Canada seem to agree on standards for organic compounds, this is where the similarity ends. As of 1990, the Canadian government had no primary drinking

water standards for radionuclides other than radium, and it had not set primary standards for volatile organic chemicals such as benzene and vinyl chloride. By 1990, the European Economic Community had set primary drinking water standards only for total pesticides and trihalomethanes. The World Health Organization (WHO) has recommended primary standards only for the organics 2,4-D, methoxychlor, and trihalomethanes, beta and alpha radionuclides, and five of the eight volatile organic chemicals regulated in the United States.

In the United States, there is evidence that primary standards regarding air and water pollutants have had some impact on air and water quality. Some of the nation's surface waters have improved since the implementation of the Clean Water Act. **Coliform bacteria** counts and **dissolved solids** have been reduced, and **dissolved oxygen** levels have increased enough to permit the reestablishment of plants and animals that had died out in polluted waters.

Air quality standards have led to an overall 20 percent reduction in emissions of particulates, sulfur dioxide, and carbon monoxide during the 1980s. Los Angeles, which suffers as much or more than any American city from air pollution, has proposed an air quality plan that would surpass the EPA's standards, requiring lifestyle changes to help combat air pollution.

Implementing laws and regulations to enforce water and air standards that will protect the health of humans will be expensive: The yearly price tag just for implementation of the Clean Air Act of 1990 is estimated to be $25 billion. But many believe the savings in health costs and environmental damage to be incalculable. *See also* Air pollution index; Air Quality Control Region; Air quality criteria; Attainment area; Nitrates and nitrites; Nonattainment area; Radioactive waste; Safe Drinking Water Act; Sewage treatment; Toxic substance

[*Linda Rehkopf*]

Resources

Books

Freeman, M. *Air and Water Pollution Regulation: Accomplishments and Economic Consequences.* Westport, CT: Greenwood Publishing Group, 1993.

Harte, John, et al. *Toxics A to Z.* Berkeley: University of California Press, 1991.

van der Leeden, F., et al. *The Water Encyclopedia.* Chelsea, MI: Lewis Publishers, 1990.

Prince William Sound

British explorer Captain James Cook made the European discovery of Prince William Sound in 1778. Today, the 25,090-square-mile (65,000-square-kilometer) passage southeast of Anchorage, Alaska, is the focal point of an ongoing political and environmental issue: whether oil can be safely transported in extreme climates without seriously threatening terrestrial and marine habitats or the recreational, agricultural, and industrial interests of the region.

Prince William Sound is bordered on the north and west by the Chugach and Kenai mountain ranges, on the east by the **Copper** River, and by the Hinchinbrook Islands on the south. Ten percent of its area is open water, with depths ranging from 492-2,952 ft (149-895 m). The remainder consists of shallow coastal waters, shoals, and reefs. The combined 2,980 miles (4,800 km) of mainland and island shoreline are home to more than 200 **species** of birds, among them approximately 3,000 bald eagles. Ten species of marine mammals, including **sea lions**, **seals**, **whales**, porpoises, and some 10,000 sea otters thrive in its bountiful waters. The natural beauty of Prince William Sound is responsible for the area's healthy tourism industry.

The Sound is world-renowned for its **salmon** fishery. More than 300 streams are used by salmon for spawning and, combined with hatchery inputs, more than one billion fry are released annually into the Sound. The herring season brings in an additional 12 million dollars to the area each year. The inhabitants of the four communities on the Sound are supported primarily by fishing and its related industries. For economic reasons, residents of the area are concerned with preserving the Sound's **natural resources**.

In the mid-1970s, the Cordova Island fishermen fought the oil industry's plan to build an oil pipeline from Prudhoe Bay in northern Alaska to Prince William Sound. The decision was left to the United States Senate which, despite warnings from scientists and environmental groups, voted to approve the proposed pipeline route into the Sound.

Natural and man-induced disasters threaten the health of the Sound. The Alaskan coast makes up the western boundary of the North American continental plate and is, therefore, subject to considerable seismic activity. On March 24, 1964, an **earthquake** centered in the mountains surrounding Prince William Sound had substantial effects on the Sound. Twenty-five years later to the day, the oil tanker ***Exxon Valdez*** wrecked on the rocks of the Sound's Bligh Reef in waters less than 39 ft (12 m) deep. Approximately 265,000 barrels (42 million liters) of heavy crude oil leaked into the Sound, one of the worst oil spills in United States history. The local **sea otter** population was nearly destroyed; many pelagic and shore bird populations were also heavily impacted. Effects of the spill and subsequent clean-up efforts on shoreline and seafloor communities are unknown, as are the long-term effects on Prince William Sound and adjacent habitats.

[*William G. Ambrose Jr. and Paul E. Renaud*]

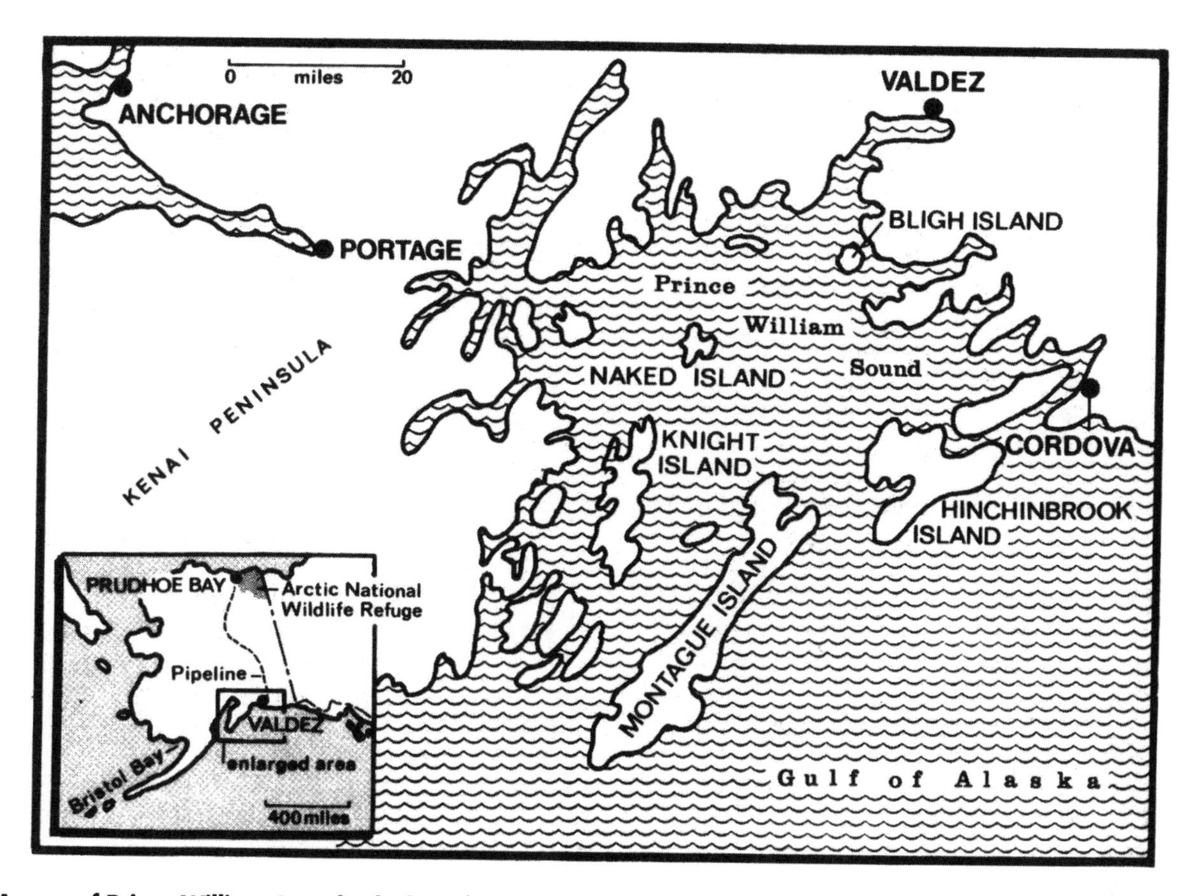

A map of Prince William Sound, Alaska. (The Conservation Fund. Reproduced by permission.)

Resources
Books

Lethcoe, N., ed. *Prince William Sound Environmental Reader, 1989: Exxon Valdez Oil Spill.* Valdez, AK: Prince William Sound Books, 1989.

Periodicals

Dold, C. A., et al. "Just the Facts: Prince William Sound." *Audubon* 91 (1989): 80.

Heacox, K. "Sound of Silence." *Buzzworm* 5 (January–February 1993): 52.

Steiner, R. "Probing an Oil-Stained Legacy." *National Wildlife* 31 (April–May 1993): 4–11.

Priority pollutant

Under the 1977 amendments to the **Clean Water Act**, the **Environmental Protection Agency** is required to compile a list of priority toxic pollutants and to establish toxic pollutant **effluent** standards. A list of 126 key water pollutants was produced by EPA in 1981, and is updated regularly, with the updates appearing in the *Code of Federal Regulations*. This list, consisting of metals and organic compounds and known as the Priority Toxic Pollutant list, also prescribes numerical **water quality standards** for each compound. State **water quality** programs must use these standards or develop their own standards of at least equal stringency.

Privatization movement

This movement, initiated and supported primarily by economists, peaked in the early 1980s. Public ownership of land was inefficient because the administration of the lands was removed from the incentives and discipline of the free market. The case for the transfer of these lands to the private sector was made most strongly for lands managed for commodities (e.g., grazing lands, mineral lands, timber lands), but some also advocated transferring **wilderness** lands to the private sector, where it, too, would be more efficiently managed. Many of the economists who supported the program were a part of a movement referred to as the New Resource Economics (NRE), which advocated an increased reliance on private property rights and the free market for managing **natural resources**. Such an approach meshed

well with the Reagan Administration's philosophy of free market economics.

The privatization idea moved from theory to practice in February 1982 at a Cabinet Council on Economic Affairs meeting with the creation of the Asset Management Program. This program was designed to identify federal property for disposal, to develop legislation needed to dispose the land, and to oversee the sale of the land. The program was formalized by President Reagan in February 1982. The Fiscal Year (FY) 1983 budget proposal called for the sale of 5 percent of the nation's land (excluding Alaska), approximately 35 million acres, over five years. The revenues projected from the program were $17 billion from FY 1983 through FY 1987, the bulk of which was to come from sales of **Bureau of Land Management** (BLM) and **Forest Service** lands.

The BLM began to develop a program for land disposal, but in July 1983, Secretary of the Interior James Watt removed Interior Department lands from the Asset Management Program. He thought that the program was a mistake and was undermining the President's support in the West that had developed through the *good neighbor* program (which had defused the **Sagebrush Rebellion**).

In the summer of 1982, the Forest Service began to identify possible lands for disposal and announced that it would seek legislative authority to dispose the land. In March 1983, the agency announced that it would seek legislative authority to dispose of up to 6 million acres of land managed by the Forest Service (3.2% of the lands in the system). At this time, they indicated the specific amounts of land under consideration for disposal in each state, with high figures of 872,054 acres in Montana and 36 percent of its land in Ohio.

Opposition to the Asset Management Program was immediate and intense. The chief opponents were environmentalists, but they were also joined by the forestry profession and many western politicians. This already strong opposition to the program intensified once the specific areas for disposal were identified. In the face of this intense opposition, the Forest Service never presented legislation to Congress to allow the sale of these lands. The attempt to put privatization into practice was aborted by early 1984.

[*Christopher McGrory Klyza*]

RESOURCES
BOOKS

Short, C. B. *Ronald Reagan and the Public Lands: America's Conservation Debate, 1979–1984.* College Station, TX: Texas A&M University Press, 1989.

Truluck, P. N., ed. *Private Rights and Public Lands.* Washington, DC: Heritage Foundation, 1983.

Probability

Probability is a quantity that describes the likelihood that an event will occur. If an event is almost certain to occur, the probability of the event is high. If the event is unlikely to occur, the probability is low. Probability is represented in **statistics** as a number between one and zero, and is derived from the ratio of the number of potential outcomes of interest to the number of all potential outcomes. For example, if there are 52 cards in a deck and a card player is equally likely to draw any card, then the probability of drawing hearts is 1/4 or 0.25%. The probability of drawing a king is 4/52 or 0.0769%. An event that has a probability of one will always occur and an event that has a probability of zero will never occur.

[*Marie H. Bundy*]

Project Eco-School

Project Eco-School (PES) is a nonprofit resource center designed to promote **environmental education** by serving as a link between schools and a vast library of environmental information. PES was founded in 1989 by Jayni Chase and her husband, comic-actor Chevy Chase, who were determined to foster greater environmental awareness by educating children.

Since its inception, PES has worked most closely with schools and children in California. In Inglewood, for instance, PES provided Worthington Elementary School with an environmental library. The organization has also been instrumental in developing consumer responsibility among students. It promoted the Zero Waste Lunch, in which students were encouraged to use reusable containers and avoid using the more wasteful, disposable, and prepackaged single-serving containers. The PES flyer "Guidelines for Packing a Zero Waste Lunch" provided tips on **recycling** and stressed the importance of our actions and choices on the **environment**.

Among PES's notable publications is*Blueprint for a Green School*, a reference book geared toward educating young people about a host of environmental issues, including recycling, **waste reduction**, and consumer alternatives. In addition,*Blueprint for a Green School* provides instruction on developing letter-writing campaigns and organizing a range of schoolchildren's activities such as field trips, community services, and fund raising.

The newsletter*Grapevine* is an important element in PES's endeavor to promote environmental awareness. A typical issue provides coverage of events and environmental activism in various schools throughout the country and reports on **environmental policy** affecting the nation and the

world. The newsletter also functions as a valuable networking tool by providing readers with coverage of recent developments in environmental laws and publications. An "Eco-Stars" segment details events relating to recently created organizations, especially those directed to students.

[*Les Stone*]

RESOURCES
ORGANIZATIONS
Project Eco-School, 881 Alma Real Drive, Suite 301, Pacific Palisades, CA USA 90272

Propellants

Propellants are used to disperse aerosols, powders and other materials in a wide variety of applications. Consumers are familiar with products such as cleaners, waxes, and spray paints that are packaged in "aerosol spray cans" that use propellants. In addition, many medical, commercial and industrial products use propellants. These products include adhesives; document preservation sprays; portable fire extinguishing equipment; insecticides; sterilants; animal repellants; medical devices such as anti-asthma inhalers and topical anesthetic applicators; lubricants, coatings and cleaning fluids used in the electrical, electronic, and aerospace industries; and blowing agents and mold release agents used in the production of foams, plastic and elastomeric materials. In many cases, the propellant pressurizes, atomizes and delivers the product as an **aerosol** spray. In other cases, the propellant itself is the sole or major "active ingredient" of the product, usually for solvent and/or cleaning applications.

Aerosol sprays and other dispersions made possible using propellants can form very small and uniform droplets with controllable properties. These characteristics are desirable or necessary in critical applications where a very thin layer or coating must be applied to a surface, e.g., as a lubricant coating in an aerospace application. Additionally, the small aerosols can be inhaled deeply into the lung and the droplet itself may evaporate quickly, depending on the product. Such characteristics help optimize the delivery of drugs using medical inhalers.

The ideal propellant should generate sufficient volume and pressure of vapor or gas for a particular application, should be safe to humans and the **environment**, inert with respect to reactions with other materials, miscible or highly soluble in the product, easy to handle, and inexpensive. Each chemical has its own physical and chemical properties that affect the selection of a propellant for a given product and application. Important properties are solvency, performance, cost, and environmental considerations. Many of these properties are well characterized. For example, the solvency of a propellant can be measured as its solubility in water or other **chemicals**. Performance properties include the amount and pressure of vapor generated from the liquid, measured as vapor pressure, the volume of vapor generated per mass or volume of the liquid, and the ratio of the volume of gas to the volume of liquid. Performance also includes the range of temperatures where the propellant will function adequately, measured as the boiling and freezing points of the propellant, and as the vapor generation rates at different temperatures. The flammability of the propellant is another often critical property, measured as the flammability limits in air and the flash point. Additional properties that may be important include the density of the liquid and gas, the critical temperature and pressure, the specific heat of gas and liquid, the heat of vaporization, the viscosity of the liquid propellant, the coefficient of liquid expansion, and the surface tension. Environmental characteristics of propellants include the toxicity (often measured as **cancer** causing potency), the **ozone** depleting potential, the greenhouse warming potential, and the reactivity (in terms of forming ground level ozone). It is the environmental considerations that have brought propellants forward in the last 20+ years as an national and international environmental issue.

Restrictions on the Use of Propellants

Prior to the late 1970s, aerosol propellants were frequently chlorofluorocarbon (CFC) propellants, and this use constituted over 50% of total U.S. CFC consumption. At the time, the most widely used CFCs were CFC-11 and CFC-12. (These chemicals were also used in refrigeration, foam blowing, and sterilization). Following concerns raised in 1974 regarding possible stratospheric ozone depletion resulting from CFCs, the U.S. **Environmental Protection Agency** (EPA) and the **Food and Drug Administration** (FDA) acted on March 17, 1978, to ban the use of CFCs as aerosol propellants in all but essential applications. This reduced aerosol use of CFCs by approximately 95% and achieved nearly a 50% reduction in the U.S. consumption of CFCs. This reduction was largely accomplished without economic penalty as consumers voluntarily responded to advertising for "ozone safe" substitutes. Also in the late 1970s, Canada and a few Nordic nations banned or restricted aerosol propellant uses, also resulting in a sharp drop in their total CFC use. Later in the 1980s, however, increases in other uses of CFCs, largely as solvents, offset the earlier decreases.

The 1978 ban specifically exempted certain products based on a determination of essentiality. Also excluded were products where the CFC itself was the active ingredient or sole ingredient in an aerosol or pressurized dispenser products, as this did not fit the ban's narrow definition of an "aerosol" propellant. These restrictions remained in effect until 1990. Title VI of the 1990 **Clean Air Act** Amendments (CAAA) included provisions relevant to use of ozone deplet-

ing chemicals used as propellants and expanded restrictions for propellants. EPA's response is the "Significant New Alternatives Policy" (SNAP) Program which includes the evaluation of alternatives to ozone-depleting substances, including an assessment of their **ozone layer depletion** potential, global warming potential, toxicity, flammability, and exposure potential (section 612 of the CAAA; 59 FR 13044).

In general, ozone-depleting substances are divided into two classes with different product bans (section 610 in the CAAA): Class I comprises CFCs, **halons**, **carbon** tetrachloride, methyl chloroform (MCF), hydrobromofluorocarbons and methyl bromide; and Class II comprises solely of **hydrochlorofluorocarbons** (HCFCs). EPA is to prohibit the sale or distribution of certain "nonessential" products (as determined by Congress and EPA) that release class I substances, mandate the phaseout of class I and class II substances (sections 604 and 605), review substitutes (section 612), and prohibit the sale of certain nonessential products made with class I and class II substances (section 610). After its review, EPA issued regulations on January 15, 1993 (58 FR 4767) which banned CFC propellants in aerosols and other pressurized dispensers (and also in some other products such as flexible and packaging foam). Exceptions were made for a number of products, including certain medical devices; lubricants for pharmaceutical and tablet manufacture; gauze bandage adhesives and adhesive removers; topical anesthetic and vapocoolant products; lubricants, coatings and cleaning fluids for electrical and electronic equipment and aircraft maintenance that contain CFC-11, CFC-12 or CFC-113; release agents for molds used to produce plastic or elastomeric materials that contain CFC-11 or CFC-113; spinnerette lubricant/cleaning sprays used to produce synthetic fibers that contain CFC-114; CFCs used as halogen **ion** sources in **plasma** etching; document preservation sprays that contain CFC-113; and red pepper bear repellant sprays that contain CFC-113. However, these products are not exempted from the phase-out requirements for CFCs.

HCFCs can make excellent propellants, but their use has been limited due to their cost and restrictions regarding the class II ozone-depleting substances. In 1993, as a result of increased taxation and CFC phase-out, HCFCs were temporarily economically viable in applications where flammability is a concern and cheaper alternatives (**hydrocarbons**) could not be used. However, section 610(d) of the CAAA prohibits the sale or distribution of aerosol or foam products that contain or are manufactured with class II substances after January 1, 1994. Again, exceptions and exclusions from this ban have been made for "essential uses," i.e., if necessary due to flammability or worker safety issues and if the only available alternative is the use of a class I substance. On December 30, 1993, EPA published a final rule (58 FR 69637) which exempted medical devices, lubricants, coatings or cleaning fluids for electrical, electronic equipment, and aircraft maintenance, mold release agents used in the production of plastic and elastomeric materials and synthetic fibers, document preservation sprays containing HCFC-141b or HCFC-22, portable fire extinguishing equipment sold to commercial users, owners of boats, noncommercial aircraft, and wasp and hornet sprays for use near high-tension power lines. However, no other exceptions for class II propellants were made since substitutes are available.

Propellant Alternatives and Substitutes

Due to the restrictions on propellants use, a large number of propellant alternatives and substitutes have been investigated. Both EPA and industry have been active in this area.

Many products can be effectively packaged, distributed and used without employing a propellant. For example, some products are now sold for direct application as liquids or use manually operated finger and trigger pumps, two-compartment aerosol mechanism, mechanical pressure dispenser systems, and nonspray dispensers (e.g., solid stick dispensers). The elimination of propellants altogether can be viable option for some products and applications. In other cases, however, this may not provide proper dispersal or accurate application of the product. Also, persons using manual pumps or sprays may become fatigued with the constant pumping motion and thus produce poor product performance. Because propellants are considered essential in a variety of uses, there remains a need to find propellants that are not environmentally damaging. Unfortunately, no propellant exists which has all favorable properties. For example, some CFC propellant substitutes, e.g., ammonia, butane and pentane, have problems with toxicity and flammability. There may be some essential applications for CFCs propellants for which no practical substitutes exist, and the use of CFC propellants in anti-asthma inhalers is a frequently cited example.

A variety of propellants are being considered or are being used as alternative propellants for class I and II controlled substances. Each alternative propellant has its own physical and chemical characteristics that influence its suitability for a given application. The primary substitutes for aerosol propellant uses of CFC-11, HCFC-22 and HCFC-142b are saturated hydrocarbons (C3-C6); dimethyl ether; compressed gases; and HFCs. A few EPA-approved alternative propellants are discussed below.

Of the hydrocarbons, butane, isobutane and propane may be used singly or in mixtures. All have low boiling points, are relatively nontoxic, inexpensive and readily available. (As with essentially any propellant, very high concentrations may result in asphyxiation because of the lack of oxygen.) However, these propellants are flammable and, for example, should not be used around electrical equipment if

sparks could ignite the hydrocarbon propellant. To reduce product flammability, hydrocarbons can be used with water-based formulations. In the United States, nearly 50% of aerosol propellants were using hydrocarbon propellants prior to 1978, and nearly 90% in 1979 as a result of the CFC ban. Propane/butane propellants have been used since 1987 in the Scandinavian market.

Dimethyl ether (DME) is a medium pressure, flammable, liquefied propellant generally used in combination with other propellants. Its properties are similar to the hydrocarbons.

Compressed gases, including **carbon dioxide**, **nitrogen**, air, and **nitrous oxide**, are used in applications where a nonflammable propellant is necessary. These gases are inexpensive, readily available, nonflammable (although certain temperatures and pressures of nitrous oxide may create a moderate explosion risk), relatively nontoxic, and industrial practices for using these substitutes are well established. Since these gases are under significantly greater pressure than CFCs and HCFCs, containers holding these gases must be larger and bulkier, and safety precautions are necessary during filling operations. Often, modifications must be made to use compressed gases as they require new dispensing mechanisms and stronger containers due to the greater pressure; their low molecular weights restrict certain applications; high pressure compressed gases dispel material faster which may waste product; compressed gases cool upon expansion; and finally, they are not well suited for applications that require a fine and even dispersion. At present, about 7-9% of the aerosol products use compressed gases, and their use is expected to grow.

Hydrofluorocarbons (HFCs) such as HFC-134a, HFC-125 and HFC-152a are partially fluorinated hydrocarbons, developed relatively recently and currently priced significantly higher than HCFC-22. HFCs are less dense than HCFC-22, but can provide good performance in applications (but not products such as noise horns, which require a more dense gas). HFC-134a and HFC-125 are nonflammable and have very low toxicity. HFC-152a is slightly flammable. All three HFCs have zero ozone depletion potential, but are potential **greenhouse gases** although their atmospheric residence times are short. Additionally, these HFCs are chemically reactive and they contribute to the formation of tropospheric ozone. In ozone nonattainment areas, state and local controls on VOCs may restrict the use of these products. HFCs may be combined with flammable propellants to reduce the flammability of the mixtures. HFC-134a and HFC-227a are possible alternative propellants in medical applications (currently using CFC-12 and CFC-114). Approvals from both FDA and EPA are needed for these applications.

[*Stuart Batterman*]

Public Health Service

The United States Public Health Service is the health component of the **U.S. Department of Health and Human Services**. It originated in 1798 with the organization of the Marine Hospital Service, out of concern for the health of the nation's seafarers who brought diseases back to this country. As immigrants came to America, they brought with them **cholera**, smallpox, and yellow fever; the Public Health Service was charged with protecting the nation from infectious diseases.

Today the Service helps city and state health departments with health problems. Its responsibilities include controlling infectious diseases, immunizing children, controlling sexually transmitted diseases, preventing the spread of tuberculosis, and operating a quarantine program.

The Centers for Disease Control in Atlanta, Georgia, is the Public Health Service agency responsible for promoting health and preventing disease.

Public interest group

Public interest groups may be defined as those groups pursuing goals the achievement of which ostensibly will provide benefits to the public at large, or at least to a broader population than the group's own membership. Thus, for example, if a public interest group concerned with **air quality** is successful in its various strategies and activities, the achieved benefit--cleaner air--is available to the public at large, not merely to the group's members. The competition of interest groups, each pursuing either its own good or its conception of the public good, has been an increasingly prominent feature of American politics in the latter half of the twentieth century.

There is no single, universally applicable definition or test of the public good, and thus there is often a great deal of disagreement about what happens to be in the public interest, with different public interest groups taking quite different positions on various issues. In any particular political controversy, moreover, there may be several quite different public interests at stake. For example, the question of whether to build a nuclear-powered generator plant may involve competing public interests in the protection of the **environment** from **radioactive waste** and other dangers, the maintenance of public safety, and the promotion of economic growth, among others.

Similarly, the question of who benefits from the activities of a public interest group can also be quite complicated. Some benefits that are generally available to the public may not be equally available or accessible to everyone. If **wilderness** preservation groups are successful, for example, in hav-

ing land set aside in, say, Maine, that land is in principle available to potential recreational users from all over the country. But people from New England will find that benefit much more accessible than people from another region.

The membership, resources, and number of active public interest groups in the United States have all increased dramatically in the previous twenty-five years. There are now well over 2,500 national organizations promoting the public interest, as determined from almost every conceivable viewpoint, in a wide variety of issue areas. Over 40 million individual members support these groups with membership fees and other contributions totalling more than $4 billion every year. Many groups find additional support from various corporations, private foundations, and governmental agencies.

This growth in the public interest sector has been more than matched by, and partly was a response to, a similar explosion in the number and activities of organized interest groups and other politically active organizations pursuing benefits available only or primarily to their own members. Public interest groups often provide an effective counterweight to the activities of these more narrowly-oriented associations.

The growth of interest group politics has not been without its negative consequences. Some critics argue that every interest group, including "public interest groups," ultimately pursues relatively narrow goals important mainly to fairly limited constituencies. A politics based on the competition of these groups may be one in which well-organized narrow interests prevail over less well-organized broader interests. It may also produce decreased governmental performance and economic inefficiencies of various types. Public interest groups, moreover, seem to draw most of their support and membership from the middle and upper economic strata, whose interests and concerns are then disproportionately influential.

[*Lawrence J. Biskowski*]

Resources

Books

Berry, J. M. *The Interest Group Society.* Boston: Little, Brown, 1984.

Cigler, A. J., and B. A. Loomis, eds. *Interest Group Politics.* Washington, DC: CQ Press, 1991.

McFarland, A. S. *Public Interest Lobbies.* Washington, DC: The American Enterprise Institute, 1976.

Schlozman, K. L., and J. T. Tierney. *Organized Interests and American Democracy.* New York: Harper and Row, 1986.

Public Interest Research Group

see **U.S. Public Interest Research Group**

Public land

Public land refers to land owned by the government. Most frequently, it is used to refer to land owned and managed by the United States government, although it is sometimes used to refer to lands owned by state governments. The U.S. government owns 262 million acres, about one eighth of the land in the country, the bulk of which is located in the western states, including Alaska. The lands are managed primarily by the **Bureau of Land Management**, the Department of Defense, the **Fish and Wildlife Service**, the **Forest Service**, and the **National Park Service**.

Public Lands Council

An important environmental debate is currently focused on the use of public lands by livestock owners, particularly owners of cattle and sheep. Historically, the federal government has sold leases and permits for grazing on **public land** to these individuals at very low prices, often a few cents per acre.

In recent years, many environmentalists have argued that cattle are responsible for the destruction of large tracts of rangeland in the western states. They believe the government should either greatly increase the rates they charge for the use of western lands or prohibit grazing entirely on large parts of it. In response, livestock owners have claimed that they practice good-land management techniques and that in many cases the land is in better condition than it was before they began using it.

The Public Lands Council (PLC) is one of the primary groups representing those who use public lands for grazing. It is a nonprofit corporation that represents approximately 31,000 individuals and groups who hold permits and leases allowing them to use federal lands in 14 western states for the grazing of livestock. Twenty-six state groups belong to the council, and in addition, the council coordinates the public lands policies of three other organizations, the National Cattlemen's Association, the American Sheep Industry Association, and the Association of National **Grasslands**.

The PLC also represents the interests of public land ranchers before the United States Congress. It lobbies and monitors Congress and various federal agencies responsible for grazing, water use, **wilderness**, **wildlife**, and other federal land management policies that are of concern to the livestock industry.

The council was founded in 1968 for the purpose of promoting principles of sound management of federal lands for grazing and other purposes. The council obtains its funds from dues collected by state organizations and by contributions from other organizations it represents. It has two classes

of membership. General members are those who belong to a state organization that contributes to the PLC or those who make individual contributions. Voting members are elected by the general membership of each state, which may have a maximum of four voting members. Voting members meet at least once each year to establish PLC policy.

The PLC maintains a 500-volume library at its offices in Washington, D.C. It also publishes a quarterly newsletter, as well as news releases on specific issues, and regular columns in various western livestock publications. The council has also sponsored workshops and seminars on issues of importance to users of federal lands.

The American Lands and Resources Foundation has been established as an arm of the council for the purpose of receiving charitable and tax-deductible contributions to be used for the education of the general public about the benefits of using federal lands for livestock grazing.

[*David E. Newton*]

Public trust

The public trust doctrine is a legal doctrine dealing with the protection of certain uses and resources for public purposes, regardless of ownership. These uses or resources must be made available to the public, regardless of whether they are under public or private control. That is, they are held in trust for the public.

The modern public trust doctrine can be traced back at least as far as Roman civil law. It was further developed in English law as "things common to all." This public trust was applied mostly to navigation: the Crown possessed the ocean, the rivers, and the lands underlying these bodies of water (referred to as *jus publicum*). These were to be controlled in order to guarantee the public benefit of free navigation.

In the United States, the public trust doctrine is rooted in *Illinois Central Railroad Co. v. Illinois* (1892). In this case, the Supreme Court ruled that a grant held by the Illinois Central Railroad of the Chicago waterfront was void because it violated the state's public trust responsibilities to protect the rights of its citizens to navigate and fish in these waters. In making this decision, the court relied on English common law and served to underscore the existence of the public trust doctrine in the United States.

As a trustee, the government (federal or state) faces certain restrictions in protecting the public trust. These include using the trust property for common purpose and making it available for use by the public; not selling the trust property; and maintaining the trust property for particular types of uses. If any of these responsibilities are violated, the government can be sued by its citizens for neglecting its trust responsibilities.

With the rise of the environmental movement, the public trust doctrine has been applied with greater frequency and to a broader array of subjects. In addition to navigable waters, the doctrine has been applied to **wetlands**, state and national parks, and fossil beds. This expanded use of the doctrine has led to increased conflict. On the ground, there has been growing conflict over the limitations of private property. Landholders who find their actions limited argue that the public trust doctrine is a way to mask indirect **takings** and argue that they deserve just compensation in return for the restrictions. Among legal scholars, there is concern that the public trust doctrine is not firmly grounded in the law and simply reflects the opinions of judges in various cases. Proponents of the public trust doctrine, however, argue that it is a useful vehicle to temper unrestricted private property rights in the United States. Indeed, it has been argued that if public trust rights exist in private property, then no takings can occur since the regulation is merely recognizing the pre-existing limitation in the property.

[*Christopher McGrory Klyza*]

Resources

Books

Plater, Z. J. B., R. H. Abrams, and W. Goldfarb, eds. *Environmental Law and Policy: Nature, Law, and Society.* New York: West Publishing, 1992.

Periodicals

Brady, T. P. "'But Most of It Belongs to Those Yet to be Born:' The Public Trust Doctrine, NEPA, and the Stewardship Ethic." *Boston College Environmental Law Review* 17 (1990): 621–46.

Kagan, D. G. "Property Rights and the Public Trust: Opposing Lakeshore Funnel Development." *Boston College Environmental Law Review* 15 (1987): 105–34.

Puget Sound/Georgia Basin International Task Force

The Puget Sound/Georgia Basin **ecosystem** consists of three shallow marine basins: the Strait of Georgia to the north, Puget Sound to the south, and the Strait of Juan de Fuca that connects this inland sea to the Pacific Ocean. In addition to open water, the ecosystem includes islands, shorelines, **wetlands**, and the watersheds of several mountain ranges whose freshwater rivers dilute the salt water. This rich and diverse ecosystem spans the United States-Canadian border and is threatened by **population growth**, urbanization, agriculture, and industry. In addition to Pacific **salmon**, valuable ground fish, and marine mammals including orca **whales**, Puget Sound/Georgia Basin has three of North America's busiest ports—Seattle, Tacoma, and

Vancouver, British Columbia (B.C.). Already an estimated 58% of the coastal wetlands of Puget Sound and 18% of the Strait of Georgia wetlands, which provide vital **habitat** to fish and birds, have been lost to development.

The Puget Sound/Georgia Basin International Task Force was created 1992 by the B.C./Washington Environmental Cooperation Council to coordinate priority environmental efforts between the state of Washington and the Canadian province of British Columbia. The task force includes representatives from the U.S. **Environmental Protection Agency** (EPA), the U.S. **Fish and Wildlife Service**, the Northwest Fisheries Science Center, the Department of **Fisheries and Oceans Canada**, and the Department of **Environment Canada**. State and provincial representation include the Washington Departments of Ecology, Natural Resources, and Fish and Wildlife and the B.C. Ministries of Water, Land and Air Protection and Sustainable Resource Management (formerly combined under the B.C. Ministry of Environment, Lands and Parks). The task force also includes representatives from the Puget Sound **Water Quality** Action Team, the Northwest Straits Commission, the Coast Salish Sea Council, and the Northwest Indian Fisheries Commission. The task force is charged with addressing marine, near-shore, and shoreline environmental issues for the entire Puget Sound/Georgia Basin shared ecosystem, as well as identifying and responding to new issues that may affect the shared waters.

The task force pursues project funding and promotes communication and cooperation among federal, state, provincial, local, and tribal/aboriginal governments and groups. In 1994, a Marine Science Panel (MSP), made up of Washington and B.C. scientists, identified priority environmental issues to be addressed by the task force's working groups. The task force cooperates with government agencies on implementation of the working groups' proposals.

The MSP recommended that fish and **wildlife management** shift its goal from maximum sustainable harvest to **species** protection. To this end, the task force's highest priority issues are protecting marine life, establishing **marine protected areas** (MPAs), and preventing near-shore and wetland habitat loss and the introduction of exotic or nonindigenous species (NIS). The Task Force Work Group on Marine Protected Areas has conducted an assessment of Puget Sound MPAs. The B.C. Nearshore Habitat Loss Work Group has devised a plan for preventing coastal habitat loss in the Georgia Basin. The Washington Aquatic Nuisance Species Coordination Committee and the B.C. Working Group on Non-Indigenous Species have worked to develop nuisance species legislation and proposals for regulating ballast water release from ships, a major source of NIS introduction. In collaboration with various agencies, the task force has launched a public NIS education program.

The medium priority recommendations of the MSP included the control of toxic waste discharges and the prevention of large **oil spills** and major diversions of freshwater for **dams** and other projects. The Washington and B.C. toxic chemical work groups have undertaken research, inventoried contaminated sites, and assessed the movement of toxic **chemicals** through the shared waters.

On May 15, 2002 the Transboundary Georgia Basin-Puget Sound Environmental Indicators Working Group released their ecosystem indicators report, concluding that population growth constitutes the primary threat to the region. During the 1990s, the region's population grew by about 20% to seven million people. It is projected to grow another 32% by 2020. Whereas most of the growth during the 1990s was in urban areas, new growth is expected to occur in more rural areas. The report found that **air pollution** from inhalable particles has declined since 1994; the amount of waste produced per person has remained about the same; and **recycling** has increased, particularly in B.C. However the loss of stream and shoreline habitats to development is endangering the region's plants and animals. In the Georgia Basin almost 35% of the freshwater fish and 12% of the reptiles are in danger of **extinction**, as are 18% of Puget Sound's freshwater fish and 25% of its reptiles. The harbor **seals** of Puget Sound are much more heavily contaminated with **polychlorinated biphenyls** (PCBs) than the Strait of Georgia seals. PCB contamination levels have remained constant for more than 14 years, despite major cleanup efforts. The harbor seals also are contaminated with other **persistent organic pollutants** including dioxins and **furans**. The report found that most of the ecosystem's protected land is in the mountains: only 1% of land below 3,000 ft (914 m) is protected.

[*Margaret Alic Ph.D.*]

Resources

Books

Mills, Mary Lou. *Strategy and Recommended Action List for Protection and Restoration of Marine Life in the Inland Waters of Washington State.* Seattle: Puget Sound/Georgia Basin International Task Force, 1999.

Other

The British Columbia Nearshore Habitat Loss Work Group. *A Strategy to Prevent Coastal Habitat Loss and Degradation in the Georgia Basin.* June 2001 [May 2002]. <www.wa.gov/pswqat/shared/pdfs/Coastal_main.pdf>.

Puget Sound/Georgia Basin International Task Force. *Pathways to Our Optimal Future: A Five-Year Review of the Activities of the International Task Force.* Puget Sound-Georgia Basin Environmental Initiative. [May 2002]. <http://www.wa.gov/puget_sound/shared/pdfs/_December_1_f.al_in_sequenc.pdf>.

"Shared Waters, Puget Sound On-Line." *Puget Sound/Georgia Basin International Task Force.* [May 2002]. <www.wa.gov/pswqat/shared/backgrnd.html>.

Transboundary Georgia Basin-Puget Sound Environmental Indicators Working Group. *Georgia Basin-Puget Sound Ecosystem Indicators Report.* May 15, 2002 [June 2002]. <wlapwww.gov.bc.ca/cppl/gbpsei/index.html>.

Washington Sea Grant Program. *Shared Waters: The Vulnerable Inland Sea of British Columbia and Washington.* Shared Waters, Puget Sound On-Line. [May 2002]. <www.wa.gov/pswqat/shared/bcwaswl.html>.

ORGANIZATIONS

Ministry of Water, Land and Air Protection, PO Box 9360 Stn Prov Govt, Victoria, BCCanada V8W 9M2 (250) 387-9422, Fax: (250) 356-6464, <http://www.gov.bc.ca/wlap>

Puget Sound/Georgia Basin International Task Force, Email: jdohrmann@psat.wa.gov, <http://www.wa.gov/puget_sound/shared/shared.html>

Puget Sound Water Quality Action Team, PO Box 40900, Olympia, WA USA 98504-0900 (360) 407-7300, Toll Free: (800) 54-SOUND, <http://www.wa.gov/puget_sound>

Pulp and paper mills

Pulp and paper mills take wood and transform the raw product into paper. Hardwood logs (beech, birch, and maple) and softwoods (pine, spruce, and fir) are harvested from managed forestlands or purchased from local farms and timberlands across the world and are transported to mills for processing. Hardwoods are more dense, shorter fibered, and slower growing. Softwoods are less dense, longer fibered, and faster growing.

Today, the process is mainly done with high tech, sophisticated machinery. Wood products, which consist of lignin (30 percent), fiber (50 percent), and other materials--carbohydrates, proteins, fats, turpentine, resins, etc., (20 percent) are transformed into paper consisting of fiber, and additives--clay, titanium dioxide, calcium carbonate, water, rosin, alum, starches, gums, dyes, synthetic polymers, and pigments. Wood is about 50 percent cellulose fiber. The structure of paper is a tightly bonded web of cellulose fibers. About 80 percent of a typical printing paper by weight is cellulose fiber. First in the process, the standard eight-foot (2.4-m) logs are debarked by tumbling them in a giant barking drum and then chipped by a machine that reduces them to half-inch chips. The chips are cooked, after being screened and steamed, in a digester using sodium bisulfite cooking liquor to remove most of the lignin, the sticky matter in a tree that bonds the cellulose fibers together. This is the pulping process.

Then the chips are washed, refined, and cleaned to separate the cellulose fibers and create the watery suspension called pulp. The pulp is bleached in a two-stage process with a number of possible **chemicals**. Those companies that choose to avoid **chlorine** bleach will use **hydrogen** peroxide and sodium hydrosulfite which yields a northern high-yield hardwood sulfite pulp. This pulp is blended with additional softwood kraft pulp after refining as part of the stock preparation process, which involves adding such materials as dyes, pigments, clay fillers, internal sizing, additional brighteners, and opacifiers.

Late in the process, the stock is further refined to adjust fiber length and **drainage** characteristics for good formation and bonding strength. The consistency of the stock is reduced by adding more water and the stock is cleaned again to remove foreign particles. The product is then pumped to the paper machine headbox.

From here, the dilute stock (99.5 percent water) flows out in a uniformly thin slice onto a Fourdrinier wire--an endless moving screen that drains water from the stock to form a self-supporting web of paper. The web moves off the wire into the press section which squeezes out more water between two press felts, then into the first drier section where more moisture is removed by evaporation as the paper web winds forward around an array of steam-heated drums. At the size press, a water-resistant surface sizing is added in an immersion bath.

From there the sheet enters a second drier section where the sheet is redried to the final desired moisture level before passing through the computer scanner. The scanner is part of a system for automatically monitoring and regulating basis weight and moisture. The paper enters the calendar stack, where massive steel polishing rolls give the sheet its final machine finish and bulking properties.

The web of paper is then wound up in a single long reel, which is cut and moved off the paper machine to a slitter/winder machine which slices the reel into rolls of the desired width and rewinds them onto the appropriate cores. The rolls are then conveyed to the finishing room where they are weighed, wrapped, labeled, and shipped.

In practice, all papers, even newsprint, are pulp blends, but they are placed in one of two categories for convenient description: groundwood and free sheet. And in practice, other pulp varieties enter into the picture. They may be reclaimed pulps such as de-inked or post-consumer waste; recycled pulps which included scraps, trim, and unprinted waste; cotton fiber pulps; synthetic fibers; and pulps from plants other than trees: bagasse, esparto, bamboo, hemp, **water hyacinth**; and banana, or rice. But the dominant raw material remains wood pulp. Paper makers choose and blend from the spectrum of pulps according to the demands on their grades for strength, cleanliness, brightness, opacity, printing, and converting requirements, aesthetics, and market price.

From cotton fiber-based sheets to the less expensive papers made from groundwood, to recycled grades manufactured with various percentages of wastepaper content, papermakers have consistently responded to the need of the marketplace. In today's increasingly environmentally conscious marketplace, papermakers are being called on to produce pulp that is environmentally friendly. Eliminating chlo-

rine from the bleaching process is a major step in eliminating unwanted **toxins**. The changeover costs money and is the source of controversy here in the United States.

However, the chlorine-free trend has taken a firm hold in Europe. All of Sweden requires its printing and writing paper mills to be chlorine-free by the year 2010. France, Germany, and several other countries have several mills that are reported to be chlorine-free and the trend is moving across Canada.

While there are growing exceptions, most North American mills still use a chlorine bleaching process to create a bright, white pulp. Why the need to eliminate the chemical?

In the bleaching process, chlorine, chlorine dioxide, and other chlorine compounds create toxic byproducts. These byproducts consist of over 1,000 chemicals, some of which are the most toxic known to man. The list includes: **dioxin** and other organochlorines compounds such as PCBs, DDT, **chlordane**, aldrin, dieldrin, **toxaphene**, chloroform, heptachlor and **furans**. These are formed by the reaction of lignin in the pulp with chlorine or chlorine-based compounds used in the bleaching sequence of all kraft pulps.

These unwanted chemicals byproducts must be discharged and end up in the **effluent**. The effluent is released into our rivers, lakes and streams and is threatening the **groundwater** and our drinking water, as well as the food chain through fish and birds. Epidemic health effects among thirteen **species** of fish and **wildlife** near the top of the **Great Lakes** food web have been identified. Not only are these toxic chemicals causing **cancer** and birth deformities in humans and wildlife, but they are very persistent, building up in our waterways and eventually into our bodies.

Dioxin traces have been found in papers and even in coffee from chlorine-bleached coffee **filters** and milk from chlorine-bleached milk cartons, as well as in women's hygiene products. Most of the paper being sold in the United States today as "dioxin free" is actually "dioxin undetectable." That is because dioxin can be measured in **parts per trillion** or parts per quintillion, but beyond that level, there are no scientific measurements sophisticated enough. Or if measurable, the process becomes very expensive. (If exposed often enough, even these minuscule quantities build up in the **environment** and in humans.) To be truly dioxin free, the paper must be made from pulps that have been bleached without chlorine or chlorine-based compounds.

The newer trend is to eliminate chlorine from the bleaching process completely. Some North American mills are turning to hydrogen peroxide, oxygen brightening, or **ozone** brightening. These compounds do not produce dioxin or other organochlorine compounds and are considered "environmentally-sound." The United States pulp and paper industry has sharply reduced its use of the chemical and plans to curtail use further during the next few years.

In part, this reduction can be traced to the increased sophistication of pulp and paper plants during the past decade. The cooking and bleaching operations have been fine-tuned. Wood chips are cooked more before they go to the bleach plant, so less bleaching is required. Also, in some plants, industry is trying chlorine dioxide as a substitute; it produces less dioxin, but still contains chlorine.

The result of these changes means an 80 percent reduction in the amount of dioxin associated with bleaching. Between 1988-1989, a total of 2.5 lb (11.1 kg) of dioxin was produced. As of 1993, that number fell below 8 ounces (22.7 g) per year. Eight ounces sound like a small amount, but scientists measure dioxin in **parts per million**, billion, trillion and quintillion, so eight ounces is still too high.

A number of lawsuits have been filed by residents living near or downstream of dioxin-contaminated pulp mills because of the health threats. The more the plaintiffs win, the sooner will the use of chlorine be eliminated completely. It is estimated that the amount of chlorine used in pulp and paper bleaching will fall from 1.4 million tons in 1990 to 920,000 tons by 1995.The eventual goal is **zero discharge**.

[*Liane Clorfene Casten*]

RESOURCES

BOOKS

Ferguson, K. *Environmental Solutions for the Pulp and Paper Industry.* San Francisco: Miller Freeman, 1991.

PERIODICALS

Jenish, D. "Cleaning Up a Chemical Soup." *Maclean's* 103 (29 January 1990): 32–4.

Purple loosestrife

Purple loosestrife (*Lythrum salicaria*) is an aggressive wetland plant **species** first introduced into the United States from Europe. It is a showy, attractive plant that grows up to 4 feet (1.2 m) in height with pink and purple flowers arranged on a spike, and it is common in shallow marshes and lakeshores all across the northern half of the United States.

Loosestrife is a perennial plant species that spreads rapidly because of the high quantity of seeds it produces (sometimes up to 300,000 seeds per plant) and the efficient dispersal of seeds by wind and water. The plant has a well-developed root system and is able to tolerate a variety of **soil** moisture conditions. This has made it an effective colonizer of disturbed ground as well as areas with fluctuating

Purpole loosestrife (*Lythrum salicaria*)**.** (Photograph by John Dudak. Phototake. Reproduced by permission.)

hydrological regimes. Because of its attractive flowers, it has also been planted as an ornamental species.

As a highly adaptive and tolerant plant, loosestrife is often able to outcompete native plant species in many environments, particularly in **wetlands** and disturbed areas. Other species often have difficulty surviving once it becomes established, and loosestrife has low value as food and nesting **habitat**. For example, the cattail plants in many wetlands of New York state have been replaced by purple loosestrife. **Wildlife** experts prize cattails as plants whose roots and tubes are of high value as food source to small mammals and rodents, and the spread of loosestrife has led to a decline in the habitat values of these wetlands. So, in spite of its pleasing appearance, New York and a number of other states have embarked on purple loosestrife eradication programs.

The most commonly used chemical method for eradicating purple loosestrife is a **herbicide** known as Rodeo--a selective herbicide that kills only dicotyledonous or broadleaf plants. When a diluted solution of Rodeo is sprayed directly on mature leaves, it disrupts protein synthesis and causes plant death within two to seven days. Physical removal of the plants by mowing early in the growing season, prior to seed set, has also been effective in keeping areas free of loosestrife. *See also* Introduced species

[*Usha Vedagiri*]

Resources
Books

Schmidt, J. C. *How to Identify and Control Water Weeds and Algae.* Milwaukee, WI: Applied Biochemists, 1987.

Thompson, D. Q. *Spread, Impact and Control of Purple Loosestrife (Lythrum salicaria) in North American Wetlands.* Washington, DC: U.S. Department of the Interior, 1987.

———. *Waterfowl Management Handbook. 13.4.11, Control of Purple Loosestrife.* Washington, DC: U. S. Fish and Wildlife Service, 1989.

PVC

see **Polyvinyl chloride**

Q

David Quaamen (1948 –)

American writer

Nature writing is a well established tradition in American literature, yet the label has in recent years come to be considered one-dimensional and even a little quaint by 'serious' writers and 'real' scientists. Many of the best nature writers are not trained ecologists and ecology is supposed to have out-grown natural history. Intellectuals belittle the popularity of nature writing and see themselves as beyond the backwoods genre.

Still, the American public continues to gain much of its understanding and awareness of natural systems, of environmental issues, and of humanity's threat to wild species and places from the best and most popular of the nature writers—Edwin Way Teale, Annie Dillard, Terry Tempest Williams...and David Quammen. Part of the exposure comes from the fact that many of these writers make a good part of their living by lecturing and 'reading' widely in popular venues. Quammen, however, considers his writing as departing "distantly from conventional nature writing, of which I have never much cared to be either a reader or an author." He started writing as a novelist, but, beginning in 1981, spent 15 years writing a column titled "Natural Acts," for Outside magazine, his job "to think about the natural world, and about human relationships with that world, in a way that would interest" the half million readers of the magazine, an assignment which he says allowed him "unimaginable freedom to explore a wide range of peculiar theories, remote places, bizarre facts, and unpopular opinions." His first collection of essays (*Natural Acts: A Sidelong View of Science and Nature*), mostly taken from his column, appeared in 1985. That collection was well received, with one reviewer suggesting that Quammen "breathes importance into the little-known nitty-gritty of biology" and in the process "describes wondrous places, people and situations."

Still, Quammen does write about nature, and though many of his essays are "very funny and very offbeat," (the word quirky is often used) they are just as frequently serious commentaries on science and its relationship to politics, philosophy, and a variety of other subjects. Perhaps then, it is more appropriate to use his own descriptor and call him a science journalist. Lee Dembart in the *Los Angeles Times* stated that "Quammen likes science for its own sake, but he also likes it for the larger truths it suggests. He works the fringes of science and draws conclusions that are universal."

Marilyn McEntyre, author of one of the most detailed looks to date at Quammen's work, is enthusiastic: Quammen's "gift to the general reader is to make science accessible, compelling, and, not least important, entertaining." Her comments underline the idiosyncrasies of Quammen's work: "Certainly Quammen is concerned with those central and urgent ethical dilemmas that define and unite what may loosely be known as the 'environmental movement,' but he also pauses to consider the ethics of domestic dog ownership," a reference to his questions about possible 'unnatural' harm to dogs from their confinement in urban and suburban areas, "daring to throw darts at one of the most sacred of American icons, 'man's best friend.'"

Born in Cincinnati, in 1948, and spending a childhood in southwest Ohio as the only boy of three children, he was the "designated mowist" of the family lawn, a repeated chore that probably had something to do with the "why lawns?" kinds of questions about human-nature relationships that appear so often in his writings . Quammen went on to earn a B.A. degree in English from Yale University in 1970, and, as a Rhodes scholar, a B. Litt. in 1973 from Oxford.

The *Song of the Dodo* (1996), one of Quammen's most recent, yet most recognized books, illustrates both his grasp of current ideas in science and the heat that he (and the scientists themselves) take when they relate those ideas to how "humans in all their variousness, regard and react to the natural world, in all its variousness." The variety in human-environment relationships is a basic theme Quammen claims for much of his writing. He built the popular book *Song of the Dodo*on the less publicly accessible island biogeography research of Robert MacArthur and E. O. Wil-

son, creating a sharp, clear depiction of the ideas that islands are bounded, isolated places vulnerable to extinctions, a vulnerability repeated as humans create "isolated, island-like enclaves" in terrestrial ecosystems, places too small, fragmented, and isolated to support diverse populations.

This mix of grave warnings of environmental degradation and species extinctions with personal anecdotes and popular travel writing draws both praise and criticism. His books are accessible to a wide audience, so the environmental issues he discusses are widely disseminated, increasing the public's awareness. Even reviews in biology journals like *BioScience* laud Quammen's island biogeography book as "a work in the finest tradition of autobiographical natural history," a history traced back to the great naturalists of the nineteenth century, including Darwin. But Quammen doesn't escape the general questions raised by critics of island biogeography theory of how different land 'islands' surrounded by land are from islands bounded by water.

Some critics think that Quammen's accessibility is based on the sugar-coating of information: "Quammen has succumbed to the fashionable conceit that 'popular' science has to be diluted with travelogue and soap biography." However, Bill McKibben, in *Audubon* magazine, echoed many others when he proclaimed *Song of the Dodo* a "masterpiece of scientific journalism...and a heroic achievement."

[*Gerald L. Young Ph.D.*]

Further Reading

Books

McEntyre, Marilyn C. "David Quammen." Vol. 2, *American Nature Writers* edited by John Elder. NY: Charles Scribner's Sons, 1996.

"Quammen, David 1948–." Vol. 79, *Contemporary Authors.* Detroit: Gale Group, 1999.

Wallace, Allison B. "Contemporary Ecophilosophy in David Quammen's Popular Natural Histories." In *The Literature of Science: Perspectives on Popular Scientific Writings* edited by Murdo W. William. Athens, GA: University of Georgia Press, 1993.

Other

"Inventory for the David Quammen Papers, 1956–1999 and undated." *Texas Archival Resources Online.* [2002]. <http://www.lib.utexas.edu/taro/tturb/00159/00159.html>.

R

Rabbits in Australia

Imported into **Australia** in the mid-nineteenth century, rabbits have overrun much of the country, causing extensive agricultural and environmental damage and demonstrating the dangers of introducing non-native **species** into an area. Before the first humans arrived in Australia, the only mammals living there were about 150 species of marsupials as well as **bats**, rats, mice, platypuses, and echidnas. The chief predator in Australia today is the dingo, a wild dog introduced about 40,000 years ago by Australia's first human settlers, the Aborigines. When the British settled Australia as a penal colony in the late 1700s, they brought a variety of pets and livestock with them, including rabbits.

The problems with rabbits began in 1859, when 12 pairs of European rabbits were released on a ranch. Since no significant numbers of natural predators were present in Australia, the rabbit population exploded within a few years. They overgrazed the grass used for sheep raising, and despite extensive efforts to reduce and control the population, by 1953 approximately 1.2 million mi^2 (3 million km^2) were inhabited by between several hundred million and over one billion rabbits.

The rabbits have caused considerable damage to Australia's **environment**, especially its native plants and **wildlife**. Their extensive burrows cause **soil erosion**, and they eat large quantities of grass and shoots of other plants, devastating native **flora** that birds, insects, and other creatures depend on for food and cover.

Repeated attempts have been made to exterminate the rabbits and limit their populations, often by fumigating and ripping up their burrows, or by using poison bait. The most effective effort undertaken involved the introduction of myxomatosis in the 1950s, a disease that is fatal to rabbits. But after an initial decline, the rabbit population began to recover, and much of Australia remained overrun by the animals.

Proponents of eradication say that the most promising new biological agent to control wild rabbits is know as RCV (rabbit calicivirus), which is thought to be a new disease, and has killed farmed rabbits in China and Europe. In its early stages, it kills over 90% of infected rabbits. In the laboratory, it has been shown to kill rabbits shortly after infection; and after 31 different native species were exposed to the **virus**, it did not appear to be capable of infecting non-target animals. But the certainty of this is disputed by other scientists, who worry that release of the virus is "an uncontrollable and unpredictable" biological experiment; that it could eventually jump the species barrier, especially if it mutates into a different form; and that rabbits will eventually develop **resistance** to the virus, as they did to myxomatosis.

The elimination of rabbits might even have adverse consequences, such as removing an important food source from eagles and other predators who have come to depend on them. Wild dogs and cats, deprived of their normal prey, may turn to increased **hunting** of kangaroos and other marsupials. On the other hand, the number of feral canines and felines might decline with the demise of an easy and abundant food source, which would benefit marsupials. So, the ultimate impact of RCV is not entirely predictable.

There is little danger of an immediate collapse of the rabbit population. The latest data estimates that the population has been reduced from about 200–300 million rabbits to 100 million.

The proliferation of rabbits in Australia has cost the government and ranchers billions of dollars. This situation remains a primary example of the harm that can result from the introduction of non-native species, no matter how seemingly harmless, into a foreign environment.

[*Lewis G. Regenstein*]

Resources

Books

Ecology of Biological Invasions. New York: Cambridge University Press, 1986.

Periodicals

Anderson, I., and R. Nowak. "Australia's Giant Lab." *New Scientist.*

Chris Pert, rabbit hunter in Western Australia. (Photograph by Ulrike Welsch. Reproduced by permission.)

Rachel Carson Council

The Rachel Carson Council focuses on the dangers of pesticides and other toxic **chemicals** and their impact on human health, **wildlife**, and the **environment**. After the 1962 publication of her classic book on pesticides, *Silent Spring*, **Rachel Carson** was overwhelmed by the public interest it generated, including letters from many people asking for advice, guidance, and information. Shortly after her death in April 1964, colleagues and friends of Carson established an organization to keep the public informed on new developments in the field of chemical contamination. Originally called the Rachel Carson Trust for the Living Environment, it was incorporated in 1965 to work for **conservation** of **natural resources**, to increase knowledge about threats to the environment, and to serve as a clearinghouse of information for scientists, government officials, environmentalists, journalists, and the public.

The Council has long warned that many pesticides, regulated by the **Environmental Protection Agency** (EPA) and widely used by homeowners, farmers, and industry, are extremely harmful. Many of these chemicals can cause **cancer**, miscarriages, **birth defects**, genetic damage, and harm to the central nervous system in humans, as well as destroy wildlife and poison the environment and **food chain/web** for years to come.

Other, less acutely toxic chemicals that are commonly used, the organization points out, can cause "delayed neurotoxicity," a milder form of nerve damage that can show up in subtle behavior changes such as memory loss, fatigue, irritability, sleep disturbance, and altered brain wave patterns. Concerning termiticides used in schools, "children are especially vulnerable to this kind of poisoning," the Council observes, "and the implications for disturbing their ability to learn are especially serious."

Through its studies, publications, and information distribution, the Council has provided data strongly indicating that many pesticides now in widespread use should be banned or carefully restricted. The group has urged and petitioned the EPA to take such action on a variety of chemicals that represent serious potential dangers to the health and lives of millions of Americans and even to **future generations**.

The Council has also expressed strong concern about, and sponsored extensive research into, the link between exposure to toxic chemicals and the dramatic increase in recent years in cancer incidence and death rates. The group's publications and officials have warned that the presence of dozens of cancer-causing chemicals in our food, air, and water is constantly exposing us to deadly carcinogens and is contributing to the mounting incidence of cancer, which eventually strikes almost one American in three, and kills over 500,000 Americans every year.

The Council publishes a wide variety of books, booklets, and brochures on pesticides and toxic chemicals and alternatives to their use. Its most recent comprehensive work, *Basic Guide to Pesticides: Their Characteristics and Hazards*

(1992), describes and analyzes over 700 pesticides. Other publications discuss the least toxic methods of dealing with pests in the home, garden, and greenhouse; non-toxic gardening; ways to safely cure and prevent lawn diseases; and the dangers of poisons used to keep lawns green.

The Council's board of directors includes experts and leaders in the fields of **environmental science**, medicine, education, law, and consumer interest, and a board of consulting experts includes scientists from many fields.

[*Lewis G. Regenstein*]

RESOURCES

ORGANIZATIONS

Rachel Carson Council, Inc., 8940 Jones Mill Road, Chevy Chase, MD USA 20815. (301) 652-1877, Fax: (301) 951-7179, Email: rccouncil@aol.com, <http://members.aol.com/rccouncil/ourpage/index.htm>

Radiation exposure

Radiation is defined as the **emission** of energy from an atom in the form of a wave or particle. Such energy is released as electromagnetic radiation or as **radioactivity**. Electromagnetic radiation includes radio waves, infrared waves or heat, visible light, **ultraviolet radiation**, x rays, gamma rays, and cosmic rays. Radioactivity, emitted when an atomic nucleus undergoes decay, usually takes the form of a particle such as an **alpha particle** or **beta particle**, though atomic decay can also release electromagnetic gamma rays.

While radiation in the form of heat, visible light, and even ultraviolet light is essential to life, the word "radiation" is often used to refer only to those emissions which can damage or kill living things. Such harm is specifically attributed to radioactive particles as well as the electromagnetic rays with frequencies higher than visible light (ultraviolet, x rays, gamma rays). Harmful electromagnetic radiation is also known as **ionizing radiation** because it strips atoms of their electrons, leaving highly reactive ions called *free radicals* which can damage tissue or genetic material.

Effects of radiation

The effects of radiation depend upon the type of radiation absorbed, the amount or dose received, and the part of the body irradiated. While alpha and beta particles have only limited power to penetrate the body, gamma rays and x rays are far more potent. The damage potential of a radiation dose is expressed in *rems*, a quantity equal to the actual dose in *rads* (units per kg) multiplied by a quality factor, called *Q*, representing the potency of the radiation in living tissue. Over a lifetime, a person typically receives 7–14 rems from natural sources. Exposure to 5–75 rems causes few observable symptoms. Exposure to 75–200 rems leads to vomiting, fatigue, and loss of appetite. Exposure to 300 rems or more leads to severe changes in blood cells accompanied by hemorrhage. Such a dosage delivered to the whole body is lethal 50% of the time. An exposure of more than 600 rems causes loss of hair, loss of the body's ability to fight infection, and results in death. A dose of 10,000 rem will kill quickly through damage to the central nervous system.

The symptoms that follow exposure to a sufficient dose of radiation are often termed "radiation sickness" or "radiation burn." Bone marrow and lymphoid tissue cells, testes and ovaries, and embryonic tissue are most sensitive to radiation exposure. Since the lymphatic tissue manufactures white blood cells (WBCs), **radiation sickness** is almost always accompanied by a reduction in WBC production within 72 hours, and recovery from a radiation dose is first indicated by an increase in WBC production.

Any exposure to radiation increases the risk of **cancer**, **birth defects**, and genetic damage, as well as accelerating the aging process, and causing other health problems including impaired immunity. Among the chronic diseases suffered by those exposed to radiation are cancer, stroke, diabetes, hypertension, and cardiovascular and renal disease.

Sources of radiation

Some 82% of the average American's radiation exposure comes from natural sources. These sources include **radon** gas emissions from underground, cosmic rays from space, naturally occurring radioactive elements within our own bodies, and radioactive particles emitted from **soil** and rocks. Man-made radiation, the other 18%, comes primarily from medical x rays and nuclear medicine, but is also emitted from some consumer products (such as **smoke** detectors and blue topaz jewelry), or originates in the production and testing of **nuclear weapons** and the manufacture of nuclear fuels.

Environmental scientists believe that radon, a radioactive gas, accounts for most of the radiation dose Americans receive. Released by the decay of **uranium** in the earth, radon can infiltrate a house through pores in block walls, cracks in basement walls or floors, or around pipes. The **Environmental Protection Agency** estimates that eight million homes in the United States have potentially dangerous levels of radon, and calls radon "the largest environmental radiation health problem affecting Americans." Inhaled radon may contribute to 20,000 lung cancer deaths each year in the United States. The EPA now recommends that homeowners test their houses for radon gas and install a specialized ventilation system if excessive levels of gas are detected.

Though artificial sources of radiation contribute only a small fraction to overall radiation exposure, they remain a strong concern for two reasons. First, they are preventable or avoidable, unlike cosmic radiation, for example. Second, while the *average* individual may not receive a significant

dose of radiation from artificial sources, geographic and occupational factors may mean dramatically higher doses of radiation for large numbers of people. For instance, many Americans have been exposed to radiation from nearly 600 nuclear tests conducted at the **Nevada Test Site**. From the early 1950s to the early 1960s, atmospheric blasts caused a lingering increase in radiation-related sickness downwind of the site, and increased the overall dose of radiation received by Americans by as much as 7%. Once the tests were moved underground, that figure fell to less than 1%.

A February 1990 study of the Windscale **plutonium** processing plant in Britain clearly demonstrated the importance of the indirect effects of radiation exposure. The study correlated an abnormally high rate of **leukemia** among children in the area with male workers at the plant, who evidently passed a tendency to leukemia to their children even though they had been receiving radiation doses that were considered "acceptable."

Recently, the EPA published their National Radon Results listing studies from 1984 to 1999. This study showed that 18 million homes have been tested for radon in the United States. In 1999 alone, there were about 1.4 million homes tested for radon. The knowledge of radon's harmful effects is on the rise (88% of Americans were aware of the harmful effects in 1999 compared to 73% in 1996) which is apparently contributing to the increase in testing.

[*Linda Rehkopf and Jeffrey Muhr*]

RESOURCES

BOOKS

Caufield, C. *Multiple Exposures.* Harper & Row, Publishers, New York, 1989.

Wagner Jr., H. N., and L. E. Ketchum. *Living With Radiation: The Risk, the Promise.* Baltimore: Johns Hopkins University Press, 1989.

PERIODICALS

Cobb Jr., C. E. "Living With Radiation." *National Geographic* 175 (April 1989): 403–437.

OTHER

Environmental Protection Agency. *National Radon Results.* June 2002 [cited July 2002]. <http://www.epa.gov/air/oarnew3.html>.

Radiation Sources

Natural Sources	
Radon gas	55%
Inside body	11%
Rocks, soil, and groundwater	8%
Cosmic rays	8%
Artificial sources	
Medical x rays	11%
Nuclear medicine	4%
Consumer products	3%
Miscellaneous*	<1%

* includes occupation exposure, nuclear fallout, and the production of nuclear materials for nuclear power and weapons

Radiation sickness

Radiation sickness, also known as acute radiation syndrome or **ionizing radiation** injury, is illness resulting from human exposure to ionizing radiation. The **radiation exposure** may be from natural sources, such as radium, or from man-made sources, such as x rays, nuclear reactors, or atomic bombs. Ionizing radiation penetrates cells of the body and ultimately causes damage to critically important molecules such as nucleic acids and enzymes. Immediate cell death may occur if the dose of ionizing radiation is sufficiently high; lower doses result in cell injury which may preclude cell replication. Tissues at greatest risk for radiation injury are those which have cells that are rapidly dividing. Blood forming cells, the lining to the gastrointestinal tract, skin, and hair forming cells are particularly vulnerable. Muscle, brain, liver, and other tissues, which have a low rate of cell division, are less so.

Epidemiological data on radiation sickness has been accumulated from the study of individual cases, as well as from the study of large numbers of afflicted individuals, such as the survivors of the atomic bombing of Hiroshima and Nagasaki and the 135,000 people evacuated from close proximity to the fire at the **Chernobyl Nuclear Power Station** in 1986.

In those who survive, radiation sickness may be characterized by four phases. The initial stage occurs immediately after exposure and is characterized by nausea, vomiting, weakness, and diarrhea. This is a period of short duration, typically one or two days. It is followed by a period of apparent recovery lasting for one to three weeks. No particular symptoms appear during this time. The third stage is characterized by fever, infection, vomiting, lesions in the mouth and pharynx, abscesses, bloody diarrhea, hemorrhages, weight loss, hair loss, bleeding ulcers, and petechiae, small hemorrhagic spots on the skin. During this phase, there is a loss of appetite, nausea, weakness, and weight loss. These symptoms are due to the depletion of cells which would normally be rapidly dividing. Bone marrow depression occurs with reduced numbers of white blood cells, red blood cells, and blood platelets. Gut cells that are normally lost

are not replaced and hair follicle cells are depressed resulting in hair loss. If death does not occur during this phase, a slow recovery follows. This is the fourth phase, but recovery is frequently accompanied by long-lasting or permanent disabilities including widespread scar tissue, cataracts, and blindness.

The toxic side effects of many **cancer** chemotherapeutic agents are similar to acute radiation sickness, because both radiation and chemotherapy primarily affect rapidly dividing cells. Little effective treatment can be administered in the event of casualties occurring in enormous numbers such as in war or a major **nuclear power** plant disaster. However, in individual cases such as in accidental laboratory or industrial exposure, some may be helped by isolation placement to prevent infection, transfusions for hemorrhage, and bone marrow transplantation. *See also* Epidemiology; Nuclear weapons; Nuclear winter; Radiation exposure; Radioactive pollution; Threshold dose

[*Robert G. McKinnell*]

Resources
Books

Guskova, A. K., et al. "Acute Effects of Radiation Exposure Following the Chernobyl Accident." In *Treatment of Radiation Injuries*, edited by D. Browne, et al. New York: Plenum Press, 1990.

Wald, N. "Radiation Injury." In *Cecil Textbook of Medicine*, edited by P. B. Beeson, et al. 15th ed. Philadelphia: Saunders, 1979.

Radioactive decay

Radioactive decay is the process by which an atomic nucleus undergoes a spontaneous change, emitting an **alpha particle** or **beta particle** and/or a **gamma ray**. Radioactive decay is a natural process that takes place in the air, water, and **soil** at all times. The decay of isotopes such as uranium-238, radium-226, radon-222, potassium-40, and carbon-14 produce radiation that poses an unavoidable and, probably, minimal hazard to human health. Scientists have also learned how to convert stable isotopes to radioactive forms. The radioactive decay of these isotopes has been added to the natural **background radiation** from naturally radioactive materials. *See also* Carbon; Radioactive fallout; Radioactivity

Radioactive fallout

In the 1930s, scientists found that bombarding **uranium** metal with neutrons caused the nuclei of uranium atoms to break apart, or fission. One significant feature of this reaction was that very large amounts of energy were released during **nuclear fission**. The first practical application of this discovery was the atomic bomb, developed by scientists working in the United States in the early 1940s.

The atomic bomb takes advantage of the energy released during fission to bring about massive destruction of property and human life. However, every bomb blast is also accompanied by another event known as radioactive fallout. The term radioactive fallout refers to all radioactive dust and particles that fall to the earth after a nuclear explosion. This combination of dust and particles contains hundreds of isotopes formed when a uranium nucleus fissions. Iodine-131 and yttrium-98 are only two of the many isotopes that are formed during an atomic bomb blast.

When these isotopes are formed, they come together as only very small particles. The force of the blast assures that large particles will not survive. In this form, the radioactive dust and particles may remain suspended in air for days, weeks, or months. Only when they have consolidated to form larger particles will they fall back to Earth.

Once they have reached the earth, these particles face different fates. Some isotopes formed during fission have very short half-lives. They will decay rapidly and pose little or no environmental threat. Others have longer half-lives and may remain in the **environment** for many years.

The components of radioactive fallout that cause the greatest concern are those that can take some role in plant or animal **metabolism**. For example, the element strontium is chemically similar to the element calcium. Strontium can replace calcium in many biochemical reactions. That explains why the following scenario is so troubling.

Strontium-90 released during a fission bomb blast falls to the earth, coats grass, and is eaten by cows. The cows incorporate strontium-90 into their milk, just as they do calcium. When growing children drink that milk, the strontium-90 is used to build bones and teeth, just as calcium is. Once incorporated into bones and teeth, however, the radioactive strontium continues to emit harmful radiation for many years.

The dangers of radioactive fallout are one of the major reasons that the United States and the former Soviet Union were able to agree in 1963 on a limited ban of **nuclear weapons** testing. In that agreement, both nations promised to stop nuclear weapons testing in the **atmosphere**, under water, and in outer space. *See also* Half-life; Radioactive decay; Radioactivity

[*David E. Newton*]

Resources
Books

Inglis, D. R. *Nuclear Energy: Its Physics and Social Challenge.* Reading, MA: Addison-Wesley, 1973.

Jagger, J. *The Nuclear Lion: What Every Citizen Should Know About Nuclear Power and Nuclear War.* New York: Plenum, 1991.

Radioactive pollution

Radioactive **pollution** can be defined as the release of radioactive substances or high-energy particles into the air, water, or earth as a result of human activity, either by accident or by design. The sources of such waste include: 1) nuclear weapon testing or detonation; 2) the *nuclear fuel cycle*, including the mining, separation, and production of nuclear materials for use in nuclear **power plants** or nuclear bombs; 3) accidental release of radioactive material from **nuclear power** plants. Sometimes natural sources of **radioactivity**, such as **radon** gas emitted from beneath the ground, are considered pollutants when they become a threat to human health.

Since even a small amount of **radiation exposure** can have serious (and cumulative) biological consequences, and since many radioactive wastes remain toxic for centuries, radioactive pollution is a serious environmental concern even though natural sources of radioactivity far exceed artificial ones at present.

The problem of radioactive pollution is compounded by the difficulty in assessing its effects. **Radioactive waste** may spread over a broad area quite rapidly and irregularly (from an abandoned dump into an **aquifer**, for example), and may not fully show its effects upon humans and organisms for decades in the form of **cancer** or other chronic diseases. For instance, radioactive iodine-131, while short-lived, may leave those who ingest it with long-term health problems. By the time a radioactive leak is revealed and a health survey is carried out, many of the individuals affected by the leak may have already moved from the area (or died without having been examined with the possible cause in mind). In addition, since most radioactive materials are under the jurisdiction of governmental agencies—usually in the secretive defense and military establishments—the dumping activities and accidental discharges that accompany nuclear materials production and bomb testing tend to remain concealed until governments are obligated to disclose them under public pressure.

Atmospheric pollution

Radioactive pollution that is spread through the earth's **atmosphere** is termed *fallout.* Such pollution was most common in the two decades following World War II, when the United States, the Soviet Union, and Great Britain conducted hundreds of **nuclear weapons** tests in the atmosphere. France and China did not begin testing nuclear weapons until the 1960s and continued atmospheric testing even after other nations had agreed to move their tests underground.

Three types of fallout result from nuclear detonations: local, tropospheric, and stratospheric. Local fallout is quite intense but short-lived. Tropospheric fallout (in the lower atmosphere) is deposited at a later time and covers a larger area, depending on meteorological conditions. Stratospheric fallout, which releases extremely fine particles into the upper atmosphere, may continue for years after an explosion and attain a worldwide distribution.

The two best known examples illustrating the effect of fallout contamination are the bombing of Hiroshima and **Nagasaki, Japan** in 1945, and the **Chernobyl Nuclear Power Station** disaster in April 1986. Within five years of the American bombing of Japan, as many as 225,000 people had died as a result of long-term exposure to radiation from the bomb blast, chiefly in the form of fallout.

The disaster at the Chernobyl Nuclear Power Station in Ukraine on April 26, 1986 produced a staggering release of radioactivity. In 10 days at least 36 million curies spewed across the world. The fallout contaminated approximately 1,000 square mi (2,590 sq km) of farmland and villages in the Soviet Union. Dangerous levels of radioactivity were reported in virtually every European country, and radioactive pollutants contaminated rainwater, pastures and food crops. Radiation alerts were posted in almost every country, children were kept indoors, and sales of milk, vegetables, and meat were banned in some areas. In the Soviet Union alone, 135,000 people were evacuated from their homes. In addition to the hundreds killed at the time of the explosion, scientists predict the eventual Soviet death toll from the Chernobyl accident may reach 200,000; the estimated **mortality** in western Europe may approach 40,000.

Pollution on land and in water

The major sources of radioactive pollution on land and water include: 1) the nuclear fuel cycle—the extraction, separation and refinement of materials for use in nuclear weapons and nuclear power—and 2) the day-to-day operations of nuclear power plants.

At every stage in the production of nuclear fuels, contaminants are left behind. The mining of **uranium**, for example, produces highly radioactive *tailings* which can be blown into the air, contaminate **soil**, or leach into bodies of water. The magnitude of radioactive pollution caused by the nuclear fuel cycle, especially in the United States, Great Britain, and the Soviet Union, has only recently been revealed. Through the years of the cold war, the extent of accidental discharges and intentional dumping carried out by government plants like Britain's Windscale and the United States' **Hanford Nuclear Reservation** remained largely unknown. Not until the late 1980s, for instance, was the legacy of flagrant pollution at Hanford exposed to public scrutiny. This legacy included the release of half a million curies of radioactive iodine into the air from 1944 to 1955, and the

release of millions of curies of radioactive material into the Columbia River, Washington from 1944 through at least the 1960s. In 1956, 450,000 gal (1.7 million l) of high-level waste were accidently spilled on the Hanford grounds. Through the 1950s, additional millions of gallons of waste were dumped into the ground. Residents in the area and along the river were never notified about any of the radioactive discharges that took place at Hanford.

In 1988, the **U.S. Department of Energy** reported that radioactive wastes from Hanford were contaminating some underground water supplies; Hanford was shut down a year later. At present, a farm of deteriorating storage tanks at Hanford containing approximately 57 million gal (216 million l) of radioactive and toxic waste is being monitored for leaks. Officials are not even sure what **chemicals** lie in some of these tanks. Federal officials estimate that cleaning up Hanford and the U.S. government's other weapons facilities (at Fernald, Ohio; Rocky Flats; Colorado; Savannah River, Georgia; and other locations) may cost over $200 billion.

Nuclear power plants also contribute to radioactive pollution. Spent nuclear fuel from these plants, a *high-level* waste, must be kept from human contact for hundreds or thousands of years, yet no completely reliable disposal method exists. At present, most high-level waste has simply been left in pools at power plant sites while the government seeks a location for permanent disposal. Most *low-level* waste (anything that is not spent fuel or transuranic waste) generated by nuclear power plants has been landfilled throughout the country. Three such **landfill** sites (Maxey Flats, Kentucky; West Valley, New York; and Sheffield, Illinois) have been shut down due to leakage of radioactive liquid into the ground.

Nuclear power plants also **discharge** wastes directly, as a result of malfunctions or intentional dumping. During the decade from 1979 to 1989, the **Nuclear Regulatory Commission** recorded 33,000 mishaps at power plants in the United States, 1,000 of which it classified as "particularly significant." Though the vast majority of these incidents resulted in no release of radioactive material, some accidents, such as the Three Mile Island partial meltdown and the Chernobyl explosion, have caused significant discharges of radioactivity into the **environment**. A sampling of incidents:

- the Vermont Yankee power plant, a power station with a poor safety record, was fined $30,000 for dumping 83,000 gal (314,000 l) of radioactive water into the Connecticut River in July 1976;
- the accident at Three Mile Island, Pennsylvania in 1979 released 2.5 million curies of radioactive noble gases and a small quantity of radioactive iodine into the atmosphere;
- in one of a series of 21 unreported leaks occurring up to March 1981, radioactive material from the Tsuruga nuclear power plant in Japan was dumped into Tsuruga Bay after warning alarms were shut off; operators of the plant later admitted that they intentionally dumped wastes regularly;
- in October 1982, heavy rains caused a flood at the Cofrentes nuclear power plant in Spain which released 154 gal (580 l) of radioactive waste;
- in December 1985, a power failure at the Rancho Seco, California plant resulted in a release of a small quantity of radiation into the air; in 1989 the NRC fined the facility $100,000 for violating waste disposal regulations from 1983 to 1986;
- at the Douglas Point, Canada, station in January 1986, it was discovered that **high-level radioactive waste** had been leaking from a spent-fuel "storage pond" at a rate of 16 gal (60 l) per hour.

The health effects of radioactive leaks are still debated. While the harmful effects of Chernobyl are probably beyond question, analysis of the Three Mile Island incident has not detected a long-term increase in cancer or diseases as a result of the discharge there. Studies conducted in England, however, reveal increased rates of **leukemia** in areas surrounding the nuclear plants at Hinkley Point, Dounreay, and Windscale.

[*Linda Rehkopf and Jeffrey Muhr*]

Resources

Books

Brill, A. Bertrand. *Low-Level Radiation Effects: A Fact Book.* New York: Society of Nuclear Medicine, 1985.

Caufield, Catherine. *Multiple Exposures.* New York: Harper & Row, Publishers, 1989.

Dresser, Peter D., ed. *Nuclear Power Plants Worldwide.* Detroit: Gale Research, 1993.

Gore, Al. *Earth in the Balance.* New York: Houghton Mifflin, 1992.

Jagger, J. *The Nuclear Lion: What Every Citizen Should Know About Nuclear Power and Nuclear War.* New York: Plenum, 1991.

Jones, R. R., and R. Southwood, eds. *Radiation and Health.* New York: Wiley, 1987.

Regenstein, Lewis. *How to Survive in America the Poisoned.* Washington, DC: Acropolis Books Ltd., 1986.

Shulman, S. *The Threat at Home: Confronting the Toxic Legacy of the U.S. Military.* Boston: Beacon Press, 1992.

Periodicals

Goldsmith, E., et al. "Chernobyl: The End of Nuclear Power?" *The Economist* 16 (1986): 138–209.

Radioactive waste

Radioactive waste is the "garbage" left as a result of the use of nuclear materials by human societies. Such waste can be categorized as low-level, intermediate-level, or high-level waste. The term *transuranic waste* is also used to describe

materials consisting of elements heavier than **uranium** in the periodic table.

The term **low-level radioactive waste** usually refers to materials that contain a small amount of **radioactivity** dispersed in a large volume of material. Such materials are produced in a great variety of industrial, medical, and research procedures. A common practice is to store these materials in sealed containers until their level of radioactivity is very low and then to dispose of them by shallow burial or in other traditional **solid waste** disposal systems.

The assumption is that the level of radiation released by these wastes is too low to cause any harmful environmental effects. That assumption has been challenged by some scientists who believe that enough is not yet known about the long-term effects of radiation. They suggest that safer methods of disposal for such wastes need to be developed.

Intermediate-level wastes, as the name suggests, contain a higher level of radioactivity than low-level wastes, but a lower level than high-level wastes. These materials cannot be discharged directly into the **environment**. An important source of such wastes is the re-processing of nuclear fuels. At one time, large quantities of intermediate-level wastes were dumped into the deepest parts of the Atlantic Ocean. That practice has been discontinued and intermediate-level wastes are now being stored on land until a permanent disposal system is developed.

High-level radioactive wastes consist of materials that contain a large amount of radioactivity that will remain at dangerous levels for hundreds or even thousands of years. These materials pose the most difficult disposal problem of all since they must be completely isolated and stored for very long periods of time. The primary sources of high-level wastes are nuclear **power plants** and research and development of **nuclear weapons**.

A number of methods for the storage of high-level wastes have been suggested. Among these are burial in large chunks of concrete, encapsulation in glass or ceramic, projection of them inside rockets into outer space, and burial in the Antarctic ice sheet. Various countries around the world have developed a variety of methods for storing their high-level wastes. In Canada, such wastes have been stored in water-filled pools for more than 25 years. France, with one of the world's largest **nuclear power** establishments, has developed no permanent storage system but plans to build a large underground vault for its wastes by the early twenty-first century.

In the United States, Congress passed the Nuclear Waste Policy Act in 1982, outlining a complete program for the construction of a high-level waste repository in the early twenty-first century. In 1987, **Yucca Mountain**, Nevada, was selected as the location for that site. Current plans call for a huge vault 1,000 ft (305 m) underground as the site for long-term, high-level waste storage at this location.

See also Nuclear fission; Nuclear Regulatory Commission (NRC); Ocean dumping; Office of Civilian Radioactive Waste Management (OCRWM); Radioactive decay; Radioactive waste management

[*David E. Newton*]

Resources

Books

Bartlett, D. L., and J. B. Steele. *Forevermore: Nuclear Waste in America.* New York: W. W. Norton, 1985.

Carter, L. J. *Nuclear Imperatives and Public Trust: Dealing With Radioactive Waste.* Baltimore: Resources for the Future, 1987.

League of Women Voters. *The Nuclear Waste Primer.* Washington, DC: League of Women Voters, 1985.

Managing the Nation's Nuclear Waste. Washington, DC: Office of Civilian Radioactive Waste Management, March 1990.

Resnikoff, M. *Deadly Defense: Military Radioactive Landfills.* New York: Radioactive Waste Campaign, 1988.

Periodicals

Hunt, C. B. "Disposal of Radioactive Wastes." *Bulletin of the Atomic Scientists* (April 1984): 44–46.

Radioactive waste management

Radioactive waste materials are produced as by-products of research, **nuclear power** generation, and **nuclear weapons** manufacture. Radioactive waste is classified by the U.S. government into five groups: high-level, transuranic (chemical elements heavier than **uranium**), spent fuel, uranium mill **tailings**, and **low-level radioactive waste**.

The management and disposal of radioactive waste receives attention at all levels of state and local governments, but the regulations sometimes conflict or are confusing. The U.S. Congress passes relevant legislation, the **Environmental Protection Agency** (EPA) sets applicable environmental standards, and the **Nuclear Regulatory Commission** (NRC) develops regulations to implement the standards. For **high-level radioactive waste**, the **U.S. Department of Energy** (DOE) is responsible for the design, construction and operation of suitable disposal facilities. And courts have recently gotten into the act, ordering states to arrange to have necessary disposal facilities designed, constructed and operated for management of low-level wastes.

The principal federal laws related to the management and disposal of radioactive waste include the Atomic Energy Act (1954), the Uranium Mill Tailings Radiation Control Act (1978), the Low-Level Radioactive Waste Policy Act (1980), the Nuclear Waste Policy Act (1982), the Low-Level Radioactive Waste Policy Amendments Act (1985), and the Nuclear Waste Policy Amendments Act (1987). In

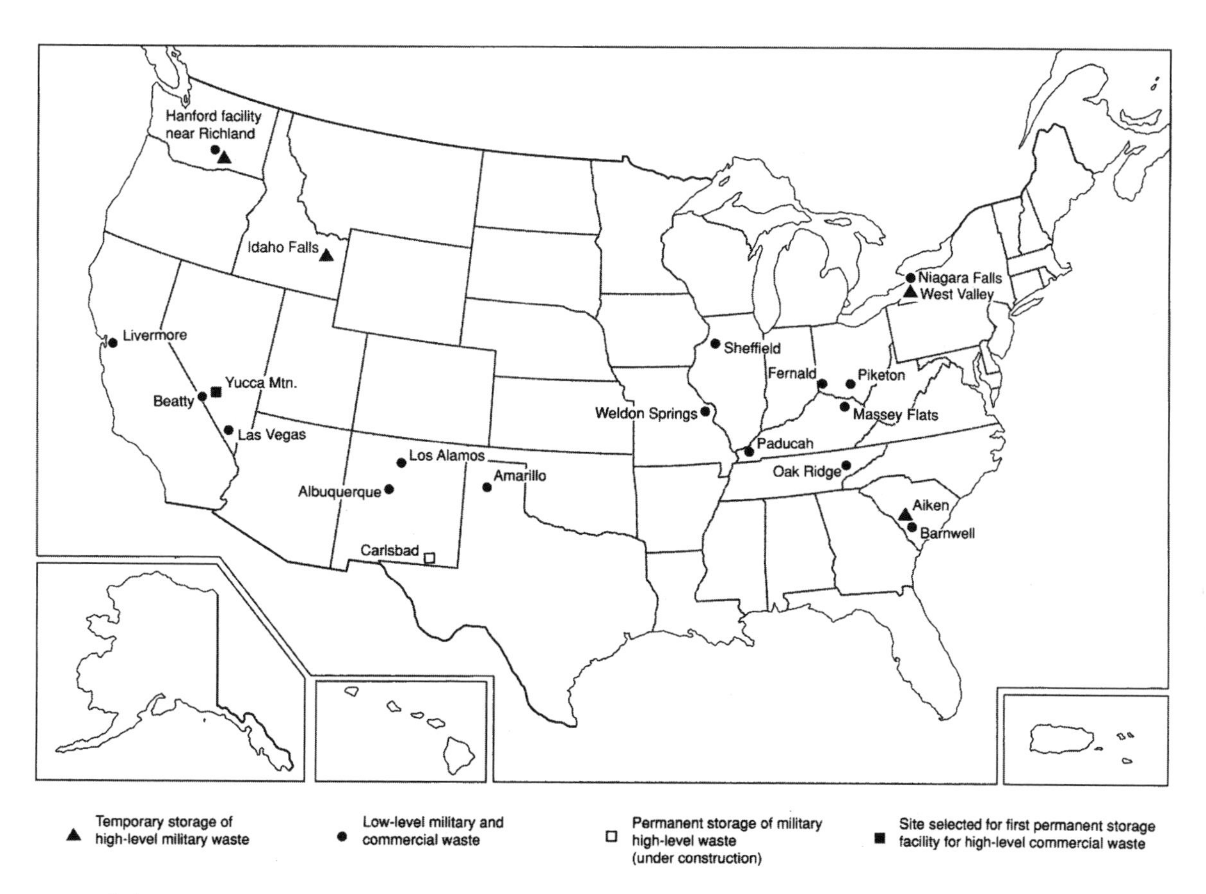

A map of the United States showing radioactive waste disposal site locations. (McGraw-Hill Inc. Reproduced by permission.)

addition, the Federal Facility Compliance Act (1992) forces the military to clean up its waste sites.

One of the more difficult aspects of the regulatory quagmire is the problem of dealing with waste, especially spent nuclear fuel and contaminated material such as worn-out reactor parts. Today, the official policy objective for nuclear waste is to dispose of it so that it will never do any appreciable damage to anyone, under any circumstances, for all time. However, for some radioactive waste, "for all time" is measured in thousands of years.

Although radioactive waste carries some hazard for many years, **radioactive decay** removes most of the hazard after a few hundred years. A well-designed waste storage system can be made safe for a long time, some scientists and policy-makers insist, provided that engineers ensure that **erosion**, **groundwater**, earthquakes, and other unpredictable natural or human activities do not breach safety barriers.

Disposal methods for radioactive wastes have varied. Over the past 50 years, low-level wastes have been flushed down drains, dumped into the ocean, and tossed into landfills. Uranium mill tailings have been mounded into small hills at sites throughout the western United States. Storage tanks and barrels at DOE sites hold millions of gallons of radioactive waste and toxic **chemicals**, the by-products of **plutonium** production for nuclear weapons.

Over 80% of the total volume of radioactive waste generated in the United States is considered low-level. There is now a shift away from the most common disposal method of shallow land burial. Because of public demand for disposal methods that provide the greatest safety and security, disposal methods now include above- and below-ground vaults and earth-mounded concrete bunkers.

However, the country's 17,000 laboratories, hospitals, and nuclear **power plants** that produce low-level radioactive byproducts could become *de facto* disposal sites. The federal low-level radioactive waste policy enacted in 1980 was designed to remedy the inequity of shipping all the nation's low-level radioactive waste to three states (Illinois, Kentucky, and New York) which never agreed to serve as the nation's sole disposal facilities. Its intent was to require every state to either build its own low-level repository or compact with other states to build regional facilities by 1986. The 1980

federal act was amended in 1985, when it became clear that the 1986 target would not be met, and the deadline was postponed to 1993. As of late 1992, however, few states had even determined sites for such facilities. And after 1996, the institutions generating low-level wastes will be held liable for all low-level wastes they produce.

The principal sources of high-level radioactive waste are nuclear power plants, and programs of the U. S. Department of Defense and DOE, especially those dealing with nuclear weapons. These wastes include spent fuel removed from nuclear plants, and fission products separated from military fuel that has been chemically processed to reclaim unused uranium and plutonium. In the United States, commercial nuclear power plant fuel removed from those facilities is stored on-site until a geologic repository is completed. The expected completion date for the repository is 2010.

Typically, waste fuel is stored at the power station in a pressurized **buffer** storage tube for 28 days, then the fuel rod is broken down to component parts. Irradiated fuel elements then are stored for 80 days in a water-filled pond to allow for further decay. For the nonreusable radioactive waste that remains, about 2.5%, there is no totally safe method for disposal.

Radioactive waste from one of the first nuclear bomb plants is so volatile that the clean-up problems have stymied the experts. The plant at the DOE's **Hanford Nuclear Reservation** near Richland, Washington, is the nation's largest repository of nuclear waste. There, 177 tanks contain more than 57 million gal (26 million l) of radioactive waste and toxic chemicals, byproducts of plutonium production for nuclear weapons. The DOE and Westinghouse Corporation (the contractor in charge of Hanford's clean-up) are still not sure exactly what mixture of chemicals and radioactive waste each tank contains, but a video taken inside one tank shows the liquid bubbling and roiling from chemical and nuclear reactions. Corrosive, highly radioactive liquids have eaten through Hanford's storage tanks and are being removed to computer-monitored, carbon-steel storage tanks. The clean-up at Hanford could cost as much as $57 billion. The only other country known to have had a waste problem as large as the one at Hanford is the former Soviet Union, where a nuclear waste dump exploded in the nuclear complex at Chelyabinsk in 1957, contaminating thousands of square miles of land.

As the military starts massive clean-up efforts, it has identified 17 nuclear facilities as among the worst of the more than 20,000 suspected toxic sites. For example, since the 1970s, about 77,000 barrels of low-level radioactive wastes have been stored at the **Oak Ridge, Tennessee** nuclear reservation. Today, some barrels are rusting and leaking **radioactivity**. Other identified radioactive waste sites include California's Lawrence Livermore National Laboratory and Sandia National Laboratory, Colorado's **Rocky Flats nuclear plant**, and New Mexico's Los Alamos National Laboratory. The Savannah River nuclear plant in South Carolina has been called especially dangerous. Both the **Savannah River site** and the Hanford site pose the risk of the kind of massive nuclear waste explosion that occurred at Chelyabinsk.

One new technology in development for the treatment of hazardous and radioactive waste is *in situ vitrification* (ISV), which utilizes electricity to melt contaminated material in place. This technology was developed primarily to treat radioactive waste, but it also has applications to hazardous chemical wastes. The end-result glass formed by ISV is unaffected by extremes of temperature, is not biotoxic, should remain relatively stable for one million years, and passes government tests that measure the speed of contaminant **leaching** over time. The process permits treatment of mixtures of various wastes including organic, inorganic, and radioactive. The fact that it is not necessary to excavate the waste prior to treatment is seen as a major advantage of the technology, since excavation along with **transportation** increases health risks. The technology is currently being used in the United States at Superfund sites.

Incineration is also used for disposal of low-level radioactive waste, but incomplete **combustion** can produce dioxins and other toxic ash and aerosols. Although finding a site for ash disposal is difficult, the ash is considered a better form for burial than the original waste. It is biologically and structurally more stable, and many of the compounds it contains are insoluble. The residual waste produced by incineration also is less susceptible to leaching by rain and groundwater.

Although progress is slow and the situation is often chaotic in the United States, the state of radioactive **waste management** in much of the rest of the world is even less advanced. The situation is particularly acute in Eastern Europe and the former Soviet Union, since these countries have generated huge quantities of radioactive waste. In many cases the information-gathering necessary to plan clean-up efforts has not begun or is barely underway, and most, if not all, of the actual clean-up remains. *See also* Eastern European pollution; Ocean dumping; Radiation exposure; Radiation sickness; Radioactive pollution; Waste Isolation Pilot Plant; Yucca Mountain, Nevada

[*Linda Rehkopf*]

Resources

Books

Deadly Defense: Military Radioactive Landfills. New York: Radioactive Waste Campaign, 1988.

Moeller, D. W. *Environmental Health.* Cambridge: Harvard University Press, 1992.

Shulman, S. *The Threat at Home: Confronting the Toxic Legacy of the U.S. Military.* Boston: Beacon Press, 1992.

The Nuclear Waste Primer. Washington, DC: League of Women Voters, 1985.

PERIODICALS

Hammond, R. P. "Nuclear Wastes and Public Acceptance." *American Scientist* 67 (March–April 1979): 146–50.

Shulman, S. "Operation Restore Earth: Cleaning Up After the Cold War." *E Magazine* 4 (March–April 1993): 36–43.

Radioactivity

In 1896, the French physicist Henri Becquerel accidentally found that an ore of **uranium**, pitchblende, emits an invisible form of radiation, somewhat similar to light. The phenomenon was soon given the name *radioactivity* and materials like pitchblende were called *radioactive.*

The radiation Becquerel discovered actually consists of three distinct parts, called alpha, beta, and gamma rays. Alpha and beta rays are made up of rapidly moving particles—helium nuclei in the case of alpha rays, and electrons in the case of beta rays. Gamma rays are a form of electromagnetic radiation with very short wavelengths.

Alpha rays have relatively low energies and can be stopped by a thin sheet of paper. They are not able to penetrate the human skin and, in most circumstances, pose a relatively low health risk. Beta rays are more energetic, penetrating a short distance into human tissue, but they can be stopped by a thin sheet of **aluminum**. Gamma rays are by far the most penetrating form of radiation, permeating wood, paper, plastic, tissue, water, and other low-density materials in the **environment**. They can be stopped, however, by sheets of **lead** a few inches thick.

Radioactivity is a normal and ubiquitous part of the environment. The most important sources of natural radioactivity are rocks containing radioactive isotopes of uranium, thorium, potassium, and other elements. The most common radioactive **isotope** in air is carbon-14, formed when neutrons from cosmic ray showers react with **nitrogen** in the **atmosphere**. Humans, other animals, and plants are constantly exposed to low-level radiation emitted from these isotopes, and they do suffer to some extent from that exposure. A certain number of human health problems—cancer and genetic disorders, for example—are attributed to damage caused by natural radioactivity.

In recent years, scientists have been investigating the special health problems related to one naturally occurring radioactive isotope, radon-226. This isotope is produced when uranium decays, and since uranium occurs widely in rocks, radon-226 is also a common constituent of the environment. Radon-226 is an alpha-emitter, and though the isotope does have a long **half-life** (1,620 years), the alpha particles are not energetic enough to penetrate the skin. The substance, however, is a health risk because it is a gas that can be directly inhaled. The alpha particles come into contact with lung tissue, and some scientists now believe that radon-226 may be responsible for a certain number of cases of lung **cancer**. The isotope can be a problem when homes are constructed on land containing an unusually high concentration of uranium. Radon-226 released by the uranium can escape into the basements of homes, spreading to the rest of a house. Studies by the **Environmental Protection Agency** (EPA) have found that as many as eight million houses in the United States have levels of radon-226 that exceed the **maximum permissible concentration** recommended by experts.

Though Becquerel had discovered radiation occurring naturally in the environment, scientists immediately began asking themselves whether it was possible to convert normally stable isotopes into radioactive forms. This question became the subject of intense investigation in the 1920s and 1930s, and was finally answered in 1934 when Irène Curie and Frèdèric Joliot bombarded the stable isotope aluminum-27 with alpha particles and produced phosphorus-30, a radioactive isotope. Since the Joliot-Curie experiment, scientists have found ways to manufacture hundreds of artificially radioactive isotopes. One of the most common methods is to bombard a stable isotope with gamma rays. In many cases, the product of this reaction is a radioactive isotope of the same element.

Highly specialized techniques have recently been devised to meet specific needs. Medical workers often use radioactive isotopes with short half-lives because they can be used for diagnostic purposes without remaining in a patient's body for long periods of time. But the isotope cannot have such a short half-life that it will all but totally decay between its point of manufacture and its point of use.

One solution to this problem is the so-called "molybdenum cow." The cow is no more than a shielded container of radioactive molybdenum-99. This isotope decays with a long half-life to produce technetium-99, whose half-life is only six hours. When medical workers require technetium-99 for some diagnostic procedure, they simply "milk" the molybdenum cow to get the short-lived isotope they need.

Artificially radioactive isotopes have been widely employed in industry, research, and medicine. Their value lies in the fact that the radiation they emit allows them to be tracked through settings in which they cannot be otherwise observed. For example, a physician might want to know if a patient's thyroid is functioning normally. In such a case, the patient drinks a solution containing radioactive iodine, which concentrates in the thyroid like stable iodine. The isotope's movement through the body can be detected by a Geiger counter or some other detecting device, and the speed as well as the extent to which the isotope is taken up by the thyroid is an indication of how the organ is functioning.

Artificially radioactive isotopes can pose a hazard to the environment. The materials in which they are wrapped, the tools with which they are handled, and the clothing worn by workers may all be contaminated by the isotopes. Even after they have been used and discarded, they may continue to be radioactive. Users must find ways of disposing of these wastes without allowing the release of dangerous radiation into the environment, a relatively manageable problem. Most materials discarded by industry, medical facilities, and researchers are **low-level radioactive waste**. The amount of radiation released decreases quite rapidly, and after isolation for just a few years, the materials can be disposed of safely with other non-radioactive wastes.

The same cannot be said for the **high-level radioactive waste** produced by nuclear **power plants** and defense research and production. Consisting of radioactive isotopes, such wastes are produced during fission reactions and release dangerously large amounts of radiation for hundreds or thousands of years.

Nuclear fission was discovered accidentally in the 1930s by scientists who were trying to produce artificial radioactive isotopes. In a number of cases, they found that the reactions they used did not result in the formation of new radioactive isotopes, but in the splitting of atomic nuclei, a process that came to be known as nuclear fission.

By the early 1940s, nuclear fission was recognized as an important new source of energy. That energy source was first put to use for destructive purposes, in the construction of **nuclear weapons**. Later, scientists found ways to control the release of energy from nuclear fission in nuclear reactors.

The most serious environmental problem associated with fission reactions is that their waste products are largely long-lived radioactive isotopes. Attempts have been made to isolate these wastes by burying them underground or sinking them in the ocean. All such methods have proved so far to be unsatisfactory, however, as containers break open and their contents leak into the environment.

The United States government has been working for more than four decades to find better methods for dealing with these wastes. In 1982, Congress passed a *Nuclear Waste Policy Act*, providing for the development of one or more permanent burial sites for high-level wastes. Political and environmental pressures have stalled the implementation of the act and a decade after its passage, the nation still has no method for the safe disposal of its most dangerous radioactive wastes. *See also* Ecotoxicology; Hazardous waste; Nuclear fusion; Nuclear winter; Radiation exposure; Radiation sickness; Radioactive fallout; Radioactive pollution; Radioactive waste management

[*David E. Newton*]

RESOURCES

BOOKS

Gofman, J. W. *Radiation and Human Health.* San Francisco: Sierra Club Books, 1981.

Inglis, D. R. *Nuclear Energy: Its Physics and Social Challenge.* Reading, MA: Addison-Wesley, 1973.

Jones, R. R., and R. Southwood, eds. *Radiation and Health.* New York: Wiley, 1987.

Wagner, H. N., and L. E. Ketchum. *Living With Radiation.* Baltimore: Johns Hopkins University Press, 1989.

Radiocarbon dating

Radiocarbon dating is a technique for determining the age of very old objects consisting of organic (carbon-based) materials, such as wood, paper, cloth, and bone. The technique is based on the fact that both stable and radioactive isotopes of **carbon** exist. These isotopes behave almost identically in biological, chemical, and physical processes.

Carbon-12, a stable **isotope**, makes up about 99% of all carbon found in **nature**. Radioactive carbon-14 is formed in the **atmosphere** when neutrons produced in cosmic ray showers react with **nitrogen** atoms.

Despite the fact that it makes up no more than 0.08% of the earth's crust, carbon is an exceedingly important element. It occurs in all living materials and is found in many important rocks and minerals, including limestone and marble, as well as in **carbon dioxide**. Carbon moves through the atmosphere, hydrosphere, lithosphere, and **biosphere** in a series of reactions known as the **carbon cycle**. Stable and radioactive isotopes of the element take part in identical reactions in the cycle. Thus, when green plants convert carbon dioxide to carbohydrates through the process of **photosynthesis**, they use both stable carbon-12 and radioactive carbon-14 in exactly the same way. Any living material consists, therefore, of a constant ration of carbon-14 to carbon-12.

In the mid-1940s, Willard F. Libby realized that this fact could be used to date organic material. As long as that material was alive, he pointed out, it should continue to take in both carbon-12 and carbon-14 in a constant ratio. At its death, the material would no longer incorporate carbon in any form into its structure. From that point on, the amount of stable carbon-12 would remain constant. The amount of carbon-14, however, would continuously decrease as it decayed by beta **emission** to form nitrogen. Over time, the ratio of carbon-14 to carbon-12 would grow smaller and smaller. That ratio would provide an indication of the length of time since the material had ceased being alive.

Radiocarbon dating has been used to estimate the age of a wide variety of objects ranging from charcoal taken from tombs to wood found in Egyptian and Roman ships. One

of its most famous applications was in the dating of the Shroud of Turin. Some religious leaders had claimed that the Shroud was the burial cloth in which Jesus was wrapped after his crucifixion. If so, the material of which it was made would have to be nearly 2,000 years old. Radiocarbon dating of the material showed, however, that the cloth could not be more than about 700 years old.

Radiocarbon dating can be used for objects up to 30,000 years of age, but it is highly reliable only for objects less than 7,000 years old. These limits result from the fact that eventually carbon-14 has decayed to such an extent that it can no longer be detected well or, eventually, at all in a sample. For older specimens, radioisotopes with longer half-lives can be used for age determination. *See also* Half-life; Radioactive decay

[*David E. Newton*]

Resources
Books

Taylor, R. E., et al., eds. *Radiocarbon Dating: An Archaeological Perspective.* Orlando: Academic Press, 1987.

———. *Radiocarbon After Four Decades: An Interdisciplinary Perspective.* New York: Springer Verlag, 1992.

Radioisotope

The term radioisotope is shorthand for radioactive **isotope**. Isotopes are forms of an element whose atoms differ from each other in the number of neutrons contained in their nuclei and, hence, in their atomic masses. Hydrogen-1, hydrogen-2, and hydrogen-3 are all isotopes of each other.

Isotopes may be stable or radioactive. That is, they may exist essentially unchanged forever (stable), or they may spontaneously emit an **alpha particle** or **beta particle** and/or a **gamma ray**, changing in the process into a new substance. Hydrogen-1 and hydrogen-2 are stable isotopes, but hydrogen-3 is radioactive.

The first naturally occurring radioisotopes were discovered in the late 1890s. Scientists found that all isotopes of the heaviest elements—uranium, radium, **radon**, thorium, and protactinium, for example—are radioactive. This discovery raised the question as to whether stable isotopes of other elements could be converted to radioactive forms.

By the 1930s, the techniques for doing so were well established, and scientists routinely produced hundreds of radioisotopes that do not occur in **nature**. As an example, when the stable isotope carbon-12 is bombarded with neutrons, it may be converted to a radioactive cousin, carbon-13. This method can be used to manufacture radioisotopes of nearly every element.

Naturally occurring radioisotopes are responsible for the existence of **background radiation**. Background radiation consists of alpha and beta particles and gamma rays emitted by these isotopes. In addition to the heavy isotopes mentioned above, the most important contributors to background radiation are carbon-14 and potassium-40.

Synthetic radioisotopes have now become ubiquitous in human society. They occur commonly in every part of **nuclear power** plant operations. They are also used extensively in the health sciences, industry, and scientific research. A single example of their medical application is **cancer** therapy. Gamma rays emitted by the radioisotope cobalt-60 have been found to be very effective in treating some forms of cancer. Other gamma-emitting radioisotopes can also be used in this procedure.

The potential for the release of radioisotopes to the **environment** is great. For example, medical wastes might very well contain radioisotopes that still emit measurable amounts of radiation. Stringent efforts are made, therefore, to isolate and store radioisotopes until their radiation has reached a safe level.

These efforts have two aspects. Some radioisotopes have short half-lives. The level of radiation they emit drops to less than 1% of the original amount in a matter of hours or days. These isotopes need only be stored in a safe place for a short time before they can be safely discarded with other solid wastes.

Other radioisotopes have half-lives of centuries or millennia. They will continue to emit harmful radiation for thousands of years. Safe disposal of such wastes may require burying them deep in the earth, a procedure that still has not been satisfactorily demonstrated. In spite of the potential environmental hazard posed by radioisotopes, they do not presently pose a serious threat to plants, animals, or humans. The best estimates place the level of radiation from artificial sources at less than five percent of that from natural sources. *See also* Half-life; Radioactive decay; Radioactive waste management; Radioactivity

[*David E. Newton*]

Resources
Books

Baker, P., et al. *Radioisotopes in Industry.* Washington, DC: U.S. Atomic Energy Commission, 1965.

Corless, W. R., and R. L. Mead. *Power from Radioisotopes.* Washington, DC: U.S. Atomic Energy Commission, 1971.

Kisieleski, W. E., and R. Baserga. *Radioisotopes and Life Processes.* Washington, DC: U.S. Atomic Energy Commission, 1967.

Radioisotope	Half-life (Years)	Radiation Emitted
Americium-241	460	
Cesium-134	2.1	
Cesium-135	2,000,000	
Cesium-137	30	Beta and gamma
Curium-243	32	
Iodine-131	0.02	Beta and gamma
Krypton-85	10	Beta and gamma
Neptunium-239	2.4	
*Plutonium-239	24,000	Gamma and neutron
Radon-226	1,600	Alpha
Ruthenium-106	1	Beta and gamma
Strontium-90	28	Beta
Technetium-99	200,000	
*Thorium-230	76,000	
Tritium	13	Beta
*Uranium-235	713,000,000	Alpha and neutron
Xenon-133	0.01	
Zirconium-93	900,000	

*Used as fuel in nuclear reactions. All others are by-products.

Some radioisotopes associated with nuclear power. (McGraw-Hill Inc. Reproduced by permission.)

Radiological Emergency Response Team

The Radiological Emergency Response Team (RERT) of the federal **Environmental Protection Agency** (EPA) responds to emergencies that involve the release of radioactive materials. The team responds to emergencies such as accidents at nuclear **power plants**, accidents involving the shipment of radioactive material, and acts of nuclear terrorism. RERT works with the EPA Superfund Program, as well as federal, state, and local agencies to develop and enforce strategic plans.

RERT is based in the EPA Office of Radiation and Indoor Air in Washington, D.C., and at two national laboratories. **Environmental monitoring** and assessment is performed by employees at the National Air and Radiation Environmental Laboratory in Montgomery, Alabama, and the Radiation and Indoor Environments National Laboratory in Las Vegas, Nevada. Approximately 75 RERT members were stationed in Washington and at the laboratories in the spring of 2002.

In an emergency, an RERT field team goes to the site where radioactive material was released. The team's duties include taking environmental measurements and doing laboratory work. In addition, RERT works with state and local authorities to protect the public from exposure to harmful radiation levels. Team equipment ranges from a mobile radiation laboratory to the personal dosimeter used to measure the radiation dose in an individual.

[*Liz Swain*]

Radionuclides

Radionuclides are radioactive elements. **Radioactivity** can be defined as the release of **alpha and beta** particles from atoms, and/or **gamma rays** that takes place when the nuclei of certain unstable substances spontaneously disintegrate. It is during this disintegration process that they emit radiation. The two main types of radiation released during such processes are termed **ionizing**, or of sufficient strength to forcibly eject electrons from their orbit around an atoms nucleus, producing ions, and **non-ionizing**, of less strength and incapable of displacing electrons and forming ions. **Ionizing radiation** is highly significant because when it occurs within the atoms of molecules in living things, it is capable of causing biological damage such as the death of cells or the unnatural reproduction of cells. We term this unnatural reproduction of cells **cancer**.

Radionuclides occur in the **environment** both naturally and as a result of human industry. Some naturally-occurring radionuclides exist all over the earth and have been present since its formation 4.5 billion years ago. But most elements can be made artificially radioactive by bombarding them with high-energy particles such as neutrons. Each radionuclides forms and behaves in unique ways, with its own method and rate of decay. Such decay is measured by what is called **half-life**, or the time it takes for a group of atoms to decay to half of their original number.

One of the substances that occurs naturally in the earth's crust is **uranium**, a radionuclide used often as fuel to produce **nuclear power**. Uranium is the heaviest element found on earth with the exception of tiny amounts of an element called neptunium. When the German chemist Martin H. Klaproth discovered uranium in 1789, he named it in honor of the recently discovered planet Uranus. The scientific community showed little interest in uranium until 1896, when Henri Becquerel discovered radioactivity, and uranium was named as one of only two known radioactive elements of the time. In 1938, scientists discovered **nuclear fission**, a reaction that produces energy. In fact, uranium can produce energy at nearly three million times that of **coal** (1 lb of uranium can produce the same amount of energy produced by 3 million lb [1.4 million kg] of coal).

The most abundant form (**isotope**) of uranium boasts a **half-life** nearly identical to that of the earth itself. This allows scientists to use the substance's disintegration to date

other geological features of the earth by comparison. However, uranium use also has many drawbacks. Most notably, use of the substance produces nuclear waste that must be carefully transported and stored. In addition, easily obtainable supplies of uranium on Earth are limited, therefore the costs of locating and refining uranium can be extremely high.

Radium occurs as a result of uranium disintegration. Marie Curie, the first woman to win a Nobel Prize, and one of the few scientists to ever receive the Nobel prize twice, along with her husband Pierre, discovered the substance. She had followed up on Becquerel's observations on uranium's radioactive properties for her doctoral dissertation in the late 1890s. The Curies began refining pitchblende, a waste ore that was commonly found around uranium mines, known to emit radiation. The Curies first discovered the radioactive salt they named polonium, and then radium, found to be thousands of times more radioactive than any other substance discovered to date.

Radium is found in water, **soil**, plants, and food, but at low concentrations. Drinking water holds the highest potential for human exposure to radium. When humans are exposed to radium orally, they are at risk for lung, bone, brain and nasal passage tumors. Chronic exposure has led to acute **leukemia** and other complications. Radium has been classified by the **Environmental Protection Agency** (EPA) as a human **carcinogen**. Today, radium is still considered among the most radioactive metals on Earth, and requires careful attention and handling. Its use has been limited in many products, but the substance still helps treat cancers through radiation therapy and aids in some forms of research. As evidence of the validity of the EPA's finding, its discoverer, Marie Curie died at the age of 67 of leukemia, caused by her prolonged exposure to radiation.

Radon is a gas that can diffuse out of uranium (from radium) in the ground from uranium and thorium present in minerals, ores and rocks. A German physics professor named Friedrich Dorn discovered radon in 1900, after he followed the experiments of Marie Curie. He found that radium emitted a radioactive gas he termed **radium emanation**. Dorns discovery was important because it showed that an element could be transmuted from metal to gas as part of the **radioactive decay** process. The hazards of radon were not discovered until late in the twentieth century, and since that time, testing houses for the gas has become an important precaution. Radon has shown the capability of seeping into **groundwater** and contaminating public drinking supplies.

As Radon is colorless and odorless, it can prove particularly dangerous. Radon exposure occurs mostly through inhalation of the gas in indoor locations such as schools, homes, or office buildings. Chronic exposure produces serious respiratory effects, and smokers are at particular risk for lung cancer, estimated at 10–20 times the risk of nonsmokers. Although hazardous, radon has been useful in predicting earthquakes. In 1918, the discovery by a group of Chinese scientists that radon levels in groundwater rise just before earthquakes led to the prediction of several earthquakes by monitoring radon concentrations in well water. Radon has also proves useful in detecting leaks, inspecting metal welds, and measuring flow rates. However, radons high risk status as a carcinogen outweighs its beneficial uses.

Plutonium is an example of an artificial radioactive substance. It was discovered around the early 1940s during experiments on nuclear fission conducted by a number of scientists. Plutonium is used in breeder reactors when uranium undergoes nuclear fission to produce energy and occurs as a waste product of uranium. Like uranium, plutonium produces large amounts of energy and its properties were put to the test in the first **nuclear weapons** test in New Mexico in 1945.

Knowledge of the properties and risks of all radionuclides have made it necessary for people all over the world to become more aware of the inherent dangers they possess and the necessary preventative measures needed for protection. People that work in factories that process uranium or with phosphate fertilizers are at increased risk of exposure to radionuclides, as are those living in close proximity to uranium mines. Tests can measure radioactivity levels in the body and elimination of radium and radon in exhaled breath. Uranium, radon, and radium levels can be measured in the urine.

One of the greatest challenges facing scientists and policymakers throughout the world is the safe transport and disposal of the **hazardous waste** resulting from radionuclides. The subject has been steeped in controversy for many years, as no place seems safe enough to handle storage of a substance that lives for millions of years. However, scientists continue to develop new methods for storage of nuclear byproducts. In early 2001, a Los Alamos National Laboratory team announced that certain ceramic materials held up against radiation damage and could potentially offer new solutions to resist **leaching** and radiation for thousands of years.

A consortium of international agencies promotes international cooperation in managing nuclear waste, since nearly every developed country has some sort of nuclear power or weaponry project in place. Most of these countries have developed plans to safely dispose of nuclear waste early in the twenty-first century. The United States leads most countries in storage efforts, with the 1982 passage of the Nuclear Waste Policy Act. The act identified objectives for developing geologic repositories for high-level nuclear waste.

The federal government identified **Yucca Mountain** in Nevada for deep underground storage of wastes to replace

storage that now lies at commercial nuclear **power plants** and research reactor sites in 43 states. The EPA continues to develop public health and environmental standards to set safe limits for the long-term storage of highly **radioactive waste**. However, a coalition of environmental groups also continues to watch and question the agency to ensure that the radiation standard remains protective enough. Like the **Waste Isolation Pilot Plant** (WIPP) storage facility in southern New Mexico, the Yucca Mountain site will no doubt remain controversial for some time.

The EPA provides printed fact sheets on radionuclides that warn of environmental and occupational exposure to the substances and acceptable levels of contact for humans. According to the agency, uranium is present in rocks and soil and throughout the environment, and although exposure can occur through air, higher levels generally occur in food or drinking water. Chronic long-term exposure to uranium and radon has been linked to both lung and kidney diseases. The EPA is currently in the process of promulgating new drinking water standards (the first update since 1976) regarding (non-Radon) radionuclides. These standards will become effective December 8, 2003, and will affect only Community Water Systems (CWSs), or those systems that serve more than 25 residents regularly all year.

[*Joan M. Schonbeck*]

RESOURCES

BOOKS

Alexander, D. E., and R. W. Fairbridge. *Encyclopedia of Environmental Science.* Dordrecht, The Netherlands: Kluwer Academic Publishers, 1999.

Clayman, Charles. *The American Medical Association Home Medical Encyclopedia.* New York: Random House, 1998.

Emsley, J. *The Elements.* New York: Clarendon Press, 1998.

OTHER

Environmental News Network. [cited July 2002]. <http://wwww.enn.com>.

ORGANIZATIONS

Alliance for Nuclear Accountability, 1914 N. 34th St., Ste. 407, Seattle, WA USA 98103 (206) 547-3175, Fax: (206) 547-7158, Email: ananuclear@earthlink.net, <http://www.ananuclear.org>

Environmental Protection Agency Office of Air and Radiation, Ariel Rios Building, 1200 Pennsylvania Ave. NW, Washington, DC USA 20460 (202) 564-7400, <http://www.epa.gov/air/oarofcs.html>

Radiotracer

A radioactive **isotope** progressing through a biological or physical system can be followed by several tracking procedures. For example, **fertilizer** containing radioactive **phosphorus** can be added to **soil**. Plants grown in this soil then take up the radioactive phosphorus just as they do non-radioactive phosphorus. If one of these plants is placed on a photographic plate, radiation from the radioactive phosphorus exposes the plate. The plant "takes its own picture," showing where the phosphorus concentrates in the plant. Radiotracers are a highly desirable research technique as they do not require the destruction of an organism for its study.

Radon

Although it has received attention as an environmental hazard only recently, radon is a naturally occurring radioactive gas that is present at low concentrations everywhere in the **environment**. Colorless and odorless, radon is a decay product of radium; radium is a decay product of the radioactive element **uranium**, which occurs naturally in the earth's crust. Radon continues to break down into products called radon progeny. Radon is measured in units called picocuries per liter (pCi/L), and it becomes a health concern when people are exposed to concentrations higher than normal background levels. Some geologic formations, such as the Reading Prong in New Jersey, are naturally very high in radon emissions.

During their normal decay process, radioactive elements emit several kinds of radiation, one of which is alpha radiation. The health effects of radon are associated with these alpha particles. These particles are too heavy to travel far and they cannot penetrate the skin, but they can enter the body through the lungs during inhalation. Studies of miners exposed to high concentrations of radon have shown an increased risk of lung **cancer**, and this is the health effect most commonly associated with radon. Background levels are usually estimated at 1 pCi/L. It is estimated that a person exposed to this concentration for 18 hours a day for five years increases their risk of developing cancer to one in 1000. At radon levels of 200 pCi/L, the increased risk of lung cancer after five years of exposure at 18 hours per day rises to 60 in 1,000. Because cancer is a disease that is slow to develop, it may take five to 50 years after exposure to radon to detect lung cancer.

In the outdoor environment, radon gas and its decay products are usually too well-dispersed to accumulate to dangerous levels. It is indoors without proper ventilation, in places such as basements and ground floors, where radon can seep from the **soil** and accumulate to dangerous concentrations. The most common methods of reducing radon buildup inside the home include installing blowers or simply opening windows. Plugging cracks and sealing floors that are in contact with soil also reduces the concentration. In the United States, environmental and public health agencies have instituted free programs to test for radon concentrations, and they also offer assistance and guidelines for reme-

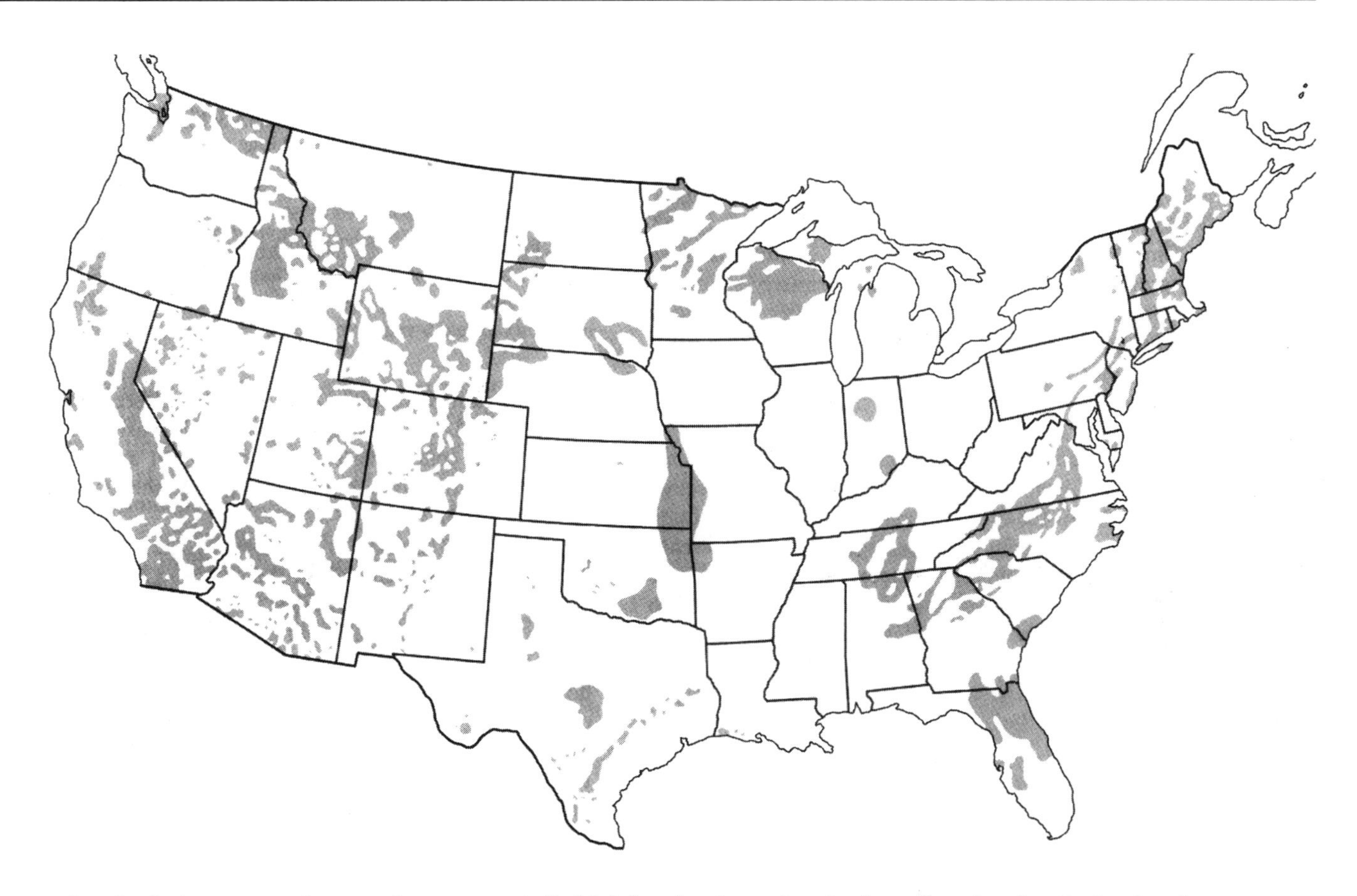

The shaded areas on the map have potentially high levels of uranium in the soil and rocks. Radon levels may be high in these areas. (McGraw-Hill Inc. Reproduced by permission.)

dying the problem. *See also* Radiation exposure; Radioactive decay; Radioactive pollution; Radioactivity

[*Usha Vedagiri*]

RESOURCES

BOOKS

Brenner, D. J. *Radon: Risk and Remedy.* Salt Lake City: W. H. Freeman, 1989.

Cohen, B. *Radon: A Homeowner's Guide to Detection and Control.* Mt. Vernon, NY: Consumer Report Books, 1988.

Kay, J. G., et al. *Indoor Air Pollution: Radon, Bioaerosols, and VOCs.* Chelsea, MI: Lewis, 1991.

Lafavore, M. *Radon: The Invisible Threat.* Emmaus, PA: Rodale Press, 1987.

Rails-to-Trails Conservancy

In 1985, the Rails-to-Trails Conservancy, a non-profit organization, was established to convert abandoned railroad corridors into open spaces for public use. During the nineteenth century, the railroad industry in the United States boomed as rail companies rushed to acquire land and assemble the largest rail system in the world. By 1916, over 250,000 mi (402,000) of track had been laid across the country, connecting even the most remote towns to the rest of the nation. However, during the next few decades, the **automobile** drastically changed the way Americans live and travel.

With Henry Ford's introduction of mass production techniques, the automobile became affordable for nearly everyone, and industry shifted to trucking for much of its overland **transportation**. The invention and widespread use of airplanes had a significant impact on the railroad industry as well, and people began abandoning railroad transportation.

As a result of these developments, thousands of miles of rail corridors fell into disuse. Rails-to-Trails Conservancy estimates that over 3,000 mi (4,828 km) of track are abandoned each year. Since much of the track is in the public domain, Rails-to-Trails strives to find uses for the land that will benefit the public. Thus Rails-to-Trails works with citizen groups, public agencies, railroads, and other concerned parties to, according to the organization, "build a

transcontinental trailway network that will preserve for the future our nation's spectacular railroad corridor system."

Some of the rails have been converted to trails for hiking, biking, and cross-country skiing as well as **wildlife** habitats. By providing people with information about upcoming abandonments, assisting public and private agencies in the effort to gain control of those lands, sponsoring short-term land purchases, and working with Congress and other federal and state agencies to simplify the acquisition of abandoned railways, Rails-to-Trails has mobilized a powerful grassroots movement across the country. Today there are over 11,500 mi (18,506 km) of converted rails-to-trails.

Rails-to-Trails also sponsors an annual conference and other special meetings and publishes materials related to railway **conservation**. Its quarterly newsletter, *Trailblazer*, is sent to members and keeps them aware of rails-to-trails issues. Many books and pamphlets are available through the group, as are studies such as *How to Get Involved with the Rails-to-Trails Movement* and *The Economic Benefits of Rails-to-Trails Conversions to Local Economies*. In addition, the Rails-to-Trails Conservancy provides legal support and advice to local groups seeking to convert rails to trails.

[*Linda M. Ross*]

Resources

Organizations

Rails-to-Trails Conservancy. 1100 17th Street, 10th Floor, NW, Washington, D.C. USA 20036. (202) 331-9696, Email: railtrails@transact.org, <http://www.railtrails.org>.

Rain forest

The world's rain forests are the richest ecosystems on Earth, containing an incredible variety of plant and animal life. These forests play an important role in maintaining the health and **biodiversity** of the planet. But rain forests throughout the world are rapidly being destroyed, threatening the survival of millions of **species** of plants and animals and disrupting **climate** and weather patterns. The rain forests of greatest concern are those located in tropical regions, particularly those found in Central and South America, and the ancient temperate rain forests along the northeastern coast of North America.

Tropical rain forests (TRFs) are amazingly rich and diverse biologically and may contain one-half to two-thirds of all species of plants and animals, though these forests cover only about 5–7% of the world's land surface. Tropical rain forests are found near the equatorial regions of Central and South America, Africa, Asia, and on Pacific Islands, with the largest remaining forest being the Amazon rain forest, which covers a third of South America.

TRFs remain warm, green, and humid throughout the year and receive at least 150 in (4 m) of rain annually, up to half of which may come from trees giving off water through the pores of their leaves in a process called **transpiration**. The tall, lush trees of the forest form a two- to three-layer closed canopy, allowing very little light to reach the ground. Although tropical forests are known for their lush, green vegetation, the **soil** stores very few nutrients. Dead and decomposing animals, trees, and leaves are quickly taken up by forest organisms, and very little is absorbed into the ground.

In 1989 *World Resources Institute* predicted that "between 1990 and 2020, species extinctions caused primarily by tropical **deforestation** may eliminate somewhere between 5–15% of the world's species...This would amount to a potential loss of 15,000 to 50,000 species per year, or 50 to 150 species per day." It is estimated that 30–80 million species of insects alone may exist in TRFs, at least 97% of which have never been identified or even discovered.

Tropical forests also provide essential winter **habitat** for many birds that breed and spend the rest of the year in the United States. Some 250 species found in the United States and Canada spend the winter in the tropics, but their population levels are decreasing alarmingly due to forest depletion.

Tropical rain forests have unique resources, many of which have yet to be utilized. Food, industrial products, and medicinal supplies are common examples. Among the many fruits, nuts, and vegetables that we use on a regular basis and which originated in tropical forests are citrus fruits, coffee, yams, nuts, chocolate, peppers, and cola. A variety of oils, lubricants, resins, dyes, and steroids are also products of the forests. Natural **rubber**, the fourth biggest agricultural export of southeast Asian nations, brings in over $3 billion a year to developing countries. The forests could also yield a sustainable supply of woods like teak, mahogany, bamboo, and others. Although only about 1% of known tropical plants have been studied for medicinal or pharmaceutical applications, these have produced 25–40% of all prescription drugs used in the United States. Some 2,000 tropical plants now being studied have shown potential as cancer-fighting agents.

Scientists studying ways to commercially utilize these forests in a sustainable, non-destructive way have determined that two to three times more money could be made from the long-term collection of such products as nuts, rubber, medicines, and food, as from cutting the trees for **logging** or cattle ranching. Perhaps the greatest value of tropical rain forests is the essential role they play in the earth's climate. By absorbing **carbon dioxide** and producing oxygen through **photosynthesis**, the forests help prevent global warming

(the **greenhouse effect**) and are important in generating oxygen for the planet. The forests help prevent droughts and **flooding**, soil **erosion** and stream **sedimentation**, maintain the **hydrologic cycle**, and keep streams and rivers flowing by absorbing rainfall and releasing moisture into the air.

Despite the worldwide outcry over deforestation, destruction is actually increasing. A 1990 study by the United Nations Food and Agriculture Organization found that tropical forests were disappearing at a rate exceeding 40 million acres (16.2 million ha) a year—an area the size of Washington state. This rate is almost twice that of the previous decade.

Timber companies in the United States and western Europe are responsible for most of this destruction, mainly through farming, cattle ranching, logging, and huge development projects. Japan, the world's largest hardwood importer, buys 40% of the timber produced, with the United States a close second. Ironically, much of the destruction of forests worldwide has been paid for by American taxpayers through such government-funded international lending and development agencies as the **World Bank**, the International Monetary Fund, and the Inter-American Development Bank, along with the United States Agency for International Development.

The release of the World Bank's new Operational Policy on Forests has been delayed since October 2001, and had not been released as of the first week in July 2002. The release has been delayed by a World Bank dispute over whether its new Forest Policy would apply to the World Bank's growing area of lending, which directly or indirectly finances logging activities. There has been widespread demand that the Forest Policy must apply to all World Bank operations that might have an impact on forests. It remains to be seen how the Operational Policy on Forests will handle this important question.

American, European, and Latin American demand for beef has contributed heavily to the conversion of rain forest to pasture land. It is estimated that one-fourth of all tropical forests destroyed each year are cut and cleared for cattle ranching. Between 1950 and 1980, two-thirds of Central America's primary forests were cut, mostly to supply the United States with beef for fast food outlets and pet food.

In Brazil and other parts of the **Amazon Basin**, cattle ranchers, plantation owners, and small landowners clear the forest by setting it on fire, which causes an estimated 23–43% increase in **carbon** dioxide levels worldwide and spreads **smoke** over millions of square miles, which interferes with air travel and causes respiratory difficulties.

Unfortunately, cleared forest that is turned into pasture land provides very poor quality soil, which can only be ranched for a few years before the land becomes infertile and has to be abandoned. Eventually **desertification** sets in, causing cattle ranchers to move on to new areas of the forest.

While huge timber and multi-national corporations have justifiably received much of the blame for the destruction of TRFs, local people also play a major role. Populations of the **Third World** gather wood for heating and cooking, and the demand brought about by their growing numbers has resulted in many deforested areas. The proliferation of coca farms, producing cocaine mainly for the American market, has also caused significant deforestation and **pollution**, as has gold prospecting in the Amazon.

The destruction of TRFs has already had devastating effects on **indigenous peoples** in tropical regions. Entire tribes, societies, and cultures have been displaced by environmental damage caused by deforestation. In Brazil fewer than 200,000 Indians remain, compared to a population of some six million about 400 years ago. Sometimes they are killed outright when they come into contact with settlers, loggers, or prospectors, either by disease or because they are shot. And those who are not killed are often herded into miserable reservations or become landless peasants working for slave labor wages.

Today, less than 5% of the world's remaining TRFs have some type of protective status, and there is often little or no enforcement of prohibitions against logging, **hunting**, and other destructive activities.

The ancient rain forests of North America are also important ecosystems, composed in large part of trees that are hundreds and even thousands of years old. Temperate rain forests (or evergreen forests) are usually composed of conifers (needle-leafed, cone-bearing plants) or broadleaf evergreen trees. They thrive in cool coastal climates with mild winters and heavy rainfall and are found along the coasts of the Pacific Northwest area of North America, southern Chile, western New Zealand, and southeast **Australia**, as well as on the lower mountain slopes of western North America, Europe, and Asia.

The rain forest of the Pacific Northwest, the largest **coniferous forest** in the world, stretches over 112,000 mi^2 (129,000 km^2) of coast from Alaska to northern California, and parts extend east into mountain valleys. The forests of the Pacific Northwest consist of several species of coniferous trees, including varieties of spruce, cedar, pine, Douglas fir, Hemlock, and Pacific yew. Broadleaf trees, such as Oregon oak, tanoak, and madrone, are also found there. Redwood tree growth extends to central California. Further south are the giant sequoias, the largest living organisms on earth, some of which are over 3,000 years old.

In some ways, the temperate rain forests of the Pacific Northwest may be the most biologically rich in the world. Although TRFs contain many more species, temperate rain

Rain forest. (Photograph by David Julian. Phototake NYC. Reproduced by permission.)

forests have far more plant matter per acre and contain the tallest and oldest trees on Earth. Over 210 species of fish and **wildlife** live in ancient forests, and a single tree can support over 100 different species of plants. One tree found in cool, moist forests is the slow-growing Pacific yew, whose bark and needles contain taxol, considered one of the most powerful anti-cancer drugs ever discovered.

Unfortunately, the **clear-cutting** of most of the ancient forests and their yew trees has caused a serious shortage of taxol. Some yew trees are unavailable for harvesting because they grow in forests protected as habitat for the endangered **northern spotted owl**. The logging of federal land under the jurisdiction of the **Bureau of Land Management** (BLM) and the U.S. **Forest Service** (USFS) has eliminated some of the last and best habitats for the Pacific yew.

The remaining ancient forest, almost all of which is now on BLM and USFS land, is being cut at a rate of 200,000 acres (81,000 ha) a year, as of 1992. At this rate it will be destroyed within less than two decades. The consequences of this destruction will include the disappearance of **rare species** dependent on this habitat, the silting of waterways and erosion of soil, and the decimation of **salmon** populations, which provide the world's richest salmon fishery, worth billions of dollars annually.

Although the timber industry claims that logging maintains jobs in the Pacific Northwest, logging **national forest** often makes no economic sense. Because of the expense of building logging roads, surveying the area to be cut, and the low price it charges for trees, the USFS often loses money on its timber sales. As a result of such "deficit" or "below cost" timber sales between 1989 and 1992, the USFS lost an average of almost $300 million a year by selling timber from national forests.

Several private **conservation** groups, such as the **Wilderness Society** and the **Sierra Club**, are working to preserve the remaining ancient forests on BLM and USFS land through federal legislation, lawsuits, and other actions. In April 1993, President Bill Clinton attended a "timber summit" in Portland, Oregon, to discuss the ancient forests, **endangered species**, and timber jobs. But the cutting of old growth forests and "below cost" timber sales have continued much as before. The Bush administration has not supported environmental issues. In 2001, President Bush triggered international outrage when he refused to agree to the Kyoto Protocol, a United Nations approved plan to preserve the **environment**.

Two of the largest remaining temperate rainforests are Alaska's 16.9 million-acre (6.8 million ha) Tongass National Forest and the Russian **Taiga**. Tongass is the last large, relatively undisturbed, temperate rain forest in the United States, stretching over 600 mi (966 km) of coast along the Alaska Panhandle. It has the highest concentration of bald eagles and grizzly bears anywhere on Earth. In addition to reducing habitat for wildlife and salmon hatching, scenic areas are being destroyed, reducing tourism and **recreation** in the area.

The huge forests in the Russian Far East and Siberia are called the Taiga. The Siberian portion is the world's largest forest. Comprising some two million mi^2 (5.1 million km^2), the Taiga is much larger than the Brazilian Amazon and would cover the entire American lower 48 states. With Russia becoming more market-oriented and in desperate need of money, some have discussed selling the logging rights to some of these forests to American, Japanese, and Korean logging companies.

Pressure from environmentalists and increasing public concern for rain forests has encouraged world leaders to consider more environmental laws. In June 1992, at the **United Nations Earth Summit** conference in Rio de Janeiro, Brazil, a set of voluntary principles to conserve the world's threatened forests were agreed upon. The document affirms the right of countries to economically exploit forests but states that this should be done "on a sustainable basis," recognizing the value of forests in absorbing carbon dioxide and slowing climate change. Initial hope that the agreement on principles might eventually be turned into a binding

international convention has evaporated, and throughout the world, forests continue to be destroyed. Some of the last remaining untouched forests are being opened up to **poaching**, logging, and other exploitation, such as the one million acre (405,000 ha) virgin Ndoke rainforest of the northern Congo Republic, a wildlife paradise full of thousand-year-old trees, along with **elephants**, leopards, gorillas, and other endangered species.

There is a small but positive sign of a much-needed change. In October 2001 Brazil suspended all trade in mahogany. The government's decision followed a two-year investigation by **Greenpeace** using ground, air, and satellite surveillance to document rampant illegal logging on Indian reservations and other protected wildlife areas. In addition, a report released in June 2002 showed that the rate of forest destruction of Brazil's Amazon jungle fell 13.4% from a five-year peak in 2000. However, the rate of destruction still deeply troubles environmentalists. *See also* Compaction; Deciduous forest; Decomposition; Migration; Old-growth forest

[*Bill Asenjo Ph.D.*]

RESOURCES
BOOKS

Caufield, C. *In the Rainforest: Report From a Strange, Beautiful, Imperiled World.* Chicago: University of Chicago Press, 1986.

Mitchell, G. J. *World on Fire: Saving an Endangered Earth.* New York: Charles Scribner's Sons, 1991.

Myers, N. *The Sinking Ark: A New Look at Disappearing Species.* Oxford: Pergamon Press, 1979.

Porritt, J. *Save the Earth.* Atlanta: Turner Publishing, 1991.

Raven, P. H. "The Cause and Impact of Deforestation." In *Earth 88: Changing Geographic Perspectives.* Washington, DC: National Geographic Society, 1988.

Repetto, R. *The Forest for the Trees? Government Policies and the Misuse of Resources.* Washington, DC: World Resources Institute, 1988.

Zuckerman, S. *Saving Our Ancient Forest.* Los Angeles: Living Planet Press, 1991.

PERIODICALS

Bugge, A. "Brazil's Amazon Destruction Down but Still Alarming." *Reuters* June 12, 2002.

Jordan, M. "Brazilian Mahogany: Too Much In Demand—Illegal Logging, Exports Are Lucrative for Criminals, Disastrous for Rain Forest." *Wall Street Journal,* November 14, 2001.

"United States Government Report Blames Humans for Global Warming." *Reuters* [cited July 2002]. <http://www.enn.com/news/wirestories/2002/06/06042002>.

"World Bank Forest Policy." *World Rainforest Movement Bulletin 55* [cited February 2002]. <http://www.wrm.org.uy>.

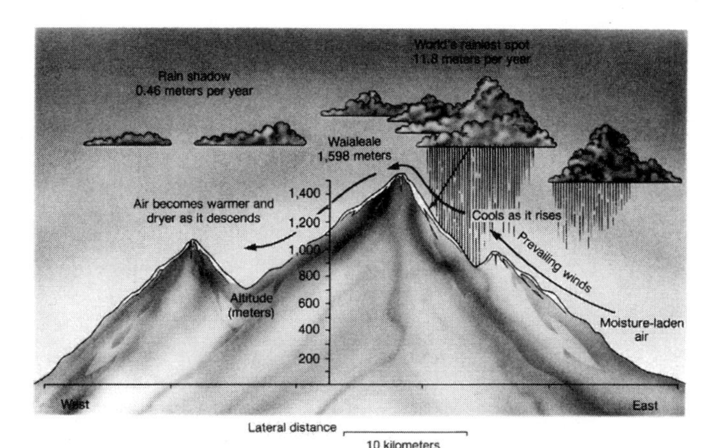

A rain shadow in Hawaii on the eastern side of Mount Waialeale. (McGraw-Hill Inc. Reproduced by permission.)

Rain shadow

A region of relative dryness found on the downwind side of a mountain range or other upland area. As an air mass rises over the upwind side of a mountain range, pressure drops and temperature falls. This causes the relative humidity of the air mass to rise. Eventually the moisture in the air condenses and precipitation occurs. An air mass that is now cooler and drier passes over the top of the mountain range. As it descends on the downwind side of the range, it warms again and its relative humidity is further reduced. This reduction in relative humidity not only prevents further rainfall, but also causes the air mass to absorb moisture from other sources, drying the **climate** on the downwind side. The ultimate result is lush forest on the windward side of a mountain separated by the summit from an **arid environment** on the downwind side. Examples of rain shadows include the arid areas on the eastern sides of the mountain ranges of western North America, and the Atacama **Desert** in Chile on the downwind side of the Andes Mountains.

Rainforest Action Network

Rainforest Action Network (RAN), founded in 1985, is an activist group that works to protect rain forests and their inhabitants worldwide. Its strategy includes imposing public pressure on those corporations, agencies, nations, and politicians whom the group believes are responsible for the destruction of the world's rain forests by organizing letter-writing campaigns and consumer boycotts. In addition to mobilizing consumer and environmental groups in the United States, RAN organizes and supports conservationists committed to **rain forest** protection around the world.

RAN's first direct action campaign was the boycott of Burger King. The fast-food restaurant chain was importing much of its beef from Central and South America, where large areas of forests have been turned into pastureland for cattle. RAN contends that "after sales dropped 12% during the boycott in 1987, Burger King canceled $35 million worth of beef contracts in Central America and announced that it had stopped importing rainforest beef...The formation of Rainforest Action Groups (RAGs) that staged demonstrations and held letter-writing parties in U.S. cities helped make the boycott and other campaigns a success." RAN states that there are now over 150 RAGs in North America alone.

RAN also works with human rights groups around the world, attempting to protect and save the cultures of **indigenous peoples** dependent on rain forests. The group helps support ecologically sustainable ways to use the rain forest, such as **rubber** tapping and the harvesting of such foods as nuts and fruits.

RAN urges the public to avoid buying tropical wood, such as rosewood and mahogany, and plywood made from **rain forest** timber. The group also recommends that people not buy rainforest beef, which is often found in fast-food hamburgers and processed beef products.

RAN is strongly pushing its boycott of the Japanese conglomerate Mitsubishi, which makes televisions, VCRs, fax machines, and stereos. RAN points out that "Mitsubishi has big **logging** operations in Malaysia, Borneo, Philippines, Indonesia, Chile, Canada, and Brazil...It is the world's number one importer of tropical timber." Other corporations that RAN has criticized for damaging rain forests include ARCO, Scott Paper, Coca-Cola, Texaco, and CONOCO, as well as such international agencies as the **World Bank** and the International Tropical Timber Organization (ITTO).

RAN operates an extensive media campaign targeted at companies and decision-makers who create policy on rainforests by running full-page advertisements in such major newspapers as *The New York Times* and the *Wall Street Journal*, RAN's publications include its quarterly *World Rainforest Report*; its monthly *Action Alerts*; *The Rainforest Action Guide*; and *The Rainforest Catalogue*, offering books, videos, bumper stickers, T-shirts, and other products promoting rainforest protection, as well as cosmetics and food made from rain forest plants.

Among the victories achieved by or with the help of RAN are the halting of a plan to cut down the one-million-acre (404,687-ha) La Mosquito Forest in Honduras—made famous by the movie *The Mosquito Coast*—which is essential to the cultures and livelihoods of some 35,000 indigenous people; forcing the World Bank to stop making loans to nations that destroy their rain forests; stopping **oil drilling** in the Ecuadorian Amazon by CONOCO and Du Pont; preventing the opening of a major road to take timber out of the Amazon to the Pacific Coast; and persuading Coca-Cola to donate land in Belize for a **nature** preserve.

Nevertheless, the destruction of the world's rain forests continues on an enormous scale. As RAN's director Randall Hayes points out: "Over 50 percent of the world's tropical rainforests are gone forever. Two-thirds of the southeast Asia forests have disappeared, mostly for hardwood shipped to Japan, Europe, and the U.S. And this destruction continues at a rate of 150 acres per minute, or a football field per second."

[*Lewis G. Regenstein*]

RESOURCES

ORGANIZATIONS

Rainforest Action Network. 221 Pine Street, Suite 500, San Francisco, CA USA 94104 (415) 398-4404, Fax: (415) 398-2732, Email: rainforest@ran.org, <http://www.ran.org>

Ramsar Convention

see **Convention on Wetlands of International Importance (1971)**

Rangelands

Concentrated in 16 western states, rangelands comprise 770 million acres (311.6 million ha) and over one-third of the land base in the United States. Rangelands are vegetated predominately by shrubs, and they include **grasslands**, **tundra**, marsh, meadow, **savanna**, **desert**, and alpine communities. They are fragile ecosystems that depend on a complex interaction of plant and animal **species** with limited resources, and many range areas have been severely damaged by **overgrazing**. Currently over 300 million acres (121 million ha) within the United States alone are classified as being in need of **conservation** treatment and management. Effective range-management techniques include **soil conservation**, preservation of **wildlife habitat**, and protection of watersheds.

Raphus cucullatus

see **Dodo**

Raprenox (nitrogen scrubbing)

A recently-developed technique for removing **nitrogen oxides** from waste gases makes use of a common, nontoxic organic compound known as cyanuric **acid**, $C_3H_3N_3O_3$. When heated to temperatures of about 660°F (345°C), cya-

nuric acid decomposes to form isocyanic acid. The acid, in turn, reacts with oxides of **nitrogen** to form **carbon dioxide**, **carbon monoxide**, nitrogen, and water. The process has been given the name of *raprenox*, which comes from the expression *rap*id *re*moval of *n*itrogen *ox*ides. In tests so far, the method has worked very well with internal **combustion** engines, removing up to 99% of all nitrogen oxides from exhaust gases. Its efficiency with gases released from smokestacks has not yet been determined. *See also* Scrubbers

Rare species

A **species** that is uncommon, few in number, or not abundant. A species can be rare and not necessarily be endangered or threatened, for example, an organism found only on an island or one that is naturally low in numbers because of a restricted range. Such species are, however, usually vulnerable to any exploitation, interference, or disturbance of their habitats. Species may also be common in some areas but rare in others, such as at the edge of its natural range.

"Rare" is also a designation that the IUCN—The World Conservation Union gives to certain species "with small world populations that are not at present 'endangered' or 'vulnerable' but are at risk. These species are usually localized within restricted geographical areas or habitats or are thinly scattered over a more extensive range." Some American states have also employed this category in protective legislation.

RDFs

see **Refuse-derived fuels**

Recharge zone

The area in which water enters an **aquifer**. In a recharge zone surface water or precipitation percolate through relatively porous, unconsolidated, or fractured materials, such as sand, moraine deposits, or cracked basalt, that lie over a water bearing, or aquifer, formation. In some cases recharge occurs where the water bearing formation itself encounters the ground surface and precipitation or surface water seeps directly into the aquifer. Recharge zones most often lie in topographically elevated areas where the **water table** lies at some depth. Aquifer recharge can also occur locally where streams or lakes, especially temporary ponds, are fed by precipitation and lie above an aquifer. Karst **sinkholes** also frequently serve as recharge conduits. A recharge zone can extend hundreds of square miles, or it can occupy only a small area, depending upon geology, rainfall, and surface **topography** over the aquifer. Recharge rates in an aquifer depend upon the amount of local precipitation, the ability of surface deposits to allow water to filter through, and the rate at which water moves through the aquifer. Water moves through the porous rock of an aquifer sometimes a few centimeters a day and sometimes, as in karst limestone regions, many kilometers in a day. Surface water can enter an aquifer only as fast as water within the aquifer moves away from the recharge zone.

Because recharge zones are the water intake for extensive underground reservoirs, they can easily be a source of **groundwater** contamination. Agricultural pesticides and fertilizers are especially common groundwater pollutants. Applied year after year and washed downward by rainfall and **irrigation** water, **agricultural chemicals** frequently percolate into aquifers and then spread through the local groundwater system. Equally serious are contaminants **leaching** from **solid waste** dumps. Rainwater percolating through **household waste** picks up dozens of different organic and inorganic compounds, and contamination appears to continue a long time: pollutants have recently been found leaching from waste dumps left by the Romans almost 2,000 years ago. Perhaps most serious are **petroleum** products, including **automobile** oil, which Americans dump or bury in their back yards at the rate of 240 million gal (910 million l) per year, or 4.5 million gal (18 million l) each week. On a more industrial scale, inadequately sealed toxic waste and radioactive materials contaminate extensive areas of groundwater when they are deposited near recharge zones. Because recharge occurs in a vast range of geologic conditions, all these contamination sources present real threats to groundwater quality. *See also* Hazardous waste; Radioactive waste

[*Mary Ann Cunningham Ph.D.*]

Resources

Books

Fetter, C. W. *Applied Hydrology*. Columbus, OH: Charles E. Merrill, 1980.

Freeze, R. A., and J. A. Cherry. *Groundwater*. Englewood Cliffs, NJ: Prentice-Hall, 1979.

Reclamation

This term has been used environmentally in two distinct ways. The more historic use refers to making land productive for agriculture. The current usage refers mostly to the restoration of disturbed land to an ecologically stable condition.

The main application for agricultural purposes is the development of **irrigation**. That was the mission given in 1902 to the federal **Bureau of Reclamation** under the **U.S. Department of Agriculture**. This agency has built a total of 180 water projects in 17 western states, including the

Hoover and Glen Canyon **Dams** in Arizona, and supplies 10 trillion gal (37.8 trillion l) of water to 31 million people every year. However, this type of reclamation has led to much environmental damage and now scientists are trying to figure out how to reverse this damage. The term reclamation has also been used to describe the system of dikes and pumps in the Netherlands to allow farming of lands below sea level.

The main thrust of reclamation today is the restoration of land damaged by human activity, especially by **strip mining** for **coal**. Surface or strip mining represented a third of coal production in 1963, increasing to 60% in 1973. This increase in production led to more damage to the land and once the environmental impact was realized, the reclamation movement began. Since 1970, over two million acres (800,000 ha) of mined lands have been restored, plus 100,000 acres (40,500 ha) of abandoned mines.

However, it wasn't until 1977 that Congress finally passed the landmark legislation, the **Surface Mining Control and Reclamation Act**. Earlier attempts had failed because of the perceived threat to jobs. This act applies only to coal mining and restricts mining in prime western farmlands or where owners of surface rights object. The act requires mine operators to 1) demonstrate reclamation proficiency; 2) restore the shape of the land to the original contour and revegetate it if requested by the landowner; 3) minimize impacts on the local **watershed** and **groundwater** and prevent **acid** contamination; and 4) pay a fee on each ton of coal mined into a $4.1 billion fund to reclaim orphaned lands left by earlier mining.

Since the act was passed, it has come under fire by both industry and environmentalists, resulting in more than 50 amendments. During the Carter administration, excessive litigation caused many delays to implementing and enforcing the regulations. When Reagan was in office, 60% of the regulations were reviewed or eliminated. The **Office of Surface Mining** Reclamation and Enforcement came under attack during the first Bush administration because of its lack of performance in enforcing the legislation. During this time 6,000 mines were abandoned without reclamation and there were many exemptions from regulations.

In 1995, several amendments were proposed to minimize the duplication of state and federal regulations. A 1997 report by the Public Employees for Environmental Responsibility claimed that continued lack of enforcement led to less than one percent of the 120,000 acres (48,500 ha) strip mined in Colorado and not one acre of over 90,000 acres (36,000 ha) of stripped Indian lands being reclaimed as of 1996. During the second Bush administration, new regulations governing mining on federal lands were established in 2001, which environmentalists claimed would reinstate dated reclamation standards that led to **pollution** of land and water.

Another type of strip mining, mountaintop removal, has recently come under attack for violating the reclamation legislation. This method involves using machinery to cut off entire mountaintops, as much as 400 ft (122 m), to reach the coal underneath. The rock and earth removed from these mountaintops are dumped into nearby streams in waste piles called valley fills. Environmentalists believe many streams are being destroyed; in West Virginia alone over 1,000 mi (1600 km) of streams have been buried. These mines violate the conditions for **buffer** zone variances (no mining activities can take place within a 100-ft (30-m) buffer zone near streams, unless environmental conditions are satisfied.) Another federal court ruling in 1999 said that valley fills also violated the **Clean Water Act**. For some mine operators who claim their sites will be used for grazing or **wildlife habitat**, improper reclamation methods prevent the sites from being used for this purpose.

Reclamation clearly adds to the cost of coal. For Appalachian coal this reaches 15% of the cost per **Btu** (British Thermal Unit). The 1977 act created a level playing field where all compete under the same rules. Costs easily involve $1,000–$5,000 per acre, but this is small compared to royalties paid the landowner. In Germany, where coal seams are very thick and land values are at a premium because of heavy population pressure, great efforts are made to reclaim the land, even up to $10,000 per acre. In terms of productive farm, grazing, or timber lands, restoration after a one-time extraction of coal allows a return to **sustainable agriculture** or forestry.

Strip mining creates four landforms which the reclamation process must address: rows of **spoil** banks, final cut canyons, high walls adjacent to the final cut, and coal-haul roads. The first two are relatively easy, but the latter two require special treatment. Reshaping the land to the original contour and keeping the **soil** in place are major challenges in hilly terrain. Operators usually find it necessary to cut into the unmined hillside to make it grade into the mined land below. **Subsidence** of **overburden** could conceivably create a cliff face along the headwall if not adequately compacted. Minimizing grading during the mining process is one way to minimize the amount of land excavation and alteration.

When possible, natural looking slopes can be achieved during the mining process by mining to the prescribed safety angles, or by the cut and fill method. This is generally the most inexpensive means of reclamation. Before mining, **topography** maps should be made of existing slopes and contours so that mining can match these as close as possible.

Besides physically reshaping the land, for reclamation to succeed the essential needs of **erosion** control, **topsoil**

replacement, and nurturing and protecting young vegetation until it can survive on its own must be met. These needs are interdependent. Vegetation is crucial for erosion control, especially as the slope gradient increases; however vegetation struggles without topsoil, critical nutrients, and protection from wildlife and livestock. One additional problem, **acid mine drainage**, is eliminated by good reclamation, as the oxidizing materials which produce the acid are buried.

Coal-haul roads are very dense from the heavy vehicles traversing them. They must be ripped up and plowed for even minimal revegetation success or, less desirable, be buried under overburden. Such access roads can cause the biggest disturbance to sites since they must be designed to meet roadway standards. These roads should thus be designed to minimize grading and require careful planning.

The one most crucial element, and most expensive, is topsoil replacement. Earthwork (backfilling and grading) accounts for up to 90% of reclamation costs. Original soils provide five major benefits: 1) a seedbed with the physical properties needed for survival; 2) a **reservoir** for needed nutrients; 3) a superior medium for water **absorption** and retention; 4) a source of native seeds and plants; and 5) an **ecosystem** where the decomposer and aerator-mixer organisms can thrive. The absence of even one of these categories often dooms reclamation efforts.

The more one studies this problem, the more important the topsoil becomes; truly it is one of the earth's most vital resources. Loose overburden is sometimes so coarse and lacking in nutrients that it can support little plant cover. A quick buildup of **biomass** is critical for erosion control, but without the topsoil to sustain both plant productivity and the **microorganisms** needed to decompose the dead biomass, any ground cover is soon lost, exposing the soil to increasingly higher erosion rates.

Some areas, such as flat lowlands, can revegetate with or without human aid. But for many lands, especially those with large fractures of rock and air voids, reclamation is practically an all or nothing venture, with little tolerance for halfway measures. Indeed, in conditions where most of the negative impacts are retained within the site, reclamation may easily worsen conditions by removing the barriers to **runoff** and **sediment**.

Stabilization of the site may also be needed to complete the mining and reclamation process. Methods include extensive grading, slope alteration, hazard removal, and soil stabilization. The latter method increases the load **carrying capacity** of soils and can be achieved by using reinforced earth or a chemical treatment. The ability to handle runoff and precipitation is improved. Runoff can also be reduced by planting vegetation on top of the slope or cut.

Much of what is needed for effective reclamation is known. What has been missing has been the will to do it and the legal clout for enforcement. A new approach holds promise to increase this willingness to reclaim the land, which is based on the emerging field of eco-asset management. Ecological resources such as forests and **wetlands** are developed and treated as financial assets to their owner. This market-based approach results in higher quality reclamation, an increase in the number of sites reclaimed, and economic benefits to property owners and other participants. Hopefully, such an approach will help persuade the mining industry that reclamation is an investment in the future. *See also* Mine spoil waste; Restoration ecology

[*Laurel M. Sheppard*]

Resources

Books

Bradshaw, A. D., and M. J. Chadwick. *The Restoration of Land: The Ecology and Reclamation of Derelict and Degraded Land.* Berkeley: University of California Press, 1980.

Law, D. L. *Mined-Land Rehabilitation.* New York: Van Nostrand Reinhold, 1984.

Meleen, N. H. *Geomorphological Perspectives on Land Disturbance and Reclamation.* Oxford Polytechnic Discussion Papers in Geography, No. 22. Oxford, England: Oxford Polytechnic, 1986.

Powell, J. W. "The Reclamation Idea." In *American Environmentalism: Readings in Conservation History*, edited by R. F. Nash. 3rd ed. New York: McGraw-Hill, 1990.

The Practical Guide to Reclamation in Utah. Division of Oil, Gas & Mining, State of Utah, Department of Natural Resources, 2001.

Williams, R. D., and G. E. Schuman, eds. *Reclaiming Mine Soils and Overburden in the Western United States: Analytic Parameters and Procedures.* Ankeny, IA: Soil Conservation Society of America, 1987.

Periodicals

Barnard, J. "Has the Bureau of Reclamation Met the Needs of the Changing West?" *Associated Press*, June 18, 2002.

"Hardrock Mining on Federal Lands." *The National Academy Press*, 2000.

Kenworthy, T. "New Mining Rules Reverse Provisions." *USA Today*, October 25, 2001.

"Mountain Top Removal Mining Called Illegal." *Environmental News Network* 1998 [cited July 2002]. <http://www.enn.com>.

Ward Jr., K. "Lawmaker Offers Mixed Reaction to Strip Mining Ruling." *Knight-Ridder/Tribune Business News*, May 15, 2002.

Other

Citizens Coal Council. [cited July 2002]. <http://www.citizencoalscouncil.org>.

Empty Promise. Public Employees for Environmental Responsibility, 1997.

Green, E., et al. *The Surface Mining Control and Reclamation At of 1977: New Era of Federal-State Cooperation or Prologue to Future Controversy?* 16 E. Min. L. Inst., Chapter 11, 1997.

Narten, P. F., et al. *Reclamation of Mined Lands in the Western Coal Region.* U.S. Geological Survey Circular 872. Alexandria, VA: U.S. Geological Survey, 1983.

Natural Strategies. October 2001 [cited July 2002]. <http://www.naturalstrategies.com/nsn-10-01.htm>.

Record of Decision

A Record of Decision (ROD) is a public document that explains which remedial alternatives will be used to clean up a Superfund site. A Superfund site is a site listed on the **National Priorities List** (NPL), which identifies sites in the United States that pose the greatest long term threat to human health and the **environment**. Placement on the NPL means that clean up of the site must follow the requirements of the **Comprehensive Environmental Response, Compensation, and Liability Act**, which outlines the process for completing site clean up.

The first step in the process is to conduct a Remedial Investigation and Feasibility Study (RI/FS). During the RI, data are collected to characterize site conditions, determine the nature of the wastes, assess risk to human health and the environment, and conduct treatability studies to evaluate the potential performance and cost of treatment technologies being considered. The FS is the mechanism used to develop, screen and evaluate alternative remedial actions. A result of the RI/FS is the development of the Proposed Plan, which is the identification of a recommended remedy. The preferred remedy is one that will be effective over both the short and long-term, reduce toxicity, mobility and/or volume of contamination, be technically and economically feasible to implement, and be acceptable to the state and the community. After a period for public comment on the RI/FS and the Proposed Plan, the ROD is prepared. The ROD has three basic components: (1) the Declaration, which is an abstract and data certification sheet for the key information in the ROD and is the formal authorizing signature page; (2) the Decision Summary, which provides an overview of site characteristics, alternatives evaluated, and the analysis of those options; it also identifies the selected remedy and explains how the remedy fulfills statutory and regulatory requirements; and (3) the Responsiveness Summary, which presents stakeholder concerns that were obtained during the public comment period about the site and preferences regarding the remedial alternatives. The Summary also explains how those concerns were addressed and how the preferences were factored into the remedy selection process.

[*Judith L. Sims*]

Resources

Books

U.S. Environmental Protection Agency. *A Guide to Preparing Superfund Proposed Plans, Records of Decision, and other Remedy Selection Decision Documents. EPA 540-R-98-031.* Washington, DC: Office of Solid Waste and Emergency Response, U.S. Environmental Protection Agency, 1999.

Other

U.S. Environmental Protection Agency. *Superfund Cleanup Process.* March 28, 2001 [cited June 2002]. <http://www.epa.gov/superfund/action/process/sfproces.htm>.

Recreation

The term recreation comes from the Latin word *recreatio*, referring to refreshment, restoration, or recovery. The modern notion of recreation is complex, and many definitions have been suggested to capture its meaning. In general, the term recreation carries the idea of purpose, usually restoration of the body, mind, or spirit. Modern definitions of recreation often include the following elements: 1) it is an activity rather than idleness or rest, 2) the choice of activity or involvement is voluntary, 3) recreation is prompted by internal motivation to achieve personal satisfaction, and 4) whether an activity is recreation is dependent on the individual's feelings or attitudes about the activity.

The terms "leisure" and "play" are often confused with recreation. Recreation is one kind of leisure, but only part of the expressive activity is leisure. Leisure may also include non-recreational pursuits such as religion, education, or community service. Although play and recreation overlap, play is not so much an activity as a form of behavior, characterized by make-believe, **competition**, or exploration. Moreover, whereas recreation is usually thought of as a purposeful and constructive activity, play may not be goal-oriented and in some cases may be negative and self-destructive.

The benefits of recreation include producing feelings of relaxation or excitement and enhancing self-reliance, mental health, and life-satisfaction. Societal benefits also result from recreation. Recreation can contribute to improved public health, increased community involvement, civic pride, and social unity. It may strengthen family structures, decrease crime, and enhance **rehabilitation** of individuals. Outdoor recreation promotes interest in protecting our **environment** and has played an important educational role. However, recreational activities have also damaged the environment. Edward Abbey, among others, have decried the tendency of *industrial tourism* to destroy wild areas and animal habitats. For example, some cite the damming of wild rivers to create lakes for boating and skiing or defacing mountain sides for ski runs and ski lifts as putting human recreation over environment.

Recreation is big business, creating jobs and economic vitality. In 2001 American consumers spent $100 billion on recreation and leisure activities. Much of these expenditures are for wildlife-related recreation. In 2001, more than 80 million Americans age 16 or older enjoyed some sort of recreation related to **wildlife** like fishing, **hunting**, birdwatching, or wildlife photography. More than 13 million

adults hunted in 2001, with overall hunting participation dropping about 7% from 1996 to 2001. Wildlife watching remained a popular outdoor activity and more than 66 milion people age 16 or older fed, photographed, or observed wildlife in 2001, spending some $40 billion on their wildlife activities.

More adults participate in walking for pleasure than any other recreational activity. Driving for pleasure, sightseeing, picnicking, and swimming are also among top recreational activities. Canoeing has been the fastest growing activity over the past 30 years.

Outdoor recreation also stimulates tourism. In 2001, almost 300 million recreation visits were made to our national parks. The United States has more than 778 million acres (314 million ha) of publicly-owned recreation lands, more than any other nation. Over one-third of the contiguous United States is public recreation land; however, these lands are not evenly distributed. Most of these lands are in the sparsely populated western states. There are 8,373 acres (3,391 ha) of public recreation lands per 1,000 people in the West, whereas there is only 294 acres (119 ha) per 1,000 people in the northeastern states. Almost 91% of the public lands area is administered by the federal government; 8% by states, 0.7% by counties, and 0.4% by municipalities. However, when number of sites, rather than acres, is considered, the recreation supply picture changes considerably. Almost 62% of all recreation sites are municipal, whereas only 1.6% of the sites are federal.

The private sector is also a major recreation supplier, particularly for certain activities. For example, there are more than 10,000 private campgrounds, 600 ski resorts, and 4,789 privately-owned **golf courses** in the United States.

Several important trends will affect the supply and demand of future recreation opportunities in the United States. Increasing populations, increasing ethnic diversity, the aging of the population, changes in leisure time, disposable income, and mobility will affect the demand for recreation. The loss of open spaces; **pollution** of our lakes, rivers, and coastlines; increasingly limited access to private lands; and increasing liability concerns will make the task of meeting future demand for recreation more difficult.

[*Ted T. Cable*]

Resources

Books

Kelly, J. R. *Recreation Business.* New York: Wiley, 1985.

Kraus, R. *Recreation and Leisure in Modern Society.* 3rd ed. Dallas, TX: Scott, Foresman, 1984.

The Report of the President's Commission on Americans Outdoors, The Legacy, The Challenge. Washington, DC: Island Press, 1987.

Other

"National Park 2001 Visitor Use Summary." *National Park Service.* [cited July 9, 2002]. <http://www2.nature.nps.gov/stats>.

U.S. Fish and Wildlife Service. *2001 National Survey of Fishing, Hunting, and Wildlife Associated Recreation.* Washington, DC: U.S. Government Printing Office, 2002.

Recyclables

Recyclables are products or materials that can be separated from the **waste stream** and used again in place of raw materials. Since colonial times, Americans have recycled a host of materials, ranging from corn husks used for mattress stuffing to old clothes used for quilts. Today, household recyclables include newspapers, mixed waste paper, glass, tin, **aluminum**, steel, **copper**, **plastics**, batteries, yard debris, wood, and used oil. Commercial recyclables include scrap metals, concrete, plastics, corrugated cardboard, and other nonferrous scrap material. The list of recyclables will likely expand as technology meets a growing demand for more **recycling** in response to increased consumer awareness and waste disposal costs, dwindling **landfill** space, and more stringent **waste management** regulations. The **Environmental Protection Agency** (EPA) has emphasized the importance of diverting recyclables from the waste stream by endorsing integrated waste management, in which **municipal solid waste** is managed according to a hierarchy of source reduction, recycling, **solid waste incineration**, and landfilling.

Programs to divert household recyclables from the waste stream are typically developed and implemented at the local or community level, while commercial recyclables are usually collected by private industry. Collection and preparation methods vary among recycling programs. Some communities have implemented curbside collection programs that require households to separate recyclables from their waste and sort them into segregated containers. Other curbside collection programs pick up recyclables separated from **household waste** but commingled together. Still other communities are responsible for picking out recyclables mixed in the waste stream. In some municipalities, recyclables are sorted and prepared for processing or market at a Materials Recovery Facility (MRF). Most MRFs rely on workers to hand-sort recyclables, though some also use sorting equipment, including magnets for removing metals and blowers for sorting plastics. Plastics are also hand-sorted according to resins identified by a voluntary coding system consisting of a triangular arrow stamp with a number in the center and letters underneath that identify the seven major types of plastic resins used in containers. Other equipment exists for size reduction (shredding or grinding), weighing, and baling recyclables.

Increasing both the type and quantity of recyclables are goals of many states. In 1999, recycling (including **composting**) diverted 64 million tons of material away from landfills and incinerators. That's nearly double the amount diverted in 1990 of 34 million tons. Batteries had the highest recovery rate of any recyclable; almost 97% of all batteries were recycled. Slightly more than 41% of all paper and paperboard products were recycled and almost 40% of all plastic drink containers made their way to recycling centers, a major improvement over the last decade or so. An estimated 27% of glass was recovered from the waste stream. Approximately 54% of aluminum packaging was recovered for recycling. Recyclables also exist in the form of durable goods, which are products that have a lifetime of over three years. These are usually bulky items, such as major appliances, which are not mixed with the rest of the waste stream. Ferrous metals can be recycled from refrigerators, washing machines, and other major appliances, known as "white goods."

Recyclables are commodities, and markets for these commodities fluctuate dramatically. The dynamic nature of markets is a key factor in whether a recyclable is actually recycled instead of being disposed in a landfill or combusted in an incinerator. Municipalities involved in recycling programs must deal with rapid shifts in the market for recyclables. For example, the demand for a particular recyclable may drop, reducing the price and forcing the municipality to pay for the material to be taken away for recycling or disposal. Consequently, recycling must compete with raw material markets, as well as waste disposal methods.

An additional advantage of recycling is that it helps reduce greenhouse gas emissions. These emissions can impact the earth's **climate**. In 1996, it was estimated that recycling of **solid waste** in the United States alone may have prevented the release of 33 million tons of **carbon** into the air, about the same amount as emitted by 25 million cars in one year. *See also* Garbage; Green packaging; Municipal solid waste; Solid waste incineration; Waste reduction

[*Marci L. Bortman*]

Resources

Periodicals

Pollock, C. S. "Building a Market for Recyclables." *World Watch* 1 (May–June 1988): 12–18.

Other

"Basic Facts, Municipal Solid Waste" *U.S. Environmental Protection Agency.* [cited July 8, 2002]. <http:www.epa.gov/epaoswer/non-hw/muncpl/facts.htm>.

U.S. Environmental Protection Agency. *Characterization of Municipal Solid Waste in the United States: 1990 Update.* Washington, DC: U.S. Government Printing Office, 1990.

U.S. Office of Technology Assessment. 1989. *Facing America's Trash: What Next for Municipal Solid.* Washington, DC: U.S. Government Printing Office, 1989.

Recycling

Recycling waste is not a new idea. Throughout history, people have disposed **garbage** in myriad ways. They fed household garbage to domestic animals. Scavengers gleaned the **waste stream** for usable items that could be fixed, then sold or traded for other goods and services. Homemakers mended clothing, and children grew up in hand-me-downs. Durable goods were just that; goods that could be reused until their durability wore out. These practices were not due to a desire to reduce the waste stream, but rather a need to produce products from all available resources. Modern society has moved away from such straightforward recycling practices, choosing instead to toss out the old and buy new goods. This throwaway society now faces a trash crisis.

The volume of **solid waste** generated in the United States has continued to increase, along with the cost of building landfills and **incineration** facilities. Recycling some of this material into new uses saves existing **landfill** space, conserves energy and **natural resources**, reduces **pollution**, and saves tax money. Recycling can provide new jobs, create new industries, and contribute to the increase in the GNP. Instead of referring to what people discard every day as garbage, advocates of recycling emphasize the need to refer to waste products as Post Consumer Materials (PCMs) and consider them renewable resources for various manufacturing processes.

The challenge of collecting adequate volumes of recyclable materials in a form ready for manufacturing into new products is a formidable task. Unlike raw materials that are extracted from the earth or manufactured in a lab, PCMs are mixed materials, and are sometimes contaminated with toxic and non-toxic residue. PCMs must be cleaned and source-separated at the individual household and business level. Once the materials are made available at the point of generation, they must be collected and transported to a collection site, usually referred to as a Materials Recycling Facility (MRF). At this location, materials are processed to make **transportation** easier and increase their value. Glass is separated by color, **plastics** are pelletized, and paper is baled. Other materials may be cleaned and reduced in volume, depending on the equipment available. The exceptions to this rule are single materials collected and transported directly to the point of use, such as newspaper collected at a drop off site and transported to a local insulation manufacturer. The existence of an MRF enhances the recycling process because it allows large volumes of materials to be amassed at a single location, making marketing of PCMs

more profitable. If manufacturers know that there are adequate volumes of high-quality raw material consistently available, they are more likely to agree to long-term purchasing contracts.

Manufacturers are being encouraged to retrofit, or retool, their manufacturing processes in order to use PCMs instead of raw materials. Many glass manufacturers now use cullet (cut glass) instead of silica and sand to make new glass containers, and some paper manufacturers have installed deinking equipment in order to manufacture paper from used newsprint instead of wood pulp. Once the initial expense of equipment change has been absorbed, such changes can reduce costs. But these are long-term investments on the part of manufacturers, and they must be ensured that recycling is a long-term commitment on the part of consumers and communities.

Recycling begins with people, and the biggest challenge for recycling lies in educating and motivating the public upon whom successful **source separation** depends. Except in rare circumstances recycling is a voluntary act, so the attitudes, knowledge, and feelings of the participants are the key to sustained recycling behavior. It has been found that people will change their behavior to protect the health of their families, the value of their property, and their own self-image, and recycling has often been justified in these terms. Once the rationale to begin recycling has been shared with community residents, a program for making recycling convenient and inexpensive must be institutionalized.

Curbside recycling is currently the most desirable recycling option, primarily because of its convenience. BioCycle Magazine reported in April of 1999 only nine states with fewer than 10 curbside recycling programs. These are usually commingled, or mixed, systems. Residents clean and prepare glass, cans, paper, and plastic, before putting them together in a container and placing them at the curb. Most waste haulers pick up the **recyclables** on the same day as the rest of the waste, separating them into bins on the truck right at the curb. In more rural areas, where homes are too far apart, recycling often takes place at a drop off center. This is usually a tractor trailer with separate bins, which is hauled directly to an MRF when it is full and replaced by an empty trailer. Research has shown that the best system is one where curbside recycling is offered, but a drop-off site is also available as a backup in case residents miss their recycling day. In some communities there is a weekly or monthly drop-off. Trucks from different industries use a convenient parking lot and residents are encouraged to bring their recyclables to company representatives. Some communities also have buy-back centers where residents can sell PCMs to a broker who in turn sells the material to manufacturers.

Many towns now have curbside recycling programs with trucks picking up recyclables on a weekly basis. (Photograph by Robert J. Huffman. Field Mark Publications. Reproduced by permission.)

In addition to residential recycling there is office or business recycling. Office paper recycling programs are usually set up so office workers separate the paper at their desks. The paper is taken regularly to a location where it is either baled or picked up loose and transported to a manufacturing plant. Offices that generate a large amount of white office paper can collect and market the paper directly to paper manufacturers. Smaller offices can cooperate with other businesses, contracting with paper manufacturers to pick up paper at a central location. Another example of business recycling is the recycling of corrugated cardboard by groceries and clothing stores. These businesses have been breaking down boxes and baling them on site for pickup by collection trucks. **Dry cleaning** establishments are currently taking back hangers and plastic bags from their customers for recycling. Businesses of many kinds have begun to realize that positive publicity can be achieved through their recycling efforts, and they have been increasingly inclined to participate.

One of the biggest challenges to recycling in the future is market development. For communities near to cities that have major manufacturers taking PCMs, marketing is not a problem. For smaller more rural communities, however, different strategies must be developed. In some cases a well run MRF that brokers recycled materials is adequate. In other cases, it may be necessary to search for local markets to utilize PCMs, such as the use of newspaper as animal bedding. Newspaper collected and processed within the community can be sold to local farmers at a lower cost less than straw or sawdust. Another approach is encouraging a niche industry to locate nearby, such as a small insulation contractor who is promised all the post-consumer newsprint from the community. Ultimately, recycling will not be successful unless consumers buy products made from PCMs. Products made from these materials are becoming easier to find, and they are more clearly marked than in the past, though some green or earth-friendly product labeling has been controversial. Recycling advocates have argued that such labeling should not only say that the product is made from recycled material, it should also note what percentage of the material is post consumer waste, as opposed to manufacturing waste.

Many states are encouraging recycling through legislation. According to the Container Recycling Institute in 1999 10 states had bottle bills requiring stores to take back beer and soft drink bottles and cans, giving consumers a refund. Certain states have banned all compostable materials from landfills in order to force **composting**. In some states, daily newspapers are required to use a certain percentage of recycled paper in order to publish. These and other legislative steps have enhanced the recycling movement nationwide. States with the highest recycling rate of 40% or higher include Maine, Minnesota, New York, South Carolina, South Dakota, and Virginia according to a report in BioCycle magazine in 1999. Alaska, Montana, and Wyoming have the lowest recycle rate of 9% or less according to the report. The majority of states (40) recycle between 10 and 39% of materials. No data was available for Idaho.

According the **Environmental Protection Agency** (EPA) in 2002, U.S. consumers recycle approximately 28% of all waste. The highest recycle rate is for steel packaging, with 57% being recycled. **Aluminum** drinking cans are recycled at a rate of 55%, compared to 40% of plastic soft drink bottles. The paper recycling rate is 42%, and 52% of appliances meet with the recycler. The EPA sponsors a free voluntary program to all U.S. organizations, called WasteWise, to encourage and track the results of corporate recycling programs.

The growing tonnage of computers, computer monitors, and television sets that overpower municipal landfills is an increasing problem that demands resolution. An estimated 70% of **heavy metals** found in third millennium landfills comes from digital hardware. Eight pounds (3.6 kg) of **lead** are contained in a single cathode ray tube (CRT) monitor, contributing to the threats of lead contamination. In May of 2002 the EPA proposed a recycling program to control this high-tech waste, including a provision to classify CRT computer monitors as reusable waste products. The necessity for these measures is evident in consideration of the 500 million computers in use worldwide, which contain an estimated 1.58 billion lb (716 million kg) of lead and 632,000 lb (287,000 kg) of **mercury**. According to EPA estimates the number of discarded computers between 2002 and 2007 might reach 250 million units in the United States alone.

Equally critical is the problem of **battery recycling**. Growing concerns over poor battery recycling practices by consumers are well founded because they are classified as toxic waste, with mercury and lead contamination the most serious problems caused by improper battery disposal. Yet rarely are batteries disposed of accordingly. Most nonrechargeable batteries are tossed in the trash where they end up in the landfills. The use of rechargeable batteries offers a very effective solution to alleviate much of this problem, because each rechargeable battery has a life cycle equal to hundreds of nonrechargeable batteries; some rechargeables have a life cycle of 1,000 recharges. Because of poor disposal practices, every rechargeable battery protects the landfill from potential contamination by hundreds of batteries.

The **National Recycling Coalition** is a nonprofit organization that provides leadership and coordination at the national level, sponsoring events, scholarships, and educational programs to encourage effective recycling practices. *See also* Container deposit legislation; Municipal solid waste; Solid waste recycling and recovery; Solid waste volume reduction; Transfer station; Waste management; Waste reduction

[*Cynthia Fridgen*]

Resources

Books

BioCycle: Journal of Composting and Recycling. JG Press.

Carless, J. *Taking Out the Trash: A No-Nonsense Guide to Recycling.* Washington, DC: Island Press, 1992.

Coming Full Circle: Successful Recycling Today. New York: Environmental Defense Fund, 1988.

Pollock, C. *Mining Urban Wastes: The Potential for Recycling.* Washington, DC: Worldwatch Institute, 1987.

Recycling and recovery, solid waste

see **Solid waste recycling and recovery**

Red tide

Red tide is the common name for phenomenon created when toxic algal blooms turn seawater red, killing marine life and making water unsuitable for human or animal use.

Red tides are caused by several **species** of dinoflagellates and diatoms, microscopic unicellular **phytoplankton** that live in cold and warm seas. A red pigment, called peridinin, which collects light during **photosynthesis**, colors the water red when large numbers of the **plankton** populate an area.

Red tides are a danger because the **toxins** released by large numbers of these plankton can paralyze fish and bioaccumulate in the tissues of shellfish and filter-feeding mollusks. Predators of the shellfish, including humans, consume the toxins causing paralysis and death.

Species of dinoflagellates that cause red tides generally belong to the "red tide genera," *Gymnodinium* and *Gonyaulax*. Though several are known to cause red tides, a single red tide is nearly always tied to a specific species. This may be the result of three factors. First, the conditions responsible for the bloom may also cause the red tide species to reproduce more rapidly than other phytoplankton, thus outcompeting them for available nutrients. Second, the toxins excreted may prevent the growth of other species. Or, behavioral differences may give them competitive advantages.

Red tides have occurred periodically through recorded history in seas around the world. The Red Sea is thought to have been named for algal blooms back in biblical times. However, in recent decades red tides have been more frequent in areas where algal blooms have never been a problem.

The coast of Chile has reported red tides since the beginning of the nineteenth century, and the Peruvian coast has recorded red tides, which they named *aguajes*, since 1828. They appear more frequently in the summer months when the water becomes warmer, and they also seem to be associated with the **El Niño** current. Coastal up-welling events are also known to be essential for the occurrence of *Gonyaulax tamarensis* blooms in the Gulf of Maine.

Red tides have been plaguing the east and west coasts of Florida on a nearly annual basis since at least 1947. Blooms on the west Florida shelf are initiated by the Loop Current, an annual intrusion of oceanic waters into the Gulf of Mexico. Here, the toxic dinoflagellate *Gymnodinium breve* produces rather predictable blooms. The red tide is transported inshore by winds, tides, and other currents where it may be sustained if adequate nutrients are available.

In 1987 and 1988, red tide spread northward from the Gulf Coast of Florida as far as North Carolina, where it caused a loss of $25 million to the shellfish industry due to brevetoxin contamination. This was the first time that *G. breve* blooms were reported so far north, and it raises the question of whether blooms will occur annually in this region. Also during the 1987–88 blooms, 700 bottlenose **dolphins** washed ashore along the United States Atlantic coast. Red tides commonly occur along the northwest coast of British Columbia, Canada, and as far north as Alaska and along the Russian coast of the Bering Sea.

Researchers blame the recent spread of red tides and the occurrence of algal blooms, which ordinarily occur in harmless low concentrations, on the increasing amount of coastal **water pollution**. Sewage and agricultural **runoff** increase concentrations of nutrients, such as **nitrogen** and **phosphorus**, in coastal waters. This increase in nutrients may create favorable conditions for the growth of red tide organisms.

In April 1992, two fishermen suffered paralytic shellfish **poisoning** (PSP) within several minutes of eating a few butter clams from Kingcome Inlet, British Columbia. In August 1992, saxitoxin, another dinoflagellate toxin associated with PSP, was found in the digestive systems of dungeness crabs caught in Alaskan waters. On Prince Edward Island, Canada, in 1987, three people died and more than 100 others became ill from domoic acid-contaminated mussels. Some of those who became ill from this toxin developed amnesic shellfish poisoning, which still causes short-term memory loss.

It is difficult to accurately predict when and where red tides may occur, and how seriously they will contaminate shellfish. Monitoring levels of toxins in shellfish meat and population sizes of the dinoflagellates is important in preventing widespread poisoning of humans. Any detectable level of brevetoxin in 3.5 oz (100 g) of shellfish meat is potentially harmful to humans. The U.S. **Food and Drug Administration** (FDA) does not inspect shellfish growing areas regularly, but relies on state agencies to monitor toxin levels. In selected states, surveillance programs have been established that randomly examine shellfish that have been harvested in natural shellfish beds. Florida is the only state that has a constant monitoring and research program for both shellfish poisoning and dinoflagellate blooms that cause poisoning. Monitoring for shellfish poisoning is costly, but the expense may increase if red tides become more frequent due to deteriorating coastal **water quality**. In April and May of 1996, 158 **manatees** in southwest Florida died from an unusual type of red tide. In China from 1997 until 1999, there were 45 red tides which wiped out 75% of the entire stock of Hong Kong's fish farms. This was an estimated $240 million loss to Hong Kong's ecomonic system. *See also* Agricultural pollution; Bioaccumulation; Competition; Marine pollution; Sewage treatment

[*William G. Ambrose and Paul E. Renaud*]

Resources

Books

Taylor, D., and H. Seliger, eds. *Toxic Dinoflagellate Blooms.* Vol. 1. New York: Elsevier North Holland, 1979.

PERIODICALS

Culotta, E. "Red Menace in the World's Oceans." *Science* 257 (1992): 1476–1477.

Konovalova, G. V. "Harmful Dinoflagellate Blooms Along the Eastern Coast of Kamchatka." *Harmful Algae News* 4 (1993): 2.

Taylor, F. J. R. "Artificial Respiration Saves Two From Fatal PSP in Canada." *Harmful Algae News* 3 (1992): 1.

Redwoods

There are three genera of redwood trees, each with a single **species**. The native range of the coast redwood (*Sequoia sempervirens*) is a narrow 450-mi (725-km) strip along the Pacific Ocean from central California to southern Oregon. The giant sequoia (*Sequoiadendron giganteum*) is restricted to about 75 groves scattered over a 260-mi (418-km) belt, nowhere more than 15 mi (24 km) wide, extending along the west slope of the Sierra Nevadas in central California. The third species, dawn redwood (*Metasequoia glyptostroboides*), was described first from a fossil and was presumed to be extinct. However, in 1946 live trees were discovered in a remote region of China. Since then, seeds have been brought to North America, and this species is now found in many communities as an ornamental planting. Unlike the coast redwoods and giant sequoias, dawn redwoods are deciduous.

Redwoods are named for the color of their heartwood and bark. The thick bark protects the trees from fires which occur naturally throughout their ranges. Their wood has a high tannin content which makes it resistant to **fungi** and insects, making redwood a particularly desirable building material. This demand for lumber was responsible for most of the destruction of the original redwood forests in North America. In 1918, the **Save-the-Redwoods League** was formed to save redwoods from destruction and to establish redwood parks. This organization has purchased and protected over 280,000 acres (113,400 ha) of redwood forests and was instrumental in the establishment of the Redwoods **National Park**. The term *Sequoia* used in the generic names of these species, and in the common name giant sequoia, honors the Cherokee Chief Sequoyah who developed an alphabet for the Cherokee language.

Coast redwoods. (Photograph by Jack Dermid. Photo Researchers Inc. Reproduced by permission.)

Coast redwoods

Coast redwoods are the tallest and one of the longest living tree species. Average mature trees are typically 200–240 ft (61–73 m) tall, although some trees exceed 360 ft (109 m). The world's tallest known tree is a coast redwood that stands 368 ft (112 m) tall on the banks of Redwood Creek in Redwood National Park. In some areas coast redwoods can live for more than 2,000 years. They are evergreen, with delicate foliage consisting of narrow needles 0.5–0.75 in (1.3–2 cm) long, growing flat along their stems and forming a feathery spray.

Coast redwoods are prolific. Their cones are about an inch long and contain from 30 to 100 seeds. They produce seeds almost every year with maximum seed production oc-

curring between the ages of 20–250 years old. They also have an advantage over other species in that new trees can sprout from the roots of damaged or fallen trees.

Unlike the other two species of redwoods, coastal redwoods cannot tolerate freezing temperatures. They thrive in areas below 2,000 ft (607 m), with summer fog, abundant winter rainfall, and moderate temperatures. Although coast redwoods are often found in mixed evergreen forest communities, they can form impressive pure stands, especially on flat, riparian areas with rich soils.

Giant sequoia

The giant sequoia, although not as tall as the coast redwood, is a larger and more long-lived species. Giant sequoias can attain a diameter of 35 ft (10.6 m); whereas the largest Coast redwood has a 22-ft (6.7-m) diameter. The largest tree species by volume, giant sequoias the trees through which roads were built and rangers' residences hollowed out. The most massive specimen, the General Sherman tree, located in Sequoia National Park, has a bole volume of 52,500 ft^3 (1,575 m^3)

The oldest giant sequoia is 3,600 years old, compared with 2,200 years old for the oldest coast redwood. This makes the giant sequoia the second oldest living thing on earth (the bristlecone pine is the most long-lived).

Giant sequoias can be found between elevations of 5,000 and 8,000 ft (1,517 and 2,427 m), growing best on mesic sites (such as bottomlands) with deep, well-drained sandy loam soil. Unlike coast redwoods, mature trees of this species cannot sprout from the roots, stumps, or trunks of injured or fallen trees (young trees can produce stump sprouts subsequent to injury). It is paradoxical that one of the largest living organisms is produced by one of the smallest seeds. Three thousand seeds would weigh only an ounce. Typically, cones bearing fertile seeds are not produced until the tree is 150–200 years old. The cones are egg-shaped and 2.0–3.5 in (5–9 cm) in length. Once cones develop they may continue to grow without releasing the seeds for another 20 years. A typical mature tree may produce 1,500 cones each year, and because they are not dropped annually, a tree may have 30,000 cones at one time, each with about 200 seeds. Two species of animals, a wood-boring beetle (*Phymatodes nitidus*) and the chickaree or Douglas squirrel (*Tamiasciurus douglasii*), play important roles in dislodging the seed from the cones. Fire is also important in seed release as the heat drys the cones. Fire has the added advantage of also preparing a seed bed favorable for germination. The light seeds are well-adapted for wind dispersal, often travelling up to a quarter mile away from the source tree.

Like the other redwoods, the wood of the giant sequoia is extremely durable. Because of this durability and the ornate designs in the wood, this species was harvested extensively. One tree can produce up to 600,000 board feet of lumber, enough to fill 280 railroad freight cars or build 150 five-room houses. Unfortunately, because of the enormous size of the trees and the brittle nature of the wood, when a tree was felled, as much as half would be wasted because of splintering and splitting. Now that virtually all giant trees have been protected from **logging**, the greatest threats from humans are **soil compaction** around their bases and the elimination of fire which allows fuels to accumulate, thereby increasing the chances of deadly crown fires. Toppling over is the most common natural cause of death for mature giant sequoia trees. Weakening of the shallow roots and lower trunk by fire and decay, coupled with the tremendous weight of the trees, results in the tree falling over. Sometimes wind, heavy snows, undercutting by streams, or water-soaked soils contribute to the toppling. In its natural range this species is now valued primarily for its aesthetic appeal.

[*Ted T. Cable*]

Resources

Books

Burns, R. M., and B. H. Honkala. *Silvics of North America.* Vol. 1, *Conifers.* Washington, DC: U.S. Dept. of Agriculture, 1990.

Dewitt, J. B. *California Redwood Parks and Preserves.* San Francisco: Save-the-Redwoods League, 1985.

Walker, L. C. *Trees: An Introduction to Trees and Forest Ecology for the Amateur Naturalist.* Englewood Cliffs, NJ: Prentice-Hall, 1984.

Refuse-derived fuels

The concept of refuse-derived fuels (RDFs) is one that has the potential for addressing two of the most troubling environmental problems in the world at the same time: **solid waste** disposal and a source of energy. The term refuse-derived fuel refers to any process or method by which waste materials are converted into a form in which they can be burned as a source of energy.

In regions of the world characterized by a throw-away ethic and swamped with essentially non-degradable materials, accumulation and disposal of solid waste continue to be a growing problem. In 1960, each American generated 2.7 lb (1.2 kg) of solid waste. This grew to 4.3 lb (1.9 kg) per person by 1990. Americans continue to generate more solid waste each day but the rate of growth has decreased. In 2002, every man, woman, and child in the United States produced an average of 4.5 lb (2.0 kg) of waste each day.

Materials are labeled wastes because they tend to have no future use once they are discarded. Yet, by their very

nature, most solid wastes are potentially valuable as fuels. A typical sample of municipal waste in the United States, for instance, may consist of about 30–40% paper, 5% textiles, 5% wood, and 20–30% organic material. All of these materials are combustible. Metals, glass, **plastics**, sand, and other non-combustible materials constitute the remaining portion of a waste sample. Each of these non-combustible materials is potentially recyclable.

Even though 70% or more of municipal wastes appear to have potential value as sources of fuel, that potential has not, as yet, been extensively developed. In most countries, solid wastes are mainly disposed of in landfills. A small fraction is incinerated or used for other purposes. However, conversion of solid wastes to a useable fuel is still an experimental process.

The primary roadblock to the commercial development of RDFs is the economic cost of preparing such fuels. Given that most RDFs produced have only about half the energy value of a typical sample of industrial **coal** and given the relatively low price of coal, there is little economic incentive for municipalities to build energy systems based on refuse-derived fuels.

Many scientists, engineers, and environmentalists see a positive future for refuse-derived fuels. They argue that as fossil fuel reserves are consumed, the cost of traditional fuels such as coal, gas, and oil will inevitably increase. In addition, RDFs tend to burn more cleanly and have a significant environmental advantage over coal and other **fossil fuels**. Research must continue, therefore, to develop more efficient and less expensive RDFs and associated processes to obtain and use them.

One of the fundamental problems in the development of refuse-derived fuels is obtaining the raw materials in a physical condition that will allow the extraction of combustible organic matter. The solid waste material entering most landfills consists of a complex mixture of substances. Some of these are combustible while others are not. Some have other commercial values while others have none. Plastics provide a convenient example that illustrates both of these aspects.

The first step in preparing wastes for the production of refuse-derived fuels is known as size reduction. In this step, waste materials are shredded, chopped, sliced, pulverized, or otherwise treated in order to break them up into smaller pieces. Most methods for separating combustible from non-combustible materials require that waste particles be small in size. Both **magnetic separation** and air classification will work only with small particles.

Size reduction also serves a number of other functions. For example, it breaks open plastic and glass bottles, releasing any contents that may still be in them. It also reduces clumping and tangling that often occurs with larger particles. Finally, when used as a fuel, particles of small size burn more efficiently than larger particles.

One indication of the importance of size reduction in the production of RDFs is the number of different machines that have been developed to accomplish this step. Bag breakers are used to tear open plastic bags and cardboard boxes. Machines called chippers reduce cut wood into small pieces. Granulators have sharp knives on rotating blades. These devices cut materials against a stationary blade, somewhat similar to a pair of scissors. Flail mills are rotary machines that swing hammers around in a circle to smash materials. Ball and rod mills consist of cylindrical drums that hold balls or rods. When the drums spin, balls or rods smash against the inside of the drum, crushing solid wastes also present in them.

Once waste materials have been reduced in size, the portion that is combustible must be removed from that which is not. Several different procedures can be used. A common process, for example, is to pass the solid wastes under a magnet. The magnet will extract all metals containing iron (ferrous) from the mixture.

Another technique is rotary screening, in which wastes are spun in a drum and sorted using screens that have openings of either 50 mm or 200 mm diameter. Sand, broken glass, dust, and other non-organic materials that do not burn tend to be removed by this procedure.

A third process is known as air classification. When solid wastes are exposed to a blast of air, various components are separated from each other on the basis of their size, shape, density, and moisture content. A number of different kinds of air classifiers exist, and all are reasonably efficient in separating organic from inorganic materials.

One type of air classifier is the horizontal model, in which size-reduced wastes are dropped into a horizontal stream of air. Various types of material are separated on the basis of the horizontal distance they travel from their point of origin. Vertical classifiers separate on a similar principle except that particles are injected into a rising stream of air. The rotary classifier consists of an inclined cylindrical column into which wastes are injected and then separated by a rotating, rising column of air.

The organic matter that remains after size reduction and separation can be used as a fuel in four distinct ways. First of all, it can be converted to other combustible forms through pyrolysis. Pyrolysis is a process in which organic materials are heated to high temperatures in an oxygen-free or oxygen-deficient **atmosphere**. The product of this reaction is a mixture of gaseous, liquid, and solid compounds. The solid compounds can be used as fuels and are collectively called char.

Two particularly valuable fuels, **methane** (a gas) and **ethanol** (ethyl alcohol, a liquid), can also be produced from

organic wastes. Scientists have long known that methane is produced when **anaerobic** bacteria act on organic compounds found in solid wastes. They have successfully extracted methane from landfills and used it as a fuel. However, they have achieved only limited success in generating methane from other refuse outside of landfills.

Some success has been obtained in the production of ethanol from wastes. This process separates materials containing cellulose from other wastes and then hydrolyzes them in an acidic medium. As of 2002, ethanol has not been generated in commercially useful amounts from solid waste streams. However, ethanol has been obtained from corn and some other crop materials. In some states, ethanol is added to **gasoline**, producing a fuel that burns more cleanly and efficiently than pure gasoline.

Some heating systems make use of the dried material produced as a result of size reduction and separation directly as a fuel. The Imperial Metal Industries (IMI) company in Birmingham, England, has operated such a plant since 1976. Solid wastes provided by the West Midlands County Council is delivered to the IMI plant where it is reduced in size, separated, dried, and then fed directly into furnaces, where it is burned. A similar plant operated by Commonwealth Edison and run on wastes from the city of Chicago was opened in 1978.

A convenient and popular method of handling RDFs is to process them as pellets or briquettes. Pellets are formed from wastes that have been size reduced, separated, and partially dried. The material is extruded through a dye or compacted into small nuggets in devices known as densifying machines. An important factor in the success of this method is keeping the wastes moist enough to stick together, but dry enough for efficient **combustion**.

Finally, research has been conducted on the use of RDFs in combustion systems, such as **fluidized bed combustion**. In this process, fuel is mixed with some material such as sand and suspended in a furnace by a stream of air. The second material makes thorough mixing of the fuel possible. The fuel is then burned in the air stream.

A considerable number of research and engineering studies have been conducted on refuse-derived fuels. As of 2002, many municipalities operate plants that generate electricity. With the help of modest tax benefits in some but not all areas, most of these facilities are commercially successful. They have reduced the volume of waste entering local landfills. In turn, the life of existing landfills has been extended and the need to open new landfills has been reduced. They generate a modest but significant proportion of electricity for customers in their municipalities.

[*L. Fleming Fallon Jr., M.D., Dr.P.H.*]

Resources

Books

Diane Publishing *Test Firing and Emissions Analysis of Densified RDF Combustion in a Small Power Boiler.* Collingdale, PA: Diane Publishing, 1994.

Hasselriis, F. *Refuse-derived Fuel Processing.* Boston: Butterworth Publishers, 1984.

Porteous, A. *Refuse Derived Fuels.* London: Applied Science Publishers, 1981.

Vesilind, P. A., J. J. Peirce, and R. Weiner. *Environmental Engineering.* 2nd ed. Boston: Butterworth Publishers, 1988.

Periodicals

Pian, C. C., and K. Yoshikawa. "Development of a High-temperature Air-blown Gasification System." *Bioresource Technology* 3, no. 79 (2001): 231–241.

Other

"Biomass." *Understanding Energy.* [cited July 2002]. <http://www.energy.-org.uk/EFBioMas.htm>.

PSC Analytical Services. [cited July 2002]. <http://www.philipanalytical.-com/company.htm>.

Ruzic, David. *Biomass.* 1998 [cited July 2002]. <http://starfire.ne.uiuc.edu/ne201/course/topics/biomass>.

Tom [Thomas Howard] Regan (1938 –)

American philosopher and animal rights activist

Regan is a well-known figure in the animal liberation movement. His *The Case for Animal Rights* (1983) is a systematic and scholarly defense of the controversial claim that animals have rights that humans are morally obligated to recognize and respect. Arguing against the views of earlier philosophers like René Descartes, who claimed that animals are machine-like and incapable of having mental states such as consciousness or feelings of pleasure or pain, Regan attempts to show that such a view is misguided, muddled, or incorrect.

Regan was born in Pittsburgh, Pennsylvania. He attended Thiel College and the University of Virginia, where he earned his doctorate in philosophy. Before publishing *The Case for Animal Rights*, Regan also edited *Animal Rights and Human Obligations* (1976) and *Matters of Life and Death* (1980).

Within the **animal rights** movement, Regan disagrees with certain philosophical arguments advanced in defense of animal liberation by, most notably, the nineteenth-century English philosopher Jeremy Bentham and the twentieth-century Australian philosopher **Peter Singer**. Both Bentham and Singer are utilitarians who believe that morality requires the pleasures of all sentient creatures—human and nonhuman alike—be maximized and their pain minimized. From a utilitarian perspective, most discussions about rights are unnecessary. Regan disagrees, arguing that sentience—the ability to feel pleasure and pain—is an inadequate basis on which to build a case for animal rights and against such

practices as meat-eating, factory farming, fur **trapping**, and the use of animals in laboratory experiments.

Regan is currently a professor at North Carolina State University where he set up the National Bioethics Institute in 1999. He was also the recipient of the 2000 Holladay Medal. Regan views the case for animal rights as an integral part of a broader and more inclusive environmental ethic. He predicts and participates in the coming of a new kind of revolution—a revolution of gentility and concern—made not by a self-centered *me-generation* but by an emerging *thee generation* of caring and concerned people from every political party, religion, and socioeconomic class. Among the indications of this forthcoming revolution is the growing popularity of and financial support for the various groups and organizations that comprise the more broadly based environmental and animal rights movements.

[*Terence Ball*]

Resources
Books

Regan, Tom. *All That Dwell Therein: Essays on Animal Rights and Environmental Ethics.* Berkeley: University of California Press, 1982.

———, and Carl Cohen. *The Animal Rights Debate.* Lanham, MD: Rowman & Littlefield, 2001.

———. *The Case for Animal Rights.* Berkeley: University of California Press, 1983.

———. *The Thee Generation: Reflections on the Coming Revolution.* Philadelphia: Temple University Press, 1991.

Regulatory review

The regulatory review process allows the executive branch of the United States federal government to ensure that the regulations drafted by different agencies contribute to the current administration's overall goals. Regulatory review is also carried on at the state level, although the organizational structure of these councils or commissions varies somewhat from state to state. In the past the regulatory review process has emphasized a **cost-benefit analysis** that has tended to delay or halt the publication of environmental regulations. This pattern shifted after 1993 to a greater emphasis on citizen participation in the process of rulemaking.

Executive agencies carry out congressionally enacted laws by drafting and enforcing regulations. The **Environmental Protection Agency** (EPA) has responsibility for writing the regulations that implement environmental laws, like the **Clean Air Act** and the Comprehensive Environmental Response, Compensation and Liability Act (Superfund). For example, when the **Resource Conservation and Recovery Act** (RCRA) called for states to protect **groundwater** from **landfill** leachate, the EPA drafted a regulation that included specifications for the kind of equipment that landfills should install to ensure that leachate would not leak.

An emerging concern in regulatory review is the spiraling costs of regulations related to the **environment**. The price tag on environmental regulation has tripled over the last quarter century, from about $80 billion in 1977 to more than $267 billion in 2000. In addition, a recent study of regulatory review found that federal agencies involved with environmental regulations received the largest increases in the federal budget in recent years, the EPA in particular receiving an increase of over $1 billion for the fiscal year 2002.

Once drafted, before being published for public comment, all regulations are sent for review to the Office of Information and Regulatory Affairs (OIRA) within the **Office of Management and Budget** (OMB). OIRA either approves, rejects, or asks for specific changes in the regulation. If a regulation is approved it can be published on the Internet or in hard copy for public comment.

This centralization of the regulatory process was accomplished by Ronald Reagan's executive orders 12291 and 12498. Executive Order 12291, issued in 1981, established a cost-benefit review process. All federal agencies must weigh costs and benefits for each "major" regulation and submit these considerations to OIRA. (A "major" regulation costs more than $100 million or is otherwise deemed major by OIRA.) Agencies cannot publish proposed or final rules until OIRA has ensured that the benefits of a regulation outweigh the costs. Executive Order 12498, issued in 1984, requires agency heads to predict their regulatory actions for the following year. All regulations are then compiled into the *Regulatory Program.* Compiling this program allows OIRA to be involved in the early planning stages of all regulations.

In September 1993, Executive Orders 12291 and 12498 were replaced by President Clinton with Executive Order 12866 on regulatory planning and review. While Executive Order 12866 left the basic structure of OIRA in place, including the figure of $100 million as the standard of a major regulation, the order was intended to harmonize federal regulations regarding the environment with related state, local, and tribal regulations and procedures. The order also allowed for negotiated rulemaking and other consensual approaches to developing regulations in appropriate situations. A third important feature of Executive Order 12866 is its emphasis on transparency; that is, federal agencies are not only required to "provide the public with meaningful participation in the regulatory process," but also to allow time for comments on proposed regulations. To facilitate citizens' participation and response, all regulatory information is to be provided to the public "in plain, understandable language."

Executive Order 12866 became the basis for the establishment of two online databases maintained by OMB for purposes of regulatory review and communication with the public. OIRA has a Regulation Tracking System, or REGS, to monitor the regulatory review process. When a proposed regulation is submitted to OIRA, the REGS system verifies the information contained in the document and assigns each submission a unique OMB tracking number. It then prints out a worksheet for review. After the worksheet has been reviewed, its contents are entered into the REGS system and maintained in its database. REGS also provides a listing, updated daily, of all pending and recently approved regulations on the OMB home page. The OIRA Docket Library maintains certain records related to proposed Federal regulatory actions reviewed by OIRA under Executive Order 12866. Telephone logs and materials from public meetings attended by the OIRA Administrator are also available in the Docket Library. A major upgrading of OIRAs information system is scheduled for the second half of 2002.

Though administrations and policy priorities change, the regulatory review process will most likely remain centralized in the Office of Information and Regulatory Affairs. This office allows the president to ensure that agencies implement legislation through regulations that reflect the central priorities of the administration. For example, Executive Order 13211, signed by President George W. Bush on May 22, 2001, requires federal agencies to submit a formal Statement of Energy Effects to OIRA along with their submissions of proposed regulations. The Statement of Energy Effects must include a description of any negative effects on the supply, distribution, or use of energy that are likely to result from a proposed regulation. These negative effects could include lowered production of **coal**, oil, gas, or electricity; increased costs; or harm to the environment. The order does not, however, change the basic process of regulatory review.

One structural modification that may be made within the next few years, however, is related to the computerization of government information. As of 2002, the federal government does not have a "seamless" system of data collection, analysis, and use. For the past several decades, government agencies have purchased computers and software systems on their own, without any attempt to coordinate their systems with those of other agencies. The lack of a unified computer language complicates the process of regulatory review as well as contributing to overall inefficiency and higher costs. In the summer of 2002, the Senate passed a bill called the E-Government Act, intended to help government agencies communicate more effectively with one another as well as make government information more accessible to citizens. The E-Government Act would allocate $345 million over a four-year period to help federal agencies upgrade their online services, and to establish an Office of Electronic Government alongside OIRA under the Office of Management and Budget.

[*Alair MacLean and Rebecca J. Frey, Ph.D.*]

RESOURCES
BOOKS

Coleman, B. *Through the Corridors of Power: A Citizen's Guide to Federal Rulemaking.* Washington, DC: OMB Watch, 1987.

PERIODICALS

Raney, Rebecca Fairley. "Changing Federal Buying Habits." *New York Times*, July 8, 2002.

OTHER

Dudley, Susan, and Melinda Warren. *Regulatory Response: An Analysis of the Shifting Priorities of the U.S. Budget for Fiscal Years 2002 and 2003.* Regulatory Report 24, Weidenbaum Center, Washington University, St. Louis, MO.

Federal Register, volume 58, Presidential Documents. *Executive Order 12866 of September 30, 1993— Regulatory Planning and Review.* Washington, DC: U.S. Government Printing Office, 1993.

Federal Register, volume 66, Presidential Documents. *Executive Order 13211 of May 22, 2001— Actions Concerning Regulations That Significantly Affect Energy Supply, Distribution, or Use.* Washington, DC: U.S. Government Printing Office, 2001.

ORGANIZATIONS

Office of Information and Regulatory Affairs (OIRA), The Office of Management and Budget, 725 17th Street, NW, Washington, DC USA 20503 (202) 395-4852, Fax: (202) 395-3888, <http://www.whitehouse.gov/omb/inforeg>

Rehabilitation

Rehabilitation is a process which is being applied more frequently to the **environment**. It aims to reverse the deterioration of a national resource, even if it cannot be restored to its original state.

Attempts to rehabilitate deteriorated areas have a long history. In England, for example, gardener and architect Lancelot "Capability" Brown devoted his life to restoring vast stretches of the English countryside that had been dramatically modified by human activities.

Reforestation was one of the common forms of rehabilitation used by Brown, and this method continues to be used throughout the world. The demand for wood both as a building material and a source of fuel has resulted in the devastation of forests on every continent. Sometimes the objective of reforestation is to ensure a new supply of lumber for human needs, and in other cases the motivation is to protect the environment by reducing land **erosion**. Aesthetic concerns have also been the basis for reforestation programs. Recently, the role of trees in managing atmospheric **carbon dioxide** and global climate has created yet another motivation for the planting of trees.

Another activity in which rehabilitation has become important is **surface mining**. The process by which **coal** and other minerals are mined using this method results in massive disruption of the environment. For decades, the policy of mining companies was to abandon damaged land after all minerals had been removed. Increasing environmental awareness in the 1960s and 1970s led to a change in that policy, however. In 1977, the United States Congress passed the **Surface Mining Control and Reclamation Act** (SMCRA), requiring companies to rehabilitate land damaged by these activities. The act worked well at first and mining companies began to take a more serious view of their responsibilities for restoring the land they had damaged, but by the mid-1980s that trend had been reversed to some extent. The Reagan and Bush administrations were both committed to reducing the regulatory pressure on American businesses, and tended to be less aggressive about the enforcement of environmental laws such as SMCRA.

Rehabilitation is also widely used in the area of human resources. For example, the spread of urban blight in American cities has led to an interest in the rehabilitation of public and private buildings. After World War II, many people moved out of central cities into the suburbs. Untold numbers of houses were abandoned and fell into disrepair. In the last decade, municipal, state, and federal governments have shown an interest in rehabilitating such dwellings and the areas where they are located. Many cities now have urban homesteading laws under which buildings in depressed areas are sold at low prices, and often with tax breaks, to buyers who agree to rehabilitate and live in them. *See also* Environmental degradation; Environmental engineering; Forest management; Greenhouse effect; Strip mining; Restoration ecology; Wildlife rehabilitation

[*David E. Newton*]

Resources

Books

Moran, J. M., M. D. Morgan, and J. H. Wiersma. *Introduction to Environmental Science.* 2nd ed. New York: W. H. Freeman, 1986.

Newton, D. E. *Land Use, A–Z.* Hillside, NJ: Enslow Publishers, 1989.

William Kane Reilly (1940 –)

American conservationist and Environmental Protection Agency administrator

Called the "first professional environmentalist" to head the **Environmental Protection Agency** since its founding in 1970, Reilly came to the agency in 1989 with a background in law and urban planning. He had been appointed to President Richard Nixon's **Council on Environmental Quality** in 1970, and was named executive director of the Task Force on **Land Use** and Urban Growth two years later. In 1973, Reilly became president of the Conservation Foundation, a non-profit environmental research group based in Washington, D.C., which he is credited with transforming into a considerable force for environmental protection around the world. In 1985, the Conservation Foundation merged with the **World Wildlife Fund**; Reilly was named as president, and under his direction, membership grew to 600,000 with an annual budget of $35 million.

William K. Reilly describing the Bush administration's clean air proposals in 1989. (Corbis-Bettmann. Reproduced by permission.)

Reilly began to cautiously criticize White House **environmental policy** during Ronald Reagan's administration, objecting to the appointments of **James G. Watt**, and his successor William P. Clark, as secretary of the interior. He published articles and spoke publicly about **pollution**, **species** diversification, **rain forest** destruction, and **wetlands** loss.

After his confirmation as EPA administrator, Reilly laid out an agenda that was clearly different from his predecessor, Lee Thomas. Reilly promised "vigorous and aggressive enforcement of the environmental laws," a tough new **Clean Air Act**, and an international agreement to reduce **chlorofluorocarbons** (CFCs) in the **atmosphere**. Calling toxic waste cleanup a top priority, Reilly also promised action

to decrease urban **smog**, remove dangerous **chemicals** from the market more quickly, encourage strict fuel efficiency standards, increase **recycling** efforts, protect wetlands, and encourage international cooperation on global warming and **ozone layer depletion**.

In March 1989, Reilly overruled a recommendation of a regional EPA director and suspended plans to build the Two Forks Dam on the South Platte River in Colorado. The decision relieved environmentalists who had feared construction of the dam would foul a prime trout stream, flood a scenic canyon, and disrupt **wildlife migration** patterns. That same year, Reilly criticized the federal government's response to the grounding of the ***Exxon Valdez*** and the oil spill in **Prince William Sound**, Alaska.

At Reilly's urging, President George Bush proposed major revisions of the Clean Air Act of 1970. The new act required public utilities to reduce emissions of **sulfur dioxide** by nearly half. It also contained measures that reduced emissions of toxic chemicals by industry and lowered the levels of urban smog.

Under Reilly's direction, the EPA also enacted a gradual ban on the production and importation of most products made of **asbestos**. He left the agency in 1992, and in 1993 became President and Chief Executive Officer of Aqua International Partners, L. P.

[*Linda Rehkopf*]

Resources
Periodicals

Williams, T. "It's Lonely Being Green: In an Environmentally Unfriendly Administration, William K. Reilly and John Turner Have Stood Alone as Conservationists." *Audubon* 94 (September–October 1993): 52–6.

Relict species

A **species** surviving from an ancient time in isolated populations that represent the localized remains of a distribution which was originally much wider. These populations become isolated through disruptive geophysical events such as **glaciation**, or immigration to outlying islands, which is not followed by reunification of the fragmented populations. The origins and relationships of several relict species are well documented, and these include local Central American avian populations which are remnants of the North American avifauna left after the last glacial retreat. The origins of other relict species are unknown because all related species are extinct. This group includes lungfishes, rhynchocephalian reptiles (genus *Sphenodon*), and the duck-billed platypus (*Ornithorhynchus anatinus*). *See also* California condor; Endangered species; Extinction; Rare Species

Religion and the environment

All of the world's major faiths have, as integral parts of their laws and traditions, teachings requiring protection of the **environment**, respect for **nature** and **wildlife**, and kindness to animals. While such tenets are well-known in such Eastern religions as Buddhism and Hinduism, there is also a largely-forgotten but strong tradition of such teachings in Christianity, Judaism, and Islam. All of these faiths recognize a doctrine of their deity's love for creation and for all of the living creatures of the world. The obligation of humans to respect and protect the natural environment and other life forms appears throughout the sacred writings of the prophets and leaders of the world's great religions.

These tenets of "environmental theology" contained in the world's religions are infrequently observed or practiced, but many theologians feel that they are more relevant today than ever. At a time when the earth faces a potentially fatal ecological crisis, these leaders say, traditional religion shows us a way to preserve our planet and the interdependent life forms living on it. Efforts are under way by representatives of virtually all the major faiths to make people more aware of the strong **conservation** and humane teachings that are an integral part of the religions of most cultures around the globe. This movement holds tremendous potential to stimulate a spiritually-based ecological ethic throughout the world to improve and protect the natural environment and the welfare of humans who rely upon it.

The Bible's ecological message

The early founders and followers of monotheism were filled with a sense of wonder, delight, and awe of the greatness of God's creation. Indeed, nature and wildlife were sources of inspiration for many of the prophets of the Bible. The Bible contains a strong message of conservation, respect for nature, and kindness to animals. It promotes a reverence for life, for "God's Creation," over which humans were given stewardship responsibilities to care for and protect. The Bible clearly teaches that in despoiling nature, the Lord's handiwork is being destroyed and the sacred trust as caretakers of the land over which humans were given stewardship is violated.

Modern-day policies and programs that despoil the land, desecrate the environment, and destroy entire **species** of wildlife are not justified by the Bible. Such actions clearly violate biblical commands to humans to "replenish the earth," conserve **natural resources**, and treat animals with kindness, and to animals to "be fruitful and multiply" and fill the earth. For example, various laws requiring the protection

of natural resources are found in the Mosaic law, including passages mandating the preservation of fruit trees (Deuteronomy 20:19, Genesis 19:23–25); agricultural lands (Leviticus 25:2–4); and wildlife (Deuteronomy 22:6–7; Genesis 9). Numerous other biblical passages extol the wonders of nature (Psalms 19, 24, and 104) and teach kindness to animals, including the Ten Commandments, which require that farm animals be allowed to rest on the Sabbath.

Christianity: Jesus' nature teachings

The New Testament contains many references by Jesus and his disciples that teach people to protect nature and its life forms. Throughout the Sermon on the Mount, Jesus used nature and pastoral imagery to illustrate his points and uphold the creatures of nature as worthy of being emulated. In stressing the lack of importance of material possessions such as fancy clothes, Jesus observed that "God so clothes the grass of the field" and cited wildflowers as possessing more beauty than any human garments ever could: "Consider the lilies of the field, how they grow; they toil not, neither do they spin. And yet I say unto you, that even Solomon in all his glory was not arrayed like one of these" (Matthew 6:28–30; Luke 12:27). In Luke 12:6 and Matthew 10:29, Jesus stresses that even the lowliest of creatures is loved by God: "Are not five sparrows sold for two pennies? And not one of them is forgotten before God."

Judaism: a tradition of reverence for nature

The teachings and laws of Judaism, going back thousands of years, strongly emphasize kindness to animals and respect for nature. Indeed, an entire code of laws relates to preventing "the suffering of living creatures." Also important are the concepts of protecting the elements of nature, and *tikkun olam*, or "repairing the world." Jewish prayers, traditions, and literature contain countless stories and admonitions stressing the importance of the natural world and animals as manifestations of God's greatness and love for His creation.

Many examples of practices based on Judaism's respect for nature can be cited. Early Jewish law prevented **pollution** of waterways by mandating that sewage be buried in the ground, not dumped into rivers. In ancient Jerusalem, dung heaps and **garbage** piles were banned, and refuse could not be disposed of near water systems. The Israelites wisely protected their drinking water supply and avoided creating hazardous and unhealthy waste dumps. The rabbis of old Jerusalem also dealt with the problem of **air pollution** from wheat chaff by requiring that threshing houses for grain be built no closer than 2 miles (3.2 km) from the city. In order to prevent foul odors, a similar ordinance existed for graves, carrion, and tanneries, with tanneries sometimes required to be constructed on the edge of the city downwind from prevailing air currents. Wood from certain types of rare trees could not be burned at all, and the Talmud cautioned that lamps should be set to burn slowly so as not to use up too much naphtha. The biblical injunction to allow land to lie fallow every seven years (Lev. 25:3–7) permitted the **soil** to replenish itself.

Islam and ecology

In the Qur'an (Koran), the holy book of the Islamic faith, scholars have estimated that as many as 750 out of the book's 6,000 verses (about one-eighth of the entire text) have to do with nature. In Islamic doctrine there are three central principles that relate to an environmental ethic, including *tawhid* (unity), *khilafa* (trusteeship), and *akhirah* (accountability). Nature is considered sacred because it is God's work, and a unity and interconnectedness of living things is implied in certain scriptures, such as: "There is no God but He, the Creator of all things" (Q.6: 102). The balance of the natural world is also described in some verses: "And the earth we have spread out like a carpet; set thereon mountains firm and immobile; and produced therein all kinds of things in due balance" (Q.15:1 9). On the earth, humans are given the role of stewards (called "vicegerents"), when the Qur'anic scripture states, "Behold, the Lord said to the angels: 'I will create a vicegerent on earth...'" (Q.2: 30). In the teachings of the Prophet Muhammad, humans are told to cultivate and care for the earth. ("Whoever brings dead land to life, that is, cultivates wasteland, for him is a reward therein"), and humans are cautioned in the Qur'an against abusing the creation: "Do no mischief on the earth after it hath been set in order, but call on Him with fear and longing in your hearts: for the Mercy of God is always near to those who do good" (Q.7: 56). Interestingly, the Qur'an also contains scriptures that caution humankind about the use of metals, perhaps foreshadowing the machine age of the future: "We bestowed on you from on high the ability to make use of iron, in which there is awesome power as well as a source of benefits for man" (Q.57.25).

Other faiths

The religions of the East have also recognized and stressed the importance of protecting natural resources and living creatures. Buddhism and Hinduism have doctrines of non-violence to living beings and have teachings that stress the unity and sacredness of all of life. Some Hindu gods and goddesses are embodiments of the natural processes of the world. Taoists, who practice a philosophy and worldview that originated in ancient China, strive not to attempt to dominate but to live in harmony with the *tao*, or the natural way of the universe. Many other religions, including the Baha'is and those of Native Americans, Amazon Indian tribes, and other indigenous and tribal peoples, stress the sanctity of nature and the need to conserve wildlife, forests, plants, water, fertile land, and other natural resources.

[*Lewis G. Regenstein and Douglas Dupler*]

RESOURCES
BOOKS

Barnhill, David L., and Roger S. Gottlieb. *Deep Ecology and World Religions.* Albany: State University of New York Press, 2001.

Berry, T. *The Dream of the Earth.* San Francisco: Sierra Club Books, 1988.

Carson, Rachel. *Silent Spring.* Cambridge, MA: Riverside Press, 1962.

Cobb, J. *Is It Too Late? A Theology of Ecology.* Environmental Ethics Books, 1995.

Fox, M. W. *The Boundless Circle: Caring for Creatures and Creation.* Wheaton, IL: Quest Books, 1996.

Gore, Al. *Earth in the Balance: Ecology and the Human Spirit.* New York: Houghton Mifflin, 1992.

Hoyt J. *Animals in Peril.* Garden City-Park, NY: Avery, 1994.

Leopold, A. *A Sand County Almanac.* New York: Oxford University Press, 1949.

Regenstein, L. G. *Replenish the Earth: The Teachings of the World's Religions on Protecting Animals and Nature.* New York: Crossroads, 1991.

Wilson, E. O. *Biophilia.* Cambridge: Harvard University Press, 1984.

ORGANIZATIONS

Harvard Center for the Study of World Religions: Religions of the World and Ecology, 42 Francis Avenue, Cambridge, MA USA 02138 (617) 495-4495, Fax: (617) 496-5411, Email: cswrinfo@hds.harvard.edu, <http://www.hds.harvard.edu/cswr/ecology/index.htm>

Remediation

The treatment of hazardous wastes in most industrialized countries is now controlled through regulations, policies, incentive programs, and voluntary efforts. However, until the 1980s, society knew little of the harmful effects of inadequate management of hazardous wastes, which resulted in many contaminated sites and polluted **natural resources**. Even in the 1970s, when environmental legislation required disposal of hazardous wastes to landfills and other systems, contamination of soils and **groundwater** continued, because these systems were often not leak proof. To meet the environmental and public health standards of the twenty-first century, these contaminated sites must continue to be rehabilitated, or remediated, so they no longer pose a threat to the public or the **ecosystem**. However, the price will be high; total costs for cleaning up the U.S. Department of Energy's (DOE) **nuclear weapons** sites alone will cost $147 billion from 1997 to 2070. Cleaning America's contaminated groundwater will cost even more: $750 billion over the next 30 years.

The Superfund program for remediation of contaminated sites was created in 1980 by the **Comprehensive Environmental Response, Compensation, and Liability Act** (CERCLA) and was amended by the Superfund Amendments and Reauthorization Act (SARA) of 1984. The Superfund program is designed to provide an immediate response to emergency situations that pose an imminent hazard, known as *removal or emergency response-actions*, and to provide permanent remedies for environmental problems resulting from past practices (i.e., abandoned or inactive waste sites), know as *remedial-response actions*. Under the Superfund program, the remedial action process is implemented in structured stages: (1) Preliminary assessment; (2) Site inspection; (3) Listing on the **National Priorities List** (NPL), a rank ordering of sites representing the greatest threats; (4) Remedial investigation/ feasibility study (RI/FS); (5) U.S. Environmental Protection Agency's **Record of Decision** (ROD); (6) Negotiation of a consent decree and remedial action plan between the U.S. **Environmental Protection Agency** and the responsible parties; and (7) Final design and construction.

Money for the Superfund came from special taxes on chemical and **petroleum** companies, which expired in 1995 and were not renewed. These taxes contributed about $1.3 billion a year prior to 1996. Since 1995, the fund has dwindled from a high of $3.6 billion to a projected $28 million in 2003. Currently, debate continues on where additional funds will come from, which may eventually be the taxpayer if special taxes are not reestablished. The Bush administration had proposed that the Superfund pay $593 million of 2002's projected $1.3 billion in cleanup costs, with the remaining $700 million coming from the Treasury. After this, the Superfund will be depleted. Money will have to come from somewhere because according to **Resources for the Future**, Superfund programs will cost $14–$16.4 billion between 2000 and 2009, with annual costs of between $1.3 billion and $1.7 billion.

In addition to the Superfund program, site remediation projects may be required for: the **Resource Conservation and Recovery Act** (RCRA) Corrective Action Program for operating treatment, storage, and disposal facilities; the underground storage tanks program established by RCRA; Federal Facility cleanup of sites operated primarily by DOE and the Department of Defense (regulated under both CERCLA and RCRA); and state-regulated, private, and voluntary programs. The cleanup of federal facilities is expected to be the most expensive, followed by the RCRA Corrective Action Program, and then Superfund. DOE must characterize, treat, and dispose of hazardous and **radioactive waste** at more than 120 sites in 36 states and territories, which is expected to cost more than $60 billion over the next 10 years.

The costs of remediation are determined by the degree of remediation required. Superfund cleanup standards require that contaminated waters be remediated to concentrations at least as clean at the Maximum Contaminant Levels (MCLs) of the **Safe Drinking Water Act** and the **water quality** criteria of the **Clean Water Act**. **Carcinogen** risks from exposures to a Superfund site should be less than one excess **cancer** in a population of one million people. Some

states require that remediation projects restore the site to an equivalent of its original condition.

Remedial techniques are divided into two basic types: on-site methods and removal methods. Most remedial techniques are used in combination (e.g., pump and treat systems) rather than singly. Generally remediation techniques that significantly reduce the volume, toxicity, or mobility of hazardous wastes are preferred for use. Off-site transport and disposal is usually the least favored option. On-site methods include containment, extraction, treatment, and destruction, while removal methods consist of excavation and **dredging**. Containment is used to contain liquid wastes or contaminated ground waters within a site; or to divert ground or surface waters away from the site. Containment is achieved by the construction of low-permeability or impermeable cutoff walls or diversions. On-site groundwater extraction is accomplished by pumping, which is usually followed by aboveground treatment (referred to as *pump and treat)*. Pumping is used to control water tables so as to contain or remove contaminated groundwater plumes or to prevent **plume** formation. **Soil** vapor extraction, by either passive or forced ventilation collection systems, is used to remove hazardous gases from the soil. Gases are collected in PVC perforated pipe **wells** or trenches and transported above ground to treatment or **combustion** units.

On-site treatment of hazardous wastes can be conducted above ground, following extraction or excavation, or below ground (*in situ*). Above ground treatment methods include the use of conventional **environmental engineering** unit processes. Below ground treatment can be nonbiological or biological. Non-biological *in situ* treatment involves the delivery of a treatment fluid through injection wells into the contaminated zone to achieve immobilization and/or **detoxification** of the contaminants. This treatment is accomplished by processes such as oxidation, reduction, neutralization, hydrolysis, precipitation, chelation, and stabilization/solidification. Biological *in situ* treatment (**bioremediation**) is accomplished by modifying site conditions to promote microbial degradation of organic compounds by naturally occurring **microorganisms**. Exogenous microorganisms that have been adapted to degrade specific compounds or classes of compounds are also sometimes pumped into ground water or applied to **contaminated soil**. Certain types of bacteria and plants containing **peptides** are also being developed for treating radioactive materials and **heavy metals**.

Finally, on-site destruction of organic hazardous wastes can be accomplished by the use of high-temperature **incineration**. Inorganic materials are not destroyed, so residues from the incineration process must be disposed of in a secure **landfill**. In situ vitrification (ISV), or thermal destruction, involves the heating of buried wastes to melting temperatures (by applying electric current to the soil) so that a chemically inert, stable glass block is produced. Nonvolatile materials are immobilized in the vitrified mass, and organic materials are pyrolyzed. Combustion off-gases are collected and treated.

Removal methods of remediation include extraction and dredging. Excavation of contaminated solid or semisolid hazardous wastes may be required if the wastes are not amenable to *in situ* treatment. Excavated materials may be treated and replaced or transported off-site. Dredging is used to remove contaminated sediments from streams, estuaries, or surface impoundments. Removal of well-consolidated sediments in shallow waters requires the use of a clamshell, dragline, or backhoe. Sediments with a high liquid content or that are located in deeper waters are removed by hydraulic dredging. Dredged materials are pumped or barged to treatment facilities.

Although not every remediation project has been a complete success, many Superfund sites have been cleaned up significantly. Of the 1,310 sites on the list, 773 sites had been cleaned up by 2001, with the remainder in various stages of cleanup. However, the rate of cleanups is slowing; about half as many will be completed during the current administration compared to the previous one. As better procedures for remediation continue to be developed and refined, perhaps this rate will increase. Nanoparticle technology is one method that shows promise for cleaning groundwater sites, since it can reduce 96% of contaminants compared to 25% with conventional methods.

[*Laurel M. Sheppard*]

Resources

Books

Blackman Jr., W. L. *Basic Hazardous Waste Management.* Boca Raton, FL: Lewis Publishers, 1993.

Freeman, H. M., ed. *Standard Handbook of Hazardous Waste Treatment and Disposal.* New York: McGraw-Hill, 1989.

Henry, J. G., and G. W. Heinke. *Environmental Science and Engineering.* 2nd ed. Upper Saddle River, NJ: Prentice Hall, 1996.

Probst, Katherine N., David M. Konishky, et al.*Superfund's Future: What Will It Cost?* Resources for the Future, 2001.

Periodicals

"Administration Accused of Slowing Superfund." *Chemical Market Report,* April 15, 2002, 4.

"DOE: Nuclear Weapon Cleanup Will Cost $147B." *USAToday.com,* August 23, 2001.

Hellprin, John. "EPA Chief Defends Halving Toxic Waste Cleanups as Superfund Money Nears Depletion." *Associated Press,* March 13, 2002.

Morgan, Dan. "Hazardous Waste Site Cleanup Delayed, EPA Inspector Reports." *Washington Post,* July 2, 2002, A02.

"New Technology Revolutionizing Ground Water Clean-Up." *PR Newswire,* March 13, 2002.

Remote sensing

see **Measurement and sensing**

Renew America

Renew America serves as a clearinghouse for information on the **environment**. Renew America distributes reports, pamphlets, and books to educators, organizations, and the general public in an attempt to fulfill its primary goal of "renewing America's community spirit through environmental success." One of Renew America's major projects is the annual *Environmental Success Index*, a compilation of over 1,400 successful environmental programs throughout the United States. The index lists organizations involved in a variety of environmental issues, including **air pollution control**, drinking water and **groundwater** protection, hazardous and solid **waste reduction** and **recycling**, **renewable energy**, and forest protection and urban forestry. The purpose of the index is to disseminate information on organizations that can function as prototypes for other programs. Renew America also publishes an annual report entitled *State of the States*, which serves as a report card on environmental policies in all 50 states. *State of the States* is compiled in cooperation with over 100 state and local environmental organizations.

In addition to disseminating published information, Renew America distributes the Renew America National Awards each year. The awards are the result of a national search conducted by the Renew America Advisory Council, which is composed of 28 environmental groups. Renew America also selects a recipient for the Robert Rodale Environmental Achievement Award. Rodale served as chair of Renew America and the Rodale Press and administered the **Rodale Institute**, a scientific and educational foundation focusing on food, health, and natural resource issues. The awards are presented at the Environmental Leadership Conference, which also features several environmentally-oriented workshops and panel discussions. Like the *Environmental Success Index*, the advisory council selects recipients from a broad range of environmental programs. Past winners have included a group that assists area residents in creating community food gardens, surfers who won a lawsuit under the federal **Clean Water Act**, an adopt-a-manatee program, and an asphalt recycling project.

Renew America also offers educational tools such as a community resource kit called *Sharing Success*, which features articles on ways to approach environmental problems and highlights special environmental groups. It was developed to meet the needs of local environmental groups and other interested activists who wanted to share successful approaches to environmental problems.

Renew America does not utilize volunteers or engage in grassroots projects itself but provides information on organizations that do. Members receive the *State of the States* report and a quarterly newsletter highlighting significant environmental issues. Board members include former congress member Claudine Schneider and actor Eddie Albert as well as officials from several prominent environmental organizations.

[*Kristin Palm*]

Resources

Organizations

Renew America, 1200 18th Street, NW, Suite 1100, Washington, D.C. USA 20036 (202) 721-1545, Fax: (202) 467-5780, Email: renewamerica@counterpart.org, <http://sol.crest.org/environment/renew_america>

Renewable energy

The earth's resources are commonly divided into renewable and **nonrenewable resources**. Some renewable resources are perpetual, meaning that they are not affected by human use, such as **solar energy** or **wind energy**. Other renewable resources are organic and inorganic materials that are replenished by physical and biogeochemical cycles. Examples of organic renewable resources are all plant and animal **species** that people use for food, building materials, drugs, leisure, and so on. Examples of inorganic renewable resources are water and oxygen, which are replenished in the hydrological and oxygen cycles, respectively. The main sources of renewable energy in the United States are **biomass** (wood and waste burned for fuel), hydroelectric power (energy produced from flowing water), geothermal sources (energy from heat sources in the earth's surface), solar (energy from the sun), and wind energy. Nonrenewable resources are those materials that are present in the earth in limited amounts (minerals) or are produced only over many millions of years (**fossil fuels**). Nonrenewable resources may be recyclable so that their usefulness to human beings can be extended, but often they are transformed during use into useless matter such as waste gas.

Taking advantage of renewable energy sources became important in the United States after the OPEC (**Organization of Petroleum Exporting Countries**) **oil embargo** during the 1970s. Until that time, **coal** and oil easily supplied nearly 90% of energy to the United States, but sudden shortages and huge price increases prompted the United States government to invest in researching alternative and renewable energy sources. The 1990s were a decade of cheap and readily available fossil fuel, and United States development of renewable energy grew relatively slowly. In the early twenty-first century renewable energy became important

again, with an energy crisis in California, the nation's most populous state, and with increasing concern over **pollution** and global warming. Political issues related to energy production, such as growing United States dependence on Middle Eastern oil reserves and the protection of **wildlife** and offshore areas that contain United States oil reserves, also brought attention and hope to renewable energy.

According to figures released by the Energy Information Administration (EIA) of the **U.S. Department of Energy** (DOE), renewable energy sources accounted for about 7% of all United States energy used in 2000. Biomass contributed about 48% of that renewable energy total. Biomass includes wood used to heat homes, as well as the waste burned in municipal waste recovery plants and converted to electricity. Other biomass sources of energy include **ethanol** and **methanol**, which are alternative fuels made from fermented plant material. Ethanol is made from corn and often added to **gasoline** for use in internal **combustion** engines. Methanol is an alternative fuel made from wood, and like ethanol, is expensive to produce and thus at a disadvantage when compared to fossil fuels and to other renewables. Another biomass energy source is **landfill** gas, which is **methane** that is captured and used for fuel as waste landfills decompose.

Hydropower accounted for 46% of United States renewable energy in 2000, and was the world's largest renewable energy source. Hydropower uses the force of dammed water to turn turbines, which produce electricity. Future United States development of hydropower is not projected to increase, as the most productive sites have already been used. Several **Third World** countries have viewed hydropower as a cheap source of future energy and are developing new projects, while environmentalists have strongly protested the damming of more rivers and the **habitat** destruction associated with large hydropower facilities.

Geothermal sources produced 5% of United States renewable energy in 2000. **Geothermal energy** is produced by converting the heat of the earth into electricity, using steam-powered turbines, or by using the heat directly for buildings and industrial purposes. The United States was the world's largest producer of geothermal power, accounting for about 44% of the world total. Like hydropower, U.S. geothermal power is not projected to grow in the future, as the most advantageous sites are already taken. Other countries including the Philippines, New Zealand, and Iceland are developing significant thermal energy resources.

Solar power accounted for 1% of United States renewable energy consumption in 2000, although the solar power industry grew by 20% for five straight years in the late 1990s. Solar energy systems include passive ones, such as greenhouses or building designs utilizing the sun's heat and light, and active systems, which use mechanical methods to convey the sun's energy. Photovoltaic conversion (PV) systems use chemical cells that convert the sun's energy to electricity. The solar energy industry illustrates how technology affects renewable energy development and use. Between 1975 and 1999, the cost of electricity produced by PV systems declined from a prohibitively expensive $100 per watt to less than $4 per watt, making the costly technology more accessible to consumers.

Wind energy accounted for less 1% of United States renewable energy consumption in 2000. Wind energy is produced when wind forces the movement of turbines on windmills, and generators convert that movement into electricity. Technology improvements helped wind energy to become the world's fastest growing alternative energy source by 2001, lowering its price and making it more competitive with other energy sources. During the 1990s, the United States lost its edge in wind energy technology, as countries such as Germany, Denmark, Spain, and Japan significantly invested in wind energy production. Global wind electric production capacity doubled between 1995 and 1998, and doubled again by 2001, reaching 23,300 megawatts in 2001.

Scientists are busy researching other renewable energy sources. Some countries are attempting to develop means to capture the energy in the ocean's waves (**tidal power**) or the energy in warm ocean water (**ocean thermal energy conversion**). Some scientists believe that **hydrogen** may be the energy source of the future, as hydrogen gas can be made from water and other compounds. However, no cheap or efficient method has yet been discovered for taking advantage of highly combustible hydrogen, the most plentiful chemical element in the universe.

Fossil fuel prices and environmental laws are predicted to affect the development and use of renewable energy in the future. Furthermore, technology improvements may make the cost of producing renewable energy competitive with fossil fuel energy. Research and development of alternative energy technology in the future may depend upon government-sponsored initiatives, as they have in the past.

[*Douglas Dupler*]

Resources

Books

Asmus, Peter. *Reaping the Wind: How Mechanical Wizards, Visionaries, and Profiteers Helped Shape Our Energy Future.* Washington, DC: Island Press, 2001.

Dupler, Douglas. *Energy: Shortage, Glut, or Enough?* Farmington Hills, MI: Gale, 2001.

Hazen, Mark E., and Michael Hauben. *Alternative Energy.* Clifton Park, NY: Delmar Learning, 2001.

Schaeffer, John, and Doug Pratt, eds. *Real Goods Solar Living Source Book: The Complete Guide to Renewable Energy Technologies and Sustainable Living.* Broomfield, CO: Real Goods, 2001.

PERIODICALS

Gelbspan, Ross. "A Modest Proposal to Stop Global Warming." *Sierra* (May/June 2001): 63.

OTHER

Energy Information Administration. *Renewable Energy Consumption by Source, 1989–2000.* May 7, 2002 [cited July 6, 2002]. <http://www.eia.doe.-gov/emeu/aer/renew.html>.

ORGANIZATIONS

Energy Information Administration, 1000 Independence Avenue, SW, Washington, DC USA 20585 (202) 586-8800, Email: infoctr@eia.doe.gov, <http://www.eia.doe.gov>

National Renewable Energy Laboratory, 1617 Cole Blvd., Golden, CO USA 80401-3393 (303) 275-3000, Email: client_services@nrel.gov, <http://www.nrel.gov>

Renewable Natural Resources Foundation

The Renewable Natural Resources Foundation (RNRF) was founded in 1972 to further education regarding renewable resources in the scientific and public sectors. According to the organization, it seeks to "promote the application of sound scientific practices in managing and conserving renewable natural resources; foster coordination and cooperation among professional, scientific, and educational organizations having leadership responsibilities for renewable resources; and develop a Renewable Natural Resources Center."

The Foundation is composed of organizations that are actively involved with renewable natural resources and related public policy, including the American Fisheries Society, the American **Water Resources** Association, the Association of American Geographers, **Resources for the Future**, the **Soil and Water Conservation Society, the Nature Conservancy**, and the **Wildlife** Society. Programs sponsored by the Foundation include conferences, workshops, summits of elected and appointed leaders of RNRF member organizations, the RNRF Round Table on Public Policy, publication of the *Renewable Resources Journal*, joint human resources development of member organizations, and annual awards to recognize outstanding contributions to the field of renewable resources.

In 1992, the RNRF organized the "Congress on Renewable Natural Resources: Critical Issues and Concepts for the Twenty-First Century." At the Congress, held in Vail, Colorado, 135 of America's leading scientists and resource professionals gathered to discuss critical natural resource issues that this nation will be facing as the twenty-first century approaches. According to the *Renewable Resources Journal*, the "synergy created by bringing together a diverse group—resource managers, policymakers, and physical, biological, and social scientists—resulted in scores of recommendations for innovative policies."

[*Linda Ross*]

RESOURCES

ORGANIZATIONS

Renewable Natural Resources Foundation, 5430 Grosvenor Lane, Bethesda, MD USA 20814-2193 (301) 493-9101, Fax: (301) 493-6148, Email: info@rnrf.org, <http://www.rnrf.org>

Reserve Mining Corporation

In 1947, the beginnings of one of the first classic cases of environmental conflict emerged when the Reserve Mining Corporation received permits to dump up to 67,000 tons per day of asbestos-like fiber **tailings** from its taconite mining operations in **Silver Bay**, Minnesota, into western Lake Superior. The Reserve case typifies the conflict that occurs when the complex problems of **pollution**, economics, politics, and law collide. After decades of investigation and lawsuits, the corporation stopped the dumping in spring 1980, but the change did not come easily for any of the participants.

Taconite is a low-grade iron ore, which can be crushed, concentrated, and pelletized to provide a good source of iron ore. Silver Bay, Minnesota, was created as a result of the mine's development by Reserve, and approximately 80% of the local labor force of 3,500 residents worked at the $350 million facility in 1980. The first permits were given to the company in 1947, on the condition that it would strictly comply with all provisions, including providing continued assurances that the discharges posed no harm to the lake or surrounding inhabitants. In 1956 and 1960, Reserve was allowed to increase its use of Lake Superior water, thus increasing its dumping of waste taconite tailings. This continued until the late 1960s, when two major signs emerged that the lake—and potentially local residents—were being harmed by the waste discharges.

The first sign was the obvious discoloration of the water on the northwest shores of Lake Superior. Up to 67,000 tons of tiny fibers resembling **asbestos**, a **carcinogen**, were dumped into the lake each day, turning the water into a green slime from Silver Bay to Duluth, Minnesota, and nearby Superior, Wisconsin. Second, several local citizens made the link between the company's **discharge** and possible harmful effects on humans when they read about a discovered connection between **cancer** and asbestos used to polish rice in Japan in the 1960s.

By June 1973, after completing several studies at the request of these local citizens, the U.S. **Environmental Protection Agency** concluded that the drinking water of Duluth and other northshore communities was contaminated with

asbestos-like fibers, which might cause cancer. The Minnesota **Pollution Control** Agency followed with the finding that a suspected source of the fibers was the Reserve Mining Corporation discharge. Duluth and other affected communities installed **filtration** systems to prevent further contamination of local drinking water supplies, and convinced state and federal agencies to join citizen groups in legal action against the company.

In a series of subsequent court appearances, Reserve continually argued that the company would be forced to close its operations if Lake Superior discharges were stopped, presumably resulting in the demise of the town of Silver Bay as well. Five years of lawsuits seemingly ended on April 20, 1974, when federal District Judge Miles Lord completed a nine-month trial and ordered the plant to cease discharges of taconite tailings into the air and water within 24 hours. Less than 48 hours later, however, the plant was ordered reopened by three judges from the Eighth U.S. Circuit Court of Appeals. While concluding that the original permits issued to Reserve Mining were a "monumental environmental mistake," they allowed the company to remain open while it altered its production process to create alternative disposal methods. In June 1974, the governors of Minnesota and Wisconsin filed an appeal of this ruling to the Supreme Court, but were rebuffed when the court refused to reinstate Judge Lord's order. Justice William Douglas' written dissent to the court's ruling stated, "I am not aware of a constitutional principle that allows either private or public enterprises to despoil any part of the domain that belongs to all of the people. Our guiding principle should be Mr. Justice Holmes' dictum that our waterways, great and small, are treasures not **garbage** dumps or cesspools."

Six years later, Reserve Mining Company and its parent owners Armco and Republic Steel were fined more than one million dollars for their permit violations and intransigence in completing disposal upgrades. Reserve switched to onland disposal, via rail and pipeline, to a 6-mi^2 (9.7-km^2) basin 40 mi (64.4 km) inland, which kept tailings covered under 10 ft (3 m) of water to prevent fiber dust from escaping. The basin operated efficiently until a production cut resulted in more water and plant **effluent** being sent to the disposal basin than was used in the plant to concentrate the crushed taconite rock into iron ore. Reserve sought a permit in October 1985 from the Minnesota Pollution Control Agency to pump water out of the disposal basin to a filtration plant on the Beaver River, which flows into Lake Superior. After several arguments from environmental groups and others, the permit was eventually issued at a lowered level from 15 to 1 million fibers per liter. Operations ultimately closed in the summer of 1986, due to reduced demand for the taconite pellets.

The mine was purchased and reopened in 1990 by Cyprus Mining of Colorado, and operated at 30% of capacity. The one-million-fiber-per-liter limit has generally been met, and thus asbestos-like fibers continue to be discharged into Lake Superior, albeit at a drastically reduced rate from those discharged for decades directly into the largest freshwater lake in the world.

[*Sally Cole-Misch*]

Resources

Books

Rousmaniere, J., ed. *The Enduring Great Lakes, A Natural History Book.* New York: W.W. Norton & Company, 1979.

Periodicals

Merritt, Grant. "The Reserve Mining Case: Promises, Promises." *Great Lakes Focus on Water Quality* 1, no. 1 (Fall 1974): 1–3.

———. "The Reserve Mining Case—20 Years Later." *Focus on International Joint Commission Activities* 19, no. 3 (November/December 1994): 1–4.

"Tailings' End." *Time,* March 31, 1980, 45.

Reservoir

A reservoir is a body of water held by a dam on a river or stream, usually for use in **irrigation**, electricity generation, or urban consumption. By catching and holding floods in spring or in a rainy season, reservoirs also prevent **flooding** downstream. Most reservoirs fill a few miles of river basin, but large reservoirs on major rivers can cover thousands of square miles. Lake Nasser, located behind Egypt's **Aswan High Dam**, stretches 310 miles (500 km), with an average width of 6 miles (10 km). Utah's Lake Powell, on the **Colorado River**, fills almost 93 miles (150 km) of canyon.

Because of their size and their role in altering water flow in large ecosystems, reservoirs have a great number of environmental effects, positive and negative. Reservoirs allow more settlement on flat, arable flood plains near the river's edge because the threat of flooding is greatly diminished. Water storage benefits farms and cities by allowing a gradual release of water through the year. Without a dam and reservoir, much of a river's annual **discharge** may pass in a few days of flooding, leaving the river low and muddy the rest of the year. Once a reservoir is built, water remains available for irrigation, domestic use, and industry even in the dry season.

At the same time, the negative effects of reservoirs abound. Foremost is the destruction of instream and stream side vegetation and **habitat** caused by flooding hundreds of miles of river basin behind reservoirs. Because reservoirs often stretch several miles across, as well as far upstream, they can drown habitat essential to aquatic and terrestrial plants, as well as extensive tracts of forest. Humans also frequently lose agricultural land and river-side cities to reservoir flooding. China's proposed **Three Gorges Dam** on the

Chang Jiang (Yangtze River) will displace 1.4 million people when it is completed; India's Narmada Valley reservoir will flood the homes of 1.5 million people. In such cases the displaced populations must move elsewhere, clearing new land to reestablish towns and farms. Water loss from evaporation and **seepage** can drastically decrease water volumes in a river, especially in hot or **arid** regions. Lakes Powell and Mead on the Colorado River annually lose 1.3 billion cubic yards (1 billion cubic m) of water through evaporation, water that both people and natural habitats downstream sorely need. Egypt's long and shallow Lake Nasser is even worse, losing as much as 20 billion cubic yards (15 billion cubic m) per year. Impounded water also seeps into the surrounding bedrock, especially in porous sandstone or limestone country. This further decreases river flow and sometimes causes slope instability in the reservoir's banks. A disastrous 1963 **landslide** in the waterlogged banks of Italy's Vaiont reservoir sloshed 192 million cubic yards (300 million cubic m) of water over the dam and down the river valley, killing almost 2,500 people in towns downstream. A further risk arises from the sheer weight of stored water, which can strain faults deep in the bedrock and occasionally cause earthquakes.

Sedimentation is another problem associated with reservoirs. Free flowing rivers usually carry a great deal of suspended **sediment**, but the still water of a reservoir allows these sediments to settle. As they accumulate in the reservoir, storage capacity decreases. The lake gradually becomes shallower and warmer, with an accompanying decrease in **water quality**. Even more important, sediment-free water downstream of the dam no longer adds sand and mud to river banks and deltas. **Erosion** of islands, banks, and deltas results, undermining bridges and walls as well as natural riverside habitat.

One of the losses felt most acutely by humans is that of scenic river valleys, gorges, and canyons. China's Three Gorges Dam will drown ancient cultural and historic relics to which travellers have made pilgrimages for centuries. In the United States, the loss of Utah's cathedral-like Glen Canyon and other beautiful or unusual environmental features are tragedies that many environmentalists still lament. *See also* Arable land; Environmental economics; Environmental policy; Riparian land; River basins

[*Mary Ann Cunningham Ph.D.*]

Resources

Books

Driver, E. E., and W. O. Wunderlich, eds. *Environmental Effects of Hydraulic Engineering Works.* Proceedings of an International Symposium Held at Knoxville, Tennessee, Sept. 12-14, 1978. Knoxville: Tennessee Valley Authority, 1979.

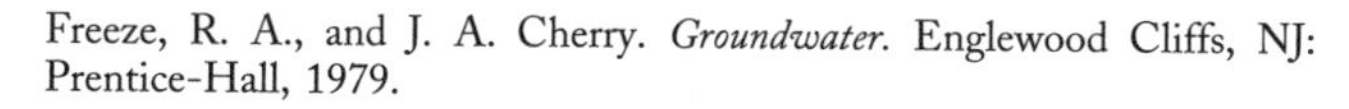
Freeze, R. A., and J. A. Cherry. *Groundwater.* Englewood Cliffs, NJ: Prentice-Hall, 1979.

Periodicals

Esteva, G., and M. S. Prakash. "Grassroots Resistance to Sustainable Development." *The Ecologist* 22 (1992): 45–51.

Residence time

Residence time is the length of time that a substance, usually a **hazardous material**, can be detected in a given **environment**. For example, airborne pollutants are carried into the **atmosphere** by natural updrafts. The concentration and size of the particulates and their residence time exerts varying degrees of influence on the solar radiation balance. Other examples include the period of time that oxidant concentrations such as PAN-type or **ozone** remain in the atmosphere, or the period of time that nuclear by-products are found in the atmosphere or in the **soil**. *See also* Hazardous waste siting; Radioactive waste

Resilience

In **ecology**, resilience refers to the rate at which a community returns to some state of development after it has been displaced from that state. Ecosystems comprised of communities with high inherent resilience are, over time, relatively stable in the face of **environmental stress**. These communities return to their original structure and functions, or similar ones, relatively quickly in response to a disturbance. Of course, there are thresholds for resilience. If a perturbation is too intense, then the inherent resilience of the community may be exceeded, and the previous state of ecological development may not be rapidly reattained, if at all. Populations of **species** that are smaller in size, poor competitors, short-lived, with short generation times and the ability to spread quickly are relatively resilient. Those that are larger in size, competitive, longer-lived, with longer generation times and high investment in offspring have a higher threshold, or **resistance**, to a stress factor, but have less resilience, and regenerate more slowly. *See also* Biosphere; Biotic community

Resistance (inertia)

In **ecology**, resistance is the ability of a population or community to avoid displacement from some state of development as a result of an **environmental stress**. Populations or communities with inherently high resistance are relatively stable when challenged by such conditions. If the stress is greater than the population threshold, though, change must occur.

In general, **species** that are larger in size, relatively competitive, longer-lived, with longer generation times and high investment in offspring are relatively resistant to intensified stresses. When these species or their communities are overcome by environmental change, however, they have little **resilience** and tend to recover slowly.

In contrast, species that are smaller in size, short-lived, highly fecund, and with shorter generation times have little ability to resist the effects of perturbation. However, these species and their communities are resilient, and have the ability to quickly recover from disturbance. This assumes, of course, that the environmental change has not been too excessive, and that the **habitat** still remains suitable for their regeneration and growth. *See also* Biotic community; Ecosystem; Population biology

Resource Conservation and Recovery Act

The Resource Conservation and Recovery Act (RCRA), an amendment to the **Solid Waste** Disposal Act, was enacted in 1976 to address a problem of enormous magnitude—how to safely dispose of the huge volumes of municipal and industrial solid waste generated nationwide. It is a problem with roots that go back well before 1976.

There was a time when the amount of waste produced in the United States was small and its impact on the **environment** relatively minor. (A river could purify itself every 10 miles [16 km].) However, with the industrial revolution in the latter part of the nineteenth century, the country began to grow with unprecedented speed. New products were developed and consumers were offered an ever-expanding array of material goods.

This growth continued through the early twentieth century and took off after World War II when the nation's industrial base, strengthened by the war, turned its energies toward domestic production. The results of this growth were not all positive. With more goods came more waste, both hazardous and nonhazardous. In the late 1940s, the United States was generating roughly 500,000 metric tons of **hazardous waste** a year. In 1985, the **Environmental Protection Agency** (EPA) estimated that 275 million tons of hazardous waste were generated nationwide.

Waste management was slow in coming. Much of the waste produced made its way into the environment where it began to pose a serious threat to ecological systems and public health. In the mid-1970s, it became clear, to both Congress and the nation alike, that action had to be taken to ensure that solid wastes were managed properly. This realization began the process that resulted in the passage of RCRA. The goals set by RCRA were: to protect human health and the environment; to reduce waste and conserve energy and **natural resources**; to reduce or eliminate the generation of hazardous waste as expeditiously as possible.

To achieve these goals, four distinct yet interrelated programs exist under RCRA. The first program, under Subtitle D, encourages states to develop comprehensive plans to manage primarily nonhazardous solid waste, e.g., **household waste**. The second program, under Subtitle C, establishes a system for controlling hazardous waste from the time it is generated until its ultimate disposal—or from "cradle-to-grave." The third program, under Subtitle I, regulates certain underground storage tanks. It establishes performance standards for new tanks and requires leak detection, prevention and corrective action at underground tank sites. The newest program to be established is the **medical waste** program under Subtitle J. It establishes a demonstration program to track medical waste from generation to disposal.

Although RCRA created a framework for the proper management of hazardous and nonhazardous solid waste, it did not address the problems of hazardous waste encountered at inactive or abandoned sites or those resulting from spills that require emergency response. These problems are addressed by the Comprehensive Environmental Response, Compensation and Liability Act (CERCLA), commonly called Superfund.

Today RCRA refers to the overall program resulting from the Solid Waste Disposal Act. The Solid Waste Disposal Act was enacted in 1965 for the primary purpose of improving solid waste disposal methods. It was amended in 1970 by the **Resource Recovery** Act and again in 1976 by RCRA. The changes embodied in RCRA remodeled our nation's solid waste management system and added provisions pertaining to hazardous waste management.

The Act continues to evolve as Congress amends it to reflect changing needs. It has been amended several times since 1976, most significantly on November 8, 1984. The 1984 amendments, called the Hazardous and Solid Waste Amendments (HSWA), expanded the scope and requirements of RCRA. Provisions resulting from the 1984 amendments are significant, since they tend to deal with the waste problems resulting from more complex technology.

The Act is a law that describes the kind of waste management program that Congress wants to establish. This description is in very broad terms—directing the EPA to develop and promulgate criteria for identifying the characteristics of hazardous waste. The Act also provides the Administrator of the EPA (or designated representative) with the authority necessary to carry out the intent of the Act—the authority to conduct inspections.

The Act includes a Congressional mandate for EPA to develop a comprehensive set of regulations. Regulations

are legal mechanisms that define how a statute's broad policy directives are to be implemented. RCRA regulations have been developed by the EPA, covering a range of topics from guidelines for state solid waste plans to a framework for the hazardous waste permit program.

RCRA regulations are published according to an established process. When a regulation is proposed it is published in the *Federal Register*. It is usually first published as a proposed regulation, allowing the public to comment on it for a period of time—normally 30 to 60 days. Included with the proposed regulation is a discussion of the Agency's rationale for the regulatory approach and an explanation, or preamble, of the technical basis for the proposed regulation. Following the comment period, EPA evaluates public comments. In addressing the comments, the EPA usually revises the proposed regulation. The final regulation is published in the *Federal Register* ("promulgated"). Regulations are compiled annually and bound in the *Code of Federal Regulations* (CFR) according to a highly structured format. The codified RCRA regulations can be found in Title 40 of the CFLRL, Parts 240-280; regulations are often cited as 40 CAR.

Although the relationship between an Act and its regulations is the norm, the relationship between HSWA and its regulations differs. HSWA is unusual in that Congress placed explicit requirements in the statute in addition to instructing the EPA in general language to develop regulations. Many of these requirements are so specific that the EPA incorporated them directly into the regulations. HSWA is all the more significant because of the ambitious schedules that Congress established to implement the Act's provisions. Another unique aspect of HSWA is that it establishes "hammer" provisions—statutory requirements that go into effect automatically as regulations if EPA fails to issue these regulations by certain dates. EPA further clarifies its regulations through guidance documents and policy.

The EPA issues guidance documents primarily to elaborate and provide direction for implementing the regulations—essentially they explain how to do something. For example, the regulations in 40 CAR Part 270 detail what is required in a permit application for a hazardous management facility. The guidance for this part gives instructions on how to evaluate a permit application to see if everything is included. Guidance documents also provide the Agency's interpretations of the Act.

Policy statements, however, specify operating procedures that *should* be followed. They are a mechanism used by EPA program offices to outline the manner in which pieces of the RCRA program are to be carried out. In most cases, policy statements are addressed to staff working on implementation. Many guidance and policy documents have been developed to aid in implementing the RCRA Program. To find out what documents are available, the Office of Solid Waste's Directives System lists all RCRA-related policy guidance and memoranda and identifies where they can be obtained.

RCRA works as follows: Subtitle D of the Act encourages States to develop and implement solid waste management plans. These plans are intended to promote **recycling** of solid wastes and require closing or upgrading of all environmentally unsound dumps. Due to increasing volumes, solid waste management has become a key issue facing many localities and states. Recognizing this, Congress directed the EPA in HSWA to take an active role with the states in solving the difficult problem of solid waste management. The EPA has been in the process of revising the standards that apply to **municipal solid waste** landfills. Current with the revision of these standards, the EPA formed a task force to analyze solid waste source reduction and recycling options. The revised solid waste standards, together with the task force findings, form the basis for the EPA's development of strategies to better regulate municipal solid waste management.

Subtitle C establishes a program to manage hazardous wastes from "cradle-to-grave." The objective of the C program is to ensure that hazardous waste is handled in a manner that protects human health and the environment. To this end, there are Subtitle C regulations regarding the generation, **transportation** and treatment, storage, or disposal of hazardous wastes. In practical terms, this means regulating a large number of hazardous waste handlers. As of June, 1989, the EPA had on record more than 7,000 treatment, storage and disposal facilities; 17,000 transporters; and about 180,000 large and small quantity generators. By February, 1992, that number had increased substantially (43,000 treatment, storage, and disposal facilities; 19,700 transporters; 238,000 large and small quantity generators).

The Subtitle C program has resulted in perhaps the most comprehensive regulations the EPA has ever developed. They first identify those solid wastes that are "hazardous" and then establish various administrative requirements for the three categories of hazardous waste handlers: generators, transporters and owners or operators of treatment, storage and disposal facilities (TSDFs). In addition, Subtitle C regulations set technical standards for the design and safe operation of TSDFs. These standards are designed to minimize the release of hazardous waste into the environment. Furthermore, the regulations for TSDFs serve as the basis for developing and using the permits required for each facility. Issuing permits is essential to making the Subtitle C regulatory program work, since it is through the permitting process that the EPA or a State actually applies the technical standards to facilities.

Subtitle I regulates **petroleum** products and hazardous substances (as defined under Superfund) stored in underground tanks. The objective of Subtitle I is to prevent leakage to **groundwater** from tanks and to clean up past releases. Under Subtitle I, the EPA developed performance standards for new tanks and regulations for leak detection, prevention, closure, financial responsibility, and corrective action at all underground tank sites. This program may be delegated to states.

Subtitle J was added by Congress when, in the summer of 1988, medical wastes washed up on Atlantic beaches, thus highlighting the inadequacy of medical waste management practices. Subtitle J instructed the EPA to develop a two-year demonstration program to track medical waste from generation to disposal in demonstration States. The demonstration program was completed in 1991 and Congress has yet to decide upon the merits of national medical waste regulation. At this point no Federal regulatory authority is administering regulations, but individual states are; some are setting up very stringent regulations on storage, packaging, transportation, and disposal.

Congress and the President set overall national direction for RCRA programs. The EPA's Office of Solid Waste and Emergency Response translates this direction into operating programs by developing regulations, guidelines and policy. the EPA then implements RCRA programs or delegates implementation to the states, providing them with technical and financial assistance.

Waste minimization or reduction is a major EPA goal. It is defined as any source reduction or recycling activity that results in either reduction of total volume of hazardous waste or reduction of toxicity of hazardous waste, or both, as long as that reduction is consistent with the general goal of minimizing present and future threats to human health and the environment. *See also* Waste reduction

[*Liane Clorfene Casten*]

Resources
Books

Hazardous Waste Management Compliance Handbook. New York: Van Nostrand Reinhold, 1992.

RCRA Deskbook. Washington, DC: Environmental Law Institute, 1991.

Other

U. S. Environmental Protection Agency. *RCRA Orientation Manual.* Washington, DC: U.S. Government Printing Office, 1990.

Resource recovery

Resource recovery is the process of recovering materials or energy from **solid waste** for **reuse**. The aim is to make the best use of the economic, environmental, and social costs of these materials before they are permanently laid to rest in a **landfill**. The **Environmental Protection Agency** (EPA) and environmentalists have set up a hierarchy for resource recovery: reduce first, then reuse, recycle, incinerate with **energy recovery**, and landfill last. Following the hierarchy will cut solid waste and reduce resources consumed in production. Solid waste managers have turned to resource recovery in an effort to cut disposal costs, and the hierarchy has become not only an important guideline but a major inspiration for local **recycling** programs.

After the industrial revolution made a consumer society possible, **garbage** was considered a resource for a class of people who made their living sorting through open dumps, scavenging for usable items and recovering valuable scrap metals. Once public health concerns forced cities to institute garbage collection and dump owners began to worry about liability for injuries, city dumps were for the most part closed to the public. Materials recovery continued in the commercial sector, however, with an entire industry growing up around the capturing of old refrigerators, junk cars, discarded clothing, and anything else that could be broken down into its raw materials and made into something else. That industry is still strong; it is represented by powerful trade associations, and accounts for three-quarters of all ocean-borne bulk cargo that leaves the Port of New York and New Jersey for foreign markets.

The Arab **oil embargo** of the 1970s focused attention on the fact that **natural resources** are limited and can be made unavailable, and environmentalists took up the cause of recycling. But only after a barge loaded with New York City garbage spent months stopping in port after port on the eastern seaboard because it was unable to find a landfill was the real problem recognized. There were too few places to put it. The places that did take garbage often leaked a toxic brew of leachate and were seldom subject to environmental requirements.

Environmental restrictions on waste disposal, however, began to change. The EPA required that landfills be lined with expensive materials; states began mandating complicated leachate collection systems to protect **groundwater**. Amendments to the **Clean Air Act** of 1990 changed the regulations for garbage incinerators and waste-to-energy plants, which capture the energy created from burning trash to run turbines or to heat and cool buildings. These facilities were now required to install **scrubbers** or other devices for cleaning emissions and to dispose of the ash in safe landfills. The result of all these changes was a dramatic increase in the cost of dumping; as the price shot up, municipal leaders took notice. Economics joined **environmentalism** to support recycling, and those charged with managing solid wastes

began to look for ways to get the most out of their trash or avoid creating it altogether.

Reduction—cutting waste by using less material to begin with—is at the top of the resource recovery hierarchy because it eliminates entirely the need for disposal, while avoiding the environmental costs of using raw materials to make replacement goods. **Waste reduction**, or source reduction, requires a change in behavior that many say is unrealistic when both the culture and the economy in the United States are based on consumption. However, manufacturers have responded to consumer demands that products last longer and use fewer resources. The durability of the average passenger tire, for example, nearly doubled during the 1970s and 1980s. Makers of **disposable diapers** advertise the fact that they have reduced diaper size so that they take up less space in landfills.

Reuse also cuts the need for raw materials. Perhaps the best example is the reusable beverage bottles that were widely used until the second half of this century. Studies have shown that reusing a glass bottle takes less energy than making a new one or melting it down to make a new one.

Recycling has become a popular way to make new use of the resources in old products. In its purist and most efficient form, recycling means turning an old object into the same kind of new object—an old newspaper into a new one, an old plastic detergent jug into a new one. But new uses for certain materials are limited, and some laws prevent recycled plastic from coming into contact with food; in such cases old items can only be made into lesser products. Recycled plastic soda bottles, for example, can be used in carpeting, television sets, and plastic lumber. The act of recycling itself also uses energy and causes **pollution**. Materials must be shipped from drop-off site to the remanufacturer and then to producer before they can be returned to the retailer. Grinding plastic into pellets takes energy, and cleaning old newsprint produces by-products that can pollute water. Another drawback to recycling is the unstable price of materials, which has caused some solid waste managers to reevaluate local programs.

After **composting**, the resource recovery hierarchy lists **incineration** with energy recovery. As the last option before landfilling, incinerators or waste-to-energy plants only burn what cannot be recycled, and they can recover one-quarter to one-third of their operating costs by selling the energy, usually in the form of electricity or steam. But incineration also produces waste ash, which can contain toxic materials such as **lead** or **mercury**. Incinerator ash often meets EPA standards for hazardous wastes, but because it mainly comes from household garbage, it is considered nonhazardous no matter what is detected, although this may change in the future. Other hazardous materials escape out the stacks of incinerators in the form of **air pollution**, despite high-tech scrubbers and other pollution controls. Environmentalists also maintain that the resources used in making materials that are incinerated are lost, despite the energy that is produced when they burn. They claim that waste-to-energy proponents have adopted the term resource recovery for the process in an effort to mask its problems and put an updated face on an industry that once was unscientific and uncontrolled.

Last on the list is landfilling. Although the United States has potential landfill space to last centuries, much of it is socially or environmentally unsound. Even with extensive liner systems, landfills are expected to leak **chemicals** into groundwater. Most communities, moreover, are unwilling to have new landfills as neighbors, and citizens groups throughout the country have demonstrated an increasing ability to keep them out through sophisticated protests. Finding space for new landfills has become a nightmare for elected officials and public works managers.

Despite their need for solutions, the resource recovery hierarchy does not always prove practical for local officials who are usually faced with tight budgets and the restrictions of democratic politics. Many of the advantages of the methods at the top of the hierarchy, such as waste reduction or recycling, are distant in either time or geography for many municipalities. A city might be hundreds of miles from the nearest plant that recycles paper and from the forests that might benefit from an overall decrease in demand for trees, for instance. In such a case, recycling paper makes little sense on a small scale. Environmentalists believe that the solution is for manufacturers and solid waste managers to make decisions together. Finding the most efficient use of resources derived from both virgin materials and recovered from wastes may have to be considered regionally or nationally to be fully realized. *See also* Container deposit legislation; Groundwater pollution; NIMBY; Plastics; Solid waste incineration; Solid waste volume reduction; Waste stream

[*L. Carol Richie*]

Resources

Books

Brown, L. R., et al. *State of the World 1991.* New York: W.W. Norton, 1991.

Periodicals

Peterson, C. "What Does 'Waste Reduction' Mean?" *Waste Age* (January 1989).

Schall, J. "Does the Hierarchy Make Sense?" *MSW Management* (January/February 1993).

Resources for the Future

In 1952 President Harry S. Truman's Materials Policy Commission, headed by William Paley of the Columbia Broad-

casting System (CBS), released its final report. The commission had been asked to examine the status of the nation's material resources, such as fuels, metals, and minerals. It concluded that the country's material base was in quite good condition, but the report maintained that further research was necessary to insure the continued use of these resources. Paley established Resources for the Future (RFF) as a nonprofit corporation to carry out this research, especially the updating of resource **statistics**.

At the same time, the Ford Foundation had created a fund to support resource **conservation** issues and to organize a national conference on the issue. Paley allowed this fund to use the name and facilities of his new corporation, and RFF hosted the Mid-Century Conference on Resources for the Future in Washington, D.C. with Ford Foundation funding.

Perhaps the major result of the conference was organizational. The Ford Foundation decided that RFF should become the base of its resource research program as an independent foundation. These organizations began in 1954 with a five-year grant to RFF of $3.4 million "aimed primarily at the economics of the nation's resource base." This has been the goal of research at RFF ever since. Most of the staff and most of the visiting scholars, as well as most of the beneficiaries of their small grants program have been economists, analyzing the importance of material resources to the United States economy, and examining a wide range of intertwined economic, resource, conservation, and environmental issues.

Resources for the Future is now organized into two divisions, the Energy and Natural Resources Division, and the Quality of the Environment Division. There are also two centers, the National Center for Food and Agricultural Policy, and the Center for Risk Management. Well-known experts on resource and environmental issues such as Marion Clawson, Hans Landsberg, John Krutilla, and Allen Kneese are long-time staff members or "fellows" of RFF. The organization also supports resource and environmental research by non-staff scholars through visiting fellowships and its small grants program, and the range of these activities exceeds RFF's original mandate. RFF publishes an annual report and a quarterly journal called *Resources*. The annual report lists the year's publications by resident staff, making it a good reference source for up-to-date materials on a wide range of resource and environmental issues. *Resources* publishes short articles, mostly by staff members or visiting fellows, and a quarterly report entitled "Inside RFF."

A scan of these two documents on any given date reveals the range of RFF interests. A recent annual report included staff publications on energy on different aspects of agriculture and agricultural policy, on various aspects connecting economics and public **forest management**, on climatic change, on risk management, cost analysis, and regulations and regulatory activities affecting resources and the environment.

[*Gerald L. Young Ph.D.*]

RESOURCES

ORGANIZATIONS

Resources for the Future, 1616 P Street NW, Washington, DC USA 20036-1400 (202) 328-5000, Fax: (202) 939-3460, <http://www.rff.org>

Respiration

The term respiration has two major definitions. On a cellular level, respiration is the chemical process by which food molecules are oxidized to release energy. In the process, **carbon dioxide** and water are also produced. Cellular respiration is, therefore, the reverse of **photosynthesis**, the process by which **carbon** dioxide and water are combined to form complex carbohydrate molecules. On an organismic level, respiration refers to the act of inhaling and exhaling. **Respiratory diseases** are among the most common of all environmental illnesses. Inhaling **particulate** matter in polluted air may result, for example, in various forms of pneumoconiosis.

Respiratory diseases

Respiratory diseases—diseases of the lungs and airways such as **asthma**, chronic **bronchitis**, **emphysema**, and lung cancer—have diverse causes. While cigarette smoking is the leading cause of most major respiratory diseases, air pollutants and workplace **toxins** can also contribute to respiratory illness.

The lungs are equipped with an elaborate defense system to repel toxins and invading organisms. Before air reaches the lungs, it passes through the nose, throat, and bronchi that are lined with mucus to trap irritants. Within the bronchi, smaller bronchial tubes and bronchioles are covered with cilia that sweep particles out. Lungs are far from vulnerable, however. The defenses can fail, leading to any number of respiratory diseases.

Cigarette smoking is the most important cause of Chronic Obstructive Pulmonary Disease (COPD), a name covering pulmonary emphysema and chronic bronchitis. These lung diseases damage the air passageways and interfere with the lung's ability to function. COPD is the fifth leading cause of death in the United States, and in a smoking, aging population, will probably not decline.

Chronic bronchitis is a persistent inflammation of the bronchial tubes. Mucus cells of the bronchial tree produce excess mucus, and the lungs may become permanently filled

with fluid. Eventually, scar tissue from infection replaces the cilia, decreasing the lungs' efficiency. With each infection, mucus clogs the alveoli, or air sacs. Little or no gas exchange occurs, and the ventilation-blood flow imbalance reduces oxygen levels and raises **carbon dioxide** levels in the blood.

Chronic bronchitis usually accompanies the development of emphysema. The destruction of too many elastic fibers in the lungs' framework and air sac walls results in hyperinflated air sacs that impair the lungs' ability to recoil during expiration. Eventually, the alveoli merge into one large air sac, and the network of capillaries in the lungs is lost. This reduces gas exchange that occurs, and stale air is trapped in the lungs.

These two COPDs are not curable, but can be treated. Patients can prevent pulmonary infections and the inhalation of harmful substances, reduce airway obstruction, improve muscle conditioning and use supplemental oxygen.

Asthma is a disease characterized by a narrowing of the airways, episodic wheezing, tightness in the chest, shortness of breath, and coughing. There are various types of asthma, including exercise-induced asthma, occupational asthma triggered by irritants in the workplace, and asthmatic bronchitis. Trigger factors can include dust, odors, cold air, **sulfur dioxide** fumes, emotional stress, upper-respiratory infection, exertion, and airborne allergens. Asthma is often preventable if trigger factors can be identified and eliminated.

Scientists are also turning their attention to health hazards posed by atmospheric acids and other air pollutants. One study at 79 southern Ontario hospitals showed a consistent association between the summer levels of atmospheric sulfates and **ozone** and hospital admissions for acute respiratory illnesses such as asthma, chronic bronchitis, and emphysema.

This "acid air" forms when emissions of sulfur dioxide and **nitrogen oxides**, mostly from coal-burning **power plants** and motor vehicles, are transformed to sulfuric and nitric acids. **Acid aerosol** concentrations tend to be higher on hot summer days. The fine acid particles can penetrate the deepest, most delicate tissues of the lungs, inflame respiratory-tract tissue, depress pulmonary function, and constrict air passages. Health effects are more pronounced when acid aerosols are accompanied by ozone in the lower **atmosphere**. Together, ozone and acid aerosols produce changes in the lungs that inhibit their ability to clear themselves of toxins and other irritants.

High levels of **air pollution** may foster respiratory-disease symptoms in otherwise healthy individuals. One study indicates that air **pollution** in Los Angeles may begin to permanently "derange" an individual's lung cellular architecture by age 14. Another study has implicated wood-burning stoves, which foul the air with tiny particulates that may cause or exacerbate outbreaks of respiratory illnesses. An environmental team studying respiratory problems at a Boston-area high school implicated the school's poorly designed ventilation system as the most likely cause of the students' high rate of respiratory illnesses.

Workplace-related respiratory diseases are on the increase. In 1990, the National Safety Council listed occupational lung diseases as the leading work-related diseases in this country. The Chicago-based National Safe Workplace Institute recently concluded that 2–4% of pulmonary disease is related to working conditions. Diseases that top the list include **asbestosis**, lung **cancer**, silicosis, occupational asthma, and **coal** workers' pneumoconiosis, commonly known as **black lung disease**.

Asbestosis is a chronic fibrotic lung disease caused by the inhalation of inert dusts. This affliction has received widespread media coverage because the majority of its victims have been subjected to long-term exposure to **asbestos**. Asbestosis causes a scarring of lung tissue that can result in serious shortness of breath. Silicosis is a disease of the lungs caused by breathing in dust that contains silica. Black lung disease is a long-term lung disease caused by the settlement of coal dust on the lungs, eventually resulting in emphysema.

Insurance data indicates that people in certain high-risk occupations, including agriculture, construction, mining, and quarrying have three to four times the average death rate for all industries, mostly from respiratory diseases.

Residential homes are also vulnerable. **Chlorine** bleach and cleaning fluids, insecticides, wood-burning fireplaces, and gas stoves that produce **nitrogen** dioxide can be toxic to the lungs. The causes of the expected 170,000 new cases of lung cancer in 1993 include exposure to **cigarette smoke**, **ionizing radiation**, **heavy metals**, and industrial carcinogens. *See also* Air quality; Automobile emissions; Fibrosis; Respiration; Sick building syndrome; Smog; Smoke; Yokkaichi asthma

[*Linda Rehkopf*]

Resources

Books

Haas, F., et al. *The Chronic Bronchitis and Emphysema Handbook.* New York: Wiley Science Editions, 1990.

Moeller, D. W. *Environmental Health.* Cambridge, MA: Harvard University Press, 1992.

Periodicals

"Experts Finger Tight Building Syndrome." *Science News* 137 (June 9, 1990): 365.

Fackelmann, K. "The High and Low of Respiratory Illness." *Science News* 137 (June 9, 1990): 365.

Raloff, J. "Air Pollution: A Respiratory Hue and Cry." *Science News* 139 (March 30, 1991): 203.

Shepherd, S. L., et al. "Is Passive Smoking a Health Threat?" *Consumers Research Magazine* 74 (October 1991): 30–1.

Restoration ecology

Ecological restoration is an attempt to reset the ecological clock and return a damaged **ecosystem** to its predisturbance state—to turn a disused farm into a **prairie** or to convert a parcel of low-lying acreage into a vigorous wetland. Precise replication of the predisturbance condition is unlikely to occur because each ecosystem is the result of a sequence of climatic and biological events unrepeatable in precisely the same order and intensity as the original sequence. However, close approximations of the predisturbance condition are often possible, with differences from the original apparent only to professionals.

Within this limitation, restorationists strive to re-build ecosystems that, if not exactly like their original predecessors, possess the qualities of a healthy ecosystem. These properties include:

- dominance of indigenous (native) **species**
- sustainability or the ability to perpetuate
- resistance to invasion by non-native or **pest** species
- the presence of healthy functions such as **photosynthesis**, **respiration**, and plant and animal reproduction.
- the ability to generate nutrients such as **nitrogen** and store them in the ecosystem
- a pattern of interaction between key species similar to the pattern found in undisturbed ecosystems, including relationships in the food chain

Regrettably, the term restoration as used in the media and by organizations responsible for ecological damage often refers to cosmetic activity—the clean up of oil, for instance, with the expectation that once a substantial amount of spilled oil has been removed, natural processes will restore the ecosystem to its former condition. Merely re-creating the form of an ecosystem without attention to whether it is functioning is not considered to be true ecological restoration. Developers also abused the term restoration when acquiring land for malls or residences in exchange for creating equivalent or larger **wetlands** elsewhere. Insead of functioning restorations, they often carry out substandard cosmetic restoration efforts.

Restoration **ecology** developed in the twentieth century as a response to **habitat** destruction brought about by industrialization, overpopulation, and agricultural mismanagement. Pioneering biologists, such as **Aldo Leopold**, began to advocate an approach to healing damaged ecosystems that involved direct human intervention.

Moreover, restoration ecology is more than just applied science that uses the insights of ecology to build ecosystems. It is becoming an essential research tool to test the **accuracy** of ecological theories through direct experience. For example, the role of fire in the maintenance of prairie ecosystems has come to be better understood through its application on restored tracts that were failing to thrive.

As habitat loss accelerates, restoration takes on increasing importance as means of preserving habitat for threatened species, rekindling lost **biodiversity**, restoring mineral balance to eroded and infertile lands, improving **water quality**, and preserving atmospheric gas balance. The task of restoration, seen as rather abstract in years past, is now viewed more concretely as the consequences of destruction of habitat begin to be understood.

Although the specific stages of a restoration project depend on the initial state of the land and the desired end, a typical restoration project begins with certain steps to create the conditions necessary for natural processes to gain a foothold. On eroded and compacted soils the first step is often to physically loosen the damaged **soil** by tearing or furrowing. This promotes water retention and creates suitable microenvironments for plant seeds. On depleted land, it also might be necessary to add **fertilizer** or organic **mulch** to provide initial replenishment of nutrients such as nitrogen and **phosphorus**.

Suitable plant seeds, preferably from indigenous species, are then introduced. These are usually key varieties called matrix species. Nurse species that create the conditions needed to hasten growth of other plants (a process called facilitation) may be introduced. Sometimes a sequence of species must be introduced in a specific order. For example, on clay mining wastes, the best growth is obtained by planting annual grasses and legumes, then perennial grasses a year later.

Often animal colonization is left to **nature**, especially if the site under restoration is near other predisturbance sites. Management of the site to monitor its progress and to remove unwanted invading species is usually carried out in succeeding months and years. If a restoration effort is successful, the site gradually returns to its natural state and the need for further intervention decreases.

Ecological restoration requires contributions from a large number of academic disciplines, although the precise mixture will vary from one restoration project to another. The term restoration ecology is misleading in suggesting that the activity is for ecologists only. All restoration projects require some degree of funding, and large-scale projects, such as surface mine **reclamation** or restoration of a wetland, river, or lake, may require substantial multi-year funding. In addition, if the restored ecosystem is to be protected from further damage, public understanding and support are necessary. As a consequence, disciplines that study the values of society and identify those that align with ecological values

are extremely important for large-scale, long-term successful restoration efforts.

Equally important is the knowledge obtained from climatologists, chemists, engineers (where restructuring is necessary), hydrologists, geologists, statisticians, forestry and **wildlife** specialists, geneticists, soils and **sediment** chemists, political scientists, attorneys, and a variety of other professions. Historians and anthropologists often have a crucial role in restoration because the history of ecological damage may be reconstructed with historic evidence, both written and through cultural and biological artifacts and relics.

The role of private citizens in a restoration effort often is pivotal. Community leaders may provide funds toward carrying out restoration, while the work of local volunteers helps to connect the community with the organizations and institutions administering a project. Community involvement also provides a project with a long-term focus months or years after the initial effort is complete. Moreover, the labor-intensive task of restoration seems to thrive when it is carried out by volunteers and concerned individuals working together with specialists.

Although underlying theory in the field of restoration ecology is still in its infancy and the precise outcome of a project is almost always uncertain, restoration efforts virtually all result in improvement to damaged environments. The condition of a restored system may be strikingly superior to the damaged condition. For example, the tidal Thames River in England had virtually no fish species in the 1950s. However, many years after **pollution** clean up and restoration, over 100 species were found in the tidal area. In the United States, **Lake Washington** in the Pacific Northwest, the Kissimmee River in Florida, the Rio Blanco in Colorado, and the Hackensack River Meadowlands in the New York metropolitan area are examples of successful ecological pollution clean-ups and restoration. Many of these efforts were citizen-initiated.

The United States Department of Agriculture **Fish and Wildlife Service** (USFWS) has undertaken a new approach to restoring waterways, lakes, and wetlands looking at the processes that impact these ecosystems from a larger **watershed** perspective. Identifying sources of pollution and **environmental stress** at the level of the watershed allows scientists to incorporate study of the surrounding land and its effects on habitat destruction in the at risk body of water. An example where watershed analysis sucessfully provided the necessary information for ecological restoration is in the restoration of stream corridors in the Whitefish Mountains of Montana.

Another success story is the restoration of the Lanphere **Dunes** Unit of the Humboldt Bay **National Wildlife Refuge** by **the Nature Conservancy** and its partners. The major challenge in the Lanphere Dunes was eradication of a 10-acre (4-ha) patch of non-native invasive grass called European beachgrass (*Ammophila arenaria*). European beachgrass is destructive to dunes because it changes that way that sand accumulates. This alters the suitability of the habitat for native plants. The elimination of the non-native grass and restoration of native plant communities was accomplished with more than 2,000 person hours per acre of volunteer labor over the course of three years. As of 1997, native plant cover had increased by almost 50%.

Similar international restorations, such as the Guanacaste dry forest in Costa Rica, show that citizens of developing countries with far fewer monetary resources the United States can also have strong involvement in ecological restoration. In a time of environmental attrition, the restoration movement plays a role in shaping the future by helping citizens develop a feeling of connection between themselves and their wild lands, while providing concrete improvements in ecological conditions.

[*John Cairns Jr. and Jeffrey Muhr and Marie H. Bundy*]

RESOURCES
BOOKS

Berger, J. J., ed. *Environmental Restoration: Science and Strategies for Restoring the Earth.* Covelo, CA: Island Press, 1989.

Bradshaw, A. D., and M. J. Chadwick. *The Restoration of Land.* Berkeley: University of California Press, 1980.

Cairns Jr., J. *Rehabilitating Damaged Ecosystems.* 2 vols. Boca Raton: CRC Press, 1988.

Ehrlich, P. R., and A. H. Ehrlich. *Healing the Planet.* New York: Addison-Wesley, 1991.

Jordan III, W. R., et al., eds. *Restoration Ecology: A Synthetic Approach to Ecological Research.* Cambridge: Cambridge University Press, 1990.

National Research Council. *Restoration of Aquatic Ecosystems: Science, Technology, and Public Policy.* Washington, DC: National Academy Press, 1991.

PERIODICALS

Burke, William K. "Return of the Native: The Art and Science of Environmental Restoration." *E Magazine*, July/August 1992.

Retention time

Retention time refers to a time frame in which **chemicals** stay in a certain location. The term is sometimes used interchangeably with residence or renewal time, which in **limnology** (aquatic **ecology**) describes the length of time a water molecule or a chemical resides in a body of water. The **residence time** of water is called the hydraulic retention time. It may range from days to hundreds of years, depending on the volume of the lake and rates of inflow and outflow, and is often used in calculations of **nutrient loading**. Retention time can also refer to the length of time needed to detoxify harmful substances or to break down hazardous chemicals in pharmaceutical and **sewage treatment** plants;

the length of time chemicals stay in a living organism; or the length of time needed to detect certain chemicals by instruments such as gas **chromatography**.

Reuse

Reuse, using a product more than once in its original form, is one of the preferred methods of solid **waste management** because it prevents materials from becoming part of the **waste stream**. It is a form of source reduction of waste.

Some products are specifically designed to be reusable while others are commonly reused as a matter of convenience. In the former instance, products such as canvas shopping bags and cloth napkins are meant to be used again and again, unlike their paper cousins. In the latter case, glass and plastic food containers are often reused for numerous household purposes, even though they are intended to be used only as food packaging.

Reuse of products is a basic component of diaper services and home bottled water delivery. Although neither of these services were originally created with **waste reduction** in mind, they do impact **solid waste** generation. A cotton diaper can be used 70 or more times, unlike a single-use disposable diaper. The three and five gallon bottles used by most bottled water services are picked up and taken back to the water plant for refilling, eliminating the need for consumers to buy individual gallon containers in supermarkets.

Deposit systems for beer and soft drink containers represent another form of product reuse. In these systems, consumers pay a nominal fee when they purchase beverages sold in specific types of containers. The fee is refunded when they return the empty containers to the retailer or other designated location. Virtually all beer and soft drinks used to be sold in returnable bottles. During the 1950s, the use of nonreturnable bottles and metal cans started to be more commonplace. By the mid-1980s, most beverages were packaged in nonreturnable bottles, cans, or plastic containers.

Nine states currently have legislation requiring deposits on beverage containers. The deposits are meant to be incentives to consumers to return the bottles. Where such legislation exists, between 70 and 90% of the targeted containers are returned. Litter control rather than solid waste reduction is sometimes the purpose of beverage **container deposit legislation**. Reduction in litter has been documented to be as high as 80%, but the impact on solid waste disposal is more difficult to calculate since beverage containers account for a relatively small portion of the waste stream.

Another type of product reuse is resale of used clothing and household items through thrift stores and garage sales. Although calculating the impact of such practices on waste generation would be extremely difficult, there is no doubt that they result in some reduction in waste disposal. *See also* Recyclables; Recycling; Waste reduction

[*Teresa C. Donkin*]

Resources
Books

Bohm, P. *Deposit-Refund Systems: Theory and Applications to Environmental, Conservation and Consumer Policy.* Baltimore: Johns Hopkins University Press, 1981.

Revegetation

see **Restoration ecology**

Rhinoceroses

Popularly called rhinos, rhinoceroses are heavily-built, thick-skinned herbivores with one or two horns on their snout and three toes on their feet. The family Rhinocerotidae includes five **species** found in Asia and Africa, all of which face **extinction**.

The two-ton, one-horned Great Indian rhinoceroses (*Rhinoceros unicornis*) are shy and inoffensive animals that seldom act aggressively. These rhinos were once abundant in Pakistan, northern India, Nepal, Bangladesh, and Bhutan. Today, there are about 2,400 Great Indian rhinos left in two game reserves in Assam, India, and in Nepal. The smaller one-horned Javan rhinoceros (*Rhinoceros sondaicus*) is the only species in which the females are hornless. Once ranging throughout southeast Asia, Javan rhinos are now on the verge of extinction, with only 60 living on reserves in Java and Vietnam.

The Sumatran rhinoceros (*Didermocerus sumatrensis*), the smallest of the rhino family, has two horns and a hairy hide. There are two subspecies—*D. s. sumatrensis* (found in Sumatra and Borneo) and *D. s. lasiotis*—found in Thailand, Malaysia, and Burma. Sumatran rhinos are found in hilly jungle terrain and once coexisted in southeast Asia with Javan rhinos. Now there are only 300 Sumatran rhinos left.

The two-horned, white, or square-lipped, rhinoceros (*Ceratotherium simum*) of the African **savanna** is the largest land mammal after the African elephant, standing 7 ft (2 m) at the shoulder and weighing more than 3 tons. White rhinos have a wide upper lip for grazing. There are two subspecies: the northern white (*C. s. cottoni*) and the southern white (*C. s. simum*). Once common in the Sudan, Uganda, and Zaire, northern white rhinos are now extremely rare, with only 40 left (28 in Zaire, the rest in zoos). Southern African white rhinos are faring somewhat better (7,500) and are the world's most common rhino.

Great Indian rhinoceros. (Photograph by Gerald Davis. Phototake. Reproduced by permission.)

The smaller two-horned black rhinoceros (*Diceros bicornis*) has a pointed upper lip for feeding on leaves and twigs. Black rhinos can be aggressive but their poor eyesight makes for blundering charges. Black rhinos (which are actually dark brown) were once common throughout sub-Saharan Africa but are now found only in Kenya, Zimbabwe, Namibia, and South Africa. Today, there are only 2,600 black rhinos left in the wild, compared to 100,000 30 years ago.

Widespread **poaching** has diminished rhino populations. The animals are slaughtered for their horns, which are made of hardened, compressed hair-like fibers. In Asia, the horn is prized for its supposed medicinal properties, and powdered horn brings $28,000 per kg. In Yemen, a dagger handle made of rhino horn can command more than $1,000. As a result, rhinos now survive only where there is strict protection from poachers. Captive breeding programs for endangered rhinos are hindered by the general lack of breeding success for most species in zoos and a painfully slow reproduction rate of only one calf every three to five years. The present world rhino population of about 16,000 is little more than half the estimated "safe" long-term survival number of 22,500.

[*Neil Cumberlidge Ph.D.*]

Resources

Books

Cumming, D. H. M., R. F. Du Toit, and S. N. Stuart. *African Elephants and Rhinos: Status Survey and Conservation Action Plan.* Gland, Switzerland: IUCN-The World Conservation Union, 1990.

Penny, M. *Rhinos, Endangered Species.* New York: Facts on File, 1988.

Periodicals

"Southern White Rhino comes back after Its Brush with Extinction." *Winston-Salem Journal*, January 23, 2000, D6.

Tudge, C. "Time to Save the Rhinoceroses." *New Scientist* 28 (September 1991): 30–5.

Other

Save the Rhino. [cited May 2002]. <http://www.savetherhino.org/index>.

Ribonucleic acid

Ribonucleic **acid** (RNA) exists as a polymer constructed of four kinds of nucleotides. Ribonucleic acids are ordinarily involved in the conversion of genetic information from DNA into protein: information flows from the genetic material via RNA for the fabrication of an organism. Ribosomes are

cytoplasmic particles structured of protein and ribosomal RNA (rRNA) and are the sites of protein synthesis. Messenger RNA (mRNA), transcribed from genomic DNA, translocates genetic information to the ribosomes. In addition, there are about 20 transfer RNAs (tRNA), which bind to specific amino acids and to particular regions of mRNA for the assembly of amino acids into proteins. The genetic material of some viruses is RNA.

Ellen Henrietta Swallow Richards (1842 – 1911)

American chemist

Ellen Swallow Richards was born in Ipswich, Massachusetts, and was a daughter of Peter and Fanny Swallow, who were school-teachers and part-time farmers. Ellen Swallow obtained a Baccalaureate (the equivalent of an undergraduate degree) from Vassar Female College, and in 1870 she became the first woman to be accepted as a student by the Massachusetts Institute of Technology (M.I.T.). Founded in Boston in 1865, MIT had rapidly become a highly regarded school of engineering and applied science. In 1873, she graduated as a Bachelor of Science in chemistry and in the same year received a Master of Arts degree from Vassar, based on a thesis involving examination of the vanadium content of an iron ore.

Ellen Swallow then spent several years studying towards a doctorate at MIT, but did not graduate, apparently because the faculty members of the department in which she was studying, all of whom were men, were unwilling to allow a woman to be the first-ever Doctor of Science graduate from their program. Ellen Swallow was undaunted by this and other forms of gender bias that were pervasive at the time. She went on to become a pioneer in the development of the new engineering field of public **sanitation**, and was a strong advocate of domestic science and home economics.

In 1875 Ellen Swallow married Robert Hallowell Richards, a professor of metallurgy and mining engineering at M.I.T. The couple shared a devotion to science and to each other, but had no children. Ellen Richards collaborated with her husband in his metallurgical and mining work, and her contributions in those fields were recognized later on, when she became the first woman to be elected to the American Institute of Mining and Metallurgical Engineers.

Ellen Richards understood that one of her major responsibilities was to foster the scientific education of American women. She was central in the establishment of a Women's Laboratory at M.I.T., as well as other facilities and programs that made important contributions towards allowing women to pursue careers in science. Some of her early research involved investigations of household and food chemistry and **food additives**. In 1882, she published *The Chemistry of Cooking and Cleaning*, and in 1885 *Food Materials and Their Adulterations*.

In 1884, Ellen Richards was appointed to a position in a new chemical laboratory at M.I.T. that specialized in the study of public sanitation. She held this position for 27 years until her death in 1911. One of her initial accomplishments was running the chemical laboratory responsible for the first survey of lakes and rivers in Massachusetts, an extensive study that became a classic in its field.

In 1890, M.I.T. established a ground-breaking program in sanitary engineering in which Ellen Richards taught the chemistry and analyses of water, sewage, and **atmosphere**. In collaboration with A. G. Woodman, in 1900 she published *Air, Water, and Food Chemistry for Colleges*, a textbook in support of the teaching of public sanitation. Her most important scientific contributions were in the field of water sanitation. She was, for example, an influential advocate of the **chlorination** of drinking water to kill pathogenic bacteria. This practice was critical in lowering the high death rates that were being caused by drinking unsanitary water in most towns and cities of the time.

After about 1890, Ellen Richard's interests increasingly focused on what was to become known as the "home economics movement." This involved the study and teaching of nutrition, food preparation, household sanitation and hygiene, and related subjects, mostly in the public school system. The home economics movement can be likened to a domestic-science literacy program for the masses, and it became an important means by which urban women learned how to run economical and healthy households. Ellen Richards wrote numerous papers and several books in this subject area, including *Home Sanitation: A Manual for Housekeepers* in 1887, *Domestic Economy as a Factor in Public Education* in 1889, and *Euthenics: The Science of Controllable Environments* in 1912. Today Ellen Richards is considered a parent of home economics in the United States. For this contribution, she is indirectly responsible for many pervasive improvements in living conditions throughout America and in other countries where home economics has become a popular subject in school curricula.

[*Bill Freedman Ph.D.*]

Resources

Books

James, E. T., J. W. James, and P. S. Boyer, eds. *Notable American Women. 1607–1950.* Cambridge, MA: Belknap Press, 1971.

Stern, M. B. "The First Woman Graduate of M.I.T.—Ellen H. Richards, Chemist." In *We the Women: Career Firsts of Nineteenth Century America.* New York: Schulte Pub. Co., 1962.

Right-to-act legislation

On September 3, 1991, 25 men and women perished in a fire in a chicken-processing plant in Hamlet, North Carolina. Workers were trapped inside the burning building because managers of the plant had illegally bolted emergency exits in order to prevent possible theft of chickens. The American public reacted with outrage to news of the fire because it was recognized that the employees' deaths could have been prevented if **Occupational Safety and Health Administration** (OSHA) standards regarding access to fire exits had been enforced. As a result, labor representatives have repeatedly called for new, more effective means for protecting Americans from the hazards of injury, illness, and death at their workplaces.

In North Carolina, the site of the 1991 poultry plant fire, Worker Right to Act (RTA) legislation has been enacted in an effort to meet those demands, and such legislation is being proposed in other states. RTA legislation is designed to give workers some of the power they need to avoid or prevent exposure to such workplace hazards as those leading to the poultry plant fire and those associated with use of toxic **chemicals**. It is important to note that the goals of RTA legislation overlap with those of **Toxics Use Reduction legislation** (TUR). TUR laws, which have been enacted in at least 26 states, require business facilities to reduce their use of toxic substances, thus protecting workers and other community residents.

Although North Carolina has adopted a number of RTA provisions through a series of separate statutes instead of in a comprehensive, single package, RTA laws are only in the proposal stage in most states in which they are being advocated. In New Jersey, an initial four-year campaign for adoption of Worker and Community RTA was unsuccessful. Advocates of RTA in that state were optimistic that future efforts will be successful and that their experiences would benefit groups and individuals working for the enactment of RTA laws in other states. Several years later, Passaic County in New Jersey passed the nation's first right-to-act law in the spring of 1999. The measure allows 25 or more neighbors and/or workers to petition the county Health Officer for creation of a Neighborhood Hazard Prevention Advisory Committee to monitor a specific facility. In addition, such committees have the authority to conduct walk-through surveys of the plant, accompanied by technical experts. If a company refuses to cooperate, the county can sue on behalf of the committee.

In Michigan, a comprehensive RTA bill was first considered in 1993 and has been reintroduced before the state legislature several times since then. Therefore, the bill proposed in Michigan is used here to illustrate the provisions of a comprehensive RTA statute. Second, some of North Carolina's RTA provisions are described to provide further examples of RTA mechanisms.

Michigan's proposed RTA law

A Worker RTA bill considered in Michigan in 1993 includes five sets of RTA protections. First, the bill mandates that worker-management committees be established at each worksite where the number of employees regularly exceeds ten. If there is a government-certified labor organization at the worksite, it will select the workers' representatives. If there is no certified labor organization, nonsupervisory employees will select their own representatives. The committee must: (1) inspect the site at least monthly for existing or potential safety, health, and environmental problems; (2) investigate accidents and exposures that have the potential to harm employees and the **environment**; and (3) conduct annual reviews. The committee's duties are cross-referenced to a proposed Toxics Use Reduction and Community Right to Act (TUR/CRTA) bill.

Second, the Michigan Worker RTA bill mandates that each employer develop and implement a worksite safety and health plan. Such plans must provide for periodic inspections of the worksite and require documentation of hazards and actions taken to correct them.

Third, there are provisions giving any employee or employee representative the right to request an inspection by the Michigan Department of Labor or the Michigan Department of Health if he or she believes that there is a violation of a standard and that that violation threatens physical harm to an employee. Before a representative of one of those agencies makes a determination as to imminent danger, an employee may choose not to perform an assigned task if the employee has a reasonable apprehension of death or serious injury and he or she reasonably believes that no less drastic alternative is available.

Fourth, the Michigan bill increases the authority of state inspectors regarding citations and penalties and authorizes employees to contest the failure of such inspectors to conduct inspections and issue citations. There are substantial penalties for willful or repeated violations of the Act. Fifth, an employer cannot **discharge** an employee or in any way discriminate against an employee or job applicant because he or she has filed a complaint under the Act or testified at a proceeding under the Act.

It is significant that the Michigan TUR bill, entitled the "Toxic Use Reduction and Community Right to Act Bill" (TUR/CRTR), also includes RTA provisions. A summary of some provisions of Michigan's TUR/CRTA bill illustrates the interrelationship between RTA and TUR laws. First, TUR/CRTA establishes a goal of a 50% reduction in toxics use and toxics waste generation over a period of five years, thus reducing workers' exposure to toxics. Second, companies and governmental bodies that must report information under the

federal Emergency Planning and Community Right to Know Act (EPCRA) must conduct audits of toxics used and generated as waste, and they must prepare plans and set goals for reducing the amounts of those toxics at their facilities. Then, annual reports must be filed, documenting progress in reaching those goals. Third, communities are granted rights to monitor business facilities through "community environmental committees." This provision is cross-referenced to workers' rights to investigate hazards under the Worker RTA bill. Also, workers, through their committees established under Worker RTA, have the opportunity to review and provide input on the facility's TUR plan before it is completed. Fourth, there are provisions giving workers and community members the right to take companies to court to compel them to comply with the TUR/CRTA law.

North Carolina's RTA statutes

North Carolina's RTA statutes include some of the same kinds of mechanisms in Michigan's Worker RTA bill, but, overall, they are not as comprehensive. For example, North Carolina requires workplace safety committees and the establishment of health and safety programs, but those requirements are imposed only on employers of 11 or more employees if those employers have a poor "experience rating" under workers' compensation laws. (A "1.5" rating or worse is the measure used.)

North Carolina's RTA statutes do include several provisions which are not in the Michigan bill. North Carolina has created a special emphasis inspection program to target employers with high rates of violations or high rates of illness, injury or death. Also, an interagency task force has been created to study and issue a report setting out a plan for reorganization of the occupational health and safety and fire safety networks within North Carolina.

Significance of RTA laws

Labor leaders and workers in Michigan and in other states continue to advocate adoption of RTA and are optimistic that it will be adopted as leaders and citizens become aware of the need for it and its benefits. RTA laws are designed to lead to better enforcement of existing OSHA standards. Under RTA, worker-management committees conduct inspections of a workplace on a regular basis instead of waiting for OSHA or its state counterpart to do so. Thus, management and workers, as a team, become the primary watchdogs for the work facility. A major reason for the lack of enforcement by OSHA and its state counterparts is their lack of funding for inspectors. Use of worker-management committees provides a means of protecting workers despite scarce government resources.

By mandating that employers prepare worksite safety and health plans and that employee representatives be included in that planning process, RTA laws are designed to prompt employers and their employees to take a proactive stance with respect to workplace hazards. Workers are included in planning because they are on the job on a day-to-day basis and are in a good position to identify hazards and recommend safer ways of working.

An important feature of RTA laws is that protections are extended to non-unionized as well as unionized workers. This is significant in view of figures showing that as of 1993 union membership in the United States had shrunk to a five-decade low—16% of the work force.

Worker RTA also recognizes that even workers who know about on-the-job hazards often lack viable alternatives for safer ways to earn a living and support their families. RTA provides the worker with mechanisms for reporting hazards as well as the right to refuse to perform an assigned task because of a reasonable apprehension of death or serious injury. Also, anti-discrimination provisions encourage workers to exercise their rights under the RTA law.

Finally, the interrelationships between the goals and provisions of RTA and TUR statutes and the fact that such laws are being supported by a coalition of labor interests and environmentalists appear to signal a shift in public policy in this country. For decades, United States laws have divided laws on health and safety regulatory authority according to site: OSHA in the workplace and the **Environmental Protection Agency** (EPA) outside the workplace. Also, existing laws have mandated different approaches to regulation of different media pursuant to such separate laws as the **Clean Air Act**, the **Clean Water Act**, and the **Occupational Safety and Health Act**. Supporters of RTA and TUR laws view workplace and environmental problems as interrelated parts of an integrated "whole." Therefore, the provisions of RTA and TUR have been drafted to reflect an **holistic approach** to regulation. New Jersey's experience, however, suggests that environmentalists, community leaders, and workers need to work together more closely to secure adequate right-to-act legislation. Many environmentalists do not fully understand the intensity of the fear of unemployment among workers, and their perception that job loss results from environmental regulation.

In terms of right-to-act legislation, Canada has progressed further than the United States. As of January 2002, Canadian workers are guaranteed the right to refuse dangerous work as well as the right to report unsafe conditions and to participate in workplace health and safety committees. It remains to be seen whether legislators and citizens throughout the United States will be convinced that enactment of RTA laws is an appropriate way to deal with hazards faced by American citizens within and outside of their workplaces. **Ralph Nader** pointed out during his presidential campaign in 2000 that not a single control standard for toxic chemicals was passed during the Clinton administration, thus reflecting the longest period of inactivity in OSHA's

history. Nader's platform called for national right-to-act legislation, defined as "an unambiguous statutory right for employees to refuse dangerous work." Although job security remains a pressing concern for American workers, the incidents that have focused public attention on workplace hazards over the last twenty years have brought about a fundamental shift in public opinion. In the words of Frances Lynn, "ordinary people are challenging the assertion that a regrettable but necessary cost of business is environmental damage and threats to human health."

[*Paulette L. Stenzel and Rebecca J. Frey, Ph.D.*]

RESOURCES
PERIODICALS

"Exits Blocked, 25 Die As Blaze Sweeps Plant." *Chicago Tribune*, September 4, 1991.

Garland, S. B. "What a Way to Watch Out for Workers." *Business Week*, September 23, 1991.

Lynn, Frances M. "Public Participation in Risk Management Decisions: The Right to Define, the Right to Know and the Right to Act." *Risk: Issues in Health and Safety* 1 (1990): 95–102.

Stenzel, P. L. "Right to Act: Advancing the Common Interests of Labor and Environmentalists." *57 Albany Law Review* (1993): 1–40.

Young, Jim. "Jersey County Gives Workers and Residents 'Right to Act' Against Corporate Polluters. *Labor Notes* May, 1999.

OTHER

Engler, Rick. *Right to Act Backgrounder.* Lawrenceville, NJ: New Jersey Work Environment Council, October 28, 2001.

Industry Canada. *A Canadian Guide to Occupational Health and Safety.* Sydney, Nova Scotia: University College of Cape Breton, 2002.

Michigan Senate Bill 946 (1992).

Nader 2000 News Room. *Statement on Occupational Safety and Health.* Issued October 31, 2000.

ORGANIZATIONS

New Jersey Work Environment Council (NJWEC), 1543 Brunswick Avenue, Lawrenceville, NJ USA 08648-4627 (609) 695-7100, Fax: (609) 695-4200, Email: rickengler@aol.com, <http://www.njwec.org>

Right-to-know

Many states and local governments, as well as the federal government, have passed legislation, referred to as right-to-know laws, that requires the release of information on the hazards associated with **chemicals** produced or used in a given facility. Most right-to-know laws address both community and employee access to information about potential hazards. Requirements of these laws usually include providing public access to information on hazardous materials present, conducting inventories or surveys, establishing record-keeping and exposure reporting systems, and complying with labeling regulations. Notification of emergency releases of hazardous substances into the **environment** is also required under right-to-know laws, such as the Federal **Emergency Planning and Community Right-to-Know Act** of 1986 (EPCRA).

Rio Conference

see **United Nations Earth Summit (1992)**

Riparian land

Riparian land refers to terrain that is adjacent to rivers and streams and is subject to periodic or occasional **flooding**. The plant **species** that grow in riparian areas are adapted to tolerate conditions of periodically waterlogged soils. Riparian lands are generally linear in shape and may occur as narrow strips of streambank vegetation in dry regions of the American Southwest or as large expanses of bottomland hardwood forests in the wetter Southeast. The ecosystems of riparian areas are generally called riparian **wetlands**. In the western United States, riparian vegetation generally includes willows, cottonwoods, saltcedar, tamarisk, and mesquite, depending on the degree of dryness.

The riparian zone of bottomland hardwood forests can be differentiated into several zones based on the frequency of flooding and degree of wetness of the soils. Proceeding away from the channel of the river, the zones may be described as follows: intermittently exposed, semipermanently flooded, seasonally flooded, temporarily flooded, and intermittently flooded. The vegetation of each zone is adapted to survive and thrive under the conditions of flooding peculiar to that zone. Due to the irregularities in **topography** and the formation of streambank levees that are normal to any landscape, it is rare that these zones always occur in the same predictable sequence.

Intermittently exposed zones have standing water present throughout the year and the vegetation grows in saturated **soil** throughout the growing season. Bald cypress and water tupelo are typical trees of this zone and have adaptations such as stilt roots and **anaerobic** root **respiration** to cope with the permanently flooded conditions. Semipermanently flooded zones have standing water or saturated soils through most of the year, and flooding duration may last more than six to eight weeks of the growing season. Black willow and silver maple are abundant tree species of this zone. Seasonally flooded zones are areas where flooding is usually present for three to six weeks of the growing season. A number of hardwood tree species thrive in this zone, including green ash, American elm, sweetgum, and laurel oak. Temporarily flooded zones have saturated soils for one to three weeks of the growing season. For the rest of the year, the **water table** will be well below the soil surface. Many oaks, such as swamp chestnut oak and water oak, as well as hickories may be found

here. Intermittently flooded zones are areas where soil saturation is rarely present and flooding occurs with no predictable frequency. This is an area that may actually be difficult to distinguish from the adjacent uplands. Many transitional and upland species such as eastern red cedar, beech, sassafras, and hop-hornbeam are common.

For all riparian wetlands in the field, these zones of moisture and vegetation gradients occur in an overlapping, intergrading fashion, and the plant species are distributed in varying degrees throughout the riparian zone. Riparian wetlands are valued for their specialized plantlife and **wildlife** values. They are recognized as interfaces where uplands and aquatic areas meet to form intermediate ecosystems that are themselves unique in their diversity, productivity, and function. They also provide significant economic benefits by minimizing flood and **erosion** damage.

[*Usha Vedagiri*]

Resources

Books

Mitsch, W. J., and J. G. Gosselink. *Wetlands.* New York: Van Nostrand Reinhold, 1986.

Riparian rights

Riparian rights are the rights of persons who own land bordering a river, bay, or other natural surface water. A riparian right entitles the property owner to the use of both the shore and the bed as well as the water upon it. Such rights, however, may not be exercised to the detriment of others with similar rights to the same watercourse. As Sir William Blackstone observed in his *Commentaries on the Laws of England* (1765–69), "If a stream be unoccupied, I may erect a mill thereon, and detain the water; yet not so as to injure my neighbor's prior mill, or his meadow; for he hath by the first occupancy acquired a property in the current."

Riparian rights, or riparianism, is a legal doctrine used in the eastern United States to govern water claims. Riparian landowners have rights to the use of the water adjoining their property, but not to **groundwater** or artificial waterways such as canals. Riparian rights were once subject to the "natural flow doctrine," which held that the right-holder could use and divert the water adjoining his or her property at will unless this use diminished the quantity or quality of the natural flow to other right-holders. Since almost any use could be said to diminish the natural flow, this doctrine has given way to the "reasonable use" rule. David H. Getches explains the "reasonable use" rules in this way: "If there is insufficient water to satisfy the reasonable needs of all riparians, all must reduce usage of water in proportion to their rights, usually based on the amount of land they own."

In the **arid** western states, **water rights** are governed by the "prior appropriation" doctrine, under which rights are derived from the first use of water rather than ownership of riparian property. If someone bought land on a river and simply let it flow past, for instance, someone else could acquire a right to that water by putting it to use. The right remains in force, furthermore, unless the right-holder abandons the use of the water. Prior appropriation rights may also be transferred, and some cities in Arizona and elsewhere in the arid Southwest have recently taken advantage of this option by buying large ranches with established water rights in order to secure water supplies for their residents.

California has given its name to a hybrid system, followed in several states, that combines features of the riparian and prior appropriation doctrines. Whatever the system, water rights are limited by the "reserved rights doctrine," intended to assure adequate water for Native American reservations and public lands.

State and federal agencies have the legal right to regulate **land use** in the protection of riparian areas. Some states have established setbacks, or **buffer** zones, along shorelines or riverfronts that regulate development activities within the area. For example, in 1996, Massachusetts enacted the Rivers Protection Act that required property owners to follow a permitting procedure before developing any part of a corridor of 200 ft (61 m) along a riverfront or stream. New Hampshire, Wisconsin, and Montana are among the other states that have implemented state regulations of activity in riparian areas. Some of these state regulatory actions have been challenged in the federal court system as violating the Fifth Amendment constitutional right of citizens to be protected from "takings," by any government entity. **Takings** law prohibits the government from appropriating property for a non-public use, or from failing to provide the owner with suitable and just compensation. The placement of development, the destruction of **wetlands**, and the amounts of fair compensation to property owners are common points of contention in court cases that challenge the government's ability to control development in riparian areas. For example, the Federal Circuit Court determined that a mining company in Florida was due compensation when it was prohibited from removing minerals from a portion, but not all of its property located in a wetland. The Court ruled that the restriction placed on the company constituted a "partial takings," because it forced the company to bear an expense that was the public's responsibility.

Other issues that have been increasingly contested in the courts are the rights of private property owners to build docks on their waterfront property. In the 2001 case in New York, *Stutchin vs. the Village of Lloyd Harbor*, the plaintiffs charged that the action by the Village to deny them the right to build a longer dock than the regulations allowed

constituted a takings because the action denied them the use of their riparian rights. Reparian rights include the right to wharf out to a navigable depth as long as navigation is not impeded. The Village was acting on the regulations developed along the guidelines of the New York Coastal Management Program, which assists communities in protection of coastal resources. In this case, a New York district court judge ruled that the municipality of Lloyd Harbor had the authority and the right to restrict the length of the private dock because the plaintiff's access to the water was not eliminated, and the local jurisdiction has the right to insure that public rights are protected. The judge saw that the construction of the planned dock would interfere with navigation and pose a threat to the **natural resources** of the area. *See also* Irrigation; Water allocation; Water conservation; Water resources; Water table draw-down

[*Richard K. Dagger and Marie H. Bundy*]

RESOURCES
BOOKS

Davis, C. *Riparian Water Law: A Functional Analysis.* Arlington, VA: National Water Commission, 1971.

Water Rights in the Fifty States and Territories. Denver, CO: American Water Works Association, 1990.

Waters and Water Rights. Charlottesville, VA: Michie, 1991.

Risk analysis

Risk is the chance that something undesirable will happen. Everyone faces personal risks daily; we all have a chance of being struck by a car or by lightning or of catching a cold. None of these are certain to happen today, but they all can and do happen occasionally, some more frequently than others. Even though all risk is unpleasant, the consequences of being struck by an **automobile** are much more serious than those of catching a cold. Most people would do more and pay more to avoid the risks they consider most serious. Thus, risk has two important components: (1) the consequences of an event and (2) its **probability**. In addition, while the threat of lightning has always been present, the possibility of being struck by a car emerged only in the last century. Modern risks are constantly evolving.

Events that challenge the health of ecological systems are also becoming apparent. Ecosystems have always faced the risk of severe damage from fire, **flooding**, and volcanoes. More recently, however, population increases, especially in cities; global climate change; **deforestation**; **acid rain**; pesticides; and sewage, **garbage**, and industrial waste disposal have all threatened ecosystems.

Healthy ecosystems supply air, water, food, and raw materials that make life possible; they also process the wastes human societies produce. For these compelling reasons, we must protect them. In the United States, the **National Environmental Policy Act**, the Toxic Substances Control Act, the **Clean Water Act**, the Federal Insecticide, **Fungicide**, and Rodenticide Act, and other legislation has been passed by Congress to protect the **environment**.

Risk analysis can determine whether proposed actions will damage ecosystems. This screening process evaluates plans that might prove destructive, and allows people to make informed decisions about which would be most environmentally sound. Proposed utilities, roads, waste-disposal sites, factories, and even new products can use risk analysis to address environmental concerns during planning stages, when changes are most easily made. The process also allows people to rank environmental problems and allocate attention, resources, and corrective efforts.

Risk analyses are made by both scholars and government decision makers. Scientists are interested in a thorough understanding of the way ecosystems function, but decision makers need quick and efficient tools for making choices.

Scholarly approaches take many forms: Synoptic surveys assess the characteristics of stressed and unstressed natural systems. Experiments determine how the whole or one part of an **ecosystem** (such as fish) will respond to stress. Extrapolations apply specific observations to other ecosystems, **chemicals**, or properties of interest by analyzing dose-response curves, establishing relationships between the molecular structure of a chemical and its likely environmental effects, or simulation of entire ecosystems on a computer. Using these tools, scientists can also measure change in ecosystems and translate the results for the general public.

A less precise method, often used by public officials and other nonscientists, is called ecosystem **risk assessment**. It uses available toxicological, ecological, geological, geographical, chemical, and sociological information to estimate possible damage. The process has three steps:

- The problem is identified. For example: could **nutrient runoff** from local agriculture affect **commercial fishing** on a nearby lake?
- Available scientific information is gathered to predict both the level of stress that could result and the likely ecosystem response. Where do the nutrients go and how are organisms exposed to them? How does increasing nutrient levels affect biological systems, especially fish?
- Risk is quantified by comparing exposure and effects data. If the predicted stress level is lower than those known to cause serious damage, risk is low. On the other hand, if the predicted stress is higher, risk is high.

Despite the reams of data available, there is never enough information about the possible effects of any stress. An assessment based squarely on facts is more reliable than

one that uses scarce, preliminary information. Anyone charged with making decisions must weigh all options.

The possibility of an undesirable occurrence, the seriousness of the consequences, and the uncertainty involved in any prediction all factor into the estimate. Alternative actions and their risks and benefits must also be considered, and the consistency of the action balanced with other societal goals. An action's risks and benefits are often not distributed evenly in society. For example, people sharing the **water table** with a proposed **landfill** may shoulder more risk, while those who use many nonrecyclable consumer products may benefit disproportionately. Who benefits and who loses can also affect decisions.

Risk predictions are similar to weather forecasts—they are based on careful observation and are useful, but they are far from perfect. They indicate useful precautions—whether these involve carrying an umbrella or treating waste before it enters the water. They help us understand ecosystems and allow us to consider the environment before potentially harmful action is taken. If ecological risks are considered when decisions are being made, ecosystems on which people depend can be protected. *See also* Environmental policy; Environmental stress; Greenhouse effect; Industrial waste treatment; Sewage treatment; Solid waste

[*John Cairns Jr. and B. R. Neiderlehner*]

RESOURCES

BOOKS

Bartell, S. M., R. H. Gardner, and R. V. O'Neill. *Ecological Risk Estimation.* Chelsea, MI: Lewis Publishers, 1992.

Ehrlich, P. R., and A. H. Ehrlich. *Healing the Planet: Strategies for Solving the Environmental Crisis.* New York: Addison-Wesley Publishing, 1992.

National Research Council. *Risk Assessment in the Federal Government: Managing the Process.* Washington, DC: National Academy Press, 1983.

PERIODICALS

Norton, S. B., et al. "A Framework for Ecological Risk Assessment at the EPA." *Environmental Toxicology and Chemistry* 11–12 (1992): 1663–672.

OTHER

Cairns Jr., J., K. L. Dickson, and A. W. Maki, eds. *Estimating the Hazard of Chemical Substances to Aquatic Life, STP 657.* Philadelphia: American Society for Testing and Materials, 1978.

Risk assessment (public health)

Risk assessment refers to the process by which the short and long-term adverse consequences to individuals or groups in a particular area resulting from the use of specific technology, chemical substance, or natural hazard is determined. Generally, quantitative methods are used to predict the number of affected individuals, morbidity or **mortality**, or other outcome measures of adverse consequences. Many risk assessments have been completed over the last two decades to predict human and ecological impacts with the intent of aiding policy and regulatory decisions. Well-known examples of risk assessments include evaluating potential effects of herbicides and insecticides, nuclear **power plants**, incinerators, **dams** (including dam failures), **automobile pollution**, **tobacco** smoking, and such natural catastrophes as volcanoes, earthquakes, and hurricanes. Risk assessment studies often consider financial and economic factors as well.

Human health risk assessments

Human health risk assessments for chemical substances that are suspected or known to have toxic or carcinogenic effects is one critical and especially controversial subset of risk assessments. These health risk assessments study small populations that have been exposed to the chemical in question. Health effects are then extrapolated to predict health impacts in large populations or to the general public who may be exposed to lower concentrations of the same chemical.

One mathematical formula that determines an individual's risk from chemical exposures is:

$$\text{Risk} = \begin{array}{c} \textit{(emissions)} \times \textit{(transport)} \times \\ \textit{(loss factor)} \times \textit{(exposure period)} \times \\ \textit{(uptake)} \times \textit{(toxicity factor)} \end{array}$$

For the case of a **hazardous waste** incinerator, emissions might be average **smoke stack emissions** of gas; the transport term represents dilution in the air from the stack to the community; the loss factor might represent chemical degradation of reactive contaminants as stack gases are transported in the **atmosphere**; the exposure period is the number of hours that the community is downwind of the incinerator; uptake is the amount of contaminants absorbed into the lung (a function of breathing rate and other factors); and toxicity is the chemical potency. Multiplying these factors indicates the **probability** of a specific adverse health impact caused by contaminants from the incinerator. In typical applications, such models give the incremental lifetime risk of **cancer** or other health hazards in the range of one in a million (equivalent to 0.000001). Cancers currently cause about one-third of deaths, thus, a one in a million probability represents a tiny increase in the total cancer incidence. However, calculated risks can vary over a large range—0.001 to 0.00000001.

While the same equation is used for all individuals, some assumptions regarding uptake and toxicity might be modified for certain individuals such as pregnant women, children, or individuals who are routinely exposed to the **chemicals**. In some cases, monitoring might be used to verify exposure levels. The equation illustrates the complexity of the risk assessment process.

Risk assessment process

The risk assessment/management procedure consists of five steps: (1) *Hazard assessment* seeks to identify causative agent(s). Simply put, is the substance toxic and are people exposed to it? The hazard assessment demonstrates the link between human actions and adverse effects. Often, hazard assessment involves a chain of events. For example, the release of **pesticide** may cause **soil** and ground **water pollution**. Drinking contaminated **groundwater** from the site or skin contact with contaminated soils may therefore result in adverse health effects. (2) *Dose-response relationships* describe the toxicity of a chemical using models based on human studies (including clinical and epidemiologic approaches) and animal studies. Many studies have indicated a threshold or "no-effect" level, that is, an exposure level where no adverse effects are observed in test populations. Some health impacts may be reversible once the chemical is removed. In the case of potential carcinogens, linear models are used almost exclusively. Risk or potency factors are usually set using animal data, such as experiments with mice exposed to varying levels of the chemical. With a linear dose-response model, a doubling of exposure would double the predicted risk. (3) *Exposure assessment* identifies the exposed population, detailing the level, duration, and frequency of exposure. Exposure *pathways* of the chemical include ingestion, inhalation, and dermal contact. Human and technological defenses against exposure must be considered. For example, respirators and other protective equipment reduce workplace exposures. In the case of prospective risk assessments for facilities that are not yet constructed—for example, a proposed hazardous waste incinerator—the exposure assessment uses mathematical models to predict emissions and distribution of contaminants around the site. Probably the largest effort in the risk assessment process is in estimating exposures. (4) *Risk characterization* determines the overall risk, preferably including quantification of uncertainty. In essence, the factors listed in the equation are multiplied for each chemical and for each affected population. To arrive at the total risk, risks from different exposure pathways and for different chemicals are added. Populations with the maximum risk are identified. To gauge their significance, results are compared to other environmental and societal risks. These four steps constitute the scientific component of risk assessment. (5) *Risk management* is the final decision-making step. It encompasses the administrative, political, and economic actions taken to decide if and how a particular societal risk is to be reduced to a certain level and at what cost. Risk management in the United States is often an adversarial process involving complicated and often conflicting testimony by expert witnesses. In recent years, a number of disputes have been resolved by mediation.

Risk management and risk reduction

Options that result from the risk management step include performing no action, product labeling, and placing regulations and bans. Examples of product labeling include warning labels for consumer products, such as those on tobacco products and cigarette advertising, and **Material Safety Data Sheets** (MSDS) for chemicals in the workplace. Regulations might be used to set maximum permissible levels of chemicals in the air and water (e.g., air and **water quality** criteria is set by the U.S. **Environmental Protection Agency**). In the workplace, maximum exposures known as Threshold Limit Values (TLVs) have been set by the U.S. **Occupational Safety and Health Administration**. Such regulations have been established for hundreds of chemicals. Governments have banned the production of only a few materials, including DDT and PCBs, and product liability concerns have largely eliminated sales of some pesticides such as Paraquat and most uses of **asbestos**.

A variety of social and political factors influence the outcome of the risk assessment/management process. Options to reduce risk, like banning a particular pesticide that is a suspected **carcinogen**, may decrease productivity, profits, and jobs. Furthermore, agricultural losses due to insects or other pests if pesticide is banned might increase malnutrition and death in subsistence economies. In general, risk assessments are most useful when used in a relative or comparative fashion, weighing the benefits of alternative chemicals or agricultural practices to another. Risk management decisions must consider what degree of risk is acceptable, whether it is a voluntary or involuntary risk, and the public's perception of the risk. A risk level of one in a million is generally considered an acceptable lifetime risk by many federal and state regulatory agencies. This risk level is mathematically equivalent to a decreased life expectancy of 40 minutes for an individual with an average expected lifetime of 74 years. By comparison, the 40,000 traffic fatalities annually in the United States represent over a 1% lifetime chance of dying in a wreck—10,000 times higher than acceptable for a chemical hazard. The discrepancy between what an individual accepts for a chemical hazard in comparison to risks associated with personal choices like driving or smoking might indicate a need for more effective communication about risk management.

Risk assessments are often controversial. Scientific studies and conclusions about risk factors have been questioned. For example, animals are often used to determine dose-response and exposure relationships. Results from these studies are then applied to humans, sometimes without accounting for physiological differences. The scientific ability to accurately predict absolute risks is also poor. The **accuracy** of predictions might be no better than a factor of 10, thus 10 to 1,000 cancers or other health hazard might be

experienced. The uncertainty might be even higher, a factor of 100, for example. Risks due to multiple factors are considered independent and additive. For instance, smoking and asbestos exposure together have been shown to greatly increase health risks than exposure to one factor alone. Conversely, multiple chemicals might inhibit or cancel risks. In nearly all cases, these factors cannot be modeled with our present knowledge. Finally, assessments often use a worst-case scenario, for example, the complete failure of a **pollution control** system, rather than a more modest but common failure like operator error.

Ecological risk assessment

Ecological risk assessments are similar to human health risk assessments but they estimate the severity and extent of ecological effects associated with an exposure to an **anthropogenic** agent or a perturbational change. Again, the risk estimate is stated in probability terms that reflect the degree of certainty. Ecological assessments tend to be more complex than human health assessments since a variety of dynamic ecological communities or systems may be involved, and these systems have important but often poorly understood interactions and feedback loops. In addition the current status and health of ecological systems must be defined by measurements and analysis before an assessment can begin. In some cases, animal **species** or ecosystems may be more sensitive than humans. Contingency or hazard assessment resembles that made for human health but focuses on low probability events such as failure of dams, **nuclear power** plants, and industrial facilities that have the potential for significant public health and welfare damage. Finally, risk reduction approaches have been suggested that shift focus from end-of-pipe controls, for instance, pollution control equipment, to preventing pollution in the first place by minimizing waste and **recycling**.

[*Stuart Batterman*]

Resources

Books

Chemical Risk: A Primer. Washington, DC: American Chemical Society, 1984.

Periodicals

Naugle, D. F., and T. K. Pierson. "A Framework for Risk Characterization of Environmental Pollutants." *Journal of the Air and Waste Management Association* 4 (1992): 1298–1307.

Other

U.S. Environmental Protection Agency. *Integrated Risk Information System Background Document.* Washington, DC: U. S. Government Printing Office, 1991.

Risk assessors

Risk assessors endeavor to define a risk that will be realized under actual or anticipated conditions by establishing the "average truth" from numerous probabilities. Assessing risk based on conclusions drawn from an infinite number of integrations cannot be expressed in terms of "safe" or "nontoxic." Such expressions imply that a chemical or an event (e.g., accident) is without risk or harm, which may be misleading. Instead a chemical or an event is ranked as minor, moderate, or high to reflect the degree of risk that it represents within a given set of parameters. Such is the role of the risk assessor to calculate the effects of variables while classifying the uncertainties and presenting risk managers with a list of choices.

Comprehending the number of variables and uncertainties associated with an event can be explored in the following example. If one were to assess the number of deaths from **cancer** following a lifetime exposure to a chemical contained in drinking water, one must examine a diverse array of interlocking factors, such as the composition of the chemical, the chemical's observable effects on animals and/ or **environment**, and whether the reported dose-exposure rates are applicable to humans. Likewise, if one were to examine the amount of chemical which can remain in the **soil** and not create **groundwater** contamination following a chemical spill, factors such as the following should be examined: the composition of the spilled chemical, its specific gravity, solubility, and viscosity; the soil's composition, **pH**, precipitation and **infiltration** rates; and hydrologic setting, the depth to the **water table** and its vertical and horizontal flow.

In addition to these factors, one must also take into account underlying uncertainties in data acquisition. For example, health databases and findings are typically derived from experimental or laboratory animal tests on a specific chemical or in connection with unrelated human studies. Therefore predictions from these studies may not be applicable to human subjects.

Sometimes lifestyle choices can also play a role in **risk assessment**. In addition to the chemical that a person is exposed to, other factors—such as smoking or exposure to another chemical substance—can have additive, accumulative, or antagonistic effects. Such anomalies require supplemental research, such as mathematical **modeling**.

Mathematical modeling mirrors the processes and interrelationships of real-life systems. The mathematical content of a model may extend to different equations or to simple look-up tables; its purpose is to reflect the required outcome of interest. For example, a health model frequently employed by the **Environmental Protection Agency** (EPA) is a Linearized Multistage Model because it yields the most

conservative risk estimates when exposure occurs in very low doses. While ecological modeling is still in its infancy (because it involves outcomes at numerous levels from a single **species** to communities of organisms), models, nonetheless, provide valuable information to risk assessors.

[*George M. Fell*]

River basins

Recognition of the river as the dominant force in forming basins may be traced back to John Playfair in 1802. In contrast to the leading opinions, Playfair observed that rivers were proportional to valley size and tributaries were accordant, neither of which would be likely unless rivers had created the basins, rather than the other way around. Now known as Playfair's law, his observation has led to extensive efforts to quantify river basin characteristics.

In 1945, R. E. Horton developed the concept of stream order, and Arthur Strahler further elaborated on the subject. The smallest tributaries are labeled "1," and when two first order streams converge, they form a second order stream ("2"), and so forth. The ratio of lower order to higher order streams remains remarkably consistent throughout a given basin. Uniformity is the important factor in all of these because it demonstrates that **drainage** network characteristics are quantitatively consistent.

Renowned geomorphologist William Morris Davis, noted in *Water, Earth, and Man* (1969) that "the river is like the veins of a leaf; broadly viewed, it is like the entire leaf." There is, however, one critical difference. In the leaf, the flow of water and nutrients is primarily from larger to smaller veins. However, in river basins all matter, good and bad, flows downstream. That is why when urban water systems use rivers, their intakes are located upstream and drainage outlets downstream. **Groundwater pollution** follows this same downward pathway, though at a far slower pace; it is one of our most serious problems.

The river basin, as part of the **hydrologic cycle**, is increasingly a technological and social system. For the United States, an estimated 10% of the national wealth is devoted to structures involving the movement of water, including **dams, irrigation** systems, water supply networks, and sewers, with increasingly sophisticated controls.

The importance of river basins is well illustrated by the **Tennessee Valley Authority** (TVA) project. Launched by the Roosevelt Administration in 1933, it was a massive economic and social effort aimed at the chronic problems of depleted soils, rampant **soil erosion**, recurrent **flooding**, and economic desolation. Industrial demand for electricity has subsequently grown so large that hydroelectric power, the initial source, now supplies less than 20% of TVA demand. *See also* Environmental economics; Sewage treatment; Topography

[*Nathan H. Meleen*]

Resources

Books

Chorley, R. J., ed. *Water, Earth, and Man: A Synthesis of Hydrology, Geomorphology, and Socio-economic Geography.* London: Methuen, 1969.

Gore, J. A., ed. *The Restoration of Rivers and Streams.* Boston: Butterworth, 1985.

Morisawa, M. *Streams: Their Dynamics and Morphology.* New York: McGraw-Hill, 1968.

Petulla, J. M. *American Environmental History: The Exploitation and Conservation of Natural Resources.* San Francisco: Boyd & Fraser, 1977.

River blindness

River blindness is a disease responsible for a high incidence of partial or total blindness in parts of tropical Africa and Central America. Also called *onchocerciasis,* the disease is caused by infection with *Onchocerca volvulus,* a thread-shaped round worm (a nematode), which is transmitted between people by the biting blackfly *Simulium.* The larvae of *Onchocerca* develop into the infective stage, called *L4,* inside the blackfly and are introduced into humans by the bite of an infected blackfly.

Adult *Onchocerca* (2.5 ft or 0.76 m in length) develop in the connective tissue under the skin of humans. The adult worms lie coiled within subcutaneous nodules several inches in diameter. The nodules are painless and cause little damage, but can be cosmetically unattractive; fortunately, they are easily removed by simple surgery. The more serious health problems associated with onchocerciasis are caused by the release of masses of early-stage larvae, known as *microfilaria,* into the host's connective tissue under the skin. The mobile larvae spread throughout the body, including the surface tissues of the eyes. It is the burrowing activities of these larvae that cause the symptoms associated with onchocerciasis—either severe *dermatitis* or blindness. Onchocerciasis can now be treated with drugs such as ivermectin, which kill the larvae, but the blindness is usually permanent.

Avoiding blackfly bites with protective clothing and skin repellents is not a practical control measure on a large scale. The major method of prevention of onchocerciasis in humans is the control of the blackfly intermediate hosts and vectors. The larvae and pupae of blackfly are strictly aquatic and are found only in fast-running water. In Kenya, the larval stages of *S. neavei* have been killed by releasing DDT into streams where the blackflies breed and evolve. However, the waterways must be treated frequently to prevent the re-establishment of blackfly populations, and there have been

A victim of river blindness is led through a sugar plantation in his village at Banfora, Burkino Faso. (United Nations. Reproduced by permission.)

serious questions raised over the safety of DDT. In West Africa, *S. damnosum* is more difficult to control, since the adults can fly over considerable distances, and can easily reinfect cleared sites from up to 60 mi (97 km) away.

Onchocerciasis has long been socially debilitating in an 11-country area of West Africa where both the flies and **parasites** are abundant. Here, more than one-fifth of all males over the age of 30 may be blind, turning productive people into long-term dependents. The presence of the disease often results in the **migration** of people away from rivers to higher ground, which they then clear for cultivation. Clearing vegetation in Africa frequently results in **soil erosion** and the formation of gullies that channel water during heavy rains. The moving water allows blackflies to breed, spreading the disease to the new area. The reappearance of river blindness results in still further human migration, until large areas of badly eroded land are left unpopulated and unproductive, and entire villages are abandoned. The disease has now been brought largely under control by a World Health Organization program begun in 1974, in which 63 million acres (25 million ha) of land have been made safe for resettlement.

[*Neil Cumberlidge*]

Resources

Books

Bullock, W. A. *People, Parasites, and Pestilence. An introduction to the Natural History of Infectious Disease.* Minnesota: Burgess Publishing Company, 1982.

Crosskey, R. W. *The Natural History of Blackflies.* New York: John Wiley and Sons, 1990.

Markell, E. K., M. Voge, and D. T. John. *Medical Parasitology.* 7th ed. Philadelphia: Saunders, 1992.

Rodger, F. C., ed. *Onchocerciasis in Zaire: a New Approach to the Problem of River Blindness.* New York: Pergamon Press, 1977.

River dolphins

River **dolphins** are cetaceans, which means that they related to **whales** and porpoises. These mammals live only in rivers

and waterways of Asia and South America. There are five **species** of river dolphins belonging to three different families. In Asia the Baiji or Chinese river dolphin (*Lipotes vexillifer*) lives in the lower part of China's Chang Jiang or Yangtze River. On the Indian subcontinent, Indus river dolphins (*Platanista indi*) originally were found throughout almost the entire length of the Indus River, but now are limited to a stretch of that river in the Sind Province of Pakistan. Ganges river dolphins (*Platanista gangetica*) are found throughout the Ganges River, although their numbers are decreasing. In South America the Amazon river dolphin (*Inia geoffrensis*), also called the boto, is found primarily in the Amazon and Orinoco Rivers of Brazil and Venezuela. The South American franciscana river dolphin (*Pontoporia blainvillei*), unlike other river dolphins, is a marine (ocean) dolphin that lives in shallow water close to shore from central Brazil south to Argentina.

The baiji live in China's Yangtze River, one of the busiest waterways in the world. The Chinese have recognized the baiji river dolphin for centuries. The name "baiji," which means "white dolphin," has been found in a Chinese dictionary from 200 B.C. This dolphin is also called the Chinese river dolphin or whitefin dolphin.

Baiji coloring ranges from pale to dark blue-gray on the back and sides. The belly is white or gray-white. The dolphin has a triangular dorsal fin and a snout that continues to grow as it ages.

Newborn calves measure 32–35 in (about 0.8 m) in length and weigh from 6 to 11 lb (2.7 to 5 kg). Adult females are somewhat larger than the males. Adult length ranges from 6.5 to 8 ft (2 to 2.4 m). Adults weigh from 220 to 355 lb (100 to 160 kg).

Indus river dolphins live in a short stretch of the Indus River in Pakistan. Most dolphins live in an area between two **dams** built during the 1930s. The dams have limited their **migration** to other parts of the river. Ganges river dolphins live in the Ganges, Meghna, and Brahmaputra rivers in western India, Nepal, Bhutan, and Bangladesh. They are also found in the Karnaphuli River in Bangladesh.

Indus and Ganges river dolphins share many similarities. They are both called "susu." The word sounds like a sneeze and resembles the sound of a dolphin breathing. They are also called blind river dolphins, because they are the only dolphins that cannot see. These dolphins navigate by echolocation, producing sound waves, then interpreting the echoes these sound waves generate.

The Indus and Ganges river dolphins range in color from gray-blue to brown. Both have a hump on their back instead of a dorsal fin. They swim on their side, and females sometimes carry their young on their back.

Newborn Indus and Ganges river dolphins weigh about 17 lb (7.7 kg) and measure 28–35 in (about 0.8 m) in length. Adults weigh 155 to 200 lb (70 to 90 kg). Their length ranges from 5 to 8 ft (1.5 to 2.5 m).

Amazon river dolphins live in rivers in Brazil, Bolivia, Colombia, Guyana, Ecuador, Peru, and Venezuela that are part of the Amazon and Orinoco watersheds. They are also called boto or pink river dolphins. Adults range in color from blue-gray to pink. The pink color becomes more intense when these dolphins are more active and dims when they are less active. The Amazon river dolphin as a hump on its back instead of a dorsal fin.

Newborn calves weigh about 15 lb (7 kg) and are about 31 in (0.8 m) long. Adults weigh from 185 to 355 lb (85 to 160kg). Adult females range in length from 5 to 7.5 ft (1.5 to 2.4 m). Male dolphin length ranges from 6.5 to 9 ft (2 to 2.7 m).

According to the Whale and Dolphin Conservation Society, Amazon river dolphins have entered local South American folklore. One legend says that male dolphins captured women and took them underwater. Some women explained their pregnancies by saying that dolphins fathered their babies.

Franciscana are classified as river dolphins, although they live in the Atlantic Ocean. They were originally thought to move from fresh water to salt water during their lifecycle, but scientists have now determined that they live their entire life in the sea. Franciscana are found in the shallow coastal waters off eastern South America, primarily in the La Plata estuary. They are also known as La Plata dolphins.

Franciscana have gray-brown backs. Their color may lighten during the winter or as they age. Some older dolphins are primarily white. In relation to their size, franciscana have the longest beaks of any dolphin.

Newborn franciscana are 28 in (0.7 m) long and weigh from 16 to 19 lb (8.6 kg). Adult length ranges from about 4 to 6 ft (1.2 to 1.8 m). Adult franciscana dolphins weigh from 65 to 115 lb (29 to 52 kg).

Declining populations

The river dolphin population is decreasing throughout the world, often through stresses caused by development. Some dolphins are killed by chemical **pollution**. Other dolphins become entangled in fishing nets and die, and **hunting** has led to declines in the populations of all river dolphins except franciscana.

The damming of rivers divides populations and prevents migration to better environments. This has caused a decline in the population of baiji, Ganges, and Indus river dolphins. The lack of adequate food supply (prey depletion), has caused declines in franciscana, baiji, and Ganges dolphins. Development has destroyed or polluted much of the **habitat** where dolphins live. Loss of habitat threatens baiji, franciscana, and Ganges river dolphins.

Baiji are the most rare cetacean. The population of baiji dolphins is in the dozens, according to the World Conservation Union (IUCN). The Chinese government declared the baiji a protected animal in 1949, and people were punished for intentionally killing baiji. However, accidents and development still caused many deaths.

In 1975, China declared the baiji a national treasure, and subject to more protection. In 1996, the IUCN rated baiji as one of the 12 most **endangered species** in the world. In 2000, fewer than 100 individuals were believed to survive in the wild. Increasing development and the building of dams along the Yangtze River is likely to kill the few animals remaining.

In 1998, the IUCN estimated that the population of Indus river dolphins was less than 1,000. In 2000, the Indus dolphin was considered critically endangered and threatened with **extinction**. In addition, in 2000 the IUCN estimated that only a few thousand wild Ganges river dolphins remained. It is also considered an endangered species.

South American river dolphin population has not been depleted as much as the river dolphin population of Asia. Although the exact population of Amazon river dolphins has not been estimated, they are believed to have to the largest population of all species of river dolphin. In 1996, the IUCN considered Amazon river dolphins to be vulnerable to population decreases because of increased dam building along the Amazon River. According to the IUCN, inadequate verifiable data exists about the population of franciscana dolphins. Lack of data has made protection and management of this dolphin population difficult.

[*Liz Swain*]

Resources

Books

Blair, Cornelia, Mark Siegal, and Nancy Jacobs, eds. *Endangered Species 1998.* Wylie, TX: Information Plus, 1998.

Carwardine, Mark, ed. *Whales, Dolphins and Porpoises.* New York: Checkmark Books, 1999.

Day, Trevor. *Oceans.* New York: Facts on File, 1999.

Organizations

American Cetacean Society, P.O. Box 1391, San Pedro, CA USA 90733-1391 (310) 548-6279, Fax: (310) 548-6950, Email: acs@pobox.com, <http://www.acsonline.org>

Whale and Dolphin Conservation Society, P.O. Box 232, Melksham, WiltshireUnited Kingdom SN12 7SB 44 (0) 1225-354333, Fax: (44) (0) 1225-791577, Email: webmaster@wdcs.org, <http://www.wdcs.org>

World Conservation Union Headquarters, Rue Mauverney 28, Gland, Switzerland 1196 41 (22) 999-0000, Fax: 41 (22) 999-0002, Email: mail@hq.iucn.org, <http://www.iucn.org>

Rivers

see **Amazon basin; Colorado River; Cuyahoga River; Hudson River; River basins; Wild river**

RMA

see **Rocky Mountain Arsenal**

RNA

see **Ribonucleic acid**

Rocky Flats nuclear plant

The production of **nuclear weapons** inherently poses serious risks to the **environment**. At any point in the production process, radioactive materials may escape into the surrounding air and water, and safe methods for the disposal of waste from the manufacturing process still have not been developed. The environmental risks posed by the production of nuclear weapons are illustrated by the history of the Rocky Flats Nuclear Munitions Plant, located 16 mi (26 km) northwest of Denver, Colorado.

Rocky Flats was built in 1952, following an extensive search for sites at which to build plants for the processing of **plutonium** metal, a critical raw material used in the production of nuclear weapons. Authorities wanted a location that was close enough to a large city to attract scientists, but far enough away to ensure the safety of city residents.

Another important factor in site selection was wind measurements. The government wanted to be sure that, in the event of an accident, radioactive gases would not be blown over heavily populated areas. The selection of Rocky Flats was justified on the basis of wind measurements made at Denver's Stapleton airport, showing that prevailing winds blow from the south in that area. Had the same studies been carried out at Rocky Flats itself, however, they would have shown that prevailing winds come from the northwest, and any release of radioactive gases would be carried not away from Denver but toward it.

Over the next four decades, this unfortunate mistake was to have serious consequences as spills, leaks, fires, and other accidents became routine at Rocky Flats. On September 11, 1957, for example, the **filters** on glove boxes caught fire and burned for 13 hours. These filters were used to prevent plutonium dust on used gloves from escaping into the outside air, but once the fire began this is exactly what happened. The release of plutonium was even accelerated when workers turned on exhaust fans to clear the plant of **smoke**. Smokestack monitors showed levels of plutonium 16,000 times greater than the maximum recommended level. Officials at Rocky Flats reportedly made no effort to notify

local authorities or residents about the accidental release of the radioactive gases.

This incident reflects the contradiction between the commitment of the United States government to the development of nuclear weapons and its concern for protecting the health of its citizens, as well as the natural environment. In 1992, a government report on Rocky Flats accused the Department of Energy (DOE) of resisting efforts by the **Environmental Protection Agency** (EPA) and state environmental agencies to make nuclear weapons plants comply with environmental laws and regulations. DOE officials defended this policy by saying that Rocky Flats was the only site in the United States at which plutonium triggers for nuclear weapons were being produced.

The Rocky Flats plant was originally operated by Dow Chemical Company. In 1975, Rockwell International Corporation replaced Dow as manager of the plant. Over the next 14 years, Rockwell faced increasing criticism for its inattention to safety considerations both within the plant and in the surrounding area.

Rockwell's problems came to a head on a June morning in 1989, when a team of 75 FBI agents entered the plant and began searching the 6,550-acre (2,653-ha) complex for evidence of deliberate violations of environmental laws. As a result of the search, Rockwell was relieved of its contract at Rocky Flats and replaced by EG&G, Inc., an engineering firm based in Wellesley, Massachusetts. The ensuing investigation of safety violations lasted over two years and in March 1992 Rockwell plead guilty to 10 crimes, five of them felonies, involving intentional violations of environmental laws. The company agreed to pay $18.5 million in fines, the second largest fine for an environmental offense in United States history.

The fine did not, however, end the dispute over the safety record at Rocky Flats. Rockwell officials claimed that the Department of Energy was also at fault for the plant's poor environmental record. The company argued that the DOE had not only exempted them from environmental compliance but had even encouraged it to break environmental laws, especially **hazardous waste** laws. The federal grand jury that investigated the Rocky Flats case agreed with Rockwell. Not only was the DOE equally guilty, the grand jury decided, but the plant's new manager, EG&G, was continuing to violate environmental laws even as the case was being heard in Denver. Members of the jury were so angry about the way the case had been handled that they wrote President Bill Clinton, asking him to investigate the government's role at Rocky Flats.

Secretary of Energy James Watkins had closed Rocky Flats for repairs in November 1989 and it remained closed during the course of the investigation. Rocky Flat's problems appeared to be over in January 1992 when EG&G announced that, after spending $50 million in repairs, the plant was ready to re-open. Within a matter of days, however, Secretary Watkins ordered that weapons production at the plant permanently cease.

When Secretary Watkins made his decision to close Rocky Flats, the area faced two environmental challenges. In the first place, the site contained the largest stockpile of weapons grade plutonium in the United States, totaling more than 14 tons of the metal. The plutonium had been left in whatever form it occurred at the time of the plant shutdown, including water solutions, partially machined parts, and raw materials. The plutonium poses a threat because of its potential for starting fires, exposing workers to radiation, and threatening nearby communities in the event of an accident at the site.

In the second environmental challenge, buildings and facilities covering a 6,500 acre (2,630 ha) region had become seriously contaminated during the plant's 40-year lifetime. While a less serious short-term danger, this extensive contamination still poses a threat that could extend over centuries without some action to deal with it.

Agencies responsible for the future of Rocky Flats—primarily the EPA, the Colorado Department of Public Health and Environment, and the Defense Nuclear Facilities Safety Board—decided on a two-prong approach to the clean-up of Rocky Flats. The first step is to deal with the immediate threat of plutonium left behind on the site by finding safe ways of safely storing the material. This step involves activities such as draining tanks and pipes that have been used for plutonium-containing solutions; venting plutonium-containing waste drums and tanks of **hydrogen** gas that has built up within them; and re-packaging containers in which plutonium has been stored, making them safer for long-term storage. The cost of this operation in the site's first full year of clean-up was $573 million and employed an estimated 9,374 workers.

The second stage of the Rocky Flats clean-up operation will involve dealing with contamination that had developed on the site over its 40-year history. For this purpose, Rocky Flats was added to the **National Priorities List** for Superfund in 1989. Among the challenges to be solved in this clean-up are **soil** and **groundwater** that have been contaminated by the burial of nuclear materials, **chemicals**, spills, and other accidents. There are also tens of thousands of cubic yards of wastes produced while the plant was in operation that must now be shipped to some permanent storage area. Originally estimated to take upwards of 70 years, a contract was signed by DOE and Kaiser-Hill on January 24, 2000 to have the Rocky Flats Environmental Technology Site (renamed soon after production closure) closed by December 15, 2006. As of 2002, the project was operating well within the six year time frame.

[*David E. Newton*]

RESOURCES
PERIODICALS

Pasternak, D. "A $200 Billion Scandal." *U.S. News & World Report* (December 14, 1992): 34–37+.

"The Rocky Flats Cover-Up, Continued." *Harper's* (December 1992): 19–23.

Schneider, K. "U.S. Shares Blame in Abuses at A-Plant." *New York Times* (March 27, 1992): A12.

Wald, M. L. "Rockwell To Plead Guilty and Pay Large Fine for Dumping Waste." *New York Times* (March 26, 1992): A1.

———. "New Disclosures Over Bomb Plant." *New York Times* (November 22, 1992): 23.

OTHER

U.S. Department of Energy. *Rocky Flats Closure Project.* [cited July 2002]. <http://www.rfets.gov>.

Rocky Mountain Arsenal

The Rocky Mountain Arsenal (RMA), a few miles northeast of Denver, was originally constructed and operated by the Chemical Corps of the United States Army. Beginning in 1942, the arsenal was the main site at which the Chemical Corps manufactured chemical weapons such as mustard gas, nerve gas, and phosgene. The Army eventually leased part of the 27 mi^2 (70 km^2) plot of land to the Shell Oil Company which produced DDT, dieldrin, **chlordane**, parathion, aldrin, and other pesticides at the site.

The presence of a chemical weapons plant has long been a source of concern for many Coloradans. In 1968, for example, a group of Denver-area residents complained that nerve gas was being stored in an open pit directly beneath one of the flight paths into Denver's Stapleton International Airport.

Indeed, the Army was well aware of the hazard posed by its RMA products. In 1961, it found that wastes from manufacturing processes were seeping into the ground, contaminating **groundwater** and endangering crops in the area. The Army's solution was to institute a new method of waste disposal. In September 1961, engineers drilled a deep well 12,045 ft (3,654 m) into the earth. The lowest 75 ft (23 m) of the well was located in a highly fractured layer of rock. The Army's plan was to dump its chemical wastes into this deep well. The unexpressed principle seemed to be out of sight, out of mind.

Fluids were first injected under pressure into the well on March 8, 1962, and pressure-injection continued over the next six months at the rate of 5.5 million gal (21 million l) per month. After a delay of about a year, wastes were once more injected at the rate of 2 million gal (7.5 million l) per month from August 1964 to February 1966. This practice was then terminated.

The reason for ending this method of waste disposal was the discovery that earthquakes had begun to occur in the Denver area at about the same time that the Army had started using its deep **injection well**. Seismologists found that the pattern of the earthquakes in the region between 1962 and 1966 closely matched the pattern of waste injection in the **wells**. When large volumes of wastes were injected into the well, many earthquakes occurred. When injection stopped, the number decreased. Scientists believe that the liquid wastes pumped into the injection well lubricated the joints between rock layers, making it easier for them to slide back and forth, creating earthquakes.

Earthquakes are hardly a new phenomenon for residents of Colorado. Situated high in the Rocky Mountains, they experience dozens each year although most are minor earthquakes. However, the suggestion that the Army's activity at RMA might be increasing the risk of earthquakes became a matter of great concern. The proximity of RMA to Denver raised the possibility of a major disaster in one of the West's largest metropolitan areas.

Faced with this possibility, the Army decided to stop using its injection well on February 20, 1966. Earthquakes continued to occur at an abnormally high rate, however, for at least another five years. Scientists hope that liquids in the well will eventually diffuse through the earth, reducing the risk of further major earthquakes.

In 1992 Congress announced plans to convert the arsenal to a **wildlife refuge**. The land around RMA had been so badly poisoned that humans essentially abandoned the area for many years. The absence of humans, however, made it possible for a number of **species** of **wildlife** to flourish.

On October 16, 1996, the U.S. Army announced its plans for the clean-up of the Rocky Mountain Arsenal. That plan dealt with three areas of environmental concern: water, structures, and **soil**. The Army declared its intent to continue its program of treating water supplies and groundwater on the site and to provide safe drinking water to neighboring off-site communities. It outlined plans to demolish all existing buildings for which no future use had been designated. These buildings along with all other contaminated materials were scheduled for disposal in new on-site capped landfills.

The most extensive clean-up efforts were to be focused on soil that had been contaminated by activities at the Arsenal. Any unexploded weapons were scheduled for removal and off-site detonation. Contaminated soils and other materials were to be buried in new waste landfills. These landfills were designed to consist of leak-proof barriers with 6 in (15 cm) concrete barrier caps and underground barrier walls to protect wildlife from their contents. Existing trenches and sewers are to be surrounded by barrier walls, covered with concrete barrier caps, and/or plugged with concrete caps, all to protect wildlife from their contents.

In addition to these physical plans, the Army announced plans for three other on-going activities at the Arsenal: (1) the discontinuation of any additional **hazardous waste** disposal activities at the Arsenal; (2) on-going medical monitoring at adjacent communities during the period of clean-up; and (3) monitoring and control of emissions and odors from the site during clean-up.

Clean-up activities at RMA are being directed by the U.S. Army under the supervision of the **Environmental Protection Agency**. In 2001, there were 31 environmental clean-up projects proposed, and 26 were underway. The Shell Chemical Company is also participating in the study of environmental problems at RMA and in efforts to deal with those problems. Until the clean-up process has been completed, the former RMA site is being managed as a wildlife refuge by the U.S. Department of the Interior's **Fish and Wildlife Service**. At the conclusion of the process, the refuge will become a **National Wildlife Refuge**.

[*David E. Newton*]

RESOURCES
PERIODICALS

"Aresenal Update." *Defense Cleanup* 12 no. 10 (March 9, 2001): 78.

Breen, B. "From Superfund Site to Wildlife Refuge." *Garbage* 4 (May-June 1992): 22.

Gascoyne, S. "From Toxic Site to Wildlife Refuge; If Approved an Ambitious Plan Would Transform a Former Chemical-Weapons Arsenal near Denver." *The Christian Science Monitor* (September 12, 1991): 10.

Rocky Mountain Institute

Founded by energy analysts Hunter and **Amory B. Lovins** in 1982, the Rocky Mountain Institute (RMI) is a nonprofit research and education center dedicated to the **conservation** of energy and other resources worldwide. According to literature published by the Institute, the Lovinses founded RMI with the intention of fostering "the efficient and sustainable use of resources as a path to global security." RMI targets seven main areas for reform: energy, water, agriculture, **transportation**, economic renewal, green development, and global security.

RMI's energy and water programs attempt to promote **energy efficiency** and the use of renewable resources. The programs take an "end-use/least-cost" approach, promoting awareness of which activities require energy, how much and what types of energy those activities require, and the cheapest way that energy can be supplied. E Source, a subsidiary of the energy program, serves as a clearinghouse for technological information on energy efficiency.

The agriculture program at RMI focuses on several conservation-based methods, including low-input, organic and alternative-crop farming, efficient **irrigation**, local and direct marketing, and the raising of extra-lean and range beef. The program is closely linked to RMI's water project.

RMI's transportation program is based on the belief that "inefficient transportation systems shape and misshape our world." The project seeks an end to a transportation-based society and a start to one that is access-based, emphasizing superefficient vehicles and a decrease in the necessity to travel over mobility.

RMI's economic renewal program seeks to create lasting economic bases in rural areas. The project works on a grassroots level and has been tested in four towns. Through workshops and workbooks based on studies of several towns, RMI hopes to promote energy use based on sustainability in the future rather than industrialism in the present.

In the area of green development, RMI is involved in cost-effective innovative construction and energy planning for new towns. RMI has conducted efficiency studies for a new building and has acted as consultants on a prototype store for a major retailer as part of this project.

The global security program is more scholarship- and theory-oriented than RMI's other projects and is based on the concept that innovation in the other six areas—energy, water, agriculture, economic renewal, transportation, and green development—will lead to new ways of thinking with regards to global security. Through scholarly exchange and analysis of security practice around the world, the staff at RMI hope to help develop a post-Cold War, cooperative approach to global security that is less reliant on military strength.

RMI's technologically advanced, energy efficient facility that houses their headquarters features a semi-tropical bioshelter among other showcases of conservation innovations. The facility stands as a visible monument to RMI's commitment to conserving the world's resources.

Hunter and Amory Lovins have received several awards for their work with the Institute, including a Mitchell Prize in 1982, a Right Livelihood award in 1983, and the Onassis Foundation's first Delphi Prize in 1989. The Delphi Prize is considered one of the top two environmental awards in the world.

[*Kristin Palm*]

RESOURCES
ORGANIZATIONS

Rocky Mountain Institute, 1739 Snowmass Creek Road, Snowmass, CO USA 81654-9199 (970) 927-3851, <http://www.rmi.org>

Rodale Institute

The Rodale Institute developed out of the efforts begun by J. I. Rodale to promote **organic gardening and farming** in the 1930s. The notion of **recycling** organic matter back into the **soil** to yield healthier and more productive crops was not widely accepted at that time, so in 1942, Rodale began publishing *Organic Gardening and Farming* magazine.

From those simple beginnings, Rodale's mission grew to become a multi-faceted organization dedicated to "improving human health through regenerative farming and organic gardening." Today, the Rodale Institute supports and publishes research to further **organic farming**, facilitates farming networks, engages in international farming programs, and publishes numerous resources for gardeners and organic farmers.

The Rodale Institute Research Center is a 333-acre (135-ha) farm in Kutztown, Pennsylvania, where organic horticulture and **sustainable agriculture** techniques are tested. The Rodale staff focuses on farming projects that help enrich and protect the world's **natural resources**. Of particular importance to the institute's gardeners are the flower, fruit, vegetable, and herb gardens that are maintained at the Research Center. Practices that are employed in these gardens include the use of beneficial insects and cover crops to reduce reliance on chemical pesticides.

Another goal of the institute is to facilitate communication between farmers and urban dwellers. This is especially important as urban communities continue to expand into the countryside. The Rodale Institute sponsors programs to help these two seemingly competing groups work together for mutual benefit. For example, Rodale has initiated a community **composting** program whereby the grass cuttings and leaves from urban areas are collected and delivered to farms, where they are composted and used to enrich the soil. This program decreases the need for **landfill** space and provides soil-enriching organic matter to farmers. Rodale Institute also encourages mutual understanding between farmers and city dwellers by hosting such events as Field Days and GardenFest.

The Rodale Institute's networking program helps farmers link up with one another to share information on sustainable farming. These farms participate in Rodale Institute's research and often experiment with alternatives to conventional farming methods. Through the publication *The New Farm*, the cooperational farmers share their experiences and questions with many farmers across the country.

Rodale has established cooperative programs with several universities around the world to further their research. Among those universities are Pennsylvania State University, Cornell University, Michigan State University, University of Padova (Italy), Northeast Forestry University (China), South China Environmental Institute, Jiangsu Academy of Agricultural Sciences (China), and the Institute for Land Improvement and Grassland Farming (Poland).

Rodale Institute's programs extend to international farmers as well. With sights set on developing long-term, preventative measures against **famine** and poverty, the institute has established programs in Africa and South America. The African program is based in Senegal and brings together farmers, villages, government agencies, and other organizations to work toward famine prevention through **conservation** and soil quality improvement. In South America, Rodale works with Guatemalan farmers to encourage sustainable farming and help preserve the disappearing forest regions. This particular program combines modern knowledge with traditional Mayan farming techniques to produce high quality yields from healthy soil.

Perhaps to the non-farmer, Rodale is best known for its publications. In addition to *Organic Gardening* and *The New Farm* magazines, Rodale publishes a large selection of instructional books for farmers, gardeners, and others interested in organic farming.

[*Linda Ross*]

Resources

Organizations

The Rodale Institute, 611 Siegfriedale Road, Kutztown, PA USA 19530-9320 (610) 683-1400, Fax: (610) 683-8548, Email: info@rodaleinst.org

Rodenticide

see **Pesticide**

RODs

see **Records of decision**

Holmes Rolston (1932 –)

American environmental and religious philosopher

Holmes Rolston has devoted his distinguished career to plausibly and meaningfully interpreting the natural world from a philosophical perspective and is regarded as one of the world's leading scholars on the philosophical, scientific, and religious conceptions of **nature**. His early work on values in nature, as well as his role as a founder of the influential academic journal *Environmental Ethics*, was critical not only in establishing but also in shaping and defining the modern field of environmental philosophy. In his 1988 book *Environmental Ethics: Values in and Duties to the Natu-*

Holmes Rolston. (Colorado State University Photographic Archives. Reproduced by permission.)

ral World, Rolston presented a philosophically sophisticated and defensible case for a value-centered ecological ethic, one which derives ethical conclusions from descriptive premises. Rolston clearly states that intrinsic values objectively exist at the **species**, **biotic community**, and individual levels in nature and that these values impose on humans certain direct obligations to nonhuman entities, such as species and ecosystems. These obligations are separate from and sometimes in conflict with those based on the instrumental value of nature, which may motivate humans to protect the **environment** for their own benefit.

Rolston is one of the most prolific writers and sought-after speakers in the field today. His work is unusually accessible to a wide audience, and he has pioneered the application of ethical theory to actual environmental problems by consulting with two dozen **conservation** and policy groups, including the United States Congress and a Presidential Commission.

Rolston came to prominence in this field in a roundabout way. Born in the Shenandoah Valley of Virginia, he studied physics as an undergraduate at Davidson College before entering theological seminary. After completing his Ph.D. in theology and religious studies at the University of Edinburgh in Scotland, Rolston spent nearly a decade as a Presbyterian pastor in rural southwest Virginia. During this time, Rolston's love for and curiosity about nature and **wilderness** grew unabated. He fed his love of these things by learning the natural history of his surroundings in splendid detail and by becoming an activist on local environmental issues. In his search for a philosophy of nature to complement his biology, Rolston entered the philosophy program at the University of Pittsburgh, where he received a master's degree in philosophy of science in 1968. He then embarked on an academic career at Colorado State University, where he currently holds the prestigious position of University Distinguished Professor.

The major theoretical innovations of Rolston's work include the reconceptualization of ethical extensionism to accommodate **intrinsic value** at all levels of nature, a task for which the ordinary methods and vocabulary of traditional ethics proved inadequate. The result is a biologically-based account of natural values which is hierarchical but nonanthropocentric. He has thus established himself as the foremost proponent and defender of intrinsic natural value theory. Rolston is often identified as the father of **environmental ethics** as a modern academic discipline. As such, he occupies a singular place of importance in modern philosophy. Rolston continues to integrate the practical and theoretical dimensions of his work, as he examines life to discover its meaning and expands the circle of moral significance to include all natural entities, processes, and systems. In addition to *Environmental Ethics*, Rolston has written several other critically acclaimed books, including *Philosophy Gone Wild*, *Science and Religion: A Critical Survey*, and *Conserving Natural Value*, and has contributed to dozen other books and professional and popular periodicals. Rolston is also an avid backpacker, accomplished field naturalist, and respected bryologist.

[*Ann S. Causey*]

Resources

Books

Rolston, Holmes. *Environmental Ethics: Duties to and Values in the Natural World.* Philadelphia: Temple University Press, 1988.

———. *Genes, Genesis, and God.* New York: Cambridge University Press, 1999.

———. *Philosophy Gone Wild: Environmental Ethics.* Buffalo, NY: Prometheus Books, 1989.

———. *Science and Religion: A Critical Survey.* Philadelphia: Temple University Press, 1987.

Periodicals

Rolston, Holmes. "Values Deep in the Woods." *American Forests* 94 (May–June 1988): 33–7.

Pierre Ronsard (1524 - 1585)

French naturalist and poet

Pierre Ronsard transformed his life from one of possible anguish into one in which he created a legacy that has lasted for centuries. Born into a noble family in near Vendome, France in September of 1524, Ronsard was the younger of their sons. His life was one of privilege—from the benefits of a classical education in his family's residence at the Chateau de la Poissoinniere to being sent away to the College of Navarre, in Paris, at the early age of nine. Before graduating from college, he left and was appointed page to the Duke of Orleans, son of King Francis I. Later, he would become page to James V, King of Scotland where he stayed for three years. While there and through time spent in England Ronsard became proficient in the English language. His travels then took him to Germany, Italy, and other countries throughout Europe.

In 1841 Ronsard was afflicted with an incurable deafness. Plans for a military career were no longer an option for him. He was only 17 and decided to retire from public life and pursue his studies again. For seven years he studied Greek at the College de Coqueret; and he became further involved with his passion of poetry. By 1550 he was awarded with the title of"Prince of Poets," successful in his desire to find a new direction for French poetry. He continued to be a favorite of royalty abroad. He received a diamond from Queen Elizabeth of England who was a relative. Mary Stuart of Scotland escaped the woes of prison life through Ronsard's poetry. The city of Toulouse in his native France presented him with a solid silver award in tribute to the goddess Minerva.

Ronsard remained in ill health while writing volumes of poetry that was revolutionary in many ways, including the words he carried into it and enhance the entire French vocabulary. Not only did he add words from Greek and Latin; he also uncovered old romantic dialects, and utilized the technical languages found in the trades, in science, and in sports. Writing for him produced harmony for others. Scholars consider him to be the best lyrical poet in France until the romantic age that arrived with the nineteenth century poets. He was a figure known and beloved throughout his country, a celebration that would rival only writer Victor Hugo nearly 300 years later. One of his many poems, *Roses* would inspire an honor bestowed on the two hundreth anniversary of his death—the Pierre Ronsard rose, a delicate pink known also as the **Eden Climber**.

During the last years of his life Ronsard became a minor monk with the Priory of Saint Cosme on an ancient island on the Loire River. The original buildings were established between the eleventh and fifteenth centuries. Ronsard's original works are contained there in the Prior's house, rebuilt over a hundred years following his death in 1585. Visitors to the priory (reservations arerequired) can visit Ronsard's tomb, walk through the gardens, see his original works, and attend festivals in his honor. His devotion to love and **nature** has given him the honor of being known as the "Father of Nature." Many present-day environmentalists credit Ronsard with beginning the movement toward preservation of lands, forests, and gardens. He carried his passion with him and gave of it so generously to all his readers, and to generations following that would enjoy the graceful harmony with which he presented it.

[*Jane Spear*]

Resources

Other

"Gardens of the Priory of Saint-Cosme." *Gardens of France (Jardins-de-France).* [cited July 2002]. <http://www.jardins-de-france.com>.

"Pierre de Ronsard." *France web.* [cited July 2002]. <http://franceweb.fr/poesie>.

"Pierre de Ronsard." *New Advent.* [cited July 2002]. <http://www.newadvent.org>.

"Pierre de Ronsard."*Sonnett Central.* [cited July 2002]. <http://www.sonnets.org>.

"Pinks." *Potee.* [cited July 2002]. <http://potee.com/rose.html>.

"Priory of Saint Cosme." *Counsel General of Loire, France.* [cited July 2002]. <http://www.cg37.fr>.

Organizations

Priory of Saint Cosme, Loire France 33-(2) 47 37 32 70, <http://www.cg37.fr/>

Theodore Roosevelt (1858 - 1919)

American politician and conservationist

Historians often cite **conservation** of **natural resources** as Theodore Roosevelt's most enduring contribution to the country. As the nation's twenty-sixth President, Roosevelt was faced with critical conservation issues and made decisive moves to promote conservation, thus becoming the national leader most clearly associated with preservation of **public land**.

Roosevelt was born into a wealthy family in New York City, and early took an interest in the outdoors, partly to compensate for **asthma** and a frail constitution. He was an avid naturalist as a child, an early interest that lasted all his life and, as Paul R. Cutright documents in detail in his *Theodore Roosevelt: The Making of a Conservationist*, Roosevelt was instrumental in creating a role model for conservationists. Certainly, much of his attention to conservation derived at least in part from his interest in the natural science that underlay those issues.

He was, for example, a life-long bird watcher and relished the fact that he could match John Burroughs's prow-

ess at identifying birds on a field walk with the naturalist-writer. Roosevelt's interest in the natural history of birds and other animals provided much of the motivation for two long and perilous trips taken after his presidency to South America and Africa, trips that compromised his health and may have shortened his life.

From a very young age, Roosevelt was also a politician. At the age of 23, he was elected to the New York State Assembly in 1881. Always contradictory, the young Roosevelt was conservative and pro-establishment but was also a reformer, anticorruption and anti-machine politics.

As governor of New York (1898–1900), Roosevelt defined and tried to act on conservation issues. In 1900, **Gifford Pinchot** helped Roosevelt formulate his message to the New York State Assembly about the need for **forest management**. The Governor also tried to outlaw the use of bird feathers for adornment. Historians claim that his actions so alarmed the Assembly that he was "manipulated" out of the governor's mansion into the vice-presidency, from there he became President after William McKinley's assassination.

Roosevelt's contribution to conservation can be divided into four categories: first, his role in setting aside and managing what are now called national forests; second, his decisive initiation of a **national wildlife refuge** system; third, his impact on transforming the **arid** lands of the American West into irrigated farmland; and, fourth, his efforts to promote natural resources nation-wide.

Roosevelt is best known for his collaboration with Pinchot in appropriating public forest lands once controlled by private interests and "reserving" them for "our people unborn." Congress passed the Forest Reserve Act (authorizing the Presidents to create forest reserves from the public domain) in 1891, well before Roosevelt took office. The three Presidents immediately before him established forest reserves of some 50 million acres (20.3 million ha). Roosevelt publicized the value of forest reserves to the people; and reorganized management of the reserves, placing them under the Bureau of Forestry (later the **Forest Service**) in the Department of Agriculture. Increased the acreage in reserves—by 150 million acres (60.8 million ha), 16 million (6.5 million ha) of which were set aside almost overnight in a famous and successful action in 1907 by which he and Pinchot worked long hours to create the reserves before the President had to sign a congressional act with a rider limiting his powers to do so. All of these actions remain controversial today. The Forest Service's timber management policies are still being criticized for catering to special interests, and not realizing Roosevelt's goal to preserve the forests for the people, not powerful private interests.

The story of Pelican Island illustrates perhaps better than any other incident in Roosevelt's administration his approach to conservation issues. Visited by naturalist friends alarmed at the decimation of birds on Florida's Pelican Island, Roosevelt asked if the law prevented him from declaring the island "a Federal Bird Reservation." Told that no such law existed, the President responded, "Very well, then I so declare it." During his tenure, Roosevelt created 50 additional **wildlife** refuges. His enthusiastic initiation of these preserves provided a base for the extensive national **wildlife refuge** system.

Roosevelt entered the Presidency with the idea that the western drylands could become productive through **irrigation**. He realized, however, that the scale of such projects prohibited private enterprise from undertaking the task. One of his first initiatives was to work with congressmen representing western states to pass the **Reclamation** Act of 1902. Sixteen projects were soon initiated in those states. Reclamation from Roosevelt's point of view—especially large dams—are considered a mixed blessing by many conservationists today.

Roosevelt, was a natural publicist and, along with Pinchot and **John Muir**, he provided an extraordinary legacy to the American people of a variety of lands and resources in public ownership. Roosevelt used the Presidency's "bully pulpit" effectively to arouse public interest in conservation issues. Harold Pinket argues that Roosevelt's main contribution to the conservation movement was "wielding his presidential prestige to craft a coalition of people with otherwise opposed perspectives on natural resources, from naturalists and civic leaders who favored preservation to utilitarian resource specialists and users." No accomplishment illustrates this better than the Governor's Conference of 1907. At this conference Roosevelt bought all the nation's governors and many other leaders together and, using his own enthusiasm for conservation, he ignited discussions, policies and actions that resonate still today at many levels of government.

Roosevelt's concern for conservation was reflected in his message to Congress, delivered two months after becoming President. It contained strong references to all the relevant issues of the time—preservation and use of forests, **soil** and **water conservation**, wildlife protection, **recreation**, and reclamation of arid lands. He is still recognized for his leadership on these areas. In addition to his concern for conservation issues Roosevelt was a widely published author. Many readers can still be entertained by vivid accounts of **hunting** trips in the west or his life as a rancher in the Dakotas.

Most environmentalists pay strong tribute to Roosevelt's accomplishments as a conservationist and his contributions to the conservation movement. However, he is not universally admired. Some view Roosevelt as an elitist while other criticize what some consider his excessive slaughter as

Theodore Roosevelt. (Archive/Corbis-Bettmann. Reproduced by permission.)

a hunter. Outright defends Roosevelt, arguing "it was only after Roosevelt put the full force of his power as President behind the conservation program that it got off the ground." Theodore Roosevelt's energy and bluster, his cultivation of well-known naturalists and conservationists, his willingness to listen to and act on their advice, his political skills in getting policies enacted—all those endure in a lasting legacy of national forests, a national wildlife refuge system, a strengthened **national park** system, and increased and ongoing awareness of the importance of protecting these for **future generations**. His words still ring true today, that "any nation which...lives only for the day, reaps without sowing, and consumes without husbanding, must expect the penalty of the prodigal." A very verbal leader, Roosevelt often had the last word: "For the people...must always include the people unborn as well as the people now alive, or the democratic ideal is not realized."

[*Gerald L. Young Ph.D.*]

Resources

Books

Brooks, P. "A Naturalist in the White House." In *Speaking for Nature.* Boston: Houghton Mifflin, 1980.

Cutright, P. R. *Theodore Roosevelt: The Making of a Conservationist.* Champaign: University of Illinois Press, 1985.

Periodicals

Ponder, S. "Publicity in the Interest of the Public: Theodore Roosevelt's Conservation Crusade." *Presidential Studies Quarterly* 20 (Summer 1990): 547–555.

Theodore Roszak (1933 –)

American social critic

Perhaps most prominently associated with the counterculture movement in the 1960s, Roszak was born in 1933. He received his B.A. from the University of California at Los Angeles in 1955 and a Ph.D. from Princeton University in 1958. Roszak began his career as an instructor in history at Stanford University and is currently professor of history at California State University, Hayward. He received a Guggenheim fellowship in 1971.

Like his mentor Lewis Mumford, Roszak combines political and cultural criticism with a thoroughgoing critique of technology and technological society. Published in 1969, his first book was an effort to understand the counterculture movement. In *The Making of a Counterculture: Reflections on the Technocratic Society and its Youthful Opposition*, Roszak criticizes consumer society, the military-industrial complex it supports, the increasing concentration of populations in unclean, unsafe, and ungovernable cities, and the technocratic and bureaucratic mentality that views such dilemmas as essentially technical problems with technical or scientific solutions. Roszak is highly critical of this rationalist point-of-view and argues that modern society should attempt to recover and return to a sense of the sacred and mysterious dimensions of human life.

In 1972, Roszak published *Where the Wasteland Ends: Politics and Transcendence in Post-Industrial Society.* Here, he offers an outline of what he terms a "visionary commonwealth," an alternative society that would check or eliminate the destructive tendencies of modern technocratic civilization. The commonwealth he describes is decentralized and small in scale. Politics is participatory, technology is appropriate and intermediate, and there is widespread experimentation with different forms of social, economic, and political organization. Roszak believes that resettling populations into such small-scale, economically self-sufficient, and politically self-governing communities cannot happen quickly. However, he contends that such a shift will happen and he argues that it should happen, if humans are to live spiritually rich and meaningful lives. Roszak maintains that the possibilities for this utopia are present as "springs" within the technocratic wasteland, waiting to be discovered and used to transform this spiritual **desert** into a humanly habitable garden of earthly delights. It is he says, "more humanly beautiful to

risk failure in searching for the hidden springs than to resign to the futurelessness of the wasteland."

Published in 1978, *Person/Planet* expands on Roszak's vision of the future of humanity and continues his critique of modern society. Roszak argues that industrial society is disintegrating in a way that, he maintains, is creative. Large, complex institutions, including government itself, are failing to attract the loyalty and allegiance they need to maintain their authority, and Roszak believes their disintegration will make his utopian commonwealth possible. For him, the needs of the individual and the needs of the planet are identical. Both flourish in an **atmosphere** of authenticity, diversity, and respect, and he argues that these are things that large industrial institutions, with their emphasis on uniformity, linearity, and wastefulness, can neither comprehend nor tolerate.

Roszak has been criticized for his romanticism and his utopianism. His attacks on science and rationalism, in particular, have been frequently condemned as vague and imprecise, and he has been accused of confusing the methodology of science with the failings of the people who employ it. However, many admire Roszak not only for his passionate prose but also for his vision of human possibilities, and his books are still frequently consulted for their images of people living reverently and responsibly in harmony with the earth.

[*Terence Ball*]

Resources
Books

Roszak, Theodore. *Person/Planet: The Creative Disintegration of Industrial Society.* Garden City, NY: Anchor Press/Doubleday, 1978.

———. *The Gendered Atom: Reflections on the Sexual Psychology of Science.* Conari Press, 1999.

———. *The Making of a Counterculture: Reflections on the Technocratic Society and its Youthful Opposition.* Garden City, NY: Doubleday, 1969.

———. *The Voice of the Earth.* New York: Simon & Schuster, 1992.

———. *Where the Wasteland Ends: Politics and Transcendence in Postindustrial Society.* New York: Doubleday, 1972.

———. M. E. Gomes, and A. D. Kanner. *Ecopsychology: Restoring the Earth, Healing the Mind.* San Francisco: Sierra Club Books, 1995.

Frank Sherwood Rowland (1927 –)

American atmospheric chemist

Rowland was born in the central Ohio city of Delaware. After a brief period of service in the United States Navy, he attended Ohio Wesleyan University where he earned his bachelor's degree in chemistry in 1948. He balanced his semi-professional baseball career with his studies and continued his education in chemistry, earning master's and doctorate degrees at the University of Chicago in 1951 and 1952, respectively. Fresh out of school, Rowland pursued a career in education and research. He taught chemistry at Princeton University from 1952 to 1956 and the University of Kansas between 1963 and 1964.

Rowland has conducted research in many areas of the chemical and radiochemical fields. In 1971, for example, to calm an alarmed public, Rowland and a team of scientists investigated the seemingly high levels of **mercury** being found in tuna and **swordfish**. They tested the tissues of museum exhibit, century-old fish and found the levels of the dangerous substance were in about the same range as those recently pulled from the water, and therefore proved the fish were not a health threat. In addition to this type of testing, he completed work for such organizations as the International Atomic Energy Administration and the United States **Atomic Energy Commission**. Rowland is probably best known, however, for his work in atmospheric and chemical kinetics and especially his investigations of **chlorofluorocarbons** (CFCs).

Inert and versatile compounds, CFCs were most often found in items like cooling devices and **aerosol** cans; CFCs are also found throughout the **atmosphere**. The inherent dangers of CFCs were unknown until Rowland teamed up with University of California associate **Mario Molina** in 1973 and found that certain atoms of those CFCs present in the atmosphere were combining with the **ozone**, causing rapid depletion of the protective ozone layer itself. Such destruction could result in drastic climatic changes, as well as increased atmospheric penetration by the sun's rays (causing a huge upswing of the occurrence of skin **cancer**).

Initially, Rowland's and Molina's theory was not readily accepted. In recent years however, the theory has come be accepted as an unpleasant fact; the importance of CFC elimination has been realized. Major industries are taking steps to address the CFC problem. Du Pont for example, a major developer and manufacturer of CFCs, has pledged to reduce its production of the compounds. The 1995 Nobel Prize in chemistry was awarded to Rowland, Molina, and Paul Crutzen.

Rowland is currently the chemistry department chairman at the University of California at Irvine, where he has been a faculty member since 1964. He maintains membership in numerous organizations, among them: the Ozone Commission of the International Association of Meteorological and Atmospheric Physics; Committee of Atmospheric Chemical and Global **Pollution**; and various committees of the United States National Academy of Science.

[*Kimberley A. Peterson*]

RESOURCES
PERIODICALS

Edelson, E. "The Man Who Knew Too Much." *Popular Science* 234 (January 1989): 60–65, 102.

Moreau, D. "Change Agents." *Changing Times* (January 1990): 99.

Rowland, F. S. "Chlorofluorocarbons and the Depletion of Stratospheric Ozone." *American Scientist* 77 (1989): 36–45.

Rubber

Rubber was first used to manufacture tires in the mid-nineteenth century. By the turn of the twenty-first century, there were about three billion waste tires stockpiled or clogging landfills in the United States alone. Worldwide, about 700 million scrap tires are generated annually. This waste rubber has become a major source of **pollution**.

Although rubber, made from latex secreted by trees, has been used for thousands of years, it wasn't until 1839, when Charles Goodyear invented vulcanization, that rubber became important for manufacturing. Vulcanization uses sulfur to cross-link the latex fibers into a rubber that is strong, flexible, durable, and resistant to heat and cold. Like natural rubber, modern synthetic rubbers are polymers (long chains of similar molecules) that are cross-linked by vulcanization. Tires consume 60–70% of all rubber produced, two-thirds of which is synthetic.

Of the 273 million waste tires generated annually in the United States, about 25% end up in landfills. There they leak pollutants into **soil** and **groundwater** and tend to rise to the surface, harming **landfill** covers. The largest tire dump in the Northeast United States holds an estimated 20–30 million tires. The majority of states now ban tire disposal in landfills and collect disposal fees on tires or require that the tires be chipped or ground before disposal. An additional 800 million tires are stockpiled in the United States, however, and many more are dumped illegally. These mountains of tires can fill up with water and become breeding grounds for rats and mosquitoes. They also can ignite, have a high heat output, and are very difficult to contain. These fires can burn for months or years, producing toxic **smoke** and oils that pollute the air, water, and soil.

Rubber recycling

Rubber is difficult to recycle because of its chemical cross-linking. Furthermore most tires contain a mixture of three or four types of synthetic rubbers, as well as natural rubber, other fibers, and steel. Nevertheless in the United States, markets now exist for 76% of newly-scrapped tires, up from 17% in 1990. About 42% of scrapped tires are used for fuel in facilities such as **pulp and paper mills** and cement kilns. As a fuel, tires are equivalent to oil and produce 25% more energy than **coal**. Because of new technologies and pollution controls, tire **combustion** now proceeds at higher temperatures with less **air pollution**.

About 33 million retread or recapped tires are sold annually in the United States. A retread tire reuses 75% of the old tire and requires 70% less oil than manufacturing a new tire.

In 2001 about 40 million scrap tires were used in the construction of playground equipment, artificial reefs, boat bumpers, crash barriers, stabilizers for slopes, and **erosion** control for **dams**. Tire material is used for **mulch**, mats, septic systems, building products, coatings and sealants, and in **hazardous waste** containers.

Many new products contain material from recycled tires. In the United States about 24.5 million scrap tires annually are ground up, cut, stamped, or punched for new products. Ground tires are used for running tracks, playgrounds, flooring, and the soles of shoes, or mixed with asphalt for road paving. Rubber crumb is used in products such as athletic turf and auto parts.

In typical rubber **recycling**, the tires are cut up into small pieces and ground, or frozen in liquid **nitrogen** and shattered or pulverized. The steel is extracted with magnets and **filters** separate the rubber from other synthetic fibers. However 15–50% of the original tire remains as a useless rubber-fiber blend that goes to the landfill. The **U.S. Department of Agriculture** is attempting to develop more efficient methods of separating the rubber from the other fibers.

The development of new tire recycling technologies is an active area of research. Ground-up rubber can be mixed with virgin rubber and vulcanized to form new cross-links, restoring its strength and elasticity. About 5% of scrap tires are recycled in this way. Improved methods for rejuvenating old rubber, without mixing it with newly-manufactured rubber, would enable tire manufacturers to increase the recycled content of new tires.

Another area of research uses heat and pressure to combine the powder from ground-up tires with powder made from asphalt. This asphalt-modified rubber is superior to asphalt for roads, construction materials, and roofing shingles. Likewise, researchers are experimenting with adding rubber crumb to fresh concrete to increase strength and durability. A composite that contains 50% rubber crumb could potentially replace **plastics** such as polyvinyl chlorides for various applications. Other researchers are studying whether the oils produced by waste tire combustion can be reprocessed into **carbon** black for use in various products.

[*Margaret Alic Ph.D.*]

Resources

Other

"Arizona State University Research Finds Recycling Cure for Used Tires." *ScienceDaily Magazine* September 13, 2001 [cited July 7, 2002]. <http://www.sciencedaily.com/releases/2001/09/010913074634.htm>.

"Recycling Research Institute." *Scrap Tire News Online.* 2002 [cited July 7, 2002]. <http://www.scraptirenews.com>.

U.S. Environmental Protection Agency. *Jobs Through Recycling.* May 31, 2002 [cited July 7, 2002]. <http://www.epa.gov/epaoswer/non-hw/recycle/jtr/comm/rubber.htm>.

U.S. Environmental Protection Agency. *Municipal Solid Waste.* May 9, 2002 [cited July 7, 2002]. <http://www.epa.gov/msw/tires.htm>.

"Umass Polymer Scientists Aiming to Turn Scrap Tires into Environmentally Friendly Products." *ScienceDaily Magazine* March 6, 2002 [cited July 7, 2002]. <http://www.sciencedaily.com/releases/2002/03/020306073739.htm>.

Organizations

International Tire & Rubber Association Foundation, Inc., PO Box 37203, Louisville, KY USA 40233-7203 (502) 968-8900, Fax: (502) 964-7859, Toll Free: (800) 426-8835, Email: itra@itra.com, <http://www.itra.com>

Recycled Materials Resource Center, 220 Environmental Technology Building, Durham, NH USA 03824 (603) 862-3957, Email: rmrc@rmrc.unh.edu, <http://www.rmrc.unh.edu>

Rubber Manufacturers Association, 1400 K Street, NW, Washington, DC USA 20005 (202) 682-4800, Email: info@rma.org, <http://www.rma.org>

U.S. Environmental Protection Agency, 1200 Pennsylvania Avenue, NW, Washington, DC USA 20460 (703) 412-9810, Toll Free: (800) 424-9346, Email: public-access@epa.gov, <http://www.epa.gov>

William Doyle Ruckleshaus (1934 –)

American former Environmental Protection Agency administrator

Ruckleshaus is known as a lawyer, a loyal member of the Republican party, and a skilled administrator who has been able to work effectively with environmentalists as well as industry representatives. His reputation also bespeaks his integrity in law enforcement and his ability to withstand pressure. However, Ruckleshaus is best known on the national level for his service as administrator of the **Environmental Protection Agency** (EPA) from 1970 to 1973 and again from 1983 to 1984.

Ruckleshaus was born on July 24, 1934 in Indianapolis, Indiana, into a renowned Republican family. He earned his bachelor of arts degree *cum laude* from Princeton University in 1957 and his law degree from Harvard University in 1960. His early career in the 1960s included the practice of law between 1960 and 1968 and service in the Indiana House of Representatives from 1967 to 1969. In 1969 he was appointed Assistant Attorney General for the United States by the newly elected President Richard Nixon. In 1970, President Nixon selected Ruckleshaus to become the first head of the recently created EPA. Under his direction, 15 environmental programs were brought together under the agency. Ruckleshaus left the EPA in 1973 to serve as acting director of the Federal Bureau of Investigation (FBI), and later that year, he was appointed Deputy Attorney General for the United States. In 1974 he resigned from that position rather than comply with President Richard Nixon's order to dismiss the special Watergate prosecutor.

From 1974 to 1976 he practiced law with the firm Ruckleshaus, Beveridge, Fairbanks and Diamond, in Washington, D.C. Ruckleshaus has been criticized for going through the "revolving door" of government. While at his law firm he was the legal representative of several companies that contested rules made by the EPA while he was its administrator. From 1975 to 1983 he was Senior Vice-President for Legal Affairs of the Weyerhaeuser Company, Tacoma, Washington, a large timber and wood products company.

Ruckleshaus was again called to head the Environmental Protection Agency in 1983. Early in 1983 the EPA came under criticism from the public and from Congress regarding allegations of mishandling of the federal Superfund program. Allegations included lax enforcement against polluters, mishandling of Superfund monies, manipulation of the Superfund for political purposes, and conflicts of interest involving ties between EPA officials and regulated businesses. The allegations and the ensuing investigation led to the resignation of 21 top EPA officials including its administrator, Anne Burford Gorsuch.

Upon Gorsuch's resignation in March 1993, President Ronald Reagan asked Ruckleshaus to serve as interim EPA administrator. He agreed to do so, serving until the appointment of his successor Lee Thomas in November of 1984. During this second period of service as the EPA's top administrator, Ruckleshaus used his experience, his skills as an administrator, and ability to work with industry and environmentalists to stabilize the EPA. He succeeded in quelling much of the criticism being leveled against the agency.

After leaving the EPA at the end of 1985, Ruckleshaus joined the firm Perkins Coie in Seattle, Washington. He has served on the boards of directors of several major corporations and on the board of advisors for the Wharton School of Business, University of Pennsylvania.

In 1989 he was named chairperson of Browning-Ferris Industries, Inc. (BFI) and, as of 1997, he continues to hold that position. BFI is the world's second largest **waste management** firm after WMX Technologies (Waste Management, Inc.) and owns and operates substantial hazardous and non-hazardous waste-disposal facilities throughout the states. When Ruckleshaus took over as chairperson, BFI had been cited in various lawsuits related to its operations. BFI has been making steady progress since 1989 in obtaining required permits. Two major lawsuits against BFI were also settled under Ruckleshaus leadership in 1996. The U.S. Justice Department's antitrust division had alleged that BFI

and WMX had used long-term contracts to keep competition out of certain states. Thus, over the past eight years, Ruckleshaus has been gaining substantial experience as the leader of a major waste disposal business.

In 1995, the **Environmental Law Institute** (ELI) honored Ruckleshaus on the twenty-fifth anniversary of the U.S. EPA by presenting him with the 1995 ELI award. The ELI presents its award annually to a lifelong leader in environmental protection who represents the highest ideals of service.

[*Paulette L. Stenzel*]

Resources
Periodicals

"EPA: Ruckleshaus Bows Out." *Newsweek* (December 10, 1984): 39.

"Government's Pollution Fighter." *New York Times* (April 12, 1973): 22.

Ivey, M. "Can Bill Ruckelshaus Clean Up Browning-Ferris' Act?" *Business Week* (October 14, 1991): 46.

King, Seth S. "Return of First E.P.A. Chief." *New York Times Biographical Service* (March 1983): 372.

Simon, R. "Mr. Clean's New Mess." *Forbes* 146 (November 26, 1990): 166–67.

Runoff

The amount of rainfall or snowmelt that either flows over the **soil** surface or that drains from the soil and enters a body of water, thereby leaving a **watershed**. This water is the excess amount of precipitation that is not held in the soil nor is it evaporated or transpired back to the **atmosphere**. Water that reaches deep **groundwater** and does not, therefore, directly flow into a surface body of water is usually not considered runoff. Runoff can follow many pathways on its journey to streams, rivers, lakes, and oceans. Water that primarily flows over the soil surface is surface runoff. It travels more quickly to bodies of water than water that flows through the soil, called subsurface flow. As a rule, the greater the proportion of surface to subsurface flow, the greater the chance of **flooding**. Likewise, the greater the amount of surface runoff, the greater the potential for soil **erosion**. *See also* Storm runoff

S

Safe Drinking Water Act (1974)

The Safe Drinking Water Act (SDWA, 1974) is the main federal law that ensures the quality of drinking water in the Unites States. When implemented, it extended coverage of federal drinking water standards to all public water supplies. Previous standards, established by the United States **Public Health Service** beginning in 1914 and administered by the United States **Environmental Protection Agency** (EPA) since its creation in 1970, had legally applied only to water supplies serving interstate carriers (e.g., planes, ships, and rail cars engaged in interstate commerce). However, many states and municipalities complied with them on a voluntary basis. Under the SDWA, public water supplies were defined as those publicly or privately owned community water systems having at least 15 connections or serving at least 25 year-round customers or non-community water supplies serving at least 25 non-residents for at least 60 days per year.

The SDWA required the EPA to promulgate primary drinking water regulations to protect public health and secondary drinking water regulations to protect the aesthetic and economic qualities of the water. The EPA was granted authority to regulate: 1) contaminants which may affect health (e.g., trace levels of carcinogenic **chemicals** which may or may not have an impact on human health); 2) compounds which react during **water treatment** to form contaminants; 3) classes of compounds (if more convenient than regulating individual compounds); and 4) treatment techniques, when it is not feasible to regulate individual contaminants (e.g., disinfection is required in lieu of standards on individual disease-causing **microorganisms**).

Recognizing the right and the responsibility of the states to oversee the safety of their own drinking water supplies, the SDWA authorized the EPA to grant primacy to states willing to accept primary responsibility for administering their own drinking water program. To obtain primacy, a state must develop a drinking water program meeting minimum federal requirements and must establish and enforce primary regulations at least as stringent as those promulgated by the EPA. States with primacy are encouraged to enforce the federal secondary drinking water regulations, but are not required to do so.

Other provisions of the SDWA authorized control of underground injection (e.g., waste disposal **wells**); required special protection of sole-source aquifers (those providing the only source of drinking water in a given area); authorized funds for research on drinking water treatment; required the EPA to conduct a rural water supply survey to investigate the quality of drinking water in rural areas; allocated funds to subsidize up to 75% of the cost of enlarging state drinking water programs; required utilities to publicly notify their customers when the primary regulations are violated; permitted citizens to file suit against the EPA or a state having primacy; and granted the EPA emergency powers to protect public health.

Dissatisfied with the slow pace at which new regulations were being promulgated by the EPA, which had in 1983 initiated a process to revise the primary and **secondary standards**, Congress amended the SDWA in 1986. The 1986 amendments required the EPA: to set **primary standards** for nine contaminants within one year, 40 more within two years, 34 more within three years, and 25 more by 1991; to specify criteria for **filtration** of surface water supplies and disinfection of **groundwater** supplies; to require large public water systems to monitor for the presence of certain unregulated contaminants; to establish programs to demonstrate how to protect sole-source aquifers; to require the states to develop well-head protection programs; and to issue, within 18 months, rules regarding injection of waste below a water source.

The 1986 SDWA amendments also prohibited the use of **lead** solder, flux, and pipe; authorized the EPA to treat Indian tribes as states, making them eligible for primacy and grant assistance; required the EPA to conduct a survey of drinking **water quality** on Indian lands; authorized the EPA to initiate enforcement action if a state fails to take appropriate action within 30 days; and increased both civil and criminal penalties for failure to comply with the SDWA.

Since 1986 was a Congressional election year and every member of Congress wanted to go on record as having voted for safe drinking water, the amendments passed unanimously. However, Congress failed to provide federal funds to assist state programs in complying with the many new provisions of the SDWA. The average annual cost per state to comply with the SDWA by the year 1995 has been estimated to be nearly $500 million.

Further amendments to SWDA were passed in 1996. The new amendments established a Drinking Water State Revolving Fund to finance state compliance costs for water treatment facilities, easier access to water quality information for consumers, and contamination prevention initiatives. The mandate for contaminant testing was changed to a risk-based prioritized system that granted the EPA the authority to decide whether or not to regulate a contaminant after completing a required review of five contaminants every five years. The amendments also called for specific risk assessments and final regulation of **radon**, **arsenic**, DBP/cryptosporidium, and sulfate.

As of 2002, the SWDA ranks the following drinking water standards as rule making priorities:

- **Arsenic–** The SWDA requires the EPA to revise the existing 50 **parts per billion** (ppb) standard for arsenic in drinking water; the EPA implemented a 10 ppb standard for arsenic in January 2001; After the Bush administration briefly withdrew the standard, the new rule became effective in February 2002. All U.S. water systems must be in compliance by 2006;
- **Ground Water Rule–** The EPA is proposing to regulate the appropriate use of disinfection in ground water and of other components of ground water systems to ensure public health protection.
- **Lead and copper–** The EPA estimates that approximately 20% of human exposure to lead is attributable to lead in drinking water.
- **Microbials and disinfection byproducts–** The EPA considers that a major challenge for water suppliers is how to balance the risks from **microbial pathogens** and disinfection byproducts.
- **MTBE–** MTBE (methyl-t-butyl ether) belongs to a group of chemicals commonly known as fuel oxygenates and has replaced lead as an octane enhancer since 1979.
- **Radionuclides–** The EPA has updated standards for **radionuclides** in drinking water.
- **Radon–** Radon is a naturally-occurring radioactive gas associated with **cancer**, and that may be found in drinking water and indoor air. The EPA has developed a regulation to reduce radon in drinking water.

[*Stephen Randtke and Paula Anne Ford-Martin*]

Resources

Books

Ingram, Colin. *The Drinking Water Book: A Complete Guide to Safe Drinking Water.* Berkeley, CA: Ten Speed Press, 1991.

Lewis, Scott A. *The Sierra Club Guide to Safe Drinking Water.* San Francisco, CA: Sierra Club Books, 1996.

National Research Council. *Setting Priorities for Drinking Water Contaminants.* Washington, DC: National Academy Press, 1999.

Subcommittee on Arsenic in Drinking Water, Committee on Toxicology, Board on Environmental Studies and Toxicology, Commission on Life Sciences, National Research Council. *Arsenic in Drinking Water* Washington, DC: National Academy Press, 1999. Available online at http://books.-nap.edu/books/0309063337/html/index.html. Accessed June 2, 2002.

Other

U.S. Environmental Protection Agency, Office of Water. *The Safe Water Drinking Act.* [cited June 2, 2002]. <http://www.epa.gov/OGWDW/sdwa/sdwa.html>.

Organizations

Water Environment Federation, 601 Wythe Street, Alexandria, VA USA 22314-1994 (703) 684-2400, Fax: (703) 684-2492, Toll Free: (800) 666-0206, Email: csc@wef.org, <http://www.wef.org>

Sagebrush Rebellion

A political movement in certain western states during the late 1970s, sparked by passage of the **Federal Land Policy and Management Act** in 1976. The federal government owns an average of 60% of the land in the twelve states that include the Rockies or lie west of them. Cattlemen, miners, loggers, developers, farmers, and others argued not only that federal ownership had an adverse impact on the economy of their states, but that it violated the principle of states' rights. This group demanded that the federal government transfer control over large amounts of this land to individual states, insisting on their right to make their own decisions about the management of both the land itself and the **natural resources**. The rebellion was defused after the election of Ronald Reagan in 1980. He appointed James Watt, who had been a leader of this movement, as Secretary of the Interior, and oversaw the institution of the so-called "good neighbor" policy for the management of federal lands. *See also* Wise use movement

Sahel

A 3,000-mile (5,000 km) band of semi-arid country extending across Africa south of the Sahara **desert**, the Sahel zone ("the shore" in Arabic) passes through Mauritania, Senegal, Mali, Burkina Faso, Niger, Chad, and the Cape Verde Islands. Similar semi-arid conditions prevail in Sudan, Ethiopia, and Somalia. These countries are among the poorest in the world. The low annual rainfall in this region is variable (4-20 in or 10-50 cm) and falls in a short, intense period in July and Au-

gust. In some years the rains fail to develop, and droughts are a common occurrence. The uncertain rainfall of the Sahel makes it generally unfavorable for agriculture.

For centuries, the indigenous nomadic Tuareg people used the Sahel in a sustainable way, constantly moving herds of camels from one grazing area to the next; they practiced little agriculture. In the nineteenth and twentieth centuries, following European colonization, herds of water-dependent, non-native cattle were introduced, which were poorly suited to the **arid** conditions of the region. The above average rainfall of the 1950s and 1960s attracted large numbers of farmers and pastoralists into the Sahel, which placed new stresses on this fragile **ecosystem**. A series of droughts of the 1970s and 1980s in the Sahel resulted in episodes of large-scale starvation.

Studies of long-term **climate** patterns show that while droughts have been common in the Sahel for at least 2,500 years, the droughts of recent years have increased in frequency and duration. Records also show that the annual rainfall has decreased and that the sands of the Sahara have shifted some 60 miles (100 km) south into the region.

The causes of these changes have been linked to the expanding human settlement of the Sahel, with consequent increased demands on the area to produce more food and more firewood. These demands were met by increases in domestic animal herds and by more intensive agriculture. This in turn led to drastic reductions in vegetation cover. Land was cleared for farming and human settlements, vegetation was overgrazed, and large numbers of trees were cut for firewood. In this way, the natural vegetation of the Sahel (sparse, coarse grasses interspersed with thorn trees and shrubs) was dramatically altered and the ecosystem degraded.

Less vegetation cover meant more **soil erosion** and less **groundwater** recharge as heavy seasonal rainstorms hit exposed ground, carrying away valuable **topsoil** in flash floods. Less vegetation also meant more soil erosion from wind and rain, as there were fewer root systems to bind the soil together. In addition, fewer plants meant that less water was released into the air from their leaves to form rain-making clouds. The net result of these processes was the trend towards less annual rainfall, more soil erosion, and **desertification**. Other reasons for **famine** and desertification in the Sahel include political events (such as prolonged civil wars) and social changes (such as the breakdown of the old sustainable tribal systems of using the land).

[*Neil Cumberlidge Ph.D.*]

Resources
Books

Brown, L. R., ed. *State of the World 1986.* New York: Norton, 1986.

Gorse, J. E., and D. R. Steeds. *Desertification in the Sahelian and Sudanian Zones of West Africa.* Washington, DC: World Bank, 1987.

Myers, N., ed. *The Gaia Atlas of Planet Management.* London: Pan Books, 1985.

Southwick, C. H. *Global Ecology.* Sunderland, MA: Sinauer, 1985.

St. Lawrence Seaway

The St. Lawrence Seaway is a series of canals, locks, and lakes giving ocean-going ships of the Atlantic access to the **Great Lakes** and dozens of inland ports, including Toronto, Cleveland, Chicago, and Duluth. A herculean engineering project built jointly by the Canadian and United States governments, the St. Lawrence Seaway has become an essential trade outlet for the Midwest. It has also become an inlet for exotic and often harmful aquatic plants and animals, from which the landlocked Great Lakes were historically protected. The seaway's canals and locks, completed in 1959 after decades of planning and five years of round-the-clock construction, bypass such drops as Niagara Falls and allow over 6,000 ships to sail in and out of the Great Lakes every year.

The initial impetus for canal construction came from the steel industries of Ohio and Pennsylvania. During the first half of this century, these industries relied on rich ore supplies from the Mesabi iron range in northern Minnesota. The end of this source was in sight by the 1930s; mining engineers pointed to remote but rich iron deposits in Labrador as the next alternative. Lacking a cheap transport method from Labrador to the Midwest, steel industry backers began lobbying for the St. Lawrence Seaway. Midwestern grain traders and manufacturers joined the effort, promoting the route as a public works and jobs project, an economic booster for inland states and provinces, and a symbol of joint United States and Canadian cooperation, power, nationalism, and progress.

Unfortunately the seaway and its heavy traffic have also brought modern problems to the Great Lakes. Aside from the **petroleum** and chemical leaks associated with active shipping routes, the seaway opened the Great Lakes to aggressive foreign animal and plant invaders. The world's largest freshwater **ecosystem**, the Great Lakes, had been isolated from external invasion by steep waterfalls and sheer distance until the canals and locks opened. Since the 1960s, an increasing number of Atlantic, European, and Asian **species** have spread through the lakes. Some arrive under their own power, but most appear to have entered with ocean-going ships. Until recently, ships arriving at Great Lakes ports commonly carried freshwater ballast picked up in Europe or Asia. When the ships reached their Great Lakes destinations they discharged their ballast and loaded grain, ore, automobiles, and other goods. A profusion of **plankton**

and larvae riding in the ballast water thus entered a new **environment**; some of them thrived and spread with alarming speed.

One of the first **introduced species** to make its mark was the lamprey (*Petromyzon marinus*), an 18–in (45.7 cm) long, eel-shaped Atlantic species that attaches itself to the side of a fish with its circular, suction-like mouth. Once attached, the lamprey uses its rasping teeth to feed on the fish's living tissue, usually killing its host. After the lamprey's arrival in the 1960s, Lake Superior's commercial whitefish and trout fisheries collapsed. Between 1970 and 1980, Great Lakes trout and **salmon** populations plummeted by 90%. The region's $2 billion per year sport fishery nearly disappeared with **commercial fishing**. Since the 1970s, innovative control methods such as carefully-timed chemical spraying on spawning grounds have reduced lamprey populations, and native game fish have shown some recovery.

However, many other invaders have reached the Great Lakes via the seaway. Tubenose gobies (*Proterorhinus marmoratus*), a bottom-dwelling fish from the Black Sea, compete for food and spawning grounds with native perch and sculpins. The prolific North European ruffes (*Gymnocephalus cernuus*, a small perch) and spiny water flea (*Bythotrephes cederstroemi*, a tiny crustacean) also compete with native fish, and having hard, sharp spines, both are difficult for other species to eat. Asian invaders include the Asiatic clam (*Corbicula fluminea*), which colonizes and blocks industrial water outlets.

Most threatening to the region's businesses and economies is the **zebra mussel** (*Dreissena polymorpha*). Originating in the Baltic region, this tiny, extraordinarily prolific bivalve colonizes and suffocates industrial water pipes, docks, hard lake-bottom surfaces, and even the shells of native clams. Ironically, the zebra mussel has also benefited the lakes by raising public awareness of exotic invasions. Alarmed industries and states, faced with the enormous cost of cleaning and replacing zebra mussel-clogged pipes and screens, have begun supporting laws forcing incoming ships to dump their ballast at sea. Although reversals of exotic invasions are probably impossible, such remedial control measures may help limit damages. *See also* Biofouling; Exotic species; Introduced species; Parasites; Water pollution

[*Mary Ann Cunningham Ph.D.*]

Resources

Books

Cunningham, W. *Understanding Our Environment: An Introduction.* Dubuque, IA: William C. Brown, 1994.

Mabec, C. *The Seaway Story.* New York: Macmillan Co., 1961.

Periodicals

Raloff, J. "From Tough Ruffe to Quagga: Intimidating Invaders Alter the Earth's Largest Freshwater Ecosystem." *Science News* 142 (July 25, 1992): 56–8.

Other

St. Lawrence Seaway: 1991 Navigation Season. Ottawa: Saint Lawrence Seaway Authority.

Kirkpatrick Sale (1937 –)

American environmental writer

Kirkpatrick Sale is an influential environmental writer whose work has focused on the threat that a growing, resource-hungry population poses to the **environment**.

Sale was born in Ithaca, New York. His father was an English professor at Cornell University who was considered something of a campus rebel. Sale attended Swarthmore College for one year before transferring to Cornell in 1955, where he majored in history and edited the student newspaper. By the time he graduated in 1958, writing had become more important to him than history and he decided to pursue a career in journalism.

Sale first worked as an editor for the *New Leader*, an important leftist journal. During the early 1960s, he spent time in Africa, which he believed would become the center of world attention, and then worked briefly at the *New York Times Magazine*. By 1968, he had given up journalism and begun a career as a freelance writer. His first book, *SDS*, dealt with the radical Vietnam-era organization, Students for a Democratic Society. Working on the book, Sale later said, "radicalized me in a way beyond where I'd been." After *SDS* was published in 1972, Sale began work on *Power Shift*, an analysis of shifting political themes in America, published in 1975.

Power Shift was published at the height of the environmental movement in the United States. Like many, Sale had grown concerned about the future of a world in which an ethic of continuous progress and development required the continued consumption of **natural resources** at a terrifying rate. *In Human Scale* (1980) and *Dwellers in the Land* (1985), Sale focused on a concept that he defined as **bioregionalism**. He used the term to describe an ethic of living within the limitations of the environment. Society, he believed, must be a part of **nature**; human ends and means must accommodate nature, not the reverse.

The approaching quincentennial celebration of Columbus' arrival in the New World gave Sale the inspiration for his next book. For more than four years, he immersed himself in rereading source documents about the discoverer and his voyages. He began to think about the ways in which the transplantation of European culture and **environmental**

ethics had transformed the New World and contributed to the modern environmental crisis. The result of this work was *Conquest of Paradise: Christopher Columbus and the Columbian Legacy*. One conclusion Sale reached in this book was that the very cultures destroyed by the European invasion, those of the Native Americans, had practiced many of the environmental concepts to which modern societies must return if the world is to survive.

Although Sale has expressed interest in writing novels, his early books have been devoted to activist topics because "the world is in such a mess, and it needs writing about." He continues: "I write my books to help save society and the planet."

[*David E. Newton*]

RESOURCES
BOOKS

Sale, K. *Dwellers in the Land: The Bioregional Vision.* San Francisco: Sierra Club Books, 1983.

———. *Human Scale.* New York: Coward, McCann & Geoghegan, 1980.

———. *Rebels Against the Future: The Luddites and Their War on the Industrial Revolution: Lessons For the Computer Age.* Addison-Wesley, 1995.

———. *The Conquest of Paradise.* New York: Knopf, 1990.

———. *The Fire of His Genius: Robert Fulton and the American Dream.* New York: Free Press, 2001.

———. *The Green Revolution: The American Environmental Movement, 1962–1992.* Hill and Wang, 1993.

PERIODICALS

Baker, J. F. "Kirkpatrick Sale." *Publishers Weekly* (October 19, 1990): 41–42.

Saline soil

Soils containing enough soluble salts to interfere with the ability of plants to take up water. The conventional measurement that determines the **salinity** of **soil** is a *deciSiemens meter*; soils are considered saline if the conductivity of their saturation extract solution exceeds 4 deciSiemens meter-1. This unit of measure closely approximates the ionic salt concentration, and it is relatively easy to evaluate. The most common salts are composed of mixtures of sodium, calcium, and magnesium with chlorides, sulfates, and bicarbonates. Other less soluble salts of calcium sulfate and calcium and magnesium carbonate may be present as well. The **pH** is commonly less than 8.5.

In saline soils, there is often what is known as a perched water table—water close to the surface of the land. This phenomenon can be caused by restricting layers of fine clay within the soil, or by the application of waters at a rate greater than the natural permeability of the ground. When the **water table** is so close to the top of the soil, water is often transmitted to the surface where evaporation occurs, leaving the ions in the water to precipitate as salts. The resulting complex of salts can be seen on the surface of the soil as a white, crust-like layer.

The surface layers of a saline soil commonly have very good soil structure, and this is important to understand if the soil is to be reclaimed or leached of the high salt concentrations. This kind of structure has relatively large pores; water can flow quickly through them, which aids in carrying away salty water, thus making it easier to flush out the soil. Artificial drains can also be installed beneath the surface to provide an outlet for water trapped within the soil, and once these drains are in place, salt concentrations can be further reduced by the application of good quality water.

Excessive **irrigation** and the use of fertilizers and animal wastes can all increase the salinity of soil. The yields of common crops and the level of agricultural production are severely reduced in areas where salts have been allowed to accumulate in the soil. **Salinization** can be so severe in some cases that only salt-tolerant crops can be grown. **Leaching** is often necessary to reduce the levels of salt and keep the soil suitable for crop production. But this process can remove other soluble components from the soil and carry them into the **waste stream**, polluting both **groundwater** and surface water. The **environmental degradation** from this form of **agricultural pollution** can be extensive, and a method needs to be developed for leaching saline soils without these consequences. If this problem cannot be solved, it will no longer be possible to use some saline soils for agriculture. As **population growth** continues and the global demand for food increases, another approach might be the development of more salt-tolerant plant **species**.

[*Royce Lambert and Douglas Smith*]

RESOURCES
BOOKS

Brady, N. C. *The Nature and Properties of Soils.* 13th ed. New York: Macmillan, 2001.

Millar, R. W., and R. L. Danahue. *Soils: An Introduction to Soils and Plant Growth.* 6th ed. Englewood Cliffs, NJ: Prentice-Hall, 1990.

Salinity

A salt is a compound of a metal with a nonmetal other than **hydrogen** or oxygen. NaCl (sodium chloride), or table salt, is the best-known example. The solubility of various salts in water at a standard temperature is highly variable. When salts are dissolved in water, the result is called a saline solution, and the salinity of the solution is measured by its ability to carry an electrical current. Salinity of water is one of the components of **water quality**. Salinity is also measured

in **soil**. Soil can become salinized from water containing sufficient salts, from the natural degradation of soil minerals, or from materials added to the soil such as **fertilizer**. Large amounts of water may be required to leach the accumulated salts from the root zone and prevent reduced plant growth or plant desiccation. *See also* Soil profile; Water quality standards

Salinization

An increase in salt content, usually of agricultural soils, **irrigation** water, or drinking water is called salinization. Salinization is a problem because most food crops, like the human body, require fresh (nonsaline) water to survive. Although a variety of natural processes and human activities serve to raise the salt contents of **soil** and water, irrigation is the most widespread cause of salinization. Almost any natural water source carries some salts; with repeated applications these salts accumulate in the soil of irrigated fields. In **arid** regions, streams, lakes, and even aquifers can have high salt concentrations. Farmers forced to use such saline water sources for irrigation further jeopardize the fertility of their fields. In coastal areas, field salinization also results when seawater floods or seeps into crop lands. This occurs when falling water tables allow sea water to seep inland under ground, or where **aquifer subsidence** causes land to settle. Salinization also affects water sources, especially in arid regions where evaporation results in concentrated salt levels in rivers and lakes. Remedies for salinization include the selection of salt-tolerant crops and flood irrigation, which washes accumulated salts away from fields but deposits them elsewhere.

Most of the world's people rely on irrigated agriculture for food supplies. Regular applications of water, either from rivers, lakes, or underground aquifers, allow grain crops, vegetables, and fruits to grow even in dry regions. California's rich Central Valley is an outstanding example of irrigation-dependent agriculture. One of North America's primary gardens, the valley is naturally dry and sun baked. Canals carrying water from distant mountains allow strawberries, tomatoes, and lettuce to grow almost year round. The cost of this miraculous productivity is a gradual accumulation of salts, which irrigation water carries onto the valley's fields from ancient sea bed sediments in the surrounding mountains. Without heavy **flooding** and washing, Central Valley soils would become salty and infertile within a few years. Some of California's soils have become salty and infertile despite flooding. Similar situations are extremely widespread and have been known since humankind's earliest efforts in agriculture. The collapse of early civilizations in Mesopotamia and the Indus Valley resulted in large part from salt accumulation, caused by irrigation, which made food supplies unreliable.

Salts are a group of mineral compounds, composed chiefly of sodium, calcium, magnesium, potassium, sulfur, **chlorine**, and a number of other elements that occur naturally in rocks, clays, and soil. The most familiar salts are sodium chloride (table salt) and calcium sulfate (gypsum). These and other salts dissolve easily in water, so they are highly mobile. When plenty of water is available to dilute salt concentrations in water or wash away salt from soil, these naturally occurring compounds have little impact. Where salt-laden water accumulates and evaporates in basins or on fields, it leaves behind increasing concentrations of salts.

Most crop plants exposed to highly saline environments have difficulty taking up water and nutrients. Healthy plants wilt, even when soil moisture is high. Leaves produced in saline conditions are small, which limits the photosynthetic process. Fruits, when fruiting is successful, are also small and few. Seed production is poor, and plants are weak. With increasing **salinity**, crop damage increases, until plants cannot grow at all. Water begins to have a negative effect on some crops when it contains 250–500 **parts per million** (ppm) salts; highly saline water, sometimes used for irrigation out of necessity, may contain 2,000–5,000 ppm or more. For comparison, sea water has salt concentrations upwards of 32,000 ppm. In soil, noticeable effects appear when salinity reaches 0.2%; soil with 0.7% salt is unsuitable for agriculture.

Another cause of soil salinization is subsidence. When water is pumped from underground aquifers, pore spaces within rocks and sediments collapse. The land then compacts, or subsides, often lowering several meters from its previous level. Sometimes this **compaction** brings the land surface close to the surface of remaining **groundwater**. Capillary action pulls this groundwater incrementally toward the surface, where it evaporates, leaving the salts it carried behind in the soil. Near coastlines such processes can be especially severe. Seawater often seeps below the land surface, especially when fresh-water aquifers have been depleted. When seawater, with especially high salt concentrations, rises to the surface it evaporates, leaving crystalline salt in the soil.

In such cases, the salinization of the aquifer itself is also a serious problem. Many near-shore aquifers are threatened today by seawater invasions. Usually seawater invasions occur when farms and cities have extracted a substantial amount of the aquifer's water volume. Water pressure falls in the fresh-water aquifer until it no longer equals pressure from adjacent sea water. Sea water then invades the porous aquifer formation, introducing salts to formerly fresh water supplies.

Rivers are also subject to salinization. Both cities and farms that use river water return their **wastewater** to the

river after use. Urban storm sewers and **sewage treatment** plants often send poor quality water back to rivers; **drainage** canals carry intensely saline **runoff** from irrigated fields back to the rivers that provided the water in the first place. When **dams** block rivers, especially in dry regions, millions of cubic meters of water can evaporate from reservoirs, further intensifying in-stream salt concentrations.

The **Colorado River** is one familiar example out of many rivers suffering from artificial salinization. The Colorado, running from Colorado through Utah and Arizona, used to empty into the Sea of Cortez south of California's Imperial Valley before human activities began consuming the river's entire **discharge**. Farms and cities in adjacent states consume the river's water, adding salts in wastewater returned to the river. In addition, the Colorado's two huge reservoirs, Lake Powell and Lake Mead, lie in one of the continent's hottest and driest regions and lose about 10% of the river's annual flow through evaporation each year. By the time it reaches the Mexican border, the river contains 850 ppm salts, too much for most urban or agricultural uses. Following a suit from Mexico, the United States government has built a $350 million **desalinization** plant to restore the river's **water quality** before it leaves Arizona. The Colorado's story is, unfortunately a common one. Similar situations abound on major and minor rivers from the Nile to the Indus to the Danube.

Salinization occurs on every occupied continent. The world's most severely affected regions are those with arid climates and long histories of human occupation or recent introductions of intense agricultural activity. North America's Great Plains, the southwestern states, California, and much of Mexico are experiencing salinization. Pakistan and northwestern India have seen losses in agricultural productivity, as have western China and inland Asian states from Mongolia and Kazakhstan to Afghanistan. Iran and Iraq both suffer from salinization, and salinization has become widespread in Africa. Egypt's Nile valley, long northern Africa's most bountiful bread basket, also has rising salt levels because of irrigation and subsidence. One of the world's most notorious case histories of salinization occurs around the **Aral Sea**, in southern Russia. This inland basin, historically saline because it lacks an outlet to the sea, is fed by two rivers running from northern Afghanistan. Since the 1950s, large portions of these rivers' annual discharge has been diverted for cotton production. Consequently, the Aral Sea is steadily drying and shrinking, leaving great wastes of salty, dried sea bottom. Dust storms crossing these new deserts carry salts to both cotton and food crops hundreds of miles away.

Avoiding salinization is difficult. Where farmers have a great deal of capital to invest, as in California's Central Valley and other major agricultural regions of the United States, irrigators install a network of perforated pipes, known as tiles, below their fields. They then flood the fields with copious amounts of water. Flooding washes excess salts through the soil and into the tiles, which carry the hypersaline water away from the fields. This is an expensive method that wastes water and produces a toxic brine that must be disposed of elsewhere. Usually this brine enters natural rivers or lakes, which are then contaminated unless their volume is sufficient to once again dilute salts to harmless levels. However this method does protect fields. More efficient irrigation systems, with pipes that drip water just near plant roots themselves may be an effective alternative that contaminates minimal volumes of water.

Water can also be purified after agricultural or urban use. Purification, usually by reverse osmosis, is an expensive but effective means of removing salts from rivers. The best way to prevent water salinization is to avoid dumping urban or irrigation wastes into rivers and lakes. Equally important is avoiding evaporation by reconsidering large dam and **reservoir** developments. Unfortunately, most societies are reluctant to consider these options: reservoirs are widely viewed as essential to national development, and wastewater purification is an expensive process that usually benefits someone else downstream.

Perhaps the best way to deal with salinization is to find or develop crop plants that flourish under saline conditions. Governments, scientists, and farmers around the world are working hard to develop this alternative. Many wild plants, especially those native to deserts or sea coasts, are naturally adapted to grow in salty soil and water. Most food plants on which we now depend—wheat, rice, vegetables, fruits—originate in nondesert, nonsaline environments. When domestic food plants are crossed with salt-tolerant wild plants, however, salt-tolerant domestics can result. This process was used to breed tomatoes that can bear fruit when watered with 70% seawater. Other vegetables and grains, including rice, barley, millet, asparagus, melons, onions, and cabbage, have produced such useful crossbreeds.

Equally important are innovative uses of plants that are naturally salt tolerant. Some salt-adapted plants already occupy a place in our diet—beets, dates, quinoa (an Andean grain), and others. Furthermore, careful allocation of land could help preserve remaining salt-free acreage. Planting salt-tolerant fodder and fiber crops in soil that is already saline can preserve better land for more delicate food crops, thus reducing pressure on prime lands and extending soil viability. *See also* Salinization of soils

[*Mary Ann Cunningham Ph.D.*]

Resources
Books

Frenkel, H., and A. Meiri, eds. *Soil Salinity: Two Decades of Research in Irrigated Agriculture.* New York: Van Nostrand Reinhold, 1985.

National Research Council. *Saline Agriculture.* Washington, DC: National Research Council, 1990.

Scabocs, I. *Salt-Affected Soils.* Boca Raton, FL: CRC Press, 1989.

Shainberg, I., and J. Shalhevet. *Soil Salinity Under Irrigation.* Berlin: Springer-Verlag, 1984.

Salinization of soils

Salinization of **soil** involves the processes of salt accumulation in the upper rooting zone so that many plants are inhibited or prohibited from normal growth. Salinization occurs primarily in the semi-arid and **arid** portions of the earth. Salinization is commonly thought to occur only in the hot climatic regions, but may be found in cooler to cold portions of the earth where precipitation is very limited. Where the annual precipitation exceeds about 20 inches (500 mm), there is usually adequate downward movement of salts through **leaching** to prohibit the development of **saline soil**. Occasionally salinization of soil will occur where there has been an inundation of land by sea water. Some areas of the world were once under sea water, but due to uplift the land is now many feet above sea level. These lands are common sources of "ancient salts" and may lead to the development of saline waters as these soils are slowly leached.

Natural sources of salts in soil are primarily from the **decomposition** of rocks and minerals through the processes of chemical **weathering**. With adequate amounts of moisture present, hydrolysis, hydration, solution, oxidation, and carbonation cause the minerals to decompose and release the ionic constituents to form various salts. The more common cations are sodium, calcium, and magnesium with lesser amounts of potassium and boron. The common anions are chloride and sulfate with lesser amounts of bicarbonate, carbonate, and occasionally nitrate.

Inherent factors of the landscape may also lead to salinization of soils. One of the more common factors is the restriction of water **drainage** within the soil, often referred to as soil permeability. Soil becomes less **permeable** because of genetic or inherited layers within the **soil profile** that have relatively high amounts of clay. Because of the extremely small sizes of the clay particles, the natural pore size is also very small, frequently being less than 0.0001 mm average diameter. The natural tortuosity and small size combine to severely reduce the rate of water movement downward which tends to increase salt buildup.

In addition, because of the soil pore size, saline waters may be transported upward through the processes of capillary action. While the rate of upward movement of water may be relatively slow, the process can deliver saline water from several inches below the surface of the land and create a zone of highly saline soil at the surface of the earth.

Salinization of soils has also occurred in areas where **irrigation** waters have been applied to lands that have not been subject to long-term natural leaching by rainfall. In several cases the application of relatively good quality water to arid soils with poor internal drainage has caused the soil to develop a higher **water table** which in turn has permitted the salts within the soil to be carried to the surface by high evaporation demand. In other cases water of relatively high salt content has been applied to soils without proper drainage, resulting in the development of saline soils. As the water is evaporated from the soil or the plant surfaces, the salts are left to accumulate in the soil.

Because of the increasing demand for food and fiber, many marginal lands are being brought into agronomic production. Proper irrigation management and drainage decisions must be considered on a worldwide and long-term basis to avoid a deleterious influence on the ability to produce on these soils and to determine the appropriate environmental considerations for disposal of salts from salinized soils. *See also* Arable land; Nitrates and nitrates; Runoff; Soil conservation; Soil eluviation; Soil illuviation

[*Royce Lambert*]

Resources

Books

Foth, H. D. *Fundamentals of Soil Science.* 7th ed. New York: Wiley, 1984.

Singer, M. J., and D. N. Munns. *Soils: An Introduction.* 5th ed. New York: Macmillan, 2001.

Other

Richards, L. A., ed. *Diagnosis and Improvement of Saline and Alkali Soils.* USDA Handbook 60. Washington, DC: U.S. Government Printing Office, 1964.

Salmon

Salmon is a popular fish for food and sport fishing. Five **species** of salmon live in the North Pacific Ocean: Pink, Sockeye, Coho, Chum, and Chinook. One species, the Atlantic salmon, lives in the North Atlantic Ocean. Two other fish species that are also members of the *Salmonidae* fish family—steelhead and sea-run cutthroat trout—live in the Pacific Northwest. The Pacific Coast salmon populations are being threatened with extirpation from much, if not all, of their range.

At the heart of the Pacific salmon species' range, and perhaps indicative of the heart of its problems, is the Columbia River basin. Covering parts of seven states and two Canadian provinces, the Columbia River system contains over 100 **dams**, 56 of which are major structures, including 19 major generators of hydroelectric power. These structures present an insurmountable obstacle for these migrating

fishes. Adult salmon, after growing and maturing in the ocean, return to the freshwater stream of their origin as they swim upstream to spawn. The adults will die shortly after this culmination of their arduous journey, and, after hatching, the young salmon—called smolts—swim downstream to the ocean to continue this life cycle.

About three-fourths of all of the population declines of salmon are directly attributable to hydroelectric dams. The dams simply do not allow a majority of these fish to successfully complete their **migration**, and many salmon die as they swim, or are swept, directly into the turbines. Fish ladders, stepped pools intended as an aid for fish to bypass the dams, enable some salmon to continue their journey, but many do not find their way through. As they move downstream, the smolts are slowed or stopped by the reservoirs created by the dams. Here they are exposed to larger populations of predators than in their natural riverine **habitat**. They are exposed to a wide variety of pathogens as well as a physical **environment** of warmer, slow moving waters, to which they are only moderately tolerant. Only 20% of the downstream migrants ever make it to the Pacific. Poor **water quality** and **nutrient** deficits in many stretches of the Columbia River also takes a toll on the young smolts.

Overfishing, both offshore and along the rivers, contributes to the decline of salmon populations. Fishery biologists have attempted to offset these losses of native stocks by releasing hatchery-raised salmon. However, interbreeding reduces the genetic hardiness of these fish. They also weaken the genetic lines of wild fish when they breed with them. Historically, when hatchery programs have increased the mixed stock fish population (i.e., wild and farmed) dramatically, fishing activity increases and a further depletion of wild salmon results.

To address overfishing issues, in 1996 Washington State began using a mass-marking program designed to enable fishermen to more easily identify hatchery chinook and coho salmon. Fish bred in a hatchery have their adipose fin (a small tail fin) removed before they are released. Anyone who catches a marked hatchery fish may keep it; wild salmon that are unmarked must be released back into the wild.

Realizing the need for a more balanced management plan for the Columbia River basin, Congress passed the Northwest Power Act in 1980. This act established the Northwest Power Planning Council (NWPPC), which was charged with the task of balancing long-term hydroelectric energy needs with minimizing the negative impact of dams on native salmon populations. However, despite modifications to water flow along the Snake and Columbia Rivers instituted by the NWPPC, the native salmon population continued to decline throughout the 1980s.

In 1991, the National Marine Fisheries Service (NMFS) began an extensive study of salmon populations in the northwestern United States. The NMFS found that 52 distinct populations (termed Evolutionarily Significant Units, or ESUs) of Pacific salmon have been identified in west coast states. That same year, Snake River stocks of sockeye salmon were first listed under the **Endangered Species Act** (ESA) as endangered. Twenty-six salmon ESUs are now listed as threatened or endangered status.

In June 2000, the NMFS adopted a rule prohibiting the killing or injuring of 14 ESUs of Pacific salmon and steelhead classified as threatened under the Endangered Species Act (ESA). This "take" rule was adopted under section 4(d) of the ESA. The rule does allow for the removal of ESA-listed salmon in association with approved programs such as scientific research and tribal fishing rights.

Various species of salmon have been added to the ESA list in the past several decades. As of May 2002, Atlantic salmon were included on the ESA with an endangered status, while chum and coho were listed as threatened. Dual status ESA species (endangered in one part of their range and threatened in another) include chinook salmon, sockeye salmon, and steelhead.

In late 2001, a U.S. district court ruled that the NMFS listing of Oregon coast coho salmon as endangered was "arbitrary and capricious" (*Alsea Valley Alliance v. Evans*). The court determined that excluding hatchery stock from the population assessment of this species, as NMFS had done, was inappropriate. This ruling could have far-reaching implications for other salmon stocks listed as endangered or threatened; after the *Alsea* ruling, six delisting petitions were filed by farming **irrigation** groups and other agencies requesting the removal of additional ESUs from endangered status. As of May 2002, NMFS was appealing the *Alsea* decision, but had also announced status reviews on fourteen endangered salmon ESUs. The Oregon coho remains on the endangered list pending the decision of the appeal.

Long-term, sustained population recovery for these ecologically, as well as economically, important salmon populations will depend on changes in both habitat and human behavior. More water is needed downstream to aid migration. This would mean less water for irrigation and for hydroelectric-generated power. Increased water flow through releases from reservoirs and spillway openings in the hydroelectric dam system has been shown to improve salmon survival rates in the Snake River. Balancing the power requirements of the Pacific Northwest with the habitat needs of salmon species will be a key part of ensuring their continued survival.

[*Eugene C. Beckham and Paula A. Ford-Martin*]

Resources

Books

Lichatowich, Jim. *Salmon Without Rivers: A History of the Pacific Salmon.* Washington, DC: Island Press, 2001.

Taylor, Joseph. *Making Salmon: An Environmental History of the Northwest Fisheries Crisis.* Seattle, WA: University of Washington Press, 2001.

Periodicals

Curtis, S. "Power Plan Trumps Salmon Recovery." *Field & Stream* 106, no.1 (May 2001): 16.

Gresh, Ted, J. Lichatowich, and P. Schoonmaker. "An Estimation of Historic and Current Levels of Salmon Production in the Northeast Pacific Ecosystem." *Fisheries* 25, no.1 (January 2000): 15–21.

Other

Northwest Fisheries Science Center, National Marine Fisheries Service. *Salmonid Travel Time and Survival Related to Flow in the Columbia River Basin.* March 2000 [cited May 2002]. <http://www.nwfsc.noaa.gov/pubs/nwfscpubs.html>.

Northwest Salmon Recovery Planning. [cited May 31, 2002]. <http://research.nwfsc.noaa.gov/cbd/trt/index.html>.

Organizations

Northwest Fisheries Science Center, NMFS, NOAA, 2725 Montlake Blvd. E, Seattle, WA USA 98112 (206) 860-3200, Email: NWFSC.Webmaster@noaa.gov, <http://www.nwfsc.noaa.gov/>

Henry S. Salt (1851 – 1939)

English writer and reformer

Although he is probably best known as the vegetarian author of *Animals' Rights Considered in Relation to Social Progress* (1892), Henry Salt was a man with many roles—teacher, biographer, literary critic, and energetic advocate of causes he grouped under the title of humanitarianism. These included pacifism and socialism as well as **vegetarianism**, anti-vivisectionism, and other attempts to promote the welfare of animals. Late in his life, Salt also advocated efforts to conserve the beauties of **nature**, especially wildflowers and mountain districts.

Salt was born in India in 1851, the son of a colonel in the Royal Bengal Artillery. When his parents separated a year later, Salt's mother moved to England, and he spent much of his childhood at the home of her well-to-do parents. Educated at Eton and Cambridge, he embarked on an apparently comfortable career when he returned to Eton as an assistant master. While teaching at Eton, however, Salt met some of the leading radicals and reformers of the day, including William Morris, John Ruskin, and George Bernard Shaw, with whom he formed a lasting friendship. During this period Salt became a vegetarian. He also read **Henry David Thoreau's** *Walden,* which inspired Salt and his wife to leave Eton and move to a country cottage in 1885 to lead a simple, self-sufficient life. For Salt, the simple life was far from dull. He became a prolific writer, although his efforts brought little financial profit. In addition to *Animals' Rights,* his works include a biography of Thoreau, studies of Shelley and Tennyson, a translation of Virgil's *Aeneid,* and an autobiography, *Seventy Years Among Savages* (1921), in which he examined English life as if he were an anthropologist studying a primitive tribe. Another book, *A Plea for Vegetarianism,* came to the attention of **Mohandas Karanchand Gandhi** when he was a student in London. Gandhi, who was raised a vegetarian, later wrote in his *Autobiography* that Salt's book made him "a vegetarian by choice." Salt apparently met Gandhi in 1891, then corresponded with him in 1929 during Gandhi's nonviolent struggle for the independence of India—a struggle inspired, in part, by Thoreau's essay, "Civil Disobedience."

Perhaps the best statement of Salt's views is the address he wrote to be read at his funeral. Declaring himself "a rationalist, socialist, pacifist, and humanitarian," Salt disavowed any belief "in the present established religion," but acknowledged "a very firm religious faith" in "a Creed of Kinship:" "a belief that in years yet to come there will be a recognition of the brotherhood between man and man, nation and nation, human and sub-human, which will transform a state of semi-savagery...into one of civilization, when there will be no such barbarity as warfare, or the robbery of the poor by the rich, or the ill-usage of the lower animals by mankind."

[*Richard K. Dagger*]

Resources

Books

Hendrick, G., and W. Hendrick, eds. *The Savour of Salt: A Henry Salt Anthology.* Fontwell, Sussex: Centaur Press, 1989.

Periodicals

Jolma, D. J. "Henry Salt and 100 Years of Animal Rights." *The Animals' Agenda* (November–December 1992): 30–2.

Salt marsh

see **Wetlands**

Salt (road)

While several **chemicals** are available for deicing winter roads, common salt (NaCl) is most frequently used. Approximately 20 billion lb (9 billion kg) of salt are used each year in the United States for treating ice and snow on roads. Calcium chloride ($CaCl_2$), potassium chloride (KCl), and urea are also available but used in smaller quantities. Common salt is preferred because it is cheaper per pound and more effective. While the price per pound of salt is cheap, about 1.5 billion dollars are spent per year on the enormous quantity used, and its distribution and application.

Salt in its solid form does not melt ice. It first must go into solution to form a brine, and the brine effects a melting of ice and snow. Ordinarily, snow is packed on the road surface by vehicular traffic; known as the "hard-pack," it forms a bond with the underlying pavement that is frequently impossible to remove with snowplows. Salt melts through the hard-pack and breaks the ice-pavement bonding. Traffic breaks the loosened ice, and snowplows are then able to remove the broken ice and packed snow. Deicing facilitates traffic movement after a snow fall and it is thought to make winter driving safer.

But there are hidden costs in the use of salt. It corrodes steel, and it is estimated that about three billion dollars is spent annually to protect vehicles from rust with corrosion-resistant coatings. Some have added to this estimate the costs of frequent washings to save vehicles from rust. The expense of salt deicing does not stop, however, with motor vehicle protection and damage. The United States has about 500,000 bridges, of which an estimated 40% are currently considered deficient. Damage to bridges comes from a variety of causes, but many experts consider the most significant agent for premature deterioration is deicing salt. It is estimated that repair and protection to damaged bridges in the United States in the next decade will cost between one-half and two-thirds of a billion dollars. Salt damage is also causing bridges in Great Britain to deteriorate more rapidly than expected and is believed to be the principal cause of bridge damage with an anticipated cost of repair in the next decade of a half billion pounds.

Salt damages roadside vegetation, and it has been shown that waters downstream from a deiced highway contain significantly more (in one case 31 times as much) chloride than waters in the same rivers upstream. Well water can similarly become contaminated with salt, and both Massachusetts and Connecticut have large numbers of **wells** with a sodium content in excess of 20 mg per liter, which is considered to be the upper limit for individuals who must control sodium intake. One area of Massachusetts has quit using deicing salt because of concern for sodium in their drinking water.

It would be useful to develop a deicer that has less impact on the economy and the **environment**. One such alternative is calcium magnesium acetate (CME) which, while initially far more costly than salt, is believed to have less potential for damaging the environment. *See also* Automobile; Groundwater pollution; Salinity; Salinization

[*Robert G. McKinnell*]

Resources

Periodicals

Boice, L. P. "Environmental Management: CMA, An Alternative to Road Salt?" *Environment* 28 (1985): 45.

"Deicing Agents: A Primer." *Public Works* (July 1991): 50–51.

Dunker, K. F., and B. G. Rabbat. "Why America's Bridges Are Crumbling." *Scientific American* 268 (March 1993): 66–72.

"Highway Deicing: Comparing Salt and Calcium Magnesium Acetate." *TR News* 163 (November–December 1992): 17–19.

Reina, P. "Salt Breaks the Back of Motorway Bridges." *New Scientist* (April 29, 1989): 30.

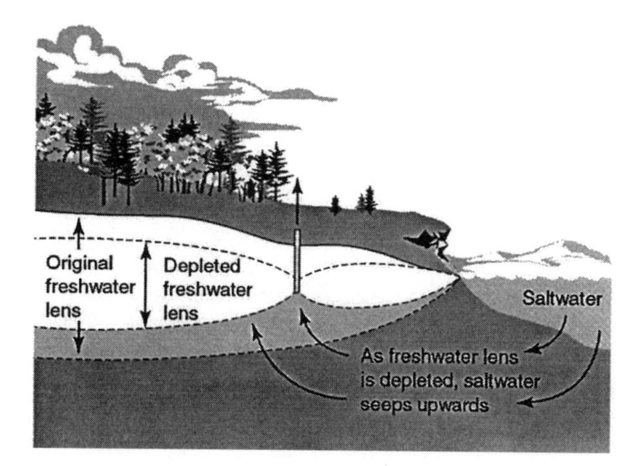

Saltwater intrusion into a coastal aquifer due to depletion of the freshwater in the aquifer. (Illustration by Hans & Cassidy.)

Salt water intrusion

Aquifers in coastal areas where fresh **groundwater** is discharged into bodies of salt water such as oceans are subject to salt water intrusion. Intrusion occurs when water usage lowers the level of freshwater contained in the **aquifer**. The natural gradient sloping down toward the ocean is changed, resulting in a decrease or reversal of the flow from the aquifer to the salt water body, which causes salt water to enter and penetrate inland. If salt water travels far enough inland well fields supplying freshwater can be ruined and the aquifer can become so contaminated that it may take years to remove the salt, even with fresh groundwater available to flush out the saline water.

Salt water intrusion can also develop where there is artificial access to salt water, such as sea level canals or **drainage** ditches. On the coastal perimeter of the United States there are a number of areas with such intrusion problems. There are five methods available to control intrusion: 1) the reduction or rearrangement of the pumping draw-down; 2) direct recharge; 3) the development of a pumping trough adjacent to the coast; 4) maintenance of a freshwater ridge above sea level along the coast; and 5) construction of artificial subsurface barriers.

Developing a pumping trough or maintaining a pressure ridge are methods that do not solve the basic problem: the overdraft or excessive withdrawal of water from the aquifer. Only when this problem is solved by reducing drawdown, or isolated by the subsurface barrier method, can the intrusion be stopped or reversed. In some areas, it is simply not economically feasible to control intrusion.

Oceanic islands, such as Hawaii, have unique problems with salt water intrusion. Most islands consist of sand, lava, coral, or limestone; they are relatively **permeable**, so freshwater is in contact with salt water on all sides. Because fresh water is supplied entirely by rainfall in these places, only a limited amount is available. A freshwater lens is formed by movement of the freshwater toward the coast. Depth to salt water at any location is a function of rainfall recharge, the size of the island, and permeability. Tidal and seasonal fluctuations may form a transition zone between freshwater and salt water.

The close proximity of salt water in oceanic islands can introduce saline water into fresh groundwater even without excessive overdraft. An island well pumping at a rate high enough to lower the **water table** disturbs the fresh-salt water equilibrium, and salt water then rises as a cone within the well. To avoid this condition, island **wells** are designed for minimum drawdown, and they skim freshwater from the top of the aquifer. In general, drawdowns of a few inches to a few feet provide plentiful water supplies.

[*James L. Anderson*]

RESOURCES

PERIODICALS

Gorrie, P. "Water From the Ground." *Canadian Geographic* 112 (September–October 1992): 68–77.

Wagman, D. "Protecting Hidden Assets: Many American Cities and Counties Are Designing Groundwater Regulations to Protect Underground Aquifers." *American City and County* 105 (March 1990): 38–41.

Sand dune ecology

Dunes are mounds of sand that have been piled by the action of winds. The sand is usually composed of bits of minerals that have been eroded from rocks, picked up by water or winds, and then re-deposited somewhere else. Typically, the sand is deposited behind some object that is a barrier to the movement of air currents, which causes the windspeed to slow suddenly so that the load of sand particles can no longer be contained against the force of gravity, and it falls to the ground. If there is a suitable source of sand, dunes may deposit along the edges of oceans, large lakes, rivers, and even in inland locations.

The mineralogical composition of sand in dunes varies greatly from place to place. The most common minerals are quartzitic (this is a silica sand), but other minerals may also be mixed in, and they may dominate the composition of the sand in some places. The White Sands National Monument in New Mexico, for example, is characterized by a sand of gypsum (calcium sulfate). The grain size of sands varies depending on the strength of the winds that delivered the sand to the location where the dune has formed—the greater the typical windspeed, the larger the sand grains can be.

Once deposited to the surface, unconsolidated sand tends to be moved about and shifted by the winds. In fact, dunes tend to "migrate" in the direction of the prevailing winds, with sand being picked up at the windward side of the dune, and deposited on the leeward side. Because of the mechanical instability of their surface young dunes initially provide rather precarious habitats for the establishment of vegetation. However, if plants do manage to establish a foothold on a dune, the progressive development of vegetation over time will help to stabilize the surface, allowing an **ecosystem** to develop through the process of **succession**.

Some of the most extensive dune systems in the world occur in deserts, including parts of the Sahara **Desert** of northern Africa, the Gobi Desert of eastern Asia, and the Great Western Desert of the United States and Mexico. Because of the aridity of those environments only a very limited amount of ecological development is possible. In places with a more moderate **climate**, however, sand dunes can be colonized by vegetation and succession can proceed to the development of tall, mixed-species forests. This type of succession is typical of many dune systems along the coasts of the oceans and many large lakes and rivers of North America.

Environmental conditions

Sandy soils are generally free-draining, so that even in regions with abundant rainfall sand-dune habitats can be quite **arid**. The dry **soil** conditions are made worse by the frequently windy conditions in dune habitats, a factor that speeds the evaporation of water from plant foliage and the soil surface. Droughty conditions are especially important on the higher parts of dunes; in low places the **water table** may come to the surface, creating locally moist or wet conditions, which are sometimes referred to as dune slacks.

In addition, most sands contain few nutrients, so they provide rather infertile conditions for plant growth. Sandy soils are also lacking in organic matter, which prevents them from holding much water or nutrients for later uptake by plants. In fact, some sand-dune habitats can be referred to as ombrotrophic, or "fed from the clouds," which means that **atmospheric deposition** is the principal source of **nutrient** inputs in support of plant productivity. In addition, the low levels of calcium concentrations make sandy soils highly

Moving sand dunes are slowly engulfing a forest in Nags Head, North Carolina. (Photograph by Jack Dermid. National Audubon Society Collection/Photo Researchers Inc. Reproduced by permission.)

vulnerable to **acidification**, a condition that can be stressful to some **species** of plants.

Vegetation of sand dunes

The species of plants that grow on sand dunes vary depending on the climate, the **stability** of the dunes, soil chemistry, and **biogeography**. Dunes of the deserts of southern California, southwestern Arizona, and nearby Mexico are sparsely vegetated with a mixture of perennial grasses, such as *Pleuraphis rigida*, perennial herbs such as the sand sagebrush (*Artemisia filifolia*), and species of shrubs, including ephedra (*Ephedra trifurca*), daleas (*Dalea* spp.), Colorado desert buckwheat (*Eriogonum deserticola*), and mesquite (*Propsis spp.*). During occasional rainy periods these sandy deserts can bloom prolifically with showy annual herbs, such as species of *Allionia*, *Boerhavia*, *Euphorbia*, *Oenothera*, and *Pectis*. Areas of gypsum-rich sand have other species that are specific to that soil type, including yucca (*Yucca elata*), ephedra (*Ephedra torreyana*), onion blanket (*Gaillardia multiceps*), and grasses such as *Bouteloua breviseta*.

In more humid climatic regimes of North America, younger sand-dune systems are initially colonized by species of dune-grasses, such as American beach grass (*Ammophila breviligulata*), sea oats (*Uniola paniculata*), sand rye (*Elymus mollis*), and beach grass (*Calamovilfa longifolia*). As the dunes stabilize, other species invade the community. These may form a **prairie** with interspersed shrub- and tree-dominated nuclei, which may eventually coalesce to form continuous forest (see the next section on dune succession). Some dune-rich areas of the east coast of the United States support forests of pines (*Pinus* spp.), oaks (*Quercus* spp.), and magnolia (*Magnolia virginiana*). Older dunes in the **Great Lakes** region tend to be covered by mixed stands of oaks, pines, and tulip-tree (*Liriodendron tulipifera*). The inland sand hills of northern Alberta are commonly occupied by stands of jack pine (*Pinus banksiana*).

Primary succession on sand dunes

Succession on recently formed sand dunes is a type of primary succession. This is because the sandy substrate that is newly exposed for ecological development has no pre-existing biological capability to initiate the succession. Succession can not begin until **microbes**, plants, and animals have invaded the site from existing ecosystems somewhere else. (Secondary succession occurs after a disturbance, but one that has allowed some of the original biota to survive, so they can play a prominent role in ecological recovery.)

Some classic studies of primary succession have involved sand-dune ecosystems. In North America, the most famous studies were conducted on systems of dune ridges at several places on Lake Michigan and Lake Huron. Dune ridges are actively developing in those places to this day. This occurs as sand is piled up onto the beach by wave action, to be moved inland by the prevailing on-shore winds and mostly deposited on the dune closest to the edge of the lake, on its leeward, or building side. This process of dune-building is an ancient phenomenon, and it has been actively occurring in the region since deglaciation, more than twelve thousand years ago.

The dune building is, however, superimposed upon another geological process—the slow rebounding in elevation of the regional ground surface, a phenomenon that is known as isostasy. This slow, uplifting process is still occurring in response to the long-ago melting of the continental glaciers that once covered virtually all of northern North America; when the glaciers were present, their enormous weight actually depressed crustal bedrocks beneath them into Earth's plastic mantle. The surface has been rebounding since deglaciation released the crust from the weight of the gigantic glaciers, at a fairly steady rate of about 1 yd (1 m) per century. Because of isostatic rebound, ancient shorelines of the Great Lakes are now higher in elevation than they used to be, and in some flatter areas well-defined dune systems formed on ancient shores can be found miles inland of the present shoreline of the lakes. These circumstances mean that dunes located further inland are older than dunes occurring closer to the modern lake shores. Because the ages of the various dunes can be dated, ecologists are able to study a series of dunes of various age and their associated communities. This so-called chronosequence technique allows the ecologists to reconstruct the basic elements of the successional process.

The first plants to colonize the embryonic dunes are annual and biennial plants of shoreline habitats, such as the American searocket (*Cakile edentula*), Russian thistle (*Salsola kali*), orache (*Atriplex patula*), and a small euphorb (*Eurphorbia polygonifolia*). Although buffeted by severe weather, these plants help in the initial stages of dune stabilization. The next invaders of the site are several species of dune grasses, particularly *Ammophila breviligulata* and *Calamovilfa longifolia*. The extensive roots and rhizomes of these grasses are important in binding the sandy substrate, and in stabilizing the lakeward dune ridges and allowing them to grow larger. Once these grasses take hold, some other foredune plants can establish and begin to grow, including the beach pea (*Lathyrus maritimus*) and evening primrose (*Oenothera biennis*). With time, a tall grass prairie develops, dominated by various species of grasses and broad-leaved herbs. The prairie becomes invaded by shade-intolerant shrub and tree species, which form forest nuclei. Eventually a climax forest develops that is dominated by oaks, pines, and tulip-tree.

[*Bill Freedman Ph.D.*]

Resources

Books

Barbour, M. G., and W. D. Billings. *North American Terrestrial Vegetation.* Cambridge, UK: Cambridge University Press, 1988.

Ranwell, D. S. *Ecology of Salt Marshes and Sand Dunes.* Chapman and Hall, London, 1972.

Periodicals

Morrison, R. G., and G. A. Yarranton. "Diversity, Richness, and Evenness during a Primary Sand Dune Succession at Grand Bend, Ontario," *Canadian Journal of Botany* 51 (1973): 2401–2411.

Olson, J. S. "Rates of Succession and Soil Changes on Southern Lake Michigan Sand Dunes." *Botanical Gazette* 119 (958): 125–170.

Sanitary landfill

see **Landfill**

Sanitary sewer overflows

Sanitary sewer overflows (Ssos) are discharges of untreated sewage from municipal sanitary sewer systems that were not designed to carry storm-water **runoff**. Almost all sewer systems experience at least occasional Ssos and they occur frequently in some systems. Ssos are a major source of **water pollution** in lakes, rivers, and streams. Although they are illegal under the **Clean Water Act**, the U.S. **Environmental Protection Agency** (EPA) estimates that at least 40,000 Ssos occur annually nationwide.

Of the approximately 50 trillion gal (189 trillion l) of raw sewage that flow daily through about 19,500 sewer systems in the United States, it is estimated that about 1.2 billion gal (4.5 billion l) were released in Ssos in 2000. This raw sewage is discharged from manholes, bypassing pump stations and treatment plants. It flows into basements, lawns, streets, parks, streams, swimming areas, and drinking water. It is estimated that sewers back up into basements 400,000 times a year in the United States.

Ssos pose a serious threat to human health and to the **environment**. Raw sewage can introduce disease-causing organisms into drinking water and swimming and fishing areas. People may swallow contaminated water or eat contaminated fish or shellfish. Diseases also can be contracted by breathing in organisms from sewage or by **absorption** through the skin. Although most such illnesses are mild

stomach upsets, there is a risk of life-threatening diseases such as **cholera**, dysentery, and hepatitis. Ssos are a major cause of beach closures. In 2000, an Sso discharged an estimated 72 million gal (273 million l) of raw sewage into the Indian River, causing drinking-water advisories and beach closures throughout much of Florida. Sewage contamination can rob water of oxygen and result in floating debris and harmful blooms of algae. Sewage from Ssos can destroy **habitat** and harm plants and animals.

Ssos often occur when excess rainfall or snowmelt in the ground enters leaky sanitary sewers that were not built to hold rainfall or which do not drain properly. Excess water also can enter sewers from roof drains that are connected to sewers, through broken pipes, or through poorly connected service lines. In some cities, as many as 60% of Ssos originate at service connections to buildings. Sometimes new subdivisions or commercial buildings are connected to systems that are too small. Ssos also can occur when pipes crack, break, or become blocked by tree roots or debris. **Sediment**, debris, oil, and grease can build up in pipes, causing them to break or collapse. Sometimes pipes settle or shift so that the joints no longer match. Pump or power failures also can cause Ssos. Many sewer systems in the United States are 30–100 years old and the EPA estimates that about 75% of all systems have deteriorated to where they carry only 50% or less of their original capacity. Ssos are also caused by improper installation or maintenance of the system and by operational errors. Some municipalities have had to spend as much as a billion dollars correcting sewer problems or have halted new development until the system is fixed or its capacity increased.

Most Ssos are avoidable. Cleaning, maintenance, and repair of systems, enlarging or upgrading sewers, pump stations, or treatment plants, building wet-weather storage and treatment facilities, or improving operational procedures may be necessary to prevent Ssos. In some cases backup facilities may be required. Many municipalities are developing flow-equalization basins to retain excessive wet-weather flow until the system can handle it. The EPA estimates that an additional 12 billion dollars will be required annually until 2020, if the nation's sewer systems are to be repaired.

In 1995 representatives from states, municipalities, health agencies, and environmental groups formed the Sso Federal Advisory Subcommittee to advise the EPA on Sso regulations. In May 1999 President Bill Clinton ordered stronger measures to prevent Ssos that close beaches and adversely affect **water quality** and public health. Each EPA region is now required to inventory Sso violations and to address Ssos in priority sewer systems. In January 2001 the EPA's Office of **Wastewater** Management proposed a new SSO Rule that will expand the Clean Water Act permit requirements for 19,000 municipal sanitary sewer collection systems to reduce Ssos. Furthermore, some 4,800 additional municipal collection systems will be required to obtain permits from the EPA. Both the public and health and community officials will have to be notified immediately of raw sewage overflows that endanger public health.

[*Margaret Alic Ph.D.*]

Resources

Books

Office of Water. *Benefits of Protecting Your Community from Sanitary Sewer Overflows.* Washington, DC: U.S. Environmental Protection Agency, 2000.

Periodicals

Bell Jr., Robert E., and Maggie L. Powell. "Pending SSO Regulations." *Water Engineering & Management* 148, no. 4 (April 2001): 30–37.

Hobbs, David V., Eva V. Tor, and Robert D. Shelton. "Equalizing Wet Weather Flows." *Civil Engineering* 69, no. 1 (January 1999): 56–9.

Whitman, David. "The Sickening Sewer Crisis: Aging Systems Across the Country are Causing Nasty—and Costly —Problems." *U.S. News & World Report* (June 12, 2000): 16–18.

Other

Beach Watch. *Frequently Asked Questions.* Office of Water, U.S. Environmental Protection Agency. April 16, 1999 [cited May 31, 2002]. <http://www.epa.gov/ost/beaches/faq.html>.

Liquid Assets 2000: Today's Challenges. U.S. Environmental Protection Agency. May 31, 2002 [cited May 31, 2002]. <http://www.epa.gov/cgi-bin/epaprintonly.cgi>.

Office of Wastewater Management. *Sanitary Sewer Overflows.* U.S. Environmental Protection Agency. March 14, 2001 [cited May 31, 2002]. <http://www.cfpub.epa.gov/npdes/home.cfm?program_id=4>.

Organizations

American Public Works Association, 2345 Grand Blvd., Suite 500, Kansas City, MO USA 64108-2641, (816) 472-6100, <http://www.pubworks.org>

U.S. Environmental Protection Agency Office of Wastewater Management, Mail Code 4201, 401 M Street, SW, Washington, DC, USA 20460 (202) 260-7786, Fax: (202) 260-6257, <http://www.epa.gov/owm/sso>

Sanitation

Sanitation can be defined as the measures, methods, and activities that prevent the transmission of diseases and ensure public health. Specifically, "sanitation" refers to the hygienic principles and practices relating to the safe collection, removal and disposal of human excreta, refuse, and **wastewater**.

For a household, sanitation refers to the provision and ongoing operation and maintenance of a safe and easily accessible means of disposing of human excreta, **garbage**, and wastewater, and providing an effective barrier against excreta-related diseases.

The problems that result from inadequate sanitation can be illustrated by the following events in history:

1700 B.C.: Ahead of his time by a few thousand years, King Minos of Crete had running water in his bathrooms in his palace at Knossos. Although there is evidence of plumbing and sewerage systems at several ancient sites, including the cloaca maxima (or great sewer) of ancient Rome, their use did not become widespread until modern times.

1817: A major epidemic of **cholera** hit Calcutta, India, after a national festival. There is no record of exactly how many people were affected, but there were 10,000 fatalities among British troops alone. The epidemic then spread to other countries and arrived in the United States and Canada in 1832. The governor of New York quarantined the Canadian border in a vain attempt to stop the epidemic. When cholera reached New York City, people were so frightened, they either fled or stayed inside, leaving the city streets deserted.

1854: A London physician, Dr. John Snow, demonstrated that cholera deaths in an area of the city could all be traced to a common public drinking water pump that was contaminated with sewage from a nearby house. Although he could not identify the exact cause, he did convince authorities to close the pump.

1859: The British Parliament was suspended during the summer because of the stench coming from the Thames. As was the case in many cities at this time, storm sewers carried a combination of sewage, street debris, and other wastes, and storm water to the nearest body of water. According to one account, the river began to "seethe and ferment under a burning sun."

1892: The comma-shaped bacteria that causes cholera was identified by German scientist, Robert Loch, during an epidemic in Hamburg. His discovery proved the relationship between contaminated water and the disease.

1939: Sixty people died in an outbreak of typhoid fever at Manteno State Hospital in Illinois. The cause was traced to a sewer line passing too close to the hospital's water supply.

1940: A valve that was accidentally opened caused polluted water from the Genesee River to be pumped into the Rochester, New York, public water supply system. About 35,000 cases of gastroenteritis and six cases of typhoid fever were reported.

1955: Water containing a large amount of sewage was blamed for overwhelming a **water treatment** plant and causing an epidemic of hepatitis in Delhi, India. An estimated one million people were infected.

1961: A worldwide epidemic of cholera began in Indonesia and spread to eastern Asia and India by 1964; Russia, Iran, and Iraq by 1966; Africa by 1970; and Latin America by 1991.

1968: A four-year epidemic of dysentery began in Central America resulting in more than 500,000 cases and at least 20,000 deaths. Epidemic dysentery is currently a problem in many African nations.

1993: An outbreak of cryptosporidiosis in Milwaukee, Wisconsin, claimed 104 lives and infected more than 400,000 people, making it the largest recorded outbreak of waterborne disease in the United States.

The problem of sanitation in developed countries, who have the luxury of adequate financial and technical resources is more concerned with the consequences arising from inadequate commercial food preparation and the results of bacteria becoming resistant to disinfection techniques and antibiotics. Flush **toilets** and high quality drinking water supplies have all but eliminated cholera and epidemic diarrheal diseases. However, in many developing countries, such as the Pacific islands, inadequate sanitation is still the cause of life or death struggles.

In 1992, the South Pacific Regional Environment Programme (SPREP) and a Land-Based Pollutants Inventory stated that "[t]he disposal of solid and liquid wastes (particularly of human excrement and household garbage in urban areas), which have long plagued the Pacific, emerge now as perhaps the foremost regional environmental problem of the decade."

High levels of fecal **coliform bacteria** have been found in surface and coastal waters. The SPREP Land-Based Sources of **Marine Pollution** Inventory describes the Federated States of Micronesia's sewage **pollution** problems in striking terms: The prevalence of water-related diseases and **water quality** monitoring data indicate that the sewage pollutant **loading** to the environment is very high. A recent waste quality monitoring study (as part of a workshop) was unable to find a clean, uncontaminated site in the Kolonia, Pohnpei, area.

Many central wastewater treatment plants constructed with funds from United States **Environmental Protection Agency** in Pohnpei and in Chuuk States have failed due to lack of trained personnel and funding for maintenance.

In addition, septic systems used in some rural areas are said to be of poor design and construction, while pour-flush toilets and latrines—which frequently overflow in heavy rains—are more common. Over-the-water latrines are found in many coastal areas, as well.

In the Marshall Islands, signs of eutrophication—excess water plant growth due to too much nutrients—resulting from sewage disposal are evident next to settlements, particularly urban centers. According to a draft by the Marshall Islands NEMS, "one-gallon blooms occur along the coastline in Majuro and Ebeye, and are especially apparent on the **lagoon** side adjacent to households lacking toilet facilities." Stagnation of lagoon waters, reef degradation, and **fish kills**

resulting from the low levels of oxygen have been well documented over the years. Additionally, red tides **plague** the lagoon waters adjacent to Majuro.

There is significant **groundwater pollution** in the Marshall Islands as well. The Marshall Island EPA estimates that more than 75% of the rural **wells** tested are contaminated with fecal coliform and other bacteria. Cholera, typhoid and various diarrheal disorders occur.

With very little industry present, most of these problems are blamed on domestic sewage, with the greatest contamination problems believed to be from pit latrines, septic tanks, and the complete lack of sanitation facilities for 60% of rural households. As is often the case, poor design and inappropriate placement of these systems are often identified as the cause of contamination problems. In fact, even the best of these systems in the most favorable **soil** conditions allow significant amounts of nutrients and pathogens into the surrounding environment, and the soil characteristics and high **water table** typically found on atolls significantly inhibit treatment. In addition, the lack of proper maintenance, due to a lack of equipment to pump out septic tanks, is likely to have degraded the performance of these systems even further.

Forty percent of the population in the Republic of Palau is served by a secondary **sewage treatment** plant in the state of Koror, which is generally thought to provide adequate treatment. However, the Koror State government has expressed concern over the possible contamination of Malakal Harbor, into which the plant discharges. Also, some low-lying areas served by the system experience periodic back-flows of sewage which run into mangrove areas, due to mechanical failures with pumps and electrical power outages. In other low-lying areas not covered by the sewer system, septic tanks and latrines are used, which also overflow, affecting marine water quality.

Rural areas primarily rely on latrines, causing localized marine contamination in some areas. Though there have been an increasing number of septic systems installed as part of a rural sanitation program funded by the United States, there is anecdotal evidence that they may not be very effective. Many of the **septic tank** leach fields may not be of adequate size. In addition, a number of the systems are not used at all, as some families prefer instead to use latrines since the actual toilets and enclosures are not provided with the septic tanks as part of the program.

Wastewater problems also result from agriculture. According to the EPA, pig waste is considered to be a more significant problem than human sewage in many areas.

Because sanitation has become a social responsibility, national, state and local governments have adopted regulations that, when followed, should provide adequate sanitation for the governed society. However, the very technologies and practices that were instituted to provide better health and sanitation now have been found to be contaminating ground and surface waters. For example, placing **chlorine** in drinking water and waste water to provide disinfection, has now been found to produce carcinogenic compounds called **trihalomethanes** and **dioxin**. Collecting sanitary waste and transporting them along with industrial waste to inadequate treatment plants costing billions of dollars has failed to provide adequate protection for public health and environmental security.

Increasingly, the solution seems to be found in methods and practices that borrow from the stable ecosystems model of **waste management**. That is, there are no wastes, only resources that need to be connected to the appropriate organism that requires the residuals from one organism as the nutritional requirement of another. New waterless **composting** toilets that destroy human fecal organisms while they produce **fertilizer**, are now the technology of choice in the developing world and have also found a growing **niche** in the developed world. Wash water, rather than being disposed of into ground and surface waters, is now being utilized for **irrigation**. The combination of these two ecologically engineered technologies provides economical sanitation, eliminates pollution, and creates valuable fertilizers and plants, while reducing the use of potable water for irrigation and toilets.

Simple hand washing is now re-emerging as the most important measure in preventing disease transmission. Handwashing breaks the primary connection between surfaces contaminated with fecal organisms and the introduction of these pathogens into the human body. The use of basic soap and water, not exotic disinfectants, when practiced before eating and after defecating may save more lives than all of the modern methodologies and technologies combined.

[*Carol Steinfeld*]

Resources

Books

Salvato, J. *Environmental Engineering and Sanitation.* 4th ed. 1992.

"Sanitation Problems in Micronesia." A report. Concord, MA: Sustainable Strategies, 1997.

Periodicals

Fair, G., J. Geyer, and A. Okun. *Water and Wastewater Engineering* 1 (1992).

Santa Barbara oil spill

On January 28, 1969, a Union Oil Company **oil drilling** platform 6 mi (10 km) off the coast of Santa Barbara, California, suffered a blowout, leading to a tremendous ecological disaster. Before it could be stopped, 3 million gal (11.4

million l) of crude oil gushed into the Pacific Ocean, killing thousands of birds, fish, **sea lions**, and other marine life. For weeks after the spill, the nightly television news programs showed footage of the effects of the giant black slick, including oil–soaked birds on the shore dead or dying.

Many people viewed the disaster as an event that gave the modern environmental movement—which began with the publication of Rachel Carson's book *Silent Spring* in 1962—a new impetus in the United States. "The blowout was the spark that brought the environmental issue to the nation's attention," Arent Schuyler, an environmental studies lecturer at the University of California at Santa Barbara (UCSB), said in a 1989 interview with the *Los Angeles Times.* "People could see very vividly that their communities could bear the brunt of industrial accidents. They began forming environmental groups to protect their communities and started fighting for legislation to protect the environment," Schuyler explained.

The background

The spill began when oil platform workers were pulling a drilling tube out of one of the **wells** to replace a drill bit. The pressure differential created by removing the tube was not adequately compensated for by pumping mud back down into the 3,500-ft-deep well (about 1,067 m). This action caused a drastic buildup of pressure, resulting in a blowout; **natural gas**, oil, and mud shot up the well and into the ocean. The pressure also caused breaks in the ocean floor surrounding the well, from which more gas and oil escaped.

Cleanup and environmental impact

It took workers 11 days to cap the well with cement. Several weeks later, a leak in the well occurred, and for months, the well continued to spew more oil. On the surface of the ocean, an 800-mi^2-long (about 1,278 km^2) oil slick formed. Incoming tides carried the thick tar-like oil onto 35 mi (56 km) of scenic California beaches—from Rincon Point to Goleta. The spill also reached the off-shore islands of Anacapa, Santa Cruz, Santa Rosa, and San Miguel.

The Santa Barbara area is known to have some "natural" **petroleum** seepages, thus allowing for some degree of adaptability in the local marine plants and animals. However, the impact of the oil spill on sea life was disastrous. Dozens of sea lions and **dolphins** were believed killed, and many were washed ashore. The oil clogged the blowholes of the dolphins, causing massive bleeding in the lungs, and suffocation. It has been estimated that approximately 9,000 birds died as the oil stripped their feathers of the natural waterproofing that kept them afloat. Multitudes of fish are believed to have been killed, and many others fled the area, causing economic harm to the region's fishermen. A great number of **whales** migrating from the Gulf of Alaska to Baja, Mexico, were forced to take a sweeping detour around the polluted water.

Although hundred of volunteers, many of them UCSB students, worked on-shore catching and cleaning birds and mammals, an army of workers used skimmers to scoop up oil from the ocean surface. Airplanes were used to drop **detergents** on the oil slick to try to break it up. The cleanup effort took months to complete and cost millions of dollars. It also had a detrimental effect on tourism in the Santa Barbara area for many months. The excessively toxic dispersants and detergents used in these sensitive habitats also caused great ecological damage to the area. In later years, **chemicals** used to control petroleum spills were far less toxic.

Legal impact

Environmentalists claimed the spill could have been prevented and blamed both Union Oil and the U.S. **Geological Survey** (USGS). By approving the drilling permit, USGS officials were waiving current federal safety regulations for the oil company's benefit. Federal and state investigations also determined that if additional steel-pipe sheeting had been placed inside the drilling hole, the blowout would have been prevented. This type of sheeting was required under the regulations waived by the USGS.

In the years following the oil spill, the state and federal governments enacted many environmental protection laws, including the California Environmental Quality Act and the **National Environmental Policy Act** of 1969. The state also banned offshore oil drilling for 16 years, established the federal **Environmental Protection Agency** (EPA) and the California Coastal Commission, and created the nation'first environmental studies program at UCSB.

[*Ken R. Wells*]

Resources

Books

Jurcik, Liz. *Black Tide: The Santa Barbara Oil Spill and Its Consequences.* New York: Delacorte Press, 1972.

Periodicals

Corwin, Miles. "The Oil Spill Heard 'Round the Country!" *Los Angeles Times* (January 28, 1989): A1.

Cowan, Edward. "Mankind's Fouled Nest." *The Nation* (March 10, 1969): 304.

"Grotesque Beauty of a Sea Shimmering With Oil." *Life* (February 21, 1969): 58–62.

"Runaway Oil Well: Will It Mean New Rules in Offshore Drilling?" *U.S. News & World Report* (February 17, 1969): 14.

Organizations

Environmental Protection Agency (EPA), 1200 Pennsylvania Avenue, NW, Washington, DC USA 20460 (202) 260-2090, Email: public-access@epa.gov, <http://www.epa.gov>

Santa Barbara Wildlife Care Network, P.O. Box 6594, Santa Barbara, CA USA 93160 (805) 966-9005, Email: sbwcn@juno.com, <http://www.silcom.com/~sbwcn/spill.htm>

Saprophyte

A saprophyte is an organism that survives by consuming nutrients from dead and decaying plant and animal material, that is, organic matter. Saprophytes include **fungi**, molds, most bacteria, actinomycetes, and a few plants and animals. Saprophytes contain no chlorophyll and are, therefore, unable to produce food through **photosynthesis**, the conversion of chemical compounds into energy when light is present. Organisms that do not produce their own food, heterotrophs, obtain nutrients from surrounding sources, living or dead. For saprophytes, the source is non-living organic matter. Saprophytes are known as **decomposers**. They absorb nutrients from forest floor material, reducing complex compounds in organic material into components useful to themselves, plants, and other **microorganisms**. For example, lignin, one of three major materials found in plant cell walls, is not digestible by plant-eating animals or useable by plants unless broken down into its various components, mainly complex sugars. Certain saprophytic fungi are able to reduce lignin into useful compounds. Saprophytes thrive in moist, temperate to tropical environments. Most require oxygen to live. North American plant saprophytes include truffles, Indian Pipe (*Monotropa uniflora*), pinedrops (*Pterospora andromedea*), and snow orchid (*Cephalanthera austinae*), all of which feed on forest floor litter in the northeast forests.

[*Monica Anderson*]

SARA

see **Superfund Amendments and Reauthorization Act (1986)**

Savanna

A savanna is a dry grassland with scattered trees. Most ecologists agree that a characteristic savanna has an open or sparse canopy with 10–25% tree cover, a dominant ground cover of annual and perennial grasses, and less than 20 in (50 cm) of rainfall per year. Greatly varied environments, from open deciduous forests and parklands, to dry, thorny scrub, to nearly pure **grasslands**, can be considered savannas. At their margins these communities merge, more or less gradually, with drier prairies or with denser, taller forests. Savannas occur at both tropical and temperate latitudes and on all continents except **Antarctica**. Most often savannas occupy relatively level, or sometimes rolling, terrain. Characteristic savanna soils are dry, well-developed ultisols, oxisols, and alfisols, usually basic and sometimes lateritic. These soils develop under savannas' strongly seasonal rainfall regimes, with extended dry periods that can last up to 10 months. Under natural conditions an abundance of insect, bird, reptile, and mammal **species** populate the land.

Savanna shrubs and trees have leathery, sometimes thorny, and often small leaves that resist **drought**, heat, and intense sunshine normally found in this **environment**. Savanna grasses are likewise thin and tough, frequently growing in clumps, with seasonal stalks rising from longer-lived underground roots. Many of these plants have oily or resinous leaves that burn intensely and quickly in a fire. Extensive root systems allow most savanna plants to exploit moisture and nutrients in a large volume of **soil**. Most savanna trees stand less than 32 ft (10 m) tall; some have a wide spreading canopy while others have a narrower, more vertical shape. Characteristic trees of African and South American savannas include acacias and miombo (*Brachystegia* spp.). Australian savannas share the African baobab, but are dominated by eucalyptus species. Oaks characterize many European savannas, while in North America oaks, pines, and aspens are common savanna trees.

Because of their extensive and often nutritious grass cover, savannas support extensive populations of large herbivores. Giraffes, **zebras**, impalas, kudus, and other charismatic residents of African savannas are especially well-known. Savanna herbivores in other regions include North American **bison** and elk and Australian kangaroos and wallabies. Carnivores—lions, cheetahs, and jackals in Africa, **tigers** in Asia, **wolves** and pumas in the Americas—historically preyed upon these huge herds of grazers. Large running birds, such as the African ostrich and the Australian emu, inhabit savanna environments, as do a plethora of smaller animal species. In the past century or two many of the world's native savanna species, especially the large carnivores, have disappeared with the expansion of human settlement. Today ranchers and their livestock take the place of many native grazers and carnivores.

Savannas owe their existence to a great variety of convergent environmental conditions, including temperature and precipitation regimes, soil conditions, fire frequency, and **fauna**. Grazing and browsing activity can influence the balance of trees to grasses. Fires, common and useful for some savannas but rare and harmful in others, are an influential factor in these dry environments. Precipitation must be sufficient to allow some tree growth, but where rainfall is high some other factors, such as grazing, fire, or soil **drainage** needs to limit tree growth. Human activity also influences the occurrence of these lightly-treed grasslands. In some regions recent expansion of ranches, villages, or agriculture have visibly extended savanna conditions. Elsewhere centuries or millennia of human occupation make natural and **anthropogenic** conditions difficult to distinguish. Because savannas are well suited to human needs, people have

occupied some savannas for tens of thousands of years. In such cases people appear to be an environmental factor, along with **climate**, soils, and grazing animals, that help savannas persist. *See also* Deforestation

[*Mary Ann Cunningham*]

RESOURCES
BOOKS

Cole, M. M. *The Savannas: Biogeography and Geobotany*. London: Academic Press, 1986.

Danserreau, P. *Biogeography: An Ecological Perspective*. New York: Ronald Press Company, 1957.

de Laubenfels, D. J. *Mapping the World's Vegetation*. Syracuse, NY: Syracuse University Press, 1975.

White, R. O. *Grasslands of the Monsoon*. New York: Praeger, 1968.

Savannah River Site

From the 1950s to the 1990s, the manufacture of **nuclear weapons** was a high priority item in the United States. The nation depended heavily on an adequate supply of atomic and **hydrogen** bombs as its ultimate defense against enemy attack. As a result, the government had established 17 major plants in 12 states to produce the materials needed for nuclear weapons.

Five of these plants were built along the Savannah River near Aiken, South Carolina. The Savannah River Site (SRS) is about 20 mi (32 km) southeast of Augusta, Georgia, and 150 mi (242 km) upstream from the Atlantic Ocean. The five reactors sit on a 310 mi^2 (803 km^2) site that also holds 34 million gal (129 million l) of high-level radioactive wastes.

The SRS reactors were originally used to produce **plutonium** fuel used in nuclear weapons. In recent years, they have been converted to the production of tritium gas, an important component of fusion (hydrogen) bombs.

For more than a decade, environmentalists have been concerned about the safety of the SRS reactors. The reactors were aging; one of the original SRS reactors, for instance, had been closed in the late 1970s because of cracks in the reactor vessel.

The severity of safety issues became more obvious, however, in the late 1980s. In September 1987, the **U.S. Department of Energy** (DOE) released a memorandum summarizing 30 "incidents" that had occurred at SRS between 1975 and 1985. Among these were a number of unexplained power surges that nearly went out of control. In addition, the meltdown of a radioactive fuel element in November 1970 took 900 people three months to clean up.

E. I. du Pont de Nemours and Company, operator of the SRS facility, claimed that all "incidents" had been properly reported and that the reactors presented no serious risk to the area. Worried about maintaining a dependable supply of tritium, DOE officials accepted du Pont's explanation and allowed the site to continue operating.

In 1988, however, the problems at SRS began to snowball. By that time, two of the original five reactors had been shut down permanently and two, temporarily. Then, in August of 1988 an unexplained power surge caused officials to temporarily close down the fifth, "K," reactor. Investigations by federal officials found that the accident was a result of human errors. Operators were poorly trained and improperly supervised. In addition, they were working with inefficient, aging equipment that lacked adequate safety precautions.

Still concerned about its tritium supply, the DOE spent more than $1 billion to upgrade and repair the K reactor and to retrain staff at SRS. On December 13, 1991, Energy Secretary James D. Watkins announced that the reactor was to be re-started again, operating at 30% capacity.

During final tests prior to re-starting, however, 150 gal (569 l) of cooling water containing radioactive tritium leaked into the Savannah River. A South Carolina water utility company and two food processing companies in Savannah, Georgia, had to discontinue using river water in their operations.

The ill-fated reactor experienced yet another set-back only five months later. During another effort at re-starting the K reactor in May 1992, radioactive tritium once again escaped into the surrounding **environment** and, once again, start-up was postponed.

By the fall of 1992, changes in the balance of world power would cast additional doubt on the future of SRS. The Soviet Union had collapsed and neither Russia nor its former satellite states were regarded as an immediate threat to U.S. security, as former Soviet military might was now spread among a number of nations with urgent economic priorities. American president George Bush and Russian leader Boris Yeltsin had agreed to sharp cut-backs in the number of nuclear warheads held by both sides. In addition, experts determined that the existing supply of tritium would be sufficient for U.S. defense needs until at least 2012.

In 2000, the K reactor was turned into a materials storage facility. The site will soon be put back into action after the delivery of 6 tons of plutonium in June of 2002. The plutonium will be held at the site until it is converted into nuclear reactor fluid by 2017. If the plutonium has not been converted, it will be transferred to an other location.

[*David E. Newton*]

RESOURCES
PERIODICALS

Applebome, P. "Anger Lingers After Leak at Atomic Site." *New York Times* (January 13, 1992): A7.

Schneider, K. "U.S. Dropping Plan to Build Reactor." *New York Times* (September 12, 1992): 5.

Sharp, Deborah. "Six Tons of Plutonium begin Journey to S.C." *USA Today* June 24, 2002 [cited July 2002]. <http://www.usatoday.com/news/nation/2002/06/25/plutonium-usat.htm>.

Sweet, W. "Severe Accident Scenarios at Issue in DOE Plan to Restart Reactor." *Physics Today* (November 1991): 78–81.

Wald, M. L. "How an Old Government Reactor Managed to Outlive the Cold War." *New York Times* (December 22, 1991): E2.

Save the Whales

Save the Whales is a non-profit, grassroots organization that was founded and incorporated in 1977. It was conceived by 14-year-old Maris Sidenstecker II when she learned that **whales** were being cruelly slaughtered. She designed a T-shirt with a logo that read "Save the Whales" and designated the proceeds to help save whales. Her mother, Maris Sidenstecker I, helped co-found the organization and still serves as its executive director.

Unlike some environmental groups, **Greenpeace** for example, Save the Whales is involved in very few direct action protests and activities. Its primary focus is education. The group's educational staff is comprised of marine biologists, environmental educators, and researchers. They speak to school groups, senior citizen organizations, clubs, and numerous other organizations. Save the Whales is opposed to commercial **whaling** and works to save all whale **species** from **extinction**. The group also sends a representative to the annual International Whaling Commission conference.

WOW!, Whales On Wheels, is an innovative program developed by Save the Whales. WOW! travels throughout California educating thousands of children and adults about marine mammals and their **environment**. WOW! features live presenters and incorporates audio and video into the program along with hands-on exhibits and marine mammal activity projects. In 1992 alone, Maris Sidenstecker II gave presentations to over 7,000 school children.

Save the Whales has also developed a regional Marine Mammal Beach Program for school aged children. It consists of an abbreviated WOW! lecture after which the participants assist in cleaning up a beach. Save the Whales has adopted Venice Beach in California and cleans it several times a year.

A five-minute educational video entitled "One Person Makes A Difference" has been produced by Save the Whales. It combines footage of the humpback whales, orcas, and gray whales in the wild which is followed by shots of the slaughter of pilot whales in the Faroe Islands. The video ends with brief interviews that advocate whale **conservation** and describe the ways in which one person can make a difference. Save the Whales is currently involved in raising funds to produce a 30-minute documentary film entitled "Barometers of The Ocean." This video will explore chemical **pollution** in the oceans and how it is affecting whale populations.

The organization also supports marine mammal research. It helps finance research on **seals and sea lions** in Puget Sound (Washington), and makes financial contributions to SCAMP (Southern California **Migration** Project) to support studies of gray whale migration. Save the Whales publishes a quarterly newsletter that is free with membership, and the organization currently has 2,000 members worldwide.

[*Debra Glidden*]

Resources

Organizations

Save The Whales, P.O. Box 2397, Venice, CA USA 90291 Email: maris@savethewhales.org, <http://www.savethewhales.org>

Save-the-Redwoods League

The Save-the-Redwoods League was founded in 1918 to protect California's redwood forests for **future generations** to enjoy. Prominent individuals involved in the formation of the League included Stephen T. Mather, first Director of the **National Park Service**; Congressman William Kent, author of the bill creating the **National Park** Service; Newton Drury, Director of the National Park Service from 1940–1951; and John Merriam, paleontologist and later president of the Carnegie Institute. The impetus for forming the organization was a trip taken in 1917 by several of these men and others during which they witnessed widespread destruction of the forests by loggers. They were appalled to learn that not one tree was protected by either the state or the federal government. Upon their return, they wrote an article for *National Geographic* detailing the devastation. Shortly thereafter they formed the Save-the-Redwoods League. One of the League's first actions was to recommend to Congress that a **Redwoods** National Park be established.

The specific stated objectives of the organization are:

- "to rescue from destruction representative areas of our primeval forests,
- to cooperate with the California State Park Commission, the National Park Service, and other agencies, in establishing redwood parks, and other parks and reservations,
- to purchase redwood groves by private subscription,
- to foster and encourage a better and more general understanding of the value of the primeval redwood or sequoia and other forests of America as natural objects of extraordinary interest to present and future generations,
- to support reforestation and **conservation** of our forest areas."

This non-profit organization uses donations to purchase redwood lands from willing sellers at fair market value. All contributions, except those specified for research or reforestation, are used for land acquisition.

The League's members and donors have given more than $70 million in private contributions since the formation of the organization. These monies have been used to purchase and protect more than 130,000 acres (52,609 ha) of redwood park land. The establishment of the Redwoods National Park in 1968, and the park's subsequent expansion in 1978, represented milestones for the Save-the-Redwoods League which had fought for such a national park for 60 years.

Mere acquisition of the redwood groves does not ensure the long-term survival of the redwoods **ecosystem**. Based on almost 70 years of study by park planners and ecologists, a major goal of the Save-the-Redwoods League is to complete each of the existing redwoods parks as ecological units along logical **watershed** boundary lines. The acquisition of these watershed lands are necessary to act as a buffer around the groves to protect them from effects of adjacent **logging** and development.

[*Ted T. Cable*]

Resources

Organizations

Save-the-Redwoods League, 114 Sansome Street, Room 1200, San Francisco, CA USA 94104-3823 (415) 362-2352, Fax: (415) 362-7017, Email: info@savetheredwoods.org, <http://www.savetheredwoods.org>

Scarcity

As most commonly used, the term scarcity refers to a limited supply of some material. With the rise of environmental consciousness in the 1960s, scarcity of **natural resources** became an important issue. Critics saw that the United States and other developed nations were using natural resources at a frightening pace. How long, they asked, could non-renewable resources such as **coal**, oil, **natural gas**, and metals last at the rate they were being consumed?

A number of studies produced some frightening predictions. The world's oil reserves could be totally depleted in less than a century, according to some experts, and scarce metals such as silver, **mercury**, zinc, and **cadmium** might be used up even faster at then-current rates of use.

One of the most famous studies of this issue was that of the **Club of Rome**, conducted in the early 1970s. The Club of Rome was an international group of men and women from 25 nations concerned about the ultimate environmental impact of continued **population growth** and unlimited development. The Club commissioned a complex computer study of this issue to be conducted by a group of scholars at the Massachusetts Institute of Technology. The result of that study was the now famous book *The Limits to Growth.*

Limits presented a depressing view of the Earth's future if population and technological development were to continue at then-current rates. With regard to non-renewable resources such as metals and **fossil fuels**, the study concluded that the great majority of those resources would become "extremely costly 100 years from now." Unchecked population growth and development would, therefore, lead to widespread scarcity of many critical resources.

The *Limits* argument appears to make sense. There is only a specific limited supply of coal, oil, chromium, and other natural resources on Earth. As population grows and societies become more advanced technologically, those resources are used up more rapidly. A time must come, therefore, when the supply of those resources becomes more and more limited, that is, they become more and more scarce.

Yet, as with many environmental issues, the obvious reality is not necessarily true. The reason is that there is a second way to define scarcity, an economic definition. To an economist, a resource is scarce if people pay more money for it. Resources that are bought and sold cheaply are not scarce.

One measure of the *Limits* argument, then is to follow the price of various resources over time, to see if they are becoming more scarce in an economic sense. When that study is conducted, an interesting result is obtained. Supposedly "scarce" resources such as coal, oil, and various metals have actually become *less* costly since 1970 and must be considered, therefore, to be less scarce than they were two decades ago.

How this can happen has been explained by economist Julian Simon, a prominent critic of the *Limits* message. As the supply of a resource diminishes, Simon says, humans become more imaginative and more creative in finding and using the resource. For example, gold mines that were once regarded as exhausted have been re-opened because improved technology makes it possible to recover less concentrated reserves of the metal. Industries also become more efficient in the way they use resources, wasting less and making what they have go further.

In one sense, this debate is a long-term versus short-term argument. One can hardly argue that the Earth's supply of oil, for example, will last forever. However, given Simon's argument, that supply may last much longer than the authors of *Limits* could have imagined twenty years ago.

[*David E. Newton*]

Resources

Books

Meadows, D. H., et al. *The Limits to Growth.* New York: Universe Books, 1972.

Ophuls, W., and A. S. Boyan Jr. *Ecology and the Politics of Scarcity Revisited.* San Francisco: W. H. Freeman, 1992.

Simon, J. L. and H. Kahn, eds. *The Resourceful Earth.* Oxford: Basil Blackwell, 1984.

———. *The Ultimate Resource.* Princeton, NJ: Princeton University Press, 1981.

Scavenger

Any substance or organism that cleans a setting by removing dirt, decaying matter, or some other unwanted material. Vultures are typical biological scavengers because they feed on the carcasses of dead animals. Certain **chemicals** can act as scavengers in chemical reactions. Lithium and magnesium are used in the metal industry as scavengers since these metals react with and remove small amounts of oxygen and **nitrogen** from molten metals. Even rain and snow can be regarded as scavengers because they wash pollutants out of the **atmosphere**.

Schistosomiasis

Human blood fluke disease, also called schistosomiasis or bilharziasis, is a major parasitic disease affecting over 200 million people worldwide, mostly those in the tropics. Although sometimes fatal, schistosomiasis more commonly results in chronic ill-health and low energy levels. The disease is caused by small parasitic flatworms of the genus *Schistosoma.* Of the three **species**, two (*S. haematobium* and *S. mansoni*) are found in Africa and the Middle East, the third (*S. japonicum*) in the Orient. *Schistosoma haematobium* lives in the blood vessels of the urinary bladder and is responsible for over 100 million human cases of the disease a year. *Schistosoma mansoni* and *Schistosoma japonicum* reside in the intestine; the former species infect 75 million people a year and the latter 25 million.

Schistosomiasis is spread when infected people urinate or defecate into open waterways and introduce parasite eggs that hatch in the water. Each egg liberates a microscopic free-living larva called the *miracidium* which bores into the tissues of a water snail of the genus *Biomphalaria*, *Bulinus*, or *Onchomelania*, the intermediate host. Inside the snail the parasite multiplies in sporocyst sacs to produce masses of larger, mobile, long-tailed larvae known as *cercariae.* The cercariae emerge from the snail into the water, actively seek out a human host, and bore deep into the skin. Larvae that reach the blood vessels are carried to the liver where they develop into adult egg-producing worms that settle in the vessels of the urinary bladder or intestine. Adult *Schistosoma* live entwined in mating couples inside the small veins of their host. Fertilized females release small eggs (0.2 mm long), at the rate of 3,500 per day, which are carried out of the body with the urine or the feces.

The symptoms of schistosomiasis correlate with the progress of the disease. Immediately after infection migrating cercariae cause itching skin. Subsequent establishment of larvae in the liver damages this organ. Later, egg release causes blood in the stool (dysentery), damage to the intestinal wall, or blood in the urine (hematuria), and damage to the urinary bladder.

Schistosomiasis is increasing in developing countries due in part to rapidly increasing human populations. In rural areas, attempts to increase food production that include more **irrigation** and more **dams** also increase the **habitat** for water snails. In urban areas the combination of crowding and lack of **sanitation** ensures that increasingly large numbers of people become exposed to the parasite.

Most control strategies for schistosomiasis target the snail hosts. One strategy kills snails directly by adding snail poisons (molluscicides) to the water. Another strategy either kills or removes vegetation upon which snails feed. Biological methods of snail control include the introduction of fish that feed on snails, of snails that kill schistosome snail hosts, of insect larvae that prey on snails, and of flukes that kill schistosomes inside the snail. Some countries, such as Egypt, have attempted to eliminate the parasite in humans through mass treatment with curative drugs including ambilar, niridazole, nicolifan, and praziquantel. Total eradication programs for schistosomiasis focus both on avoiding contact with the parasite through education, better sanitation, and on breaking its life cycle through snail control and human treatment.

[*Neil Cumberlidge Ph.D.*]

Resources
Books

Basch, P. F. *Schistosomes: Development, Reproduction, and Host Relations.* New York: Oxford University Press, 1991.

Bullock, W. L. *People, Parasites, and Pestilence: An Introduction to the Natural History of Infectious Disease.* Minneapolis: Burgess Publishing Company, 1982.

Malek, E. A. *Snail-Transmitted Parasitic Diseases.* Boca Raton: CRC Press, 1980.

Markell, E. K., M. Voge, and D. T. John. *Medical Parasitology.* 7th ed. Philadelphia: Saunders, 1992.

Ernst Friedrich Schumacher (1911 – 1977)

German/English economist

E. F. Schumacher combined his background in economics with an extensive background in theology to create a unique

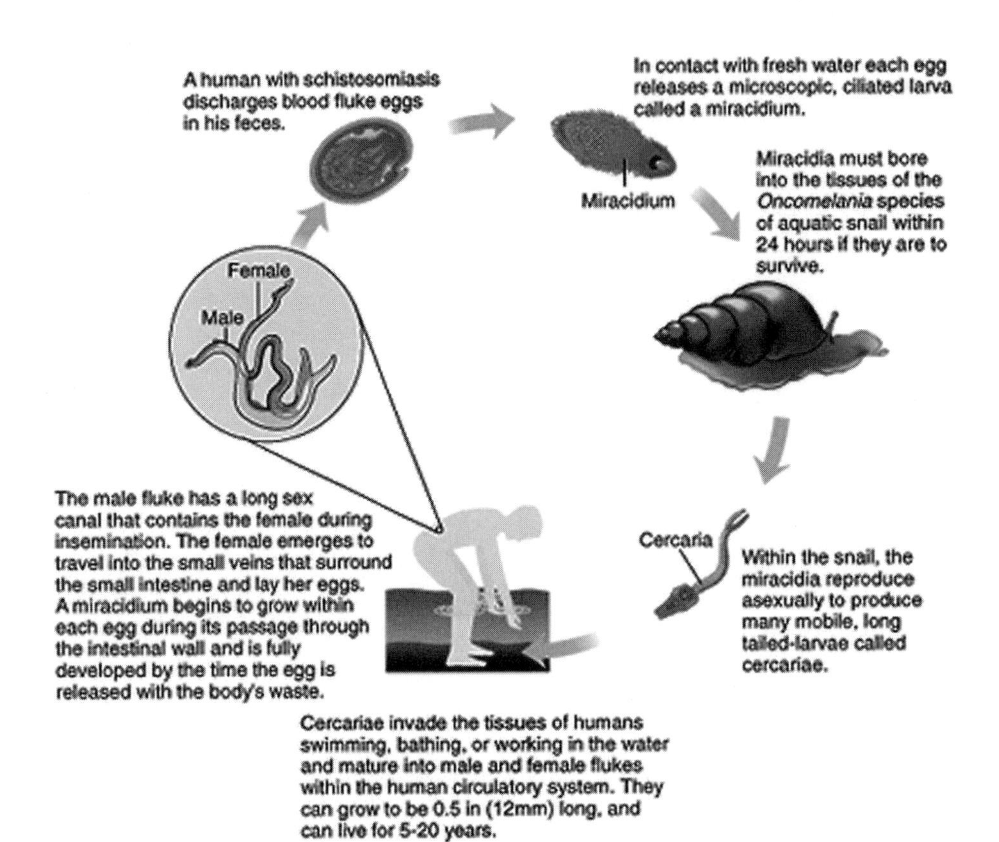

A human with schistosomiasis discharges blood fluke eggs in his feces.
In contact with fresh water each egg releases a microscopic, ciliated larva called a miracidium.
Miracidium
Miracidia must bore into the tissues of the *Oncomelania* species of aquatic snail within 24 hours if they are to survive.
Female
Male
The male fluke has a long sex canal that contains the female during insemination. The female emerges to travel into the small veins that surround the small intestine and lay her eggs. A miracidium begins to grow within each egg during its passage through the intestinal wall and is fully developed by the time the egg is released with the body's waste.
Cercaria
Within the snail, the miracidia reproduce asexually to produce many mobile, long tailed-larvae called cercariae.
Cercariae invade the tissues of humans swimming, bathing, or working in the water and mature into male and female flukes within the human circulatory system. They can grow to be 0.5 in (12mm) long, and can live for 5-20 years.

The life cycle of the blood fluke. This parasite causes the disease schistosomiasis in humans. (Illustration by Hans & Cassidy.)

strategy for reforming the world's socioeconomic systems, and by the early 1970s he had risen nearly to folk hero status.

Schumacher was born in Bonn, Germany. He attended a number of educational institutions, including the universities of Bonn and Berlin, as well as Columbia University in the United States. He eventually received a diploma in economics from Oxford University, and emigrated to England in 1937. During World War II, Schumacher was forced to work as a farm laborer in an internment camp. By 1946, however, he had become a naturalized British citizen, and in that same year he accepted a position with the British section of the Control Commission of West Germany. In 1953 he was appointed to Great Britain's National Coal Board as an economic advisor, and in 1963 he became director of **statistics** there. He stayed with the National Coal Board until 1970.

Schumacher recognized the contemporary indifference to the course of world development, as well as its implications. He began to seek alternative plans for **sustainable development**, and his interest in this issue led him to found the Intermediate Technology Development Group in 1966. Today, this organization continues to provide information and other assistance to **less developed countries**. Schumacher believed that in order to build a global community that would last, development could not exploit the **environment** and must take into account the sensitive links between the environment and human health. He promoted research in these areas during the 1970s as president of the Soil Association, an organization that advocates **organic farming** worldwide.

Schumacher also combined his studies of Buddhist and Roman Catholic theology into a philosophy of life. In his highly regarded book, *Small is Beautiful: Economics as if People Mattered* (1973), he stresses the importance of self-reliance and promotes the virtues of working with **nature** rather than against it. He argues that continuous economic growth is more destructive than productive, maintaining that growth should be directly proportionate to human need. People in society, he continues, should apply the ideals of **conservation** (durable-goods production, **solar energy**, **recycling**) and appropriate technology in their lives whenever possible.

In his *A Guide For the Perplexed* (1977), published in 1977, Schumacher is even more philosophical. He attempts to provide personal rather than global guidance, encouraging self-awareness and urging individuals to embrace what he termed a "new age" ethic based on Judeo-Christian principles. He also wrote various pamphlets, including: *Clean Air and Future Energy*, *Think About Land*, and *Education for the Future*.

Schumacher died in 1977 while traveling in Switzerland from Lausanne to Zurich.

[*Kimberley A. Peterson*]

Resources

Books

Schumacher, E. F. *A Guide for the Perplexed.* New York: Harper & Row, 1977.

———. *Small is Beautiful: Economics As If People Mattered.* London: Blond and Briggs, 1973.

Periodicals

Fraker, S., and G. C. Lubenow. "Mr. Small." *Newsweek* (March 28, 1977): 18.

"Why It Is Important to Think Small." *International Management* (August 1977): 18–20.

Albert Schweitzer (1875 – 1965)

German theologian, musicologist, philosopher, and physician

Albert Schweitzer was an individual of remarkable depth and diversity. He was born in the Upper Alsace region of what is now Germany. His parents, Louis (an Evangelical Lutheran pastor and musician) and Adele cultivated his inquisitive mind and passion for music. He developed a strong theological background and, under his father's tutelage, studied piano, eventually acting as substitute musician at the church.

As a young adult, Schweitzer pursued extensive studies in philosophy and theology at the University of Strasbourg, where he received doctorate degrees in both fields (1899 and 1900 respectively). He continued to further his interests in music. While on a fellowship in Paris (researching Kant), he also studied organ at the Sorbonne. He began a career as an organist in 1893 and was eventually renowned as expert in the area of organ construction and as one of the finest interpreters and scholars of Johann Sebastian Bach. His multitudinous published works include the respected *J. S. Bach: the Musician Poet* (1908).

In 1905, after years of pursuing various careers—including minister, musician, and teacher—he determined to dedicate his work to the benefit of others. In 1913 he and his wife, Helene, traveled to Lambarene in French Equatorial Africa (now Gabon), where they built a hospital on the Ogowe River. At the outbreak of World War I, they were allowed to continue working at the hospital for a time but were then sent to France as prisoners of war (both were German citizens). They were held in an internment camp until 1918. After their release, the Schweitzers remained in France and Albert returned to the pulpit. He also gave organ concerts and lectures. During this time in France he wrote "Philosophy of Civilization" (1923), an essay in which he described his philosophy of "reverence for life." Schweitzer felt that it was the responsibility of all people to "sacrifice a portion of their own lives for others." The Schweitzers returned to Africa in 1924, only to find their hospital overgrown with jungle vegetation. With the assistance of volunteers, the facility was rebuilt, using typical African villages as models.

Awarded the Nobel Peace Prize in 1952, Schweitzer used $33,000 of the Prize money to establish a leper colony near the hospital in Africa. It was not until late 1954 that he gave his Nobel lecture in Oslo. He took advantage of the opportunity to object to war as "a crime of inhumanity." Three more years passed before Schweitzer sent out an impassioned plea to the world in his "Declaration of Conscience," which he read over Oslo radio. He called for citizens to demand a ban of **nuclear weapons** testing by their governments. His words ignited a series of arms control talks among the superpowers that began in 1958 and, five years later, resulted in a limited but formal test-ban treaty.

Schweitzer received numerous awards and degrees. Among his long list of published works are *The Quest of the Historical Jesus* (1906), *The Mysticism of Paul the Apostle* (1931), *Out of My Life and Thought* (1933, autobiography), and *The Light Within Us* (1959). He died at the age of 90 at Lambarene.

[*Kimberley A. Peterson*]

Resources

Books

Bentley, J. *Albert Schweitzer: The Enigma.* New York: Harper Collins, 1992.

Miller, D. C. *Relevance of Albert Schweitzer at the Dawn of the Twenty-First Century.* Lanham: University Press of America, 1992.

Schweitzer, A. *Out of My Life and Thought.* New York: Henry Holt, 1990.

Periodicals

Negri, M. "The Humanism of Albert Schweitzer." *Humanist* 53 (March–April 1993): 26–31.

Scientific Committee on Problems of the Environment

The Scientific Committee on Problems of the Environment

(SCOPE) is a worldwide group of 22 international science unions and 40 national committees that work together on issues pertaining to the environment. SCOPE was founded in 1969 and today is headquartered in Paris, France. SCOPE stands as a permanent committee of the International Council for Science.

SCOPE describes its organization as "an interdisciplinary body of natural and social science expertise focused on global environmental issues, operating at the interface between scientific and decision-making instances" and "a worldwide network of scientists and scientific institutions developing syntheses and reviews of scientific knowledge on current or potential environmental issues." The committee aims to determine not only how human activities impact the environment, but also how environmental changes impact people's health and welfare.

With its broad international nature, SCOPE acts as an independent source of information for governments and nongovernmental entities around the world by offering research and consulting expertise on environmental topics. In past years, the committee has gathered scientists to produce reports including possible effects of nuclear war, **biosphere** programs to keep the earth inhabitable, and **radioactive waste**.

SCOPE's scientific program focuses on three main areas: managing societal and **natural resources**, **ecosystem** processes and **biodiversity**, and health and the environment. Managing societal and natural resources involves those projects that are founded on scientific research but that can be applied in a practical manner to help sustain the biosphere. In other words, the committee wants to ensure that Earth can continue to produce the fuel, food, and other natural resources needed to support the population. As of early 2002, nearly a dozen projects involving biosphere development, urban **waste management**, agriculture, **conservation**, and others fell under the spectrum of this SCOPE task.

Ecosystem processes and biodiversity projects focus on how ecosystems interact with human activities. These projects look at how activities such as mining and movement of substances such as **nitrogen** in rivers will affect the earth's land and water. Projects also look at changes in Earth's **climate** and the impact that those changes may have on future **ecology**.

The committee's health and environment projects develop methods to assess chemical risks of various activities to humans, plants, animals, and habitats. For example, the committee is working on a project involving the study of effects of **mercury**, particularly in aquatic environments. Another major project studies **radioactivity** at nuclear sites. Called RADSITE, the project aims to review radioactive wastes generated in development of **nuclear weapons**.

In addition to its own work and affiliation with the International Council for Science, SCOPE works in partnership with several international bodies including the United Nations Education, Social and Culture Organization (UNESCO), and the International Human Dimensions Program on Global Environmental Change (IHDP). SCOPE is a unique organization in its international scientific approach to, and cooperation on, environmental issues.

[*Teresa G. Norris*]

Resources

Periodicals

O'Riordan, T. "A New Science for a New Age; *New Scientist* 133 (January 25, 1992): 14.

Organizations

Scientific Committee on Problems of the Environment (SCOPE), %1, blvd. De Montmorency F-75016, Paris, France 33-1-45250498, Fax: 33-1-42881466, Email: secretariat@icsu-scope-org, <http://www.icsu-scope.org>

Scientists' Institute for Public Information

Working to bridge a gap they perceive between scientists and the media, the Scientists' Institute for Public Information (SIPI) was established in 1963 to disseminate expert information on science and technology to journalists through a variety of means.

SIPI's best-known program is the Media Resource Service (MRS), which was founded in 1980. The MRS serves as a referral service for journalists seeking information from scientists, engineers, physicians, and policymakers. The MRS maintains a database listing more than 30,000 experts who are available for and willing to comment on a variety of topics. The service is free to any media outlet. The honorary chair of the MRS advisory committee is the highly regarded newscaster Walter Cronkite. The MRS is funded by such noteworthy news organizations as CBS Inc., the Scripps Howard Foundation, Time Warner, The Washington Post Company, the Associated Press, the National Broadcasting Company, and The New York Times Company Foundation.

In addition to the MRS, SIPI operates the Videotape Referral Service (VRS), another free resource service which aids broadcast journalists in finding videotapes to accompany science- and technology-related stories. The VRS also provides a list of videotapes for an annual SIPI conference called "TV News: The Cutting Edge," a meeting of scientists, television news directors, and science reporters. The VRS also publishes a monthly newsletter featuring topical listings

of story ideas and a current catalogue of available videotapes on science-related issues.

An outgrowth of the MRS, SIPI also operates the International Hot Line in connection with its Global Change program. The hotline provides assistance to journalists worldwide by referring them to scientists and environmental experts. The Global Change program holds briefings to update the media on current international scientific and environmental issues as well.

SIPI also organizes roundtable discussions, seminars, and symposia for scientists and national journalists. These programs have focused on such issues as nuclear waste disposal, military technology and budget priorities, and human gene therapy. SIPI has also developed smaller-scale versions of these programs for state and regional press associations and journalism schools.

In addition, SIPI sponsors the Defense Writers Group. This group is made up of members of the Pentagon press corps who gather to discuss views with defense experts. In addition, SIPI publishes a newsletter addressing current issues in science policy and featuring reviews of media coverage of science and technology.

[*Kristin Palm*]

SCOPE

see **Scientists' Committee on Problems of the Environment**

Scotch broom

Scotch broom (*Cytisus scoparius*) is a member of the pea family that is native to southern Europe and northern Africa. It can grow up to 10 ft (3 m) in height. Scotch broom's yellow flowers are shaped like peas, and its leaves each have three leaflets.

During the 1800s, Scotch broom was introduced into the United States as an ornamental plant. In addition, people used the plant for sweeping and make beer and tea with the seeds. During the mid-1800s, Scotch broom was planted in the western United States to stabilize roads and prevent **soil erosion**. By the end of the twentieth century, the nonnative plant was classified as an invasive weed in the Pacific Northwest and in parts of the northeastern United States and southeastern Canada.

Scotch broom is a hearty plant that grows quickly. Its seeds can last 80 years. Thick fields of Scotch broom can crowd out vegetation that is native to an area. Spread of the plant can reduce the amount of grazing **habitat** for animals. Broom can block pathways used by **wildlife**. In addition, broom can be a fire hazard.

Methods of controlling the spread of Scotch broom include removing the plant by its roots, spraying **herbicide**, and controlled burning to destroy seeds.

[*Liz Swain*]

Scrubbers

Scrubbers are **air pollution control** devices that help cleanse the emissions coming out of an incinerator's **smoke** stack. Hot exhaust gas comes out of the incinerator duct and scrubbers help to wash the **particulate** matter (dust) resulting from the **combustion** out of the gas. High efficiency scrubbers literally scrub the smoke by mixing dust particles and droplets of water (as fine as mist) together at a very high speed. Scrubbers force the dust to move like a bullet fired at high velocity into the water droplet. The process is similar to the way that rain washes the air.

Scrubbers can also be used as absorbers. **Absorption** dissolves material into a liquid, much as sugar is absorbed into coffee. From an **air pollution** standpoint, absorption is a useful method of reducing or eliminating the **discharge** of air contaminants into the **atmosphere**. The gaseous air contaminants most commonly controlled by absorption include **sulfur dioxide**, **hydrogen** sulfide, hydrogen chloride, **chlorine**, ammonia, **nitrogen oxides**, and light **hydrocarbons**.

Gas absorption equipment is designed to provide thorough contact between the gas and liquid solvent in order to permit interphase diffusion and solution of the materials. This contact between gas and liquid can be accomplished by dispersing gas in liquid and visa versa. Scrubbers help wash out polluting **chemicals** from the exhaust gas by facilitating the mixture of liquid (solvent) and gas together.

A number of engineering designs serve to disperse liquid. These include a packed tower, a spray tower or spray chamber, venturi absorbers, and a bubble tray tower.

The most appropriate design for **incineration** facilities is the packed tower—a tower filled with one of many available packing materials. The packing material should provide a large surface area and, for good fluid flow, should be shaped to give large void space when packed. It should also be chemically inert and inexpensive. It must be designed so as to expose a large surface area and be made of materials, such as stainless steel, ceramic or certain forms of plastic.

Packing materials come in various manufactured shapes. They may look like a saddle, a thick tube, a many-faceted star, a scouring pad, or a cylinder with a number of holes carved in it. Packing may be dumped into the column at random or stacked in some kind of order. Randomly dumped packing has a higher gas pressure drop across the bed. The stacked packings have an advantage of lower pres-

sure and higher possible liquid throughout, but the installation cost is much higher because they are packed by hand.

Rock and gravel have also been used as packing materials but are usually considered too heavy. They also have small surface areas, give poor fluid flow and at times are not chemically inert.

The liquid is introduced at the top of the tower and trickles down through to the bottom. Since the effectiveness of a packed tower depends on the availability of a large, exposed liquid film, poor liquid distribution that prevents a portion of the packing from being irrigated renders that portion of the tower ineffective. Poor distribution can result from improper introduction of the liquid at the top of the tower or using the wrong rate of liquid flow. The liquid rate must be sufficient to wet the packing but not flood the tower. A liquid rate of at least 800 pounds of liquid per hour per square foot of tower cross-section is typical.

While the liquid introduced at the top of the tower trickles down through the packing, the gas is introduced at the bottom and passes upward through the packing. This process results in the highest possible efficiency. Where the gas stream and solvent enter at the top of the column, there is initially a very high rate of absorption that constantly decreases until the gas and liquid exit in equilibrium.

Scrubbers are used in **coal** burning industries that generate electricity. They can be used in high sulfur coal emissions because high sulfur coal emits high levels of sulfur dioxide. Scrubbers are also utilized in industrial chemical manufacturing as an important operation in the production of a chemical compound. For example, one step in the manufacture of hydrochloric **acid** involves the absorption of hydrochloric acid gas in water. Scrubbers are used as a method of recovering valuable products from gas streams—as in **petroleum** production where natural **gasoline** is removed from gas streams by absorption in a special hydrocarbon oil. *See also* Air pollution; Flue-gas scrubbing; Tall stacks

[*Liane Clorfene Casten*]

RESOURCES

BOOKS

Schiffner, K. C. *Wet Scrubbers.* Chelsea, MI: Lewis, 1986.

PERIODICALS

"Fume Scrubbers Benefit Environment and Manufacturing." *Design News* 48 (August 24, 1992): 28–9.

"Scrubbing Emissions." *Environment* 31 (March 1989): 22.

Sea cows

see **Manatees**

Sea level change

For at least tens of thousands of years, changes in sea level seem to be a natural part of the earth's **environment**. Up until a few hundred years ago, these changes, both up and down, occurred due to land movement, ice melting from glaciers, and an increase or decrease in the amount of water trapped in the polar ice caps. In most cases, the changes were very gradual in human terms. But in the past several decades, many scientists have become alarmed at the rapid increase in ocean levels.

The reason is because the earth is heating up, which causes sea water to expand in volume and ice caps to melt. Over the past 100 years, scientists have measured a mean sea level rise of about 4 in (10 cm). They blame it on an average increase of 1.8°F (1°C) in world-wide surface temperatures of the planet. With the advent of satellites and other technology in the last half of the twentieth century, scientists have been able to more precisely measure the ocean levels. They have found that sea levels are rising at a rate of about 0.4–1.2 in (1–2 cm) per year. A 1-ft (30-cm) rise in sea level would place at least 100 ft (30 m) of beach width underwater.

Why the earth is heating up is the subject of much discussion and disagreement among scientists. Some believe it is due to the heavy burning of **fossil fuels**, such as **coal** and oil, which causes increases the amounts of certain gases, particularly **carbon dioxide**, in the **atmosphere**. Other scientists believe the current global warming is mostly a natural phenomenon and a part of the normal cycle of the planet's environment.

Scientists are studying the ice cap on **Antarctica** to determine if, in fact, the earth's **climate** is warming due to the burning of fossil fuels. The global warming hypothesis is based on the atmospheric process known as the **greenhouse effect**, in which **pollution** prevents the heat energy of the earth from escaping into the outer atmosphere. Global warming could cause some of the ice caps to melt, raising the sea level and **flooding** many of the world's largest cities, including New York, Los Angeles, Tokyo, and London, and other lowland areas. Nearly half of the world's population live in coastal areas. Because the polar regions are the engines that drive the world's weather system, this research is essential to identify the effect of human activity on these regions.

Many scientists are concerned about the increasing levels of **carbon** dioxide in Earth's atmosphere. With more carbon dioxide in the atmosphere, they say, more heat will be trapped. Earth's annual average temperature will begin to rise. Some estimates suggest that a doubling of atmospheric carbon dioxide will result in an increase of 4.5°F (2.5°C) in the planet's annual average temperature.

While that number may seem small, it could have disastrous effects on the world's economies. One result might be the melting of Earth's ice caps at the North and South poles, with a resulting increase in the volume of the ocean's water. Were that to happen, many of the world's largest cities, those located along the edge of the oceans, might be flooded. Some experts predict dramatic changes in climate that could turn currently productive croplands into deserts, and deserts into productive agricultural regions.

As with many environmental issues, experts tend to disagree about one or more aspects of anticipated climate change. Some authorities are not convinced that the addition of carbon dioxide to the atmosphere will have any significant long-term effects on Earth's average annual temperature. Others concede that Earth's temperature may increase, but that the changes predicted are unlikely to occur. They point out that other factors, such as the formation of clouds, might counteract the presence of additional carbon dioxide in the atmosphere. They warn that nations should not act too quickly to reduce the **combustion** of fossil fuels since that will cause serious economic problems in many parts of the world. They suggest it would be prudent to wait for a while to see if greenhouse factors really are beginning to change.

[*Ken R. Wells*]

Resources

Books

Douglas, B. C. ed, et al. *Sea Level Rise: History and Consequences.* San Diego: Academic Press, 2000.

Periodicals

Laber, Emily. "Meltdown." *The Sciences* (July 1999): 6.

Middleton, Nick. "The Heat is On." *Geographical* (January 2000): 44.

Moore, Curtis A. "Awash in a Rising Sea."*International Wildlife* (January–February 2002).

Spalding, Mark. "Danger on the High Seas." *Geographical* (February 2002): 15–16.

Foley, Grover. "The Threat of Rising Seas." *The Ecologist* (March–April 1999): 76–79.

Organizations

National Oceanic and Atmospheric Administration, 14th St. and Constitution Ave. NW, Room 6013, Washington, DC USA 20230 (202)482-6090, Fax: (202)482-3154, Email: answers@noaa.gov, <http://www.noaa.gov>

Sea lions

see **Seals and sea lions**

Sea otter

The sea otter (*Enhydra lutris*) is found in coastal marine waters of the northeastern Pacific Ocean, ranging from California to as far north as the Aleutian Islands. Sea otters spend their entire lives in the ocean and even give birth while floating among the kelp beds. Their ability to use tools, often "favorite" rocks, to open clam shells and sea urchins is well-known and fascinating since few other animals are known to exhibit this behavior. Their playful, curious nature makes them the subjects of many **wildlife** photographers but also has aided in their demise. Many are injured or killed by ship propellers or in fishing nets.

A female sea otter eating squid. (Photograph by Karl W. Kenyon, National Audubon Society Collection. Photo Researchers Inc. Reproduced by permission.)

Some sea otters are killed for their highly valued fur. Their thick hair traps air, insulating the otter from the cold water in which it lives. Before they were placed under international protection in 1924, 800,000 to one million sea otters were slaughtered for their pelts, eliminating them from large portions of their original range. Despite **poaching**, which remains a problem, they are slowly returning through relocation efforts and natural **migration**. Their proximity to human settlement, however, still poses a problem for their continued survival.

Pollution, especially from **oil spills**, is deadly to the sea otter. The insulating and water-repellant properties of their fur are inhibited when oil causes the fine hairs to stick together, and otters die from hypothermia. Ingestion of oil during grooming does extensive, often fatal, internal damage.

In 1965, an oil spill near Great Sitkin Island, Alaska, reduced the island's otter population from 600 to six. In 1989, the oil tanker **Exxon Valdez** spilled over 11 million gal (40 million l) of crude oil into **Prince William Sound**, Alaska, leading to the nearly complete elimination of the Sound's once thriving sea otter population. In California, fears that a similar incident could destroy the sea otter population there have led to relocation efforts.

Sea otters have few natural enemies, but they were extensively hunted by Aleuts and later by Europeans. Sea otters were hunted to **extinction** around several islands in Alaska, an event that led to studies on the importance of sea otters in maintaining marine communities. Attu Island, one of the islands that has lost its otter population, has high sea urchin populations that have, through their grazing, transformed a kelp forest into a "bare" hard ground of coralline and green algae. Few fish or abalone are present in these waters anymore. On nearby Amchitka Island, otters are present in densities of 7.7–11.6 per mi^2 (20–30 per km^2) and forage at depths up to 22 yd (20 m). In this area, few sea urchins persist and dense kelp forests harbor healthy fish and abalone populations. These in turn support higher-order predators such as **seals** and bald eagles.

Effects of sea otter foraging have also been documented in soft-bottom communities, where they reduce densities of sea urchins and clams. In addition, disturbance of the bottom **sediment** leads to increased predation of small bivalves by sea stars. Otters' voracious appetite for invertebrates also brings them into conflict with people. Fishermen in northern California blame sea otters for the decline of the abalone industry. Farther south, residents of Pismo Beach, an area noted for its clam industry, are exerting pressure to remove otters. Sea urchin and crab fishermen have also come into conflict with these competitors. It remains a challenge for fishermen, environmentalists, and regulators to arrive at a mutually agreeable management policy that will allow successful coexistence with sea otters. The sea otter census of 2001 counted only 2,161 otters in California, less than 6,000 in Alaska, 2,500 in Canada, 555 in Washington, and about 15,000 in Russia. They are considered endangered by the IUCN.

[*William G. Ambrose Jr. and Paul E. Renaud*]

Resources

Books

Sumick, J. L. *An Introduction to the Biology of Marine Life.* 5th ed. Dubuque, IA: W. C. Brown, 1992.

Periodicals

Brazil, Eric. "Annual Census Begins in State for Nearly Extinct Sea Mammal. *San Francisco Chronicle* (May 18, 2001): A3.

Kvitek, R. G., et al. "Changes in the Alaskan Soft-Bottom Prey Communities Along a Gradient of Sea Otter Predation." *Ecology* 73 (1992): 413–28.

"Northern Sea Otters may be Declared Endangered." *The Grand Rapids Press* (November 12, 2000): A3.

Raloff, J. "An Otter Tragedy." *Science News* 143 (1993): 200–202.

Other

Help Save the Sea Otters. [cited May 2002]. <http://www.saveseaotters.org>.

Friends of the Sea Otter. [cited May 2002]. <http://www.seaotters.org>.

Sea Shepherd Conservation Society

The Sea Shepherd Conservation Society was founded in 1977 by Paul Watson, one of the founding members of **Greenpeace**, as an aggressive direct action organization dedicated to the international conservation and protection of marine **wildlife** in general and marine mammals in particular. The society seeks to combat exploitative practices through education, confrontation, and the enforcement of existing laws, statutes, treaties, and regulations. It maintains offices in the United States, Great Britain, and Canada, and has an international membership of about 15,000.

Sea Shepherd regards itself virtually as a police force dedicated to ocean and marine life conservation. Most of its attention over the years has been devoted to the enforcement of the regulations of the International Whaling Commission (IWC), which makes policies for signatory states on **whaling** practices but does not itself have powers of enforcement. The stated objective of the society has been to harass, interfere with, and ultimately shut down all continuing illegal whaling activities.

Called a "samurai conservation organization" by the Japanese media, Sea Shepherd often walks a thin line between legal and illegal tactics. The society operates two research ships, the *Sea Shepherd* and the *Edward Abbey*, and has been known to ram illegal or pirate whaling ships and to sabotage whale processing operations. All crew members are trained in techniques of "creative non-violence:" They are forbidden to carry weapons or explosives or to endanger human life and are enjoined to accept all moral responsibility and legal consequences for their actions.

Crew members also pledge never to compromise on the lives of the marine mammals they protect. The Society has documented on film illegal whaling operations in the former Soviet Union and presented this evidence to the IWC, despite being chased back to United States waters by a Soviet frigate and helicopter gunships. Moreover, members are not at all squeamish about the destruction of weapons, ships, and other property used in the slaughter of marine wildlife. In 1979, Sea Shepherd hunted down and rammed the pirate whaler *Sierra*, eventually putting it out of business. Publicity over the *Sierra* operation motivated the arrest of two other pirate whalers in South Africa. The next year Sea Shepherd was involved in the sinking of two Spanish whalers

that had flagrantly exceeded whale quotas set by the IWC. In 1986, Sea Shepherd was involved in the sinking of two Icelandic whalers (half the Icelandic whaling fleet) in Reykjavik harbor and also managed to damage the nearby whale-processing plant. Seeking publicity, crew members demanded to be arrested for their actions, but Iceland refused to charge them. Indeed, Sea Shepherd claims that in all of its operations it has never caused nor suffered an injury, nor have any of its crew members been convicted in criminal proceedings.

Typically, Sea Shepherd invites members of the news media along to document and publicize the destructive and exploitative practices it opposes. Such documentary footage has been shown on major television networks in the United States, Britain, Canada, **Australia**, and Western Europe. This publicity played an important role in increasing public awareness of marine conservation issues and in mobilizing public opinion against the slaughter of marine mammals. Sea Shepherd helped to bring about the end of commercial seal-killing in Canada and in the Orkney Islands, Scotland.

Highly successful in its efforts against outlaw whalers, Sea Shepherd continues to conduct research on conservation and **pollution** issues and to monitor national and international law on marine conservation issues. Its members are working to establish a wildlife sanctuary in the Orkney Islands. Its present campaign is focused primarily against drift net fishing in the North Pacific, in support of a United Nations call for a complete international ban on drift-net fishing.

[*Lawrence J. Biskowski*]

Resources

Organizations

Sea Shepherd Conservation Society, 22774 Pacific Coast Hwy., Malibu, CA USA 90265 (310) 456-1141, Toll Free: (310) 456-2488, <http://www.seashepherd.com>

Sea turtles

Sea turtle populations have dramatically declined in numbers over the past half century. Green sea turtles (*Chelonia mydas*), hawksbills (*Eretmochelys imbricata*), Kemp's Ridleys (*Lepidochelys kempii*), loggerheads (*Caretta caretta*), and leatherbacks (*Dermochelys coriacea*) have all had their numbers decimated by human activity. The decline has been caused by several factors, including the development of a highly industrialized fishery to meet the demand for seafood on a worldwide basis. The most economical fishing method involves pulling multiple nets underwater for extended periods of time, and any air-breathing animals, such as sea turtles, which get caught in the net are usually drowned before they are hoisted on board.

Loggerhead sea turtle returns to sea after laying eggs on a Florida beach. (Corbis-Bettmann. Reproduced by permission.)

In the United States, this problem has led to the introduction of the **turtle excluder device** (TED), which must be placed on each tow net used by commercial shrimpers and fishermen. These cage-like devices have a slanted section of bars which allow fish and shellfish into the net but deflect turtles. These highly controversial devices have been mandatory for less than a decade, but there are some indications that they are saving thousands of turtles per year.

As significant as the impact of **commercial fishing** may seem, it does little to sea turtle populations compared to losses incurred at the earliest stages of the turtles' life history. In the late 1940s, along an isolated beach near Tamaulipas, Mexico, an extremely dense assemblage of sea turtles were observed digging out nests and laying eggs at the beach. So many females were present that they were seen crawling over one another and digging out the nests of others in order to lay their own eggs. At this location alone, the sea turtle population was estimated in the millions. In the early 1960s, scientists realized that the turtles found at Tamaulipas were a distinct **species**, Kemp's Ridley, and that they nested nowhere else in the world; but by that time the population had declined to only a few hundred turtles.

Threats to the survival of newly hatched sea turtles have always been enormous; crows, gulls, and other predators attack them as they scurry seaward, and they are prey for waiting barracudas and jacks as they reach water. Other animals raid their nests for the eggs, and humans are among these nest predators, collecting the eggs for food. Sea turtles concentrate their numbers in small nesting locations such as Tamaulipas in order to greatly outnumber their natural predators, thus allowing for the survival of at least a few individuals to perpetuate the species. However, this congre-

A green sea turtle. (Photograph by Dr. Paula A. Zahl. Photo Researchers Inc. Reproduced by permission.)

gating behavior has contributed to their demise, because it has made human predation easier and more profitable.

Adult turtles are harvested as a protein source in many **Third World** countries, and many turtles are also subjected to increasing levels of **marine pollution**. Both of these factors have contributed to the sharp decline in their population. Public awareness and **conservation** efforts may keep sea turtles from **extinction**, but it is not clear whether species will be capable of rebounding from the decimation that has already taken place.

[*Eugene C. Beckham*]

RESOURCES
BOOKS

Bjorndal, K. A., ed. *Biology and Conservation of Sea Turtles.* Washington, D.C.: Smithsonian Institution Press, 1981.

Carr, Archie. *So Excellent a Fish: Tales of Sea Turtles.* New York: Scribners, 1984.

National Research Council. *Decline of the Sea Turtles: Causes and Prevention.* Washington, DC: National Academy Press, 1990.

PERIODICALS

Ezell, C. "Turtle Recovery Could Take Many Decades." *Science News* 142 (August 22, 1992): 118.

"Sea Turtle Nesting Begins Soon." *The Florida Times Union* (April 13, 2002): L–13.

Stolzenburg, W. "Requiem for the Ancient Mariner." *Sea Frontiers* 39 (March–April 1993): 16–18.

OTHER

National Marine Fisheries Service. [cited May 2002]. <http://www.nmfs.noaa.gov>.

Seabed disposal

Over 70% of the earth's surface is covered by water. The coastal zone—the boundary between the ocean and land—is under the primary influence of humans, while the rest of the ocean remains fairly remote from human activity. This remoteness has in part led scientists and policy makers to examine the deep ocean, particularly the seabed, as a potential location for waste disposal.

Much of the deep ocean seabed consists of abyssal hills and vast plains that are geologically stable and have sparse numbers of bottom-dwelling organisms. These areas have been characterized as oozes, hundreds of meters thick, that are in effect "deserts" in the sea. Other attributes of the deep ocean seabed that have led scientists and policy makers to consider the sea bottom as a repository for waste include the immobility of the interstitial pore water within the **sediment**, and the tendency for ions to adsorb or stick to the sediment, which limits movement of elements within the waste. Another important factor has been the lack of known commercial resources such as **hydrocarbons**, minerals, or fisheries.

The deep seabed has been studied as a potential disposal option specifically for the placement of **high-level radioactive waste**. Investigations on the feasibility of disposing radioactive wastes in the seabed were carried out for over a decade by a host of scientists from around the world. In 1976, the Organization for Economic Cooperation and Development (OECD) and the Nuclear Energy Agency coordinated research at the international level and formed the Seabed Working Group. Members of the group included Belgium, Canada, France, the Federal Republic of Germany, Italy, Japan, the Netherlands, Switzerland, United Kingdom, United States, and the Commission of European Communities. The Seabed Working Group focused its investigation on two sites in the Atlantic Ocean, Great Meteor East and the Southern Nares Abyssal Plain, and one site in the Pacific Ocean, known as E2. The Great Meteor East site lies between 30.5°N and 32.5°N, 23°W and 26°W, approximately 1,865 mi (3,000 km) southwest of Britain. The Southern Nares Abyssal Plain site lies between 22.58°N and 23.17°N, 63.25°W and 63.67°W, approximately 375 mi (600 km) north of Puerto Rico. Site E2 in the Pacific Ocean lies between 31.3°N and 32.67°N, 163.42°E and 165°E, and is approximately 1,240 mi (2,000 km) east of Japan.

This working group pursued a multidisciplinary approach to studying the deep-ocean sediments as a potential disposal option for high-level **radioactive waste**. High-level radioactive waste consists of spent nuclear fuel or byproducts from the reprocessing of spent nuclear fuel. It also includes transuranic wastes, a byproduct of fuel assembly, weapons fabrications, and reprocessing operations, and **uranium** mill **tailings**, a byproduct of mining operations. **Low-level radioactive waste** is legally defined as all types of waste that do not fall into the high-level radioactive waste category. They are made up primarily of byproducts of nuclear reactor operations and products that have been in contact with the use of radioisotopes. Low-level radioactive wastes are characterized as having small amounts of **radioactivity** that do not usually require shielding or heat-removing equipment.

One proposal to dispose of high-level radioactive waste in the deep seabed involved the enclosure of the waste in an insoluble solid with metal sheathing or projectile-shaped canisters. When dropped overboard from a ship, the canisters would fall freely to the ocean bottom and bury themselves 33–44 yd (30–40 m) into the soft sediments of the seabed. Other proposals recommend drilling holes in the seabed and mechanically inserting the canisters. After emplacement of the canisters, the holes would then be plugged with inert material.

The 1972 international **Convention on the Prevention of Marine Pollution by Dumping of Waste and Other Matter**, commonly called the London Dumping Convention, prohibits the disposal of high-level radioactive wastes. In the United States, the **Marine Protection, Research and Sanctuaries Act** of 1972 also bans the ocean disposal of high-level radioactive waste. However, Britain has recently considered using the continental shelf seabed for disposal of low- and intermediate-level radioactive waste. The European countries discontinued **ocean dumping** of low-level radioactive waste in 1982. In the United States, the ocean disposal of low-level radioactive waste ceased in the 1970s. Between 1951 and 1967, approximately 34,000 containers of low-level radioactive waste were dumped in the Atlantic Ocean by the United States.

Another important aspect in the debate about seabed disposal is the risk to humans, not only because of the potential for direct contact with wastes but also because of the possibility of accidents during transport to the disposal location and contamination of fishery resources. When comparing seabed disposal of high-level radioactive waste to land disposal, for example, the **transportation** risks may be higher because travel to a site at sea would likely be longer than travel to a location on land. Increased handling by personnel substantially increases the risk. Also, there is a statistically greater risk of accidents at sea than on land when transporting anything, especially radioactive wastes, although such an accident at sea would probably pose less risk to humans.

It is also a concern that the seabed **environment** may be more inhospitable than a land site to the metal canisters containing the radioactive waste, because corrosion is more rapid due to the salts in marine systems. In addition, it is uncertain how fast **radionuclides** will be transported away from the site. The heat associated with the decay of high-level radioactive waste may cause convection in the sediment pore waters, resulting in the possibility that the dissolved radioactive material will diffuse to the sediment-water interface. Predictions from calculations, however, indicate that convection may not be significant. According to laboratory experiments that simulate subseabed conditions, it would take roughly 1,000 years for radioactive waste buried at a depth of 33 yd (30 m) to reach the sediment-water interface. Other technical considerations adding to the uncertainty of the ultimate fate of buried radioactive waste in the seabed involve possible **sorption** of the radionuclide cations to clay particles in the sediment, and possible uptake of radionuclides by bottom dwelling organisms.

The Seabed Working Group concluded from their investigation that seabed disposal of high-level radioactive waste is safe. Compared to land disposal sites, the predicted doses of possible **radiation exposure** are lower than published radiological assessments. However, there are political concerns over deep-ocean seabed disposal of wastes. Deep-ocean disposal sites would likely be in international waters. Therefore, international agreements would have to be reached, which may be very difficult with countries without a **nuclear power** industry, particularly for disposal of radioactive waste.

In 1991, the Woods Hole Oceanographic Institution held a workshop to discuss the research required to assess the potential of the abyssal ocean as an option for disposal of sewage **sludge**, incinerator ash, and other high volume benign wastes. The disposal technology considered at this workshop entailed employing an enclosed elevator from a ship to emplace the waste at or close to the seabed. One issue raised at the workshop was the need to investigate the incidence of benthic storms that may occur along the deep ocean seabed. These benthic storms, also called turbidity flows, are currents with high concentrations of sediment that can stir up the sea bottom, erode the seabed, and redistribute sediment further downstream.

Woods Hole held a follow-up workshop in 1992. Included were a broader array of scientists and representatives from environmental organizations, and these two groups disagreed over the use of the ocean floor as a waste-disposal option. The researchers supported the consideration and study of the seabed and the ocean in general as sites for

disposal of wastes. The environmentalists did not support ocean disposal of wastes. The environmentalists' view is consistent with the law passed in 1988, the **Ocean Dumping Ban Act**, which prohibits the dumping of sewage sludge and industrial waste in the marine environment. In 1993, a ban on the dumping of any radioactive materials into the sea was put into effect at the London Convention and will be enforced until 2018. *See also* Convention on the Law of the Sea; Dredging; Hazardous waste siting; Marine pollution; Ocean dumping; Radioactive pollution

[*Marci L. Bortman*]

Resources

Books

Chapman, N. A., and I. G. McKinley. *The Geological Disposal of Nuclear Waste.* New York: Wiley, 1987.

Freeman, T. J., ed. *Advances in Underwater Technology, Ocean Science and Offshore Engineering.* Vol. 18, *Disposal of Radioactive Waste in Seabed Sediments.* Boston: Graham & Trotman, 1989.

Krauskopf, K. B. *Radioactive Waste Disposal and Geology.* New York: Chapman and Hall, 1988.

Murray, R. L. *Understanding Radioactive Waste.* Ed. Judith A. Powell. Columbus, OH: Battelle Press, 1989.

Periodicals

Spencer, D. W. "The Ocean and Waste Management." *Oceanus* 33 (Summer 1990): 7–23.

Seabrook Nuclear Reactor

Americans once looked to nuclear energy as the nation's great hope for power generation in the twenty-first century. Today, nuclear **power plants** are regarded with suspicion and distrust, and new proposals to construct them are met with opposition. Perhaps the best transition in the perception of **nuclear power** is the debate that surrounded the construction of the Seabrook Nuclear Reactor in Seabrook, New Hampshire.

The plant was first proposed in 1969 by the Public Service Company of New Hampshire (PSC), an agency then responsible for providing 90% of all electrical power used in that state. The PSC planned to construct a pair of atomic reactors in marshlands near Seabrook in order to ensure an adequate supply of electricity in the future.

Residents were not enthusiastic about the plan. The marshlands and beaches around Seabrook have long been a source of pride to the community, and in March of 1976, the town voted to oppose the plant. Townspeople soon received a great deal of support. Seabrook is only a few miles north of the Massachusetts border, and residents and government officials from that state joined the opposition against the proposed plant. In addition, an umbrella organization of 15 anti-nuclear groups called the Clamshell Alliance was formed to fight the PSC plan.

The next 12 years were characterized by almost nonstop confrontation between the PSC and its supporters and Clamshell Alliance and other groups opposed to the proposal. Hardly a month passed during the 1970s and 80s without news of another demonstration or the arrest of someone protesting construction. The issues became more complex as economic and technical considerations changed during this time. The demand for electricity, for example, began to drop instead of increasing, as the PSC had predicted, and at least three other utilities that had agreed to work with PSC on construction of the plant withdrew from the program. The total cost of construction also continued to rise. When first designed, construction costs were estimated at $973 million for both reactors. Only one reactor was ever built, and by the time that it was finally licensed in 1990, total expenditures for it alone had reached nearly $6.5 billion.

The decision to build at Seabrook eventually proved to be a disaster financially and from a public relations stance for PSC. The company's economic woes peaked in 1979 when the courts ruled that PSC could not pass along additional construction costs at Seabrook to its customers. Over the next decade, the company fell into even more difficult financial straits, and on January 28, 1988, it filed for bankruptcy protection. The company promised that its action was not the end for the Seabrook reactor and maintained that the plant would eventually be licensed.

A little more than two years later, the **Nuclear Regulatory Commission** (NRC) granted a full-power operating license to the Seabrook plant. The decision was received enthusiastically by the utility companies in the New England Power Pool, who believed that Seabrook's additional capacity would reduce the number of power shortages experienced by consumers in the six-state region.

Private citizens and government officials were not as enthusiastic. Consumers faced the prospect of higher electrical bills to pay for Seabrook's operating costs, and many observers continued to worry about potential safety problems. Massachusetts attorney general, James Shannon, for example, was quoted as saying that Seabrook received "the most legally vulnerable license the NRC has ever issued."

As of July 2002, Seabrook was operating at 100% power and providing electricity for over one million homes. *See also* Electric utilities; Energy policy; Nuclear fission; Radioactive waste management; Three Mile Island Nuclear Reactor

[*David E. Newton*]

Seabrook Nuclear Reactor (Seabrook, New Hampshire). (Photograph by Tom Pantages. Reproduced by permission.)

RESOURCES
PERIODICALS

Shulman, S. "Embattled Seabrook Wins License at Last." *Nature* (March 8, 1990): 96.

Wasserman, H. "Clamshell Alliance: Getting It Together." *Progressive* (September 1977): 14–18.

———. "Clamshell Reaction: Protest against Nuclear Power Plant at Seabrook, N.H. by Clamshell Alliance." *Nation* (June 18, 1977): 744—749.

———. "Nuclear War by the Sea." *Nation* (September 11, 1976): 203–205.

Seals and sea lions

Members of the suborder Pinnipedia, seals and sea lions are characterized by paddle-like flippers on a pair of limbs. Most pinnipeds are found in boreal or polar regions and are the most important predators in many high latitude areas, feeding primarily on fish and squid. Seals and sea lions catch their prey during extended dives of up to 25 minutes at depths of 2,625 ft (800 m) or more. Pinniped biologists are interested in the physiological changes these animals undergo during their dives. Known as mammalian dive response, a combination of reduced heart rate and a lowered core body temperature enables these warm-blooded animals to complete such dives.

Seals and sea lions must return to beaches and ice floes each year to give birth. Here they raise their pups in large congregations known as colonies. In some **species** the larger, more aggressive males will form polygamous mating groups, or harems, and claim and defend territories in these limited breeding areas. Males that do not get to mate form "bachelor male" groups in areas less suitable for weaning pups. Males of most species of pinnipeds do not mate before age four but may acquire harems when they attain large size and fighting experience. Some individual seals have been known to live as long as 46 years.

Because of the commercial value of their pelts, many species of seals and sea lions are threatened by **hunting**. One of the largest pinnipeds, the Stellar sea lion (*Eumetopias jubatus*) has been steadily decreasing in numbers and is now listed as endangered by the IUCN. Within 15 years, the new pup count at the Marmot Island Rookery (the largest known sea lion rookery in the world) decreased from an

A Hawaiian monk seal. (Photograph by Jim Leupold. U.S. Fish & Wildlife Service. Reproduced by permission.)

average of 6,700 to only 804. The northern or Pribilof fur seal (*Callorhinus ursinus*), which had been heavily exploited for its fur and in steady decline since the mid-1950s, is now listed as vulnerable. In California, the Guadeloupe fur seal (*Arctocephalus townsendi*) is also listed as vulnerable. The Monk seal (*Monachus monachus*) in the Mediterranean is listed as critically endangered.

These animals are in danger from natural and man-induced pressures, and the consequences of their demise are unknown. Other than hunting, perceived competition with fishermen, **pollution**, and **habitat** destruction are threatening pinnipeds. Despite the sanctions afforded by the **Marine Mammals Protection Act** (1972), fishers continue to kill seals that are "interfering" with fishing operations. Recent revisions to the act require some vessels to include observers who can monitor compliance with the law. Some environmentalists contend that it is the fisher who "interferes" with the seals. The California Department of Fish and Game estimates that 2,000 seals die each year from accidental entanglement in fishing nets.

In the summer of 1988, 18,000 harbor seals washed up on European shores. They were discovered to be suffering from an acute viral infection, possibly due to an immune-system breakdown. Chemical pollutants, mainly polychlorinated biphenyl (PCB), have been implicated in reduced immune function in marine mammals, and levels of PCBs have been constantly increasing in many coastal waters. A connection has also been made between mass seal die-offs and increasing temperature. Four of the six recorded mass mortalities have occurred in the past 12 years, a period that includes some of the warmest weather in the twentieth century. Other pollutants, such as **heavy metals**, may also affect seal health or reproductive ability. **Oil spills** also poison seal food sources and reduce the insulation ability of the seals' fur.

Finally, the reduction of habitat for feeding and reproduction has not been investigated thoroughly. Reputedly caused by reductions in prey-fish stocks due to **overfishing** and by development of coastal shorelines, this reduction may negatively affect pinniped populations.

[*William G. Ambrose Jr. and Paul E. Renaud*]

Resources
Books

Sumick, J. L. *An Introduction to the Biology of Marine Life.* 5th ed. Dubuque, IA: W. C. Brown, 1992.

Periodicals

Gentry, R. "Seals and Their Kin." *National Geographic* 171 (1987): 474–501.

Hersh, S. "Saga of North Sea Seals (Phocine Distemper Virus)." *Sea Frontiers* 36 (1990): 55.

Rosenberg, S. "Sea Lions of Monterrey." *Sea Frontiers* 35 (1989): 97–103.

Stirrup, M. "A Sea Lion Mystery." *Sea Frontiers* 36 (1990): 46–53.

Stolzenberg, W. "Seals Under Siege: A Heated Warning." *Science News* 138 (1990): 84.

Paul Bigelow Sears (1891 - 1990)

American ecologist and conservationist

Born in Bucyrus, Ohio, Paul Sears obtained bachelor's degrees in zoology and economics, then earned a Ph.D. in botany from the University of Chicago. He spent most of his career as professor or head of various botany departments. In these positions, Sears researched changes in native **flora** as a result of human activities, conducted pioneering studies of fossil pollen, and studied the relationship between vegetation and climatic change. A respected and influential ecologist, he served as president of the **Ecological Society of America** (1948) and received the ESA's "Eminent Ecologist Award" in 1965. He spent the last ten years of his academic career as chair of the graduate program in **conservation** at Yale University and retired in 1960.

Sears was one of the few biological ecologists interested in **human ecology**, writing cogently and consistently in a field that he saw as a problem in synthesis. In his 1957 Condon Lecture at the University of Oregon, titled "The Ecology of Man," he mandated "serious attention to the ecology of man" and demanded "its skillful application to human affairs."

He considered ecology a "subversive subject," arguing that if it were taken seriously, it would "endanger the assumptions and practices accepted by modern societies, whatever their doctrinal commitments." But Sears was a optimist and believed that scientists would eventually agree because the nature of their work mandated that they have "confidence that the world hangs together."

As a conservationist, Sears believed that one of the basic lessons ecology teaches is that materials cycle and recycle through natural systems. Thus, he became a strong advocate of intensive **recycling** by human societies. He also taught that a return to greater use of human muscle power would be healthy for people because it would promote fitness and **energy conservation**, as well as an impact on the **biosphere**.

Sears was also one of the few prominent ecologists to successfully write for popular audiences. At least one of his popular books, *Deserts on the March*, first published in 1935, has become a minor American classic, reprinted in a fourth edition in 1980. The title provides an apt summary: in it, Sears documents the mistakes American farmers made in creating conditions that led to the disastrous **Dust Bowl**. This book had a major influence on the **soil conservation** movement in the United States.

Throughout his life, Sears believed that the best way to solve ecological problems was to teach every person about their own immediate **environment**: "Each of us can begin quite simply by learning to look about himself, wherever he may be." The touchstone of education, especially a scientific one, should be "the final ability to read and enjoy the landscape. While there is life there is hope, but only for the enlightened."

[*Gerald L. Young Ph.D.*]

Resources
Books

Sears, P. B. *Deserts on the March.* 4th ed. Norman: University of Oklahoma Press, 1980.

———. "The Processes of Environmental Change by Man." In *Man's Role in Changing the Face of the Earth*, edited by W. L. Thomas Jr. Chicago: The University of Chicago Press, 1956.

———. *Where There is Life: An Introduction to Ecology.* New York: Dell, 1970.

Noah Seattle (1786 - 1866)

Duwamish chief

Noah Seattle (or See-athl) was a chief of the Duwamish or Suquamish tribe, one of the Salish group of the Northwest Coast of North America. Born in the Puget Sound area in 1786, Seattle lived there until his death on June 7, 1866. He was baptized a Roman Catholic about 1830 and is buried in the graveyard of the Port Madison Catholic Church. Ironically, Native Americans were banned by law from living in Seattle, Washington, the city named after him, one year after the chief's death.

By most accounts, Seattle was a great orator and a skilled diplomat. Although he never fought in a war against white people, he was a warrior with a reputation for daring raids on neighboring tribes. Seattle owned eight Native American slaves, but freed them after President Abraham Lincoln's Emancipation Proclamation. He was the first to sign the Port Elliott Treaty negotiations of 1855, which surrendered most Native American lands in the Puget Sound area for white settlement. He gave two short speeches on that occasion, which are preserved in the National Archives

in Washington, D.C. Both speeches encouraged others to sign the agreement and to cooperate with the United States authorities.

Seattle has become famous in recent years as the author of one of the most widely quoted pieces of environmental literature in the world. Among some familiar passages are: "How can you buy or sell the sky, the warmth of the land? The earth is our mother. This we know, the earth does not belong to man. Man belongs to the earth. Whatever befalls the earth, befalls the sons of the earth. Man did not weave the web of life; he is merely a strand in it. Whatever he does to the web, he does to himself."

Generally called the speech, letter, or lament of Chief Seattle, this text exists in many forms. It has been set to music, called the *Fifth Gospel* in religious services, and used to sell everything from toilet tissue to recyclable plastic bags. A version was used by artist Susan Jeffers in 1991 to accompany drawings in a best selling children's book called, *Brother Eagle, Sister Sky.* This piece is a poetic, moving, environmentally sensitive work that supports beliefs about the reverence of Native Americans for the earth. Unfortunately, there is no evidence that Chief Seattle ever said any of the wise words attributed to him.

Historian Rudolf Kaiser has traced the origins and myths surrounding the Chief Seattle text. The first report of what we now know as Seattle's speech appeared in an article by Dr. Henry A. Smith published in the *Seattle Sunday Star* on October 29, 1887, as part of a series of old pioneer reminiscences. Smith was recalling remarks made by the Chief in 1854 on the arrival of Governor Stevens to the territory. Although 33 years had passed since the event, Smith wrote that he clearly remembered the "grace and earnestness of the sable old orator."

As Smith reconstructed it, the speech was dark and gloomy, the farewell of a vanishing race. "Your God loves his people and hates mine...It matters but little where we pass the remainder of our days. They are not many...Some grim Nemesis of our race is on the redman's trail, and wherever he goes he will still hear the sure approaching footsteps of the fell destroyer." This version was mostly a justification for displacement of the native peoples by the pioneers. Smith remembered the Chief saying that the President's offer to buy Native American lands was generous "for we are no longer in need of a great country." This version has no mention of the web of life or other ecological concepts.

In 1969 Smith's stuffy Victorian prose was translated into modern English by poet William Arrowsmith. Two year later, Ted Perry wrote the script for a film called *Home* produced by the Southern Baptist Convention. He used some quotes from Arrowsmith's translation together with a great deal of imagery, symbolism, and sentiments of 1970s **environmentalism**. Perry is the source of 90% of what we now know as Seattle's speech. Over Perry's objections, the film's editors attributed the text to Chief Seattle to make it seem more authentic.

The piece has since assumed a life of its own. Some obvious inconsistencies exist, such as "hearing the lovely cry of the whippoorwill" or having "seen a thousand rotting buffaloes on the **prairie**, left by the white man who shot them from a passing train." Chief Seattle lived his whole life in the Puget Sound area and died 13 years before the railroads reached the West Coast. He never heard a whippoorwill or saw a buffalo. Many people who use the speech are unperturbed by evidence that it is untrue. It so perfectly supports their view of Native Americans that they believe some Native American should have said these things even if Seattle did not.

This story is a good example of a myth that persists because it fits our preconceived views. In Chief Seattle, we seem to find an unsophisticated primitive who spoke in poetic, beautiful language and prophesied all the evils we now experience. Indigenous cultures probably contain much ecological wisdom that can be taught to others, but care must be taken not to blindly accept mythological texts such as this.

[*William P. Cunningham Ph.D.*]

Resources
Books

Kaiser, R. "Chief Seattle's Speech(es): American Origins and European Reception—Almost a Detective Story." In *Indians in Europe*, edited by C. F. Feest. Gottingen, 1985.

Secchi disk

A simple instrument used to measure the transparency of surface water bodies such as lakes and reservoirs. A Secchi disk is a metal, plastic, or wooden disk that is 8–12 in (20–30 cm) in diameter. It may be white or have alternating black and white quadrants. A long calibrated rope is attached to the center of the disk. Researchers use the disk by lowering it into the water, usually from the shaded side of a boat, and measuring the depth at which it just disappears from sight. The disk is then lowered a little more and then raised until it reappears. The average of the depths at which the disk disappears and then reappears is known as the Secchi depth. The amount of **sediment**, algae, and other solids in the water affects clarity and depth of light penetration and thus, Secchi depth. Scientists use this measure of transparency as a simple indication of **water quality**, often by comparing Secchi depth measurements over a period of time to track changes in the clarity of a particular water body.

Testing water clarity with a secchi disk. (Photograph by W.A. Banaszewski. Visuals Unlimited. Reproduced by permission.)

Second World

The former Soviet Union and those (formerly) communist states that were politically and economically allied with it are often informally called the Second World. The term developed after World War II to distinguish communist bloc states from powerful capitalist states (**First World**) and from smaller, nonaligned developing states (**Third World**). These terms developed as a part of theories that the world's countries together made up a single dynamic system in which these three sectors interacted. Although the term is often used to indicate countries and regions of intermediate economic strength, its proper meaning has to do with political economy, or the politics of communist-socialist economics. In recent years, with the shrinking role and then disintegration of the communist bloc, the term has become much less common than those of First and Third Worlds.

Secondary recovery technique

The term secondary recovery technique refers to any method for removing oil from a **reservoir** after all natural recovery methods have been exhausted. The term has slightly different meanings depending on the stage of recovery at which such methods are used.

The oil trapped in an underground reservoir is typically mixed with water, **natural gas**, and other gases. When a well is sunk into the reservoir, oil may flow up the well pipe to the earth's surface at a rate determined by the concentration of these other substances. If the gas pressure is high, for example, the oil may be pushed out in a fountain-like gusher.

Flow out of the reservoir continues under the influence of a number of natural factors, such as gravity, pressure of surrounding water, and gas pressure. Later, flow is continued by means of pumping. All such recovery approaches that depend primarily on natural forces are know as primary recovery techniques.

Primary recovery techniques normally remove no more than about 30% of the oil in a reservoir. **Petroleum** engineers have long realized that another fraction of the remaining oil can be forced out by fluid injection. The process of fluid injection involves the drilling of a second hole into the reservoir at some distance from the first hole through which oil is removed. Some gas or liquid is then pumped down into the second hole, increasing pressure on the oil remaining in the reservoir. The increased pressure forces more oil out of the reservoir and into the recovery pipe.

The single most common secondary recovery technique is water **flooding**. When water is pumped into the second well, it diffuses out into the oil reservoir and tends to displace oil from the particles to which it is absorbed. This process forces more of the residual oil up into the recovery pipe.

Water flooding was used as early as 1900, but did not become legal until 1921. A common practice was to drill a series of **wells**, some of which were still producing and others of which employed water injection. As the former became exhausted, they were converted to water injection wells and another group of producer wells were drilled. The process was repeated until all available oil was recovered from the field.

Recently, more sophisticated approaches to fluid injection have been developed from a reservoir. The two fluids that have been used most extensively in these approaches are liquid **hydrocarbons** and **carbon dioxide**. The principle behind liquid hydrocarbons is to find some material that will mix completely with oil and then push the oil-mixture that is formed out of the reservoir into the recovery pipe. A commonly used hydrocarbon for this process is liquified petroleum gas (LPG), which is completely miscible with oil.

Since LPG is fairly expensive, only a small volume is actually used. It is pumped down into the reservoir and followed by a "pusher gas." The pusher gas, often **methane**, is inexpensive and can be used in larger volume. The pusher gas forces LPG into the reservoir where it (the LPG) mixes with residual oil.

This system has worked well in the laboratory, but not so well in actual practice. The LPG has a tendency to get lost in the reservoir to an extent that it does not effectively remove very much residual oil.

The most effective fluid now available for injection appears to be **carbon** dioxide. A mixture of carbon dioxide and water is pumped down into the reservoir and followed by an injection of pure water that drives the carbon dioxide-water mixture through the reservoir. As carbon dioxide comes into contact with oil, it dissolves in the oil, causing it to expand and break loose from surrounding rock. The oil-carbon dioxide-water is then pumped out of the recovery pipe where the carbon dioxide is removed from the mixture and re-used in the next recovery pass.

The carbon dioxide process has been effective in removing oil after water flooding has already been used and only 25% of the oil in a reservoir still remains. In most cases, however, it is more efficiently used with reservoirs containing a larger fraction of residual oil.

Fluid injection is one type of secondary recovery technique. Another whole group of methods can also be used to extract the remaining oil from a reservoir. If these methods are employed after fluid injection has been tried, they are often referred to as tertiary recovery techniques. If they are used immediately after primary recovery, they are known as secondary recovery techniques. A whole set of recovery techniques can be called by different names, therefore, depending on the stage at which they are used. It is becoming more common today to refer to *any* method for removing the residual oil from a reservoir as an enhanced recovery technique.

Another technique for removing residual oil from a reservoir makes use of surfactants. A surfactant is a substance whose molecules are attracted to water at one end and oil at the other end. The most familiar surfactants are probably the soaps and **detergents** found in every home.

If surfactants are injected into an oil reservoir, they will form emulsions between the oil and water in the reservoir. The oil is essentially washed off particles of rock in the reservoir the way grease is washed off a pan by a household detergent. The emulsion that is formed is then pushed through the reservoir and out the producer pipe by a flood of water pushed down the injection pipe.

The surfactant method works well in the laboratory, although it has been less successful in the field. Surfactants tend to adsorb on rock particles and get left behind as the water pushes forward. Methods for overcoming this problem are now being explored.

One of the fundamental problems with recovering residual oil in a reservoir is that oil droplets often have difficulty in squeezing through the small openings between adjacent rock particles. The use of surfactants is one way of helping the oil particles slip through those openings more easily. Another approach is to increase the temperature of the oil in the reservoir, thereby reducing its viscosity (tendency to flow). As it becomes less viscous, the oil can more easily force its way through pores in the reservoir.

One of the earliest applications of this principle, the steam soak method, was first used in Venezuela in 1959. In this method, steam is injected into one part of the reservoir and the producer pipe is closed off. After a few days, the pipe is reopened, and the loosened oil flows out. This process is repeated a few times before a change is made in the method and steam is piped in continuously while the producer pipe remains open.

Steam injection works especially well with heavy oils that are not easily displaced by other secondary recovery techniques. It is now used commercially in a number of fields, primarily in Venezuela and California.

A dramatic form of secondary recovery is *in situ* (in place) **combustion**. The principle involved is fairly simple. A portion of the oil in the reservoir is set on fire. The heat from that fire then warms the remaining residual oil and reduces its viscosity, forcing it up the producer pipe.

In practice, the fire can sometimes be made to ignite spontaneously simply by pumping air down the **injection well**. In some cases, however, the oil must actually be ignited at the bottom of the well. The temperature produced in this process may reach 650–1200°F (350–650°C) and the region of burning oil may creep through the rock at a speed of 1–12 in (3–30 cm) per day. As the fire continues, air, and usually water, are continually pumped into the injection well to keep the combustion zone moving. Under the best of

circumstances, *in situ* combustion has recovered up to half of all the oil remaining in a reservoir.

Research has demonstrated that each recovery method is suitable for particular kinds of reservoirs. Oil viscosity, rock porosity, depth of the reservoir, and amount of oil remaining in the reservoir are all factors in determining which method to use. To date, however, the only method that has proved to be practical in actual field situations is steam injection.

[*David E. Newton*]

RESOURCES
BOOKS

Dickey, P. A. *Petroleum Development Geology*. 3rd ed. Tulsa, OK: PennWell Books, 1986.

Secondary standards

Pollution control levels that affect the welfare of plants, animals, buildings and materials. Secondary standards may relate to either **water quality** or **air pollution** under the control of either the **Safe Drinking Water Act** or the **Clean Air Act**. These so-called "welfare effects" are thought not to affect human health but primarily crops, livestock, vegetation, and buildings. Secondary standards can refer to effects to which some monetary value could be ascribed, such as damage to materials, **recreation**, **natural resources**, community property, or aesthetics. **Primary standards**, on the other hand, refer to **pollution** levels that affect human health.

Second-hand smoke

see **Cigarette smoke**

Secure landfill

see **Landfill**

Sediment

A mixture of sand, **silt**, clay, and perhaps organic components. **Soil** eroded from one location and deposited in another is identified as sediment. The sedimentary fraction has the ability to carry not only the mineral (sand, silt, and clay) and organic (**humus**) components, but also other components that may be attached such as **nitrogen** compounds, herbicides, and pesticides. These riders are of high concern to those involved in environmental studies. Products applied to the soil in one location and beneficial to that system may be transported to other locations where the effect is detrimental to the **habitat** of other life forms. Care must be exercised: 1) when applying supplementary items to the soil, and 2) to develop systems that keep sediment from finding its way into the streams and water bodies.

Sedimentation

The deposition of material suspended in a liquid. Sedimentation is normally considered a function of water deposition of the finer **soil** separates of sand, **silt**, and clay, but it may also include organic debris. Sometimes this is referred to as the **siltation** process, although there may be other fractions of material present other than silt. The term can be applied to wind-transported sediments as well. Sedimentation can be both harmful and beneficial. River and stream channels, reservoirs, and other water bodies may be degraded because of the deposition of **sediment** materials. Many of the most important food- and fiber-producing soils of the world have been developed from the deposition of fine particulates by both water and wind. In some cases, large topographic land forms are the result of long-term sedimentation processes.

Seed bank

A seed bank is the reservoir of viable seeds present in a plant community. Seed banks are evaluated by a variety of methods. For some **species**, it is possible to make careful, direct counts of viable seeds. In most cases, however, the surface substrate of the **ecosystem** must be collected and seeds encouraged to germinate by exposure to light, moisture, and warmth. The germinating seedlings are then counted and, where possible, identified to species.

In most cases, the majority of seeds are found in surface layers. For example, the organic-rich forest floor contains almost all of the forest's seed bank, with much smaller numbers of seeds present in the mineral **soil**.

The seeds of some plant species can be remarkably long-lived, extending the life of the seed bank. For example, in northeastern North America, the seeds of pin cherry (*Prunus pensylvanica*) and red raspberry (*Rubus idaeus*) can persist in the forest floor for perhaps a century or longer. This considerably exceeds the period of time that these ruderal species are present as mature, vegetative plants during the initial stages of post-disturbance forest **succession**. However, because these species maintain a more-or-less permanent presence on the site through their persistent seed bank, they are well placed to take advantage of temporary opportunities of resource availability that follow disturbance of the stand by **wildfire**, windstorm, or harvesting.

The seeds of many other plant species have only an ephemeral presence in the seed bank. In addition to some tropical species whose seeds are short-lived, many species in temperate and northern latitudes produce seeds that cannot survive exposure to more than one winter. This is a common trait in many grasses, asters, birches, and most conifers, including pines, spruces, and fir. Often these species produce seeds that disperse widely, and can dominate the short-lived seed banks during the autumn and springtime. Species with an ephemeral presence in the seed bank must produce large numbers of well-dispersed seeds each year or at least frequently, if they are to successfully colonize newly disturbed sites and persist on the landscape.

Although part of the plant community, seed banks are much less prominent than mature plants. In some situations, however, individual plants in the seed bank can numerically dominate the total-plant density of the community. For example, in some cultivated situations the persistent seed bank can commonly build up to tens of thousands of seeds per square meter and sometimes densities which exceed 75,000 seeds per square meter. Even natural communities can have seed banks in the low tens of thousands of seeds per square meter. However, these are much larger than the densities of mature plants in those ecosystems.

The seed bank of the plant community is of great ecological importance because it can profoundly influence the vigour and species composition of the vegetation that develops after disturbance.

[*Bill Freedman Ph.D.*]

Resources

Books

Harper, J. L. *Population Biology of Plants.* San Diego: Academic Press, 1977.

Grime, P. *Plant Strategies and Vegetation Processes.* New York: Wiley, 1979.

Seed dressings

see **Methylmercury seed dressings**

Seepage

The process by which water gradually flows through the **soil**. Seepage is the cause of a variety of environmental problems. For example, pesticides used on a farm may enter **groundwater** and be transported by seepage to a human water supply. In some cases, toxic or radioactive wastes stored in sealed tanks underground have gotten into water supplies by seepage after the tanks have rusted and broken apart. *See also* Water quality

Selection cutting

A harvesting method that removes mature trees individually or in small groups. The resulting gaps in the canopy allow understory trees to develop under the protection of the remaining overstory. Among all regeneration methods selection cuttings offer seedlings the most protection against sun and wind. They also protect against the **erosion** of **soil** and maintain aesthetic value in forests. On the negative side, selection **logging** can damage the remaining trees and requires highly skilled labor and more expense. Because most of the overstory remains, selection cutting favors shade-tolerant **species** and is not appropriate when the goal is growth of light-demanding species. *See also* Clear-cutting

Sellafield (U.K.)

see **Windscale (Sellafield) plutonium reactor**

Sense of place

Every adult seems to remember a special place from their past: a place of refuge as a child; sites of family vacations; a grandparent's farm; somewhere shared with a loved one at a special time. Many people, though often with some embarrassment, will confess to a favorite place in the present where they can go to be alone, to gather with friends, or spend leisure time. Places we can identify with—and that feel special. A literature about that identification with place—that sense of place—has burgeoned in the last two decades of the twentieth century.

Yet, contemporary literature, both fictional and factual, is packed with documentation of alienation, rootlessness, displaced people, loss of connection to any particular place, loss of identity and character in suburb and city. In 1957, in *America as a Civilization*, Max Lerner documented the loss of community and the quest to regain it, claiming "this is what I call the problem of place in America, and unless it is somehow resolved, American life will become more jangled and fragmented than it is, and American personality will continue to be unquiet and unfulfilled." As the century ends, an increasing number of scholars, thinkers, and writers are advocating a reinhabitation of place, a **reclamation** of the spirit and sense of place as a solution to numerous contemporary problems, including **environmental degradation**, sustainability, and social alienation.

Sense of place as a phrase has at least two meanings. First, the particular characteristic of a place that makes it what it is. For example, though few people have visited **Antarctica**, most have some sense, an image in their mind's eye, of what that continent is like. That image may be

realistic, or unrealistic, or may be dramatically simplified, but it will usually be based on physical characteristics that the place actually does have.

The second meaning is the particular sense that individuals have of places they know by experience; we all have a sense of many places that we have visited, but a sense of the same place is experienced in many ways by many different people. Even after a visit to Antarctica, for example, people will return with a variety of different views, depending on their reasons for being there, on how long they stayed, and on how much they know about the place. An explorer will carry one view, a natural scientist studying how organisms adapt to extreme environments will bring back a quite different view, and a tourist yet another viewpoint. Yi-Fu Tuan even claimed that "sense of place...implies a certain distance between self and place that allows the self to appreciate a place."

For many, a third meaning is the only one of consequence: that one can gain a sense of place only from being or becoming deeply involved with a place and by coming to know that one place and its inhabitants intimately. This is the meaning implicit in the claim that modern Americans must regain a sense of place to counteract their mobility and alienation from **environment**. The poet Gary Snyder refers to this when he claims that "there are many people on the planet now who are not 'inhabitants.' He teaches that spirit of place is accessed only through knowledge gained by direct experience in a specific locale; "Know the plants" has become almost his mantra. When you really know the plants, you are beginning to get a sense of the place, a sense of what is possible and a sense of how to live there in harmony with the site and setting.

Interwoven with the concept of place is the sense of home. We all have a sense of many places (in the first two meanings of the phrase), but a feeling of home is always associated with a sense of that specific place where home is thought to be. Home can be a dwelling, a town, even a state or a nation (in the sense of "homeland")—or all of these—but it always is a place, at one or several scales. Most of us live in some degree of intimacy with the place we call home; even if that is restricted to a dwelling, it is a place we know and are to some degree comfortable with. Beyond the dwelling, and even our identity with region and nation, to recognize what it means to say "the earth is our home" might change the way we behave toward the world on which we live.

Scholars and students agree that sense of place in all three meanings is best developed by increasing one's knowledge of the locality in question. A full sense of the third meaning can only be developed by really getting to know one place well. That usually means living there, becoming engaged with the surroundings, getting involved with the community, making enough of a commitment of time to get beyond the surface. It takes real effort to get to know the plants (even in cities, to know what the trees and shrubs are in the parks and on the streets, and to know why those particular **species** are there rather than others), to know, in rural areas, what crops are planted and why, to know the climatic patterns and the reasons for those patterns. If such a sense of place could be enhanced and strengthened, this might change the way people use energy in their homes and cars, interact with other parts of the environment, and consider how connected each of us really is to the particular places where we live.

Some of the best American literature focuses on the complex relationships of an individual writer with a specific place. For example, Henry David Thoreau's *Walden Pond* trumpeted the personal significance and the transcendence of one small place. More recently, Aldo Leopold's struggle to reclaim a ravaged farm in the sand counties of Wisconsin has become a metaphor for attempts to understand the importance to human lives of being immersed in the workings of a place of one's own. Annie Dillard's mystical tome on Tinker Creek has, as a modern day Walden, helped many better understand the spirit of place. And, the "Refuge" Terry Tempest Williams found on the shores of the Great Salt Lake has generated empathy for the human condition and for the natural world with which it is so entwined.

If, says the poet Wendell Berry, you don't know where you are, you don't know who you are. People who have a better sense of who they are by knowing more about the places where they are, start paying attention to the impact those places have on their lives. Recognizing that place and identity are related, that place and well being cannot be disconnected, should lead to taking better care of those places.

[*Gerald L. Young Ph.D.*]

Resources

Books

Swan, J., and R., eds. *Dialogues With the Living Earth: New Ideas on the Spirit of Place From Designers, Architects, and Innovators.* Wheaton, IL: Quest Books-Theosophical Publishing House, 1996.

Tall, D. *From Where We Stand: Recovering a Sense of Place.* Baltimore: The Johns Hopkins University Press, 1993.

Periodicals

Cuba, L., and D. M. Hummon. "A Place to Call Home: Identification With Dwelling: Community, and Region," *The Sociological Quarterly* 34, no. 1 (Spring 1993): 111–131.

Septic tank

Nearly 20 million homes, which include almost 30% of the population of the United States, dispose of their **waste-**

water through an on-site disposal system. The most commonly used type of system is the septic tank, which is an individual treatment system that uses the **soil** to treat small wastewater flows. The system is usually used in rural or large lot settings where centralized wastewater treatment is impractical. Septic tank systems are designed specifically for each site, using standardized design principles that are usually state-regulated. Septic tank systems commonly contain three components: the septic tank, a distribution box, and a drainfield, all of which are connected by conveyance lines.

The septic tank serves to separate solids from the liquids in the wastewater. All sources of wastewater, including those from sinks, baths, showers, washing machines, dishwashers, and **toilets**, are directed into the septic tank, since any of these waters can contain disease-causing **microorganisms** or environmental pollutants. The size of the septic tank varies depending on the number of bedrooms in the home, but an average tank holds 1,000 gal (3,790 l) of liquid. Wastewater in a septic tank is treated by **anaerobic** bacteria that digest organic materials, while encouraging the separation of solid materials from the wastewater. The solids accumulate and remain in the septic tank in the form of **sludge**, which collects at the bottom of the tank, and also in the form of scum, which floats on the top of the wastewater. Periodically (for example, every two or three years) the indigestible sludge and scum (referred to as septage) are removed from the tank by pumping and are disposed of in a septage disposal system (like a municipal **sewage treatment** system). Unfortunately, many homeowners do not properly maintain septic systems by pumping them out as frequently as every two to three years and often wait until there is a back up in the system or some other type of problem. Periodic pumping is designed to prevent the solids from leaking out of the septic tank in the wastewater **effluent**. The effluent from the septic tank is a cloudy liquid that still contains many pollutants (including **nitrogen** compounds, suspended solids, and organic and inorganic materials) and microorganisms (including bacteria and viruses, some of which that may be potentially pathogenic), which require further treatment.

Treatment of the wastewater effluent from the septic tank is continued by transporting the wastewater by gravity to a soil **absorption** field through a connecting pipe. The absorption field is also referred to as the soil drainfield or the **nitrification** field. The absorption field consists of a series of underground perforated pipes covered with soil and turf, which may be connected in a closed loop system. The wastewater enters a constructed gravel bed (the trench) through perforations in the pipe, where it is stored before entering the underlying unsaturated soil. The wastewater is treated as it trickles into and through the soil by **filtration** and **adsorption** processes as well as by **aerobic** degradation processes before the wastewater enters the ground water. Filtration removes most of the suspended solids and may also remove microorganisms. Adsorption is the process by which pollutants and microorganisms are attracted to and held on the surfaces of soil particles, thus immobilizing them. Adsorption attracts such nutrients such as **phosphorus** and some forms of nitrogen (mostly ammonium [NH^4]) and is most effective when fine-textured soil is used as the adsorption medium. However, soils with a very fine texture, such as soils high in clay, may have too low permeability to allow much wastewater to pass through the soil. Microbial degradation results in the removal of many remaining nutrients and organic materials. If the volume and type of soil underlying a soil absorption system are adequate, most pollutants (with the exception of nitrate nitrogen) should be removed before the wastewater reaches the **groundwater**.

Estuaries (bays, harbors, etc.) often experience nitrogen **loading** from septic systems along the shore (within 656 mi [200 m] of the shore or greater in sandy soil environments). Typically on-site disposal systems such as septic tanks cannot remove more than 50–60% of the nitrogen (mostly in the form of nitrate) and it ends up in the nearby coastal waterway, which is a problem because many of the estuaries around the world are experiencing problems with eutrophication as a direct result of overloading of nitrogen.

Although some difficulties can arise with septic systems, there are some simple practices that can prevent common problems. For example, **groundwater pollution** and surfacing of untreated or poorly filtered effluent from a septic tank system can be prevented by ensuring that excessive amounts of water are not allowed to enter or flood the drainfield. Reduced production of wastewater, or **water conservation**, is recommended to prevent system overload. Water from roof drains, basement sump pump drains, and other rain water or surface water **drainage** systems should be directed away from the absorption field. The functioning of the system can also be damaged by the addition of such materials as coffee grounds, wet-strength towels, **disposable diapers**, facial tissues, cigarette butts, and excessive amounts of grease, which can clog the inlet to the septic tank, or if carried out of the septic tank, may impede drainage of wastewater in the soil absorption field. The septic tank should be pumped more frequently if a **garbage** disposal is used.

Groundwater **pollution** can also be caused by the addition of hazardous **chemicals** to the septic tank system, which may be transported through the system to the ground water without removal or treatment in the system. Hazardous chemicals may be found in such commonly used products as pesticides, solvents, latex paint, oven cleaners, **dry cleaning** fluids, motor oils, or degreasers.

Siting requirements for a soil absorption system depend on the amount of daily sewage flow and site conditions

that affect the ability of the soil to absorb, treat, and dispose of septic tank effluent without creating a public health hazard or contamination of ground or surface waters. If a proposed site is located on a gently sloping surface that is not susceptible to **flooding**, and has at least 6 ft (1.8 m) of well-drained, **permeable** soil, with a low content of coarse fragments, only a minimum area is required for installation of the absorption drainfield. However, area requirements increase as slope increases and soil permeability, depth of suitable soil, and depth to groundwater decrease. However, at some sites, the soil type, depth, or site **topography** may not be suitable for the use of the conventional soil absorption drainfield, and modifications or additions to the conventional system may be required.

Siting requirements usually also include that sufficient area be reserved for installation of a repair system if the original system fails. This additional area should meet all requirements of the original soil absorption system and should be kept free of development and traffic. The disposal site should also be located at safe distances from ground water supply sources, **wetlands**, lakes, streams, drain tile, and escarpments where **seepage** may occur, as well as set back from buildings and roads that may interfere with the proper operation of the system.

A common problem encountered in drainfields is excessive development of a clogging mat at the interface between the gravel bed in the absorption trench and the underlying soil due to the accumulation of organic materials and the growth of microorganisms. The development of a clogging mat is a natural process and at some sites, such as those with sandy soils, may be desirable to slow the movement of the water through the sandy materials to allow for treatment of the wastewater to occur. However, excessive development of a clogging mat may result in formation of anaerobic soil conditions, which are less conducive to degradation of the **organic waste** materials, as well as in surfacing of the effluent or backing up of the wastewaters into the residence.

At sites with limiting features or where problems with the excessive development of clogging mats occur, landowners may modify or enhance the conventional soil absorption drainfield to increase its performance. Examples of such alterations include:

- Alternating drainfields: The wastewater effluent from the septic tank is directed to two or more separate drainfields; a section of the drainfield receives effluent for six to 12 months and then is allowed to rest for a similar period of time. The clogging mat that forms at the soil/trench interface dries and is oxidized during this dormant period, thus increasing the expected life span of the section and restoring aerobic soil conditions.
- Pressure distribution: A pressure head is created within the distribution pipe system in the drainfield. This is usually achieved by using a dosing tank and a pump or a siphon and yields uniform distribution of the wastewater throughout the system. The pressure distribution system differs from a conventional system because approximately the same amount of effluent flows out of each hole in the distribution pipes, rather than a concentrated amount of effluent flowing by gravity into a few localized areas. The effluent is discharged periodically to the drainfield so that a dose/rest cycle is maintained, and in turn allows for the wastewater to be absorbed into the underlying soil before additional effluent is added to the drainfield. The dose/rest cycle may also slow down the formation rate of the clogging mat that naturally occurs over time. Most commonly, landowners use these types of pressure distribution systems (which also employ vegetation to help evaporate liquids): (a) low-pressure subsurface pipe distribution system, which consists of a network of small-diameter perforated plastic pipes buried in narrow, shallow trenches; (b) mound system, in which wastewater is pumped to perforated plastic pipe that is placed in a vegetated sand mound constructed above the natural surface of the ground; and (c) **evapotranspiration** bed, in which a vegetated sand bed is lined with plastic or other waterproof material.

A wide variety of onsite septic systems exist from which to select the most appropriate for a specific site. The primary criterion for selection of an appropriate system is the ability of the system to protect public health and to prevent **environmental degradation** at the specific site.

Experimental septic treatment systems now run in private solar-powered greenhouses that house complete ecosystems of plants and fish to create a natural cleansing action that can process wastewater into clean drinking water in as little as 12 days. Systems were cropping up sparingly by the end of the 1990s, often in milder climates where the solar source is plentiful. One self-contained eco-system proved capable of servicing an entire small town at a development cost as low as $300,000. The potential for **third world** implementation is under consideration, and prototype systems developed in the early part of the decade were implemented even in the harsh New England **climate**, where **water pollution** is particularly severe.

Serious ecological threats in the northeast led to a series of comprehensive changes to the Massachusetts State Environmental Code (Title V) beginning in 1995. The state set restrictions on the siting and maintenance of septic tanks and systems and imposed requirements for state-certified testing of systems whenever a property title transfers. The code changes were prompted by dangerous levels of water pollution making it unsafe to swim and fish in many of the state's lakes and bays and contributing to **toxins** in the drinking water. The legislation requires that detailed engineering plans be included as part of construction permits.

Initial inspection of existing systems is required combined with regular inspections of certain private and public systems. Follow-up inspections are mandated coincident with any changes to building occupancy. Permissible septic system additives are clearly defined.

Some developers hope to eliminate the septic tank altogether with high-tech waste systems allowing an entire sewer system to be housed inside a single-family dwelling, including a 91-gal (344-l) wastewater tank that operates on 240 volts of electricity. The system, with an ability to pump wastewater vertically or as far as 2 mi (3.2 km) horizontally, can easily follow the contour of the landscape. The system is easier and cheaper to build because it eliminates the deep digging associated with septic tank system installation and offers the advantages of a completely contained sewer system.

[*Judith Sims*]

Resources
Books

Design Manual: Onsite Wastewater Treatment and Disposal Systems. Cincinnati: U.S. Environmental Protection Agency, Municipal Environmental Research Laboratory, 1980.

National Small Flows Clearinghouse. *So...Now You Own a Septic Tank.* Morgantown, WV: West Virginia University, 1990.

Northern Virginia Planning District Commission. *A Reference Guide for Homeowners: Your Septic System.* Morgantown, WV: West Virginia University, National Small Flows Clearinghouse, 1990.

Small Wastewater Systems: Alternative Systems for Small Communities and Rural Areas. Washington, DC: U.S. Environmental Protection Agency, Office of Water Program Operations, 1980.

Serengeti National Park

Serengeti National Park lies in northern Tanzania between Lake Victoria and the East African Rift Valley. It was established in 1929 (and expanded in 1940) to protect 5,600 square miles (14,500 sq. km) of the Serengeti plains **ecosystem**. This vast park spans an area twice the size of **Yellowstone National Park** and supports over 94 **species** of mammals, 400 species of birds, and includes the spectacular **migration** routes of the largest herds of grazing animals to be seen anywhere in the world.

Each year, migrating herds move clockwise around the Park, constantly seeking better feeding grounds. Changing water availability is the key factor in the annual migrations, which correlate closely with the local cycles of rainfall. At the right time of year, visitors to the Park can see hundreds of thousands of migrating herds of wildebeest (*Connochates taurinus*), running in winding lines several miles long. The wildebeest are accompanied by herds of zebra (*Equus burchelli*) and Thomson's gazelle (*Gazella thomsoni*).

During the peak dry season (August to November), the grazers congregate in the Serengeti's northern extension, moving south east when the first storms appear in December. The greatest concentration of animals occurs in the short grass pastures of the eastern Serengeti from December to May when millions of the Park's migratory grazers assemble. They are accompanied by packs of nomadic predators such as lion (*Panthera leo*) and hyenas (*Crocuta crocuta*). As the dry season progresses (June and July) and the grazing gets worse, the massive herds move to the wooded savannas of the western corridor of the Park. By mid-August, when this food supply is exhausted, the herds move back into the northern plains, crossing also into the adjacent Masai Mara National Reserve in Kenya.

The non-migrant inhabitants of the Serengeti plains include ostrich (*Struthio camelus*), impala (*Aepyceros melanious*), topi (*Damalicus korrigum*), buffalo (*Syncercus kaffer*), giraffe (*Giraffa camelopardalis*), Grant's gazelle (*Gazella granti*), leopard (*Panthera pardus*), cheetah (*Acinonyx jubatus*), hunting dogs (*Lycaeon pictus*), and jackals (*Canis adustus*).

The three principle species of migratory grazers—zebra, wildebeest, and Thomson's gazelle—do not compete directly for food, and it is common to find them grazing together. **Zebras** eat the upper parts of the grass shoots, exposing the softer leaf bases for the wildebeest. Grazing wildebeest in turn expose the herb layer beneath the grass, which is eaten by gazelles. Balanced populations of grazers actively maintain the **stability** of grassland ecosystems: too little grazing will allow woody vegetation to grow, while too much grazing will turn grassland into **desert**.

Today, the Serengeti National Park is under pressure from the rapidly growing human population outside the Park. Domestic cattle herders and farmers operate inside Park boundaries, competing with the **wildlife** for food, water, and land, while well-armed poachers kill game for meat, horns, and tusks. Human ancestors (*Australopithecus robustus* and *Homo habilis*) once hunted game on the Serengeti plains, as evidenced by the finds of the Louis and Mary Leakey at Olduvai Gorge, which lies close to the Park's eastern entrance.

[*Neil Cumberlidge Ph.D.*]

Resources
Books

A Field Guide to the National Parks of East Africa. London: J. G. Williams, Collins, 1981.

Schaller, George B. *Golden Shadows, Flying Hooves.* Chicago: University of Chicago Press, 1970.

Zebras and wildebeasts at Ngorongoro Crater, Serengeti National Park, Tanzania. (Photograph by Carolina Biological Supplies. Phototake. Reproduced by permission.)

Seveso, Italy

Accidents in which large quantities of dangerous **chemicals** are released into the **environment** are almost inevitable in the modern world. Toxic chemicals are produced in such large volumes today that it would be a surprise if such accidents were never to occur. One of the most infamous accidents of this kind occurred at Seveso, Italy, a town near Milan, on July 10, 1976.

The Swiss manufacturing firm of Hoffman-LaRoche operated a plant at Seveso for the production of hexachlorophene, a widely used disinfectant. One of the raw materials used in this process is 2,4,5-trichlorophenol (2,4,5-TCP). At one point in the operation, a vessel containing 2,4,5-TCP exploded, releasing the chemical into the **atmosphere**. A cloud 100–160 ft (30–50 m) high escaped from the plant and then drifted downwind. It eventually covered an area about 2,300 ft (700 m) wide and 1.2 m (2 km) long.

Although 2,4,5-TCP is a skin irritant, it was not this chemical that caused concern. Instead, it was an impurity in 2,4,5-TCP, a compound called 2,3,7,8-tetrachlorodibenzo-p-dioxin, that caused alarm. This compound, one of a family known as dioxins, is one the most toxic chemicals known to science. It occurs as a by-product in many manufacturing reactions in which 2,4,5-TCP is involved. Experts estimated that 7–35 lb (3–16 kg) of **dioxin** were released into the atmosphere as a result of the Seveso explosion.

People living closest to the Hoffman-LaRoche plant were evacuated from their homes and the area was closed off. About 5,000 nearby residents were allowed to stay, but were prohibited from raising crops or farm animals.

Damage to plants and animals in the exposed area was severe. Thousands of farm animals died or had to be destroyed. More than 2.5 tons (225 kg) of **contaminated soil** were removed before planting could begin again. Short- and long-term effects on human health, however, were relatively modest. In the months following the accident, 176 individuals were found to have chloracne, an inflammation of the skin caused by chlorine-based chemicals. An additional 137 cases of the condition were found in a follow-up survey six months after the accident.

Other health problems were also detected in the human population. About 8% of the exposed population had enlarged livers and a few residents showed signs of minor nerve damage. Some people claimed that exposed women had higher rates of miscarriage and of deformed children, but local authorities were unable to substantiate these claims. No human lives were lost in the accident.

[*David E. Newton*]

Resources

Books

Harrison, R. M., ed. *Pollution: Causes, Effects, and Control.* Cambridge, Royal Society of Chemistry, 1990.

Periodicals

Walsh, J. "Seveso: The Questions Persist Where Dioxin Created a Wasteland." *Science* (September 9, 1977): 1064–1067.

———. "Reporter at Large: Dioxin Pollution of Seveso." *New Yorker* (September 4, 1978): 34–36+.

Whiteside, T. "Reporter at Large: TCDD Explosion at Icmesa Chemical Plant." *New Yorker* (July 25, 1977): 41+.

Sewage treatment

Sewage is **wastewater** discharged from a home, business, or industry. Sewage is treated to remove or alter contaminants in order to minimize the impact of discharging wastewater into the **environment**. The operations and processes used in sewage treatment consist of physicochemical and biological systems.

The concerns of those involved in designing sewage treatment systems have changed over the years. Originally, the **biochemical oxygen demand** (BOD) and total suspended solids (TSS) received most of the attention. This was primarily because excessive BOD and TSS levels could cause severe and readily apparent problems, such as oxygen deficits that led to odors and **fish kills**, and **sludge** deposits that suffocated benthic organisms. By removing BOD and TSS, other contaminants were also removed and other benefits were realized; so even today, some **discharge** permits contain only limits for BOD and TSS. However, many permits now contain limits on other contaminants as well, and these limits, as well as other requirements, are constantly changing.

Among the first contaminants to be added to the requirements for discharge permits were nutrients. The most commonly regulated nutrients are **phosphorus** and **nitrogen**. Originally removing phosphorus and nitrogen could only be done through expensive, advanced methods. But scientists have recently discovered ways to accomplish enhanced removals of nutrients in conventional biological treatment plants with relatively minor operational and structural adjustments.

The most recently regulated pollutants are toxicants. There are regulations for specific toxic agents, and there are the generic-type regulations, which specify that the toxicity to certain test organisms should not exceed a certain level. For example, the wastewater discharged from a particular municipality may be restricted from killing more than 50% of the *Ceriodaphnia* in an aquatic toxicity test. The municipality would not need to determine what is causing the toxicity, just how to minimize its effects. Efforts to understand the causes of toxicity are referred to as toxicity reduction evaluations. The generic limit can therefore sometimes turn into a more specific standard, in the view of the municipality or industry, when the identity of a toxicant is determined; the general regulatory limit might remain, but treatment personnel are more cognizant of the role that a certain pollutant plays in overall, **effluent** toxicity.

The systems used to treat sewage can be divided into stages. The first stage is known as preliminary treatment. Preliminary treatment includes such operations as flow equalization, screening, comminution (or grinding), grease removal, flow measurement, and grit removal. Screenings and grit are taken to a **landfill**. Grease is directed to sludge handling facilities at the plant.

The next stage is primary treatment, which consists of gravity settling to remove suspended solids. Approximately 60% of the TSS in a domestic wastewater is removed during primary settling. Grease that floats to the surface of the **sedimentation** tank is skimmed off and handled along with the sludge (known as primary sludge) collected from the bottom of the tank.

The next stage is secondary treatment, which is designed to remove soluble organics from the wastewater. Secondary treatment consists of a biological process and secondary settling. There are a number of biological processes. The most common is **activated sludge**, a process in which **microbes**, also known as **biomass**, are allowed to feed on organic matter in the wastewater. The make-up and dynamics of the biomass population is a function of how the activated sludge system is operated. There are many types of activated sludge systems that differ based on the time wastewater remains in the biological reactor and the time microbes remain there. They also differ depending on whether air or oxygen is introduced, how gas is introduced, and where wastewater enters the biological reactor, as well as the number of tanks and the mixing conditions.

There are also biological treatment systems in which the biomass is attached. Trickling **filters** and biological towers are examples of systems that contain biomass adsorbed to rocks plastic. Wastewater is sprayed over the top of the rocks or plastic and allowed to trickle down and over

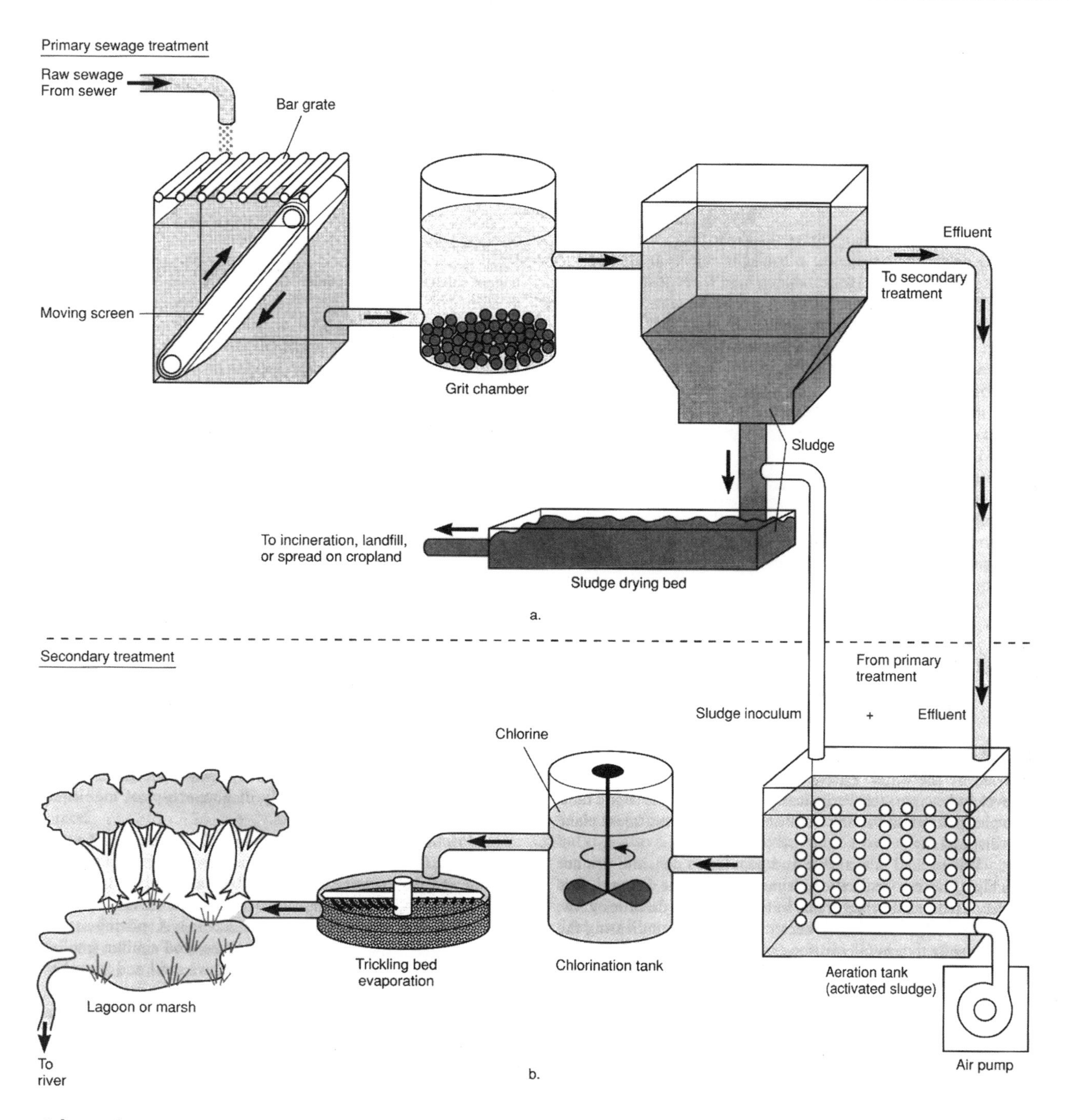

Schematic representation of a sewage treatment system. (McGraw-Hill Inc. Reproduced by permission.)

the attached biomass, which removes materials from the waste through **sorption** and biodegradation. A related type of attached-growth system is the rotating biological contactor, where biomass is attached to a series of thin, plastic wheels that rotate the biomass in and out of the wastewater.

It is important to note that each of the above biological systems is **aerobic**, meaning that oxygen is present for the microbes. **Anaerobic** biological systems are also available in both attached and suspended growth configurations. Examples of the attached and suspended growth systems are,

respectively, anaerobic filters and upflow anaerobic sludge blanket units.

The end-products of aerobic and anaerobic processes are different. Under aerobic conditions, if completely oxidized, organic matter is transformed into products that are not hazardous. But an anaerobic process can produce **methane** (CH_4), which is explosive, and ammonia (NH_3) and **hydrogen** sulfide (H_2S), which are toxic. There are thus special design considerations associated with **anaerobic systems**, though methane can be recovered and used as a source of energy. Some materials are better degraded under anaerobic conditions than under aerobic conditions. In some cases, the combination of anaerobic and **aerobic systems** in a series provides better and more economical treatment than either system could alone. Many substances are not completely mineralized to the end-products mentioned above, and other types of intermediate metabolites can be considered in selecting a biological process.

Biomass generated during biological treatment is settled in secondary clarifiers. This settled biomass or secondary sludge is then piped to sludge-management systems or returned to the biological reactor in amounts needed to maintain the appropriate biomass level. The hydraulic detention time of secondary clarifiers is generally in the area of two hours.

As mentioned above, biological systems are designed on the basis of hydraulic **residence time** and sludge age. In a conventional activated-sludge system, sewage is retained in the reactor for about five to seven hours. Biomass, due to the **recycling** of sludge from the secondary clarifier, remains in the reactor, on average, for about ten days.

Disinfection follows secondary clarification in most treatment plants. Disinfection is normally accomplished with **chlorine**. Due to the potential environmental impact of chlorine, most plants now dechlorinate wastewater effluents before discharge.

Some facilities use another stage of treatment before disinfection. This stage is referred to as tertiary treatment or advanced treatment. Included among the more commonly used advanced systems are **adsorption** to activated **carbon**, **filtration** through sand and other media, **ion exchange**, various membrane processes, nitrification-denitrification, coagulation-flocculation, and fine screening.

The treatment systems used for municipal sewage can be different from the systems used by industry, for industrial wastes can pose special problems which require innovative applications of the technologies available. Additionally, industrial wastes are sometimes pretreated before being discharged to a sewer, as opposed to being totally treated for direct discharge to the environment. *See also* Aerobic sludge digestion; Anaerobic digestion; Bioremediation; Industrial waste treatment; Sludge treatment and disposal; Waste management

[*Gregory D. Boardman*]

Resources
Books

Metcalf and Eddy, Inc. *Wastewater Engineering Treatment, Disposal, and Reuse.* Revised by G. Tchobanoglous and F. Burton. 3rd ed. New York: McGraw-Hill, 1991.

Peavy, H. S., D. R. Rowe, and G. Tchobanoglous. *Environmental Engineering.* New York: McGraw-Hill, 1985.

Viessman, W., and M. J. Hammer. *Water Supply and Pollution Control.* 4th ed. New York: Harper and Row, 1985.

Shade-grown coffee and cacao

Coffee (*Coffea arabica*) and cacao (*Theobroma cacao*) are important agricultural crops in the developing world that have been traditionally grown under a light canopy of rainforest trees created by thinning the original rainforest. In the 1970s, growers in South America began changing their planting practices, setting out sun-tolerant varieties of these plants in uniform rows. Coffee and cacao grown in full sun require more pesticides and **fertilizer** in order to thrive. In addition, when rainforest is cleared in order to grow these crops in sun, animals and native plants lose their **habitat**, and **biodiversity** (the number of different plant an animal **species** in a habitat) decreases.

Coffee and cacao are grown in places that would normally support tropical rainforests with enormous biodiversity. The original technique of thinning the forest and growing these crops in shade under a light canopy of trees reduced biodiversity. Completely clearing the forest and growing a single crop in full sun severely reduces biodiversity. When coffee and cacao are grown in shade, far less of the original habitat is destroyed.

Destruction of tropical rainforests due to **logging**, subsistence cultivation (**slash and burn agriculture**), and cash-crop cultivation, including the growing of coffee and cacao, has become of increasing concern to ecologists worldwide because of its far-reaching effects. In the late 1990s, ecologists in the United States recognized that **deforestation** in the tropics was destroying the winter habitat of many migrating songbirds as well as native birds. One solution, among many, was to advocate for the return of traditional agricultural methods for South American coffee and cacao. Combined efforts of **conservation** groups, farmers, and coffee buyers working across international borders led to a growing market in shade-grown coffee in the early 2000s. Advocating for shade-grown cacao has met with a more limited success.

The coffee plant is a low shrub or tree originating in Africa in the high forests of Ethiopia and the Sudan. Colonists introduced the plant to Central and South America in the 1700s. Coffee developed into an increasingly important **cash crop** in the nineteenth century. It grew well in the higher elevations of Central and South America, where moist conditions and the cloud and tree cover of the rainforests allowed it to flourish. Coffee production grew two- to three-fold in parts of Central America between 1870 and 1910, as coffee became a valued commodity worldwide. By 2000, over 40% of cropland in Central America was planted with coffee, and coffee was second only to oil as Central America's most valuable export.

Until the 1970s, coffee was produced mostly by small farmers who grew the crop under the forest canopy. Farmers began planting coffee in rows in full sun in the 1970s. United States Agency for International Development (USAID) and local government groups encouraged farmers to apply **monoculture** plantation farming to their coffee farms, setting the plants out in deforested fields and treating them with chemical pesticides and fertilizers. Almost 70% of Columbia's coffee plantations were transformed to full-sun plantations between the 1970s and early 1990s. In Mexico, Costa Rica, and other parts of Central America, some 30–40% of coffee production shifted to full-sun fields.

In many cases, sun-grown plants produced larger crops than shade-grown plant. Yet this method had hidden costs, because the plants soon depleted nutrients from the **soil** and required chemical fertilizer to continue to produce high yields. Coffee plants grown in full sun also do not live as long as plants grown in shade. While a shade-grown coffee shrub can survive for 80–100 years, plants grown in full sun live only about 15 years.

Clearing land for full-sun coffee production (and to plant other agricultural crops and for logging) had devastating effects on all parts of the **ecosystem**. For example, full-sun coffee plantations were found to support fewer than ten percent of the species of birds found in the rainforest.

Conservationists in the United States began recording a decline in the population of migratory songbirds such as the northern oriole (*Icterus galbula*) in the 1980s. By the 1990s some scientists suspected the falling songbird population in the United States was due in part to loss of habitat in the birds' winter homes in Central and South America. This loss of habitat was wide-ranging, with full-sun plantations making up only one small part of the problem.

Conservation groups began promoting shade-grown coffee in the late 1990s as a way to conserve rainforest habitat. One of the biggest coffee buyers in the United States, Starbucks, began purchasing shade-grown coffee from Mexico in 2000 to support the rainforest conservation effort. A growing market for gourmet coffee in the United States, made it economically feasible for some farmers to return to traditional shade-grown crops.

Cacao production has followed a similar path. Cacao is native to South America, and grows in similar conditions to coffee. The conversion to full-sun cacao plantations began in the 1970s. In Ecuador, a prime producer of cacao, collapse of the market led farmers to replant with quick-growing, sun-tolerant hybrids. Like coffee, cacao grown on plantations without forest shade depletes the soil, requiring the addition of chemical fertilizers. Furthermore, monoculture fields diminish biodiversity and increase susceptibility to disease and pests. Some growers' cooperatives, in league with North American conservation groups, began returning to shade-grown cacao in the late 1990s. To date, shade-grown cacao has lacked the substantial market shade-grown coffee has found, yet by 2000, hundreds of small cacao growers in Ecuador had converted to traditional shade cultivation.

[*Angela Woodward*]

Resources

Periodicals

Chou, Sophie. "Coffee and Deforestation." *World Watch* (March/April 1998): 9.

"Finally, the Message Is Clear: Less Pollution, More Profits." *In Business* 23, no. 4 (September/October 2001): 30.

Goldberg, Carey. "Songbirds' Plight Starts a Buzz in Coffee Circles." *New York Times* (July 27, 1997): 12.

Hull, Jennifer Bingham. "Can Coffee Drinkers Save the Rainforest?" *Atlantic Monthly* (August 1999): 19–21.

LaFranchi, Howard. "Made in the Shade: Java that Saves Forests." *Christian Science Monitor* 92, no. 69 (March 2, 2000): 1.

Rice, Robert A. "A Place Unbecoming: The Coffee Farm of Northern Latin America." *Geographical Review* 89, no. 4 (October 99): 554–580.

"Shade-Grown Cocoa Returns to Ecuador." *Candy Industry* 165, no. 6 (June 2000): 20.

Organizations

Rainforest Alliance, 65 Bleecker Street, New York, NY USA 10012 (212) 677-1900, Email: canopy@ra.org, <http://www.rainforestalliance.com>

Shadow pricing

A practice employed by some economists that involves putting a price on things that normally do not have a market value. For example, **pollution** degrades common property and the natural heritage of present and **future generations**, but because it is considered a "free" good, no one "pays" for it at the marketplace. Yet pollution has a cost even if the market fails to explicitly take into account this cost and assign it to responsible parties. Thus pollution is perceived by most economists as a defect that undermines otherwise efficient markets. To correct this defect, economists often advocate the use of some type of shadow pricing for pollu-

tion. Such prices are set by the "shadow" procedure of asking people what they would be willing to pay for breathing clean air, watching **whales** migrate, or preventing the **extinction** of a particular **species** of plant or animal. Hypothetical or shadow prices for these "commodities" can thereby be estimated. Some environmentalists and economists, however, criticize shadow pricing for falsely attempting to assign a monetary value to things that are invaluable precisely because they (like beauty, love, and respect) are not and cannot be bought, sold, or traded in markets. To assign a shadow price, these critics claim, is to make the mistake of assuming that nothing has **intrinsic value** and that the worth of everything can be measured and reduced to its utilitarian or instrumental value.

Shanty towns

The United Nations estimates that at least one billion people—20% of the world's population—live in crowded, unsanitary slums of the central cities and in the vast shanty towns and squatter settlements that ring the outskirts of most **Third World** cities. Around 100 million people have no home at all. In Bombay, India, for example, it is thought that half a million people sleep on the streets, sidewalks, and traffic circles because they can find no other place to live. In São Paulo, Brazil, at least three million "street kids" who have run away from home or have been abandoned by their parents live however and wherever they can. This is surely a symptom of a tragic failure of social systems.

Slums are generally legal but otherwise inadequate multifamily tenements or rooming houses, either custom built for rent to poor people or converted from some other use. The *chals* of Bombay, for example, are high-rise tenements built in the 1950s to house immigrant workers. Never very safe or sturdy, these dingy, airless buildings are already crumbling and often collapse without warning. Eighty-four percent of the families in these tenements live in a single room; half of those families consist of six or more people. Typically they have less than 22 ft^2 (2 m^2) of floor space per person and only one or two beds for the whole family. They may share kitchen and bathroom facilities down the hall with 50–75 other people. Even more crowded are the rooming houses for mill workers, where up to 25 men sleep in a single room only 75 ft^2 (7 m^2). Because of this overcrowding, household accidents are a common cause of injuries and deaths in cities of developing countries, especially to children. Charcoal braziers or kerosene stoves used in crowded homes are a routine source of fires and injuries. With no place to store dangerous materials beyond the reach of children, accidental poisonings and other mishaps are a constant hazard.

Shanty towns are created when people move onto undeveloped lands and build their own houses. Shacks are built of corrugated metal, discarded packing crates, brush, plastic sheets, or whatever building materials people can scavenge. Some shanty towns are simply illegal subdivisions where the landowner rents land without city approval. Others are spontaneous or popular settlements or squatter towns where people occupy land without the owner's permission. Sometimes this occupation involves thousands of people who move onto unused land in a highly organized, overnight land invasion, building huts and laying out streets, markets, and schools before authorities can evict them. In other cases, shanty towns gradually appear.

Called *barriads*, *barrios*, *favelas*, or *turgios* in Latin America, *bidonvillas* in Africa, or *bustees* in India, shanty towns surround every megacity of the developing world. They are not an exclusive feature of poor countries, however. Some 200,000 immigrants live in the *colonias* along the southern Rio Grande in Texas. Only 2% have access to adequate **sanitation**. Most live in conditions as poor as those of any city of a developing nation. Smaller enclaves of the poor and dispossessed can be found in most American cities.

The problem is magnified in **less developed countries**. Nouakchott, Mauritania, the fastest growing city in the world, consists almost entirely of squatter settlements and shanty towns. It has been called "the world's largest refugee camp." About 80% of the people in Addis Ababa, Ethiopia, and about 70% of those in Luanda, Angola, live in these squalid refugee camps. Two-thirds of the population of Calcutta live in shanty towns or squatter settlements and nearly half of the 19.4 million people in **Mexico City, Mexico**, live in uncontrolled, unauthorized shanty towns and squatter settlements. Many governments try to clean out illegal settlements by bulldozing the huts and sending riot police to drive out the settlers, but the people either move back or relocate to another shanty town elsewhere.

These popular but unauthorized settlements usually lack sewers, clean water supplies, electricity, and roads. Often the land on which they are built was not previously used because it is unsafe or unsuitable for habitation. In **Bhopal, India**, and Mexico City, for example, squatter settlements were built next to deadly industrial sites. In such cities as Rio de Janiero, Brazil; La Paz, Bolivia; Guatemala City, Guatemala; and Caracas, Venezuela, they are perched on landslide-prone hills. In Bangkok, Thailand, thousands of people live in shacks built over a fetid tidal swamp. In Lima, Peru; Khartoum, Sudan; and Nouakchott, shanty towns have spread onto sandy deserts. In Manila in the Phillipines, 20,000 people live in huts built on towering mounds of **garbage** amidst burning industrial waste in city dumps.

Living in tents on the streets of New Delhi, India. (Visuals Unlimited. Reproduced by permission.)

Few developing countries can afford to build modern waste treatment systems for their rapidly growing cities, and the spontaneous settlements or shanty towns are the last to be served. The **World Bank** estimates that only 35% of urban residents in developing countries have satisfactory sanitation services. The situation is especially desperate in Latin America, where only 2% of urban sewage receives any treatment. In Egypt, Cairo's sewer system was built about 50 years ago to serve a population of two million people. It is now being overwhelmed by more than 11 million inhabitants. Only 217 of India's 3,119 towns and cities have even partial sewage systems and **water treatment** facilities. These systems serve less than 16% of India's 200 million urban residents. In Colombia, the Bogota River, 125 mi (200 km) downstream from Bogota's five million residents, still has an average fecal bacterial count of 7.3 million cells per liter, more than 7,000 times the safe drinking level and 3,500 times higher than the limit for swimming.

Some 400 million people, or about one-third of the population in developing world cities, do not have safe drinking water, according to the World Bank. Where people must buy water from merchants, it often costs 100 times as much as piped city water and may not be safe to drink after all. Many rivers and streams in Third World countries are little more than open sewers, and yet, they are all that poor people have for washing clothes, bathing, cooking, and, in the worstcases, for drinking. Diarrhea, dysentery, typhoid, and **cholera** are epidemic diseases in these countries, and infant **mortality** is tragically high.

A striking aspect of most shanty towns is the number of people selling goods and services of all types on the streets or from small stands in informal markets. Food vendors push carts through crowded streets; children dart between cars selling papers or cigarettes; curb-side mechanics make repairs using primitive tools and ingenuity. Nearly everything city residents need is available on the streets. These individual entrepreneurs are part of the informal economy: small-scale family businesses in temporary locations outside the control of normal regulatory agencies. In many developing countries, this informal sector accounts for 60–80% of the economy.

Governments often consider these independent businesses to be backward and embarrassing, a barrier to orderly development. It is difficult to collect taxes or to control these activities. In many cities, police drive food vendors, beggars, peddlers, and private taxis off the streets at the same time that they destroy shanties and squatter settlements.

Recent studies, however, have shown that this informal economy is a vital, dynamic force that is more often positive than negative. The sheer size and vigor of this sector means that it can no longer be ignored or neglected. The informal economy is often the only feasible source of new housing, jobs, food distribution, trash removal, **transportation**, or **recycling** for the city. Small businesses and individual entrepreneurs provide services that people can afford and that cities cannot or will not provide.

The businesses common to the informal sector are ideal in a rapidly changing world. They tend to be small, flexible, and labor intensive. They are highly competitive and dynamic, avoiding much of the corruption of developing nations' bureaucracies. Government leaders beginning to recognize that the informal sector should be encouraged rather than discouraged are making microloans and assisting communities with self-help projects. When people own their houses or businesses, they put more time, energy, and money into improving and upgrading them.

[*William P. Cunningham Ph.D.*]

Resources

Books

Hardoy, J. E., and D. E. Satterthwaite. "Third World Cities and the Environment of Poverty." In *Global Possible*, edited by R. Repetto. Washington, DC: World Resources Institute, 1985.

Livermash, R. "Human Settlements." In *World Resources 1990–1991*. Washington, DC: World Resources Institute, 1990.

PERIODICALS

"The World's Urban Explosion." *Unesco Courier* (March 1985): 24–30.

Sharks

Sharks are eight orders of cartilaginous fishes in the class Elasmobranchii. Sharks first appear in the fossil record of about 430 million years, during the Silurian Period. The Silurian is sometimes known as "Age of Fishes," because this is when the first kinds of fish-like animals appeared and then rapidly radiated into a great diversity of forms. Today, there are about 400 living **species** of sharks, divided into eight orders and 30 families. New species continue to be discovered as marine biologists begin to explore the relatively unknown abyssal waters of the oceans.

The living orders of sharks are:

- Squatiniformes, including 13 species of angelsharks, with a flat body, extremely wide and elongate pectoral fins, two dorsal fins, no anal fin, and a body length up to 8 ft (2.4 m); an example is the Pacific angelshark (*Squatina californica*).
- Pristiophoriformes, including five species of sawsharks, with a narrow cylindical body, two dorsal fins, no anal fin, a long blade-like snout edged with needle-like teeth, and a body length up to 6 ft (1.8 m); an example is the longnose sawshark (*Pristiophorus cirratus*).
- Squaliformes, including more than 80 species of dogfish sharks, with a cylindrical body, two dorsal fins, no anal fin, a moderately long snout, and a body length up to 23 ft (7.0 m); examples are the Greenland shark (*Somniosus microcephalus*) and spiny dogfish (*Squalus acanthias*).
- Hexanchiformes, including five species of six-gill and seven-gill sharks, with a cylindrical body, one dorsal fin, an anal fin, and a body length up to 16.5 ft (5.0 m); an example is the six-gill shark (*Hexanchus griseus*).
- Carcharhiniformes, including almost 200 species of ground sharks, with a cylindrical body, two dorsal fins, an anal fin, a moderately long snout, a nictitating eye membrane, and a body length up to 25 ft (7.6 m); examples are the tiger shark (*Galeocerdo cuvieri*) and scalloped hammerhead (*Sphyrna lewini*).
- Lamniformes, including 16 species of mackerel sharks, with a cylindrical body, two dorsal fins, an anal fin, a moderately long snout, and a body length up to 45 ft (14 m); examples are the megamouth shark (*Megachasma pelagios*), pelagic thresher shark (*Alopias pelagicus*), and basking shark (*Cetorhinus maximus*).
- Orectolobiformes, including 33 species of carpet sharks, with a cylindrical body, two spineless dorsal fins, an anal fin, a short snout, and a body length up to 45 ft (14 m); examples are the nurse shark (*Ginglymostoma cirratum*), great white shark (*Carcharodon carcharias*), and whale shark (*Rhincodon typus*; the world's largest fish).
- Heterodontiformes, including eight species of bullhead sharks, with a cylindrical body, two spined dorsal fins, no anal fin, a short snout, teeth adapted for either gripping or crushing, and a body length up to 66 in (1.7 m); an example is the horn shark (*Heterodontus francisci*).

Species of sharks vary greatly in body design. In general, however, they have a relatively streamlined body shape, which allows them to swim without using much energy. The **conservation** of energy is important to sharks because they do not appear to sleep and most species must swim continuously to maintain a flow of water over their gills, which is necessary for the exchange of respiratory oxygen and **carbon dioxide**. As sharks swim, water passes through their mouth, over the gills, and then exits the body through lateral gill slits located behind the head.

Sharks do not have true bones; rather, their skeleton is made up of a softer material known as cartilage. The center of their larger vertebrae, however, may contain a bone-like deposit of calcium phosphate. Sharks do have hard teeth, and to a lesser degree hard scales and spines, which preserve well in the fossil record and are the main indicators of their presence in ancient times. Sharks grow new teeth throughout their life, replacing the ones that break or wear down with use. Sharks also have rigid fins, and they do not have an air bladder. Sharks have extremely acute senses, including those of vision, hearing (or vibration), smell, taste, and electro-perception.

Sharks appear to be highly resistant to infections, cancers, and circulatory diseases, and they can often recover from severe injuries. Sharks are slow-growing animals that can live for a long time, and most species appear to take decades to reach sexual maturity. Almost all species are ovoviviparous, meaning the eggs are retained inside the body of the mother until they hatch, and so the young are born "alive," or as fully formed miniatures of the adult form. The **fecundity** of sharks is much less than that of most other fish, and ranges from as few as two to as many as about 135 pups per reproductive event.

All species of sharks are predators or scavengers of other sea creatures. However, the size range of the prey varies enormously among species of sharks. The largest species are the whale shark and the basking shark, which can attain a length of up to 45 ft (14 m) and may weigh as much as one ton (1 tonne). However, these enormous animals filter-feed on such relatively tiny prey as **zooplankton**, small fish, and fish eggs and larvae. The smallest species of shark is the spined pygmy shark (*Squaliolus laticaudus*), which only grows to a length of about 10 in (25 cm). Some sharks are fierce predators of large fish, marine mammals, and other relatively big animals. Almost any shark longer than about 6 ft (2 m)

A hammerhead shark swimming just below the water surface. (Photograph by John Bortniak, NOAA Corp. National Oceanic and Atmospheric Administration.)

is a potential danger to swimming or surfing humans. The species most often linked with attacks on people are the great white shark, the tiger shark, and the bull shark (*Carcharhinus leucas*). These are all widely distributed species that are adapted to feeding on **seals** or **sea turtles**, but may also opportunistically attack humans.

According to data compiled by the International Shark Attack File, in the year 2001 there were about 76 unprovoked shark attacks on humans world-wide. This included 54 attacks in U.S. waters, mostly occurring off beaches in Florida. (This tally does not include "provoked" attacks, as might happen when a fisher is attempting to remove an entangled shark from a net or hook, or scavenging damage done to an already-drowned human.) Only five of those unprovoked attacks, however, resulted in fatalities (three were in the United States). This relatively low, 7% fatality rate is thought to occur because, from the perspective of the shark, an attack on a human is usually a "mistake" that happens when a swimmer or surfer is confused with a more usual prey animal, such as a seal. In such cases, the shark usually breaks off the attack before lethal damage.

Although there is always a risk of being attacked by a large shark while swimming or surfing in waters that they frequent, the actual **probability** of this happening is extremely small. In the United States, for example, many more human fatalities are caused by bee and wasp stings and by venomous snakebites. Even lightning strikes cause about 30 times as many deaths per year as do shark attacks. In fact, it is much more dangerous to drive to and from a beach in the United States, compared to the risk of a shark attack while swimming there.

Although it is a fact that very few humans are hurt by sharks each year, the reverse is definitely not true. Sharks are extremely vulnerable to fishing pressure because of their slow growth rate, length maturation period, and low birth rate. Consequently, **commercial fishing** practices are devastating populations of many species of sharks throughout the oceans of the world. In waters of the United States, for

example, fishing **mortality** of sharks in recent years has been equivalent to about 18-thousand tons (20-thousand tonnes) per year. This is much more than the maximum sustainable yield calculated by fishery scientists, which is equivalent to about 9–11-thousand tons (10–15-thousand tonnes) per year. Because of this intense **overfishing**, there has been a catastrophic decline in the numbers of sharks in U.S. territorial waters, by as much as 80% since the 1970s and 1980s. A similar decrease is also occurring in many other regions of the world, and for the same reasons.

The excessive killing of sharks is occurring for several reasons. Some species of sharks, such as the spiny dogfish, are commercial species caught for their meat. Larger species of sharks may be caught in large numbers solely for their fins, which are cut off and sold as a delicacy in certain Asian markets, particularly in China. Shark "finning" is a widely illegal but common practice. It is also extremely wasteful as the de-finned sharks are thrown back into the sea, either as dead carcasses or as live fish doomed to starve to death because they are incapable of swimming properly. As many as 30–100 million sharks may be subjected to this finning practice each year. There are also economically significant recreational fisheries for sharks, involving sportsmen catching them as trophies.

In addition to sharks being targets of commercial and recreational fisheries, immense numbers are also incidentally caught as so-called "bycatch" in fisheries targeted for other species of fish. This is especially the case in long-line fisheries for large tuna and **swordfish**. This open-ocean fishing method, mostly practiced by boats fishing for Japan, Korea, and Taiwan, involves setting tens of miles of line containing thousands of baited hooks. This is a highly non-selective practice that kills enormous numbers of species that are not the intended target of the fishery, including many sharks. Some fishery biologists believe that the **bycatch** mortality of sharks, about 12-million animals per year, may be at least half the size of the commercial fishery of these animals. This excessive bycatch of sharks, almost all of which is discarded at sea as "waste," is further contributing to the devastation of their populations. The World Conservation Union (IUCN) lists numerous species of sharks as being at risk of endangerment or **extinction**.

In a few places, sharks are contributing to the local economy in ways that do not require the destruction of these marvelous animals. This is happening through **ecotourism** involving recreational diving with sharks in non-threatening situations. For instance, waters off Cocos Island, Costa Rica, sustain large numbers of whitetip reef sharks (*Triaenodon obesus*), hammerheads (*Sphyrna zygaena*), and whale sharks (*Rhincodon typus*), and this is a local attraction for scuba-diving tourists.

[*Bill Freedman Ph.D.*]

Resources

Books

Compagno, L. J. V. *Sharks of the World.* New York: UN Food and Agricultural Organization and UN Development Programme, 1984.

Springer, V. G., and J. P. Gold. *Sharks in Question: The Smithsonian Answer Book.* Washington, DC: Smithsonian Institution Press, 1989.

Stevens, J. D., ed. *Sharks.* New York: Facts on File Publications, 1987.

Other

Public Broadcasting Service. *The World of Sharks.* [cited July 2002]. <http://www.pbs.org/wgbh/nova/sharks/world>.

Shark Specialist Group. [cited July 2002].<http://www.flmnh.ufl.edu/fish/Organizations/SSG/SSGDefault.html>.

The Pelagic Shark Research Foundation. [cited July 2002]. <http://www.pelagic.org>.

The University of Florida. *The International Shark Attack File.* [cited July 2002]. <http://www.flmnh.ufl.edu/fish/Sharks/ISAF/ISAF.htm>.

Sheet erosion

see **Erosion**

Paul Howe Sheppard (1925 - 1996)

American environmental theorist

A scientist, teacher, and environmental theorist, Shepard moved beyond the factual realm of biological science to expand upon the nature of human behavior and its roots in the natural world. He was a prolific writer who have had a profound influence on a variety of contemporary thinkers.

In his introduction to *The Subversive Science: Essays Toward an Ecology of Man*, edited with Daniel McKinley, Shepard discussed the reluctance of modern (as opposed to primitive) people to view humanity as one element of the whole of creation. Instead, humanity is wrongly seen as the singular focus or the intended culmination of creation, thereby denying the role played by all other life forms in the **evolution** of humankind. Shepard suggested this need to dominate has led to physical and mental separation from the whole, resulting in innumerable problems for humankind, among them **environmental degradation**. He believed that people should simultaneously appreciate the integrity of every being as well as the relatedness of all things, acquiring a world view that is inclusive and holistic rather than exclusive and superficially hierarchical: "Without losing our sense of a great human destiny and without intellectual surrender, we must affirm that the world is a being, a part of our own body."

Shepard continued to assess and emphasize the relationships between other living things and human development, particularly the role played by animals. In *The Others: How Animals Made Us Human*, he discusses the paradox of people who both love and kill or hunt animals, and contrasts

them with those who would protect the "others" of the natural world from humans in the name of "animal rights:" "In the perspective of the enormous history of life and the role of animals in human evolution for a million years, I feel only disconnected by the precept of untouchability...Great naturalists and primal peoples were motivated not by the ideal of untouchability but by a cautious willingness to consume and be consumed, both literally and in a mythic sense."

Shepard was born on July 12, 1925, in Kansas City, Missouri; his father was a horticulturist. During World War II he served in the U.S. Army from 1943–46, in the European theater. He received his B.A. from the University of Missouri (1949) and his M.S. in **conservation** (1952). His Ph.D. from Yale (1954) was based on an interdisciplinary study of **ecology**, landscape architecture, and art history. Shepard was a Fulbright senior research scholar in New Zealand in 1961, and later received both a Guggenheim fellowship to research the cultures of hunting-gathering peoples (1968–69) and a Rockefeller fellowship in the humanities (1979).

For 21 years Shepard taught concurrently at Pitzer College and Claremont Graduate School (both in Claremont, California), serving as the Avery Professor of Natural Philosophy and **Human Ecology** at Claremont from 1973 until his retirement in 1994. Earlier he had taught at other schools, including Knox College, Dartmouth College, and Smith College. He published numerous books, was a regular contributor to *Landscape* and *North American Review*, and also wrote for *BioScience*, *Perspectives in Biology and Medicine*, and *School Science and Mathematics*.

[*Ellen Link*]

Resources

Books

Contemporary Authors. Vol. 10. Detroit: Gale Research, 1983.

Shepard, Paul. *Environmental: Essays on the Planet as a Home*. Boston: Houghton, 1971.

———. *Man in the Landscape: A Historic View of the Esthetics of Nature*. New York: Knopf, 1967.

———. *Nature and Madness*. San Francisco: Sierra Club Books, 1982.

———. *The Others: How Animals Made Us Human*. Washington, D.C.: Island Press, 1996.

———. *The Tender Carnivore and the Sacred Game*. New York: Scribner, 1973.

———. *Thinking Animals: Animals and the Development of Human Intelligence*. New York: Viking Press, 1978.

———, and B. Sanders. *The Sacred Paw: The Bear in Nature, Myth and Literature*. New York: Viking Press, 1983.

———, and D. McKinley, eds. *The Subversive Science: Essays Toward an Ecology of Man*. Boston: Houghton Mifflin, 1969.

———, ed. *The Only World We've Got: A Paul Shepard Reader*. San Francisco: Sierra Club Books, 1996.

Oelschlaeger, M., ed. *The Company of Others: Essays in Celebration of Paul Shepard*. Durango, Co: Kivaki Press, 1995.

Periodicals

Pace, E. "Paul Shepard, Professor and Author, 71." *New York Times*, July 22, 96: A15(L).

Shifting cultivation

Shifting cultivation refers to a practice whereby a tract of land is alternately used for crop production and then allowed to return to native vegetation for a period of years. Typically, the land is cleared of vegetation, crops are grown for two or three years, and then the land abandoned for a period of 10 or more years. To facilitate land clearing prior to cultivation, the vegetation is cut and the debris burned. The practice is also called slash-and-burn agriculture or swidden agriculture.

Shifting cultivation is most common in the tropics where farming techniques are less technologically advanced. The soils are usually low in plant nutrients. For two or three years following land clearing, the nutrients brought to, or near, the **soil** surface by deep rooting trees, shrubs, and other plants support cultivated crops. With native management, the available nutrients are removed by the cultivated crops or leached so that after a few years the soil will support only minimal plant growth. It is then allowed to return to native vegetation which slowly concentrates the nutrients in the surface soil again. After a period of years the cycle repeats. **Erosion** is often severe on sloping land during the cultivated phase. On some soils, particularly in the tropics, the soil structure becomes massive and hard. *See also* Leaching; Soil compaction; Soil organic matter

[*William E. Larson*]

Resources

Books

Ramakrishnan, P. S. *Shifting Agriculture and Sustainable Development: An Interdisciplinary Study from North-Eastern India*. Park Ridge, NJ: Parthenon Publishing Group, 1992.

Vasey, D. E. *An Ecological History of Agriculture*. Ames, IA: Iowa State University Press, 1992.

Periodicals

Monastersky, R. "Legacy of Fire: The Soil Strikes Back." *Science News* 133 (April 9, 1988): 231.

Shoreline armoring

Shoreline armoring is the construction of barriers or structures for the purpose of preventing coastal **erosion** and/or manipulating ocean currents. Although they may prevent the deterioration of the immediate shoreline, these man-made structures almost always exacerbate erosion problems

at nearby, downstream beaches by changing wave patterns and water flow.

Armoring may also promote the loss of shoreline vegetation, which in turn can degrade the **ecosystem** that seabirds and other beach and sea-dwelling life depend upon. By altering water flow and erosion and deposition and creating physical barriers, some types of shoreline armoring prevent **sea turtles** from reaching their nesting sites on shore. Finally, shoreline structures can block sunlight imperative to aquatic plants such as **eelgrass**, a prime **habitat** for herring and other marine life.

Shoreline erosion is a natural process determined by a complex array of environmental causes. Weather changes, tidal currents, and sea level adjustments all impact erosion over time. Catastrophic weather events such as hurricanes, **tsunamis**, and tropical storms can completely change the shape and nature of the coast in one fell swoop. Beach location, currents, and other conditions may also create depositional shorelines, where sand and **sediment** collect rather than erode. Armoring is a human attempt to minimize the impact of these natural forces on an isolated section of coastal shoreline. Because the coastal ecosystem is extensive and interconnected, these attempts ultimately affect other nearby, unprotected locations.

Common types of shoreline armoring include

- *Groins.* Long barriers that run perpendicular from the shoreline and are designed to trap sand and deposit it on the adjacent beach.
- *Jetties.* Similar to groins, jetties are rock structures that are built out from the shoreline. They are often erected for the purpose of keeping ship or boating channels clear of sediment build up by directing currents appropriately.
- *Seawalls.* These concrete or rock walls act as a barrier between beach and ocean; but increase neighboring beach erosion by deflecting normal wave patterns.
- *Bulkheads.* Like a seawall, a bulkhead is designed to separate the also promotes erosion further downstream.
- *Revetments.* A small seawall barrier, often constructed from rocks or boulders.
- *Breakwaters.* Breakwaters are offshore structures that absorb much of the force of waves before they reach the shoreline. They sometimes serve a dual purpose as navigational aids in boating channels.
- *Rip Rap.* Rip rap is a rock armoring sometimes used on river beds as well as coastal slopes to deter erosion.
- *Docks.* Although they are not built for the purpose of preventing coastal erosion, boat dock structures do impact marine life by blocking sunlight crucial for vegetation. Pilings, or posts, for docks can also disrupt sea-floor sediments. Depending on their size, structure, and location, they can also influence natural water flow patterns.

Alternatives to shoreline armoring include maintaining or planting native coastal shrubs, trees, and other plants that have an established root system to help prevent erosion. **Beach renourishment**, the process of replacing eroded beach with dredged offshore sand, is also used for erosion management. However, renourishment is cost-prohibitive in many cases, and may have a negative impact on marine **flora** and **fauna**. It also requires ongoing maintenance, as erosion is a perpetual process. Finally, the relocation of structures placed at risk by severe erosion, sometimes referred to as a retreat strategy, allows the natural processes of shoreline **evolution** to continue while preserving the public interest.

Shoreline armoring is becoming more widely recognized as an environmental hazard rather than a help. In South Carolina, state legislation entitled the 1988 Beachfront Management Act established a retreat rather than armor strategy for dealing with shoreline changes. The Act states that "[t]he use of armoring in the form of hard erosion control devices such as seawalls, bulkheads, and rip rap to protect erosion-threatened structures adjacent to the beach has not proven effective. These armoring devices have given a false sense of security to beachfront property owners. In reality, these hard structures, in many instances, have increased the vulnerability of beachfront property to damage from wind and waves while contributing to the deterioration and loss of the dry sand beach which is so important to the tourism industry." South Carolina no longer allows the construction of shoreline armoring within its coastal setback area, nor the rebuilding of existing armoring that is more than two-thirds damaged.

[*Paula Anne Ford-Martin*]

Resources

Books

Dean, Cornelia. *Against the Tide: The Battle for America's Beaches.* New York: Columbia University Press, 2001.

Periodicals

Dunn, Steve, et al. "Coastal Erosion: Evaluating the Risk." *Environment* 42, no. 7 (September 2000): 36 (10).

Other

South Carolina General Assembly. *Beachfront Management Act of 1988.* S.C. Code of Regulations. Chapter 30, Section 48-39-320(B). 1998 [cited June 6, 2002]. <http://www.lpitr.state.sc.us/coderegs/c030.htm#30-21>.

Sick Building Syndrome

Sick Building Syndrome (SBS) is a term applied to a building that makes its occupants sick because of indoor air pollutants. **Indoor air quality** (IAQ) problems fall into three categories: SBS, building-related illnesses, and **multiple chemical sen-**

sitivity. Of the three, SBS accounts for 75% of all IAQ complaints.

Indoor air is a health hazard in 30% of all buildings, according to the World Health Organization. The **Environmental Protection Agency** (EPA) lists IAQ fourth among top **environmental health** threats. The problem of SBS is of increasing concern to employees and occupational health specialists, as well as landlords and corporations who fear the financial consequences of illnesses among tenants and employees. **Respiratory diseases** attributed to SBS account for about 150 million lost work days each year, $59 billion in indirect costs, and $15 billion in medical costs.

Sick building syndrome was first recognized in the 1970s around the time of the energy crisis and the move toward **conservation**. Because heating and air conditioning systems accounted for a major portion of energy consumption in the United States, buildings were sealed for **energy efficiency**. Occupants depend on mechanical systems rather than open windows for outside air and ventilation. A tight building, however, can seal in and create contaminants. Common complaints of SBS include headaches, fatigue, cough, sneezing, nausea, difficulty concentrating, bleary eyes, and nose and throat irritations. Symptoms are caused by a range of contaminants, including volatile organic compounds (VOC), which are **chemicals** that turn to gas at room temperature and are given off by paints, adhesives, caulking, vinyl, telephone cable, printed documents, furniture, and solvents. Most common VOCs are **benzene** and chloroform, both of which may be carcinogens. Formaldehyde in building materials also is a culprit.

Biological agents such as viruses, bacteria, fungal spores, algae, pollen, mold, and dust mites add to the problems. These are produced by water-damaged carpet and furnishing or standing water in ventilation systems, humidifiers, and flush **toilets**.

Carbon dioxide levels increase as the number of people in a room increases, and too much can cause occupants to suffer hyperventilation, headaches, dizziness, shortness of breath, and drowsiness, as does **carbon monoxide** and the other **toxins** from **cigarette smoke**.

Schoolchildren are considered more vulnerable to SBS because schools typically have more people per room breathing the same stale air. Their size, childhood allergies, and asthmas add to their vulnerability.

Sick buildings can be treated by updating and cleaning ventilation systems regularly and using air cleaners and **filtration** devices. Also, plants spaced every 100 ft^2 (9.3 m^2) in offices, homes, and schools have been shown to filter out pollutants in recycled air.

A simple survey of the indoor **environment** can detect many SBS problems. Check each room for an air source; if windows cannot be opened, every room should have a supply vent and exhaust vent. Clean the vents. Check to see if air is circulating by placing a strip of tissue at each vent opening. The tissue should blow out at a supply vent and blow in at an exhaust vent. Move partitions, file cabinets, and boxes away from vents. Supply and exhaust vents should be more than a few feet apart. Dead spaces where air stagnates and pollutants build up should be renovated. Move printing and copying machines away from people and give those machines adequate exhaust. Check that the ventilation system operates fully in every season and whenever people are in the building.

The EPA enforces tough laws on outdoor **air pollution**, but not for indoor air except for some smoking bans. Yet almost every pollutant, according to the EPA, is at higher levels indoors than out. Help in detecting and correcting sick building syndrome is available from the **National Institute of Occupational Safety and Health**, the federal agency responsible for conducting research and making recommendations for safe and healthy work standards. *See also* Occupational Safety and Health Administration (OSHA); Occupational Health and Safety Act

[*Linda Rehkopf*]

RESOURCES

BOOKS

Kay, J. G., et al. *Indoor Air Pollution: Radon, Bioaerosols, and VOCs.* Chelsea, MI: Lewis, 1991.

Samet, J. M., and J. D. Spangler. *Indoor Air Pollution: A Health Perspective.* Baltimore, MD: Johns Hopkins University Press, 1991.

PERIODICALS

Soviero, M. M. "Can Your House Make You Sick?" *Popular Science* (July 1992): 80.

Sierra Club

The Sierra Club is one of the nation's foremost **conservation** organizations and has worked for over 100 years to preserve "the wild places of the earth." Founded in 1892 by author and **wilderness** explorer **John Muir**, who helped lead the fight to establish **Yosemite National Park**, the group's first goal was to preserve the Sierra Nevada mountain chain. Since then, the club has worked to protect dozens of other national treasures.

The preserve of Mount Rainier was one of the Sierra Club's earliest achievements, and in 1899 Congress made that area into a **national park**. The group also helped to establish Glacier National Park in 1910. The Sierra Club supported the creation of the **National Park Service** in 1916, and in 1919 began a campaign to halt the indiscriminate cutting of redwood trees.

The club has helped secure many conservation victories. They worked to create such national parks as Kings

Canyon, Olympic, and Redwood, national seashores such as Point Reyes in California and Padre Island in Texas, as well as the Jackson Hole National Monument. The club also campaigned to expand Sequoia and Grand Teton national parks. In the 1960s, the Sierra Club helped to secure such legislative victories as the **Wilderness Act** in 1964, the establishment of the National Wilderness Preservation System, and the expansion of the Land and **Water Conservation** Fund in 1968.

By 1970, the Sierra Club had 100,000 members, with chapters in every state, and the group took advantage of growing public support for the **environment** to accelerate progress towards conserving America's natural heritage. The **National Environmental Policy Act** was passed by Congress that year, and the **Environmental Protection Agency** (EPA) was created. Later, the club helped defeat a proposal to build a fleet of polluting Supersonic Transports, and they organized the Sierra Club Legal Defense Fund. In 1976, the club's lobbying efforts sped passage of the **Bureau of Land Management** (BLM) Organic Act, which increased governmental protection for an additional 459 million acres (185 ha).

One of the most important victories for the Sierra Club came in 1980, when a year-long campaign culminated in passage of the Alaska National Interest Conservation Act, establishing 103 million acres (41.6 million ha) as either national parks, monuments, refuges, or wilderness areas. Superfund legislation was also enacted to clean up the nation's abandoned toxic waste sites.

The decade of the 1980s, however, was a difficult one for conservationists. With James Watt as Secretary of Interior under President Ronald Reagan, and Ann Gorsuch Burford as EPA administrator, the Sierra Club was placed in a defensive position. The group focused mainly on preventing environmentally destructive projects and legislation—for example, blocking the MX missile complex in the Great Basin (1981), preventing weakening of the **Clean Air Act**, and stopping BLM from dropping 1.5 million acres (607,030 ha) from its wilderness inventory in 1983. Despite government interference, pressure from the public and from Congress helped the club continue its record of positive accomplishments, including the designation of 6.8 million acres (2.7 million ha) of wilderness in 18 states (1984), new wilderness designations in Alabama, Oklahoma, and Washington, and the addition of 40 rivers to the National Wild and Scenic River System.

In 1990, after years of grassroots lobbying, a compromise Clean Air Act was reauthorized, strengthening safeguards against **acid rain** and **air pollution**. Current projects include protecting the last remaining ancient forests of the Pacific Northwest; preventing oil and gas drilling in the 1.5-million-acre (607,030-ha) Arctic **National Wildlife Refuge** in Alaska; securing wilderness and park areas in California, Colorado, Idaho, Montana, Nebraska, North Carolina, South Dakota, New Mexico, and Utah; and combating global warming and the depletion of the world's protective **ozone** layer.

In 2001, the Sierra Club began its hundred and tenth year of work to protect the environment. By 2002, it had grown to 700,000 members and had 58 chapters across the United States, with an annual budget of $38 million. Having become so large and influential, the Sierra Club is now considered one of the "big ten" American conservation organizations. An extensive professional staff is required to operate this complex organization, and members tend to have little influence over club policy at the national level. Some radical activists have criticized mainline organizations of this kind for being too conservative, too comfortable in their relationship to established powers, and too willing to compromise basic principles in order to maintain power and prestige. Supporters of the club argue that a spectrum of environmental organizations is desirable and that different organizations can play useful roles.

[*Lewis G. Regenstein*]

Resources

Organizations

Sierra Club, 85 Second St., Second Floor, San Francisco, CA USA 94105-3441 (415) 977-5500, Fax: (415) 977-5799, Email: information@sierraclub.org, <http://www.sierraclub.org>

Silent Spring

see **Carson, Rachel**

Silt

A **soil** separate consisting of particles of a certain equivalent diameter. The most commonly used size for silt is from 0.05 to 0.002 mm equivalent diameter. This is the size used by the Soil Science Society of America and the **U.S. Department of Agriculture**, but others recognize slightly different equivalent diameters. As compared to clay (less than 0.002 mm), the silt fraction is less reactive and has a low cation exchange capacity. Because of its size, which is intermediate between clay and sand, silt contributes to formation of desirable pore sizes, and the **weathering** of silt minerals provides available plant nutrients. Wind-blown silt deposits are referred to as "loess." *See also* Soil conservation; Soil consistency; Soil profile

Siltation

The process or action of depositing **sediment**. Sediment is composed of solid material, mineral or organic, and can be of any texture. The material has been moved from its site of origin by the forces of air, water, gravity, or ice and has come to rest on the earth's surface. The term siltation does not imply the deposition of **silt** separates, although the sediment deposited from **erosion** of agricultural land is often high in silt because of the sorting action into **soil** separates during the erosion process.

Silver Bay

Silver Bay, on the Minnesota shore of Lake Superior, became the center of **pollution control** lawsuits in the 1970s when cancer-causing asbestos-type fibers, released into the lake by a Silver Bay factory, turned up in the drinking water of numerous Lake Superior cities. While **pollution** lawsuits have become common, Silver Bay was a landmark case in which a polluter was held liable for probable, but not proven, **environmental health** risks. **Asbestos**, a fibrous silicate mineral that occurs naturally in rock formations across the United States and Canada, entered Lake Superior in the waste material produced by Silver Bay's **Reserve Mining Corporation**. This company processed taconite, a low-grade form of iron ore, for shipment across the **Great Lakes** to steel-producing regions. Fibrous asbestos crystals removed from the purified ore composed a portion of the plant's waste **tailings**. These tailings were disposed of in the lake, an inexpensive and expedient disposal method. For almost 25 years the processing plant discharged wastes at a rate of 67,000 tons per day into the lake.

Generally clean, Lake Superior provides drinking water to most of its shoreline communities. However, water samples from Duluth, Minnesota, 50 mi (80.5 km) southwest of Silver Bay, showed trace amounts of asbestos-like fibers as early as 1939. While the term "asbestos" properly signifies a specific long, thin crystal shape that appears in many mineral types, both the long fibers and shorter ones, known as "asbestos-like" or "asbestiform," have been linked to **cancer** in humans.

Early incidences of asbestos-like fibers in drinking water probably resulted from nearby mining activities, but fiber concentrations suddenly increased in the late 1950s when Reserve Mining began its tailing **discharge** into the lake. By 1965 asbestiform fiber concentrations had climbed significantly, and municipal water samples in the 1970s were showing twice the acceptable levels defined by the federal **Occupational Safety and Health Administration** (OSHA). Lawsuits filed against Reserve Mining charged that the company's activity endangered the lives of the region's residents.

Although industrial discharge often endangers communities, the Silver Bay case was a pivotal one because it was an early test of scientific uncertainty in cases of legal responsibility. Cancer that results from exposure to asbestos fibers appears decades after exposure, and in a large population an individual's **probability** of death may be relatively small. Asbestos concentrations in Lake Superior water also varied, depending largely upon weather patterns and city **filtration** systems. Finally, it was not entirely proven that the particular type of fibers released by Reserve Mining were as carcinogenic as similar fibers found elsewhere. In such circumstances it is difficult to place clear blame on the agency producing the pollutants. The 200,000 people living along the lake's western arm were clearly at some risk, but the question of how much risk must be proven to close company operations was difficult to answer. Furthermore, the Reserve plant employed nearly all the breadwinners from nearby towns. Plant closure essentially spelled death for Silver Bay.

In 1980 a federal judge ordered the plant closed until an on-land disposal site could be built. Reserve Mining did construct on-shore tailings ponds, which served the company for several years until economic losses finally closed the plant in the late 1980s.

[*Mary Ann Cunningham Ph.D.*]

Resources

Periodicals

Carter, L. J. "Pollution and Public Health: Taconite Case Poses Major Test." *Science* 186 (October 4, 1974): 31–6.

Sigurdson, E. E. "Observations of Cancer Incidence Surveillance in Duluth, Minnesota." *Environmental Health Perspectives* 53 (1983): 61–7.

Peter Alfred David Singer (1946 –)

Australian philosopher and animal rights activist

Philosopher and leading advocate of the animal liberation movement, Singer was born in Melbourne, **Australia**. While teaching at Oxford University in England, Singer encountered a group of people who were vegetarians not because of any personal distaste for meat, but because they felt, as Singer later wrote, that "there was no way in which [maltreatment of animals by humans] could be justified ethically." Impressed by their argument, Singer soon joined their ranks. Out of his growing concern for the rights of animals came the book *Animal Liberation*, a study of the suffering we inflict upon animals in the name of scientific experimentation and food production. *Animal Liberation* caused a sensation when it was published in 1972 and soon became a major manifesto

of the growing animal liberation movement in North America, Australia, England, and elsewhere.

As a utilitarian, Singer—like his nineteenth-century forebear and founder of **utilitarianism**, the English philosopher Jeremy Bentham—believes that morality requires that the total amount of happiness be maximized and pain minimized. Or, as the point is sometimes put, we are morally obligated to perform actions and promote policies and practices that produce "the greatest happiness of the greatest number." But, Singer says, the creatures to be counted within this number should include all sentient creatures, animals as well as humans.

To promote only the happiness of humans and to disregard the pains of animals Singer calls speciesism—the view that one **species**, *Homo sapiens*, is privileged above all others. Singer likens **speciesism** to sexism and racism. The idea that one sex or race is innately superior to another has been discredited. The next step, Singer believes, is to recognize that all sentient creatures—human and nonhuman alike—deserve moral recognition and respect. Just as we do not eat the flesh or use the skin of our fellow humans, so, Singer argues, should we not eat meat or wear fur from animals. Nor is it morally permissible for humans to kill animals, to confine them, or to subject them to lethal laboratory experiments.

Although Singer's conclusions are congruent with those of **Tom Regan** and other defenders of **animal rights**, the route by which he reaches them is quite different. As a utilitarian, Singer emphasizes sentience, or the ability to experience pleasure and pain. Regan, by contrast, emphasizes the **intrinsic value** or inherent moral worth of all living creatures. Despite their differences, both have come under attack from the fur industry, defenders of "factory farming," and advocates of animal experimentation. Singer remains a key figure at the center of this continuing storm.

[*Terence Ball*]

Resources
Books

Ball, T., and R. Dagger. "Liberation Ideologies." In *Political Ideologies and the Democratic Ideal.* New York: Harper-Collins, 1991.

Singer, Peter. *A Darwinian Left: Politics, Evolution, and Cooperation.* New Haven, CT: Yale University Press, 2000.

———. *Animal Liberation.* 2nd ed. New York: Random House, 1990.

———. *Ethics into Action: Henry Spim and the Animal Rights Movement.* Latham, MD: Rowman & Littlefield, 1998.

———. *Practical Ethics.* New York: Cambridge University Press, 1979.

———. *Writings on an Ethical Life.* New York: Ecco Press, 2000.

———, and Helga Kuhse, eds. *Bioethics: An Anthology.* Malden, MA: Blackwell Publishers, 1999.

———, and T. Regan, eds. *Animal Rights and Human Obligations.* Englewood Cliffs, NJ: Prentice-Hall, 1976.

Sinkholes

Sinkholes are one of the main landforms in karst **topography**, so named for the region in Yugoslavia where solution features such as caves, caverns, disappearing streams and hummocky terrain predominate. Karst features occur primarily in limestone but may also occur in dolomite, chert, or even gypsum (Alabaster Caverns in Oklahoma).

As the name implies, sinkholes are depressions formed by solution enlargement or the **subsidence** of a cavern roof. Subsidence may occur slowly, as the cavern roof is gradually weakened by solution, or rapidly as the roof collapses. Several of the latter occurrences have gained widespread coverage because of the size and amount of property damage involved.

An often-described sinkhole formed during May 1981 in Winter Park, Florida, swallowing a three-bedroom house, half a swimming pool, and six Porsches in a dealer's lot. The massive "December Giant" occurred near Montevallo, Alabama, and measured 400 ft (122 m) wide by 50 ft (15 m) deep. A nearby resident reported hearing a roaring noise and breaking timbers, as well as feeling earth tremors under his house.

Cenotes is the Spanish name for sinkholes. One sacred cenote at the Mayan city of Chichen Itza in Yucatan, Mexico, was known as "the Well of Sacrifice." Archaeologists postulate that, to appease the gods during a **drought**, human sacrifices were cast into the water 80 ft (24 m) below, followed by a showering of precious possessions from onlookers. Since most of the gold and silver objects from the New World were melted down, these sacrifices are now highly prized artifacts from pre-Columbian civilizations.

Although occurring naturally, sinkhole formation can be intensified by human activity. These sinkholes offer an easy pathway for injection of contaminated **runoff** and sewage from septic systems into the **groundwater**. Because karst landscapes have extensive underground channels, the polluted water often travels considerable distances with little **filtration** or chemical modification from the relatively inert limestone. Therefore, the most serious hazard posed by sinkholes is the access they provide for turbid, polluted surface waters. This allows bacteria to thrive, so testing of spring water that emerges within or below karst regions is vital.

Sinkholes also pose special problems for construction of highways, reservoirs, and other massive objects. Fluctuating water levels weaken the overlying rock when the **water table** is high, but remove support when water levels are low. Thornbury (1954) described the problems resulting from efforts by Bloomington, Indiana, to build a water supply **reservoir** on top of karst topography. Much valuable water escaped through channels in the limestone beneath the dam. This structure eventually was abandoned and a new reservoir constructed in a region composed of relatively impervious siltstone below the limestone.

Anthropogenic (human-caused) sinkholes form as a result of mine subsidence, often catastrophically. Collapsing mine tunnels within the 1,000-mi (1,600-km) labyrinth in the historic Tri-State Mines of Kansas, Missouri, and Oklahoma have created scenarios very similar to the Winter Park, Florida, example. Subsidence of the strata overlying underground **coal** mines is another rich source of these anthropogenic sinkholes.

[*Nathan H. Meleen*]

Resources
Books

Coates, D. R., ed. *Environmental Geomorphology.* Binghampton, NY: State University of New York, 1971.

Keller, E. A. *Environmental Geology.* 4th ed. Columbus, OH: Charles E. Merrill Publishing Co., 1985.

Thornbury, W. D. *Principles of Geomorphology.* New York: John Wiley & Sons, 1954.

Periodicals

"Into the Well of Sacrifice," *National Geographic Magazine* (October 1961): 540–561.

Site index

A means of evaluating a forest's potential to produce trees for timber. Site refers to a defined area, including all of its environmental features, that supports or is capable of supporting a stand of trees. Site index is an indicator used to predict timber yield based on the height of dominant and co-dominant trees in a stand at a given index age, usually 50 or 100 years. Normally, site index curves are prepared by graphically plotting projected tree height as a function of age. *See also* Forest management

Site remediation

see **Hazardous waste site remediation**

Skidding

A technique in forest harvesting by which logs or whole trees are dragged over the ground, as opposed to being lifted in the air, to a landing, where they are loaded on trucks for transport to a mill. The logs may be dragged by mechanical means, such as a crawler tractor or rubber-tired skidder, or by draft animals. Skidding normally disturbs the ground surface and forms "skid trails" which may be used once or reused in subsequent harvests. **Soil** in skid trails, especially on steep land, may channel water flow during rainfall and snow melt and be susceptible to **erosion**.

SLAP suits

see **Strategic lawsuits to intimidate public advocates**

Slash

The waste material, consisting of limbs, branches, twigs, leaves, or needles, left after forest harvesting. **Logging** slash left on the ground can protect the **soil** from raindrop impacts and **erosion**, and it can decompose to make **humus** and recycle nutrients. Unusually large amounts of slash, which may present a fire hazard or provide favorable **habitat** for harmful insects or disease organisms, usually must be reduced by burning or mechanical chopping.

Slash and burn agriculture

Also known as swidden cultivation or **shifting cultivation**, slash-and-burn agriculture is a primitive agricultural system in which sections of forest are repeatedly cleared, cultivated, and allowed to regenerate over a period of many years. This kind of cultivation was used in Europe during the Neolithic period, and it is still widely used by **indigenous peoples** and landless peasants in the tropical rain forests of South America.

The plots used in slash-and-burn agriculture are small, typically 1–1.5 acres (0.4–0.6 hectare). They are also polycultural and polyvarietal; farmers plant more than one crop on them at a time, and each of these crops may be grown in several varieties. This helps control populations of agricultural pests. The cutting and burning involved in clearing the site releases nutrients which the cultivated crops can utilize, and the fallow period, which usually lasts at least as long as 15 years, allows these nutrients to accumulate again. In addition to restoring fertility, re-growth protects the **soil** from **erosion**.

Families and other small groups practicing slash-and-burn agriculture generally clear one or two new plots a year, working a number of areas at various stages of cultivation at a time. These plots can be close to each other, even interconnected, or spread out at a distance through the forest, designed to take advantage of particularly fertile soils or to meet different needs of the group. As the nutrients are exhausted and productivity declines, the areas cleared for slash-and-burn agriculture are rarely simply abandoned; the fallow period begins gradually, and **species** such as fruit trees are still cultivated as the forest begins to reclaim the open spaces. The forest may contain originally cultivated species that still yield a harvest many years after the plot has been overgrown.

Jungle in Guatemala is burned to clear land for raising corn and cattle. (Photograph by George Holton. Photo Researchers Inc. Reproduced by permission.)

Although this system of agriculture was practiced for thousands of years with relatively modest effects on the **environment**, the pressures of a rapidly growing population in South America have made it considerably less benign. **Population growth** has greatly increased the number of peasants who do not own their own land; they have been forced to migrate into the rain forests, where they subsist practicing slash-and-burn agriculture. In Brazil, the number of farmers employing this system of agriculture has increased by more than 15% a year since 1975. A recently released report by the United Nations Population Fund has emphasized the destruction this system can cause when practiced on such a large scale and identified it as a threat to species diversity. *See also* Agricultural revolution; Agricultural technology; Agroecology

[*Douglas Smith*]

Sludge

A suspension of solids in liquid, usually in the form of a liquid or a **slurry**. It is the residue that results from **wastewater** treatment operations, and typical concentrations range from 0.25 to 12% solids by weight. An estimated 8.5 million dry tons of municipal sludge is produced in the United States each year.

The volume of sludge produced is small compared with the volume of wastewater treated. The cost of sludge treatment, however, is estimated to be from 25 to 40% of the total cost of operating a wastewater-treatment plant. In the design of **sludge treatment and disposal** facilities, the term sludge refers to primary, biological, and chemical sludges, and excludes grit and screenings. Primary sludge results from primary **sedimentation**, while the sources of biological and chemical sludges are secondary biological and chemical settling and the processes used for thickening, digesting, conditioning, and dewatering the sludge from the primary and secondary settling operations.

Primary sludge is usually gray to dark gray in color and has a strong offensive odor. Fresh biological sludge—activated sludge, for example—is brownish and has a musty or earthy odor. Chemical sludge can vary in color depending on composition and may have objectionable odor. There are

several sludge treatment options available. These operations and processes are principally intended for size reduction, grit removal, moisture removal, and stabilization of the organic matter in the sludge. Examples of these sludge processing methods are sludge grinding and **cyclone** degritter for the preliminary operations, chemical conditioning and heat treatment, centrifugation and vacuum **filtration** for dewatering operations, **aerobic** or **anaerobic digestion**, **composting**, and **incineration** for sludge stabilization.

Other than its organic content, sewage sludge is known to contain some essential **soil** nutrients, such as **phosphorus** and **nitrogen**, that make it suitable for use as **fertilizer** or soil conditioner after further processing. Sludge may also be disposed of in landfills. Landfilling and incineration, however, are strictly disposal methods and, unlike land application, do not recycle sludge.

Raw sludge contains potentially harmful pathogens, **heavy metals** and toxic organics; it is regulated by the **Environmental Protection Agency** (EPA). The EPA has recently proposed new regulations establishing management practices and other requirements for the disposal of sewage sludge. *See also* Aerobic sludge digestion; Contaminated soil; Industrial waste treatment; Sewage treatment; Waste management

[*James W. Patterson*]

Resources

Books

Sundstrom, D. W., and H. E. Klei. *Wastewater Treatment.* Englewood Cliffs, NJ: Prentice-Hall, 1979.

Weber Jr., W. *Physicochemical Processes for Water Quality Control.* New York: Wiley-Interscience, 1972.

Periodicals

Hasbach, A. C., ed. "Putting Sludge to Work." *Pollution Engineering* (December 1991).

Other

Process Design Manual for Sludge Treatment and Disposal. Washington, DC: U.S. Environmental Protection Agency, 1979.

Sludge digestion

see **Aerobic sludge digestion**

Sludge treatment and disposal

The proper treatment and disposal of **sludge** require knowledge of the origin of the solids to be handled, as well as their characteristics and quantities. The type of treatment employed and the method of operation determine the origin of sludge. In **sewage treatment** plants, sludge is produced by primary settling, which is used to remove readily settleable solids from raw **wastewater**. Biological sludges are produced by treatment processes such as **activated sludge**, trickling filter, and rotating biological contractors. Chemical sludges result from the use of **chemicals** to remove constituents through precipitation; examples of precipitates that are produced by this process include phosphate precipitates, carbonate precipitates, hydroxide precipitates, and polymer solids.

Sludge is characterized by the presence or absence of organic matter, nutrients, pathogens, metals and toxic organics. These characteristics are an important consideration for determining both the type of treatment to be used and the method of disposal after processing. According to the **Environmental Protection Agency** (EPA), the typical chemical compositions of untreated and digested primary sludge include solids, grease and fats, protein, **nitrogen**, **phosphorus**, potash, iron, silica, and **pH**.

Municipal sludge is a high-volume waste. Typical volumes for raw primary sludge range from 2,950 to 3,530 gallons per million gallons of wastewater treated. Volumes for trickling filter **humus** range from 530 to 750 gallons per million, while the volumes for activated sludge are much higher, from 14,600 to 19,400 gallons per million. It is possible to calculate the quantities of sludge theoretically, but the EPA recommends that treatment operations use pilot plant equipment to make these calculations whenever possible. In developing a treatment and disposal system, the EPA also recommends that large wastewater treatment plants adopt a methodical approach "to prevent cursory dismissal of options." For small plants, with a capacity of less than a million gallons a day, the task of determining an operating procedure is often shorter and less complex.

Sludge treatment and disposal generally include several unit processes and operations, which fall under the following classifications: thickening, stabilization, disinfection, conditioning, dewatering, drying, thermal reduction, miscellaneous processes, ultimate disposal, or **reuse**.

Thickening is a volume-reducing process in which the sludge solids are concentrated to increase the efficiency of further treatment. It has recently been reported that sludge with 0.8% solids thickened to a content of 4% solids yields a five-fold decrease in sludge volume. Thickening methods commonly employed are gravity thickening, **flotation** thickening, and centrifuge.

Sludges are stabilized to eliminate offensive odors and reduce toxicity. A stable sludge has been defined as "one that can be disposed of without damage to the **environment**, and without creating nuisance conditions." In sludges, toxicity is characterized by high concentrations of metals and toxic organics, as well as by high oxygen demand, abnormally high or low pH levels, and unsafe levels of pathogenic **micro-**

organisms. There are a variety of technologies available for stabilizing toxic sludge; these include lime stabilization, heat treatment, and biological stabilization, which consists of **aerobic** or **anaerobic digestion** and **composting**.

Sludge that has been stabilized may also be disinfected in order to further reduce the level of pathogens. There are several methods of sludge disinfection: thermal treatment such as pasteurization, chemical treatment, and irradiation. This process is important for the reuse and application of sludge on land.

Dewatering is used to achieve further reductions in moisture content. It is a process designed to reduce moisture to the point where the sludge behaves like a solid; at the end of this process, the concentration of solids in sludge is often greater than 15%. Dewatering can include a number of unit operations: Sludge can be dried in drying beds or lagoons, filtered through a vacuum filter, a filter press, or a strainer, and separated in machines such as a solid-bowl centrifuge. It can be determined whether a sludge will settle in a centrifuge by testing it in a test-tube centrifuge, where the concentration of cake solids are determined as a function of centrifugal acceleration. The Capillary Suction Time (CST) and the specific **resistance** to **filtration** are important parameters for the filterability of sludge.

Sludge is conditioned to prepare it for other treatment processes; the purpose of sludge conditioning is to improve the effectiveness of dewater and thickening. The methods most commonly employed are the addition of organic materials such as polymers, or the addition of inorganic materials such as **aluminum** compounds. Heat treatment is also used; in this treatment, temperatures range from 356–392°F (180–200°C) for a period of 20–30 minutes.

After thickening and dewatering, further reduction of moisture is necessary if the sludge is going to be incinerated or processed into, starved air **combustion**, co-disposal, and wet oxidation. These processes have a number of advantages. They can achieve maximum volume reduction; they can destroy toxic organic compounds, and they also produce heat energy which can be utilized. But sludge combustion cannot be considered an ultimate or long-term disposal option because of the residuals it produces, such ash and air emissions, which may have detrimental effects on the environment. Landfills are another disposal option, but limitations on space as well as regulatory constraints restrict their long-term feasibility. Reuse is probably the best solution for the long-term management of sludge; the most feasible beneficial use option will probably be either land application, land **reclamation**, or raw material recovery. Examples include the conversion of sludge into commercial **fertilizer**, fuel, and building products.

Whatever the selected array of sludge treatment and disposal measures are, the main factors that influence the choice will always be cost-effectiveness, public health, and environmental protection. *See also* Activated sludge; Municipal solid waste composting; Solid waste; Waste management

[*James W. Patterson*]

Resources

Books

Vesilind, P. A. *Treatment and Disposal of Wastewater Sludges.* Ann Arbor, MI: Ann Arbor Science Publishers, 1979.

Periodicals

Hasbach, A. C. "Putting Sludge to Work." *Pollution Engineering* (December 1991).

Other

Process Design Manual for Sludge Treatment and Disposal. Washington, DC: U.S. Environmental Protection Agency, 1979.

Slurry

A thin mixture of water and fine, insoluble materials suitable for pumping as a liquid. Pipelines are the cheapest and most efficient means of moving materials long distances. In Arizona, the Black Mesa slurry pipeline carries eight tons of powdered **coal** per minute to the electrical power plant near Page on Powell **Reservoir**. A much longer pipeline has been proposed for shipping subbituminous coal in Wyoming to Midwestern cities. Water supply is a crucial factor in these **arid** regions lying in the Rocky Mountain **rain shadow** zone.

Small quantity generator

Small quantity generators (SQGs) are facilities that do not generate more than 2,205 lb (1,000 kg) of regulated **hazardous waste** or more than 2.2 lb (1 kg) of acute hazardous waste in a calendar month, or do not store more than 13,228 lb (6,000 kg) of hazardous waste in any 180 day period. If the facility has to transport the waste for more than 200 mi (322 km) to dispose of it, then the period that waste can be stored may be increased to 270 days, without the operator of the site losing the small quantity generator designation. If a site generates 220 lb (100 kg) or less of hazardous waste in a calendar year, then the site is designated

as a conditionally exempt small quantity generator or (CESQG).

[*Marie H. Bundy*]

Smart growth

Smart growth is a relatively new movement in the United States, at least by name, a movement promoted since the early 1990s as a new way to direct growth and development, especially in urban areas, away from sprawl and toward urban centers of various kinds. There seems to be no universally accepted definition of smart growth, and that alone indicates problems in agreement about what it is, what its goals are, the nature and extent of its benefits and impacts, and to what degree it can be implemented.

Disagreements seem to be more at the level of specific issues rather than broad goals, however, and a fairly simple if still broad and somewhat vague definition can be readily distilled from a rapidly increasing literature. A composite definition suggests that smart growth is growth management rather than "control of growth," and probably no one would disagree with that; it is inward development rather than outward development, i.e., it is intended to limit sprawl and to fill in or reclaim already developed areas rather than expand into new ones; and it is high density rather than low density. According to the **Environmental Protection Agency** (EPA), "smart growth is development that serves the economy, community, and the environment."

Its advocates see it as a replacement for the "failed" environmental movement of the 1960's, but others describe it as an attempt to gloss over growth, to hide continued dominance by the marketplace in uncoordinated decisions on human settlement patterns, or to avoid facing the economic and political difficulties of the often better environmental alternative of no growth. Still others claim that too many smart growth advocates are just the opposite, ignorant of and naive about those same market factors. Though backed by individuals and groups that represent all spectrums of political thought, smart growth is criticized by skeptics from an equally broad range of viewpoints. Smart growth as concept, as philosophy, and as policy, smart growth when implemented, rejected or delayed, all remain controversial. The advocates of smart growth do seem to agree on ten principles, listed here essentially as they are stated in a 2002 EPA document titled *Getting to Smart Growth*. The benefits foreseen by its proponents are evident in the principles as stated. 1) Mix land uses rather than the separation mandated by zoning codes in many communities in the United States. Mix commercial with residential, high density with low, pedestrian and bike access with other means of transport. 2) Take advantage of compact building design to create smaller footprints, to increase density, and to reduce the amount of space required to accommodate increases in population. 3) Create a range of housing opportunities and choices to increase neighborhood diversity in multiple ways and better accommodate the housing needs of a diverse population. 4) Create walkable neighborhoods by making them pedestrian-friendly and scaled to people on foot rather than to people in cars. 5) Foster distinctive, attractive communities with a strong **sense of place** by introducing more distinctiveness and uniqueness to residential and retail areas, for example, through less reliance on large malls that are indistinguishable from one city to the next in every region of the country. Create structures that reflect local and regional differences in **climate**, vegetation, and landforms. 6) Preserve open space, farmland, natural beauty, and critical environmental areas to enhance quality of life and increase local recreational opportunities. 7) Strengthen and direct development towards existing communities through less reliance on new, low-density dispersed developments outside urban cores and close-in suburbs. 8) Provide a variety of **transportation** choices, especially by creating transit and non-motorized alternatives to the **automobile**, in the process reducing traffic congestion. 9) Make development decisions predictable, fair, and cost effective. 10) Encourage community and stakeholder collaboration in development decisions.

Given the **visibility** of arguments in recent years against sprawl, these principles seem like apple pie and motherhood; who could be against them? But smart growth has many critics. Numerous individuals and groups can and do take issue, especially with specific principles when local implementation is actually attempted on the ground. Some of the statements are viewed more critically than others, and the movement in general also has its skeptics. Critics, especially those without a NIMBY interest, are as quick as proponents to point out good elements in the smart growth movement, most of them embodied in the ten stated principles. Such elements as in-fill, high density, neighborhood development and identity, and brownfield redevelopment, are widely supported, though not universally.

A problem pointed out over and over again by critics of smart growth is that the **reclamation** and filling in of already developed areas increases their attractiveness, raises their value and the cost of living there, i.e., the areas are gentrified and the people who did live in the reclaimed areas are displaced. Minorities and the poor are squeezed out and their difficulties in finding acceptable accommodations is traced to declining stocks of affordable housing. Much of the debate over smart growth centers on whether or not it is the cause of increases in the price of housing. Critics say yes, land prices do increase, affordable housing is torn down and replaced by more expensive dwellings, and yes, these

changes are directly attributable to reclamation efforts implemented in the name of smart growth or growth management.

Proponents argue that growth management is not the cause of declines in the availability of affordable houses, but that increases are the result of conventional marketing dynamics and growth in population, employment, and income, especially during the boom of the 1990s. They further argue that smart growth initiatives are much less expensive, for individuals and families as well as for local governments, because they decrease investments in infrastructure, especially that related to automobile use, and that such savings can then be passed on to decrease housing costs.

Among the problems of implementing smart growth is the tight hold many Americans have on the American Dream: detached houses that "return people to nature" are not located in the central parts of cities and certainly not in **brownfields**. And critics are quick to see the irony in the coincidence of interest in a smart growth movement with reports in the daily papers like the following two illustrations: first, increases in "Supersize Suburbs," "big-house communities" in which smaller families live in hypertrophy homes: "in the country's long pursuit of ever more elbow room...the American house has been swelling for decades." In some suburban counties, "more than two-thirds of the houses have nine rooms or more..." The mega-mansions are literally built on green "fields" because many owners don't want trees and vegetation blocking the "full effect" the houses have on passers-by. Surveys show that many of these monster homes are occupied by an average of slightly more than three people. A typical response when this trend is questioned: "We can afford it, so why not?"

A second illustration is the realization by a big timber company in Florida that its extensive timber holding in the state are "too valuable for trees alone." The company's plans for one stretch of 40 mi (64 km) of beaches on the panhandle "are so sweeping that it hopes to reduce the "redneck riviera" image of the area by calling it "Florida's Great Northwest." Some 20 developments are in various stages of planning, with more than 10,000 large, high-end homes permitted so far, aimed at affluent buyers from Atlanta and Birmingham. The chairman of the company calls the expected transformation "regional place making." Locals shrug with "we've got to take the hand we're dealt" statements and environmentalists concede that a company so powerful will prevail, so they try to work with it to minimize the impacts, recognizing that "it's absolutely clear that the Florida Panhandle will change" fundamentally, hopefully "in a way that creates livable communities and protects the natural system, and with [this company] we have at least a shot at doing that."

These two illustrations are drawn from high-growth areas, with little indication that such principles as compact building or neighborhood diversity or people-scale neighborhoods not dependent on cars or less reliance on new low-density developments outside urban areas or widely based stake-hold collaboration are driving many land-use decisions in the United States, at least on private land. The question isn't one of smart growth being or not being a policy in these areas, but that the attitudes and actions in such illustrations make smart growth as a solution to settlement problems in the nation as a whole problematic at best.

One especially severe criticism is that nowhere in the 10 principles is there even a hint at any real attempt to actually limit growth, or even delay it, that smart growth just accommodates growth rather than addressing it as a problem. Critics point out that the capability of local or regional environmental systems—notably water supplies—to support growth is not mentioned in any of the principles, or anywhere in the smart growth philosophy. Growth is still growth and calling it "smart" does not help answer the question of ultimate limits, of how to plan for no growth. Even its critics admit, however, that smart growth creates better management and planning alternatives than totally unregulated sprawl.

[*Gerald L. Young Ph.D.*]

Resources

Periodicals

Chen, Donald D. T. "The Science of Smart Growth." *Scientific American* 283, no. 6 (December 2000): 84–91.

Daniels, Tom. "Smart Growth: A New American Approach to Regional Planning." *Planning Practice & Research* 16, no. 3–4 (2001): 271–179.

Downs, Anthony. "What Does 'Smart Growth' Really Mean?" *Planning* 67, no. 4 (April 2001): 20–25.

Pelley, Janet. "Building Smart-Growth Communities." *Environmental Science & Technology* 33, no. 1 (January 1999): 28–32.

Staley, Samuel R., and Leonard C. Gilroy. "Why 'Smart Growth' Isn't Smart." *Consumers' Research* 85, no. 1 (January 2002): 10–13.

Other

Environmental Protection Agency. *Getting to Smart Growth: 100 Policies for Implementation.* [cited July 2002]. <http://www.smartgrowth.org/pdf/gettosg.pdf.>.

Smelter

Smelters are industrial facilities that are used to treat metal ores or concentrates with heat, **carbon**, and oxygen in order to produce a crude-metal product, which is then sent to a refinery to manufacture pure metals.

In many cases, smelters process sulfide minerals, which yield gaseous **sulfur dioxide**, a significant waste product. Other smelters, including some that process iron ores, do not treat sulfide minerals. Similarly, secondary smelters used for **recycling** metal products, such as used **automobile** batteries, do not emit sulfur dioxide. However, all smelters emit

metal particulates to the **environment**, and unless this is prevented by **pollution control** devices, these emissions can cause substantial environmental damages.

The earliest large industrial smelting technique involved the oxidation of sulfide ores using roast beds, which were heaps of ore piled upon wood. The heaps were ignited in order to oxidize the sulfide-sulfur to sulfur dioxide, thereby increasing the concentration of desired metals in the residual product. The roast beds were allowed to smoulder for several months, after which the crude-metal product was collected for further processing at a refinery. The use of roast beds produced intense, ground-level plumes filled with sulfur dioxide, acidic mists, and metals, which could devastate local ecosystems through direct toxicity and by causing **acidification**.

More modern smelters emit pollutants to the **atmosphere** through tall smokestacks. These can effectively disperse emissions, so that local **air quality** is enhanced, and damages are greatly reduced. However, the actual acreage of land affected by the contamination is enlarged because the tall smokestacks broadcast emissions over a greater distance and **acid rain** may be spread over a larger area as well.

Emissions of toxic **chemicals** can be reduced at the source. Metal-containing particulates can be controlled through the use of electrostatic precipitators or baghouses. These devices can remove particulates from flue-gas streams, so that they can be recovered and refined into pure metals, instead of being emitted into the atmosphere. Electrostatic precipitators and baghouses can often achieve particle-removal efficiencies of 99% or better.

It is more difficult and expensive to reduce emissions of gases such as sulfur dioxide. Existing technologies usually rely on the reaction of sulfur dioxide with lime or limestone to produce a **slurry** of gypsum (calcium sulfate) that can be disposed into landfills or sometimes manufactured into products such as wallboard. It is also possible to produce sulfuric **acid** by some flue-gas desulfurization processes. At best, removal efficiencies for sulfur dioxide are about 95%, and often considerably less.

The types of smelters that do not treat sulfide metals, such as iron ore smelters, logically do not emit sulfur dioxide. They do, however, spew metal particulates into the atmosphere. For example, a facility that has been operating for centuries at Gusum, Sweden, has caused a significant local **pollution** with **copper** and **lead**, the toxic effects of which have damaged vegetation. The surface organic matter of sites close to the Gusum facility has been polluted with as much as 2% each of zinc and copper. Secondary smelters also do not generally emit toxic gases, but they can be important sources of metal particulates. This has been especially well documented for secondary lead smelters, many of which are present in urban or suburban environments. The danger from these is that people can be affected by lead in their environment, in addition to long-lasting ecological effects. For example, lead concentrations in **soil** as large as less than 5% dry weight were found in the immediate vicinity of a battery smelter in Toronto, Ontario. In this case, people were living beside the smelter. Garden soils and vegetables, house dust, and human tissues were all significantly contaminated with lead in the vicinity of that smelter. Similar observations have been made around other lead-battery smelters, including many that are situated dangerously close to human habitation.

One of the best-known case studies of environmental damage caused by smelters concerns the effects of emissions from the metal processing plants around **Sudbury, Ontario**. This area has a long history as a mining community. For many years, roast beds were the primary metal-processing technology used at Sudbury. In fact, in its heyday, up to 30 roast beds were operating in Sudbury. Unfortunately, this process saturated the air and soil with sulfur dioxide, **nickel**, and copper. Local ecosystems were devastated by the direct toxicity of sulfur dioxide and, to a lesser extent, the metals. In addition, **dry deposition** of sulfur dioxide caused a severe acidification of lakes and soil. This caused much of the plant life to die, which in turn started soil **erosion**. Naked bedrock was exposed, and then blackened and pitted by reaction with the sulfurous plumes and acidic mists.

When the devastating consequences of the roast bed method became clear, the government prohibited their use. The processors turned to new technology in 1928 when they began construction of three smelters with **tall stacks**. Since these emitted pollutants high into the atmosphere, they showed substantial improvements in local air quality. However, the damage to vegetation continued; lakes and soils were still being acidified, and toxic contaminants were spread over an increasingly large area.

Over time, well-defined patterns of ecological damage developed around the Sudbury smelters. The problems that had occurred in the roast beds were being repeated on a large scale. The most devastated sites were close to the smelters, and had concentrations of nickel and copper in soil in the thousands of **parts per million**; they were very acidic with resulting **aluminum** toxicity, and frequently fumigated by sulfur dioxide. Such toxic sites were barren, or at most had very little plant cover. The few plants that were present were usually specific ecotypes of a few widespread **species** that had evolved a tolerance to the toxic effects of nickel, copper, and acidity. Aquatic lake habitats close to the smelters were similarly affected. These waterbodies were acidified by the dry deposition of sulfur dioxide, and had large concentrations of soluble nickel, copper, aluminum, and other toxic metals. Of course, the plant and animal life of these lakes was highly impoverished and dominated by

life forms that were specifically adapted to the toxic stresses associated with the metals and acidification.

It is well recognized that increased linear distance from the **point source** of toxic chemicals reduces deposition rates and toxic stress. Correspondingly, the pattern of environmental pollution and ecological damage around the Sudbury areas decreases more-or-less exponentially with increasing distance from the smelters. In general, it is difficult to demonstrate damages to terrestrial communities beyond 9–12 miles (15–20 km), although contamination with nickel and copper can be found at least 62 miles (100 km) away. In comparison, lakes that are deficient in plant life have little acid-neutralizing capacity and can be shown to have been damaged by the dry deposition of sulfur dioxide at least 25–30 miles (40–48 km) from Sudbury.

To improve regional air quality, the world's tallest smokestack, at 1,247 feet (380 m), was constructed at the largest of the Sudbury smelters. This "superstack" allowed for an even wider dispersion of smelter emissions. At the same time, a smelter was closed down, and steps were taken to reduce sulfur dioxide emissions by installing desulferization, or **flue-gas scrubbing**, equipment and by processing lower-sulfur ores. In aggregate, these actions resulted in a great improvement of air quality in the vicinity of Sudbury.

The improvement of air quality allowed vegetation to regenerate over large areas, with many existing species increasing in abundance and new species appearing in the progressively detoxifying habitats. The revegetation has been actively encouraged by re-seeding and liming activities along roadways and other amenity areas where soil had not been eroded. Aquatic habitats have also considerably detoxified since 1972. Lakes near the superstack and the closed smelter have become less acidic, metal concentrations have decreased, and the plants and animals have increased **biomass**.

However, there is important controversy about contributions that the still-large emissions of sulfur dioxide from Sudbury may be making towards the regional acid rain problem. The superstack may actually exacerbate regional difficulties because of the wide dispersal of its emissions of sulfur dioxide. It is possible that height of the superstack may be decreased in order to increase the local dry deposition of emitted sulfur dioxide and reduce the long range dispersion of this acid-precursor gas.

The Sudbury scenario is by no means an unusual one. Many other smelters have substantially degraded their surrounding environment. The specifics of environmental pollution and ecological damage around particular smelters depend on the intensity of toxic stress and the types of ecological communities that are being affected. Nevertheless, broadly similar patterns of ecological change are observed around all point sources of toxic emissions, such as smelters.

In most modern smelters, a reduction of emissions at the source can eliminate the most significant of the ecological damages that have plagued older smelters. The latter were commissioned and operated under social and political climates that were much more tolerant of environmental damages than are generally considered to be acceptable today. *See also* Air pollutant transport; Electrostatic precipitation; Water pollution

[*Bill Freedman Ph.D.*]

Resources

Books

Freedman, B. *Environmental Ecology.* San Diego: Academic Press, 1989.

———, and T. C. Hutchinson. "Sources of Metal and Elemental Contamination of Terrestrial Ecosystems." In *Metals in the Environment.* Vol. 2. Edited by N. W. Lepp. London: Applied Science Publishers, 1981.

Nriagu, J., ed. *Environmental Impacts of Smelters.* New York: Wiley, 1984.

Robert Angus Smith (1817 – 1884)

Scottish chemist

Smith has two claims to fame. Through his studies of **air pollution** he was, in 1852, the discoverer of **acid rain**, and his appointment as Queen Victoria's first inspector under the Alkali Acts Administration of 1863 made him the prototype of the scientific civil servant. He was one of the earlier scientists to study the chemistry of air and **water pollution** and among the first to see that such study was important in identifying and controlling environmental problems caused by industrial growth in urban centers. He was also interested in public health, disinfection, peat formation, and antiquarian subjects. Although his work has sometimes been dismissed as pedestrian, Smith's pioneering studies of the chemistry of atmospheric precipitation, published in 1872 in the book *Air and Rain*, were far ahead of their time and a major contribution to a new discipline that he called "chemical climatology."

[*Eville Gorham Ph.D.*]

Resources

Periodicals

Gorham, E. "Robert Angus Smith, F.R.S., and 'Chemical Climatology.'" *Notes and Records of the Royal Society of London* 36 (1982): 267–72.

MacLeod, R. M. "The Alkali Acts Administration, 1863–84: The Emergence of the Civil Scientist." *Victorian Studies* 9 (1965): 85–112.

Smog over New York City. (Photograph by Yoav Levy. Phototake. Reproduced by permission.)

Smog

A term chosen by the Glasgow public health official Des Voeux at the beginning of the twentieth century to describe the smoky fogs that characterized coal-burning cities of the time. The word is formed by adding the words **smoke** and fog together and has persisted as a description of this type of urban **atmosphere**. It has more and more been used to describe **photochemical smog**, the **haze** that became a characteristic of the **Los Angeles Basin** from the 1940s. "Smog" is sometimes even used to describe **air pollution** in general, even where there is no reduction in **visibility** at all. However, the term is most properly used to describe the two distinctive types of **pollution** that dominated the atmospheres of late nineteenth century London, England, known as winter smog, and twentieth century Los Angeles, called summer smog.

The city of London burned almost 20 million tons of **coal** annually by the end of the nineteenth century. Although industrialized, much coal was burned in domestic hearths, and the smoke and **sulfur dioxide** produced barely rose from the chimneys above the housetops. Only a few rather inaccurate measurements of the pollutant concentrations in the air were made in the last century, although they hint at concentrations much higher than what we might expect in London today.

The smogs of nineteenth-century London took the form of dense, vividly colored fogs. The smog was frequently so dense that people became lost and had to be lead home by linksmen. It is said that visibility became so restricted that fingers on an outstretched arm were invisible. The fog that rolled over window sills and into rooms became such an integral part of what we know as Victorian London that almost any Sherlock Holmes story mentions it.

With London's high humidity and incipient fog, smoke particles from coal-burning formed a nucleus for the condensation of vapor into large fog droplets. This water also serves as a site for chemical reactions, in particular the formation of sulfuric **acid**. Sulfur dioxide dissolved in fog droplets, perhaps aided by the presence of alkaline material such as ammonia or coal ash. Once in solution the sulfur was oxidized, a process often catalyzed by the presence of dissolved metallic ions such as iron and manganese. Dissolution and oxidation of sulfur dioxide gave rise to sulfuric acid droplets, and it was sulfuric acid that made the smog so damaging to the health of Londoners.

London's severe smogs occurred throughout the last decades of the nineteenth century. Detective writer Robert Barr even published *The Doom of London* at the turn of the century, which saw the entire population of London eliminated by an apocalyptic fog. Many residents of Victorian London recognized that the fogs increased death rates, but the most infamous incident occurred in 1952, when a slow-moving anti-cyclone stalled the air over the city. On the first morning the fog was thicker than many people could ever remember. By the afternoon people noticed the choking smell in the air and started experiencing discomfort. Those who walked about in the fog found their skin and clothing filthy after just a short time. At night the treatment of respiratory cases was running at twice its normal level. The situation continued for four days.

It was difficult to describe exactly what had happened, because primitive air pollution monitoring equipment could not cope with high and rapidly changing concentrations of pollutants, but it has been argued that for short periods the smoke and sulfur dioxide concentrations may have approached ten thousand micrograms per 1.3 yd^3 (1 m^3). Today, in a relatively healthy city, the desired maximum is about a hundred micrograms per 1.3 yd^3 in short-term exposures.

Normal death rates were exceeded by many thousands through the four-day period of the fog. Public feelings ran high, and the United Kingdom government, barraged with questions, set up an investigative committee. The Beaver Committee report eventually served as the basis for the UK **Clean Air Act** of 1956. This law was gradually adopted through many towns and cities of the UK and has been seen by many as a model piece of legislation. Although it is true that the classic London smog has gone, it is far from clear the extent to which this change came about through legislation rather than through broader social developments, such as the use of electricity in homes (although the act did encourage this).

Photochemical smog is sometimes called summer smog, because unlike the classical London type smog, it is more typical in summer at many localities, often because it requires long hours of sunshine to build up. When photochemical smogs were first noticed in Los Angeles, people believed them to be much the same as the smogs of London and Pittsburgh. Early attempts at control looked largely at local industry emissions. The **automobile** was eliminated as a likely cause because of low concentrations of sulfur in the fuel and the fact that only minute amounts of smoke were generated.

It was some time before the biochemist **Arie Jan Haagen-Smit** recognized that damage to crops in the Los Angeles area arose not from familiar pollutants, but from a reaction that took place in the presence of **petroleum** vapors and sunlight. His observations focused unwelcome attention on the automobile as an important factor in the generation of summer smog.

The Los Angeles area proved an almost perfect place for generating smogs of this type. It had a large number of cars, long hours of sunshine, gentle sea breezes to back the pollutants up against the mountains, and high level inversions preventing the pollutants from dispersing vertically.

Studies through the 1950s revealed that the smog was generated through a photolytic cycle. Sunlight split **nitrogen** dioxide into nitric oxide and atomic oxygen that could subsequently react and form **ozone**. This was the key pollutant that clearly distinguished the Los Angeles smogs from those found in London. Although the highly reactive ozone reacts rapidly with nitric oxide, converting it back into nitrogen dioxide, organic radicals produced from petroleum vapor react with nitric oxide very quickly. The nitrogen dioxide is again split by the sunlight, leading to the formation of more ozone. The cycle continues to build ozone concentrations to higher levels throughout the day.

The nitrogen oxide-ozone cycle is just one of many processes initiated in a smog of this kind. The photochemically active atmosphere contains a great number of reactive molecular fragments that lead to a range of complex organic compounds. Some of the hydrocarbon molecules of petroleum vapor are oxidized to aldehydes or **ketones**, such as acrolein or formaldehyde, which are irritants and suspected carcinogens. Some oxygenated fragments of organic molecules react with the **nitrogen oxides** present in the atmosphere. The best-known product of these reactions is **peroxyacetyl nitrate**, often called PAN, one of a class of nitrated compounds causing eye irritation experienced in summer smogs.

The reactions that were recognized in the Los Angeles smog are now known to occur over wide areas of the industrialized world. The production of smog of this kind is not limited to urban or suburban areas, but may occur for many hundreds of miles to the lee of cities using large quantities of liquid fuel. The importance of **hydrocarbons** in sustaining the processes that generate photochemical smog has given rise to control policies that recognize the need to lower the **emission** of hydrocarbons into the atmosphere, and hence the emphasis of the use of catalytic converters and low volatility fuels as part of **air pollution control** strategies.

See also Alternative fuels; Environmental policy; Fossil fuels; Respiratory diseases

[*Peter Brimblecombe*]

RESOURCES
BOOKS

Findlayson-Pitts, B. J., and J. N. Pitts. *Atmospheric Chemistry.* New York: Wiley, 1986.

OTHER

Exhausting Our Future: An Eighty-Two City Study of Smog in the 80s. Washington, DC: U.S. Public Interest Research Group, 1989.

Smoke

Air pollutant, usually white, grey, or black in appearance, that arises from **combustion** processes. Benjamin Franklin recognized that smoke was the product of inefficient combustion and argued that one should "burn one's own smoke." Smoke generally forms when there is insufficient oxygen to oxidize all the vaporized fuel to **carbon dioxide**, so much remains as **carbon** or soot. Smoke is largely soot, although it also contains small amounts of **fly ash**. Although domestic smoke and poorly constructed chimneys had long been annoying to city dwellers, it was the development of the steam engine that drew particular attention to the need to get rid of unwanted smoke. The first half of the nineteenth century witnessed an awakening interest on the part of local administrators in Europe and North America to combat the smoke problem.

Smoke has long been recognized as more than just an aesthetic nuisance. It has been proven to have a broad range of effects on human health and well-being. Early laws were preoccupied with reducing black smoke, and various tests and "smoke shades" were created to test the color of smoke in the **atmosphere**. For example, the Ringelmann Chart showed a set of shades on a card, which was compared to the color of smoke in the atmosphere. Today it has been shown that the color of suspended smoke particles in cities have gradually changed as **coal** is used less frequently as an energy source. In many cities the principal soiling agent in the air is now diesel smoke. Diesel smoke blackens buildings and even gets inside houses, and studies show that diesel smoke contains a range of carcinogens.

While in many of the major cities smoke production has been reduced, coal, peat, and wood continue to be burnt in large quantities. In these locations, often in developing countries, smoke control represents considerable challenge as environmental agencies remain underfunded. *See also* Cigarette smoke; Smog; Tobacco

[*Peter Brimblecombe*]

RESOURCES
BOOKS

Brimblecombe, P. *The Big Smoke.* London: Methuen, 1987.

Smoking

see **Cigarette smoke; Respiratory diseases; Tobacco**

Snail darter

Most new **species** are discovered, then described in the scientific literature with little fanfare, and most are then known only to a relatively small group of specialists. This was not the case with the snail darter (*Percina tanasi*), a small member of the freshwater fish family of perches, *Percidae.* The snail darter's discovery cast it in the limelight of a highly controversial, environmental battle over the impoundment of the Little Tennessee River by the **Tellico Dam**. Because its discovery coincided with the enactment of the **Endangered Species Act**, its only known **habitat** was the free-flowing channel of the Little Tennessee River, and it was perceived as a means of successfully challenging the completion of this **Tennessee Valley Authority** project.

Two ichthyologists at the University of Tennessee, Drs. David Etnier and Robert Stiles, discovered the snail darter in the Little Tennessee River in August of 1973. After catching several specimens of these three-inch creatures, they returned to Knoxville to examine their find. By that fall, after careful comparison with other members of the genus, it was clear that they had found a new species of fish.

Because this darter was not known from any other location, Dr. Etnier submitted a status report on the species to the U.S. **Fish and Wildlife Service** the following year. The darter's existence was being threatened by the completion of the Tellico Dam, which would ultimately eliminate the free-flowing, clear, riverine habitat needed for its survival. The darter was recommended for listing as an **endangered species** in 1975, and in January 1976 its official scientific description was published. It was now the snail darter.

The notoriety this fish was beginning to receive did not go unnoticed by the TVA. They began transplanting snail darters to the Hiwassee River in 1975 and, by early 1976, had moved over 700 to the new location. All of this was done in secrecy, neither the Fish and Wildlife Service nor the appropriate Tennessee state agencies were notified. When the snail darter was designated an endangered species, the court battles began. Injunctions to halt completion of the Tellico Dam were granted and overturned all the way to the Supreme Court, where the justices ruled in favor of the snail darter. However, the High Court left an opening for the U.S. Congress to step in and exempt Tellico Dam from the Endangered Species Act. That is just what Congress did, and in January 1980 the gates closed and the **reservoir** behind Tellico Dam began to fill,

Snail darter. (Photograph by J. R. Shute. Visuals Unlimited. Reproduced by permission.)

thus sealing the fate of the Little Tennessee River and its darter population.

An additional population of snail darters was discovered in the Hiwassee River, and this population seems to be thriving. The snail darter, so named because of the principle component of its diet, continues its existence and is thought to number upwards of 100,000. It was moved from the endangered listing to one of threatened by the Department of the Interior (DOI) in 1984. Dr. Etnier believes that now the snail darter should be removed from the list of threatened species. The species' scientific name, *Percina tanasi*, is in reference to the ancient Cherokee Village and Native American burial ground of Tanasi, from which the state got its name, and which, now, lies at the bottom of Tellico's reservoir.

Snail darters live for an average of two years, with a maximum recorded longevity of four years. These darters reach sexual maturity at one year and migrate from their downstream, slackwater habitat to the sand and gravel shoals upstream, where they will eventually spawn. These clear, shallow shoal areas represent the habitat needed by the snail darters and their mollusk prey for survival.

[*Eugene C. Beckham*]

Resources

Books

Ono, R., J. Williams, and A. Wagner. *Vanishing Fishes of North America.* Washington, DC: Stone Wall Press, 1983.

Page, L. *Handbook of Darters.* Neptune City, NJ: TFH Publications, 1983.

Periodicals

Etnier, D. "*Percina (Imostoma) tanasi*, A New Percid Fish From the Little Tennessee River, Tennessee." *Proceedings of the Biological Society of Washington* 88 (1976): 469–488.

Other

Mansfield, Duncan. "Snail Darter is No Longer in Danger of Extinction." *Appalachian Focus Environmental News* December 11, 2000 [cited May 2002]. <http://www.appalachianfocus.org/_enviro1/00000011.htm>.

Snow leopard

The snow leopard (*Uncia uncia*), also known as the ounce, or irbis, is a large cat that ranges over highland habitats in Central Asia. It occurs from Turkistan and northern Afghanistan in the western part of its range, through the Kashmir region of Pakistan and India, to southern Mongolia, extreme southwestern China, and Tibet. The snow leopard is sometimes referred to as *Panthera uncia*, which indicates affinity with other large cats, such as the tiger, African lion, and mountain lion, which are also named in the genus *Panthera*.

The snow leopard has an adult body length of 29–51 in (75–130 cm) and a tail of 27–39 in (70–100 cm). It stands 20–25 in (50–65 cm) at the shoulder, and has a body weight of 77–155 lb (35–70 kg). Its pelage is characterized by dense, long hair and a woolly underfur. The underbelly is whitish, while the upper parts of the animal are a creamy yellowish color, patterned by darker rings. Following a gestation period of about 93–103 days, snow leopards give birth in a rocky lair. Usually two cubs are born, but there can be as many as five. Cubs reach sexual maturity at an age of two years in captivity, but probably later in the wild. They have lived as long as 17–19 years in captivity, but the life span would be considerably less in the wild.

During the summer the snow leopard generally occurs in alpine **tundra** habitats, including meadows and rocky places above the forested tree line but below the snow line. During the summer snow leopards venture to 19,685 ft (6,000 m). During winter they occur as low as 5,000 ft (1,500 m), and in harsh winters they may occur in montane (or sub-alpine) forests. Because of the sparse populations of their natural prey, snow leopards have large home ranges, typically amounting to several square miles. The ranges of adjacent animals can overlap to a substantial degree.

Snow leopards are solitary predators. They typically rest during the day and feed during the early morning or late afternoon on wild prey, such as ibex, sheep, musk deer, boar, hare, rabbit, pika, marmot, ptarmigan, and pheasant. Because of expanding human populations and agriculture within the range of snow leopards, they are increasingly feeding on domestic animals, such as goats, sheep, dogs, and chickens.

The global population of wild snow leopards is only about 3,500–7,000 individuals. Another 600–700 animals occur in zoos. Because of its small and decreasing populations the snow leopard is an **endangered species**. Although the

snow leopard is officially protected throughout its range, it is nevertheless widely killed as a **pest** where it intersects with livestock-raising mountain people. It is also widely poached as a trophy and, more importantly, for its beautiful and valuable fur, which can be readily purchased in some rural marketplaces within the range of the **species**, even though its **hunting** is illegal. Most hunting of snow leopards is conducted using leg-hold traps or deep pits baited with a live goat or sheep.

[*Bill Freedman Ph.D.*]

RESOURCES
BOOKS

Grzimek, B., ed. *Grzimek's Encyclopedia of Mammals.* London: McGraw Hill, 1990.

Nowak, R. M. *Walker's Mammals of the World.* 6th ed. Baltimore: Johns Hopkins University Press, 1999.

Gary Sherman Snyder (1930 –)

American writer and poet

Snyder was born in San Francisco but grew up in the Northwest, learning about **nature** and life in cow pastures and second-growth forests. He earned his B.A. in anthropology at Reed College in Portland and spent some time at other universities, learning Asian languages and literatures. During the 1950s, Snyder became a part of the Beat Movement in San Francisco along with such noted figures as Jack Kerouac and Allen Ginsberg. While there he supported himself by working at a variety of odd jobs before leaving to study Zen Buddhism in Japan where he remained for nearly a decade.

Best known as a Pulitzer-prize-winning poet (for *Turtle Island*, 1974; *No Nature*, 1992), Snyder is also an elegant essayist whose collections include *The Practice of the Wild* (1990). Snyder is often described as *the* ecological poet, producing work that demonstrates a deep understanding of the subject. Through his poetry Snyder teaches ecology—as science, philosophy, and world view—to a wide audience.

Snyder's ecological principles center on eating and being eaten, house-keeping principles, the quest for community, identification with and fitting into locale and place, and understanding connective cycles, relationships and interdependence. He does not, for example, condemn **hunting** or taking life, but instead explains it as necessary and sacred: "there is no death that is not somebody's food, no life that is not somebody's death." But he urges his reader to treat the taking of life and the life ingested with respect. Because eating involves consuming other lives, Snyder believes, "eating is truly a sacrament." For Snyder, the primary ethical teaching of all times and places is "cause no unnecessary harm."

Gary Snyder. (Photograph by Chris Felver. Reproduced by permission.)

He declares that "there are many people on the planet now who are not 'inhabitants,'" meaning they do not identify with or know anything about the place in which they live. He emphasizes that "to know the spirit of a place is to realize that you are a part of a part and the whole is made of parts, each of which is whole. You start with the part you are whole in." He notes that "our relation to the natural world takes place in a *place*, and it must be grounded in information and experience"; basically, you "settle in and take responsibility and pay attention."

In *Myths & Texts*, Snyder notes that "As a poet I hold the most archaic values on earth. They go back to the upper Paleolithic: the fertility of the **soil**, the magic of animals, the power-vision in solitude, the terrifying initiation and rebirth...the common work of the tribe." He is an activist who does not seek to return to an unclaimable past, but who tries to raise awareness of how humans are connected to each other, to other life-forms, and to the earth. Currently, Snyder is an English professor at the University of California.

[*Gerald L. Young Ph.D.*]

RESOURCES
BOOKS

Halper, J., ed. *Gary Snyder: Dimensions of a Life.* San Francisco: Sierra Club Books, 1991.

Murphy, P. D. *Critical Essays on Gary Snyder.* Boston: G. K. Hall, 1990.

Paul, S. *In Search of the Primitive: Rereading David Antin, Jerome Rothenberg, and Gary Snyder.* Baton Rouge: Louisiana State University Press, 1986.

Snyder, G. *Myths & Texts.* New York: New Directions, 1978.

———. *No Nature.* New York: Pantheon Books, 1992.

———. *The Practice of the Wild.* San Francisco: North Point Press, 1990.

———. *Turtle Island.* New York: New Directions, 1974.

PERIODICALS

Martin, J. "Speaking for the Green of the Leaf: Gary Snyder Writes Nature's Literature." *CEA Critic: An Official Journal of the College English Association* 54 (Fall 1991): 98–109.

Social ecology

Social **ecology** has many definitions. It is used as a synonym for **human ecology**, especially in sociology; it is considered one form of ecological psychology; and it is the name chosen for the mix of approaches taught at the University of California, Irvine, as well as the radical revisionism of **Murray Bookchin**.

In her 1935 critique of the ecological approach in sociology, Milla Alihan clearly preferred the label "social ecology." In a book on urban society in the mid-forties, Gist and Halbert discussed the social ecology of the city, and Ruth Young examined the social ecology of a rural community in the journal *Rural Sociology* in the early 1970s. A bibliography for the Council for Planning Librarians went so far as to define social ecology as a "subfield of sociology that incorporates the influence of not only sociology but economics, biology, political science, and urban studies," and Mlinar and Teune, in their *Social Ecology of Change*, clearly considered it a part of sociology.

The Program in Social Ecology at the University of California, Irvine has been defined by Binder as "a new context for psychology," but it includes major subprograms in community psychology, urban and regional planning, **environmental health**, human ecology, criminal justice, and educational policy. He concluded by defining social ecology as the study of "the interaction of man with his **environment** in all of its ramifications." In a later article, Binder and others divided the ecological psychology into an ecological approach, an environmental approach, and social ecology itself, which they then enlarged into "a new departure in environmental studies" at Irvine.

Bookchin is equally ambitious, claiming to have "formulated a discipline unique to our age," and defining social ecology as integrating "the study of human and natural ecosystems through understanding the interrelationships of culture and nature." For Bookchin, social ecology "advances a critical, holistic world view and suggests that creative human enterprise can construct an alternative future: reharmonizing people's relationship to the natural world by reharmonizing their relationship with each other." He goes on to claim that "this interdisciplinary approach draws on studies in the natural sciences, feminism, anthropology, and philosophy to provide a coherent critique of current anti-ecological trends and a reconstructive, communitarian, technical, and ethical approach to social life." Social ecology "not only provides a critique of the split between humanity and **nature**; it also poses the need to heal them." Bookchin established an Institute for Social Ecology in Plainfield, Vermont, which organizes conferences, publishes a journal, and offers, in conjunction with Goddard College, a graduate program leading to a Master of Arts degree in Social Ecology.

Social ecology, then, is defined in a variety of ways by different individuals. In general use, the term remains ambiguous. *See also* Bioregionalism; Ecology; Economic growth and the environment; Environmental economics; Environmental ethics; Environmentalism

[*Gerald L. Young Ph.D.*]

RESOURCES

BOOKS

Alihan, M. A. *Social Ecology: A Critical Analysis.* New York: Cooper Square Publishers, 1964.

Bookchin, M. "What is Social Ecology?" In *The Modern Crisis.* Philadelphia: New Society Publishers, 1986.

———. *The Philosophy of Social Ecology: Essays on Dialectical Naturalism.* Montreal: Black Rose Books, 1990.

Emery, R. E., and E. L. Trist. *Towards a Social Ecology: Contextual Appreciation of the Future in the Present.* New York: Plenum Publishing Corporation, 1973.

Socially responsible investing

Socially responsible investing allows people to use financial assets to influence how corporations and governments address a variety of social problems. Stocks, bonds, mutual funds, special lending programs, and other financial tools can be chosen and tailored by investors to favor particular social issues, including tobacco-related health problems, civil rights, environmental destruction, violence, community development, and other social concerns. SRI is sometimes referred to as ethical investing, social investing, mission-based investing, and natural investing.

For hundreds of years religious groups have practiced socially responsible investing. John Wesley, the founder of the Methodist Church, believed that the use of money is an important lesson from the New Testament. The Quakers in early America and other Christian groups practiced a form of SRI called avoidance investing, when they avoided investing in companies that supported activities to which they were opposed on religious grounds, such as slavery, gambling, alcohol, **tobacco**, and weapons production.

The rise of the modern corporation brought with it new forms of investment, and the growing corporate economy allowed more people to own shares, or stocks, in corporations. In 1934, the U.S. Congress created the Securities and Exchange Commission (SEC), which would oversee the trading of financial instruments such as stocks. Congress mandated that all owners, or stockholders, should be able to vote on company concerns when it stated that, "fair corporate suffrage is an important right that should be attached to every equity security bought on a public exchange."

The political activism of the 1960s, when people strongly protested issues concerning civil rights, women's rights, war, and **environmental degradation**, had rippling effects upon corporation activity and the development of SRI. In the 1970s, church groups including the National Council of Churches, which had stock holdings in major corporations, began actively using shareholders' resolutions to address social concerns. As an offshoot of the National Council of Churches, an important organization in SRI was formed in 1972, the Interfaith Center on Corporate Responsibility (ICCR), which brought together religious groups to increase shareholder power. One issue in particular had a profound effect on the development of SRI—the policy of apartheid in South Africa, which was the forced and often violent segregation of blacks from whites. In the 1980s, many shareholders demanded investment tools that utilized divestment, which eliminated from stock collections all companies that had South African holdings. Shareholders in some companies also demanded disinvestment, or the withdrawal of corporate activities from South Africa.

Increased concerns about environmental problems have led to the development of environmental investing tools. As early as 1970, an environmental mutual fund called the Pax World Fund was offered, which used **pollution control** as one of its standards for selecting companies. (Mutual funds are professionally managed collections of stocks, shares of which can be purchased and traded.) In March 1989 the ***Exxon Valdez*** oil spill in Alaska prompted further development of environmental investing. The Coalition for Environmentally Responsible Economies (CERES) developed the **Valdez Principles**, a list of 10 environmental principles related to corporate activity. Companies can be rated on how well they adhere to the principles, and shareholders have also created resolutions asking companies to adopt the principles for their operations. These principles include protection of the environment, sustainable use of natural resources, reduction and disposal of wastes, wise use of energy, risk reduction, marketing of safe products, damage compensation, disclosure, environmental management, and self-assessment.

In the free market society of the United States, corporations are set up by law to be responsible to their shareholders, and the market is governed as little as possible. In a capitalistic system, maximizing the profits for shareholders is the main goal of the corporation. As efficiently as this system has worked to create a materialistically wealthy society, the profit-motive corporate structure has also contributed to some social and environmental problems that were growing as capitalism moved into the new millennium. For instance, energy companies can maximize their profits by producing and selling as much energy as possible. In the early twenty-first century, the problems of global warming and **air pollution** were receiving increased attention, problems that are directly related to the burning of **fossil fuels**, from which most energy is produced. Thus, the shareholders of an energy company might benefit from increased production of energy, while society as a whole may face new problems from such activity. Energy companies powerfully oppose new governmental regulations such as air **pollution** controls, and there can be conflicts between private and public interest. That is, corporations desire a market free from government interference, while the government is the main tool the public has to regulate large corporations. Some people, wishing to affect change from within, become what are known in SRI as activist shareholders, and utilize shareholders' rights. Activist shareholders are often in the minority, and it may take many millions of dollars worth of stock to be able to pass shareholders' resolutions in large companies. In the early twenty-first century, shareholders in some of the world's largest energy companies were initiating resolutions on air pollution, alternative energy, **sustainable development**, and global warming, for instance.

Socially responsible investing relies upon three main strategies to further its social and environmental goals: screening, shareholder advocacy, and community investing. Positive screening uses various criteria to select companies that have excellent track records, while negative screens eliminate companies with poor records from investment possibilities. Tobacco is the most commonly used screening criteria, and other popular screens include environmental protection, human rights, employment equality, gambling, alcohol, and weapons production. Other commonly used screens have been labor relations, animal testing, community investing, and community relations, while specialty screens include issues such as executive compensation, abortion and birth control, and international labor standards, among others.

Shareholder advocacy utilizes the legal rights of shareholders as owners of companies. Shareholders can influence corporate operations by direct communication with management as well as by shareholder resolutions. These resolutions, when filed, get published in a company's proxy statement and are voted on by all shareholders during the company's annual meeting. Although they rarely receive a majority vote,

which would mandate their adoption, proxy resolutions can initiate the process of change.

Community investing is direct investment into underprivileged communities that have failed to receive adequate assistance from traditional financial institutions. This type of investing encourages economic development by providing start-up loans to small businesses, affordable housing, employment programs, and financial assistance for low-income people, for example. Community investing utilizes community development banks and credit unions that are set up specifically for low- and middle-income communities. Community loan funds and venture capital funds are other community investing tools that provide funding to underdeveloped areas.

According to the Social Investment Forum, a nonprofit industry group, screening strategies accounted for about 61% of SRI funds in 2001, shareholder advocacy was utilized by 13%, screening and advocacy together accounted for 26%, while community investing garnered less than 1% of total SRI funds.

In the new millennium, SRI has grown to be a significant and influential part of the investment world. From a 1985 total of $65 billion, SRI funds had grown to over $2 trillion by 2001, according to the Social Investment Forum. The 2001 total represented 13% of the $16 trillion that was in professionally managed funds, or one in every eight investment dollars. Furthermore, SRI funds grew at a faster rate (1.5%) than the rate for all professionally managed funds, meaning that socially and environmentally conscious companies were positioned strongly in the economy of the 2000s. Industry analysts have noted that socially responsible firms tend to be well managed, adaptive, and technologically adept. In 2001, there were 230 mutual funds that were using social and environmental screening for various causes. There were funds that emphasized alternative energy, sustainable development, and workers' rights, and other funds that avoided **nuclear power**, weapons building, gambling, and pornography. There were funds for fundamental Christians, Catholics, and a fund specifically tailored for Muslims.

Shareholder advocacy actions were on the increase in 2001, due in part to new reports on global warming and concerns over corporate ethics brought about by the Enron energy scandal. Since the 1980s, institutions, including religious groups, mutual funds, foundations, pension funds for public and private organizations, and others, have become the largest shareholders in the United States. When these institutions come together they create powerful shareholder associations, such as the Council on Institutional Investors which controlled over $1.5 trillion in assets in 2001 and has its own corporate governance guidelines. Shareholder advocacy for social and environmental issues had some successes in the late 1990s and early 2000s. Shareholders in 2000 persuaded Home Depot to gradually stop using old-growth timber, and investors convinced General Electric to spend up to $250 million to clean the Housatonic River in Massachusetts from PCB contamination. Furthermore, shareholders convinced Texaco, General Motors, Ford, and DaimlerChrysler to end their memberships in the Global Climate Coalition, an industry group that maintains that global warming is not valid and opposes actions to curb it. In 2001, 262 social policy resolutions were presented to 177 companies, and 45 of these resolutions garnered more than 10% of shareholders' votes.

[*Douglas Dupler*]

Resources

Books

Brill, Jack A., and Alan Reder. *Investing from the Heart: The Guide to Socially Responsible Investing and Money Making.* New York: Crown, 1992.

Domini, Amy L. *Socially Responsible Investing: Making a Difference and Making Money.* Dearborn, MI: Dearborn Trade, 2001.

Hawkin, Paul, Amory Lovins, and L. Hunter Lovins. *Natural Capitalism: Creating the Next Industrial Revolution.* Boston: Little Brown, 1999.

Kinder, Peter, Steven D. Lydenberg, and Amy L. Domini. *Social Investment Almanac.* New York: Henry Holt, 1992.

Periodicals

Bogo, Jennifer. "Business Savvy: Making Room on the Shelves for a New Generation of Greener Goods." *E Magazine* (July/August 2000): 26–33.

Rose, Sarah. "Responsible Investing: Is It Living Up to Its Promise?" *Sierra* (November/December 2000): 26.

Social Investment Forum. "2001 Report on Socially Responsible Investing Trends in the United States." Washington, D.C.: Social Investment Forum. [cited July 2002]. <http://www.socialinvest.org>.

Other

Co-op America Home Page. [cited July 2002]. <http://www.coopamerica.org>.

Shareholder Action Network Home Page. [cited July 2002]. <http://www.shareholderaction.org>.

Social Investment Forum Home Page. [cited July 2002]. <http://www.socialinvest.org>.

Society for Conservation Biology

The idea of **conservation biology** as a scientific discipline with a specific application to society-at-large was brought to life in 1978 by organizers Michael Soule and Bruce Wilcox. They formalized their interest in a new **conservation** discipline at the first International Conference of Conservation Biology in San Diego. Between 1978 and 1985, the ideals of conservation biology survived through some growing pains, most importantly expanding from the idea of several single disciplines looking at a problem to a unified examination of a problem including the insights of many disciplines.

The core of that interdisciplinary approach formed a foundation from which, in 1985, the Society for Conservation Biology was born. The Society is, in its own words, "...an international professional organization dedicated to promoting the scientific study of the phenomena that affect the maintenance, loss, and restoration of biological diversity. The Society's membership is comprised of a wide range of people interested in the conservation and study of biological diversity; resource managers, educators, government and private conservation workers, and students."

What is unusual for a professional scientific group is that the Society was formed "to develop the scientific and technical means for the protection, maintenance, and restoration of life on this planet—its **species**, its ecological and evolutionary processes, and its particular and total environment." Thus, a mission is an inherent part of the Society mandate. That mission can be very broadly interpreted to include political, economic, and cultural ramifications, which are not commonly the arena for professional scientists.

To undertake the above mission the SCB promotes research, disseminates information, encourages communication and collaboration, educates, and recognizes outstanding contributions in the field. Part of that is done through a scientific journal, *Conservation Biology*, a newsletter, the development of local chapters, a Policy and Resolutions Committee, and the independent actions of the members.

The ideals noted above are based on the Society's official statements, but they undervalue the Society. It has grown to be a true meeting ground for scientists and non-scientists, and has maintained open and honest debate on vital topics, such as the roles of science and advocacy, that are touchstones for the new generation of conservation scientists. The message that comes through from the overall body of the Society's work is that careful science and compassion are not antithetical, rather they work together to promote **nature** in a natural way.

The journal *Conservation Biology* is the center of discussion and dissemination of thought for conservationists worldwide. Its overall quality, maintained by the highest editorial principles, has grown from a few articles in a quarterly issue to 300–400 page editions published six times per year. In each volume philosophical treatments are found alongside the latest molecular technique for recognizing levels of genetic variability.

The Society meets annually in a conference setting, complete with technical paper presentations, research posters, workshops, and a series of conservation awards. They occasionally meet in conjunction with other scientific groups and explore particular thematic topics in the Annual Meeting.

[*David Duffus*]

RESOURCES

ORGANIZATIONS

Society for Conservation Biology, 4245 N. Fairfax Drive, Arlington VA USA 22203 (703) 276-2384, Fax: (703) 995-4633, Email: information@conservationbiology.org, <http://conbio.net/scb>

Society of American Foresters

Established in 1900 by **Gifford Pinchot**, considered to be America's first scientifically trained forester, the Society of American Foresters (SAF) is a scientific and educational organization representing the forestry profession in the United States. With a membership of over 18,700 in 1997, it is the largest professional society for foresters in the world. SAF's mission is to "advance the science, education, technology, and practice of forestry; to enhance the competency of its members; to establish professional excellence; and to use the knowledge, skills, and **conservation** ethic of the profession to ensure the continued health and use of forest ecosystems and the present and future availability of forest resources to benefit society." Although the focus of this organization is on forestry, the SAF addresses issues and concerns of a variety of natural resource professionals.

To be a professional member of SAF, one must hold a bachelor's or higher degree in forestry from an SAF-accredited or candidate institution, or hold a comparable degree in a closely related field of **natural resources** and have at least three years of forestry-related experience. There is a code of ethics for members that emphasizes stewardship of the land and that advocates land management that is consistent with ecologically sound principles. Members are expected to perform services consistent with their knowledge and skills, and for which they are qualified by education or experience.

The SAF provides an array of services for its members and for the forestry community as a whole, including the publication of the *Journal of Forestry*, the SAF's chief publication, three regional technical journals, a newsletter, and *Forest Science*, a quarterly research journal. Technical expertise is provided through the organization's 28 working groups that cover a broad range of subdisciplines and through supporting continuing education programs. The SAF accredits forestry programs in universities, ensuring that curriculum meet standards considered to be needed by forestry graduates. It provides background and key testimony to natural resource policy makers on forestry issues, including timber harvesting methods, **forest management**, clean water legislation, and **wetlands**. For example, many of the recommendations of the SAF were incorporated into the **Endangered Species** Conservation and Management Act of 1995.

Environmental issues, particularly during the 1980s and 1990s, have stimulated a diversity of opinions and views

about the profession of forestry and the role of the SAF. The spotted owl issue, the role of **clear-cutting** in forest management, the management and protection of **old growth forests**, and other forestry issues have been debated within the SAF just as they have been debated in society as a whole. Many views are represented by members of the SAF, including those with a forest products orientation and those promoting an ecosystems-based management approach that places high values on forest attributes such as **biodiversity**. Some members advocate the return to the basics of conservation, emphasizing both production and environmental protection. Clearly there is a broader view of the profession of forestry within SAF and perhaps a greater diversity of opinions than in the past. With this diversity, the SAF will continue to play a major role in how future forests are managed, particularly on public lands.

[*Kenneth N. Brooks*]

Resources

Organizations

Society of American Foresters, 5400 Grosvenor Lane, Bethesda, MD USA 20814 (301) 897-8720, Fax: (301) 897-3690, Email: safweb@safnet.org, <http://www.safnet.org>

Sociobiology

A perennial debate about human nature, pitting the idea that behavior is culturally conditioned against the notion that human actions are innately controlled, was rekindled in 1975 when **Edward O. Wilson's** *Sociobiology: The New Synthesis* introduced a new scientific discipline.

Sociobiology is defined by Wilson as "the systematic study of the biological basis of all social behavior" and can be described as a science that applies the principles of evolutionary biology to the social behavior of organisms, analyzing data from a number of disciplines, especially ethology (the biological study of behavior), **ecology**, and **evolution**. It employs two chief postulates. The first is the *interaction principle*, the idea that the action of any organism, including man, arises as a fusion of genotype and learning or experience. The second is the *fitness maximization* principle, the notion that all organisms will attempt to behave in a manner that passes copies of their genes to **future generations**.

Sociobiology provides several ways to understand the social behavior of an organism, including those acts termed "altruistic." First, social behaviors can be seen as the genetically conditioned attempts of an organism to save copies of its genes carried in others, especially close relatives. Such acts can also be viewed as a form of "genetic reciprocity" in which an organism helps another and the helped organism (or others) repays the help at some time in the future, thereby enhancing the fitness of each of the organisms involved. Finally, cooperative actions can be understood as the result of natural selection operating at the level of groups, in which "altruistic" groups pass their genes on more successfully than "selfish" groups.

In *Sociobiology*, Wilson, an entomologist specializing in the biology of ants, develops these concepts using the example of social insects. He shows how their cooperative and "altruistic" behaviors, never fully considered by traditional evolutionary biology, can be understood in terms of genetic fitness and natural selection. But the crucial aspect of Wilson's book is his assertion, within a single chapter, that such genetic theories can be applied to the social behavior of primates and man as well as to "lower" animals. Indeed, Wilson contends that scientists can shrink "the humanities and social sciences to specialized branches of biology" and predicts that to maintain the human **species** indefinitely, we will have to "drive toward total knowledge, right down to the levels of the neuron and gene," statements in which he seems to suggest that human beings can eventually be explained totally in "mechanistic terms."

Broadly, the premise of Wilson's argument finds its origins in the ideas of René Descartes and the mechanists of the eighteenth century such as Julien Offroy de La Mettrie, whose 1748 book, *L'homme-machine*, helped define the modern biological study of human beings. Wilson argues that if humans are indeed biological machines, then their actions in the realm of society should not be excluded from examination under biological principles, and dramatically defines his goal by stating that "when the same parameters and quantitative theory are used to analyze both termite colonies and troops of rhesus macaques, we will have a unified science of sociobiology."

Wilson's book, especially his statements in the final chapter, generated a storm of controversy. Some feared that a genetic approach to human behavior could be used by racists (and eugenicists) as a scientific base for their arguments of racial superiority, or that delving too deeply into human genetics might lead to manipulation of genes for undesirable purposes. Well-known evolutionary biologists such as Stephen Jay Gould and Richard Lewontin criticized Wilson for stepping over the line between science and speculation. Gould quite openly declared his unhappiness with the final chapter of *Sociobiology*, stating that it "is not about the range of potential human behavior or even an argument for the restriction of that range from a much larger total domain among all animals. It is, primarily, an extended speculation on the existence of genes for specific and variable traits in human behavior—including spite, aggression, xenophobia, conformity, homosexuality, and the characteristic behavior differences between men and women in Western society." Lewontin argued that sociobiology "is too theoreti-

cally impoverished to deal with real life," that "by its very nature, sociobiological theory is unable to cope with the extraordinary historical and cultural *contingency* of human behavior, nor with the diversity of individual behavior and its development in the course of individual life histories."

Feminists have also taken up the cudgel against sociobiology, asserting that it implicitly sanctions existing social behavior with a latent determinism. If, for example, repressive acts are defined as the result of direct genetic influence, the basis for their moral evaluation disappears.

Wilson has responded to his critics energetically and condemns those using sociobiological arguments to support societal evils by equating "what is" with "what should be." Wilson says: "The 'what is' in human nature is to a large extent the heritage of a Pleistocene hunter-gatherer existence. When any genetic bias is demonstrated, it cannot be used to justify a continuing practice in present and future societies. Since most of us live in a radically new **environment** of our own making, the pursuit of such a practice would be bad biology; and like all bad biology, it would lead to disaster. For example, the tendency under certain conditions to conduct warfare against competing groups might well be in our genes, having been advantageous to our Neolithic ancestors, but it could lead to global suicide now. To rear as many healthy children as possible was long the road to security; yet with the population of the world brimming over, it is now the way to environmental disaster." He goes on to suggest that "our primitive old genes will therefore have to carry the load of much more cultural change in the future" and, indeed, that our "genetic biases can be trespassed" by thinking and decision-making.

Sociobiology seems to have survived its vilification by critics. Indeed, Wilson's vigorous advocacy of a new synthesis between the social and biological sciences has helped bring into the open a debate about the relationship of biology and society that will probably bear a great deal of scientific fruit.

[*Gerald L. Young and Jeffrey Muhr*]

Resources

Books

Wilson, Edward O. *On Human Nature.* Cambridge, MA: Harvard University Press, 1978.

Wilson, Edward O. *Sociobiology: The New Synthesis.* Cambridge, MA: Harvard University Press, 1975.

Periodicals

Lieberman, L., L. T. Reynolds, and D. Friedrich. "The Fitness of Human Sociobiology: The Future Utility of Four Concepts in Four Subdisciplines." *Social Biology* 39 (Spring-Summer 1992): 158–169.

Wolfe, Alan. "Social Theory and the Second Biological Revolution." *Social Research* 57 (Fall 1990): 615–648.

Soil

Soil is the unconsolidated mineral material on the immediate surface of the earth that serves as a natural medium for the growth of land plants. Soil is found on all surfaces except on steep, rugged mountain peaks and areas of perpetual ice and snow.

Soil is related to the earth much as the rind is related to an orange. But this rind of the earth is far less uniform than the rind of the orange. Soil is deep in some places and shallow in others. Soil can be red in Georgia and black in Iowa. It can be sand, loam, or clay. Soil is the link between the rock core of the earth and the living things on its surface. Soil is the foothold for the plants we grow and the foundation for the roads we travel and for the buildings we live in.

Soil consists of mineral and organic matter, air, and water. The proportions vary, but the major components remain the same. Minerals make up 50% of an ideal soil while air and water make up 25% each. Every soil occupies space. Soil extends down into the planet as well as over its surface. Soil has length, breadth, and depth. The concept that a soil occupies a segment of the earth is called the "soil body." A single soil in a soil body is referred to as a "pedon." The soil body is composed of many pedons and is thus called a "polypedon."

Every soil has a profile or a succession of layers (horizons) in a vertical section down into the non-soil zone referred to as the parent material. Parent materials can be soft rock, glacial drift, wind blown sediments, or alluvial materials. The nature of the **soil profile** is important for determining a soil's potential for root growth, storage of water, and supply of plant nutrients.

Soil formation proceeds in steps and stages, none of which is distinct. It is impossible to determine where one step or stage in soil formation begins and another ends. The two major steps in the formation of soils are the accumulation of soil parent materials and the differentiation of horizons in the profile. Soil horizons develop due to the processes of additions, losses, translocations, and transformations. These processes act on the soil parent material to produce soil horizons. The intensity of any one process will vary from location to location due to five soil-forming factors:

- Soil forms from the parent materials, and they can be rock, sand, glacial till, loess, alluvial sediments, or lacustrine clays.
- Topography is the position on the landscape where the parent material is located. Positions can be summits, sideslopes, and footslopes. **Erosion** and the depth to the **water table** will influence soil formation as a result of topography.
- **Climate** acts on the parent material and determines the rate of rainfall and **evapotranspiration**, which influence the amount of **leaching** in the soil profile. The greater the leaching, the faster soil horizons develop.

- Biotic factors include vegetation and animals. Animals are important mixers of the soils and can destroy soil horizons. Vegetation influences soils by determining the amount of organic matter incorporated into the soils. Soils formed under **prairie** grasses will have a thick black surface layer, compared to soils formed under forests where the surface layer is thin with a light-colored leached zone under it.
- Time refers to the number of years that the parent material has been acted on by climate and vegetation. The older a soil is, the greater the **horizon** development. Young soils will have very minimal horizon development. The age of a soil is thus dependent on its development and not on the total number of years. Therefore, developed soils in the Midwest may be 10,000 years old, while young soils in the valleys of California will also be 10,000 years old. Developed soils on high terraces in California can be as old as one million years.

Because the soil-forming factors vary from location to location, soils are different across the landscape. The ability to interpret where the soils will change has allowed soil scientists to make a map of these changes. The soil map is thus an interpretation of the soils that occur on a landscape. The soil map is published on a county-by-county basis and is called a **soil survey**. Soil surveys are very useful documents in planning **land use** according to the soil's potential.

Soil is an absolutely essential natural resource but one that is both limited and fragile. The soil is often overlooked when **natural resources** are listed. Throughout history the progress of civilizations has been marked by a trail of wind-blown or water-washed soils that resulted in barren lands. Continuing to use the soil without appropriate **soil conservation** management is very destructive to the **environment**. Protecting the quality of our nation's **topsoil** is largely within human control. To many soil scientists, saving our soil is much more important than saving oil, **coal**, or **natural gas** resources.

[*Terrence H. Cooper*]

Resources

Books

Miller, R. W., and R. L. Donahue. *Soils: An Introduction to Soils and Plant Growth.* Englewood Cliffs, NJ: Prentice-Hall, 1990.

Other

Stefferund, A., ed. *Soils: 1957 Yearbook of Agriculture.* Washington, DC: U.S. Government Printing Office, 1957.

Soil and Water Conservation Society

Established in 1945, the **Soil** and **Water Conservation** Society (SWCS) is a nonprofit organization that advocates the "protection, enhancement and wise use of soil, water and related natural resources." The organization was formed during an era of public concern about soil **erosion** and water resource management following the **dust bowl** period of the 1930s in the United States. Improved farm tillage practices and an improved **conservation** ethic for **land use** were key objectives of the SWCS and became focal points for the organization. From its inception, the SWCS has been an advocate of **soil conservation** and improved use of soil and **water resources**, with its primary audience being professionals in agriculture.

In recent years, the SWCS has expanded its conservation focus to a broader natural resource perspective that goes beyond agriculture, although environmental issues associated with agriculture are still of significant interest. As a result, the SWCS membership is now made up of conservation professionals from many natural resource and agricultural disciplines. Individual and business membership in the SWCS exceeds 11,000, making it one of the largest multidisciplinary organizations for conservation professionals in the United States.

The SWCS actively promotes education, with the objective of a membership that can play leadership roles in dealing with environmental issues. Working groups within the SWCS develop policy statements on a wide range of environmental topics, including **water quality**, **wetlands**, **biodiversity**, **floodplain** management and **sustainable agriculture**. Through education programs and professional meetings, members of the society are challenged to set ethical standards that recognize the interdependence of people and their **environment**. Emphasis on sustainable land use practices and a sustainable environment have emerged as themes for its membership in the 1990s.

Educational programs have targeted policy makers and the general public alike. In 1995, the SWCS published a survey of 17,000 landowners who participated in the federal **Conservation Reserve Program** (CRP), which illustrated what farmers planned to do with their CRP land after their contracts expired. The intent was to inform policy makers about the environmental benefits of the CRP program and the consequences if this program was terminated. Other education programs focus on farmers, such as a White Paper on **conservation tillage** that stressed the environmental benefits of no-till farming. This had an important effect on farmers' attitudes toward no-till farming, assessing both the environmental benefits and costs of no-till farming.

Following the floods in the upper Mississippi River Valley in 1993, the SWCS supported Internet-linked communications with members about farm practices and their effects on the great flood. This program led to an 80% increase in farmer participation in the emergency Wetlands Reserve Program. The SWCS also played a key role in

promoting conservation language and funding in the 1996 federal farm bill.

In 2002, an initiative was established to examine the conservation implications of **climate** change. This project was set up after the National Climatic Data Center commented on the 20% increase of precipitation intensity.

Continuing education programs on the latest developments in land and water management are carried out through workshops, field trips, conferences, and symposiums. The SWCS now sponsors a program for certifying professionals in soil erosion and **sediment** control. As part of its educational mission, the SWCS publishes the *Journal of Soil and Water Conservation*, a bimonthly publication that reports the results of new natural resource and agricultural research.

[*Kenneth N. Brooks*]

RESOURCES

ORGANIZATIONS

Soil and Water Conservation Society, 7515 NE Ankeny Road, Ankeny, IA USA 50021-9764 (515) 289-2331, Fax: (515) 289-1227, Email: webmaster@swcs.org, <http://www.swcs.org>

Soil, buried

see **Buried soil**

Soil compaction

The forcing together of **soil** particles under pressure, usually by foot or vehicle traffic. **Compaction** decreases soil porosity and increases **bulk density**. The degree of compaction is determined by the amount of pressure applied and soil characteristics, including clay, water, and organic matter contents. Although compaction sometimes can be beneficial by improving seed and root contact with the soil and by increasing the soil's ability to hold water, most compaction is detrimental to plants and soil animals. It can destroy soil structure, decreasing water intake, **percolation**, gas exchange, and biological activity, while increasing water **runoff**, **erosion**, and **resistance** to root penetration. Soil compaction often is a serious problem in agricultural fields, forests, range lands, lawns, and **golf courses**. Conversely, soil compaction is necessary for most construction purposes.

Soil conservation

Soil conservation is the protection of soil against excessive loss of fertility by natural, chemical, or artificial means. It encompasses all management and land-use methods protecting soil against degradation, focusing on damage by **erosion** and **chemicals**. Soil conservation techniques can be divided into six categories, crop selection and rotation, **fertilizer** and lime application, **tilth**, residue management, contouring and strip cropping, and mechanical (e.g., **terracing**).

While the potential dangers of chemical degradation and soil erosion were recognized as early as the American Revolution, it was not until the early 1930s that soil conservation became a familiar term. The soil conservation movement was a result of the droughts during the 1930s, the effects of water erosion, the terrific dust storms created by wind erosion in the Great Plains, and by the urging of **Hugh Hammond Bennett.**

Dr. Bennett, a soil scientist from North Carolina, recognized the erosion damage to previously **arable land** in the Southeast, Midwest, and elsewhere in the 1920s and 1930s. In 1929, he published a bulletin entitled "Soil Erosion, A National Menace" and started a successful personal campaign to get federal support, beginning with a $160,000 appropriation by Congress to initiate a national study. In 1933, the **U.S. Department of the Interior** named Bennett as head of the Soil Erosion Service, which conducted soil erosion control demonstrations nationwide. In 1935, the **Soil Conservation Service**, led by Bennett, was established as a permanent agency of the **U.S. Department of Agriculture**.

Soil degradation problems addressed include **soil compaction**, **salinity** build-up, and excessive soil acidity. Because soil and water are so intimately related, the program also deals with **water conservation** and **water quality**. Under M. L. Wilson, then Assistant Secretary of Agriculture, the Soil Conservation Service established conservation districts guided by elected officials assisted by Soil Conservation Service personnel. Currently there are more than 3,000 districts in the United States.

Sharing the cost of conservation became federal policy with the passage of the Soil Conservation and Domestic Allotment Act of 1936. The act funded the shift of croplands to "soil-building" crops and established soil conservation practices on croplands and **grasslands**. The Great Plains Conservation Program, enacted by Congress in 1956, sought to shift some of the highly **erodible** land from cropland to grassland. The Water Bank and Experimental Rural Clean Waters Programs were an attempt to resolve disputes over **drainage** of "potholes" in the Midwest and Great Plains and demonstrate the influence of soil and water conservation practices on water quality.

As early as the 1930s, the federal government began to purchase "submarginal" lands outright. The Conservation Reserve segment of the Soil Bank (1956-1960) bought substandard farmland to conserve soil and alleviate surplus crop production. The current **Conservation Reserve Program**, authorized in a 1985 farm bill, paid farmers to convert land from cropland to grassland or trees under a long-term lease.

Currently the "sodbuster," "swampbuster," and conservation compliance programs attempt to force farmers to comply with soil and water conservation programs to be eligible for other government programs, such as price supports.

The role of the Soil Conservation Service has changed over the years, but its central mission is still to provide technical information for good **land use**. Today the Soil Conservation Service is highly concerned with environmental problems, water quality, wetland preservation, and prime farm land protection, as well as urban concerns related to their mission. The soil conservation movement has spawned a number of professional societies, including the **Soil and Water Conservation Society** and the World Association of Soil and Water Conservation. *See also* Conservation tillage; Contour plowing; Dust Bowl; Environmental degradation; Soil organic matter; Strip-farming; Sustainable agriculture

[*William E. Larson*]

Resources
Books

National Academy of Sciences. *Soil Conservation.* 2 vols. Washington, DC: National Academy Press, 1986.

Wilson, G. F., et al. *The Soul of the Soil: A Guide to Ecological Soil Management.* 3rd ed. Agaccess, 1996.

Yudelman, M., et al. *New Vegetative Approaches to Soil and Water Conservation.* Washington, DC: World Wildlife Fund, 1990.

Soil Conservation Service

Soil erosion is an age-old problem recorded in many documents of human civilizations. In the United States, Simms traces concern for soil erosion, and attempts to combat it, back to the very first European settlers. **U.S. Department of Agriculture** (USDA) bulletins on soil erosion date back at least to "Washed Soils: How to Prevent and Reclaim Them," dated 1894.

The contemporary **Soil Conservation** Service (SCS) was predated by the Bureau of Soils in the USDA, later called the Bureau of Chemistry and Soils. The catalyst for an action agency was **Hugh Hammond Bennett**, a soils scientist who went to work for the Bureau of Soils in 1903. From his urging, the USDA published the classic circular on "Soil Erosion, A National Menace," in 1928. Further pressure by Bennett resulted in the establishment of a temporary agency, the Soil Erosion Service, in the **U.S. Department of the Interior** in 1933, what Simms describes as "the first national action program of soil **conservation** anywhere in the world." Bennett was the agency's first director.

Hugh Bennett pushed his agenda hard, and his "primary tenet of soil conservation" can still be seen in the activities of the service today: "No single practice will suffice. A physical inventory of the land of the farm should be made, and each should be used in accordance with its adaptability and treated in accordance with its needs."

In 1935 President Franklin D. Roosevelt signed an executive order transferring the Soil Erosion Service to the USDA. The Department of Agriculture immediately consolidated all of its activities bearing on erosion control into the agency. Also in 1935 the Soil Conservation Act was passed and signed into law, establishing a permanent Soil Conservation Service in the USDA, with Hugh Bennett as the first Director. In 1994 The Soil Conservation Service became the Natural Resources Conservation Service (NRCS).

Scanning lists of publications from the Natural Resources Conservation Service shows work on a surprising range of contemporary conservation concerns. As one might expect, topping the list are numerous soil surveys of various counties in the United States: a **soil survey** of Esmeralda County, New York in 1991, one of the Asotin County area in Washington, and one of Crockett County, Tennessee. Less expected is a paper on **salinity** control in Colorado. Also included are status reports of various NRCS programs: status of water quantity and **water quality** programs, and the situation regarding **wetlands** and riparian programs. The NRCS also publishes a wide variety of maps, depicting, for example, major watersheds in Michigan, land resource regions and major land resource areas, and soil erosion in Colfax County, Nebraska.

The Service provides appropriate databases for computer research, including a pesticides properties database for environmental decision-making, state soil geographic databases, and a computer design for grassed waterways. The NRCS provides local landowners a "how-to" video (1990) on "Better Land, Better Water." The United States still has severe erosion problems—Steiner notes that "only 3.23% of the cropland in the United States was considered adequately protected by the NRCS in 1983." But, this and other conservation problems are not nearly as severe as they would be without Bennett's wisdom, foresight, and tenacity. He set the stage for 50 years of research and action by the national network of Natural Resources Conservation Service offices and agents.

[*Gerald L. Young Ph.D.*]

Resources
Books

Simms, D. H. *The Soil Conservation Service.* New York: Praeger, 1970.

Steiner, F. R. *Soil Conservation in the United States: Policy and Planning.* Baltimore: Johns Hopkins University Press, 1990.

PERIODICALS

Helms, D. "Conserving the Plains: The Soil Conservation Service in the Great Plains." *Agricultural History* 64 (Spring 1990): 58–73+.

Soil consistency

The manifestations of the forces of cohesion and adhesion acting within the **soil** at various water contents, as expressed by the relative ease with which a soil can be deformed or ruptured. Consistency states are described by terms such as friable, soft, hard, or very hard. These states are assessed by thumb and thumbnail penetrability and indentability, or more quantitatively, by Atterberg limits, which consist of liquid limit, plastic limit, and plasticity number. Atterberg limits are usually determined in the laboratory and are expressed numerically. *See also* Soil compaction; Soil conservation; Soil texture

Soil, contaminated

see **Contaminated soil**

Soil eluviation

When water moves through the **soil**, it also moves small colloidal-sized materials. This movement or **leaching** of materials like clay, iron, or calcium carbonate is called eluviation. The area where the materials have been removed is the zone of eluviation and is called the E **horizon**. Zones of eluviation contain fewer nutrients for plant growth. E horizons are often found in forested soils.

Soil illuviation

When water moves through the **soil**, it moves small colloidal-sized particles with it. These particles of clay, iron, **humus**, and calcium carbonate will be deposited in zones below the surface or in the **subsoil**. The zones are called illuvial zones and the process is referred to as illuviation. Illuvial zones are most often referred to as B horizons, and the subscript with the B designates the kind of material translocated (i.e., B_t = clay accumulation).

Soil liner

One of the requirements of modern sanitary landfills is that a **soil** liner be placed on top of the existing soil. Liners are needed to prevent the penetration of **landfill** leachate into the soil. Without a liner, leachate could move through the soil and contaminate the **groundwater**. Soil liners can be made of clay, plastic, **rubber**, blacktop, or concrete.

Soil loss tolerance

Soil loss tolerance is the maximum average annual soil removal by **erosion** that will allow continuous cropping and maintain soil productivity (T). It is occasionally defined as the maximum amount of soil erosion offset by the maximum amount of soil development while maintaining an equilibrium between soil losses and gains. T is usually expressed in terms of tons per acre or tons per hectare. Because T values are difficult to quantify, they are usually inferred by human judgment rather than scientific analysis. In determining T values, the depth of the soil to consolidated material or depth to an unfavorable **subsoil** is considered. T values are an expression of concern for plant growth but may not adequately reflect environmental concerns.

Soil organic matter

Additions of plant debris to soils will initiate the build up of organisms that will decompose the plant debris. After **decomposition** the organic material will be called organic matter or **soil humus**. This decomposed plant debris will be in the form of very small particles and will coat the sand, **silt**, and clay particles making up the mineral soil particles, thus making the soil black. The more organic matter accumulating in a soil, the darker the soil will become. Continued additions of organic matter are important to create a soft, tillable soil that is conducive to plant growth. Organic matter is important in soils because it will add nutrients, store **nitrogen** and other positive cations, and create a stronger soil aggregate that will withstand the impact of raindrops and thus prevent water **erosion**.

Soil profile

The layers in the **soil** from the surface to the **subsoil**. The soil profile is the collection of soil **horizons**. Soil profiles will vary from location to location because of the five soil-forming factors: **climate**, vegetation, **topography**, soil parent material, and the length of time the soil has been **weathering**. By looking at a soil's profile many interpretations can be made about **land use** and the suitability of the soil for a specific use.

Soil, saline

see **Saline soil**

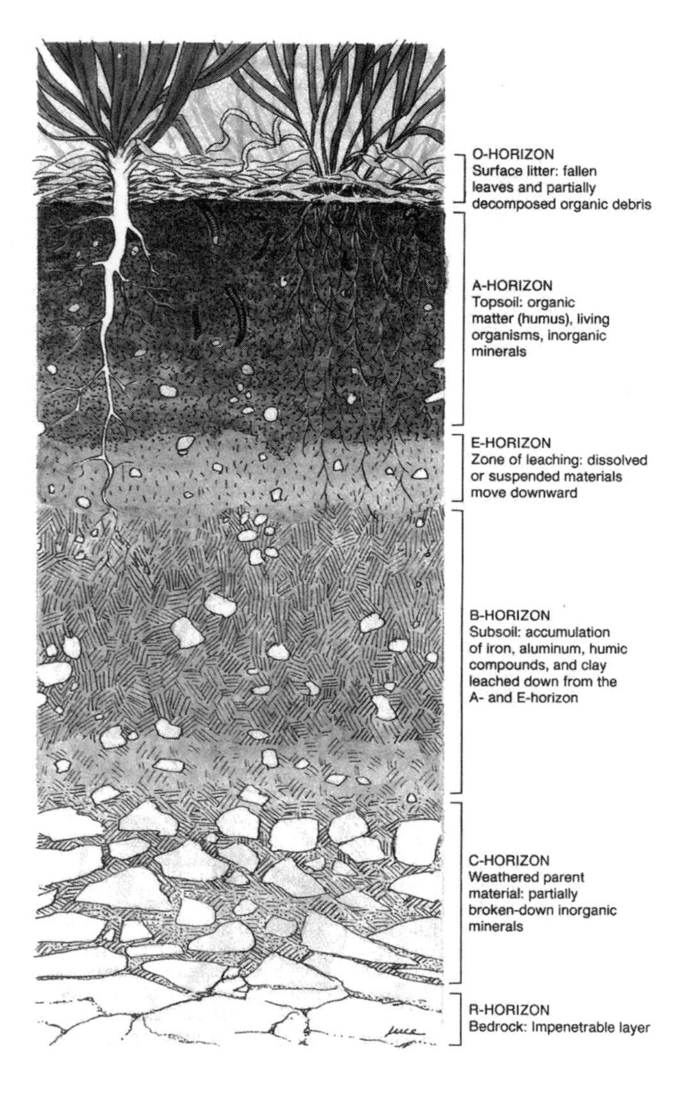

A soil profile showing the horizons in a cross-section of soil. (McGraw-Hill Inc. Reproduced by permission.)

Soil survey

A **soil** survey is a combination of field and laboratory activities intended to identify the basic physical and chemical properties of soils, establish the distribution of those soils at specific map scales, and interpret the information for a variety of uses.

There are two kinds of soil surveys, general purpose and specific use. General purpose surveys collect and display data on a wide range of soil properties which can be used to evaluate the suitability of the soil or area for a variety of purposes. Specific use soil surveys evaluate the suitability of the land for one specific use. Only physical and chemical data related to that particular use are collected, and map units are designed to convey that specific suitability.

A soil survey consists of three major activities: research, mapping, and interpretation. The research phase involves investigations that relate to the distribution and performance of soil; during the mapping phase, the land is actually walked and the soil distribution is noted on a map base which is reproducible. Interpretation involves the evaluation of the soil distribution and performance data to provide assessments of soil suitability for different kinds of **land use**.

During the research phase of the survey, soil scientists first establish which soil properties are important for that type of survey. They then establish field relationships between soil properties and landscape features and determine the types of soils to be mapped, preparing a map legend. In this phase, scientists also evaluate land productivity and recommend management practices as they relate to the mapping units proposed in the map legend.

Mapping is perhaps the most widely recognized phase of a soil survey. It is conducted by evaluating and delineating the soils in the field at a specified map intensity or scale. **Soil profile** observations are made at three levels of detail. Representative profiles are taken from soil pits, generally with laboratory corroboration of the observations made by soil scientists. Intermediate profiles are taken from soil pits, roadside excavations, pipelines or other chance exposures, and some sampling and description does occur at this level. Soil-type identifications are taken from auger holes or small pits, and only brief descriptions are made without laboratory confirmation.

With information gathered from sampling, as well as from the earlier research phase, soil scientists draw soil boundaries on an aerial photograph. These delineations or map units are systematically checked by field transects—straight-line traverses across the landscape with samples taken at specified intervals to confirm the map-unit composition.

Field research and the process of mapping produce a range of information that requires interpretation. Interpretations can include discussions of land use potential, management practices, avoidance of hazards, and economic evaluations of soil data. The interpretation of the information gained during a soil survey is based on crop yield estimates and soil response to specific management. Crop yields are estimated in the following ways: by comparison with data from experimental sites on identified soil types; by field experiments conducted within the survey area; from farm records, demonstration plots or other farm system studies; and by comparison of known crop requirements with the physical and chemical properties of soils. Soil response involves the evaluation of how the soils will respond to changes

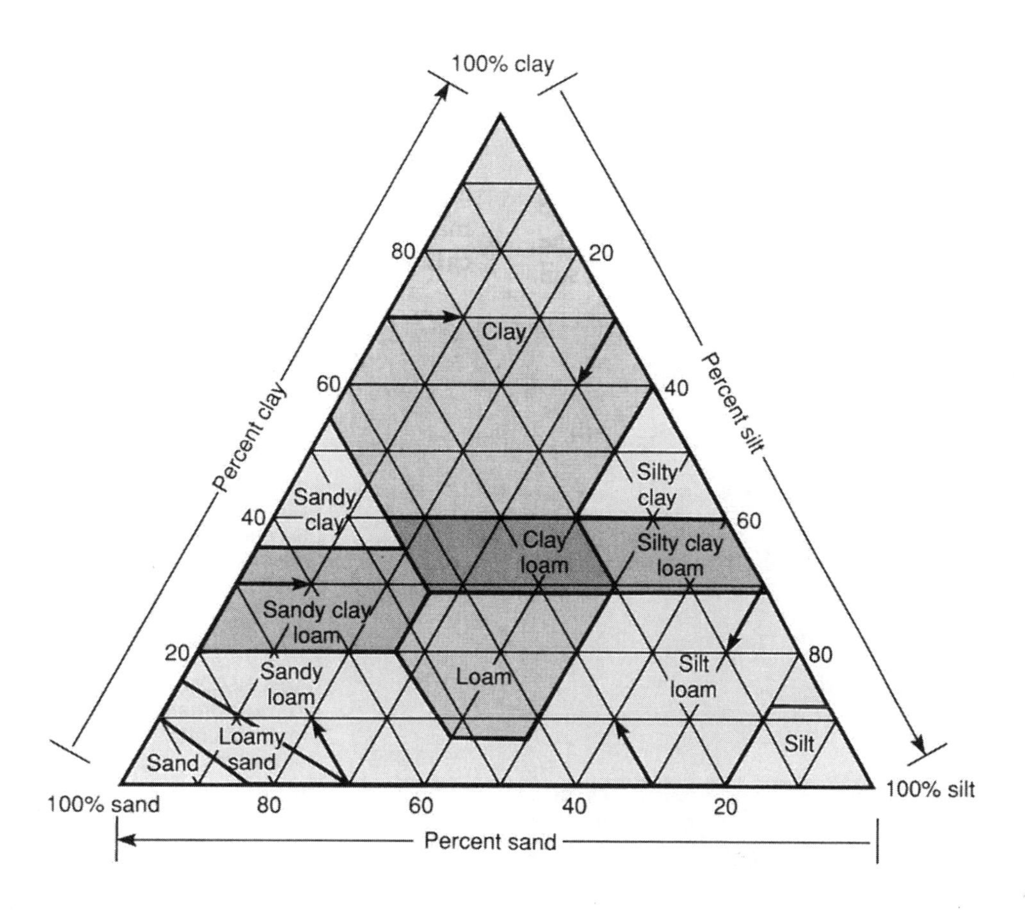

Soil texture depends on the relative proportions of sand, silt and clay in a soil, as represented in this diagram. (McGraw-Hill Inc. Reproduced by permission.)

in use or management, such as **irrigation**, **drainage**, and land **reclamation**. Part of this evaluation includes an assessment of hazards that may result from changes such as **erosion** or **salinization**.

In general purpose surveys, which are the kind conducted in the United States, soils are mapped according to their properties on the hypothesis that soils which look and feel alike will behave the same, while those that do not will respond differently. A soil survey attempts to delineate areas that behave differently or will respond differently to some specified management, and the mapping units provide the basis for locating and predicting these differences. *See also* Contaminated soil; Soil conservation; Soil Conservation Service; Soil texture; U. S. Department of Agriculture

[*James L. Anderson*]

Resources

Books

Dent, D., and A. Young. *Soil Survey and Land Evaluation.* Boston: Allen and Unwin, 1981.

Soil texture

The relative proportion of the mineral particles that make up a **soil** or the percent of sand, **silt**, and clay found in a soil. Texture is an important soil characteristic because it influences water **infiltration**, water storage, amount of **aeration**, ease of tilling the soil, ability to withstand a load, and soil fertility. Textural names are given to soil based on the percentage of sand, silt, and clay. For example, loam is a soil with equal proportions of sand, silt, and clay. It is best for growing most crops.

Solar cell

see **Photovoltaic cell**

Solar constant cycle

The solar constant is a measure of the amount of radiant energy reaching the earth's outer **atmosphere** from the sun.

More precisely, it is equal to the rate at which solar radiation falls on a unit area of a plane surface at the top of the atmosphere and oriented at a perpendicular distance of 9.277 x 10^7 mi (1.496 X 10^8 km) from the sun. That distance is the average (mean) distance between earth and sun during the course of a single year.

Measuring the solar constant has historically been a difficult challenge since there were few good methods for measuring energy levels at the top of the atmosphere. A solution to this problem was made possible in 1980 with the launching of the Solar Maximum (Solar Max) Mission spacecraft. After repair by the *Challenger* astronauts in 1984, Solar Max remained in orbit, taking measurements of solar phenomena until 1989. As a result of the data provided by Solar Max, the solar constant has now been determined to be 1.96 calories per 0.3 in^2 (2 cm^2) per minute, or 1,367 watts per 3.3 ft (1 m). This result confirms the value of 2.0 calories per cm^2 per minute long used by scientists.

In addition to obtaining a good value for the solar constant, however, Solar Max made another interesting discovery. The solar constant is not, after all, really constant. It varies on a daily, weekly, monthly, and yearly basis, and probably much longer. The variations in the solar constant are not large, averaging at a few tenths of a percent.

Scientists have determined the cause of some variability in the solar constant and are hypothesizing others. For example, the presence of sunspots results in a decrease in the solar constant of a few tenths of a percent. Sunspots are regions on the sun's surface where temperatures are significantly lower than surrounding areas. Since the sun presents a somewhat cooler face to the earth when sunspots are present, a decrease in the solar constant is not surprising.

Other surface features on the sun also affect the solar constant. Solar flares, for example, are outbursts or explosions of energy on the surface that may last for many weeks. Correlations have been made between the presence of solar flares and the solar constant over the twenty-seven-day period of the sun's rotation.

A longer-lasting effect is caused by the eleven-year sunspot cycle. As the number of sunspots increase during the cycle, so does the solar constant. For some unknown reason, this long-term effect is just the opposite of that observed for single sunspots.

Scientists believe that historical studies of the solar constant may provide some clues about changes in the earth's **climate**. The nineteenth-century British astronomer E. Walter Maunder pointed out that sunspots were essentially absent during the period of 1645 to 1715, a period now called the Maunder minimum. The earth's climate experienced a dramatic change at about the same time, with temperatures dropping to record lows. These changes are now known as the Little **Ice Age**. It seems possible that further study of the solar constant cycle may explain the connection between these two phenomena.

[*David E. Newton*]

Resources

Books

Chorlton, Windsor, and the editors of Time-Life Books. *Ice Ages*. Alexandria, VA: Time-Life Books, 1983.

Hudson, Hugh S. "Solar Constant." In *McGraw-Hill Encyclopedia of Science & Technology*. 7th ed. New York: McGraw-Hill, 1992.

Solar detoxification

Solar energy is being investigated as a potential means to destroy environmental contaminants. In solar **detoxification**, photons in sunlight are used to break down contaminants into harmless or more easily treatable products. Solar detoxification is a "destructive" technology—it destroys contaminants as opposed to "transfer-of-phase" technologies such as activated **carbon** or air stripping, which are more commonly used to remove contaminants from the **environment**.

In a typical photocatalytic process, water or **soil** containing organic contaminants is exposed to sunlight in the presence of a catalyst such as titanium dioxide or humic and fulvic acids. The catalyst absorbs the high energy photons, and oxygen collects on the catalyst surface, resulting in the formation of reactive **chemicals** referred to as hydroxyl free-radicals and atomic oxygen (singlet oxygen). These reactive chemicals transform the organic contaminants into degradation products, such as **carbon dioxide** and water. Solar detoxification can be accomplished using natural sunlight or by using inside solar simulators or outside solar concentrators, both of which can concentrate light 20 times or more. The effectiveness of solar detoxification in both water and soil is affected by **sorption** of the toxic compounds on sediments or soil and the depth of light penetration. Solar detoxification of volatilized toxic compounds may also occur naturally in the **atmosphere**.

For solar detoxification to be successfully accomplished, toxic chemicals should be converted to thermodynamically stable, non-toxic end products. For example, chlorinated compounds should be transformed to carbon dioxide, water, and hydrochloric **acid** through the general sequence:

organic pollutants → aldehydes → carboxylic acids → carbon dioxide

Photocatalytic degradation such as this usually results in complete mineralization only after prolonged irradiation. If intermediates formed in the degradation pathway are non-toxic, however, the reaction does not have to be driven completely to carbon dioxide and water for acceptable detox-

ification to have occurred. In the solar detoxification of **pentachlorophenol** (PCP) and 2,4-dichlorophenol, for example, toxic intermediates were detected; however, with extended exposure to sunlight, the compounds were rendered completely nontoxic, as measured by **respiration** rate measurements in **activated sludge**.

Many applications of solar detoxification implemented at ambient temperatures have only 90–99% efficiency and do not completely mineralize the compounds. The rate of some photolytic reactions can be increased by raising the temperature of the reaction system, but the production of stable reaction intermediates (some of which may be toxic) is reduced. The changes in reaction rate have been attributed to a combination of a thermally induced increase in the photon-absorption rate, an increase in the quantum yield of the primary photoreaction, and the initiation of photoinduced radical-chain reactions.

Research conducted at the Solar Energy Research Institute (SERI), a laboratory funded by the **U.S. Department of Energy**, has demonstrated that **chlorinated hydrocarbons**, such as trichloroethylene (TCE), trichloroethane (TCA), and **vinyl chloride**, are vulnerable to photocatalytic treatment. Other toxic chemicals shown to be degraded by solar detoxification include a textile dye (Direct Red No. 79), pinkwater, a munitions production waste, and many types of pesticides, including chlorinated cyclodiene insecticides, triazines, ureas, and dinitroaniline herbicides. In conjunction with **ozone** and **hydrogen** peroxide, both of which are strong oxidants, ultraviolet light has been shown to be effective in oxidizing some refractory chemicals such as methyl ethyl ketone, a degreaser, and **polychlorinated biphenyls** (PCBs). *See also* Oxidizing agent

[*Judith Sims*]

Resources
Books

Mill, T., and W. Mabey. "Photodegradation in Water." In *Environmental Exposure from Chemicals*, edited by W. B. Neely and G. E. Blau. CRC Press, Boca Raton, FL: 1985.

Periodicals

Al-Ekabi, H., et al. "Advanced Technology for Water Purification by Heterogenous Photocatalysis." *International Journal of Environment and Pollution* 1, nos. 1–2 (1991): 125–136.

Manilai, V. B., et al. "Photocatalytic Treatment of Toxic Organics in Wastewater: Toxicity of Photodegradation Products." *Water Research* 26, no. 8 (1992): 1035–1038.

Stephenson, F. A. "Chemical Oxidizers Treat Wastewater." *Environmental Protection* 3, no. 10 (1992): 23–27.

Other

Solar Energy Research Institute, Development and Communications Office. *Solar Treatment of Contaminated Water*. SERI/SP-220-3517. Goldon, CO: Solar Heat Research Division, Solar Energy Research Institute, 1989.

Solar energy

The sun is a powerful fusion reactor, where **hydrogen** atoms fuse to form helium and give off a tremendous amount of energy. The surface of the sun, also known as the photosphere, has a temperature of 6,000 K (10,000°F [5,538°C]). The temperature at the core, the region of **nuclear fusion**, is 36,000,000°F (20,000,000°C). A ball of **coal** the size of the sun would burn up completely in 3,000 years, yet the sun has already been burning for three billion years and is expected to burn for another four billion. The power emitted by the sun is 3.9 x 10^{26} watts.

Only a very small fraction of the sun's radiant energy, or insolation, reaches the earth's **atmosphere**, and only about half of that reaches the surface of the earth. The other half is either reflected back into space by clouds and ice or is absorbed or scattered by molecules within the atmosphere. The sun's energy travels 93,000,000 mi (149,667,000 km) to reach the earth's surface. It arrives about 8.5 minutes after leaving the photosphere in various forms of radiant energy with different wave lengths, known as the electromagnetic spectrum.

Solar radiation, also known as solar flux, is measured in Langleys per minute. One Langley equals one calorie of radiant energy per square centimeter. It is possible to appreciate the magnitude of the energy produced by the sun by comparing it to the total energy produced on earth each year by all sources. The annual energy output of the entire world is equivalent to the amount the sun produces in about five billionths of a second. Solar radiation over the United States each year is equal to 500 times its energy consumption.

The solar energy that reaches the surface of the earth and enters the biological cycle through **photosynthesis** is responsible for all forms of life, as well as all deposits of fossil fuel. All energy on earth comes from the sun, and it can be utilized directly or indirectly. Direct uses include passive solar systems such as greenhouses and atriums, as well as windmills, hydropower, and the burning of **biomass**. Indirect uses of solar energy include photovoltaic cells, in which semiconductor crystals convert sunlight into electrical power, and a process that produces methyl alcohol from plants.

Wind results from the uneven heating of the earth's atmosphere. About 2% of the solar energy which reaches the earth is used to move air masses, and at any one time the kinetic energy in the wind is equivalent to 20 times the current electricity use. Due to mechanical losses and other factors, windmills cannot extract all this power. The power produced by a windmill depends on the speed of the wind and the effective surface areas of the blades, and the maximum extractable power is about 60%.

A solar furnace in Albuquerque, New Mexico. (Photograph by Mark Antman. Phototake. Reproduced by permission.)

The use of the water wheel preceded windmills, and it may be the most ancient technology for utilizing solar energy. The sun causes water to evaporate, clouds form upon cooling, and the subsequent precipitation can be stored behind **dams**. This water is of high potential energy, and it is used to run water wheels or turbines. Modern turbines that run electric generators are approximately 90% efficient, and in 1998 hydroelectric energy produced about 4% of the primary energy used in the United States.

A passive solar heating system absorbs the radiation of the sun directly, without moving parts such as pumps. This kind of low-temperature heat is used for space heating. Solar radiation may be collected by the use of south-facing windows in the Northern Hemisphere. Glass is transparent to visible light, allowing long-wave visible rays to enter, and it hinders the escape of long-wave heat, therefore raising the temperature of a building or a greenhouse.

A thermal mass such as rocks, brine, or a concrete floor stores the collected solar energy as heat and then releases it slowly when the surrounding temperature drops. In addition to collecting and storing solar energy as heat, passive systems must be designed to reduce both heat loss in cold weather and heat gain in hot weather. Reductions in heat transfer can be accomplished by heavy insulation, by the use of double glazed windows, and by the construction of an earth brim around the building. In hot summer weather, passive cooling can be provided by building extended overhangs or by planting deciduous trees. In dry areas such as in the southwestern United States or the Mediterranean, solar-driven fans or evaporative coolers can remove a great deal of interior heat. Examples of passive solar heating include roof-mounted hot-water heaters, solarian glass-walled rooms or patios, and earth-sheltered houses with windows facing south.

A simple and innovative technology in passive systems is a design based on the fact that at depth of 15 ft (4.6 m) the temperature of the earth remains at about 55°F (13°C) all year in a cold northern **climate** and about 67°F (19°C) in warm a southern climate. By constructing air intake tubes at depths of 15 ft, the air can be either cooled or heated to reach the earth's temperature, proving an efficient air conditioning system.

Active solar systems differ from passive systems in that they include machinery, such as pumps or electric fans, which lowers the net energy yield. The most common type of active systems are photovoltaic cells, which convert sunlight into direct current electricity. The thin cells are made from semiconductor material, mainly silicon, with small amounts of gallium arsenide or **cadmium** sulfide added so that the cell emits electrons when exposed to sunlight. Solar cells are connected in a series and framed on a rigid background. These modules are used to charge storage batteries aboard boats, operate lighthouses, and supply power for emergency telephones on highways. They are also used in remote areas not connected to a power supply grid for pumping water either for cattle or **irrigation**.

Both passive and active solar systems can be installed without much technical knowledge. Neither produces **air**

pollution, and both have a very low environmental impact. But a well-designed passive system is cheaper than active one and does not require as much operative maintenance. *See also* Alternative energy sources; Energy and the environment; Energy policy; Passive solar design

[*Douglas Smith*]

RESOURCES
BOOKS

Balcomb, J. D. *Passive Solar Building.* Cambridge: MIT Press, 1992.

Hedger, J. *Solar Energy–The Sleeping Giant: Basics of Solar Energy.* Deming: Akela West Publishers, 1993.

PERIODICALS

Brown, L. R., et al. "A World Fit to Live In." *UNESCO Courier* (November 1991): 28–31.

Peck, L. "Here Comes the Sun." *Amicus Journal* 12 (Spring 1990): 27–32.

Solar Energy Research, Development and Demonstration Act (1974)

Following the outbreak of the Arab-Israeli war in 1973, Arab oil-producing states imposed an embargo on oil exports to the United States. The embargo lasted from October 1973 to March 1974, and the long gas lines it caused highlighted the United States' dependence on foreign **petroleum**. Congress responded by enacting the **Solar Energy** Research, Development and Demonstration Act of 1974.

The act stated that it was henceforth the policy of the federal government to "pursue a vigorous and viable program of research and resource assessment of solar energy as a major source of energy for our national needs." The act's scope embraced all energy sources which are renewable by the sun—including solar thermal energy, photovoltaic energy, and energy derived from wind, sea thermal gradients, and **photosynthesis**. To achieve its goals, the act established two programs: the Solar Energy Coordination and Management Project and the Solar Energy Research Institute.

The Solar Energy Coordination and Management Project consisted of six members, five of whom were drawn from other federal agencies, including the National Science Foundation, the Department of Housing and Urban Development, the **Federal Power Commission**, NASA, and the **Atomic Energy Commission**. Congress intended that the project would coordinate national solar energy research, development, and demonstration projects, and would survey resources and technologies available for solar energy production. This information was to be placed in a Solar Energy Information Data Bank and made available to those involved in solar energy development.

Over the decade following the passage of the act in 1974, the United States government spent $4 billion on research in solar and other **renewable energy** technologies. During the same period, the government spent an additional $2 billion on tax incentives to promote these alternatives. According to a **U.S. Department of Energy** report issued in 1985, these efforts displaced petroleum worth an estimated $36 billion.

Despite the promise of solar energy in the 1970s and the fear of reliance on foreign petroleum, spending on renewable energy sources in the United States declined dramatically during the 1980s. Several factors combined during that decade to weaken the federal government's commitment to solar power, including the availability of inexpensive petroleum and the skeptical attitudes of the Reagan and Bush administrations, which were distrustful of government-sponsored initiatives and concerned about government spending.

To carry out the research and development initiatives of the Solar Energy Coordination and Management Project, the act also established the Solar Energy Research Institute, located in Golden, Colorado. In 1991, the Solar Energy Research Institute was renamed the National Renewable Energy Laboratory and made a part of the national laboratory system. The laboratory continues to conduct research in the production of solar energy and energy from other renewable sources. In addition, the laboratory studies applications of solar energy. For example, it has worked on a project using solar energy to detoxify **soil** contaminated with hazardous wastes.

[*L. Carol Ritchie*]

RESOURCES
PERIODICALS

"Federal R&D Funding for Solar Technologies." *Solar Industry Journal* (First Quarter 1992).

"Photovoltaics." *Solar Industry Journal* (First Quarter 1992).

OTHER

Solar Energy Research, Development, and Demonstration Act of 1974, Pub. L. No. 93-473, 88 Stat. 1431 (1974), codified at 42 U.S.C. 5551, et seq. (1988).

Solid waste

Solid waste is composed of a broad array of materials discarded by households, businesses, industries, and agriculture. The United States generates more than 11 billion tons (10 billion metric tons) of solid waste each year. The waste is composed of 7.6 billion tons (6.9 billion metric) of industrial nonhazardous waste, 2–3 billion tons (1.8–2.7 billion metric) of oil and gas waste, over 1.4 billion tons (1.3 billion metric)

of mining waste, and 195 million tons (177 billion metric) of **municipal solid waste**.

Not all solid waste is actually solid. Some semi-solid, liquid, and gaseous wastes are included in the definition of solid waste. The **Resource Conservation and Recovery Act** (RCRA) defines solid waste to include **garbage**, refuse, **sludge** from municipal **sewage treatment** plants, ash from solid waste incinerators, mining waste, waste from construction and demolition, and some hazardous wastes. Since the definition is so broad, it is worth considering what the act excludes from regulations concerning solid waste: untreated sewage, industrial **wastewater** regulated by the **Clean Water Act**, **irrigation** return flows, nuclear materials and by-products, and hazardous wastes in large quantities.

RCRA defines and establishes regulatory authority for **hazardous waste** and solid waste. According to the act, some hazardous waste may be disposed of in solid waste facilities. These include hazardous wastes discarded from households, such as paint, cleaning solvents, and batteries, and small quantities of hazardous materials discarded by business and industry. Some states have their own definitions of solid waste which may vary somewhat from the federal definition. Federal oversight of solid-waste management is the responsibility of the **Environmental Protection Agency** (EPA).

Facilities for the disposal of solid waste include municipal and industrial landfills, industrial surface impoundments, and incinerators. Incinerators that recover energy as a by-product of waste **combustion** are called **resource recovery** or waste-to-energy facilities. Sewage sludge and agricultural waste may be applied to land surfaces as fertilizers or **soil** conditioners. Other types of **waste management** practices include **composting**, most commonly of separated organic wastes, and **recycling**. Some solid waste ends up in illegal open dumps.

Three quarters of industrial nonhazardous waste comes from four industries: iron and steel manufacturers, **electric utilities**, companies making industrial inorganic **chemicals**, and firms producing **plastics** and resins. About one-third of industrial nonhazardous waste is managed on the site where it is generated, and the rest is transported to off-site municipal or industrial waste facilities. Although surveys conducted by some states are beginning to fill in the gaps, there is still not enough landfills and surface impoundments for industrial solid wastes. Available data suggest there is limited use of environmental controls at those facilities, but there is insufficient information to determine the extent of **pollution** they may have caused.

The materials in municipal solid waste (MSW) are discarded from residential, commercial, institutional, and industrial sources. The materials include plastics, paper, glass, metals, wood, food, and **yard waste**; the amount of each material is evaluated by weight or volume. The distinction between weight and volume is important when considering such factors as **landfill** capacity. For example, plastics account for only about 8% of MSW by weight, but more than 21% by volume. Conversely, glass represents about 7% of the weight and only 2% of the volume of MSW.

MSW has recently been the focus of much attention in the United States. Americans generated 4.5 lb (2 kg) per day of MSW in 2000, an increase from 4.3 lb (1.95 kg) per day 1990, 4.0 lb (1.8 kg) per day in 1980, and 2.7 lb (1.2 kg) per day in 1960. This increase has been accompanied by tightening federal regulations concerning the use and construction of landfills. The expense of constructing new landfills to meet these regulations, as well as frequently strong public opposition to new sites for them, have sharply limited the number of disposal options available, and the result is what many consider to be a solid waste disposal crisis. The much-publicized "garbage barge" from Islip, New York, which roamed the oceans from port to port during 1987 looking for a place to unload, has become a symbol of this crisis.

The disposal of MSW is only the most visible aspect of the waste disposal crisis; there are increasingly limited disposal options for all the solid waste generated in America. In response to this crisis, the EPA introduced a waste management hierarchy in 1989. The hierarchy places source reduction and recycling above **incineration** and landfilling as the preferred options for managing solid waste.

Recycling diverts waste already created away from incinerators and landfills. Source reduction, in contrast, decreases the amount of waste created. It is considered the best waste management option, and the EPA defines it as reducing the quantity and toxicity of waste through the design, manufacture, and use of products. Source reduction measures include reducing packaging in products, reusing materials instead of throwing them away, and designing products to be long lasting. Individuals can practice **waste reduction** by the goods they choose to buy and how they use these products once they bring them home. Many businesses and industries have established procedures for waste reduction, and some have reduced waste toxicity by using less toxic materials in products and packaging. Source reduction can be part of an overall industrial pollution prevention and waste minimization strategy, including recapturing process wastes for **reuse** rather than disposal.

Some solid wastes are potentially threatening to the **environment** if thrown away but can be valuable resources if reused or recycled. Used motor oil is one example. It contains **heavy metals** and other hazardous substances that can contaminate **groundwater**, surface water, and soils. One gallon (3.8 L) of used oil can contaminate 1 million gal (3.8 million L) of water, but these problems can be

avoided and energy saved if used oil is rerefined into motor oil or reprocessed for use as industrial fuel. Much progress remains to be made in this area: of the 200 million gal (757 million L) of used oil generated annually by people who change their own oil, only 10% is recycled.

Another solid waste with recycling potential are used tires, which take up a large amount of space in landfills and cause uneven settling. Tire stockpiles and open dumps can be breeding grounds for mosquitoes, and are hazardous if ignited, emitting noxious fumes that are difficult to extinguish. Instead of being thrown away, tires can be shredded and recycled into objects such as hoses and doormats; they can also be mixed in road-paving materials or used as fuel in suitable facilities. Whole tires can be retreaded and reused on automobiles, or simply used for such purposes as artificial reef construction. Other solid wastes with potential for increased recycling include construction and demolition wastes (building materials), household appliances, and waste wood.

The EPA recommends implementation of this waste management hierarchy through integrated waste management. The agency has encouraged businesses and communities to develop systems where components in the hierarchy complement each other. For example, removing **recyclables** before burning waste for **energy recovery** not only provides all the benefits of recycling, but it reduces the amount of residual ash. Removing recyclable materials that are difficult to burn, such as glass, increases the **Btu** value of waste, thereby improving the efficiency of energy recovery.

Some state and local governments have chosen to conserve landfill space or reduce the toxicity of waste by instituting bans on the burial or burning of certain materials. The most commonly banned materials are automotive batteries, tires, motor oil, yard waste, and appliances. Where such bans exist, there is usually a complementary system in place for either recycling the banned materials or reusing them in some way. Programs such as these usually compost yard waste and institute separate collections for hazardous waste.

Perhaps because of the solid waste disposal crisis, there have been recent changes in solid waste management practices. About 55.3% of MSW was buried in landfills in 2000, down from 67% in 1990 and 81% in 1980. Recovery of materials for recycling and composting also increased, with 64 million tons of material diverted away from landfills and incinerators in 1999, up from 34 million tons in 1990. *See also* Refuse-derived fuel; Sludge treatment and disposal; Solid waste incineration; Solid waste landfilling; Solid waste recycling and recovery; Solid waste volume reduction; Source separation

[*Teresa C. Donkin and Douglas Smith*]

Resources

Books

Blumberg, L., and R. Grottleib. *War on Waste—Can America Win Its Battle With Garbage?* Covelo, CA: Island Press, 1988.

Neal, H. A., and J. R. Schubel. *Solid Waste Management and the Environment: The Mounting Garbage and Trash Crisis.* Englewood Cliffs, NJ: Prentice-Hall, 1987.

Robinson, W. D., ed. *The Solid Waste Handbook.* New York: Wiley, 1986.

Underwood, J. D., and A. Hershkowitz. *Facts About U. S. Garbage Management: Problems and Practices.* New York: INFORM, 1989.

Periodicals

Glenn, J. "The State of Garbage in America." *Biocycle* (May 1992).

Other

U.S. Environmental Protection Agency. *Characterization of Municipal Solid Waste in the United States: 1992 Update (Executive Summary).* Washington, DC: U. S. Government Printing Office, July 1992.

U.S. Environmental Protection Agency. *Solid Waste Disposal in the United States.* Washington, DC: U.S. Government Printing Office, 1989.

Solid waste incineration

Incineration is the burning of waste in a specially designed **combustion** chamber. The idea of burning **garbage** is not new, but with the increase in knowledge about toxic **chemicals** known to be released during burning, and with the increase in the amount of garbage to be burned, incineration now is done under controlled conditions. It has become the method of choice of many **waste management** companies and municipalities.

According to the **Environmental Protection Agency** (EPA), there are 135 operational waste combustion facilities in the United States. About 120 of them recover energy, and in all, the facilities process about 14.5% of the nation's 232 million tons (210.5 million metric tons) of **municipal solid waste** produced each year.

There are several types of combustion facilities in operation. At incinerators, mixed trash goes in one end unsorted, and it is all burned together. The resulting ash is typically placed in a **landfill**.

At **mass burn** incinerators, also known as mass burn combustors, the heat generated from the burning material is turned into useable electricity. Mixed garbage burns in a special chamber where temperatures reach at least 2000°F (1093 °C). The byproducts are ash, which is landfilled, and combustion gases. As the hot gases rise from the burning waste, they heat water held in special tubes around the combustion chamber. The boiling water generates steam and/or electricity. The gases are filtered for contaminants before being released into the air.

Modular combustion systems typically have two combustion chambers: one to burn mixed trash and another to heat gases. Energy is recovered with a heat recovery steam

generator. Refuse-derived fuel combustors burn presorted waste and convert the resulting heat into energy. In all, incinerators in this country in 1992 typically consumed 80,100 tons (72,730 metric tons) of waste each day, and generated the annual equivalent of 16.4 million megawatts of usable power, equal to roughly 30 million barrels of oil.

One major problem with incineration is **air pollution**. Even when equipped with **scrubbers**, many substances, some of them toxic, are released into the **atmosphere**. In the United States, emissions from incinerators are among targets of the **Clean Air Act**, and research continues on ways to improve the efficiency of incinerators. For now, however, many environmentalists protest the use of incinerators.

Incinerators burning municipal solid waste produce the pollutants **carbon monoxide**, **sulfur dioxide**, and particulates containing **heavy metals**. The generation of pollutants can be controlled by proper operation and by the proper use of air **emission** control devices, including dry scrubbers, electrostatic precipitators, fabric **filters**, and proper stack height.

Dry scrubbers wash **particulate** matter and gases from the air by passing them through a liquid. The scrubber removes **acid** gases by injecting a lime **slurry** into a reaction tower through which the gases flow. A salt powder is produced and collected along with the **fly ash**. The lime also causes small particles to stick together, forming larger particles that are more easily removed.

Electrostatic precipitators use high voltage to negatively charge dust particles, then the charged particulates are collected on positively charged plates. This device has been documented as removing 98% of particulates, including heavy metals; nearly 43% of all existing facilities use this method to control air **pollution**.

Fabric filters or **baghouses** consist of hundreds of long fabric bags made of heat-resistant material suspended in an enclosed housing which filters particles from the gas stream. Fabric filters are able to trap fine, inhalable particles, up to 99% of the particulates in the gas flow coming out of the scrubber, including condensed toxic organic and heavy metal compounds. Stack height is an extra precaution taken to assure that any remaining pollutants will not reach the ground in a concentrated area.

Using the **Best Available Control Technology** (BAT), the National Solid Wastes Management Association states more than 95% of gases and fly ash are captured and removed. However, such state-of-the-art facilities are more the exception than the rule. In a study of 15 mass burn and refuse-derived fuel plants of varying age, size, design, and control systems, the environmental research group **INFORM** found that only one of the 15 achieved the emission levels set by the group for six primary air pollutants (dioxins and **furans**, particulates, **carbon** monoxide, sulfur dioxide, **hydrogen** chloride, and **nitrogen oxides**). Six plants did not meet any. Only two employed the safest ash management techniques, and the EPA is still trying to define standards for air emissions from municipal incinerators.

Ash is the solid material left over after combustion in the incinerator. It is composed of noncombustible inorganic materials that are present in cans, bottles, rocks and stones, and complex organic materials formed primarily from carbon atoms that escape combustion. Municipal solid waste ash also can contain **lead** and **cadmium** from such sources as old appliances and car batteries.

Bottom ash, the unburned and unburnable matter left over, comprises 75–90% of all ash produced in incineration. Fly ash is a powdery material suspended in the **flue gas** stream and is collected in the **air pollution control** equipment. Fly ash tends to have higher concentrations of certain metals and organic materials, and comprises 10–25% of total ash generated.

The greatest concern with ash is proper disposal and the potential for harmful substances to be released into the **groundwater**. Federal regulations governing ash are in transition because it is not known whether ash should be regulated as a hazardous or nonhazardous waste as specified in the **Resource Conservation and Recovery Act** (RCRA) of 1976.

EPA draft guidelines for handling ash include: ash containers and transport vehicles must be leak-tight; **groundwater monitoring** must be performed at disposal sites; and liners must be used at all ash disposal landfills.

While waste-to-energy plants can decrease the volume of solid waste by 60–90% and at the same time recover energy from discarded products, the cost of building such facilities is too high for most municipalities. The $400 million price tag for a large plant is prohibitive, even if the revenue from the sale of energy helps offset the cost. But without a strong market for produced energy, the plant may not be economically feasible for many areas.

The Public Utility Regulatory Policy Act (PURPA), enacted in 1979, helps to ensure that small power generators, including waste-to-energy facilities, will have a market for produced energy. PURPA requires utilities to purchase such energy from qualifying facilities at avoided costs, that is, the cost avoided by not generating the energy themselves. Some waste-to-energy plants are developing markets themselves for steam produced. Some facilities supply steam to industrial plants or district heating systems.

Another concern over the use of incinerators to solve the garbage dilemma in the United States is that materials incinerated are resources lost—resources that must be recreated with a considerable effort, high expense, and potential environmental damage. From this point of view, incineration represents a failure of waste disposal policy. Waste disposal,

many believe, requires an integrated approach that includes reduction, **recycling**, **composting**, and landfilling, along with incineration.

[*Linda Rehkopf*]

Resources

Books

Clarke, M. *Improving Environmental Performance of MSW Incinerators.* New York: INFORM, 1988.

Other

Denison, R. A., and J. Ruston, eds. *Recycling and Incineration: Evaluating the Choices.* New York: Environmental Defense Fund, 1990.

Solid waste landfilling

The **Resource Conservation and Recovery Act** (RCRA) defines **solid waste** as **garbage**, refuse, **sludge** from **sewage treatment** plants, ash from incinerators, mining waste, construction and demolition materials. It also includes some small quantities of **hazardous waste**. There are several acceptable methods for disposing of these wastes. These include **incineration**, **composting**, **recycling**, industrial surface impoundments, and landfills. In terms of personal inconvenience, disposing of waste in landfills is easier than the other methods. Waste can have any shape or condition prior to being placed into a **landfill**. This is different from composting and incineration. Waste to be composted must be **biodegradable** and have maximal surface area to hasten the process of composting. Waste to be incinerated must be combustible and be reduced into small pieces to maximize surface area and promote burning. Waste to be disposed of in a landfill is usually broken down with heavy machinery, then compacted and placed into specially prepared sites. Formerly, waste was dumped into unused sand or gravel pits. In 2002, most waste is placed into sanitary landfills. These are specially prepared sites into which drains and special linings have been installed prior to the placement of waste. In sanitary landfills, waste is compacted each day and covered with a layer of dirt to decrease odor and discourage flies and other organisms such as rats, mice and birds that can transmit disease.

In 1960, each American generated 2.7 lb (1.2 kg) of solid waste. This grew to 4.3 lb (1.9 kg) per person by 1990. Americans continue to generate more solid waste each day but the rate of growth has decreased. In 2000, each person generated 4.5 lb (2.0 kg) each day. The population also continues to increase. The net effect is to create increased amounts of municipal waste. The supply of available landfill space is rapidly decreasing. The attitude of **not in my backyard** (NIMBY) further slows the development and construction of new landfills. With fewer landfills, the cost of sanitary waste disposal has dramatically increased. The composition of waste dumped into landfills is also important in terms of capacity and useful life. For instance, **plastics** account for 8% of **municipal solid waste** by weight, but more than 21% by volume. In an attempt to conserve land space and reduce other, long term problems, some municipalities have banned the deposition of certain materials such as car batteries, used tires, motor oil, **yard waste**, and appliances.

In 1980, 81% of solid waste was buried in landfills; in 1990 the amount had decreased to 67%. In 2000, the amount of solid waste put into landfills again increased to approximately 75%. This decrease through 1990 was the result of a concentrated effort by federal and local organizations to address problems associated with landfills. Experts differ as to the reasons for the increase. Many attribute it to decreased participation in recycling programs. In addition, contamination of surface and ground waters near landfills has been reported for the past two decades. Because of these and other health-related problems associated with landfill use, the **Environmental Protection Agency** (EPA) continues to recommend source reduction, recycling, and incineration as the preferred **waste management** solutions. Placing solid waste in landfills is the least desirable method for disposing of solid waste. However, it is an acceptable alternative that is far superior to unrestricted dumping.

[*L. Fleming Fallon Jr., M.D., Dr.P.H.*]

Resources

Books

Kreith, F. *Handbook of Solid Waste Management.* New York: McGraw Hill, 2002.

McDougall, F. R. *Integrated Solid Waste Management: A Life Cycle Inventory.* 2nd ed. Oxford, U.K.: Blackwell, 2001.

Thomas-Hope, E. *Solid Waste Management.* Kingston, Jamaica: University Press of the West Indies, 2000.

Periodicals

Fabbricino, M. "An Integrated Programme for Municipal Solid Waste Management." *Waste Management Research* 19, no. 5 (2001): 368–379.

Wilson, E. J. "Life Cycle Inventory for Municipal Solid Waste Management." *Waste Management Research* 20, no. 1 (2002): 16–22.

Youcai Z., C. Zhugen, S. Qingwen, and H. Renhua. "Monitoring and Long-term Prediction of Refuse Compositions and Settlement in Large-scale Landfill." *Waste Management Research* 19, no. 2 (2001): 160–168.

Other

"The Basics of Landfills." *Environmental Justice Advocates.* April 9, 2001 [cited July 2002]. <http://www.ejnet.org/landfills>.

Paddock, Todd. "Looking into Landfills." *Academy of Natural Sciences.* June 1989 [cited July 2002]. <http://www.acnatsci.org/research/kye/landfills.html>.

"Solid Waste." *Annenberg Foundation.* 1997–2002 [cited July 2002]. <http://www.learner.org/exhibits/garbage/solidwaste.html>.

U.S. Environmental Protection Agency. "Landfills." *Industrial Water Pollution Controls.* May 14, 2002 [cited July 2002]. <http://www.epa.gov/ost/guide/landfills>.

U.S. Environmental Protection Agency. "MSW Disposal." *Municipal Solid Waste.* July 10, 2002 [cited July 2002]. <http://www.epa.gov/epaoswer/non-hw/muncpl/disposal.htm>.

Solid waste management

see **Waste management**

Solid waste recycling and recovery

Recycling is the recovery and **reuse** of materials from wastes. **Solid waste** recycling refers to the reuse of manufactured goods from which resources such as steel, **copper**, or **plastics** can be recovered and reused. Recycling and recovery is only one phase of an integrated approach to solid **waste management** that also includes reducing the amount of waste produced, **composting**, incinerating, and landfilling.

Municipal solid waste (MSW) comes from household, commercial, institutional, and light industrial sources, and from some hospital and laboratory sources. In 2000 the United States produces nearly 232 million tons (210.5 million metric tons) of MSW per year, almost 4.5 lb (2 kg) per resident per day. The percentages of MSW generated in this country include paper and paperboard, 38.1%; yard wastes, 12.1%; metals, 7.8%; glass, 5.5%; **rubber**, textiles, leather and wood, 11.9%; food wastes, 10.9%; plastics, 10.5%; and other, 3.2%.

Recycling is a significant way to keep large amounts of solid waste out of landfills, conserve resources, and save energy. As of 2000, Americans recovered, recycled, or composted 30.1% of MSW, incinerated 14.5%, and landfilled 55.3%.

The technology of recycling involves collection, separation, preparing the material to buyer's specifications, sale to markets, processing, and the eventual reuse of materials. Separation and collection is only the first step; if the material is not also processed and returned to commerce, then it is not being recycled. In many parts of the country, markets are not yet sufficiently developed to handle the growing supply of collected material.

Intermediate markets for recyclable materials include scrap dealers or brokers, who wait for favorable market conditions in which to sell their inventory. Final markets are facilities where recycled materials are converted to new products, the last phase in the recycling circle.

The materials recycled today include **aluminum**, paper, glass, plastics, iron and steel, scrap tires, and used oil. Aluminum, particularly cans, is a valuable commodity. By the late 1980s, over 50% of all aluminum cans were recycled. Recycling aluminum saves a tremendous amount of energy: it takes 95% less energy to produce an aluminum can from an existing one rather than from ore. Other aluminum products that are recycled include siding, gutters, door and window frames, and lawn furniture.

Over 40% of the paper and paperboard used in the U.S. is collected and utilized as either raw material to make recycled paper, or as an export to overseas markets. Recycled paper shows up in newsprint, roofing shingles, tar paper, and insulation. Other recyclable paper products include old corrugated containers, mixed office waste, and high-grade waste paper. Contaminants must be removed from paper products before the remanufacture process can begin, however, such as food wastes, metal, glass, rubber, and other extraneous materials.

The market for crushed glass, or cullet, has increased. Recycled glass is used to make fiberglass and new glass containers. About 1.25 million tons (1.14 million metric tons) of glass is recycled annually in the United States.

Three types of plastic are successfully being recycled, the most common being PET (polyethylene terephthalate), or soft drink containers. Recycled PET is used for fiberfill in sleeping bags and ski jackets, carpet backing, **automobile** bumpers, bathtubs, floor tiles, and paintbrushes. HDPE plastic (high density polyethylene) is used for milk jugs and the bottoms of soft drink bottles. It can be recycled into trash cans and flower pots, among other items. **Polystyrene** foam is crushed into pellets and turned into plastic lumber for benches and walkways. Commingled plastics are recycled into fence posts and park benches.

Iron and steel are the most recycled materials used today. In 1987, 51 million tons (46 million metric tons) were recycled, more than twice the amount of all other materials combined. The material is remelted and shaped into new products.

More than one billion discarded tires are stockpiled in the United States, but scrap tires can be shredded and used for asphalt-rubber or retreading; are incinerated for fuel; or used to construct artificial marine reefs.

Used oil is a valuable resource, and of the 1.2 billion gal (4.5 billion L) generated annually, two-thirds is recycled. The rest, about 400 million gal (1.5 billion L), is disposed of or dumped. About 57% of used oil is reprocessed for fuel, 26% is refined and turned into base stock for use as lubricating oil, and about 17% is recycled for other uses.

Composting is the **aerobic** biological **decomposition** of **organic waste** materials, usually lawn clippings. Com-

posting is not an option for a major portion of the solid **waste stream**, but is an important component of the **resource recovery** program.

Recycling collection methods vary, but curbside collection is the most popular and has the highest participation rates. It is also the most expensive way for municipalities to collect **recyclables** in their communities. Collection centers do not yield as many recyclables because residents must do the sorting themselves, but centers offer the most affordable method of collection.

Precycling is an option that is gaining widespread recognition in this country. Basically, precycling refers to the consumer making environmentally sound choices at the point of purchase. It includes avoiding products with extra packaging, or products made to satisfy only short-term needs, such as disposable razors.

Resource recovery or materials recovery is the recycling of waste in an industrial setting. It does not involve recycling consumer waste or municipal solid waste, but includes reprocessed industrial material that, for whatever reason, is not able to be used as it was initially intended. Some consumer groups are pressing for government guidelines on labeling packaging or products "reprocessed" as opposed to "recycled."

While the **Environmental Protection Agency** (EPA) insists that no single alternative to the municipal solid waste problem should be relied upon, its generally accepted hierarchy of waste management alternatives is 1) source reduction and 2) reusing products. Waste that is not generated never enters the waste stream.

If recycling is to be used as a genuine MSW management alternative rather than a "feel good" way to conserve resources, then materials must be recovered and made into new products in large quantities. For some materials, however, an insufficient market exists, so communities must pay to have some recyclable materials taken away until a market is developed. Recycling programs depend on the will of the community to follow through, and in many areas, response is weak and enforcement lacking. However, dwindling **landfill** space in the 1990s may force communities to mandate recycling programs.

Recent EPA regulations seriously affected the number of operable landfills. The requirements include installing liners, collecting and treating liquids that leach, monitoring **groundwater** and surface water for harmful **chemicals**, and monitoring the escape of **methane** gas. These regulations will increase the number of corporate-run landfills, but the cost of building and maintaining a landfill that adheres to the regulations will top $125 million. The end cost to consumers to have trash hauled away may also force many **garbage** makers to become reducers, reusers, and recyclers.

[*Linda Rehkopf*]

Resources

Books

Kharbanda, O. P., and E. A. Stallworthy. *Waste Management: Towards a Sustainable Society.* Westport, CT: Auburn House/Greenwood, 1990.

Robinson, W. D., ed. *The Solid Waste Handbook.* New York: Wiley, 1986.

Periodicals

Franklin, W. E., and M. A. Franklin. "Recycling." *The EPA Journal* (July-August 1992): 7.

Other

Solid Waste Recycling: The Complete Resource Guide. Washington, DC: Bureau of National Affairs, 1990.

Solid waste volume reduction

Solid waste volume reduction can take place at several points in the **waste management** process. Solid waste volume reduction can take the form of **precycling** or **reuse** behavior on the part of consumers. This behavior reduces solid waste at the source and prevents materials from ever entering the **waste stream**. Precycling on the part of consumers is the best initial activity to reduce the volume of solid waste. Reuse is also a measure that prevents or delays the **migration** of materials to the **landfill**. Once the decision is made that a product is no longer useful and needs to be discarded, there are several management techniques that can be used.

Recycling diverts large volumes of materials from the waste stream to a manufacturing process. Glass, plastic, paper, cardboard, and other clean, source-separated materials are well-suited for use in the manufacture of new products. As markets become more readily available, the volume of materials that end up in landfills will be reduced even more. Such troublesome wastes as tires, appliances and construction debris are targets of serious recycling efforts because they are large-volume items, take up a lot of space in landfills, and do not provide good fuel for waste-to-energy facilities in their discarded state. Once **recyclables** have been separated from the waste stream, solid waste can be further reduced in volume through several methods.

Compaction of waste materials after **source separation** is useful for two reasons. Compaction prepares waste for efficient transport by truck, boat or rail car to landfills or other waste disposal facilities. Compacted waste takes up less space in a landfill, thereby extending the life of the landfill. In some cases, compacted waste can be stored for later disposal. It must be understood that the compaction of waste should only be done after all recyclables have been removed since this process contaminates post-consumer materials and makes it almost impossible to recover them for a manufacturing process.

Incineration has long been accepted as a waste volume reduction technique. The burning of solid waste can reduce the volume by 95%. In the early 1920s, incineration was used to reduce waste materials to what was thought of as a harmless ash. We now know that contaminants can become concentrated in the ash, qualifying it as a **hazardous waste**. It wasn't until the mid-1970s that waste-to-energy facilities using incineration were considered a viable option, as a consequence of the added benefit of energy supply. Most states now require that the residual ash from burning **municipal solid waste** be landfilled in a monofill. A monofill is a landfill compartment that can only be used for incinerator ash.

Composting is another volume reduction technique that can divert large volumes of waste material from the waste stream. Leaves, grass clippings and tree prunings can be reduced in volume through an active composting process both in the backyard and in municipal composting yards. Individual homeowners can compost in their back yards by using simple rounds of wire fencing placed in a shady, cool corner. The composting materials need to be managed through the addition of moisture, temperature monitoring, and frequent turning, until they become a rich **soil** for the garden. Municipal yards perform the same process on a much larger scale and then use the resulting soil in city parks or give it back to the homeowners for personal use. These solid waste volume reduction techniques for organics in the waste stream are simple and low cost and can be very effective.

Methods for encouraging solid waste volume reduction at the individual household level consist of various incentive programs. The most common technique is a system of volume-based user fees. With this method, homeowners are given free bags into which they can put anything that is designated by the city as a recyclable material. Distinctly different bags are sold at the local grocery store or city hall. These bags usually sell for somewhere between $2 and $5 a package and must be used for all materials destined for disposal. Homeowners who want to reduce their **garbage** disposal costs would try to reduce the number of purchased bags they need to use. They can do this through a concerted effort of solid waste volume reduction. Careful decisions at the supermarket to reduce packaging and single-use items, reuse of packaging and containers, careful source separation of recyclables, and composting of all organically based materials can reduce the volume of solid waste considerably and reduce the cost of garbage service. An even greater effort can be launched by dedicated individuals by taking used clothing to consignment shops or donating it to organizations for distribution to needy populations, taking clothes hangers back to cleaning establishments, and requesting that junk mail no longer be sent to their households. Of course, there is also the ongoing debate on using **disposable diapers**, which greatly add to the volume of the waste stream.

Businesses that contribute to the reduction of solid waste are those that have been set up to take back, repair, and reuse products and materials that would otherwise end up in the waste stream. Small appliance repair shops can recondition appliances and resell them. Businesses that repair wood pallets make a relatively new contribution to the effort to reduce the waste volume. Tires are being chipped for fuel in waste-to-energy incinerators, and are also being pulverized for use in soils for playing fields. All of these volume reduction techniques have a significant effect on how we live and how we do business. The goal of solid waste volume reduction is to view the waste stream as a resource to be "mined," leaving only those items behind that have truly outlived their usefulness. *See also* Solid waste incineration; Yard waste

[*Cynthia Fridgen*]

Resources

Books

Blumberg, L., and R. Gottlieb. *War on Waste: Can America Win Its Battle With Garbage?* Covelo, CA: Island Press, 1989.

Kharbanda, O. P., and E. A. Stallworthy. *Waste Management: Toward a Sustainable Society.* Westport, CT: Auburn House/Greenwood, 1990.

Noyes, R., ed. *Pollution Prevention Technology Handbook.* Park Ridge, NJ: Noyes Press, 1993.

Periodicals

Porter, J. W., and J. Z. Cannon. "Waste Minimization: Challenge for American Industry." *Business Horizons* 35 (March-April 1992): 46–9.

Solidification of hazardous materials

Solidification refers to a process in which waste materials are bound in a solid mass, often a monolithic block. The waste may or may not react chemically with the agents used to create the solid. Solidification is generally discussed in conjunction with stabilization as a means of reducing the mobility of a pollutant. Actually, stabilization is a broad term which includes solidification, as well as other chemical processes that result in the transformation of a **toxic substance** to a less or non-toxic form.

Experts often speak of the technologies collectively as solidification and stabilization (S/S) methods. Chemical fixation, where chemical bonding transforms the toxicant to non-toxic form, and encapsulation, in which toxic materials are coated with an additive are processes referred to in discussions of S/S methods. There are currently about 40 different

vendors of S/S services in the United States, and though the details of some processes are privileged, many fundamental aspects are widely known and practiced by the companies.

S/S technologies try to decrease the solubility, the exposed surface area, and/or the toxicity of a **hazardous material**. While the methods also make wastes easier to handle, there are some disadvantages. Certain wastes are not good candidates for S/S. For example, a number of inorganic and organic substances interfere with the way that S/S additives will perform, resulting in weaker, less durable, more **permeable** solids or blocks. Another disadvantage is that S/S often double the volume and weight of a waste material, which may greatly affect **transportation** and final disposal costs (not considering potential costs associated with untreated materials contaminating the **environment**). S/S additives, such as encapsulators, are available that will not increase the weight and volume of the wastes so dramatically, but these additives tend to be more expensive and difficult to use.

Methods for S/S are characterized by binders, reaction types, and processing schemes. Binders may be inorganic or organic substances. Examples of inorganic binders which are often used in various combinations include cements, lime, pozzolans, which react with lime and moisture to form a cement such as **fly ash**, and silicates. Among the organic types generally used are epoxies, polyesters, asphalt, and polyolefins (e.g. polyethylene). Organic binders have also been mixed with inorganic types, e.g., polyurethane and cement. The performance of a binder system for a given waste is evaluated on a case-by-case basis; however, much has been learned in recent years about the compatibility and performance of binders with certain wastes, which allows for some intelligent initial decisions related to binder selection and processing requirements.

Among the types of reactions used to characterize S/S are **sorption**, pozzolan, pozzolan-portland cement, and thermoplastic microencapsulation. Sorption refers to the addition of a solid to sorb free liquid in a waste. Activated **carbon**, gypsum, and clays have been used in this capacity. Pozzolan reactions typically involve adding fly ash, lime, and perhaps water to a waste. The mixture of fly ash, lime, and water form a low-strength cement that physically traps contaminants. This system is very alkaline and therefore may not be compatible with certain wastes. For example, a waste containing high amounts of ammonium ions would pose a problem because, under highly alkaline conditions, the toxic gas ammonia would be released. Also, sodium borate, carbohydrates, oil, and grease are known to interfere with the process.

The pozzolan-portland cement process consists of adding a pozzolan, often fly ash, and a portland cement to a waste. It may be necessary to add water if enough is not present in the waste. The resulting product is a high-strength matrix that primarily entraps contaminants. The performance of the system can be enhanced through the use of silicates to prevent interference by metals, clays to absorb excessive liquids and certain ions, surfactants to incorporate organic solvents, and a variety of sorbents that will hold on to toxicants as the solid matrix forms. Care is needed in selecting a sorbent because, for example, an acidic sorbent might dissolve a metal hydroxide, thereby increasing the mobility of the metal, or result in the release of toxic gases such as **hydrogen** sulfide or hydrogen cyanide. Borates, oil, and grease can also interfere with this process.

Thermoplastic encapsulation is accomplished by blending a waste with materials such as melted asphalt, polyethylene, or wax. The technique is more difficult and costly than the other methods introduced above because specialized equipment and higher temperatures are required. At these higher temperatures it is possible that certain hazardous materials will violently react. Additionally, it is known that high salt levels, certain organic solvents, and grease will interfere with the process.

There are basically four categories of processing schemes for S/S. For in-drum processing, S/S additives and waste are mixed and allowed to solidify in a drum. The drum and its contents are then disposed. In-plant processing is a second category which refers simply to performing S/S procedures at an established facility. The facility might have been designed by a company for their own wastes or as a S/S plant which serves a number of industries.

A third category is mobile-plant processing, in which S/S operations are moved from site to site. The fourth category is *in-situ* processing which involves adding S/S additives directly to a **lagoon** or **contaminated soil**.

As may be inferred from the above discussion, the goals of S/S operations are to remove free liquids from a waste, generate a solid matrix that will reliably contain hazardous materials, and/or create a waste that is no longer hazardous. The first goal is important because current regulations in the **Resource Conservation and Recovery Act** (RCRA) stipulate that free liquids are not to be disposed of in a **landfill**. The third goal is obviously important because disposing of a **hazardous waste**, called delisting, is much more costly and time-consuming than disposing of a regular waste.

Wastes are deemed to be hazardous on the basis of four characteristic tests and a series of listings. The tests are related to the ignitability, reactivity, corrosiveness, and extraction procedure (EP) toxicity of the waste. It is possible that S/S procedures delist a waste in any of the four test characteristics. The processes may also chemically transform a substance listed as hazardous waste by the **Environmental Protection Agency** (EPA) into a non-hazardous chemical.

However, S/S techniques generally delist a waste through effecting a change in the results of the EP toxicity test, related to the goal of proper containment. If a waste can be contained well, it may pass the extraction test.

The extraction procedure has changed somewhat in recent years. Originally, the **solid waste** to be tested was stirred in a weakly acidic solution overnight, and then the supernatant was tested for certain inorganic and organic agents. Recently, the test has been replaced by what is known as the Toxicity Characteristic Leaching Procedure (TCLP). Shortcomings of the original extraction procedure were recognized some years ago, as Congress directed the EPA to develop a more reliable, second generation test through the Hazardous and Solid Waste Amendments of 1984. Work on the TCLP test actually began in 1981. In the TCLP, a solid waste is again suspended in an acidic solution, but the method of contacting the liquid with the solid has been changed. The suspension is now placed in a container that revolves about a horizontal axis, tumbling solids amidst the extracting solution. The extraction is allowed to proceed for 18 hours, after which time the solution is filtered and tested for a variety of organic and inorganic substances. The number of compounds tested under the TCLP is greater than the number measured for the original EP test. Failing the TCLP test dictates that a waste is hazardous and must be managed as such. *See also* Hazardous Materials Transportation Act (1975); Storage and transport of hazardous materials

[*Gregory D. Boardman*]

Resources

Books

Freeman, H. M. *Standard Handbook of Hazardous Waste Treatment and Disposal.* New York: McGraw-Hill, 1989.

Martin, E. J., and J. H. Johnson. *Hazardous Waste Management Engineering.* New York: Van Nostrand Reinhold, 1987.

Wentz, C. A. *Hazardous Waste Management.* New York: McGraw-Hill, 1989.

Other

U.S. Environmental Protection Agency. "Background Document: Toxicity Characteristic Leaching Procedure." Report No. PB87-154886, Washington, DC: 1986.

U.S. Environmental Protection Agency. "Guide to the Disposal of Chemically Stabilized and Solidified Waste." Report No. SW-872, Washington, DC: 1980.

Sonic boom

When an object moves through a fluid, it displaces that fluid in the form of a shock wave. The path left by a speedboat in water is an example of a shock wave. A sonic boom is a special kind of shock wave produced when an object travels though air at a speed greater than the speed of sound (1,100 ft/sec [335 m/sec] at sea level). Supersonic aircraft, such as the *Concorde*, produce a sonic boom when they fly faster than the speed of sound. A number of adverse environmental effects have been attributed to sonic booms from supersonic airplanes. These include the breaking of windows and the frightening of animals and people.

Sorption

Term generally referring to either **adsorption** or **absorption**. Adsorption is the process by which one material attaches itself to the surface of a second material. Some air pollutants settle out and coat the outer surface of a building, for example. Absorption is the process by which one material soaks into a second material. Liquid air pollutants may actually soak into the surface of a building. The term sorption is most commonly used when the exact process involved, adsorption or absorption, is unknown. Various sorption methods are employed to reduce pollutants from contaminated air and water. *See also* Pollution control

Source reduction

see **Solid waste**

Source separation

Source separation is the segregation of different types of **solid waste** at the location where they are generated (a household or business). The number and types of categories into which wastes are divided usually depends on the collection system used and the final destination of the wastes. The most common reason for separating wastes at the source is for **recycling**. **Recyclables** that are segregated from other trash are usually cleaner and easier to process. Yard wastes are often separated so they may be composted or used as **mulch**. Some experimental municipal recycling projects also require homeowners to separate household compostibles such as food scraps, coffee grounds, bones, and **disposable diapers**. Some studies suggest that as much as 30% of **household waste** may be compostible; another 40% may be recyclable.

Separate collection of household trash, recyclables, and **yard waste** is gaining popularity in the United States. In some communities source separation is mandated, while in others it is voluntary. Many cities provide residents with recycling bins to be filled with recyclables and placed next to **garbage** cans on collection day. Source-separated yard waste is usually placed in plastic bags or bundled if it is bulky, like tree trimmings. In areas where curbside collection of recyclables and yard waste is not available, residents often

take these source-separated wastes to drop-off centers, or sell recyclables to buy-back facilities. For source-separated recycling programs to be successful, citizen participation is essential. Incentives to increase participation, such as reduced trash collection charges for recyclers, are sometimes implemented.

Household recyclables that are source separated from trash can either be commingled (all recyclables mixed together in one container) or segregated into individual containers for each material (i.e., glass, newspaper, **aluminum**). Commingled recyclables are eventually separated manually, mechanically, or by some combination of both at transfer stations or materials recovery facilities. In some cases, commingled recyclables are manually separated at the curbside by the collection crew. Recyclables that residents have separated into individual containers are usually collected in trucks with compartments for each material. The collected materials are then processed further at materials-recovery facilities or other types of recycling plants.

Many businesses also separate their solid wastes. This can be as simple as placing recycling bins next to soda-vending machines in employee cafeterias or more complex separation systems on assembly lines. One of the most prevalent wastes from the commercial sector is corrugated cardboard (13% of **municipal solid waste** generated). Once it becomes contaminated by other wastes, it may not be suitable for recycling. Some businesses find it easier and more economical to separate and bale corrugated cardboard for recycling because this can reduce their waste-disposal costs.

Source-separation programs can reduce the undesirable effects of landfills or incinerators. For instance, batteries and household **chemicals** can increase the toxicity of **landfill** leachate, air emissions from incinerators, and incinerator ash. In addition, some potentially noncombustible wastes, such as glass, can reduce the efficiency of incinerators. Reducing the volume of residual ash is another incentive for diverting wastes from **incineration**.

Recyclables and special wastes can be retrieved from the **waste stream** without source separation programs. Many communities find it more convenient or economical to separate wastes after collection. In these programs, recyclables and special wastes are manually or mechanically separated at transfer stations or materials-recovery facilities. Separating recyclables in this way may require more labor and higher energy costs, but it's more convenient for residents since it requires no extra effort beyond regular trash disposal procedures.

Source separation may be only one part of an overall community recycling program. These, in turn, are components of more comprehensive waste-management strategies. To reduce the environmental impact of waste disposal, the **Environmental Protection Agency** (EPA) encourages communities to develop strategies to decrease landfill use and lower the risks and inefficiencies of incineration. **Waste reduction** and recycling are considered to be the most environmentally beneficial methods to manage waste.

[*Teresa C. Donkin*]

RESOURCES
OTHER

U.S. Congress, Office of Technology Assessment. *Facing America's Trash: What Next for Municipal Solid Waste.* Washington, DC: U.S. Government Printing Office, October 1989.

U.S. Environmental Protection Agency, Solid Waste and Emergency Response. *Characterization of Municipal Solid Waste in the United States: 1990 Update.* Washington, DC: U.S. Government Printing Office, June 1990.

U.S. Environmental Protection Agency, Solid Waste and Emergency Response. *Decision-Makers Guide to Solid Waste Management.* Washington, DC: U.S. Government Printing Office, November 1989.

South

The **United Nations Earth Summit**, held in Rio de Janeiro in June of 1992, illuminated some major differences among the nations of the world, differences that are summarized in the terms North and South. The latter term refers to nations of the Southern Hemisphere—nations that have significantly different environmental concerns from those of their northern neighbors. These concerns arise primarily from rapid **population growth**; low levels of technological and industrial development; and generally difficult living conditions.

The distinction between North and South did not originate with the Rio conference. It was popularized by a report delivered in 1980 to the Secretary-General of the United Nations by a commission headed by Willy Brandt (1913-1992), the former Chancellor of West Germany. The report, which was entitled *To Ensure Survival— Common Interests of the Industrial and Developing Countries*, is sometimes referred to as the Brandt Report. It recommended a marked rise in development assistance from the developed countries to the countries of the South. The rich countries were to increase their official development assistance (ODA) to 0.7% of their gross national product (GNP) by 1985, and to 1% by 2000. The Brandt Report was followed by a second study in 1983, *Common Crisis North–South: Cooperation for World Recovery*. This report predicted conflict and catastrophe if the imbalance between North and South were not corrected.

An important conclusion that was drawn from the Rio conference is that the marked differences between North and South must somehow be reduced if global problems are ever to be solved. Twenty years after the Brandt Report, however, the results are discouraging. Instead of an increase

in ODA, there has been a marked decrease, in terms of both absolute dollar amounts and GNP percentages. As of 2002, financing for the development of the South is in serious crisis. ODA contributions by all OECD countries fell from a high of $59.6 billion (in US dollars) in 1994 to an estimated $56.0 billion in 1999. ODAs share of the North's GNP fell accordingly from 0.30–0.24%. Policy recommendations that were made in 2001 included a proposal to finance the costs of development by an international tax on energy consumption and **carbon dioxide** emissions.

[*Rebecca J. Frey Ph.D.*]

Resources

Books

Kaul, Inge, Isabelle Grunberg, and Marc A. Stern. *Global Public Goods: International Cooperation in the Twenty-First Century.* New York: Oxford University Press, 1999.

Reid, David. *Sustainable Development: An Introductory Guide.* London, UK: Earthscan, 1995.

Other

Martens, Jens. *Overcoming the Crisis of ODA—The Case for a Global Development Partnership Agreement.* Presentation to the Civil Society Hearings at the United Nations, November 7, 2000.

Organizations

Dante B. Fascell North–South Center, University of Miami., 1500 Monza Avenue, Coral Gables, FL USA 33146 (305) 284-6868, Fax: (305) 284-6370, Email: nscenter@miami.edu, www.miami.edu/nsc

Spaceship Earth

Spaceship Earth is a metaphor which suggests that the earth is a small, vulnerable craft in space.

Adlai Stevenson used the metaphor in his presidential campaign speeches during the 1950s, but it is not clear who originated it. Perhaps R. Buckminster Fuller was most responsible for popularizing it; he wrote a book entitled *Operating Manual for Spaceship Earth* and was described by one biographer as the ship's "pilot." Fuller was impressed by how negligible the craft was in the infinity of the universe, how fast it was flying, and how well it had been "designed" to support life.

Fuller took the metaphor further, noting that what interested him about the earth was "that it is a mechanical vehicle, just as is an automobile." He noted that people are quick to service their automobiles and keep them in running condition but that "we have not been seeing our Spaceship Earth as an integrally-designed machine which to be persistently successful must be comprehended and serviced in total." He did observe one difference between the spaceship and a car: there is no owner's manual for the earth. The lack of operating instructions was significant to Fuller because it has forced humans to use their intellect; but he also maintained that "designed into this Spaceship Earth's total wealth was a big safety factor" which allowed the support system to survive human ignorance until that intellect was sufficiently developed. It was Fuller's lifelong quest to persuade humans to use their intellect and become good pilots and mechanics for Spaceship Earth. He was an optimist, and believed that humans are all astronauts—"always have been, and so long as we exist, always will be"—and as such can learn the mechanics of the system well enough to operate the vehicle satisfactorily.

Political scientist Barbara Ward borrowed the phrase from Fuller to claim that "planet earth, on its journey through infinity, has acquired the intimacy, the fellowship, and the vulnerability of a spaceship." She claimed this image to be "the most rational way of considering the whole human race today." Humans must begin to see humanity "as the ship's crew of a single spaceship on which all of us, with a remarkable combination of security and vulnerability, are making our pilgrimage through infinity." Ward used the metaphor to argue for the reality of global community: "This is how we have to think of ourselves. We are a ship's company on a small ship. Rational behavior is the condition of survival." The rational behavior she advocated was building the institutions, the laws, the habits, and the traditions needed to get along together in the world.

Nigel Calder appreciated the value of the spaceship metaphor and described it in more specific terms: "Whether its watchkeepers were **microbes** or dinosaurs, the Earth system of rocks, air, water and life worked like the life-support system of a manned spacecraft." He went on to suggest that "the gas and water tanks of Spaceship Earth are the air and the oceans" and that "the sun is the spaceship's main power supply." Calder used the metaphor as an introduction to a detailed consideration of what he describes as a new "Earth-system science," moving from there to a depiction of the globe as a total system.

Kenneth Boulding, an economist, has made essentially the same point, describing an inevitable transition from a "cowboy economy" where support systems are open with no linkage between inputs and outputs, to a "spaceman" economy where "the earth has become a single spaceship, without unlimited reservoirs of anything, either for extraction or for **pollution**, and in which, therefore, man must find his place in a cyclical ecological system which is capable of continuous reproduction of material form even though it cannot escape having inputs of energy."

The purpose of the spaceship metaphor is to persuade people that the earth has limits and that humans must respect those limits. It provides a modern, new-age image of a small, comprehensible system, which many people can understand. In that sense, the metaphor helps people understand their

relationship to the **environment** by depicting a system, as Ward notes, that is small enough to be vulnerable and needs to be cared for if it is to sustain life.

But the image can also be delusive, even specious, with negative implications not generally recognized. A spaceship is an artifact, a structure of human creation. Some have argued that depicting the earth as such is seductive, but borders on the arrogant by implying, as Fuller seems to, that humans can completely "control" the operations of the earth. Furthermore, spaceships are commonly thought of as small and crowded, with a life-support system devoted exclusively to human inhabitants. Since humans are using more and more of the planet's resources for their own benefit, and the pressure of an increasing human population is extinguishing other life-forms, some environmentalists argue that the spaceship metaphor should be used with caution. *See also* Environmental ethics; Green politics

[*Gerald L. Young Ph.D.*]

Resources

Books

Boulding, K. "The Economics of the Coming Spaceship Earth" and "Spaceship Earth Revisited." In *Valuing the Earth: Economics, Ecology and Ethics.* Cambridge, MA: The MIT Press, 1993.

Calder, N. *Spaceship Earth.* New York: Viking Penguin, 1991.

Fuller, R. B. *Operating Manual for Spaceship Earth.* New York: Simon and Schuster, 1969.

Ward, B. *Spaceship Earth.* New York: Columbia University Press, 1966.

Spanish flu

see **Flu pandemic**

Spawning aggregations

Spawning aggregation is defined as a group of fish of the same **species** that are gathered together for the purpose of spawning—releasing sperm or eggs for the purpose of reproduction. The fish population that is together at this time is significantly greater than during periods the fish are not reproducing. For fish whose **habitat** is stable, drawing aggregations revolves around a relatively small area. In transient populations the individuals might travel for days or weeks in order to reach the aggregation site.

The U.S. **Fish and Wildlife Service** (FWS) oversees the recovery of species listed under the **Endangered Species Act**. For fish, that goal is tended through the National Fish Hatchery System—coordinating efforts between hatcheries and fisheries management. As a result, national fish restoration programs have returned abundance to the populations of various species, including the **Great Lakes** lake trout, the Atlantic coast striped bass, Atlantic **salmon**, and Pacific salmon. Much of the success has come through the agency's captive propagation programs. And, according to the FWS, "The success of captive propagation for recovery depends upon a number things, including careful genetics planning and management, concurrent habitat restoration, thorough evaluation studies—and funding. Propagation of imperiled fish species is often more than twice as costly as rearing non-native game fish due to genetic analyses, special diet requirements, and rearing conditions that enhance survival in the wild, along with rigorous monitoring and evaluation studies." A critical factor in the reproduction of **endangered species** is the understanding of spawning aggregations.

In various coastal systems in the United States and throughout the world, fisheries management relies on the knowledge of spawning aggregation locations. It is crucial to understanding how to protect the populations, as well as key to maintaining the ecological balance of marine life.

Reef fish are more vulnerable to disruption of spawning aggregations because many species only aggregate for brief time periods. If reef fish spawning aggregations are fished during the activity, then their populations are depleted due to unsuccessful reproduction. There are many reef fish species that reproduce in unprotected federal waters of the U.S. Caribbean, the Gulf of Mexico, and the South Atlantic. These species include the black grouper, cubera snapper, gag grouper, gray snapper, jewfish, Nassau grouper, red hind, scamp, and yellowfin grouper.

Two examples of decline in particular fish populations, according to Mark W. Sprague of the Department of Physics at East Carolina University in Greenville, North Carolina, in (2000) are the weakfish (*Cynoscion regalis*), and the red drum (*Sciaenops ocellatus*). It was his theory that using fish sounds could assist researchers in identifying their spawning activity. In the abstract presented for his presentation at the conference Sprague presented this preview of his presentation. "Weakfish and red drum, both members of family Sciaenidae, use their swim bladders to produce species-specific sounds associated with spawning activity. Large spawning aggregations of these fish can produce sound levels as high as 145 decibels (re: 1[mu]Pa), and these sounds will be presented for each species. Water depth, bottom type and contour, sound-speed gradient, and water current all affect the propagation of the fish sounds through the water. Measurements of these factors and the sound level of the fish calls are used to obtain an approximate range to the spawning aggregation." Sprague's work was supported by the North Carolina Division of Marine Fisheries and the U.S. Fish & Wildlife Service.

Some debate exists among those who fish commercially regarding whether the federal and state regulations for spawning aggregations are overly conservative. An example

of one such debate appeared in *Fishermen's Voice Monthly Newspaper* a periodical of the fishing industry in Maine. The article recounted an aspect of the spawning issues in August 2000. The new plan unveiled there, "calls for a series of three rolling closures of one month duration. Flexible starting dates and options for extending the closures are designed to protect the fish only when they are spawning. Less astringent tolerance levels in the amended plan allow fishing within the closed areas provided the fish landed are less than 20 percent Stage V or Stage VI spawners." The purpose of regulation over spawning areas is intended to protect the spawning populations and therefore protect future populations so that **overfishing** does not become a problem.

The **conservation** of fish and the protection of marine wildlife species, such as **sharks** and **whales**, depends on the management of spawning aggregations. The debate over how that is best done will continue well into the twenty-first century.

[*Jane E. Spear*]

Resources

Periodicals

Zeller, Dirk. "Spawning Aggregations: Patterns of Movement of the Coral Trout *Plectropomus leopardus* (Serranidae) as Determined by Ultrasonic Telemetry." *Marine Ecology* (February 12, 1998).

Other

Darwin Initiative. *Conversation of Whale Sharks & Fish Spawning Aggregations in Belize.* [cited March 25, 2002]. <http://www.ayork.ac.uk>.

Fish Base. *Global Information System on Fishes.* [cited June 2002]. <http://www.fishbase.org>.

"Herring." *Fishermen's Voice* August 2000 [cited July 2002]. <http://www.fishermensvoice.com>.

Johannes. R. E. "MPA Perspective: Indo-Pacific Should Protect More Reef Fish Spawning Aggregation Sites." *MPA News.* March 2000 [cited July 2002]. <http://www.depts.washington.edu/mpanews>.

U.S. Fish and Wildlife Service. "National Fish Hatchery System." [June 2002]. <http://www.fisheries.fws.gov>.

Organizations

U. S. Fish and Wildlife Service, Washington, D.C. , <http://www.fws.gov>

Special use permit

An authorization from the appropriate governmental agency to another agency or to a private operator to use publicly owned land for certain purposes. The Secretary of Agriculture and the **Forest Service** issue permits for special uses not explicitly covered by timber, mining, or grazing laws or regulations on lands in the **national forest** system. Uses must be carefully controlled and be compatible with forest policy as well as with other uses. Commercial users pay a fair market fee to mine stone and gravel, operate ski areas, conduct sporting and planned **recreation** events, and construct rights-of-way, pipelines, power lines, and microwave stations. Other special uses include archeological explorations and leases for public buildings and summer homes.

Species

A species is a group of closely related, physically similar beings that can interbreed freely. In practice, the dividing lines between species or between species and subspecies are sometimes unclear.

In sorting out species, biologists search for easily recognized diagnostic characteristics. For example, it is easier to recognize a giraffe by its long neck than by its blood proteins. However, species differ from one another not just by conspicuous features. Members of a species share a common **gene pool** and have a common geographic range, **habitat**, and similar characteristics ranging from the biochemical and morphological to the behavioral.

In taxonomy, the hierarchy of biological classification, species is the category just below genus. The major groups in the classification hierarchy are kingdom, phylum (for animals) or division (plants), class, order, family, genus, and species. Taxonomically, a species is designated in italics by the genus name followed by its specific name, as in *Felis domesticus*, or domestic cat. New species are named and existing species names are altered according to the International Rules of Botanical Nomenclature, the International Rules of Zoological Nomenclature, and the International Bacteriological Code of Nomenclature.

Isolating mechanisms prevent closely related species living in the same geographic area from mating with each other. These are called sympatric species, as opposed to allopatric species, which are closely related species living in different geographical areas. Sympatric species are very common. Bullfrogs, green **frogs**, wood frogs, and pickerel frogs, all members of the genus *Rana*, may be found in or near the same pond. They are prevented from mating by reproductive isolating mechanisms.

Allopatric speciation allows new species to arise from one or more preexisting species, which is a long, slow process. Most allopatric species evolve when a small population becomes physically isolated from the main part of the species, and gene flow between them stops. Genetic differences spread and accumulate in the small, isolated group. This process is most likely to occur at the border of a species, where environmental conditions exceed the range of tolerance of members of the species and there is a barrier to the species' spread. The barrier could be a mountain range, a **desert**, forest, large body of water and so on.

Organisms that are part of a region's natural **flora** and **fauna** are called native or indigenous species. Those

associated with the mature or climax form of community are called climax species. Fragments of a climax species that remain after a major disturbance, such as a forest fire, are referred to as relic species.

Nonnative organisms are called introduced or **exotic species**. They can wreak havoc when brought deliberately or by accident into a new area where they have no natural enemies or where they compete aggressively with other organisms in the **environment**. A classic example is the introduction of the **zebra mussel** (*Dreissena polymorpha*) into the **Great Lakes**, where it reproduces wildly and has clogged **water treatment** facilities. In south Florida, the Australian pine (*Casuarina equisetifolia*) has overtaken many coastal areas because it is extremely tolerant of salt spray and otherwise poor beach soils. Its thick foliage and quick rate of multiplication allows the Australian pine to compete with native species for light, food, and space.

All of the plants and animals found in a particular area, whether native or introduced, are called resident species. Where new species adapt well so that they do not need special help to perpetuate, they are said to be acclimatized or naturalized. Invader species are not original to a region but arrive after an area is disturbed, such as by fire or **overgrazing**. Native species that are the first to recolonize a disturbed area are called pioneer species. Several pioneer species quickly reappeared on and around Mount Saint Helens after the volcano's 1980 eruption.

Species that are especially adapted to a highly variable, unpredictable, or transient environment are called opportunistic species. Within a particular area, dominant species are those that because of their number, coverage, or size, strongly influence the conditions of other species in the area. For example, certain trees in **old-growth forests** in the Pacific Northwest are vital to the success of other species in the area. Destruction of the forests by **logging** also causes the decline of other species, such as the spotted owl (*Strix occidentalis*). In other areas, plants that are sought after by grazing animals are called ice cream species.

So-called indicator species, or indicator organisms, are watched for warning signs that their ecosystems are ill. The decline of migratory songbirds around the world is an indication that nesting and breeding grounds of the songbirds, particularly rain forests, are in danger. Other indicator species are amphibians. Some scientists believe that the decline of many frog and toad populations is a result of environmental **pollution**. Since amphibians have highly **permeable** skins, they are particularly vulnerable to **environmental degradation** such as **pesticide** pollution. The depletion of the **ozone** layer, which allows more **ultraviolet radiation** to penetrate the **atmosphere**, has been blamed for the decline in other frog populations. Increased ultraviolet radiation is believed to be responsible for the destruction of frog eggs, which are deposited in shallow water. When the striped bass population in **Chesapeake Bay** declined in 1975, it indicated pollution problems in the water. An indicator species also may be used to show the stage of development, or regeneration, of an **ecosystem**. For example, **land use** managers may watch plant species to assess the level of grazing use in rangeland.

If the survival of other species depends on the survival of a single species, that single species is called a **keystone species**. Where a species plays a role in more than one ecosystem, it is said to be a mobile link species.

A species population has a dimension in space, called a range, and a dimension in time. The population extends backwards in time, merging with other species populations much like the branches of a tree. The population has the potential to extend forward in time, but various factors may prevent the perpetuation of the species. When the continued existence of a species is in question, it is regarded as an **endangered species**. This causes great concern to conservationists, especially when the species is the only representative of its genus or family, as is the case with the **giant panda** (*Ailuropoda melanoleuca*). A species found in a very restricted geographic range is endemic or narrowly endemic, and is at special risk of becoming extinct. **Extinction** results from an imbalance between a species and the environment in which it is living.

Many species exist today in name only. On a 1991 United States **Fish and Wildlife Service** report on species, the following were removed from the endangered species list because they all became extinct since the 1980 report was issued: the Amistad gambusia fish (*Gambusia amistadensis*), the Tecopa pupfish (*Cyprinodon nevadensis calidae*), Sampson's pearly mussel (*Epioblasma sampsonii*), the blue pike (*Stizostedion vitreum*), and the dusky seaside sparrow (*Ammodramus maritimus nigrescens*).

Species richness varies with the type of region, and it is especially diverse in areas of warm or wet places, such as tropical forests. Many biologists are concerned that the unprecedented destruction of tropical forests is forcing to extinction species that have not even been discovered. When a species is lost, it is lost forever, and with it the tremendous potential, especially in medicine, of that species' contributions to the world.

[*Linda Rehkopf*]

Resources

Books

Villee, C., et. al. *General Zoology*. New York: Saunders College Publishing, 1984.

Species diversity

see **Biodiversity**

Speciesism

This term, popularized by author **Peter Singer** in his book *Animal Liberation* (1975), refers to a human attitude of superiority over other creatures and a tendency among humans to place the interests of their own **species** above all others. As Singer puts it, speciesism "is as indefensible as the most blatant racism. There is no basis for elevating membership of one particular species into a morally crucial characteristic. From an ethical point of view, we all stand on an equal footing—whether we stand on two feet, or four, or none at all. That is the crux of the philosophy of the animal liberation movement...Just as we have progressed beyond the blatantly racist ethic of the era of slavery and colonialism, so we must now progress beyond the speciesist ethic of the era of factory farming, seal hunting...and the destruction of wilderness." Opposition to speciesism has become one of the foundations of the modern **animal rights** movement. *See also* Environmental ethics

Spectrometry, mass

see **Mass spectrometry**

Speyeria zerene hippolyta

see **Oregon silverspot butterfly**

Spoil

Normally used to describe **overburden** removed during **strip mining**. These operations leave long rows of intermixed rock and **soil**, commonly described as spoil banks. Often composed of loose shale and placed downhill from the final cut, they block outflow, forming lakes, and in level terrain store large amounts of **groundwater**. Prior to the 1977 enactment of the **Surface Mining Control and Reclamation Act**, these man-made landforms were often abandoned, creating a distinctive landscape. Efforts are now made to reclaim these aptly-named "orphan" lands. However, unless properly done, **reclamation** may exacerbate **erosion**; in such cases they are best left undisturbed. Trees often lead the **succession**, perhaps hastening the day when effective reclamation might be feasible.

Spotted owl

see **Northern spotted owl**

SSOs

see **Sanitary sewer overflows**

Stability

The term "stability" refers to the tendency of an individual organism, a community, a population, or an **ecosystem** to maintain a more or less constant structure over relatively long periods of time. Stability does not suggest that changes do not occur, but that the net result of those changes is nearly zero. A healthy human body is an example of a stable system. Changes are constantly taking place in a body. Cells die and are replaced by new ones. Chemical compounds are manufactured in one part of the body and degraded in another part. But, in spite of these changes, the body tends to look very much the same from day to day, week to week, and month to month.

The same is true with groups of individuals. A **prairie** grassland may look the same from year to year, even though the individual plants that make up the community change. Many ecosystems exhibit mosaic stability. For instance, a forest with openings caused by windthrow or fire at different times regrows trees of the same **species**, so that the character of the forest remains unchanged.

Stability is threatened by various kinds of stress. For example, lack of food can threaten the stability of an individual person, and fire can endanger the stability of a grassland community. Both organisms and communities have a rather remarkable **resilience** to such threats, however, and are often able to return to their original structure. Given good food once again, a person can recover her or his health and structure, just as a grassland tends to regrow after a fire.

The biotic and abiotic elements of an ecosystem cause stress on each other which can threaten the stability of each. An unusually long dry spell, for example, can cause plants to die and endanger the biotic community's stability. In turn, animals may multiply in number and physically degrade the **soil** and water in their **habitat**.

Because of the magnitude of the stress they place on other communities, humans are often the most serious threat to the stability of an ecosystem. An **old-growth forest** that has been clear-cut may, in theory, restore itself eventually to its original structure. But the time frame needed for such a restoration involves hundreds or thousands of years. For all practical purposes, therefore, the community of organisms in the region has ceased to exist as such.

One of the important mechanisms by which ecosystems maintain stability is negative feedback. Suppose, for example, that favorable weather conditions cause an unusually large growth of plants in a valley. Animals that live on those plants, then, will also increase in number and reproduce

in greater quantity than in less abundant years. The increased population of animals will place a severe stress on the plant community if and when weather conditions return to normal. With decreased food supplies per individual organism, more animals will die off and the population will return to its previous size.

The factors that contribute to the stability of an ecosystem are not well-understood. At one time, most scholars accepted the hypothesis that diversity was an important factor in maintaining stability. From a common sense perspective, it is reasonable to assume that the more organisms and more kinds of organisms in a system, the greater the distribution of stress and the more stable the system as a whole will be.

Some early experiments appeared to confirm this view, but those experiments were often conducted with artificial or only simplified systems. Ecologists are increasingly beginning to question the relationship between **biodiversity** and stability, at least partly because of examples from the real world. Some highly complex systems, such as **tropical rain forests**, have been less successful in responding to external stress than have simpler systems, such as **tundra**.

The term "stability" has a more specialized meaning in atmospheric studies. There, it also refers to a condition in which vertical layers of air at different temperatures are in equilibrium with each other. *See also* Population biology

[*David E. Newton*]

Resources
Books

Clapham Jr., W. B. *Natural Ecosystems.* New York: The Macmillan Company, 1973.

Suthers, R. A., and R. A. Gallant. *Biology: The Behavioral View.* Lexington, MA: Xerox College Publishing, 1973.

Stack emissions

Stack emissions are those gases and solids that come out of the **smoke** stack after the **incineration** process. Incinerators can be designed to accept wastes of any physical form, including gases, liquids, solids, sludges, and slurries. Incineration is primarily for the treatment of wastes that contain organic compounds. Wastes with a wide range of chemical and physical characteristics are considered suitable for burning. Most of these wastes are by-products of industrial manufacturing and chemical production processes, or result from the clean-up of contaminated sites.

There is a great deal of controversy about the content of incinerator stack emissions. The **Environmental Protection Agency** (EPA) supports incineration as a **waste management** tool and claims that these emissions are not dangerous. In an official publication, the EPA has stated: "Incinerator **emission** gases are composed primarily of two harmless inorganic compounds, **carbon dioxide** and water. The type and quantity of other compounds depends on the composition of the wastes, the completeness of the **combustion** process, and the **air pollution control** equipment with which the incinerator is equipped. These compounds include organic and inorganic compounds contained in the original waste and organic and inorganic compounds created during combustion."

Contrary to the EPA, many environmentalists believe that burning **hazardous waste**, even in "state-of-the-art" incinerators, releases far more **heavy metals**, unburned wastes, dioxins, and new **chemicals** formed during the incineration process (PICs) than is healthy for the **environment** or humans. In a report published in 1990, *Playing with Fire*, **Greenpeace** disagrees strongly about the toxic materials emitted from incinerator stacks.

The report argues that metals are not destroyed during incineration; in fact, they are often released in forms that are far more dangerous than the original wastes. At least 19 metals have been identified in the air emissions of hazardous waste incinerators. An average-sized commercial incinerator burning hazardous waste with an average metals content emits these metals into the air at the rate of 204 lb (92.6 kg) per year and deposits another 670,000 lb (304,180 kg) of metals per year in its residual ashes and liquids, which must be properly disposed of in landfills designed for that purpose.

The consequences to human health are significant. **Cancer**, **birth defects**, reproductive dysfunction, neurological damage, and other health effects are known to occur at very low exposures to many of the metals, organochlorines, and other pollutants released by waste-burning facilities. Increased cancer rates, respiratory ailments, reproductive abnormalities, and other health effects have been noted among people living near some waste-burning facilities, according to scientific studies in other countries and surveys conducted by community groups and local physicians in the United States.

Hazardous waste incinerators in the United States are producing at least 324 million lb (147 million kg) per year of ash residues. These ashes, which are buried in landfills, are contaminated by PICs, many of which are more toxic than the original waste. The ashes also contain increased concentrations of heavy metals, often in more leachable forms than in the original wastes.

Trial burns are used to determine an incinerator's destruction and removal efficiency (DRE). Under current federal regulations, an incinerator of general hazardous waste must, during the trial burn, demonstrate a DRE of 99.99% with just a few chemicals to be tested—perhaps one or two.

Many environmentalists consider these standards unsatisfactory. In the unlikely event that this standard could be met at all times with all wastes burned throughout the lifetime of an incinerator, a hazardous waste incinerator of average size (70 million lb [32 million kg] per year) would still be emitting 7,000 lb (3,178 kg) per year of unburned wastes. With corrections and accidents, emissions may be as high as 700,000 lb (317,800 kg) per year. Environmentalists also argue that DRE addresses only the stack emissions of the few chemicals selected for the trial burn, and that it does not reflect other substances that are released. These include unburned chemicals other than those selected for the trial burn, heavy metals, or newly formed PICs. In addition, the present method for determining DREs does not account for the retention within the combustion system of the chemicals selected for the trial burn and their continued release for hours, or even days, after stack gas sampling has ceased. *See also* Hazardous waste site remediation; Hazardous waste siting; Heavy metals precipitation; NIMBY (Not In My Backyard); Toxic use reduction legislation

[*Liane Clorfene Casten*]

Resources
Books

Costner, P., and J. Thornton. "Playing With Fire: Hazardous Waste Incineration." 2nd ed. Washington, DC: Greenpeace, 1993.

Stakeholder analysis

Stakeholder analysis is an approach to making policy decisions, primarily in business, which is based on identifying and prioritizing the interests different groups have in an institution. A stakeholder analysis examines the stakes which shareholders, suppliers, customers, employees, or communities may have in a particular issue.

A stakeholder analysis is often used to develop generic strategies; these are meant to be broad descriptions of what the corporation stands for and how it intends to mediate between competing stakeholder concerns. Many firms have made public statements concerning these strategies. Hewlett-Packard has said that it is dedicated to the dignity and worth of its individual employees. Aetna Life and Casualty claims that tending to the broader needs of society is essential to fulfilling its economic role. Perhaps the most famous example of a stakeholder strategy is Johnson and Johnson's: "We believe our first responsibility is to the doctors, nurses and patients, to mothers and all others who use our products and services."

People both inside and outside the business world consider the stakeholder analysis an effective method for uniting business strategy with social responsibility. This approach can be used to provide a variety of business goals with an ethical basis, and it has played a role in developing the marketing of environmentally safe products, known as **green marketing**. While many environmentalists consider this kind of analysis a step in the right direction, some have argued that public statements of strategies can be misleading, and that companies can be more concerned with their image and their relationship with the media than they are with social responsibility. *See also* Green advertising and marketing

[*Alfred A. Marcus and Douglas Smith*]

Resources
Books

Marcus, A. A. *Business and Society*. Homewood, IL: Irwin Press, 1992.

Statistics

Statistics is the mathematical science of collecting, organizing, summarizing, and interpreting information in a numerical form. There are two main branches of statistics. Descriptive statistics summarizes particular data about a given situation through the collection, organization, and presentation of those data. Inferential statistics is used to test hypotheses, to make predictions, and to draw conclusions, often about larger groups than the one from which the data have been collected.

Statistics enables the discernment of patterns and trends, and causes and effects in a world that may otherwise seem made up of random events and phenomena. Much of the current debate over issues like global warming, the effects of industrial **pollution**, and **conservation** of **endangered species** rests on statistics. Arguments for and against the human impact on the **environment** can be made or broken based on the quality of statistics supporting the argument, the interpretation of those data, or both. Providing accurate statistics to help researchers and policy-makers come to the most sound environmental conclusions is the domain of statistical **ecology** in particular, and **environmental science** in general.

Basic concepts

Statistical analysis begins with the collection of data—the values and measurements describing an event or phenomenon. Researchers collect two types of data: qualitative and quantitative. Qualitative data refers to information that cannot be ascribed a numerical value but that can be counted.

For example, suppose a city wants to gauge the success of its curbside **recycling** program. Researchers might survey residents to better understand their attitudes toward recycling. The survey might ask whether or not people actually

recycle, or whether they feel the program is adequate. The results of that survey—the number of people in the city who recycle, the number who think the program could be improved, and so on—would provide quantitative data on the city's program. Quantitative data, then, refers to information that can be ascribed numeric values—for example, a study of how many tons each of glass and paper get recycled over a certain period of time. A more detailed breakdown might include data on the recycling habits of individual residents by measuring the amount of recycled materials people place in curbside bins.

Sorting through and weighing the contents of every recycling bin in a town would be not only time consuming but expensive. When researchers are unable to collect data on every member of a population or group under consideration, they collect data from a sample, or subset, of that population. In this case, the population would be everyone who recycles in the city.

To bring the scope of the study to a manageable size, the researchers might study the recycling habits of a particular neighborhood as a sample of all recyclers. A certain neighborhood's habits may not reflect the behavior of an entire city, however. To avoid this type of potential bias, researchers try to take random samples. That is, researchers collect data in such a way that each member of the population stands an equal chance of being selected for the sample. In this case, investigators might collect data on the contents of one bin from every block in the city.

Researchers must also define the particular characteristics of a population they intend to study. A measurement on a population that characterizes one of its features is called a parameter. A statistic is a characteristic of or fact about a sample. Researchers often use statistics from a sample to estimate the values of an entire population's parameters when it is impossible or impractical to collect data on every member of the population. The aim of random sampling is to help make that estimate as accurate as possible.

Descriptive statistics

Once the data have been collected, they must be organized into a readily understandable form. Organizing data according to groups in a table is called a frequency distribution. Graphic representations of data include bar charts, histograms, frequency polygons (line graphs), and pie charts.

Measures of central tendency, or the average, provide an idea of what the "typical" value is for a group of data. There are three measures of central tendency: the mean, the median, and the mode. The mean, which is what most people refer to when they use the term "average," is derived by adding all the values in a data set and then dividing the resulting sum by the number of values. The median is the middle value in a data set when all the numbers are arranged in ascending or descending order. When there is an even number of values, the median is derived by calculating the mean of the two middle values. The mode is the value that most often occurs in a set of data.

Although averages describe a "typical" member of a group or set of data, it can also be helpful to know about the exceptions. Statisticians have therefore devised several measures of variability—the extent to which data fluctuate between different measures. The range of data set is the difference between the highest and lowest values. Deviation is the difference between any one measure and the mean.

Range and deviation provide information on the variability of the individual members of a group, but there are also ways to describe the variability of the group as a whole if, for example, a statistician wants to compare the variability of two sets of data. Variance is derived from squaring the deviations of a set of measures and then calculating the mean of those squares. Standard deviation is the most common statistic used to describe a data sets variability because it can be expressed in the same units as the original data. Standard deviation is derived by calculating the square root of the variance.

Inferential statistics

Inferential statistics is largely concerned with predicting the probability—the likelihood (or not)—of certain outcomes, and establishing relationships or links between different variables. Variables are the changing factors or measurements that can affect the outcome of a study or experiment.

Inferential statistics is particularly important in fields such as ecological and environmental studies. For example, there are **chemicals** contained in **cigarette smoke** that are considered to be carcinogenic. Researchers rely in no small part on the methods of inferential statistics to justify such a conclusion.

The process begins by establishing a statistical link. To use a common example, health experts begin noticing an elevated incidence of lung **cancer** among cigarette smokers. The experts may suspect that the cause of the cancer is a particular chemical (if there are 40 suspected chemical carcinogens in the **smoke**, each chemical must be evaluated separately) in the cigarette smoke. Thus, there is a suspected association, or possible relationship, between a chemical in the smoke and lung cancer.

The next step is to examine the type of correlation that exists, if any. Correlation is the statistical measure of association, that is, the extent or degree to which two or more variables (a potential chemical **carcinogen** in cigarette smoke and lung cancer, in this case) are related. If statistical evidence shows that lung cancer rates consistently rise among a group of smokers who smoke cigarettes containing the suspected chemical compared with a group of nonsmokers

(who are similar in other ways, such as age and general health), then researchers may say that a correlation exists.

Correlation does not prove a cause and effect relationship, however. The reason, in part, is the possible presence of confounders—other variables that might cause or contribute to the observed effect. Therefore, before proposing a cause-effect relationship between a chemical in cigarette smoke and lung cancer, researchers would have to consider whether other contributing factors (confounders)—such as diet, exposure to environmental **toxins**, stress, or genetics — may have contributed to onset of lung cancer in the study population. For example, do some smokers also live in homes containing **asbestos**? Are there high levels of naturally occurring carcinogens such as **radon** in the work or home environment?

Teasing out the many possible confounders in the real world can be extremely difficult, so although statistics based on such observations are useful in establishing correlation, researchers must find a way to limit confounders to better determine whether a cause-effect relationship exists. Much of the available information on environmentally related causes and effects is verified with data from lab experiments; in a lab setting, variables can be better controlled than in the field.

Statistically and scientifically, cause and effect can never be proved or disproved 100%. Researchers test hypotheses, or explanations for observed phenomena, with an approach that may at first appear backwards. They begin by positing a null hypothesis, which states that the effects of the experiment will be opposite of what is expected. For example, researchers testing a chemical (called, for example, chemical x) in cigarette smoke might start with a null hypothesis such as: "exposure to chemical x does not produce cancer in lab mice." If the results of the experiment *disprove* the null hypothesis, then researchers are justified in advancing an alternative hypothesis. To establish that there is an effect, an experiment of this nature would rely on comparing an experimental group (mice exposed to chemical x, in this case) with a control group—an unexposed group used as a standard for comparison.

The next step is to determine whether the results are statistically significant. Researchers establish a test or experiments *P* value, the likelihood that the observed results are due to chance. Frequently, the results of an experiment or test are deemed statistically significant if the *P* value is equal to or less than 0.05. A *P* value of 0.05 means there are five or fewer chances in 100 that the observed results were due to random processes or statistical variability. In other words, researchers are 95% sure they have documented a real cause and effect.

Other important considerations include whether the results can be confirmed and are reliable. Findings are considered confirmed if another person running the same test or experiment can produce the same results. Reliability means that the same results can be reproduced in similar studies.

[*Darrin Gunkel*]

Resources

Books

Cohn, Victor. *News and Numbers.* Ames, IA: Iowa State University Press, 1989.

Graham, Alan. *Statistics.* Lincolnwood, IL: NTC/Contemporary Publishing, 1999.

Jaisingh, Lloyd R. *Statistics for the Utterly Confused.* New York: McGraw-Hill, 2000.

Slavin, Stephen. *Chances Are: The Only Statistics Book You'll Ever Need.* Lanham, MD: Madison Books, 1998.

Other

Elementary Concepts in Statistics. 1984–2002 [cited July 6, 2002]. <http://www.statsoft.com/textbook/stathome.htm>.

Introduction to Statistics. 1997–2002 [cited July 6, 2002]. <http://writing.colostate.edu/references/research/stats/pop2a.cfm>.

Steady-state economy

Steady-state economics is a branch of economic thinking which applies the perspectives of steady-state systems developed in thermodynamic physics to economic analysis. This direction in economics is largely associated with the work of Herman Daly, who has written the classic work in the field, *Steady-State Economics* (1977). While its impact has been relatively minor within the discipline of economics itself, the concept of steady-state economics has gathered a significant audience among life scientists and within the larger environmental movement.

Steady-state economics is closely related to **sustainable development**, and many consider it as but one component of the larger issue of sustainability. The constraints imposed by the laws of physics determine, in Daly's terminology, the "ultimate means" or the "ultimate supply limit" beyond which no measure of human development can make better use of resources and energy. One of the fundamental criticisms of traditional economic thinking made by steady-state economics is founded on this concept of absolute limits: Standard economics implicitly assumes that it is always possible to cope with **population growth** and resource shortages by technical advances and the substitution of one resource for another. This is an assumption which steady-state economics views as not only wrong, but dangerously misleading.

For Daly, the greater danger of this perspective is that it is ultimately a form of hubris, of acting as though we can

surpass the limitations of the physical world and attain the freedom of the divine. This is the classic sin of pride, and Daly sees its corrective as the corresponding virtue of humility. Science, he claims, "sees man as a potentially infallible creator whose hope lies in his marvelous scientific creativity." He contrasts this with the view of steady-state economics, which "conceives of man as a fallen creature whose hope lies in the benevolence of his Creator not in the excellence of his own creations." In Daly's view, it is only when we are humble that we are able to see human life and the entire evolutionary process in which it is embedded as a gift bestowed upon us by God, not something we have made. For Daly, this gift of the **evolution** of life is a minimum definition of the "Ultimate End," whose preservation and further development must be the goal of all our actions. The "Ultimate End" is fostering the continuance of the evolutionary process, and the "ultimate means" is determined by the laws of physics; they both define boundaries only within which is it possible to have a steady-state economy and a sustainable society.

Daly offers three large-scale social institutions for the United States to help make a steady-state economy a reality. The first of these is a socially determined limit on the national population, with licenses issued to each person allocating exactly the number of births required to maintain **zero population growth** (approximately 2.1 births per female). These licenses could be purchased or otherwise transferred between individuals, so that those wanting no children could transfer their licenses to those wishing more than their allotment. The second institution would stabilize the stock of human artifacts and would maintain the resources needed to maintain and replace this stock at levels which do not exceed the physical limits of the **environment**. A set of marketable quotas for each resource would be the primary mechanism to attain this goal. The third institution would be a set of minimum and maximum limits on personal income and a maximum cap on personal wealth. The first two institutions are designed to structure population and economic production within the fundamental thermodynamic limits or "ultimate means." The third is the extension into human society of the moral boundaries set by the goal of preserving and fostering life—in this case to ensure that all people in the steady-state economy have access to society's resources. *See also* Bioregionalism; Carrying capacity; Deep ecology; Family planning; Growth limiting factors; Sustainable agriculture

[*Eugene R. Wahl*]

Resources

Books

Boulding, K. "The Economics of the Coming Spaceship Earth." In *Environmental Quality in a Growing Economy*, edited by H. Jarrett. Baltimore: Johns Hopkins University Press, 1966.

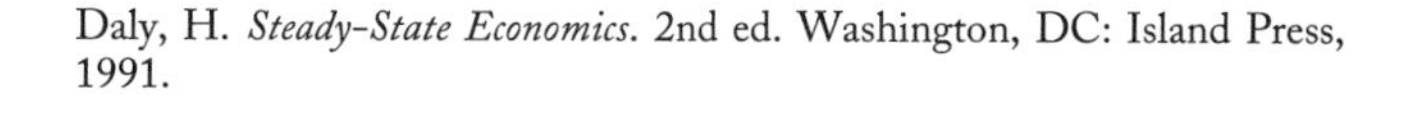

Daly, H. *Steady-State Economics*. 2nd ed. Washington, DC: Island Press, 1991.

Wallace Stegner (1909 – 1993)

American writer

Wallace Stegner was an American novelist, historian, biographer, and teacher. Widely regarded as the dean of western writers, Stegner evoked a vivid sense of the western United States as a place and of the intimate relationship of the people with that place. Among his best known novels are *Big Rock Candy Mountain* (1943), *Angle of Repose* (1971), *The Spectator Bird* (1976), and *Crossing to Safety* (1987). His works of nonfiction include *Beyond the Hundredth Meridian* (1954), *Wolf Willow* (1963), *The Sound of Mountain Water* (1969), and *Where the Bluebird Sings to the Lemonade Springs: Living and Writing in the West* (1992).

Stegner was born on February 18, 1909, on a farm outside Lake Mills, Iowa, the second son of George and Hilda Paulson Stegner. His father, a restless and rootless risk-taker, moved the family to North Dakota, then to Washington state, and then to Saskatchewan, Montana, and Utah. His family was so poor that he was sent for a time to an orphanage. Stegner's early education was spotty at best. He was an avid hunter and outdoorsman who enjoyed the company of Native Americans, cowboys, miners, Mormons, and others who eked out a precarious living in the west. In these hardscrabble early years, Stegner met the people and endured the experiences that were later to reappear in fictional form in his novels and short stories.

At age sixteen, Stegner enrolled at the University of Utah, where he majored in English. In 1930, he began graduate work at the University of Iowa, where he earned a master's degree. Two years later, he returned to the University of Utah, where he received his Ph.D. in 1935. He went on to teach at several universities, including Harvard, Wisconsin, and Stanford, where he founded the creative writing program that he directed until 1971. Among his students and the recipients of Stanford's Wallace Stegner Writing Fellowship were **Wendell Berry** and **Edward Abbey**. An ardent outdoorsman and conservationist, Stegner was active in the **Sierra Club** and fought successfully to save Echo Park from being dammed. In later years he said that his greatest regret was not having worked harder to prevent the damming of Glen Canyon.

Stegner was a pioneer in the **environmental education** of American citizens and their elected representatives. His new medium was the large coffee-table book featuring photographs of stunning seldom-seen places that were under threat from mining, **logging**, or development interests. Interspersed among pictures by photographers such as **Ansel**

Adams were short essays by Stegner and others about the fragility and beauty of these wild places and the threats they faced. Published by the Sierra Club and distributed free to Senators and Congressmen, these books proved to be powerful weapons in the ongoing effort to protect the natural **environment** from the predations of developers.

Many of Stegner's works of fiction also deal with environmental themes and problems—the rush to develop the West, the desire to get rich by riding roughshod over fragile environments and ecosystems, the wanton indifference to the peoples and the history of a place, and the ravages wrought by this onslaught. Arrayed against these forces are the quieter forces of history and memory, of respect for a region and its past, and the desire to preserve and pass these on to **future generations**.

Wallace Stegner died on April 12, 1993, from a heart attack resulting from a serious **automobile** accident two days earlier. In accordance with his wishes, his ashes were scattered on the western slope of Baker Hill in Vermont.

[*Terence Ball*]

RESOURCES

BOOKS

Benson, J. J. *Wallace Stegner: His Life and Work.* New York: Viking Press, 1996.

Rankin, C. E., ed. *Wallace Stegner: Man and Writer.* Albuquerque: University of New Mexico Press, 1996.

Stochastic change

Any change that involves random behavior. The term "stochastic" is derived from the Greek *stokhiastikos*, meaning to guess at random. Stochastic changes obey the laws of **probability**. A familiar example of such a change is a series of card games. The way that cards fall in each hand is a random event, but well-known mathematical laws can describe the likelihood of various possibilities. For instance, if a traffic engineer wants to study traffic patterns at a busy downtown intersection, the engineer can use stochastic principles to analyze the traffic flow, since the movement of vehicles through that intersection is likely to occur at random. Some natural processes, such as genetic drift, also can best be characterized as the result of a series of stochastic changes.

Storage and transport of hazardous material

Hazardous materials consist of numerous types of explosive, corrosive, and poisonous substances. The materials might be used as a reactant in an industrial process, as an additive in **water treatment** (acids and bases, for example), or as a source of fuel or energy (**gasoline** and nuclear materials). These materials might also be produced, wasted, or contaminated and thus require management as a **hazardous waste**. In all cases, it is necessary to understand the characteristics of the chemical so that appropriate containers and labels for storage and **transportation** are used.

The U.S. Department of Transportation (DOT) and the **Environmental Protection Agency** (EPA) have issued regulation on how hazardous materials are to be stored, labeled, and transported. The DOT was actually the first to address these issues; the EPA simply adopted many of their regulations, and the agency continues to work closely with the DOT to minimize confusion, conflicts, and redundancy. The DOT was originally given authority to regulate the transportation of hazardous materials in 1966, but it was not until 1975 that the **Hazardous Materials Transportation Act** (HMTA) was passed, giving the DOT broad authority over all aspects of transporting hazardous materials. In 1980, the EPA adopted the DOT regulations, and the DOT amended their policies to make them more appropriate for hazardous wastes. The **Nuclear Regulatory Commission** (NRC) is also involved in overseeing the transportation of nuclear materials.

In 1976, the EPA promulgated the **Resource Conservation and Recovery Act** (RCRA), which dealt with many hazardous **waste management** issues, including storage and transportation. RCRA stipulated that the EPA was to adopt transportation regulations that were consistent with the HMTA, but hazardous waste storage and transportation posed some additional problems. For example, it was necessary to define the term "hazardous waste" and to develop a rigorous system to track the waste from generation ("cradle") to ultimate disposal ("grave"). The RCRA and later amendments (Hazardous and **Solid Waste** Amendments of 1984) distinguished between small and large quantity generators and established different regulations for them. Large quantity generators (LQG) are prohibited from storing hazardous wastes for more than 90 days. Small quantity generators (SQG) can store wastes for 180 or 270 days, depending upon how far they have to ship the wastes for disposal.

Transportation of hazardous wastes to disposal sites is carefully monitored; this is done by using a form known as a manifest. The manifest contains the name and EPA identification number of the generator, the transporter, and the treatment disposal facility (TSD facility). The manifest also provides a description of the waste and documents each step in the transportation network; generators that have TSD facilities on-site need not prepare a manifest.

The storage and transportation of hazardous materials is closely regulated. Despite these regulations, however, acci-

dents do occur, and research is continuing in the development of better shipping and storage containers, more reliable means and routes for shipments, more accurate risk assessments, and improved systems for accident prevention and **remediation** of spills. *See also* Hazardous Substances Act; Hazardous waste siting; NIMBY (Not In My Backyard); Toxic substance; Toxic Substances Control Act; Toxic use reduction legislation

[*Gregory D. Boardman*]

RESOURCES
BOOKS

Freeman, H. M. *Standard Handbook of Hazardous Waste Treatment and Disposal.* New York: McGraw-Hill Book Company, 1989.

Martin, E. J., and J. H. Johnson. *Hazardous Waste Management Engineering.* New York: Van Nostrand Reinhold, 1987.

Wentz, C. A. *Hazardous Waste Management.* New York: McGraw-Hill, 1989.

Storm King Mountain

One of the most important cases in **environmental law** involved the proposed construction in 1963 of a huge hydroelectric plant by Consolidated Edison on the shores of the **Hudson River** at the foot of Storm King Mountain. The plant was designed to generate electricity for New York City during its periods of peak demands. Objections to the planned construction were raised by the Scenic Hudson Preservation Conference, which claimed that the plant would seriously damage the area's natural beauty. After nearly a decade of hearings and court cases, Consolidated Edison received permission to build the plant. In the process, however, the right of citizen groups to argue the value of aesthetic considerations in such cases was affirmed. *See also* Electric utilities

Storm runoff

The amount of water that flows into streams and rivers soon after a rainfall, causing the stream to rise above its stable condition. Sometimes called stormflow, quickflow, or direct **runoff**, this flow of water occurs relatively quickly to channels and causes water levels to rise, peak, and then recede as the storm water drains from the **watershed** following the storm. Storm runoff is the sum of precipitation falling directly on the channel, overland flow, and subsurface flow. **Groundwater** generally takes a long period of time to contribute to streamflow and does not appreciably affect streamflow rise immediately after the storm; therefore, it is not considered a part of stormflow. When water levels rise above the banks of a stream or river, storm runoff is considered to cause a flood. In technical **hydrology** terms, stormflow is the portion of the hydrograph that is above base flow.

Storm sewer

Modern stormwater management consists of two components: the major **drainage** system comprising overland flow and retention facilities including ponds, playing fields, parking lots, underground reservoirs, and similar devices; and the minor drainage system consisting of storm sewers. The major system is designed to handle major regional storms which have an infrequent **probability** of occurrence. The minor system uses storm sewers to quickly drain the rainfall from yards, sidewalks, and streets after a rainfall event or during snow melt. Storm sewers are the single largest cost in servicing land for housing. Storm sewers typically **discharge** into surface waters such as creeks, rivers, or lakes without treatment. *See also* Storm runoff

Strategic Lawsuits Against Public Participation

Strategic Lawsuits Against Public Participation (SLAPP suits) are lawsuits initiated by interested parties, most often businesses, that seek to counteract criticism from individuals, businesses, and nonprofit organizations that the litigant believes to be harmful to their operations—often those in the past, present, and future possible detrimental events. The lawsuits often have arisen in response to environmental activists who have protested against those they deem as polluters, or otherwise harmful to the **environment** and human health. Primarily these lawsuits are designed to discourage pubic activism simply by threatening to financially impair those engaging in such activities.

According to the **Environmental Law** and Policy Center, (ELPC) serving Illinois, Indiana, Michigan, Minnesota, Ohio, and Wisconsin, in its January 2001 *ELPC News* "SLAPP suits, unfortunately, are becoming a popular weapon among industry lawyers and conservative ideologues, who tend to be hostile to environmental activism. Often the claims embodied in SLAPP suits are legally tenuous and ultimately collapse under judicial scrutiny; but that doesn't mean they're not successful. Even without obtaining a favorable ruling, the business plaintiffs who file SLAPP suits can deter grassroots activists by subjecting them to the financial and emotional harassment of a potentially punitive court ruling." Often, such suits discourage any future challenges to environmental and **pollution** law violations. It should be noted that ELPC provides assistance to environmental organizations in matters of litigation.

During the late 1990s, one of the most famous SLAPP suits involved television personality, Oprah Winfrey. The uproar followed a guest appearance on her show by Howard Lyman of the **Humane Society of the United States** during which he said that, "mad cow disease would make **AIDS** look like the common cold." The show aired on April 16, 1996, only a month after the British government admitted that the disease was a problem in England. After hearing the evidence presented, Winfrey added her own remark during that same show saying that, "It has just stopped me cold from eating another burger." The broadcast seemed to directly affect a dramatic reduction in cattle futures on the Chicago Mercantile Exchange. The meat industry, represented in the lead by the National Cattlemen's Beef Association (NCBA) were outraged enough to file a $2 million lawsuit against Lyman, and Winfrey, despite her offer to do an interview with the NCBA's policy director. It was Paul Engler, a beef feed-lot operator who filed the suit. This particular form of lawsuit instituted by members of the food and agriculture industry came to be known as a *food disparagement* suit—one category of SLAPP. Winfrey took her television production to Texas during the court proceedings, airing her show daily from a local auditorium, and was ultimately successful in winning the case.

The first food disparagement law was passed in Colorado in 1991 at the behest of apple growers using **Alar**, a chemical sprayed on apples while still on the tree to allow them to ripen longer. Due to a public campaign with vocal celebrities airing their complaints against the carcinogenic dangers of Alar, the industry stopped using the chemical but proceeded to sue the *Columbia Broadcasting System (CBS)* and others for their negative publicity regarding the industry's practice. The American Civil Liberties Union attempted to challenge a similar law passed in Georgia, but the state court upheld it on a technicality in 1995. As of 1998, 13 states had already adopted food disparagement laws, and 14 other states were considering them.

In 1999 the Oregon House of Representatives overwhelmingly passed a bill that would stop SLAPP suits. According to a 1998 article from *New Jersey Land Use Law* a law school study of 228 SLAPP suits found that the SLAPP targets prevailed in approximately 77% of all cases; but those that targeted smaller, less-affluent groups were significantly more successful.

[*Jane E. Spear*]

Resources

Books

Meadows, Donella H. "A SLAPP in the Face." *Valley Advocate*. Easthampton, MA: 1998.

Periodicals

Ivins, Molly. "SLAPP Lawsuits Bruise the Face of Free Speech." *The Seattle Times* (January 24, 2000).

Tanner, Linda. "When Citizens SLAPP Back!" *National Wildlife* 4, no. 4 (June/July 1994): 12.

Other

American Civil Liberties Union Massachusetts. *SLAPPing Back at SLAPP Suits.* August 13, 2001 [cited July 2002]. <http://www.aclu-mass.org>

Community Lawyer. *Citizens Win SLAPP Suit.* (June 2002). <http://www.communitylawyer.org>

Counter Punch Organization. *SLAPP Suits in California Forests.* January 2001 [cited July 2002]. <http://www.counterpunch.org/pipermail/counterpunch-list>.

Environmental Law & Policy Center. "Guarding Against SLAPP Lawsuits." *ELPC News* January 2001 [cited July 2002]. <http://www.elpc.org>.

Environmental News Network. *Greenpeace Takes on Oil Companies over Spill Cleanup in Alaska.* May 13, 2000 [cited July 2002]. <http://www.enn.com/news>.

Environmental News Network. *Greens Slap USFWS with Lawsuits in Florida.* December 21, 2000 [cited July 2002]. <http://www.enn.com/news>.

Environmental News Network. *Mad Cowboy: The Beef Industry Strikes Back.* February 25, 1998 [cited July 2002]. <http://www.enn.com/news>.

Environmental News Network. *Mexico's Fledgling Ecotourism Industry Struggles to Survive.* April 26, 2002 [cited July 2002]. <http://www.enn.com/news>.

Hudson, David. *Anti-SLAPP Measure Easily Clears Oregon House.* May 17, 1999 [cited July 2002]. <http://www.casp.net/or-1999html>.

New Jersey Land Use Law. *SLAPP Suits.* March 2, 1998 [cited July 2002]. <http://www.nj-laanduselaw.com>.

Organizations

Environmental Law & Policy Center, 35 East Wacker Drive, #1300, Chicago, IL USA 60601, (312) 673-6500, Fax: (312) 795-3730, <http://www.elpc.org>

U.S. Environmental Protection Agency, 1200 Pennsylvania Avenue, N.W., Washington, D.C. USA 20460, (202) 260-2090, <http://www.epa.gov>

Strategic minerals

American industry has a voracious appetite for minerals. The manufacture of a typical **automobile**, for example, requires not only such familiar metals as iron, **copper**, **lead**, and **aluminum**, but also such less-familiar metals as manganese, platinum, molybdenum, and vanadium. For the time being, the United States has an abundant supply of many critical minerals. The country as of 2002 is essentially self-sufficient in such major metals as iron, copper, lead, and aluminum. In each case, we import less than a quarter of the metals used in industrial production.

There are some minerals, however, that do not occur naturally to any considerable extent in the United States. For example, the United States has essentially no reserves of columbium (niobium), strontium, manganese, tantalum, or cobalt; sheet mica; or bauxite ore. To the extent that these minerals are important in various industrial processes, they are regarded as critical or strategic minerals. Some

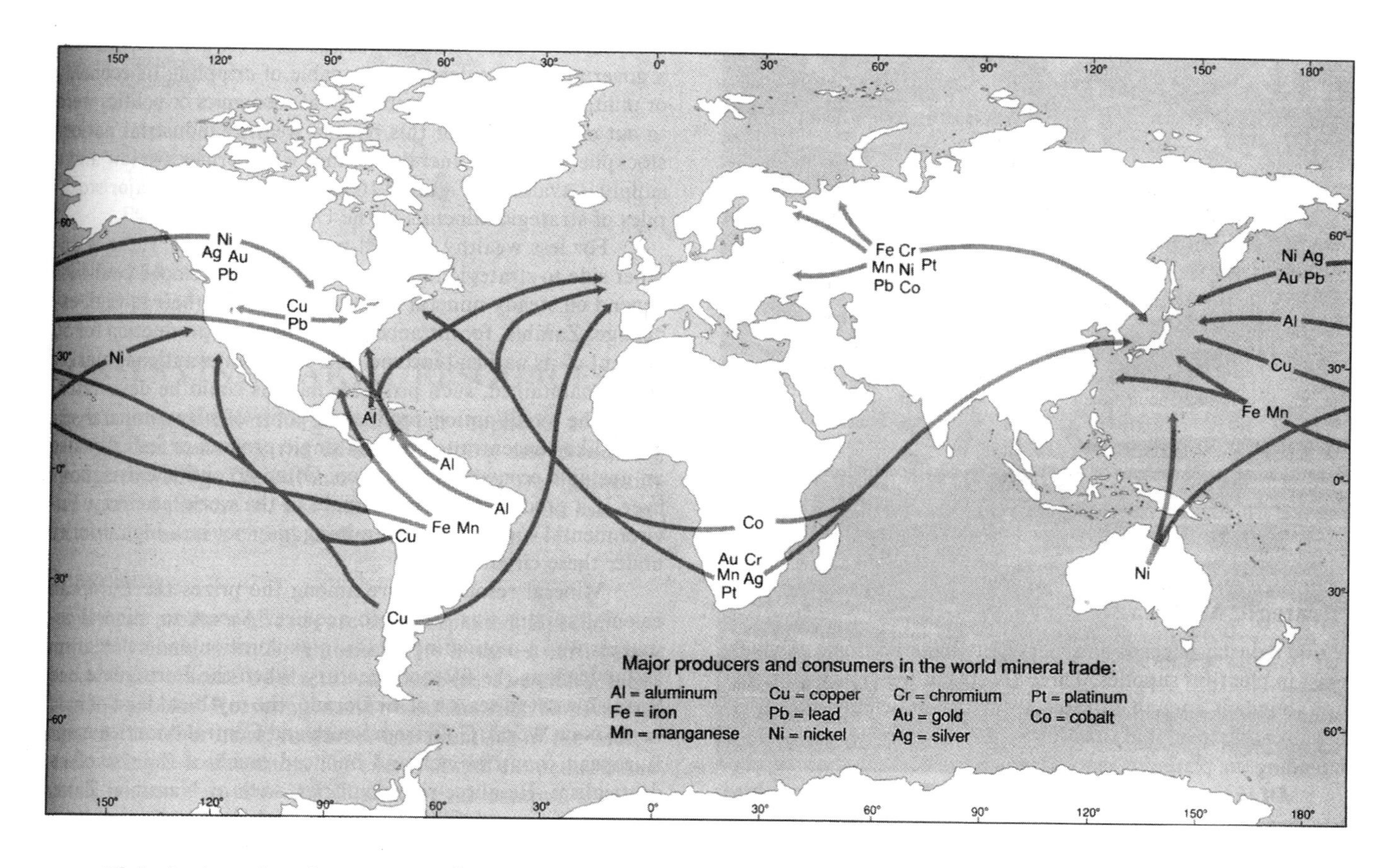

Global mineral trade. (McGraw-Hill Inc. Reproduced by permission.)

examples of strategic minerals are tin, silver, cobalt, manganese, tungsten, zinc, titanium, platinum, chromium, bauxite, and diamonds. The United States must import at least half the amount of each of these minerals that it uses each year.

Ensuring a constant and dependable supply of strategic minerals is a complex political problem. In some cases, the minerals we need can be obtained from friendly nations with whom we can negotiate relatively easily. Canada, for example, supplies a large part of the **nickel**, columbium, gallium, tantalum, **cadmium**, and cesium used by American industry. Nickel may be obtained from Norway; cobalt and antimony from Belgium; and fluorspar from Italy.

Other nations on whom we depend, however, are less friendly, less dependable, or less stable. The southern African nations of Zaire, Zambia, Zimbabwe, Botswana, and South Africa, for example, are major suppliers of such strategic minerals as chromium, gold, platinum, vanadium, manganese, and diamonds. Africa also has large deposits of copper, cobalt and chromium. When these nations experience political unrest, supplies of these minerals may become scarce. A report issued by the United States Agency for International Development (USAID) in November 2000 underscored the point that ongoing American investment in Africa is necessary because of our need for the continent's strategic minerals.

Since 1980, the United States government has tried to protect American industry, especially defense industries, from the danger of running out of strategic minerals. Some analysts have feared that political factors might result in the loss of certain minerals that are needed by industry, particularly those used in the manufacture of military hardware. In 1984, the United States Congress established the National Critical Materials Council (NCMC) to advise the President on issues involving strategic minerals. The Council monitored domestic and international needs and trends for nearly a decade to ensure the nation's access to a dependable supply of strategic minerals. In 1993, the NCMC was placed under the supervision of the newly established National Science and Technology Council.

A similar monitoring function has been performed by the Defense Logistics Agency (DLA), a division of the Department of Defense dating back to World War II. The DLA maintains and administers the Defense National Stockpile Center, or DNSC, which sells and maintains strategic and critical materials in order to reduce the country's dependence on foreign sources of supply. The DNSC is presently located in Fort Belvoir, Virginia.

In addition to political unrest, however, the supply of strategic minerals depends on humankind's general management of the earth's **nonrenewable resources**. Of the minerals in common use in manufacturing and industry, 80 are in abundant supply. There are, however, about 18 minerals that are expected to be in short supply around the globe by 2015; these include gold, silver, **mercury**, lead, sulfur, tin, tungsten, and zinc. It is estimated that the earth's stores of these minerals will be 80% depleted by 2040. Strategies for lowering present levels of use in the developed nations include **recycling**; inventing new substitute materials and finding new uses for old ones; decreasing the size of products; and extending the lifespan of items made from strategic minerals. A recent report to the Canadian government from the assistant deputy minister for the minerals and metals sector of **natural resources** stated that competition among nations for strategic minerals has already begun to transform the manufacturing processes used by Canadian industries as of 2002.

[*David E. Newton and Rebecca J. Frey Ph.D.*]

Resources

Books

Anderson, Ewan W., and Liam D. Anderson. *Strategic Minerals: Resource Geopolitics and Global Geo-Economics.* New York: John Wiley and Sons, 1997.

Collins, John M. *Military Geography for Professionals and the Public.* Washington, DC: National Defense University Press, 2001.

Strategic Minerals: A Resource Crisis. Washington, DC: Council on Economics and National Security, 1981.

U.S. Bureau of Mines. *The Domestic Supply of Critical Minerals.* Washington, DC: U S. Government Printing Office, 1983.

Periodicals

Homer-Dixon, Thomas F. "Environmental Scarcity, Mass Violence, and the Limits to Ingenuity." *Current History* 95 (November 1996): 359–365.

Other

Derryck, Vivian L. *Remarks to the Foreign Service Association of Northern California.* Presentation given in San Francisco, CA, November 3, 2000.

Kelly, Thomas, David Buckingham, and Carl DiFrancesco, et al. *Historical Statistics for Mineral Commodities in the United States.* United States Geological Survey Open-File Report 01-006, June 2002.

Minerals and Metals Sector of Natural Resources Canada. *Focus 2006: A Strategic Vision for 2001–2006.* Ottawa, Ontario: Public Works and Government Services Canada, May 2002.

Organizations

U.S. Department of the Interior, U.S. Geological Survey, Minerals Information, 988 National Center, Reston, VA USA 20192 Toll Free: (888) ASK-USGS (888-275-8747), Email: Contact person: Joseph Gambogi, jgambogi@usgs.gov, www.minerals.usgs.gov/minerals/

Stratification

The process by which a region is divided into relatively distinct, nearly horizontal, layers. Sedimentary rock often consists of various strata because the rock was laid at different times when different materials were being deposited. Lakes and oceans also consist of strata where plant and animal life may be very different depending on the temperature and amount of light available. The **atmosphere** is divided into strata that differ in density, temperature, chemical composition, and other factors. In the lower atmosphere, temporary stratification, such as temperature inversions, may occur, often resulting in severe **pollution** conditions.

Stratosphere

A layer of the **atmosphere** that lies between about 7 and 31 mi (11 and 50 km) above the earth's surface, bounded at the bottom by the **tropopause** and at the top by the stratopause. Scientists became aware of the presence of the stratosphere with observations of high level dust after the eruption of **Krakatoa** in 1883. However, the real discovery of the stratosphere had to await Teisserenc de Bort's work of 1900.

The stratosphere is characterized by temperatures that rise with height. As a result, the air is very stable, not mixing much vertically, and allowing distinct layers of air, or strata, to form. There are also persistent regular strong winds, of which the best known are the intense western winds during the winter called the polar night jet stream.

The air in the stratosphere has much the same composition as the lower atmosphere except for a higher proportion of **ozone**. **Absorption** of incoming **solar energy** by this ozone makes the upper parts of the stratosphere warm and sets up the characteristic temperature gradient. The relatively high ozone concentrations are maintained by photochemical reactions. It has become clear in more recent years that other chemical reactions involving **nitrogen** and **chlorine** are also important in maintaining the ozone balance of the upper atmosphere. This balance can be easily disturbed through the input of additional **nitrogen oxides** and halogen compounds (from CFCs, or **chlorofluorocarbons**). Transfer of gases across the tropopause into the stratosphere is rather slow, but some gases, such as CFCs and **nitrous oxide**, are sufficiently long-lived in the **troposphere** to leak across into the stratosphere and cause **ozone layer depletion**. Large volcanic eruptions can have sufficient force to drive gases and particles into the stratosphere where they can also disturb the ozone balance. High-flying aircraft represent another source of **pollution** in the stratosphere, but the lack of development of a supersonic passenger fleet has meant that this contribution has remained fairly small.

The stratosphere is extremely cold (about -112°F [-80°C]) in places, so there is relatively little water. Nevertheless, nacreous or mother-of-pearl clouds, although not frequently observed, have long been known. More recently there has been much interest in polar stratospheric clouds which had hitherto received little attention. Stratospheric cloud particles can be water-containing sulfuric **acid** droplets or solid nitric acid hydrates. Studies of the Antarctic ozone hole have revealed that these clouds are likely to play an important role in the depletion of ozone in the polar stratosphere. *See also* Acid rain; Cloud chemistry; Ozonation; Stratification; Volcano

[*Peter Brimblecombe Ph.D.*]

RESOURCES

BOOKS

Chamberlain, J. W., and D. M. Hunten. *Theory of Planetary Atmospheres.* New York: Academic Press, 1987.

Stray voltage

see **Electromagnetic field**

Stream channelization

The process of straightening or redirecting natural streams in an artificially modified or constructed stream bed. Channelization has been carried out for numerous reasons, most often to drain **wetlands**, direct water flow for agricultural use, and control **flooding**. While this process makes a stream more useful for human activities, it tends to interfere with natural river habitats and to destabilize stream banks by destroying riparian vegetation. When annual flood patterns are disrupted, fertilizing **sediment** is no longer deposited on river banks and excessive sediment accumulation can occur downstream. Perhaps most importantly, wetland **drainage** and the removal of instream obstacles such as rocks, fallen trees, shallow backwaters, and sand bars eliminate feeding and reproductive habitats for fish, aquatic insects, and birds.

Stringfellow Acid Pits

Aerospace, electronics, and other high-technology businesses expanded rapidly in California during the 1950s, bringing **population growth** and rapid economic progress. These businesses also brought a huge volume of toxic wastes and the problem of safely disposing of them. A modern-day reminder of those years is the Stringfellow Acid Pits located near the Riverside suburb of Glen Avon, 50 mi (80 km) east of Los Angeles.

The Acid Pits, also known as the Stringfellow Quarry Waste Pits, are located on a 20-acre (8-ha) site in Pyrite Canyon above Glen Avon. In the mid-1950s, a number of high-tech companies began to dump their hazardous wastes into the canyon. No special precautions were taken in the dumping process; as one observer noted, the companies got rid of their wastes just as cavemen did: "They dug a hole and dumped it in."

Over the next two decades, more than 34 million gal (129 million L) of waste were disposed of in a series of pan-shaped reservoirs dug into the canyon floor. The wastes came from more than a dozen of the nation's most prominent companies, including McDonnell-Douglas, Montrose Chemical, General Electric, Hughes Aircraft, Sunkist Growers, Philco-Ford, Northrop, and Rockwell-International. The wastes consisted of a complex mixture of more than 200 hazardous **chemicals**. These included hydrochloric, sulfuric, and nitric acids; sodium hydroxide; trichlorethylene and methylene chloride; **polychlorinated biphenyls** (PCBs); a variety of pesticides; volatile organic compounds (VOCs); and **heavy metals** such as **lead**, **nickel**, **cadmium**, chromium, and manganese.

By 1972, residents of Glen Avon had begun to complain about health effects caused by the wastes in the Stringfellow Pits. They claimed that some chemicals were evaporating and polluting the town's air, while other chemicals were **leaching** out of the dump and contaminating the town's **drinking-water supply**. People attributed health problems to chemicals escaping from the dump; these problems ranged from nose bleeds, emotional distress, and insomnia to **cancer** and genetic defects. Medical studies were unable to confirm these complaints, but residents continued to insist that these problems did exist.

In November 1972, James Stringfellow, owner of the pits, announced that he was shutting them down. However, his decision did not solve the problem of what to do with the wastes still in the pit. Stringfellow claimed his company was without assets, and the state of California had to take over responsibility for maintaining the site.

The situation at Stringfellow continued to deteriorate under state management. During a March 1978 rainstorm, the pits became so badly flooded that officials doubted the ability of the existing **dams** to hold back more than 8 million gal (30.3 million L) of wastes. To prevent a possible disaster, they released nearly 1 million gal (3,785 million L) of liquid wastes into flood control channels running through Glen Avon. Children in nearby schools and neighborhoods, not knowing what the brown water contained, waded and played in the toxic wastes.

When the **Comprehensive Environmental Response, Compensation and Liability Act** (Superfund) was passed in 1980, the Stringfellow Pits were named the most polluted waste site in California. The pits became one of first targets for **remediation** by the **Environmental Protection Agency** (EPA), but this effort collapsed in the wake of a scandal that rocked both the EPA and the Reagan administration in 1983. EPA administrators Rita Lavelle and Anne McGill Burford were found guilty of mishandling the Superfund program, and were forced to resign from office along with 22 other officials.

During the early 1990s, citizens of Glen Avon finally began to experience some success in their battle to clean up the pits. The EPA had finally begun its remediation efforts in earnest, and residents won judgments of more then $34 million against Stringfellow and four companies that had used the site. In 1993, residents initiated the largest single civil suit over toxic wastes in history. The suit involved 4,000 plaintiffs from Glen Avon and 13 defendants, including the state of California, Riverside County, and a number of major companies. *See also* Contaminated soil; Groundwater pollution; Hazardous waste site remediation; Hazardous waste siting; Storage and transport of hazardous materials

[*David E. Newton*]

Resources
Books

Brown, M. H. *Laying Waste: The Poisoning of America by Toxic Chemicals.* New York: Pantheon Books, 1980.

Periodicals

Gorman, T. "A Tainted Legacy: Toxic Dump Site in Riverside County Has Sparked the Nation's Largest Civil Suit." *Los Angeles Times* (January 10, 1993): A3.

Madigan, N. "Largest-Ever Toxic-Waste Suit Opens in California." *New York Times* (February 5, 1993): B16.

Mydans, S. "Settlements Reached on Toxic Dump in California." *New York Times* (December 24, 1991): A11.

Strip-farming

In the United States, **soil conservation** first became an important political issue in the 1930s, when President Franklin D. Roosevelt led a campaign to study the loss of valuable **topsoil** because of **erosion**. It soon became clear that one source of the problem was the fact that farmers tended to plow and plant their fields according to property lines, which usually formed squares or rectangles. As a result, furrows often ran up and down the slope of a hill, forming a natural channel for the **runoff** of rain, and each new storm would wash away more fertile topsoil. This kind of erosion resulted not only in the loss of **soil**, but also in the **pollution** of nearby waterways.

The **Soil Conservation Service** was founded in 1935 as a division of the **U.S. Department of Agriculture**, and one of its goals was the development of farming techniques that would reduce the loss of soil. One such technique was strip-farming, also known as strip-cropping. Strip-farming involves the planting of crops in rows across the slope of the land at right angles to it rather than parallel to it. On gently sloping land, soil **conservation** can be achieved by plowing and planting in lines that simply follow along the slope of the land rather than cutting across it, a technique known as **contour plowing**. On very steep slopes, a more aggressive technique known as **terracing** is used. Strip-farming is a middle point between these two extremes, and it is used on land with intermediate slopes.

In strip-farming, two different kinds of crops are planted in alternate rows. One set of rows consists of crops in which individual plants can be relatively widely spaced, such as corn, soybeans, cotton, or sugar beets. The second set of rows contains plants that grow very close together, such as alfalfa, hay, wheat, or legumes. As a result of this system, water is channeled along the contour of the land, not down its slope. In addition, the closely planted crops in one row protect the exposed soil in the more widely spaced crops in the second row. Crops such as alfalfa also slow down the movement of water through the field, allowing it to be absorbed by the soil.

The precise design of a strip farm is determined by a number of factors, such as the length and steepness of the slope. The crops used in the strip, as well as the width of rows, can be adjusted to achieve minimal loss of soil. Under the most favorable conditions, soil erosion can be reduced by as much as 75% through the use of this technique. *See also* Agricultural pollution; Agricultural technology; Soil compaction; Soil loss tolerance

[*Lawrence H. Smith*]

Resources
Books

Enger, E. D., et al. *Environmental Science: The Study of Relationships.* 2nd ed. Dubuque, IA: W. C. Brown, 1986.

Moran, J. M., M. D. Morgan, and J. H. Wiersma. *Introduction to Environmental Science.* 2nd ed. New York: W. H. Freeman, 1986.

Petulla, J. M. *American Environmental History.* San Francisco: Boyd & Fraser, 1977.

Strip mining

This technique is used for near-surface, relatively flat sedimentary mineral deposits. How deeply the mining can occur

is essentially determined by the combination of technological capabilities and the economics involved. The latter includes the current value of the mineral, contractual arrangements with the landowner, and mining costs, including **reclamation**. Strip mining is used for mining phosphate **fertilizer** in Florida, North Carolina, and Idaho, and for obtaining gypsum (mainly for wallboard) in western states.

However, the most common association of strip mining is with **coal**. The examples of decimated land in Appalachia have motivated calls for prevention, or at least major efforts at reclamation. Strip mining for coal comprises well over half of the land that is strip-mined, which totaled less that 0.3% of land in the United States between 1930 and 1990. This is far less land than the amount lost to agriculture and urbanization. However, in agriculturally rich areas like Illinois and Indiana there is a growing concern over the one-time disruption of land for mineral extraction, compared to the long term use for food production.

Strip mining has occurred mainly in the Appalachian Mountains and adjacent areas, the Central Plains from Indiana and Illinois through Oklahoma, and new mines for subbituminous coal in North Dakota, Wyoming, and Montana. Important mining is also carried out on Hopi and Navajo lands, notably Black Mesa in northeastern Arizona.

Despite the small amount of land used in strip mining, the process radically alters landforms and ecosystems where it is practiced. Depending on state laws, mining landscapes prior to 1977 were often left as is, dubbed "orphan lands." The 1977 act required the land to be restored as closely as possible to the original condition. This is a nearly impossible task, especially when one considers the reconstruction of the preexisting **soil** conditions and **ecosystem**. Even so, reclamation is a vital first step in the healing process. Generally, the steeper the terrain, the greater the impact on the landforms and river systems, and the more difficult the reclamation.

Detailed economic planning precedes any strip mining effort. Numerous cores are drilled to determine the depth, thickness, and quality of the coal, and to assess the difficulty of removing the **overburden**, which consists of **topsoil** and rock above the resource. If caprock is encountered, expensive and time-consuming blasting is required, a frequent occurrence in the United States. Economic analysis then determines the area and depth of profitable overburden removal. Finally, contracts must be negotiated with landowners; strip mines commonly end abruptly at property lines.

Two kinds of earth removal equipment are typically used: a front-loading bucket (the classic steam shovel), or a dragline bucket that pulls the material toward the operator. Power shovels and draglines built prior to World War II generally have bucket capacities of 30–50 yd^3 (23–38 m^3). Post-World-War-II equipment may have a capacity up to 200 yd^3 (153 m^3). A new development, encouraged by the 1977 reclamation law, is the combination of dozers and scrapers (belly loaders) more commonly seen on road-building or construction sites.

After the overburden is removed, mining begins. The process is conducted in rows, creating long ridges and valleys in the countryside that resemble a washboard. Coal extraction follows behind power shovels, leaving a flat, canyon-like cut. Upon completion of a row, the shovel starts back in the opposite direction, placing the new overburden in the now-empty cut.

In hilly terrain, only a few cuts are all that is usually profitable because the depth of overburden increases rapidly into the hillside. Since the worst complications as a result of strip mining occur on hillsides, the environmental price for a limited amount of coal is very high. Hillside mining such as this is called "contour mining," in contrast to "area mining" on relatively flat terrain. In the latter, the number of rows are limited mostly by contractual arrangements. Consequently, the main difference between area and contour strip mining are the number of rows and the steepness of the terrain.

Both types of strip mining leave behind four basic land configurations: 1) **spoil** bank ridges; 2) a final-cut canyon often partially filled by a lake; 3) a high headwall marking the uphill end of the mining; and 4) coal-haul roads, usually at the base of the outermost spoil bank and through gaps in the spoil-bank rows left for this purpose. In some orphan lands, wilderness-like conditions prevail, where trees populate the spoil banks and aquatic ecosystems thrive in the final-cut lake. Left alone by man, these may afford a surprisingly rich **habitat** for **wildlife**, especially birds. Deer thrive in some North Dakota abandoned mines.

Reclamation of area mining is relatively simple compared to contour strip mining. Prior to mining, the topsoil is removed and stockpiled. The overburden from the initial cut may be used to fill in the final cut, and the top part of the headwall is sometimes cut down to grade into the spoil. The spoil banks are leveled and the topsoil replaced; fertilization and replanting, usually with grasses or trees for **erosion** control, and subsequent monitoring of revegetation efforts, complete the process. In large operations, the leveling and replanting coincide with mining, which is ideal since this rapidly rebuilds the vegetation cover.

Reclamation of contour mining presents far greater difficulties, primarily because of the slope angles encountered. Research in Great Britain revealed that even well-vegetated slopes were producing 50 to 200 times as much **sediment** as similar, undisturbed slopes. Furthermore, the greater slope angles allow much more of the sediment to reach the channel below, where it eventually flows into streams and rivers.

Another problem for orphan lands in hilly terrain is the ecological island left when hills are completely enclosed by high headwalls. This is not unlike the ecological islands created in the southwestern United States from **climate** changes and vertical zonation of vegetation. Though far more recent, ecologists hope these "orphan islands" will allow interesting case studies of genetic isolation.

The long-term effect of strip mining has been the subject of research in Kentucky, Indiana, and Oklahoma. For over a decade the United States **Geological Survey** studied Beaver Creek Basin, Kentucky, obtaining valuable data before and after contour mining. Their findings were published in a 1970 report authorized by C.R. Collier and others. As expected, mining left a degraded landscape, and resulted in much greater **runoff**, sediment production, and **water quality** problems. By contrast, area mining in Indiana trapped vast quantities of **groundwater** within the loosened soil, reducing peak discharges, extending base flow, and yielding water of acceptable quality.

In Oklahoma, a study conducted by Nathan Meleen dealt with a mix of both flat and hilly terrain. The findings were published as a doctoral dissertation in 1977 by Clark University. Area mining, Meleen found, produced far more benign impacts than did contour stripping. The worst conditions were encountered where contour mining was adjacent to streams below, especially around the end of ridges. The short distances and relatively steep gradients gave ideal conditions for sediment and **acid drainage** into the channels below. Summer runoff decreased while winter runoff increased. The huge holes left by these unreclaimed operations acted like reservoirs during the drier summer months, but in winter when wetter conditions prevailed, they yielded more outflow than before mining, because of altered **infiltration** rates.

Strangely enough, a 1971 Oklahoma law produced effects similar to what the Geological Survey found in Kentucky. The incomplete reclamation and lack of sediment retention or topsoil replacement created ideal erosion conditions, with rates approaching 13% sediment by weight. This focuses the crucial role of topsoil replacement and rapid revegetation as the preeminent needs in reclamation. Efforts to revegetate some orphan lands where topsoil replacement is impossible will only result in worse conditions, especially downstream. Given the fact that most impacts from area mining are retained onsite, and that orphan lands possess great potential for **recreation**, especially fishing and hiking, at least some should be left undisturbed. *See also* High-grading; Mine spoil waste; Surface mining

[*Nathan H. Meleen*]

RESOURCES

BOOKS

Caudill, H. M. *Night Comes to the Cumberlands.* Boston: Atlantic-Little, Brown, 1963.

Collier, C. R., et al. *Influences of Strip Mining on the Hydrologic Environment of Parts of Beaver Creek Basin, Kentucky, 1955–66.* USGS Professional Paper 427-C. Washington, DC: U. S. Government Printing Office, 1970.

Doyle, W. S. *Strip Mining of Coal: Environmental Solutions.* Park Ridge, NJ: Noyes Data Corp., 1976.

Landy, M. K. *The Politics of Environmental Reform: Controlling Kentucky Strip Mining.* Baltimore: Johns Hopkins University Press, 1976.

OTHER

U.S. Department of the Interior. "Surface Mining and Our Environment: A Special Report to the Nation." Washington, DC: U. S. Government Printing Office, 1967.

U.S. Environmental Protection Agency. "Erosion and Sediment Control: Surface Mining in the Eastern U.S.: Planning." Washington, DC: U. S. Government Printing Office, 1976.

Strix occidentalis caurina

see **Northern spotted owl**

Strontium 90

A radioactive **isotope** of strontium, produced during **nuclear fission**. The isotope was of great concern to environmental scientists during the period of atmospheric testing of **nuclear weapons**. Strontium 90 released in these tests fell to the Earth's surface, adhered to grass and other green plants, and was eaten by cows and other animals. Since strontium is chemically similar to calcium, it follows the same metabolic pathways, ending up in an animal's milk. When a child drinks this milk, strontium 90 becomes incorporated into their bones and teeth. With a **half-life** of about 28 years, strontium 90 continues to emit radiation throughout the individual's lifetime.

Student Environmental Action Coalition

Founded in 1990, the Student Environmental Action Coalition (SEAC) focuses on social justice and on environmental and political issues throughout the United States. The mission of SEAC is to unite youth that have exhibited leadership in their community and to create an environmental movement. It has activists that build allegiances among local, regional, national, and international communities engaged in environmental struggles.

The unique aspect of SEAC membership is that there is not a national office that dictates an agenda to the local groups. Instead, local groups coordinate a National Council each year to run SEAC. The local groups are composed of schools or neighborhood clubs, with each choosing its own problem areas to focus on. Some topics that are frequently

addressed include combating racism and sexism, and developing a sustainable living style.

Skills in each local group are developed at conferences held throughout the year. These conferences are led by students and consist of workshops on the strategies, lectures, and publicity for grassroots organizing.

In 1998, SEAC was forced to cut one of its key funding programs which proved disastrous for the group. Fortunately, they were able to reopen, although with a much smaller staff and reduction of size and publication of its magazine, *Threshold*. Now that SEAC relies primarily on donations and membership dues, the group has continued to expand and work on projects such as the Clean Energy Campaign.

[*Nicole Beatty*]

Resources

Organizations

Student Environmental Action Coalition, P.O. Box 31909, Philadelphia, PA USA 19104-0609 (215) 222-4711, Fax: (215) 222-2896, Email: seac@seac.org, <http://www.seac.org>

Styrene

Styrene is an oily organic liquid with an aromatic odor that is used as a building block for polymers in the manufacture of **plastics**, resins, coatings, and paints. Short-term health effects of styrene exposure include nervous system effects such as depression, loss of concentration, weakness, fatigue, and nausea. Potential long-term effects include liver and nerve tissue damage. Styrene has been designated as a possible human **carcinogen**. The drinking water standard (Maximum Contaminant Level, or MCL) for styrene is 0.1 **parts per million** (ppm). Styrene when released into water rapidly evaporates or is degraded by **microorganisms**. It does not bind to soils and may leach to ground water. However, its rapid degradation minimizes its **leaching** potential. Styrene does not tend to accumulate in aquatic life. It is also found in the air and in the microgram/cubic meter range.

[*Judith L. Sims*]

Resources

Other

U.S. Environmental Protection Agency. *Consumer Factsheet on: Styrene.* May 22, 2002 [cited June 23, 2002]. <http://www.epa.gov/safewater/dwh/c-voc/styrene902.html>.

Organizations

The Styrene Information and Research Center, 1300 Wilson Boulevard, Suite 1200, Arlington, Virginia USA 22209, (703) 741-5010, Fax: (703) 741-6010), <http://www.styrene.org>

Subbituminous coal

see **Coal**

Submerged aquatic vegetation

Submerged aquatic vegetation consists of a taxonomically diverse group of plants that lives entirely beneath the water surface. This diverse group of aquatic plants includes **species** of angiosperm vascular plants, mosses, and liverworts, and macroalgae (seaweeds). Their underwater growth habit separates them from other kinds of aquatic plants that are free-floating, have floating leaves, or are emergent above the water surface.

Almost all species of submerged aquatic plants live in freshwater ponds, lakes, rivers, streams, and **wetlands**, and shallow marine waters. They can occur in ponds with an acidic **pH** less than 4 or in alkaline waterbodies with pH greater than 10, but they tend to be most rich in species at pHs of 6 to 8. Only a few angiosperm species occur in **brackish** estuarine or marine habitats, including the **eelgrass** (*Zostera marina*), widgeon grass (*Ruppia maritima*), and turtle-grasses (*Thalassia* species). No aquatic mosses or bryophytes occur in saline waters.

Macroalgae, commonly known as seaweeds, are also classified as submerged aquatic vegetation though they are not biologically part of the kingdom Plantae. They are members of the kingdom Protista and live within the marine **environment**. They, too, are beneficial to the aquatic **habitat**.

Both grasses and macroalgae absorb nutrients which can be a source of pollutants in aquatic habitats. Submerged vegetation also is a major food source for many different species of organisms including waterfowl, **sea turtles**, and **manatees**. These plants and seaweeds add oxygen to the water, and grass roots help stabilize shorelines against **erosion**. Additionally, submerged aquatic vegetation provides shelter and food for organisms. For example, crustaceans, especially the blue crab, and juvenile or larval fish use submerged vegetation as protective nurseries and a means to hide from predators. Some organisms like barnacles and bryozoans attach themselves to plant surfaces to live. Other species use submerged vegetation as place to lay eggs and hatch new offspring.

Though communities of submerged aquatic vegetation help improve **water quality**, their own survival depends on maintaining good water quality. These plants and macroalgae are sensitive to environmental changes brought about by agricultural **runoff**, industrial waste, and global warming. Increased temperature, **salinity**, and depth of coastal waters in particular can contribute to marked habitat change for these species, thus threatening their distribution and abundance.

Submerged aquatic plants are most abundant in relatively shallow water, where they have access to enough sunlight to engage in **photosynthesis** at a high enough rate to survive. Waterbodies with poor **visibility** may support few or no submerged aquatic plants, and only at very shallow depths. Waterbodies may have poor visibility because of an excessive abundance of **phytoplankton**, or they may be highly turbid because of suspended clays, or they may be brown-colored because of dissolved organic matter leached from nearby bogs or peaty **soil**. Waterbodies with exceptionally clear water may have submerged aquatic plants growing on the bottom as deep as about 19 feet (6 m). Shallow, moderately fertile, mesotrophic or eutrophic waterbodies may support especially large populations of submerged aquatic plants, where they may even be regarded as nuisance "weeds".

Some of the most widespread and familiar species of angiosperm submerged aquatic plants include the tape-grass or wild celery (*Vallisneria americana*), the waterweed (*Elodea canadensis*), the ribbon-leaf pondweed (*Potamogeton epihydris*), the slender water-nymph (*Najas gracillima*), the pipewort (*Eriocaulon septangulare*), the greater bladderwort (*Utricularia vulgaris*), and the lake quillwort (*Isoetes lacustris*). Examples of bryophytes that are submerged aquatic plants include an aquatic peatmoss (*Sphagnum macrophyllum*), an aquatic moss (*Fontinalis antipyretica*), and an aquatic liverwort (*Ricciocarpus natans*).

A few species of submerged aquatic plants are used as ornamentals in aquaria and outdoor water-gardens. There are even some clubs of aficionados of the aesthetics of these cultivated aquatic plants. Examples include the Amazon sword (*Echinodorus amazonicus*), the tropical hornwort (*Ceratophyllum submersum*), the temple plant (*Hygrophila corymbosa*), the Java moss (*Vesicularia dubyana*), and a tropical aquatic liverwort (*Riccia fluitans*). Submerged aquatic plants are also used as **ecosystem** components in the constructed wetlands that are sometimes used to treat human sewage and other wastes.

A few submerged aquatic plants, particularly several non-native invasive species, can be sufficiently abundant that they are considered serious weeds because of their effects on the use of waterbodies for **recreation** and **transportation**. Some of the most important invasive weeds of this kind in North America are the Eurasian water-milfoil (*Myriophyllum spicatum*), the Brazilian waterweed (*Egeria densa*), and the hydrilla (*Hydrilla verticillata*). Attempts are sometimes made to reduce infestations of these species using mechanical harvesters or herbicides.

[*Bill Freedman Ph.D.*]

Resources
Books

Borman, S., R. Korth, and J. Temte. *Through the Looking Glass: A Field Guide to Aquatic Plants.* Madison, WI: University of Wisconsin Press, 1997.

Crow, G. E. and C. B. Hellquist. *Aquatic and Wetland Plants of Northeastern North America.* Madison, WI: The University of Wisconsin Press, 2000.

Fassett. N. C. *A Manual of Aquatic Plants.* Madison, WI: University of Wisconsin Press, 1957.

Subsidence

To subside is to sink or fall. Subsidence is commonly associated with the lowering of the earth's surface due to actions that have occurred below the surface. Sometimes this is a natural phenomenon, such as the solubilizing and removal of minerals by water. When the underground support system is removed in the process, the surface of the land sinks to a new level. This often leads to a special topographic form known as Karst **topography**. Human activities that lead to the extraction of ores, minerals, and **fossil fuels** often lead to a weakened mineral structural support and subsidence of the surface of the earth. In the **arid** portions of the earth, extraction of water from sub-surface aquifers has led to subsidence of the earth's surface as well.

Subsoil

The portion of the **soil** that is below the surface. The subsoil is often referred to as the B **horizon**. Subsoils are generally lower in organic matter, lighter in color, denser, and often have a higher clay content than surface soils. Subsoils are generally not as productive as surface soils for plant growth. Thus, if surface soils are eroded, future productivity is reduced when the subsoil is used for crop production.

Succession

Succession is the gradual transformation or creation of a **biological community** as new **species** move into an area and modify local environmental conditions. Primary succession occurs when plant and animal species colonize a previously barren area, such as a new volcanic island, a sand dune, or recently glaciated ground. In these cases every living thing, from **soil** bacteria and **fungi** to larger plants and animals, must arrive from some adjacent **habitat**. Secondary succession is the development of communities in an area that has been disturbed by fire, hurricanes, field clearing, tree felling, or some other process that removes most plants and animals. Intermediate successional communities are known as "seral stages" or "seres."

As early successional species become established, they alter their **environment** and make it more habitable for later seral stages. Usually a disturbed or barren area has low soil **nutrient** levels, intense sunlight, and no protection from violent weather. Because precipitation quickly runs off the bare ground, little moisture is available for plant growth. Species that can survive under such harsh conditions have little **competition**, and they spread quickly. As they grow and thicken, these plants add organic matter to the soil, which aids moisture retention and helps soil bacteria to grow. As soil nutrients and moisture increase, larger shrubs and perennial plants can take root. The shade of these larger species weakens and eventually eliminates the original pioneering species, but it cools the local environment, further improving moisture availability and allowing species that demand relative environmental **stability**, such as woodland species, to begin moving in. Eventually, shade-tolerant species of plants will come to dominate the area that sun-loving plants had first colonized.

One of the most well-documented examples of primary succession occurred in 1883 when the **volcano** on the Indonesian island of **Krakatoa** erupted, destroying most of the island and its life forms, and leaving a new island of bare volcanic rock and ash. Within a few years several species of grasses, ferns, and flowering shrubs had managed to arrive, carried by wind, water, or passing birds from islands 25 mi (40 km) or more distant. Just 40 years later, more than 300 plant species were growing, and in some areas over 12 in (30 cm) of soil had developed: soil bacteria, insects, and **decomposers** had managed to reach the island and were turning fallen organic **detritus** into soil. Over the decades, the total number of species gradually stabilized, but the character of the community continued to change as incoming species replaced earlier arrivals. On Krakatoa succession happened with unusual speed because the **climate** is warm and humid and because the fresh volcanic ash made a nutrient-rich soil.

Early successional species, those most able to quickly establish a foothold, are known as pioneer species. Usually pioneer species are opportunists, able to find nourishment and survive under a great variety of conditions, quick to grow, and able to produce a great number of seeds or offspring at once. Pioneer species have very effective means of dispersing seeds or young, and their seeds can often remain dormant in soil for some time, sprouting only after conditions become suitable. Dandelions are well-known pioneers because they can quickly produce a seed head with hundreds of seeds. Each tiny seed has a lightweight structure that allows the wind to carry it long distances. Dandelions are quick to invade a lawn because of this effective seed-dispersal tactic, because they can survive under sunny or shady conditions, and because they grow and reproduce quickly. In open sunshine, they are competitive enough to establish themselves despite the presence of a thick mat of turf grass.

Later successional species tend to be more shade tolerant, slower to grow and reproduce, and longer lived than pioneer species. Their larger seeds do not disperse as easily (compare the size of an acorn or walnut with that of a dandelion seed), and their seeds cannot remain dormant for very long before they lose viability. Seedlings of some late successional species, such as the Pacific Northwest hemlock, require shade to survive, and most require considerable moisture and soil nutrients.

We usually think of succession occurring after a catastrophic environmental disturbance, but in some cases the gradual environmental changes of succession proceed in the absence of disturbance. Two outstanding examples of this are bog succession and the invasion of prairies by shrubs and trees.

In cool, moist, temperate climates, plant succession often turns ponds into forest through bog succession. Water-loving plants, mainly rush-like sedges and sphagnum moss, gradually creep out from the pond's edges. Often these plants form a floating mat of living and dead vegetation. Other plants—cranberries, Labrador tea, bog rosemary—become established on top of the mat. Organic detritus accumulates below the mat. Eventually the bog becomes firm enough to support black spruce, tamarack, and other tree species. As it fills in and dries, the former pond slowly becomes indistinguishable from the surrounding forest.

Many **grasslands** persist only in the presence of occasional wildfires. During wet decades or when human activity prevents fires, woody species tend to creep in from the edges of a **prairie**. If sufficient moisture is available and if fires do not return, open grassland can give way to forest. Fire suppression has aided the advance of forests in this way across much of the United States, Canada, and Mexico. Grazing or browsing animals can also be important forces in maintaining biological communities. Adding or removing grazers can initiate successional processes.

Longer, slower disturbances than fire or field clearing can also initiate succession. Over the course of centuries, climate change can cause significant alteration in the character of biomes. Minor variations in rainfall or temperature ranges can alter community structure for decades or centuries. The end of the last glacial period about 10,000 years ago allowed **tundra**, then grasslands, then temperate forests to advance northward across North America. Even geologic activity, such as mountain building or changes in sea level, has caused succession. More recently, human introductions of species from one continent to another have caused significant restructuring of some biological communities.

Turnover from one community to another may occur in just a few years or decades, as it did on Krakatoa. Stages

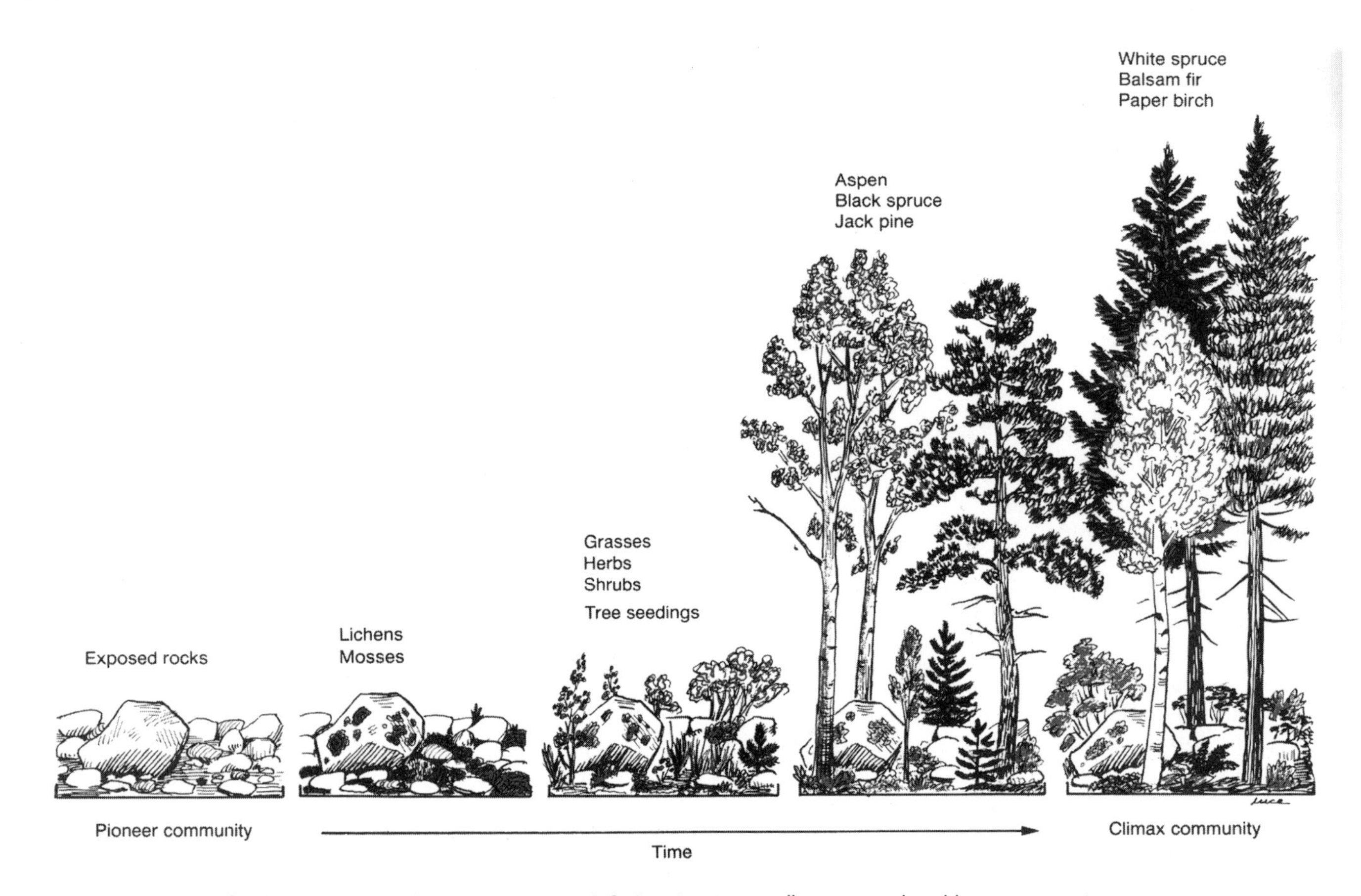

Five stages of primary succession on a terrestrial site. (McGraw-Hill Inc. Reproduced by permission.)

in bog succession may take a hundred years or more. Often, succession continues for centuries or millennia before a stable and relatively constant biological community emerges. This is especially likely in areas whose climate is moist and mild enough to support the high species diversity of temperate or **tropical rain forests**. However, simpler ecosystems such as arctic tundra may take centuries to recover from disturbance because low temperatures and short summers make growth rates almost universally slow.

Generally, successional processes are understood to lead ultimately to the emergence of a **climax** community, a group of species perfectly suited to a region's climate, soil types, and other environmental conditions. Climax communities are said to exist in equilibrium in their environments; where early successional species groups tend to facilitate the development of later stages, climax communities, sometimes called mature communities, tend to have self-perpetuating characteristics or mechanisms that help maintain local environmental conditions and help the community to persist. In the 1930s, F. E. Clements, one of the best-known plant ecologists in the United States, identified a list of just 14 climax communities that he claimed were the final result of succession for all the various sets of environmental conditions in the country.

More recent evaluations of climax communities, however, have concluded that climax forests are often a patchwork of stable and unstable areas. In an **old-growth forest** in the Pacific Northwest, hemlock and spruce may compose the local climax community, but every time a large tree falls, it creates a sunny opening that makes way for other seral species to grow for a time. Occasional fires disturb larger patches of forest, and succession starts again from the beginning in disturbed areas. The entire **biome** is more a shifting and changing mosaic than a uniform and unchanging forest. Often, slight gradations in soil types, slope, exposure, and moisture cause a variety of stable communities to blend in an area. The term "polyclimax community" was developed to recognize the complexity of such assemblages.

In some cases, climax communities may be difficult to distinguish at all. For instance, in a biome regularly disturbed by fire, "climax" species usually require fire to aid seed germination and eliminate competitors. Definitions here become somewhat blurred: Is a climax community one

The area around Sudbury, Ontario, was deforested by emissions from nickel and copper smelting. (Photograph by A. J. Copley. Visuals Unlimited. Reproduced by permission.)

that exists without disturbance? Is fire a disturbance or a maintenance factor? Even if a very old forest community is considered, on a time scale of thousands of years, it may be difficult to determine whether what is observed today is really a permanent climax community or just a very long-standing seral stage. Despite these questions, the concept of an ultimate climax community is central to the idea of succession.

Usually climax communities are considered more diverse than intermediate communities. Because climax communities can remain stable for a long time, species can diversify and develop specialized niches. This specialization can lead to the coexistence of a great number of species, as is the case in tropical rain forests, where thousands of species may share just a few acres of forest. In some cases, however, intermediate seral stages actually have greater diversity because they contain elements of multiple communities at once. *See also* Biodiversity; Biological fertility; Ecological productivity; Growth limiting factors; Introduced species; Restoration ecology

[*Linda Rehkopf*]

Resources

Books

Brown, J. H., and A. C. Gibson. *Biogeography*. St. Louis: Mosby, 1983.

Ricklefs, R. E. *Ecology*. 3rd ed. New York: W. H. Freeman and Co., 1990.

Sudbury, Ontario

The town of Sudbury, Ontario, has been the site of a large metal mining and processing industry since the latter part of the nineteenth century. The principal metals that have been sought from the Sudbury mines are **nickel** and **copper**. Because of Sudbury's long history of mining and processing, it has sustained significant ecological damage and has provided scientists and environmentalists with a clear case study of the results.

Mining and processing companies headquartered at Sudbury began by using a processing technique that oxidized sulfide ores using roast beds, or heaps of ore piled upon wood. The heaps were ignited and left to smolder for several months, after which cooled metal concentrate was collected

and shipped to a refinery for processing into pure metals. The side effect of roast beds was the intense, ground-level plumes of **sulfur dioxide** and metals, especially nickel and copper, they produced. The **smoke** devastated local ecosystems through direct phytotoxocity and by causing **acidification** of **soil** and water. After the vegetative cover was killed, soils eroded from slopes and exposed naked granitic-gneissic bedrock, which became pitted and blackened from reaction with the roast-bed plumes.

After 1928, the use of roast beds was outlawed, and three smelters with **tall stacks** were constructed. These emitted pollutants higher into the **atmosphere**, but some local vegetation damage was still caused, lakes were acidified, and toxic contaminants were spread over an increasingly larger area.

Over the decades, well-defined patterns of ecological damage developed around the Sudbury smelters. The most devastated sites occurred closest to the sources of **emission**. They had large concentrations of nickel, copper, and other metals, were very acidic with resulting toxicity from soluble **aluminum** ions, and were frequently subjected to toxic fumigations by sulfur dioxide. Such sites had very little or no plant cover, and the few **species** that were present were usually physiologically tolerant ecotypes of a few widespread species.

Ecological damage and environmental contamination lessened with increasing distance from the **point source** of emissions. Obvious damage to terrestrial ecosystems was difficult to detect beyond 10–12 mi (15–20 km), but contamination with nickel and copper could be observed much farther away. **Oligotrophic** lakes with clear water, low fertility, and little buffering capacity, however, were acidified by the **dry deposition** of sulfur dioxide at least 25–31 mi (40–50 km) from Sudbury.

In 1972, a very tall, 1,247-ft (380-m) "superstack" was constructed at the largest of the Sudbury smelters. The superstack resulted in an even greater dispersion of **smelter** emissions. This, combined with closing of another smelter, and reduction of emissions by **flue gas** desulfurization and the processing of lower-sulfur ores, resulted in a substantial improvement of local **air quality**. Consequently, a notable increase in the cover and species richness of plant cover close to the Sudbury smelters has occurred, a process that has been actively encouraged by revegetation activities along roadways and other amenity areas where soil remained. Lakes close to the superstack and the closed smelter have also become less acidic, and their biota has responded accordingly. There is controversy, however, about the contributions that the still large emissions of sulfur dioxide may be making towards the regional **acid rain** problem. It is possible that height of the superstack will be decreased to reduce the longer-range transport of emitted sulfur dioxide. *See also* Air pollution; Contaminated soil; Water pollution

[*Bill Freedman Ph.D.*]

Resources
Books

Freedman, B. *Environmental Ecology*. San Diego: Academic Press, 1989.

Nriagu, J., ed. *Environmental Impacts of Smelters*. New York: J. Wiley, 1984.

Sulfate particles

Sulfate particles are sub-micron sized, sulfur-containing airborne particles. Most sulfate is a secondary pollutant, formed by the oxidation in the **atmosphere** of **sulfur dioxide** gas. Sulfur dioxide is emitted largely in fossil fuel **combustion**, particularly from **power plants** burning **coal**. A small fraction (generally well under 10%) of sulfur is emitted as primary sulfate at the combustion source. The use of coal-cleaning, **scrubbers**, and low sulfur coal have reduced sulfur dioxide emissions in the United States, and thus airborne sulfate concentrations have decreased.

In the atmosphere, sulfur dioxide (SO_2) emissions are slowly transformed to sulfate (SO_4) at a rate of 0.1–5% per hour, with the rate increased by higher temperatures, sunshine, and the presence of oxidants. Further reactions with water vapor may produce sulfuric **acid** (H_2SO_4), a corrosive acid which is injurious to ecosystems and humans, and also ammonium sulfate $(NH_4)_2SO_4$, which is particularly effective at impairing **visibility**. The atmospheric **residence time** for sulfate particles is long, 2–10 days, which permits transport on a regional or continental scale across hundreds or thousands of miles. Because sulfate particles are relatively soluble, precipitation effectively washes them out, resulting in **acid rain**. Of the various sizes of aerosols, sulfate particles are in the accumulation mode, and typically there might be about 10,000 particles per 0.06 in^3 (1 cm^3) in an urban area.

In addition to problems of acid rain and visibility degradation, sulfate particles can cause a number of health problems. Community epidemiological studies report associations of annual and multi-year average concentrations of PM_{10} (particulates), $PM_{2.5}$ (fine **particulate**), and sulfates with health effects that include premature **mortality**, increased respiratory symptoms and illness (e.g., **bronchitis** and cough in children), and reduced lung function. The risks associated with long-term exposures, although highly uncertain, appear to be larger than those associated with short-term exposures. Other analyses have shown statistically significant associations between sulfate and other particulate air pollutants with total and cardiopulmonary mortality. Ad-

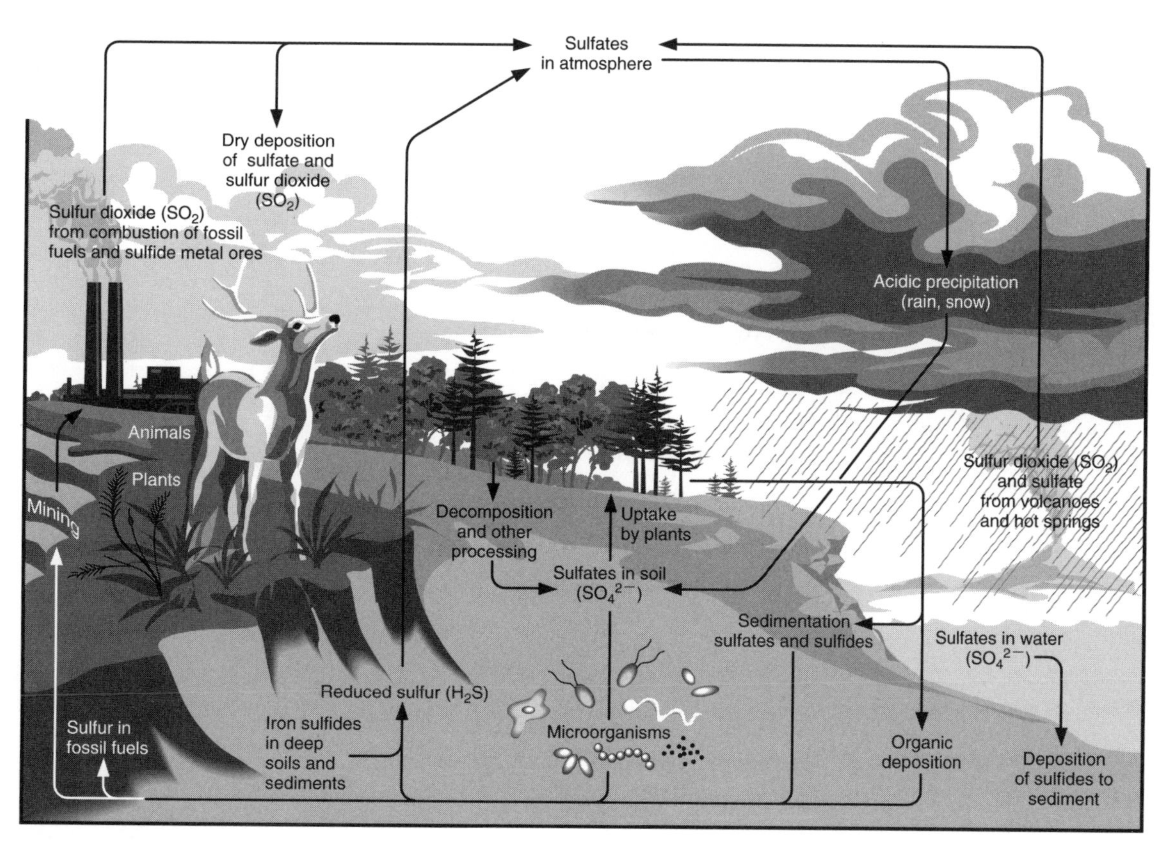

The sulfur cycle. (Illustration by Hans & Cassidy.)

ditionally, animal studies suggest that concentrations approaching ambient levels of ammonium sulfate and nitrate can cause morphometric changes that could lead to a decrease in compliance or a "stiffening" of the lung.

Because of the long time and distance scales required for transformation, and because sulfate particles are small, which allows them to remain airborne for several days, sulfate concentrations tend to be uniformly distributed across broad regions. In the eastern United States, sulfate particles constitute half or more of fine fraction particle concentrations (also called $PM_{2.5}$. Thus, much of the sulfate in an urban area or **airshed** arises from distant or "regional" sources. If at high levels, this "background" component can pose a dilemma for **air quality** management, since **emission** reductions from local sources do not control the problem. Instead, transboundary agreements such as the Canada-United States Air Quality Agreement, which was signed March 13, 1991, have been used to monitor and control emissions from distant sources that cause most of the emissions.

Governmental regulations in the United States do not directly address sulfate; however, the **National Ambient Air Quality Standards** do set limits on particulate concentrations (PM_{10}), and regulations on fine particulate ($PM_{2.5}$) have been proposed. In addition to these ambient standards, regulations limit emissions of sulfur dioxide at sources (in the New Source Performance Standards), and the degradation of existing air quality levels (Prevention of Significant Deterioration). In Europe, deposition of acidic compounds, largely from atmospheric sulfate, is controlled by setting critical loads that depend on the capacity of a region to withstand acid inputs without lake and **soil acidification**.

[*Stuart Batterman*]

Sulfur cycle

The series of chemical reactions by which sulfur moves through the earth's **atmosphere**, hydrosphere, lithosphere, and **biosphere**. Sulfur enters the atmosphere naturally from volcanoes and hot springs and through the **anaerobic** decay of organisms. It exists in the atmosphere primarily as **hydrogen** sulfide and **sulfur dioxide**. After conversion to sulfates in the air, sulfur is carried to the earth's surface by precipitation. There it is incorporated into plants and animals who

return sulfur to the earth's crust when they die. Through their use of **fossil fuels**, humans have a large effect on the sulfur cycle, approximately tripling the amount of the element returned to the atmosphere.

Sulfur dioxide

Along with sulfur trioxide, one of the two common oxides of sulfur existing in the **atmosphere**. Sulfur dioxide is produced naturally in the atmosphere by the oxidation of **hydrogen** sulfide, a gas released from volcanoes and hot springs. About 22 million tons (20 million metric tons) of sulfur dioxide is released annually into the atmosphere by human activities, primarily during the **combustion** of **fossil fuels**. The amount of **anthropogenic** sulfur dioxide is roughly equal to that formed naturally from hydrogen sulfide. Sulfur dioxide reacts with water in clouds to form **acid rain**, causes damage in plants, and is responsible for a variety of respiratory problems in humans.

Superconductivity

In 1911, Dutch physicist Heike Kamerlingh-Onnes discovered that some materials, when cooled to very low temperatures—within a few degrees of absolute zero—become superconductive, losing all **resistance** to the flow of electric current. Potentially, that discovery had enormous practical significance because a large fraction of the electrical energy that flows through any appliance is wasted in overcoming resistance. Kamerlingh-Onnes's discovery remained a laboratory curiosity for over 70 years, however, because the low temperatures needed to produce superconductivity are difficult to achieve. Then, in 1985, scientists discovered a new class of compounds that become superconductive at much higher temperatures (about -74°F [-170°C]). The use of such materials in the manufacture of electrical equipment promises to greatly increase the efficiency of such equipment.

Superfund

see **Comprehensive Environmental Response, Compensation, and Liability Act; Superfund Amendments and Reauthorization Act (1986)**

Superfund Amendments and Reauthorization Act (1986)

Congress initially authorized **Comprehensive Environmental Response, Compensation, and Liability Act** (CERCLA) legislation in 1980 to clean up abandoned dump sites in the United States that contained **hazardous waste**. The activities mandated under CERCLA were to be administered by the **Environmental Protection Agency** (EPA). The program was reauthorized in 1986 by the Superfund Amendment and Reauthorization Act (SARA) and more commonly referred to as simply the Superfund. Several provisions of CERCLA were changed or clarified.

Superfund sites identified in the original legislation were virtually ignored during the early 1980s. Several key EPA officials resigned after they were charged with mismanaging the monies allocated by the original legislation. The EPA attempted to speed cleanup of contaminated sites, but progress was still too slow for critics, some members of Congress and many individual citizens. When the program expired in September 1985, the cleanup activities at more than 200 sites were delayed for lack of funds. Concern about hazardous waste sites continued. This pressure on Congress was sufficient to facilitate reauthorization of CERCLA.

The Superfund was originally financed by a tax on receipt of hazardous waste, and by a tax on domestic refined or imported crude oil and **chemicals**. The SARA reauthorization increased funding from $1.6 billion to $8.5 billion over five years. It also authorized the use of contributions from potentially responsible parties (persons who had created the environmental hazards or who currently owned the land on which former dump sited were located). SARA declined to place the full financial burden of cleanup on oil and chemical companies. As of 2002, funding is obtained from a broad-based combination of business and public contributions.

SARA emphasizes the importance of remedial actions, specifically those that reduce the volume, toxicity, or mobility of hazardous substances, pollutants and contaminants. Targets for long-term remedial actions are listed on the **National Priorities List**. This listing is revised each year.

As of 2002, more than 1,300 sites around the nation have been placed on the National Priorities List. These sites are considered to be the worst in the country. A total of 40,000 uncontrolled waste sites have been reported to U.S. federal agencies. As of June 2002, work has been completed on a total of 812 sites in the National Priorities List. Factors used to rank the severity of reported sites include the type, quantity, and toxicity of the substance(s) found at the site, as well as the number of people likely to be exposed, the pathways of exposure, and the vulnerability of the **groundwater** supply at the site. If a site poses immediate threats such as the risk of fire or explosion, the EPA may initiate short-term actions to remove those threats before actual cleanup begins.

Critics charge that the number of hazardous waste sites nationwide is still underreported. Many states have

developed their own programs to supplement the federal Superfund.

Under SARA guidelines, the **Agency for Toxic Substances and Disease Registry** performs health assessments at Superfund sites. This program, administered by the **Centers for Disease Control and Prevention**, also lists hazardous substances found on sites, prepares toxicological profiles, identifies gaps in research on health effects, and publishes findings. *See also* Chemical spills; Emergency Planning and Community Right-to-Know Act; Hazardous material; Hazardous Substances Act; Toxic substance; Toxic Substances Control Act; Toxic use reduction legislation

[*L. Fleming Fallon Jr., M.D., Dr.P.H.*]

Resources

Books

Church, T. W. *Cleaning Up the Mess: Implementation Strategies in the Superfund Program.* Washington, DC: Brookings Institute, 1993.

Fogelman, V. M. *Hazardous Waste Cleanup, Liability, and Litigation: A Comprehensive Guide to Superfund Law.* Westport, CT: Greenwood Publishing Group, 1992.

Hutchins, D. C., and R. J. Moriarty. *Walking By Day.* Solon, OH: CPR Prompt Corp, 1998.

Meyers, R. A., and D. K. Dittrick. *The Wiley Encyclopedia of Environmental Pollution and Cleanup.* New York: Wiley, 1999.

Pearson, Eric. *Environmental and Natural Resources Law.* Albany, NY: Matthew Bender & Company, 2001.

Periodicals

Au, W. W., and H. Falk. "Superfund Research Program—Accomplishments and Future Opportunities." *International Journal of Hygiene and Environmental Health* (2002): 165–168.

Lindell, M. K., and R. W. Perry. "Community Innovation in Hazardous Materials Management: Progress in Implementing SARA Title III in the United States." *Journal of Hazardous Materials* (2001): 169–194.

Wentsel, R. S., B. Blaney, L. Kowalski, D. A. Bennett, P. Grevatt, and S. Frey. "Future Directions for EPA Superfund Research." *International Journal of Hygiene and Environmental Health* (2002): 161–163.

Other

Agency for Toxic Substances and Disease Registry. *Superfund Reauthorization.* [cited July 2002]. <http://www.atsdr.cdc.gov/superfnd.html>.

Superfund Basic Research Program. [cited July 2002]. <http://benson.niehs.nih.gov/sbrp>.

U.S. Environmental Protection Agency. <http://www.epa.gov/superfund/> and <http://www.epa.gov/superfund/sites/> and <http://es.epa.gov/oeca/spfund/> and <http://www.epa.gov/superfund/sites/npl/npl.htm> and <http://www.epa.gov/superfund/sites/npl/info.htm>.

Surface mining

Surface mining techniques are used when a vein of **coal** or other substance lies so close to the surface that it can be mined with bulldozers, power shovels, and trucks instead of using deep shaft mines, explosive devices, or **coal gasification** techniques. Surface mining is especially useful when the rock contains so little of the ore being mined that conventional techniques, such as tunneling along veins, cannot be used. Surface mining removes the earth and rock that lies above the coal or mineral seam and places the **overburden** off to one side as **spoil**. The exposed ore is removed and preliminary processing is done on site or the ore is taken by truck to processing plants. After the mining operations are complete, the surface can be recontoured, restored, and reclaimed.

Surface mining already accounts for over 60% of the world's total mineral production, and the percentage is increasing substantially. Many factors contribute to the popularity of surface mining. The lead time for developing a surface mine averages four years, as opposed to eight years for underground mines. Productivity of workers at surface mines is three times greater than that of workers in underground mining operations. The capital cost for surface mine development is between 20 and 40 dollars per annual ton of salable coal, and the start-up expenses for underground mining operations are at least twice that, averaging around 80 dollars per annual ton.

The preliminary stages of mine development involve gathering detailed information about the potential mining site. Trenching and core drilling provide information on the coal seam as well as the overburden and the general geological composition of the site. Analysis of the drill core from the overburden is an important step in preventing environmental hazards. For example, when the shale between coal seams and the underlying strata is disturbed by mining, it can produce **acid**. If this instability in the strata is detected by analysis of the drill cores, acid **runoff** can be avoided by appropriate mine design. The data obtained from the preliminary drilling is plotted on topographic maps so that the relationship between seams of ore and the overlying terrain are clearly visible.

After test results and all other relevant data have been obtained, the information is entered into a computer system for analysis. Various mine designs and mining sequences are tested by the computer models and evaluated according to technical and economic criteria in order to determine the optimal mine design. Satellite imagery and aerial photographs are usually taken during the exploratory stages. These photographs serve as an excellent visual record of the actual environmental conditions prior to the mining operation, and they are frequently utilized during the **reclamation** process.

There are several types of surface mining, and these include area mining, contour mining, auger mining, and open-pit mining. Area mining is used predominantly in the Midwest and western mountain states, where coal seams lie horizontally beneath the surface. Operations begin near the coal outcrop—the point where the ore lies closest to the surface. Large stripping shovels or draglines dig long parallel

trenches; this removes the overburden, leaving the ore exposed. The overburden excavated from the trench is thrown into the previous trench, from which coal has already been extracted. The process is similar to a farmer plowing a field in furrows. Because of the steep slopes and rugged terrain frequently encountered in area mining operations, the ore is usually recovered by small equipment such as front end bucket loaders, bulldozers, and trucks.

Area mining is also practiced in Appalachia, where many of the coal seams lie under mountains or foothills. This type of area mining is commonly known as mountaintop removal. The mountains mined using this technique generally have long ridges with underlying deposits. A cut is made parallel to the ridge, and subsequent cuts are made parallel to the first; this results in the entire top of the mountain being leveled off and flattened out. When these areas are reclaimed, the most frequent designated uses are grazing lands or development sites.

Contour mining is used primarily at the point where the seam lies closest to the surface, on steep inclines such as in the Appalachian mountains. In these areas, coal usually lies in flat continuous beds, and the excavation continues along the side of the mountain. This produces a long, narrow trench, with highwall extending along the trench—the contour line for horizontal coal seams. Recent federal regulations have outlawed many contour mining practices, such as leaving exposed highwall and spoil on the mountainside.

The auger process is used primarily in salvage operations. When it is no longer economically profitable to remove the overburden in the highwall, augering techniques are utilized to recover additional tonnage. The auger extracts coal by boring underneath the final highwall. Currently, auger mining only accounts for 4% of surface-mined coal production in the United States.

Open-pit mining is used primarily in western states, where coal seams are at least 100 ft (30 m) thick. The thin overburden is removed and taken away from the site by truck, leaving the exposed coal seam. This type of mine operation is very similar to rock quarry operations.

The equipment used in surface mining ranges from bulldozers, front-end bucket loaders, scrapers, and trucks to gigantic power shovels, bucket-wheel excavators, and draglines. Over the past ten years, technological development has concentrated on mechanization and development of heavy-duty equipment. Due to economic factors, equipment manufacturers have focused on improving performance of existing equipment instead of developing new technologies. Since the recession of the 1980s, bucket capacity and the size of conventional machines have increased. Concentration of production technology has created mining systems that incorporate mining equipment with continuous **transportation** systems and integrated computers into all aspects of the industry.

Surface mining can have severe environmental effects. The process removes all vegetation, destroying microflora and **microorganisms**. The **soil**, **subsoil**, and strata are broken and removed. **Wildlife** is displaced, **air quality** suffers, and surface changes occur due to oxidation and topographic changes.

Hydrology associated with surface mining has a major effect on the **environment**. Removal of overburden may change the **groundwater** in numerous ways, including **drainage** of water from the area, altering the direction of **aquifer** flow, and lowering the water tables. It also creates channels that allow contaminated water to mingle with water of other aquifers. Acid water runoff from the mining operations can contaminate the area and other water sources. In addition to being highly acidic, the runoff from mining operations also contains many other trace elements that adversely affect the environment.

Federal regulations require that **topsoil** be redistributed after the mining is complete, but many people do not consider this requirement sufficient. When soil is removed, the soil structure breaks down and compacts, preventing normal organic matter from getting into the soil. Microorganisms are destroyed by the changes in the soil and the lack of organic components. The rate of **erosion** in mining areas is also greatly increased due to the lack of native vegetation.

The removal of vegetation and overburden at the mining site displaces all wildlife, and a large portion of it may be completely destroyed. Some forms of wildlife, such as birds and game animals, may get out of the area safely, but those that hibernate or burrow usually die as a result of the mining. Ponds, streams, and swamps are routinely drained before mining operations commence and all aquatic life in the region is destroyed.

In addition to the short-term environmental effects, surface mining also has long-term impact on the **flora** and **fauna** within the region of the mine. Salts, **heavy metals**, acids, and other minerals exposed during removal of overburden suppress growth rate and productivity. Due to changes in soil composition, many native **species** of plants are unable to adjust. The loss of vegetation means a loss of feeding grounds, which in turn disrupts **migration** patterns. Displaced species encroach on neighboring ecosystems, which may cause overpopulation and disruption of adjacent habitats. *See also* Surface Mining Control and Reclamation Act

[*Debra Glidden*]

An open-pit copper mine. (Photograph by Paul Logsdon. Phototake. Reproduced by permission.)

RESOURCES

PERIODICALS

Chironis, N. "With Dozers Bigger Is Better." *Coal Age* 91 (July 1986): 56.

Sanda, A. "Draglines Dominate Big Surface Mines." *Coal Age* 96 (July 1991): 30.

Schmidt, B. "GAO to Interior: Get Tough on Mining." *American Metal Market* 96 (December 21, 1988): 2.

Singhal, R. K. "In Pit Crushing and Conveying Systems." *World Mining Equipment* 10 (January 1986): 24.

"Surface Mining Costs Decline in the Midwest." *Mining* 91 (May 1986): 10.

Surface Mining Control and Reclamation Act (1977)

This act set minimum federal standards for surface **coal** mining and **reclamation** of mining sites. The act requires that: (1) mine operators demonstrate reclamation capability; (2) previous uses must be restored and the land reshaped to its original contour, with **topsoil** replacement and replanting; (3) mining be prohibited on prime western agricultural land, with farmers and ranchers holding veto power; (4) the hydrologic **environment** must be protected, especially from **acid drainage**; and (5) a $4.1 billion fund be established to reclaim abandoned strip mines. States have enforcement responsibility, but the **U.S. Department of the Interior** steps in if they fail to act.

Surface-water quality

see **Water quality**

Survivorship

The term "survivorship" describes the likelihood that an organism will remain alive from one time period to the next. For example, the survivorship of human males in a given country to age 20 might be around 96.5%, while survivorship to age 70 might be 55%.

Age-specific **mortality** can be illustrated in the form of a *survivorship curve* that usually yields one of three patterns of survivorship: mortality concentrated at the end of the maximum life span; a constant **probability** of death for every age group; or high early mortality followed by high

survival. Survivorship can also be depicted with a *life table* showing the number of survivors from a given starting population at specific intervals.

Suspended solid

Refers to a solid that is suspended in a liquid. Most treatment facilities for both municipal and industrial wastewaters must meet **effluent** standards for total suspended solids (TSS). A typical TSS limit for a secondary **wastewater** treatment plant is 20 mg/L. However, some industries may have permits that allow them to **discharge** much more, for example, 500 mg/L. The test for TSS is commonly performed by filtering a known amount of water through a pre-weighed glass-fiber or 0.45 micron (μ) porosity filter. The filter with the solids is then dried in an oven (217°–221°F [103°–105°C]) and weighed. The amount of dried solid matter on the filter per amount of water originally filtered is expressed in terms of mg/L. Solids which pass through the filter in the filtrate are referred to as **dissolved solids**. Settleable solids (i.e., those that settle in a standard test within 30 minutes) are a type of suspended solids, but not all suspended solids are settleable solids.

Sustainable agriculture

Because of concerns over pesticides and **nitrates** in **groundwater**, **soil erosion**, **pesticide** residues in food, **pest** resistance to pesticides, and the rising costs of purchased inputs needed for conventional agriculture, many farmers have begun to adopt alternative practices with the goals of reducing input costs, preserving the resource base, and protecting human health. This is called sustainable agriculture.

Many of the components of sustainable agriculture are derived from conventional agronomic practices and livestock husbandry. Sustainable systems more deliberately integrate and take advantage of naturally occurring beneficial interactions. Sustainable systems emphasize management, biological relationships such as those between the pest and predator, and natural processes such as **nitrogen fixation** instead of chemically intensive methods. The objective is to sustain and enhance, rather than reduce and simplify, the biological interactions on which production agriculture depends, thereby reducing the harmful off-farm effects of production practices.

Examples of practices and principles emphasized in sustainable agriculture systems include:

- Crop rotations that mitigate weed, disease, insect, and other pest problems; increase available soil **nitrogen** and reduce the need for purchased fertilizers; and, in conjunction with **conservation tillage** practices, reduce soil erosion.
- **Integrated pest management** (IPM) that reduces the need for pesticides by crop rotations, scouting weather monitoring, use of resistant cultivars, timing of planting, and biological pest controls.
- Soil and **water conservation** tillage practices that increase the amount of crop residues on the soil surface and reduce the number of times farmers have to till the soil.
- Animal production systems that emphasize disease prevention through health maintenance, thereby reducing the need for antibiotics.

Many farmers and people of rural communities are starting to explore the possibilities of systems of sustainable agriculture. The term "systems" is used because there is no one single way to farm sustainably. The possible ways are as numerous as farmers and potential farmers.

The first aspect of sustainable agriculture is the understanding that a respect for life, in its various forms, is not only desirable but necessary to human survival. A second aspect requires that the farming system not put life in jeopardy, and that its methods not deplete the soil or the water or place farmers in situations where they themselves are depleted, either in numbers or in the quality of their lives.

Another aspect of sustainable agriculture recognizes that farming families are an essential part of a sustainable system. Farmers are the systems' stewards, or caregivers. As stewards, they know their land better than anyone else and are equipped to shoulder the challenge of developing a sustainable system on that land. In a sustainable system, farmers ideally move toward less dependence on off-farm purchased inputs and more toward natural or organic materials. This is accomplished by gaining knowledge about the intricate biological and economic workings of the farm. Lastly, sustainable agricultural systems require the support of consumers as well; they can give support, for example, by selectively buying food raised in close proximity to a buyer's local market.

[*Terence H. Cooper*]

Resources

Books

National Research Council Board on Agriculture. *Alternative Agriculture.* Washington, DC: National Academy Press, 1989.

Periodicals

Thesing, C. "What Is Sustainable Agriculture?" *The Land Stewardship Letter* 10 (1992): 13–14.

Sustainable architecture

Sustainable architecture refers to the practice of designing buildings which create living environments that work to minimize the human use of resources. This is reflected both in a building's construction materials and methods and in its use of resources, such as in heating, cooling, power, water, and **wastewater** treatment.

The operating concept is that structures so designed "sustain" their users by providing healthy environments, improving the quality of life, and avoiding the production of waste, to preserve the long-term survivability of the human **species**.

Hunter and **Amory Lovins** of the **Rocky Mountain Institute** say the purpose of sustainable architecture is to "meet the needs of the present without compromising the ability of **future generations** to meet their own needs."

The term, however, is a broad one, and is used to describe a wide variety of aspects of building design and use. For some, it applies to designing buildings that produce as much energy as they consume. Another interpretation calls for a consciousness of the spiritual significance of a building's design, construction, and siting. Also, some maintain that the buildings must foster the spiritual and physical well-being of their users.

One school of thought maintains that, in its highest form, sustainable architecture replicates a stable **ecosystem**. According to noted ecological engineer David Del Porto, a building designed for sustainability is a balanced system where there are no wastes, because the outputs of one process become the inputs of another. Energy, matter, and information are cascaded through connected processes in cyclical pathways, which by virtue of their efficiency and interdependence yield the matrix elements of environmental and economic security, high quality of life, and no waste. The constant input of the sun replenishes any energy lost in the process.

Sustainability, as it relates to resources, became a widely used term with **Lester Brown's** book, *Building a Sustainable Society,* and with the publishing of the International Union on the Conservation of Nature's "World Conservation Strategy" in 1980.

Sustainability then came to describe a state whereby natural renewable resources are used in a manner that does not eliminate or degrade them or otherwise diminish their renewable usefulness for future generations, while maintaining effectively constant or non-declining stocks of **natural resources** such as **soil**, **groundwater**, and **biomass** (**World Resources Institute**).

Before "sustainable architecture," the term "solar architecture" was used to express the architectural approach to reducing the consumption of natural resources and fuels by capturing **solar energy**. This evolved into the current and broader concept of sustainable architecture, which expands the scope of issues involved to include water use, **climate** control, food production, air purification, **solid waste reclamation**, wastewater treatment, and overall **energy efficiency**. It also encompasses building materials, emphasizing the use of local materials, renewable resources and recycled materials, and the mental and physical comfort of the building's inhabitants. In addition, sustainable architecture calls for the siting and design of a building to harmonize with its surroundings.

The United Nations lists the following five principles of sustainable architecture:

- *Healthful interior environment.* All possible measures are to be taken to ensure that materials and building systems do not emit toxic substances and gasses into the interior **atmosphere**. Additional measures are to be taken to clean and revitalize interior air with **filtration** and plantings.
- *Resource efficiency.* All possible measures are to be taken to ensure that the building's use of energy and other resources is minimal. Cooling, heating, and lighting systems are to use methods and products that conserve or eliminate energy use. Water use and the production of wastewater are minimized.
- *Ecologically benign materials.* All possible measures are to be taken to use building materials and products that minimize destruction of the global **environment**. Wood is to be selected based on non-destructive forestry practices. Other materials and products are to be considered based on the toxic waste output of production. Many practitioners cite an additional criterion: that the long-term environmental and societal costs to produce the building's materials must be considered and prove in keeping with sustainability goals.
- *Environmental form.* All possible measures are to be taken to relate the form and plan of the design to the site, the region, and the climate. Measures are to be taken to "heal" and augment the **ecology** of the site. Accommodations are to be made for **recycling** and energy efficiency. Measures are to be taken to relate the form of building to a harmonious relationship between the inhabitants and **nature**.
- *Good design.* All possible measures are to be taken to achieve an efficient, long-lasting, and elegant relationship of area use, circulation, building form, mechanical systems, and construction technology. Symbolic relationships with appropriate history, the Earth, and spiritual principles are to be searched for and expressed. Finished buildings shall be well built, easy to use, and beautiful.

The NMB Bank headquarters in Amsterdam, the Netherlands, is an example of sustainable architecture. Constructed in 1978, this approximately 150,000-ft^2 (45,500-

m^2) complex is a meandering S-curve of 10 buildings, each offering different orientations and views of gardens. Constructed of natural and low-polluting materials, the buildings feature organic design lines, indoor and outdoor gardens, passive solar elements, heat recovery, water features, and natural lighting and ventilation. Built for an estimated 5% more than a conventional office building, the NMB building's operating costs are only 30% of those of a conventional building. Another example is the Solar Living Center in Hopland, California, which employs both passive and photovoltaic solar elements, as well as ecological wastewater systems. The rice straw bale and cement building is constructed around a solar calendar.

Sustainable architecture as a movement

Some maintain that sustainability, as it relates to architecture, refers to a process and an attitude or viewpoint. Sustainability is "a process of responsible consumption, wherein waste is minimized, and buildings interact in balanced ways with natural environments and cycles, balancing the desires and activities of humankind within the integrity and **carrying capacity** of nature, and achieving a stable, long-term relationship within the limits of their local and global environment." (Rocky Mountain Institute.)

However, sustainable architecture does not necessarily mean a reduction in material comfort. Sustainability represents a transition from a period of degradation of the natural environment (as represented by the industrial revolution and its associated unplanned and wasteful patterns of growth) to a more humane and natural environment. It is doing more with less.

Proponents of sustainable architecture occasionally debate the broader applications of the term. Some say that sustainable buildings should generate more energy over time (in the form of power, etc.) than was required to construct, fabricate their materials, operate, and maintain them. This is also referred to as "regenerative architecture," which John Tillman Lyle sums up in his book, *Regenerative Design for Sustainable Development*, as "living on the interest yielded by natural resources rather than the capital." Others simply see it as an approach to making buildings less consumptive of natural resources.

Spiritual aspects of sustainable architecture

A spiritual viewpoint is that sustainable architecture is "stewardship," a recognition and celebration of the human environment as a vital part of the larger universe and of humankind's role as caretakers of the earth. Viewed in this way, resources are regarded as sacred. Another perspective is that the creation of a building in the likeness of a living system is somewhat religious, as a divine entity creates a living order.

Although the term communicates slightly different meanings to various audiences, it nevertheless serves as a consciousness-raising focus for creating greater concern for the built environment and its long-term viability. Rather than representing a return to subsistence living, buildings designed for sustainability aim to improve the quality and standards of living. Sustainable architecture recognizes people as temporary stewards of their environments, working toward a respect for natural systems and a higher quality of life.

[*Carol Steinfeld*]

Resources

Books

Barnett, D., and W. Browning. *A Primer on Sustainable Building.* Colorado: Rocky Mountain Institute, 1995.

Dimensions of Sustainable Development. Washington, DC: World Resources Institute, 1990.

Lyle, J. *Regenerative Design for Sustainable Development.* New York: John Wiley & Sons, Inc., 1994.

Nebel, B. J. *Environmental Science: The Way the World Works.* 3rd ed. New York: Prentice Hall, 1990.

Orr, D. W. *Ecological Literacy.* New York: State University of New York Press, 1992.

World Resources 1992–93: A Guide to the Global Environment. New York: Oxford University Press, 1992.

Other

Kremers, J. "Sustainable Architecture." *Architronic* December 1996 [cited July 2002] <http://architronic.saed.kent.edu>.

Sustainable biosphere

The **biosphere** is the region of the earth that supports life; it includes all land, water, and the thin layer of air above the earth. A sustainable biosphere is one with the continuing ability to support life.

Some scientists have suggested that the oceans and **atmosphere** in the biosphere adjust to ensure the continuation of life—a theory known as the **Gaia hypothesis**. British scientist **James Lovelock** first proposed it in the late 1970s. He argued that in the last 4,500 years of the earth's approximately five billion years of existence, humans have been placing increasingly greater stress on the biosphere. He called for a new awareness of the situation and insisted that changes in human behavior were necessary to maintain a biosphere capable of supporting life. Maintaining a sustainable biosphere involves the integrated management of land, water, and air.

The land management practices that many believe are necessary for sustaining the biosphere require **conservation** and planting techniques that control **soil erosion**. These techniques encourage farmers to use organic fertilizers instead of synthesized ones. Recycled manure and composted plant materials can be used to put nutrients back into the

soil instead of increasing **pollution** in the **environment**. Crop rotation reduces crop loss to insects and also increases soil fertility. Monitoring **pest** populations helps determine the best times as well as the best methods for eliminating them, and using pesticides only when needed lowers pollution of soil and water and prevents insects from developing resistances.

China provides one model of good land management. Through techniques of **sustainable agriculture** which emphasize **organic farming**, China has met (though not exceeded) the nutritional needs of most of its people. On Chinese farms, **methane** digesters recycle animal wastes, by-products, and general refuse; the gas produced from the digesters is in turn used as an energy source. Biological controls are also used to reduce the number of harmful insects and weeds. Crops are selected for their ability to grow in particular locations, and **strip-farming** and **terracing** help conserve the soil. Crops that put certain nutrients in the soil are planted in alternate rows with crops that deplete those nutrients. Forests, important for **watershed** and as a fuel source, are maintained. Fish farming, or **aquaculture**, is widely practiced, and the fish raised on these farms produce about 10 times more animal protein than livestock raised on the same amount of land. In China, agriculture relies upon human labor rather than more expensive and environmentally harmful machinery, which has the added advantage of keeping millions of people gainfully employed.

Another important aspect of land management is **biodiversity**. Each organism is specially suited for its **niche**, yet all are interconnected, each depending on others for nutrients, energy, the purification of wastes, and other needs. Plants provide food and oxygen for animals, sustain the hydrocycle, and protect the soil. Animals help pollinate plants and give off the **carbon dioxide** that plants need to generate energy. **Microorganisms** recycle wastes, which returns nutrients to the system. The activity of many of these organisms is directly useful to humans. Additionally, having a broad range of organisms means a broad source of genetic material from which to create new, as well as better, crops and livestock. However, a reduction in the number of **species** reduces biological diversity. **Hunting** can destroy species, as can eliminating habitats by cutting down forests, draining **wetlands**, building cities and polluting excessively.

The water portion of the biosphere has two interconnected parts. Freshwater lakes and rivers account for about 3% of the earth's water, while the oceans contain the other 97%. Oceans are important to the biosphere in a number of ways. They absorb some of the gases in the air, regulating its composition, and they influence **climate** all over the earth, preventing drastic weather changes. Additionally, the ocean contains a myriad of plant and animal species, which provide food for humans and other organisms and aid in maintaining their environment.

However, according to many scientists, exploitation of the oceans must be curtailed to ensure that they remain viable. Their argument is that species are too often harvested until they are destroyed, creating an imbalance in the **ecosystem**, and that fishing quotas that do not allow populations to be depleted faster than they can be replaced should be set and enforced. Rather than harvesting immature fish, the **commercial fishing** industry should release them, thus keeping the species population in equilibrium. Pressure must also be taken off individual species, and interactions in the ocean ecosystem should be examined. A strategy designed to consider these factors would benefit the whole ocean community as well as humans. Pollution of the ocean should also be reduced. Pollution can take the form of **garbage**, sewage, toxic wastes, and oil. In excess, these harm the organisms of the ocean, which in turn harms humans. Disposal of these substances should be controlled through, though not limited to, **sewage treatment**, **discharge** restrictions and improved dispersal techniques.

Although freshwater is a renewable resource, care needs to be taken to keep it clean. Wise water management, which includes monitoring water supplies and **water quality**, can ensure that everyone's needs are met. Agriculture and commercial fishing require water, and manufacturing industries use water in many of their processes. Households consume water in cooking, cleaning, and drinking, among other activities. Without water, certain recreations such as water skiing and swimming would be impossible to enjoy. **Wildlife** and fish also need water to survive.

Pollutants reduce and damage the freshwater supply. Pollutants include **chemicals**, sewage, toxic waste, hot water, and many other organic and inorganic materials. Waste **leaching** from landfills contaminates **groundwater**, and surface water is corrupted by wastes dumped into the water by polluted groundwater flowing into the stream or lake, or by polluted rain. Like oceans, freshwater can break down small amounts of pollutants, but large amounts place great pressure on the system. Disposal of potential pollutants requires monitoring and treatment. A useful strategy is to balance the demands of humans with the needs of the ecosystem, emphasizing conservation and pollution reduction.

As with land and water, the atmosphere has been adversely affected by pollution. Pollution in the air causes such phenomena as **smog**, **acid rain**, **atmospheric inversion**, and **ozone layer depletion**. Smog lowers the amount of oxygen in the air, which can cause health problems. **Acid** rain destroys plants, animals, and freshwater supplies. Atmospheric inversion exists when **air pollution** is unable to escape because a layer of warm air covers a layer of cooler air, trapping it close to the ground. Certain compounds, such

as chlorofluoromethanes, thin the **ozone** layer by combining with ozone to split it apart. A thinner ozone layer provides less protection from the sun's **ultraviolet radiation**, which promotes skin **cancer** in humans and reduces the productivity of some crops. Reducing or eliminating harmful emissions can also control air pollution.

Sustainable development is one approach to keeping the biosphere healthy. This method attempts to increase local food production without increasing the amount of land taken. It involves **nature** conservation and **environmental monitoring**, and it advocates encouraging and training local communities to participate in maintaining the environment. The goal is to balance human needs with environmental needs, and proponents of this view maintain that economic growth depends on renewable resources, which in turn depends on permanent damage to the environment being kept at a minimum.

The **Ecological Society of America** published the Sustainable Biosphere Initiative (SBI) in the early 1990s. This document was an attempt to define ecological priorities for the twenty-first century, and it is based on the realization that research in applied **ecology** is necessary for better management of the earth's resources and of the systems that support life. The SBI calls for basic ecological research and emphasizes the importance of education and policy and management decisions informed by such research.

The SBI sets out criteria for determining which research projects should be pursued, and it proposes three top research priorities: global change, biodiversity, and sustainable ecological systems. Global change research focuses on the causes and effects of various changes in all aspects of the biosphere (air, water, land). Biodiversity research focuses on both naturally occurring and human-caused changes in species. It also studies the consequences of those changes. Research on sustainable ecological systems exams the pressures placed on the biosphere, methods of correcting the damage and management of the environment to maintain life.

Creating a sustainable biosphere requires a cohesive policy for reducing consumption and seeking nonmaterial means of satisfaction. Many believe that such a cohesive policy would provide a much-needed focus for the many diverse environmental interests currently vying for public support.

[*Nikola Vrtis*]

RESOURCES

BOOKS

Myers, N., ed. *GAIA: An Atlas of Planet Management.* Garden City, NY: Anchor Press/Doubleday, 1984.

PERIODICALS

"Biosphere Reserves." *Futurist* (May-June 1992): 31–32.

Lubchenco, J., et al. "The Sustainable Biosphere Initiative: An Ecological Research Agenda." *Ecology* 72 (1991): 371–412.

Risser, P. G., J. Lubchenco, and S. A. Levin. "Biological Research Priorities–A Sustainable Biosphere." *Bioscience* (October 1991): 625–627.

Sustainable development

Sustainable development is a term first introduced to the international community by ***Our Common Future***, the 1987 report of the World Commission on **Environment** and Development, which was chartered by the United Nations to examine the planet's critical social and environmental problems and to formulate realistic proposals to solve them in ways that ensure sustained human progress without depleting the resources of **future generations**. This Commission—which was originally chaired by **Gro Harlem Brundtland**, Prime Minister of Norway, and was consequently often called the Brundtland Commission—defined sustainable development as "meeting the needs of the present without compromising the ability of future generations to meet their own needs." The goal of sustainable development, according to the Commission, is to create a new era of economic growth as a way of eliminating poverty and extending to all people the opportunity to fulfill their aspirations for a better life.

Economic growth, in this view, is the only way to bring about a long-range transformation to more advanced and productive societies and to improve the lot of all people. As former President John F. Kennedy said: "A rising tide lifts all boats." But economic growth is not sufficient in itself to meet all essential needs. The Commission, which was actually named the Commission on Sustainable Development, or CSD, pointed out that we must make sure that the poor and powerless will get their fair share of the resources required to sustain that growth. The CSD stated that "an equitable distribution of benefits requires political systems that secure effective citizen participation in decision making and by greater democracy in international decision making."

The concept of "sustainable" growth of any sort is regarded as an oxymoron by some people because the availability of **nonrenewable resources** and the capacity of the **biosphere** to absorb the effects of human activities are clearly limited and must impose limits to growth at some point. Nevertheless, supporters of sustainable development maintain that both technology and social organization can be managed and improved to meet essential needs within those limits.

To clarify this concept, a discussion of what is development and what is meant by "sustainable" is necessary. Development is a process by which something grows, matures,

improves, or becomes enhanced or more fruitful. Organisms develop as they progress from a juvenile form to an adult. The plot in a novel develops as it becomes more complex, clearer, or more interesting. Human systems develop as they become more technologically, economically, or socially advanced and productive. In human history, social development usually means an improved standard of living or a more gratifying way of life.

Something is sustainable if it is permanent, enduring, or can be maintained for the long-term. Sustainable development, then, means improvements in human well-being that can be extended or prolonged over many generations rather than just a few years. The hope of sustainable development is that if its benefits are truly enduring, they may be extended to all humans rather than just the members of a privileged group.

Some development projects have been viewed by some as environmental, economic, and social disasters. Large-scale hydropower projects--for instance in the James Bay region of Quebec or the Brazilian Amazon--were supposed to be beneficial but displaced indigenous people, destroyed **wildlife**, and poisoned local ecosystems with acids from decaying vegetation and **heavy metals** from flooded soils released through **leaching**. Similarly, introduction of "miracle" crop varieties in Asia and huge grazing projects in Africa financed by international lending agencies crowded out wildlife, diminished the diversity of traditional crops, and destroyed markets for small-scale farmers.

Other development projects, however, worked more closely with both **nature** and local social systems. Projects such as the Tagua Palm Nut project sponsored by **Conservation International** in South America encouraged native people to gather natural products from the forest on a sustainable basis and to turn them into valuable products (in this case "vegetable ivory" buttons) that could be sold for good prices on the world market. Another exemplary local development project was the "microloan" financing pioneered by the *grameen* or village banks of Bangladesh. These loans, generally less than $100, allowed poor people who usually would not have access to capital to buy farm tools, a spinning wheel, a three-wheeled pedicab, or some other means of supporting themselves. The banks were not concerned merely with finances, however; they also offer business management techniques, social organization, and education to ensure successful projects and loan repayment.

Still, critics of sustainable development argue that the term is self-contradictory because almost every form of development requires resource consumption. The limits and boundaries imposed by natural systems, they argue, must at some point make further growth unsustainable. Using ever-increasing amounts of goods and services to make human life more comfortable, pleasant, or agreeable must inevitably interfere with the survival of other **species** and eventually of humans themselves in a world of fixed resources.

While it is probably true that traditional patterns of growth that require ever-increasing consumption of resources are unsustainable, development could be possible if it means: (a) finding more efficient and less environmentally stressful ways to provide goods and services; and (b) living with reduced levels of goods and services for those of us who are relatively rich so that more can be made available to the impoverished majority.

Some economists argue that the physical limits to growth are somewhat vague, not because the laws of physics are inaccurate, but because substituting one resource for another in producing goods and services allows a tremendous flexibility in the face of declining abundance of some resources. For example, the extremely rapid adoption of **petroleum** products as sources of energy and raw materials in the early twentieth century is a good case study of how quickly new discoveries and technical progress can change the resource picture.

From this perspective, the decline of known petroleum reserves and the effects of gases from the **combustion** of petroleum products on possible global warming—while serious problems—are also signals to humans to look for substitutes for these products, to explore for unknown reserves, and to develop more efficient and possibly altogether new ways of performing the tasks for which oil is now used. Teleconferencing, for instance, which allows people at widely distant places to consult without traveling to a common meeting site, is a good example of this principle. It is far more energy-efficient to transmit video and audio information than to move human bodies.

The issues of distributive justice or fairness between "developed" and "developing" societies—concerned with who should bear the burden of sustaining future development and how resources will be shared in development—raise many difficult questions. Similarly, the technical problems of finding new, efficient, non-destructive ways of providing goods and services require entirely new ways of doing things. The range of theoretical understanding, personal and group activity, and policy and politics encompassed by this project is so vast that it might be considered the first time humans have tried to grasp all their activities and the workings of nature in its entirety as one integrated, global system.

The first planet-wide meeting to address these global issues was the **United Nations Conference on the Human Environment** which took place in 1972. It was followed by the **United Nations Earth Summit** held in June, 1992, in Rio de Janeiro, Brazil. The scope of sustainability issues taken up at the Earth Summit is in itself remarkable. Agenda 21, a 900-page report drafted in preparation for the conference, included proposals for allocation of international aid

to alleviate poverty and improve **environmental health**. Other recommendations included providing **sanitation** and clean water to everyone, reducing indoor **air pollution**, meeting basic healthcare needs for all, reducing **soil erosion** and degradation, introducing sustainable farming techniques, providing more resources for **family planning** and education (especially for young women), removing economic distortions and imbalances that damage the environment, protecting **habitat** and **biodiversity**, and developing non-carbon energy alternatives.

In June 1997, a follow-up Earth Summit was held to adopt a comprehensive program to further implement Agenda 21. Those attending developed a multi-year program of work that addressed themes and economic sectors each year. Examples include global attention to freshwater management, oceans and seas, agriculture and forests, the **atmosphere** and energy. The CSD is made up of elected members from Member States of the United Nations, and its 2001 chair was Professor Bedrich Moldan of the Czech Republic.

See also Amazon basin; Greenhouse effect

[*Eugene R. Wahl and E. Shrdlu*]

Resources

Books

Leonard, H. J., et al. *Environment and the Poor: Development Strategies for a Common Agenda.* New Brunswick: Transaction Books for the Overseas Development Council, 1989.

MacNeill, J., et al. *Beyond Interdependence: The Meshing of the World's Economy and the Earth's Ecology.* New York: Oxford University Press, 1991.

Ramphal, S. *Our Country, The Planet.* Covelo, CA: Island Press, 1992.

World Bank. *World Development Report: Development and the Environment.* New York: Oxford University Press, 1992.

World Commission on Environment and Development. *Our Common Future.* New York: Oxford University Press, 1987.

Other

"About Commission on Sustainable Development." *United Nations Sustainable Development* [cited July 9, 2002]. <http:www.un.org/esa/sustdev/csdgen.htm>.

Organizations

United Nations Commission on Sustainable Development, United Nations Plaza, Room DC2-2220, New York, NY, USA 10017, 212-963-3170, Fax: 212-963-4260, Email: dsd@un.org, <www.un.org/esa/sustdev/>

Sustainable forestry

Sustainability may be defined in terms of sustaining biophysical properties of the forest, in terms of sustaining a flow of goods and services from the forest, or a combination of the two. A combined definition follows: sustainable forests are able to provide goods and services to the present without impairing their capacity to be equally or more useful to **future generations**. The goods and services demanded of the forest include wood of specified quality; **habitat** for **wildlife**, fish, and invertebrates; recreational, aesthetic, and spiritual opportunities; sufficient water of appropriate quality; protection (e.g., against floods and **erosion**); preservation of natural ecosystems and their processes (i.e. allowing large forested landscapes to be affected only by natural dynamics); and the preservation of **species**.

To provide in perpetuity for future demands, the productivity, diversity, and function of the forest must be maintained and enhanced. A listing of key biophysical properties follows, along with examples of current threats to their maintenance: 1) **Soil** productivity (reduced by erosion and **nutrient** depletion from overharvesting). 2) **Biomass** (degraded in quantity and quality by overharvesting and destructive **logging** practices). 3) Climatic **stability** (possibly threatened by **emission** of **greenhouse gases**). 4) Atmospheric quality (lowered by **ozone** and **sulfur dioxide**). 5) Ground and surface waters (altered by **deforestation** or **drainage**). 6) Diversity of plants, animals, and ecosystems (reduced by deforestation, fragmentation, and replacement of complex natural forests with even-aged forests harvested at young ages).

Forest sustainability also depends on the maintenance and improvement of socioeconomic factors. These include human talent and knowledge; infrastructure (e.g., roads); and political, social, and economic institutions (e.g., marketing systems, stable political systems, international cooperation, and peace). Also essential are technological developments that permit wiser use of the forest and substitute ways to meet human needs. Perhaps the major threat to forest sustainability arises from an ever-increasing population, especially in developing countries, combined with the ever-increasing material demands of the developed nations.

Forests have been managed as a renewable and sustainable resource by stable indigenous tribes for many generations. In recent years, some governments have made efforts in this direction as well. Managing any forest for sustainability is a complex process. The spatial scale of sustainability is often global or regional, crossing land ownerships and political boundaries. The time horizons for sustainability by far exceeds those used for "long-term" planning by business or government.

Some people have even questioned the feasibility of sustainable forestry. Optimists believe technological innovation and changed value systems will enable us to achieve sustainability. Others feel demand will outstrip supply, leading to spiralling forest degradation, and have called for conservative and careful management of existing forests.

[*Edward Sucoff*]

RESOURCES
BOOKS
Kimmins, H. *Balancing Act.* UBI Press, 1992.
World Resource Institute. *World Resources 1992-1993.* New York: Oxford Press, 1992.

Swamp

see **Mangrove swamp; Wetlands**

Swimming advisories

Swimming advisories are warnings to the public that contact with beach water could cause an illness. Advisories may be issued after water monitoring, testing that reveals potentially harmful bacteria levels. Water monitoring assesses levels of *E. coli* and fecal coliform contamination in recreational waters. Local authorities restrict use of water for recreational activities when the bacteria level exceeds standards for safe usage.

A beach may be closed when the level of water contamination poses a definite health risk. Mildly polluted water can cause conditions such as a headache, sore throat, or vomiting. Highly polluted water can cause hepatitis, **cholera**, and typhoid fever.

Swimming advisories are usually issued by governmental agencies and apply to public beaches at the ocean and waterways such as bays, lakes, and rivers. Advisories may be issued for public pools, too. However, not all states had programs to monitor **water quality** and issue public advisories as of May of 2002.

According to the U.S. **Environmental Protection Agency** (EPA), state and local governments issued nearly 4,000 beach closings and swimming advisories in 1995. Those actions involved the ocean, bays, and the **Great Lakes**. In 2000, there were at least 11,270 beach closings and advisories in the United States, according to the **Natural Resources Defense Council**. NRDC also reported 48 extended closings and advisories that lasted from six to 12 weeks. Also reported were 50 permanent closings and advisories that lasted for more than 12 weeks.

NRDC is an environmental group that has surveyed beaches since 1990. Each July, the group issues "Testing the Waters," a state-by-state assessment of beaches. The survey covers ocean, bay, and Great Lakes beaches. The report was based on information from local, county, state and federal agencies including the EPA.

According to NRDC, 11 states and Guam monitored water quality at all beaches at least once a week. Regular monitoring of all beaches was performed in California, Connecticut, Delaware, Hawaii, Illinois, Indiana, New Hampshire, New Jersey, North Carolina, Ohio, and Pennsylvania. Monitoring of some beaches was performed at least weekly in Alabama, Florida, Georgia, Maine, Maryland, Massachusetts, Michigan, Minnesota, Mississippi, New York, Rhode Island, South Carolina, Texas, Virginia, Wisconsin, Puerto Rico, and the Virgin Islands.

In states where monitoring improved, there was more awareness about unhealthy water quality. More monitoring led to additional advisories and closures. The 2000 NRDC report ranked California at the top of the state list in terms of advisories and closures. Nearly one-third of California advisories and closures occurred in San Diego and Los Angeles counties.

An example of procedures for swimming advisories can be found in a San Diego County report for 2001. For the county, 2001 was the first full year that San Diego followed regulations in the state Beach Safety Bill. In 2001, water contamination caused 1,301 advisory and closure days, according to a report from the San Diego County Department of Health.

An advisory was triggered when bacterial levels exceeded state standards. Sources of bacteria could include human or **animal waste**, **soil**, or decaying plant material. During an advisory, signs on the beach warn, "Contact With This Water May Cause An Illness."

Beach closings can be caused by sewage spills. These pose the greatest health risk to swimmers, and beach signs warned, "Contaminated Water — Keep Out."

San Diego County also listed 72-hour general advisories. The three-day advisories occurred after rainfall measuring 0.2 in (0.5 cm) or more. Rainfall washes pollutants off areas like streets. Pollutants traveled to the ocean through storm drains. Advisories were posted on the County Web page and phone tape, and in the local newspaper.

San Diego was typical of beach cities in terms of what caused **water pollution** and illnesses. Sources of water **pollution** include water **runoff**, malfunctioning **sewage treatment** plants, and the waste and litter left by people.

People are at risk when they swim or come into contact with polluted water. The risk of illness increases when a person swallows water. Furthermore, direct exposure to water sometimes causes skin and eye infections, according to the EPA.

The EPA advised the public to visit beaches that are regularly monitored. Groups including the EPA provide beach monitoring information on their Web sites. The EPA also cautioned people against swimming near sewage **discharge** pipes and to avoid swimming in urban beaches after a heavy rainfall.

Groups campaigning for national water monitoring policy include **American Oceans Campaign**, an organiza-

tion founded by actor Ted Danson. In 2002, American Oceans Campaign and NRDC were among the groups that wanted the Bush administration to implement federal **water quality standards** that President Bill Clinton approved before leaving office in 2000.

The Beaches Environmental Assessment and Coastal Health (BEACH) Act of 2000 authorized $30 million for beach monitoring grants. States with coasts and bays and those containing the Great Lakes are eligible for grants allocated through the EPA. States can use the grants to set up or improve programs to monitor water quality. Grant money can also be used for programs to inform the public about water pollution. In 2002, President George W. Bush budgeted $2 million for BEACH grants.

[*Liz Swain*]

Resources

Organizations

American Oceans Campaign., 600 Pennsylvania Avenue SE, Suite 210, Washington, D.C. USA 20003, (202) 544-3526, Fax: (202)544-5625, Email: info@americanoceans.org, <http://www.americanoceans.org>

Environmental Protection Agency Office of Water (4101M), 1200 Pennsylvania Avenue, N.W., Washington, D.C. USA 20460, (202) 566-0388, Fax: (202) 566-0409, Email: OW-GENERAL@epa.gov, <http://www.epa.gov/waterscience/beaches/>

Natural Resources Defense Council., 40 West 20th Street, New York, NY USA 10011, (212) 727-2700, Fax: (212) 727-1773, Email: nrdcinfo@nrdc.org, <http://www.nrdc.org>

Swordfish

The swordfish, *Xiphias gladius*, is classified in a family by itself, the Xiphiidae. This family is part of the Scombroidea, a subfamily of marine fishes that includes tuna and mackerel.

Tuna and mackerel are two of the fastest-swimming creatures in the ocean. Swordfish rival them, as well as mako **sharks**, with their ability to reach speeds approaching 60 mph (96.5 kph) in short bursts. Their streamlined form and powerful build account for their speed, as well as the fact that their torpedo-shaped body is scaleless and smooth, thus decreasing surface drag. Swordfish have a tall, sickle-shaped dorsal fin that cuts through the water like a knife, and their long pectoral fins, set low on their sides, are held tightly against their body while swimming. Swordfish lack pelvic fins, and they have reduced second dorsal and anal fins which are set far back on the body, adding to the streamlining. It has been suggested that their characteristic "sword"—a broad, flat, beak-like projection of the upper jaw—is the ultimate in streamlining. The shape of their bill is one of the characteristics that separates them from the true billfish, which include sailfish and marlin.

Swordfish live in temperate and tropical oceans and are particularly abundant between 30 and 45 degrees north latitude. They are only found in colder waters during the summer months. Although fish in general are cold-blooded vertebrates, swordfish and many of their scombroid relatives are warm-blooded, at least during strenuous activity, when their blood will be several degrees higher than the surrounding waters. Another physiological **adaptation** is an increased surface area of the gills, which provides for additional oxygen transfer and thus allows swordfish to swim faster and longer than other **species** of comparable size.

Swordfish are valuable for both sport fishing and **commercial fishing**. The size of swordfish taken commercially is usually under 250 lb (114 kg), but some individuals have been caught that weigh over 1,000 lb (455 kg). The seasonal **migration** of this species has meant that winter catches are limited, but long-lining in deep waters off the Atlantic coast of the United States has proven somewhat successful in that season.

Swordfish have only become popular as food over the past 50 years. Their overall commercial importance was reduced in the 1970s, however, when relatively high levels of **mercury** were discovered in their flesh. Because of this discovery, restrictions were placed on their sale in the United States, which had a negative effect on the fishery. Mercury contamination was thought to be solely the result of dramatic increase in industrial **pollution**, but recent studies have shown that high levels of mercury were present in museum specimens collected many years ago. Although still a critical environmental and health concern, industrial sources of mercury contamination have apparently only added to levels from natural sources. Today, swordfish is considered somewhat of a luxury food item, sold at fine dining establishments.

Like any delicacy, swordfish entrees catch a tasty price for the establishments that serve them, but at what cost to the **environment**? In 1999 alone, global fisheries caught roughly 55,000 tons of swordfish. About 40% of the all swordfish caught comes from the Pacific Ocean. California fisheries account for about 10% of the Pacific proportion netted each year. As mentioned, until the middle of this century, most of the swordfish caught were relatively large in size. By 1995, however, about 58% of the Atlantic swordfish caught were immature, weighing less than half the average adult mass. Juvenile fish are now caught more frequently because fishing techniques have become very efficient, leading to **overfishing**.

According to the **Natural Resources Defense Council**, an active **conservation** organization, two-thirds of the swordfish caught by United States fishermen are juvenile. In 1963, the average weight of swordfish caught was 250 lb (114 kg). The average weight of swordfish caught in 1999 was 90 lb (41 kg). Among conservation scientists, it is generally agreed that the minimum weight a female swordfish must attain for reproductive success is roughly 150 lb (68

kg). Standards set by the International Commission for the Conservation of Atlantic Tunas (ICCAT), however, set a minimum weight limit at 50 lb (23 kg), well below reproductive weight. The ICCAT, a regulatory council consisting of 36 nations that governs global swordfish and tuna fish catches, has limited overfishing by setting quotas for its member countries. Unfortunately, scientists believe that swordfish populations are still in danger because most swordfish are not reaching reproductive age before being caught.

Government agencies have warned fisheries that swordfish populations are in danger. The United States Marine Fisheries Service reports that the global swordfish numbers have decreased by 70% since the 1960s. Also, it is believed that if the present rate of decline continues, swordfish will become commercially extinct (too few in number to be profitable) within a decade. Despite this, excessive catching of swordfish remains a problem. In an effort to limit the rate and extent of swordfish decline, a boycott and marine conservation campaign was launched. In 1998, the swordfish became the official symbol of global overfishing for the National Coalition for Marine Conservation, representing dozens of **endangered species**. That year, environmental organizations such as the Natural Resource Conservation Council and Seaweb created a partnership with prominent chefs at famous fine dining establishments across the nation. Participating restaurants, leading the boycott, removed swordfish from their menus while the environmental awareness groups spread the word about swordfish decline. Restaurants in Dallas, Philadelphia, Boston, New York City, and Baltimore were among the cities that led the boycott.

Resteraunts also took swordfish off the menu due to high levels of mercury. The **Food and Drug Administration** has set the level of 1 part per million (ppm), but some samples of swordfish have come up as triple that regulated amount. Due to these large amounts, the FDA recommends that women who are or may become pregnant and small children avoid eating swordfish. Although it is too soon to tell, this may have a large impact on the current status of the fish.

In 1999, Carl Safina, director of the Living Oceans Program of the **National Audubon Society**, published the book *Song for the Blue Ocean.* In response to the dramatic changes in fish populations that were being observed, his book urged readers to adopt an interest in the oceans and the growing threat to their ecological **stability**. Still regarded by many to be an unchangeable expanse and infinite source of goods, Safina alerts readers of the threat of human progress to the ocean, the world's largest **ecosystem**. Inspiring activism, and some controversy, Safina's book exposes how human activity affects even the ocean's inhabitants. Among the list of species discussed as being in peril is the swordfish. With efforts such as the restaurant boycott, federal conservation regulation improvements, and social interventions such as the book *Song for the Blue Ocean*, it is hoped that swordfish populations will make a steady come back within the next decade. Their fate, however, ultimately depends upon consumers, who control demand.

[*Eugene C. Beckham*]

Resources
Books

Burton, R., C. Devaney, and T. Long. *The Living Sea: An Illustrated Encyclopedia of Marine Life.* New York: G. P. Putnam's Sons, 1974.

Hoese, H., and R. Moore. *Fishes of the Gulf of Mexico: Texas, Louisiana, and Adjacent Waters.* College Station: Texas A & M University Press, 1977.

McClane, A. *Field Guide to Saltwater Fishes of North America.* New York: Holt, Rinehart, & Winston, 1978.

Nelson, J. *Fishes of the World.* New York: Wiley, 1976.

Wheeler, A. *Fishes of the World: An Illustrated Dictionary.* New York: MacMillan, 1975.

Symbiosis

In the broad sense, symbiosis means simply "living together"—the union of two separately evolved organisms into a single functional unit regardless of the positive or negative influence on either **species**.

Symbiotic relationships fall into three categories. **Mutualism** describes the condition in which both organisms benefit from the relationship, **commensalism** exists when one organism benefits and the other is unaffected, and parasitism describes a situation in which one organism benefits while the other suffers. Symbionts may also be classified according to their mode of life. Endosymbionts carry out the relationship within the body of one species, the host. Exosymbionts are attached to the outside of the host in a variety of ways, or are unattached, seeking contact for specific purposes or at particular times. Similarly, symbionts may either be host-specific, co-evolved to one species in particular, or a generalist, able to make use of many potential partners. In either case, the relationship usually involves coevolutionary levels of integration, where partners influence each other's fitness.

Symbiosis generally requires either behavioral or morphological **adaptation**. Morphological adaptation ranges from the development of special organs for attachment, color patterns that signal partners to chemical cues, and intercellular alterations to accommodate partners. Behavioral adaptations include a range of body postures and displays that indicate readiness to accommodate symbionts.

Symbiosis is fundamental to life. The various forms of relationships between species—principally symbiosis,

A Javan moray eel with cleaner wrasses on the Great Barrier Reef, Australia. This is an example of a symbiotic relationship. (Photograph by Fred McConnaughey. Photo Researchers Inc. Reproduced by permission.)

competition, and predation—create the structure of ecosystems and produce co-evolutionary phenomena which shape species and communities. At the most fundamental level, the transition from non-nucleated prokaryotic cells to eukaryotic cells, which is the evolutionary link to multicellular organisms, is generally believed to be by symbiosis. In higher organisms, 90% of land plants use some sort of association between roots and mycorrhizal **fungi** to provide **nutrient** uptake from **soil**. Almost 6,000 species of fungi and about 300,000 land plants can form mycorrhizal associations. In the next step of the terrestrial food chain, almost all herbivorous species of insects and mammals use symbionts in the gut to digest cellulose in plant cell walls and allow nutrient assimilation. Thus, the energetics of life on earth, and certainly that of higher life forms, are largely dependent on symbiotic relationships. *See also* Mycorrhiza; Parasites

[*David A. Duffus*]

Resources
Books

Margulis, L., and R. Fester, eds. *Symbiosis As a Source of Evolutionary Innovation.* Cambridge, MA: MIT Press, 1991.

Synergism

Synergism is a term used by toxicologists to describe the phenomenon in which a combination of **chemicals** has a toxic effect greater than the sum of its parts.

Malathion and Delnav are two **organophosphate** insecticides that were tested separately for the number of fish deaths they would cause at certain concentrations. When the two chemicals were combined, however, their toxicity was significantly greater than the individual tests would have led scientists to expect. The toxicity of the mixture, in such cases, is greater than the total toxicity of the individual chemicals.

In the manufacture of pesticides, toxic chemicals are deliberately combined to produce a synergistic effect, using an additive known as a "potentiator" or "synergist" to enhance the action of the basic active ingredient. Piperonyl butoxide, for example, is added to the insecticide rotenone to promote synergistic effects.

The study of unintended synergistic effects, however, is still in its early stages, and the majority of documented synergistic effects deal with toxicity to insects and fish. Labo-

ratory studies on the effects of **copper** and **cadmium** on fish have established that synergistic interactions between multiple pollutants can have unanticipated effects. Both copper and cadmium are frequently released into the **environment** from a number of sources, without any precautions taken against possible synergistic impacts. Scientists speculate that there may be many similar occurrences of unpredicted and currently unknown synergistic toxicities operating in the environment.

Synergism has been cited as a reason to make environmental standards, such as the standards for **water quality**, more stringent. Current **water quality standards** for metals and organic compounds are derived from toxicological research on their individual effects on aquatic life. But pollutants discharged from a factory could meet all water quality standards and still harm aquatic life or humans because of synergistic interactions between compounds present in the **discharge** or in the receiving body of water. Many scientists have expressed reservations about current standards for individual chemicals and concern about the possible effects of synergism on the environment. But synergistic effects are difficult to isolate and prove under field conditions.

[*Usha Vedagiri*]

RESOURCES

BOOKS

Connell, D. W., and G. J. Miller. *Chemistry and Ecotoxicology of Pollution.* New York: Wiley, 1984.

Rand, G. M., and S. R. Petrocelli. *Fundamentals of Aquatic Toxicology.* Bristol, PA: Hemisphere Publishing Corporation, 1985.

Synthetic fuels

Synthetic fuels are gaseous and liquid fuels produced synthetically, primarily from **coal** and **oil shale**. They are commonly referred to "synfuels."

The general principle behind coal-based synfuels is that coal can be converted to gaseous or liquid forms that are more easily transported and that burn more cleanly than coal itself. In either case, **hydrogen** is added to coal, converting some of the **carbon** it contains into ozone-depleting **hydrocarbons**. Coal liquefaction involves the conversion of solid coal to a petroleum-like liquid. The technology needed for this process is well known and has been used in special circumstances since the 1920s. During World War II, the Germans used this technology to produce **gasoline** from coal. In the Bergius process used in Germany, coal is reacted with hydrogen gas at high temperatures (about 890°F/475°C) and pressures (200 atmospheres). The product of this reaction is a mixture of liquid hydrocarbons that can be fractionated, as is done with natural **petroleum**.

The processes of **coal gasification** are similar to those used for liquefaction. In one method, coal is first converted to coke (nearly pure carbon), which is then reacted with oxygen and steam. The products of this reaction are **carbon monoxide** and hydrogen, both combustible gases. The heat content of this gas is not high enough to justify transporting it great distances through pipelines, and it is used, therefore, only by industries that actually produce it.

A second type of gas produced from coal, synthetic **natural gas** (SNG), is chemically similar to natural gas. It has a relatively high heat content and can be economically transmitted through pipelines from source to use-site.

For all their appeal, liquid and gaseous coal products have many serious disadvantages. For one thing, they involve the conversion of an already valuable resource, coal, into new fuels with 30–40% less fuel content. Such a conversion can be justified only on environmental grounds or on the basis of other economic criteria.

In addition, coal liquefaction and gasification are water-intensive processes. In areas where they can be most logically carried out, such as the western states, sufficient supplies of water are often not available.

Another approach to the production of synfuels involves the use of oil shale or **tar sands**. These two materials are naturally occurring substances that contain petroleum-like oils trapped within a rocky (oil shale) or sandy (tar sand) base. The oils can be extracted by crushing and heating the raw material. Refining the oil obtained in this way results in the production of gasoline, kerosene, and other fractions similar to those collected from natural petroleum.

A fifth form of synfuel is obtained from **biomass**. For example, human and animal wastes can be converted into **methane** gas by the action of **anaerobic** bacteria in a digester. In some parts of the world, this technology is well-advanced and widely used. China, India, and Korea all have tens of thousands of small digesters for use on farms and in private homes.

During the 1970s, the United States government encouraged the development of all five synfuel technologies. In 1980, the United States Congress created the United States Synthetic Fuels Corporation to distribute and manage $88 billion in grants to promote the development of synfuel technologies. The timing of this action could not have been worse, however. Within two years, the administration of President Ronald Reagan decided to place more emphasis on traditional fuels by deregulating the price of oil and natural gas and by extending tax breaks to oil companies. It also withdrew support from research on synfuels.

At the same time, worldwide prices for all **fossil fuels** dropped dramatically. There seemed to be no indication that synfuels would compete economically with fossil fuels in the foreseeable future. Unable to deal with all these setbacks,

the United States Synthetic Fuels Corporation went out of business in 1992. *See also* Alternative fuels

[*David E. Newton*]

RESOURCES
BOOKS

McGraw-Hill Encyclopedia of Science & Technology. 7th ed. New York: McGraw-Hill, 1992.

PERIODICALS

Douglas, J. "Quickening the Pace in Clean Coal Technology." *EPRI Journal* (January-February 1989): 4—15.

Moore, T. "How Advanced Options Stack Up." *EPRI Journal* (July-August 1987): 4–13.

Peterson, I. "Squeezing Oil Out of Stone." *Science News* 124 (December 3, 1983): 362–364.

Shepard, M. "Coal Technologies for a New Age." *EPRI Journal* (January-February 1988): 5–17.

Systemic

Refers to a general condition of process or system, rather than a condition of one of the component pieces. For example, when a plant wilts in response to **drought**, the entire organism is responding to the effect of loss. Communities of plants and animals and their abiotic **environment** comprise an **ecosystem**. Whole ecosystems may respond to an event, such as changes in local **climate**, by undergoing a systemic change—for example, the composition and abundance of **species** may be altered, and the functional ecological processes of those organisms would be rearranged.

T

Taiga

Taiga is a generic term for a type of conifer-dominated boreal forest found in northern environments. The word was first used to describe dense forests of spruce (especially *Picea abies*) in northern Russia. It has been extended to refer to boreal forests of similar structure in North America but dominated by other conifer **species** (especially *Picea mariana* and *Picea glauca*). Broad-leafed tree species are uncommon in taiga, although species of poplar (*Populus* spp.) and birch (*Betula* spp.) are present. Taiga environments are characterized by cool and short growing seasons. Plant roots can only exploit a superficial layer of seasonally thawed ground, the active layer, situated above permanently frozen substrate, or **permafrost**.

Tailings

Tailings are produced when metallic ores are ground into a fine powder to free the metal-bearing mineral. Its maximum particle size of about 0.08 in (2 mm) is small enough to retain water and support plant growth. Fine dust is one potential hazard, especially if trapped in the lungs of miners. The greatest revegetation difficulties come from metals in lethal concentrations and deficiencies of critical nutrients, especially **nitrogen** and **phosphorus**. Some old tailings have remained barren for many years, with consequent high **erosion** rates. However, some metalophytes are able to survive on these wastes. Curiously, some plants have populations which can also survive on these tailings.

Tailings pond

Refers to the water stored on-site to wash ore and receive waste **tailings** as a residue. This provides both a water supply and a **sediment** trap. Two serious environmental problems ensue. The water leaches out metals and other dangerous elements that were formerly trapped underground in a reducing (non-oxidizing) **environment**; acidic wastes amplify the **leaching**. And though usually stable, some **dams** have burst during flood stages, unleashing a torrent of toxic materials and posing a combined hazard of mudflow and **flooding**.

Takings

The term "takings" refers to the practice of taking, or otherwise restricting the use of, private property through the governmental power of eminent domain, environmental regulation, or other restrictions. The power of eminent domain—and the property owner's right to compensation—is stated in the Bill of Rights (Fifth Amendment to the U.S. Constitution): "nor shall private property be taken for public use, without just compensation." Likewise, the Fourteenth Amendment stipulates that no state may "deprive any person of life, liberty, or property, without due process of law; nor deny to any person within its jurisdiction the equal protection of the laws."

Critics of environmental regulation contend that government cannot constitutionally interfere with a property owner's freedom to do as he or she wishes with his or her own property without compensating the owner. For example, a farmer who wishes to extend or improve his land by draining a swamp or other wetland cannot be forbidden to do so without adequate compensation. A rancher who is forbidden by law to shoot or trap **wolves** that threaten his cattle can claim compensation for livestock losses due to wolf predation. To "take" or otherwise regulate or restrict the use of their property without full and fair compensation would constitute a violation of their constitutionally guaranteed rights under the Fifth and Fourteenth Amendments.

Many environmentalists maintain that the high cost of such compensation would mean the effective end of many environmental regulations regarding **wetlands**, woodlands, and other habitats on private property. They contend that the Constitution does not construe the concept of "takings" as broadly as their critics do. There have long been restric-

White spruce-willow taiga in Alaska. (Photograph by Charlie Ott. Photo Researchers Inc. Reproduced by permission.)

tions on what one may do to or with his or her property. There are, for example, restrictions on erecting billboards on private property along public highways. **Automobile** salvage yards can be required to erect walls to shut out the unsightly view. In neither case is the owner to be compensated for loss, cost, or inconvenience. In these instances, the public interest in avoiding eyesores overrides the private interests of the property owners. So too, environmentalists contend, does the public interest in a clean, safe, and sustainable natural **environment** empower the federal, state, or local government to enact and enforce laws restricting what property owners may do with or to **wildlife** and the **habitat** that sustains them.

The legal and political battles over "takings" continues to be waged in the courts of law and public opinion. Some critics of environmental protection (most notably, libertarians associated with the Cato Institute) contend that the Constitution does not go far enough in guaranteeing the protection of private property rights and should therefore be further amended. Some environmentalists claim that the Constitution should, on the contrary, be amended to extend protection to animals, ecosystems, and **future generations**. It seems plausible to predict that these battles will be waged well into the twenty-first century.

[*Terence Ball*]

Resources
Books

Epstein, R. A. *Takings: Private Property and the Power of Eminent Domain.* Cambridge, MA: Harvard University Press, 1985.

Tall stacks

The waste gases that escape from factories or **power plants** running on **fossil fuels** typically contain a variety of pollutants, including **carbon monoxide**, particulates, **nitrogen oxides**, and sulfur. The first of these pollutants, **carbon** monoxide, tends to dispense rapidly once it is released to the **atmosphere** and seldom poses a serious threat to human health or the **environment**. The same cannot be said of the last three categories of pollutants, all of which pose a threat to the well-being of plants, humans, and other animals.

During the 1960s, humans became increasingly aware of the range of hazards, posed by pollutants escaping from smokestacks. One of the most common events of the time was for residents in the area immediately adjacent to a factory or plant to find their homes and cars covered with fine dust from time to time. They soon learned that the dust was **fly ash** released from the nearby smokestacks. Particulates (fine particles of unburned carbon) also tend to settle out fairly rapidly and cover objects with a fine black powder.

It was clear that visible pollutants, such as fly ash and particulates, offered a hint that invisible pollutants, such as oxides of **nitrogen** and sulfur, were also accumulating in areas close to factories and power plants. As the health hazards of these pollutants became better-known, efforts to bring them under control increased.

Certain technical methods of **air pollution** control—electrostatic precipitation and scrubbing, for example—were reasonably well-known. However, many industries and utilities preferred not to make the significant financial investment needed to install these technologies. Instead, they proposed a simpler and less expensive solution: the construction of taller smokestacks.

The argument was that, with taller smokestacks, pollutants would be carried higher into the atmosphere. They would then have more time and space to disperse. Residents close to the plants and factories would be spared the fallout from the smokestacks. The "tall stack" solution met the requirements of the 1970 **Clean Air Act** as a "last-ditch alternative" to gas-cleaning technologies.

As a result, companies began building taller and taller smokestacks. The average stack height increased from about 200 ft (61 m) in 1956 to over 500 ft (152 m) in 1978. The tallest stack in the United States in 1956 was under 600 ft (182 m) tall, but a decade later, some stacks were more than 1,000 ft (305 m) high.

At first, this effort appeared to be a satisfactory way of dealing with air pollutants. The concentration of particulates and oxides of nitrogen and sulfur close to plants did indeed, decrease after the installation of tall stacks.

But new problems soon began to appear, the most serious of which was **acid deposition**. The longer oxides of sulfur and nitrogen remain in the air, the more likely they are to be oxidized to other forms, sulfur trioxide and nitrogen dioxide. These oxides, in turn, have the tendency to react with water droplets in the atmosphere forming sulfuric and nitric acids, respectively. These acids eventually fall back to earth as rain, snow, or some other form of precipitation. When they reach the ground, they then have the potential to damage plants, buildings, bridges, and other objects.

The mechanism by which stack gases are transported in the atmosphere is still not completely understood. Yet it is now clear that when they are emitted high enough into the atmosphere, they can be carried hundreds or thousands of miles away by prevailing winds. The very method that reduces the threat of air pollutants for communities near a factory or power plant, therefore, also increases the risk for communities farther away. *See also* Acid rain; Air quality; Criteria pollutant; Emission; Scrubbers; Smoke; Sulfur dioxide

[*David E. Newton*]

Resources
Periodicals

"The Dirtiest Half-Dozen." *Time* 139 (June 8, 1992): 31.

Harris, J. "How to Sell Smoke." *Forbes* 145 (June 11, 1990): 204–205.

"New Clues to What An Incinerator Spews." *Science News* 140 (September 21, 1991): 189.

Talloires Declaration

The Talloires Declaration was created in October 1990 in Talloires, France. Jean Mayer, then president of Tufts University in Boston, called together 22 other university presidents and chancellors from universities throughout the world in order to express their concerns regarding the state of the world as well as prepare a document that identified key environmental actions that institutions of higher learning must take to prepare.

According to their organization, University Leaders for a Sustainable Future (ULSF), "Recognizing the shortage of specialists in environmental management and related fields, as well as the lack of comprehension by professionals in all fields of their effect on the **environment** and public health, this gathering defined the role of the university in the following way: "Universities educate most of the people who develop and manage society's institutions. For this reason, universities bear profound responsibilities to increase the awareness, knowledge, technologies, and tools to create an environmentally sustainable future.'"

The actions outlined are listed as follows:

- increase awareness of environmentally **sustainable development**
- create an institutional culture of sustainability
- educate for environmentally responsible citizenship
- foster environmental literacy for all
- practice institutional ecology
- involve all stakeholders
- collaborate for interdisciplinary approaches
- enhance capacity of primary and secondary schools
- broaden service outreach nationally and internationally
- maintain the movement

By May 2002, the declaration had been signed by 288 institutions, including 17 in Africa; 99 in Canada and the United States; 108 in Latin America and the Caribbean; 33 in Asia, the Middle East, and the South Pacific; and 31 in Europe and Russia. ULSF has reported that the document has provided inspiration for other similar official declarations in locations throughout the world, including Halifax, Nova Scotia, Canada; and, Kyoto, Japan—home of the monumental **Kyoto Treaty**.

Programs have been developed at universities and colleges throughout the United States. One exemplary program was established at Ball State University in Muncie, Indiana. In an introduction to the Ball State plan, associate professor James Elfin of the Ball State Department of Natural Resources and Environmental Management, stated that, "If colleges and universities are leaders in the quest for knowledge, why can't they be leaders on the path to a sustainable future?...Academic institutions are regarded—rightly or wrongly—as vanguards for what is possible and encuring. They are perceived to be on the forefront of what society can attain. As an institution, the university endures over a long span of time—if successful, it can be imagined to last forever. In an earlier edition of *The Declaration*...architect William McDonough was asked how long sustainability will take. His reply: "It will take forever. That's the point."' Ball State outlined an extensive plan with a list of Action Items, practical ways in which to implement a "greening" of the campus.

The list of 10 items are as follows:

- audits (three of them: one of campus environment, another of faculty expertise, and a third of needs of K–12 schools across the state of Indiana)
- campus bikeway system
- developing an **environmental education** resource center (in the meantime, a separate effort on campus has resulted in the creation of a field station and environmental education center, FSEEC)
- green issues awareness (or, "green facts" that would be distributed via various media on and off campus)
- retrofit or reconfiguration of the Central (District) Heating Plant
- adopting hybrid electric (or other energy-efficient) vehicles for the campus motor pool
- campus-wide **recycling** program (one has existed, but is not effectively run nor fully integrated with all campus operations)
- sustainability trunk show (to spread the word to larger publics outside the university community)
- incorporation of sustainability within the university core curriculum (or, general education program that is required by all students)

[*Jane E. Spear*]

Resources

Other

Environmental News Network. *The Worldwatch Report: Campus Environmental Programs Spur Action.* June 1, 1998 [cited June 2002]. <http://www.enn.com/features>.

Safe Climate. *University of Buffalo: How to Shrink the University's Carbon Footprint While Reducing the Cost of Education.* 1996 [cited June 2002]. <http://www.safeclimate.net/index>.

Organizations

University Leaders for a Sustainable Future, 2100 "L" Street, NW, Washington, DC USA, 20037, 202-778-6133, Fax: 202-778-6138, <http://www.ulsf.org>

Arthur G. Tansley (1871 – 1951)

British botanist

Arthur G. Tansley was a highly influential British botanist and ecologist. He was educated at University College, London, and Trinity College of Cambridge University. He taught at University College until 1906, and then moved to Cambridge. In 1927, He became a Professor of Botany at Oxford, and he retired as a Professor Emeritus in 1937. Much of his retirement was spent working for the **conservation** of **nature** in Britain, largely by public advocacy and by working with the **Nature Conservancy**, of which he was the first Chairman.

In 1902, Tansley founded the influential botanical journal, *New Phytologist*, and was its editor for thirty years. He was a founder of the British Ecological Society in 1913, and was the editor of its key periodical, the *Journal of Ecology*, for twenty years. Tansley's classic book, *The British Islands and Their Vegetation*, was published in 1939.

Tansley was highly influential in the development of the field of ecology. He was the first person to use the word "ecosystem" to describe a basic unit of nature. He coined the word as a reduction of the phrase "ecological system" (the word "system" was borrowed from its prevalent use at the time in the field of physics). The new word was initially used at a conference at Ithaca, New York, in 1929. It was first published in 1935 in a highly influential paper in which Tansley discussed and clarified key terms and concepts in ecology. His ideas about "systems" were important in helping ecologists to understand that organisms and their communities are profoundly influenced by many non-living environmental factors, and vice versa (i.e., organisms and their communities also influence environmental factors). Much of the

power of the **ecosystem** concept is inherent in its consideration of the complex, often unpredictable behavior of systems. The subsequent emergence of the field of systems ecology drew heavily on physics, mathematics, and computer science, as well as the more traditional ecological disciplines of plant and animal ecology. Prior to Tansley's ideas about ecosystems, ecologists mostly studied populations of plant and animal **species**, or their communities, but with relatively little regard for the influence of and interaction with environmental factors.

Modern ecologists now consider ecosystems as consisting of all of the biotic (plants, animals, and **microorganisms**) and abiotic (**climate**, **soil**, and other non-living environmental factors) elements of a specified area or volume. As such, an ecosystem can be relatively small (e.g., contained within a **clod** of earth or an aquarium) or as large as the entire **biosphere** of the Earth. Moreover, the state of an ecosystem at any point in time (for example, during **succession** following disturbance) represents the present balance between biological and abiotic influences.

Many ecologists also believe that ecosystems do not just consist of the sum of the qualities of their individual parts. Rather, there are also "emergent" properties that cannot be easily predicted, but may nevertheless be extremely important in the **stability** and **resilience** of the ecosystem. For example, some particularly important species, known as **keystone species**, may have a highly disproportionate influence on the integrity of their ecosystem, which could not be predicted based only on their relative contribution of **biomass** or productivity. Some top predators, for instance, play a key role in setting the structure of their entire food web, while certain microorganisms (such as **nitrogen** fixers) have an inordinate influence on overall productivity. Interestingly, in his 1935 paper, Tansley objected to some of the original, emerging ideas about holism in ecology; nevertheless, as evidence of emergent properties has accumulated, holistic ideas have taken root in the theory and practice of ecology.

Tansley is also known for his criticisms of the views of the influential American ecologist, **Frederic Clements** (1874-1945). Clements believed that plant communities were highly integrated and "organismic" in their qualities, and not merely a sum of their individual species. He also suggested that succession, or the process of development of plant communities after disturbance, should proceed to a highly predictable and stable **climax** community determined by regional conditions, particularly climate. This view was heavily criticized by another American ecologist, **Henry Gleason** (1882-1975), who argued that plant species respond to succession and environmental gradients in an individualistic manner, rather than in a strongly integrated, or organismic way. Tansley played an important role in this debate, and favored the individualistic hypothesis as being most realistic.

Tansley also performed important research that showed that plant species are capable of growing over a wider range of environmental conditions than they actually manage to exploit in nature. His experiments showed that **competition** is the key factor restricting species to a relatively narrow range of **habitat** conditions. This work influenced the development of the ecological concept of the **niche**. Modern textbooks define the fundamental niche as the inherent tolerance of a species to environmental extremes (such as high and low temperatures), and the realized niche as the range of environmental conditions that a species actually exploits in nature, being mediated by competition.

[*Bill Freedman Ph.D.*]

Resources

Books

McIntosh, R. P. *The Background of Ecology: Concept and Theory.* Cambridge, UK: Cambridge University Press, 1985.

Sheail, J. *Seventy-five Years in Ecology: The British Ecological Society.* Oxford, UK: The British Ecological Society, 1988.

Tansley, Arthur G., ed. *The British Islands and Their Vegetation.* Cambridge, UK: Cambridge University Press, 1939.

Tansley, Arthur G. *"On Competition Between Galium saxatile and Galium sylvestre on Different Kinds of Soils." Journal of Ecology 5 (1917): 173–179.*

Tansley, Arthur G. *"The Use and Abuse of Vegational Concepts and Terms."* Ecology 16 (1935): 284–307.

Tar sands

Oil sands, formerly referred to as tar sands, are sandy substrates that contain large deposits of bitumin, a kind of fossil fuel. Bitumen is a thick, viscous, black, sticky, tar-like form of liquid **petroleum**, but its physical consistency is comparable to that of cold molasses. Oil sands occurring at or near the surface can be mined using open-pit techniques. The bituminous materials are then extracted by heating and the **hydrocarbons** refined and manufactured into a synthetic petroleum product. Once the readily extractable, surface deposits are extracted, deeper bitumen may be mined using steam assisted, *in situ* techniques that decrease viscosity and allow the bitumen to flow so it can be pumped to the surface.

Mined oil sand typically contains 10-12% bitumen, 4-6% water, and 80-85% mineral sand and clays. About 2.2 tons (2 tonnes) of oil sand must be processed to produce one barrel (0.16 m^3) of liquid petroleum. About 75% of the bitumen present can typically be recovered from mined oil sand. After processing, the sand waste and any unextracted bitumen are eventually used to back-fill and contour mined-out areas, which are then seeded with vegetation to reclaim a stable plant cover.

Most economically recoverable oil sand occurs at or fairly near the ground surface. Initially it is mined in enormous open pits, using extremely large trucks and industrial shovels, among the biggest of their kind in the world. After the open-pit mining, remaining deposits deeper than about 250 feet (75 m) and up to 1,300 ft (400 m) or deeper are recovered using so-called *in situ*, or in-site practices. Various techniques are used for *in situ* extraction of bitumen and heavy crude oil. They include the injection of steam, solvents, or **carbon dioxide** to reduce the viscosity of the heavy fluids so they can flow and be collected and pumped to the surface. The largest *in situ* recovery plant in Alberta is located at Cold Lake, where steam injection is used to enhance the ability of the bitumen to flow. This allows it to be pumped to the surface, where it is treated with liquid natural-gas hydrocarbon condensate to reduce its viscosity. The treated bitumen is then shipped by pipeline to a specialized refinery complex where a relatively light synthetic petroleum is manufactured from the heavy, bituminous, crude fossil fuel.

The world's largest resources of oil sand occur in northern Alberta, Canada, and in Venezuela, in northern South America. In Alberta, there are about 54-thousand mi^2 (141-thousand km^2) of oil-sand deposits. This region contains a total estimated resource of 1.6 trillion barrels (0.25 trillion m^3) of bitumen, of which about 311 billion barrels (49 billion m^3) are thought to be potentially recoverable using known mining and extraction technologies. Once processed, this bitumen could yield an estimated 189 billion barrels (30 billion m^3) of synthetic petroleum.

The first pilot plant to process oil-sand bitumen was constructed in Alberta in 1962. The first full-scale facility was developed between 1964 and 1967, with an initial capacity of 31,000 barrels per day (4.9-thousand m^3/day). During the 1990s, extremely large investments were made to mine and process the oil-sand deposits of Alberta. Between 1996 and 2000, the total investment in development of this resource was $11 billion (Can.), and it was $4.2 billion in 2000 alone. In the year 2000, 66,000 people were directly employed in the oil-sand industry, and the Government of Alberta received royalties of $711 billion. In 2001, oil sands were yielding about one-third of all the petroleum produced in Canada, and continued investment is expected to raise this to 50% by 2005. That would be equivalent to about 10% of the total production of petroleum in North America. Clearly, oil sands are a key component of the resource of liquid **fossil fuels** in North America, and a critical element contributing to the domestic energy security on the continent.

The mining and processing of oil sands results in important environmental impacts, although the most intense effects are restricted to the immediate vicinity of the industrial operations. For example, the development of mine sites, refineries, and pipelines results in severe direct damage to local natural habitats. In the case of open-pit mine development, the **soil** and other **overburden** removed from the site are generally stockpiled for use as a top-dressing during the **reclamation** of the pit after it is mined out. The original surface materials are conserved for this use because they provide a much more hospitable material for plant establishment and growth than does processed oil-sand waste, which is a highly mineral and infertile substrate. The reclamation usually attempts to re-create a facsimile of the original contours of the landscape, and is intended to establish a self-maintaining vegetation with at least as high a level of productivity as what occurred prior to the mining. The reclamation effort is typically designed to develop grassland, forest, or wetland habitats, or a mixture of these. Often, the intention is to establish vegetation dominated by native **species** of plants. The resulting habitats may then be attractive to indigenous species of birds, mammals, and other animals.

The process by which bitumen is extracted from oil sand and then manufactured into synthetic petroleum results in some emissions of **sulfur dioxide**, a toxic gas, into the **atmosphere**. The permissible amounts of emissions are regulated within levels intended to prevent local vegetation damage. This is achieved by the use of scrubbing technologies that remove most sulfur dioxide from the flue gases before they are vented to the ambient **environment**. The mining and processing processes also require extremely large inputs of energy to drive their processes. Industrial engineers have been working to make the processes run more efficiently, and they anticipate achieving a 45% reduction of **carbon** dioxide emissions per barrel by 2010, compared to 1990 levels. This will result in large savings in process costs, while also reducing the emissions of carbon dioxide, an important greenhouse gas, to the atmosphere.

[*Bill Freedman Ph.D.*]

Resources

Books

Fitzgerald, J. *Gold with Grit: The Alberta Oil Sands.* Sidney, BC: Gray's Publishing, 1978.

Lyttle, R. B. *Shale Oil and Tar Sands: The Promises and Pitfalls.* Danbury, CT: Impact Books, 1982.

Perrini, E. M. *Oil from Shale and Tar Sands.* Park Ridge, NJ: Noyes Data Corporation, 1975.

Other

Government of Alberta. *Oil Sands.* 2002 [cited July 2002]. <http://www.energy.gov.ab.ca/com/Sands/default.htm>.

Target species

The object of a **hunting**, fishing, or collecting exercise, or an extermination effort aimed at destroying a particular orga-

nism. This term is often used to describe the intended victim of an application of **pesticide** or **herbicide**. Unfortunately, many non-target **species** can be affected by so-called targeted activities. For example, Pacific net fishermen target the yellowfin tuna, but also snare and kill non-target species such as **dolphins**. Pesticides applied to crops to kill insects (the target species) can enter the **food chain/web**, become biologically magnified, and kill non-target species such as fish-eating birds or birds of prey. *See also* Biomagnification

Taylor Grazing Act (1934)

This act established federal management policy on public grazing lands, the last major category of public lands to be actively managed by the government. The delay in its passage was due to a lack of interest (the lands were sometimes referred to as "the lands no one wanted") and due to competition between the Agriculture Department and the Interior Department over which department would administer the new program.

The purpose of the Taylor Grazing Act was to "stop injury to the public grazing lands...to provide for their orderly use, improvement and development...[and] to stabilize the livestock industry dependent on the public range." To achieve these purposes, the Secretary of the Interior was authorized to establish grazing districts on the public domain lands. The lands within these established districts were to be classified for their potential use, with agricultural lands to remain open for homesteading. In 1934 and 1935, President Franklin D. Roosevelt issued Executive Orders that withdrew all remaining public lands for such classification, an action that essentially closed the public domain. The Secretary was authorized to develop any regulations necessary to administer these grazing districts, including the granting of leases for up to ten years, the charging of fees, the undertaking of range improvement projects, and the establishing of cooperative agreements with grazing landholders in the area.

Another important feature of the law called upon the Secretary of the Interior to cooperate with "local associations of stockmen" in the administration of the grazing districts, referred to by supporters of the law as "democracy on the range" or "home rule on the range." This was formalized in the creation of local advisory boards. The law created a Division of Grazing (renamed the Grazing Service in 1939) to administer the law, but for a number of reasons, the agency was ineffective. It became quite dependent on the local advisory boards and was often cited as an example of an agency captured by the interests it was supposed to be controlling. In 1946, the agency merged with the General Land Office to create the **Bureau of Land Management**.

Also included in the bill was the phrase "pending final disposal," which implied that these grazing lands would not necessarily be retained in federal ownership. This phrase was included because many thought the lands would eventually be transferred to the states or the private sector and because it lessened opposition to the law. The uncertainty introduced by this phrase made management of the grazing lands more difficult, and it was not eliminated until passage of the **Federal Land Policy and Management Act** in 1976.

The grazing fees charged to ranchers were initiated at a low level, based on a cost-of-administration approach rather than a market level approach. This has led to lower grazing fees on public lands than on private lands. Typically, three-fourths of the fees go to the local advisory boards and are used for range improvement projects.

[*Christopher McGrory Klyza*]

Resources

Books

Dana, S. T., and S. K. Fairfax. *Forest and Range Policy.* 2nd ed. New York: McGraw-Hill, 1980.

Foss, P. O. *Politics and Grass.* Seattle: University of Washington Press, 1960.

Peffer, E. L. *The Closing of the Public Domain: Disposal and Reservation Policies, 1900–1950.* Stanford: Stanford University Press, 1951.

Technological risk assessment

see **Risk assessment**

TED

see **Turtle excluder device**

Tellico Dam

In an effort to put people back to work during the Great Depression, several public works and **conservation** programs were set in motion. One of these, the **Tennessee Valley Authority** (TVA), was established in 1933 to protect the Tennessee River basin, its water, **soil**, forests, and **wildlife**. TVA initiated many projects to meet these agency goals, but also to provide employment to this impoverished area and to slow the emigration of the region's youth. Many of the projects completed and operated by TVA were **dams** and reservoirs along the Tennessee River and its tributaries. In 1936 one such project site was identified on the lower end of the Little Tennessee River. This proposal for Tellico Dam was reviewed, but it was abandoned because the cost was too high and the returns too low. The project was reconsidered in 1942, and plans were drawn up for its construction; however, World War II halted its development.

Two decades passed before the Tellico Dam project became active again. By this time, there was some opposition to the dam from local citizens, as well as from the Tennessee State Planning Commission. Opposition grew through the mid-1960s and congressional hearings were held to study the economic and environmental factors of this project. Supreme Court Justice William O. Douglas even visited the site to lend his support to the Eastern Band of the Cherokee Indian Nation, whose land would be inundated by Tellico's **reservoir**. Congress, however, approved the project in 1966 and authorized funds to begin construction the following year.

After construction started, opponents of Tellico Dam requested that TVA prepare an **Environmental Impact Statement** (EIS) to be in compliance with the **National Environmental Policy Act** of 1969. TVA refused because Tellico Dam was already authorized and under construction before the Act was passed. Local farmers and landowners, whose land would be affected by the reservoir, joined forces with conservation groups in 1971 and filed suit against TVA to halt the project because there was no impact statement. The court issued an injunction, and construction of Tellico Dam was halted until the EIS was submitted. TVA complied with the order and submitted the final EIS in 1973. That same year the **Endangered Species Act** was signed into law, and two ichthyologists from the University of Tennessee made a startling discovery in a group of fishes they collected from the Little Tennessee River.

They discovered a new **species** of fish, the **snail darter**, later scientifically described and named *Percina tanasi*. Its scientific name honors the ancient Cherokee village of Tanasi, a site that was threatened by Tellico Dam. The state of Tennessee also takes its name from this ancient village. The snail darter's common name refers to the small mollusks that comprise the bulk of its diet. In January 1975, the U.S. **Fish and Wildlife Service** was petitioned to list this new species as endangered. Foreseeing a potential problem, TVA transplanted snail darters into the nearby Hiwassee River without consulting with the Fish and Wildlife Service or the appropriate state agencies. Two more transplants of specimens in 1975 and 1976 brought the number of snail darters in the Hiwassee to over 700. Meanwhile, in October 1975, the snail darter was placed on the federal **endangered species** list, and the stretch of the Little Tennessee River above Tellico Dam was listed as **critical habitat**.

TVA worked feverishly to complete the project, in obvious violation of the Endangered Species Act, and also prepared for the ensuing court battles. A suit was filed for a permanent injunction on the project, but this was denied. An appeal was immediately made to the U.S. Court of Appeals, which overturned the decision and issued the injunction. TVA then appealed this decision to the Supreme Court, which, to the surprise of many, upheld the Court of Appeals ruling, but with a loophole. The Supreme Court left an opening for the U.S. Congress to come to the aid of Tellico Dam. In 1979, an amendment was attached to energy legislation to exempt Tellico Dam from all federal laws, including the Endangered Species Act. President Carter reluctantly, because of the amendment, signed the bill. In January 1980, the gates closed on Tellico Dam, **flooding** over 17,000 acres (6,885 ha) of valuable agricultural land, the homes of displaced landowners, the Cherokee's ancestral burial grounds at Tanasi, and the riverine **habitat** for the Little Tennessee River's snail darters.

Fortunately, Tellico Dam did not cause the **extinction** of the snail darter. During the 1980s, snail darter populations were found in several other rivers in Tennessee, Georgia, and Alabama, and the transplanted population in the Hiwassee River continues to thrive.

[*Eugene C. Beckham*]

Resources
Books

Ono, R., J. Williams, and A. Wagner. *Vanishing Fishes of North America.* Washington, DC: Stone Wall Press, 1983.

Wheeler, W. B. *TVA and the Tellico Dam: 1936–1979: A Bureaucratic Crisis in Post-Industrial America.* Knoxville: University of Tennessee Press, 1986.

Temperate rain forest

A temperate rain forest is an evergreen, broad-leaved, or **coniferous forest** which generally occurs in a coastal **climate** with cool to warm summers, mild winters, and year-round moisture abundance, often as fog. Broad-leaved temperate forests are found in areas including western Tasmania, southeastern **Australia**, New Zealand, Chile, southeastern China, and southern Japan. They often have close evolutionary ties to tropical and subtropical forests.

Temperate conifer rain forests are more cold-tolerant than broad-leaved rain forests and are rich in mosses while lacking tree ferns and vines. The original range of the temperate conifer rain forest included portions of Great Britain, Ireland, Norway, and the Pacific Coast of North America. The Pacific Northwest (PNW) rain forest extends from Northern California to the Gulf of Alaska and is the most extensive temperate rain forest in the world. The eastern boundary of the PNW forest is sometimes set at the crest of the most western mountain range and sometimes extended further east to include all areas in the maritime climatic zone, which has mild winters and only moderately dry summers. These forests are dominated by Douglas fir (*Pseudotsuga menziesii*), western hemlock (*Tsuga heterophylla*), sitka spruce (*Picea sitchensis*), western red cedar (*Thuja plicata*), coastal **redwoods** (*Sequoia sempervirens*), and associated hardwoods

and conifers. In the absence of harvesting, these forests are regenerated by small wind storms and very infrequent catastrophic disturbances.

The PNW forests are noteworthy in many ways. Of all the world's forests, they have the tallest trees, at 367 ft (112 m), and areas with the highest **biomass** (4,521 Mt per ha). On better sites the dominant **species** typically live 400 to 1,250 years or more and reach 3–13 ft (1–4 m) in diameter and 164–328 feet (50–100 m) in height, depending on site and species.

The natural temperate rain forest is generally less biodiverse than its tropical counterpart, but still contains a rich **flora** and **fauna**, including several species that are unique to its **ecosystem**. Preservation of temperate rain forests is a global **conservation** issue. Humans have virtually eliminated them from Europe and left only isolated remnants in Asia, New Zealand, and Australia. In North America, sizeable areas of natural rain forest still exist, but those not specifically reserved are predicted to be harvested within 25 to 50 years.

The same pressures that previously converted most of the natural temperate rain forests to plantations and other land uses still threaten the remaining natural forests. There is the desire to harvest the massive trees and huge timber volumes before they are lost to insects or wind blow. There is also the desire to replace slow-growing natural forests with younger forests of fast-growing species. Those who favor preservation cite the connection between intact rain forests and high salmon/trout production. They note that the replacement forests, being simpler in structure and species composition, cannot offer the same biological diversity and ecosystem functioning. Also extolled are the aesthetic, spiritual, and scientific benefits inherent in a naturally functioning ecosystem containing ancient massive trees. *See also* Biodiversity; Clear-cutting; Deciduous forest; Deforestation; Forest Service; Tropical rain forest

[*Edward Sucoff and Klaus Puettmann*]

Resources

Books

Adam, P. *Australian Rainforests.* Oxford: Clarendon Press, 1992.

Franklin, J. F. "Pacific Northwest Forests." In *North American Terrestrial Vegetation*, edited by M. G. Barbour and W. D. Billings. Cambridge: Cambridge University Press, 1988.

Tennessee Valley Authority

The idea for the Tennessee Valley Authority (TVA) emerged with the election of Franklin D. Roosevelt as President of the United States, and it became one of the major symbols of his "New Deal" policies designed to rescue the country from the social and economic problems of the Depression of the 1930s. TVA activities have included improved navigation, flood control, production of electricity (and its distribution to small towns and rural areas), **fertilizer** production, **soil conservation** and stabilization, reforestation, improved **transportation** facilities, and recreational sites.

The TVA was established when Roosevelt signed the Implementing Act in May of 1933. Within a few months of its passage, the Norris Dam and the town of Norris, Tennessee were under construction and the controversy began. Early opponents labeled it "socialistic" and un-American, contrary to the American ideal of private enterprise. The most enduring criticism has been that TVA is too single-minded in its emphasis on power production, especially after hydro-capacity was exhausted and the agency turned first to **coal** and then to **nuclear power** to fuel its **power plants**. TVA has most recently been criticized for being too concerned about profit-making and too little concerned about a project's environmental impacts, such as **air pollution** created by generating plants and effects on **endangered species**.

The Tennessee Valley Authority has, however, effected change in the intended region. Nearly all of the region's farms now have electricity, for example, compared to only 3% in 1933, and the area has been growing economically. It is now closer to the nation's norms for employment, income, purchasing power, and quality of life as measured by jobs, electrification, and recreational facilities. A clean-air plan was initiated in 1998 to cut 170,000 tons of toxic emissions from TVA's coal-fired plants yearly. *See also* Tellico Dam

[*Gerald R. Young Ph.D.*]

Resources

Books

Creese, W. L. *TVA's Public Planning: The Vision, The Reality.* Knoxville: University of Tennessee Press, 1990.

Nurick, A. J. *Participation in Organizational Change: The TVA Experiment.* New York: Praeger Publishers, 1985.

Organizations

Tennessee Valley Authority, 400 W. Summit Hill Dr., Knoxville, TN USA 37902-1499, (865) 632-2101, Email: tvainfo@tva.gov

Teratogen

An environmental agent that can cause abnormalities in a developing organism resulting in either fetal death or congenital abnormality. The human fetus is separated from the mother by the placental barrier, but the barrier is imperfect and permits a number of chemical and infectious agents to pass to the fetus. Well known teratogens include (but are

not limited to) alcohol, excess vitamin A and retinoic **acid**, the rubella **virus**, and high levels of **ionizing radiation**. Perhaps the best known teratogenic agent is the drug thalidomide, which induced severe limb abnormalities known as phocomelia in children whose mothers took the drug. *See also* Birth defects; Environmental health; Mutagen

Terracing

A procedure to reduce the speed at which water is removed from the land. Water is directed to follow the gentler slopes of the terrace rather than the steeper natural slopes. Terracing is usually recommended only for intensively used, eroding cropland in areas of high-intensity rainfall. Terraces are costly to construct and require annual maintenance, but are feasible where **arable land** is in short supply or where valuable crops can be grown. In order to become self-sufficient in food production, ancient civilizations in Peru often constructed terraces in very steep, mountainous areas. Today, in Nepal, people living in the foothills of the Himalayas use terraces in order to have enough land available for food production.

Territorial sea

Territorial sea is the part of the ocean that a nation controls. For centuries, countries had different opinions about who owned the seas. In 1609, Dutch statesman Hugo Grotius said that all nations should have access to the oceans. Most countries wanted territorial rights.

Methods for measuring boundaries included a 1702 recommendation from Cornelius van Bynkershonk. He proposed that a nation had the right to the area that could be protected by cannons located on land.

During the twentieth century, an international law defined territorial sea and three other coastal zones. The 1994 United Nations **Convention on the Law of the Sea** treaty defined territorial sea as the area extending up to 12 nautical miles (13.8 statute miles) from a nation's shoreline.

A nation's territorial sea consists of the ocean surface, the airspace above the sea, and the seabed below. Within this area, the nation has the exclusive rights to **natural resources**, as well as rights for fishing, shipping, and navigation.

The U.N. had discussed coastal boundaries since 1958, and nations met numerous times before approving the treaty in 1994. The United States claimed its 12-mile territorial sea in 1988 when President Ronald Reagan signed proclamation 5928.

[*Liz Swain*]

Territoriality

The attempt by an individual organism or group of organisms to control a specified area. The area or territory, once controlled, is usually bound by some kind of marker (such as a scent or a fence). Control of territory usually means defense of that territory, primarily against other members of the same **species**. This defense, which may or may not be aggressive, typically involves threats, displays of superior features (e.g., size or color), or displays of fighting equipment (e.g., teeth, claws, antlers). Actual physical combat is relatively rare. A songbird establishes its territory by vigorous singing and will chase intruders away during the mating and nesting season. A leopard (*Panthera pardus*) marks the boundaries of its territory with urine and will defend this area from other leopards of the same sex. Territoriality is found in many organisms, probably including humans, and it serves several purposes. It may provide a good nesting or breeding site and a sufficient feeding or **hunting** area to support offspring. It may also protect a female from males other than her mate during the mating season.

Tertiary sewage treatment

see **Sewage treatment**

Tetrachloroethylene

Tetrachloroethylene is a manufactured chemical used for **dry cleaning** and metal degreasing. It is also used in the manufacture of other **chemicals** and for textile finishing and dyeing. Tetrachloroethylene has been produced commercially since the early 1900s. At room temperature, tetrachloroethylene is a nonflammable colorless liquid. It evaporates easily and has a sharp, sweet odor similar to chloroform, which most people can smell at a level of about one part per million (ppm). Other names for tetrachloroethylene are perchloroethylene, PCE, tetrachloroethene, and "perc".

Results of animal studies using high levels of tetrachloroethylene indicated that it can cause liver and kidney damage. Exposure to very high levels of tetrachloroethylene was toxic to the unborn pups of pregnant rats and mice. The offspring of rats exposed to high levels of tetrachloroethylene during pregnancy exhibited changes in behavior. Since tetrachloroethylene has been shown to cause liver tumors in mice

and kidney tumors in male rats, it is considered to be a potential human **carcinogen**.

Exposure to high concentrations of tetrachloroethylene, especially in poorly ventilated areas, can cause nose and throat irritation and dizziness, headache, sleepiness, confusion, nausea, difficulty in speaking and walking, unconsciousness, and death. Severe skin irritation may result from repeated or extended skin contact with tetrachloroethylene.

The **Occupational Safety and Health Administration** (OSHA) has set a limit of 100 ppm for an eight-hour workday over a 40-hour workweek. The **National Institute for Occupational Safety and Health** (NIOSH) recommends that because tetrachloroethylene is a potential carcinogen, levels in the workplace should be kept as low as possible. Since tetrachloroethylene is considered a hazardous air pollutant under the U.S. **Clean Air Act**, a federal rule went into effect in 1996 for dry cleaning establishments that regulates operating procedures and equipment type and use, with the goal of reducing tetrachloroethylene emissions from those facilities. However, tetrachloroethylene does not contribute to stratospheric **ozone** depletion and has been approved by the U.S. **Environmental Protection Agency** (EPA) for use as a replacement for ozone-depleting solvents.

Tetrachloroethylene can be released into air and water through dry cleaning and industrial metal cleaning or finishing activities. **Water pollution** can result from tetrachloroethylene **leaching** from vinyl pipe liners. Tetrachloroethylene released into **soil** will readily evaporate or may leach slowly to groundwater, for it biodegrades slowly in soil. When released into water bodies, tetrachloroethylene will primarily evaporate and has little potential for accumulating in aquatic life.

The Maximum Contaminant Level (MCL), which is an enforceable standard for tetrachloroethylene in drinking water, has been set at 5 **parts per billion** (ppb) by the EPA. This is the lowest level of tetrachloroethylene, given present technologies and resources, that a **water treatment** system can be expected to achieve. In contrast to the enforceable level of tetrachloroethylene is the desired level in drinking water, which is referred to as the Maximum Contaminant Level Goal (MCLG). This been set at zero because of possible health risks. If levels of tetrachloroethylene are too high in a drinking water source, it can be removed through the use of granular activated **carbon filters** and packed tower **aeration**, which is a way to remove chemicals from ground water by increasing the surface area of the contaminated water that is exposed to air. Tetrachloroethylene wastes are classified as hazardous under the **Resource Conservation and Recovery Act** (RCRA) and must be stored, transported, and disposed of according to RCRA requirements.

[*Judith L. Sims*]

Resources

Books

Leeds, Michelle S. *Perchloroethylene (Carbon Dichloride, Tetrachloroethylene, Drycleaner, Fumigant): Effects on Health and Work.* Washington, DC: Abbe Publishing Association, 1995.

Other

Halogenated Solvents Industry Alliance. *Perchloroethylene: White Paper.* November 1999 [cited June 22, 2002]. <http://www.hsia.org/white_papers/perc.htm>.

U.S. Environmental Protection Agency. *Consumer Factsheet on: Tetrachloroethylene.* May 22, 2002 [cited June 22, 2002]. <http://www.epa.gov/safewater/dwh/c-voc/tetrachl.html>.

U.S. Environmental Protection Agency. *Rule and Implementation Information for Perchloroethylene Dry Cleaning Facilities.* June 10, 2002 [cited June 23, 2002]. <http://www.epa.gov/ttn/atw/dryperc/dryclpg.html>.

Tetraethyl lead

Tetraethyl **lead** is an organometallic compound with the chemical formula $(C_2H_5)_4Pb$. In 1922, automotive engineers found that the addition of a small amount of this compound to **gasoline** improves engine performance and reduces knocking. Knocking is a physical phenomenon that results when low-octane gasoline is burned in an internal **combustion** engine. Until 1975, tetraethyl lead was the most common additive used to reduce knocking in motor fuels. This additive presents an environmental hazard, however, since lead is expelled into the **environment** during operation of **automobile** engines using leaded fuels. Given the growing concern about the health effects of lead, the compound has now been banned for use in gasolines in the United States.

See also Air pollution; Air pollution index; Emission; Gasohol

The Coastal Society

see **Coastal Society, The**

The Cousteau Society

see **Cousteau Society, The**

The Global 2000 Report

Released in July 1980, this landmark study warned of grave consequences for humanity if changes were not made in **environmental policy** around the globe. Prepared over a three year period by the President's **Council on Environmental Quality** (CEQ) in cooperation with the U.S. Department of State and other federal agencies, this was the first comprehensive and integrated report by the United States or any other government projecting long-term environmental, re-

source, and population trends. It was the subject of extensive publicity, attention, and debate, influencing political leaders and policy makers the world over.

In announcing release of the report, CEQ warned that "U.S. Government projections show that unless the nations of the world act quickly and decisively to change current policies, life for most of the world's people will be more difficult and more precarious in the year 2000."

Specifically, the study states that "If present trends continue, the world in 2000 will be more crowded, more polluted, less stable ecologically, and more vulnerable to disruption than the world we live in now...For hundreds of millions of the desperately poor, the outlook for food and other necessities of life will be no better. For many, it will be worse."

Among the report's findings and conclusions were the following:

- The world will add almost 100 million people a year to its population, which will grow from 4.5 billion in 1980 to 6 billion in 2000;
- Billions of tons and millions of acres of cropland are being lost each year to **erosion** and development, and **desertification** is claiming an area the size of Maine each year;
- The planet's genetic resource base is being severely depleted, and between 500,000 and 2 million plant and animal species--15 to 20 percent of all **species** on the earth--could be extinguished by the year 2000;
- Periodic and severe water shortages will be accompanied by a doubling of the demand for water from 1971 levels; increased burning of **coal** and other **fossil fuels** will cause **acid** rain-induced damage to lakes, crops, forests, and buildings, and could lead to catastrophic **climate** change (global warming) "that could have highly disruptive effects on world agriculture."
- Depletion of the stratospheric **ozone** layer by industrial **chemicals** (**chlorofluorocarbons**) could cause serious damage to food crops and human health.

The report noted the ongoing worldwide efforts to protect and replant forests, conserve energy, promote **family planning** and birth control programs, prevent **soil** erosion and desertification, and find alternatives to the present reliance on toxic pesticides and nonrenewable, polluting energy sources such as **petroleum** and coal. But the study emphasized that "Encouraging as these developments are, they are far from adequate to meet the global challenges projected in this study. Vigorous, determined new initiatives are needed if worsening poverty and human suffering, **environmental degradation**, and international tension and conflicts are to be prevented."

Some skeptics criticized the report's pessimistic tone and dire warnings as exaggerated and overblown, and its recommendations later were largely ignored by the Reagan and Bush administrations. However, it is now apparent that its major points were not only well founded but may turn out to be far too conservative, rather than radical and alarmist. *See also* Deforestation; Drought; Energy conservation; Environmental policy; Gene pool; Greenhouse effect; Ozone layer depletion; Pollution control; Population growth; Sustainable agriculture; Water allocation

[*Lewis G. Regenstein*]

Resources

Books

Council on Environmental Quality. *The Global 2000 Report to the President.* 3 vols. Washington, DC: Government Printing Office, 1980.

The Nature Conservancy

see **Nature Conservancy, The**

The Ocean Conservatory

see **Ocean Conservatory, The**

Thermal plume

Water used for cooling by **power plants** and factories is commonly returned to its original source at a temperature greater than its original temperature. This heated warm water leaves an outlet pipe in a stream-like flow known as a thermal **plume**. The water within the plume is significantly warmer than the water immediately adjacent to it. Thermal plumes are environmentally important because the introduction of heated water into lakes or slow-moving rivers can have adverse effects on aquatic life. Warmer temperatures decrease the solubility of oxygen in water, thus lowering the amount of **dissolved oxygen** available to aquatic organisms. Warmer water also causes an increase in the **respiration** rates of these organisms, and they deplete the already-reduced supply of oxygen more quickly. Warmer water also makes aquatic organisms more susceptible to diseases, **parasites**, and toxic substances. *See also* Thermal pollution

Thermal pollution

The **combustion** of **fossil fuels** always produces heat, sometimes as a primary, desired product, and sometimes as a secondary, less-desired by-product. For example, families burn **coal**, oil, **natural gas**, or some other fuel to heat their homes. In such cases, the production of heat is the object of burning a fuel. Heat is also produced when fossil fuels

are burned to generate electricity. In this case, heat is a by-product, not the main reason that fuels are burned.

Heat is produced in a number of other common processes. For example, electricity is also generated in nuclear **power plants**, where no combustion occurs. The decay of organic matter in landfills also releases heat to the **atmosphere**.

It is clear, therefore, that a vast array of human activities result in the release of heat to the **environment**. As those activities increase in number and extent, so does the amount of heat released. In many cases, heat added to the environment begins to cause problems for plants, humans, or other animals. This effect is then known as thermal **pollution**.

One example of thermal pollution is the development of urban heat islands. An **urban heat island** consists of a dome of warm air over an urban area caused by the release of heat in the region. Since more human activity occurs in an urban area than in the surrounding rural areas, the atmosphere over the urban area becomes warmer than it is over the rural areas.

It is not uncommon for urban heat islands to produce measurable **climate** changes. For example, the levels of pollutants trapped in an urban heat island can reach 5–25% greater than the levels over rural areas. Fog and clouds may reach twice the level of comparable rural areas, wind speeds may be reduced by up to 30%, and temperatures may be 0.9–3.6°F (0.5°–2°C) higher than in surrounding rural areas. Such differences may cause both personal discomfort and, in some cases, actual health problems for those living within an urban heat island.

The term thermal pollution has traditionally been used more often to refer to the heating of lakes, rivers, streams, and other bodies of water, usually by electric power-generating plants or by factories. For example, a one-megawatt **nuclear power** plant may require 1.3 billion gal (4.9 billion L) of cooling water each day. The water used in such a plant has its temperature increased by about 30.6°F (17°C) during the cooling process. For this reason, such plants are usually built very close to an abundant water supply such as a lake, a large river, or the ocean.

During its operation, the plant takes in cool water from its source, uses it to cool its operations, and then returns the water to its original source. The water is usually recycled through large cooling towers before being returned to the source, but its temperature is still likely to be significantly higher than it was originally. In many cases, an increase of only a degree or two may be "significantly higher" for organisms living in the water.

When heated water is released from a plant or factory, it does not readily mix with the cooler water around it. Instead, it forms a stream-like mass known as a **thermal plume** that spreads out from the outflow pipes. It is in this thermal **plume** that the most severe effects of thermal pollution are likely to occur. Only over an extended period of time does the plume gradually mix with surrounding water, producing a mass of homogenous temperature.

Heating the water in a lake or river can have both beneficial and harmful effects. Every **species** of plant and animal has a certain range that is best for its survival. Raising the temperature of water may cause the death of some organisms, but may improve the environment of other species. Pike, perch, walleye, and small mouth bass, for example, survive best in water with a temperature of about 84°F (29°C), while catfish, gar, shad, and other types of bass prefer water that is about 10°F (5.5°C) warmer.

Spawning and egg development are also very sensitive to water temperature. Lake trout, walleye, Atlantic **salmon**, and northern pike require relatively low temperatures (about 48°F/9°C), while the eggs of perch and small mouth bass require a much higher temperature, around 68°F (20°C).

Clearly, changes in water temperature produced by a nuclear power plant, for example, is likely to change the mix of organisms in a waterway.

Of course, large increases in temperature can have disastrous effects on an aquatic environment. Few organisms could survive an accident in which large amounts of very warm water were suddenly dumped into a lake or river. This effect can be observed especially when a power plant first begins operation, when it shuts down for repairs, and when it restarts once more. In each case, a sudden change in water temperature can cause the death of many individuals and lead to a change in the make-up of an aquatic community. Sudden temperature changes of this kind produce an effect known as thermal shock. To avoid the worst effects of thermal shock, power plants often close down or re-start slowly, reducing or increasing temperature in a waterway gradually rather than all at once.

One inevitable result of thermal pollution is a reduction in the amount of **dissolved oxygen** in water. The amount of any gas that can be dissolved in water varies inversely with the temperature. As water is warmed, therefore, it is capable of dissolving less and less oxygen. Organisms that need oxygen to survive will, in such cases, be less able to survive.

Water temperatures can have other, less-expected effects as well. As an example, trout can swim less rapidly in water above 66°F (19°C), making them less efficient predators. Organisms may also become more subject to disease in warmer water. The bacterium *Chondrococcus columnaris* is harmless to fish at temperatures of less then 50°F (10°C). Between temperatures of 50°–70°F (10°–21°C), however, it is able to invade through wounds in a fish's body. It can even attack healthy tissue at temperatures above 70°F (21°C).

The loss of a single aquatic species or the change in the structure of an aquatic community can have far-reaching effects. Each organism is part of a **food chain/web**. Its loss may mean the loss of other organisms farther up the web who depend on it as a source of food.

The water heated by thermal pollution has a number of potentially useful applications. For example, it may be possible to establish aquatic farms where commercially desirable fish and shellfish can be raised. The Japanese have been especially successful in pursuing this option. Some experts have also suggested using this water to heat buildings, remove snow, fill swimming pools, use for **irrigation**, de-ice canals, and operate industrial processes that have modest heat requirements.

The fundamental problem with most of these suggestions is that waste heat has to be used where it is produced. It might not be practical to build a factory close to a nuclear power plant solely for the purpose of using heat generated by the plant. As a result, few of the suggested uses for waste heat have actually been acted upon.

There are no easy solutions to the problems of thermal pollution. To the extent that industries and utilities use less energy or begin to use it more efficiently, a reduction in thermal pollution should result as a fringe benefit.

The other option most often suggested is to do a better job of cooling water before it is returned to a river, lake, or the ocean. This goal could be accomplished by enlarging the cooling towers that most plants already have. However, those towers might have to be as tall as a 30-story building, in which case they would create new problems. In cold weather, for example, the water vapor released from such towers could condense to produce fog, creating driving hazards over an extended area.

Another approach is to divert cooling water to large artificial ponds where it can remain until its temperature has dropped sufficiently. Such cooling ponds are in use in some locations, but are not very attractive alternatives because they require so much space. A one-megawatt plant, for example, would require a cooling pond with 1,000-2,000 acres (405-810 ha) of surface area. In many areas, the cost of using land for this purpose would be too great to justify the procedure.

Some people have also used the term thermal pollution to describe changes in the earth's climate that may result from human activities. The large quantities of fossil fuels burned each year release a correspondingly large amount of **carbon dioxide** to the earth's atmosphere. This **carbon** dioxide, in turn, may increase the amount of heat trapped in the atmosphere through the **greenhouse effect**. One possible result of this change could be a gradual increase in the earth's annual average temperature. A warmer climate might, in turn, have possibly far-reaching and largely unknown effects on agriculture, rainfall, sea levels, and other phenomena around the world. *See also* Alternative energy sources; Industrial waste treatment

[*David E. Newton*]

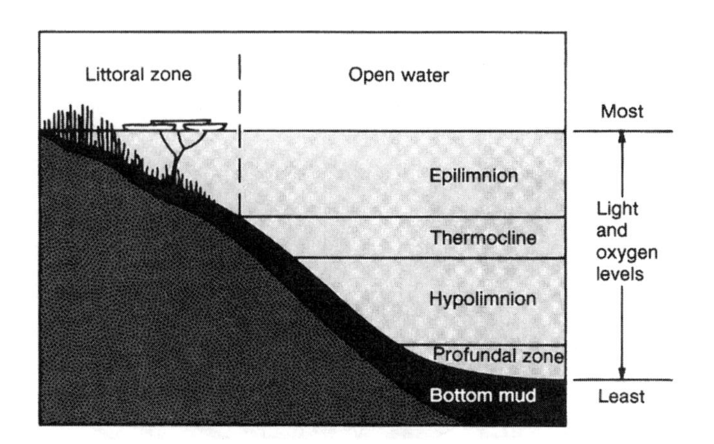

Thermal stratification of a deep lake. (McGraw-Hill Inc. Reproduced by permission.)

Resources

Books

Harrison, R. M., ed. *Pollution: Causes, Effects, and Control.* Cambridge: Royal Society of Chemistry, 1990.

Langford, T. E., ed. *Ecological Effects of Thermal Discharges.* New York: Elsevier Science, 1990.

Periodicals

Hudson, J., and J. B. Cravens. "Thermal Effects." *Water Environment Research* 64 (June 1992): 570–81.

Thermal stratification (water)

The development of relatively stable, warmer and colder layers within a body of water. Thermal **stratification** is related to incoming heat, water depth, and degree of water column mixing. Deep, large lakes such as the **Great Lakes** receive insufficient heat to warm the entire water column and lack adequate physical turnover or mixing of the water for uniform temperature distribution. Thus, they have an upper layer of water that is warmed by surface heating (epilimnion) and a lower layer of much colder water (hypolimnion), separated by a layer called the **thermocline** in which the temperature decreases rapidly with depth. Both daily and seasonal variations in heat input can promote thermal stratification. Stratification in summer followed by mixing in the fall is a phenomenon commonly observed in temperate lakes of moderate depths (over 33 ft [10 m]).

Thermocline

A thermocline is a zone of rapid temperature change with depth in a body of water. It is the boundary between two layers of water that have different temperatures, in a lake, **estuary**, or an ocean. The thermocline is marked by a dramatic change in temperature, where the water temperature changes at least one Celsius degree with every meter of depth. Because of density differences associated with a change in temperature, thermoclines can prevent mixing of nutrients from deep to shallow water, and can therefore cause the surface waters of some lakes and ocean to have very low **primary productivity**, even when there is sufficient light for **phytoplankton** to grow. In other areas, the thermocline can prevent mixing of oxygen-rich surface waters with bottom waters in which oxygen has been depleted as a result of high rates of **decomposition** related to eutrophication. In temperate freshwater lakes, the thermocline is disrupted in fall when the surface waters become denser as decreasing air temperatures cool the surface of the lake. Cooler water is denser and sinks, thereby causing warmer bottom water to rise to the surface and mix nutrients throughout the water column.

[*Marie H. Bundy*]

Thermodynamics, Laws of

One way of understanding the **environment** is to understand the way matter and **energy flow** through the natural world. For example, it helps to know that a fundamental law of **nature** is that matter can be neither created nor destroyed. That law describes how humans can never really "throw something away." When wastes are discarded, they do not just disappear. They may change location or change into some other form, but they still exist and are likely to have some impact on the environment.

Perhaps the most important laws involving energy are the laws of thermodynamics. These laws were first discovered in the nineteenth century by scientists studying heat engines. Eventually, it became clear that the laws describing energy changes in these engines apply to all forms of energy.

The first law of thermodynamics says that energy can be changed from one form to another, but it can be neither created nor destroyed. Energy can occur in a variety of forms such as thermal (heat), electrical, magnetic, nuclear, kinetic, or chemical. The conversion of one form of energy to another is familiar to everyone. For example, the striking of a match involves two energy conversions. In the first conversion, the kinetic energy involved in rubbing a match on a scratch pad is converted into a chemical change in the match head. That chemical change results in the release of chemical energy that is then converted into heat and light energy.

Most energy changes on the earth can be traced back to a single common source: the sun. The following describes the movement of energy through a common environmental pathway, the production of food in a green plant: **solar energy** reaches the earth and is captured by the leaves of a green plant. Individual cells in the leaves then make use of solar energy to convert **carbon dioxide** and water to carbohydrates in the process known as **photosynthesis**. The solar energy is converted to a new form, chemical energy, that is stored within carbohydrate molecules.

"Stored" energy is called potential energy. The term means that energy is available to do work, but is not currently doing so. A rock sitting at the top of a hill has potential energy because it has the capacity to do work. Once it starts rolling down the hill, it uses a different type of energy to push aside plants, other rocks, and other objects.

Chemical energy stored within molecules is another form of potential energy. When chemical changes occur, that energy is released to do some kind of work.

Energy that is actually being used is called kinetic energy. The term kinetic refers to motion. A rock rolling down the hill converts potential energy into the energy of motion, kinetic energy. The first law of thermodynamics says that, theoretically, all of the potential energy stored in the rock can be converted into kinetic energy, without any loss of energy at all.

One can follow, therefore, the movement of solar energy through all parts of the environment and show how it is eventually converted into the chemical energy of carbohydrate molecules, then into the chemical energy of molecules in animals who eat the plant, then into the kinetic energy of animal movement, and so on.

Environmental scientists sometimes put the first law into everyday language by saying that "there is no such thing as a free lunch." By that expression, they mean that in order to produce energy, energy must be used. For many years, for example, scientists have known that vast amounts of oil can be found in rock-like formations known as **oil shale**, but the amount of energy needed to extract that oil by any known process is much greater than the energy that could be obtained from it.

Scientists apply the first law of thermodynamics to an endless variety of situations. A nuclear engineer, for example, can calculate the amount of heat energy that can be obtained from a reactor using nuclear materials (nuclear energy) and the amount of electrical energy that can be obtained from that heat energy.

However, in all such calculations, the engineer also has to take into consideration the second law of thermody-

namics. This law states that in any energy conversion, there is always some decrease in the amount of usable energy.

A familiar example of that law is the incandescent light bulb. Light is produced in the bulb when a thin wire inside is heated until it begins to glow. Electrical energy is converted into both heat energy and light energy in the wire.

As far as the bulb is concerned, the desired conversion is electrical energy to light energy. The heat that is produced, while necessary to get the light, is really "waste" energy. In fact, the incandescent lightbulb is a very inefficient device. More than 90% of the electrical energy that goes into the bulb comes out as heat. Less than 10% is used for the intended purpose, making light.

Examples of the second law can be found everywhere in the natural and human-made environment. And, in many cases, they are the cause of serious problems.

If one follows the movement of solar energy through the environment again, the amazing observation is how much energy is wasted at each stage. Although green plants do convert solar energy to chemical energy, they do not achieve 100% efficiency. Some of the solar energy is used to heat a plant's leaf and is converted, therefore, to heat energy. As far as the plant is concerned, that heat energy is wasted. By the time solar energy is converted to the kinetic energy used by a school child in writing a test, more than 99% of the original energy received from the sun has been wasted.

Some people use the second law of thermodynamics to argue for less meat-eating by humans. They point out how much energy is wasted in using grains to feed cattle. If humans would eat more plants, they say, less energy would be wasted and more people could be fed with available resources.

The second law explains other environmental problems as well. In a **nuclear power** plant, energy conversion is relatively low, around 30%. That means that about 70% of the nuclear energy stored in radioactive materials is eventually converted not to electricity, but to waste heat. Large cooling towers have to be built to remove that waste heat. Often, the waste heat is carried away into lakes, rivers, and other bodies of water. The waste heat raises the temperature of this water, creating problems of **thermal pollution**.

Scientists often use the concept of entropy in talking about the second law. Entropy is a measure of the disorder or randomness of a system and its surroundings. A beautiful glass vase is an example of a system with low entropy because the atoms of which it is made are carefully arranged in a highly structured system. If the vase is broken, the structure is destroyed and the atoms of which it was made are more randomly distributed.

The second law says that any system and its surroundings tends naturally to have increasing entropy. Things tend to spontaneously "fall apart" and become less organized. In some respects, the single most important thing that humans do to the environment is to appear to reverse that process. When they build new objects from raw materials, they tend to introduce order where none appeared before. Instead of iron ore being spread evenly through the earth, it is brought together and arranged into a new **automobile**, a new building, a piece of art, or some other object.

But the apparent decrease in entropy thus produced is really misleading. In the process of producing this order, humans have also brought together, used up, and then dispersed huge amounts of energy. In the long run, the increase in entropy resulting from energy use exceeds the decrease produced by construction. In the end, of course, the production of order in manufactured goods is only temporary since these objects eventually wear out, break, fall apart, and return to the earth.

For many people, therefore, second law of thermodynamics is a very gloomy concept. It suggests that the universe is "running down." Every time an energy change occurs, it results in less usable energy and more wasted heat.

People concerned about the environment do well, therefore, to know about the law. It suggests that humans think about ways of using waste heat. Perhaps there would be a way, for example, of using the waste heat from a nuclear power plant to heat homes or commercial buildings. Techniques for making productive use of waste heat are known as **cogeneration**.

Another way to deal with the problem of waste heat in energy conversions is to make such conversions more efficient or to find more efficient methods of conversion. The average efficiency rate for **power plants** using **fossil fuels** is only 33%. Two-thirds of the chemical energy stored in **coal**, oil, and **natural gas** is, therefore, wasted. Methods for improving the efficiency of such plants would obviously provide a large environmental and economic benefit.

New energy conversion devices can also help. Fluorescent light builds, for example, are far more efficient at converting electrical energy into light energy than are incandescent bulbs. Experts point out that simply replacing existing incandescent light bulbs with fluorescent lamps would make a significant contribution in reducing the nation's energy expenditures. *See also* Alternative energy sources; Energy and the environment; Energy flow; Environmental science; Pollution

[*David E. Newton*]

Resources

Books

Joesten, M. D., et al. *World of Chemistry*. Philadelphia: Saunders, 1991.

———. *Living in the Environment*. 7th ed. Belmont, CA: Wadsworth, 1992.

Miller Jr., F. *College Physics*. 6th ed. New York: Harcourt Brace Jovanovich, 1987.

Miller Jr., G. T. *Energy and Environment: The Four Energy Crises*. 2nd ed. Belmont, CA: Wadsworth, 1980.

Thermoplastics

A thermoplastic is any material that can be heated and cooled a number of times. Some common thermoplastics are **polystyrene**, polyethylene, the acrylics, the polyvinyl **plastics**, and polymeric derivatives of cellulose. Thermoplastics are attractive commercial and industrial materials because they can be molded, shaped, extruded, and otherwise formed while they are molten. A few of the products made from thermoplastics are bottles, bags, toys, packing materials, food wrap, adhesives, yarns, and electrical insulation. The ability to be reshaped is also an environmental benefit. Waste thermoplastics can be separated from other solid wastes and recycled by reforming them into new products. *See also* Recyclables; Recycling; Solid waste; Solid waste recycling and recovery; Solid waste volume reduction

Thermosetting polymers

Thermosetting polymers are compounds that solidify, or "set," after cooling from the molten state and then cannot be remelted. Some typical thermosetting polymers are the epoxys, alkyds, polyurethanes, **furans**, silicones, polyesters, and phenolic **plastics**. Products made from thermosetting polymers include: radio cases, buttons, dinnerware, glass substitutes, paints, synthetic **rubber**, insulation, and synthetic body parts. Because they cannot be recycled and do not readily decompose, thermosetting polymers pose a serious environmental hazard. They contribute significantly, therefore, to the problem of **solid waste** disposal and, in some cases, pose a threat to **wildlife** who swallow or become ensnared in plastic materials. *See also* Solid waste incineration; Solid waste recycling and recovery; Solid waste volume reduction

Third World

The unofficial but common term "Third World" refers to the world's less wealthy and **less developed countries**. In the decades after World War II, the term was developed in recognition of the fact that these countries were emerging from colonial control and were prepared to play an independent role in world affairs. In academic discussions of the world as a single, dynamic system, or world systems theory, the term distinguishes smaller, nonaligned countries from powerful capitalist countries (the **First World**) and from the now disintegrating communist bloc (the **Second World**). In more recent usage, the term has come to designate the world's less developed countries that are understood to share a number of characteristics, including low levels of industrial activity, low per capita income and literacy rates, and relatively poor health care that leads to high infant **mortality** rates and short life expectancies. Often these conditions accompany inequitable distribution of land, wealth, and political power and an economy highly dependent on exploitation of **natural resources**.

Third World pollution

As the countries of the **Third World** struggle with **population growth**, poverty, famines, and wars, their residents are discovering the environmental effects of these problems, in the form of increasing air, water, and land **pollution**. Pollution is almost unchecked in many developing nations, where Western nations dump toxic wastes and untreated sewage flows into rivers. Many times, the choice for Third World governments is between poverty or poison, and basic human needs like food, clothing, and shelter take precedence.

Industrialized nations often dump wastes in developing countries where there is little or no environmental regulation, and governments may collect considerable fees for accepting their **garbage**. In 1991, *World Watch* magazine reported that Western companies dumped more than 24 million tons (22 million metric tons) of **hazardous waste** in Africa alone during 1988.

Companies can also export industrial hazards by moving their plants to countries with less restrictive **pollution control** laws than industrialized nations. This was the case with Union Carbine, which moved its chemical manufacturing plant to **Bhopal, India**, to manufacture a product it was not allowed to make in the United States. As Western nations enact laws promoting environmental and worker safety, more manufacturers have moved their hazardous and polluting factories to **less developed countries**, where there are little or no environmental or occupations laws, or no enforcement agencies. Hazardous industries such as textile, **petrochemical**, and chemical production, as well as smelting and electronics, have migrated to Latin America, Africa, Asia, and Eastern Europe.

For example, IBM, General Motors, and Sony have established manufacturing plants in Mexico, and some of these have created severe environmental problems. At least 10 million gal (38 million L) of the factories' raw sewage is discharged into the Tijuana River daily. Because pollution threatens San Diego beaches, most of the cleanup is paid for by the United States and California governments. Although consumers pay less for goods from these companies, they

are paying for their manufacture in the form of higher taxes for environmental cleanup.

Industries with shrinking markets in developed countries due to environmental concerns have begun to advertise vigorously in the Third World. For example, DDT production, led by United States and Canadian companies, is at an all-time high even though it is illegal to produce or use the **pesticide** in the United States or Europe since the 1970s. DDT is widely used in the Third World, especially in Latin America, Africa, and on the Indian subcontinent.

Industrial waste is handled more recklessly in underdeveloped countries. The New River, for example, which flows from northern Mexico into southern California before dumping into the Pacific Ocean, is generally regarded as the most polluted river in North America due to lax enforcement of environmental standards in Mexico.

Industrial pollution in Third World countries is not their only environmental problem. Now, in addition to worrying about the environmental implications of **deforestation**, **desertification**, and **soil erosion**, developing countries are facing threats of pollution that come from development, industrialization, poverty, and war.

The number of gasoline-powered vehicles in use worldwide is expected to double to one billion in the next 40 years, adding to **air pollution** problems. Mexico City, for example, has had air pollution episodes so severe that the government temporarily closed schools and factories. Much of the auto-industry growth will take place in developing countries, where the **automobile** population is rapidly increasing.

In the Third World, the effects of **water pollution** are felt in the form of high rates of death from **cholera**, typhoid, dysentery, and diarrhea from viral and bacteriological sources. More than 1.7 billion people in the Third World have an inadequate supply of safe drinking water. In India, for example, 114 towns and cities dump their human waste and other untreated sewage directly into the Ganges River. Of 3,119 Indian towns and cities, only 209 have partial **sewage treatment**, and only eight have complete treatment.

Zimbabwe's industrialization has created pollution problems in both urban and rural areas. Several lakes have experienced eutrophication because of the **discharge** of untreated sewage and industrial waste. In Bangladesh, degradation of water and soil resources is widespread, and flood conditions result in the spread of polluted water across areas used for fishing and rice cultivation. Heavy use of pesticides is also a concern there.

In the Philippines, air and soil pollution poses increasing health risks, especially in urban areas. Industrial and toxic waste disposals have severely polluted 38 river systems.

Widespread poverty and political instability has exacerbated Haiti's environmental problems. Haiti, the poorest country in the Western Hemisphere, suffers from deforestation, land degradation, and water pollution. While the country has plentiful **groundwater**, less than 40% of the population has access to safe drinking water.

As 117 world leaders and their representatives met at the **United Nations Earth Summit** in 1992, some of these concerns were addressed as poorer Third World countries sought the help of richer industrialized nations to preserve the **environment**. Almost always, environmental cleanup in the Third World is an economic issue, and the countries cannot afford to spend more on it.

Another issue affecting Third World pollution control is the role of transnational corporations (TNCs) in global environmental problems. A Third World Network economist cited TNCs for their responsibility for water and air pollution, toxic wastes, hazardous **chemicals** and unsafe working conditions in Third World countries. None of the Earth Summit documents, however, regulated transnational corporations, and the United Nations has closed its Center for Transnational Corporations, which had been monitoring TNC activities in the Third World.

There is a growing environment awareness in Third World nations, and many are trying to correct the problems. In Madras, India, sidewalk vendors sell rice wrapped in banana leaf; the leaf can be thrown to the ground and is consumed by one of the free-roaming cows on Madras' streets. In Bombay, India, tea is sold in a brown clay cup which can be crushed into the earth when empty. The Chinese city of Shanghai produced all its own vegetables, fertilizes them with human waste, and exports a surplus.

Environmentalists worldwide are calling for a strengthened **United Nations Environment Programme** (UNEP) to enact sanctions and keep polluters out of the Third World. It could also enforce the "polluter pays" principle, eventually affecting Western governments and companies that dump on the Third World. Already, UNEP and the **World Bank** provide location advice and environmental **risk assessment** when the host country is not able to do so, and the World Health Organization and the International Labor Organization provide some guidance on occupational health and safety to developing countries. *See also* Drinking-water supply; Environmental policy; Environmental racism; Flooding; Hazardous material; Industrial waste treatment; Marine pollution; Watershed management

[*Linda Rehkopf*]

Resources

Periodicals

Barber, B. "Lessons from the Third World." *Omni* (January 1992): 25.

Collins, C., and C. Darch. "Summit Sets a Rich Table, but Africa Gets the Crumbs." *National Catholic Reporter* 29 (July 3, 1992): 12.

Kumar, P. "Stop Dumping on the South." *World Press Review* (June 1992): 12–13.

LaDou, J. "Deadly Migration: Hazardous Industries' Flight to the Third World." *Technology Review* (July 1991): 47.

"Pollution and the Poor." *Economist* 322 (February 15, 1992): 18–19.

Lee M. Thomas (1944 –)

American former Environmental Protection Agency administrator

If the 1960s and 1970s have become known as decades of growing concern about environmental causes, the 1980s will probably be remembered as a decade of stagnation and retreat on many environmental issues. Presidents Ronald Reagan and George Bush held very different beliefs about many environmental problems than did their immediate predecessors, both Democrat and Republican. Reagan and Bush both argued that environmental concerns had resulted in costly overregulation that acted as a brake on economic development and contributed to the expansion of government bureaucracy.

One of the key offices through which these policies were implemented was the **Environmental Protection Agency** (EPA). In 1985, President Reagan nominated Lee M. Thomas to be Administrator of that agency. Thomas had a long record of public service before his selection for this position. Little of his service had anything to do with environmental issues, however. Thomas began his political career as a member of the Town Council in his hometown of Ridgeway, South Carolina. He then moved on to a series of posts in the South Carolina state government.

The first of these positions was as executive director of the State Office of Criminal Justice Programs, a post he assumed in 1972. In that office, Thomas was responsible for developing criminal justice plans for the state and for administering funds from the Law Enforcement Assistance Administration.

Thomas left this office in 1977 and spent two years working as an independent consultant in criminal justice. Then, in 1979, he returned to state government as director of Public Safety Programs for the state of South Carolina. In addition to his responsibilities in public safety, Thomas served as chairman of the Governor's Task Force on Emergency Response Capabilities in Support of Fixed Nuclear Facilities. The purpose of the task force was to assess the role of various state agencies and local governments in dealing with emergencies at nuclear installations in the state.

Thomas's first federal appointment came in 1981, when he was appointed executive deputy director and associate director for State and Local Programs and Support of the Federal Emergency Management Agency (FEMA). His responsibilities covered a number of domestic programs, including Disaster Relief, **Floodplain** Management, **Earthquake** Hazard Reduction, and Radiological Emergency Preparedness. While working at FEMA, Thomas also served as chairman of the U.S./Mexican Working Group on Hydrological Phenomena and Geological Phenomena.

In 1983, President Reagan appointed Thomas assistant administrator for **Solid Waste** and Emergency Response of the EPA. In this position, Thomas was responsible for two of the largest and most important of EPA programs, the **Comprehensive Environmental Response Compensation and Liability Act** (Superfund) and the **Resource Conservation and Recovery Act.** Two years later, Thomas was confirmed by the Senate as administrator of the EPA, a position he held until 1989, when he became the Chairman and Chief Executive Officer for Law Companies Environmental Group Inc. Thomas then was employed as the Senior Vice President, Environmental and Government Affairs for the Georgia-Pacific Corporation. He currently is the President of Building Products and Distribution.

[*David E. Newton*]

Resources

Periodicals

Thomas, Lee M. "The Business Community and the Environment: An Important Partnership." *Business Horizons* (March–April 1992): 21–24.

———. "Trends Affecting Corporate Environmental Policy: The Real Estate Perspective." *Site Selection and Industrial Development* 35 (October 1990): 1(1183)–3(1185).

Henry David Thoreau (1817 – 1862)

American writer and natural philosopher

Thoreau was a member of the group of radical Transcendentalists who lived in New England, especially Concord, Massachusetts, around the mid-nineteenth century. He is known worldwide for two written works, both still widely read and influential today: *Walden*, a book, and a tract entitled "Civil Disobedience." All of his works are still in print, but most noteworthy is his 14-volume *Journal*, which some critics think contains his best writing. Contemporary readers interested in **conservation**, **environmentalism**, **ecology**, natural history, the human **species**, or philosophy can gain great understanding and wisdom from reading Thoreau.

Today, Thoreau would be considered not only a philosopher, a humanist, and a writer, but also an ecologist (though that word was not coined until after his death). His status as a writer, naturalist, and conservationist has been secure for decades; in conservation circles, he is recognized as what *Backpacker* magazine calls one of the "elders of the tribe."

Trying to trace any idea through Thoreau's work is a complicated task, and that is true of his ecology as well. One straightforward example of Thoreau's sophistication as an ecologist is his essay on "The **Succession** of Forest Trees." Thoreau found the same unity in **nature** that present-day ecologists study, and he often commented on it: "The birds with their plumage and their notes are in harmony with the flowers." Only humans, he felt, find such connections difficult: "Nature has no human inhabitants who appreciate her." Thoreau did appreciate his surroundings, both natural and human, and studied them with a scientist's eye. The linkages he made showed an awareness of **niche** theory, hierarchical connections, and trophic structure: "The perch swallows the grub-worm, the pickerel swallows the perch, and the fisherman swallows the pickerel; and so all the chinks in the scale of being are filled."

Much of Thoreau's philosophy was concerned with **human ecology**, as he was perhaps most interested in how human beings relate to the world around them. Often characterized as a misanthrope, he should instead be recognized for how deeply he cared about people and about how they related to each other and to the natural world. As Walter Harding notes, Thoreau believed that humans "Antaeus-like, derived [their] strength from contact with nature." When Thoreau insists, as he does, for example, in his journal, that "I assert no independence," he is claiming relationship, not only to "summer and winter...life and death," but also to "village life and commercial routine." He flatly asserts that "we belong to the community." Present-day humans could not more urgently ask the questions he asked: "Shall I not have intelligence with the earth? Am I not partly leaves and vegetable mold myself?"

The essence of Thoreau's message to present-day citizens of the United States can be found in his dictum in *Walden* to "simplify, simplify." That is a straightforward message, but one he elaborates and repeats over and over again. It is a message that many critics of today's materialism believe American citizens need to hear over and over again. Right after those two words in his chapter on "What I Lived For" is a directive on how to achieve simplicity. "Instead of three meals a day, if it be necessary eat but one; instead of a hundred dishes, five; and reduce other things in proportion," he asserts. The message is repeated in different ways, making a major theme, especially in *Walden*, but also in his other writings: "A man is rich in proportion to the number of things he can afford to let alone." Thoreau's preference for simplicity is clear in the fact that he had only three chairs in his house: "one for solitude, two for friendship, three for society." Thoreau is often quoted as stating "I would rather sit on a pumpkin and have it all to myself than be crowded on a velvet cushion."

Hundreds of writers have joined Thoreau in censuring the materialist root of current environmental problems, but reading Thoreau may still be the best literary antidote to that materialism. Consider the stressed commuter/city worker who does not realize that the "cost of a thing is the amount of...life which is required to be exchanged for it, immediately or in the long run." In the pursuit of fashion, ponder his admonition to "beware of all enterprises that require new clothes."

Thoreau firmly believed that the rich are the most impoverished: "Give me the poverty that enjoys true wealth." The enterprises he thought important were intangible, like being present when the sun rose, or, instead of spending money, spending hours observing a heron on a pond. Working for "treasures that moth and rust will corrupt and thieves break through and steal...is a fool's life." As Thoreau noted, too many of us make ourselves sick so that we "may lay up something against a sick day." To him, most of the luxuries, and many of the so-called comforts of life, are not only dispensable, "but positive hindrances to the elevation of mankind." For most possessions, Thoreau's forthright answer was "it costs more than it comes to."

Thoreau was a humanist, an abolitionist, and a strong believer in egalitarian social systems. One of his criticisms of materialism was that, in the race for more and more money and goods, "a few are riding, but the rest are run over." He recalled that, before the modern materialist state, it was less unfair: "In the savage state every family owns a shelter as good as the best." Thoreau was anti-materialistic and believed that the relentless pursuit of "things" divert people from the real problems at hand, including destruction of the **environment**: "Our inventions are wont to be pretty toys, which distract our attention from serious things." In this same vein, he claimed that "the greater part of what my neighbors call good I believe in my soul to be bad."

In modern life, stress is a major contributor to illness and death. And, much of that stress is generated by the constant acquisitive quest for more and more material goods. Thoreau asks "why should we be in such desperate haste to succeed and in such desperate enterprises?" He notes that "from the desperate city you go into the desperate country" with the result that "the mass of men lead lives of silent desperation." This passage was written in the nineteenth century, but still echoes through the twenty first.

Simplification of lifestyle is now widely taught as a practical antidote to the environmental and personal consequences of the materialist cultures of the urban/industrial twentieth century. And that kind of simplification is central to Thoreau's thought. But readers must remember that Thoreau was unmarried, childless, and often dependent on friends and family for room and board. Thoreau lived on and enjoyed the land around Walden Pond, for example,

Henry David Thoreau. (The Library of Congress.)

but he did not own it or have to pay taxes or upkeep on it; he "borrowed" it from his friend Ralph Waldo Emerson (1803–1882). It is not realistic for most people to try to emulate Thoreau directly or to take his suggestions literally. And he agreed, saying "I would not have anyone adopt my mode of living on any account." Still, he did figure out that by working only six weeks a year, he could support himself and so free the other forty-six to live as he saw fit.

Often characterized as an impractical dreamer, Thoreau's wisdom is repeated widely today, a wisdom that is commonly of direct applicability and down to earth. **Aldo Leopold**, for example, is credited with the axiom that wood you cut yourself warms you twice, but it can be found almost a hundred years earlier in Thoreau's chapter on "House-Warming." In 1850, he was an advocate of **national forest** preserves, writing eloquently on the subject in his essay "Chesuncook" in *The Maine Woods*. He also predicted the devastation wreaked on the shad runs by **dams** built on the Concord River.

Thoreau was first and finally a writer. His greatest contribution remains that his writing still raises the consciousness of the reader, causing people who come in contact with his work to be more aware of themselves and of the world around them. As he said "only that day dawns to which we are awake." A raised consciousness means an increase in humility because, as Thoreau teaches, "the universe is wider than our views of it." He remained convinced "that to maintain one's self on this earth is not a hardship but a pastime."

[*Gerald L. Young*]

Resources

Books

Fritzell, P. A. "Walden and Paradox: Thoreau as Self-Conscious Ecologist." In *Nature Writing and America: Essays Upon a Cultural Type*. Ames: Iowa State University Press, 1990.

Johnson Jr., W. C. "Thoreau's Language of Ecology." In *Wilderness Tapestry: An Eclectic Approach to Preservation*, edited by S. I. Zeveloff, L. M. Vause, and W. H. McVaugh. Reno: University of Nevada Press, 1992.

Thoreau, H. D. *Walden*. New York: A. L. Burt, 1902.

Threatened species

see **Endangered species**

Three Gorges Dam

When finished in 2009, the Three Gorges Dam on the Chang Jiang (Yangtze River) will be the largest dam in the world. Spanning 1.2 mi (2.0 km), and standing 600 ft (175 m) above normal river level, the dam will create a **reservoir** more than 400 mi (644 km) long. The project was designed to control floods, ease navigation, and provide badly needed electricity for China's heartland. The 26 giant turbines in the dam's powerhouse are expected to generate 18,200 megawatts of electricity—equal to 10% of the country's current supply—to support industrialization and modernization in the region. The reservoir will flood the scenic Three Gorges, one of China's most historically significant areas. It also will displace at least 1.3 million people and will flood 150 towns and cities, and more than 1,300 villages.

Proponents of the project argue that China needs the project's electricity for modernization. With dangerous rapids drowned under the reservoir, ocean-going ships will be able to sail all the way to Chongking, nearly 2,000 mi (3,200 km) inland from the ocean. The water stored in the reservoir also will make possible a long-discussed plan to build aqueducts to carry water from southern China to the dry plains around Beijing. Furthermore, annual **flooding** on the river caused 300,000 deaths in the twentieth century. In 1998, the worst flood in history drove 56 million people from their homes. Planners expect the dam to reduce these floods and eliminate untold misery for the 300 million people who live in the Yangtze River Valley.

The Three Gorges Dam was first proposed in 1919 by Dr. Sun Yat-sen, a revolutionary and founder of the Chinese republic. Later it was championed by Mao Tsetung (or Zedong) who led the country from 1949 to 1976.

Mao saw this project as a way to instill national pride and demonstrate China's modernization. Although Mao ordered a full-scale survey and project design in 1955, prohibitive costs together with doubts about the safety and feasibility of the dam have slowed construction. In 1992, Premier Li Peng pushed through the final vote to authorize the project. The first stage of construction was completed in 2002, and flooding of the reservoir will begin in 2003. The water is expected to reach its maximum height by 2009.

Environmentalists criticize the dam on the grounds that it will reduce fish stocks, eliminate 78,000 acres (32,000 ha) of important agricultural lands, and threaten habitats of critically **endangered species** such as the Yangtze river dolphin, Chinese sturgeon, finless porpoise, and Yangtze alligator. The reservoir will also fill one of China's most important historical and cultural regions, sinking 8,000 archeological sites and historic monuments. Currently, more than 250 billion gallons (nearly a trillion liters) of untreated sewage is dumped into the Yangtze every year. Planners claim that the new cities into which residents are being relocated will have better **sanitation** than the old river towns. How sewage is handled in these new towns, and how much contamination leeches out of abandoned structures in the reservoir remains to be seen. Environmentalists worry that the reservoir will become a stagnant cesspool, dangerous to both aquatic life and to the millions of people who depend on the river for their drinking water.

Sediment and **silt** accumulation is another problem to be solved, since tremendous volumes of sediment will accumulate as the silt- and sand-laden Yangtze River slows in the reservoir. Elsewhere in China silt and sand accumulation has decreased reservoir storage capacity nationwide by 14%. Simply controlling river-bottom gravel may require **dredging** as much as 200,000 m^2 of material each year. To reduce sediment accumulation, dam operators plan to let spring floods, which carry much of the annual sediment load, flow through silt doors at the bottom of the dam. They hope this flow will scour out the bottom of the reservoir.

Geologists worry about catastrophic dam failure because the dam is built over an active seismic fault. Engineers assure us that the dam can withstand the maximum expected **earthquake**, but China has a poor record of dam safety. More than 3,200 Chinese **dams** have failed since 1949, a failure rate of 3.7% compared to a 0.6% failure rate for the rest of the world. Probably the worst series of dam failures in world history occurred in Henan Province in 1975, when heavy **monsoon** rains caused 62 modern dams to fall like a line of dominoes. Some 230,000 people died in the massive flooding that followed. Even if the dam is able to withstand earthquakes, giant waves spawned by upstream landslides could easily cause a calamitous dam failure. In 1986, a **landslide** just a few miles upstream from the dam site dumped 15 million cubic meters of rock and **soil** into the river. Witnesses reported an 260 ft (80 m) wave rolling down the river. If a wave that size hits the upstream side of the dam with a full reservoir, one geologist predicts a flood "of biblical proportions" as a wall of water races downstream through lowlands where hundreds of millions of people live.

The Chinese government is installing an early warning system to predict landslides, and has banned timber cutting and farming on steep upstream hillsides in an effort to control both landslides and sediment loads in the river. Critics claim that there would be safer and cheaper ways to store water and generate electricity than a single huge dam. Original estimates were that the Three Gorges Dam would cost $11 billion. By 2002, the costs for construction, relocation, and landscape stabilization had risen to $75 billion, and the project is not yet finished. A series of relatively small dams on tributary streams might have been much cheaper and less disruptive than the current project.

In 2002, the first stage of the dam was completed and water started filling the reservoir. By late 2003, the water is expected to reach a height of 438 ft (135 m) above normal river level. Most of the new cities above the final 175 m watermark have already been finished, and relocation of residents is well underway. Destruction and decontamination of cities that will be flooded has also begun.

[*William P. Cunningham Ph.D.*]

Resources

Books

Dai Qing. *The River Dragon Has Come!* Armonk, New York: M. E. Sharpe, 1998.

Hessler, Peter. *River Town: Two Years on the Yangtze.* New York: HarperCollins, 2001.

Leopold, Luna. *Sediment Problems at Three Gorges Dam.* Berkely, CA: International Rivers Network, 1996.

Ryder, Grainne, ed. *Damming the Three Gorges: What the Dam Builders Don't Want You to Know.* Toronto: Probe International, 1990.

Periodicals

Sullivan, L. R. "The Three Gorges Project: Dammed If They Do?" *Curent History* 94 (1995): 266–70.

Three Mile Island Nuclear Reactor

The 1970s were a decade of great optimism about the role of **nuclear power** in meeting world and national demands for energy. Warnings about the declining reserves of **coal**, **petroleum**, and **natural gas**, along with concerns for the environmental hazards posed by **power plants** run on **fossil fuels**, fed the hope that nuclear power would soon have a growing role in energy production. Those expectations were suddenly and dramatically dashed on the morning of March 28, 1979.

An aerial view of Three Mile Island Nuclear Reactor in Harrisburg, Pennsylvania. (AP/Wide World Photos. Reproduced by permission.)

On that date, an unlikely sequence of events resulted in a disastrous accident at the Three Mile Island (TMI) Nuclear Reactor at Harrisburg, Pennsylvania. As a result of the accident, radioactive water was released into the Susquehanna River, radioactive steam escaped into the **atmosphere**, and a huge bubble of explosive **hydrogen** gas filled the reactor's cooling system. For a period of time, there existed a very real danger that the reactor core might melt.

News of the accident produced near panic among residents of the area. Initial responses by government and industry officials downplayed the seriousness of the accident and, in some cases, were misleading and self-serving.

The first official study of the TMI accident was carried out by the President's Commission on the Accident at Three Mile Island. Chaired by John G. Kemeny, then president of Dartmouth College, the Commission attempted to reconstruct the events that led to the accident. It found that the accident was initiated during a routine maintenance operation in which a water purifier used in the system was replaced. Apparently air was accidentally introduced into the system along with the purified fresh water.

Normally the presence of a foreign material in the system would be detected by safety devices in the reactor. However, a series of equipment malfunctions and operator errors negated the plant's monitoring system and eventually resulted in the accident.

At one point, for example, operators turned off emergency cooling pumps when faulty pressure gauges showed that the system was operating normally. Also, tags which hung on water pumps indicating that they were being repaired blocked indicator lights showing an emergency condition inside the reactor.

The Kemeny Commission placed blame for the TMI accident in a number of places. They criticized the nuclear industry and the **Nuclear Regulatory Commission** for being too complacent about reactor safety. They suggested that workers needed better training. The Commission also found that initial reactions to the accident by government and industry officials were inadequate.

No one was killed in the TMI accident, but a 1990 study showed that there were no statistically significant in-

creases in cancers for residents downwind of the accident. A study completed in the mid-1990s, however, refutes those findings stating that "cancer incidence, specifically, lung **cancer** and **leukemia**, increased more following the TMI accident in areas estimated to have been in the pathway of radioactive plumes than in other areas." This study also indicates that the radiation levels may have been higher than originally reported, and that cancer rates (lung/leukemia) were 2–10 times higher than upwind rates. Cleanup operations at the plant took over six years to complete, and with a $1 billion price tag, cost more than the plant's original construction.

Probably the greatest effect of the TMI accident was the nation's loss of confidence in nuclear power as a source of energy. Since the accident, not one new nuclear power plant has been ordered, and existing plans 65 others were eventually canceled. *See also* Carcinogen; Radiation exposure; Radioactive waste; Radioactivity

[*David E. Newton*]

Resources
Books

Dresser, P. D. *Nuclear Power Plants Worldwide.* Detroit, MI: Gale Research, 1993.

Ford, D. *Three Mile Island: Thirty Minutes to Meltdown.* New York: Penguin, 1983.

Gray, M., and I. Rosen. *The Warming: Accident at Three Mile Island.* New York: W.W. Norton, 1982.

Moss, T. H., and D. L. Sills, eds. *The Three Mile Island Nuclear Accident: Lessons and Implications.* New York: Academy of Sciences, 1981.

Periodicals

Booth, W. "Post-Mortem on Three Mile Island." *Science* (December 4, 1987): 1342–1345.

Stranahan, S. Q. "Three Mile Island: It's Worse Than You Think." *Science Digest* (June 1985): 54–57+.

Other

President's Commission on the Accident at Three Mile Island. *Report of the President's Commission on the Accident at Three Mile Island.* Washington, DC: U.S. Government Printing Office, 1979.

Threshold dose

In radiology, the threshold dose is the smallest dose of radiation that will produce a specified effect. In the larger context of toxic exposure, threshold dose refers to the dose below which no harm is done. There is a threshold below which relatively little damage occurs from exposure, and above which the damage increases dramatically. For noncarcinogenic **toxins**, there does seem to be a safe dose for some substances, a threshold dose below which no harm is done. With carcinogens, however, just one change in the genetic materials of one cell may be enough to cause a malignant transformation that can lead eventually to **cancer**. Although there is some evidence that repair mechanisms or surveillance by the immune system may reduce the incidence of some cancers, it is generally considered prudent to assume that no safe or threshold dose exists for carcinogens.

Tidal power

In looking for **alternative energy sources** to meet future needs, some common physical phenomena are obvious candidates. One of these is tidal power. Twice each day on every coastline in the world, bodies of water are pulled onto and off of the shore as a result of gravitational forces exerted by the moon and sun. Only on ocean coasts is this change large enough to notice, however, and therefore, to take advantage of as an energy source.

The potential of tidal power as an energy source is clearly demonstrated. Pieces of wood are carried onto a beach and then off again every time the tide comes in or goes out. In theory, the energy that moves this wood could also push against a turbine blade and turn a generator.

In fact, the number of places on the earth where tides are strong enough to spin a turbine is relatively small. The simple back-and-forth movement seen on any shoreline does not contain enough energy by itself. Geographical conditions must concentrate and focus tidal action in a limited area. In such places, tides do not move in and out at a leisurely pace, but rush in and out with the force of a small river.

One of the few commercial tidal power stations in operation is located at the mouth of La Rance River in France. Tides at this location reach a maximum of 44 ft (13.5 m). Each time the tide comes in, a dam at the La Rance station holds water back until it reaches its maximum depth. At that point, gates in the dam are opened and water is forced to flow into the river, driving a turbine and generator in the process. Gates in the dam are then closed, trapping the water inside the dam. At low tide, the gates open once again, allowing water to flow out of the river, back into the ocean. Again the power of moving water is used to drive a turbine and generator.

Hence the plant produces electricity only four times each day, during each of two high tides and each of two low tides. Although it generates only 250 megawatts daily, the plant's 25% efficiency rate is about equal to that of a power plant operating on **fossil fuels**.

One area where tides are high enough to make a power plant feasible is on Canada's **Bay of Fundy**. Experiments suggest that plants with capacities of 1–20 megawatts could be located at various places along the bay. So far, however, the cost of building such plants is significantly greater than

the cost of building conventional **power plants** of similar capacity. Still, optimists suggest that serious development of tidal power could provide up to a third of the electrical energy now obtained from hydropower worldwide at some time in the future.

Despite the drawbacks and expense of tidal power, China is one country that has used it extensively. Since energy needs are often modest in China, low capacity plants are more feasible. As of the late 1980s, therefore, the Chinese had built more than 120 tidal power plants to provide electricity for small local regions. *See also* Dams; Energy and the environment; Wave power

[*David E. Newton*]

Resources

Periodicals

Greenburg, D. A. "Modeling Tidal Power." *Scientific American* 257 (November 1987): 128–135.

Holloway, T. "Tidal Power." *Sea Frontiers* 35 (March-April 1989): 114–49.

"The Potential of Tidal Power—Still All At Sea." *New Scientist* 126 (May 19, 1990): 52.

Webb, J. "Tide of Optimism Ebbs Over Underwater Windmill." *New Scientist* 138 (April 24, 1993): 10.

Tigers

Tigers (*Panthera tigris*) are the largest living members of the family Felidae, which includes all cats. Siberian tigers (*P. t. altaica*) are the largest and most massive of the eight recognized subspecies. They normally reach a weight of 660 lb (300 kg), with a record male that reached 845 lb (384 kg). Several of the subspecies have had their populations totally decimated and are probably extinct, mostly through direct human actions. The species' range, overall, has been greatly reduced in historic times. Currently, tigers are found in isolated regions of India, Bangladesh, Nepal, Bhutan, Southeast Asia, Manchuria, China, Korea, Russia, and Indonesia. Tigers are designated "endangered" by the U. S. **Fish and Wildlife Service** and by **IUCN—The World Conservation Union**. They are also listed in Appendix I of the **Convention on International Trade in Endangered Species of Wild Fauna and Flora**.

Unlike their relatives, the lion and the cheetah, tigers are not found in open habitats. They tend to be solitary hunters, stalking medium- to large-sized prey (such as pigs, deer, antelope, buffalo, and gaur) in moderately dense cover of a variety of forest habitats, including **tropical rain forest**, moist coniferous and deciduous forests, dry forests, or mangrove swamps. Within their **habitat** both males and females establish home ranges that do not overlap with members of their own sex. Home ranges average 8 mi^2 (21 km^2) for females, but vary from 25–40 mi^2 (65–104 km^2) for males depending on prey availability and to allow for the inclusion of several females in his range. Tigers that live in areas of prime habitat raise more offspring than can establish ranges within that habitat, therefore, several are forced to the periphery to establish territories and live. This creates an important condition in that this series of peripheral individuals helps promote genetic mixing in the breeding population.

Tigers are often labeled "maneaters." Although most tigers shy away from humans, some have been provoked into attack. Others, having been encountered unexpectedly, attack people as a defense, and a very few are thought to hunt humans consciously. An estimated 60–120 people fall victim to tigers each year. Tigers' typical prey includes larger mammals such as deer and buffalo, and the cats will actively search for this prey, instead of waiting in ambush. Tigers hunt alone, and, even though they are highly skilled predators, are rarely successful more than once in every 15 attempts. With a **scarcity** of habitat for large prey and reduced cover, many tigers will opportunistically attack domesticated livestock, thus, themselves becoming targets of humans.

Population pressures from humans, habitat loss, **poaching**, and **overhunting** have led to the **extinction** or probable extinction of four subspecies and large reductions in the populations of the other four subspecies of tiger. The Balinese tiger (*P. t. balica*) has been extinct for several decades. The South China tiger (*P. t. amoyensis*), which was the target of an extermination campaign, may be extinct in the wild. There has been only one unconfirmed report of the Caspian tiger (*P. t. virgata*) in the last 30 years, thus it is probably extinct. The Javan tiger (*P. t. sondiaca*) is also probably extinct, as its population was reduced to four or five individuals, and there has not been a confirmed sighting since the late 1970s. Little is known of the population levels of the Indochinese tiger (*P. t. corbetti*) but the number is thought to be between 1,000 and 1,500, with 60 in captivity. The Sumatran tiger (*P. t. sumatrae*) has a population in the wild of about 500 individuals. The Siberian tiger, whose population dropped to 20–30 individuals in 1940, has a population today of over 500 in the wild. The Bengal tiger (*P. t. tigris*) has been the target of renewed poaching efforts since mid-1990. The bones of these tigers bring a handsome price on the black market. Tiger bones are ground up, dissolved in a liquid, and used in Chinese medicine. Early estimates are that 500 or more Bengal tigers have been poached in the last three years for this purpose.

[*Eugene C. Beckham*]

Resources

Books

Grzimek, B., ed. *Grzimek's Encyclopedia of Mammals.* Vol. 4. New York: McGraw-Hill, 1990.

Nowak, R. M. *Walker's Mammals of the World.* 6th ed. 2 vols. Baltimore: Johns Hopkins University Press, 1999.

PERIODICALS

Angier, Natalie. "The Cat Comes Back." *New York Times Upfront* (January 31, 2000): 24.

Luoma, J. R. "The State of the Tiger." *Audubon* 89 (1987): 61–63.

"Man Saving Endangered Tigers." *Xinhua News Agency* (March 15, 2001).

OTHER

The Tiger Foundation. [cited May 2002]. <http://www.tigers.ca>.

Tilth

Tilth is the ability of the **soil** to facilitate tillage, resist weeds, and allow plants to take root. It can also refer to the act of tilling, or cultivating soil. Some have related soil tilth to the physical condition of the soil created by integrating the effects of all physical, chemical, and biological processes occurring within a soil matrix. Soil tilth is usually used as a general descriptive term (good, moderate, or poor) rather than a precisely defined scientific quotient. *See also* Agricultural chemicals; Arable land; Soil fertility

Timberline

The elevational or latitudinal extent of forests. Above upper timberline, either at high elevations or at the Arctic or Antarctic limits, atmospheric and **soil** temperatures are too cold for forest development. Below lower timberline, conditions are too dry to support forests. Timberlines are more common and pronounced in the western than in the eastern United States because of more extreme topographic and climatic conditions in the West.

Times Beach

More than 2,000 residents of this Missouri town, located about 30 mi (48 km) southwest of St. Louis, were evacuated after it was contaminated with **dioxin**. In 1971, chemical wastes containing dioxin were mixed with oil and sprayed along the streets of Times Beach to keep down the dust. The spraying was done by Russell M. Bliss, whose company collected and disposed of waste oils and **chemicals** from service stations and industrial plants. Much of this toxic waste oil was sprayed on roads throughout Missouri.

Shortly after the spraying around Times Beach, horses on local farms began dying mysteriously. At one breeding stable, from 1971 to 1973, 62 horses died, as well as several dogs and cats. **Soil** samples were sent to the Centers for Disease Control (CDC) in Atlanta for analysis, and after three years of testing, the agency determined that dangerous levels of dioxin were present in the soil.

A decade after the spraying, the town experienced the worst **flooding** in its history. On December 5, 1982, the Meramec River flooded its banks and inundated the town, submerging homes and contaminating them and almost everything else in Times Beach with dioxin. This compound binds tightly to soil and degrades very slowly, so it was still present in significantly high levels ten years after being used on the roads as a dust suppressant. Following a CDC warning that the town had become uninhabitable, most of the families were temporarily evacuated. The U.S. **Environmental Protection Agency** (EPA) eventually agreed to buy out the town for about $33 million and relocate its inhabitants.

At the time, the CDC considered soil dioxin levels of one part per billion (ppb) and above to be potentially hazardous to human health. Levels found in some areas of Times Beach reached 100 to 300 ppb and above. Dioxin is considered a very toxic chemical. Exposure to this chemical has been linked to **cancer**, miscarriages, genetic mutations, liver and nerve damage, and other health effects, including death, in humans and animals.

Indeed, the contamination seemed to take a serious toll on the health of Times Beach residents. Town officials claimed that virtually every household in Times Beach experienced health disorders, ranging from nosebleeds, depression, and chloracne (a severe skin disorder) to cancer and heart disease. Almost all of the residents tested for dioxin contamination showed abnormalities in their blood, liver, and kidney functions.

By 1983, federal and state officials had located about 100 other sites in Missouri where dioxin wastes had been improperly dumped or sprayed, with levels of the compound reaching as high as 1,750 ppb in some areas.

A decade after the evacuation of Times Beach, debate over dioxin's dangers continues, and some CDC officials now say that the agency overreacted and that the town should not have been abandoned. But by then, Times Beach had become a household name, joining **Love Canal**, New York, and **Seveso, Italy**, on the list of municipalities that were ruined by toxic chemical contamination. *See also* Carcinogen

[*Lewis G. Regenstein*]

RESOURCES

PERIODICALS

"Dioxin Cleanup: Status and Opinions." *Science News* 141 (January 11, 1992): 30.

Felton, E. "The Times Beach Fiasco." *Insight* 7 (August 12, 1991): 12–19.

Gorman, C. "The Double Take on Dioxin." *Time* 138 (August 26, 1991): 52.

Kemezis, P. "Times Beach Cleanup Begins." *ENR* 225 (August 2, 1990): 37–8.

Tipping fee

The fee charged by the owner or operator of a **landfill** for the acceptance of a unit weight or volume of **solid waste** for disposal, usually done by the truckload. The tipping fee is passed back along the chain of waste acceptor to hauler to generator in the form of fees or taxes. Tipping fees rise as the volume of available landfill space is depleted, or as it becomes harder to open new landfills due to public opposition and stricter environmental regulations. *See also* Garbage; Municipal waste; Transfer station; Waste management; Waste stream

Tobacco

Tobacco (*Nicotiana tabacum*) is an herbaceous plant cultivated around the world for its leaves, which can be rolled into cigars, shredded for cigarettes and pipes, processed for chewing, or ground into snuff. Tobacco leaves are the source of commercial nicotine, a component of many pesticides. The tobacco plant is fast-growing with a stem from 4–8 ft (1–3 m) in height.

Native to the Americas, tobacco was cultivated by Native Americans, and Christopher Columbus found them using it in much the same manner as today. American Indians believed it to possess medicinal properties, and it was important in the ceremonies of the plains tribes.

Tobacco was introduced into Europe in the mid-1500s on the basis of its purported medicinal qualities. Tobacco culture by European settlers in America began in 1612 at Jamestown, and it soon became the chief commodity exchanged by colonists for articles manufactured in Europe.

The leading tobacco-growing countries in the world today are China and the United States, followed by India, Brazil, and Turkey, as well as certain countries in the former Soviet Union. Although about one-third of the annual production in the United States is exported, the country also imports about half as much tobacco as it exports. The leading tobacco-growing state is North Carolina, followed by Kentucky, South Carolina, Tennessee, Virginia, and Georgia.

Nicotine occurs in tobacco along with related alkaloids and organic acids such as malic and citric. Nicotine content is determined by the **species**, variety, and strain of tobacco; it is also affected by the growing conditions, methods of culture and cure, and the position on the plant from which the leaves are taken. Tobacco is high in ash content, which can range from 15–25% of the leaf. Flue-cured tobacco is rich in sugar, and cigar tobaccos are high in nitrogenous compounds but almost free of starch and sugars.

Most tobacco products are manufactured by blending various types of leaves, as well as leaves of different origins, grades, and crop years. Cigarette manufacturers usually add sweetening preparations and other flavorings, but the preparation of tobaccos for pipe smoking and chewing is as varied as the assortment of products themselves. Snuff is made by fermenting fire-cured leaves and stems and grinding them before adding salts and other flavorings. Cigars are made by wrapping a binder leaf around a bunch of cut filler leaf and overwrapping with a fine wrapper leaf.

In the United States and elsewhere, stems and scraps of tobacco are used for nicotine extractions. They can also be ground down and made into a reconstituted sheet in a process like papermaking, which is used as a substitute cigar binder or wrapper or cut to supplement the natural tobacco in cigarettes. *See also* Agricultural chemicals; Cigarette smoke; Respiratory diseases

[*Linda Rehkopf*]

Resources

Books

Chandler, W. U. *Banishing Tobacco.* Washington, DC: Worldwatch Institute, 1986.

Periodicals

Heise, L. "Unhealthy Alliance: With U.S. Government Help, Tobacco Firms Push Their Goods Overseas." *World Watch* (September-October 1988): 19–28.

Other

National Academy of Sciences. *Environmental Tobacco Smoke: Measuring Exposures and Assessing Health Effects.* Washington, DC: National Academy Press, 1986.

Toilets

The origin of the indoor toilet for the disposal of human wastes goes far back in history. Archaeologists found in the palace of King Minos on Crete an indoor latrine that had a wooden seat and may have worked like a modern flush toilet; they also discovered a water-supply system of terra cotta pipes to provide water for the toilet. Between 2500 and 1500 B.C., cities in the Indus Valley also had indoor toilets that were flushed with water. The **wastewater** was carried to street drains through brick-lined pits. In 1860, Reverend Henry Moule invented the earth closet, a wooden seat over a bucket and a hopper filled with dry earth, charcoal, or ashes. The user of the toilet pulled a handle to release a layer of earth from the hopper over the wastes in the bucket. The container was emptied periodically. During the eighteenth and nineteenth centuries in Europe, human wastes were deposited in pan closets or jerry pots. After use, the pots were emptied or concealed in commodes. The contents of the jerry pots were often collected by nearby farmers who used the wastes as organic **fertilizer**. However, as cities

grew larger, **transportation** of the wastes to farms became uneconomical, and the wastes were dumped into communal cesspits or into rivers. The flush toilet common in use today was supposedly invented by Thomas Crapper in the nineteenth century; Wallace Rayburn wrote a biography of Crapper, titled *Flushed with Pride*, in 1969.

The development of the flush toilet was primarily responsible for the development of the modern sanitary system, consisting of a maze of underground pipes, pumps, and centralized treatment systems. Modern sanitary systems are efficient in removing human and other wastes from human dwellings but are costly in terms of capital investment in the infrastructure, operational requirements, and energy requirements. The treated wastewater is usually disposed of in rivers and lakes, sometimes causing adverse impacts upon the receiving waters.

Sanitary systems require an abundant supply of water, and the flush toilet is responsible for the largest use of water in the home. Each flush of a conventional water-carriage toilet uses between 4–7 gal (15–26 L) of water, depending on the model and water supply pressure. The average amount of water used per flush is 4.3 gal (16 L). Since each person flushes the toilet an average of 3.5 times per day, the average daily flow per person is approximately 16 gal (60 L) for a yearly flow of 5,840 gal (22,104 L).

To reduce the volume of water used for flushing, a variety of devices are available for use with a conventional flush toilet. These devices include:

- Tank insert - a displacement device placed in storage tank of conventional toilets to reduce the volume (but not the height) of stored water.
- Dual flush toilet - devices used with conventional toilets to enable the user to select from two flush volumes, based on the presence of solid or liquid waste materials.
- Water-saving toilet - variation of conventional toilet with redesigned flushing rim and priming jet that allows the initiation of the siphon flush in a smaller trapway with less water.
- Pressurized and compressed air (assisted flush toilet) - variation of conventional toilet designed to utilize compressed air to aid in flushing by propelling water into the bowl at increased velocity.
- Vacuum-assisted flush toilet - variation of conventional toilet in which the fixture is connected to a vacuum system that is used to assist a small amount of water in flushing.

In addition to modifications to conventional flush toilets, non-water carriage toilets are available to reduce the amount of water required. They are also used for disposing of toilet wastes. Types of non-water carriage toilet systems include:

- Composting toilet - self-contained units that accept toilet wastes and utilize the addition of heat in combination with **aerobic** biological activity to stabilize human excreta; larger units may accept other organic wastes in addition to toilet wastes and requires periodic disposal of residuals.
- Incinerating toilet - small self-contained units that utilize a burning assembly or heating element to volatilize the organic components of human waste and evaporate the liquids; requires periodic disposal of residuals.
- Oil-recycle toilets - self-contained unit that uses a mineral oil to transport human excreta from a toilet fixture to a storage tank; oil is purified and reused for flushing; requires removal and disposal of excreta from storage tank periodically (usually annually).

The wastes from toilets are referred to as "blackwater." If the wastes from toilets are segregated and handled separately using alternative non-water carriage toilets from the wastewaters generated from other fixtures in the home (referred to as "graywater"), significant quantities of pollutants, especially suspended solids, **nitrogen**, and pathogenic organisms, can be eliminated from the total wastewater flow. Graywater, though it still may contain significant numbers of pathogenic organisms, may be simpler to manage than total residential wastewater due to a reduced flow volume. *See also* Municipal solid waste; Sewage treatment

[*Judith Sims*]

Resources

Periodicals

Love, S. "An Idea in Need of Rethinking: The Flush Toilet." *Smithsonian* 6 (1975): 61–66.

Rodale, R. "Goodbye to the Flush Toilet." *Compost Science* 12 (1971): 24–25.

Other

U.S. Environmental Protection Agency. *Design Manual: Onsite Wastewater Treatment and Disposal Systems.* Cincinnati, OH: Municipal Environmental Research Laboratory, U.S. Environmental Protection Agency, 1980.

Tolerance level

Tolerance level refers to the maximum allowable amount of chemical residue, such as a **pesticide**, legally permitted in food. Tolerance levels are determined by government agencies such as the **Environmental Protection Agency** (EPA) and the **Food and Drug Administration** (FDA), and are based on the results of testing, primarily animal testing. This testing determines the dosage level at which few or no effects are observed. This dose is then adjusted to a human equivalency dose with a margin of error built in for added safety.

While pesticides play an important role in modern society, and are used for a variety of purposes, not until

recent decades was it clear how dangerous and persistent many pesticides are. The EPA monitors only a small fraction of pesticides, and in many cases their probable effects on the **environment** are unknown. Pesticide tolerance levels are particularly important worldwide, since pesticides are found in the tissues of people living even in very remote areas of the world. For example, as a result of environmental contamination, the concentration of pesticides in human breast milk has, at certain times in some areas, exceeded the tolerance level for cow's milk.

To establish tolerance levels, scientists conduct a **risk assessment**, or an evaluation of the hazards to the environment, including human health, from exposure to the substance. Data on the toxicity of the substance are combined with data about exposure. The calculation assesses the theoretical maximum residue contribution, or the amount of a chemical that would be present in the average daily diet if all foods treated by that chemical had amounts at the tolerance level. The highest concentration allowable is called the maximum acceptable tolerance concentration. For drinking water, tolerance levels are called the National Primary Drinking Water Requirements. For **air pollution** standards, tolerance levels are called the permissible exposure level.

Environmentalists who criticize the government's tolerance levels charge that standards are not safe enough either for the general population or for workers exposed to **toxins**. Critics also claim that tolerance levels do not take into account either the cumulative effect of residues from a variety of sources in the environment, the duration of exposure, or the unanticipated effects of two or more **chemicals** combined. Tolerance levels also are criticized for not recognizing threats to at-risk populations such as children, the elderly, and pregnant women. **Animal rights** activists have criticized the use of animals to determine tolerance levels, and advocate the use of computer models or *in vitro* testing (testing on tissue cultures) to make determinations of toxicity or safety of chemicals. *See also* Pesticide residue; Toxic substance

[*Linda Rehkopf*]

RESOURCES
BOOKS

Harte, John, et al. *Toxics A to Z.* Berkeley: University of California Press, 1991.

Toluene

Toluene is a potentially hazardous liquid that occurs naturally in crude oil, and can be obtained from the tolu tree. Toluene is used in manufacturing paints, adhesives, and **rubber**, as well as in the leather-making and printing industries. People can be exposed to toluene by breathing **automobile** exhaust, fumes from paints, kerosene, and heating oils, or by drinking well-water that is contaminated with toluene. At high doses, toluene may causes dizziness, unconsciousness, and possibly death. High doses may cause kidney damage, while low doses may cause nausea, hearing and vision problems, disorientation, and fatigue. Toluene in air is very reactive so it increases **photochemical smog** and leads to the production of peroxybenzoylnitrate, a potent eye irritant. The **Environmental Protection Agency** (EPA) does not classify toluene as a **carcinogen**, although the EPA has set a limit of 1 milligram per liter for safe drinking water.

[*Marie H. Bundy*]

Top predator

see **Predator-prey interactions**

Topography

The relief or surface configuration of an area. Topographic studies are valuable because they show how lands are developed and give insight into the history and relative age of mountains or plains. Topographic features are developed by physical and chemical processes. Physical processes include the relatively long-term tectonic actions and continental movement that lead to subduction of lands in some cases or to the development of high elevation mountains. Earth surface forms are usually altered more quickly by the action of water, ice, and wind, leading to the development of deep canyons, leveling of mountains, and filling of valleys. Chemical processes include oxidation, reduction, carbonation, solution, and hydrolysis. These reactions lead to the alteration of organic and mineral materials that also influence the topographic forms of the earth.

Topsoil

The upper portion of the **soil** that is used by plants for obtaining water and nutrients, often referred to as the *horizon*. Higher levels of organic matter in topsoil cause it to be darker and richer than **subsoil** and give it greater potential for crop production. The loss of topsoil is a critical problem worldwide. The net effect of this widespread topsoil **erosion** is a reduction in crop production equivalent to removing about 1% of the world's cropland each year. Soil material that can be purchased to add to existing surface soils is also referred to as topsoil. This kind of topsoil may or may not be from the surface of a soil. Often it is a dark-colored soil that is high in organic matter and in some cases may be more organic soil than mineral soil.

Tornado and cyclone

A tornado is a vortex or powerful whirling wind, often visible as a funnel-shaped cloud hanging from the base of a thunderstorm. It can be very violent and destructive as it moves across land in a fairly narrow path, usually a few hundred yards in width. Wind speeds are most often too strong to measure with instruments and are often estimated from the damages they cause. Winds have been estimated to exceed 350 mph (563 kph). Very steep pressure gradients are also associated with tornados and contribute to their destructiveness. Sudden changes in atmospheric pressure taking place as the storm passes sometimes cause walls and roofs of buildings to explode or collapse.

In some places such storms are referred to as cyclones. Tornados most frequently occur in the United States in the central plains where maritime polar and maritime tropical air masses often meet, producing highly unstable atmospheric conditions conducive to the development of severe thunderstorms. Most tornados occur between noon and sunset, when late afternoon heating contributes to atmospheric instability.

Torrey Canyon

The grounding in 1967 of the supertanker *Torrey Canyon* on protruding granite rocks near the Scilly Isles off the southwest coast of England introduced the world to a new hazard: immense **oil spills**, especially from supertankers. It exposed current technology's inability to handle such massive quantities of spilled oil and the need for ship design that would help prevent them.

Although oil spills are tragic, each incident provides insight into the progress being made to prevent them. The *Torrey Canyon* accident pointed out needed research in the area of oil spills. These include the efficacy of storm waves, oil-consuming **microbes**, and time to heal the **environment**.

This wreck demonstrated the inadequacies of existing international marine law, especially questions of responsibility and liability. The international aspects of the *Torrey Canyon* case were complex. This American-owned vessel was registered in Liberia, sailed with a crew of mixed nationals, and grounded in United Kingdom waters, also contaminating the Brittany coast of France. The British government set precedent by ordering the bombing of the wreck in a futile effort to torch the remaining oil.

Efforts to salvage the spilled oil quickly proved useless, as the escaping oil rapidly thinned and became widespread. It was later learned that the most volatile parts of the oil evaporated within several days. Fortunately, this amounted to a large percentage of the total spill.

A number of crucial lessons were learned as a result of abatement efforts. Dispersants, which break the oil into tiny droplets, were inadequate and applied too little and too late. Once "mousse" (the term for water encased in the oil) forms under wave action, little can be done to break up this ecologically hazardous muck. One of the first actions in response to the 1993 grounding of the *Braer* near the Shetland Islands was the aerial spraying of dispersants on the rapidly thinning oil.

Use of **detergents** on the intertidal zone of England's resort beaches proved deadly to grazing organisms. The most effective treatment of oil-tainted water and beaches was discovered to be nature's storm action combined with metabolic breakdown by **microorganisms**. Efforts to clean oiled seabirds proved largely futile, as they succumbed to hypothermia, stress, and **poisoning** from ingested oil.

The French discovered an effective emergency tactic to lessen the damage. They used straw and other absorbent materials to sop up the incoming oil. They also protected a 100-year-old research section at Roscoff by making a long boom out of burlap stuffed with straw and wood cuttings, an operation dubbed "Big Sausage." To compensate for inadequate anchoring, students physically held the boom in place while the tide rolled in, with new troops periodically relieving those chilled in the cold water. *See also Amoco Cadiz*, Clean Water Act; *Exxon Valdez*

[*Nathan H. Meleen*]

Resources

Books

Cowan, E. *Oil and Water: The Torrey Canyon Disaster.* Philadelphia: Lippincott, 1968.

Petrow, R. *In the Wake of Torrey Canyon.* New York: David McKay, 1968.

Williams, A. S. *Saving Oiled Seabirds.* Rev. ed. Washington, DC: American Petroleum Institute, 1987.

Periodicals

Walsh, J. "Pollution: The Wake of the *Torrey Canyon. Science* 160 (April 12, 1968): 167–169.

Other

Office of Technology Assessment. *Coping With Oiled Environments.* Washington, DC: U.S. Government Printing Office, 1989.

Toxaphene

Toxaphene was once the nation's most heavily used **pesticide** and accounted for one-fifth of all pesticide use in the United States. It was used mainly on cotton and dozens of food crops and on livestock to kill **parasites**. It is severely toxic to fish and **wildlife** and has been implicated in massive kills of fish, ducks, pelicans, and other waterfowl. Because of this, its ability to cause cancerous tumors and genetic

abnormalities in animals, and fears that it may be similarly dangerous to humans, all sale and use of toxaphene have been banned except for already existing stocks. However, this chlorinated hydrocarbon pesticide is extremely persistent. Years after application, it has been found in fish, water, wildlife, and the **food chain/web**, posing a continuing potential threat to the **environment** and human health.

Toxic Chemicals Release Inventory (EPA)

see **Toxics Release Inventory (EPA)**

Toxic substance

Toxic substances are materials that are poisonous to living organisms. There are hundreds of thousands of artificial and natural toxic substances, also known as **toxins**, that are found in solid, liquid, or gaseous form. Toxic substances damage living tissues or organs by interfering with specific functions of cells, membranes, or organs. Some destroy cell membranes, others prevent important cell processes from occurring. Many cause cells to mutate, or make mistakes when they replicate themselves. Some of the most common and dangerous **anthropogenic** (human-made) toxic substances in our **environment** are **chlorinated hydrocarbons**, including DDT (**dichlorodiphenyl-trichloroethane**) and polychlorinated biphenyl (PCB). Many of these are produced by **pesticide** manufacturers and other chemical industries specifically because of their ability to kill pests. **Petroleum** products, produced and used in oil refining, **plastics** manufacturing, industrial solvents, and household cleaning agents, are also widespread and highly toxic agents. **Heavy metals**, including **cadmium**, chromium, **lead**, **mercury**, and **nickel**, and radioactive substances such as **uranium** and **plutonium** are also dangerous toxic agents. Once a toxic substance is released into the environment, plants may absorb it along with water and nutrients through their roots or through pores or tissues in their leaves and stems. Animals, including humans, take up environmental toxic substances by eating, drinking, breathing, absorbing them through the skin, or by direct transmission from mother to egg or fetus.

The toxicity (the potential danger) of all toxic materials depends upon dosage. Some toxins are deadly in very small doses; others can be tolerated at relatively high levels before an observable reaction occurs. All chemical substances can become toxic in high enough concentrations, but even very toxic **chemicals** may cause no reaction in very small amounts. If you eat an extremely large dose of table salt (sodium chloride), you will suffer severe reactions, possibly even death. On the other hand, in moderate doses table salt is not toxic. On the contrary, it is essential for your body to continue functioning normally. Likewise, more unusual trace elements such as selenium are important to us in extremely minute amounts, but high concentrations have been observed to cause severe **birth defects** and high **mortality** among birds.

One of the most important characteristics that determines a substance's toxicity is the way the substance moves through the environment and through our bodies. Most chemicals and minerals move most effectively when they are dissolved by water or by an oil-based liquid. Generally compounds of mineral substances, including sodium chloride, selenium, zinc, **copper**, lead, and cadmium, dissolve best in water. Organic chemicals (those containing **carbon**), including chlorinated **hydrocarbons**, **benzene**, **toluene**, chloroform, and others, dissolve most readily in oily solvents, including **gasoline**, **acetone**, and the fatty tissues in our bodies. In the environment, inorganic substances are often mobilized when ground is disturbed and watered, as in the case of irrigated agriculture or mining, or when waste dumps become wet and their soluble contents move with **runoff** into **groundwater** or surface water systems. Animals and plants readily take up these substances once they are mobilized and widely distributed in natural water systems. Organic chemicals mainly move through the environment when human activity releases them through pesticide spraying, by allowing aging storage barrels to leak, or by accidental spills. Natural surface and groundwater systems further distribute these compounds. Animals that ingest these compounds also store and distribute toxic organic substances in their bodies.

Once we ingest or breathe a toxic substance, its mobility in our body depends upon its solubility and upon its molecular shape. As they enter our bodies through the tissues lining our lungs or intestines or more rarely through our skin, fat soluble compounds can be picked up and stored by fatty tissues, or lipids, in our cells. Proteins and enzymes in our blood, organs, tissues, and bones recognize and bond with molecules whose shape fits those proteins and enzymes. In most cases our bodies have mechanisms to break down, or metabolize, foreign substances into smaller components. When possible, our bodies metabolize foreign substances and turn them into simpler, water soluble compounds. These are easier for our bodies to eliminate by excretion in feces, urine, sweat, or saliva.

Not all substances are easy to metabolize, however. Some persistent compounds such as DDT simply accumulate in tissues, a process known as **bioaccumulation**. If gradual accumulation goes on long enough, toxic dosages are reached and the animal will suffer a severe reaction or death. Furthermore, the by-products (metabolites) of some substances are more dangerous than the original toxin and more difficult to excrete. Toxic metabolites accumulate in

our tissues along with persistent toxins. Although accumulation can occur in bones, fat reserves, blood, and many organs, the most common locations for toxins to accumulate is in the liver and kidneys. These organs have the primary responsibility of removing foreign substances from the blood stream, so any substance that the body cannot excrete tends to collect here.

The parts of our bodies that are most easily damaged by exposure to toxic substances, however, are areas where cells and tissues are reproducing and growing. Brain cells, bones, and organs in children are especially susceptible, and improper cell replication is magnified as growth proceeds. Serious developmental defects can result—a widespread example is the accumulation of lead in children's brains, causing permanent retardation. In adults, any tissue that reproduces, repairs, or replaces itself regularly is likely to exhibit the effects of toxicity. Linings of the lungs and intestines, bone marrow, and other tissues that regularly reproduce cells can all develop defective growths, including **cancer**, when exposed to toxic substances. An individual's susceptibility to a toxic substance depends upon exposure—usually workers who handle chemicals are the first to exhibit reactions—and upon the person's **genetic resistance**, age, size, gender, general health, and previous history of exposures.

Toxins that cause an immediate response, usually a health crisis occurring within a few days, are said to have **acute effects**. Subacute toxic effects appear more gradually, over the course of weeks or months. **Chronic effects** may begin more subtly, and they may last a lifetime. General classes of toxic substances with chronic effects include the following:

- Neurotoxins, which disable portions of the nervous system, including the brain. Because nerves regulate body functions and because nerve cells are not replaced after an individual reaches maturity, damage is especially critical.
- Mutagens, which cause genetic alterations so that cells are improperly reproduced. These can lead to birth defects or tumors. Compounds that specifically affect embryos are called teratogens.
- Carcinogens cause cancer by altering cell reproduction and causing excessive growth, which becomes a tumor.
- Tumerogens are substances that cause tumors, but if tumors are benign, they are not considered cancerous.
- Irritants, which damage cells on contact and also make them susceptible to infection or other toxic effects.

Natural toxins occur everywhere, and many are extremely toxic. Some, such as peroxides and nitric oxide, even occur naturally within our bodies. Many people avoid potatoes, peppers, tomatoes, and other members of the nightshade family because they are sensitive to the alkaloid *solanine* that they contain. Mushrooms and molds contain innumerable toxic substances that can be lethal to adults in minute quantities. Ricin, a protein produced by castor beans, can kill a mouse with a dose of just three ten-billionths (3/10,000,000,000) of a gram. Ricin is one of the most toxic organic substance known. Usually we encounter these natural substances in small enough doses that we do not suffer from them, and in many cases our bodies are equipped to metabolize and eliminate them. Except for toxic minerals and heavy metals released by agriculture and mining, most people rarely worry too much about naturally occurring toxic substances, even though they include some of the most toxic agents known.

The most problematic environmental toxic substances are anthropogenic materials that are produced in large quantities for industrial processes and home use. These are dangerous first because most of them are organic and thus bond readily with our tissues and second because their production and distribution occurs rapidly and is poorly controlled. Every year hundreds of new substances are developed, but the testing process is time-consuming and expensive. The National Institute of Safety and Health has listed 99,585 different toxic and hazardous substances, but there are over 700,000 different chemicals in commercial use, and of these only 20% have been thoroughly tested for toxic effects. One-third have not been tested at all. Once these substances are manufactured, their sale, **transportation**, and especially disposal are often inadequately monitored. Of the more than over 292 million tons (265 million metric) of toxic and hazardous materials produced every year the United States, about 220 million tons (200 million metric) are used or disposed of properly by **recycling**, chemical conversion to non-toxic substances, **incineration**, or permanent storage. About 66 million tons (60 million metric) are inappropriately disposed of, mostly in landfills, with non-toxic **solid waste**. Of course, significant amounts of substances that are properly used also freely enter the environment—including pesticides applied to fields, benzene, gasoline, and other volatile organic compounds that evaporate and enter the **atmosphere**, and paints and solvents that evaporate or gradually break down after use.

National efforts to control toxic substances in the United States began with the **National Environmental Policy Act**, passed by Congress in 1969. This act required the establishment of standardized rules governing the identification, testing, and regulation of toxic and hazardous substances. The law setting out those rules finally appeared in the 1976 **Resource Conservation and Recovery Act** (RCRA). This law sets up standard policy for regulating toxic and hazardous substances from the time of production to disposal. Efforts to deal with new and historic environmental problems associated with toxic substances began with the 1980 **Comprehensive Environmental Response,**

Compensation, and Liability Act (CERCLA). This act provided for setting up a multi-billion dollar Superfund to pay clean up costs at the hundreds of abandoned toxic waste sites around the country. Unfortunately, Superfund money is proving woefully inadequate in meeting real cleanup expenses.

International trade in toxic waste remains a dire problem that has not yet been adequately addressed. In many developing countries pesticides, organic solvents, heavy metals associated with mining, and other noxious materials are regularly released to the environment. Systems for monitoring their release are often completely non-existent. Workers handling these substances frequently suffer terrible diseases and pass on genetic disabilities to their children. Because many of the most dangerous toxins are produced in the United States and Europe for sale to developing regions, wealthier nations stand in a position to force some efforts at regulation. Thus far, however, neither regulation nor worker education has been an international priority.

[*Mary Ann Cunningham Ph.D.*]

Resources

Books

Cunningham, W. P. *Environmental Science: A Global Concern*. Dubuque, IA: William C. Brown, 1992.

Kamrin, M. A. *Toxicology*. Chelsea, MI: Lewis Publishers, 1988.

Toxic Substances Control Act (1976)

The Toxic Substances Control Act (TSCA), enacted in 1976, is a key law for regulating toxic substances in the United States. The Act authorizes the **Environmental Protection Agency** (EPA) to study the health and environmental effects of **chemicals** already on the market and new chemicals proposed for commercial manufacture. If the EPA finds these health and environmental effects to pose unreasonable risks, it can regulate, or even ban, the chemical(s) under consideration.

The initiative to regulate toxic substances began in the early 1970s. In 1971, the **Council on Environmental Quality** produced a report on toxic substances. It concluded that existing health and environmental laws were not adequately regulating such substances. The report recommended that a new comprehensive law to deal with toxic substances be enacted. Among the problem substances that had helped to focus national concern on toxic substances were **asbestos**, **chlorofluorocarbons** (CFCs), **Kepone** (a **pesticide**), **mercury**, polychlorinated biphenyl (PCB)s, and **vinyl chloride**. Research proved that these substances caused significant health and environmental problems, yet they were unregulated.

Congressional debate on regulating toxic substances began in 1971. The Senate passed bills in 1972 and 1973 that were strongly supported by environmentalists and labor, but the House approach to the issue was more limited and had the support of the chemical industry. The key difference between the two approaches was how much pre-market review would be required before new chemicals could be introduced and marketed. Although none of the key stakeholders, the chemical industry, environmentalists, or labor, was entirely happy with the compromise bill of 1976, all three groups supported it.

The approach adopted in TSCA is pre-market notification, rather than the more rigorous pre-market testing that is required before new drugs are introduced on the market. When a new chemical is to be manufactured, or an existing chemical is to be used in a substantially different way, the manufacturer of the chemical must provide the EPA with pre-manufacture notification data at least 90 days before the chemical is to be commercially produced. The EPA then examines this data and determines if regulation is necessary for the new chemical or new use of an existing chemical. The agency can require additional testing by manufacturers if it deems the existing data insufficient. Until such data is available, the EPA has the option to ban or limit the manufacture, distribution, use, or disposal of the chemical. For new chemicals or new applications of existing chemicals, producers must demonstrate that the chemicals do not pose unreasonable risks to humans or the **environment**. The EPA can act quickly and with limited burden to prevent such new chemicals from being produced or distributed.

Between July 1979, when TSCA went into effect for new chemicals, and April 2002, the EPA received 23,486 new chemical notices. The agency determined that no action was necessary for nearly 90% of these chemicals. Of the 2,702 chemicals that required further action, 917 were never produced commercially due to EPA concerns. The remaining 1,785 chemicals were controlled through formal actions or negotiated agreements.

As of mid-2002, there were nearly 75,000 chemicals within the purview of TSCA (including 5,000 new chemicals introduced in the past decade). Although the law requires that the EPA examine each of these chemicals for its potential health and environmental risk, the volume of chemicals and lack of EPA resources has made such a task virtually impossible. Through 2000, the agency had examined almost 700 existing chemicals. Of the compounds on this list, the EPA determined that 370 chemicals required no further action. This was due to the fact that the chemical was no longer being manufactured, it was under study by another federal agency or industry, adequate data on the chemical existed, or exposure to the chemical was limited. The EPA

has issued rules to regulate 112 of these chemicals. The remaining chemicals are in various stages of testing and review.

TSCA created an Interagency Testing Committee (ITC) to help the EPA set priorities. The ITC designates which existing chemicals should be examined first. Once designated by the ITC, the EPA has one year to study the chemical and issue a ruling. The agency initially had difficulty meeting this one-year limit. In fact, the EPA was sued over its failure to meet the deadline and worked under a court schedule to test these identified chemicals. The record of the EPA has improved but the agency still struggles to meet deadlines imposed by ITC due to budgetary and personnel shortages.

The EPA also identifies specific chemicals already in use for review. Unlike the policy with new chemicals, the EPA must demonstrate that existing chemicals warrant toxicity testing. Furthermore, if the EPA desires to regulate the chemical once it has been tested, it must pursue another complex course of action. By design, the regulation of existing chemicals is a lengthy and complicated process.

There are several options available to the EPA if it concludes that a new or existing chemical presents "an unreasonable risk of injury to health or the environment." The EPA can: 1) prohibit the manufacture of the chemical by applying a rule or obtaining a court injunction; 2) limit the amount of the chemical produced or the concentration at which the chemical is used; 3) impose a ban or limit uses of the chemical; 4) regulate the disposal of the chemical; 5) require public notification of its use; and 6) require labeling and record keeping for the chemical. The EPA is required to use the least burdensome of these regulatory approaches for chemicals already in use.

There are several other important components to TSCA. First, the law requires strict reporting and record keeping by the manufacturers and processors of the chemicals regulated under the law. In addition to keeping track of the chemicals, these records also include data on environmental and health effects. Second, TSCA is one of the few federal environmental laws that require environmental and health benefits to be balanced by economic and societal costs in the regulatory process. The act states that toxic substances should only be regulated when an unreasonable risk to people or the environment is present. Although "unreasonable risk" is not clearly defined in the language of the act, it does incorporate a concern for balancing costs and benefits.

Third, PCBs were singled out in the TSCA legislation. Although the manufacture of PCBs was prohibited by 1979, fewer than 1% of all PCBs in use up to that time were phased out. Despite the ban, a significant PCB problem still remains. One major problem is how to process the millions of pounds of PCBs abandoned by firms that went bankrupt in the 1980s. There is also concern about the safety of disposing of PCBs. Environmentalists argued that PCB treatment required new legislation. In response, the EPA decided to issue stricter regulation under the TSCA to ensure the safe the **transportation** and final disposal of PCBs.

Fourth, TSCA contained provisions for private citizens, as opposed to the U.S. government, to file lawsuits. Under the TSCA rules, private citizens could sue companies for violation of the law. They could also sue the EPA for failure to implement TSCA in a timely manner. Fifth, certain materials are not covered by TSCA. These include ammunition and firearms; nuclear materials; **tobacco**; and chemicals used exclusively in cosmetics, drugs, food and **food additives**, and pesticides. These substances are regulated by other laws.

Although the authorization of TSCA expired in 1983 and Congress has not acted to re-authorize the law, it has continued to fund the program. The scope of the law was expanded in 1986 by an amendment requiring asbestos hazards to be reduced in schools, and in 1988 by an amendment providing grants and technical assistance to state programs that reduce indoor **radon** and assigning the regulation of genetically engineered organisms to TSCA. A 1990 amendment required accreditation of persons who inspect for asbestos-containing material in school, public, and commercial buildings, while the Residential Lead-Based Paint Hazard Reduction Act of 1992 introduced new requirements for the reduction of hazards associated with the inspection and abatement of lead-based paints. Reauthorization legislation has been introduced in the past decade. However, it has never gotten beyond consideration by congressional committees. Many experts attribute this trend to Republican control of congress since 1994.

Evaluations of the law have revealed three chief problems. First, the amount of data generated on chemicals is less than expected. A 1984 study by the **National Academy of Sciences** reported that only 4% of the major chemicals in the TSCA inventory had studies showing data on possible chronic toxicity. Toxicity data was absent for 78% of the chemicals produced at over 1 million lb (454,000 kg) per year and for 76% of chemicals that were produced at less than 1 million lb per year. The lack of data is thought to be due to early uncertainty over how the EPA would implement the law and the requirement that any regulation must follow rule-making procedures. Such rule-making procedures are slow and costly; the procedures associated with implementing the rules can sometimes cost more than the actual toxicity testing.

Second, in terms of existing chemicals, as of 2002, a total of 50 have been banned in the United States. Most of these are pesticides. A notable EPA action was the ban on the use of chlorofluorocarbons (CFCs) as **aerosol propellants**.

This occurred in March 1978. Part of the reason for a lack of action may be due to the costs involved in implementing such a complex regulatory program. It is also difficult to determine unreasonable risk in light of inadequate data and more general uncertainties. Third, most pre-market notifications for new chemicals are not accompanied by test data. Rather, the EPA identifies potentially harmful chemicals by comparing the new chemical to existing chemicals for which data exists. If the EPA were to require more data for most new chemicals, in many cases the new chemicals might not be produced since the test costs would exceed the expected profit.

The chemical industry and environmentalists have reached conflicting conclusions about the TSCA legislation. The chemical industry has argued that the EPA has required more testing than is scientifically needed and that the regulatory approach for new chemicals is overly burdensome. Environmentalists have criticized the EPA for its slow progress in examining existing chemicals, for requiring no health data for new chemicals, and for withholding much of the data for new chemicals to comply with the desire for confidentiality requested by chemical manufacturers. *See also* Chronic effects; Radioactive waste; Risk analysis

[*L. Fleming Fallon Jr., M.D., Dr.P.H.*]

Resources

Books

Kreith, F. *Handbook of Solid Waste Management.* New York: McGraw Hill, 2002.

McDougall, F. R. *Integrated Solid Waste Management: A Life Cycle Inventory.* 2nd ed. Oxford, U.K.: Blackwell, 2001.

Meyers, R. A., D. K. Dittrick. *The Wiley Encyclopedia of Environmental Pollution and Cleanup.* New York: Wiley, 1999.

Pearson, Eric. *Environmental and Natural Resources Law.* Albany, NY: Matthew Bender & Company, 2001.

Thomas-Hope, E. *Solid Waste Management.* Kingston, Jamaica: University Press of the West Indies, 2000.

Periodicals

Gori, G. B. "The Costly Illusion of Regulating Unknowable Risks." *Regulatory Toxicology and Pharmacology* 34, no. 3 (2001): 205–212.

Landrigan, P. J. "Risk Assessment for Children and other Sensitive Populations." *Annals of New York Academy of Science* 865 (1999): 1–9.

Lee, C. C., G. L. Huffman, Y. L. Mao. "Regulatory Framework for the Thermal Treatment of Various Waste Streams." *Journal of Hazardous Materials* 76, no. 1 (2000): 13–22.

Other

U.S. Department of Energy. *EH-41 Environmental Law Summary: Toxic Substances Control Act.* April 1997 [cited July 2002]. <http://tis.eh.doe.gov/oepa/law_sum/TSCA.HTM>.

U.S. Department of Energy. *DOE Environmental Guidance - Toxic Substances Control Act.* March 1998 [cited July 2002]. <http://tis.eh.doe.gov/oepa/guidance/tsca.htm>.

U.S. Environmental Protection Agency. <http://www.epa.gov/region5/defs/html/tsca.htm> and <http://www.epa.gov/opptintr/biotech/> and <http://es.epa.gov/oeca/main/strategy/tsca.html> and <http://www.epa.gov/oppfead1/international/us-unlist.htm>.

U.S. Department of the Interior. *Toxic Substances Control Act of 1986.* <http://www.usbr.gov/laws/tsca.html>.

Toxic waste

see **Hazardous waste**

Toxics Release Inventory (EPA)

A public database containing the total amounts of toxic **chemicals** that are routinely released into the air, water, and **soil** each year from about 30,000 industrial, particularly manufacturing, facilities in the United States. Firms that use or emit any of about 667 listed chemicals in quantities over specified thresholds are required to provide the U.S. **Environmental Protection Agency** (EPA), states, and the public with estimates of the amount of each chemical stored or used at the firm, the amount emitted, including permitted releases, and other information. This inventory is required under the Superfund Amendments and Reauthorization Act of 1986 and has been compiled annually since 1987. It excludes, however, federal facilities; oil, gas, and mining industries; agricultural activities; and small firms. The most recent inventory lists about 20 billion lb (9.1 billion kg) of emissions annually, roughly 5% of the United States total toxic emissions as estimated by the U.S. Office of Technology Assessment. The inventory has not been verified using independent sources. The Toxics Release Inventory has helped to support new legislation (e.g., **pollution** prevention programs), provide data for agency projects (**cancer** studies); track toxic chemical estimates; regulate toxic chemicals; and screen environmental risks.

Resources

Organizations

Toxics Release Inventory, , (202) 566-0250, Email: tri.us@epa.gov, <http://www.epa.gov/tri>

Toxics use reduction legislation

In recent years, such disasters resulting from toxic **chemicals** as those at **Love Canal**, New York, and in **Bhopal, India**, have increased awareness of the hazards associated with their use. In response to such concerns, "right-to-know" statutes and regulations have been enacted on both the state and federal levels. In 1990, Congress passed the **Pollution Prevention Act**, but the provisions of this legislation are relatively limited when compared to the toxics use reduction (TUR) statutes that have been enacted in at least 26 states

since 1989. In the 1990s, similar legislation was passed in Great Britain and other European countries.

State TUR statutes, sometimes called **pollution** prevention statutes, are designed to motivate businesses to reduce their use of toxic chemicals. Most such statutes set specific overall goals, such as a 50% reduction in the use of toxic chemicals, with that reduction being phased in over a specified number of years. A comprehensive TUR statute usually covers planning requirements, reporting requirements, and protection of trade secrets. It encourages worker and community involvement, provides technical assistance and research, institutes enforcement mechanisms and penalties for non-compliance, and designates funding.

The Massachusetts Toxics Use Reduction Act (MTURA) is considered one of the strongest existing TUR laws, and it is often used as a model for such legislation in other states. The stated goals of the MTURA are to reduce toxic waste by 50% statewide by 1997, while continuing to sustain and promote the competitive advantages of Massachusetts' businesses. Companies subject to MTURA are called large quantity toxics users, and each of these are now required to develop an inventory of the toxic chemicals flowing both in and out of every production process at its facilities. The company must then develop a plan for reducing the use of toxic chemicals in each of these production processes. The inventory and a summary of this plan must be filed with a designated state agency, and the company must submit an annual report for each **toxic substance** manufactured or used at that facility. To accommodate concerns about trade secrets, the MTURA allows companies to report amounts of a chemical substance using an index/matrix format, instead of absolute amounts. In other states, TUR laws address such concerns by allowing companies to withhold information they believe would reveal trade secrets except on court order.

Plans filed under the MTURA are not available to the public, but any 10 residents living within 10 mi (16.09 km) of a facility required to prepare such a plan may petition to have the Massachusetts Department of Environmental Protection examine both the plan and the supporting data. Summaries of these plans must be filed every two years and these are available to Massachusetts residents, as well as the annual reports these companies must issue. Large quantity toxics users must notify their employees of new plans as well as updating existing plans, and management is required to solicit suggestions from all employees on options for reducing the use of toxic substances.

To encourage toxics use reduction, MTURA has established the Office of Toxics Use Reduction Assistance Technology, which provides technical assistance to industrial toxics users. The name of the office has since been changed to the Office of Technical Assistance for Toxics Use Reduction, or OTA. The act also established a Toxics Use Reduction Institute (TURI) at the University of Lowell, which develops training programs, conducts research on toxics use reduction methods, and provides technical assistance to individual firms seeing to adopt pollution prevention techniques. A recent example of the Institute's publications is a module on wet cleaning techniques as a replacement for standard **dry cleaning**, intended for businesses in the garment care industry. In 1995, TURI established a Toxics Use Reduction Networking (TURN) grant program that funds model projects intended to increase citizen participation in toxics use reduction. The grant program funds two basic types of projects, community awareness projects and municipal integration projects. In 2002, TURI broadened its grant program to include projects related to strategic TUR planning at the community level.

Enforcement of MTURA is overseen by the Administrative Council on Toxics Use Reduction, but there are also provisions in the act for court action by groups of ten or more Massachusetts citizens. Civil penalties up to $25,000 for each day of the violation can be assessed, and for willful violations a court can impose fines between $2,500 and $25,000 per violation or imprisonment for up to one year, or both. Administration of MTURA is funded through a toxics users fee imposed on companies subject to the act. Fees are determined according to the number of employees at each facility and the number of toxic substances reported by the facility.

Those drafting TUR laws in other states have chosen a variety of mechanisms to achieve their goals. One issue on which states differ is whether the reduction of toxics, which focuses on pollution prevention, should be the sole emphasis of the statute, or whether the statute should include other objectives. These other objectives are usually means of **pollution control**, which can include **waste reduction**, waste minimization, and what is called "out-of-process" **recycling**, which occurs when chemical wastes are taken from the production site, transported to recycling equipment, and then returned. Analysts refer to those statutes that promote toxics-use reduction exclusively or almost exclusively as having a pure focus. Those statutes that explicitly combine toxics-use reduction with pollution control are labeled as having a mixed focus. Statutes in Massachusetts are categorized as pure, while toxics use reduction legislation in Oregon, for example, is considered to have a mixed focus.

The Oregon and Massachusetts statutes are considered by many to be the strongest toxics use reduction legislation in the United States, and most states have adopted less stringent provisions. For example, the United States General Accounting Office issued a report in June of 1992 that lists only ten states out of twenty-six with such laws, as having

legislation that "clearly promotes" programs to reduce the use of toxic chemicals.

Over the past three decades, Congress has enacted various laws aimed at pollution control. These include the **Resource Conservation and Recovery Act** (1976), the Toxic Substances Control Act (1976), and the Superfund Amendments and Reauthorization Act (1986). Pollution-control laws such as these are primarily considered source-reduction laws. These laws are designed to reduce waste after it has been generated, while TUR laws are designed to restrict **hazardous waste** before generation, by reducing or eliminating the toxic chemicals that enter the production processes. Some federal legislation is, however, moving in the direction of a TUR strategy.

TUR laws have continued the goals of **right-to-know** legislation by moving away from reactive enforcement of environmental laws toward hazard prevention. A recent example of this change in direction is the relationship between toxics use reduction and **cancer** prevention. The Cancer Prevention Coalition stated in 1999 that phasing out the use of known carcinogenic substances in industry through toxics use reduction legislation is an important measure in reversing the current high rates of cancer incidence. The Coalition noted that TUR laws exemplify the "precautionary principle of risk prevention" in contrast to a strategy of risk management.

TUR laws are also different in their "multimedia" approach to regulation. Previous environmental and occupational and health and safety statutes have divided enforcement between various agencies, such as the **Environmental Protection Agency** (EPA), the **Occupational Safety and Health Administration**, and the Consumer Product Safety Commission. Even under the jurisdiction of a single agency, moreover, current environmental laws often require different approaches to regulation of toxics depending on where they are found—air, water, or land. In contrast, TUR laws require comprehensive reports and plans for the reduction of toxics discharged innto all media. TUR laws have been supported by coalitions of environmentalists and labor representatives precisely because of this comprehensive multimedia approach.

Reactions by industry, however, have been mixed. In some instances, businesses have saved money once they began purchasing substitute chemicals, or when changing production processes improved efficiency. But some firms have hesitated to make the capital investments needed to change their industrial processes because the cost-benefit ratio of such changes remains uncertain. Some firms have judged the costs of alternative chemicals or processes to be prohibitively expensive, and in other cases alternative technologies or less hazardous chemicals have not been developed and are simply not available at any price.

The EPA has found, however, that some companies do not take advantage of available technology for reducing or eliminating toxic chemicals because they are unaware of its existence. The Toxics Use Reduction Institute has been a pioneer in making such information available to companies within the Commonwealth of Massachusetts. The Office of Pollution Prevention and Toxics (OPPT), a subagency of the EPA, has sponsored the Green Chemistry Program since 1995. The Green Chemistry Program awards grants for research in green chemistry, promotes partnerships with industry in developing green chemistry technologies, and works with other federal agencies in building green chemistry principles into their operations. OPPT also supports several research centers representing industrial as well as academic and government concerns. In addition to TURI, these centers include the **Emission** Reduction Research Center at the New Jersey Institute of Technology and the Center for Process Analytical Control at the University of Washington.

On the other hand, some industries have been working since the late 1990s for repeal or abolition of TUR legislation. The Massachusetts Chemical Technology Alliance, for example, is opposed to the Commonwealth's TURA statutes on the ground that compliance is ineffectual as well as too costly. Other industry-sponsored groups maintain that the law puts Massachusetts companies at a competitive disadvantage. It is likely that controversy over the need for and effectiveness of TUR legislation in the United States will continue for the foreseeable future. *See also* Chemical spills; Emergency Planning and Community Right-to-Know Act; Environmental liability; Hazardous material; Hazardous Substances Act; International trade in toxic wastes; NIMBY; Waste management

[*Paulette L. Stenzel and Rebecca J. Frey, Ph.D.*]

Resources

Books

Epstein, Samuel S., MD. *The Politics of Cancer Revisited.* East Ridge Press, 1998.

Periodicals

DeVito, Michael J. "Toxic Reporting: Help or Hindrance?" *Chemistry and Industry* (January 5, 1998): 36.

Pelley, Janet. " Toxic Chemical Use Reporting Debated on State and National Levels." *Environmental Science and Technology* 32 (1998): 9–12.

Other

Toxics Use Reduction Institute. *Six Years of Progress in Toxics Use Reduction.* Lowell, MA: TURI, 1996.

Organizations

Massachusetts Toxics Use Reduction Institute, University of Massachusetts Lowell, One University Avenue, Lowell, MA USA 01854-2866 (978) 934-3275, Fax: (978) 934-3050, <htpp://www.turi.org>

Office of Technical Assistance for Toxics Use Reduction, Commonwealth of Massachusetts, Executive Office of Environmental Affairs, 251 Causeway

Street, Suite 900, Boston, MA USA 02114-2136 (617) 626-1060, Fax: (617) 626-1095, <http://www.state.ma.us/ota>

Toxins

Toxins are **chemicals** or physical agents that exert a toxic effect on living organisms. Toxic means poisonous: that is, causing a reaction with cellular components that disrupts essential metabolic processes. At some level of exposure, all chemicals, whether natural or synthetic, are toxic. All can either cause death or damaging effects soon after exposure, or can cause some other disease (such as **cancer** or **birth defects**) after longer-term exposure.

Although many people think toxins are mainly pesticides or industrial chemicals, they also include the poisons of marine animals, spiders, snakes, plants, and the extremely toxic botulinum toxins that can kill a human being with a single minuscule dose. Toxins can exert their effects on many different organs. The nervous system, the brain, the lungs, the skin, and the eyes are only some of the organs that can be damaged by toxins. Toxicologists use reports, **epidemiology**, and laboratory studies to characterize both the lethal doses of toxins and the doses of certain chemicals that can cause disease over the long term. Environmental laws regulate exposures to certain human-made toxins.

Trace element/micronutrient

A micronutrient is an element that plants need in small quantities for growth and **metabolism**. Common micronutrients are iron, **copper**, zinc, boron, molybdenum, manganese, and **chlorine**. Some plants may benefit from small amounts of sodium, silicon, and/or cobalt as well. Additional micronutrients may be added to the list as scientists are better able to detect minute quantities and plant requirements. Micronutrients often act as catalyst in plant chemical reactions. Historically, when **animal waste** was used as **fertilizer**, there were few micronutrient deficiencies detected. However, with the use of large amounts of chemical fertilizer, higher yields, and use of **monoculture**, micronutrient deficiencies in crop plants are becoming more evident. Several plant disorders such as beet canker, cracked stem of celery, and stem end of russet in tomatoes can be related to the lack of certain micronutrients. On the other hand, excess of micronutrients in plants may lead to plant toxicity.

Trade in pollution permits

Trade in **pollution** permits augments the traditional approach to environmental regulation by using market principles to control pollution. Since its inception, the program has been criticized as unfair and unfeasible. Yet the concept of trading pollution permits continues to spread.

Most environmental laws limit the amount of waste or pollution each regulated facility can emit to air, water, or land. These limits are then written into permits. Regulators monitor the facility to make sure the permits are followed. Any facility exceeding the permitted level of emissions may be fined or otherwise penalized. This method of controlling pollution is known as command and control.

In 1990, Title IV of the **Clean Air Act** Amendments became the first federal law to codify a market-based approach to **pollution control**. The title regulated **sulfur dioxide** emissions in an effort to reduce **acid rain**. It set a goal of a 10 million lb (4.5 million kg) reduction in **power plants** emissions of sulfur dioxide. The law granted each power plant the right to a certain level of pollution. Companies that emit less than they are allowed can sell the remainder of their pollution allowance to other companies.

Title IV redefined pollution as a commodity, like pork bellies or soybean futures. One of the largest trade in pollution transactions took place between two utilities, the **Tennessee Valley Authority** (TVA) and Wisconsin Power and Light. The TVA, one of the largest emitters of sulfur dioxide, paid several million dollars to one of the cleanest utilities in the country, Wisconsin Power, for the right to emit an additional 10,000 lb (4,500 kg) of the compound. Despite its obvious success, financial analysts and utility industry executives are still skeptical about the fledgling market's future.

If the market does become established, critics fear that trades like the one between the TVA and Wisconsin Power will only become more common. Because the market maximizes profit, certain utilities that are already highly polluting may spend money buying the right to pollute rather than to clean up their production processes. Other, cleaner utilities may continue to provide the extra pollution allowances. Certain parts of the country will have more pollution than others.

Proponents of the market-based approach respond that simply reaching the goal of reducing sulfur dioxide emissions by 10 million lb (4.5 million kg) will benefit the country as a whole. They also see an advantage in making pollution costly, for then businesses will have an economic interest in reducing it.

The debate about the feasibility and fairness of pollution trading continues. Other regional and local authorities, such as the Southern California **Air Quality** Management District, have considered adopting such measures for their jurisdictions. *See also* Agricultural pollution; Air pollution; Air pollution control; Air quality; Environmental economics; Green politics; Green tax; Industrial waste treatment; Marine pollution; Pollution control costs and benefits; Radioactive waste management; Sewage treatment; Solid waste vol-

ume reduction; Solid waste recycling and recovery; Toxic use reduction; Waste management; Waste reduction; Water pollution

[*Alair MacLean*]

Resources

Periodicals

Allen, F. E. "Tennessee Valley Authority Is Buying Pollution Rights from Wisconsin Power." *Wall Street Journal* (May 11, 1992): A12.

Mann, E. "Market-Driven Environmentalism: The False Promise." *GREEN: Grantmakers Network on the Economy and the Environment* 1 (Winter 1992): 1–3.

Portney, P. "Market-Driven Environmentalism: Tomorrow's Success." *GREEN: Grantmakers Network on the Economy and the Environment* 1 (Winter 1992): 1, 3–5.

Tragedy of the commons

A term referring to the theory that, when a group of people collectively own a resource, individuals acting in their personal self-interest will inevitably overtax and destroy the resource. According to the commons theory, each individual gains much more than he or she loses by overusing a commonly held resource, so its destruction is simply an inevitable consequence of normal and rational behavior. In the study of economics, this idea is known as the free rider problem. Human **population growth** is the issue in which the commons idea is most often applied: each individual gains personal security and wealth by producing many children. Even though each additional child taxes the global community's food, water, energy, and material resources, each family theoretically gains more than it loses for each additional child produced. Although the theory was first published in the nineteenth century, Garrett Hardin introduced it to modern discussions of population growth in a 1968 article published in the journal *Science*. Since that time, the idea of the tragedy of the commons has been a central part of population theory. Many people insist that the logic of the commons is irrefutable; others argue that the logic is flawed and the premises questionable. Despite debates over its validity, the theory of the commons has become an important part of modern efforts to understand and project population growth.

In its first published version, the idea of the commons was a scenario mapped out in mathematical logic and concerning common pasture land in an English village. In his 1833 essay, William Forster Lloyd described the demise of a common pasture through overuse. Up to that time, many English villages had a patch of shared pasture land, collectively owned, on which villagers could let their livestock graze. At the time of Lloyd's writing, however, many of these commons were being ruined through **overgrazing**. This destruction, he proposed, resulted from unchecked population growth and from the persistence of collectively held, rather than private, lands. In previous ages, said Lloyd, wars and plagues kept human and animal populations well below the maximum number the land could support. By the nineteenth century, however, England's population was climbing. More people were looking for room to graze more cattle, and they used common pasture because the resource was essentially free. Free use of a common resource, Lloyd concluded, leads directly to the ruin of that resource.

In his 1968 article, Garrett Hardin applied the same logical argument to a variety of natural amenities that we depend upon. Cattle, **grazing on public lands** in the American West, directly consume a common pasture owned by all Americans. Normal pursuit of increased capital leads ranchers to add as many cattle as they can, and much of the country's public lands are now severely denuded, gullied, and eroded from overgrazing. National parks such as Yosemite and Yellowstone are commonly held lands to which all Americans have free access. Each individual naturally wishes to maximize her or his vacation time, so that the collective result is congestion and **pollution** in the parks. The oceans represent the ultimate world commons. Private corporations and individual countries maximize their profit by catching as many fish as possible. In **whaling**, we have seen this practice effectively eliminate some **species** within a few decades. Other fisheries currently stand in danger of the same end.

Extrapolating his reasoning to arguments for population control, Hardin pointed out that if each family produces as many children as it can, then the inevitable collective result will be wholesale depletion of food, clean water, energy resources, and living space. Worldwide, quality of life will then diminish. Because we cannot increase the world's supply of water, energy, and space, argues Hardin, the only way to avoid this chain of events is to prevent population increases. Ideally, we need worldwide agreements to moderate individual childbearing activity, so that the cumulative demand on world resources will not exceed resource availability.

Critics dispute Hardin's thesis on a number of levels. The most important objection is that Lloyd and Hardin condemned the idea of the commons but that their examples did not involve real community resources. A commons, maintained over generations by a group of people for the benefit of the group, lasts because members of the group accept a certain amount of self-restraint in the interest of the community and because they know that self-restraint ensures the resource's survival. Peer pressure, respect for elders and taboos, fear of recrimination from neighbors, and consideration of neighbors' needs all reinforce individual restraint in a commons. All of these restraints rely upon a stable social structure and an understanding of commons management. In a stable society, individual survival depends

upon the prosperity of one's neighbors, and every parent has an interest in ensuring that **future generations** will have access to necessary resources. Many such commons exist around the world today, especially in traditional villages in the developing world, where generational rules dictate the protection of grazing lands, **water resources**, fields, and forests.

In Lloyd's case study, industrialization, widespread eviction of tenant farmers to make way for sheep, and the introduction of capitalism had all disrupted the traditional social fabric of English villages. Uprooted families moving from village to village did not observe traditional rules. More importantly, many common pastures were being privatized by large landowners, leaving fewer acres of **public land** for village livestock to squeeze onto. The problem described in Lloyd's essay was, in fact, one of uncontrolled free access to a resource whose collective ownership had broken down. Likewise, ocean fisheries have generally been uncontrolled, unowned resources with no collectively enforced rules of restraint. American public grazing lands have rules of restraint, but they fail because of poor enforcement and inadequate development.

Many people also object to the implication of commons logic that a system of private property is superior to one of collectively held property. In criticizing village commons, Lloyd defended the actions of powerful landholders who could, on a whim, evict a community of tenant farmers. Commons logic holds that private landowners make better guardians of resources because they can see that their personal interest is directly served by careful management. However, evidence abounds that private owners frequently exploit and destroy resources much faster than groups do. Nineteenth century landowners, after removing farmers who had husbanded their land for generations, overstocked the country with sheep and clear-cut forests for lumber. Gullied pastures, lost **wildlife habitat**, and other environmental costs resulted. In the American timber industry, virtually all private old-growth forests have been cleared, their capital liquidated. Publicly held old growth, because it is expected to serve other needs beyond lumber production (wildlife habitat, **recreation**, **watershed** for communities, biological resources), has survived better than private forests. As a community, the people of the United States have imposed some rules and limitations on forest use that private landholders have not had. In a capitalist society, private landowners often benefit most by quickly liquidating **natural resources**, which allows them to move on to another area and another resource.

Where resources are precious or irreplaceable, collective guardianship is often the only way to ensure their maintenance. In villages from India to Zaire to the Philippines, community councils monitor the use of forests, water supplies, and crop lands. In developed countries, common and highly valued resources such as schools, public roadways, and parks have long been monitored and controlled by rules that restrain individual behavior and prevent destruction of those resources. Generally, everyone agrees not to block roads, burn schools, or vandalize parks because these resources benefit them all in some way. At the same time, no private individual could adequately maintain such resources. Collective responsibility and respect are necessary.

Finally, some critics argue that it is simplistic to assume that depletion results from population size, rather than uneven distribution of resources. Garrett Hardin chooses to direct his commons logic at large numbers of people using limited amounts of resources without considering the amount of resources used per person. Residents of the world's poorer regions counter that each of their children uses only 5% of the resources that an American child uses. The world can afford a great number of these children, they argue, and it could afford even more of them if people in wealthy countries drove fewer cars, ate less beef, and polluted less of the world's air and water. The deduction that breeding causes shortages, not rates of consumption, is hotly contested by those whose survival depends on the labor of their children.

Defenders of poorer segments of the world's population point out that social **stability**, including a fair distribution of resources, is more likely than population control to deter a global tragedy of the commons. As long as war and **famine** exist, and as long as the rich continue to exploit the labor and resources of the poor (an example of which is American support of the Guatemalan military structure so that we might buy coffee for three dollars a pound and bananas for 50 cents), reliable social structures cannot be rebuilt and responsible guardianship of the world's resources cannot be reestablished.

In the end, arguments for and against commons logic are based upon examples and interpretation. Logical proof, which the scenario set out to establish, remains elusive because the real world is very complex and because one's acceptance of both premises and conclusions depends upon one's political, economic, and social outlook.

[*Mary Ann Cunningham Ph.D.*]

Resources

Periodicals

Hardin, Garrett. "The Tragedy of the Commons." *Science* 162 (December 13, 1968): 1243–48.

"Whose Common Future?" *Ecologist* 22 (July/August 1992).

Overgrazing by domestic animals destroys the grassland and contributes to the tragedy of the commons. (©Richard Dibon-Smith, National Audubon Society Collection. Photo Researchers Inc. Reproduced by permission.)

Trail Smelter arbitration

The Trail **Smelter** arbitration of 1938 and 1941 was a landmark decision about a dispute over **environmental degradation** between the United States and Canada. This was the first decision to recognize international liability for damages caused to another nation, even when no existing treaty created an obligation to prevent such damage.

A tribunal was set up by Canada and the United States to resolve a dispute over timber and crop damages caused by a smelter on the Canadian side of the border. The tribunal decided that Canada had to pay the United States for damages, and further that it was obliged to abate the **pollution**. In delivering their decision, the tribunal made an historic and often-cited declaration: "Under the principles of international law, as well as of the law of the United States, no State has the right to use or permit the use of its territory in such a manner as to cause injury by fumes in or to the territory of another or the properties or persons therein, when the case is of serious consequence and the injury is established by clear and convincing evidence..." The case was landmark because it was the first to challenge historic principles of international law, which subordinated international environmental duty to nationalistic claims of sovereignty and free-market methods of unfettered industrial development. The Trail Smelter decision has since become the primary precedent for international **environmental law**, which protects the **environment** through a process known as the "web of treaty law." International environmental law is based on individual governmental responses to discrete international problems, such as the Trail Smelter issue. Legal decisions over environmental disputes between nations are made in reference to a growing body of treaties, conventions, and other indications of "state practices."

The Trail Smelter decision has shaped the core principle underlying international environmental law. According to this principle, a country which creates **transboundary pollution** or some other environmentally hazardous effect is liable for the harm this causes, either directly or indirectly, to another country. A much older precedent for this same principle is rooted both in Roman Law and Common Law: *sic utere ut alienum non laedas*—use your own property in

such a manner as not to injure that of another. Prior to the twentieth century, this principle was not relevant to international law because actions within a nation's borders rarely conflicted with the rights of another. *See also* Acid rain; Environmental Law Institute; Environmental liability; Environmental policy; United Nations Earth Summit

[*Kevin Wolf*]

Russell Eroll Train (1920 -)

American environmentalist

Concern about environmental issues is a relatively recent phenomenon worldwide. Until the 1960s, citizens were not interested in air and **water pollution**, waste disposal, and **wetlands** destruction.

An important figure in bringing these issues to public attention was Russell Eroll Train. He was born in Jamestown, Rhode Island, on June 4, 1920, the son of a rear admiral in the United States Navy. He attended St. Alban's School and Princeton University, from which he graduated in 1941. After serving in the United States Army during World War II, Train entered Columbia Law School, where he earned his law degree in 1948.

Train's early career suggested that he would follow a somewhat traditional life of government service. He took a job as counsel to the Congressional Joint Committee on Revenue and Taxation in 1948 and five years later, became clerk of the House Ways and Means Committee. In 1957, Train was appointed a judge on the Tax Court of the United States.

This pattern was disrupted, however, because of Train's interest in **conservation** programs. In 1961, he founded the African Wildlife Leadership Foundation and became its first head. He gradually began to spend more time on conservation activities and finally resigned his judgeship to become president of the Conservation Foundation.

Train's first environmental-related government appointment came in 1968, when President Lyndon Johnson asked him to serve on the National Water Commission. The election of Republican Richard Nixon late that year did not end Train's career of government service, but instead provided him with even more opportunities. One of Nixon's first actions after the presidential election was his appointment of a 20-member inter-governmental task force on **natural resources** and the **environment**. The task force's report criticized the government's failure to fund anti-pollution programs adequately, and it recommended the appointment of a special advisor to the President on environmental matters.

In January 1969, Nixon offered Train another assignment. The President had been sharply criticized by environmentalists for his appointment of Alaska Governor Walter J. Hickel as Secretary of the Interior. To blunt that criticism, Nixon chose Train to serve as Under Secretary of the Interior, an appointment widely praised by environmental groups. In his new position, Train was faced with a number of difficult and controversial environmental issues, the most important of which were the proposed **Trans-Alaska pipeline** project and the huge new airport planned for construction in Florida's **Everglades National Park**. When Congress created the **Council on Environmental Quality** in 1970 (largely as a result of Train's urging), he was appointed chairman by President Nixon. The Council's first report identified the most important critical environmental problems facing the nation and encouraged the development of a "strong and consistent federal policy" to deal with these problems.

Train reached the pinnacle of his career in September 1973, when he was appointed administrator of the federal government's primary environmental agency, the **Environmental Protection Agency** (EPA). During the three and a half years he served in this post, he frequently disagreed with the President who had appointed him. He often felt that Nixon's administration tried to prevent the enforcement of environmental laws passed by Congress. Some of his most difficult battles involved energy issues. While the administration preferred a *laissez-faire* approach in which the marketplace controlled energy use and prices, Train argued for more controls that would help conserve energy resources and reduce air and water **pollution**.

With the election of Jimmy Carter in 1976, Train's tenure in office was limited. He resigned his position at EPA in March 1977 and returned to the Conservation Foundation.

[*David E. Newton*]

Resources

Books

Schoenbaum, E. W. "Russell E. Train." In *Political Profiles*. New York: Facts on File, 1979.

Periodicals

Durham, M. S. "Nice Guy in a Mean Job." *Audubon* (January 1974): 97–104.

Trans-Alaska pipeline

The discovery in March 1968 of oil on the Arctic slope of Alaska's Prudhoe Bay ignited an ongoing controversy over the handling of the Arctic slope's abundant energy resources. Of all the options considered for transporting the huge quantities found in North America's largest field, the least hazardous and most suitable was deemed a pipeline to the ice-free southern port of Valdez.

Plans for the pipeline began immediately. Labeled the Trans-Alaskan Pipeline System (TAPS) its cost was estimated at $1.5 billion, a pittance compared to the final cost of $7.7 billion. The total development cost for Prudhoe Bay oil was likely over $15 billion, the most expensive project ever undertaken by private industry. Antagonists, aided by the nascent environmental movement, succeeded in temporarily halting the project. Legislation that created the **Environmental Protection Agency** (EPA) and required environmental impact statements for all federally-related projects added new, critically important requirements to TAPS. Approval came with the 1973 Trans-Alaskan Authorization Act, spurred on by the 1973 OPEC embargo, the world's first severe energy crisis. Construction quickly resumed.

The pipeline is an engineering marvel, having broken new ground in dealing with **permafrost** and mountainous Arctic conditions. The northern half of the pipeline is elevated to protect the permafrost, but river crossings and portions threatened by avalanches are buried for protection. A system was developed to keep the oil warm for 21 days in the event of a shutdown, to prevent TAPS from becoming the "world's largest tube of chapstick." Especially notable is Thompson Pass near the southern end, where descent angles up to 45 degrees severely taxed construction workers, especially welders.

Everything about TAPS is colossal: 799 mi (1,286 km) of 48-in (122-cm) diameter vanadium alloy pipe; 78,000 support columns; 65,000 welds; 15,000 trucks; and peak employment of more than 20,000 workers. TAPS has an operations control center linked to each of the 12 pump stations, with computer controlled flow rates and status checks every 10 seconds.

The severe restrictions that the enabling act imposed have paid off in an enviable safety record and few problems. The worst problem thus far was caused by local sabotage of an above-ground segment. In spite of its good record, however, TAPS remains controversial. As predicted, it delivers more oil than West Coast refineries can handle, and efforts continue to allow exports to Japan.

In 2001, the Secure America's Future Energy (SAFE) Act, based on the President's National Energy Policy, stated among its goals the 2004 renewal of the existing TAPS lease, along with the construction of a new pipeline to transport **natural gas** from Alaska to the 48 contiguous states. Another SAFE goal, the development of the **Arctic National Wildlife Refuge** for **oil drilling**, is still heavily backed in government despite a resounding Senate rejection of the project in 2002. Both utilitarian conservationists and altruistic preservationists are at loggerheads over energy development in Alaska, while many native Inuits consider this an opportunity to solidify their growing involvement in the American economy. *See also* Oil drilling; Oil embargo

[*Nathan H. Meleen*]

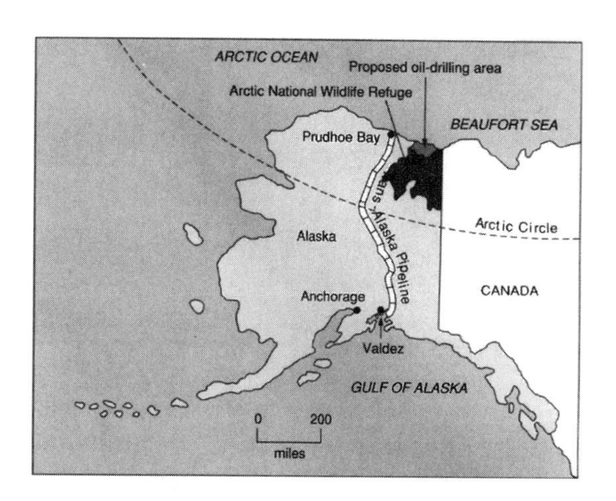

The route of the Trans-Alaska pipeline. (McGraw-Hill Inc. Reproduced by permission.)

RESOURCES

BOOKS

Dixon, M. *What Happened to Fairbanks? The Effects of the Trans-Alaska Oil Pipeline on the Community of Fairbanks, Alaska.* Boulder: Westview Press, 1978.

Roscow, J. P. *800 Miles to Valdez: The Building of the Alaska Pipeline.* Englewood Cliffs, NJ: Prentice Hall, 1977.

PERIODICALS

Hodgson, B. "The Pipeline: Alaska's Troubled Colossus." *National Geographic* 150 (November 1976): 684–707.

Lee, D. B. "Oil in the Wilderness: An Arctic Dilemma." *National Geographic* 174 (December 1988): 858–871.

Trans-Amazonian highway

The Trans-Amazonian highway begins in northeast Brazil and crosses the states of Para and Amazonia. The earth road, known as BR-230 on travel maps, was completed in the 1970s during the military regime that ruled Brazil from 1964 until 1985. The highway was intended to further **land reform** by drawing landless peasants to the area, especially from the poorest regions of northern Brazil. More than 500,000 people have migrated to Transamazonia since the early 1970s. Many of the colonists feel that the government enticed them there with false promises.

The road has never been paved, so it is nothing but dust in the dry season and an impassable swamp during the wet season. Farmers struggle to make a living, with the highway as their only means of transporting produce to

market. When the rains come, large segments of the highway wash away entirely, leaving the farmers with no way to transport their crops. Small farmers live on the brink of survival. Farmer José Ribamar Ripardo says, "People grow crops only to see them rot for lack of transport. It's really an animal's life." To survive in Amazonia, the colonists say, they require a paved road to transport their produce.

The goal of repairing and paving the highway is viewed with disfavor by many environmentalists. Roads through Amazonia are perceived as being synonymous with destruction of the rain forests. Many environmental groups fear that improved roads will bring more people into the area and lead to increased devastation.

The Movement For Survival, spearheaded by José Geraldo Torres da Silva, claims that "If farmers could have technical help to invest in **nature**, they will be able to support their families with just one third of their land lots, avoiding **deforestation**, the indiscriminate killing of numerous animal and plant species." Farmers in the region say that they can survive on the land base that they have already acquired. They have proposed preserving some natural vegetation by growing a mixture of **rubber** and cacoa trees, which grow best in shaded areas, so they will not have to devastate the forest. Efforts are currently underway to establish extractivist preserves to harvest rubber, brazil nuts, and other native products. Such proposals for the use of the forest are a viable option, but they are not enough. The environmentally friendly plans must also take into account fluctuations in the market place. Brazil nuts, for instance, are harvested in the unspoiled forest. When nut prices fall on the international market, the gatherers must have other means of income to fall back on, without having to relocate to a different part of the country.

Many residents of the area fear that if small farmers do not have a reliable road to get produce to market, they will have to leave the region. There is a real danger that their lots will be sold to cattle ranchers, loggers, and investors. Small farmers have a track record of utilizing the land in ways that are more environmentally sound than those who follow in their wake.

[*Debra Glidden*]

Resources

Periodicals

Babbitt, B. E. "Amazon Grace." *New Republic* 202 (25 June 1990): 18–19.

Fearnside, P. M. "Rethinking Continuous Cultivation in Amazonia: The 'Yurimaguas Technology' May Not Provide the Bountiful Harvest Predicted by its Originators." *Bioscience* 37 (March 1987): 209.

Néto, R. B. "The Transamazonian Highway." *Buzzworm* 4 (November-December 1992): 28–29.

Simpson, J. "To the Beginning of the World." *World Monitor* 6 (January 1993): 34–41.

Vesilind, P. J. "Brazil Moment of Promise and Pain." *National Geographic* 171 (March 1987): 372–373.

Transboundary pollution

The most common interpretation of transboundary **pollution** is that it is pollution not contained by a single nation-state, but rather travels across national borders at varying rates. The concept of the global commons is important to an understanding of transboundary pollution. As both population and production increase around the globe, the potential for pollution to spill from one country to another increases. Transboundary pollution can take the form of contaminated water or the deposition of airborne pollutants across national borders. Transboundary pollution can be caused by catastrophic events such as the Chernobyl nuclear explosion. It can also be caused by the creeping of industrial **discharge** that eventually has a measurable impact on adjacent countries. It is possible that pollution can cross state lines within a country and would indeed be referred to as transboundary pollution. This type of case is seldom held up as a serious policy problem since national controls can be brought to bear on the responsible parties and problems can be solved within national borders. It is good to understand how interstate environmental problems might develop and to have knowledge as to those regulatory units of national government that have jurisdiction.

Federalism is important in issues of national environmental pollution largely because pollution can spread across several states before it is contained. Within the United States, it is the **Environmental Protection Agency** (EPA) that writes the regulations that are enforced by the states. The general rule is that states may enact environmental regulations that are more strict than what the EPA has enacted, but not less strict. In some cases interstate compacts are designated by the EPA to deal with issues of **pollution control**. An example of this is the regional compacts for the management of low level **radioactive waste** materials. These interstate compacts meet and make policy for the siting and management of facilities, the transport of materials and the long term planning for adequate and safe storage of **low-level radioactive waste** until its **radioactivity** has been exhausted. Other jurisdictions such as this in the U.S. deal with transboundary issues of **air quality**, **wildlife management**, fisheries, **endangered species**, **water quality**, and solid **waste management**. Federalism has been manifested in specific legislation. The **National Environmental Policy Act** (NEPA) is an attempt to clarify and monitor environmental quality regardless of state boundaries. NEPA provides for a council to monitor trends. The **Council on Environmental Quality** is a three-member council appointed by the President of the United States to collect data

at the national level on environmental quality and management, regardless of jurisdiction. This picture at a national level is vastly complicated when moved to a global level.

There is increasing demand for dialogue and support for institutions that can address impacts on the global commons. The issue of "remedies" for transboundary pollution is usually near the top of the agenda. Remedies frequently take the form of payments to rectify a wrong, but more often take the form of fines or other measures to assure compliance with best practices. The problem with this approach is that charges have to be high enough to offset the cost of controlling "creeping" pollution that might cross national borders or the cost of monitoring production processes to prevent predictable disasters. **Marine pollution** is an excellent example of a transboundary pollution problem that involves many nation-states and unlimited point sources of pollution. Marine pollution can be the result of on-shore industrial processes that use the ocean as a waste disposal site. Ships at sea find the ocean just too convenient as a sewer system, and surface ocean activities such as **oil drilling** are a constant source of pollution due to unintentional discharges. Accidents at sea are well-documented, the most famous in recent years being the ***Exxon Valdez*** oil spill off of Alaska in **Prince William Sound** in 1989. Earlier, in 1969 an international conference on marine pollution was held. As a result of this conference two treaties were developed and signed that would respond to oil pollution on the high seas. These treaties were not fully ratified until 1975. The Convention on the Prevention of Marine Pollution by Dumping of Wastes and Other Matter and The Convention for the Prevention of Marine Pollution by Dumping from Ships and Aircraft cite specific pollutants that are considered extremely detrimental to marine environments. These pollutants are organohalogenic compounds and mercury, as well as slightly less harmful compounds such as lead, arsenic, copper, and pesticides. The specificity of those treaties that deal with ocean environments indicate a rising commitment to the control of transboundary pollution carried by the ocean's currents and helped to resolve the question of remedies in the *Valdez* catastrophe.

Acid rain is a clear example of how air currents can carry destructive pollution from one nation-state to another, and indeed, around the globe. Air as a vector of pollution is particularly insidious because air patterns, although known, can change abruptly, confounding a country's attempt to monitor where pollutants come from. **Acid deposition** has been an international issue for many years. The residuals from the burning of **fossil fuels** and the byproducts from radiation combine in the upper **atmosphere** with water vapor and precipitate down as **acid** rain. This precipitation is damaging to lakes and streams as well as to forests and buildings in countries that may have little sulfur and **nitrogen** generation of their own. The negative justice of this issue is that it is the most industrialized nations that are producing these pollutants due to increased production and too often the pollution rains down on developing countries that have neither the resources nor the technical expertise to clean up the mess. Only recently, due to international agreements between nation-states, has there been significant reduction in the deposition of acid rain. The **Clean Air Act** Amendments of 1990 in the United States helped to reduce emissions from coal-burning plants. In Europe, sulfur emissions are being reduced due to an agreement among eight countries. The European Community in 1992 agreed to implement **automobile emission standards** similar to what the United States agreed to in the early 1980s. In 1994 all large automobiles in Europe were required to reduce emissions with the installation of catalytic converters. These measures are all aimed at reducing transboundary pollution due to air emissions.

Probably the most famous case of transboundary river pollution is that of the Rhine River with its point of incidence in Basel, Switzerland. This was a catastrophe of great proportions. The source of pollution was an explosion at a chemical plant. In the process of putting out the fire, great volumes of water were used. The water mixed with the **chemicals** at the plant (mercury, insecticides, fungicides, herbicides and other **agricultural chemicals**), creating a highly toxic discharge. This lethal mixture washed into the Rhine River and coursed its way to the Baltic Sea, affecting every country along the way. Citizens of six sovereign nations were affected and damage was extensive. There was further consternation when compensation for damage was found to be difficult to obtain. Although the chemical company volunteered to pay some compensation it was not enough. It was possible that each affected nation-state could try the case in their own national court system, but how would collection for damages take place? "In transboundary cases, if there is no treaty or convention in force, by what combination of other international law principles can the rules of liability and remedy be determined?" It becomes problematic to institute legal action against another country if there is no precedent. Since there is no history of litigation between nation-states in cases of environmental damage there is no precedent to determine outcome. Historically, legal disputes have been resolved (albeit, extremely slowly) through negotiations and treaties. There is no "polluter pays" statute in international law doctrine.

The authority that can be brought to bear on issues of transboundary pollution is weak and confusing. The only mechanism that is currently available to compensate injury is private litigation between parties and that is difficult due to a lack of agreements to enforce civil judgments. If, in fact, the polluter is the national government, as in the case

of the Chernobyl disaster, it may be nearly impossible to gain compensation for damages.

Transboundary pollution takes many forms and can be perpetrated by both private industry and government activities. It is difficult if not impossible to litigate on an international scale. It is in fact sometimes difficult to determine who the polluter is when the pollution is "creeping" and not a catastrophic incidence. Airborne transboundary pollution is extremely difficult to track due to the pervasive nature of the practices that produce it. Global solutions to transboundary pollution can only be successful if all nations agree to implement controls to reduce known pollutants and to take responsibility for accidents that damage the environmental quality of other nations.

[*Cynthia Fridgen*]

Resources
Books

Buck, S. J. *Understanding Environmental Administration and Law*. Washington, DC: Island Press, 1996.

Plater, Z. J. B., R. H. Abrams, and W. Goldfarb. *Environmental Law and Policy: Nature, Law, and Society*. St. Paul: West Publishing Co., 1992.

World Wildlife Fund. "Choosing a Sustainable Future: The Report of the National Commission on the Environment." Washington, DC: Island Press, 1993.

Periodicals

Bernauer, T., and P. Moser. "Reducing Pollution of the River Rhine: The Influence of International Cooperation in The Journal of Environment & Development." *Sage Periodicals Press* 5, no. 4 (December 1996).

Transfer station

The regional disposal of **solid waste** requires a multi-stage system of collection, transport, consolidation, delivery, and ultimate disposal. Many states and counties have established transfer stations, where solid waste collected from curbsides and other local sources by small municipal **garbage** trucks is consolidated and transferred to a larger capacity vehicle, such as a refuse transfer truck, for **transportation** to a disposal facility. Typically, municipal garbage trucks have a capacity of 20 cubic yd (15.3 cubic m) and refuse transfer trucks may have a capacity of 80–100 cubic yd (61–76.4 cubic m). The use of transfer stations reduces hauling costs and promotes a regional approach to **waste management**.
See also Garbage; Landfill; Municipal waste; Tipping fee; Waste stream

Transmission lines

Transmission lines are used to transport electricity from places where it is generated to places where it is used. Almost all electricity in North America is generated in fossil-fueled, nuclear-fueled, or hydroelectric generating stations. These are located some distance away from the factories, businesses, institutions, and homes where the electricity is actually used, in some cases hundreds of miles away, so that the electricity must be transmitted from the generating stations to these diverse locations.

Transmission lines are strung between tall, well-spaced towers and are linear features that appropriate long, narrow areas of land. Most transmission lines carry a high voltage of alternating current, typically ranging from about 44 kilovolts (kV) to as high as 750 or more kV (some transmission lines carry a direct current, but this is uncommon). Transmission lines typically feed into lower-voltage distribution lines, which typically have voltage levels less than about 35 kV and an alternating current (in North America) of 60 Hertz (Hz; this is equivalent to 60 cycles of positive to negative per second), and is usually 50 Hz in Europe.

Electrical fields are generated by transmission lines (and by all electrical appliances), with the strength of the field being a function of the voltage level of the current being carried by the powerline. The flow of electricity in transmission lines also generates a magnetic field. Electric fields are strongly distorted by conducting objects (including the human body), but magnetic fields are little affected and freely pass through **biomass** and most structures. Electric and magnetic fields both induce extremely weak electrical currents in the bodies of humans and other animals. These electrical currents are, however, several million times weaker than those induced by the normal functions of certain cells in the human body.

Transmission lines are controversial for various reasons. These include their poor aesthetics, the fact that they can destroy and fragment large areas of natural lands or take large areas out of other economically productive land-uses, and the belief of many people that low-level health risks are associated with living in the vicinity of these structures.

Aesthetics of transmission lines

Transmission lines are very long, tall, extremely prominent linear features. Transmission lines have an unnatural appearance and their very presence disrupts the visual aesthetics of natural landscapes, as viewed from the ground or the air. As such, transmission lines represent a type of "visual pollution" that detracts from otherwise pleasing natural or pastoral landscapes. These aesthetic damages are an important environmental impact of almost all rural transmission lines. Similarly, above-ground transmission lines in urban and suburban areas are not regarded as having good aesthetics.

Damages to natural values

Apart from difficult crossings of major rivers and mountainous areas, transmission lines tend to follow the

shortest routes between their origin and destination. Often, this means that intervening natural areas must be partially cleared to develop the right-of-way for the transmission line. This can result in permanent losses of natural **habitat**, as typically happens when forests are cleared to develop a powerline right-of-way. Moreover, it is not feasible to allow trees to regenerate beneath a transmission line because they can interfere with the operation and servicing of the powerline.

For these reasons, vegetation is cleared beneath and to the sides of transmission lines (for a width of about one tree-height). This can be done by periodically cutting shrubby vegetation and young trees or by the careful use of herbicides, which can kill shrubs and trees while allowing the growth of grasses and other herbaceous plants.

These sorts of management practices result in the conversion of any original, natural habitats along a transmission right-of-way into artificial habitats. The ecological effects include a net loss of natural habitats and fragmentation of the remainder into smaller blocks. In addition, roads associated with the construction and maintenance of transmission lines may provide relatively easy access for hunters, anglers, and other outdoor recreationalists to previously remote and isolated natural areas. This can result in increased stress for certain wild **species**, especially hunted ones, as well as other ecological damages.

Some studies of transmission lines have found that hypothesized ecological damages did not occur or were unimportant. For example, naturalists suggested that the construction of extensive hydroelectric transmission lines in the boreal forest of northern Quebec would impede the movements of woodland caribou. In fact, this did not occur, and caribou were sometimes observed to use the relatively open transmission corridors during their migrations and as resting places. Similar observations have been made elsewhere for moose, white-tailed deer, and other ungulates. Predators of these animals, such as **wolves** and coyotes, will also freely move along transmission corridors, unless they are frequently disturbed by hunters, recreationalists, or maintenance crews.

Certain birds of prey, particularly osprey, may use transmission poles or pylons as a platform upon which to build their bulky nests, which may be used for many years. In some cases the birds are considered a management problem, especially if their nests get large enough to potentially short out the powerlines or if the parent birds aggressively defend their nests against linesmen attempting to repair or maintain the transmission line. Fortunately, it is relatively easy to move the nests during the nonbreeding season and to place them on a "dummy" pole located for the purpose beside the transmission line. In most cases, the ospreys will readily use the relocated nest in subsequent years.

Transmission lines also pose lethal risks for certain kinds of birds, especially larger species that may inadvertently collide with wires, severely injuring themselves. There have also been cases of raptors and other large birds being electrocuted by settling on transmission lines, particularly if they somehow span adjacent wires with their wings.

Disruption of land-use

Some economically important land-uses can occur beneath transmission lines, for example, forestry and some types of agriculture. Other land-uses, however, are not compatible with the immediate proximity of high-voltage transmission lines, particularly residential land-uses. In cases of land-use conflicts, opportunities are lost to engage in certain economically productive uses of the land, a context that detracts from benefits that are associated with the construction and operation of the transmission line.

Many people, including some highly qualified scientists, believe that low-level health risks may be associated with longer-term exposures to the electric and magnetic fields that are generated by transmission lines. These increased risks mean that people living or working in the vicinity of these industrial structures may have an increased risk of developing certain diseases or of suffering other damages to their health.

Although people are routinely exposed to electromagnetic fields through the operation of electrical appliances in the home or at work, those exposures are typically intermittent. In contrast, continuous electromagnetic fields are generated by transmission lines, so longer-term exposures can be relatively high. It must be understood, however, that the scientific knowledge in support of the low-level health risks associated with transmission lines is incomplete and equivocal and therefore highly controversial.

In particular, some studies have suggested that long-term exposure to electromagnetic fields generated by high-voltage transmission lines may be associated with an elevated incidence of certain types of cancers. The strongest suggestions have been for increased risks of childhood **leukemia**. There is weaker evidence of increased risks of cancers of the lymphatic and nervous systems and of adult leukemia. It must be remembered, however, that not all epidemiological studies of transmission lines have reported these statistical relationships and that the increased risks are rather small when they are found. Studies have also been made of possible increases in the incidences of migraine headaches, mental depression, and reproductive problems associated with longer-term exposures to electromagnetic fields near transmission lines. The results of these studies are inconsistent and equivocal.

Some researchers who have assessed the medical problems potentially associated with high-voltage transmission lines have concluded that it would be prudent to not have people living in close proximity to these industrial structures. Even though there is no strong and compelling evidence that

medical problems are actually occurring, the precautionary approach to environmental management dictates that the potential risks should be avoided to the degree possible. It would therefore be sensible for people to avoid living within about 54.5 yd (50 m) or so of a high-voltage transmission line, and they should avoid frequently using the right-of-way as travel corridors.

[*Bill Freedman Ph.D.*]

Resources

Books

Carpenter, D. O., and S. Ayrapetyan. *Biological Effects of Electric and Magnetic Fields.* Vols. 1 & 2. San Diego: Academic Press, 1994.

Levallois, P., and D. Gauvin. *Risks Associated with Electromagnetic Fields Generated by Electrical Transmission and Distribution Lines.* Great Whale Environmental Assessment: Background Paper No. 9, Part 1. Montreal, PQ: Great Whale Public Review Support Office. 1994.

Periodicals

Alonso, J. C., J. A. Alonso, and R. Munoz-Pulido. "Mitigation of Bird Collisions with Transmission Lines through Ground-wire Marking." *Biology Conservation* 67 (1994): 129–134.

Coleman, M. P., et al. "Leukemia and Residence Near Electricity Transmission Equipment: A Case-control Study." *British Journal of Cancer* 60 (1989): 793–798.

Feychting, M., and A. Ahlbom. "Magnetic Fields and Cancer in Children Residing Near Swedish High-voltage Power Lines." *American Journal of Epidemiology* 138 (1993): 467–481.

London, S. J., et al. "Exposure to Residential Electric and Magnetic Fields and Risk of Childhood Leukemia." *American Journal of Epidemiology* 134 (1991): 923–927.

Steenhof, K., M. A. Kochert, and J. A. Roppe. "Nesting by Raptors and Common Ravens on Electrical Transmission Line Towers." *Journal of Wildlife Management,* 57 (1993): 271–281.

Transpiration

Transpiration is the process by which plants give off water vapor from their leaves to the **atmosphere**. The process is an important stage in the water cycle, often more important in returning water to the atmosphere than is evaporation from rivers and lakes. A single acre of growing corn, for example, transpires an average of 3,500 gal (13,248 l) of water per acre (0.4 ha) of land per day. Transpiration is, therefore, an important mechanism for moving water through the **soil**, into plants, and back into the atmosphere. When plants are removed from an area, soil retains more moisture and is unable to absorb rain water. As a consequence, **runoff** and loss of nutrients from the soil is likely to increase. *See also* Erosion; Flooding; Soil conservation; Soil fertility

Transportation

For several million years, humans got to where they wanted to go by one means: walking. This, of course, greatly limited the distance of travel and the amount of items that could be transported. Today, transportation is accomplished in the air, on land, and in water. It ranges from carts pulled by horses or oxen and dirt roads in Africa to the Concorde supersonic airplane that can travel from Paris to New York in four hours.

The first changes in transportation came about five to six thousand years ago in three areas: the introduction of boats, the domestication of wild horses, and the invention of the wheel. Horse domestication started about 6,000 years ago, probably occurring in several different parts of the world at about the same time. Riding astride the horse may have begun in Turkestan before 3,000 B.C. At about the same time, many historians believe the wheel was invented in Mesopotamia, a development that drastically changed the way people moved about. It was not long before the wheel found its way onto platforms, forming carts that could be drawn by horses or oxen. Covered wagons and carriages soon followed. It is difficult to say when boats became a means of transportation but ships can be traced back at least to the ancient Mesopotamian, Greek, Roman, and Egyptian empires

In ancient Egypt, the main transportation corridor was the Nile River, where ships carried people and trading goods through the empire. On land, transportation was more difficult. Ordinary people traveled by foot, while high nobles were carried in chairs or covered litters. Over times, these became more elaborate, requiring as many as 28 people to carry them. Lower ranking nobles often rode on a chair fastened on the backs of donkeys. Horse-drawn chariots first appeared in Egypt about 3,500 years ago.

Wheeled vehicles were first used primarily as hearses for the great and as military adjuncts until they gradually came to be used more for carriers of goods. Even during the Classical Age in Greece, chariots were declining in importance for warfare, and they were finally used only for sport. In early times carts with shafts for single animals were preferred in outlying countries where roads were poor, but with improvement in roads came heavy wagons and the use of several animals for power.

Ships took on an ever increasingly important role in transportation starting about A.D. 1400, when ships of exploration sailed from Europe to Asia and the New World. By the seventeenth century, ships regularly carried European settlers to North America. These settlers soon began build their own transportation system, using boats on rivers and coastal waterways and horse and cart by land.

From 1600 to 1754 travel between the American colonies was accomplished most quickly and easily by boat. Everywhere in colonial America settlements sprang up first near navigable rivers. Two types of river vessel were common along colonial rivers. Dugout canoes carried small cargoes, while long (up to 40 ft [12 m]), flat-bottomed boats handled larger loads. River travel everywhere in America was slow and dangerous. Nevertheless, rivers served colonists everywhere as routes for their goods, travel, and communication.

Land travel between seventeenth-century colonies could be difficult as well. The number and condition of roads varied widely from colony to colony, depending heavily on the density of settlement and the support provided by the various colonial legislatures. In many places roadbeds were poor and bridges few. By the 1760s one of the longest roads, the Great Wagon Road, stretched nearly 800 mi (1,287 km) along old Indian trails through western Pennsylvania and Virginia's Shenandoah Valley to Georgia. The longest road in North America was the Camino Real, which connected Mexico City to Santa Fe, an 1,800-mi (2,897-km) trip that took wagon trains six months to complete.

Aside from walking and riding a horse, a person in eighteenth-century America had other means of traveling. Farmers used two-wheeled carts while Indian traders frequently had packhorses. During this era, the Conestoga wagon came to the forefront. A high-wheeled vehicle with a canvas cover, it had a curved bottom in order to keep its load from shifting. As may be expected, the Conestoga quickly became popular. By the 1770s more than 10,000 were in use in Pennsylvania while in the South Carolina backcountry there were an estimated 3,000. Several stagecoach lines operated between all the major cities in the north. The most heavily traveled route, between New York and Philadelphia, was served twice weekly by a stagecoach.

The 1800s saw the advent of two more means of transportation: the steamboat and the railway. Hundreds of steamboats once navigated rivers in the eastern half of the United States railways caught on throughout the world and are the primary form of inner-city **mass transit** in many countries, such as India. In the twentieth century, two more forms of transportation came into existence: the **automobile** and airplane. The electric streetcar also gained in popularity, finding itself in many major American cities, including San Francisco, Los Angeles, and New Orleans.

In many parts of the world today, mass transit systems are an important component of a nation's transportation system. Where people can not afford to buy automobiles, they depend on bicycles, animals, or mass transit systems such as bus lines to travel within a city and from city to city.

In the United States the automobile is the primary form of transportation. It is less important in most other parts of the world. Mass transportation, or mass transit, does play a role in the United States. One could argue that stagecoaches were the first mass transit vehicles in the country, since they could hold eight to 10 passengers, a big improvement over one person on a horse or several people in a small horse-drawn buggy. In the twentieth and twenty-first centuries, the horse has been replaced with the automobile. Today, mass transit options include airplanes, buses, trolleys, rail and light rail, and subways. The world's first subway opened in London in the late nineteenth century, soon followed by America's first subway in New York.

Most of these forms of transportation brought with them problems, including harming the **environment**. Electric trains, trolleys, and streetcars are essentially pollution-free but it takes addition **power plants** to run them, plants that often use highly polluting fuels such as **coal** and oil. Autos, buses, and airplanes are mostly oil-dependent. It was not until the 1970s that the public began to be concerned with the toxic effects of emissions from these transportation conveyances. Since then, airplanes, buses, and autos have increasingly become less polluting. This, however, has been offset by an increase in their numbers.

Mass transit usage reached its peak in the United States in the 1940s and 1950s. During the 1960s mass transit systems and their ridership declined drastically even as urban populations grew. There was a brief resurgence in the 1970s as the price of **gasoline** skyrocketed. In 1972, the country embarked on a round of new mass transit systems when the Bay Area Rapid Transit (BART), opened in the San Francisco Bay Area. BART was followed in the next two decades by new subway, bus, light rail, and trolley systems in Washington, D.C., San Diego, Atlanta, Baltimore, Los Angeles, San Jose, California, and other urban areas. At about the same time, the United States Congress gave the nation's inner-city passenger rail system, Amtrak, a new lease on life. Amtrak proved to be a huge success among inner-city passengers, but the federal government has never maintained the consistent support of the system it showed during the aftermath of the **petroleum** price hikes of the 1970s.

Over the next several decades, urban planners see little likelihood of entirely new types of mass transportation. Rather, they expect improvements over existing bus, trolley, and light rail systems. And at least one idea that has been around for 50 years is getting a new look. When the monorail was introduced at Disneyland in California in 1955, it was hailed as the clean, fast, and efficient transit system of the future. Up until now, it never caught on, other than a very short single-line monorail that opened in Seattle in 1962. But urban planners are taking a fresh look at monorails, which are attractive because they could be built above existing highways. Las Vegas has begun building a $650 million monorail. The four-line will parallel the heavily congested

main thoroughfare, the Strip, carrying an estimated 19 million passengers when it opens in 2004.

[*Ken R. Wells*]

Resources

Books

Lewis, David, and Fred Laurence Williams. *Policy and Planning as Public Choice: Mass Transit in the United States.* Burlington, VT: Ashgate Publishing Co., 1999.

Periodicals

Brovarski, Edward. "Getting Around in the Old Kingdom." *Calliope* (September 2001): 19.

Gould, Lark Ellen. "Making Tracks: Las Vegas is Preparing for a Well-Needed Monorail System That Will Move People From Hotel to Hotel." *Travel Agent* (March 4, 2002): SS6–SS7.

National Geographic Society. "Cleaner Living and Driving." *National Geographic* (May 2001): XIV.

National League of Cities. "Technology Improves City Transportation Flow, Safety." *Nation's Cities Weekly* (July 3, 2000): 5.

Reynolds, Francis D. "The Transportation System of The Future." *The Futurist* (September 2001): 44.

Organizations

U.S. Department of Transportation, 400 7th Street NW, Washington, DC USA 20590 202-366-4000, Email: dot.comments@ost.dot.gov, <http://www.transportation.gov>

Trash

see **Solid waste**

Treespiking

see **Monkey-wrenching**

Tributyl tin

Tributyl tin (TBT) is a synthetic organic tin compound that is primarily used as an additive to paints. In this application, TBT acts as a biocide or compound that kills plants such as algae, fungus, mildew, and mold that grow in or on the surface of the coating. These antifouling paints, as they are called, are applied to the hulls of ships to prevent sea life (e.g., barnacles and algae) from attaching to the hull. Growth of organisms on the hulls of ships causes ships to slow down and increases fuel consumption due to increased drag. Ships that are covered with these organisms can also transport non-native, or invasive, **species** around the world, which can disrupt an **ecosystem** and cause reductions in biological diversity. TBT is also used in lumber preservatives as an industrial biocide.

TBT-based paints were introduced more than 40 years ago. The first TBT anti-fouling paints, known as Free Association Paints (FAPs), where the TBT is mixed into the paint, prevented fouling by uncontrolled **leaching** of TBT into the **environment**. Starting 20 years ago, FAPS were gradually phased out and replaced with TBT Self Polishing Copolymer Paints, in which the TBT is bound in the paint matrix. This type of formulation allows for TBT to be released slowly and uniformly.

TBT, also known as organotin, is a widespread contaminant of water and sediments in harbors, shipyards, and waterways. TBT is persistent in marinas, estuaries, and other waters where circulation is poor and flushing irregular. In the natural aquatic environment, degradation of TBT through biological processes is the most important pathway for the removal of this toxic compound from the water column. At levels of TBT contamination typical of contaminated surface waters, the **half-life** is six days to four months. Half-life refers to the amount of time it takes for one-half of the TBT to be eliminated from the environment through natural means.

TBT has been shown to have significant environmental impact and is considered to be an endocrine disruptor that affects the immune and reproductive systems. In molluscs, TBT can be found in concentrations that are up to 250,000 times higher than in the surrounding **sediment** or seawater. Contaminated molluscs had deformed shells, slow growth rates, and poor reproduction, with eggs and larvae often dying. Exposure to TBT has resulted in sex change and infertility in at least 50 species of snails. In the dogwhelk, which is a small marine snail, TBT causes the female snail to grow a penis that blocks the oviduct and thus prevents reproduction. This abnormality is referred to as "imposex," and its development has been linked to the concentrations of TBT in tissue. Because of this association, imposex in the dogwhelk has been used as an indicator of TBT **pollution**. TBT has also been detected in large predators such as **sharks**, **seals**, and **dolphins**, indicating that TBT can accumulate in the food chain, as larger animals eat smaller animals contaminated with TBT. Currently, not enough data are available to help researchers determine whether TBT is a human **carcinogen**, so the U.S. **Environmental Protection Agency** (EPA) has assigned it to the "cannot be determined" category.

Environmental concerns over the effects of TBT-based paints has led to the regulatory control of TBT throughout the world. In 1988, the United States enacted the Organotin Antifouling Paint Control Act that restricts the use of TBT-based antifoulants to ships larger than 25 meters in length or those with **aluminum** hulls. The Act also restricted the release rate of TBT from the antifouling paints. In 1990, the International Maritime Organization (IMO) of the United Nations' Marine Environment Protection Committee adopted a resolution that recommended

that countries adopt measures to restrict the use and release rate of TBT-based antifouling paints. Since that recommendation, many countries have prohibited the use of TBT antifouling paints on small craft. New Zealand and Japan banned the use of TBT anti-fouling paints on all ships. In 1998, the IMO agreed to ban the application of TBT-based paints by 2003 and their presence on ships by 2008. Due to the ensuing reductions in the use of TBT-based paints, both TBT contamination and its effects on organisms have been decreasing worldwide.

Less toxic alternatives to TBT are being developed quickly. In the animal and plant worlds, researchers have found over 50 naturally-occurring substances that have the same effects as TBT and other antifouling paints, keeping organisms such as algae and some invertebrates from becoming laden with other forms of algae and sea life. In addition, other alternatives to TBT have been developed, and include epoxy-copper flake paints, which may protect boats for up to 10 years, but cannot be used on ships with aluminum hulls. Ceramic-epoxy coats are also available, and can be used on fiberglass and metal hulls. Other alternativess to TBT are foulant-release coatings that contain silicone elastomers that last as long as the copolymers in TBT. Finally, researchers are studying eletrical current systems that can sterilize the hulls of ships for up to four years as another alternative.

[*Judith L. Sims*]

Resources

Books

de Mora, S. J., ed. *Tributyl Tin: Case Study of an Environmental Contaminant.* Cambridge, UK: Cambridge University Press, 1997.

St. Louis, Richard. *Tributyl Tin: The Case for Virtual Elimination in Canada.* Toronto, Canada: World Wildlife Fund Canada, 1999.

Other

"Antifoulants." *Organotin Environmental Programme (ORTEP).* Association, The Hague, The Netherlands. [cited June 21, 2002]. <http://ortep-a.org/pages/antifoulants.htm>.

Trihalomethanes

Trihalomethanes (THMs) are organic **chemicals** composed of three halogen atoms (primarily **chlorine** and **bromine**) and one **hydrogen** atom bound to a single **carbon** atom. They are commonly used as solvents in a variety of applications, and trichloromethane (or chloroform) was once used by medical doctors as an anesthetic. In 1974, Johannes Rook of the Netherlands discovered the presence of THMs in chlorinated drinking water. He and other researchers subsequently demonstrated that minute quantities of THMs are formed during **water treatment**. Chlorine, when added to disinfect the water, reacts with naturally occurring organic matter present in the water. **Groundwater** can also be contaminated with THMs if people use them improperly, release them accidentally, or dispose of them unsuitably. Since high dosages of chloroform had been found to produce **cancer** in mice and rats, the United States **Environmental Protection Agency** (EPA) amended the drinking water standards in 1979 to limit the concentration of THMs to a sum total of 100 parts ber billion, or 0.1 mg/l. In 1998, a more stringent limit of 0.08 mg/l was introduced. *See also* Carcinogen; Drinking-water supply; Water quality standards

Trophic level

One way of analyzing the biological relationships within an **ecosystem** is to describe who eats whom within the system, also called a functional analysis. Each feeding level in an ecosystem is called a trophic level. In the **grasslands**, for example, plants are considered primary producers, forming the first trophic level. The second trophic level consists of primary consumers, such as deer, mice, seed and fruit-eating birds, and other animals, depending completely on the primary producers for their food. Carnivores and predators, such as hawks, are the secondary consumers. Often, the same **species** may fit into several categories. Bears, for example, are considered both primary and secondary consumers because they feed on plant matter as well as meat. Bacteria and **fungi** that decompose dead organic matter are called the **decomposers**. Thus, on the basis of food source and feeding behavior, a complex **food chain/web** exists within any ecosystem and every species belongs to one or more of several trophic levels.

In **environmental science**, the concept of trophic levels is often used to assess the potential for transfer of pollutants through an ecosystem. Since each trophic level is dependent on all the other levels, positive or negative changes in the composition or abundance of any one trophic level will ultimately affect all other levels. In ecosystems that normally have stable, complex trophic levels within the food web, **pollution** can lead to fluctuations and simplification of the trophic levels. Contaminants that are taken up by plants from the **soil** may be transferred to primary and secondary consumers through their feeding patterns. This is known as trophic level transfer.

A classic example of trophic level transfer was the release of DDT in the **environment**. DDT is an insecticide commonly used in the United States during the 1950s and early 1960s. DDT ran off treated fields into lakes and rivers, where it accumulated in the fatty tissues of primary consumers such as fish and shellfish. The **chemicals** were then

transferred into secondary consumers, such as eagles, which fed on the primary consumers. While the concentrations of DDT were rarely high enough to kill the birds, it did cause them to lay eggs with thin shells. The thin eggshells led to decreased hatching success and thus caused a decline in the eagle population.

Trophic level analysis is a commonly used method of environmental assessment. A pollutant or disturbance is assessed in terms of its effects on each trophic level. If significant amounts of nutrients are brought into a lake receiving **fertilizer runoff** from fields, a spurt in the growth of algae (primary producers) in the lake may be triggered. However the increased growth might actually be dominated by certain algal groups, such as blue-green algae, which may not constitute desirable food sources for the **zooplankton** and fish (primary consumers) which normally depend on algae for food. In this case, even though the environmental conditions might appear to stimulate increased growth in one trophic level, the nature of the change does not necessarily prove advantageous to other trophic levels within the same system. *See also* Agricultural pollution; Algal bloom; Aquatic weed control; Balance of nature; Bald eagle; Decomposition; Environmental stress; Predator control; Predator-prey interactions

[*Usha Vedagiri*]

Resources

Books

Connell, D. W., and G. J. Miller. *Chemistry and Ecotoxicology of Pollution.* New York: Wiley, 1984.

Smith, R. L. *Ecology and Field Biology.* New York: Harper and Row, 1980.

Tropical rain forest

The richest and most productive biological communities in the world are in the tropical forests. These forests have been reduced to less than half of their former extent by human activities and now cover only about 7% of Earth's land area. In this limited area, however, is about two-thirds of the vegetation mass and about half of all living **species** in the world.

The largest, lushest, and most biologically diverse of the remaining tropical moist forests are in the **Amazon Basin** of South America, the Congo River basin of central Africa, and the large islands of southeast Asia (Sumatra, Borneo, and Papua, New Guinea). Whereas the forests of mainland southeast Asia, western Africa, and Central America are strongly seasonal, with wet and dry seasons, the South American and central African forests are true rain forests. Rainfall is generally over 160 in (406 cm) per year and falls at a relatively even rate throughout the year. It is said that such rainforests "make their own rain," because about half the rain that falls in the forests comes from condensation of water vapor released by **transpiration** from the trees themselves. Rain forests at lower elevations are hot and humid year-round. At higher elevations, tropical mountains intercept moisture-laden clouds, so the forests that blanket their slopes are cool, wet, and fog-shrouded. They are aptly and poetically called "cloud forests." Tropical forests are generally very old. Unlike temperate rain forests, they have not been disturbed by **glaciation** or mountain-building for hundreds of millions of years. This long period of **evolution** under conditions of ample moisture and stable temperatures has created an incredible diversity of organisms of amazing shapes, colors, sizes, habits, and specialized adaptations.

Habitats in a tropical **rain forest** are stratified into three to five distinct layers from ground level to the tops of the tallest trees. Hundreds of tree species grow together in lush profusion, their crowns interlocking to form a dense, dappled canopy about 120 ft (37 m) above the forest floor. These unusually tall trees are supported by relatively thin trunks reinforced by wedge-shaped buttresses that attach to a thick mat of roots just under the **soil** surface. A few emergent trees rise above the seemingly solid canopy into a world of sunlight, wind, and open space. Numerous species of birds, insects, reptiles, and small mammals live exclusively in the forest canopy, never descending below the crowns of the trees.

The forest understory is composed of small trees and shrubs growing between the trunks of the major trees, as well as climbing woody vines (lianas) and many epiphytes—mainly orchids, bromeliads, and arboreal ferns—that attach themselves to the trees. Some of the larger trees may support 50–100 different species of epiphytes and an even larger population of animals that are specialized to live in the many habitats they create. These understory layers are a world of bright but filtered light abuzz with animal activity.

By contrast, the forest floor is generally dark, humid, quiet, and rather open. Few herbaceous plants can survive in the deep shade created by the layered canopy of the forest trees and their epiphytes. The most numerous animals are ants and termites that scavenge on the **detritus** raining down from above. A few rodent species gather fallen fruits and nuts. Rare predators such as leopards, jaguars, smaller cats, and large snakes hunt both on the ground and in the understory.

A tropical rain forest may produce as much as 90 tons (81.6 metric tons) of **biomass** per acre (0.4 ha) per year, and one might think that the soil that supports this incredible growth is rich and fertile. However, the soil is old, acidic, and nutrient-poor. Ages of incessant tropical rains and high temperatures have depleted minerals, leaving an iron- and

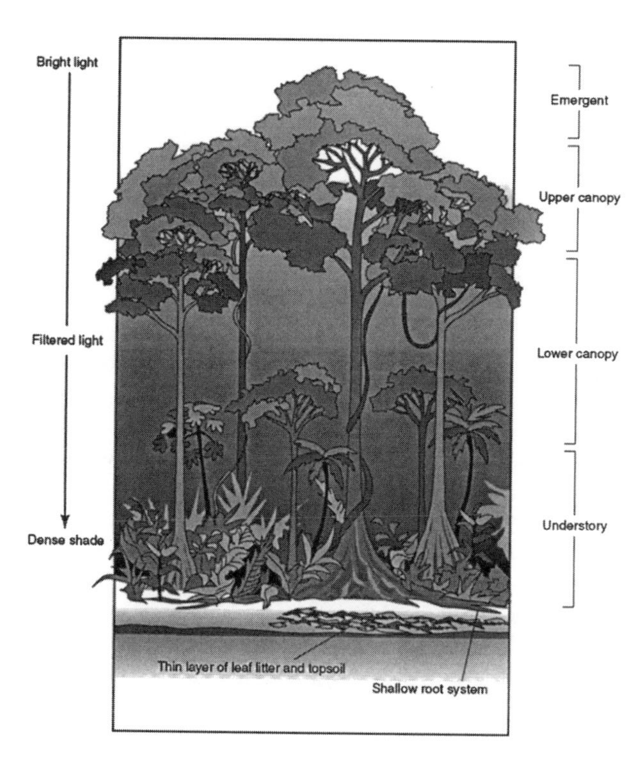

The layered communities of a tropical rain forest are directly related to the gradual lessening of light, from the brightness of the canopy to the dense shade of the forest floor. (Illustration by Hans & Cassidy.)

aluminum-rich podzol. Tropical forests have only about 10% of their organic material and nutrients in the soil, compared to boreal forests, which may have 90% of their organic material in litter and sediments.

The interactions of **decomposers** and living plant roots in the soil are, literally, the critical base that maintains the rain forest **ecosystem**. Tropical rain forests are able to maintain high productivity only through rapid **recycling** of nutrients. The constant rain of detritus and litter that falls to the ground is quickly decomposed by populations of **fungi** and bacteria that flourish in the warm, moist **environment**. Some of these decomposers have symbiotic relationships with the roots of specific trees. Trees have broad, shallow root systems to capitalize on this surface **nutrient** source; an individual tree might create a dense mat of superficial roots 328 ft (100 m) in diameter and 3 ft (0.9 m) thick. In this way, nutrients are absorbed quickly and almost entirely and are reused almost immediately to build fresh plant growth, the necessary base to the trophic pyramid of this incredible ecosystem. *See also* Biodiversity

[*William P. Cunningham Ph.D.*]

RESOURCES
BOOKS

Caufield, C. *In the Rainforest.* New York: Knopf, 1985.

Hecht, S., and A. Cockburn. *Fate of the Forest.* New York: Harper Collins, 1991.

Myers, N. *The Primary Source: Tropical Forests and Our Future.* New York: Norton, 1992.

Revkin, A. *The Burning Season: The Murder of Chico Mendes and the Fight for the Amazon Rain Forest.* New York: Houghton Mifflin, 1991.

PERIODICALS

Repetto, R. "Deforestation in the Tropics." *Scientific American* 262 (April 1990): 36–42.

Tropopause

The tropopause is the upper boundary of the **troposphere**, a layer of the earth's **atmosphere** near the ground. In the troposphere, the temperature generally decreases with increasing altitude, with restricted exceptions called inversions. However, at a height of about 6 mi (10 km) at the poles 9 mi (15 km) at the equator, the temperature abruptly becomes constant with increasing altitude. This isothermal region is called the **stratosphere**, and the interface between it and the troposphere is called the tropopause. Mixing of air across the tropopause is slow, occurring on a time scale of weeks on the average, while tropospheric mixing is more rapid.

The existence of the stratosphere is largely caused by the **absorption** of **solar energy**, mostly ultraviolet light, by oxygen to form **ozone**. The balance of the heating is caused by absorption of other parts of the solar spectrum by other trace gases. *See also* Ultraviolet radiation

Troposphere

From the Greek word *tropos*, meaning turning, troposphere is the zone of moisture-laden storms between the surface and the **stratosphere** above. Because ice crystals must form before precipitation can begin, the troposphere rises to 10–12 mi (16–19 km) over the equator but grades downward to 5–6 mi (8–9 km) over the poles. It is marked by a sharp drop in temperature vertically, averaging 3.5 degrees per 1,000 ft (305 m) because of exponentially decreasing density of air molecules. Actual "lapse" rates vary enormously, ranging from inversions, where temperatures rise and trap pollutants, to steep lapse rates with warm surface air topped by very cold polar air; the latter produces dangerous storms.

Tsunamis

Tsunamis are large seismic sea waves that can cause major destruction in coastal regions. A tsunami (Japanese for "wave in bay") is caused by underwater seismic activity, such as an **earthquake**.

While tsunamis are commonly called "tidal waves," this is an erroneous term; these potentially catastrophic waves have nothing to do with the tides. Tides are the up and down movements of the sea surface at the shore, caused by the gravitational attraction of the moon and sun on our marine waters. Tsunamis are caused by the movements of Earth's crustal plates. Tides rarely cause major damage unless they are associated with a storm, while tsunamis can cause major loss of life and property.

It has been known for several hundred years that tsunamis are caused by seismic movements of the ocean floor. This occurs most commonly during submarine earthquakes, underwater landslides, and perhaps volcanoes, all of which release large amounts of energy. The sudden movement of the earth's crust caused by an underwater earthquake, for example, displaces or moves the water above it. This causes a high-energy wave to form, which is passed rapidly through the water.

Tsunamis are very long waves, with a period (the time for one complete wave to pass a fixed point) ranging from six to 60 minutes. These waves typically travel 450 mi/h (200 m/s or 720 km/h). Therefore, an earthquake in the Gulf of Alaska could result in a tsunami hitting Hawaii less than five hours later.

It is almost impossible to feel a tsunami out at sea in deep water, however, the form of the wave changes when it reaches shallow water. Since the water is shallower, the bottom of the wave begins to "feel" the ocean bottom. The friction that results causes the wave to slow down from about 450 mi/h (200 m/s) in very deep ocean water to 49 mi/h (22 m/s) in water 164 ft (50 m) deep. While the front part of the wave has been reduced in speed, the part at sea is still moving in quickly. As a result, the energy of the wave is compressed. As the wave enters shallow water, like that in a bay, the crest rises. It quickly builds up vertically as the wave moves onto the shore. This wall of water can be more than 100 ft (30.5 m) high, in extreme cases. Since gravity is acting on this huge wall of water, it cannot support itself and crashes or breaks onto the land, similar to a normal breaker in the surf zone of a beach. However, the huge amounts of energy released by a breaking tsunami are many times greater and more destructive than an ordinary breaker at the beach, and the tsunami can literally destroy anything in its path.

In Japan, where some of the most destructive tsunamis have occurred, there have been cases in which whole fishing villages were devastated. However, the fishermen, who were at sea plying their trade, did not feel the wave, which passed right under them. They did not discover the disaster until they returned home and found their homes and villages destroyed. Because these villages were often located within shallow bays, and the fishermen, being at sea, did not experience the wave, they assumed that the tsunami arose within the bay. Therefore, these waves were called tsunamis or "wave in bay."

While scientists are not yet able to predict submarine seismic activity with much **accuracy**, they can easily measure such events when they occur. Scientists use this information about the causes of tsunamis to predict when these destructive sea waves will occur. This is extremely important in reducing loss of life and property. After the destructive 1946 tsunami that hit Hawaii, a group of tsunami early warning stations was set up to monitor seismic activity throughout the Pacific Ocean. The geographical and administrative center of this monitoring system is in Honolulu, Hawaii. When an earthquake, underwater **volcano**, or **landslide** is sensed, its location is pinpointed. If a wave is generated and a change in the water height is measured at a nearby tide-measuring station, scientists can then accurately calculate the speed of the wave to determine when the wave will make landfall. The appropriate agencies can be alerted, and if necessary, evacuations and other preparations can be made. This early warning system has been very successful in reducing damage caused by tsunamis. For example, there were no deaths from a tsunami in Hawaii in 1957 because of an early warning, even though the tsunami was over 26 ft (8 m) tall.

Before the warning systems existed, the first indication of an approaching tsunami was the rapid movement of water in a bay out to sea. This exposed areas of the bay bottom that were rarely or never exposed. The water that rushed offshore rose to build the huge crest of the wave that would crash down a few minutes later.

Despite the success of the early warning systems, there are some problems. For example, not all seismic activity generates tsunamis; they more commonly result from shallow focus earthquakes, where the actual point of crustal movement is closer to the surface. It is during these earthquakes that major crustal movement is most likely. Deep focus earthquakes, which can be very strong but often result in less crustal movement, are less likely to trigger tsunamis. It has been estimated that only one out of ten large underwater earthquakes causes damage. In addition, the chances of a tsunami hitting any one spot directly and causing major damage are relatively small because the energy in the form of the tsunami is not passed along equally in all directions. Finally, there may be other factors that reduce or enhance chances of a tsunami striking. For example, major tsunamis are rare in regions with wide continental shelves, such as

the Atlantic coast of the United States. A wide continental shelf is thought to both reflect the wave (with the energy being sent back out to sea) and absorb some of the energy of the wave through friction as it drags along the bottom. Thus, the early warning system, while essential, often gives false alarms.

Tsunamis are more common in the Pacific Ocean than other oceans of the world, primarily because there is so much seismic activity at the perimeter of the Pacific Ocean, where crustal plates meet. Thus, this region is where some of the world's most damaging tsunamis have occurred. For example, one of the most dramatic and destructive tsunamis occurred on August 27, 1883, when the volcanic island of **Krakatoa**, located in the Pacific Ocean, between Sumatra and Java, exploded and disappeared. The wall of water reached over 98 ft (30 m) in height and left catastrophic damage to coastal areas in the Sundra Strait in its wake. Over 36,000 people lost their lives. The energy from the tsunami was still measurable after it crossed the Indian Ocean, moved around the southern part of Africa, and headed north through the Atlantic Ocean into the English Channel. In addition to the tsunami, the sound of the explosion was heard 3,000 mi (4,827 km) away, and the dust that entered the **atmosphere** caused unusual sunsets for almost a year.

In 1896 there was a major tsunami in Japan that killed 27,000 people along the coast. In 1964, there was an earthquake in Alaska and the resulting tsunami caused major damage in some ports such as Kodiak and Seward. In addition, the tsunami traveled to the south, where four and a half hours later, despite warnings, it killed additional people and did major damage in Crescent City, California. A total of 119 people died and damage to property amounted to over $100 million.

Tsunamis also occur along the chain of Caribbean islands and in the Mediterranean. Both of these places, like the Pacific, are at the edges of Earth's crustal plates where earthquakes and other seismic activity are common.

[*Max Strieb*]

Resources
Books

Knauss, J. A. *Introduction to Physical Oceanography*. Englewood Cliffs, New Jersey: Prentice-Hall, Inc., 1978.

Neshyba, S. *Oceanography: Perspectives on a Fluid Earth*. New York: John Wiley and Sons, 1987.

Tumors

see **Cancer; Neoplasm**

Tundra

Tundra is a generic term for a type of low-growing **ecosystem** found in climatically stressed environments. Latitudinal tundra occurs in the Arctic and Antarctic, environments with cool and short growing seasons. Altitudinal tundra occurs at the top of mountains, where the growing season is cool, although it can be long. A major environmental factor affecting tundra communities is the availability of moisture. Wet meadows are dominated by hydric sedges and grasses, while mesic sites are dominated by dwarf shrubs and herbaceous **species** and dry sites by cushion plants and **lichens**.

Turbidity

Turbidity is a characteristic of water that describes the amount of suspended solids in the water. Suspended solids can be **phytoplankton**, **sediment**, or **detritus**. **Anthropogenic** causes of turbidity include **dredging** activities, **runoff** from agricultural and urban areas, and shoreline **erosion**. Highly turbid water can prevent light from reaching plants on the bottom or phytoplankton in the water column, and can therefore reduce the amount of **primary productivity** in an aquatic system. High concentrations of suspended solids can settle onto **submerged aquatic vegetation** (SAV) and can smother shellfish beds and fish spawning grounds. Turbidity is measured as total suspended solids as milligrams of solids per liter, or with an instrument called a nephelometer, which measures the amount and angle of light scattering that is caused by particles suspended in the water.

[*Marie H. Bundy*]

Turnover time

Turnover time refers to the period of time during which certain materials remain within a particular system. For example, the protein that we get from food is broken down by enzymes in our bodies and then resynthesized in a different form. The time during which any one protein molecule survives unchanged in the body is called the turnover time for that molecule. Each component of an environmental system also has a turnover time. On an average, for example, a molecule of water remains in the **atmosphere** for a period of 11.4 days before it falls to the earth as precipitation. *See also* Chemicals

Turtle excluder device

Sea turtles of several **species** are often accidentally caught in a variety of fishing gear in many areas of the world, including the southwest Atlantic and the shallow waters of the Gulf of Mexico. Up to 12,500 turtles died annually as a result of entanglement in shrimp trawl fishery alone. It became a concern for the **commercial fishing** industry and environmentalists alike since almost all sea turtles are endangered, and the turtle excluder device (TED) was developed in the mid-1980s in an effort to prevent turtles from entering nets designed to catch other marine animals.

Turtle excluder device (TED) designs began as barrier nets preventing turtles from being caught in the main net, then modified to gate-like attachments. Three main within-net designs are currently used. The model designed by the National Marine Fisheries Service (NMFS) consists of an addition in the throat of the trawl net, where the diameter narrows toward the end where the catch is held. The addition includes a diagonal deflator grid, that, once encountered by a turtle, forces the animal upwards and out through a door in the top of the net. The trap door is called a by-catch reduction device or BRD. A second grid deflects other fish out another BRD in the side of the net. Several other TEDs are approved that are lighter and less expensive than the NMFS design, all designed to eject turtles and other by-catch through either the top or bottom of the net. Few shrimp are lost in this process, while the capture of turtles and other by-catch is reduced significantly.

Although TEDs are required by law on Mexican shrimp fleets in the Pacific and the Gulf of Mexico, convincing United States fishing boats to adopt their use has been a struggle. A law mandating the use of TEDs on shrimp trawlers operating between March 1 and November 30 was to have gone into effect in 1988, but the State of Louisiana obtained an injunction against the order. Continued challenges brought by Louisiana and Florida, as well as political lobbying and civil disobedience in the fishing community, delayed implementation even further. In March 1993 the final United States regulations were announced, making TEDs mandatory in inshore waters, and in all waters by December 1994. TEDS became mandatory for Australian prawn trawlers in 2000, and tests conducted in 2001 show that **bycatch** is reduced by more than 90%. An unforeseen financial advantage was obtained by the use of TEDs in the Australian prawn industry because the devices are believed to reduce physical damage to the prawns from larger, heavier animals that crush the catch during capture and sorting. Even with the required TEDs, some 50,000 turtles are caught annually during shrimp trawling.

[*David A. Duffus and Marie H. Bundy*]

Resources

Books

Hillestad, H. O., et al. "Worldwide Incidental Capture of Sea Turtles." In *Biology and Conservation of Sea Turtles,* edited by K. A. Bjorndal. Washington, DC: Smithsonian Institution Press, 1981.

Seidel, W. O. and C. McVea. "Development of a Sea Turtle Excluder Shrimp Trawl in the Southeast U.S. Penaeid Shrimp Trawl." In *Biology and Conservation of Sea Turtles,* edited by K. A. Bjorndal. Washington, DC: Smithsonian Institution Press, 1981.

Periodicals

Hinman, K. "The Real Cost of Shrimp on Your Plate." *Sea Frontiers* 38 (February 1992): 14–19.

Williams, T. "The Exclusion of Sea Turtles." *Audubon* 92 (January 1990): 24–30.

2,4-D

One of the nation's most popular weed killers, the **herbicide** 2,4-D (also known as 2,4-dichlorophenoxyacetic **acid**) has been widely used by homeowners, timber companies, government agencies, farmers, and power companies to eliminate unwanted vegetation from lawns, **golf courses**, forests, **rangelands**, rights-of-way, pastures, highways, and even farmlands. Scientists and environmentalists have warned for years of the chemical's toxic effects, and Rachel Carson's classic book *Silent Spring* described its dangers to human health and the **environment**. Subsequent studies have linked 2,4-D to cancers, miscarriages, and **birth defects** in animals and humans who have been exposed to it. **Agent Orange**, a defoliant used during the Vietnam War, was a 50/50 mixture of 2,4-D and a similar herbicide, **2,4,5-T**. For years environmentalists have urged that 2,4-D be banned or strictly controlled, but the U.S. **Environmental Protection Agency** (EPA) has so far not acted to do so.

2,4,5-T

A broad-leaf **herbicide**, now banned for use in the United States. Its full name is 2,4,5-trichlorophenoxyacetic **acid**. After postemergence treatment it is readily absorbed by foliage and roots and translocated throughout the plant. This herbicide gained much notoriety during and after the Vietnam War because it was a component in the defoliant **Agent Orange**, which has been implicated in **cancer** occurrence in some war veterans. The **carcinogen** 2,3,7,8-TCDD is formed as a byproduct in the manufacture of 2,4,5-T. The **Environmental Protection Agency** (EPA) restricted the use of 2,4,5-T in 1971, and suspended usage in 1979 following concerns that it caused miscarriages in women living in areas where application of 2,4,5-T had occurred.

U

Ublic health risk assessment

see **Risk assessment**

Ultraviolet radiation

Radiation of the sun including ultraviolet A (UV-A, 320-400 nanometers) and ultraviolet B (UV-B, 280-320 nanometers). Exposure to UV-A radiation, which is utilized in tanning booths, damages dermal elastic tissue and the lens of the eye, and causes **cancer** in hairless mice. Exposure to UV-B induces breaks and other mutations in DNA and is associated with basal and squamous cell carcinoma as well as melanoma. The **ozone** layer of the earth's **atmosphere** provides protection from ultraviolet radiation, but this protective layer is becoming depleted due to the release of **chlorofluorocarbons** and other causes. *See also* Ozone layer depletion; Radiation exposure

Uncertainty in science, statistics

Uncertainty in **statistics** is measured by the amount of error in an estimate of the mean or average value of a population. The sample mean is the average of a group of measurements or parameters taken from the population, and the standard deviation of the mean provides an estimate of the uncertainty that any one measurement will represent the true mean of the population. In **environmental science**, it is often very difficult to make enough measurements of interacting processes so that uncertainty in the estimate of the mean values of important parameters is low enough to prove whether hypotheses are correct or not. Baysian statistical analysis is a statistical method that estimates the **probability** that a hypothesis is true, then modifies and updates the probability as more studies are conducted and more information becomes available.

[*Marie H. Bundy*]

Underground storage tank

see **Leaking underground storage tank**

Union of Concerned Scientists

Founded in 1969 by faculty and students at the Massachusetts Institute of Technology, the Union of Concerned Scientists (UCS) promotes four core objectives: nuclear arms reduction, a rational national security policy, safe **nuclear power**, and energy reform. UCS utilizes research, public education, and lobbying and litigation to achieve these goals. UCS also operates a speakers' bureau and publishes various informational materials, including a quarterly magazine addressing various energy and nuclear power issues.

UCS has engaged in a variety of programs targeting a wide range of people. One of the organization's first large-scale programs was the 1971 **Atomic Energy Commission** hearings, in which UCS revealed major weaknesses in nuclear plant system designs. Throughout the 1980s, UCS sponsored an annual Week of Education focusing on the nuclear arms race. The program was targeted at college students throughout the United States. In 1984 the group conducted a study that revealed technical mistakes, high costs, and other flaws in the United States Strategic Defense Initiative (SDI). In 1989 the organization operated a Voter Education Project which resulted in a television special focusing on what Americans expected from their next president. The show was broadcast in both the United States and the Soviet Union. That same year, UCS participated in the International Scientific Symposium on a Nuclear Test Ban. The organization also conducts research for the U.S. **Nuclear Regulatory Commission**.

UCS has also been involved in projects of a more political nature. Most notable are its lobbying and litigation programs which lobby Congress on various issues. The organization also provides expert testimony to congressional committees and participates in legislative coalitions. In 1991 legal action brought by UCS resulted in the closing of the Yankee Rowe Nuclear Power Plant in western Massachusetts, the oldest commercial nuclear power plant in the United States. While the plant closed voluntarily, UCS was instrumental in pressuring the plant to shut down. UCS litigators have also targeted the Kozloduy nuclear reactors in Bulgaria, citing safety concerns. Legislative programs have focused on the United States' B-2 Stealth bomber and SDI. The organization claimed a victory at the end of 1991 when the United States Congress voted to cut back on both the B-2 and SDI projects.

While the core of UCS consists of field experts, the group also maintains a strong volunteer base. UCS is made up of three core groups: the Scientists' Action Network, the Legislative Alert Network, and the Professionals' Coalition for Nuclear Arms Control. The Scientists' Network focuses on public education and lobbying, the Legislative Network informs UCS members of upcoming elections and laws, and the Professionals' Coalition promotes arms control education and legislation. The Scientists' Action Network relies on volunteers to carry out its programs on the local level. Likewise the Legislative Alert Network utilizes volunteer support in letter-writing and phone-call campaigns to members of Congress. The Professionals' Coalition for Nuclear Arms Control, too, works with physicians, lawyers, and design professionals throughout the United States to promote arms control education and legislation.

[*Kristin Palm*]

Resources

Organizations

Union of Concerned Scientists, 2 Brattle Square, Cambridge, MA USA 02238-9105 (617) 547-5552, Fax: (617) 864-9405, Email: ucs@ucsusa.org, <http://www.ucsusa.org>

United Nations Commission on Sustainable Development

A committee of 53 nations created by the United Nations General Assembly for the purpose of implementing recommendations made during the 1992 **United Nations Earth Summit** in Rio de Janeiro. Its overriding goal is to develop economies around the world while preserving the **environment** and existing **natural resources**. Major issues of interest include **water quality**, **desertification**, forest and **species** protection, toxic **chemicals**, and atmospheric and oceanic **pollution**.

Chaired by Malaysian representative Razalie Ismail, the commission endeavors to put into action the specific policies of *Agenda 21*, a United Nations plan aimed at stopping—or at least slowing down—global **environmental degradation**. The United States has taken a strong leadership role in supporting the commission's objectives. Under President Bill Clinton and Vice President Albert Gore, the **President's Council on Sustainable Development** was created to balance environmental policies with sound economic development within this country.

With the formation of the commission, several developing countries have raised concerns that the group will withhold financial aid to them based on their environmental practices. Likewise, the more developed countries (MDC) worry that money and technology allotted to the **less developed countries** (LDC) will go toward other needs rather than towards protecting the environment. The commission has no legal authority to enforce any of its policies but relies on publicity and international pressure to achieve its goals.

Resources

Organizations

Secretariat of the United Nations Commission on Sustainable Development, United Nations Plaza, Room DC2-2220, New York, NY USA 10017 (212) 963-3170, Fax: (212) 963-4260, Email: dsd@un.org, <http://www.un.org/esa/sustdev/csd.htm>

United Nations Conference on Environment and Development

see **United Nations Earth Summit (1992)**

United Nations Conference on the Human Environment (1972)

Held in Stockholm in 1972, the United Nations Conference on the Human Environment was the first global environmental conference and the precursor to the 1992 **United Nations Earth Summit** in Rio de Janeiro, Brazil.

The purpose of the conference was not to discuss scientific or technological approaches to environmental problems but to coordinate international policy. A 27-nation committee held four meetings in the two years preceding the conference, and in the months leading up to it they issued a report calling for "a major reorientation of man's values and redeployment of his energies and resources." In their preliminary report, the committee emphasized their political priorities: "The very nature of environmental prob-

lems—that is to say, their intricate interdependence—is such as to require political choices."

There was a remarkable lack of divisiveness, once the conference began, on most issues under consideration. The Soviet Union and the Eastern bloc did not attend because East Germany, which was not a member of the United Nations, had been denied full representation. But 114 of the 132 member countries of the United Nations were there, and the sessions were distinguished by what the *New York Times* called a "groundswell of unanimity." A number of resolutions passed without a dissenting vote.

Many believe the most important result of the conference was the precedent it set for international cooperation in addressing **environmental degradation**. The nations attending agreed that they shared responsibility for the quality of the environment, particularly the oceans and the **atmosphere**, and they signed a declaration of principles, after extensive negotiations, concerning their obligations. The conference also approved an environmental fund and an "action program," which involved 200 specific recommendations for addressing such problems as global climatic change, **marine pollution**, **population growth**, the dumping of toxic wastes, and the preservation of **biodiversity**. A permanent environmental unit was established for coordinating these and other international efforts on behalf of the environment; the organization that became the United Nations Environmental Programme was formally approved by the General Assembly later that same year and its base established in Nairobi, Kenya. This organization has not only coordinated action but monitored research, collecting and disseminating information, and it has played an ongoing role in international negotiations about environmental issues.

The conference in Stockholm accomplished almost everything the preparatory committee had planned. It was widely considered successful, and many observers were almost euphoric about the extent of agreement. In a speech to the nations gathered in Stockholm, the anthropologist Margaret Mead called the event "a revolution in thought fully comparable to the Copernican revolution by which, four centuries ago, men and women were compelled to revise their whole sense of the earth's place in the cosmos. Today we are challenged to recognize as great a change in our concept of man's place in the biosphere."

There were, however, some dissenting voices. Shirley Temple Black and others formally protested the fact that women were not fully represented at the gathering; only 11 delegations of the 114 nations represented included even one woman. A large "counterconference" was held in Stockholm, consisting of a number of scientific and political organizations, and environmentalists such as **Barry Commoner** argued that the official conference, though valuable, had failed to address the subjects that were most important to solving the current environmental crisis, particularly poverty and what he called "ecologically sound ways of producing goods."

[*Douglas Smith*]

United Nations Division for Sustainable Development

This United Nations division serves as the secretariat, or the research staff, of the UN Commission on **Sustainable Development**, whose representatives meet periodically to develop policies for sustainable development in 40 countries. The division comprises 30 program officers working in specific sectors: health, human settlements, freshwater resources, toxic **chemicals**, hazardous wastes, solid wastes, and radioactive wastes. According to André Vasilyev, First Officer, it prepares studies and analytical reports and conducts research to support the work of the Commission. It also measures the progress of and makes recommendations for development programs. The division's stated mission is to contribute to sustainable development worldwide by working to implement Agenda 21, the Rio Declaration on Environment and Development, the Forest Principles (a United Nations statement of principles for a global consensus on the management, **conservation**, and sustainable development of all types of forests), and the Global Programme of Action for Sustainable Development of Small Island Developing States (SIDS). The division carries out multi-year work programs on sustainable development indicators, changing consumption and production patterns, and the transfer of environmentally sustainable technology.

The Commission was established in 1992 as an adjunct to Agenda 21, the plan of action for environmental sustainability adopted at the United Nations Conference on Environment and Development (UNCED) in Rio de Janeiro. Its charge is to ensure that decisions made at UNCED are carried out at national, regional and international levels, and to further international cooperation and decision-making on these issues.

The full Commission meets once a year. Ad hoc working groups meet periodically to address specific issues, including trade and environmental development, consumption patterns, financial resources, and technology transfer. Open-ended working groups address such issues as integrated management of land resources, forests, combating **desertification**, sustainable mountain development, **sustainable agriculture**, rural development, and biological diversity.

In recent years, the Commission has examined numerous strategies to promote sustainable development: trading debt for sustainable development, creating environmental

user charges for air transport, implementing the Convention on Biological Diversity and the Convention to Combat Desertification, and developing tools for integrated land management. In the spring of 1997, the full commission met to conduct a five-year review of the results of UNCED.

[*Carol Steinfeld*]

Resources

Organizations

Division for Sustainable Development/DESA, United Nations Plaza, Room DC2-2220, New York , NY USA 10017 (212) 963-3170, Fax: (212) 963-4260, Email: dsd@un.org, <http://www.un.org/esa/sustdev/dsdgen.htm>

United Nations Earth Summit (1992)

For 12 days in June 1992, more than 35,000 environmental activists, politicians, and business representatives, along with 9,000 journalists, 25,000 troops, and uncounted vendors, taxi drivers, and assorted others converged on Rio de Janeiro, Brazil for the United Nations Conference on **Environment** and Development. Known as the Earth Summit, this was the largest environmental conference in history; in fact, it was probably the largest non-religious meeting ever held. Like a three-ring environmental circus, this conference brought together everyone from pin-striped diplomats to activists in bluejeans and indigenous Amazonian people in full ceremonial regalia. One cannot say yet whether the conventions and treaties discussed at this summit meeting will be effective, but they have the potential to make this the most important environmental meeting ever held.

The first **United Nations Conference on the Human Environment** met in Stockholm in 1972, exactly 20 years before the Rio de Janeiro meeting. Called by the industrialized nations of Western Europe primarily to discuss their worries about transboundary **air pollution**, the Stockholm conference had little input from **less developed countries** and almost no representation from non-governmental organizations (NGOs). Some major accomplishments came out of this conference, however, including the **United Nations Environment Programme**, the Global **Environmental Monitoring** System, the Convention on International Trade in **Endangered Species** (CITES), and the World Heritage Biosphere Reserve Program, which identifies particularly valuable areas of biological diversity. A companion book to the Stockholm Conference entitled *Only One Earth* was written by René Dubos and Barbara Ward.

In 1983 the United Nations established an independent commission to address the issues raised at the Stockholm conference and to propose new strategies for global environmental protection. Chaired by Norwegian Prime Minister **Gro Harlem Brundtland**, the commission spent four years in hearings and deliberations. A significantly greater voice for the developing world was heard as it became apparent that environmental problems affected the poor more than the rich. The commission's final report, published in 1987 as ***Our Common Future***, is notable for coining the term "**sustainable development**" and for linking environmental problems to social and economic systems.

In 1990 preparations for the Earth Summit began. Maurice Strong, the Canadian environmentalist who chaired the Stockholm conference, was chosen to lead once again. A series of four working meetings called PrepComs were scheduled to work out detailed agendas and agreements to be ratified in Rio de Janeiro. The first PrepCom met in Nairobi, Kenya, in August 1990. The second and third meetings were in Geneva, Switzerland, in March and August 1991. The fourth and final PrepCom convened in New York City, in March 1992. Twenty-one issues were negotiated at these conferences including **biodiversity**, **climate** change, **deforestation**, **environmental health**, marine resources, **ozone**, poverty, toxic wastes, and urban environments. Notably, population crisis was barely mentioned in the documents because of opposition by religious groups.

Intense lobbying and jockeying for power marked the two-year PrepCom process. As the date for the Rio de Janeiro conference neared, it appeared that several significant treaties would be ratified in time for presentation to the world community. Among these were treaties on climate change, biological diversity, forests, and a general **Earth Charter**, which would be an environmental bill of rights for all people. A comprehensive 400-page document called *Agenda 21* presented a practical action plan spelling out policies, laws, institutional arrangements, and financing to carry out the provisions of these and other treaties and conventions. Chairman Strong estimated that it would take $125 billion per year in aid to help the poorer nations of the world protect their environment.

In the end, however, the United States refused to accept much of the PrepCom work. During PrepCom IV in New York City, for instance, 139 nations voted for mandatory stabilization of **greenhouse gases** at 1990 levels by the year 2000, laying the groundwork for what promised to be the showcase treaty of the Earth Summit. Only the United States delegation opposed it, but after behind-the-scenes arm twisting and deal-making, the targets and compulsory aspects of the treaty were stripped away, leaving only a weak shell to take to Rio de Janeiro. Many environmentalists felt betrayed. Similarly, the United States, alone among the industrialized world, refused to sign the biodiversity treaty, the forest protection convention, or the promise to donate 0.7% of Gross Domestic Product to less developed countries for environmental protection. The nation's excuse was that

these treaties were too restrictive for American businesses and might damage the American economy. The United States did, subsequently, sign the biodiversity treaty on June 4, 1993.

Many environmentalists went to Rio de Janeiro intending to denounce United States intransigence. Even **Environmental Protection Agency** chief William Reilly, head of the United States delegation, wrote a critical memo to his staff saying that the Bush administration was slow to engage crucial issues, late in assembling a delegation, and unwilling to devote sufficient resources to the meeting. *Newsweek* magazine entitled one article about the summit: "The Grinch of Rio," saying that to much of the world, the Bush administration represented the major obstacle to environmental protection.

Not all was lost at the Rio de Janeiro meeting, however. Important contacts were made and direct negotiations begun between delegates from many countries. Great strides were made in connecting poverty to environmental destruction. Issues such as **sustainable development** and justice had a prominent place on the negotiating table for the first time. Furthermore, this meeting provided a unique forum for discussing the disparity between the rich industrialized northern nations and the poor, underdeveloped southern nations. Many bilateral—nation to nation—treaties and understandings were reached.

Perhaps more important than the official events at the remote and heavily guarded conference center was the "shadow assembly," or **Global Forum** of NGOs, held in Flamingo Park along the beach front of Guanabara Bay. Eighteen hours a day, the park pulsed and buzzed as thousands of activists debated, protested, traded information, and built informal networks. In one tent, a large TV screen tracked nearly three dozen complex agreements being negotiated by official delegates at Rio Centrum. In other tents, mini-summits discussed alternative issues such as the role of women, youth, **indigenous peoples**, workers, and the poor. Specialized meetings focused on topics ranging from sustainable energy to endangered **species**. In contrast to Stockholm, where only a handful of citizen groups attended the meetings and almost all were from the developed world, more than 9,000 NGOs sent delegates to Rio de Janeiro. There were over 700 from Brazil alone. The contacts made in these informal meetings may prove to be the most valuable part of the Earth Summit.

[*William P. Cunningham Ph.D.*]

Resources

Periodicals

Haas, P. M., et al. "Appraising the Earth Summit: How Should We Judge UNCED's Success?" *Environment* 34 (October 1992): 6.

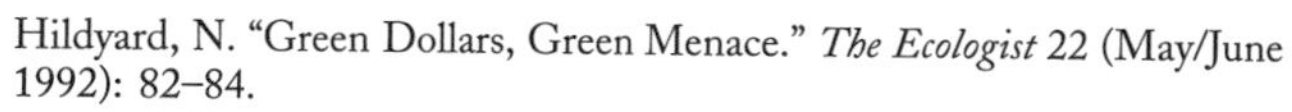
Hildyard, N. "Green Dollars, Green Menace." *The Ecologist* 22 (May/June 1992): 82–84.

Hinrichsen, D. "The Rocky Road to Rio." *The Amicus Journal* 14 (Winter 1992): 15.

Other

French, H. *After the Earth Summit: The Future of Environmental Governance.* World Watch Paper 107. Washington, DC: World Watch Institute, 1992.

United Nations Environment Programme

Formed in 1973 after the **United Nations Conference on the Human Environment**, the UN Environment Programme (UNEP) coordinates environmental policies of various nations, nongovernmental organizations, and other UN agencies in an effort to protect the environment from further degradation. Sometimes dubbed the "environmental conscience of the UN," UNEP describes its goal as follows: "to protect the environment by distributing education materials and by serving as a coordinator and catalyst of environmental initiatives."

Some of UNEP's major areas of concern are the **ozone** layer, waste disposal, toxic substances, water and **air pollution**, **deforestation**, **desertification**, and energy resources. The ways in which UNEP addresses these concerns can be best seen in the various programs it oversees or administers. The **Earthwatch** program—which includes Global **Environmental Monitoring** System (GEMS); **INFOTERRA**; and **International Register of Potentially Toxic Chemicals** (IRPTC)—monitors environmental problems worldwide and shares its information database with interested individuals and groups. UNEP's Regional Seas Program seeks to protect the life and integrity of the Caribbean, Mediterranean, South Pacific, and the Persian Gulf. The **International Geosphere-Biosphere Programme** (IGBP) promotes awareness and preservation of biogeochemical cycles. The Industry and Environment Office (IEO) advises industries on environmental issues such as **pollution** standards, industrial hazards, and clean technology.

In addition to administering these programs, UNEP has also sponsored several key environmental policies in the last few decades. For example they initiated the **Convention on International Trade in Endangered Species of Wild Fauna and Flora** (CITES), **Convention on the Conservation of Migratory Species of Wild Animals**, and the Vienna Convention for the Protection of the Ozone Layer. Most recently, UNEP was responsible for planning the 1992 **United Nations Earth Summit** in Rio de Janeiro, Brazil.

UNEP is headquartered in Nairobi, Kenya. The organization is headed by Mostafa Kamal Tolba, who assumed the position of Executive Director in 1976 after the departure of Maurice Strong.

Resources
Organizations

United Nations Environment Programme, United Nations Avenue, Gigiri, PO Box 30552, Nairobi, Kenya (254-2) 621234, Fax: (254-2) 624489/90, Email: eisinfo@unep.org, <http://www.unep.org>

Upwellings

Upwellings are highly productive areas along the edges of continents or continental shelves where waters are drawn up from the ocean depths to the surface. Rich in nutrients, these waters nourish algae, which in turn support an abundance of fish and other aquatic life. The most common locations for upwellings are the western edge of continents, such as Peru and southern California in the Pacific, northern and southwestern Africa in the Atlantic, and around **Antarctica**.

In order for upwellings to occur, there must be deep currents flowing close to the continental margin. There must also be prevailing winds that push the surface waters away from the coast—as the surface waters move offshore, the cold, nutrient-rich bottom waters move up to replace them. The Pacific upwelling near Peru is the result of the northward flow of the cold Humboldt current and the prevailing offshore wind pattern. Every seven to ten years, there is a shift in the prevailing wind pattern off the shore of Peru, a condition known as **El Niño**. The prevailing wind in the eastern Pacific shifts direction, causing the cold Humboldt current to be replaced by warm equatorial waters. This prevents upwelling, and the **ecological productivity** of the ocean in this area falls dramatically.

In oceans, the majority of nutrients sink to the sea bed, leaving the surface waters poor in nutrients. In marine ecosystems, the primary producers are microscopic planktonic algae known as **phytoplankton**, and these are only found in surface waters, where there is enough sunlight for **photosynthesis**. The lack of nutrients limits their growth and thus the productivity of the entire system, and most open oceans have the same low levels of ecological productivity as **tundra** or **desert** ecosystems. In upwellings, the water is cold and oxygen poor but it carries a tremendous amount of **nitrates** and **phosphates**, which fertilize the phytoplankton in the surface waters. These producers increase dramatically, showing high levels of net **primary productivity**, and they become the foundation of a biologically rich and varied oceanic food web, teeming with fish and bird life.

The high productivity of the upwelling area off Peru supports one of the richest sardine and anchovy fisheries in the world. In addition, the large population of seabirds associated with the Peruvian upwelling deposit equally large quantities of phosphate and **nitrogen** rich droppings called **guano** on rocky islands and the mainland. Until oil-based fertilizers were developed, Peruvian guano was the principle source for agricultural fertilizers.

In the Antarctic upwelling zone, the nutrient-rich waters support food webs at the base of which are massive populations of large marine shrimp, called **krill**. These invertebrates support a wide array of squid, **whales**, **seals**, penguins, cormorants, boobies, and sea ducks. Commercial fisheries in the Antarctic upwelling zone are now exploiting the krill harvest, an act which endangers all of the members of this unique and complex cold-water food web. *See also* Commercial fishing

[*Neil Cumberlidge Ph.D.*]

Resources
Books

Ainley, D. G., and R. J. Boekelheide. *Seabirds of the Farallon Islands: Ecology, Dynamics, and Structure of an Upwelling-System Community.* Stanford: Stanford University Press, 1990.

Glantz, M. H., and J. D. Thompson. *Resource Management and Environmental Uncertainty: Lessons from Coastal Upwelling Fisheries.* New York: Wiley, 1981.

Uranium

Uranium is a dense, silvery metallic element with the highest atomic mass of any naturally occurring element. It is the forty-seventh most abundant element in the earth's crust. All of the isotopes of uranium are radioactive, which accounts for a portion of the **background radiation** that is a natural part of the **environment**. In the 1930s, scientists discovered that one **isotope** of uranium, uranium-235, could be fissioned, or split. This information brought about a revolution in human society. The potential for **nuclear fission** is demonstrated by the fact that it has made possible not only the world's most destructive weapons, but also the generation of energy in a new way. Regardless of the purpose for which it is used, nuclear fission has created severe waste disposal problems which the world is struggling to solve. *See also* High-level radioactive waste; Nuclear power; Nuclear weapons; Radioactive waste; Radioactivity

Urban contamination

In urban areas where there has been extensive reshaping of the land surface, disturbance from building and road construction, land-filling, and additions of a variety of **chemicals** and minerals, there are many sources, causes, and types of urban contamination. Urban contamination is of concern from two perspectives: when new land is developed that was previously under agriculture or other rural **land use**, and when the central core industrial areas are redeveloped.

Sources of contamination can be summarized in three very broad categories: agricultural, urban, and industrial.

Agricultural-sourced contamination affects urban areas in several ways. Use of some pesticides in agricultural or gardening situations has increased by significant amounts the levels of **copper**, **arsenic**, and **mercury** in soils. Long-term use of **phosphorus** fertilizers can lead to a buildup of cadmium—a contaminant in many fertilizers—in soils. Synthetic organic pesticides used in agriculture can also contribute to contamination. Organochlorine insecticides are persistent and accumulate in food chains. Examples of these types of compounds are DDT, aldrin, and dieldrin. Use of these insecticides has been discontinued since their negative impacts on the **environment** have been well documented. However, these compounds do not readily decompose and are very persistent in soils and **sediment**. Former agricultural areas that have become urbanized have residues of these chemicals and others that result in a risk for contamination of water supplies as well as soils. This is especially true for very localized areas where pesticides were stored, spilled, or that were regularly used to clean or load spray equipment.

Urban activities past and present also contribute to contamination. The most common and largest volume of contamination sources comes from construction or demolition debris. Much of the debris is made up of nonhazardous materials; however, some of the debris may contain **asbestos**, high levels of trace metals, or other problem-causing substances. Asbestos is of particular concern because it can cause **cancer** in humans when ingested through air or water.

Long-term use of lead-based paints and **gasoline** has led to considerable contamination. In 1976, the amount of lead used in gasoline in the United States was approximately 210 million tons (190.5 million metric tons). This has dropped to less than 50 million tons (45.4 metric tons) per year since the use of unleaded gasoline has replaced leaded fuel. Air emissions and deposition of lead from **automobile** exhaust near roads and highways have contaminated numerous urban areas.

Disposal of domestic wastes containing trace metals and other hazardous materials by land application or in landfills are also large sources of contamination. Smokestacks from **incineration** plants for wastes often contain high levels of trace metals. Polyaromatic **hydrocarbons** (PAHs) have been deposited in urban areas as the result of **combustion** of **fossil fuels** (**coal**, peat, etc.) for generating electrical power. Other urban activities that can lead to contamination include land application of soot, coal ash, sewage **sludge**, and corrosion from metals and buildings.

In addition to agricultural and general urban sources, industrial activities such as mining and smelting contribute significantly to localized **soil** metal contamination. Atmospheric transport of the emissions can contaminate surrounding areas. Many industries have contributed a number of different kinds of toxic materials to the urban environment and as those industries close or move away they have left the areas contaminated, making it difficult to use the sites for other purposes without expensive clean-up. Examples of toxic materials that are found in former industrial sites include asbestos, chromium, arsenic, and boron. Asbestos is found most often in areas that were formerly railroad lands, shipyards, or asbestos factories. Leather tanning industries or areas used for battery production leave residues of chromium and arsenic.

Locations near chemical manufacturing companies or storage and distribution centers are also places where there is considerable contamination. One of the extreme examples where such a site was redeveloped after which major problems arose is **Love Canal** in the state of New York. Up to 248 different industrial chemicals were deposited and buried at the site before 1953. People living there following redevelopment were evacuated after many serious health problems were reported and confirmed.

Many toxic solvents are used in chemical, paint, and dye industries, and frequently these solvents can be found in soils of former industrial sites. In the Silicon Valley of California, solvent leakages into soils from computer chip manufacturing have resulted in both soil and **groundwater** contamination. Trichloroethylene and chloroethylene are common solvents used in paper making, metal plating, electrical engineering, laundries, dry cleaners, and as degreasing agents. These solvents have caused not only soil contamination, but also groundwater contamination in numerous locations. These sites are very difficult and expensive to clean up after the industries leave.

Contamination by power stations arises from pulverized fuel or **fly ash**, which is alkaline (**pH** 10-12) and contains soluble salts, in particular boron, **aluminum**, copper, and arsenic. The ash results from burning coal and is extracted from the chimneys to prevent **air pollution**. Its extremely fine particle size makes it difficult to handle and dispose. Demolition debris from these sites often contains asbestos residues, trace metals, combustible materials, oil spillages, and **polychlorinated biphenyls** (PCBs).

This urban contamination is a major problem worldwide in industrialized countries. In the Netherlands alone there have been confirmed at least 10,000 contaminated sites, with 50% of them in residential areas, with total clean up costs estimated to be $25 billion. Western Germany has identified over 50,000 potentially contaminated sites. The **Environmental Protection Agency** (EPA) in the United States has identified 32,000 potentially **hazardous waste** sites, with an estimated clean-up cost of $750 billion. There can be large impacts on the growth of plants and on the health of humans and animals that come in contact with

the contamination at the site through either the air, water, or soil. There are three main pathways by which the contaminants can enter the plants: direct uptake from soil, vapor uptake through the air, and **sorption** on plant surfaces. For organic types of contaminants the predominant pathways are through the roots or air vapors. In general, organic contaminants (including pesticides) can affect plant growth resulting in everything from depressed growth and reproduction rates to death.

The ability of the plant to take up metals and transmit them to animals or humans that eat the plants varies by plant and by metal. Since there are likely to be high concentrations of lead either from **petroleum** products or paint in contaminated sites, there is concern for backyard vegetable gardens or urban gardens where redevelopment has taken place. There are similar concerns for other metals such as zinc and **cadmium**. These metals are usually associated with electroplating industries or battery production and where municipal waste sludges have been land applied. High concentrations of trace metals can cause visible injury to plants and inhibit their growth.

The high concentrations of trace metals in many urban soils in inner-city and highly industrialized areas gives rise to concerns about potential human health effects. Lead is a trace metal of particular concern, and two predominant pathways for the uptake of lead have been identified in humans: soil and dust ingestion, and consumption of home grown foods. Depending on the extent of exposure, individuals may develop symptoms of acute or chronic lead **poisoning**. Symptoms or results of exposure include severe **anemia**, acute nervousness, kidney damage, irreversible brain damage, or death. Preschool children and women of child-bearing age are especially at risk. It has also been reported that people exposed to lead and cadmium in drinking water suffered from increased chronic kidney disease, skin cancer, heart disease, and anemia.

It is estimated that greater than 60% of human cancer is caused by contacting environmental chemicals containing PAHs. However, these compounds can be ingested through contaminated drinking water, but dietary intake of plant-based food is the largest single route of exposure.

Urban development and various industrial activities have resulted in extensive contamination in most industrial countries. When urban redevelopment or expansion is considered, the potential impacts of this contamination on plants, animals, and humans needs to be evaluated. Before the development occurs, contaminated sites need to be cleaned up, which can be very expensive. Urban populations that live in or near these sites already may be exposed to a wide range of contaminants that can have both acute and chronic health effects. As we push to revitalize our cities, this is an issue that will become increasingly important.

[*James L. Anderson*]

Resources

Books

Air Pollution, People and Plants. St. Paul, MN: Sagar Krupa, The American Phytopathological Society, 1977.

Berrow, M. L. "An Overview of Soil Contamination Problems." In *Chemicals in the Environment*, edited by J. M. Lester. Lisbon: 1986.

Thornton, I. "Metal Contamination of Soils in Urban Areas." In *Soils in the Urban Environment*, edited by P. Bullock, and P.J. Gregory. Boston: Blackwell Scientific Publications, 1991.

Urban design and planning

Urban design can be expressed as the three-dimensional relationship between buildings, open spaces, and streets within a given area. This visual expression of urban zones (combined with a specific plan that details where and what kind of development should occur) can determine the character of the community.

Currently, prevalent thinking on the subject of urban design is that new communities should be compact, mixed in both housing and commercial development, affordable, and pedestrian—a move back to living in compact, relatively self-sufficient communities and neighborhoods. It is a move back to a traditional community design in which communities and neighborhoods have affordable housing, jobs, and retail markets within easy walking distance for its residents.

This concept from the past is being promoted as the urban design and planning wave of the future. It addresses some of the ills that were brought about by the rapid suburban expansion after World War II: segregation of housing from workplace and retail areas, high housing costs, traffic congestion, long commutes, decay of community life, and negative environmental consequences (including **air pollution** and **water pollution**, inefficient use of energy and materials, etc.).

Proponents suggest that the move back to traditional urban design is good not only from the standpoint of promoting lost community values, but also positive for the **environment**. In fact, they submit that the environmental factors (such as the desire for clean air) will move the United States toward adopting this "new" approach. As the country addresses the issues of environmental quality, problems such as traffic congestion and affordable housing will become increasingly important to community design and planning, and will aid in the continuing **evolution** of urban design.

Urban ecology

Urban **ecology**, simply put, is the study of life in the city. In the late nineteenth century, biologists and zoologists created the field of ecology, seeking to understand the complex

relationships between organisms and their **environment**. The term *ecosystem* started to be used to refer to a community of organisms and its associated environment when functioning as an ecological unit. Since then, the study of ecosystems has largely been associated with scientific descriptions of pristine locales, typically in remote regions where human presence was minimal to non-existent. In contrast, urban ecology is the study of urban ecosystems. In general, urban **biodiversity** is heavily dominated by alien **species**. There are exceptions, for instance suburban communities that retain parts of the pre-existing natural **habitat**, but overall, urban ecosystems are fundamentally **anthropogenic**, meaning that they are man-made.

Evidence of human action on the environment goes back thousands of years. Human hunters in North America played a role in the **extinction** of large mammals as far back as 12,000–5,000 years ago. The development of agricultural societies led to side effects including **soil erosion**, disease, and **deforestation** before modern times. In some ancient cities, ecological degradation was severe, and ruined cities and cultures have left behind ample evidence for modern day researchers. Over thousands of years, humanity has played a major role in changing the ecosystems in which they have lived, and in modern times this process has become prolific and widespread.

In the new millenium, ecologists have little doubt that humanity's collective will is the dominant force in the welfare and outcome of the global environment. With nearly half of the world's population living in urban areas, and almost 80% of people in developed nations living in cities, mankind consumes a vast amount of resources, most of which comes from ecosystems outside the city. Large amounts of renewable and non-renewable resources are being transported from outlying ecosystems into urban ecosystems, while huge quantities of human produced pollutants are being created by urban ecosystems and spread across the spectrum of ecosystems. For this reason, urban ecologists stress the importance of understanding, in ecological terms, how the activities of people affect the total ecological spectrum from pristine **wilderness** to urban areas.

Ecologists define an urban **ecosystem** as a dependent ecosystem, meaning that it depends on other ecosystems and outside energy and resources to function. On the other hand, a natural ecosystem generally has an even balance between its energetic inputs and outputs. William Rees, a researcher at the University of British Columbia, created an analytical tool to measure just how dependent the urban ecosystem is. Called the "ecological footprint," it roughly measures how much land is required to maintain a city's activities. Rees found that Vancouver, British Columbia, for instance, had a footprint 180 times its own size in 1996, meaning that much more land, in terms of extractable resources, is required to support its inhabitants.

Another aspect of the urban process being studied is the urban fringe. The fringe is the expansion point where human activity, building, and structure meet and alter other existing ecological or natural settings. Also known as **urban sprawl**, this process, according to *Bioscience* magazine, chewed up an estimated 13 million acres (5.3 million ha) of land in the United States between 1970 and 1980, and another 9 million acres (3.6 million ha) in the 1980's. Additionally, development of farmland and forest during the economically booming 1990's reached a record pace in America, accelerating to a rate of 3.2 million acres (1.3 million ha) per year by some estimates.

This rapid development has profoundly changed the ecosystems involved and a thriving branch of urban ecology studies how patterns of urban development alter the ecological composition and organization of species. For example, the change in predator and prey relationships caused by human activity has altered the **wildlife** population in several habitats. Some animals have become more familiar and comfortable with the urban environment, and human-to-animal encounters have increased. Deer have been seen walking casually through suburban neighborhoods across the United States, feeding on saplings and freshly planted gardens, and traffic incidents involving them have increased. In Washington, D.C. deer have been seen a short distance from the White House. A **coyote** was found in an elevator in a Seattle federal building. Many coyote, whose numbers have doubled since 1850, have been frequently sighted looking for food in suburbs and urban areas. Another result of human interference with natural habitats is the changing predation habits among certain animals. For example, in California, foxes have begun killing rare bird species. Concern among environmentalists and ecologists continues to grow as the phenomenon of urban growth and development increases. Other fields of study include the mechanisms of **carbon** storage in the urban forest and cooling of the **atmosphere** by urban **evapotranspiration**.

Scientists understand that traditional ecological theory, developed without humans in mind, is not sufficient for properly understanding urban systems. Instead of considering humans as being outside or separate from the natural world, ecologists include humans as a factor affecting the natural world's **evolution**. Together with an existent natural landscape, the forces of human policies and economics have the greatest impact in how an urban ecosystem is created and maintained, as well as how plant and wildlife fare. When looking at cities, ecologists might include data such as species diversity, population sizes, and **energy flow**. Urban ecologists would further include qualities having to do with human perceptions and institutions, including cultural resources,

educational opportunities, **recreation**, wealth, city design and aesthetics, and community health. Features of the urban ecosytem include, for instance, money (a property of energy in the urban ecosystem), downtown high-rises, social institutions, industrial developments, commercial shopping developments, residential developments of all cultural and economic types, open space and recreational parks, **automobile** traffic, pedestrian traffic, **air pollution** and toxic contaminants, existing natural landforms, and adaptive plants and wildlife. The large number of factors influencing the urban environment shows the need for the inclusion of other sciences in order to understand the urban area's complexity.

In the United States, the National Science Foundation's Long-Term Ecological Research (LTER) project has the goal of creating a full-scale, interdisciplinary urban ecology research program that addresses the field's complex needs. The project, begun in the 1990's, is using two sites for research, in Phoenix, Arizona and Baltimore, Maryland. Interdisciplinary teams of the project's scientists are working to better understand the urban environment. This study provides an example of how an urban ecology analysis is conducted. First, researchers focus on the urban **topography**, and more precisely analyze a **watershed** (defined as an area that is being drained, or feeding a body of water) as a centerpoint. Scientists then begin to link up the ecological relationships of vegetation and wildlife communities. After examining the ecological relationships, the next step is to understand the human socio-cultural characteristics of the watershed. Different areas along watersheds have different land values. Urban ecologists study the relationship between the natural features of the watershed and how humans interact, according to different sociological characteristics such as age, gender, ethnic identity, and social status. Significant patterns can often be seen between social factors and a particular urban area's biophysical (natural) characteristics. For instance, studies have found that wealthy city dwellers tend to live in areas that are environmentally healthier and less polluted than those lived in by poorer city dwellers. The LTER project is comprehensively addressing issues like this, using a variety of scientific methods.

The science of urban ecology is also part of a growing awareness that the health of the overall environment depends upon the health of cities. In addition to the scientific study of ecology, this awareness is also associated with the more activist areas of "ecological urbanism" and "new urbanism," as well as "green living" and "sustainable development" trends in urban planning. The aims of ecological urbanism and new urbanism are to create justice and equity in social and **environmental policy**. These movements also strive to reform urban landscapes and urban planning methods to create sustainable, healthy cities by reducing pollutants and waste, while increasing **conservation** and the accessibility of shared resources, to insure humanity's survival and quality of life. Urban ecology contributes the resources of knowledge and information to these social movements, and to those who seek a deeper understanding of the relationship between humanity and the natural world, helping them see their city as part of a living ecosystem with valuable resources that promote better health and quality of life.

[*Douglas Dupler*]

Resources

Books

Beatley, Timothy, and Kristy Manning. *The Ecology of Place: Planning for Environment, Economy, and Community.* Washington, DC: Island Press, 1997.

Breuste, Jurgen, Hildegard Feldman, and Ogarit Uhlman. *Urban Ecology.* Berlin, New York: Springer-Verlag, 1998.

Bridgman, Howard A., Robin Warner, and John Dodson. *Urban Biophysical Environments.* New York: Oxford University Press, 1995.

Matthews, Anne. *Wild Nights: Nature Returns to the City.* New York: North Point Press, 2001.

Rocky Mountain Institute. *Green Development: Integrating Ecology and Real Estate.* New York: John Wiley & Sons, 1998.

Other

The Baltimore Ecosystem Project. [cited July 2002]. <http://www.ecostudies.org/bes>.

Organizations

The Center for Urban Ecology, 4598 MacArthur Blvd., NW., Washington, DC USA 20007-4227 202-342-1443, Fax: 202-282-1031, <http://www.nps.gov/cue/cueintro.html>

Urban Ecology, 414 13th Street, Suite 500, Oakland, CA 94612 510-251-6330, Fax: 510-251-2117, Email: urbanecology@urbanecology.org, <http://www.urbanecology.org/>

Urban heat island

An urban area where the temperature is commonly 5–10°F (-15–-12°C) warmer than the surrounding countryside. Effects include reduced heating but higher air conditioning bills. Within cities, this warmth causes spring to arrive one to two weeks earlier and fall foliage to appear one to two weeks later. The chief cause is the high energy use within cities, which ultimately ends up in the **atmosphere** as waste heat. Also important, however, are the darker surfaces within cities, which absorb more heat, the relative absence of vegetative cooling through **transpiration** because of fewer trees, and less wind because of building obstructions.

Urban runoff

Urbanization causes fundamental changes in the local **hydrologic cycle**, mainly increased speed of water movement through the system, and degraded **water quality**. They are

expressed through reduced **groundwater** recharge, faster and higher **storm runoff**, and factors that affect aquatic ecosystems, particularly **sediment**, **dissolved solids**, and temperature. The resultant problems have encouraged municipalities to reduce negative impacts through storm water management.

Important research on these issues was spearheaded by the U.S. **Geological Survey** (USGS) during the early phases of the current environmental revolution. A sampling of titles used for the USGS Circular 601 series indicates the scope of these efforts: *Urban Sprawl and Flooding in Southern California* (C601-B); *Flood Hazard Mapping in Metropolitan Chicago* (C601-C); *Water as an Urban Resource and Nuisance* (C601-D); *Sediment Problems in Urban Areas* (C601-E); and *Extent and Development of Urban Flood Plains* (C601-J). Also relevant are *Washington D.C.'s Vanishing Springs and Waterways* (C752), and *Urbanization and Its Effects on the Temperature of the Streams on Long Island, New York* (USGS Professional Paper 627-D).

Most of these works cite a 1968 publication written by Luna Leopold, *Hydrology for Urban Land Planning*. Describing a research frontier, Leopold anticipated many of the concerns currently embodied in stormwater management efforts. He identified four separate but interrelated effects of **land use** changes associated with urban **runoff**: 1) changes in peak flow characteristics, 2) changes in total runoff, 3) changes in water quality, and 4) aesthetic or amenities issues.

Urbanization transforms the physical **environment** in a number of ways that affect runoff. Initially construction strips off the vegetation cover, which results in significantly increased **erosion**. Local comparisons with farms and woodlands show that sediment from construction and highway sites may increase 20,000–40,000 times. Furthermore, as slope angles steepen, the erosion rate increases even faster; and above a 10% slope (10 ft [3 m] rise in 100 ft [30 m]) no restraints remain to hold back this sediment. On a larger scale, a Maryland study comparing relatively unurbanized and urbanized basins found a fourfold erosion increase.

Pavement and rooftops cover many **permeable** areas so that runoff occurs at a greater rate. Storm sewers are built which by their very nature speed up runoff. Estimates range from a two- to six-fold increase in runoff amounts from fully urbanized areas. Even more important, however, is that peak **discharge** (the key element in **flooding**) is higher and comes more quickly. This makes flash floods more likely, and increases the frequency of runoff events that exceed bank capacity.

This increased flooding causes much channel erosion and altered geometry, as the channels struggle to reach a new equilibrium. Less water infiltrates into groundwater reservoirs, which diminishes the base flow (the **seepage** of groundwater into humid-region channels) and causes springs to dry up. The **urban heat island** effect, combined with reduced baseflow, exposes aquatic ecosystems to higher heat stress and reduced flow in summer, and to colder temperatures in winter. Summer is especially difficult, as rainwater flows from hot streets into channels exposed to direct sunlight. As a consequence, sustaining life becomes more difficult, and sensitive organisms must adapt or die.

Water quality changes increase **cultural eutrophication** (**water pollution** caused by excessive plant nutrients), as **fertilizer** and **pesticide** residues wash from lawns and gardens to join the oil, **rubber**, pet manure, brake lining dust, and other degrading elements comprising urban **drainage**. When these are combined with the **effluent** from industrial waste and treated sewage, especially **heavy metals**, the riverine environment becomes a health concern.

Under **arid** conditions, the damage from mudflows may be increased because the land that once absorbed the sliding mud is now covered with streets and buildings. This situation can be seen in southern California, where hillside development has increased the threat of landslides.

Increasing attention is now being given to stormwater management around the United States. For example, Seattle, Washington has focused on water quality, whereas Tulsa, Oklahoma is primarily concerned with flooding. Both cities have responded to serious local problems.

By the 1950s, Seattle's **Lake Washington** and the adjacent water of Puget Sound had become so polluted that beaches had to be closed. **Sewage treatment** plants around Lake Washington and raw sewage dumping in Puget Sound were the fundamental causes. The residents of the area voted to form a regional water-quality authority to deal with the problems. As a result, all sewage treatment was removed from Lake Washington, and dumping of raw sewage into Puget Sound was halted. The environmental result was dramatic. As nitrate and phosphate levels dropped sharply, cultural eutrophication in these waters was greatly reduced, and beaches were reopened. The regional agency continues to seek better ways to improve the water quality of stormwater runoff.

Tulsa's efforts are a response to a series of devastating floods. Between 1970 and 1985, Tulsa led the country in numbers of federal flood declarations; these floods caused 17 deaths and $300 million in damages. Conditions in Tulsa are such that 6.3 in (16 cm) of rain within six hours is sufficient to produce what hydrologists call a 100-year flood. In May 1984, 10 in (25 cm) of rain fell during a seven-hour period, which was a rainfall frequency of about 200 years.

A leader in flood control, Tulsa has taken a multifaceted approach to stormwater management. During the 1970s, then Congressman James R. Jones sponsored legislation to buy out and tear down houses in the severely flood-

prone Mingo Creek Basin; some had qualified seven times for flood disaster aid, with payments far exceeding market value. In a joint venture, Tulsa and the **Army Corps of Engineers** have spent $60 million and $100 million respectively to channelize Mingo Creek in its lower reaches and create detention ponds in the middle portions of the basin. Expected benefits include a $26.9 million reduction in annual flood damage. Continuing flood control efforts have incorporated the use of a state-of-the-art computerized flood warning system.

Recreational possibilities are being exploited in Tulsa's overall stormwater management program. Access roads are being used as running tracks; athletic fields double as detention reservoirs during flood conditions. Since existing parks have been modified to meet the flood program needs, the facilities have benefitted from the improvements, and the water authority leaves the maintenance to the parks department. Using existing parks also eliminates the need to buy out houses and disrupt existing neighborhoods.

Although the Tulsa plan has been successful, more heavily developed floodplains require different measures. Often land costs are so high that it may be cheaper to spend money on flood-proofing or tearing down existing buildings. In any event, urban runoff is an expensive problem to solve. *See also* Heavy metals and heavy metal poisoning; Industrial waste treatment; Land-use control; Water treatment

[*Nathan H. Meleen*]

Resources
Books

Howard, A. D., and I. Remson. *Geology in Environmental Planning.* New York: McGraw-Hill, 1978.

Mrowka, J. P. "Man's Impact on Stream Regimen and Quality." In *Perspectives on Environment,* edited by I. R. Manners and M. W. Mikesell. Washington, DC: Association of American Geographers, 1974.

Other

Leopold, L. B. *Hydrology for Urban Land Planning: A Guidebook on the Hydrologic Effects of Urban Land Use.* USGS Circular 554. Washington, DC: U.S. Government Printing Office, 1968.

Urban sprawl

Today, most American cities are characterized by decaying central downtowns from which residents and businesses have fled to low-density suburbs spreading out around a network of increasingly congested freeways. This development pattern consumes open space, wastes resources, and leaves historic central cities with a reduced tax base and fewer civic leaders living or working in downtown neighborhoods. Streets, parks, schools, and civic buildings fall into disrepair at the same time that these facilities are being duplicated at great expense in new suburbs. The poor who are left behind when the upper and middle classes abandon the city center often can't find jobs where they live and have no way to commute to the suburbs where jobs are now located. The low-density development of suburbs is racially and economically exclusionary because it provides no affordable housing and makes it impractical to design a viable public transit system.

At the same time that inner cities are hollowing out, the amenities that people have moved to the outskirts in search of prove to be fleeting. Hoping to find a open space, opportunities for outdoor **recreation**, access to wild **nature**, scenic views, and a rural ambience, they find, instead, sprawling developments based on only a few housing styles. The checkerboard layout of nearly identical lots with little public space is permitted—or even required—by local zoning and ordinances, consumes agricultural land and fragments **wildlife habitat**. To make house construction more efficient, land in large housing developments generally is bulldozed, removing vegetation or other landmarks that might make a neighborhood recognizable.

Traffic becomes increasingly congested as more and more cars clog streets and freeways, driving the much longer distances to jobs and shopping required by dispersed living patterns. In Los Angeles, for example, which has the worst congestion in the United States, the average speed in 1982 was 58 mph (93 km/hr) and the average driver spent less than four hours per year in traffic jams. In 2000, the average speed in Los Angeles was only 35.6 mph (57.3 km/hr) and the average driver spent 82 hours per year waiting for traffic. Although new automobiles are much more efficient and cleaner operating than those of a few decades ago, the fact that we drive so much further today and spend so much more time idling in stalled traffic means that we burn more fuel and produce more **pollution** than ever before. Thus, the poor urban **air quality** people fled to the country to escape follows them and is made worse by the greater distances they drive every day.

Altogether, traffic congestion is estimated to cost the United States $78 billion per year in wasted time and fuel. Some people argue that the existence of traffic jams in cities shows that more freeways are needed. Often, however, building more traffic lanes simply encourages more people to drive further than they did before. Rather than ease congestion and save fuel, more freeways can exacerbate the problem.

As Maryland Governor Parris N. Glendening said: "In its path, sprawl consumes thousands of acres of forests and farmland, woodlands and **wetlands**. It requires government to spend millions extra to build new schools, streets and water and sewer lines." Christine Todd Whitman, former New Jersey Governor and Head of the **Environmental Protection Agency** (EPA), has said, "Sprawl eats up our

open space. It creates traffic jams that boggle the mind and pollute the air. Sprawl can make one feel downright claustrophobic about our future." While there is no universally accepted definition of urban sprawl, it generally includes the characteristics:

- unlimited outward urban expansion
- low-density residential and commercial development
- leapfrog growth that consumes farmland and natural areas
- fragmentation of power among many small units of government
- dominance of freeways and private automobiles
- no centralized planning or control of land-uses
- widespread strip-malls and "big-box" shopping centers
- great fiscal disparities among localities
- reliance on deteriorating older neighborhoods for low-income housing
- decaying city centers as new developments occurs in previously rural areas

Among the alternatives to unplanned sprawl and wasteful resource use proposed by urban planners are **smart growth** and **conservation** design. These new approaches make efficient and effective use of land resources and existing infrastructure by encouraging development that avoids costly duplication of services and inefficient **land use**. They aim to provide a mix of land uses to create a variety of affordable housing choices and opportunities. They also attempt to provide a variety of **transportation** choices including pedestrian friendly neighborhoods. This approach to planning also seeks to maintain a unique **sense of place** by respecting local cultural and natural features.

By making land use planning open and democratic, smart growth and conservation design strive to make urban expansion fair, predictable, and cost-effective. All stakeholders are encouraged to participate in creating a vision for the city and to collaborate rather than confront each other. Goals are established for staged and managed growth in urban transition areas with compact development patterns. This approach is not opposed to growth. It recognizes that the goal is not to block growth but to channel it to areas where it can be sustained over the long term. It tries to enhance access to equitable public and private resources for everyone and to promote the safety, livability, and revitalization of existing urban and rural communities.

Rather than abandon the cultural history and infrastructure investment in existing cities a group of architects and urban planners is attempting to redesign metropolitan areas to make them more appealing, efficient, and livable. European cities such as Stockholm, Sweden, Helsinki, Finland, Leichester, England, and Neerlands, the Netherlands, have a long history of innovative urban planning. In the United States, Andres Duany, Elizabeth Plater-Zyberk, Peter Calthorpe, and Sym Van Der Ryn have been leaders in what is sometimes called the new urbanist movement or neotraditionalist approach. These designers attempt to recapture some of the best features of small towns and the best cities of the past. They are designing urban neighborhoods that integrate houses, offices, shops, and civic buildings. Ideally, no house should be more than a five-minute walk from a neighborhood center with a convenience store, a coffee shop, a bus stop, and other amenities. A mix of apartments, townhouses, and detached houses in a variety of price ranges insures that neighborhoods will include a diversity of ages and income levels.

Where building new neighborhoods in rural areas is necessary, conservation design, cluster housing, or open space-zoning preserves at least half of a subdivision as natural areas, farmland, or other forms of open space. Among the leaders in this design movement are landscape architects Ian McHarg, Frederick Steiner, and Randall Arendt. They have shown that people who move to the country don't necessarily want to own a vast acreage or to live miles from the nearest neighbor; what they most desire is long views across an interesting landscape, an opportunity to see wildlife, and access to walking paths through woods or across wildflower meadows.

By carefully clustering houses on smaller lots, a conservation subdivision can provide the same number of buildable lots as a conventional subdivision and still preserve 50–70% of the land as open space. This not only reduces development costs (less distance to build roads, lay telephone lines, sewers, power cables, etc.) but also helps foster a greater sense of community among new residents. Walking paths and recreation areas get people out of their houses to meet their neighbors. Homeowners have smaller lots to care for and yet everyone has an attractive vista and a feeling of spaciousness.

Some good examples of this approach are Farmview near Yardley, Pennsylvania, and Hawknest in Delafield Township, Wisconsin. In Farmview, 332 homes are clustered in six small villages set in a 414 acre (160 ha) rural landscape, more than half of which is dedicated as permanent farmland. House lots and villages were strategically placed to maximize views, helping the development to lead its county in sales for upscale developments. Hawksnest is situated in dairy-farming country outside of Waukesha, Wisconsin. Seventy homes are situated amid 180 acres (70 ha) of meadows, ponds, and woodlands. Restored **prairie**, neighborhood recreational facilities, and connections to a national scenic trail have proved to be valuable marketing assets for this subdivision.

[*William P. Cunningham Ph.D.*]

Resources

Books

Arendt, Randall G. *Conservation Design for Subdivisions: A Practical Guide to Creating Open Space Networks.* Covelo, CA: Island Press, 1996.

Beatly, T. *Green Urbanism: Learning from European Cities.* Covelo, CA: Island Press, 2000.

Benfield, F. Kaid, Donald D. T. Chen, and Matthew D. Raimi, eds. *Once There Were Greenfields: How Urban Sprawl Is Undermining Americas's Environment, Economy, and Social Fabric.* New York: Natural Resources Defense Council, 1999.

Calthorpe, Pete, and William Fulton. *The Regional City: Planning for the end of Sprawl.* Covelo, CA: Island Press, 2000.

Duany, Andres, Jeff Speck, and Elizabeth Plater-Zyberk. *Smart Growth Manual.* New York: McGraw-Hill Co., 2002.

McHarg, Ian. *Design with Nature.* New York: John Wiley & Sons, 1995.

Ursus arctos

see **Grizzly bear**

U.S. Department of Agriculture

The U.S. Department of Agriculture (USDA) had its origin in the U.S. Patent Office, one of the first federal offices. In 1837, an employee of the Patent Office, Henry L. Ellsworth, began to distribute seeds to American farmers that he had received from overseas. By 1840, Ellsworth had obtained a grant of $1,000 from Congress to establish an Agricultural Division within the Patent Office. This division was charged with collecting **statistics** on agriculture in the United States and carrying out research, as well as distributing seeds.

Over the next two decades, the Agricultural Division continued to expand within the Patent Office, until Congress created the Department of Agriculture on May 15, 1862. The officer in charge of the department was initially called the Commissioner of Agriculture, but the Department was raised to cabinet level on February 9, 1889, and the Commissioner was renamed the Secretary of Agriculture.

The USDA today is a mammoth executive agency that manages dozens of programs. Its overall goals include the improvement and maintenance of farm incomes and the development of overseas markets for domestic agricultural products. The department is also committed to reducing poverty, hunger, and malnutrition; protecting **soil**, water, forests, and other **natural resources**; and maintaining standards of quality for agricultural products. The activities of the USDA are subdivided into seven major categories: Small Community and Rural Development, Marketing and Inspection Services, Food and Consumer Services, International Affairs and Commodity Programs, Science and Education, Natural Resources and **Environment**, and Economics.

Small Community and Rural Development oversees many of the programs which provide financial assistance to rural citizens. It administers emergency loans as well as loans for youth projects, farm ownership, rural housing, **watershed** protection, and flood prevention. It also underwrites federal crop insurance, and the Rural Electrification Administration is part of this division.

Marketing and Inspection Services is responsible for all activities relating to the inspection and maintenance of health standards for all foods produced in the United States. The Agricultural Cooperative Service, which many consider one of the USDA's most important functions, is part of the division. Food and Consumer Services helps educate consumers about good nutritional practices and provides the means by which people can act on that information. The division administers the Food Stamp program, as well as the School Breakfast, Summer Food Service, Child Care Food, and Human Nutrition Information programs.

The main functions of the International Affairs and Commodity Programs are to promote the sale and distribution of American farm products abroad and to maintain crop yields and farm income at home. The Science and Education division consists of a number of research and educational agencies including the **Agricultural Research Service**, the Extension Service, and the National Agricultural Library.

Some of the USDA's best known services are located within the Natural Resources and Environment Division. The **Forest Service** and **Soil Conservation Service** are the two largest of these. The Economics Division is responsible for collecting, collating, and distributing statistical and other economic data on national agriculture. The Economic Research Service, National Agricultural Statistics Service, Office of Energy, and World Agricultural Outlook Board are all part of this division.

[*Lawrence H. Smith*]

Resources

Books

The United States Government Manual, 1992/93. Washington, DC: U.S. Government Printing Office, 1992.

Organizations

U.S. Department of Agriculture, Washington, D.C. USA 20250 Email: vic.powell@usda.gov, <http://www.usda.gov>

U.S. Department of Commerce

see **National Oceanic and Atmospheric Administration (NOAA)**

U.S. Department of Defense

see **Army Corps of Engineers**

U.S. Department of Energy

For most of its history, the United States has felt little concern for its energy needs. The country has had huge reserves of **coal**, **petroleum**, and **natural gas**. In addition, the United States had always been able to buy all the additional **fossil fuels** it needed from other nations. As late as 1970, automotive and home heating fuels sold for about $0.20–$0.30 per gal ($0.05–$0.08 per l).

That situation changed dramatically in 1973 when members of the **Organization of Petroleum Exporting Countries** (OPEC) placed an embargo on the oil it shipped to nations around the world, including the United States. It took only a few months for the United States and other oil-dependent countries to realize that it was time to rethink their national energy strategies. The late 1970s saw, therefore, a flurry of activity by both the legislative and executive arms of government to formulate a new **energy policy** for the United States.

Out of that upheaval came a number of new laws and executive orders including the **Energy Reorganization Act** of 1973, the Federal Non-Nuclear Energy Research and Development Act of 1974, the Energy Policy and **Conservation** Act of 1976, the **Energy Conservation** and Production Act of 1976, the National Energy Act of 1978, and President Jimmy Carter's National Energy Plan of 1977.

One of the major features of both legislative and executive actions was the creation of a new Department of Energy (DOE). DOE was established to provide a central authority to develop and oversee national energy policy and research and development of energy technologies. The new department replaced or absorbed a number of other agencies previously responsible for one or another aspect of energy policy, including primarily the Federal Energy Administrations and the **Energy Research and Development Administration**. Other agencies transferred to DOE included the **Federal Power Commission**, the five power administrations responsible for production, marketing and transmission of electrical power (Bonnevile, Southeastern, Alaska, Southwestern, and Western Area Power Administrations), and agencies with a variety of other functions previously housed in the Departments of the Interior, Commerce, Housing and Urban Development, and the Navy.

DOE's mission is to provide a framework for a comprehensive and balanced national energy plan by coordinating and administering a variety of Federal energy functions. Among the Department's specific responsibilities are research and development of long-term, high-risk energy technologies, the marketing of power produced at Federal facilities, promotion of energy conservation, administration of energy regulatory programs, and collection and analysis of data on energy production and use. In addition, the department has primary responsibility for the nation's **nuclear weapons** program.

DOE is divided into a number of offices, agencies, and other divisions with specific tasks. For example, the Office of Energy Research manages the Department's programs in basic energy sciences, high energy physics, and **nuclear fusion** energy research. It also funds university research in mathematical and computational sciences and other energy-related research. Another division, the Energy Information Administration, is responsible for collecting, processing, and publishing data on energy reserves, production, demand, consumption, distribution, and technology.

[*David E. Newton*]

Resources

Other

The United States Government Manual, 1992/93. Washington, DC: U.S. Government Printing Office, 1992.

Organizations

U.S. Department of Energy, 1000 Independence Ave., SW, Washington, D.C. USA 20585 Fax: (202) 586-4403, Toll Free: (800) dial-DOE, , <http://www.energy.gov>

U.S. Department of Health and Human Services

The U.S. Department of Health and Human Services (HHS) is responsible for the welfare, safety, and health of United States citizens. Among other programs, HHS administers drug safety standards, prevents epidemics, and offers assistance to those who are economically disadvantaged.

The **Public Health Service**, a division of HHS, helps state and city governments with health problems. The service studies ways of controlling infectious diseases, works to immunize children, and operates quarantine programs. The agency also operates the Centers for Disease Control (CDC), where most of the nation's health problems are studied. These include occupational health and safety, the dangers of cigarette smoking, and childhood injuries, as well as **communicable diseases** and the epidemic of urban violence. In addition to investigating these problems, the CDC is charged with making policy suggestions on their management. HHS also administers the Social Security and Medicare programs, as well as the Head Start Program.

New lead-content standards for paint and other consumer items grew out of CDC studies. The agency conducts energy related epidemiological research for the **U.S. Department of Energy**, including studies on **radiation exposure**. Through the CDC, the HHS runs the National Center for Health **Statistics**, an **AIDS** Hotline, an international disaster

relief team, and programs to monitor influenza epidemics worldwide.

The **Food and Drug Administration** is another agency of HHS. It is charged with responding to new drug research by drug development companies and approving new drugs for use in the United States. Along with the **U.S. Department of Agriculture**, the FDA is responsible for maintaining the safety of the nation's food and drug supply. HHS also administers the **Agency for Toxic Substances and Disease Registry**, which carries out the health-related responsibilities of the Superfund legislation.

[*Linda Rehkopf*]

Resources
Other

The United States Government Manual, 1992/93. Washington, DC: U.S. Government Printing Office, 1992.

Organizations

U.S. Department of Health and Human Services, 200 Independence Avenue, SW, Washington, D.C. USA 20201 (202) 619-0257, Toll Free: (877) 696-6775, Email: HHS.Mail@hhs.gov, <http://www.hhs.gov>

U.S. Department of State

see **Bureau of Oceans and International Environmental and Scientific Affairs (OES)**

U.S. Department of the Interior

The U.S. Department of the Interior was founded by an Act of Congress on March 3, 1849. A wide variety of functions were assigned to the "Home Department" as it was called at the time, including administration of the General Land Office which passed large tracts of western lands to homesteaders (294 million acres/119 million ha), to railroads (94 million acres/38 million ha), and to colleges and universities. In the twentieth-first century, the Department of the Interior has become the custodian of public lands and **natural resources**, and is now America's primary governmental **conservation** agency. The **National Park Service, Bureau of Land Management**, **Bureau of Reclamation**, **Fish and Wildlife Service**, **Geological Survey**, **Office of Surface Mining**, **Reclamation** and Enforcement, and Bureau of Mines are part of the Department. In addition the Department is responsible for American Indian reservation communities and for people living in island territories administered by the United States.

The Department's mandate is very broad, including such duties as: 1) administration of over 507 million acres (205 million ha) of federal public lands; 2) development and conservation of mineral and **water resources**; 3) conservation and utilization of fish and **wildlife** resources; 4) coordination of federal and state **recreation** programs; 5) administration and preservation of America's scenic and historic areas; 6) reclamation of western **arid** lands through **irrigation**; and 7) management of hydroelectric power systems.

Although many Department activities and programs have generated environmental controversy over the years, it is probably these last two activities that have been most contentious. In pursuit of these mandates, the Bureau of Reclamation, working with the **Army Corps of Engineers**, has constructed massive **dams** on rivers in the West. These dams are designed to provide water storage and electricity for western cities and farms, as well as to reduce downsteam **flooding**. However, these dams also have had severe impacts on ecosystems, significantly altering and, at times, devastating riverine habitats.

On balance, however, the programs and practices of the Department of the Interior probably do more environmental good than harm. Among the Department's positive actions and accomplishments during the past several years are: 1) establishment of 23 new refuges and waterfowl production areas in Florida and in Louisiana; 2) initiation of "Refuges 2003," a Fish and Wildlife Service effort aimed at planning management programs and policies for the next 10–15 years on the nation's 472 national wildlife refuges; 3) implementation of the North American **Wetlands** Conservation Act to restore declining waterfowl populations and to conserve wetlands; 4) protection of 1,800 **species** on the List of Endangered and Threatened Species; 5) development and beginning implementation of a mitigation and enhancement plan to restore California's Kesterson **Reservoir**, which was closed due to high concentrations of selenium; 6) tripling the Coastal Barrier Resources System to 1.25 million acres (500,000 ha) over 1,200 mi (1,931 km) of shoreline including the Florida Keys, **Great Lakes**, Puerto Rico and the Virgin Islands; 7) Establishment of a new Office of **Surface Mining** Reclamation and Enforcement office in Ashland, Kentucky, for rapid response to abandoned mine land problems in need of reclamation on an emergency basis; 8) participation in filing 120 natural resource damage assessments in the period 1988–1991, the largest being the assessment of the ***Exxon Valdez*** accident in **Prince William Sound**, Alaska. The $1.1 billion settlement of this accident enabled the Federal Government to proceed with cleanup and restoration efforts.

[*Malcolm T. Hepworth*]

Resources
Other

The United States Government Manual, 1992/93. Washington, DC: U.S. Government Printing Office, 1992.

Organizations

U.S. Department of the Interior, 1849 C. Street NW, Washington, D.C. USA 20240 (202) 208-3100, <http://www.doi.gov>

U.S. Public Interest Research Group

Formally founded in 1983 by consumer advocate **Ralph Nader**, the Public Interest Research Group (PIRG) is an outgrowth of Nader's Center for the Study of Responsive Law. PIRG aims to heighten consumer awareness and focuses on such environmental issues as clean air, toxic waste cleanup, protection of the **atmosphere**, **pesticide** control, and solid **waste reduction**. While PIRG did not convene as a national organization until 1983, its roots date back to 1970, when Nader began to establish state PIRGs throughout the United States. The national umbrella organization was founded to lobby for the state units.

PIRG has been vital in bringing about several important environmental regulations. In 1986, the organization led an aggressive campaign that resulted in the strengthening of the federal Superfund program, a $9-billion venture that identifies and provides for the cleanup of sites contaminated by **hazardous waste**. That same year, PIRG was successful in influencing the United States legislature to pass the **Safe Drinking Water Act** (SDWA). The SDWA imposes limits and prohibitions on the amounts and types of **chemicals** allowable in drinking water supplies and provides for regular testing to assure that the limits imposed by the SDWA are being met. In 1987, PIRG proved instrumental in the strengthening of the **Clean Water Act** which requires cleanup of United States waterways. The CWA also imposes limits and prohibitions on chemicals which are discharged into the water system. Several state PIRGs have also met with success when suing major polluters. As the result of PIRG legal action, many polluters have been ordered to pay fines and clean up contaminated areas.

PIRG also educates the general public on various issues. Its current projects are outlined in a quarterly newsletter, *Citizen Agenda*, and some of its key concerns are detailed in such reports as *Toxic Truth and Consequences: The Magnitude of and the Problems Resulting from America's Use of Toxic Chemicals*; *Presumed Innocent: A Report on 69 Cancer-Causing Chemicals Allowed in Our Food*; and *As the World Burns: Documenting America's Failure to Address the Ozone Crisis.*

Ongoing PIRG projects address similar issues. Although successful with the Superfund program, the organization continues to work for the cleanup of toxic waste and actively supports the notion that "polluters pay"—requiring those who cause contamination to fund the subsequent cleanup. PIRG also conducts research into **energy efficiency** in an effort to curb **carbon monoxide** emissions. Through its lobbying efforts, the group also supports legislation requiring a ban on carcinogenic pesticides in food, preventing **groundwater** contamination, halting **garbage incineration**, and initiating bottle-recycling programs in all 50 states. The organization is also involved in a Clean Water Campaign, seeking to further strengthen state and federal water regulations.

PIRG achieves these goals through a balance of professional and volunteer action. PIRG volunteers are utilized in fundraising efforts, letter-writing campaigns, and election drives. Members were once asked to send **aluminum** cans to Congress to show support of PIRG's bottle-recycling campaign.

[*Kristin Palm*]

Resources

Organizations

U.S. Public Interest Research Group, 215 Pennsylvania Avenue, Washington, D.C. USA 20003, E-mail: webmaster@pirg.org, <http://pirg.org>

Used oil recycling

Used oil recycling is a procedure that involves reprocessing used motor oil so that it can be used again. Motor oil can be recycled indefinitely because the lubricant does not wear out. Recycled oil is cleaned of contaminants such as dirt, water, used additives, and fuel. Used oil may also contain toxic substances such as **lead**, **benzene**, zinc, and **cadmium**.

Recycling saves oil and helps the **environment**. According to the United States **Environmental Protection Agency** (EPA), 2.5 qt (2.4 l) of rerefined oil can be processed from 1 gal (3.8 l) of used oil. If not recycled, the gallon of lubricant from an oil change can ruin 1 million gal (3.8 million l) of fresh water, representing the annual water supply for 50 people. One pint (0.5 l) of used oil can create an approximately 1 acre (0.4 ha) oil slick on the surface of water, according to the California Integrated **Waste Management** Board. Crankcase oil accounts for more than 40% of the total oil **pollution** in harbors and waterways in the United States.

There are several markets for recycled oil. According to the American Petroleum Institute (API), a trade organization that joins with government agencies to promote the recycling of used motor oil from cars, trucks, motorcycles, boats, recreational vehicles, and lawn mowers, 75% of recycled oil is processed for use in industry, and 11% is used in space heaters in automotive bays and municipal garages. The remaining 14% is rerefined, returned to its virgin state, and again used as motor oil. The API reported that more than 640 million gal (2,422 million l) of motor oil were sold in

1997. People who changed their own oil bought 345 million gal (1,306 million l).

The API's Recycled Oil Web site includes instructions about collecting oil and where to take it for recycling. Oil should be drained from the car engine when it is warm (so that **sludge** flows out smoothly) into a pan that holds twice as much the crankcase. The drained oil should be poured through a funnel into a clean plastic bottle that has a tightly closing lid; the API recommends using a milk jug and cautions against using bottles that held bleach, cleaners, or automotive fluids such as anti-freeze (residue in those containers will contaminate the oil). The used oil should then be taken to a collection station. The API Web site has links to recycling information in every state and some Canadian provinces. Recycling services are provided by states, cities, and private companies. In addition, oil may be accepted at some service stations and oil change shops.

The next step in the recycling process is the collection of used oil by transporters who vacuum oil from service bays and collection center storage containers. The collected oil is tested for hazardous components. Oil that passes the test is mixed into a holding tank and taken by tanker truck to a recycler or a **transfer station**. At the transfer station, oil is held until it is taken to processors, rerefiners, or burners for heating.

Processors treat about 750 million gal (2,839 million l) of used oil each year, according to the API. About 43% of reprocessed oil is used by asphalt plants, 14% by industrial boilers at factories, 12% by utility boilers at electric **power plants**, 12% by steel mills, 5% by cement and lime kilns, 5% by marine boilers, 4% by **pulp and paper mills**, and less than 1% by commercial boilers that generate heat for offices, schools, and other facilities. The remaining 5% is marketed for other uses.

Rerefined oil is produced by a process that involves cleaning the lubricant of contaminants. After vacuum distillation, oil is hydrotreated to eliminate any remaining **chemicals**, then combined with additives to make virgin oil. Consumers who purchase rerefined oil should check the product for the API certification mark.

Used oil designated for specially designed space heaters helps to reduce heating costs at service bays and garages (it is not recommended for home use). In the United States, 75,000 space heaters are fueled by approximately 113 million gal (428 million l) of used oil each year.

Statewide recycling programs include Alabama's Project R.O.S.E. (Recycled Oil Saves Energy). It started in 1977 and is one of the nation's oldest programs. Project accomplishments include the collection of 9 million gal (34 million l) of used oil in 1993. In 1995 California's program was expanded to include an oil filter recycling program. More than 20 million **filters** are recycled annually in the state; recycling 1 ton (907 kg) of used filters yields 1,700 lb (771 kg) of steel and up to 60 gal (227 l) of oil.

[*Liz Swain*]

Resources

Organizations

American Petroleum Institute, 1220 L Street, NW, Washington, DC USA 20005-4070 (202) 682-8000, <http://www.api.org>

USLE

see **Erosion**

Utilitarianism

According to the ethical theory of utilitarianism, an action is right if it promises to produce better results than—or maximize the expected utility of—other action possible in the circumstances. Although there are earlier examples of utilitarian reasoning, the English philosopher Jeremy Bentham (1748–1832) gave utilitarianism its first full formulation. "Nature has placed mankind under two sovereign masters," Bentham declared, "pain and pleasure. It is for them alone to point out what we ought to do, as well as to determine what we shall do." The ethical person, then, will act to increase the amount of pleasure (or utility) in the world and decrease the amount of pain by following a single principle: promote the greatest happiness of the greatest number.

As later utilitarians discovered, this principle is not as straightforward as it seems. The very notion of happiness is problematical. Are all pleasures intrinsically equal, as Bentham suggested? Or are some inherently better or "higher" than others, as John Stuart Mill (1806–1873), the English philosopher and political economist, insisted? A related problem is the difficulty—some say the impossibility—of making interpersonal comparisons of utility. If people want to promote the greatest happiness (or good or utility) of the greatest number, they need a way to measure utility. But there is no ruler that can assess the utility of various actions in the way that weight, height, or distance are quantified.

Another problem is the ambiguity of "the greatest happiness of the greatest number." Stressing "the greatest number" implies that the utilitarian should bow to the majority. But "the greatest happiness" may mean that the intense preferences of the minority may override an equivocal majority. Utilitarians have typically taken the second tack by reformulating the principle as "maximize aggregate utility."

Either interpretation seems to call for an ever-increasing population. The more people there are, the more happiness there will be. This has led some utilitarians to argue that people should try to promote average rather than total

utility. Since the average amount of happiness could decline in an overcrowded world even as the total amount increased, the "average utilitarian" could consistently argue for population control.

Finally, who counts when happiness or utility is calculated? Everyone counts equally, Bentham said, but does "everyone" include only those people living in this place at this time? The expansive view is that "everyone" embraces all those who may be affected by one's actions, even people who may not yet be born. If so, should today's people count the preferences of **future generations** equally with those presently living? For that matter, should they restrict their concern to people? According to Bentham, the test of inclusion should not be: "Can they reason? nor Can they talk? but, Can they suffer?" This emphasis on sentience has led some contemporary utilitarians to advocate "animal liberation," including **vegetarianism**, as part of a consistent attempt to promote utility. *See also* Animal rights; Environmental ethics; Intergenerational justice

[*Richard K. Dagger*]

Resources

Books

Bentham, J. *An Introduction to the Principles of Morals and Legislation.* New York: Hafner, 1948 (originally published 1789).

Mill, J. S. "Utilitarianism." In *Utilitarianism, Liberty, Representative Government.* New York: Dutton, 1951.

Smart, J. J. C., and B. Williams. *Utilitarianism: For and Against.* Cambridge: Cambridge University Press, 1973.

V

Vadose zone

The unsaturated zone between the land surface and the **water table**. The vadose zone (from the Latin *vadosus*, meaning shallow) includes the soil-water zone, intermediate vadose zone, and capillary fringe. The pore space contains air, water, and other fluids under pressure, which is less than atmospheric pressure. Thus, the water is held to the **soil** particles by forces that are greater than the force of gravity. Saturated zones, such as perched **groundwater** aquifers, may exist in the vadose zone and water pressure within these zones is greater than atmospheric pressure. *See also* Recharge zone; Zone of saturation

Valdez Principles

In March of 1989, the ***Exxon Valdez*** ran aground in **Prince William Sound** in Alaska, spilling 11 million gal (41 million L) of crude oil. During the international outcry over the environmental consequences of the spill, environmentalists criticized a number of the structural features of the **petroleum** industry and the operational practices of the supertanker transport of oil.

In this climate, new approaches were suggested to motivate not only oil companies but all industries to support the protection of the **environment**. The Valdez Principles are perhaps the most important of the approaches suggested during this period. They were developed by the Coalition for Environmentally Responsible Economics (CERES), which was a consortium of 14 environmental groups and the Social Investment Forum, an organization of 325 socially concerned bankers, investors, and brokers. CERES was founded by Boston money manager Joan Bavaria, and it has the support of several major environmental groups, including the **Sierra Club** and the **National Wildlife Federation**.

The Valdez Principles were modelled on the Sullivan Principles, which had been developed to discourage investment in South Africa as a protest against apartheid. Members of CERES controlled $150 billion in both pension and mutual funds. The goal of the Valdez Principles was to reward the behavior that was environmentally sound and punish behavior that was not by investing or withholding funds controlled by CERES members. Corporations were also asked to sign the Valdez Principles in the hopes that the financial incentives provided by CERES would encourage companies to develop environmentally sound practices.

The Valdez Principles support a wide range of environmental issues. Protection of the **biosphere** is one of its objectives, and it encourages industries to minimize or eliminate the **emission** of pollutants. The principles are also devoted to protecting **biodiversity** and insuring the **sustainable development** of land, water, forests, and other **natural resources**. The principles advocate the use of **recycling** whenever possible, support safe disposal methods, and encourage the use of safe and sustainable energy sources. **Energy efficiency** is also a goal, as well as the marketing of products that have minimal environmental impact. The principles also call for corporations to have at least one board member qualified to represent environmental interests and a senior executive responsible for environmental affairs. Other goals include damage compensation, disclosure of accidents and hazards, and the creation of independent environmental audit procedures. Many major corporations have indicated an interest in these principles, but few have signed them. Some executives have observed that many aspects of the Valdez Principles are already required by government regulations and internal policies.

Perhaps then the most significant feature of the Valdez Principles is not what they have accomplished but the circumstances of their origin. A major disaster often engenders a social and political climate that is, at least temporarily, receptive to reform. In this case it was a rare coalition between financial investors and environmental groups. Many have argued that the principles had unrealis-

tic goals, but some see in their development hope for the future.

[*Usha Vedagiri and Douglas Smith*]

Resources

Periodicals

Ohnuma, K. "Missed Manners." *Sierra* 75 (March-April 1990): 24–26.

"The Valdez Principles." *Audubon* 91 (November 1989): 6.

Vapor recovery system

At one time, vapors that are potentially harmful to the **environment** or to human health were simply allowed to escape into the **atmosphere**. Each time an **automobile** consumer filled the gas tank, for example, **gasoline** vapors escaped into the air. As the hazards of these emissions became more evident, the use of closed systems of pipes, valves, compressors, and other components became more prevalent. In most systems, those vapors are compressed to a liquid and returned to their original source. Vapor recovery systems not only protect the environment, but conserve resources that might otherwise be wasted.

Vascular plant

A plant that possesses specialized conducting tissue (xylem and phloem) for the purpose of transporting water and solutes within the root-stem-leaf system. Although aquatic forms exist, most vascular plants are terrestrial and include herbs, shrubs and trees. The development of vascular transport allowed plants to become larger, leading to their eventual domination of terrestrial ecosystems. The uptake and translocation of soluble pollutants from the **soil** into the plant occurs through the roots and the conducting tissue. Adverse effects of such **pollution** on vascular plants include stunted growth and death of all or part of the plant.

Vector (Mosquito) Control

The largest group of animals on the face of the planet, the arthropods (meaning "jointed feet"), includes a very highly successful class of organisms, Insecta, the insects. It is estimated that insect **species** outnumber all other known species of animals and plants on earth combined. One kind of insect, the mosquito, a two-winged insect of the family Culicidae, has particular importance with respect to the activities of humankind. Mosquitoes are a type of fly that have adaptive traits that make them very successful, and very devastating pests. Most importantly, female mosquitoes of key species must consume blood from vertebrates to provide nourishment for egg production. Female mosquitoes will ingest blood-meals from amphibians, birds, or mammals. In the process of consuming blood to provide the protein necessary for egg development, the mosquito becomes an important, and dangerous, transmitter of disease. When an organism is capable of transmitting a disease, it is said to be a *vector* of that disease. Mosquitoes are perhaps the most important arthropod vectors of human disease in the world.

Mosquitoes of various species can transmit viral, protozoan, or bacterial organisms that cause disease in human beings. Viruses (complex constructions of protein and DNA or RNA) are, by definition, dependent upon living cells for replication and transmission. Some protozoans, single celled organisms possessing nuclei, have become dependent upon other organisms for their survival. Likewise, some bacteria, single-celled organisms without nuclei, cannot survive outside of a particular host organism. Mosquitoes act as vectors, or vehicles for such disease-causing organisms to jump from host to host to complete life-cycles in order to continue survival.

Mosquito vectors transmit viral strains and protozoal **parasites** that cause immense human suffering worldwide. Globally relevant viral diseases transmitted by mosquito vectors include Dengue fever and Yellow fever. Recently, in the United States, West Nile **virus**, Eastern Equine Encephalitis (EEE), La Crosse Encephalitis, and St. Louis Encephalitis are important arthropod-borne viruses, or "arboviruses." The largest impact that mosquito vector species have on humans is in the transmission of **malaria**, a disease caused by protozoan infection of red blood cells that results in massive, cyclic bouts of very high fever and in many cases, death. Vector control, or the management of disease-spreading species of mosquitoes, is a large concern among global communities. Once thought to be a historic problem, renewed interest has emerged concerning vector-borne disease as a potential threat to contemporary and future civilization.

Global mosquito control consumes tremendous financial resources and is costly in terms of time, effort, and environmental impact. Different parts of the world attempt vector control by various methods, but the most effective way of controlling mosquitoes is a combination of scientific knowledge and common sense. **Integrated pest management** combines the use of biological or chemical products, knowledge of the mosquito's **habitat**, and information about the mosquito's life cycle to control the potentially dangerous pests. Larviciding is a successful component of mosquito control. Larviciding uses an insecticide to spray areas of stagnant where mosquitoes lay their eggs and the larvae grows. Adulticiding is ground spraying for adult mosquitoes and is applied by spraying a fog of insecticide around the premises where they live. Common sense methods of mos-

quito control are emptying dishes or birdbaths where water collects, keeping grass mowed and bushes trimmed, and filling in low-lying areas so that water is not left to stagnate there.

Even with global vector control, malaria is reemerging in most countries where the disease is endemic. Malaria is caused by any of four species of protozoan parasites. *Plasmodium falciuparum* and *Plasmodium vivax* are the most deadly. The parasite is transmitted by a particular genus of mosquito, *Anopheles*. For many years, malaria was a problem in the United States. In the 1940s, a massive malaria vector control program was initiated that used immense quantities of the potent **pesticide**, DDT, to kill the mosquitoes that transmit malaria. DDT is an effective agent in the control of mosquito vectors of disease, but two problems have plagued its use. The first is its persistence in the **environment** and harmful effect on fish and **wildlife** and its implication as a mammalian **neurotoxin** that could potentially affect human beings. The second concern is the emergence of DDT-resistant strains of mosquitoes. Despite this, DDT is still used in many underdeveloped areas of the world to control mosquito infestations that pose a threat to human health and wellbeing. In the United States, the Centers for Disease Control (CDC) monitors and surveys for malaria and other vector-borne diseases that endanger the population. The CDC also advises U.S. citizens on appropriate drug therapies to utilize while visiting countries where disease-bearing vector species of mosquitoes are prevalent. In the year 2000, 246 cases of imported malaria were reported to the CDC.

One effort to control vector spread of malaria has been the Roll Back Malaria initiative which seeks to reduce the world's malaria burden by one half by the year 2010. One aspect has been the distribution of insecticide-treated mosquito nets in areas where malaria vectors are especially problematic for pregnant women and children, such as the country Nairobi. As part of the effort, 60,000 insecticide-treated mosquito nets were distributed in 2002 in cooperation with the United Nations Children's Fund. This form of vector control incorporates physical barriers with chemical insecticides and social education programs to effect change in the prevalence of this devastating disease.

Mosquito control, however, involves many diseases other than malaria, however. Dengue fever is a viral infection transmitted by mosquito vectors that causes paroxysmal joint and muscle pain, headache, vomiting, and rash. Commonly, because of its symptoms, Dengue fever is confused with malaria or typhoid fever, a bacterial infection. Dengue fever is a potential threat in Africa, China, Southeast Asia, the Middle East, Central and South America and the Caribbean. A historically notorious arbovirus, Dengue fever is returning. According to the CDC Division of Vector-Borne Infectious Disease, 1997 was the year in which Dengue fever was the most important mosquito-borne viral disease affecting humans. At that time, an estimated 2.5 billion people lived in areas of risk, making Dengue fever a mosquito (vector) control problem en par with malaria. Four major reasons for the emergence of Dengue fever as an important mosquito-borne disease have been identified. First, sufficient mosquito control efforts are practically absent in areas where the risk of contracting Dengue fever is most acute. Second, uncontrolled population expansion in such areas has led to a deficit in effective sewer and **waste management** systems, the result of which only contributes o the spread of vector-borne disease. Demographic changes leading to inadequate water management systems have exacerbated this problem. Thirdly, accessible and rapid **transportation** via airplanes had created a new, swift means for spread of the Dengue vector mosquito. Stowaway infected mosquitoes spread disease to new areas with newfound speed. Finally, limited financial resources have forced nations challenged by such vector-borne disease to choose between effective vector control and other, greatly needed endeavors. Until and unless Dengue and other vector-borne diseases are recognized as threats, funding for vector control programs may remain secondary to more pressing social dilemmas in undeveloped nations. While no Dengue vaccine exists, vigorous research employing **biotechnology** continues.

More recently, mosquito-borne illnesses have been witnessed in North America. West Nile virus has surfaced in America, formerly only attributed to central Asia, Europe, Africa, and the Middle East. By 2001, the West Nile virus had been documented by the CDC to be present in Alabama, Arkansas, Connecticut, Delaware, Florida, Georgia, Illinois, Indiana, Iowa, Kentucky, New York, Ohio, Pennsylvania, Wisconsin, and other mid-Atlantic states. As of 2001, 149 cases of West Nile virus illnesses had been reported to the CDC, which includes 18 deaths. The West Nile virus is a variety of **pathogen** called a flavovirus that can cause encephalitis, coma, and death.

Two additional arbovirus pathogens transmitted by mosquito vectors in the United States are LaCrosse Encephalitis and Eastern Equine Encephalitis. The LaCrosse virus belongs top the Bunyviridae subgroup and has approximately 70 cases of infection in the United States annually. While fatality rate of infection with LaCrosse is less than 1%, hospitalization and neurological complications that occur with infection are serious. Children under the age of 16 years are particularly at risk of contracting LaCrosse Encephalitis in endemic areas. A similar vector-borne disease is Eastern Equine Encephalitis (EEE). The manifestations of EEE range from influenza-like symptoms to severe swelling of central nervous tissue, coma, and death. The etiologic agent for EEE is a family of mosquito-borne viruses called

the Togaviridae. Since 1964, there have been approximately 153 confirmed cases of EEE in the United States.

Vector mosquito control in the United States remains a priority in many urban areas. Mosquito control is most often accomplished by chemical insecticide spraying of select areas where known vector breeding areas exist. This approach is viewed by some experts to be reactive rather than proactive. Still, potent chemical means of mosquito control remain the most cost-effective alternative for vector disease transmission in this country and abroad. However, much research is underway to find alternative methods of controlling dangerous mosquito populations. For example, now standard biotechnological techniques such as gene **cloning** and transposon insertion are presenting new and exciting ways of genetically altering mosquito vectors to become immune to the disease-causing **microorganisms**, thus rendering them mere pests, and not potential transmitters of life-threatening disease.

[*Terry Watkins*]

Vegan

The strictest type of vegetarians, refraining from eating not only meat and fish, but also eggs, dairy products, and all other food containing or derived from animals, often including honey. A vegan diet contains no cholesterol and very little fat, so it is quite healthy as long as sufficient nutrients are obtained, especially vitamins B_{12} and D, calcium, and quality protein. Most vegans also avoid wearing or using animal products of any kind, including fur, leather, and even wool. Vegans are motivated primarily by a humane and ethical concern for the welfare of animals and by a desire to avoid unhealthy food products, as well as by the realization that raising livestock and other food animals damage the natural **environment** and contributes, through its wastefulness and lack of efficiency, to world hunger. *See also* Animal rights; Environmental ethics; Vegetarianism

Vegetarianism

Vegetarians refrain from consuming animals and animal products, including meat, poultry, and fish. Lacto-ovo vegetarians will eat eggs, cheese, yogurt, and other dairy products. However, total vegetarians (vegans) avoid these animal products completely, including foods such as honey. On the whole, vegetarians emphasize the impact of dietary choice on health, on the fate of animals and the planet, and on humanity, including **future generations**.

Vegetarians avoid meat for many reasons, including concerns for animals, the **environment**, general health, and worldwide food shortages. Some cultures and religions, such as Buddhism and Hinduism, also advocate vegetarianism. Vegetarians emphasize that the overwhelming majority of food animals are raised on "factory farms," where they spend their entire lives in cramped, overcrowded conditions, lacking sunshine, exercise, and the ability to engage in natural behavior.

A vegetarian diet can also be healthier than a meat-centered one. Meat contains high amounts of saturated fat and cholesterol, which—in excess amounts—contributes to heart disease, **cancer**, and other degenerative diseases. While poultry and fish are lower in fat and cholesterol than red meat, they also carry health risks. Chicken is a major source of salmonella contamination and other dangerous bacteria, and fish (especially shellfish) are often heavily contaminated with pesticides, **heavy metals**, and toxic **chemicals**.

Vegetarians often cite the massive environmental damage caused by raising food animals. Consider the following illustrations. Livestock occupy and graze on half of the world's land mass. Cattle alone use a quarter of the earth's land. This can result in **water pollution**, clearing of forests, **soil erosion**, and **desertification**. More food could be produced and more people could be fed if resources were not used to produce meat. For example, the amount of land required to feed one meat-eater could theoretically feed 15 to 20 vegetarians. One acre (0.4 ha) of agricultural land can produce about 165 lb (75 kg) of beef or 20,000 lb (9,072 kg) of potato.

Moreover, the world's cattle eat enough grain to feed every human on earth, and most of that grain is wasted. A cow must eat 16 lb (7.3 kg) of grain and soybeans to produce 1 lb (0.45 kg) of feedlot beef—a 94% waste of food. A pig requires 7.5 lb (3.4 kg) of protein to produce 1 lb (0.45 kg) of pork protein. Ninety-five percent of all grain grown in the United States is used to feed livestock, as is 97% of all legumes (beans, peas, and lentils), and 66% of the fish caught in American waters. Over half of all the water used in the United States goes for livestock production, and it takes 100–200 times more water to produce beef than wheat.

[*Lewis G. Regenstein*]

Resources

Books

Amato, P. R., and S. A. Partridge. *The New Vegetarians: Promoting Health and Protecting Life.* New York: Plenum, 1989.

Brown, E. H. *With the Grain: The Essentially Vegetarian Way.* New York: Carroll and Graf, 1990.

Giehl, D. *Vegetarianism: A Way of Life.* New York: Harper & Row, 1979.

Mitra, A. *Food for Thought: The Vegetarian Philosophy.* Willow Springs, MO: Nucleus, 1991.

Robbins, J. *Diet for a New America.* Walpole, NH: Stillpoint Publishing, 1987.

Vladímir Ivanovich Vernadsky (1862 – 1945)

Russian mineralogist, geochemist, and biochemist

Vernadsky was born on March 12, 1863, in St. Petersburg, Russia. His father, Ivan, was a professor of political economy and edited a liberal journal that barely escaped the Tsarist regime's censorship. Vernadsky's mother, Anna Petrovna Konstantinovich, was a teacher of singing who was neither as intellectual nor as politically inclined as her husband. When Vernadsky was five, the family moved to the more provincial town of Kharkov, where he received an introduction to **nature** and astronomy from his uncle. At the age of 13, Vernadsky moved with his family back to St. Petersburg, where he attended a classical gymnasium. Because a classical Russian education in this era did not include the sciences, Vernadsky and his friends were forced to form a study group of their own.

In 1881 Vernadsky entered the physics and math departments at St. Petersburg University. Although it was the custom for men of his class to study abroad, Vernadsky remained close to home to help care for his father, who had suffered a stroke the previous spring. At St. Petersburg, Vernadsky studied with Dmitri Ivanovich Mendeleev, who derived the periodic table of the elements, chemist Aleksandr Butlerov, and mineralogist V. V. Dokuchaev. He published two scientific articles during his undergraduate years, one on mineral analysis and the other on the **prairie** rodent. Vernadsky's undergraduate thesis on isomorphism so impressed his professors that they urged him to pursue an academic career. That same year he joined an underground committee on literacy, which wrote and distributed reading materials for the common people. Through the committee, Vernadsky met Natalia Egorovna Staritskaya. The two began dating, and when Vernadsky was appointed curator of the university's mineralogical collection in 1886, they were married.

Vernadsky was not to enjoy the peaceful existence of the newlywed for long. Russia in the late 1880s was in turmoil and few places were more tumultuous than the university campuses. After a group of students there were found guilty of a plot to kill Alexander III, the Tsarist government considered St. Petersburg University a hotbed of radicalism. Administrators at the state-run university targeted students and faculty suspected of rebellious feelings towards the autocracy. The 25-year-old Vernadsky was among the suspects, not because of any radical activities but because of his decision not to study abroad, a decision which, according to administrators, branded him an avowed rabble-rouser. Vernadsky's father-in-law, a well-respected government official, appealed his ouster and the government decided to allow Vernadsky to continue his association with the university as long as he now sought that international education. As soon as his first child, George was born, Vernadsky began studying at the University of Naples.

Soon after his arrival in Italy, Vernadsky realized that the Naples department no longer led the field, so he transferred to Munich to study with the German crystallographer Paul Groth. In 1889 Vernadsky transferred to Paris's Mining Academy, where, under the guidance of Henri Le Châelier, he chose polymorphism—the ability of some chemical compounds to assume different forms—as the topic for his master's thesis. Whereas it was previously believed that the aluminosilicate minerals which make up most of the earth's crust were silicic **acid** salts, Vernadsky showed them to have a different structure, with **aluminum** that is chemically analogous to silicon. He proposed the theory of the kaolin nucleus, a structure which is made up of two aluminum, two silicon, and seven oxygen atoms, and which forms the basis of many minerals. The theory has since been confirmed, and is considered essential to an understanding of minerals.

Vernadsky started lecturing at Moscow University in 1891, the year he received his master's degree. Like many intellectuals of his time and place, he found himself balancing academic interests with political ones. In 1897, Vernadsky earned a doctoral degree with his dissertation on crystalline matter, qualifying him for a full professorship. The following year his daughter Nina was born.

The first decade of the twentieth century proved a productive one for Vernadsky. His new approach of combining geologic interests with other scientific fields, such as chemistry and biology, attracted supporters. By 1901, when he created the Mineralogical Circle at Moscow University, he had a devoted cadre of students and colleagues who formed a scientific school that was heavily influenced by the latest theories in chemistry and evolutionary biology. He also maintained an interest in politics, helping to found the Union of Liberation, a group that sought to end the Russian autocracy peacefully. In 1902 he published a summary of his political views, disguised as science, in *On a Scientific World View.* The next year, he published his first scientific book, *Fundamentals of Crystallography.*

When the universities again erupted in turmoil in 1905, Vernadsky operated a lab until the university closed. Caught up in the fervor of the times, he helped organize the Constitutional Democratic party, the largest opposition party to pose candidates for the nation's newly created Duma. His political work did not deter him from amassing scientific honors, however. In 1906 Vernadsky was elected as an adjunct member of the Academy of Sciences and appointed director of St. Petersburg's Mineralogical Museum; two years later he melded his interests with his appointment to the Agrarian Commission of the State Council.

After the university riots of 1905, the campus situation calmed down somewhat, allowing the faculty to return to teaching and research. For Vernadsky, that meant expanding the field of mineralogy to include evolutionary concerns. As he explained in volume 2 of *Izbrannye sochinenia*, his version of mineralogy held that "mineralogy, like chemistry, must study not only the products of chemical reactions but also the very processes of reaction." He was particularly interested in paragenesis, or the way in which essential minerals formed. By studying the many layers of the earth's crust in this manner, Vernadsky hoped to be able to piece together some of the planet's evolutionary history. In 1911 strikes interrupted his work once more. In retaliation for the liberal faculty's support of miscreant students, the government fired three professors. Twenty-eight percent of the faculty—including Vernadsky—resigned in protest. Soon after resigning, Vernadsky was expelled from the state council.

Loss of both these positions meant free time for Vernadsky to pursue his scientific interests. He moved to St. Petersburg, where he took a job as a scientific administrator and continued research on the distribution of rare elements such as cesium, rubidium, scandium, and indium. He made one expedition per year to remote areas in the Russian empire to catalog the nation's resources. During World War I, Vernadsky spearheaded a movement to conserve Russian **natural resources**, culminating in the formation of the Commission for the Study of the Natural Productive Forces of Russia in 1915. When the Tsarist regime collapsed in 1917, Vernadsky became involved in politics again, joining a campaign to persuade Russians to take pride in and preserve their culture. Vernadsky served briefly as the government's assistant minister in charge of universities and institutions, until the October revolution, which ushered in a government whose politics did not mesh with Vernadsky's. In 1919, he moved to Kiev to found and become president of the Ukrainian Academy of Sciences. When the Red Terror of 1919 drove him into hiding, Vernadsky spent his time in isolation, developing the blueprints for a new field he called **biogeochemistry**. As he saw it, this new discipline studied the nexus between geology, chemistry, and biology, determining how prevalent life was, the rates at which different forms of life multiply, and the processes and speed of **adaptation**.

Vernadsky left Russia in 1921 for France, where he worked with Marie Curie. During this period, he coalesced his thoughts on the interconnection of all living and nonliving beings on Earth into a book entitled *The Biosphere*. After four years in the West, Vernadsky's wife lobbied to move there permanently, rather than returning to Soviet Russia. But the government was eager to attract prominent scholars such as Vernadsky and offered him a chair in the academy with the promise of time and funding for his own work, an offer that no Western institution matched. In 1926 he founded and headed the Commission on the History of Knowledge of the Academy of Sciences of the USSR and lay the basis for the organization that later became the Vernadsky Institute of Geochemistry and Analytical Chemistry.

During the Stalin years, Vernadsky continued working, relatively uninterrupted. His earlier years of protest against the Tsar, combined with his image as an elder statesman of science, protected him from the government persecution that many of his colleagues experienced. In 1935 he began writing philosophical essays on the nature of the world. In 1940 his years of interest in **radioactivity** culminated in the creation of the **Uranium** Institute. During World War II, he argued for Russia to develop its atomic energy program. His wife died in 1943 while the couple was evacuated from their home. Vernadsky passed away two years later on January 6, 1945, a few months after suffering a cerebral hemorrhage.

Resources
Books

Balandin, R. K. *Outstanding Soviet Scientists: Vladímir Vernadsky.* Translated by Alexander Repyev. Moscow: Mir Publishers, 1982.

Bailes, Kendall. *Science and Russian Culture in an Age of Revolutions: V. I. Vernadsky and His Scientific School, 1863–1945.* Indiana University Press, 1990.

Izbrannye sochinenia (Selected works). Six volumes. Moscow, 1954–1960.

Notable Scientists: From 1900 to the Present. Farmington Hills, MI: Gale Group, 2002.

Vernadsky, Vladímir. *Biosfera* (The biosphere). Leningrad, 1926.

Victims' compensation

Traditionally, legal remedies for environmental problems have been provided by common law (judge-created law developed through private lawsuits). Common law provides remedies including compensation to victims injured by another's negligence. For example, if appropriate care is not taken in the disposal of a **toxic substance** and this substance enters a farm pond, killing or injuring the farmer's livestock, the farmer can sue the polluter for damages.

United States proposals for reforming victims' compensation fall into two general categories: 1) an approach that combines administrative relief with common law tort (a tort is a wrong actionable in civil court) reform; and 2) proposals that provide administrative relief, but eliminate tort remedies. The first approach was developed by a study group consisting of 12 attorneys designated by the American Bar Association, American Trial Lawyer's Association, the Association of State Attorneys General, and the American Law Institute. In 1980, Congress asked this study group to consider the hazardous substance personal injury problem in conjunction with the Comprehensive Environmental Re-

sponse, Compensation and Monitoring Act (Superfund). The study group recommended a "two-tier" approach. The first tier, which would be the primary remedy for injured persons, would consist of administrative relief. This part of the system would operate in a manner similar to workmen's compensation. Within three years after the discovery of an injury or disease, an applicant would make a claim based on proof of exposure, existence of the disease or injury, and compensable damages. The applicant, without having to show fault, would receive medical costs and two-thirds of earnings minus the amounts that could be obtained from other government programs. The money for this victims' compensation fund would come from industry sources through a tax on hazardous activities or some other means of eliciting contributions.

Most claims would be dealt with through the administrative system without resort to the courts. However, the second tier in the program would preserve existing tort law. Plaintiffs who chose this option would have to submit to the costs and delays of legal proceedings. However, a plaintiff able to win in the courts would have the right to collect unlimited damages, including full loss of earnings and compensation for pain and suffering.

A second proposal that combines administrative relief with traditional tort remedies originated at the **Environmental Law Institute**. Its director, Jeffrey Trauberman, published an extensive article that appeared in the *Harvard Environmental Law Review* in 1983 proposing a "Model Statute." The approach was different from the Attorneys' Study Group because it emphasized common law tort reform as the primary remedy, with the victims' compensation fund serving as "a residual or secondary source of compensation" in instances when the responsible party could not be identified, had become insolvent, or had gone out of business.

The major common law problem that Trauberman addressed was that of causation. Traditionally, the courts have been reluctant to accept probabilistic evidence as proof of causation. Trauberman, however, argued that evidence from **epidemiology**, animal and human toxicology, and other sources on the "frontiers of scientific knowledge" should be admitted. When a plaintiff was unable to demonstrate a "substantial" case of harm, Trauberman would permit "fractional recovery." By "fractional recovery" he meant that if a **hazardous waste** dump increased the total number of cancers in an area from 8–10%, then the increased incidence of **cancer** brought on by the dump was 25% in these cases, victims should be able to recover 25% of their costs. Recovery for pain and suffering, which was not available to fund claimants, would be available through litigation. These balanced proposals are to be distinguished from bills introduced in the U.S. Congress that would create an administrative compensation system but preclude tort remedies.

In contrast to the U.S., Japan has adopted an approach that combines administrative relief with tort justice. The 1973 Law for the Compensation of Pollution-Related Health Injury, which replaced a simpler 1969 law, established an administrative system to oversee compensation payments. Upon certification by a council of medical, legal, and other experts, victims of designated diseases are eligible for medical expenses, lost earnings, and other expenses, but they receive no allowance for non-economic losses such as pain and suffering. Companies pay the entire cost of this compensation. There has been an administrative review system, but in no case does this system prohibit recourse to the courts. In the United States, the institutions responsible for developing victims' compensation policy have not, as yet, forged such a comprehensive policy. *See also* Ashio, Japan; Itai-Itai disease; Minamata disease; Yokkaichi asthma

[*Alfred A. Marcus*]

Resources

Books

Marcus, A. A. *Business and Society.* Homewood, IL: Irwin Press, 1992.

Video display terminal (VDT)

see **Electromagnetic field**

Vinyl chloride

A colorless, flammable, and toxic liquid when under pressure, but a gas under ordinary conditions. Polymerized vinyl chloride, called **polyvinyl chloride**, is a ubiquitous plastic produced in enormous quantities. While vinyl chloride has been detected at some waste sites, exposure has been much greater in the factories that produced it. However, significant efforts were exerted by industry to reduce occupational exposure. Chronic exposure resulted in "vinyl chloride disease" with liver, nerve, and circulatory damage. Epidemiological studies associated **cancer** of the liver (particularly angiosarcoma) and possibly the brain with occupational exposure, and thus vinyl chloride is considered to be a human **carcinogen**.

Virus

A virus is a submicroscopic particle that contains either RNA (**ribonucleic acid**) or DNA (deoxyribonucleic **acid**). Viruses are not capable of performing metabolic functions outside of a host cell upon which the virus depends for replication. Viruses are found in the **environment** in a wide range of sizes, chemical composition, shape, and host cell specificity. Viruses can cause disease or genetic damage to host cells

and can infect many living things, including plants, animals, bacteria, and **fungi**.

Bacteriophages are viruses which use bacteria as hosts. Of particular interest in the aquatic environment are coliphages, viruses that infect ***Escherichia coli***, a bacteria that commonly grows in the colons of mammals. *E. coli* is an important bacterial indicator of fecal pollution of water. However, coliphages tend to survive much longer in the environment than *E. coli*, and their detection in water in the absence of *E. coli* tends to be a more sensitive indicator of former fecal contamination than coliform bacteria.

Human intestinal viruses are the most commonly encountered viruses in **wastewater** and water supplies since they are shed in large numbers by humans (10^9 viruses per gram of feces from infected individuals) and are largely unaffected by wastewater treatment before **discharge** to the environment. Viruses cannot replicate in the environment, but can survive for long periods of time in surface water and **groundwater**. Viruses are difficult to isolate from the environment and, once collected, are difficult to culture and identify because of their small size, numerous types, low concentrations in water, association with suspended particles, and the limitations in viral identification methods.

There are more than 100 types of known enteric viruses, and there are many others yet to be found. Enteric viruses include polio viruses, coxackieviruses A and B, echoviruses, and probably hepatitis A virus. Waterborne transmission of the polio virus in developed countries is rare. Of more consequence are the coxackieviruses and hepatitis A virus, with hepatitis A virus being a leading etiological agent in waterborne disease. Other viruses of concern include the gastroenteritis virus group, a poorly understood family of viruses which are probably a subset of the enteric viruses. The important members of this group include the Norwalk agent, rotaviruses, coronaviruses, caliciviruses, the W agent, and the cockle virus.

Preventing the transmission of viruses through water supplies depends upon adequate chemical disinfection of the water. **Resistance** of viruses to disinfectants is due largely to their biological simplicity, their tendency to clump together or aggregate, and protection afforded by association with other forms or organic material present. Proper **chlorination** of water is usually sufficient for inactivation of viruses. **Ozone** has been used for inactivation of viruses in drinking water because of its superior virucidal properties. France has been a leader in the use of ozone for inactivation of waterborne viruses.

[*Gordon R. Finch*]

Resources

Books

Feachem, R. G., et al. *Sanitation and Disease. Health Aspects of Excreta and Wastewater Management.* New York: Wiley, 1983.

Greenberg, A. E., L. S. Clesceri, and A. D. Eaton, eds. *Standard Methods for the Examination of Water and Wastewater.* 18th ed. Washington, DC: American Public Health Association, American Water Works Association, Water Environment Federation, 1982.

Periodicals

Foliguet, J. M., P. Hartemann, and J. Vial. "Microbial Pathogens Transmitted by Water." *Journal of Environmental Pathology, Toxicology, and Oncology* 7 (1987): 39–114.

Visibility

Visibility generally refers to the quality of vision through the **atmosphere**. Technically, this term denotes the greatest distance in a given direction that an observer can just see a prominent dark object against the sky at the **horizon** with the naked eye. Visibility is a measure commonly used in the observance of weather in the United States. In this context, surface visibility refers to visibility determined from a point on the ground, control tower visibility refers to visibility observed from an airport control tower, and vertical visibility refers to the distance that can be seen vertically into a ground-obscuring medium such as fog or snow.

Visibility is affected by the presence of aerosols or **haze** in the atmosphere. In the ideal case of a black target on a white background, the target can be seen at close range because it reflects no light to the eye, while the background reflects a great deal of light to the eye. As the distance between the observer and the target object increases, the light from the white background is scattered by the intervening particles, blurring to some degree the edge between the target and the surroundings, since light from the background is now scattered into the line of sight to the black target. In addition, the whiteness of the background is decreased by the scattering of light out of the direct line from the background to the eye, and by the **absorption** of some of the light by dark particles, principally **carbon** (soot). Finally, particles in the line between the target and the eye scatter light to the eye, decreasing the blackness of the target as seen from the distance. At any distance greater than the visibility, then, all these actions degrade the contrast between the object and its background so that the eye can no longer pick out the object.

Some light is scattered even by air molecules, so visibility through the atmosphere is never infinite. It can be great enough to require correction for the curvature of the earth, and the consequent fact that atmospheric density is less (fewer scattering molecules per unit of distance) at the end of the sight path than at the location of the observer.

Air pollutants now produce a haze that circles the globe, significantly reducing visibility even in places as remote as the Arctic. In areas of industrial concentration, haze may consistently reduce visibility by as much as 40%. In

addition, gaseous molecules scatter light; and some gases, most notably **nitrogen** dioxide, absorb light in the visible range, thus changing the apparent color of the background. The effects of **air pollution** on visibility are quite well understood, unlike some other effects of this **pollution** on human health and the **environment**. *See also* Arctic haze; Photochemical smog; Smog

[*James P. Lodge Jr.*]

Resources
Books

Middleton, W. E. K. *Vision Through the Atmosphere.* Toronto: University of Toronto Press, 1952.

Periodicals

Lodge Jr., J. P., et al. "Non-Health Effects of Airborne Particulate Matter." *Atmospheric Environment* 15 (1981): 449–458.

William Vogt (1902 – 1968)

American ecologist and ornithologist

An ecologist and ornithologist whose work has influenced many disciplines, William Vogt was born in 1902 in Mineola, New York. Convalescing from a childhood illness, Vogt became a voracious reader, consuming the works of poets, playwrights, and naturalists. One naturalist in particular, Ernest Thompson Seton, greatly influenced him, cultivating his interest in ornithology. Vogt graduated with honors in 1925 from St. Stephens (now Bard) College in New York, where he had studied romance languages, edited the school literary magazine, and won prizes for his poetry.

After receiving his degree, Vogt worked for two years as assistant editor at the New York Academy of Sciences, and he went on to act as curator of the Jones Beach State Bird Sanctuary from 1933 to 1935. As field naturalist and lecturer at the National Association of Audubon Societies from 1935 to 1939, he edited *Bird Lore* magazine and contributed articles to other related professional periodicals. Perhaps his most important efforts while with the association were his compilation of *Thirst for Land*, which discussed the urgent need for **water conservation**, and his editing of the classic *Birds of America.*

Vogt's chief accomplishment, however, was to make the world more fully aware of the imbalanced relationship between the rapidly growing world population and the food supply. He developed a strong interest in Latin America in the late 1930s. During World War II he acted as a consultant to the United States government on the region, and in 1942 he traveled to Chile to conduct a series of climatological studies, during which he began to realize the full scope of the depletion of **natural resources** and its consequences for world population. As he complied the results of his studies, Vogt became increasingly interested in the relationship between the **environment** and both human and bird populations. He was appointed chief of the **conservation** section of the Pan American Union in 1943, and he remained at this post until 1950, working to disseminate information on and develop solutions for the problems he had identified.

Vogt's popularity soared in 1948 when he published *Road to Survival*, which closely examined the discrepancy between the world population and food supplies. A best-seller, the book was eventually translated into nine languages. It was also a major influence on Paul R. Ehrlich, author of *The Population Bomb* and a key proponent of **zero population growth** theories.

Following the publication of *Road to Survival*, Vogt won Fulbright and Guggenheim grants to study population trends and problems in Scandinavia. Soon after returning to the United States, he was appointed national director of Planned Parenthood Federation of America, an organization wholly concerned with limiting **population growth**.

During the last seven years of his life, Vogt again turned his attentions to conservationism, and he became secretary of the Conservation Foundation, an organization created to "initiate and advance research and education in the entire field of conservation," which was recently absorbed by the **World Wildlife Fund**. Vogt died in New York City on July 11, 1968.

[*Kimberley A. Peterson*]

Resources
Books

Squire, C. B. *Heroes of Conservation.* New York: Fleet Press, 1974.
Vogt, W. *Road to Survival.* New York: W. Sloane Associates, 1948.

Volatile organic compound

In **environmental science**, the term usually refers to the **hydrocarbons**, especially those found in **air pollution**. In 1988, the five most abundant of these hydrocarbons in urban air were isopentane, n-butane, **toluene**, propane, and ethane; each of these compounds has unburned **gasoline** as its primary source. In the presence of sunlight, hydrocarbons react with **ozone**, **nitrogen oxides**, and other components of polluted air to form compounds that are hazardous to plants, animals, and humans. In 1995, volatile organic compounds ranked third behind **carbon monoxide** and **particulate** matter (which includes dust, **smoke**, soot, and chemical liquid droplets from various sources) and just above sulfur and **nitrogen** oxides in annual pollutant emissions in the United States. *See also* Air pollution control; Air quality

Volcano

Volcanoes have been called the thermostat of the planet. They wreak havoc, but also spawn far-ranging benefits for **soil** and air. Some earth scientists now say that the vast swath of destruction from a volcanic eruption can be a source of creation.

Most land volcanoes erupt along plate edges where ocean floors plunge deep under continents and melting rock rises to the surface as magma. The earth's fragmented crust pulls apart and the edges grind past or slide beneath each other at a speed of up to 8 in (20 cm) per year. But just as our blood carries nutrients that feed our body parts, volcanoes do the same for the skin of the earth.

Magma contains elements required for plant growth, such as **phosphorus**, potassium, magnesium and sulfur. When this volcanic material is blasted out as ash, the fertilization process that moves the nutrients into the soil can occur within months. Java, one of the most volcano-rich spots on the earth, is one of the world's most fertile areas.

Magma also yields energy: it heats the underground water that is tapped by **wells** to warm most of the homes in Iceland. Natural steam drives turbines that provide 7% of New Zealand's electric power, and it accounts for 1% of the United States' energy needs.

Atmospheric after-effects of a volcanic eruption can last for years, as the eruption of **Mount Pinatubo** in the Philippines on June 15, 1991, has shown. While water vapor is the main gas in magma, there are smaller amounts of **hydrogen** chloride, **sulfur dioxide**, and **carbon dioxide**. Sulfur dioxide blasted 25 mi (40 km) into the **stratosphere** after Mount Pinatubo erupted, combined with moisture to create a thin **aerosol** cloud that girdled the globe in 21 days. Scientists calculated that 2% of the earth's incoming sunlight was deflected, leading to slightly lower temperatures on worldwide average. These light sulfur dioxides can circle the globe for years and possibly damage the **ozone** layer.

Recently, scientists mapping the sea floor in the South Pacific have found what they call the greatest concentration of active volcanoes on earth. More than 1,000 seamounts and volcanic cones, some as high as 7,000 ft (2,135 m) with peaks 5,000 ft (1,525 m) beneath the ocean surface, are located in an area the size of New York state. One potential benefit of eruptions is that they generate new mineral deposits, including **copper**, iron, sulfur, and gold. The discovery is likely to intensify debate over whether volcanic activity could change water temperatures enough to affect weather patterns in the Pacific. Scientists speculate that periods of extreme volcanic activity underwater could trigger **El Niño**, a weather system that alters weather patterns around the world. *See also* Geothermal energy; Ozone layer depletion; Plate tectonics

[*Linda Rehkopf*]

Resources

Periodicals

Findley, R. "Mount St. Helens Aftermath: The Mountain That Was and Will Be." *National Geographic* 180 (December 1991): 713.

Grove, N. "Volcanoes: Crucibles of Creation." *National Geographic* 182 (December 1992): 5–41.

Powell, C. S. "Greenhouse Gusher." *Scientific American* 265 (October 1991): 20.

"Volcano Could Cool Climate, Reduce Ozone." *Science News* 140 (July 6, 1991): 7.

Dr. Richard Albert Vollenweider (1922 –)

Swiss limnologist

Vollenweider is one of the world's most renowned authorities on eutrophication, the process by which lakes mature and are gradually converted into swamps, bogs, and finally meadows. Eutrophication is a natural process that normally takes place over hundreds or even thousands of years, but human activities can accelerate the rate at which it occurs. The study of these human effects has been an important topic of environmental research since the 1960s. Richard A. Vollenweider was born in Zurich, Switzerland, on June 27, 1922. He received his diploma in biology from the University of Zurich in 1946 and his Ph.D. in biology from the same institution in 1951. After teaching at various schools in Lucerne for five years, in 1954 he was appointed a fellow in **limnology** (the study of lakes) at the Italian Hydrobiological Institute in Palanza, Italy. A year later, he accepted a similar appointment at the Swiss-Swedish Research Council in Uppsala. Vollenweider has also worked for two international scientific organizations, the United Nations Economic, Scientific, and Cultural Organization (**UNESCO**), and the Organization for Economic Cooperation and Development (OECD). From 1957 to 1959, he was stationed in Egypt for UNESCO, working on problems of lakes and fisheries. Between 1966 and 1968, he was a consultant on **water pollution** for OECD in Paris.

In 1968, Vollenweider moved to Canada to take a position as chief limnologist and head of the fisheries research board of the Canada Centre for Inland Waters (CCIW). In 1970 he was promoted to chief of the Lakes Research Division and in 1973 to the position of senior scientist at CCIW.

Perhaps the peak of Vollenweider's career came in 1978 when he was awarded the Premio Internazionale Cervia

Ambiente by Italian environmentalists for his contributions to research on the **environment**. The award was based largely on Vollenweider's work on eutrophication of lakes and waterways in the Po River region of Italy. Although the area was one of the few in Italy with sophisticated water and **sewage treatment** plants, there was abundant evidence of advanced eutrophication.

Vollenweider found a number of factors contributing to eutrophication in the area. Most importantly the treatment plants were not removing **phosphorus**, which is a major contributor to eutrophication. In addition, pig farming was widely practiced in the area, causing huge quantities of phosphorus contained in pig wastes to enter the surface water.

Vollenweider recommended a wide range of changes to reduce eutrophication. These included the addition of tertiary stages in sewage treatment plants to remove phosphorus, special treatment of wastes from pig farms, and agreements from manufacturers of **detergents** to reduce the amount of phosphorus contained in their products.

[*David E. Newton*]

Resources

Periodicals

Davey, T. "CCIW Scientist Wins Top Italian Environmental Award." *Water and Pollution Control* (January 1979): 23.

Volume reduction, solid waste

see **Solid waste volume reduction**

W

War, environmental effects of

War and similar conflicts have long been an activity of humans, and is referred to in the earliest historical records. Wars involving hunter-gatherer and early agricultural cultures included clashes between clans and village groups. This sometimes led to the deaths of some participants, as has been observed up to the present century in such places as New Guinea and Borneo. In marked contrast, modern warfare involving advanced technological societies can wreak a truly awesome destruction—about 84 million people have been killed during the wars of the twentieth century.

The conflict that has been the most destructive to human life was World War II (1939-1945), during which about 38 million people were killed. World War I (1914-1918) resulted in about 20 million deaths, the Korean War (1950-1953) about 3 million, and the Vietnam War (1961-1975) about 2.4 million.

In addition to having awful consequences for people and their civilizations, modern warfare also causes terrible environmental damages. These include the destruction caused by conventional weapons, effects of the military use of poisonous gases and herbicides, and **petroleum** spills. In addition, the potential consequences of nuclear warfare are horrific—a nuclear holocaust could kill billions of people and might also result in **climate** changes that would cause a collapse of biospheric processes.

Destruction by conventional warfare

Enormous amounts of explosive munitions are used during modern wars. An estimated 23.1 million tons (21 million metric tons) of explosives were expended during World War II, 36% of that by United States forces, 42% by the Germans, and 22% by other combatants. About 3.3 million tons (3 million metric tons) were used during the Korean War, 90% of that by U.S. and Allied forces. About 15.4 million tons (14 million metric tons) were used during the Vietnam War, 95% of that by U.S. and South Vietnamese forces. During the Gulf Conflict, several months of bombardment by U.S.-led coalition forces resulted in 88,000 tons (80,000 metric tons) of explosives being dropped on Iraq. Although not well-documented, the explosion of such enormous quantities of munitions during these wars caused great damages to people, buildings, and ecosystems.

Much of World War I, for example, was waged in an area of coastal lowland in Belgium and France known as the Western Front. This was a terrible theater of war, which involved virtually static confrontations between huge armies that fought back and forth over a well-defended landscape webbed with intricate trenchworks. Territorial gains could only be made by extraordinarily difficult frontal assaults, which resulted in an enormous waste of men and material. Battles were preceded by intense artillery bombardments, which devastated the agricultural lands and woodlands of the battlefields.

One observer described the ruined terrain of an area of Belgium known as Flanders in this way: "In this landscape nothing existed but a measureless bog of military rubble, shattered houses, and tree stumps. It was pitted with shell craters containing fetid water. Overhead hung low clouds of **smoke** and fog. The very ground was soured by poison gas." (cited in Freedman, 1995). Another passage described the destruction of a forest (or copse) by an artillery bombardment: "When a copse was caught in a fury of shells the trees flew uprooted through the air like a handful of feathers; in a flash the area became, as in a magicians trick, as barren as the expanse around it." These were common observations, and they typified the damages caused by explosions, machines, and the mass movements of men determined to kill each other.

About half of the munitions used during the Vietnam War was delivered by aircraft, half by artillery, and less than 1% from offshore ships. United States forces dropped about 20 million aerial bombs of various sizes, fired 230 million artillery shells, and used more than 100 million grenades and additional millions of rockets and mortar shells. Of course, these caused enormous physical destruction to the built **environment** of cities and towns, and to agricultural and natural environments. During 1967 and 1968, for exam-

ple, about 2.5 million craters were formed by 500- and 750-lb (227- and 337-kg) bombs dropped in saturation patterns by high-flying B-52 bombers. Each plane sortie produced a bombed-out area of about 161 acres (65 ha). Eventually about 21.6 million acres (8.1 million ha) or 11% of the landscape of Indochina was affected in this way (this includes Laos and Vietnam). Craters were about 16 yd (15 m) wide and 13 yd (12 m) deep, and they usually filled with fresh water. In addition, the explosions often started forest or grassland fires, which caused extensive secondary damages.

Some animals have been brought to the brink of **extinction** through warfare. For instance, the last wild Pere David's deer (*Elaphurus diavidianus*) were killed during the Boxer War of 1898-1900 in China. The European **bison** (*Bison bonasus*) was almost rendered extinct by **hunting** during World War I to provide meat for troops. More recently, warfare has led to lawlessness in much of Africa, allowing well-armed gangs of poachers to cause white rhinos (*Ceratotherium simum*), black rhinos (*Diceros bicornis*), **elephants** (*Loxodontia africana*), and other **species** to become critically endangered over much of their range. These animals are hunted for their horns, tusks, and other valuable body parts.

Although the effects of war on wild animals have mostly been damaging, there have been a few exceptions. Usually this happened because of decreased exploitation—men were too busy trying to kill each other to bother with hunting other species. For example, the abundance of gamebirds in Britain increased markedly during both World Wars because of decreased hunting pressures. So did fish stocks in the North Atlantic, because fishing boats were subject to attacks.

The legacy of unexploded munitions

Many aerial bombs and artillery shells do not explode and cause a lingering hazard on the landscape. These problems are made much worse by the use of explosive mines in warfare, because few of these buried devices are recovered after the hostilities cease, and they continue to be an explosive hazard for people and domestic and wild animals for decades. Modern anti-personnel mines are extremely difficult to find, largely because they can contain as little as one gram of metal, which makes it hard to detect them magnetically.

Chemical weapons in warfare

Antipersonnel chemical warfare occurred on a large scale during World War I, when more than 220.5 million lb (100 million kg) of lethal agents were used. These were devastating lung poisons such as **chlorine**, phosgene, trichloromethyl chloroformate, and chloropicrin, and the dermal agent known as mustard gas. These poisons caused about 1.3 million casualties, including 100,000 deaths.

Lethal gases were also used during the Iran-Iraq War of 1981-1987, mostly by Iraqi forces. The most famous incident involved the Kurdish town of Halabja in northern Iraq, which had rebelled against the central government and was aerially gassed with the nerve agents sabin and tabun, causing about 5,000 deaths.

The nonlethal "harassing agent" CS (or o-chlorobenzolmalononitrile) was used by the U.S. military during the Vietnam War. About 20 million lb (9 million kg) of CS was sprayed over more than 2.5 million acres (one million ha) of South Vietnam, rendering treated places uninhabitable by humans, and likely wild animals, for up to 45 days.

In addition, extensive areas of Vietnam were treated with herbicides by U.S. forces to deprive their enemy of forest cover and food production. More than 3.5 million acres (1.4 million ha) were sprayed at least once, equivalent to about one-seventh the area of South Vietnam. Most of the treated area was mangrove or upland forest, with about 247,000 acres (100,000 ha) being cropland. The most commonly used **herbicide** was a 50:50 mixture of **2,4,5-T** plus **2,4-D**, known as **Agent Orange**. More than 55 million lb (25 million kg) of 2,4-D and 47 million lb (21 million kg) of 2,4,5-T were sprayed in this military program. Because the intent was to achieve a longer-term **defoliation**, the application rates were relatively high, equivalent to about 10 times the rate normally used in forestry.

The herbicide spraying caused great ecological damages, with effects so severe that critics of the practice labeled it "ecocide," i.e., the intentional use of anti-environmental actions over a large area, carried out as a tactical component of a military strategy. The damages included loss of agricultural lands for many rural people, as well as extensive **deforestation**, which caused severe but almost undocumented effects on the native **biodiversity** of Vietnam.

Because the 2,4,5-T used in the herbicide spraying was contaminated by a **dioxin** chemical known as TCDD (2,3,7,8-tetrachlorodibenzo-*p*-dioxin), there was also a great deal of controversy about the potential effects on humans. As much as 357 lb (162 kg) of TCDD was sprayed with herbicides onto Vietnam. Although there have been claims of damages caused to exposed populations, including U.S. military personnel, the scientific studies have not demonstrated convincing linkages. Although the subject remains controversial, it seems likely that the specific effects of TCDD were small in comparison with the overall ecological effects of the use of herbicides in the Vietnam War.

Effects of nuclear warfare

Nuclear weapons have an enormous capability for destruction of humans, their civilization, and natural ecosystems. This was recognized by John F. Kennedy during a speech to the United Nations in 1961: "Mankind must put an end to war, or war will put an end to mankind."

The world's nuclear arsenal peaked during the 1980s, when the explosive yield of all weapons was about 18,000 megatons (Mt) of TNT-equivalent. This was more than

1,000 times larger than the combined yield of all conventional explosives used during World War II (6.0 Mt), the Korean War (0.8 Mt), and the Second Indochina War (4.1 Mt). The late-1980s nuclear arsenal was equivalent to 3–4 tons (2.7–3.6 metric tons) of TNT per person on Earth.

Thankfully, the cessation of the Cold War has led to substantial reductions in the world's nuclear weaponry, which has declined from a total explosive yield of about 18,000 Mt in the 1980s to about 8,000 in the mid-1990s. There were about 64,000 strategic and tactical devices in 1983, but 27,000 in 1993, and planned reductions to 11–19 thousand in 2003.

Nuclear weapons have twice been used in warfare. This involved bombs dropped by U.S. forces on Japan, in an action that brought World War II to an earlier end than would have occurred otherwise. The first bomb had an explosive yield equivalent to 0.015 Mt of TNT and was dropped on Hiroshima on August 6, 1945, and the second (0.021 Mt) was dropped on Nagasaki a few days later. The Hiroshima bomb killed about 140,000 people, while the Nagasaki device killed 74,000. The combined effects of blast and heat destroyed about two-thirds of the buildings in Hiroshima, and one-fourth of those in Nagasaki. Although enormous in comparison with conventional bombs, these nuclear devices were small in comparison to the typical yield of modern strategic warheads, which average about 0.6 Mt and range to 6 Mt.

Not surprisingly, people are concerned about the likely consequences of a large-scale nuclear exchange. One commonly used scenario of a nuclear war is for a limited exchange of 5,000–6,000 Mt, most of which would be exploded in the Northern Hemisphere. The predicted effects on people include the deaths of about 20% of the world's population, including as much as 75% of the population of the United States. The principal cause of death would be thermal radiation, but effects of blast, fire, and **ionizing radiation** would also be important. Virtually all survivors would be injured by these same forces. Such an overwhelming loss of human life, coupled with the physical devastation, would transform civilization.

It has also been suggested that a nuclear war could result in a global deterioration of climate, which would cause additional severe damages. The climatic effects might be caused by various influences of nuclear explosions, but especially the injection of large quantities of sooty smoke, inorganic particulates, and gases into the **atmosphere**. These could interfere with the planet's **absorption** of **solar energy** and its re-emission of infrared energy. These potential consequences of nuclear warfare have been studied using computer models of the type that are used to model the effects of emissions of **carbon dioxide** and **methane** on Earth's **greenhouse effect**. In general, the predictions are that nuclear war could cause a global cooling, or a "nuclear winter," which would cause widespread damages to agriculture and natural ecosystems. The longer-term climatic damages would add to the enormous destructions that were caused by blast, thermal radiation, fire, and ionizing radiation within a short time of the nuclear exchange.

Humans have probably engaged in warfare throughout their history, and regrettably we appear likely to continue to do so into the future. Fortunately, the catastrophic effects of modern technological warfare are being increasingly recognized, and many nations have been making progress in avoiding conflicts, particularly amongst the world's most heavily armed nations. Because nuclear war has especially horrific consequences, recent treaties on non-proliferation of nuclear weapons and nuclear arms reductions provide real prospects in this regard.

[*Bill Freedman Ph.D.*]

Resources

Books

Freedman, B. *Environmental Ecology.* 2nd ed. San Diego: Academic Press, 1995.

Sivard, R. L. *World Military and Social Expenditures, 1996.* Washington, DC: World Priorities, 1996.

Westing, A. H., *Explosive Remnants of War. Mitigating the Environmental Effects.* London: Taylor & Francis, 1985.

Westing, A. H., ed. *Herbicides in War: The Long-term Ecological Consequences.* London: Taylor & Francis, 1984.

Periodicals

Barnaby, F. "The Environmental Impact of the Gulf War." *Ecologist* 21 (1991): 166–172.

Grover, H. D., and M. A. Harwell. "Biological Effects of Nuclear War. I. Impact on Humans." *Bioscience* 35 (1985): 570–575.

Grover, H. D., and M. A. Harwell. "Biological Effects of Nuclear War. II: Impact on the Biosphere." *Bioscience* 35 (1985): 576–583.

Grover, H. D., and G. F. White. "Toward Understanding the Effects of Nuclear War." *Bioscience* 35 (1985): 52–556.

Warner, F. "The Environmental Consequences of the Gulf War." *Environment* 33, no. 5 (1991): 7–26.

Warbler

see **Kirtland's warbler**

Waste exchange

Getting rid of industrial waste poses many problems for those who generate it. With environmental restraints enforced by fines, companies must minimize waste at its source, recycle it within the company, or transport it for off-site **recycling**, treatment, or disposal.

The concept of waste exchange, begun in Canada in the 1980s, involves moving one institution's overstock, obsolete, damaged, contaminated, or post-dated materials to another site that might be able to use it. Waste exchange companies sprang up to meet this need, but the need for effective and rapid communication between these companies soon became apparent, because much of the waste involved in the program had to be dealt with on a timely basis. Hazardous wastes, for example, must be disposed of within 90 days.

In 1991, the **Environmental Protection Agency** (EPA) gave a $350,000 grant to the Pacific Materials Exchange in Spokane, Washington, to develop a free computer online network. Servicing about 30 waste exchange companies in the United States and Canada, the National Materials Exchange Network acts like "an industrial dating service," according to director Robert Smee. The service is easily accessible by an 800 number [800-858-6625] to anyone with a computer and a modem. Smee estimates that the computer network in its first year of operation saved companies $27 million in disposal fees.

While protecting the identity of companies generating waste, the network publishes a want list and an available source list of laboratory **chemicals**, paints, acids, and other wastes. Some are hazardous, but others are innocuous, such as one company's scrap wood, which was bought by another to be ground up for air freshener. *See also* Industrial waste treatment; Solid waste; Waste management

[*Stephanie Ocko*]

Resources

Periodicals

Manning, S. "Waste Exchanges: Why Dump When You Can Deal?" *PEM: Plant Engineering & Maintenance* (June 1990): 32–41.

Schwartz, E. I. "A Data Base That Truly Is 'Garbage In, Garbage Out'." *Business Week* (September 17, 1990): 92.

Waste Isolation Pilot Plant

Developing a safe and reliable method for disposing of radioactive wastes is one of the chief obstacles to broader applications of **nuclear power**. Nearly a half century after the world's first nuclear reactor was opened, the United States still had no permanent method for the isolation and storage of wastes that may remain dangerously radioactive for thousands of years. Scientific disputes, technical problems, and political controversies have slowed the pace at which waste disposal systems can be studied and built. The history of the Waste Isolation Pilot Plant (WIPP), located near Carlsbad, New Mexico, is an example of how difficult the solution to this challenge can be.

WIPP was designed by the **U.S. Department of Energy** (DOE) in the 1970s to test methods for isolating and storing low- and intermediate-level and transuranic radioactive wastes. (Transuranic wastes are produced mainly by **nuclear weapons** plants and consist of clothing, debris, tools, and other disposable items that have been contaminated with radioactive wastes.) Researchers decided that the most promising disposal system was to seal the wastes in steel containers and bury these in deep caves built into natural salt beds. They knew that salt beds could absorb the heat produced by **radioactive waste**, and that the beds were usually located in earthquake-free zones. In addition, salt bed caves were attractive because scientists believed they were dry, which prevented wastes from **leaching** out of their tanks. Salt would also tend to creep into openings, thus sealing the drums for thousands of years.

Between the 1970s and 1990s, the DOE spent more than $1 billion building huge caves 2,100 ft (640 m) underground near Carlsbad. The plan was to bury 800,000 drums of nuclear waste and study its behavior over a number of years. It was soon discovered, however, that some salt beds contain layers of brine (salt water), indicating that such beds are not always dry. Salt in the caves has begun to "creep," or slowly move, as expected. Concern about the possible damages caused by the storage of radioactive materials became so intense that today the WIPP is regulated as carefully as a nuclear power plant, providing regular reports to 28 different governmental agencies.

Controversy created a standstill at WIPP and prevented waste drums from being buried there for over two decades. In 1991 Secretary of Energy James Watkins announced that the DOE could wait no longer, and wastes that had been stored at 10 sites around the nation for 20 years awaiting disposal were to be shipped to WIPP. Environmentalists and some government officials in New Mexico reacted strongly to the announcement. They pointed out that Congress was required to give specific approval before any wastes could actually be buried at WIPP. A federal judge ruled that Watkins could not carry out his plan until Congress acted, and the experimental tests at WIPP were once again put on hold.

In the late 1990s, after much deliberation and further testing, Congress approved WIPP for nuclear waste storage. The facility received its first shipment of waste on March 26, 1999. Under congressional mandate, WIPP will only receive transuranic waste and no commercial or high-level nuclear waste. In 2000, more than 99% of existing U.S. transuranic waste was being temporarily stored in drums on nuclear defense sites at 23 locations, including ones in California, Colorado, Idaho, Illinois, Nevada, New Mexico, Ohio, Tennessee, South Carolina, and Washington. At first, the DOE authorized WIPP to only receive waste from Los

Alamos National Laboratory in New Mexico, from the Idaho National Engineering and Environmental Laboratory, and from Rocky Flats, a former nuclear weapons plant near Denver, Colorado. By 2002 the facility had begun receiving waste from other areas, including South Carolina's **Savannah River Site**.

On April 6, 2002, WIPP received its first shipment of waste under the Central Characterization Project (CCP) from the Savannah River Site. CCP is a program designed to make the cleanup of transuranic wastes more efficient, safe, and cost-effective. Characterization involves using a mobile **loading** system to check and approve drums of waste before loading them into specialized shipping containers for transfer to WIPP.

WIPP is projected to house up to 46.7 million gal (177 million l) of transuranic waste, which will remain radioactive for more than 10,000 years. By 2008, the number of shipments is estimated to grow to nearly 1,400 per year, declining again as the facility fills up. By 2035 the facility is projected to have received almost 37,000 truckloads of nuclear waste, barring operational or legal changes. Shipments are moved only in good weather and at night, when traffic is lighter to avoid accidents, and routed around major cities. Shipments are tracked by satellite for safety.

At the beginning of the twenty-first century, about 61 million Americans lived within 50 mi (80 km) of a military nuclear waste storage site. By the time WIPP has been in operation for 10 years, that number is projected to drop to four million. By May 2002, a total of 814 shipments had been sent to the storage facility, totaling 23,852 containers.

[*David E. Newton and Douglas Dupler*]

Resources

Books

Gerrard, Michael B. *Fairness in Toxic and Nuclear Waste Siting.* Cambridge, MA: MIT Press, 1994.

Rahm, Dianne, ed. *Toxic Waste and Environmental Policy in the Twenty-first Century United States.* Jefferson, NC: McFarland & Co., 2002.

Shrader-Frechette, K. S. *Burying Uncertainty: Risk and the Case Against Geological Disposal of Nuclear Waste.* Berkeley: University of California Press, 1994.

Streissguth, Thomas, ed. *Nuclear and Toxic Waste.* San Diego, CA: Greenhaven Press, 2001.

Other

"WIPP Receives Waste Characterized With Mobile System." *DOE News* April 12, 2002 [cited July 6, 2002]. <http://www.wipp.carlsbad.nm.us/pr/2002/WIPPReceivesCCPMobile.pdf>.

Organizations

Department of Energy, Carlsbad Field Office, PO Box 3090, Carlsbad, NM 88221, (505) 234-7352, <http://www.wipp.carlsbad.nm.us>

Waste management

The way that we manage our waste materials is a sign of the times. In our agrarian past, throwing **garbage** out into the street for roving pigs to eat seemed like a perfectly reasonable method of waste management, since most of what we threw away was organic material that increased the bulk of the pig and eliminated residuals we did not want. As manufacturing became a larger part of industry and materials had potential for **recycling** we saw the advent of the another kind of **scavenger**, the "junk man," who pulled from the **waste stream** those materials that had value. This practice increased during the Great Depression and even more so during World War II. During the war there were shortages of metals and cloth as many raw materials were diverted from domestic use to war needs. People got into the habit of **conservation**. Things were not thrown away unless they had no further use either as is or in a remanufactured state. The situation has changed in modern times.

As Jim Hightower points out in the foreword to *War on Waste* (1989), "Every year we toss into city dumps 50 million tons of paper, 41 million tons of food and **yard waste**, 13 million tons of metals, 12 million tons of glass, and 10 million tons of plastic." In addition, he points out that the **Environmental Defense** Fund has calculated that, as a nation, we throw away enough iron and steel to supply domestic auto-makers continuously; enough glass to fill the twin towers of New York City's World Trade Center every two weeks; enough **aluminum** to rebuild our commercial air fleet every three months; and enough office and writing paper to build a 12-ft (3.7 m) Great Wall coast to coast every year. The examples go on and on. The opportunity to manage this tremendous resource through careful waste management practices is urgent. Waste is indeed a renewable resource. This renewable resource, if managed well, could provide us with the ability to reserve our natural non-renewable resources until we really need them.

Most states and major metropolitan areas have a hierarchy of waste management options that guide state and local planning. Although there are refinements of this list in some areas, the standard hierarchy, in order of preference, is: reduce, **reuse**, recycle, compost, waste-to-energy (**incineration**), and **landfill**. Looking at each of these options individually gives a better understanding of an integrated waste management system.

Managing waste through a process of reduction is almost always the first line of defense and viewed as most desirable by waste management professionals. Reduction is an attempt not to generate waste in the first place. This means making certain consumer decisions that eliminate the potential for waste generation. Buying produce in bulk so that packaging is kept at a minimum is one simple consumer

WIPP FACILITY AND STRATIGRAPHIC SEQUENCE

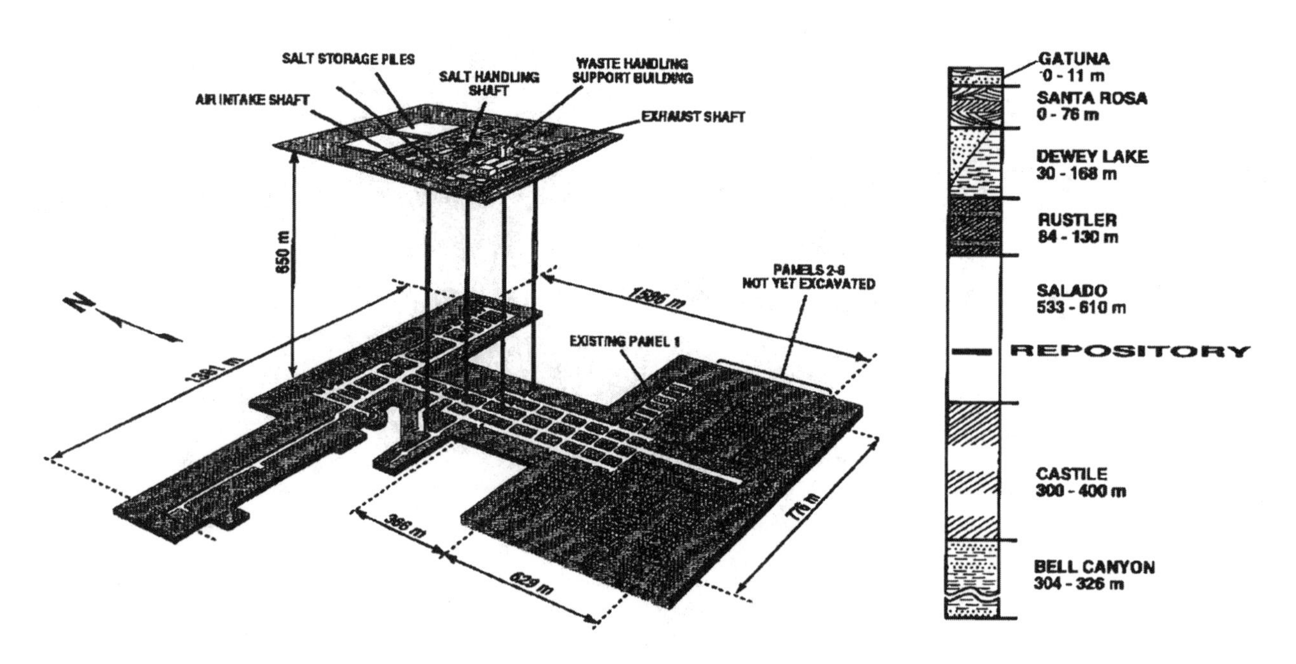

Schematic representation of the Waste Isolation Pilot Plan. (U. S. Department of Energy.)

decision that can have a profound effect, since over 40% of the waste stream is packaging material. To further reduce waste, consumers can bypass disposable items, such as disposable razors and cameras, and have small appliances repaired rather than buy new ones. Using cloth dishtowels and diapers eliminates paper towels and **disposable diapers** from the waste stream.

Reuse is often the second line of defense. This is the small area between avoiding waste and recycling post-consumer materials. A creative way to support reuse of materials is to make them available to those who can use them. Some cities have put in place a center where containers and various post-consumer materials that have been cleaned and source-separated are available to elementary school teachers for art projects. Another example of reuse is consignment shops. These stores will take clothing and small household items on consignment from people and then sell them and divide the profit with the owner.

Recycling is on the increase. Goals of recycling 15%, 25%, and even 50% of waste material have been called for in cities all across the United States. Many believe recycling is a major waste management solution. Industries in the United States are able to recycle quite a variety of materials. Paper, glass, metals, **plastics**, yard and **food waste**, motor oil, batteries, tires, asphalt, car bumpers, and all manner of scrap metal can be recycled. There is truly little that cannot be recycled, but in some cases, industry is not ready. Manufacturing equipment has not yet been retrofitted to accommodate the heterogenous nature of the waste stream. Also, communities are a dispersed source of "raw" material for manufacturers, and the cost of gathering sufficient tonnage of glass, tin, or other "raw" material for manufacturing is prohibitive. Probably the single most restrictive obstacle to full scale recycling is human behavior. As we look at recycling as a closed-loop waste management process we realize that we must source-separate and clean post-consumer waste materials, then we must get them to the point where they can be used and made into a new product. To close the loop, someone must buy that product. All of these activities demand a behavioral change on the part of consumers and manufacturers.

Composting is the biological breakdown of organic matter under **aerobic** conditions. Composting is seen as a major waste management option for large and small commu-

nities. The most common form of composting is the layering of leaves, grass and twigs and sometimes kitchen waste in a backyard compost bin. This very simple technology can also be done at the municipal level, as a community picks up leaves and other yard waste at the curb and trucks it to a central location where it is composted in long tent-shaped piles of refuse called windrows, or inside buildings where mechanical systems speed the compost process. Like recycling, composting requires a behavioral change on the part of homeowners. People must prepare the material and get it to the curb on the appropriate day. Some drawbacks to composting are that the process can generate unpleasant odors if the compost pile is not managed well. This means turning the pile frequently and maintaining aerobic conditions so that **decomposition** proceeds rapidly.

Incineration is not really a new waste management option. Incineration of **solid waste** was practiced in the early 1900s. However, not until the early 1970s did we began to look seriously at **solid waste incineration** with **energy recovery** as a major goal. There really are two types of waste-to-energy systems. One is a **mass burn** technology which does very little to the waste stream before it reaches the furnace. The other system is a refuse-derived fuel (RDF) in which waste is shredded before being delivered to the furnaces. Drawbacks to this system are ash disposal and control of air emissions. Concerns over dioxins in **stack emissions** has caused many supporters to think twice about waste-to-energy. In addition, the issue of where to dispose of the sometimes toxic ash residue must be dealt with.

Landfills are still the most common waste management option. Between 75% and 85% of the nation's waste still goes to landfills. Three types of landfills are constructed for different types of waste materials.

Type III landfills are the least expensive to build and can accommodate construction debris and other inert material. Most common are sanitary Type II landfills. These are constructed according to the criteria in subtitle D of the **Resource Conservation and Recovery Act** (RCRA). Newer **municipal solid waste** landfills are usually equipped with synthetic liners, leachate collection systems, monitoring **wells** and in some cases **methane** wells to draw off the gas that collects in landfills. Type II landfills will take any waste except **hazardous waste**. Hazardous waste can only be disposed of in a Type I landfill. These landfills are constructed in a more rigorous manner and are used primarily for contained **chemicals** and such waste as **asbestos**.

The biggest challenge surrounding modern landfills is the issue of siting. **Not in my backyard** (NIMBY) is a well-known syndrome. Current landfill construction has slowed and many of the old landfills have closed or are slated for closure. There are currently about 3,000 Type II landfills in the continental United States. As more landfills close, the life span of those remaining is shortened. Large multinational companies in the waste management business are attempting to site large regional landfills. In some cases, large companies are offering very attractive incentive packages to communities that are willing to site a landfill within their jurisdictions. Whatever method of disposal is used, the process of managing waste must bring into play a cooperative effort between manufacturers, merchants, citizens, and waste management experts if it is to be successful. *See also* Renewable resources; Source separation

[*Cynthia Fridgen*]

RESOURCES

BOOKS

Blumberg, L., and R. Gottlieb. *War on Waste.* Covelo, CA: Island Press, 1989.

Kharbanda, O. P., and E. A. Stallworthy. *Waste Management: Toward a Sustainable Society.* Westport, CT: Auburn House/Greenwood, 1990.

Underwood, J. D., et al. *Garbage: Practices, Problems, and Remedies.* New York: INFORM, 1988.

Waste reduction

Waste reduction aims to reduce environmental **pollution** by minimizing the generation of waste. It is also often an economically viable option because it requires an efficient use of raw materials. Waste-reduction methods include modifying industrial processes to reduce the amount of waste residue, changing raw materials, or **recycling** or reusing waste sources.

It is more difficult to alter existing industrial processes than it is to incorporate waste-reduction technologies into new operations. Large-scale changes in production equipment are essential to achieving waste reduction. Proper handling of materials and **fugitive emissions** reduction, as well as plugging leaks and preventing spills, all help in waste reduction. Process equipment must be checked on a regular basis for corrosion, vibrations, and leaks. Increases in automation and the prevention of vapor losses also help reduce waste generation.

Changes in what an industry puts into a process can be used to reduce the amount of waste generated. One example is the substitution of raw materials such as water to clean a part instead of solvents. End-products can also be modified to help reduce waste. Another aspect of waste reduction is the control of fugitive emissions by placing a floating roof on open tanks, for example. Other waste-reduction options include the installation of condensers, automatic tank covers, and increasing tank heights. But these and other decisions on waste reduction very much depend on the size and structure of a company. *See also* Industrial

waste treatment; Recyclables; Reuse; Toxic use reduction legislation; Waste management; Waste stream

[*James W. Patterson*]

Resources

Books

Robinson, W. D., ed. *The Solid Waste Handbook.* New York: Wiley, 1986.

Underwood, J. D., et al. *Garbage: Practices, Problems, and Remedies.* New York: INFORM, 1988.

Other

Waste Minimization: Environmental Quality With Economic Benefits. Washington, DC: U.S. Environmental Protection Agency, Office of Solid Waste and Emergency Response, 1987.

Waste stream

Streams of gas, liquid, or solids that are the by-products of a treatment operation or industrial process. The term has also been used to describe the flow of **solid waste** from various sources, including individual homes. The term is often used in **pollution** prevention strategies and **industrial waste treatment** initiatives. Many industries are now required to pretreat their wastes and seek ways to minimize waste production.

Wastewater

Water used and discharged from homes, commercial establishments, industries, **pollution control** devices, and farms. The term wastewater is now more commonly used than sewage, although for the most part, the terms are synonymous. Domestic or sanitary wastewater refers to waters used during the course of residential life or those discharged from restrooms. Industrial wastewaters refer to those generated by an industry. Municipal wastewaters are waters used by a municipality, so they would likely include both sanitary and industrial wastewaters. Combined wastewater refers to a mixture of sanitary or municipal wastewaters and stormwaters created by rainfall.

Wastewater treatment

see **Sewage treatment**

Water allocation

Agriculture and fishing have different needs for water, as do manufacturing industries, cities, and **wildlife**. Water allocation is the process of distributing water supplies to meet the various requirements of a community.

Determining how to allocate water supplies requires the consideration of certain factors, including the source of the water and methods for obtaining it. The cost of the water supply and **water treatment** systems are also taken into account, and the intended uses are reviewed. Water use is classified by whether it is an instream or a withdrawal use and by whether it is a consumptive or nonconsumptive use. Instream uses include navigation, hydroelectric power, and fish and wildlife habitats, while withdrawal uses remove water from the source. Consumptive uses make the water unavailable, either through evaporation or **transpiration**, or through incorporation into products or saltwater bodies. Water withdrawn but not consumed can be treated and returned to the water supply for **reuse** downstream.

The demand for water is determined by combining the sum of the amount of water required for instream needs with the sum of the amount of water needed for withdrawal uses. This sum is compared to the amount in the water supply. If it is equal to or greater than the sum, then the supply will meet the demand, but if this is not the case, measures to reduce consumption or increase the water supply need to taken.

About 70% of water supplies worldwide go to food production. Industry uses 23% and cities use about 7% of water supplies. In the western United States, for example, about 85% of the water supplies are currently used by farming operations at government-subsidized rates. Cities and industry as well as populations of fish and wildlife split the remaining 15%. Some politicians and environmentalists consider this division unfair, and assert that water allocation should be controlled by environmental awareness. They suggest seeking new ways of getting more water to more people while protecting wildlife and **wetlands**.

Government agencies own most of the water supplies in the western United States, and they own about 65% of the supplies in the east. Those who control the water supplies govern its use, though regulations on ownership vary by state. In western states, supplies are on a first-owned, first-served basis, even if others later run short. The distribution of supplies and shortages is shared more equally in the east.

Some water allocation programs look toward **conservation** to meet water needs. Certain ordinances, in California for example, restrict and limit water usage. One method of enforcing compliance that has been used there is publishing the names of people who do not abide by the ordinances to embarrass them. Another method uses a special section of the police force for checking water use. These officers give first-time offenders information on **water conservation**. They fine repeat offenders.

Wells, **dams**, reservoirs, conveyance systems, and artificial ponds physically control water supplies. Wells draw out **groundwater**. The other constructions manage surface

water. Some areas use **desalinization** techniques to get fresh water from seawater.

A well taps groundwater through a hole sunk to an **aquifer**. An aquifer is a **permeable** layer of rock containing water trapped by layers of impermeable rock. If the pressure is sufficient, the water will come to the surface under its force, a type of well known as an **artesian well**. If the pressure is not sufficient for an artesian well, a pump is used to mechanically drag the water to the surface.

Dams are constructed to halt some or all of a river's flow. The region behind the dams, where water collects, is called an impounding, or water collecting, **reservoir**. Each dam is designed for a specific use, though dam types may be combined depending on the river and the community's needs. Storage dams hold water during times of surplus for periods of deficit, such as holding the spring **runoff** for the summer dry season. Along with providing a water supply, storage dams can improve (or harm) habitats for fish and wildlife, create electricity, and help in **irrigation** and flood control. Diversion dams affect the course of a river, preventing it from flowing forward so that a conveyance system can redirect it to irrigate fields or to fill a distant storage reservoir. These are often used in the western United States, resulting in damage to downstream areas. Lack of water can destroy fisheries and wildfowl habitats, and shorelines and deltas erode when no **sediment** comes downstream to replace what the ocean washes away. Detention dams control floods and trap sediment. Floods are controlled to prevent destruction of communities and habitats downstream from the dam. Trapping sediment causes the downstream portion of the river to be cleaner.

Conveyance systems divert streams so they flow where humans most need them. Conveyance systems include ditches, canals, and aqueducts (also known as transmission stations). Artificial ponds store water that is pumped in using a conveyance system. Aqueducts consist of closed conduits or pipes. A closed system helps to prevent contamination. A pump system helps the flow when the water level is not high enough for gravity to take over. Consumer demand for water varies by day and by season, and enclosed ground reservoirs and water towers may be used to store treated water and meet peak demands.

Only about 3% of the earth's water is fresh, and of that, 23% is available for use as surface water and groundwater. The rest is in the unusable form of glacial ice. The amount of freshwater is not increasing with the global population, and about a third of the people in the world now live in countries with water problems. These facts make the equitable allocation of water more difficult, and the need for wise water management more acute. Worldwide shortages also make it important for water management to control the quality of water. Farming runoff, urbanization, and industry all reduce **water quality**, which causes health and environmental problems, including algal blooms caused by nitrogen-rich fertilizers, and **acid rain**. One survey done by the **Environmental Protection Agency** (EPA) found that over half of the rivers, lakes, and streams in use in the United States have been nearly or totally destroyed by **pollution**. Pollutants have also made it difficult and expensive to recover and purify groundwater. Purification treatments in general can be expensive or difficult to obtain, and this causes poor water quality in many parts of the world.

Competing needs have also created severe problems; including lowered water pressure, decreased stream flow, land **subsidence** and increased **salinity**. Additionally, **overgrazing**, **deforestation**, loss of water-retaining **topsoil**, and **strip mining** decrease the amount of available water.

Treating sewage before returning it to the water supply and properly disposing of **chemicals** can improve water quality and increase the amount of usable water available. Reducing consumption can prevent water shortages. Households and industries can lower their water consumption by employing devices that decrease the amount of water used in, for example, showers, **toilets**, and dishwashers. Farmers can lower their water consumption by growing drought-resistant crops instead of ones requiring large quantities of water, such as alfalfa, and by using more efficient watering techniques. These methods not only stabilize water supplies, they also reduce the cost to the consumer. Another method for use in water shortages is desalinization of sea water. Desalinization plants remove the salt from sea water and purify it. Although such plants can be expensive, the expense can prove to be less than attempting to pipe fresh water in from sources already strained. *See also* Aquifer restoration; Drinking-water supply; Groundwater pollution; Safe Drinking Water Act; Water quality standards

[*Nikola Vrtis*]

Resources

Periodicals

"California Drought Cops." *Newsweek* (April 30, 1990): 27.

Davis. P. A. "Senate Energy Keeps Spigot on for California Agribusiness." *Congressional Quarterly Weekly Report* (March 21, 1992): 723.

Kent, M. M. "New Report Studies Population, Water Supply." *Population Today* (December 1992): 4–5.

O'Reilly, B. "Water: How Much Is There and Whose Is It?" *Fortune* (March 25, 1991): 12.

"The U.S. No Water to Waste." *Time* (20 August 1990): 61.

Water conservation

Seventy-one percent of the earth's surface is covered by water—an area called the hydrosphere, which makes up all

of the oceans and seas of the world. Only 3% of the earth's entire water is freshwater. This includes Arctic and Antarctic ice, **groundwater**, and all the rivers and freshwater lakes. The amount of usable freshwater is only about 0.003% of the total. To put this small percentage in perspective, if the total water supply is equal to one gallon, the volume of the usable freshwater supply would be less than one drop. This relatively small amount of freshwater is recycled and purified by the **hydrologic cycle**, which includes evaporation, condensation, precipitation, **runoff**, and **percolation**. Since most of life on earth depends on the availability of freshwater, one can say "water is life."

Worldwide, agricultural **irrigation** uses about 80% of all freshwater. Cooling water for electrical **power plants**, domestic consumption, and other industry use the remaining 20%. This figure varies widely from place to place. For example, China uses 87% of its available water for agriculture. The United States uses 40% for agriculture, 40% for electrical cooling, 10% for domestic consumption, and 10% for industrial use.

Water **conservation** may be accomplished by improving crop water utilization efficiency and by decreasing the use of high-water-demanding crops and industrial products. The table shows the amount of water, in pounds, required to produce one pound of selected crops and industrial products (one gal = 7.8 lb).

Freshwater sources are either surface water (rivers and lakes) or groundwater. Water that flows on the surface of the land is called surface runoff. The relationship between surface runoff, precipitation, evaporation, and percolation is shown in the following equation: Surface runoff = precipitation - (evaporation + percolation): When surface runoff resulting from rainfall or snowmelt is confined to a well-defined channel it is called a river or stream runoff

Groundwater is surface water that has permeated through the **soil** particles and is trapped among porous soils and rock particles such as sandstone or shale. The upper **zone of saturation**, where all pores are filled with water, is the **water table**. It is estimated that the groundwater is equal to 40 times the volume of all earth's freshwater including all the rivers and freshwater lakes of the world.

The movement of groundwater depends on the porosity of the material that holds the water. Most groundwater is held within sedimentary aquifers. Aquifers are underground layers of rock and soil that hold and produce an appreciable amount of water and can be pumped economically.

Water utilization efficiency is measured by water withdrawal and water consumption. Water withdrawal is water that is pumped from rivers, reservoirs, or groundwater **wells**, and is transported elsewhere for use. Water consumption is water that is withdrawn and returned to its source due to evaporation or **transpiration**.

Water consumption varies greatly throughout the world. A conservative figure for municipal use in the United States is around 150 gal (568 L) per person per day. This includes home use for bathing, waste disposal, and landscape in addition to commercial and industrial use. The total water demand per person is around 4,500 gal (17,000 L) per person per day when one accounts for the production of food, fiber, and fuel. The consumptive use world wide is considerably less than that for the United States.

According to the United Nations World Health Organization (WHO), 5 gal (18 L) per person per day is considered a minimum water requirement. The majority of the people in the undeveloped world are unable to obtain the five gallon per day minimum requirement. The WHO estimates that nearly two billion people in the world risk consuming contaminated water. Waterborne diseases such as polio, typhoid, dysentery, and **cholera** kill nearly 25 million people per year. In order to meet this demand for freshwater, conservation is an obvious necessity.

Since irrigation consumes 80% of the world's usable water, improvements in agricultural use is the logical first step in water conservation. This can be accomplished by lining water delivery systems with concrete or other impervious materials to minimize deep percolation or by using **drip irrigation** systems to minimize evaporation losses. Drip irrigation systems have been successfully used on fruit trees, shrubs, and landscape plants.

Subsurface irrigation is an emerging technology with extremely high water utilization efficiency. Subsurface irrigation uses a special drip irrigation tubing that is buried 6–8 in (15–20 cm) underground with 12–24 in (26–50 cm) between lines. The tubing contains emitters, or drip outlets, that deliver water and dissolved nutrients at the plant's root zone at a desired rate. In addition to water conservation, subsurface irrigation has several advantages that overhead sprinklers do not: no overwatering, no disease or **aeration** problems, no runoff or **erosion**, no weeds, and no vandalism. Subsurface irrigation in California has been used on trees, field crops, and lawns with up to 50% water savings.

Xeriscape, the use of low water consuming plants, is a most suitable landscape to conserve water, especially in dry, hot urban regions such as the Southwestern United States, where approximately 50% of the domestic water consumption is used by lawns and non-drought tolerant landscape. Plants such as cacti and succulents, ceanothus, arctostaphylos, which is related to foothill manzanita, trailing rosemary (*Rosemarinus officinale*), and white rock rose (*Cistus cobariensis*) adapt well to hot, dry climates and help conserve water.

In addition to improving irrigation techniques, water conservation can be accomplished by improving domestic use of water. Such a conservation practice is the installation

of the ultra-low-flush (ULF) **toilets** in homes and commercial buildings. A standard toilet uses 5–7 gal (19–26 L) of water per flush, while the ultra-low-flush toilet uses 1.5 gal (5.7 L). Research in Santa Monica, California, shows that replacing a standard toilet with an ULF saves 30–40 gal (114–151 L) of water per day, which is equivalent to 10,000–16,000 gal (37,850–60,500 L) per year.

Another way to conserve the freshwater supply is to extract freshwater from sea water by **desalinization**. Desalinization, the removal of soluble salts and other impurities from seawater by distillation or reverse osmosis (RO), is becoming an increasingly acceptable method to provide high quality pure water for drinking, cooking, and other home uses. It is estimated that the 1993 world production of desalinated water is about 3.5 billion gal (13 billion L) per day. Most desalinated water is produced in Saudi Arabia, Persian Gulf Nations, and, more recently, in California. The cost of desalinated water depends upon the cost of energy. In the United States, it is about three dollars per thousand gallons, which is four to five times the cost paid by urban consumers and over 100 times the cost paid by farmers for irrigation water. The idea of using desalinated water for irrigation is, currently, cost prohibitive.

Water has played a vital role in the rise and fall of human cultures throughout history. The availability of usable water has always been a limiting factor for a region's ecological **carrying capacity**. It is important that humans learn to live within the limit of available **natural resources**. Conservation of water alone will not extend the natural carrying capacity for an indefinite period of time. Since the supply of available and usable water is finite, the consumption per person must be reduced. A permanent solution to the water shortage problem can be accomplished by living within the **ecosystem** carrying capacity or by reducing the number of consumers through effective control of **population growth**.

[*Muthena Naseri*]

Resources

Books

Buzzelli, B. *How to Get Water Smart: Products and Practices for Saving Water in the Nineties.* Santa Barbara: Terra Firma Publishing, 1991.

Clarke, R. *Water: The International Crisis.* Cambridge: MIT Press, 1993.

Yudelman. M., et al. *New Vegetative Approaches to Soil and Water Conservation.* Washington, DC: World Wildlife Fund, 1990.

Periodicals

Postel, S. "Plug the Leak, Save the City." *International Wildlife* 23 (January-February 1993): 38–41.

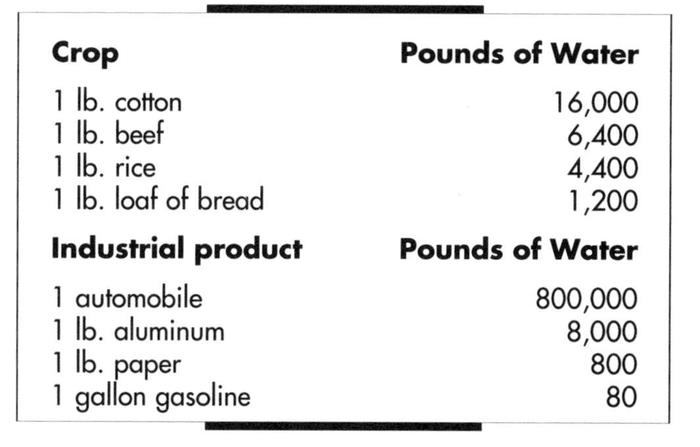

Crop	Pounds of Water
1 lb. cotton	16,000
1 lb. beef	6,400
1 lb. rice	4,400
1 lb. loaf of bread	1,200
Industrial product	**Pounds of Water**
1 automobile	800,000
1 lb. aluminum	8,000
1 lb. paper	800
1 gallon gasoline	80

Water cycle

see **Hydrologic cycle**

Water diversion projects

Water diversion projects include the construction of **dams**, levees, pumping stations, **irrigation** canals, or any other manmade structure that modifies the natural flow of a waterway. Diversion projects may be developed for purposes of hydroelectric power generation, farm irrigation, consumer and industrial water supply, and flood control. Throughout human history, communities have altered river systems for their own advantage. There is evidence that large-scale projects dating over 2000 years ago existed in China, the Mediterranean, and the Middle East.

Water diversion globally

In the early twenty-first century, water is one of the commodities deemed most important to the well-being of nations. As of 2002, the world population of 5.6 billion is growing at an annual rate of approximately 90 million, and the global demand for water is expected to rise by 2–3% annually in the decades to come.

Globally, of the more than 45,000 large dams ([49.2 ft; 15 m] or more in height, according to the International Commission of Large Dams), China has the greatest number with approximately 24,671 large dams, followed by the United States and India with 6,575 and 4,291 respectively. Approximately two-thirds of world's existing large dams are located in developing countries.

Although clean freshwater is a valuable commodity and is short supply in many regions of the world, issues surrounding its diversion and use are often the subject of heated debate. One large-scale dam project that has been hotly debated is the Mekong River project in Southeast Asia. Nations of the region want to control and develop the river

basin; environmentalists want to protect it; and caught in the middle is the river itself and the millions who depend on it for their livelihood.

The fields and numerous tributaries that feed into the Mekong River are home to more than 50 million people. Although most of the Mekong basin remains undeveloped, more than 200 dams have been proposed for the Mekong and its many tributaries. In addition to irrigating rice fields in **arid** regions, the diverted water for the first time offers the opportunity for local inhabitants to produce a variety of crops, thereby fueling economic growth and development.

Yet tens of thousands of families will be forcibly relocated; fisheries will collapse as upstream access to millions of fish is blocked; farmland will be flooded; and development will further restrict the river, leading to potential worsening of floods. Like other major water diversion projects around the world, the Mekong project illustrates some of the profound social, economic, and environmental effects of damming and redirecting the world's rivers.

Water diversion in the United States

Most large-scale water diversion projects in the United States are located along river systems in the western half of the country, where the dry, arid terrain is dependent on water diversion from high capacity rivers—like the Colorado, Columbia, Missouri, and Snake—for municipal and irrigation water supplies. The Hoover Dam, Grand Coulee Dam, Columbia River Basin Project, and scores of other dams and diversion projects are built and maintained by the U.S. **Bureau of Reclamation** (USBR), which was established by the Department of the Interior under the National **Reclamation** Act of 1902.

Water diversion practices and management in the Western states developed by necessity. Low precipitation levels in the western states meant that pioneer farmers had to find alternate methods of irrigating their crops. The USBR was so named because it was charged with the task of "reclaiming" the dry **desert** landscape for farming and homesteading through irrigation.

The Bureau's first major diversion project, the Hoover Dam and adjacent **reservoir** Lake Mead, was authorized in 1928 and completed in 1936. In addition to providing water for agriculture, community, and industrial use, the dam also generates 4 billion kilowatt-hours of hydroelectric power annually for California, Arizona, and Nevada.

As of May 2002, USBR managed 348 reservoirs that hold a capacity of 245 million acre-ft (302.2 billion m^3) of water. The Bureau is the largest wholesale water supplier in the United States, supplying irrigation water for 10 million acres of farmland and 31 million people across the West. In addition, USBR generates over 42 billion kilowatt hours of hydroelectric power annually from its 58 **power plants**. Along with the U.S. **Army Corps of Engineers**, the Bureau is responsible for the majority of national water projects.

Environmental impact

Although they occupy a smaller area compared to land and oceans, freshwaters are home to a relatively high proportion of **species** (animals, plants, and **microorganisms**), with more per unit area than any other **environment** (10% more than land, and 15% more than oceans). However, some areas, such as the Amazon, Congo, Nile, and Mekong basins are densely populated with species and are called "hotspots."

In the United States, legislation of water diversion and reclamation programs initially focused on agricultural development, job creation, and power generation. Eventually, the long-term environmental impact of some of these big river projects became apparent as western **wetlands** disappeared and scores of migratory **salmon** and other **wildlife** were designated endangered and threatened under the **Endangered Species Act** (ESA).

Diversion projects prevent migrating fish from returning to their native upstream spawning grounds. The California Department of Fish and Game estimates that two-thirds of the winter-run chinook salmon that spawn in the Sacramento River do not make it to the Pacific. Water diversion channels can be difficult to navigate, and water diversion pumps that feed the California Water Project kill and maim a large percentage of the migrating population each year. In addition, diversions for irrigation and residential water diversion lowers the water level of the river, slowing the progress of the salmon and increasing river water temperature to inhospitable levels. Despite the integration of fish lifts and fish ladders into modern dams, populations of native fish stocks are still greatly depleted by dam structures.

In addition to the problem of depleted fish stocks, migrating fish such as salmon and trout transport nutrients from oceans to rivers. When this fish population declines, so also does a significant amount of **nitrogen** and **phosphorus**, which is typically released into rivers by decomposing carcasses. This **nutrient** loss affects the **ecosystem** as well as the fish supply. A renewed interest has been placed on modifying diversion projects to recover river ecosystems and **biodiversity**.

The impact of water diversion products on habitats in and around rivers, wetlands, and streams can be profound. The U.S. Army Corps of Engineers estimates that there are over 77,000 dams of significant size over 6 ft (1.8 m) in the United States. Levees, dams, **shoreline armoring**, and other diversion structures change the very geography, geology, and **hydrology** of the river, influencing flow levels, altering channels, and preventing natural **erosion**. These structures also block the natural flow of **silt** and sediments. The natural **floodplain** is frequently impacted, resulting in a two-fold effect: destroying wetland ecosystems and elimi-

nating a recharging source for the **groundwater** table. Eventually, structures put in place to prevent flood damage may exacerbate the problem. Dams, levees, and other diversional devices prevent water from heavy rainfalls from naturally dispersing onto floodplains and wetlands.

Water diversion projects also impact the **atmosphere**. Studies of large reservoirs created behind hydroelectric dams have suggested that decaying vegetation, submerged by **flooding**, may give off quantities of greenhouse gases—such as **methane** gas—equivalent to those from other sources of electricity. Thus, hydroelectric facilities that flood large areas of land may be significant contributors to global warming. The **water quality** of many reservoirs also poses a health hazard due to new forms of bacteria that grow in many of the rivers that have been diverted by dams.

In addition to giving off **greenhouse gases**, bacteria present in decaying vegetation can also change **mercury**, present in rocks underlying a reservoir, into a water-soluble form. The mercury accumulates in the bodies of fish and poses a health hazard to both wildlife and humans who depend on these fish as a food source. Methylmercury, the most toxic form of the element, affects the immune system, alters genetic and **enzyme** systems, and damages the nervous system. Human exposure to methylmercury is almost entirely by eating contaminated fish and wildlife.

A rather surprising global impact of water diversion projects is that the weight of the world's collective reservoirs is speeding up the earth's rate of rotation and is altering the shape of earth's magnetic field, according to NASA geophysicist Dr. Benjamin Fong Chao.

Human impact

Estimates suggest that approximately 40–80 million people have been physically displaced by dams worldwide. Millions of people living downstream from dams, particularly those who rely on natural floodplain function and fisheries, have suffered serious damage to their livelihoods, and the future productivity of their resources has been put at risk. Many of the displaced are not recognized as such, and therefore are not resettled or compensated. When compensation is provided, it is often inadequate. When the physically displaced are acknowledged, many are not included in resettlement programs. Moreover, those who are resettled seldom have their livelihoods restored; resettlement programs usually focus on physical relocation rather than the economic and social needs of the displaced.

Indigenous and tribal peoples and vulnerable ethnic minorities have suffered disproportionate levels of displacement and negative impacts on not only their livelihood, but also their cultural and spiritual existence. Large dams have had significant adverse effects on cultural heritage through the loss, submergence, and degradation of plant and animal remains, burial sites, and archaeological monuments. The social, economic, and political status of minority groups often restricts their capacity to assert their interests and rights in land and **natural resources**, and restricts their role in decision-making that affects them.

In addition, affected indigenous populations living near reservoirs, as well as displaced people and downstream communities, have often faced adverse health conditions from environmental change and social disruption. Gender gaps have widened in these communities, and women have often borne a disproportionate share of the social costs while reaping few of the benefits.

Policies and initiatives

Opposition to diversion projects and their impact on the environment began to pick up momentum in the 1970s. The **National Environmental Policy Act** of 1970 (NEPA) required thorough study of the environmental impacts of proposed water projects, and inspired greater public scrutiny of the effects of major water projects on the environment. In 1978, President Jimmy Carter announced a new policy of cutting federal funding to new water projects, and developed a "hit list" of dams and irrigation projects.

The financial approach to water management evolved as well. In 1982, the Reclamation Reform Act was signed into law, changing the acreage requirements for irrigation water subsidies in an effort to get large corporate farms to pay a fair market price for their water use. The **Water Resources** Development Act (WRDA)—legislation that is reauthorized by Congress every two years to finance U.S. Army Corps of Engineers' water projects—contained amendments in 1986 that required local interests to share the construction costs of water resource projects built by the Corps.

Environmentalists charge that the USBR and the Army Corps have a history of developing water projects for financial gain regardless of the cost to the environment. At the heart of the debate is the fundamental difference in the way natural water sources are viewed—as a fiscal commodity or as a natural resource. As Marc Reisner, author of the critically acclaimed *Cadillac Desert*, states: "In the East, to 'waste' water is to consume it needlessly or excessively. In the West, to waste water is not to consume it—to let it flow unimpeded and undiverted down rivers. ... To easterners, 'conservation' of water usually means protecting rivers from development; in the West, it means building dams." Striking a balance between these two viewpoints to maintain both our riparian ecosystem and our Western communities and businesses, is the challenge facing American government, environmentalists, and other stakeholders.

Because irrigation for farming is a major consumer use of diverted water, improving growing techniques and choosing water-friendly crops may be one solution to minimizing water need. The agricultural industry can also take

steps to meet irrigation needs in a environmentally responsible manner by ensuring that adequate fish screens are in place at diversion points and agreeing to lower water delivery rates at crucial spawning times to maintain river flow for migratory fish. Finally, **conservation** and restoration of agricultural wetlands through government plans such as the Wetlands Reserve Program can help solve long-term flooding problems on farms and restore the habitats of native species.

The WRDA of 1999 enacted an Army Corps of Engineers initiative known as "Challenge 21" focused on replacing dams and other diversionary techniques for flood control with non-structural solutions and environmental restoration programs. Specifically, the initiative calls for less environmentally-invasive programs of preventing flood damage, including federally-funded voluntary buyouts of property on flood plains and restoration of altered flood plain ecosystems.

International agreements and organizations have also established standards for minimizing the negative impact of human activities on biodiversity. Legal coalitions include the World Charter for Nature, the Convention on Biological Diversity, and Agenda 21; international organizations such as the **World Bank**, the World Business Council on **Sustainable Development**, and The World Conservation Union (IUCN), have contributed to the development of accepted standards. These standards involve conservation of species and ecosystems, the recovery of degraded ecosystems, the conservation of ecological functions or processes, securing of adequate information for decision-making, and the adherence to high standards for environmental impact assessments.

[*Paula Anne Ford-Martin*]

Resources

Books

Committee on Missouri River Ecosystem Science; Water, Science, and Technology Board; Division of Earth and Life Studies; National Research Council. *The Missouri River Ecosystem: Exploring the Prospects for Recovery.* Washington DC: National Academy Press; In press, 2002. Prepublication copy available online at <http://books.nap.edu/books/0309083141/html/R1.html>

MacDonnell, Lawrence. *From Reclamation to Sustainability: Water, Agriculture, and the Environment in the American West.* Boulder: University Press of Colorado, 1999.

McNully, Patrick. *Silenced Rivers: The Ecology and Politics of Large Dams.* 2nd edition. London: Zed Books, 2001.

Reisner, Marc. *Cadillac Desert: The American West and Its Disappearing Water.* 2nd edition. New York: Penguin, 1987.

Periodicals

Bricker, Jennie, and David Filippi. "Endangered Species Act Enforcement and Western Water Law." *Environmental Law* 30, no.4 (Fall 2000): 735.

Robbins, Elaine. "Damning Dams." *emagazine.com (The Environmental Magazine)* X, no. 1 (January-February 1999): 16 [cited June 26, 2002]. <http://www.emagazine.com>.

Other

LakeNet. *Strategies and Solutions for Managing Water Diversion.* May 15, 2002 [cited June 4, 2002]. <http://www.worldlakes.org/Strategies%20and%20Solutions.htm>.

Stein, Jeff, et al. National Wildlife Federation & Taxpayers for Common Sense. *Troubled Waters: Congress, the Corps of Engineers, and Wasteful Water Projects* Washington, DC: NWF/Taxpayers for Common Sense, 2000. <http://www.nwf.org/greeningcorps/report.html>.

Organizations

American Rivers, 1025 Vermont Ave., N.W. Suite 720, Washington, DC USA 20005 (202) 347-7550, Fax: (202) 347-9240, Email: amrivers@amrivers.org, <http://www.amrivers.org>

International Rivers Network, 1847 Berkeley Way, Berkeley, CA USA 94703 (510) 848-1155, Fax: (510) 848-1008, Email: info@irn.org, <http://www.irn.org>

U.S. Bureau of Reclamation, Freedom of Information Act (FOIA) Office, PO Box 25007, D-7924, Denver, CO USA 80225-0007 (303) 445-2048, Fax: (303) 445-6575, Toll Free: (800) 822-7646, Email: borfoia@usbr.gov, <http://www.usbr.gov>

Water Environment Federation

Dedicated to "the preservation and enhancement of **water quality** worldwide," the Water Environment Federation (WEF) is an international professional organization of water quality experts. WEF is made up of 38,000 water professionals, including engineers, biologists, government officials, treatment plant managers and operators, laboratory technicians, equipment manufacturers and distributors, and educators.

WEF was founded in 1928 as the Federation of Sewage Works Associations and served as the publisher of the *Sewage Works Journal.* The organization has also been known as the **Water Pollution** Control Federation. Its scope has broadened significantly since its inception, and WEF currently publishes a variety of materials relating to the water industry. WEF also sponsors seminars, conferences, and briefings; offers technical training and education; produces informational videos for students; and provides input on environmental legislation and regulations to government bodies.

The publications issued by WEF cover a multitude of topics and are geared toward a variety of readers. Technical publications, including manuals developed by WEF's Technical Practice Committee, brief water-quality professionals on design, operation, and management innovations in the field. Periodicals and newsletters address such issues as industry trends, research findings, water analysis topics, job safety, regulatory and legislative developments, and job openings in the profession.

WEF's technical training and educational programs are geared toward both professionals and students. Audiovisual training courses and study guides are available on such topics as **wastewater** treatment, waste stabilization ponds,

and wastewater facility management. The federation also distributes health and safety videos developed by municipalities. For students, WEF offers *The Water Environment Curriculum*, a series of videos and guides outlining water quality issues geared toward children in fifth through ninth grades. *The Water Environment Curriculum* includes lessons on surface water, **groundwater**, wastewater, and **water conservation**. WEF also publishes guides on water quality issues for the general public.

WEF's major event is its annual conference, the largest water quality and **pollution control** exposition in North America. The conference features technical sessions, expert speakers, and exhibits run by companies in the water quality industry. The federation also holds specialty conferences on such topics as toxicity, landfills, and surface water quality. Two other noteworthy briefings are WEF's Pacific Rim Conference and the Washington Briefing. The Pacific Rim Conference gathers representatives from Pacific Rim countries to discuss environmental problems in that area. The Washington Briefing convenes government officials and water quality experts to discuss current issues affecting the industry.

In addition to its publications, educational tools, and conferences, WEF sponsors an affiliate, the Water Environment Research Foundation, which works to develop and implement water quality innovations. The research foundation's programs strive to find ways to enhance water quality, develop broad-based approaches to water quality that encompass all aspects of the environment, use new technology, and promote interaction among individuals involved in the water quality profession.

WEF is also associated with Water Quality 2000, a cooperative effort involving a variety of professional and scientific organizations, industry representatives, environmental groups, academic institutions, and government officials working on **water quality standards** and goals for the United States for the twenty-first century.

[*Kristin Palm*]

RESOURCES

ORGANIZATIONS

Water Environment Federation, 601 Wythe Street, Alexandria, VA USA 22314-1994 (703) 684-2452, Toll Free: (800) 666-0206, Email: csc@wef.org, http://www.wef.org

Water hyacinth

The water hyacinth (*Eichhornia crassipes*) is an aquatic plant commonly found in the southern United States, including Florida, Texas, the Gulf Coast, and California. As an **introduced species**, it has spread rapidly and is now viewed as a noxious weed. Like other aquatic weeds, water hyacinth can grow rapidly and clog waterways, damaging underwater equipment and impairing navigational and recreational facilities. It is also viewed as a weed in many tropical parts of the world.

Water hyacinth. (Photograph by Ray Coleman. Photo Researchers Inc. Reproduced by permission.)

Water hyacinths can be either free-floating or rooted, depending on the depth of the water. The plant height may vary from a few inches to 3 ft (0.9 m). The leaves, growing in rosettes, are glossy green and may be up to 8 in (20 cm) long and 6 in (15 cm) wide. The showy, attractive flowers may be blue, violet, or white and grow in spikes of several flowers. The leaf blades are inflated with air sacs, which enable the plants to float in water. The seeds are very long-lived. Studies show that seeds that have been buried in mud for many years remain viable under the right conditions, leading to reinfestation of lakes long after chemical treatment.

Weed harvesting methods to control water hyacinth growth are expensive and not very effective. The weight and volume of the harvested material usually adds greatly to the costs of hauling and disposal. Attempts to use the harvested material as **soil fertilizer** or enhancer have failed because of the low mineral and nutritive content of the plant's tissue. Chemical methods to control water hyacinth usually require

careful and repeated applications. Commonly used herbicides include Rodeo and Diquat.

In some situations, aquatic weeds serve beneficial purposes. Aquatic plants can be integral components of **wastewater** treatment, particularly in warm climates. Wastewater is routed through lagoons or ponds with dense growths of water hyacinth. In waters containing excess nutrients (**nitrates** and **phosphates**) and turbidity, water hyacinth plants have been shown to improve **water quality** by taking up the excess nutrients into the plant and by settling the suspended solids which cause turbidity. They also add oxygen to the water through **photosynthesis**. Oxygen is an essential requirement for the **decomposition** of wastewater, and oxygen deficiency is a common problem in waste treatment systems. Water hyacinth has been used successfully in California and Florida, and more states are exploring the possibility of using water hyacinth in waste treatment systems.

[*Usha Vedagiri*]

Resources

Books

Hammer, Donald A. *Constructed Wetlands for Waste Treatment.* Chelsea, MI: Lewis Publishers, Inc., 1989.

Schmidt, James C. *How to Identify and Control Water Weeds and Algae.* Milwaukee: Applied Biochemists, Inc., 1987.

Water pollution

Among the many environmental problems that offend and concern us, perhaps none is as powerful and dramatic as water **pollution**. Ugly, scummy water full of debris, **sludge**, and dark foam is surely one of the strongest and most easily recognized symbols of our misuse of the **environment**.

What is pollution? The verb "pollute" is derived from the Latin *polluere*: to foul or corrupt. Our most common meaning is to make something unfit or harmful to living things, especially by the addition of waste matter or sewage. A broader definition might include any physical, biological, or chemical change in **water quality** that adversely affects living organisms or makes water unsuitable for desired uses.

Paradoxically, however, a change that adversely affects one organism may be advantageous to another. Nutrients that stimulate oxygen consumption by bacteria and other **decomposers** in a river or lake, for instance, may be lethal to fish but will stimulate a flourishing community of decomposers. Whether the quality of the water has suffered depends on your perspective. There are natural sources of water contamination, such as poison springs, oil seeps, and **sedimentation** from **erosion**, but most discussions of water pollution focus on human-caused changes that affect water quality or usability.

The most serious water pollutants in terms of human health worldwide are pathogenic organisms. Altogether, at least 25 million deaths each year are blamed on these water-related diseases, including nearly two-thirds of the mortalities of children under five years old. The main source of these pathogens is from untreated or improperly treated human wastes. In the more developed countries, **sewage treatment** plants and other **pollution control** techniques have reduced or eliminated most of the worst sources of pathogens in inland surface waters. The United Nations estimates that 90% of the people in high-income countries have adequate sewage disposal, and 95% have clean drinking water.

For poor people, the situation is quite different. The United Nations estimates that three-quarters of the population in less-developed countries have inadequate **sanitation**, and that less than half have access to clean drinking water. Conditions are generally worse in remote, rural areas where sewage treatment is usually primitive or nonexistent, and purified water is either unavailable or too expensive to obtain. In the 33 poorest countries, 60% of the urban population have access to clean drinking water but only 20% of rural people do.

This lack of pollution control is reflected in surface and **groundwater** quality in countries that lack the resources or political will to enforce pollution control. In Poland, for example, 95% of all surface water is unfit to drink. The Vistula River, which winds through the country's most heavily industrialized region, is so badly polluted that more than half the river is utterly devoid of life and unsuited even for industrial use. In Russia, the lower Volga River is reported to be on the brink of disaster due to the 300 million tons (272.2 million metric tons) of **solid waste** and 20 trillion L (5 trillion gal) of liquid **effluent** dumped into it annually.

The less-developed countries of South America, Africa, and Asia have even worse water quality than do the poorer countries of Europe. Sewage treatment is usually either totally lacking or woefully inadequate. Low technological capabilities and little money for pollution control are made even worse by burgeoning populations, rapid urbanization, and the shift of heavy industry from developed countries where pollution laws are strict to **less developed countries** where regulations are more lenient.

In Malaysia, 42 of 50 major rivers are reported to be "ecological disasters." Residues from palm oil and **rubber** manufacturing, along with heavy erosion from **logging** of tropical rain forests, have destroyed all higher forms of life in most of these rivers. In the Philippines, domestic sewage makes up 60–70% of the total volume of Manila's Pasig River. Thousands of people use the river not only for bathing and washing clothes, but also as their source of drinking and cooking water. China treats only 2% of its sewage. Of

78 monitored rivers in China, 54 are reported to be seriously polluted. Of 44 major cities in China, 41 use "contaminated" water supplies, and few do more than rudimentary treatment before it is delivered to the public.

Pollution control standards and regulations usually distinguish between point and nonpoint pollution sources. Factories, **power plants**, sewage treatment plants, underground **coal** mines, and oil **wells** are classified as point sources because they **discharge** pollution from specific locations, such as drain pipes, ditches, or sewer outfalls. These sources are discrete and identifiable, so they are relatively easy to monitor and regulate. It is generally possible to divert effluent from the waste streams of these sources and treat it before it enters the environment.

In contrast, nonpoint sources of water pollution are scattered or diffused, having no specific location where they discharge into a particular body of water. Nonpoint sources include **runoff** from farm fields, **golf courses**, lawns and gardens, construction sites, logging areas, roads, streets, and parking lots. Multiple origins and scattered locations make this pollution more difficult to monitor, regulate, and treat than point sources.

Desert soils often contain high salt concentrations that can be mobilized by **irrigation** and concentrated by evaporation, reaching levels that are toxic for plants and animals. Salt levels in the San Joaquin River in central California rose about 50% between 1930 to 1970 as a result of agricultural runoff. **Salinity** levels in the **Colorado River** and surrounding farm fields have become so high in recent years that millions of acres of valuable croplands have had to be abandoned. The United States is building a huge desalination plant at Yuma, Arizona, to reduce salinity in the river. In northern states, millions of tons of sodium chloride and calcium chloride are used to melt road ice in the winter. The corrosive damage to highways and automobiles and the toxic effects on vegetation are enormous. **Leaching** of road salts into surface waters has a similarly devastating effect on aquatic ecosystems.

Acids are released by mining and as by-products of industrial processes, such as leather tanning, metal smelting and plating, **petroleum** distillation, and organic chemical synthesis. Coal mining is an especially important source of **acid** water pollution. Sulfides in coal are solubilized to make sulfuric acid. Thousands of miles of streams in the United States have been poisoned by acids and metals, some so severely that they are essentially lifeless.

Thousands of different natural and synthetic organic **chemicals** are used in the chemical industry to make pesticides, **plastics**, pharmaceuticals, pigments, and other products that we use in everyday life. Many of these chemicals are highly toxic. Exposure to very low concentrations can cause **birth defects**, genetic disorders, and **cancer**. Some synthetic chemicals are resistant to degradation, allowing them to persist in the environment for many years. Contamination of surface waters and groundwater by these chemicals is a serious threat to human health.

Hundreds of millions of tons of hazardous organic wastes are thought to be stored in dumps, landfills, lagoons, and underground tanks in the United States. Many, perhaps most, of these sites are leaking toxic chemicals into surface waters or groundwater or both. The **Environmental Protection Agency** (EPA) estimates that about 26,000 **hazardous waste** sites will require cleanup because they pose an imminent threat to public health, mostly through water pollution.

Although the oceans are vast, unmistakable signs of human abuse can be seen even in the most remote places. **Garbage** and human wastes from coastal cities are dumped into the ocean. **Silt**, fertilizers, and pesticides from farm fields smothered coral reefs, coastal spawning beds, and overfertilize estuaries. Every year millions of tons of plastic litter and discarded fishing nets entangle aquatic organisms, dooming them to a slow death. Generally coastal areas, where the highest concentrations of sea life are found and human activities take place, are most critically affected.

The amount of oxygen dissolved in water is a good indicator of water quality and of the kinds of life it will support. Water with an oxygen content above 8 **parts per million** (ppm) will support game fish and other desirable forms of aquatic life. Water with less than 2 ppm oxygen will support only worms, bacteria, **fungi**, and other decomposers. Oxygen is added to water by diffusion from the air, especially when turbulence and mixing rates are high, and by **photosynthesis** of green plants, algae, and cyanobacteria. Oxygen is removed from water by **respiration** and chemical processes that consume oxygen.

In spite of the multitude of bad news about water quality, some encouraging pollution control stories are emerging. One of the most outstanding examples is the Thames River in London. Since the beginning of the Industrial Revolution, the Thames had been little more than an open sewer, full of vile and toxic waste products from domestic and industrial sewers. In the 1950s, however, England undertook a massive cleanup of the Thames. More than $250 million in public funds plus millions more from industry were spent to curb pollution. By the early 1980s, the river was showing remarkable signs of rejuvenation. Oxygen levels had rebounded and some 95 **species** of fish had returned, including the pollution-sensitive **salmon**, which had not been seen in London for 300 years. With a little effort, care, and concern for the environment, similar improvements can develop elsewhere.

[*William P. Cunningham*]

Pollutant	Source	Effects
Nutrients, including nitrogen compounds	Fertilizers, sewage, acid raid from motor vehicles and power plants	Creates algae blooms, destroys marine life
Chlorinated hydrocarbons: pesticides, DDT, PCBs	Agricultural runoff, industrial waste	Contaminates and harms fish and shellfish
Petroleum hydrocarbons	Oil spills, industrial discharge, urban runoff	Kills or harms marine life, damages ecosystems
Heavy metals: arsenic, cadmium, copper, lead, zinc	Industrial waste, mining	Contaminates and harms fish
Soil and other particulate matter	Soil erosion from construction and farming; dredging, dying algae	Smothers shellfish beds, blocks light needed by marine plant life
Plastics	Ship dumping, household waste, litter	Strangles, mutilates wildlife, damages natural habitats

Major pollutants of coastal waters around the United States. (Beacon Press. Reproduced by permission.)

Resources

Books

Mayberck, M. *Global Freshwater Quality: A First Assessment.* Oxford, UK: Blackwell Reference, 1990.

Other

Protecting Our Future Today. Washington, DC: Environmental Protection Agency, 1991.

Water quality

Water quality encompasses a wide range of water characteristics, including biological, chemical, and physical descriptions of water clarity or contamination. Water quality assessment is typically based on the examination of certain attributes, conditions or properties of a lake, river, bay, **aquifer**, or other water body. The most important of these attributes are pollutants that demand oxygen or cause disease, nutrients that stimulate excessive plant growth, synthetic organic and inorganic **chemicals**, mineral substances, sediments, radioactive substances, and temperature. Since the 1950s, routine tests for water quality have evaluated temperature, turbidity, color, odor, total solids after evaporation, hardness (**pH**), and concentrations of **carbon dioxide**, iron, **nitrogen**, chloride, active **chlorine**, **microorganisms**, **coliform bacteria**, and amorphous matter. In recent years increasing contamination and growing public concern over water quality have led to the addition of a number of additional parameters, including algal growth, **chemical oxygen demand**, and the presence of **hydrocarbons**, metals, and other toxic substances.

Water quality is evaluated through a set of samples taken from a cross-section of a body of water. Because quality conditions change continually, each set of samples is understood to represent conditions at the time of sampling. Long-term water quality monitoring requires periodic data collection at regular sampling stations. In order to ensure that quality measurements are consistent at different times and locations, standards are legally set for various water uses and contaminant levels. There are two general types of standards for measuring water quality. "Water quality based standards" are set to ensure that a water body is clean enough for its expected uses, e.g., fishing, swimming, industrial use, or drinking. "Technology based standards" are set for **waste-**

water entering a water body so that overall water conditions remain acceptable. These standards have been developed worldwide. In the United States, the **Environmental Protection Agency** (EPA) is responsible for developing **water quality standards** and criteria for surface, ground, and marine waters.

Water quality standards, ambient conditions of a water body that meet expected uses, are based upon water quality criteria. These criteria designate acceptable levels of specific pollutants, clarity, or oxygen content. Usually acceptable levels of these criteria are measured in **parts per million** (ppm) of each contaminant or **nutrient**. These measurements allow scientists to assess whether or not a body of water meets the prescribed water quality standards. If standards are not met, then local governing bodies or industries must develop and implement cleanup strategies that target specific criteria. In the United States and many other countries, national water quality criteria have been developed for most conventional, toxic, and non-conventional pollutants. Conventional pollutants include suspended solids, **biochemical oxygen demand** (BOD), pH (acidity or alkalinity), fecal coliform bacteria, oil, and grease. Toxic priority pollutants include metals and organic chemicals. Non-conventional pollutants are any other contaminants that harm humans or marine resources and require regulation.

Drinking water must meet especially high standards to be safe for human consumption, and its quality is strictly monitored in the United States and most heavily populated regions of the world. Most governments have a legal responsibility to maintain acceptable drinking water quality. Until 1974, efforts to maintain acceptable levels of drinking water quality in the United States were limited to preventing the spread of contagious diseases. The **Safe Drinking Water Act** of 1974 expanded the government's regulatory role to cover all substances that may adversely affect human health or a water body's odor or appearance. This act established national regulations for acceptable levels of various contaminants.

Both point and nonpoint sources of **pollution** affect water quality. Point sources are discrete locations that **discharge** pollutants, mainly industrial outflow pipes or **sewage treatment** plants. Nonpoint sources are more diffuse, and include **storm runoff** and **runoff** from farming, **logging**, construction, and other **land use** activities.

In the United States, over 50% of the population relies on underground sources for drinking water. Because **groundwater** receives an enormous range of agricultural, industrial, and urban pollutants, groundwater quality is an issue of growing concern. Preventing **groundwater pollution** is especially important because remedial action in an aquifer is expensive and technologically difficult. Major sources of groundwater pollution include nitrate contamination from septic systems, organic pollutants from leaking fuel tanks and other storage containers, and an array of contaminants leached from landfills.

In rivers, the most extensive causes of water quality impairment are usually **siltation**, nutrient concentration (nitrogen and **phosphorus**), fecal coliform bacteria, and low **dissolved oxygen** levels caused by high organic content (sewage, grass clippings, pasture and **feedlot runoff**, etc.). Agricultural runoff including pesticides, fertilizers, and sediments is the largest source of river pollution, followed by municipal sewage discharge. Estuaries, rich ecosystems of mixed salt and fresh water where rivers enter a sea or ocean, often have serious water quality problems. Estuarine contaminants are usually the same as those in rivers—high organic content and low oxygen, disease-causing pathogens, organic chemicals, and municipal sewage discharge.

In most lakes, water quality is affected primarily by nitrogen and phosphorus **loading**, siltation and low dissolved oxygen. Like rivers, lakes suffer from agricultural runoff, **habitat** modification, storm runoff, and municipal sewage **effluent**. Lakes especially suffer from eutrophication (the excessive growth of aquatic algae and other plants) caused by high levels of nutrients. This burst of growth, also called an **algal bloom**, feeds on concentrated nutrients. Thick mats of algae and other aquatic plants clog water systems, increase turbidity, and suffocate aquatic animals so that an entire **ecosystem** is disrupted. Of the lakes assessed by the U.S. Environmental Protection Agency's 1988 National Water Quality Inventory, one-third were affected by eutrophication. In some lakes, especially the **Great Lakes**, the presence of persistent toxic substances is an additional concern. Most of these **toxins** originate in the industrial regions that surround these lakes or in the heavy shipping traffic between lakes.

Water quality degradation affects the **stability** of aquatic ecosystems as well as human health. Much is still unknown about the effects of specific pollutants on **ecosystem health**. However, fish and other aquatic organisms exposed to elevated levels of pollutants may have lowered reproduction and growth rates, diseases, and, in severe cases, high death rates. *See also* Cultural eutrophication

[*Marci L. Bortman Ph.D.*]

Resources

Books

Becker, C. D., ed. *Water Quality in North American River Systems.* Columbus, OH: Battelle Press, 1992.

Cotruvo, J. A., and E. Bellack. "Maintaining Drinking Water Quality." In *Perspectives on Water*, edited by D. H. Speidel, et al. New York: Oxford University Press, 1988.

Eckenfelder Jr., W. W. *Principles of Water Quality Management.* Boston: CBI Publishing, 1980.

OTHER

U.S. Environmental Protection Agency. *National Water Quality Inventory 1988 Report to Congress.* Washington, DC: U.S. Government Printing Office, 1990.

U.S. Environmental Protection Agency. *The Quality of Our Nation's Water.* Washington, DC: U.S. Government Printing Office, 1990.

Water quality standards

The development of **water quality** standards is a process first mandated by the Water Quality Act of 1965 and continued by requirements in the Federal **Water Pollution** Control Act of 1972 (PL 92-500), with amendments in 1977, 1982, and 1987 (collectively referred to as the **Clean Water Act**). The process is used to establish standards for stream water quality by taking into account the use and value of a stream for public water supplies, propagation of fish and **wildlife**, recreational purposes, as well as agricultural, industrial, and other legitimate uses. Water quality standards are enforceable by law and are applicable to all navigable waters. The goals of water quality standards are to protect public health and the **environment**, and to maintain a standard of water quality consistent with its designated uses. Water quality standards provide the "teeth" for water quality legislation and also the yardstick by which performance may be evaluated.

To establish water quality standards for a water body, officials (1) determine the designated beneficial water use; (2) adopt suitable water quality criteria to protect and maintain that use; and (3) develop a plan for implementing and enforcing the water quality criteria. Both uses and criteria constitute water quality standards, and water quality is evaluated based on how well the designated uses are supported.

Appropriate water use is designated by analyzing the existing use of the water as well as the potential to attain other uses based on an assessment of the physical, chemical, biological, and hydrological characteristics of the water and the economic cost and impact of achieving particular uses. The Clean Water Act requires that, whenever possible, water quality standards should ensure the protection and propagation of fish, shellfish, and wildlife and should provide for **recreation** in and on the stream. States have primary responsibility for designating stream segment uses, so stream uses may vary from state to state. However, stream use as designated by one state must not result in the violation of another state's use of the same stream.

Water quality criteria are most often expressed as numeric constituent concentrations or levels, but may be narrative statements if numeric values are not available or known. Each criterion is based on scientific information available concerning the effect of the pollutant on human health, aquatic life, and aesthetics.

Before the passage of the Clean Water Act, water **pollution control** efforts were considered successful if they achieved water quality standards. However, under the act, the accepted measurements of successful water **pollution** control is whether the **effluent** from point sources, specifically Publicly Owned Treatment Works (POTWs), meets technology-based effluent standards. These standards are based on what can be done with available technology rather than what is required to achieve water quality standards.

Water quality standards were retained, however, as part of the overall strategy to control water pollution. Rather than using water quality standards as the highest goal for determining water quality, the state and federal authorities consider water quality control standards to be the lowest acceptable level of water quality. In addition, point sources may be subjected to more stringent requirements than the use of **Best Available Control Technology** (BAT) if necessary to meet water quality standards. For example, in stream segments that do not meet water quality standards, states can establish a "total maximum daily load" of pollutants that will achieve water quality standards and assign a permissible share to individual dischargers. State water quality programs must also include provisions to prevent degradation of existing water quality when necessary to maintain existing uses and certain high quality waters.

The attainment of water quality standards is also affected by control of pollution from nonpoint sources, such as agricultural practices that result in the addition of sediments, nutrients, pesticides, and other contaminants to water bodies. The 1987 amendments to the Clean Water Act required states to identify water not meeting its standards due to **nonpoint source** pollution, identify general and specific nonpoint sources causing the problems, and develop management plans for the control of the these sources.

In 1988, state governments compiled a water quality report for the **Environmental Protection Agency** (EPA). The report indicated that of 519,412 river miles (835,734 km) assessed (out of a total of 1.8 million mi [2.9 million km] in the United States) 10% did not support their designated uses, 20% partially supported their uses, and 70% fully supported their uses. For lakes, of 16,313,962 acres (6,525,585 ha) assessed (out of a total of 39,400,000 acres [15,760,000 ha] in the United States), 10% did not support their designated uses, 17% partially supported their uses, and 74% fully supported their uses. *See also* Agricultural pollution; Agricultural runoff; Criteria pollutant; Marine pollution; Thermal pollution

[*Judith Sims*]

RESOURCES
BOOKS

Abron, L. A., and R. A. Corbitt. "Air and Water Quality Standards." In *Standard Handbook of Environmental Engineering*, edited by R. A. Corbitt. New York: McGraw-Hill, 1990.

OTHER

The Universities Council on Water Resources. "The Clean Water Act." *Water Resources Update* 84. Carbondale, IL: Southern Illinois University, 1991.

Water reclamation

The term "water reclamation" can be defined in different ways. In the early and mid-1900s, water **reclamation** referred primarily to supplying potable water to **arid** areas where water was not readily available, through the development of water projects such as **dams** and canals. Water reclamation can also mean collecting used water (i.e., **wastewater**) and treating it for **reuse** for **irrigation** and other purposes.

The federal agency primarily responsible for water reclamation projects that provided a source of water to arid lands is the **Bureau of Reclamation**, in the Department of the Interior. Formed in 1902 as the Reclamation Service and later renamed the Bureau of Reclamation in 1923, the agency's mission was "the construction and maintenance of irrigation works for the storage, diversion, and development of water for the reclamation of arid and semi-arid lands." This federal agency built dams, reservoirs, and other water development projects in the 17 western states of the United States. The mission of the Bureau of Reclamation in the early 1900s was to serve the irrigation needs of farms of 160 acres (65 ha) or less. By the 1930s the bureau's constituency expanded to larger farms, towns, and cities. The **Army Corps of Engineers** and the Department of Agriculture **Soil Conservation Service** (now called the **Natural Resources Conservation** Service) are also responsible for providing irrigation water in addition to providing flood control, **drainage**, and navigation.

During the early 1900s, water reclamation projects were a means of providing an opportunity for people to settle in the western United States. The projects provided inexpensive, government-subsidized water used for **transportation**, farming, industry, and by municipalities for drinking water and electricity. During the Depression era of the 1930s, the Bureau of Reclamation and the Public Works Administration worked together on water projects, which became an important part of the nation's economic recovery program.

In 1959, a Senate Select Committee on National **Water Resources** was formed to determine whether more water projects were necessary. After two years of investigations, the committee highlighted three main concerns: possible water shortages in the future, decreasing **water quality**, and possible **flooding**. Some of the recommendations made by the committee to address these concerns included increased regulation of stream flows by reservoirs, improved **groundwater** use, improvements in water use efficiency, and increased use of desalination and **weather modification**.

As a follow-up to the recommendations by the Select Committee on Natural Water Resources, President Kennedy created a Water Resources Council headed up by the Secretaries of the Interior, Health, Education and Welfare, and the Army to make biennial nationwide water project studies and to develop uniform regulations for evaluating water projects. With the advent of the environmental movement during the 1960s and the lowered unemployment relative to the 1930s, water reclamation projects became less of a national priority. By the 1970s, President Carter proposed that the costs of water projects be shared by the local water users. President Carter also proposed that water reclamation projects be evaluated based on economic, environmental, and safety criteria. During President Reagan's tenure, the criteria used to evaluate the feasibility of water projects were relaxed; however, President Reagan continued support of local cost sharing.

Over the years, there have been numerous water reclamation projects. A year after the formation of the Bureau of Reclamation, the agency was working on six projects. By 1907, it was participating in the development of 27 water projects. From 1905-1991, the Bureau of Reclamation built over 493 reservoirs and diversion dams, 7,670 mi (12,341 km) of irrigation canals, 1,440 mi (2,317 km) of pipelines and tunnels, and 53 hydroelectric **power plants**. All the larger rivers in the U.S. have been altered in some way by water reclamation projects. For example, the Columbia River has 28 dams. Some of the larger water reclamation projects include the Central Arizona Project (a massive canal system that diverts over 1.2 million acre-feet [1.5 billion m^3] of water per year from the **Colorado River**), Bonneville Dam in Oregon, Grand Coulee Dam in Washington, Marshall Ford Dam in Texas, Teton Dam in Idaho, Garrison Dam in North Dakota, **Tellico Dam** in Tennessee, Hoover Dam in Nevada, **Glen Canyon Dam** in Arizona, Imperial Dam in California, and the Central Valley Project in California. Costs of these projects were large. For example, the California Central Valley Project cost over $950 million. The farmers, large and small, have paid only a small portion of overall costs of water reclamation projects. An estimated 4% of the total cost of the California Central Valley Project, equal to about $38 million, was repaid by the farmers who benefitted from the water project.

Important **wildlife habitat** and wildlife were lost as a result of these efforts to create massive dams. Dam building along the Colorado River has led to the decline of eight **species** of fish that are now considered Endangered. Habitat loss can result from flooding areas to create reservoirs or from cutting off water and nutrients from naturally occurring periodic flooding. During dam construction, stream-side habitat communities can be altered and destroyed from diverting the course of the river and excavation and **dredging** activities related to dam building. The **ecology** of these

areas typically changes from that of a river to that commonly found along lakes and reservoirs.

Another ecological change occurs from dams cutting off **sediment** suspended in rivers from flowing downstream. As a result, sediment builds up in dams, reducing their volume and lowering their life expectancy. For example, the volume of Lake Waco in Texas was reduced by half over a 34-year period. When dams trap **silt** and sediment, **erosion** can result down stream. Below the Hoover Dam, millions of cubic yards of sediment were eroded over a 100 mi (160 km) stretch of the Colorado River.

Water reclamation projects continue on a much smaller scale today. Another method in which to obtain water for irrigation purposes is using water twice—once for domestic purposes and a second time for irrigation of lands such as farms and **golf courses**. Reclaiming used water occurs in the U.S. and throughout the world. Israel reclaims over 70% of its used water, or waste water, to irrigate over 47,000 acres (19,000 ha) of farm land. By the end of the 1990s, Israel plans to supply over 16% of its total water needs by reclaiming waste water. The city of Los Angeles, California, reuses its treated waste water to recharge its underground aquifers, which it taps for drinking water. In Arizona, the city of Tucson plans to reuse water to meet over 19% of its total water needs while Phoenix receives over 70,640 ft^3 (2,000 m^3) of fresh water from the Bureau of Reclamation in exchange for every 105,960 ft^3 (3,000 m^3) of reclaimed water used by the nearby irrigation district. The city of St. Petersburg, Florida, has a unique system in which it provides two sources of water; one source is potable water used for drinking, cooking, washing, etc. and the other source of water is reclaimed water from wastewater that is reused for irrigating parks, median strips along roads, and residential lawns. Because the treated wastewater contains nutrients such as **nitrogen** and **phosphorus**, less **fertilizer** is needed when used for irrigation.

Securing sufficient amounts of water while sustaining a healthy **environment** can be accomplished in many ways using all types of water reclamation projects including waste water reuse. **Water conservation**, efficiency of use, and reuse will likely be playing larger roles in ensuring adequate water supplies in the future.

[*Marci L. Bortman*]

Resources

Books

Gottlieb, R. *A Life of Its Own, The Politics and Power of Water.* New York: Harcourt Brace Jovanovich, 1988.

Hunt, C. E., and V. Huser. *Down by the River, the Impact of Federal Water Projects and Policies on Biological Diversity.* Washington, DC: Island Press, 1988.

Outwater, A. *Water, A Natural History.* New York: Basic Books, 1996.

Postel, S. *Last Oasis: Facing Water Scarcity.* The Worldwatch Environmental Alert Series. New York: W.W. Norton & Company, 1992.

Water resources

Water resources represent one of the most serious environmental issues of the twentieth century. Water is abundant globally, yet millions of people have inadequate fresh water, and droughts **plague** rich and poor countries alike. **Water pollution** compounds the **scarcity** of usable fresh water and threatens the health of millions of people each year. At the same time, the ravages of **flooding** and catastrophes associated with excessive rainfall or snowmelt cause death and destruction of property. To cope with these problems, water resource management has become a high priority worldwide.

Concerns about water scarcity seem unwarranted, considering that there are 326 million mi^3 (1.36 billion km^3) of water on earth. Although 99% of this water is either unusable (too salty) or unavailable (ice caps and deep **groundwater** storage), the remaining 1% can meet all future water needs. The total volume of fresh water on earth is not the problem. Scarcity of fresh water results from the unequal distribution of water on earth and water **pollution** that renders water unusable.

Fresh water is not always available where or when it is needed the most. In Lquique, Chile, for example, no rainfall was measured for 14 consecutive years, from 1903 to 1918. At the other extreme, Mount Waialeale, Kauai, Hawaii, averages 460 in (1,168 cm) of rainfall per year. Even such extremes in annual precipitation do not tell the complete story of water scarcity. For example, in the tropics, storms in the rainy season can deposit from 16 to over 39 in (40.6 to 99 cm) of rainfall in two to three days, often resulting in widespread flooding. However, the same areas can experience negligible rainfall for two to three months during the dry season.

The demand for water varies worldwide, with wealthier countries consuming an average of about 264,000 gallons (1,000 m^3) per person, per year. In contrast, over one-half of the population in the developing countries of North Africa and the Middle East live on a fraction of this amount. **Population growth** and droughts have forced people to use water more efficiently. However, **water conservation** alone will not solve water scarcity problems.

Increasing demands for water and the disparity of water distribution have sparked the imagination of hydrologists and engineers. Grandiose schemes have emerged, including towing icebergs from polar regions to the lower latitudes, **desalinization** of ocean water, cloud seeding, and transporting water thousands of kilometers from water-rich to water-poor regions. Although many of these ideas are

fraught with environmental, economic, and political concerns, they have placed water supply issues in the international spotlight.

Historically, more conventional, large public investment water development schemes have been a vehicle for economic development. As a result, about 75% of the fresh water used worldwide is used for **irrigation**, with 90% used for this purpose in some developing countries. Much of this water is being wasted. Poorer countries do not have the technology to irrigate efficiently, using twice as much water to produce one-half the crop yields of wealthier countries.

Multi-purpose **reservoir** projects that generate hydroelectric power, control flood, provide recreational benefits, and supply water have become commonplace worldwide. Hydroelectric power has fostered both municipal and industrial development. Since 1972, 45% of the **dams** constructed with **World Bank** support were primarily for hydropower, in contrast to 55% for irrigation and/or flood control. Although the pace of dam construction has slowed worldwide, large projects are still being constructed. For example, the proposed **Three Gorges Dam** on the Yangtze River, China, will cost at least $10 billion and will be the largest hydroelectric power project in the world. It will generate one-eighth of the total electrical power that was generated in China in 1991.

Dams and water transfer systems constructed for economic benefit often carry with them environmental costs. **Wildlife habitat** is flooded by reservoir pools, and streamflow volumes and patterns are altered, affecting aquatic ecosystems. The modification of flow by dams in the Columbia River of the United States has either eliminated or severely reduced natural spawning runs of **salmon** in much of the river. The reservoir created by the Three Gorges Dam will flood one of China's most scenic spots and will force over one million people to relocate. Already there are concerns that high rates of **soil erosion** and **sediment** deposition behind the dam will threaten the usefulness of the project. The relocation of people in response to water resource development can also bring unwanted environmental consequences. In developing countries, relocated people seek new land to cultivate, graze livestock, or build new villages, often accelerating **deforestation** and land degradation.

Small-scale projects offer environmentally attractive alternatives to large dams and reservoirs. Mini-hydroelectric power projects, which do not necessarily require dams, provide electricity to many small villages. Rainfall harvesting methods in drylands provide fresh water for drinking, livestock, and crops. Flood plain management is environmentally and economically preferable to large flood control dams in most instances.

Groundwater provides large volumes of fresh water in many regions of the world. A proposed $25 billion irrigation project in Libya would pump nearly 980 million yd^3 (750 million m^3) of groundwater per year in the Sahara to be transported 1,180 mi (1,900 km) north. Groundwater supplies in the Sahara are thought to be nearly 20 trillion yd^3 (15 trillion m^3). However, there is concern that such pumping will drop water levels 8–20 in (20–50 cm) per year, eventually causing water that is too salty for irrigation to intrude. Groundwater mining occurs if water is pumped faster than it is replenished. Such mining is occurring in many parts of the world and cannot provide sustainable sources of water.

Water requirements for urban areas pose formidable challenges in the coming years. With 90% of the population growth expected to occur in urban areas over the next few decades, **competition** for fresh water between rural agricultural areas and cities will become severe. With urban areas able to pay higher prices for water than rural areas, legal and political battles over **water rights** and subsidy issues will likely ensue. In many cities, however, finding sufficient sources of fresh water will be difficult. For example, Mexico City has a rapidly growing population of over 22 million as of 2002. Fresh water from lakes and springs have already been exhausted. Groundwater is being mined three times faster than the rate of recharge. Also, a daily **discharge** of 29 million lb (13 million kg) of sewage plus industrial wastes threaten groundwater quality. Furthermore, groundwater pumping is causing land to subside around the city. Unfortunately, many other large cities around the world are faced with equally challenging problems of water supply and pollution.

Flooding and droughts are global concerns that **lead** to **famine** in the poorer countries and will persist until solutions are found. Flood avoidance or flood plain management are not commonplace in many such countries because people have no alternative places to live and farm. The recurring floods and devastation in Bangladesh, and the droughts of North Africa emphasize the dilemma of the poorest of countries.

In contrast to problems of water quantity, water pollution accounts for over 5 million deaths per year, largely the result of drinking water that is contaminated by human and livestock waste. Inadequate **sewage treatment** and a lack of alternative fresh water supplies can be blamed. These problems can be solved but will require that **water quality** issues become a high priority in terms of policies, institutions, and financial support.

Water pollution plagues rich and poor countries alike. Municipalities and industries pollute streams and groundwater with toxic **chemicals**, sewage, disease-causing agents, oil, **heavy metals**, **thermal pollution**, and **radioactive waste**. Polluted water from municipalities and industries

can and should be treated before it is discharged. Environmental laws and enforcement can ensure that this happens.

Agricultural development has made significant strides in solving food problems in the world, but it has also contributed to water pollution. Water supplies have become polluted as a result of heavy **fertilizer** and **pesticide** use. The focus on sustainable agricultural practices that do not rely so heavily upon fertilizers and pesticides offer some promise. Sensible use of chemicals and the adoption of "best management practices" also offer a means of meeting food requirements of growing populations without contaminating water supplies.

Just as food security issues attracted global attention through the 1950s to 1970s, similar efforts will be needed to provide for water security in the twenty-first century. Long-term, holistic, and interdisciplinary solutions are needed. The crises management approach of the past—that of attempting to deal with droughts, floods, and contaminated water only when they are upon us—cannot continue.

We currently have the technology to solve most water resource problems. Implementing solutions are often constrained by inadequate policies, institutions, financial resources, and political instability. In many ways, people management is more critical than water resource management. With human populations continuing to grow, water scarcity, flooding, and disease and death from contaminated drinking water will become even more pronounced.

[*Kenneth N. Brooks*]

Resources

Books

Brooks, K. N., et al. *Hydrology and the Management of Watersheds.* Ames, IA: Iowa State University Press, 1991.

Jackson, I. J. *Climate, Water and Agriculture in the Tropics.* 2nd ed. Essex, England: Longman Scientific & Technical, 1989.

Opportunities in the Hydrologic Sciences. Committee on Opportunities in the Hydrologic Sciences, Water Sciences Technology Board. National Research Council. Washington, DC: National Academy Press, 1991.

Van der Leeden, F., F. L. Troise, and D. K. Todd. *The Water Encyclopedia.* 2nd ed. Chelsea, MI: Lewis Publishers, 1990.

Periodicals

Anton, D. "Thirsty Cities." *IDRC Reports* 18, no. 4.

"The First Commodity." *The Economist* 322 (1992): 11–12.

Winkler, P. "Qaddafi Challenges the Desert." *World Press Review* (April 1992): 47.

Water rights

Pure, potable water is an essential resource, second only to breathable air in importance for life on earth, but it has other valuable attributes as well. It powers industry, provides **recreation**, irrigates agriculture, and is a highway for ships. It also takes many forms, from oceans, rivers, and lakes to **groundwater** and even atmospheric moisture. Given this vital significance, water rights have been contended throughout history.

One area of frequent dispute is jurisdiction. If a water right is a right to use water, then who has the authority to grant or withhold that right? The obvious answer is the government that has jurisdiction over the territory in which the water is found. However, water pays no attention to boundaries. A river that begins in one jurisdiction flows into another. Problems arise when people in the first jurisdiction deprive people in the other of a sufficient quantity or quality of water from the river. Such a situation forced the United States, which uses the **Colorado River** to irrigate several western states, to build a plant to remove excess salt from the Colorado before it flows into Mexico.

Impossible as it seems, jurisdiction over the water cycle has also been litigated. From atmospheric moisture comes rain and snow, which replenish groundwater and form streams that flow into rivers, which in turn flow into lakes and oceans. Surface water thus forms an interconnected system. International law has focused on rights to "international **drainage** basins" and, more recently, "international **water resources** systems" that include atmospheric and frozen water.

When jurisdiction is not at issue, water rights are usually established in one of three ways. The first was set out in the Institutes of Justinian (533-34 A.D.), which codified Roman law. This holds that flowing water is a *res communes omnium*—a thing common to all, at least when it is navigable. It cannot be owned or used exclusively by anyone. The second and third approaches have been developed in the United States, where the doctrines of **riparian rights** and prior appropriation predominate in the eastern and western states respectively. In a riparian system, landowners whose land borders a waterway have a special, if not exclusive, right to the water. In the **arid** western states, however, water rights derive from first use, not ownership, as in the "first in time, first in right" rule. *See also* Environmental law; Hydrologic cycle; Irrigation; Land-use control; Riparian land

[*Richard K. Dagger*]

Resources

Books

Caponera, D. A., ed. *The Law of International Water Resources.* Rome: Food and Agricultural Organization of the United Nations, 1980.

Goldfarb, W. *Water Law.* 2nd ed. New York: Lewis, 1988.

Matthews, O. P. *Water Resources, Geography and Law.* Washington, DC: Resource Publications in Geography, 1984.

Sax, J. *Water Law: Cases and Commentary.* Boulder: Pruett Press, 1965.

Water table

The top or upper surface of the saturated zone of water that occurs underground, except where it is confined by an overlying impermeable layer. This surface separates the zone of aeration—the layer of **soil** or rock that is unsaturated most of the time—from the **zone of saturation** below. The water table is the level at which water rises in a well that is situated in a **groundwater** body but not confined by a layer of **impervious material**. In **wetlands**, the water table can be at the soil surface or a few centimeters below the soil surface, but in drylands it can be several hundred meters below the soil surface. Perched water tables occur where there is a shallow impervious layer (such as a clay layer) that prevents water from percolating downward to the more persistent, regional groundwater zone. As a result, water tables can occur at different depths within a region. Generally, if water tables are close to the soil surface, groundwater is susceptible to **pollution** from human activities. *See also* Percolation

Water table draw-down

A measurement of how a particular water level is affected by the withdrawal of **groundwater**. It is made by establishing the vertical distance at a given point between the original water level and the pumping water level. The **water table** of an unconfined **aquifer** or the potentiometric surface of a confined aquifer is lowered when a well is pumped, and an inverted cone of depressions develops around the pumping well.

Changes of drawdown at different distances from the well are shown with a drawdown curve. Specific capacity is the rate of **discharge** of water from the well divided by the drawdown within the well. The drawdown within a well divided by the discharge rate of water from the well—inverse of specific capacity—is called specific drawdown.

Water treatment

Water treatment—or the purification and **sanitation** of water—varies as to the source and kinds of water. Municipal waters, for example, consist of surface water and **groundwater**, and their treatment is to be distinguished from that of industrial water supplies.

Municipal water supplies are treated by public or private water utilities to make the water potable (safe to drink) and palatable (aesthetically pleasing) and to insure an adequate supply of water to meet the needs of the community at a reasonable cost. Except in exceedingly rare instances, the entire supply is treated to drinking **water quality** for three reasons: it is generally not feasible to supply water of more than one quality; it is difficult to control public access to water not treated to drinking water quality; and a substantial amount of treatment may be required even if the water is not intended for human consumption.

Raw (untreated) water is withdrawn from either a surface water supply (such as a lake or stream) or from an underground **aquifer** (by means of **wells**). The water flows or is pumped to a central treatment facility. Large municipalities may utilize more than one source and may have more than one treatment facility. The treated water is then pumped under pressure into a distribution system, which typically consists of a network of pipes (water mains) interconnected with ground-level or elevated storage facilities (reservoirs).

As it is withdrawn from the source, surface water is usually screened through steel bars, typically about 1 in (2.54 cm) thick and about 2 in (5.08 cm) apart, to prevent large objects such as logs or fish from entering the treatment facility. Finer screens are sometimes employed to remove leaves. If the water is highly turbid (cloudy or muddy), it may be pretreated in a large basin known as a presedimentation basin to allow time for sand and larger **silt** particles to settle out.

All surface waters have the potential to carry pathogenic (disease-causing) **microorganisms** and must be disinfected prior to human consumption. Since the adequacy of disinfection cannot be assured in the presence of turbidity, it is first necessary to remove the suspended solids causing the water to be turbid. This is accomplished by a sequence of treatment processes that typically includes coagulation, flocculation, **sedimentation**, and **filtration**.

Coagulation is accomplished by adding chemical coagulants, usually **aluminum** or iron salts, to neutralize the negative charge on the surfaces of the particles (suspended solids) present in the water, thereby eliminating the repulsive forces between the particles and enabling them to aggregate. Coagulants are usually dispersed in the water by rapid mixing. Other **chemicals** may be added at the same time, including powdered activated **carbon** (to absorb taste- and odor-causing chemicals or to remove synthetic chemicals); chemical oxidants such as **chlorine**, **ozone**, chlorine dioxide, or potassium permanganate (to initiate disinfection, to oxidize organic contaminants, to control taste and odor, or to oxidize inorganic contaminants such as iron, manganese, and sulfide); and **acid** or base (to control **pH**).

Coagulated particles are aggregated into large, rapidly settling "floc" particles by flocculation, accomplished by gently stirring the water using paddles, turbines, or impellers. This process typically takes 20 to 30 minutes. The flocculated water is then gently introduced into a sedimentation basin, where the floc particles are given about two to four hours

to settle out. After sedimentation, the water is filtered, most commonly through 24–30 in (61–76 cm) of sand or anthracite having an effective diameter of about 0.02 in (0.5 mm). When the raw water is low in turbidity, coagulated or flocculated water may be taken directly to the **filters**, bypassing sedimentation; this practice is referred to as direct filtration.

Once the water has been filtered, it can be satisfactorily disinfected. Disinfection is the elimination of pathogenic microorganisms from the water. It does not render the water completely sterile but does make it safe to drink from a microbial standpoint. Most water treatment plants in the United States rely primarily on chlorine for disinfection. Some utilities use ozone, chlorine dioxide, chloramines (formed from chlorine and ammonia), or a combination of chemicals added at different points during treatment. There are important advantages and disadvantages associated with each of these chemicals, and the optimum choice for a particular water requires careful study and expert advice.

Chemical disinfectants react not only with microorganisms but also with naturally occurring organic matter present in the water, producing trace amounts of contaminants collectively referred to as disinfection byproducts (DBPs). The most well-known DBPs are the **trihalomethanes**. Although DBPs are not known to be toxic at the concentrations found in drinking water, some are known to be toxic at much higher concentrations. Therefore, prudence dictates that reasonable efforts be made to minimize their presence in drinking water.

The most effective strategy for minimizing DBP formation is to avoid adding chemical disinfectants until the water has been filtered and to add only the amount required to achieve adequate disinfection. Some DBPs can be minimized by changing to another disinfectant, but all chemical disinfectants form DBPs. Regardless of which chemical disinfectant is used, great care must be exercised to ensure adequate disinfection, since the health risks associated with pathogenic microorganisms greatly outweigh those associated with DBPs.

There are a number of other processes that may be employed to treat water, depending on the quality of the source water and the desired quality of the treated water. Processes that may be used to treat either surface water or groundwater include: 1) lime softening, which involves the addition of lime during rapid mixing to precipitate calcium and magnesium ions; 2) stabilization, to prevent corrosion and scale formation, usually by adjusting the pH or alkalinity of the water or by adding scale inhibitors; 3) activated carbon **adsorption**, to remove taste- and odor-causing chemicals or synthetic organic contaminants; and 4) **fluoridation**, to increase the concentration of fluoride to the optimum level for the prevention of dental cavities.

Compared to surface waters, groundwaters are relatively free of turbidity and pathogenic microorganisms, but they are more likely to contain unacceptable levels of dissolved gases (**carbon dioxide**, **methane**, and **hydrogen** sulfide), hardness, iron and manganese, volatile organic compounds (VOCs) originating from **chemical spills** or improper waste disposal practices, and **dissolved solids** (**salinity**).

High-quality groundwaters do not require filtration, but they are usually disinfected to protect against contamination of the water as it passes through the distribution system. Small systems are sometimes exempted from disinfection requirements if they are able to meet a set of strict criteria. Groundwaters withdrawn from shallow wells or along riverbanks may be deemed to be "under the influence of surface water," in which case they are normally required by law to be filtered and disinfected.

Hard groundwaters may be treated by lime softening, as are many hard surface waters, or by **ion exchange** softening, in which calcium and magnesium ions are exchanged for sodium ions as the water passes through a bed of ion-exchange resin. Groundwaters having high levels of dissolved gases or VOCs are commonly treated by air stripping, achieved by passing air over small droplets of water to allow the gases to leave the water and enter the air.

Many groundwaters—approximately one quarter of those used for public water supply in the United States—are contaminated with naturally occurring iron and manganese, which tend to dissolve into groundwater in their chemically reduced forms in the absence of oxygen. Iron and manganese are most commonly removed by oxidation (accomplished by **aeration** or by adding a chemical oxidant, such as chlorine or potassium permanganate) followed by sedimentation and filtration; by filtration through an adsorptive media; or by lime softening.

Groundwaters high in dissolved solids may be treated using reverse osmosis, in which water is forced through a membrane under high pressure, leaving the salt behind. Membrane processes are rapidly evolving, and membranes suitable for removing hardness, dissolved organic matter, and turbidity from both ground and surface waters have recently been developed.

Industrial water treatment differs from municipal treatment in the specified quality of the treated water. Many industries use water supplied by a local municipality, while others secure their own source of water. Those securing water from a private source often treat it using the same processes used by municipalities. However, industries must often provide additional treatment to provide water suitable for their special needs, which may include process water, boiler feed water, or cooling water.

Process water is water used by an industry in a particular process or for a group of processes. The quality of water required depends on the nature of the process. For example, water used to make white paper must be free of color. In some instances, water of relatively poor quality may be acceptable, e.g., water used to granulate steelmaking slags, while other uses require water of the very highest purity, e.g., ultrapure water used in the manufacture of silicon chips.

Water used in boilers for thermoelectric or **nuclear power** generation must be very low in dissolved solids and must be treated to prevent both corrosion and scale formation in the boilers, turbines, and condensers. High purity process waters and boiler feed waters are typically produced using ion-exchange demineralization and special filters designed to remove sub-micron sized particles.

Cooling water may be used once (single-pass system) or many times (closed-loop system). Cooling water used only once may receive little treatment, typically continuous or intermittent disinfection to reduce slime growths and perhaps stabilization to control corrosion and scale formation. Water entering a closed-loop system may not only be stabilized and disinfected, but may also be treated to remove nutrients (**nitrogen** and **phosphorus**) or dissolved solids. Additional treatment is usually provided in the loop to remove dissolved solids that accumulate as a result of evaporation. *See also* Safe Drinking Water Act

[*Stephen J. Randtke*]

Resources

Books

Peavy, H. S., D. R. Rowe, and G. Tchobanoglous. *Environmental Engineering.* New York: McGraw-Hill, 1985.

Water Treatment Plant Design. 2nd ed. New York: McGraw-Hill, 1990.

Other

"Recommended Standards for Water Works." Great Lakes Upper Mississippi River Board of State Public Health and Environmental Managers. Albany, NY: Health Education Services, 1992.

Waterkeeper Alliance

The Waterkeeper Alliance is an umbrella organization consisting of 86 Waterkeeper programs (as of 2002) located throughout North and Central America. The mission of the Alliance is to restore degraded waters and to protect aquatic environments and is based on the philosophy that protection of a community's resources requires the daily vigilance of its citizens.

The Waterkeeper concept started in 1966 on the **Hudson River** in New York, when a coalition of commercial and recreational fishermen was formed to protect the Hudson from **pollution**. The coalition constructed a boat to patrol the river; in 1983 this coalition, calling itself the Riverkeepers, using funds from successful lawsuits, hired a full-time public advocate and began filing lawsuits against municipal and industrial polluters. By 1998, after over 100 successful legal actions were filed against Hudson River polluters, resulting in $1 billion in **remediation** costs, the Hudson River was once again a productive **ecosystem**. The Hudson River program was modeled after the British Isle riverkeepers, or game wardens, who were responsible for protection of private trout and **salmon** streams from **poaching** for estates, manors, and private fishing clubs.

New ecosystem protection programs, with names such as Lakekeeper, Baykeeper, Coastkeeper, as well as Riverkeeper, have been established, based on the Hudson River model. In 1992, these programs joined to form the National Alliance of River, Sound, and Baykeepers, which was renamed the Waterkeeper Alliance in 1999. Responsibilities of the Alliance include overseeing new Waterkeeper programs, licensing the use of Waterkeeper names, organizing conferences, working on issues common to Waterkeeper programs, and serving as a clearinghouse and networking center for information and strategy exchange among the programs. In 2002, the Waterkeeper Alliance was working with local advocates to establish Keeper programs in Belize, the Czech Republic, Italy, Poland, and the Philippines. The president of the Waterkeeper Alliance is Robert F. Kennedy, Jr., who also serves as the chief prosecuting attorney for the Hudson Riverkeepers.

A Waterkeeper program may utilize a variety of methods to protect a water resource, including **water quality** monitoring, participation in planning organizations, public education, development of technical solutions and remedies to problems, and litigation to ensure enforcement of environmental regulations and laws. Since most environmental laws contain citizen suit provisions, the ability of citizens to prosecute polluters is an important tool of Keeper organizations. Another essential component of a Waterkeeper program is the employment of a "Keeper," who is a full-time, privately funded, non-governmental ombudsman who acts as a public advocate for the identified water body. The Keeper is aided by a network of local fishermen, environmental experts, and concerned citizens, who act as a "neighborhood" watch program. Most Waterkeeper programs patrol and monitor their water bodies regularly, by wading with hip boots or using boats ranging in size from canoes to research vessels. Every water body has its own set of challenges, and each Waterkeeper program develops unique strategies to address those challenges.

Waterkeeper Alliance members are concerned with many different types of water issues. The Waterkeepers of Wisconsin are working to prevent water bottling companies from using spring water. Spring water is a high quality source

of surface water, as it is cold, oxygenated, and drinkable. If spring water is used commercially, the Wisconsin Waterkeepers are concerned that surface water, without recharge by spring water, will become polluted as a larger proportion of recharge would be **runoff** from fields, barnyards, and roads.

The California Coastkeeper Alliance (CCKA), a group composed of the Baja California Coastkeeper, San Diego Baykeeper, Orange County Coastkeeper, Santa Monica Baykeeper, Ventura Baykeeper, and Santa Barbara Channelkeeper, is participating in a Regional Kelp Restoration Project in coordination with the National Oceanic and Atmospheric Administration's Community Based Restoration Program. Kelp forests provide homes and shelters for a wide diversity of ocean organisms, including many commercial and sport fishing **species**. Kelp is also harvested for the extraction of algin, which is used as a thickening, stabilizing, and smoothing agent in a variety of commercial products. However, nearly 80% of the kelp canopy area along the southern California coast was lost between 1967 and 1999. Kelp losses are attributed intensified winter storms, low nutrients, and warmer sea surface temperatures associated with El Niño events. Increased turbidity, **sedimentation** and pollution from **wastewater** discharges, increasing human populations, coastal development, and **storm runoff** negatively affect the ability of kelp to recover naturally. Also over-fishing of species that prey on sea urchins also affects kelp forests, for without predators, sea urchin populations expand; sea urchins using kelp as a food source can decimate entire forests, creating "urchin barrens." The CCKA restoration project involves growing kelp in the Kelp **Mariculture** Laboratory, planting juvenile plants, and maintaining kelp forests. Another component of the project is providing portable kelp nurseries to schools throughout southern California to teach students about the importance of kelp to the coastal ecosystem.

In 2002, the Cook Inletkeeper of Alaska joined with other environmental activist groups in a citizen's lawsuit under the United States **Clean Water Act** to force the United States Department of the Army and the Department of Defense to address pollution and safety hazards associated with past and present bombing of Eagle River Flats. More than 10,000 unexploded bombs and munitions have contaminated Eagle River Flats. The munitions release **chemicals** such as RDX, 2,4-DNT, toxic metals, and other explosive and propellant compounds that present a danger to **wildlife** and people. The U.S. **Environmental Protection Agency** placed Fort Richardson on the National Priorities (Superfund) List of polluted sites in 1994, but the area has not yet been remediated.

The New Jersey Baykeeper is sponsoring an oyster gardening project where volunteers are growing oysters at their own private piers and marinas. Upon reaching sexual maturity, the oysters are transplanted to estuaries, where they serve as spawning stocks to increase the wild population. The Baykeeper provides seed oysters to volunteers, conducts workshops to teach volunteers how to grow oysters, and places oysters in the estuaries. Oyster populations in the New Jersey and New York areas have decreased due to overharvesting, pollution, **siltation**, and disease.

The Columbia Riverkeeper has been working with other groups of activists in a fight to remove the Condit Dam on the White Salmon River in southern Washington. It is estimated that removal of the Condit Dam would result in recovery of more than 4,000 salmon and steelhead in the White Salmon River per year. The Dam, owned by PacificCorps, is 90 years old, cracking, and produces only 14 megawatts of energy. In 1999, PacificCorps agreed to remove the Condit Dam by 2007. However, the agreement must be approved by the Federal Environmental Regulatory Commission, and county commissioners have suggested that the local county government operate the dam in place of PacificCorps.

The Hudson Bay Riverkeeper is working with government, business, civic, and environmental leaders to shut down the Indian Point Nuclear Power Plant in order that a full review of the plant's operating safety procedures and its emergency preparedness plans. Indian Point is located 30 mi (48 km) from New York City, with 20 million people (20% of the population of the United States) living within 50 mi (80 km) radius of the plant. It is a potential terrorist target due to its proximity to major financial centers, major **transportation** centers, and drinking water reservoirs for Westchester County and New York City, and due to its inventory of highly radioactive materials. In 2000 Indian Point was given a "red" designation by the United States **Nuclear Regulatory Commission**, giving it the highest **risk assessment**. The rating was based on operators' failure to detect flaws in a steam generator tube before a radiation leak. In December 2001, Indian Point received a notice of "substantial" safety concerns from the U.S. Nuclear Regulatory Commission when four control room crews failed to pass annual re-qualification tests.

The San Diego Baykeeper has been focusing on shipyard pollution since its formation in 1995. San Diego Bay, with 56% of its sediments acutely toxic to marine organisms, was ranked in 1996 as the second most toxic of 18 United States bays studied by the National Oceanic Atmospheric Administration. For ten years, the Bay has been posted with warnings to avoid eating bay fish because of elevated levels of **mercury**, **arsenic**, and **polychlorinated biphenyls** (PCBs). Major contributors of contamination to the bay include shipbuilding and repair facilities that serve the United States Navy and commercial oil tankers, dry cargo

carriers, ferries, and cruise lines. The Baykeeper's first activities focused on preventing ongoing and future pollution by suing shipyards over chronic stormwater permit violations. Additional activities now include campaigns to challenge and/or force shipyards, local and state regulators, and elected officials to remediate existing contaminated sediments.

The Santa Monica Baykeeper refers to itself as a "citizen" park ranger for Santa Monica Bay, San Pedro Bay, and adjacent coastal waters and watersheds. The Baykeeper maintains a 24-hour Pollution Hotline, recruits members and volunteers to serve as an "observer flotilla" and environmental watch prograram, provides a public awareness program that includes lectures and displays, publicizes violator "busts," tests the bay's water and sediments and provides the results to the public and government agencies, and participates in legal prosecution of violators.

The Waterkeeper Alliance has launched a national campaign in the United States to focus on protecting waterways from pollution caused by Confined Animal Feeding Operations, or CAFOs. Activities include filing lawsuits against animal producers, alleging that they have polluted waterways, working with farmers to reduce chemical use and improve animal environments, and educating consumers.

[*Judith L. Sims*]

Resources

Books

Kennedy Jr., Robert F., and John Cronin. *The Riverkeepers.* New York: Simon and Schuster, 1999.

Periodicals

Rosenblatt, Roger. "Heroes for the Planet: In Search of the Beauty and Mystery of Home." *Time.com* August 2, 1999 [cited July 2002]. <http://www.time.com/time/reports/environment/heroes/heroesgallery/0,2967,kennedy,00.html>.

Other

The Hudson Riverkeepers. *Videotape.* Outside Television, 1998.

The Waterkeepers. *Videotape. Outside Television.* 2000.

Organizations

Waterkeeper Alliance, 78 North Broadway, Bldg. E, White Plains, NY USA 10603 (914)422-4410, Fax: (914)422-4437, Email: info@waterkeeper.org , <http://www.waterkeeper.org>

Waterless toilets

see **Toilets**

Waterlogging

Saturation of **soil** with water, usually through **irrigation**, resulting in a condition under which most crop plants cannot grow. Although naturally wet soils are extremely common, human-caused waterlogging occurring on active croplands has created the most concern. Where fields are irrigated regularly, some means of **drainage** is usually necessary to let excess water escape, but if drainage is inadequate, water gradually accumulates, filling pore spaces in the soil. Once the oxygen in these pores is displaced by water, plant roots and soil **microorganisms** die. Waterlogging frequently occurs where **flooding** has been used. There are two main reasons for flooding: it is a simple, low-technology method of irrigation and in some cases it reduces harmful mineral concentrations in soils. Ideally, flood waters seep through the soil and into **groundwater** or stream channels, but in many agricultural regions, impermeable layers of clay block downward drainage so that irrigation water collects in the soil. Farmers who can afford it avoid this problem by installing some means of field drainage, such as perforated pipes below ground. However, the expense of this procedure is prohibitive in many regions. *See also* Soil conservation; Soil eluviation

Watershed

A catchment or **drainage** basin that is the total area of land that drains into a water body. It is usually a topographically delineated area that is drained by a stream system. **River basins** are large watersheds that contribute to water flow in a river. The watershed of a lake is the total land area that drains into the lake. In addition to being hydrologic units, watersheds are useful units of land for planning and managing multiple **natural resources**. By using the watershed as a planning unit, management activities and their effects can be determined for the land area that is directly affected by management. The hydrologic effects of land management downstream can be evaluated as well. Sometimes **land use** and management can alter the quantity and quality of water that flows to downstream communities. By considering a watershed, many of these environmental effects can be taken into consideration. *See also* Watershed management

Watershed management

Watershed management is planning, guiding, and organizing **land use** so that desired goods and services are produced from a watershed without harming **soil** productivity and **water resources**. Goods and services produced from watersheds include food, forage for livestock and **wildlife**, wood and other forest products, outdoor **recreation**, wildlife **habitat**, scenic beauty, and water. Essential to watershed management is the recognition that production must be accompanied by environmental protection.

The specific objectives of watershed management depend on human needs in a particular area and can include: (1) the **rehabilitation** of degraded lands; (2) the protection of soil and water resources under land use systems that produce multiple products of the land; and (3) the enhancement of water quantity and quality. Rehabilitating degraded lands is sometimes mistakenly thought of as the only role of watershed management. Rehabilitation requires that both the productivity and hydrologic function of degraded lands be restored. This usually entails the construction of engineering structures, such as gully control **dams**, followed by vegetation establishment, protection, and management, all of which are needed to achieve long-term healing of the landscape.

Most fundamental to watershed management is the prevention of land and water resource degradation in the first place. To achieve this goal, land use must adhere to **conservation** practices that avoid land degradation. The greatest potential for degradation arises from road construction, mining, crop cultivation, **logging**, and **overgrazing** by livestock in steep terrain. When management guidelines are not followed, soils erode and land productivity is diminished. The loss of soil and vegetative cover reduces the effectiveness of watersheds in moderating the flow of water, **sediment**, and other waterborne substances. As a result, damage to aquatic ecosystems and human communities can occur in areas that are positioned downslope or downstream. To achieve sustainable land use, the development of and adherence to land use guidelines and conservation practices must become commonplace.

Preventing degradation of **wetlands** and **riparian land** is of particular environmental concern. These soil-vegetation communities require special management and protection because the wet soils are susceptible to excessive **erosion**. Furthermore, riparian vegetation provides valuable wildlife habitat and plays a critical role in protecting **water quality**.

In some parts of the world, watershed management can be aimed at enhancing water resources. In some instances vegetative cover can be altered to increase water yield or to change the pattern of water flow for beneficial purposes. No matter what the specific objectives, watershed management recognizes that human use of land is usually aimed at producing a variety of goods and services, of which water is one product. By following the principles of soil and **water conservation**, land and **natural resources** can be managed for sustainable production with environmental protection. *See also* Sustainable agriculture; Sustainable development

[*Kenneth N. Brooks*]

Resources

Books

Black, P. E. *Watershed Hydrology.* Englewood Cliffs: Prentice Hall, 1991.

Brooks, K. N., et al. "Watershed Management: A Key to Sustainability." In *Managing the World's Forests*, edited by N. P. Sharma. Dubuque, IA: Kendall/Hunt, 1992.

Satterlund, D. R. and P. W. Adams. *Wildland Watershed Management.* 2nd ed. New York: Wiley, 1992.

Periodicals

Brooks, K. N., et al. *Hydrology and the Management of Watersheds.* Ames, IA: Iowa State University Press, 1991.

James G. Watt (1938 –)

American former Secretary of the Interior

James Gaius Watt was born in Lusk, Wyoming, and raised in nearby Wheatland, a town one of his critics described as "a place frozen in amber decades ago—the 1890s plus electricity and TV." As President Ronald Reagan's first Secretary of the Interior, Watt was heavily criticized for his position on **conservation**. He remains an icon of what environmentalists oppose. His memory was recalled during George W. Bush's administration, when his former staffer, Gale Norton, became Secretary of the Interior.

Watt came to the **U.S. Department of the Interior** from a position as founding president of the Mountain States Legal Foundation (MSLF) in Denver, a conservative group that acted on the behalf of oil, timber, development, and mineral corporations. Started with money from Joseph Coors, of Coors Brewery, the MSLF took donations from some 175 corporations. MSLF was closely identified with an anti-environmental movement called Wise Use. The group promoted so-called "takings" legislation, which asked that private and corporate landholders be compensated by the government for having to comply with environmental laws, such as **habitat** protection. Watt initiated aggressive legal tactics at MSLF, filing lawsuits and pursuing pro-business legislation on the **environment** in a way that hadn't been done before. MSLF enraged environmentalists and attracted Ronald Reagan, who plucked its president to head his Interior Department. Within months of Watt's appointment, environmentalists began to criticize his conservative policies. Watt maintained his actions were for the good of the people and the free enterprise system.

During his tenure as Secretary of the Interior, Watt cut funds for environmental programs, such as those protecting **endangered species**, and reorganized the department to put less regulatory power on the federal level. He favored the elimination of the Land and **Water Conservation** Fund, which increased the land holdings of national forests, national **wildlife** refuges, and national parks and made matching grants to state governments to do the same. Watt also favored opening extensive shorelands and **wilderness** areas for oil and gas leases, speeding the sale of public lands to private interests and doing so at bargain prices. Watt also

loosened regulations on oil and mineral resource extraction companies.

Watt's stay was marked by heavy and repeated criticisms of his policies, even by fellow Republicans. In April of 1981, the **Sierra Club** and others began a "Dump Watt" petition drive and in October of that year presented Congress with 1.1 million signatures supporting his dismissal. Even the relatively conservative **National Wildlife Federation** called for Watt's removal within six months of his taking office, stating that Watt "places a much higher priority on development and exploitation than on conservation." Watt's actions and derogatory remarks about Senate members earned him a number of enemies, which hampered his ability to carry out his policies. He resigned his post on October 9, 1983, after he made remarks about a Senate advisory panel consisting of "a black, a woman, two Jews and a cripple." Outrage at his bigotry led the Senate to draft a resolution calling for his dismissal, but Watt stepped down before he was forced out. In 1996 Watt pleaded guilty to a minor charge of attempting to sway a grand jury investigation in the 1980s, and he was fined and given five years probation. Watt's legacy was revived in the early 2000s when George W. Bush nominated Gale Norton as Secretary of the Interior. Norton had worked under Watt at the Mountain States Legal Foundation and was seen as having similar stances and goals as her predecessor. A spokesperson for the environmental group National Resource Defense Council called Norton's nomination "...dèjá vu all over again..." implying environmentalists would have to fight the battles Watt had ignited in the 1980s a second time. Watt himself was delighted with the nomination of Norton, and seemed to feel the president's choice showed that Watt's ideas of 20 years earlier were still current.

[*Gerald L. Young Ph.D.*]

Resources

Periodicals

Bratton, S. P. "The Ecotheology of James Watt." *Environmental Ethics* 5 (Fall 1983): 225–236.

Coggins, G. C., and Nagel, D.K. "Nothing Beside Remains: The Legal Legacy of James G. Watt's Tenure as Secretary of the Interior on Federal Land Law and Policy." *Boston College Environmental Affairs Law Review* 17 (Spring 1990): 473-550.

Hissom, Doug. "No Flash with Norton." *Shepherd Express Metro* 22, no. 3 (January 18, 2001).

Soraghan, Mike. "Watt Applauds Bush Energy Strategy." *Denver Post* (May 16, 2001): A1.

Organizations

Mountain States Legal Foundation, 707 17th Street, Suite 3030, Denver, CO USA 80202 (303) 292-2021, Fax: (303) 292-1980, <http://www.mountainstateslegal.com>.

Wave power

The oceans are an enormous **reservoir** of energy. One form in which that energy appears is waves. The energy in waves is derived from the **wind energy** that generates them. Since wind currents are produced by **solar energy**, wave energy is a renewable source of energy.

Humans have invented devices for capturing the power of waves at least as far back as the time of Leonardo da Vinci. The first modern device for generating electricity from wave power was patented by two French scientists in 1799. In the United States, more than 150 patents for wave power machines have been granted.

Harnessing the energy of wave motion presents many practical problems. For example, while the total amount of wave energy in the oceans is very great, the quantity available at any one specific point is usually quite small. For purposes of comparison, an wave that is 8 ft (2.4 m) tall contains the same potential energy as a hydroelectric dam 8 ft high.

One technical problem inventors face, therefore, is to find a way to magnify the energy of waves in an area. A second problem is to design a machine that will work efficiently with waves of different sizes. Over a period of days, weeks, or months, a region of the sea may be still, it may experience waves of moderate size, or it may be hit by a huge storm. A wave power machine has to be able to survive and to function under all these conditions.

Wave power has been seriously studied as an alternative energy source in the United States since the early 1970s. An experimental device constructed at the Scripps Institute of Oceanography, for example, consisted of a buoy to which was attached a long pipe with a trap near its top. As the buoy moved up and down in the waves, water entered the pipe and was captured in the trap. After a certain number of waves had occurred, enough water had been captured to drive a small turbine and electrical generator. One of the first Scripps devices consisted of a pipe 8 in (20 cm) in diameter and 320 ft (98 m) long. It was able to generate 50 watts of electricity. Scripps researchers hoped eventually to construct a machine with a pipe 15 ft (4.5 m) in diameter and 300 ft (91.5 m) long. They expected such a machine to produce 300 kilowatts of energy in 8-ft (2.4 m) waves.

Research on wave power in the United States essentially died out in the early 1980s, as did research on most other forms of **alternative energy sources**. That research has continued in other countries, however, especially in Japan, Great Britain, and Norway. The Norwegians have had the greatest success. By 1989, they had constructed two prototype wave machines on the coast west of Bergen. The machines were located on the shore and operated by using air pumped into a large tower by the rise and fall of waves. The compressed air was then used to drive a turbine and

generator. Unfortunately, one of the towers was destroyed by a series of severe storms in December 1989. The Norwegians appear to be convinced about the potential value of wave power, however, and their research on wave machines continues.

Great Britain has had a more checkered interest in wave power. One of the most ambitious wave machines, known as Salter's Duck, was first proposed in the early 1970s. In this device, the riding motion of waves is used to force water through small pipes. The high-pressure water is then used to drive a turbine and generator.

The British government was so impressed with the potential of wave power that it outlined plans in April 1976 for a 2,000-**megawatt** station. Only six years later, the government changed its mind, however, and abandoned all plans to use wave power. **Fossil fuels** and **nuclear power** had, meanwhile, regained their position as the major—and perhaps only—energy sources in the nation's future.

Then, in 1989, the British government changed its mind yet again. It announced a new review of the potential of wave power with the possibility of constructing plants off the British coast. One of the designs to be tested was a modification of a Norwegian device, the Tapchan (for "tapered channel"). The Tapchan is designed to be installed on a shoreline where waves can flow into a large chamber filled with air. As waves enter the chamber, they compress the air, which then flows though a valve and into a turbine. The compressed air rotates the turbine and drives a generator. The prototype for this machine was installed at Islay, Scotland, in the early 1990s.

Wave power has many obvious advantages. The raw materials (water and wind) are free and abundant, no harmful pollutants are released to the **environment**, and land is not taken out of use. However, the technology for using wave power is still not well-developed. It can be used only along coasts, and is still not economically competitive with traditional energy sources. *See also* Alternative energy sources; Hydrogeology; Land use; Renewable resources

[*David E. Newton*]

Resources
Books

Charlier, R. H. *Tidal Energy.* New York: Van Nostrand Reinhold, 1982.

Periodicals

Fisher, A. "Wave Power." *Popular Science* (May 1975): 68–73+.

"Norway Waves Goodbye to Wave Power Machine." *New Scientist* (January 14, 1989): 31.

Ross, D. "On the Crest of a Wave." *New Scientist* (May 19, 1990): 50–52.

Hersh, S. L. "Waves of Power." *Sea Frontiers* (February 1991): 7.

Weather

see **Climate; Meteorology; Milankovitch weather cycles; Weather modification**

Weather modification

When Mark Twain said that everybody talks about the weather, but nobody does anything about it, he was wrong. In fact, the rain falling on his California roof at that moment might have been generated by secret **chemicals** being diffused into clouds by a hired rainmaker named Charles Hatfield. Hatfield got rich selling rain to farmers in the San Joaquin Valley until he was run out of the state of California by angry San Diegoans who accused him of triggering a flood.

Experiments in rainmaking flourished in the early 1900s in American farmlands where **drought** meant not only hunger but poverty. A little like snake-oil salesmen, early rainmakers sold their ability to make it rain, but it was always ambiguous: when it worked, they were paid. More often, it was hard to tell if they had performed their promised service. Lawsuits were abundant when rain intended for an **arid** area fell across the statelines or caused floods, or when barley growers, pressured by beer companies, paid for rain that coincidentally wiped out other crops of neighboring farmers.

Controlling weather raises myriad questions. Is there an accurate scientific way of measuring human intervention in weather? Who is legally responsible if it rains in the wrong place? Does the community feel that augmenting rainfall is good for everyone? If it rains too much, can the rainmaker be sued? How much does local intervention affect global **climate**? Should regulations be in the hands of the state, the federal government, or a world organization?

For better or worse, people can modify weather both intentionally and inadvertently. Weather responds sensitively to any changes in the global **atmosphere** because it is a complex collection of energy systems powered by the sun. Modest rises in sea surface temperature in the Atlantic Ocean, for example, brought about by ocean circulation shifts or global warming, suppresses rainfall in the African **Sahel** and contributes to drought.

Sensitive climate changes are also caused by such human activities as cutting down forests or failing to let farmfields lie fallow, which, with **overgrazing** by cattle, sheep, and goats, cause **desertification**, a condition that results when **topsoil** blows away and exposes bare, unplantable land. This increases **albedo**, the reflective quality of the surface of the earth, sends back solar radiation into space, and lowers temperature. With less heat rising, fewer clouds are formed, and rainfall is reduced.

Cities, with their clustered buildings and canyons of thoroughfares, absorb infrared heat and inadvertently modify weather because their shape alters the flow of winds. Because they are localized islands of heat, cities increase cloudiness. Aerosols, or microscopic dust particles, given off in industrial **smoke**, bond with water vapor and create city **haze** and **smog**. When the aerosols contain **sulfur dioxide** and **nitrogen oxides**, they cause **acid rain**. Increased urban traffic raises levels of **carbon monoxide** and **carbon dioxide**. In the sky, jet trails contribute to the formation of clouds.

Fossil fuels, which are ancient organic matter, release CO_2 when they are burned. This collects in the greenhouse band, a protective shield that circles the earth. Naturally composed of CO_2, **methane**, chlorofluorocarbons (CFCs), **nitrous oxide**, and water vapor, the greenhouse layer processes infrared heat sent back into space by earth and regulates the temperature of the earth. When it is too full to allow infrared heat from earth to pass through into space, the temperature rises on earth, affecting local, regional, and global weather.

Intentional weather modification involves taking advantage of the energy contained within weather systems and turning it toward a specific goal. To "make" rain, a scientist mimics the natural process by introducing extra water droplets or ice crystals in clouds. However, he needs the right cloud shapes with the right internal temperature and the right winds, headed in the direction of his target.

Rainmaking became a serious science in 1950 when physicist Bernard Vonnegut at General Electric devised a way to vaporize silver iodide to let it rise on heated air currents into clouds where it solidified and bonded onto water droplets to create ice crystals. Previous attempts at rainmaking involved dropping dry ice (solid CO_2) onto clouds from planes, but this was expensive. Vonnegut chose silver iodide because its molecular structure most closely matches that of ice crystals.

In California, where the Southern California Edison Company regularly sends out planes to seed rain clouds over the dry San Joaquin Valley farmland, silver iodide is shot from rockets mounted on the leading edge of the wings. It is also vaporized into clouds from ground generators at higher altitudes in the Sierra Mountains. In rainmaking projects, the purpose is to avoid droughts, increase food productivity, and augment water supplies for drinking or hydroelectric plants. But gathering accurate data on successful seeding and subsequent precipitation has been difficult. Currently, most scientists agree with a longterm analysis that seasonal cloud seeding has increased precipitation by at least 10%, possibly as much as 20%. Clouds, which are ever-moving collections of water vapor, regulators of heat, and generators of tremendous internal winds, remain mysterious. Yet they are major players in earth's climate.

Other weather modification projects include dissipating cold fogs, done routinely at major airports around the world. In the former Soviet Union, damaging hailstorms were successfully broken up to protect ripening crops. However, **statistics** from attempts at hail suppression in the United States have been inconclusive, and research is ongoing. In the 1950s and 1960s, scientists experimented with seeding hurricanes to diminish the storms' severity and alter their path. Similarly, attempts were made to "explode" tornadoes by firing artillery into the oncoming storms. In both cases, natural energies far exceeded any attempts at control.

"We don't have the capability to turn the weather around," said Bill Blackmore of **National Oceanic and Atmospheric Administration** (NOAA)'s Weather Modification Reporting Program. "If we could modify the weather a hundred percent then we could predict the weather a hundred percent. What we need is a lot more understanding of its complexity."

NOAA funds the Federal State Coop Program, a six-state research group. The Atmospheric Modification Program at NOAA's Wave Propagation Lab in Boulder, Colorado, coordinates and evaluates state projects. Research there and at the Institute for Atmospheric Science at the South Dakota School of Mines and Technology involves doing remote sensing of clouds, computer modelling of clouds, and releasing tracers in convective clouds to better understand the dynamics of thunderstorms.

A new way of collecting rainwater is cloud "milking." Researchers have been collecting fog on the mountains of Chile by stringing 50 nylon mesh nets—39 ft (11.8 m) long by 13 ft (4 m) wide—at regular intervals on the mountainside. As the windblown fogs hit the net, they trap water particles. These are then collected into containers. On average, the system "milks" 2,500 gal (9,475 L) of drinking water a day.

Most rainmaking activities in the United States take place in the western states and are sponsored by water departments or districts and conducted by private and commercial companies. The mistakes made earlier in the history of altering the weather have been dealt with by regulations in each state. Internationally, the World Meteorological Organization (WMO) oversees weather modification, and the Treaty of War and Environmental Weather, signed at the Geneva Arms Limitation Talks in 1977, forbids uncontrolled military weather modification.

In 1971, the United States created Public Law 92-205, which requires states to file all weather modification activity with the NOAA's Weather Modification Reporting Program. Typically, about a dozen states file annually.

Two private organizations, the American Meteorological Society in Boston, Massachusetts, and the Weather Modification Association of Fresno, California, keep records

on weather modification. The *Journal of Weather Modification* is an annual publication of the Institute for Atmospheric Science at the South Dakota School of Mines and Technology. *See also* Deforestation; Greenhouse effect; Ozone; Ozone layer depletion

[*Stephanie Ocko*]

Resources
Books

Arnett, D. S. *Weather Modification by Cloud Seeding.* New York, Academic Press, 1980.

Breuer, G. *Weather Modification: Prospects and Problems.* New York: Cambridge University Press, 1979.

Periodicals

"Planned and Inadvertent Weather Modification." *Bulletin of the American Meteorological Society* (March 1992): 331–337.

Strauss, S. "To Catch a Cloud." *Technology Review* (May-June 1991): 18–19.

Other

Blackmore, W. H. *A Summary of Weather Modification Activities Reported in the United States During 1991.* Silver Spring, MD: National Oceanic and Atmospheric Administration, 1991.

Weathering

Weathering refers to the group of physical, chemical, and biological processes that change the physical and chemical state of rocks and soils at and near the surface of the earth. Weathering is primarily a result of climatic forces. Because the effects of **climate** occur at the earth's surface, the intensity of weathering decreases with depth, with most of the effects exhibited within the first meter of the surface. The most important climatic force is water, as it moves in and around rocks and **soil**.

Physical weathering is the disintegration of rock into smaller pieces by mechanical forces concentrated along rock fractures. Abrasion of rocks occurs when wind or water carry particles that wear away rocks. Physical weathering due to frost is referred to as frost shattering or frost wedging. Because water expands when it freezes, it can break rocks apart from the inside when it seeps into cracks in a rock or soil. The specific volume (volume/unit mass) of water increases by 9% during freezing, which produces a stress that is greater than the strength of most rocks. Frost action is the most common physical weathering process, as frost is widespread throughout the world. Frost even occurs in the tropics at high elevations, and as a weathering force, is most effective in coastal arctic and alpine environments, where there are hundreds of frost (freeze-thaw) cycles per year.

Exfoliation is the breaking off of rocks in curved sheets or slabs along joints that are parallel to the ground surface. Exfoliation occurs when a rock expands in response to the removal of adjacent rock. Most commonly the release of stress upon a rock occurs when overlying rock is eroded away (i.e., when the pressure of deep burial is removed). The rock breaks apart along expansion fractures that increase in spacing with depth.

Another type of physical weathering is salt wedging. Most water as it moves through the earth contains dissolved salts; in some areas the salt content may be high, with possible sources being seawater or chemical weathering of marine sediments. As the saline water moves into rock fractures and subsequently evaporates, salt crystals form. As the process continues, the crystals grow until they act as a wedge and crack and break the rocks. Salt wedging most commonly occurs in dry landscapes where the **groundwater** is near the surface.

Hydration is a physical process that also results in weathering. Soil aggregates and fine-grained rocks can disintegrate due to wetting and drying cycles and the expansion and contraction associated with the cycles. Also air that is drawn into pores under dry conditions and then trapped as water returns to the soil or rock can cause fracturing.

Thermal weathering is another physical process. Repeated daily heating and cooling of rock results in expansion during heating and contraction during cooling. Different materials expand and contract at different rates, resulting in stresses along mineral boundaries.

Chemical weathering of rocks or soils occurs through chemical reactions when rocks or soils react with water, gases, and solutions. During these chemical reactions, minerals are added or removed or are decomposed into other materials such as **clay minerals**.

Carbon dioxide, a chemical weathering agent, dissolves in rain and forms a weak carbonic **acid**. This weak acid, through the process of carbonation, can dissolve rocks such as limestone and feldspar. Carbonation of limestone can result in the formation of karst **topography** that may include caves, disappearing streams, springs, and **sinkholes**.

In chemical oxidation weathering, rocks are transformed through reactions with oxygen dissolved in water. Iron, often found in silicate minerals, is the most commonly oxidized mineral element, when ferrous iron (Fe^{+2}) is oxidized to ferric iron (Fe^{+3}). Color changes often indicate when oxidation has occurred, such as the "rusting" seen with the oxidation of iron. Other readily oxidized minerals include magnesium, sulfur, **aluminum**, and chromium.

Hydrolysis is the most common weathering process, where mineral cations in a rock or soil mineral are replaced by **hydrogen** (H^+) ions. Pure water is a poor hydrogen donor, but **carbon** dioxide dissolved in water, which produces carbonic acid, acts as a source of hydrogen ions. Weathering products formed include clay minerals.

Biological weathering occurs when organisms aid in the breakdown of rocks and minerals. Plants such as **lichens** and mosses produce a weak acid that dissolves geological materials. Plant roots growing in the cracks of rocks, through the process of root pry, can make the crack larger and may loosen other types of materials.

[*Judith L. Sims*]

Resources

Books

Ollier, Cliff, and Colin Pain. *Regolith, Soils, and Landforms.* New York: John Wiley & Sons Ltd., 1996.

Rolls, David, and Will J. Bland. *Weathering: An Introduction to the Basic Principles.* London: Edward Arnold, 1998.

Spickert, Diane Nelson, and Marianne D. Wallace. *Earthsteps: A Rock's Journey Through Time.* Golden, CO: Fulcrum Publishing, 2000.

WEE

see **Erosion**

Wells

A well is a hydraulic structure for withdrawal of **groundwater** from aquifers. A well field is an area containing two or more wells. Most wells are constructed to supply water for municipal, industrial, or agricultural use. However, wells are also used for **remediation** of the subsurface (extraction wells), recording water levels and pressure changes (observation wells), water-quality monitoring and protection (monitoring wells), artificial recharge of aquifers (injection wells), and the disposal of liquid waste (**deep-well injection**). Vacuum extraction system is a new technology for removing volatile contaminant from the unsaturated zone, in which vapor transport is induced by withdrawing or injecting air through wells screened in the **vadose zone**.

Well construction consists of several operations: (1) drilling; (2) installing the casing; (3) installing a well screen and filter pack; (4) grouting; and (5) well development.

Various well-drilling technologies have been developed because geologic conditions can range from hard rock such as granite to soft, unconsolidated geologic formation such as alluvial sediments.

Selection of a drilling method also depends on the type of the well that will be installed in the borehole, such as a water-supply well or a monitoring well. The two most widely used drilling methods are cable-tool and rotary drilling. The cable-tool percussion method is a relatively simple drilling method developed in China more than 4,000 years ago. Drilling rigs operate by lifting and dropping a string of drilling tools into the borehole. The drill bit crushes and loosens rock into small fragments that form a **slurry** when mixed with water. When the slurry has accumulated so the drilling process is significantly slowed down, it is removed from the borehole with a bailer. In rotary drilling, the borehole is drilled by a rotating bit, and the cuttings are removed from the borehole by continuous circulation of drilling fluid. Boreholes are drilled much faster with this method, and at a greater depth than with the cable-tool method. Other drilling methods include air drilling systems, jet drilling, earth augers, and driven wells.

Though well design depends on hydrogeologic conditions and the purpose of the well, every well has two main elements: the casing and the intake portion or screen. A filter pack of gravel is often placed around the screen to assure good porosity and hydraulic conductivity. After placing the screen and the gravel filter pack, the annular space between the casing and the borehole wall is filled with a slurry of cement or clay. The last phase in well construction is well development. The objective is to remove fine particles around the screen so hydraulic efficiency is improved.

A well is fully penetrating if it is drilled to the bottom of an **aquifer** and is constructed in such a way that it withdraws water from the entire thickness of the aquifer.

Wells are also used for conducting tests to determine aquifer and well characteristics. During an aquifer test, a well is pumped at a constant **discharge** rate for a period of time, and observation wells are used to record the changes in hydraulic head, also known as drawdown. The radius of influence of a pumping well is the radial distance from the center of a well to the point where there is no lowering of the **water table** or potentiometric surface (the edge of its cone of depression). The collected data are then analyzed to determine hydraulic characteristics. A pumping test with a variable discharge is often used to determine the capacity and the efficiency of the well. A slug test is a simple method for estimating the hydraulic conductivity of an aquifer, a rapid water level change is produced in a piezometer or monitoring well, usually by introducing or withdrawing a "slug" of water. The rise or decline in the water level with time is monitored. The data can be analyzed to estimate hydraulic conductivity of the aquifer.

The predominant tool for extracting vapor or contaminated groundwater from the subsurface is a vertical well. Howerver, in many situations where environmental remediation is necessary, a horizontal well offers a better choice, considering aquifer geometry, groundwater flow patterns, and the geometry of contaminant plumes. Extraction of contaminated groundwater is often more efficient with horizontal wells; a horizontal well placed through the core of a **plume** can recover higher concentrations of contaminants at a given flow rate than a vertical well. In other cases, horizontal wells may be the only option, as contaminants

are often found directly beneath buildings, landfills, and other obstacles to remedial operations. *See also* Aquifer restoration; Drinking-water supply; Groundwater monitoring; Groundwater pollution; Water table; Water table draw-down

[*Milovan S. Beljin*]

Resources

Books

Campbell, M., and J. H. Lehr. *Water Well Technology.* New York: McGraw-Hill, 1973.

Driscoll, F. G. *Groundwater and Wells.* St. Paul: Johnson Filtration Systems, 1986.

Nielsen, D. A., and A. I. Johnson. *Ground Water and Vadose Zone Monitoring.* Philadelphia: ASTM, 1990.

Adam Werbach (1972 –)

American environmentalist

Adam Werbach was born in Tarzana, California, the second son of a psychiatrist father and therapist mother. His strong environmental conscience was nurtured by his parents, who were both activists in the **Sierra Club**. Werbach became a Sierra Club member himself at age 13 and formed the Sierra Student Coalition (SSC) at age 19, composed of 300,000 members. He served two years on the Sierra Club's Board of Directors as a member of the Club's national membership committee and volunteer development committee. Then on May 18, 1996, at the age of only 23, Werbach was elected president of the Sierra Club.

It was **David Ross Brower**, 60 years older than Werbach, who ran the campaign for Werbach to be elected president by the 15-member board, which in turn is elected by the Sierra Club membership at large. As the youngest president ever to **lead** the Sierra Club, Werbach works with the executive director and other staff to manage an organization of about 600,000 members and a $44 million annual budget. The average age of Sierra Club members is around 47, emphasizing the fact that younger generations are currently under-represented in the environmental fields. One of Werbach's goals is to recruit this younger group by appealing to what interests them—the present and how it will affect their personal future.

Werbach believes: "The **environment** is the primary issue that prompts this generation, my generation, to take social and political action. Our job is to get the word out to them and to give them a place to act on their anxieties and convictions. My goal is to make that place the Sierra Club."

As a high school student, Werbach founded and served as the first director of the Sierra Club's national student program, the Sierra Student Coalition, which has trained, registered, and involved thousands of students in all states with Sierra Club **conservation** campaigns. He also organized a conference of environmental youth leaders from 20 countries for the first *World Youth Leadership Camp* in 1996. During this same year he earned a B.A. from Brown with a double major in Political science and Modern culture and media.

Werbach has many hobbies, including music, and has toured the United States, Europe, and Asia, singing baritone and playing the guitar with a men's vocal group at Brown University. He has also written several journal articles and a novel entitled *Whirled*, and has worked on several films concerning both natural- and socially-constructed environments. His book *Act Now, Apologize Later* was published in 1998 and prompted others of his generation to become much more aware of the world around them. Werbach also currently runs a cable access show, *The Thin Green Line*, which focuses on the environment.

[*Nicole Beatty*]

Wet scrubber

Wet **scrubbers** are devices used to remove pollutants from flue gases. They consist of tanks in which flue gases are allowed to mix with liquid. If the pollutant to be removed is soluble in water, water alone can be used as the scrubbing agent. However, most scrubbers are used to remove **sulfur dioxide**, which is not sufficiently soluble in water. Thus, the liquid used in such cases is one that will chemically react with the sulfur dioxide. A solution of sodium carbonate is such a liquid. When sulfur dioxide reacts with sodium carbonate, it forms sodium sulfite, which can be drawn off at the bottom of the tank. *See also* Flue-gas scrubbing

Wetlands

During the last four decades, several definitions of the term "wetland" have been offered by different sources. Today's legal and jurisdictional delineations were published in the *Corps of Engineers Wetlands Delineation Manual* and revised in 1989. It states that wetlands are "those areas that are inundated or saturated by surface or ground water at a frequency and duration sufficient to support, and that under usual circumstances support a prevalence of vegetation typically adapted for life in saturated **soil** conditions."

For an area to be a wetland, it must have certain **hydrology**, soils, and vegetation. Vegetation is dominated by **species** tolerant of saturated soil conditions. They exhibit a variety of adaptations that allow them to grow, compete,

and reproduce in standing water or waterlogged soils lacking oxygen. Soils are wet or have developed under permanent or periodic saturation. The **hydrologic cycle** produces **anaerobic** soils, excluding a strictly upland plant community.

There are seven major types of wetlands that can be divided into two major groups: coastal and inland. Coastal wetlands are those that are influenced by the ebb and flow of tides and include tidal salt marshes, tidal freshwater marshes, and mangrove swamps. Salt marshes exist in protected coastlines in the middle to high latitudes. Plants and animals in these areas are adapted to **salinity**, periodic **flooding**, and extremes in temperature. These marshes are prevalent along the eastern and Gulf coasts of the United States as well as narrow belts on the west coast and along the Alaskan coastline.

Tidal freshwater marshes occur inland from the tidal salt marshes and host a variety of grasses and perennial broad-leaved plants. They are found primarily along the middle and south Atlantic coasts and along the coasts of Louisiana and Texas.

Mangrove swamps occur in subtropical and tropical regions of the world. Mangrove refers to the type of salt-tolerant trees that dominate the vegetation of this wetland. These wetlands are only found in a few places in the United States; the largest areas are found in the southern tip of Florida.

Inland wetlands, which constitute the majority of wetlands in the United States, occur across a variety of climatic zones. They can be divided into four types: northern **peatlands**, southern deep water swamps, freshwater marshes, and riparian ecosystems.

Freshwater marshes represent a variety of different inland wetlands. They have shallow water, peat accumulation, and grow cattails, arrowheads, and different species of grasses and sedges. Major freshwater marshes include the Florida **Everglades**, **Great Lakes** coastal marshes, and areas of Minnesota and the Dakotas.

Southern deepwater swamps are freshwater woody wetlands with standing water for most of the growing season. The most recognizable type of vegetation is cypress. They are either fed by rainwater or occur in alluvial positions that are annually flooded. These wetlands are found in the southeast United States.

Northern peatlands consist of deep accumulation of peat. Primary locations are Minnesota, Wisconsin, Michigan, areas of the Northeast that have been affected by the last **glaciation**, and some mountain and coastal bays in the southeast. Bogs are marshes or swamps that lack contact with local **groundwater** and are acidified by organic acids from plants. They are noted for **nutrient** deficiency and waterlogged conditions with vegetation adapted to conserve nutrients in this **environment**.

Riparian forested wetlands, occurring along rivers and streams, are occasionally flooded but generally dry for a large part of the growing season. The most common type of wetland in the United States, they are often productive because of the periodic addition of nutrients with **sediment** deposited during floods.

Wetlands are valuable in several ways. Because of their appearance and **biodiversity** alone, wetlands are a valuable resource. Many types of **wildlife**, including **endangered species** such as the **whooping crane** and the alligator, inhabit or use wetlands. Over 50% of the 800 species of protected migratory birds rely on wetlands. Wetlands are valuable for **recreation**, attracting hunters of ducks and geese. Over 95% of the fish and shellfish that are taken commercially depend on wetland **habitat** in their life cycles.

Forest wetlands are an important source of lumber. Other wetland vegetation, such as cattails or woody shrubs, could someday be harvested for energy production. Peat is used in potted plants and as a soil amendment, particularly to grow grass sod.

Wetlands intercept and store storm waters, reducing the peak **runoff** and slowing stream discharges, reducing flood damage. In coastal areas, wetlands act as buffers to reduce the energy of ocean storms before they reach more populated areas and cause severe damage. Although most wetlands do not, some may recharge underlying groundwater. Wetlands can improve surface **water quality** by the removal of nutrients and toxic materials as water runs over or through it. Most importantly, wetlands may play a significant role in the global cycling of **nitrogen**, sulfur, **methane**, and **carbon dioxide**.

The current movement of **conservation** has encouraged the "reconstruction" of wetlands that have been destroyed through a no-net-loss policy. Wetlands are restored to protect coastlines, improve water quality, and replace lost habitat. *See also* Commercial fishing; Convention on the Conservation of Migratory Species of Wild Animals; Convention on Wetlands of International Importance; Riparian land; Soil eluviation

[*James L. Anderson*]

Resources

Books

Kusler, J. A., and M. E. Kentula, eds. *Wetland Creation and Restoration: The Status of the Science.* Covelo, CA: Island Press, 1990.

Mitsch, W. J., and J. G. Gorselink. *Wetlands.* New York: Van Norstrand Reinhold, 1993.

Williams, M., ed. *Wetlands: A Threatened Landscape.* Cambridge, MA: Basil Blackwell, 1990.

Other

A Citizen's Guide to Protecting Wetlands. Washington, DC: National Wildlife Federation, 1989.

Wetlands with blue geese and snow geese. (Photograph by Judd Cooney. Phototake. Reproduced by permission.)

IUCN Environmental Law Centre staff, eds. *The Legal Aspects of the Protection of Wetlands.* Gland, Switzerland: IUCN—The World Conservation Union, 1989.

Wetlands Convention

see **Convention on Wetlands of International Importance (1971)**

Whale strandings

An unusual **species** of mammals, **whales** are classified in the order Cetacea, the same order that includes **dolphins** and porpoises. Whales are warm blooded, breathe air and have lungs, bear live young and nurse them on milk. But unlike other mammals, they live completely in the water. That's why ancient civilizations believed whales were fish until the Greek philosopher Aristotle noted that both whales and dolphins breathed through blowholes and delivered live babies instead of laying eggs.

There are two suborders of whales that have evolved differently over time. Baleen whales (suborder Mysticeti) are named after the Norse word for "grooved," because the 10 species have large grooves or pleats on their throats and bellies. Whales in this suborder lack teeth and feed mostly on small fish and **plankton**. Yet even with this relatively small-sized diet, they can grow extremely large. The blue whale, the largest species on record, can reach lengths of more than 100 ft (30 m) and weigh over 150 tons. Other baleen species include the grey whales, minke whales and humpbacks.

Toothed whales (suborder Odontocoti) are typically smaller and faster moving species, including the orca, narwhal, beluga, and the smaller dolphins and porpoises. These whales use their speed and agility to capture prey; the orca often feeds on marine mammals and birds. The majority of toothed whales feed mainly on fish and squid.

In order to find prey in dark or murky waters, toothed whales depend on a sense called echolocation. In fact, whales generally have good vision but it is limited to 45 ft (13.7 m). Their sense of hearing is more remarkable and water is an excellent conductor of sound. Echolocation works by bouncing signals off of objects ahead. Whales can then locate prey and navigate through water, judging water depth and

shoreline. Toothed whales have more refined echolocation systems, while the echonavigation abilities of baleen whales are believed to be more rudimentary.

Whale strandings occur when whales swim or float to shore and cannot remove themselves. The mammals are then stuck in the shallow water. Usually, the cause of strandings is not known, but some causes have been identified. Whales may come to shallow water or to shore due to starvation, disease, injuries or other traumas, or exposure to **pollution**. Most stranded animals are found dead or die quickly after they are found. However, there have been cases in which stranded whales have been successfully moved to a **rehabilitation** facility, treated, and then released to the wild. Sea World facilities in Texas and Florida provide rescue and rehabilitation for stranded whales as do other organizations such as **Greenpeace** and the Atlantic Large Whale Disentanglement Network.

In recent years, underground explosions and military sonar tests may have caused otherwise unexplainable whale strandings off the coasts of Greece and the Bahamas. Over the course of two days in 1996, 12 whales beached on the coast of Greece and eventually died. These sorts of mass strandings are quite rare, and although the exact cause was not identified, it was discovered that the North Atlantic Treaty Organization (NATO) was testing an experimental sonar system in the area around the same time. No scientific connection could be proved, but no other physical explanation for the whales' beachings could be found.

In March 2000, 16 whales became stranded on two beaches in the Bahamas. Necropsies (animal autopsies) were performed on six of the seven that died and no signs of disease, **poisoning** or malnutrition were evident. However, the United States Navy had been performing underwater sonar experiments nearby that emit loud blasts underwater. An auditory specialist involved in the necropsies reported finding hemorrhages in or around the whales' ears. If the whales lost their echolocation capabilities, they would not be able to note the approaching shoreline, possibly explaining the mass stranding.

Environmentalists and scientists struggle to explain and lessen occurrence of whale strandings. Whales' social behavior is such that many species travel in strongly bonded groups. The urge to avoid separation from one another may be stronger than that of avoiding the fatal risk of stranding alongside one whale that is sick or injured and seeking shallow water.

A variety of other causes may bring about whale strandings. Pollution causes illnesses in whales that are unusual to their species and damage their nervous and immune systems. A group of killer whales stranded off the coast of British Columbia in the 1990s revealed the highest levels of **mercury** ever recorded for cetaceans.

Even weather patterns and water temperature can lead to whale strandings. Unfortunately, many strandings go undiscovered and the whales cannot be saved. What's more, the beached whales are usually not discovered in time for a useful necropsy so the strandings' cause is not determined. African stranding coordinators have been selecting samples from stranded whales for 50 years and have yet to identify any particular pattern that explains the reason for the phenomenon. One theory is that the continent's sandy sloping beach is more difficult to detect by the whales' sonar system than a more-defined, rocky coastline. Efforts to protect whales from **hunting**, polluting of their waters and strandings are increasing around the world.

[*Teresa G. Norris*]

Resources
Books

Greenaway, T. *Whales.* Austin, TX: Raintree, Steck-Vaughn, 2001.

Periodicals

Milius, S. "Whales Stranded During Military Test."*Science News* 153 (March 21, 1998): 184.

Thurston, H. "Poisoned Seas: The Cause of Whale Strandings?" *Canadian Geographic* 115 (January-February 1995): 68.

Organizations

Save the Whales, PO Box 3650, Georgetown Sta., Washington, DC USA 20007 (202) 337-2332, Fax: (202) 338-9478, Email: awi@animalwelfare.com, http://www.awionline.org/whales/indexout.html

Great Whales Foundation, PO Box 6847, Malibu, CA USA 90264 (310) 317-0755, Fax: (310) 317-1455, Toll Free: (800) 421-WAVE, Email: whales@elfi.com

Whales

Whales are aquatic mammals of the order Cetacea. The term is now applied to about 80 **species** of baleen whales and "toothed" whales, which include **dolphins**, porpoises, and non-baleen whales, as well as extinct whales. Cetaceans range from the largest known animal, the blue whale (*Balaenoptera musculus*), at a length up to 102 ft (31 m) to the diminutive vaquita (*Phoceona sinus*) at 5 ft (1.5 m).

Whales evolved from land animals and have lived exclusively in the aquatic **environment** for at least 50 million years, developing fish-like bodies with no rear limbs, powerful tails, and blow holes for breathing through the top of their heads. They have successfully colonized the seas from polar regions to the tropics, occupying ecological niches from the water's surface to ocean floor.

Baleen whales, such the right whale (*Balaena glacialis*), the blue whale, and the minke whale (*Balaenoptera acutorostrata*), differ considerably from the toothed whales in their morphology, behavior, and feeding **ecology**. To feed, a

Humpback whale breaching. (Visuals Unlimited. Reproduced by permission.)

baleen whale strains seawater through baleen plates in the roof of its mouth, capturing **plankton** and small fish. Only the gray whale (*Eschrichtius robustus*) sifts ocean sediments for bottom-dwelling invertebrates. Baleen whales migrate in small groups and travel up to 5,000 mi (8,000 km) to winter feeding grounds in warmer seas.

The toothed whales, such as the killer whale (*Orcinus orca*) and the pilot whale (*Globicephala* spp.), feed on a variety of fish, cephalopods, and other marine mammals through a variety of active predatory methods. They travel in larger groups that appear to be matriarchal. Some are a nuisance to **commercial fishing** because they target catches and damage equipment.

Large whales have virtually no natural predators besides humans, and nearly all baleen whales are now listed as **endangered species**, mostly due to commercial **whaling**. In the southern hemisphere, the blue whale has been reduced from 250,000 at the beginning of the century to its current level of a few hundred. The International Whaling Commission (IWC), which has been setting limits on whaling operations since its inception in 1946, has little power over whaling nations, such as Japan and Norway, who continue to catch hundreds of whales a year under an exemption allowing whaling for scientific research.

[*David A. Duffus*]

Resources

Books

Baker, M. L. *Whales, Dolphins, and Porpoises of the World.* Garden City, NY: Doubleday, 1987.

Ellis, R. *Men and Whales.* New York: Knopf, 1991.

Evans, G. H. *The Natural History of Whales and Dolphins.* New York: Facts on File, 1987.

U.S. National Marine Fisheries Service, Humpback Whale Recovery Team. *Final Recovery Plan for the Humpback Whale (Megaptera novaeangliae).* Silver Spring, MD: U.S. National Marine Fisheries Service, 1991.

U.S. National Marine Fisheries Service, Right Whale Recovery Team. *Final Recovery Plan for the Northern Right Whale (Eubaleana glacialis).* Silver Spring, MD: U.S. National Marine Fisheries Service, 1991.

Whaling

Although subsistence whaling by aboriginal peoples has been carried on for thousands of years, it is mainly within about

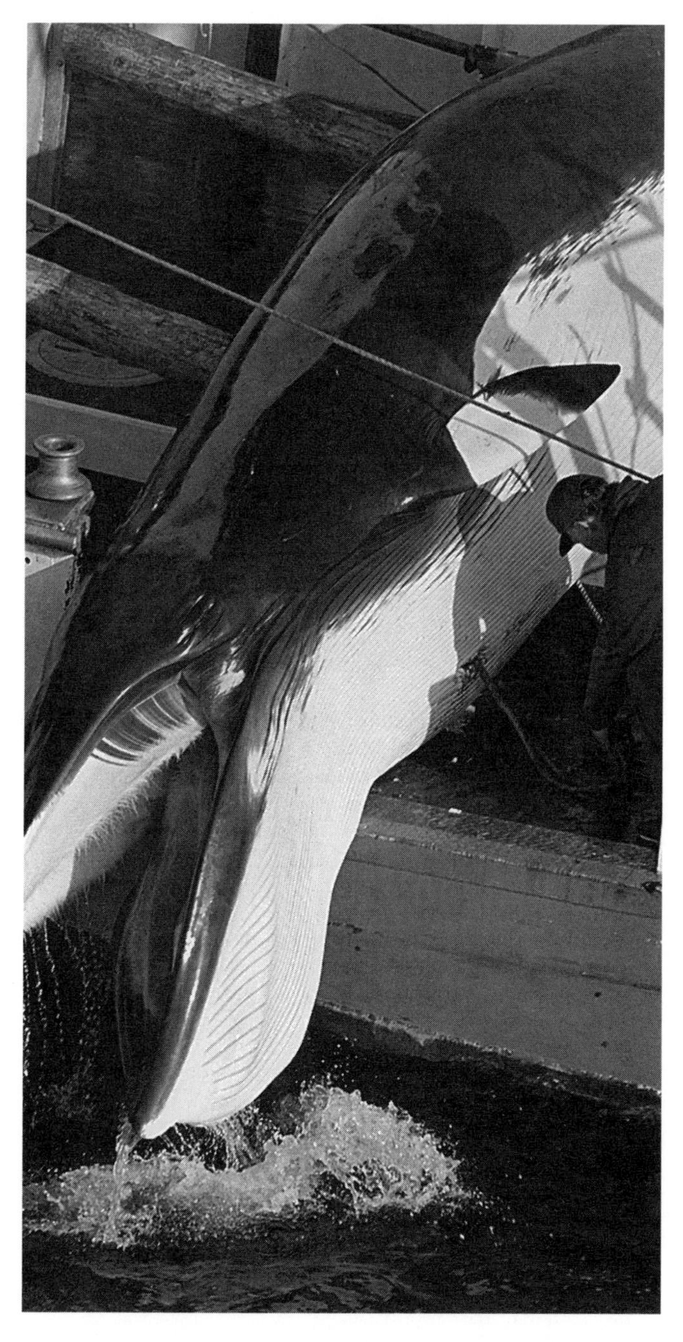

A dead whale being pulled on board a whaling vessel. (Greenpeace Photo. Reproduced by permission.)

the last thousand years that humans have pursued **whales** for commercial gain. The history of whaling may be divided into three periods: the historical whaling era, from 1000 A.D. to 1864-1871; the modern whaling era, from 1864-1871 to the 1970s; and the decline of whaling, from the 1970s to the present.

The Basques of northern Spain were the earliest commercial whalers. Concentrating on the capture of right whales (*Baleana glacialis*), Basque whaling spread over most of the northern Pacific Ocean as local populations dwindled from **overhunting**. Like many whales that were later hunted to near **extinction**, the right whale was a slow-moving and coastal **species**.

Commercial whaling is considered to have begun when the Basques took their whaling across the Atlantic to Newfoundland and Labrador in about 1530, where between 25,000 and 40,000 whales were taken over the next 80 years. The search for the bowhead whale (*Baleana mysticetus*) in the Arctic Ocean and the sperm whale (*Physeter catadon*) in the Atlantic and Pacific provided useful whale oil, waxes, and whalebone (actually the baleen from the whale's upper jaw). The oil proved to be an excellent lubricant and was used as fuel for lighting. Waxes from body tissues made household candles. A digestive chemical was employed as a fixative in perfumes. Baleen served the same purposes as many **plastics** and light metals would today, and was used in umbrella ribs, corset stays, and buggy whips.

The first species targeted were slow swimmers that stayed close to the coasts, making them easy prey. Whalers used sail and oar-powered vessels and threw harpoons to capture their prey, then dragged it back to the mainland. As technology improved and the slow-swimming whales began to disappear, whalers sought the larger and faster-swimming whales.

The historical whaling period ended for several reasons. At the end of the nineteenth century, **petroleum** was discovered to be a good substitute for whale oil in lamps. Also, the whales that were so easily caught were becoming scarce. The technology required to take advantage of larger and faster whales was first used by a Norwegian sealing captain, Svend Foyn. Between 1864 and 1871, he combined the steam-powered boat, cannon-fired harpoon, grenade-tipped harpoon head, and a **rubber** compensator to absorb the shock on harpoon lines to catch the whales. A steam-powered winch brought the catch in. Although the American whaler, Thomas Welcome Roys, was responsible for much of the development of the rocket harpoon, it was Foyn and the Norwegians who packaged the technology that would dominate whaling for the next century.

Modern whaling expanded in two sequences. In the earlier period, whaling was dominated by the spread of whaling stations in the European Arctic and around Iceland, Greenland, and Newfoundland. At the same time, it spread on the Pacific coast of Canada and the United States, around South Africa, **Australia**, and most significantly, **Antarctica**. Before 1925, whaling was tied to shore processing stations and could still be regulated from the shore. After 1925, that system broke down as the stern-slipway floating factory was developed, making it unnecessary for whalers to come ashore.

During the modern whaling period, many populations were brought near extinction as no international quotas or regulations existed. Sperm whales once numbered in the millions, and between 1804 and 1876, United States whalers alone killed an estimated 225,000. The gray whale (*Eschrichtius robustus*) has disappeared in the North Atlantic due to early whaling, although Pacific populations have rebounded significantly. The blue whale (*Balaenoptera musculus*), the largest mammal on earth, was preferred by whalers for its size after improved technology enabled them to be captured. Though protected since 1966, the blue whale has been slow to regain its numbers, and there may be less than 1,000 of these creatures left in the world. Also slow to recover has been the fin whale (*Balaenoptera physalus*), which was hunted intensively after blue whales became less numerous.

As more whales were hunted and populations diminished, the need for an international regulatory agency became apparent. The **International Convention for the Regulation of Whaling** of 1946 formed the International Whaling Commission (IWC), consisting of 38 member nations for this purpose, but the group was largely ineffectual for about 20 years. Growing environmental and political pressure during the 1960s and 1970s resulted in the establishment of the New Management Procedure (NMP) that scientifically assessed whale populations to determine safe catch limits. In 1982 IWC decided to suspend all commercial whaling as of 1986, to reopen in 1996, or when populations had rebounded enough to maintain a sustained yield.

However, as of 1993, some whaling nations, Japan and Norway in particular, threatened to leave the commission and resume commercial whaling. Iceland has already left the association. Meanwhile, the IWC is looking forward to new projects, including the protection of **dolphins** and porpoises.

Today, whaling is permitted by aboriginal groups in Canada, the United States, the Caribbean, and Siberia. Unregulated "pirate whalers" continue to kill and market whale meat, and scientific whaling continues to supply meat products primarily to the Japanese market. At the same time, various scientific specialty groups are working on comprehensive population assessments.

Because of migratory habits and the difficulty in sighting deep ocean whales, it is difficult to accurately estimate their population levels. In many cases, it is impossible to ascertain whether or not a species is in danger. It is clear, though, that the world's whales cannot sustain **hunting** at anywhere near the rates they had been harvested in the past.
See also American Cetacean Society; Environmental law; Migration

[*David A. Duffus*]

Resources
Books

Credlund, A. G. *Whales and Whaling.* New York: Seven Hills Books, 1983.

Tonnessen, J. N., and A. O. Johnsen. *The History of Modern Whaling.* London: Hurst and Co., 1982.

Periodicals

Holy, S. J. "Whale Mining, Whale Saving." *Marine Policy* (July 1985): 192–213.

M'Gonigle, R. M. "The 'Economizing' of Ecology: Why Big, Rare Whales Still Die." *Ecology Law Quarterly* 9 (1980): 119–237.

Gilbert White (1720 - 1793)

English naturalist

Gilbert White was born in 1720 in the village of Selborne, Hampshire, England. The eldest of eight children, he was expected to attend college and join the priesthood. He earned his bachelor's degree at Oxford in 1743 and his master's degree in 1746. A fair student, White was well known for his rambunctious and romantic exploits. White developed a fuller sense of religious identity after college. He was ordained a priest three years after leaving Oxford and was subsequently assigned to a parish near his family home in Hampshire, known as "The Wakes." His religious perspective played an important role throughout the rest of his life, and this is reflected in the beauty and gentleness of his writings.

It was at The Wakes that White began studying **nature** and recording his observations in letters and daily diary notations, which also provide an interesting look at various aspects of local life. He was highly susceptible to carriage sickness and rarely traveled, and his writings therefore focused solely on his immediate environment—the gardens in the village of Selborne. He organized his observations in what he called his "Garden Kalendar," and in 1789, these notes, letters, and memos were combined to comprise *The Natural History and Antiquities of Selborne*, now considered a classic of English literature.

White's ability to write in a clear and unpretentious poetic style is evident throughout the book. He closely watched leaf warblers, cuckoos, and swallows, and unlike his contemporaries, he describes not just the anatomy and plumage of the birds, but their habits and habitats. He was the first to identify the harvest mouse, Britain's smallest mammal, and sketched its physical traits as well as noting its behaviors. *The Natural History* also presents descriptive passages of insect biology and wild flowers White found at Selborne.

After a brief illness, White died alone in his family home at the age of 72.

[*Kimberley A. McGrath*]

RESOURCES
BOOKS

Jenkins, A. C. *The Naturalists: Pioneers of Natural History.* New York: Mayflower Books, 1978.

Lockley, R. M. *Gilbert White.* London: H. F. and G. Witherby, 1954.

White House Office on Environmental Policy

see **Council on Environmental Quality**

Lynn Townsend White Jr. (1907 – 1987)

American historian and writer

For most readers interested in environmental issues, White is known by only one article: "The Historical Roots of Our Ecological Crisis," first published in *Science* in 1967, and widely reprinted. That article, like most of White's work, grew out of his professional interest in medieval technology, including technology's role in "dominating nature." Born in San Francisco, schooled at Stanford and the Union Theological Seminary, White received a doctorate in history from Harvard University in 1934. He taught at Princeton and Stanford Universities, spent 15 years as the President of Mills College (1943–1958), and retired from his position of university professor of history at the University of California in Los Angeles in 1974.

White received most of the honors his profession could bestow. He was a founding member of the Society for the History of Technology, served as its president, and was also elected president of the History of Science Society, president of the American Historical Association, and a Fellow of the American Academy of Arts and Sciences. Perhaps the best introduction to his work can be obtained from two collections of articles, *Machina ex Deo: Essays in the Dynamism of Western Culture* (1968) and *Medieval Religion and Technology: Collected Essays* (1978).

White's article on the ecological crisis traced that crisis back to "modern science [as] an extrapolation of natural theology." The modern technology that emerged from that science "is at least partly to be explained as an Occidental voluntarist realization of the Christian dogma of man's transcendence of, and rightful mastery over, nature." White then concluded that, since modern science and technology have led us into an ecological crisis, "Christianity bears a huge burden of guilt." Numerous articles over many years either picked up on White's argument or took issue with it, a debate White extended by "Continuing the Conversation" in 1973, and a debate that continues today.

Although White's article received widespread attention, his more significant contributions may have been achieved through his extensive research on technology and its relationship to culture and society. Of particular importance has been his establishment of connections between religion and the way a culture perceives technology. Such perceptions have profound implications for human relationships to the **environment**. White also embraced the cause of women's rights, using his presidency at Mills College as a pulpit to advance those rights, especially in higher education. Feminist issues, such as reproduction or status, also have profound implications for human-environment relationships, and his writings in this area are still worth reading today.

[*Gerald L. Young Ph.D.*]

RESOURCES
PERIODICALS

Eckberg, D. L., and T. J. Blocker. "Varieties of Religious Involvement and Environmental Concerns: Testing the Lynn White Thesis." *Journal for the Scientific Study of Religion* 28 (December 1989): 509–517.

Hall, B. S. "Lynn Townsend White, Jr. (1907–1987)." *Technology and Culture* 30 (January 1989): 194–213.

Shaiko, R. G. "Religion, Politics, and Environmental Concern: A Powerful Mix of Passions." *Social Science Quarterly* 68 (June 1987): 244–262.

White Jr., L. T. "The Historical Roots of Our Ecological Crisis." *Science* 155 (March 10, 1967): 1203–7.

Whooping crane

The whooping crane (*Grus americana*) has long been considered the symbol for **wildlife conservation** in the United States. This large, white, wading bird of the family Gruidae is our tallest North American bird, standing nearly 5 ft (1.5 m) tall and having a wingspan of 7.6 ft (2.3 m). Whooping cranes have been threatened with **extinction** since the twentieth century. **Overhunting** in the latter part of the nineteenth and early twentieth centuries, as well as **habitat** loss—primarily due to the conversion of **prairie wetlands** into agricultural land—have been major contributors to this decline. In modern times, the number one cause of death in fledged birds has been collision with high power lines.

In 1937 the federal government established the Arkansas **National Wildlife Refuge** on the south Texas coast, which is the wintering grounds for the whooping crane, in order to protect this species' dwindling population. At the time the refuge was established, the population was at an all time low of 12 birds with a probable founding group of only six to eight birds. There were 31 birds found during a 1950

Whooping crane. (Photograph by Robert J. Huffman. Field Mark Publications. Reproduced by permission.)

census and 36 were located in 1960. The year 1961 marked the first captive breeding program for the whooping crane.

Whooping cranes usually reach maturity at four years of age. After that they establish a lifelong mate with whom they will typically nest once a year, although a pair will skip a nesting season if resources are scarce, or, in some cases for no reason whatsoever. Even though two eggs are usually laid, the parents will only raise one of the chicks. With this in mind, part of the captive breeding program involved the removal of one of the eggs from the whooping cranes' nest for mechanical incubation. In 1967, 50 eggs were removed for incubation at the Patuxent Wildlife Research Center in Laurel, Maryland. There were 48 birds at the end of 1967, the same year the whooping crane was listed as a federally **endangered species**.

Several years of egg removal and artificial incubation followed. Beginning in 1975, eggs from whooping crane nests in their primary nesting area, Wood Buffalo National Park, as well as captive-bred eggs, were placed in the nests of sandhill cranes of Grays Lake, Idaho. This cross-fostering experiment was done to get the more numerous sandhill cranes to help raise whooping cranes to adulthood, and thus increase the population at a faster rate. The experiment has been successful and by 1977, the whooping crane population increased to 120 birds. Today there are nearly 400 in the wild and in captivity.

A flock of 14 captive-raised birds was reintroduced to the Kissimmee Prairie in Florida during the winter of 1993. Nineteen more were released within the year, however, 70% of these released birds were lost due to bobcat predation. There were over 60 whooping cranes in this area by 2000 with indications of hatchlings. An additional facility in Calgary, Canada, has accepted over 30 birds since 1992 in order to establish another captive flock. Another approach to reinforce the population is to reestablish a second migratory route. Ten hatchlings were raised in captivity and followed an ultralight aircraft from Wisconsin to Florida to spend the winter. Five of the cranes returned back to Wisconsin unassisted April 18, 2002.

[*Eugene C. Beckham*]

Resources
Books

Ehrlich, Paul. *Birds In Jeopardy*. Stanford: Stanford University Press, 1992.

Periodicals

May, Peter, and David Henry. "A Whooping Crane Reintroduction Project on the Canadian Prairies: Identifying Relevant Issues Using Expert Consultation." *Endangered Species Update* 12, no. 7: (1995).

Vergano, Dan. "Endangered Cranes Set to Begin Migration." *USA Today* (October 11, 2001).

Other

Whooping Crane Eastern Partnership. *Experimental Flock of Whooping Cranes Return to Central Wisconsin.* April 18, 2002 [cited May 2002]. <http://www.bringbackthecranes.org/media/nr-4-19-02.htm>.

Wild and Scenic Rivers Act (1968)

The Wild and Scenic Rivers Act was passed in 1968, during the same week as the National Trails System Act, in the shadow of the **Wilderness Act** (1964), and in the aftermath of a heated controversy involving **dams** in the Grand Canyon. The main focus of the law is to prevent designated rivers from being dammed.

The act establishes three categories of rivers: wild, scenic, and recreational. Wild rivers are completely undeveloped and accessible only by trail. Scenic rivers are mainly undeveloped but are accessible by roads. Recreational rivers have frequent road access and may have been developed in some way. As was the case in the Wilderness Act, there would be no transfer of lands from one agency to another after designations were made. That is, for example, designated rivers in national forests are managed by the **Forest Service**, and designated rivers under **Bureau of Land Management** jurisdiction are to be managed by the BLM. Even state agencies could manage designated rivers. The law provides for the preservation of land at least 0.25 mi (0.4 km) wide on both sides of designated rivers. The act emphasizes

the use of easements on private lands, rather than acquisition, to achieve this purpose.

The first wild rivers bill, based primarily on studies done by the Agriculture and Interior Departments, was introduced in 1965 by Senator Frank Church of Idaho. The main controversy of the bill focused on which rivers to protect. In order to lessen opposition, only rivers with the support of home state senators would be immediately designated. The Senate passed the bill in early 1966. A wild rivers bill was not seriously considered in the House that year, so the process began again in 1967. The Senate passed a modified version of Church's bill in August 1967. There was some concern that the House would not pass a bill before the end of the session, but under the leadership of Representative John Saylor of Pennsylvania, it did so in September 1968. The House and Senate worked out a compromise bill in conference, and the law was signed by President Lyndon Johnson in October.

Due to the compromise involving home state approval for each designated river, only eight rivers were designated immediately by the act, totaling 773 mi (1,244 km). Twenty-seven rivers were designated for further study. The law allows for the study and potential inclusion of rivers not listed in the original act. Additionally, rivers could be added if states request the federal government to designate a river in the state as part of the system (this has happened for two rivers in Ohio and one in Maine). This system encompasses rivers throughout the country. For example, the first group of designated rivers were in California, Idaho, Minnesota, Missouri, New Mexico, and Wisconsin. As of 2002, 10,955 mi (17,630 km) of 156 rivers had been designated under the act.

[*Christopher McGrory Klyza*]

Resources

Books

Allin, C. W. *The Politics of Wilderness Preservation.* Westport, CT: Greenwood Press, 1982.

Dana, S. T., and S. K. Fairfax. *Forest and Range Policy.* 2nd ed. New York: McGraw-Hill, 1980.

Palmer, T. *Endangered Rivers and the Conservation Movement.* Berkeley: University of California Press, 1986.

The Klamath River in northern California, designated a wild rive under the Wild and Scenic Rivers Act. (Visuals Unlimited. Reproduced by permission.)

Wild river

By the mid-1960s, many rivers in the United States had been dammed or otherwise manipulated for flood control, **recreation**, and other water development projects. There was a growing concern that rivers in their natural state would soon disappear. By 1988, for example, roughly 17% (600,000 mi [965,400 km]) of all the previously free-running rivers in the United States had been trapped behind 60,000 **dams**.

In 1968, Congress passed the **Wild and Scenic Rivers Act**, establishing a program to study and protect outstanding, free-flowing rivers. Federal land management agencies such as the **Forest Service** and the **National Park Service** (NPS) were directed to identify rivers on their lands for potential inclusion in the National Wild and Scenic Rivers System. In 1982, the NPS published the National Rivers Inventory (NRI), a list of 1,524 river segments eligible under the act, though budgetary constraints prevented them from including all eligible rivers.

Eligible river segments must be undammed and have at least one outstanding resource—a **wildlife habitat**, or other recreational, scenic, historic, or geological feature. Rivers can be added to the national system through an act of Congress or by order of the **U.S. Department of the Interior** upon official request from an individual state. Congress intended all types of free-flowing rivers to be included—remote rivers as well as those that flow through urban areas—provided they meet the established criteria. Both designated rivers and rivers under study for inclusion in the system receive numerous protections. The act prohibits the building of hydroelectric or other water development projects and limits mineral extraction in a designated river corridor.

The act also mandates the development of a land management plan, covering an average of 320 acres (129 ha) per river mile (1.6 km) and roughly 0.25 mi (0.4 km) on either side of the river, which must include measures to conserve the riverside land and resources. If the land is federally owned, the responsible agency is required to specify allowable activities on the land, depending on whether the river is classified as wild, scenic, or recreational. If it is privately owned, federal and state agencies will coordinate with local governments and landowners to specify appropriate land uses within the corridor using zoning and other ordinances.

Identifying a potential wild and scenic river often stirs controversy within local communities, usually among riparian landowners concerned that such a designation will curtail the use of their property. There is no federal power to zone private land, and although the act allows federal agencies to purchase land in a designated corridor, there are strict limitations. Generally, these agencies prefer to assist state, local, and private interests in developing a cooperative plan to conserve the river's resources.

Aware of the importance of riverways for local economic development, and concerned about the issue of property rights, Congress made the Wild and Scenic Rivers Act flexible. Riparian landowners are not forced to move from their land, and designation does not affect existing land uses along the river, such as farming, mining, and **logging**. Designation can lead to some restrictions on new development, but most development will be allowed as long as it occurs in a manner that does not adversely affect the character of the river. These questions are addressed on a site-specific basis in the management plan, which is developed by all affected and interested parties, including landowners.

As of 2002, 156 rivers are designated, for a total 10,955 mi (17,630 km). Some of the segments included are the American and Klamath rivers in California, the Rio Grande in Texas, the upper and middle Delaware in New York, New Jersey and Pennsylvania, and the Bluestone in West Virginia. *See also* Ecotourism; Federal Land Policy and Management Act; Land-use control; Riparian land; Riparian rights

[*Cathryn McCue*]

Resources
Books

Coyle, K. J. *The American Rivers Guide to Wild and Scenic River Designation: A Primer on National River Conservation.* Washington, DC: American Rivers, 1988.

Wilderness

Wilderness is land that humans neither inhabit nor cultivate. Through the ages of western culture, as the human relation to land has changed, the meaning and perception of wilderness also has changed. At first, wilderness was to be either conquered or shunned. At times, it was the place for contrition or banishment, as in the biblical account of the Israelites condemned to wander 40 years in the wilderness. To European settlers of North America, wilderness was the untamed land entered only by the adventurous or perhaps the foolhardy. But the wilderness also held riches, making it new land to be exploited, tamed, and ultimately managed. Few saw wilderness as having value in its own right.

The idea of wilderness as land deserving of protection and preservation for its own sake is largely a product of late nineteenth and twentieth century North American thought. Rapidly expanding cultivation and industry not only created wealth, but it also increased the nonconsumptive, intrinsic values of wilderness. Gradually, people began to perceive the wilderness as a land of enjoyment and welcome solitude through intimacy with **nature**. For some, it became an important link to their cultural past, providing assurance that some part of the earth would be left in its primeval condition for **future generations**, and to many their image of the wilderness is as important as its physical reality.

This relatively new attitude toward wild lands has been fostered by scientific considerations. Such lands can hold a tremendous store of unadulterated native genetic material that may be important in maintaining diversity within and among **species**. Wilderness also can support nondestructive, unobtrusive research projects, which serve as references from which to gauge ecological effects in other areas.

The value of wilderness was promoted from roughly the mid-nineteenth century to mid-twentieth century by

many influential people, including George Catlin, **Henry David Thoreau**, **John Muir**, and most notably **Aldo Leopold** and **Robert Marshall**. Their ideas were ultimately incorporated into the platforms of two major organizations: the **Wilderness Society** and the **Sierra Club**. With the creation of these groups, a formal movement had begun to convince the public and lawmakers that wilderness should be preserved and protected. A significant event in this movement occurred in 1951 at the Sierra Club's second Wilderness Conference when Howard Zahniser of the Wilderness Society proposed the idea of a federal wilderness protection bill. Zahniser's work came to fruition 13 years later, four months after his death.

In 1964, the United States Congress passed the **Wilderness Act**, establishing the National Wilderness Preservation System (NWPS). The act grew out of a concern that an expanding population, with its accompanying settlement and mechanization, would leave no lands in the United States or its possessions in their natural condition. Congress intended to preserve areas of federal lands "to secure for the American people of present and future generations the benefits of an enduring resource of wilderness." The act defines wilderness as follows: "A wilderness, in contrast with those areas where man and his own works dominate the landscape, is hereby recognized as an area where the earth and its community of life are untrammeled by man, where man himself is a visitor who does not remain." Further, wilderness is to "be protected and managed so as to preserve its natural conditions," unimpaired for future use and enjoyment.

The National Wilderness Preservation System began with 54 wilderness areas totaling a little over 9 million acres (3.6 million ha), which were administered by the **Forest Service** in the **U.S. Department of Agriculture**. In the first three decades after passage of the Wilderness Act, the system has grown to nearly 500 units covering almost 95 million acres (38.5 million ha), about the size of Montana. These units are administered by the Forest Service, and by the **National Park Service**, **Fish and Wildlife Service**, and **Bureau of Land Management**, in the **U.S. Department of the Interior**. The NWPS increased to about 100 million acres (40.5 million ha), roughly the size of California, following the passage of the California Desert Protection Act of 1994.

The years of greatest growth for the wilderness system were from 1978 to 1984. In 1980, 83 units comprising more than 61 million acres (24.7 million ha) were added to this system; most of this land (56 million acres [22.7 million ha] in 35 units) was added in Alaska with the passage of the Alaska National Interest Land Conservation Act. All but six states, in the northeast and midwest, have wilderness areas. The western United States, with about 20% of the nation's population (11% in California) has almost 95% of NWPS lands. However, the largest tracts of wilderness are in Alaska, which contains nearly two-thirds of the wilderness system acreage.

Although the Wilderness Act generally defines the minimum size for wilderness as 5,000 acres (2,025 ha), it permits land "of sufficient size as to make practicable its preservation and use in an unimpaired condition." Consequently, wilderness areas range in size from the nearly 9-million-acre (3.6-million-ha) Wrangell–Saint Elias Wilderness in Alaska to the 6-acre (2.4-ha) Pelican Island Wilderness in Florida.

In addition to designated wilderness areas in the NWPS, Congress has established the National Scenic Trails System and the Wild and Scenic Rivers System. Also, several states have designated their own form of wilderness areas, the most notable of which is the Adirondack Forest Preserve in northern New York. These state systems, which are of various types and purity, help broaden the diversity of protected ecosystems and their allowable uses.

Despite the wilderness movement, not everyone agrees that wilderness preservation is a good idea. Opponents object to the restrictions imposed by the Wilderness Act, arguing that they "lock up" huge tracts of land, greatly limiting their use and value to society. The act prohibits roads, use of motorized vehicles or equipment, mechanical means of transport, structures, and commercial enterprises, including timber harvesting. Some low-impact uses, such as hiking, **hunting**, and fishing are allowed, as are limited livestock grazing and mining.

Before passage of the Wilderness Act, the debate over wilderness concerned whether or not it should be preserved. However, after 1964 the debate shifted to two major questions: How much is enough, and how should wilderness be managed? The first question was nearly settled, at least in law, with passage of the California Desert Protection Act. This was probably the last sizable addition to the wilderness areas in the United States. The second question may seem oxymoronic, but it presents a substantial challenge to the federal agencies charged with overseeing the health and welfare of these areas. As population pressures increase, decisions become more difficult. For example, wilderness and resource experts must determine an acceptable level of grazing; they must also decide what role fire should play; and, perhaps most importantly, the amount of recreational activity to be allowed.

These management issues are dynamic, and the concept of wilderness likely will change in the future as society's values change. Moreover, decisions regarding management of adjacent lands will influence the management of wilderness areas. Finally, ecosystems themselves change—that which is preserved today will not be what exists tomorrow, as

fires, storms, volcanic eruptions, and ecological **succession** reshape the landscape.

Concerns, challenges, and emotional debate over preservation of native ecosystems have arisen in other parts of the world, including Europe and the **Amazon Basin**. Although international concepts of wilderness are different from that in the United States, other countries are studying the American model for possible **adaptation** to their own situations. Apart from certain differences, concepts of wildernesses around the globe have some strong commonalities: Wilderness is the antithesis of industry; it exists in the body of the earth and in the mind of humanity; and in wilderness lies the ballast of civilization. *See also* Adirondack Mountains; Biodiversity; Frontier economy; National park; National forest; National wildlife refuge; Old-growth forest; Overgrazing; Wild and Scenic Rivers Act; Wild river; Wilderness Study Area; Wildfire; Wildlife; Wildlife management

[*Ronald D. Taskey*]

Resources

Books

Hendee, J. C., G. H. Stankey, and R. C. Lucas. *Wilderness Management.* Golden, CO: North American Press, 1990.

Periodicals

Journal of Forestry 91 (February 1993).

Other

U.S. Forest Service. *The Principal Laws Relation to Forest Service Activities.* Agriculture Handbook No. 453. Washington, DC: U.S. Government Printing Office, 1978.

Wilderness Act (1964)

The Wilderness Act of 1964 established the National Wilderness Preservation System, an area of land that now encompasses over 95 million acres (38.5 million ha). According to the law, **wilderness** is "an area where the Earth and its community of life are untrammeled by man, where man himself is a visitor who does not remain...land retaining its primeval character and influence, without permanent improvements or human habitation, which is protected and managed so as to preserve its natural conditions." This law represents the core idea of preservation: the need to preserve large amounts of land in its natural condition. This idea, which arose in the late 1800s, is in stark contrast to the desire to control **nature** and exploit the resources of nature for economic gain.

The early preservation movement focused on protecting natural areas through their designation as national parks. The birth of the wilderness system, however, can be traced back to the period around 1920. In 1919, Arthur Carhart, a landscape architect, recommended that the area around Trapper's Lake in the White Mountain **National Forest** of Colorado be kept roadless, a recommendation that was followed. A few years later, he recommended that a roadless area be established in the Superior National Forest of Minnesota, a recommendation that was also followed (this area was later named the **Boundary Waters Canoe Area**). In 1921, **Aldo Leopold**, then a **Forest Service** employee in New Mexico, proposed that large areas of land be set aside as wilderness in the national forests. For Leopold, wilderness would be an area large enough to absorb a two-week pack trip, "devoid of roads, artificial trails, cottages, or other works of man." He recommended that such an area be established in the Gila National Forest in New Mexico. The next year, Leopold's superiors acted on his recommendation and established a 574,000-acre (232,400-ha) wilderness area.

The first attempt to establish a more general wilderness policy by the Forest Service came in 1929 when the agency issued Regulation L-20. This directed the Chief of the Forest Service to establish a series of "primitive areas" in which primitive conditions would be maintained. Instructions on implementing this regulation, however, indicated that timber, forage, and **water resources** could still be developed in these areas. This effort was plagued by unclear directives and the lack of support of many foresters, who favored the use of resources, not their preservation.

The issue of Forest Service wilderness designation increased in importance in the 1930s, primarily due to the concern of the Forest Service that all of its scenic areas would be transferred to the **National Park Service** for management. **Robert Marshall**, head of the **Recreation** and Lands Division in the Forest Service, led the fight within the agency to increase wilderness designation on Forest Service lands. He and his supporters argued that increased wilderness designation was wise because: 1) it would demonstrate to Congress that the Forest Service could protect its scenic resources and that transfer of these lands to the **National Park** Service was unnecessary; 2) it would be several years before these resources might be needed, so they could be protected until they were needed; and 3) wilderness and its recreation, research, and preservation uses were suitable for national forest lands. These arguments led to the adoption of the U-Regulations in 1939, which were more precise and restrictive than Regulation L-20.

Three new wild land classifications were established by these regulations. "Wilderness" areas were to be at least 100,000 acres (40,500 ha) in size, designated by the Secretary of Agriculture, and contain no roads, motorized **transportation**, timber harvesting, or occupation. "Wild" areas would be from 5,000 to 100,000 acres (2,025 to 40,500 ha) in size and were to be under the same management restrictions

as wilderness areas with the exception that they would be designated by the Chief of the Forest Service. Lastly, the regulations created "recreation" areas, which were to be of 100,000 or more acres and were to be managed "substantially in their natural condition." In such areas, road-building and timber harvesting, among other activities, could take place at the Chief's discretion. Lands that had been classified as "primitive" under the L-20 Regulation were to be reviewed and re-classified under the U-Regulations.

The wilderness movement lost much momentum inside the Forest Service with the death of Marshall in 1939. Following World War II, preservationist groups began to discuss the need for statutory protection of wilderness due to controversies over the re-classification of primitive lands and the desire for stronger protection. The proposal for such legislation was first made in 1951 by Howard Zahniser, director of the **Wilderness Society**, at the Sierra Club's Biennial Wilderness Conference. This legislative proposal was put on hold during the battle over the proposed Echo Park dam, but in 1956 preservationists convinced Senator Hubert Humphrey to introduce the first wilderness bill. This original bill, drafted by a group of preservation and **conservation** groups, would apply not only to national forests, but also to national parks, national monuments, **wildlife** refuges, and Indian reservations. Within these areas, no farming, **logging**, grazing, mining, road building, or motorized vehicle use would be allowed. No new agency would be created; rather existing agencies would manage the wilderness lands under their jurisdiction. All Forest Service wilderness, wild, and recreation lands would be immediately designated as wilderness, as well as 49 national parks and 20 wildlife refuges. The expansion (or reduction) of the National Wilderness Preservation System would rest primarily with the executive branch: it would make recommendations that would take effect unless either chamber of Congress passed a motion against the designation within 120 days.

By 1964, when the Wilderness Act was passed, this original bill had changed substantially. Sixty-five different bills had been introduced and nearly 20 congressional hearings were held on these bills. The chief opposition to the wilderness system came from commodity and development interests, especially the timber, mining, and livestock industries. The wilderness bill met with greater success in the Senate, where a bill was passed in 1961. The House proved a more difficult arena, as Wayne Aspinall of Colorado, chair of the Interior and Insular Affairs Committee, opposed the wilderness bill. In the end, the preservationists had to compromise or risk having no law at all. The final law allowed established motorboat and aircraft use to continue; permitted control of fires, insects, and diseases; allowed established grazing to continue; allowed the President to approve water developments; and allowed the staking of new mineral claims under the Mining Law of 1872 through December 31, 1983, and development of legitimate claims indefinitely. The act designated 9.1 million acres (3.7 million ha) as wilderness immediately (Forest Service lands classified as wilderness, wild, or canoe). National park lands and wildlife refuges would be studied for potential designations, and primitive lands were to be protected until Congress determined if they should be designated as wilderness. The act also directed that future designations of wilderness be made through the legislative process.

Since the passage of the Wilderness Act, most of the attention on expanding the wilderness system has focused on Forest Service lands and Alaska lands. In 1971, the Forest Service undertook the Roadless Area Review and Evaluation (RARE I) to analyze its holdings for potential inclusion in the wilderness system. It surveyed 56 million acres (22.7 million ha) in a process that included tremendous public involvement. Based on the review, the agency recommended 12 million additional acres (4.9 million ha) be designated as wilderness. This recommendation did not satisfy preservationists, and the **Sierra Club** initiated legal action challenging the RARE I process. In an out-of-court settlement, the Forest Service agreed not to alter any of the 56 million acres (23 million ha) under study and to undertake a **land use** plan and **environmental impact statement**.

In response to RARE I, Congress passed the Eastern Wilderness Act (1975), which created 16 eastern wilderness areas and directed the Forest Service to alter its methods so that it considered areas previously affected by humans. In 1977, the Forest Service launched RARE II. Over 65 million acres (26.3 million ha) of land were examined, and the Forest Service recommended that 15 million acres (6 million ha) be declared wilderness. Preservationist groups were still not satisfied but did not challenge this process. Based upon these recommendations, additional wilderness has been designated on a state-by-state basis. The Forest Service recommendations have been the starting ground for the political process typically involving commodity groups opposed to more wilderness and preservationist groups favoring designations beyond the Forest Service's recommendations. Thus far, RARE II additions have been made for all but a few states. The completion of RARE II designations will not mean the end of the designation process, however, as preservationists will continue to fight for more wilderness land. The Alaska National Interest Conservation Land Act (1980) led to more than a doubling of the wilderness system. The act designated 57 million acres (23 million ha) as wilderness, primarily in national parks and wildlife refuges.

As of 1992, 95.3 million acres (38.6 million ha) of land are designated as wilderness. The Forest Service manages 386 areas of 34 million acres (13.8 million ha), the National Park Service 41 areas of 39.1 million acres (15.8

million ha), and the **Fish and Wildlife Service** 75 areas of 20.6 million acres (8.3 million ha). The lands managed by the **Bureau of Land Management** (BLM) were not included in the Wilderness Act, but the **Federal Land Policy and Management Act** of 1976 directed the agency to review its lands for wilderness designation. The BLM now manages 66 areas of 1.6 million acres (648,000 ha), but this figure will increase as the wilderness review and designation process is completed. Ironically, as more lands are designated wilderness by law, *de facto* wilderness in the nation has been steadily declining.

[*Christopher McGrory Klyza*]

Resources
Books

Allin, C. W. *The Politics of Wilderness Preservation.* Westport, CT: Greenwood Press, 1982.

Dana, S. T., and S. K. Fairfax. *Forest and Range Policy.* 2nd ed. New York: McGraw-Hill, 1980.

Nash, R. *Wilderness and the American Mind.* 3rd ed. New Haven, CT: Yale University Press, 1982.

Wilderness Society

This national **conservation** organization focuses on protecting national parks, forests, **wildlife** refuges, seashores, and **recreation** areas and lands administered by the Interior Department's **Bureau of Land Management** (BLM), which total almost one million mi^2 (2.6 million km^2). The Society has played a leading role on many environmental issues in the last half of the century involving public lands, including protecting 84 million acres (34 million ha) of land as **wilderness** areas since 1964.

The Society was founded in 1935 by naturalist **Aldo Leopold** and other conservationists, in part to formulate and promote a **land ethic**, a conviction that land is a precious resource "to be cherished and used wisely as an inheritance." One of the Society's proudest accomplishments was the passage of the landmark **Wilderness Act** (1964), which recognized for the first time that the nation's wild areas had a value and integrity that was worthy of protection and should remain places where "man is a visitor who does not remain."

The Society has played an important role in other major legislative victories, including passage of the National **Wild and Scenic Rivers Act** (1968); the National Trails System Act (1968); the **National Forest Management Act** (1976); the Omnibus Parks Act (1978); the Alaska National Interest Conservation Land Act (1980); the Land and **Water Conservation** Fund Reauthorization (1989); and the Tongass Timber Reform Act (1990).

The Society's agenda for the future is to prevent **oil drilling** in Alaska's **Arctic National Wildlife Refuge**; to double the size of the National Wilderness System by 2015; to eliminate **national forest** timber sales that lose money for the government; and to expand the number of wild and scenic rivers.

One of the group's major projects is to save the last remaining old-growth forests in the Pacific northwest area of the United States, some 90% or more of which have already been cut. These ancient forests, found in Washington, Oregon, and northern California, harbor some of the world's oldest and largest trees, as well as such **endangered species** as the **northern spotted owl**. Some of these trees, such as the western hemlock, Douglas fir, cedar, and Sitka spruce, were alive when the Magna Carta was signed over 700 years ago; some now tower 250 ft (76 m) in the air. These forests are home to over 200 **species** of wildlife. A single tree can harbor over 100 different species of plants, and a stand of trees can provide **habitat** for over 1,500 species of invertebrates. The Society is fighting plans by the U.S. **Forest Service** to allow the **logging**, often by **clear-cutting**, of most of the remaining ancient forest. The Society estimates that these forests are being cut at a rate of 170 acres (69 ha) a day—the equivalent of 129 football fields every 24 hours.

The Wilderness Society stresses professionalism and has a full-time staff of over 130 people, including ecologists, biologists, foresters, resource managers, lawyers, and economists. Its main activities consist of public education on the need to protect public lands; testifying before Congress and meeting with legislators and their staff, as well as with federal agency officials, on issues concerning public lands; mobilizing citizens across the nation on conservation campaigns; and sponsoring workshops, conferences, and seminars on conservation issues. The Society publishes a quarterly magazine, *Wilderness*, as well as a variety of brochures, fact sheets, press releases, and member alerts warning of impending environmental threats.

[*Lewis G. Regenstein*]

Resources
Organizations

The Wilderness Society, 1615 M St, NW, Washington, D.C. USA 20036, Toll Free: (800) THE-WILD, Email: member@tws.org, <http://www.wilderness.org>

Wilderness Study Area

A Wilderness Study Area is an area of **public land** that is a candidate for official Wilderness Area designation by the United States Government. Federally recognized Wilderness Areas, created by Congress and included in the National Wilderness Preservation System (NWPS), are legally pro-

tected from development, road building, motorized access, and most resource exploitation as dictated by the **Wilderness Act** of 1964.

The process of identifying, defining, and confirming the areas included in the NWPS is, however, long and slow. In preparation for Wilderness Area designation, states survey their public lands and identify Wilderness Study Areas (WSAs). These areas are parcels of undeveloped and undisturbed land that meet **wilderness** qualifications and on which Congress can later vote for inclusion in the NWPS. Out of over 300 million acres (121.4 million ha) of federally held roadless areas, about 23 million acres (9.3 million ha), comprising 861 different study areas, were identified as WSAs between 1976 and 1991. Of these WSAs, only a small percentage were finally recommended to Congress for Wilderness Area status.

Nearly all WSAs are on lands currently administered by the **Bureau of Land Management** (BLM) or the **Forest Service**, and nearly all are in western states. Relatively little undisturbed land remains in eastern states, and most eastern Wilderness Areas were established relatively quickly in 1975 by the Eastern Wilderness Bill. But at the time of this bill adequate surveys of the West had not been completed. The Wilderness Act had ordered western surveys back in 1964, but initial efforts were hasty and incomplete, and the West was excluded from the 1975 Eastern Wilderness Bill.

In 1976 Congress initiated a second attempt to identify western wilderness areas with the **Federal Land Policy and Management Act**. According to this act, each state was to carefully survey its BLM, Forest Service, and **National Park Service** lands. All roadless areas of 5,000 acres (2,000 ha) or more were to be identified and their wilderness, recreational, and scenic assets assessed. Each state legislature was then required to create a policy and management plan for these public lands, identifying which lands could be considered Wilderness Study Areas and which were best used for development, **logging**, or intensive livestock grazing. This public lands survey was known as the Second Roadless Area Review and Evaluation (RARE II). Although RARE II recommendations from business-minded BLM and Forest Service administrators did not satisfy most conservationists, the survey's WSA lists represented a substantial improvement over previous wilderness identification efforts.

As potential Wilderness Areas, WSAs require a number of basic "wilderness" attributes. As stressed in the 1964 Wilderness Act, these areas must possess a distinctive "wilderness character" and allow visitors an "unimpaired" wilderness experience. Set aside expressly for their wildness, these areas' intended use is mainly low-impact, short-term **recreation**. Ideally, human presence should be invisible and their impact negligible. Roadless areas are the focus of wilderness surveys precisely because of their relative inaccessibility. However, because all federal laws involve compromise, wilderness designation does not exclude livestock grazing, an extremely common use of public lands in the West. Mineral exploration and exploitation, a potentially disastrous activity in a wilderness, is also legal in Wilderness Areas.

Long and bitter conflicts between **conservation** and development interests have accompanied every state wilderness bill. Some western state legislatures, steered by business interests, have recommended less than ten percent of their roadless areas for inclusion in the NWPS. Once recommendations are made, further study can continue on each WSA until Congress votes, usually considering just a few areas at a time, to include it in the NWPS or to exclude it from consideration. *See also* Ecosystem; Environmental policy; Habitat; Land-use control

[*Mary Ann Cunningham Ph.D.*]

RESOURCES
BOOKS

Allin, C. W. *The Politics of Wilderness Preservation.* Westport, CT: Greenwood Press, 1982.

PERIODICALS

Stegner, P. "Backcountry Pilgrimage." *Wilderness* 49 (1986): 12–17+.

OTHER

U S. Forest Service. *Final Environmental Statement: Roadless Area Review and Evaluation.* Washington, DC: U.S. Department of Agriculture, 1979.

Wildfire

Few natural forces match fire for its range of impact on the human consciousness, with a roaring forest fire at one extreme and a warming and comforting campfire or cooking flame at the other. Along with earth, water, and air, fire is one of the original "elements" once thought to comprise the universe, and it has frightened and fascinated people long before the beginning of modern civilization. In **nature**, fire both destroys and renews.

Fire is an oxidation process that rapidly transforms the potential energy stored in chemical bonds of organic compounds into the kinetic energy forms of heat and light. Like the much slower oxidation process of **decomposition**, fire destroys organic matter, creating a myriad of gases and ions, and liberating much of the **carbon** and **hydrogen** as **carbon dioxide** and water. A large portion of the remaining organic matter is converted to ash which may go up in the **smoke**, blow or wash away after the fire, or, like the **humus** created by decomposition, be incorporated into the **soil**.

Although fire often is considered bad and was once thought to destroy natural ecosystems, modern scientists have recognized the importance of fire in ecological **succes-**

sion and in sustaining certain types of plant communities. Fires can help maintain seral successional stages, prolonging the time for the community to reach climax stage. Some ecosystems depend on recurring fire for their sustainability; these include many prairies, the **chaparral** of mediterranean climatic regions, pine savannas of the Southeastern United States, and long-needle pine forests of the American West. Fire controls competing vegetation, such as brush in **grasslands**, prepares new seed beds, and kills harmful insects and disease organisms. Nonetheless, diseases sometimes increase after fire because of increased susceptibility of partially burned, weakened trees.

The effects of fire on an **ecosystem** are highly variable, and they depend on the nature of the ecosystem, the fire and its fuel, and weather conditions. In climatic regions where natural fires occur seasonally, grasslands usually are the first ecosystems to burn because the fuel they supply has a large surface area to volume ratio, which allows it to dry rapidly and ignite easily. Grassland fires tend to burn quickly, but they release little energy compared to fires with heavier fuel types. As a result, the effect on soil properties normally is minor and short-lived. The intense greening of a grassland as it recovers from a burn is due largely to the flush of nutrients released from mature and dead plants and made available to new growth. Grassland fires have been set intentionally for many generations in the name of forage improvement.

Naturally caused brushland fires, including the infamous chaparral fires of the southwestern United States, usually start somewhat later in the season than the first grass fires, and they normally have more intense, longer lasting impacts. These fires burn very rapidly but with far more thermal output than grassland fires, because fuel **loading** is five to fifty times greater.

The season for forest fires normally begins somewhat later than that for grass or brush fires. The fuel in a mature forest, which may be 100 times greater than that of a medium density brushland, requires more time to dry and become available for **combustion**. When the forest burns, the ground may or may not be intensely heated, depending on the arrangement of fuels from the ground to the forest canopy. Under a hot burn with the heavy ground fuel found in some forests, heat can penetrate mineral soil to a depth of 12 in (30 cm) or more, significantly altering the physical, chemical, and biological properties of the soil. When the soil is heated, water is driven out; soil structure, which is the small aggregations of sand, **silt**, clay, and organic matter, may be destroyed, leaving a massive soil condition to a depth of several inches.

Forest and brushland soils often become hydrophobic, or water repellent. A hydrophobic layer a few inches thick commonly develops just below the burned surface. This condition is created when the fire's heat turns organic matter into gas and drives it deeper into the soil, where it then condenses on cooler particle surfaces. Under severe conditions, water simply beads and runs off this layer, like water applied to a freshly waxed car. Soil above the hydrophobic layer is highly susceptible to sheet and rill **erosion** during the first rains after a fire. Fortunately, it is soon broken up by insects and burrowing animals, which have survived the fire by going underground; they penetrate the layer, allowing water to soak through it.

Forest fires decrease soil acidity, often causing **pH** to increase by three units (e.g., from 5 to 8) before and after the fire. Normally, conditions return to prefire levels in less than a decade. Fires also transform soil nutrients, most notably converting nutritive **nitrogen** into gaseous forms that go up in the smoke. Some of the first plants to recolonize a hotly burned area are those whose roots support specialized bacteria that replenish the nutritive nitrogen through a process called **nitrogen fixation**. Large amounts of other nutrients, including **phosphorus**, potassium, and calcium, remain on the site, contributing to the so-called "ashbed effect." Plants that colonize these fertile ashbeds tend to be more vigorous than those growing outside of them. When heating has been prolonged and intense in areas such as those under burning logs, stumps, or debris piles, soil color can change from brown to reddish. Fires hot enough to cause these color changes are hot enough to sterilize the soil, prolonging the time to recovery.

In the absence of heavy ground fuels, so much of the energy of an intense forest fire may be released directly to the **atmosphere** that soils will be only moderately affected. This was the case in the great fires at **Yellowstone National Park** in 1988, after which soil scientists mapped the entire burn area as low or medium intensity with respect to soil effects. Although soils on some sites did suffer intense heating, these were too small and localized to be mapped or to be of substantial ecological consequence, and most of the areas recovered quickly after the fires.

Although fire is vital to the long-term health and sustainability of many ecosystems, wildfires take numerous human lives and destroy millions of dollars of property each year. Controlling these destructive fires means fighting them aggressively. Fire suppression efforts are based on the fact that any fire requires three things: heat, fuel, and oxygen. Together, these make up the three legs of the so–called "fire triangle" known to all fire fighters. The strategy in all fire fighting is to extinguish the blaze by breaking one of the legs of this triangle. An entire science has developed around fire behavior and the effects of changing weather, **topography**, and fuels on that behavior.

Competing conceptions of the costs and benefits of wildfires have led to conflicting fire management and suppression objectives, most notably in the national parks and

A stand of trees bursts into flames in Grand Teton National Forest in 1988. (Corbis-Bettmann. Reproduced by permission.)

national forests. The debate over which fires should be allowed to burn and which should be suppressed undoubtedly will continue for some time. *See also* Biomass; Biomass fuel; Deforestation; Ecological productivity; Forest management; Forest Service; Old-growth forest; Soil survey

[*Ronald D. Taskey*]

Resources

Books

Barbour, M. G., J. H. Burk, and W. D. Pitts. *Terrestrial Plant Ecology.* Menlo Park, CA: Benjamin/Cummings, 1980.

Other

Greater Yellowstone Pos-Fire Resource Assessment Committee, Burned Area Survey Team. *Preliminary Burned Area Survey of Yellowstone National Park and Adjoining National Forests.* 1988.

Lotan, J. E., et al. *Effects of Fire on Flora.* Forest Service General Technical Report WO–16. Washington, DC: U. S. Government Printing Office, 1981.

National Wildfire Coordinating Group. *Firefighters Guide.* NFES 1571/ PMS 414-1. Boise: Boise Interagency Fire Center, 1986.

Wells, C. G., et al. *Effects of Fire on Soil.* Forest Service General Technical Report WO–7. Washington, DC: U. S. Government Printing Office, 1979.

Wildlife

It was once customary to consider all undomesticated **species** of vertebrate animals as wildlife. Birds and mammals still receive the greatest public interest and concern, consistently higher than those expressed for reptiles and amphibians. Most concern over fishes results from interest in sport and commercial value. The tendency in recent years has been to include more life-forms under the category of wildlife. Thus, mollusks, insects, and plants are all now represented on national and international lists of threatened and **endangered species**.

People find many reasons to value wildlife. Virtually everyone appreciates the aesthetic value of natural beauty or artistic appeal present in animal life. Giant pandas (*Ailuropoda melanoleuca*), bald eagles (*Haliaetus leucocephalus*), and infant harp **seals** (*Phoca groenlandica*) are familiar examples of wildlife with outstanding aesthetic value. Wild species offer recreational value, the most common examples of which are sport **hunting** and bird watching.

Less obvious, perhaps, is ecological value, resulting from the role an individual species plays within an **ecosys-**

tem. Alligators (genus *Alligator*), for example, create depressions in swamps and marshes. During periods of droughts, these "alligator holes" offer critical refuge to water-dependent life forms. Educational and scientific values are those that serve in teaching and learning about biology and scientific principles.

Wildlife also has a utilitarian value that results from its practical uses. Examples of utilitarian value range from genetic reservoirs for crop and livestock improvement to diverse biomedical and pharmaceutical uses. A related category, commercial value, includes such familiar examples as the sale of furs and hunting leases.

Shifts in human lifestyle have been accompanied by changes in attitudes toward wildlife. Societies of hunter-gatherers depend directly on wild species for food, as many plains Indian tribes did on the **bison**. But as people shift from hunting and gathering to agriculture, wildlife comes to be viewed as more of a threat because of potential crop or livestock damage. In modern developed nations, people's lives are based less on rural ways of life and more on business and industry in cities. Urbanites rarely if ever feel threatened economically by wild animals. They have the leisure time and mobility to visit wildlife refuges or parks, where they appreciate seeing native wildlife as a unique, aesthetic experience. They also sense that wildlife is in decline and therefore favor greater protection.

The most obvious threat to wildlife is that of direct exploitation, often related to commercial use. Exploitation helped bring about the **extinction** of the **passenger pigeon** (*Ectopistes migratorius*), the great auk (*Pinguinus impennis*), Stellar's sea cow (*Hydrodamalis gigas*), and the sea mink (*Mustela macrodon*), as well as the near extinction of the American bison (*Bison bison*). In the late nineteenth and early twentieth century, state and federal laws were passed to help curb exploitation. These were successful for the most part, and they continue to play a crucial role in **wildlife management**.

Introductions of **exotic species** represent another threat to wildlife. Insular or island-dwelling species of wildlife are especially vulnerable to the impacts of exotic plants and animals. Beginning in the seventeenth century, sailors deliberately placed goats and pigs on ocean islands, intending to use their descendants as food on future voyages. As the exotic populations grew, the native vegetation proved unable to cope, creating drastic **habitat** changes. Other species, such as rats, mice, and cats, jumped ship and devastated island-dwelling birds, which had evolved in the absence of mammalian predators and had few or no defenses.

Pollution is yet another threat to wildlife. Bald eagles, ospreys (*Pandion haliaetus*), peregrine falcons (*Falco peregrinus*), and brown pelicans (*Pelecanus occidentalis*) experienced serious and sudden population declines in the 1950s and 1960s. Studies showed that these fish eaters were ingesting heavy doses of pesticides, including DDT. The pesticides left the shells of their eggs so thin that they cracked under the weight of incubating parents, and numbers declined due to reproductive failure. Populations of these birds in the United States rebounded after regulatory laws curbed the use of these pesticides. However, thousands of other **chemicals** still enter the air, water, and **soil** every year, and the effects of most of them on wildlife are unknown.

By far the most critical threat to wildlife is habitat alteration. Unfortunately, it is also more subtle than direct exploitation, and thus often escapes public attention. Human activities are altering some of the most biologically rich habitats in the world on a scale unprecedented in history. Tropical rain forests, for example, originally covered only about 7% of the earth's land surface, yet they are thought to contain half the planet's wild species. Other rich habitats undergoing rapid changes include tropical dry forests and coral reefs. As extensive areas of natural habitat are irrevocably changed, many of the native species that once occurred there will become extinct, even those with no commercial value.

Any species of wild animal has a set of habitat requirements. These begin with food requirements, adequate amounts of available food for each season. Cover requirements are structural components that are used for nesting, roosting, or watching, or that offer protection from severe weather or predators. Water is habitat requirement that affects wildlife directly, by providing drinking water, and indirectly, by influencing local vegetation. The final habitat requirement is space. Biologists can now calculate a minimum area requirement to sustain a particular species of a given size.

Even in the absence of human activities, populations of wild animals change as a result of variations in birth and death rates. When a population is sparse relative to the number that can be supported by local habitat conditions, birth rates tend to be high. In such circumstances, natural **mortality**, including predation, disease, and starvation, tends to be low. As populations increase, birth rates decline and death rates rise. These trends continue until the population reaches **carrying capacity**, the number of animals of a particular species that can be sustained within a given area.

Carrying capacity, though, is difficult to define in practice. Variations in winter severity or in summer rainfall can, between years, alter the carrying capacity for a particular area. In addition, carrying capacity changes as forests grows older, **grasslands** mature, or **wetlands** fill in through natural **siltation**. Despite these limitations, the concept of carrying capacity illustrates an important biological principle: living wild animals cannot be stockpiled beyond the practical limits that local habitat conditions can support.

While all populations vary, some undergo extreme fluctuations. When they occur regularly, such fluctuations are called cycles. Cyclical populations fall into two categories, the three- to four-year cycle typical of lemmings and voles, and the eight- to 11-year cycle known in snowshoe hares (*Lepus americanus*) and lynx of the Western Hemisphere (*Lynx canadensis*). The mechanisms that keep cycles going are complex and not completely understood, but the existence of cycles is widely accepted. Extreme populations fluctuations that occur at irregular intervals are called population irruptions. Local populations of deer tend to be irruptive, suddenly showing substantial changes at unpredictable intervals.

There are more species of wild plants and animals in tropical rain forests than on arctic tundras. Such patterns illustrate variations in the complexities of life-forms due to climatic conditions. Measurement of **biodiversity** usually focuses on a particular group of organisms such as trees or birds. The most basic indication of species richness is the total number of species in a certain area or habitat. A measure of species evenness is more valuable because it indicates the relative abundance of each species. As diversity includes richness and evenness, an area with many species uniformly distributed would have a high overall diversity.

Biodiversity is important to wildlife **conservation** as well as to basic **ecology**. Comparisons of species diversity patterns indicate the extent to which natural conditions have been affected by human activities. They also help establish priorities for acquiring new protected areas. *See also* Predator-prey interactions; Wildlife refuge; Wildlife rehabilitation

[*James H. Shaw*]

Resources
Books

Caughley, G. *Analysis of Vertebrate Populations.* New York: John Wiley & Sons, 1977.

Kellert, S. *Trends in Animal Use and Perception in 20th Century America.* Washington, DC: U.S. Department of the Interior, Fish and Wildlife Service, 1981.

Matthiessen, P. *Wildlife in America.* 2nd ed. New York: Viking Books, 1987.

McCullough, D. *The George Reserve Deer Herd.* Ann Arbor, MI: University of Michigan Press, 1979.

Wildlife management

Wildlife laws are enforced by state and federal agencies, usually known as wildlife departments or fish and game commissions. In the United States, principal responsibility for laws protecting migratory **species** and threatened and **endangered species** rests with the U.S. **Fish and Wildlife Service**.

Various nongovernmental organizations (NGOs) have grown in membership and influence since the 1960s. NGOs include private **conservation** groups such as the **National Audubon Society**, the **National Wildlife Federation**, and the **World Wildlife Fund**. Although they have no direct legal powers, NGOs have achieved considerable influence over wildlife management through fund raising, lobbying, and other kinds of political action, as well as by filing lawsuits concerning wildlife-related issues.

Depletion of native wildlife during the late nineteenth century in the United States and Canada led to passage of federal and state laws regulating the harvest and possession of resident species. Congress passed the Lacey Act in 1900, which made interstate shipment of wild animals a federal offense if taken in violation of state laws, and so curbed market **hunting**. A few years later, the Migratory Bird Treaty Act was passed, establishing federal authority over migratory game and insectivorous birds. This treaty was signed by Great Britain on behalf of Canada in 1916 and became effective two years later.

Other protective measures soon outlawed market hunting and regulated sport hunting. Hunters are now required to purchase hunting licenses, the proceeds of which help pay salaries of game wardens, and game harvests are now regulated chiefly through hunting seasons and bag limits. These are imprecise tools, but the resulting harvests are generally conservative and excessive ones are rare.

In 1973, the **Endangered Species Act** conferred federal protection on listed species, resident as well as migratory. Three years later, the United States became one of the first nations to sign the Convention on International Trade in Endangered Species (CITES), in which more than 100 nations pledged to regulate commercial trade in threatened species and to effectively outlaw trade in endangered ones.

Although laws regulating the taking or possession of wild animals are essential, they are not sufficient by themselves. Wildlife populations can only be sustained if adequate measures are taken to manage the habitats they require. **Habitat** management can take three basic approaches: preservation, enhancement, and restoration. Preservation seeks to protect natural habitats that have not been significantly affected by human activity. The principal tactic of preservation is exclusion of mining, grazing by livestock, timber harvesting, oil and gas exploration, and environmentally damaging recreational practices. Federally-designated **wilderness** areas and national parks use habitat preservation as their principal strategy.

Habitat enhancement is designed to increase both wildlife numbers and **biodiversity** through practices that improve food production, cover, water, and other habitat components. Enhancement differs from preservation because improvements are often achieved through artificial

means, such as nest boxes, the cultivation of food plants, and the installation of devices to regulate water levels. Most state wildlife management areas and many national wildlife refuges practice enhancement.

Habitat restoration seeks to reestablish the original conditions of an area that have been substantially altered by human use, such as farming. Most **land use** practices result in a simpler **biological community**, so many of the original species must be reintroduced. Although not a precise science, in the future this kind of restoration will likely be more widely used to expand rare types of habitat. Costa Rica is using restoration to create more tropical dry forests, and **The Nature Conservancy** in the United States has used it to reestablish tallgrass **prairie** habitat.

Although the strategies used in wildlife management usually involve regulation of harvests and manipulation of habitats, the objectives of most management plans are defined in terms of population levels. Most game management, for example, seeks to increase production of game. Many people assume that game production rises with game populations. In reality, game production is maximized by improving habitat conditions so as to raise the **carrying capacity** for a particular game species. Rather than let populations reach that, modern wildlife managers harvest enough game to keep the population below it, and this practice results in healthier populations with lower rates of natural **mortality** and higher rates of reproduction. Hunters may take more animals from a population managed at a population level below carrying capacity than they can from one managed at it.

Some wildlife populations are managed to minimize the damage they can inflict upon crops or livestock. Some animal-damage control is done by reducing the wildlife population, usually through **poisoning**. But just as game populations become more productive when reduced below carrying capacity, so animals that inflict crop or livestock damage often compensate for population reductions with higher rates of reproduction, as well as higher survival rates for their young. This compensatory response to control, together with increased public pressures against uses of poisons, has led to refinements in animal-damage control. The trend in recent years has been toward methods that prevent damage, such as special fencing or other protective measures, and away from widespread population reductions. When lethal measures are taken, wildlife managers have tended to endorse selective removal of offending individuals, particularly in large carnivorous species.

For **rare species**, those endangered, threatened, or declining, managers attempt to increase numbers, establish new populations, and above all, preserve each rare species' **gene pool**. Legal protection and habitat improvement are important means for managing rare species, just as they are for game animals; in addition, rare species are helped by captive propagation done under cooperative agreements between zoos and the U.S. Fish and Wildlife Service. Although expensive, captive propagation can increase numbers quickly by using new technologies such as artificial insemination and embryo transfers. Once the numbers in captivity have been built up, some of the population can then be reintroduced into suitable habitat. Wildlife managers prepare wild animals bred in captivity for reintroduction through **acclimation**, training, and by selecting favorable age and sex combinations.

Rare species often face long-term problems resulting from reductions in their gene pools. When any species is reduced to a small fraction of its original numbers, certain genes are lost because not enough animals survive to ensure their perpetuation. This loss is known as the bottleneck effect. Rare species typically exist in small, isolated populations, and they lose even more of their gene pools through the **inbreeding** that results. The **Florida panther** (*Puma concolor coryi*), a nearly extinct subspecies of cougar, exhibits loss of size, vigor, and reproductive success, clear indications of a depleted gene pool, the result of the bottleneck effect and inbreeding.

Managers try to minimize the loss of gene pools by striving to prevent severe numerical reductions from occurring in the first place. They also use population augmentation to systematically transfer selected individuals from other populations, wild or captive, to encourage outbreeding and can replenish local gene pools.

Modern wildlife managers tend not to concentrate on perpetuating one species, focusing instead on maintaining entire communities with all their diverse species of plant and animal life. Maintenance of biodiversity is important because wild species exist as ecologically interdependent communities. If enough of these communities can be maintained on a sufficiently large scale, the long-term survival of all of its species may be assured.

From a practical standpoint, there are simply not enough resources or time to develop management plans for each of the earth's 4,100 species of mammals or 9,000 species of birds, not to mention the other, more diverse life-forms. The tedious process of listing threatened and endangered species is slow and often achieved only after the species is on the brink of **extinction**. Recovery becomes expensive and despite heroic efforts, may still fail.

If biodiversity management succeeds, it may reduce the loss of wild species through proactive, as opposed to reactive, measures. Many North American songbirds, particularly those of the eastern woodlands and the prairie regions, are experiencing steady declines. Even familiar game species, including the northern bobwhite quail (*Colinus virginianus*) and several species of duck, are becoming less abundant. The causes are complex and elusive, but they probably relate

to our failure to stabilize habitat conditions on a sufficient scale. If wildlife managers can reverse this trend, they may stop and ultimately reverse these declines. *See also* Defenders of Wildlife; Migration; Overhunting; Predator control; Restoration ecology; Wildlife refuge; Wildlife rehabilitation

[*James H. Shaw*]

Resources

Books

Kohm, K. A., ed. *Balancing on the Brink of Extinction: The Endangered Species Act and Lessons for the Future.* Washington, DC: Island Press, 1991.

Matthiessen, P. *Wildlife in America.* 2nd ed. New York: Viking Books, 1987.

Shaw, J. H. *Introduction to Wildlife Management.* New York: McGraw-Hill Book Co., 1985.

Soule, M., and B. A. Wilcox, eds. 1980. *Conservation Biology.* Sunderland, MA: Sinauer Associates, 1980.

Wildlife refuge

When President **Theodore Roosevelt** issued an executive order making Pelican Island, Florida, a federal bird reservation, he introduced the idea of a **national wildlife refuge**. Nearly 90 years later, Roosevelt's action has expanded to a system of more than 400 national **wildlife** refuges with a combined area of over 90 million acres (36 million ha).

Roosevelt's action was aimed at protecting birds that used the islands. Market hunters had relentlessly killed egrets, herons, and other aquatic birds for their feathers or plumes to adorn women's fashions. The first national wildlife refuge, like other **conservation** practices of the early twentieth century, emerged in reaction to unregulated market **hunting**. Pelican Island became a sanctuary where wild animals, protected from gunfire, would multiply and then disperse to repopulate adjacent countryside.

States followed suit, establishing protected areas known variously as **game preserves**, sanctuaries, or game refuges. They were expected to function as breeding grounds to repopulate surrounding countryside. Although protected populations usually persisted within the boundaries of such refuges, they seldom repopulated areas outside those boundaries. Migratory waterfowl enjoyed some success in repopulating because they could fly over unsuitable habitats and find others that met their needs. Nonmigratory or resident **species** often found themselves surrounded by inhospitable tracts of farmlands, highways, and pastures. Dispersal chances were limited. Resident populations, particularly those of deer and other large herbivores, built up to unhealthy levels, depleting natural foods, becoming infested with heavy parasite loads, and leaving the animals susceptible to starvation. In short, refuges proved to be too small and too scattered to compensate for changes in the overall landscapes imposed by expanding human populations.

Yet elements of the refuge idea proved to be sound, even critical, under some conditions. The U.S. **Fish and Wildlife Service** administers the National Wildlife Refuge System and, in keeping with its legal responsibilities, designed the system to provide crucial **habitat** reserves for migratory waterbirds. Roughly 80% of the refuges in this system were established to provide breeding grounds, wintering grounds, or **migration** stopover sites for aquatic migrants. The task of providing **critical habitat** for threatened and **endangered species**, another area of federal commitment, has been added to this mission.

As wildlife researchers learned more about habitat needs, survival rates, and the dispersal abilities of birds and mammals, wildlife managers began to redefine refuges more in terms of meeting habitat needs than in providing sanctuary from hunters. Without considering habitat quality both on the refuge itself and along its boundaries, no level of protection from gunfire could ensure perpetuation of native wildlife.

Pelican Island is a true island off the Florida coast; other refuges are, or soon will be, habitat islands surrounded by a sea of farms, ranches, forest plantations, shopping malls, and suburbs. There is increasing evidence that even those refuges providing the highest quality habitat may simply be too small to sustain many of their wild species under conditions of isolation.

But how much is enough? How large must a habitat refuge be to ensure that its populations will survive? **Nature** provides a clue through an experiment that began 10,000–11,000 years ago at the end of the last **ice age**. As the glacial ice melted, sea levels rose. Coastal peninsulas in the Caribbean and elsewhere became chains of islands as the rising waters covered lower portions of the peninsulas.

The original peninsulas presumably had about the same number of vertebrate species as did the adjacent mainland. By comparing the numbers of birds, mammals, reptiles, and amphibians present on the islands within historical times with those on adjacent mainlands, biologists have found that the islands contain consistently fewer species. Furthermore, the number of species that a given island maintained was correlated directly to island size and inversely to distance from the mainland.

The most likely explanation for this pattern is that the smaller the island, the less likely that species will survive in isolation over long periods of time. The greater the distance, the less likely that wild animals will re-colonize the island by making it across from the mainland.

Should we consider this analogy realistic? After all, most refuges are surrounded by land, not water. Several studies from national parks in the American West, areas protected from both hunting and habitat alterations, have compared the number of large mammal species surviving in the parks. These studies show consistently that the bigger the park, the greater the number of surviving species of large mammals.

Large carnivores such as **wolves**, bears, lions, **tigers**, and jaguars need the largest expanses of protected areas. Because they attack livestock (and occasionally humans), large carnivores have long been subjected to persecution. However, there are biological reasons for their rarity as well. Any given level in the **food chain/web** can exist at only a tiny fraction of the abundance of the level beneath it. It may take 100 or more moose (*Alces alces*) to support a single wolf (genus *Canis*) or 1,000 Thompson's gazelles (*Gazella thomsonii*) to sustain one lion (*Panthera leo*). These conditions mean that large carnivores occur at low levels of abundance and range over large areas. Simulation models that take into account the areas needed for long-term perpetuation of lions or bears suggest that even the largest national parks and other protected areas may be too small. Either the protected areas themselves will need to be expanded substantially or the populations will have to be aided by reintroduction of individuals from outside the protected zones.

Although large carnivores remain controversial, they also offer popular appeal to a growing segment of the public. Refuges, parks, and other protected sites large enough to support a population of tigers or wolves, even for a short time, will almost certainly be large enough to meet the needs of other wild species as well. Thus, Project Tiger in India helped sambar (*Cervus unicolor*) and chital deer (*Axis axis*) populations by defining areas as tiger preserves.

In the United States and Canada, people tend to think of national parks as scenic areas for tourism and **recreation**. But national parks also act as refuges by protecting wildlife from direct exploitation and by attempting to preserve original habitat conditions. The **grizzly bear** (*Ursus arctos*)once ranged from the western prairies to the Pacific Ocean. By the end of World War II, the only grizzly bear populations in the lower 48 states occurred in and around two large national parks: **Yellowstone National Park** and Glacier. Had it not been for these parks, the grizzly bear would have become completely extinct in the United States, exclusive of Alaska, during the first half of the twentieth century.

Worldwide standards for refuges, parks, and other protected areas have been developed by the United Nations (UN), the **IUCN—The World Conservation Union**, and the IUCN's World Commission on Protected Areas. These agencies compile listings every five years according to three general criteria. To be included, an area must be at least 2,500 acres (1,000 ha) with the exception of offshore islands of at least 250 acres (100 ha), have effective legal protection and adequate *de facto* protection, and be managed by the highest appropriate level of government.

The 2000 tally by the IUCN/UN lists over 30,000 protected areas with a total area of 5.1 million sq mi (13.2 million sq km), about 9.5% of the earth's land surface, or roughly the size of China and India combined. Of the 193 major habitat types, 183 are represented by at least one protected area.

The number of internationally recognized areas of protected habitats grew from about 1,800 in 1970 to nearly 7,000 by 1990 and 30,000 by 2000. This expansion resulted from improved cooperation between nations and from the growing realization that without protected areas, the earth's wildlife would be in peril. During the next quarter century, the list of protected areas will likely continue to grow, especially in developing nations. They will become increasingly valued for tourism and for national prestige. The effectiveness of refuges in the more distant future will depend on **population growth** and on how quickly people shift from extractive to sustainable **land use** practices. The current tendency to view habitat protection as the antithesis to economic development will fade as people recognize the importance, both ecologically and economically, of maintaining the world's wildlife.

[*James H. Shaw*]

Resources

Books

Shafer, C. L. *Nature Reserves: Island Theory and Conservation Practice.* Washington, DC: Smithsonian Institution Press, 1990.

Western, D., and M. C. Pearl. *Conservation for the Twenty First Century.* New York: Oxford University Press, 1989.

World Resources Institute. *World Resources 1992–93.* New York: Oxford University Press, 1992.

Other

1990 United Nations List of National Parks and Protected Areas. Gland, Switzerland: IUCN—The World Conservation Union, 1990.

Wildlife rehabilitation

Wildlife rehabilitation is the practice of saving injured, sick, or orphaned wildlife by taking them out of their **habitat** and nursing them back to health. Rehabilitated animals are returned to the wild whenever possible. Although precise estimates of the numbers of wild animals rehabilitated are impossible to obtain, the practice seems to be expanding in nations such as the United States and Canada. The National Wildlife Rehabilitators Association (NWRA), an international group formed in the early 1980s, estimates that its members treat about 500,000 wild animals annually, with at least twice that number of telephone inquiries.

Wildlife rehabilitation differs from **wildlife management** in the sense that rehabilitators focus their attention on the survival of individual animals, often without regard

to whether it is a member of a **rare species**. Wildlife managers are principally concerned with the health and well-being of wildlife populations and rarely concentrate on individuals.

Wayne Marion, formerly of the University of Florida, has summarized the advantages and disadvantages of wildlife rehabilitation. Rehabilitation facilities give the public a place to bring injured or orphaned wild animals while giving veterinary students the opportunity to treat them. They also save the lives of many animals that would otherwise die; experience gained through the handling and care of common **species** may be put to use when rarer animals need emergency care. But rehabilitation, according to Marion, is expensive, and common species constitute the large majority of animals saved. Rehabilitated wildlife also lose their fear of humans and risk catching diseases from other animals.

Besides routine care for injured or orphaned animals, rehabilitation groups often organize rescues of wild animals following catastrophes. When the oil tanker ***Exxon Valdez*** ran aground in 1989, spilling 11 million gallons of crude oil into **Prince William Sound**, the International Bird Rescue Research Center (IBRRC) directed a rescue operation involving 143 boats. Over 1,600 birds representing 71 species were rescued alive. Survival rates varied by species, but overall roughly half of the birds were eventually returned to the wild. Rehabilitators also recovered 334 live sea otters from areas affected by the spill, cleaned and treated them, and returned 188 to the wild.

Wildlife rehabilitation should become more integrated with conventional wildlife management in the future, especially when catastrophes threaten **endangered species**. Meanwhile, an important role of rehabilitators is in education. Rehabilitated animals who have lost all fear of humans or who are permanently injured cannot be returned to the wild. Such individuals can be used to teach school children and civic groups about the plight of wildlife by bringing rare animals, such as the **bald eagle** or **peregrine falcon** to display at talks. *See also* Game animal; Nongame wildlife; Nongovernmental organizations

[*James H. Shaw*]

Resources

Periodicals

Maki, A. W. "The Exxon Valdez Wildlife Rescue and Rehabilitation Program." *North American Wildlife and Natural Resources Conference Transactions* 55 (1990): 193–201.

Marion, W. R. "Wildlife Rehabilitation: Its Role in Future Resource Management." *North American Wildlife and Natural Resources Conference Transactions* 54 (1989): 476–82.

Willingness to pay

see **Shadow pricing**

Edward O. Wilson Jr. (Photograph by John Chase/ Harvard. Reproduced by permission.)

Edward Osborne Wilson (1929 –)

American zoologist and behavioral and evolutionary biologist

Edward Osborne Wilson was born on June 10, 1929, in Birmingham, Alabama, and is one of the foremost authorities on the **ecology**, systematics, and **evolution** of the ant.

Wilson developed an interest in **nature** and the outdoors at an early age. Growing up in rural, south Alabama, near the border of Florida, young Wilson earned the nickname "Snake" Wilson by having collected most of the 40 snake **species** found in that part of the country. He began studying ants after an accident impaired his vision and has been studying them for over 50 years, having progressed from amateur to expert with relative swiftness. His formal training in the biological sciences resulted in his receiving the Bachelor of Science degree in 1949 and Master of Science in 1950 from the University of Alabama. He received his Ph.D. in 1955 from Harvard University.

Wilson's work on ants has taken him all over the world on numerous scientific expeditions. He has discovered several new species of these social insects, whose current tally of named species numbers nearly 9,000. Wilson estimates, however, that there are probably closer to 20,000 species in the world, and those species comprise over a million billion

individuals. Much of his work on this fascinating group of insects have been synthesized and published in a monumental book he co-authored with Bert Hölldobler in 1990, simply entitled *The Ants*.

Besides hundreds of scientific papers and articles, Wilson has published several other books that have been the focus of much attention in the scientific community. In 1967, Robert H. MacArthur, a renowned ecologist, and Wilson published *The Theory of Island Biogeography*, a work dealing with island size and the number of species that can occur on them. The book also examines the evolutionary equilibrium reached by those populations, the implications of which have been more recently applied to the loss of species through tropical **deforestation**. In 1975, Wilson published his most controversial book, *Sociobiology: The New Synthesis*, in which he discussed human behavior based on the social structure of ants. His book *On Human Nature*, also relating **sociobiology** and human evolution, earned him a Pulitzer Prize for General Nonfiction in 1979. *Biophilia* won him popular acclaim in 1984, as have his more recent books dealing with the threats to the vast diversity of species on earth, *Biodiversity*, published in 1988, and *The Diversity of Life*, published in 1992. Wilson again won the Pulitzer for General Non-fiction in 1991 for his work *The Ants*. He was also awarded the 1997 **Earthwatch** Global Citizen Award, the 1998 Zoological Society of San Diego **Conservation** Medal and 100 Champions of Conservation, 20th Century by the **National Audubon Society** (1998). Wilson was named Humanist of the Year in 1999 by the American Humanist Association.

[*Eugene C. Beckham*]

Resources

Books

Hölldobler, B., and E. Wilson. *The Ants*. Cambridge: Belknap Press, 1990.

Wilson, Edward O. *Biophilia*. Cambridge: Harvard University Press, 1984.

———. *The Diversity of Life*. Cambridge: Harvard University Press, 1992.

———. *Insect Societies*. Cambridge: Belknap Press, 1971.

———. *Sociobiology: The New Synthesis*. Cambridge: Belknap Press, 1975.

Periodicals

Lessen, D. "Dr. Ant." *International Wildlife* 21 (1991): 30–34.

Wind energy

The ultimate source of most energy used by humans is the sun. **Fossil fuels**, for example, are substances in which **solar energy** has been converted and stored as plants or animals that have died and decayed. When energy experts look for **alternative energy sources**, they often search for new ways to use solar energy. Wind energy, although not technically classified as solar energy, is produced by the sun, and when converted accounts for approximately 1–2% of the total energy of the sun.

When sunlight strikes the earth, it heats objects such as land, water, and plants, but it heats them differentially. Dark-colored objects absorb more heat than light-colored ones and rough surfaces absorb more heat than smooth ones. As these materials absorb more or less heat, they also transmit that heat to the air above them. When air is warmed, it rises into the air until it reaches approximately 6 mi (10 km) and then spreads toward the North and South Poles. The rising air is replaced by cooler air, leaving a low pressure area that attracts winds from the north and south. This movement of air results in winds, and their direction is affected by the rotation of the earth. For these reasons, some types of **topography** and some geographic regions are more likely to experience windy conditions than others, and from the standpoint of energy production, these areas are reserves from which solar energy can be extracted.

Devices used for capturing the energy of winds are known as windmills. The wind strikes the blades of a windmill, causing them to turn, and the solar energy stored in wind is converted to the kinetic energy of moving blades. That energy, in turn, is used to turn an axle, to power a pump, to run a mill, or to perform some other function. The amount of power that can be converted by a windmill depends primarily on two factors: the area swept out by the windmill blade and the wind speed. One goal in windmill design, therefore, is to produce a machine with very large blades. But, the power generated by a windmill depends on the cube of the wind speed, and so in any operating windmill, natural wind patterns are by far the most important consideration.

Farmers have been using windmills to draw water from **wells** and to operate machinery for centuries, and the technology has been traced back to the seventh century A.D. In the early twentieth century, windmills were a familiar sight on U.S. farms. More than six million of them existed, and some were used to generate electricity on a small scale. As other sources of power became available, however, that number declined until the late 1970s, when no more than 150,000 were still in use.

In the 1970s, windmills were not considered part of the nation's energy equation. Fossil fuels had long been cheap and readily available, and **nuclear power** was widely regarded as the next best alternative. The 1973 **oil embargo** by the **Organization of Petroleum Exporting Countries** (OPEC) and the Three Mile Island nuclear disaster in 1979 changed these perceptions and caused experts to begin thinking once more about wind power as an alternative energy source.

Wind turbines at Tehachapi Pass, California. (Source unknown. U.S. Department of Energy, Washington, DC.)

During the 1970s, both governmental agencies and private organizations stepped up their research on wind power. The National Aeronautics and Space Administration (NASA), for example, designed a number of windmills, ranging from designs with huge, broad-blade fans, to ones with thin, airplane-like propellers. The largest of these machines consists of blades 300 ft (100 m) long and weighing 100 tons, which rotate on top of towers 200 ft (60 m) tall. At winds of 15–45 mi per hour (25–70 km per hour) these windmills can generate 2.5 megawatts (MW) of electricity. Under the best circumstances, modern windmills achieve an energy-conversion efficiency of 30%, comparable or superior to any other form of energy conversion.

As a result of changing economic conditions and technological advances, windmills became an increasingly popular energy source for both individual homes and large energy-producing corporations. With dependable winds of 10–12 mi per hour (16–20 km per hour), a private home can generate enough electricity to meet its own needs. The 1978 Public Utilities Regulatory Policies Act requires **electric utilities** to purchase excess electricity from private citizens who produce more than they can use by alternative means, such as wind power. "Wind farms," consisting of hundreds of individual windmills, are able to generate enough electricity to meet the needs of small communities. By the late 1980s, more than 70 wind farms had begun operation in Vermont, New Hampshire, Oregon, Montana, Hawaii, and California.

North America has large wind reserves. Twelve states—North Dakota, South Dakota, Texas, Kansas, Montana, Nebraska, Wyoming, Oklahoma, Minnesota, Iowa, Colorado, and New Mexico—contain as much as 90% of the U.S. wind potential. It is estimated that North Dakota alone has enough wind to supply 37% of the electricity used in the entire United States in 1999. The largest number of wind farms in the United States are found in California (about 95% of U.S. capacity), and the greatest concentration of them are located at Altamont Pass east of San Francisco. The wind farms at Altamont Pass date to the early 1980s, when a state study declared it the most promising site for windmills, and a 25% state and federal tax credit encouraged 11 different developers to build 3,800 windmills there. By the time those credits expired, wind power had proven itself to be an inexpensive source of electricity for northern California. In 1993, a single company, U.S. Windpower, oper-

ated 4,200 windmills at Altamont Pass, dependably generating 420 megawatts of electrical energy. By 1995 California produced enough wind power to supply all of San Francisco's residents with electricity. Technology improvements have reduced the cost of wind power, and also have increased the efficiency and durability of wind power stations. From about 25 cents per kilowatt-hour (KWH) in 1980, wind energy costs declined to about 5 cents per KWH by 1995 at sites with strong winds, making it competitive with fossil fuel energy.

By the mid-1990s, wind energy had become the fastest growing source of energy in the world. Global wind electric capacity doubled between 1995 and 1998, and doubled again by 2001, reaching 23,300 megawatts in 2001. The 2001 world wind energy total was enough to power about 10 million households in industrialized countries. Since the early 1990s, the U.S. share of world wind power has declined, as countries including Japan, Germany, Denmark, Spain, and India dramatically increased their wind power production. The European Wind Energy Association estimates that the **European Union** will reach 60,000 MW in wind power production by 2010.

Cheap fossil fuel prices and changing government tax breaks for **renewable energy** caused wind energy production to stagnate or decline in the United States in the mid- to late 1990s. However, the year 2001 saw record U.S. wind installations, and the U.S. share of global capacity rose for the first time since 1986. New installations include ones in Texas (900 MW), in Oregon and Washington along the Columbia River (1,300 MW), Pennsylvania, and Palm Springs, California. A 3,000 MW wind farm was being planned in South Dakota, and the nation's first offshore wind farm was being designed for operation off the coast of Cape Cod, Massachusetts.

A trend in wind power is offshore production, particularly in Europe. Offshore wind farms are being explored due to fewer available land sites, as well as higher and more consistent winds offshore. Engineers are studying means of capturing offshore wind power. In coastal urban areas, wind farms could be built offshore on floating platforms. The electricity generated could be sent back to land on large cables or used on the platforms to generate **hydrogen** from seawater, which could then be shipped to land and used as a fuel.

Like any energy source, wind power has some disadvantages, and these are some of the reasons why wind power has not already become more popular. The most obvious disadvantage is that it can be used only in locations that have enough wind over an extended part of the day. Storage of energy is also a problem. Wind is strongest in spring and autumn when power is least needed and weakest in summer and winter when demand is greatest. Other drawbacks include controversies over the appearance of wind farms and the claim that they spoil the natural beauty of an area. Concerns have also been raised about the noise they produce and the electromagnetic fields they create. Finally, wind farms cause the death of untold thousands of birds each year, because birds use the same windy passages in their flight patterns that are good for energy production.

Around the world, wind power will become increasingly important as concerns increase over global warming and the **pollution** caused by the burning of fossil fuels, and as technology makes it cheaper and more available. By 2001 the United States had lost the technological edge in wind energy to other countries that have made wind energy a higher priority in their national energy plans. Wind energy technology has often been funded by governments, as the United States did in the 1970s, and shown favorable technology gains. During the 1980s, the Reagan and Bush administrations both took a different view about the development of alternative energy sources. Both presidents believed that the federal government should have little or no role in this type of research, and comparatively little progress was made for a decade in the development of U.S. wind power.

[*David E. Newton and Douglas Dupler*]

Resources

Books

Asmus, Peter. *Reaping the Wind: How Mechanical Wizards, Visionaries, and Profiteers Helped Shape Our Energy Future.* Washington, DC: Island Press, 2001.

Gipe, Paul. *Wind Energy Basics: A Guide to Small and Micro Wind Systems.* White River Junction, VT: Chelsea Green Pub. Co., 1999.

Other

Danish Wind Industry Association. *Guided Tour on Wind Energy.* April 17, 2002 [cited July 5, 2002]. <http://www.windpower.org/tour/index.htm>.

Energy Information Administration. *Energy Consumption by Source, 1989–2000.* May 7, 2002 [cited July 5, 2002]. <http://www.eia.doe.gov/emeu/aer/overview.html>.

Sawin, Janet. *Losing the Race for Wind Power: How the U.S. Can Retake the Lead and Solve Global Warming.* Greenpeace USA. March 2002 [cited July 5, 2002]. <http://www.greenpeaceusa.org/climate/pdfs/windreport.pdf>.

Organizations

American Wind Energy Association, 122 C Street, NW, Suite 380, Washington, DC USA 20001 (202) 383-2500, Fax: (202) 383-2505, Email: windmail@awea.org, <http://www.awea.org>

European Wind Energy Association, Rue de Trone 26, B-1000 Brussels, Belgium +32 2 546 1940, Fax: +32 2 546 1944, Email: ewea@ewea.org, <http://www.ewea.org>

Windscale (Sellafield) plutonium reactor

The Windscale nuclear reactor was built in the 1940s near

Sellafield, a remote farm area of northern England, to supply **nuclear power** to the region. This early reactor was designed with a large graphite block in which cans containing the **uranium** fuel were embedded. The graphite served to slow down fast-moving neutrons produced during **nuclear fission**, allowing the reactor to operate more efficiently.

Graphite behaves in a somewhat unusual way when bombarded with neutrons. Water, its modern counterpart, becomes warmer inside the reactor and circulates to transfer heat away from the core. Graphite, on the other hand, increases in volume and begins to store energy. At some point above 572°F (300°C), it may then suddenly release that stored energy in the form of heat.

A safety system that allowed this stored energy to be released slowly was installed in the Windscale reactor. On October 7, 1957, however, a routine procedure designed to release energy stored in the graphite cube failed, and a huge amount of heat was released in a short period of time. The graphite moderator caught fire, uranium metal melted, and radioactive gases were released to the **atmosphere**. The fire burned for two days before it was finally extinguished with water.

Fortunately, the area around Windscale is sparsely populated, and no immediate deaths resulted from the accident. However, quantities of radiation exceeding safe levels were observed shortly after the accident in Norway, Denmark, and other countries east of the British islands. British authorities estimate that 30 or more **cancer** deaths since 1957 can be attributed to **radioactivity** released during the accident. In addition, milk from cows contaminated with radioactive iodine-131 had to be destroyed. The British government eventually decided to close down and seal off the damaged nuclear reactor.

Four decades after the accident, Sellafield is still in the news. In 1991, the **Radioactive Waste Management** Advisory Committee recommended that Sellafield be chosen as the site for burying Britain's high-, low-, and intermediate-level radioactive wastes. A complex network of tunnels 2,500 ft (800 m) below ground level would be ready to accept wastes by the year 2005, according to the committee's plan.

Although some environmental groups object to the plan, many citizens do not seem to be concerned. In spite of the high levels of radiation buried in the old plant, Sellafield has become one of the most popular vacation spots for Britons. *See also* High-level radioactive waste; Liquid metal fast breeder reactor; Low-level radioactive waste; Radiation exposure; Radiation sickness

[*David E. Newton*]

Resources

Books

Dresser, P. D., ed. *Nuclear Power Plants Worldwide.* Detroit, MI: Gale Research, 1993.

Periodicals

Dickson, D. "Doctored Report Revives Debate on 1957 Mishap." *Science* (February 5, 1988): 556–557.

Goldsmith, G., et al. "Chernobyl: The End of Nuclear Power?" *The Economist* 16 (1986): 138–209.

Herbert, R. "The Day the Reactor Caught Fire." *New Scientist* (October 14, 1982): 84–87.

Howe, H. "Accident at Windscale: World's First Atomic Alarm." *Popular Science* (October 1958): 92–95+.

Pearce, F. "Penney's Windscale Thoughts." *New Scientist* (January 5, 1988): 34–35.

Urquhart, J. "Polonium: Windscale's Most Lethal Legacy." *New Scientist* (March 31, 1983): 873–875.

Winter range

Winter range is an area that animals use in winter for food and cover. Generally, winter range contains a food source and thermal cover that together maintain the organism's energy balance through the winter, as well as some type of protective cover from predators. Although some **species** of animals have special adaptations, such as hibernation, to survive winter climates, many must migrate from their summer ranges when conditions there become too harsh. Elk (*Cervus elaphus*) inhabiting mountainous regions, for example, often move from higher ground to lower in the fall, avoiding the early snow cover at higher elevations. Nothern populations of caribou or reindeer (*Rangifer tarandus*) often travel over 600 mi (965 km) between their summer ranges on the **tundra** and their winter ranges in northern woodlands. Still more extreme, the summer and winter ranges of some animals are located on different continents. North American birds known as **neotropical migrants** (including many species of songbirds) simply fly to Central or South America in the fall, inhabiting winter ranges many thousands of miles from their summer breeding grounds.

WIPP

see **Waste Isolation Pilot Plan**

Wise use movement

The wise use movement was developed in the late 1980s as a response to the environmental movement and increasing government regulation. A grassroots environmental movement came about because of perceived impotence of federal regulatory agencies to either regulate the continuing flow of untested toxics into the **environment** or clean up the massive mountain of accumulating waste through the Superfund law. The wise use movement has begun to address these same issues, but basically from a financial standpoint.

Recognizing that cleanup is far more costly than anticipated and persisting in the belief that **public land** should be available for business use, the wise use movement has begun to garner a growing constituency. One of the major new themes, for example, is that low exposures to **chemicals** and radiation are not harmful to humans or ecosystems, or at least not harmful enough to warrant the billions of dollars needed to protect and clean up the environment. Some companies, rather than changing manufacturing processes or doing research on less toxic chemicals, have chosen to continue doing business as usual and are among the leaders of the wise use movement.

The oil, mining, ranching, fishing, farming, and **off-road vehicles** industries—which are most affected by wetland regulation and restrictions on land use—also form a constituency for the wise use movement. The fight for control over land is an old one. Traditionally, timber and mining companies have sought unrestricted access to public lands. Environmental groups such as the **Wilderness Society**, the **National Audubon Society**, the **Sierra Club**, and **the Nature Conservancy** have fought to restrict access. This controversy has been in open debate since at least 1877. At that time, Carl Schurz, then Secretary of the Interior, proposed the idea of national forests, which would be rationally managed instead of exploited. Today, the debate between the environmental and wise use movements represents little more than the longstanding controversy over the best use of public lands—the 29.2% of the total area of the United States owned by the federal government. At present, the wise use movement is most active in the western states with regard to the debate over **land use**, but it is moving into the East as well, where the movement is championed by developers who want to abolish **wetlands** regulations.

At this point, several thousand small groups and countless individuals identify to some degree with the wise use movement. They claim they are the only true environmentalists and label traditional environmentalists "preservationists who hate humans." The wise use aim is to gut all environmental legislation on the theory that regulation has ruined America by curtailing the rights of property owners. Many wise use advocates avoid complexity by simply denying the existence of many widely-accepted theories. For example, some wise use leaders insist that the **ozone layer depletion** problem was manufactured by the National Aeronautics and Space Administration (NASA) and isn't a real threat.

The philosophy of the wise use movement is based on a book by Ron Arnold and Alan Gottlieb, *The Wise Use Agenda* (1988). The movement took a major step forward after a conference held in Reno, Nevada in August of 1988, sponsored by the Center for the Defense of Free Enterprise. Funding for the conference came from large corporations along with a number of right-wing business, political, and religious organizations. The conference was attended by roughly 300 people from across the United States and Canada, representing those industries that feel most threatened by current regulation. These people became the activist founders of the wise use movement. Calling themselves the "new environmentalists," they moved on to organize grassroots support.

The wise use movement has developed a 25-point agenda, seeking to foster business use of **natural resources**. Wise use goals are considered environmentally damaging and are opposed by the traditional environmentalist movement. The wise use movement pursues the development of **petroleum** resources in the **Arctic National Wildlife Refuge** in Alaska. It advocates **clear-cutting** of **old-growth forest** and replanting of public lands with baby trees, the latter at government expense. It aims to open public lands, including **wilderness** and national parks, to mining and **oil drilling**. It seeks to rescind all federal regulation of those **water resources** originating in or passing through the states, favoring state regulation exclusively. The wise use movement further advocates the use of national parks for recreational purposes and a stop to all regulation that may exclude park visitors for protective purposes. It opposes any further restrictions on **rangelands** as livestock grazing areas. It advocates the prevention and immediate **extinction** of all wildfires to protect timber for commercial harvesting.

The above is but a sampling of the agenda of wise use groups in the United States and abroad. The movement is using established corporate structures as a base, which provide training and support to activists. Corporations are now being joined by timber and **logging** associations, chambers of commerce, farm bureaus, and local organizations. The wise use movement is growing because of grassroots support. Small farmers and ranchers and small mining and logging operations have come under tremendous financial pressure. Resources are dwindling, and costs are going up. With increased mechanization, small business owners and their livelihoods are threatened. To give one example, government scientists, after conducting five separate studies, recommended that timber harvests in the Pacific Northwest's ancient forests be reduced by 60% from what they were in the mid-1980s. Loggers would not be able to cut more than two billion board feet of wood a year from national forests in Oregon and Washington. That is substantially below the five billion-plus board feet the industry harvested on those lands annually from 1983 to 1987, before the dispute over protection of the **northern spotted owl** (*Strix occidentalis caurina*) and old-growth forests wound up in court. The wise use movement is one effort to rally behind the people who risk the loss of work, and who may see **environmentalism** as their enemy.

In a strange irony, the grassroots activists of both the environmental movement and the wise use movement have much in common. The rank and file in the wise use movement represent the same kinds of concerns for environmental justice. Both sides see their well-being threatened, whether in terms of property values, livelihoods, or health, and have organized in self-defense. *See also* Environmental ethics

[*Liane Clorfene Casten and Marijke Rijsberman*]

Resources

Books

Gottlieb, A. M., and R. Arnold. *The Wise Use Agenda.* Bellevue, WA: Free Enterprise Press, 1989.

Other

Mendocino Environmental Center Newsletter 12 (Summer/Fall 1992).

Rachel's Hazardous Waste News (Environmental Research Foundation). Nos. 332, 335.

Abel Wolman (1892 – 1989)

American engineer and educator

Born June 10, 1892 in Baltimore, the fourth of six children of Polish-Jewish immigrants, Wolman became one of the world's most highly respected leaders in the field of sanitary engineering, which evolved into what is now known as **environmental engineering**. His contributions in the areas of water supply, water and **wastewater** treatment, public health, nuclear reactor safety, and engineering education helped to significantly improve the health and prosperity of people not only in the United States but also around the world.

Wolman attended Johns Hopkins University, earning a bachelor's degree in 1913 and another bachelor's in engineering in 1915. He was one of four students in the first graduating class in the School of Engineering. In 1937, having already made major contributions in the field of sanitary engineering, he was awarded an honorary doctorate by the school. That same year he helped establish the Department of Sanitary Engineering in the School of Engineering and the School of Public Health, and served as its Chairman until his retirement in 1962. As a professor emeritus from 1962 to 1989, he remained active as an educator in many different arenas.

From 1914 to 1939, Wolman worked for the Maryland State Department of Health, serving as Chief Engineer from 1922 to 1939. It was during his early years there that he made what is regarded as his single most important contribution. Working in cooperation with a chemist, Linn Enslow, he standardized the methods used to chlorinate a municipal **drinking-water supply**.

Although **chlorine** was already being applied to drinking water in some locations, the scientific basis for the practice was not well understood and many utilities were reluctant to add a poisonous substance to the water. Wolman's technical contributions and his persuasive arguments regarding the potential benefits of **chlorination** encouraged many municipalities to begin chlorinating their water supplies. Subsequently, the death rates associated with water-borne **communicable diseases** plummeted and the average life span of Americans increased dramatically. He assisted many other countries in making similar progress.

During the course of his long and illustrious career, spanning eight decades, Wolman held over 230 official positions in the fields of engineering, public health, public works, and education. He served as a consultant to numerous utilities, state and local governments and agencies, and federal agencies, including the U.S. **Public Health Service**, the National Resources Planning Board, the **Tennessee Valley Authority**, the **Atomic Energy Commission**, the U.S. **Geological Survey**, the **National Research Council**, the National Science Foundation, the Department of Defense, the Army, the Navy, and the Air Force.

On the international scene, Wolman served as an advisor to more than 50 foreign governments. For many years he served as an advisor to the World Health Organization (WHO), and he was instrumental in convincing the agency to broaden its focus to include water supply, **sanitation**, and sewage disposal. He also served as an advisor to the Pan American Health Organization.

Wolman was an active member of a broad array of professional societies, including the **National Academy of Sciences**, the National Academy of Engineering, the American Public Health Association, the American Public Works Association, the American Water Works Association, the **Water Pollution** Control Federation, and the American Society of Civil Engineers. His leadership in these organizations is exemplified by his service as President of the American Public Health Association in 1939 and the American Water Works Association in 1942.

Known as an avid reader and a prolific writer, Wolman authored four books and more than 300 professional articles. For 16 years (from 1921 to 1937) he served as editor-in-chief of the *Journal of the American Water Works Association.* He also served as Associate Editor of the *American Journal of Public Health* (1923–1927) and editor-in-chief of *Municipal Sanitation* (1929–1935).

Wolman was the recipient of more than 60 professional honors and awards, including the Albert Lasker Special Award (American Public Health Association, 1960), the National Medal of Science (presented by President Carter, 1975), the Tyler Prize for Environmental Achievement (1976), the Environmental Regeneration Award (Rene

Abel Wolman. (The Ferdinand Hamburger Jr. Archives of Johns Hopkins University. Reproduced by permission.)

Dubos Center for Human Environments, 1985), and the Health for All by 2000 Award (WHO, 1988). He was an Honorary Member of 17 different national and international organizations, some of which named prestigious awards in his honor.

He was greatly admired for his outstanding integrity and widely known for the help and encouragement he gave to others, for his keen mind and sharp wit (even at the age of 96), for his willingness to change his mind when confronted with new information, and for his devotion to his family. His wife of 65 years, Anne Gordon, passed away in 1984. Wolman's son, Gordon, is Chairman of the Department of Geography and Environmental Science at Johns Hopkins University.

[*Stephen J. Randtke*]

Resources

Books

Wolman, Abel. *Water, Health and Society: Selected Papers,* edited by G. F. White. Ann Arbor, MI: Books on Demand.

Periodicals

National Academy of Engineering of the United States of America. *Memorial Tributes* 5. Washington, DC: National Academy Press, 1992.

ReVelle, C. "Abel Wolman, 1892–1989." *EOS* 70 (29 August 1989).

APWA Reporter 56 (October 1989): 24–5.

Wolves

Persecuted by humans for centuries, these members of the dog family *Canidae* are among nature's most maligned and least understood creatures. Yet they are intelligent, highly evolved, sociable animals that play a valuable role in maintaining the **balance of nature**.

Fairy tales such as "Little Red Riding Hood" and "The Three Little Pigs" notwithstanding, healthy, unprovoked wolves do not attack humans. Rather, they avoid them whenever possible. Wolves do prey on rabbits, rodents, and especially on hoofed animals like deer, elk, moose, and caribou. By seeking out the slowest and weakest animals, those that are easiest to catch and kill, wolves tend to cull out the sick and the lame, very old and young, and the unwary, less intelligent, biologically inferior members of the herd. In this way, wolves help ensure the "survival of the fittest" and prevent overpopulation, starvation, and the spread of diseases in the prey **species**.

Wolves have a disciplined, well-organized social structure. They live in packs, share duties, and cooperate in **hunting** large prey and rearing pups. Members of the pack, often an extended family composed of several generations of wolves, appear to show great interest in and affection for the pups and for each other, and have been known to bring food to a sick or injured companion. It is thought that the orderly and complex social structure of wolf society, especially the submission of members of the pack to the leaders, made it possible for early humans to socialize and domesticate a small variety of wolf that evolved into today's dogs. The famous howls in which wolves seem to delight appear to be more than a way of establishing territory or locating each other. Howling seems to be part of their social culture, often done seemingly for the sheer pleasure of it.

Nevertheless, few animals have withstood such universal, intense, and long–term persecution as have wolves, and with little justification. Bounties on wolves have existed for well over 2,000 years and were recorded by the early Greeks and Romans. One of the first actions taken by the colonists settling in New England was to institute a similar system, which was later adopted throughout the United States. Sport and commercial **hunting and trapping**, along with federal **poisoning** and **trapping** programs—referred to as "predator control"—succeeded in eliminating the wolf from all of its original range in the contiguous 48 states excepting Minnesota, until its later reintroduction.

The **U.S. Department of the Interior**, at the urging of conservationists, has undertaken efforts to reestablish wolf populations in suitable areas, like **Yellowstone National Park**. In 1995, the 14 wolves were released to roam the park. The group adapted extremely well and multiplied quickly. In 1997, nearby ranchers, concern that the wolves could become

a threat to their livestock, filed a lawsuit to block further wolf releases. However, in 2000, a federal court upheld the releases. By 2002, there were more than 150 wolves in the park.

American timber wolves continued to be hunted and trapped, legally and illegally, in their remaining refuges. Alaska has periodically allowed and even promoted the aerial hunting and shooting of wolves and has proposed plans to shoot wolves from airplanes in order to increase the numbers of moose and caribou for sport hunters. One such proposal announced in late 1992 was postponed and then cancelled after **conservation** and **animal rights** groups threatened to launch a tourist boycott of the state.

In Minnesota, wolves are generally protected under the federal **Endangered Species Act**, but **poaching** persists because many consider wolves to be livestock killers. In 1978, the wolves were reclassified from endangered to threatened, which afforded them less protection. However, the U.S. **Fish and Wildlife Service** (FWS) adopted a recovery plan in 1978 with the goal of increasing the Minnesota wolf population to 1,400 by 2000. By 1999, the population was estimated at 2,445 by the FWS although several **wildlife** groups believed the number was much less. Wolves were also successfully reintroduced into Wisconsin and Michigan, where the population was estimated by the FWS at more than 100 in 1994. As of 2002, the FWS was considering delisting the Minnesota wolf from protection under the Endangered Species Act.

In Canada, they are frequently hunted, trapped, poisoned, and intentionally exterminated, sometimes to increase the numbers of moose, caribou, and other game animals, especially in the provinces of Alberta and British Columbia, and the Yukon Territory. The maned wolf, (*Chryocyon brachyurus*) is considered endangered throughout its entire range of Argentina, Bolivia, Brazil, Peru, and Uruguay. In Norway, there were only 28 wolves in the wild as of 2002, according to the **World Wildlife fund**. The number in Sweden was less than 100.

The most common type of wolf is the gray wolf (*Canis lupus*) that includes the timber wolf and the Arctic-dwelling **tundra** wolf. The FWS lists the gray wolf as an endangered species throughout its former range in Mexico and the continental United States, except in Minnesota, where it is "threatened." As many as 1,300 wolves may remain in the wilds of Minnesota, 6,000–10,000 in Alaska, and thousands more in Canada. A population ranging from one or two dozen also lives on Isle Royale, Michigan.

The red wolf (*Canis rufus*), a smaller type of wolf found in the southeastern United States, was nearly extinct in the wild when in 1970 the FWS began a recovery program. With less than 100 red wolves in the wilds of Texas and Louisiana, 14 were captured and became part of a captive breeding program. By 2002, about 100 red wolves had been reintroduced into the wilds in North Carolina, but are still listed as Critical by the **IUCN—The World Conservation Union**. A federal appeals court helped the effort when in 1970 it upheld FWS rules banning the killing of red wolves that wander onto privately–owned land.

Red wolf, Smoky Mountains National Park. (Photography by Tim Davis. Photo Researchers Inc. Reproduced by permission.)

In 1998, the FWS reintroduced the Mexican wolf (*Canis lupus baileyi*), a sub-species of the gray wolf, into the Apache **National Forest** in Arizona after it had been gone from the wild for 17 years. Two wolf families were released but the reintroduction suffered setbacks when five of the wolves were shot and killed by ranchers. The remaining wolves were recaptured in 2002 and relocated to the Gila National Forest and Gila **Wilderness** Area of New Mexico. The FWS hopes to have reestablished 100 Mexican wolves in the area by 2005. Less than 200 of the wolves survive in captivity. The Mexican wolf was declared an **endangered species** in 1976.

[*Ken R. Wells*]

Resources
Books

Greenberg, Daniel A. *Wolves.* New York: Benchmark Books, 2002.

Martin, Patricia A. Fink. *Gray Wolves.* New York: Children's Press, 2002.

Mech, L. David. *The Wolves of Minnesota: Howl in the Heartland.* Stillwater, MN: Voyageur Press, 2000.

Periodicals

Clugston, Michael. "Intruding on Wild Lives." *Canadian Geographic* (November–December 2001): 38.

"Defenders Applauds Decision to Translocate Mexican Wolves." *US Newswire* (March 22, 2000).

Holloway, Marguerite. "Wolves at the Door: Can We Learn to Dance With Wild Things Again?" *Discover* (June 2000): 58.

Jones, Karen. "Fighting Outlaws, Returning Wolves." *History Today* (March 2002): 38–41.

Knight, Deborah. "Uneasy Neighbors: Humans Evicted Wolves from Yellowstone and Coyotes Moved In. Now the Wolves are Back." *Animals* (Spring 2002): 6–11.

Organizations

International Wolf Center, 1396 Highway 169, Ely, MN USA 55731 (218) 365-4695, Email: wolfinfo@wolf.org, <http://www.wolf.org>

Woodpecker

see **Ivory-billed woodpecker**

George Masters Woodwell (1928 –)

American ecologist

A highly respected but controversial **biosphere** ecologist and biologist, Woodwell was born in Cambridge, Massachusetts. With his parents, who were both educators, Woodwell spent most of his summers on the family farm in Maine, where he was able to learn firsthand about biology, **ecology**, and the **environment**.

Woodwell graduated from Dartmouth College in 1950 with a bachelor's degree in zoology. Soon after graduation, he joined the Navy and served for three years on oceanographic ships. After returning to civilian life, he took advantage of a scholarship to pursue graduate studies and earned both a master's degree (1956) and a doctorate (1958) from Duke University. He began teaching at the University of Maine and was later appointed a faculty member and guest lecturer at Yale University. Throughout most of the 1960s and the early 1970s, Woodwell worked as senior ecologist at the Brookhaven National Laboratory. It was there that he conducted the innovative studies on environmental **toxins** which earned him a reputation as a nonconformist ecologist. In 1985, he founded the Woods Hole Research Center, a leading facility for ecological research.

A pioneer in the field of biospheric **metabolism**, Woodwell has worked to determine the effects of various toxins on the environment. One of the many studies he conducted examined the effects of **radiation exposure** on forest ecosystems. He used a 14-acre (5.7 ha) combination pine and oak forest as his testing area, and found that the time needed to destroy an **ecosystem** is far less than that which is necessary to rejuvenate it. He continues to investigate the effects of nuclear emissions on the environment.

Beginning in the 1950s, Woodwell worked with the Conservation Foundation—now part of the World **Wildlife** Fund—in investigating the **pesticide** DDT. Woodwell and his colleagues were the first to study the catastrophic effects of the chemical on wildlife and, in 1966, were the first to take legal action against its producers. DDT was banned in the United States in 1972.

Much of Woodwell's work has focused on the hazards of the **greenhouse effect**. He has provided valuable information for hearings on environmental issues, and has been instrumental in developing similar data for the United States and foreign government agencies. Not only is Woodwell concerned about radiation, **chemicals**, and global warming, but like **Paul Ehrlich**, he warns that **population growth** must be kept in balance with any ecosystem development.

While he does publish articles in professional journals, Woodwell produces a number of works for more publications with a broader audience. He considers it imperative that the general population play a key role in saving the planet, and his work has appeared in periodicals such as *Ecology*, *Scientific American*, and the *Christian Science Monitor*.

Woodwell maintains membership in the **National Academy of Sciences** and is a past president of the **Ecological Society of America**. He is also a member of the **Environmental Defense** Fund and **Natural Resources Defense Council**. In 2001, Woodwell was awarded the Volvo Environmental Prize.

[*Kimberley A. Peterson*]

Resources
Books

Gareffa, P. M., ed. *Contemporary Newsmakers—1987 Cumulation.* Detroit, MI: Gale Research, 1988.

Woodwell, George M., ed. *Earth in Transition: Patterns and Processes of Biotic Impoverishment.* New York: Cambridge University Press, 1991.

———, et al. *Ecological and Biological Effects of Air Pollution.* New York: Irvington, 1973.

Periodicals

Grady, D., and T. Levenson. "George Woodwell: Crusader for the Earth." *Discover* (May 1984): 44–6+.

Houghton, R. A., and G. M. Woodwell. "Global Climatic Change." *Scientific American* 260 (April 1989): 36–44.

Woodwell, George M. "On Causes of Biotic Impoverishment." *Ecology* 70 (February 1989): 14–15.

World Bank

Affiliated with the United Nations, the World Bank was formally established in 1946 to finance projects to spur the

economic development of member nations, most notably those in Europe and the **Third World**. The bank has been strongly criticized by environmental and human rights groups in recent years for funding Third World projects that destroy rain forests, damage the **environment**, and harm villagers and **indigenous peoples**.

The bank is headquartered in Washington, D.C. It is administered by a board of governors and a group of executive directors. The board of governors is composed principally of the world's finance ministers and convenes annually. There are 21 executive directors who carry out policy matters and approve all loans. The World Bank usually makes loans directly to governments or to private enterprises with their government's guarantee when private capital is not available on reasonable terms. Generally, the Bank lends only for imported materials and equipment and services obtained from abroad. The interest rate charged depends primarily on the cost of borrowing to the Bank. The subscribed capital of the bank exceeds $30 billion, and the voting power of each nation is proportional to its capital subscription.

The 1970s represented a period of extensive growth for the World Bank, both in the volume of lending and the size of its staff. The staff was pressured to disperse money quickly, and little real support for grassroots development ever materialized within the institution. Local populations were hardly ever consulted and rarely taken into account for the majority of the enterprises the bank funded during this period. As a result, billions of dollars of bank money was managed ineffectively by domestic public institutions that were often unrepresentative. The stated purpose of the loans was to promote modernization and open economics, but the lending network further supported the elite in the Third World and prevented the poor from playing a meaningful role in the policies that were so markedly reshaping their environments.

The international debt crisis during this decade also changed the bank, beginning its evolution into a debt-management institution when it lifted the restrictions on the non-project lending that had been established by its founders. In an effort to expedite its loan development, the bank underwent a major reorganization in 1986. It consolidated its activities under four senior vice presidents, but many consider the net results of these changes to be far from successful.

The most notable critics of the World Bank have been in the global environmental movement, but defects within the bank's operations complex have largely restricted their campaigns from effecting major changes. The bank has repeatedly withheld the majority of the information it generates in the preparation and implementation of projects, and many environmentalists continue to charge that it has become an institution fundamentally lacking the accountability and responsibility necessary for **sustainable development**.

The way the World Bank handled the controversial Narmada River Sandar Sarovar Dam project in India has been called a case in point. According to the **Environmental Defense** Fund, the history of bank involvement with the project reveals an institution whose primary agenda is to transfer money quickly, no matter what the cost in "systematic violation of its own environmental, economic and social policies and deception of its senior management and Board of Executive Directors." United States Executive Director E. Patrick Coady accused bank management and staff of a cover-up in relation to the Sandar Sarovar Dam, and he directly challenged the bank's credibility, noting that "no matter how egregious the situation, no matter how flawed the project, no matter how many policies have been violated, and no matter how dear the remedies prescribed, the bank will go forward on its own terms." Despite support from Germany, Japan, Canada, **Australia**, and the Scandinavian countries, Coady's accusations were ignored. Bank management continued to finance the dam until early 1993, when criticism of the project became so intense that India decided to finance the project itself. There was no plan to resettle the 250,000 people who would be displaced.

Many argue that the failures of policies at the World Bank are illustrated by other case studies in Brazil, Costa Rica, and Ghana. In Brazil, International Monetary Fund (IMF) stabilization programs have hit the poor particularly hard and depleted the country's **natural resources**. Although these structural programs were designed to manage Brazil's tremendous foreign debt, critics argue they have been socially, economically, and environmentally catastrophic. The "economic miracle" of the 1970s never materialized for Brazil's masses, despite the fact that the economy grew more than any other country in the world between 1955 and 1980. Due largely to ill-conceived IMF and World Bank advice, as well as ineffective leadership by the Brazilian government, development decisions in Brazil have leaned toward extravagant and poorly implemented projects. Many major projects have failed miserably, only adding to lost investments, and the environmental devastation caused by such projects has been enormous, with much of it irreversible, involving massive destruction of the **tropical rain forest**.

The Tucurui Dam project, for example, was part of the Brazilian government's plan to build some 200 new hydroelectric power **dams**, most of them in the Amazonian region. This project has destroyed thousands of acres of tropical **rain forest** and degraded **water quality** in the **reservoir** and downstream, further lowering the income and affecting the health of communities below the dam. For environmentalists, Tucurui is representative of the problems inherent in building large dams in tropical regions: "Opening of forest areas leads to **migration** of landless people, **deforestation** of reservoir margins, **erosion** and **siltation** causing destruction of power-generating equipment and reduction of its useful lifespan; spread of waterborne disease to human and **wildlife** populations; and the permanent destruction of fish and wildlife habitats, eliminating previously existing local economic activity and unknown numbers of **species** of plants and animals."

Ghana is usually touted by the World Bank and IMF as a particularly successful example of structural adjustment programs in Africa, but many believe that a closer examination reveals something different. The goal of the current adjustment program was to reduce the fiscal deficit, inflation, and external deficits by reducing domestic demand. A large portion of this plan was "to reverse the decline in agricultural production, restore overseas confidence in the Ghanian economy, increase foreign-exchange earnings, restore acceptable living standards, control inflation, reform prices, and reestablish production incentives for cocoa." Although cocoa production had increased 20% by 1988 as a result of incentives in agricultural sector reform, with the income for cocoa production increasing by as much as 700% in some cases, cocoa farmers make up only 18% of Ghana's farming population. Furthermore, in recent decades a growing inequity within the cocoa-producing population has been documented. Half of the land cultivated for cocoa today is owned by the top 7% of Ghana's cocoa producers, while 70% own farms of less than 6 acres (2.4 ha).

The timber industry in Ghana was also identified by the IMF and World Bank as an additional source of foreign exchange, and this led to the steady destruction of Ghana's forests. If timber production is maintained at its current rate and without appropriate environmental controls, many environmentalists and others believe that the Ghanian countryside will be stripped bare by the year 2000. The fishing industry is also threatened. Rising costs have caused the price of fish to increase while real wages have fallen. Ghanians receive 60% of their protein from the products of ocean fisheries, and decreased fish consumption in considered one of the leading factors in the rise of malnutrition in the country.

While mounting public pressure might cause the bank to reconsider some of its programs, reforms are possible only if supported by those nations controlling more than half of the Bank's voting shares. And no matter what direction is taken, many experts believe that internal contradictions will continue to trouble the bank for years, further lowering the morale of the staff. As long as the bank continues to place priority on its debt-management functions, it will likely fail to change its policy of making large-scale loans for questionable and destructive projects. Overall quality may be further diminished as developing nations continue to give up natural resources for the rapid earning of foreign exchange. Debt management could then force the bank to support large-scale, poorly-organized, and poorly-supervised projects, while the staff would be pressed harder to address wider development problems that could continue to be exacerbated by other aspects of bank lending.

The World Bank's dismal record of ignoring the interests of the poor and the environment clearly raises fundamental questions about the future of the institution. Environmentalists, human rights groups, and other critics of the Bank have offered the following suggestions for improving its operations. While there are obvious reasons for keeping some information confidential, they argue, the bank has abused this right. Greater freedom of information is necessary because participation with population groups in the development process is impossible without public access. Critics also argue for the establishment within the Bank of an independent appeals commission, maintaining that there is a clear need for a body that would hear and act on complaints of environmental and social abuses. It has been suggested that this commission could be made up of environmentalists, academics, church representatives, human rights groups, and others who would encourage the bank to emphasize sustainable development and eliminate environmentally destructive projects. And finally, it has been widely suggested that project quality should be the first priority of the World Bank, with the United Nations taking the lead in demanding this. In short, many believe the bank should no longer be allowed to function as a money-moving machine to address macro-economic imbalances. *See also* Economic growth and the environment; Environmental economics; Environmental policy; Environmental stress; Green politics; Greens; United Nations Earth Summit; United Nations Environment Programme

[*Roderick T. White Jr.*]

Resources

Books

Le Prestre, P. G. *The World Bank and the Environmental Challenge.* London: Associated University Presses, 1989.

Periodicals

Payer, C. *The World Bank: A Critical Analysis.* New York: Monthly Review Press, 1982.

World Commission on Environment and Development

see ***Our Common Future* (Brundtland Report)**

World Conservation Strategy

The World Conservation Strategy (WCS): Living Resource Conservation for Sustainable Development is contained in a report published in 1980 and prepared by the International Union for Conservation of Nature and Natural Resources (now called **IUCN—The World Conservation Union**). Assistance and collaboration was received from the **United Nations Environment Programme** (UNEP), the **World Wildlife Fund** (WWF), the Food and Agriculture Organization of the United Nations (FAO), and the United Nations Educational, Scientific and Cultural Organization (**UNESCO**).

The three main objectives of the WCS are: (1) to maintain essential ecological processes and life-support systems on which human survival and development depend. Items of concern include **soil** regeneration and protection, the **recycling** of nutrients, and protection of **water quality**; (2) to preserve genetic diversity on which depend the functioning of many of the above processes and life-support systems, the breeding programs necessary for the protection and improvement of cultivated plants, domesticated animals, and **microorganisms**, as well as much scientific and medical advance, technical innovation, and the security of the many industries that use living resources; (3) to ensure the sustainable utilization of **species** and ecosystems which support millions of rural communities as well as major industries.

The WCS believes humans must recognize that the world's **natural resources** are limited, with limited capacities to support life, and must consider the needs of **future generations**. The object, then, is to conserve the natural resources, sustain development, and to support all life. Humans have great capacities for the creation of wants or needs and also have great powers of destruction and annihilation. Human action has global consequences, and thus global responsibilities are crucial. The aim of the WCS is to provide an intellectual framework as well as practical implementation guidelines for achieving its three primary objectives.

The WCS has been endorsed by numerous leaders, organizations, and governments, and has formed the basis for preparation of National Conservation Strategies in over fifty countries.

The World Conservation Monitoring Centre (WCMC) mission is to support conservation and sustainable development by providing information on the world's **biodiversity**. It is a joint venture between the three main cooperators of WCS, IUCN, UNEP, and WWF.

The WCS has been supplemented and restated in a document called *Caring for The Earth: A Strategy for Sustainable Living*, published in 1991. This document restates current thinking about conservation and development and suggests practical actions. It establishes targets for change and urges a concerted effort in personal, national, and international relations. It stresses measuring achievements against the objectives of actions.

The WCS of 1980 and the 1991 update have done much to bring attention to the need for sustainable management of the world's natural resources. It outlines problems, suggests needed changes, and stresses the need to quantitate the progress in meeting the needs of a sustainable world. *See also* Environmental education; Environmental ethics; Environmental monitoring; Sustainable biosphere

[*William E. Larson*]

Resources
Other

Caring for the Earth: A Strategy for Sustainable Living. Gland, Switzerland: IUCN—The World Conservtion Union, 1991.

World Conservation Strategy. Gland, Switzerland: International Union for the Conservation of Nature and Natural Resources, 1980.

World Resources Institute

An international environmental and resource management policy center, the World Resources Institute researches ways to meet human needs and foster economic growth while conserving **natural resources** and protecting the **environment**. WRI's primary areas of concern are the economic effects of environmental deterioration and the demands on **energy and the environment** posed by both industrial and developing nations. Since its inception in 1982, WRI has provided governments, organizations, and individuals with information, analysis, technical support, and policy analysis on the interrelated areas of environment, development, and resource management.

WRI conducts various policy research programs and operates the Center for International Development and Environment, formerly the North American arm of the International Institute for Environment and Development. The institute aids governments and nonfederal organizations in developing countries, providing technical assistance and policy recommendations, among other services. To promote public education of the issues with which it is involved, WRI also publishes books, reports, and papers; holds briefings, seminars, and conferences; and keeps the media abreast of developments in these areas.

WRI's research projects include programs in biological resources and institutions; economies and population; **climate**, energy, and **pollution**; technology and the environment; and resource and environmental information. In collaboration with the Brookings Institution and the Santa Fe Institute, WRI is also involved in a program called the 2050 Project, which seeks to provide a sustainable future for

coming generations. As part of the program, studies will be conducted on such topics as food, energy, **biodiversity**, and the elimination of poverty.

The Center for International Development and Environment assists developing countries assess and manage their natural resources. The center's four main programs are: natural resources management strategies and assessments; natural resource information management; community planning and **nongovernmental organization** support; and sectoral resource policy and planning.

WRI's other programs are equally innovative. As part of the Biological Resources and Institutions project, WRI has developed a Global Biodiversity Strategy in collaboration with the World Conservation Union and the **United Nations Environment Programme**. The strategy, developed in 1992, outlines 85 specific actions required in the following decade to slow the decline in biodiversity worldwide. The program has also researched ways to reform forest policy in an attempt to halt **deforestation**.

The program on climate, energy, and pollution strives to develop new and different **transportation** strategies. In so doing, WRI staff have explored the use of hydrogen- and electric-powered vehicles and proposed policies that would facilitate their use in modern society. The program also researches renewable **alternative energy sources**, including solar, wind, and **biomass** power.

WRI is funded privately, by the United Nations, and by national governments. It is run by a 40-member international Board of Directors.

[*Kristin Palm*]

Resources

Organizations

World Resources Institute, 10 G Street, NE, Suite 800, Washington, D.C. USA 20002 (202) 729-7600, Fax: (202) 729-7610, Email: front@wri.org, <http://www.wri.org>

World Trade Organization (WTO)

In December, 1999, the streets of Seattle, Washington, filled with billowing clouds of tear gas and pepper spray as squadrons of police in full riot gear skirmished with surging masses of protestors in the most confrontational political demonstrations in the United States in nearly three decades. The angry crowds were there to confront delegates from 135 member nations of the World Trade Organization (WTO) who were meeting to hammer out an agenda for the next round of negotiations to regulate international trade.

Few people in America had ever heard of the WTO before the historic protest in Seattle, and yet, this exclusive body has power that affects us all. Created in 1995 by an international treaty, the WTO is the successor to the General Agreement on Tariffs and Grade (GATT), established at the end of the World War II to eliminate tariffs and trade barriers. Both GATT and WTO are part of the Bretton Woods system (named after the location in New Hampshire where the system was established in 1944) that includes the **World Bank** group and the International Monetary Fund (IMF). Where GATT was limited to considering economic issues, however, the scope of the WTO has been expanded to "noneconomic trade barriers" such as food safety laws, quality standards, product labeling, workers rights, and environmental protection standards. With legal standing equivalent to the United Nations, the WTO operates largely in secret. When considering trade disputes, it meets behind closed doors based on confidential evidence.

WTO judges are trade bureaucrats, usually corporate lawyers with ties to the industries being regulated. There are no rules against conflicts of interest, nor are there requirements that judges know anything about the culture or circumstances of the countries they judge. No appeal of WTO rulings is allowed. A country that loses a trade dispute has three options: (1) amend laws to comply with WTO rules, (2) pay annual compensation—often millions of dollars—to the complainants, or (3) face nonnegotiable trade sanctions. Critics claim that the WTO always serves the interest of transnational corporations and the world's richest countries.

Among the most controversial issues brought up in this round of WTO negotiations are agricultural subsidies, child labor laws, occupational health and safety standards, protection of intellectual property, and environmental standards. Environmentalists, for example, were outraged by a 1998 WTO ruling that a U.S. law prohibiting the import of shrimp caught in nets that can entrap **sea turtles** is a barrier to trade. The United States must either accept shrimp regardless of how they are caught, or face large fines. Some other WTO rulings that overturn environmental or consumer safety laws require Europeans to allow importation of U.S. hormone-treated beef, Americans must accept tuna from Mexico that endangers **dolphins**, and the U. S. **Environmental Protection Agency** (EPA) cannot bar import of low-quality **gasoline** that causes excessive **air pollution**. In some pending cases, Denmark wants to ban 200 **lead** compounds in consumer products; France wants to prohibit **asbestos**; and several countries want to eliminate electronic devices containing lead, **mercury**, and **cadmium**. Under current WTO rules, all these cases probably will be ruled illegal.

More than 50,000 people came to Seattle from all over the world to show their displeasure with the WTO. French farmers, Zapatista rebels from Mexico, Tibetan refugees, German anarchists, First Nations people from Canada, labor unionists, environmentalists in turtle suits, **animal rights**

activists, and organic farmers were among those who filled the streets. They held teach-ins, workshops, and peaceful marches with thousands of participants (sometimes described as "turtles and teamsters"), but what got headline coverage was civil disobedience including blocking streets, unfurling banners from atop giant construction cranes, and, for a few of the most radical anarchists, breaking windows, setting fires in dumpsters, and looting stores. These destructive acts were condemned by mainstream groups, but got most of the press anyway.

Unprepared for such a massive protest movement, the police reacted erratically. Ordered to avoid confrontation on the first day of the protests, the police stood by while a small contingent of black-hooded anarchists smashed windows and vandalized property. The next day, stung by criticism of being too soft, the police used excessive force to clear the streets, firing **rubber** bullets and tear gas indiscriminately, spraying innocent bystanders with pepper spray, and clubbing nonviolent groups engaged in passive sit-ins. The mayor declared a civil emergency and a 24–hour curfew in the area around the Civic Center. Eventually, the National Guard was called in to assist the thousands of city police. As often happens in confrontations, positions harden and violence begets more violence.

Most of the people protesting in Seattle agreed that the current WTO represents a threat to democracy, quality of life, **environmental health**, social justice, labor rights, and national sovereignty. Underneath these complaints is a broader unease about trends towards globalization and the power of transnational corporations. Although the diverse band of protestors shared many concerns, many disagreed about the best solutions to these problems and how to achieve them. While many claimed they wanted to shut down the WTO, others actually want a stronger trade organization that can enforce rules to protect workers, environmental quality, and **endangered species**.

In the end, the delegates adjourned without agreement on an agenda for the "Millennium Round" of the WTO. Developing countries, such as Malaysia, Brazil, Egypt, and India, refused to allow labor conditions into the debate. Major agricultural exporters such as the United States, Canada, Argentina, and **Australia**, continued to demand an end to crop subsidies and protective policies. Japan and the **European Union** (EU), on the other hand, maintain that they have a right to preserve small, family farms, rural lifestyles, and traditional methods of food production against foreign competition. Developing countries insist that protection of their **environment** and **wildlife** is no one's business but their own.

Following Seattle in 1999, activists have demonstrated against the effects of globalization at a number of world governance meetings. The most violent of these occurred in July 2001, when 100,000 protestors converged on a meeting of the Group of Eight Industrialized Nations in Genoa, Italy. As was the case in Seattle, the vast majority of the demonstrators were peaceful and non-violent, but a small group of radicals attacked police and vandalized property. The police responded with what many observers considered excessive force, killing one man and injuring hundreds of others. Outrage at police behavior spread across Europe as live television showed unprovoked attacks on peaceful marchers and innocent bystanders.

In the aftermath of Genoa, both protestors and government officials began to re-examine their strategies for future meetings. Leaders of many community groups question whether they should take part in mass demonstrations, because of both the personal danger and the negative image resulting from association with marauding vandals. They began to reflect on other ways to carry out their goals while avoiding the violence that marred previous demonstrations. Government officials announced that future meetings would be held in remote, inaccessible locations that limit public participation. The 2001 meeting of the WTO, for example, was held in Qatar, an authoritarian country that strictly forbids any form of public demonstration. Those who weren't official delegates to the meeting weren't even allowed into the country, perpetuating the image of the WTO as a secretive and high-handed organization.

While the location and tactics of the debate about globalization has changed, the basic problems remain. As Renato Ruggiero, the former director-general of the WTO once said, "We are no longer writing the rules of interaction among separate national economies. We are writing the constitution of a single global economy." The politicians and transnational corporations who currently control much of direction of global governance dismiss their critics as an irrelevant collection of environmental extremists and bleeding-heart social activists who know nothing about economics or practical politics. On the other hand, even World Bank president, James D. Wolfensohn admits that "at the level of people, globalization isn't working." We clearly need a way to engage governments and others in a dialogue on how we will organize global trade in an increasingly interconnected world.

Interestingly, the organizational power of protest groups in Seattle and elsewhere shows the growing internationalism of social movements and grass-roots organizations. Internet technology, the declining cost of travel, and rising educational levels have dramatically extended their capacity to both think and act globally. Those who protest globalization and argue for traditional, time-honored ways of doing things are themselves using the technology and power of global organization. Somehow we need to find a way to get beyond "No." How do we want to govern ourselves? How

will we respect local autonomy and culture, and still enjoy the benefits that come from the increased flow of goods and services across international borders?

[*William P. Cunningham Ph.D.*]

Resources

Books

Gallagher, Peter. *Guide to the WTO and Developing Countries.* Boston, MA: Kluwer Law International, 2000.

Howse, Robert. "Eyes Wide Shut in Seattle: The Legitimacy of the World Trade Organization." In *The Legitimacy of International Institutions.* United Nations University Press, 2000.

Von Moltke, Konrad. *Trade and the Environment: The Linkages and the Politics.* Winnipeg, Canada: International Institute for Sustainable Development, 2000.

Periodicals

Cohn, Marjorie, ed. "Human Rights and Wrongs." *Guild Practitioner* 57 (2000): 121.

Kovel, Joel. "Beyond the World Trade Organization." *Synthesis/Regeneration* 21 (2000): 6–9.

Other

"A Citizen's Guide to the World Trade Organization: Everything You need to know to Fight for Fair Trade." Working group on the WTO. July 1997 [cited July 9, 2002]. <http://www.citizen.org>.

Anderson, Sarah, and John Cavanagh. "The World Trade Organization" *Foreign Policy in Focus.* The Institute for Policy Studies.1997 [cited July 9, 2002]. <http://www.foreignpolicy-infocus.org/briefs/vol2/v2n14wto.html>.

Murphy, Sophia. "Managing the Invisible Hand: Markets, Farmers and International Trade" 2002. *Institute for Agriculture and Trade Policy.* [cited July 9, 2002]. <http://www.tradeobservatory.org>.

Seattle Weekly Editors. "Answering WTO's Big Questions: In a Nutshell, What was WTO Seattle?" *Seattle Weekly.* August 3–9, 2000 [cited July 9, 2002]. <http://www.seattleweekly.com/features/0031/news-editors.shtml>.

World Wildlife Fund

The World Wildlife Fund (WWF) is an international **conservation** organization founded in 1961, and known internationally as the World Wide Fund for Nature. The World Wildlife Fund acts with other U.S. organizations in a network to conserve the natural **environment** and ecological processes essential to life. Particular attention is paid to **endangered species** and to natural habitats important for human welfare. In hundreds of projects conducted or supported around the world, WWF helps protect endangered wildlife and habitats and helps protect the earth's **biodiversity** through fieldwork, scientific research, institutional development, wildlife trade monitoring, public policy initiatives, technical assistance and training, **environmental education**, and communications.

WWF has articulated nine goals that guide its work:

- to protect habitat
- to protect individual **species**
- to promote ecologically sound development
- to support scientific investigation
- to promote education in developing countries
- to provide training for local wildlife professionals
- to encourage self-sufficiency in developing countries
- to monitor international wildlife trade
- to influence public opinion and the policies of governments and private institutions

Toward these goals, WWF monitors international trade in wild plants and animals through its TRAFFIC (Trade Records Analysis of Flora and **Fauna** in Commerce) program, part of an international network in cooperation with **IUCN—The World Conservation Union**. TRAFFIC focuses on the trade regulations of the **Convention on International Trade in Endangered Species of Wild Fauna and Flora** (CITES), tracking and reporting on traded wildlife species; helping governments comply with CITES provisions; developing training materials and enforcement tools for wildlife-trade enforcement officers; pressing for stronger enforcement under national wildlife trade laws; and seeking protection for newly threatened species.

Other recent WWF projects have included programs to save the African elephant, especially through banning ivory imports under CITES legislation. The organization also has undertaken projects to preserve tropical rain forests; to identify conservation priorities in the earth's biogeographical regions; to halt overexploitation of renewable resources; and to create new parks and wildlife preserves to conserve species before they become endangered or threatened. Based on WWF's conservation efforts, The Charles Darwin Foundation for the Galapagos Islands awarded WWF an institutional seat on their Board of Directors.

WWF also funded a study of more than 2,000 projects or activities reviewed by the U.S. **Fish and Wildlife Service** between 1986 and 1991. The study found that only 18 of these projects were blocked or withdrawn because of detrimental effects to human populations. The study also showed that the Northwest lost more timber jobs to automation of timber cutting and milling, increased exports of raw logs, and a shift of the industry to the Southeastern United States, than it may lose from the listing of the **northern spotted owl** (*Strix occidentalis caurina*) as an endangered species.

WWF has been credited with saving some 30 endangered animal species from **extinction**, most notably the **giant panda**, which has become the group's logo. Using **Geographic Information Systems** satellite technology, WWF has identified regions where conservation of distinct animal and plant species are most needed. By overlaying these areas with existing parks and projects, WWF is able to see where

it needs to concentrate conservation programs. The organization publishes a wide variety of materials, including the periodicals *WWF Letter*, *TRAFFIC*, and *Tropical Forest Conservation*. Booklets produced include "Speaker and News Media Sourcebook: A Guide to Experts in Domestic and International Environmental Issues," for news media, policymakers, and organizations. Educational materials, research papers, and books jointly published include *The Gaia Atlas of Future Worlds: Challenge and Opportunity in an Age of Change* (by Norman Myers); *Options for Conservation: The Different Roles of Nongovernmental Conservation Organizations* (by Sarah Fitzgerald); *WWF Atlas of the Environment* (By Geoffrey Lean, et al.); and *The Official World Wildlife Fund Guide to Endangered Species of North America*. In 1996 the WWF produced a nationwide environmental program, Windows on the Wild, for educators concerning the challenges of global conservation.

[*Linda Rehkopf*]

Resources

Organizations

World Wildlife Fund-U.S., 1250 12th Street, NW, Washington, D.C. USA 20037, <http://www.worldwildlife.org>

Worldwatch Institute

Worldwatch Institute, a research organization based in Washington, D.C., compiles and publishes information on worldwide environmental problems; it also suggests solutions and alternative courses of action. The Institute was founded in 1975 by **Lester R. Brown**, with financial assistance from the Rockefeller Brothers Fund. Brown is now its president and director of research; chairman of the board is former Secretary of Agriculture Orville Freeman. Membership is $25 annually. The fee helps support the Institute's research and publications, which members receive, but there are no avenues for membership participation within the Institute.

The Institute has published the *State of the World* report annually since 1984; sales in recent years reached $200,000. It is issued in 26 languages and is used as a textbook in over 600 colleges in the United states alone. The publication examines such topics as global warming, water and **air quality**, and the environmental impact of social policies. In 1990, *State of the World* was produced as a 10-part series on public television, a joint venture with the producers of *Nova*. The 1992 edition describes "a planet at risk" and warned that "the policy decisions we make during this decade will determine" the quality of life for **future generations**.

Six to eight other Worldwatch papers are published each year, and over a hundred monographs have been issued on specific subjects. *Worldwatch* magazine is published in several languages, and the *Environmental Alert* series targets particular environmental issues.

[*Lewis G. Regenstein and Amy Strumolo*]

Resources

Organizations

Worldwatch Institute, 1776 Massachusetts Ave., NW, Washington, D.C. USA 20036-1904 (202) 452-1999, Fax: (202) 296-7365, Email: worldwatch@worldwatch.org, <http://www.worldwatch.org>

WTO

see **World Trade Organization**

Charles Frederick Wurster (1930 –)

American environmental scientist

Dr. Charles Wurster, a founding trustee of **Environmental Defense** (formerly the Environmental Defense Fund, EDF), was instrumental in banning the **pesticide** DDT. An emeritus professor at the Marine Sciences Research Center at the State University of New York at Stony Brook, Wurster is an expert on the environmental effects of toxic **chemicals**.

Born in Philadelphia, Pennsylvania, Wurster's interest in birds was evident at a young age. He attended Quaker schools—Germantown Friends and Haverford College—receiving his bachelor's of science degree in 1952. He earned a master of science degree in chemistry from the University of Delaware in 1954 and a Ph.D. in organic chemistry from Stanford University in 1957. Subsequently Wurster spent a year in Innsbruck, Austria, as a Fulbright Fellow. From 1959 to 1962, he worked as a research chemist at the Monsanto Research Corporation.

In 1963, as a research associate in biology at Dartmouth College in New Hampshire, Wurster demonstrated that the ubiquitous pesticide DDT was killing the campus birds. DDT sprayed on elm trees killed 70% of the robins in just two months. Myrtle warblers, who were not even on the campus at the time of the spraying, also were killed, as were all of the chipping sparrows. Furthermore DDT was not saving the trees from fatal blight. Following a two-year battle by Wurster and his colleagues, the local spraying of DDT was halted.

After moving to Stony Brook in 1965 as an assistant professor of biology, Wurster continued his studies on the harmful effects of DDT. Among other results, he found that high concentrations of DDT in Long Island osprey were associated with thin eggshells and poor reproduction. Much of his work was published in the prestigious journal *Science*.

At Stony Brook, Wurster joined the Brookhaven Town Natural Resources Committee, a group of scientists, lawyers, and citizens. In one of the first-ever legal actions to protect the **environment**, the committee forced Suffolk County to stop spraying DDT for mosquito control in the marshes of Long Island. In 1967 the committee incorporated as the EDF. The EDF's first major victory came with the 1972 nationwide ban on DDT, with Wurster in charge of the scientific arguments. Wurster is credited with creating a network of scientific expertise within the EDF that lent credibility to the organization. In 1973, while on sabbatical in New Zealand, Wurster helped found the Environmental Defence Society (EDS), modeled on the EDF.

Wurster served as Associate Professor of Environmental Sciences at the Marine Sciences Research Center from 1970 until 1994 when he became an emeritus professor. His research has focused on DDT, **polychlorinated biphenyls** (PCBs), and other **chlorinated hydrocarbons**, and their effects on birds and **phytoplankton** (marine plants at the base of the food web). He also has studied the **mortality** of diving birds caught on the longlines used for tuna fishing. Wurster's interest in the environmental sciences and public policy has continued. He was instrumental in the banning of the pesticides Dieldrin and Aldrin. In 1990 Wurster assisted the U.S. Department of Justice in its legal case against the Montrose Chemical Corporation, formerly the world's largest producer of DDT. The EDF first sued Montrose in 1970 to stop it from dumping DDT into the Santa Monica Bay. In 2000 Montrose, now owned by other companies, was finally ordered to pay for the clean-up of 17 mi^2 (44 km^2) of ocean floor. A fellow of the American Association for the Advancement of Science, Wurster was a director of **Defenders of Wildlife** from 1975 to 1984 and a trustee of the **National Parks and Conservation Association** from 1970 to 1979. He continues as a director of the EDS.

From its beginnings as a small group meeting at the Brookhaven National Laboratory, Environmental Defense has grown into one of world's largest and most influential environmental organizations, with 400,000 members and a staff of 250 that includes scientists, economists, and lawyers. Wurster continues to work with the EDF program for the establishment of marine preserves in the United States and in the open ocean and he leads ecological tours for the EDF.

[*Margaret Alic Ph.D.*]

Resources

Periodicals

Mosser, J. L., N. S. Fisher, and C. F. Wurster. "Polychlorinated Biphenyls and DDT Alter Species Composition in Mixed Cultures of Algae." *Science* 176 (1972): 533–5.

Wurster, C. F. "Beetles and Dieldrin." *Science* 163 (1969): 229.

Wurster, C. F. "DDT and Robins." *Science* 159 (1968): 1413–4.

Wurster, C. F. "DDT Reduces Photosynthesis by Marine Phytoplankton." *Science* 159 (1968): 1474–5.

Other

Bowman, Malcolm J. "Charles Wurster: Environmental Hero and Advocacy Pioneer." *EDS News*, January 2002 [cited June 10, 2002]. <www.eds.org.nz/News_Vol_4.htm>.

Organizations

Dr. Charles F. Wurster, Marine Sciences Research Center, State University of New York at Stony Brook, Stony Brook, NY USA 11794-5000 (631) 632-8738, Email: Charles.F.Wurster@stonybrook.edu,

Environmental Defence Society, PO Box 95 152 Swanson, Aukland, New Zealand 1008 (64-9) 810-9594, Fax: (64-9) 810-9120, Email: manager@eds.org.nz, <http://www.eds.org.nz>

Environmental Defense, 257 Park Avenue South, New York, NY USA 10010 (212) 505-2100, Fax: (212) 505-2375, Toll Free: (800) 684-3322, Email: members@environmentaldefense.org, <http://www.environmentaldefense.org>

WWF

see **World Wildlife Fund**

X

X ray

Discovered in 1895 by German physicist Wilhelm K. Roentgen, x rays are a form of electromagnetic radiation, widely used in medicine, industry, metal detectors, and scientific research. X rays are most commonly used by doctors and dentists to make pictures of bones, teeth, and internal organs in order to find breaks in bones, evidence of disease, and cavities in teeth.

Since x rays are a form of **ionizing radiation**, they can be very dangerous. They penetrate into, and are absorbed by, plants and animals and can age, damage, and destroy living tissue. They can also cause skin burns, genetic mutations, **cancer**, and death at high levels of exposure. The effects of ionizing radiation tend to be cumulative, and every dose adds to the possibility of further damage.

Some authorities feel that people should try to minimize their exposure to such radiation and avoid being x-rayed unless absolutely necessary. This is especially true of pregnant women, since studies show a much higher rate of childhood **leukemia** and other diseases among children who were exposed to x rays in utero. Ironically, fear of malpractice suits has prompted many doctors to increase the number of x rays performed while examining patients for disease. *See also* Radiation exposure

Xenobiotic

Designating a foreign and usually harmful substance or organism in a biological system. Xenobiotic, derived from the Greek root *xeno*, meaning "stranger" or "foreign," and *bio*, meaning "life," describes some toxic substances, **parasites**, and symbionts. Food, drugs, and poisons are examples of xenobiotic substances in individual organisms, and their toxicity is linked to the level of consumption. In communities or **species**, xenobiosis happens when two distinct species, such as different kinds of ants, share living space like nests. At the **ecosystem** level, toxic waste, when bioaccumulated in the **food chain/web**, is xenobiotic. *See also* Bioaccumulation; Hazardous waste; Symbiosis

Xeriscaping

see **Arid landscaping**

Xylene

The term "xylene" refers to any of three **benzene** derivative isomers that share the same chemical formula, $C_6H_4(CH_3)_2$, but differ in their molecular structure. They are useful as solvents, as additives to improve the **octane rating** of aviation fuels, and as raw materials in the manufacture of fibers, films, dyes, and other synthetic products. Because of their high volatility they are classified as aromatic **hydrocarbons** (volatile compounds containing only **carbon** and **hydrogen** atoms). Xylene fractions were first isolated from **coal** tar in the mid-nineteenth century in Germany. Coal tar is the thick liquid product of the carbonization, or destructive distillation, of coal, a process in which coal is heated without air to temperatures above 1,600°F (862°C). Benzene and **toluene**, two aromatic hydrocarbons similar to xylene, are formed in the same process. Coal tar remained the source for xylene until industrial demand outgrew the supply. Later, techniques were developed to obtain xylene from **petroleum** refining, and today petroleum is a major source.

The three isomeric xylenes are classified by the arrangement of the methyl (CH_3) groups substituting for hydrogen atoms on the benzene ring. In each case, the methyl groups replace two of the hydrogens attached to the six-carbon benzene ring, but the structure and chemical properties of each isomer depend on which hydrogens are replaced. Ortho-xylene (o-xylene) has methyl groups on two adjacent carbons in the benzene ring structure, while in meta-xylene (m-xylene) the methyl-containing carbons are separated by a hydrogen-containing carbon. In para-xylene (p-xylene), two hydrogen-bearing carbons separate the methyl-con-

taining carbons, placing the methyl groups at opposite ends of the molecule. Ortho-xylene, the least volatile of the three forms, serves as raw material for the manufacture of coatings and **plastics**. Para-xylene is used to manufacture polyesters. Meta-xylene, less useful than the other forms, is used to make coatings, plastics, and dyes. A commercial xylene mixture in the form of a colorless, flammable, non-viscous, toxic liquid is used as a solvent for lacquers and **rubber** cements. Emulsified xylene has been used as an economical and effective way of controlling aquatic weeds in **irrigation** systems, but concentrations must be carefully controlled to avoid harming plants and fish exposed to the residues.

Xylene's volatility leads to its easy escape into the **atmosphere**. Unless care is taken, its fumes soon permeate the air in laboratories where it is used. Although its solubility in water is quite limited, xylene (along with benzene and toluene) is classified as one of the primary components of the "water-soluble" fraction of petroleum. The widespread use of xylene in manufacturing has resulted in significant releases of the chemical into the air and water of the **environment**. Xylene's toxicity in water is influenced by **salinity**, temperature, and the presence of other toxic materials. It affects cell permeability and acts as a **neurotoxin** for fish and other animals. At high concentrations of xylene, fish exhibit a series of behavioral changes including restlessness (rapid and erratic swimming), loss of equilibrium, paralysis, and death. Although each of the three forms of xylene can be detoxified by **microorganisms**, para-xylene is more difficult to detoxify, and therefore a more persistent threat.

[*Douglas C. Pratt*]

Resources

Books

Buikema, L., and A. C. Hendricks. *Benzene, Xylene, and Toluene in Aquatic Systems: A Review.* Washington, DC: American Petroleum Institute, 1980.

Hancock, E. G., ed. *Toluene, the Xylenes, and Their Industrial Derivatives.* Amsterdam; New York: Elsevier, 1982.

Parmeggiani, L., ed. *Encyclopaedia of Occupational Health and Safety.* 3rd ed. Geneva: International Labour Office, 1983.

Walsh, D. F., ed. *Residues of Emulsified Xylene in Aquatic Weed Control and Their Impact on Rainbow Trout, Salmo gairdneri.* Denver, CO: U.S. Dept. of the Interior [Bureau of Reclamation, Engineering and Research Center, Division of General Research, Applied Sciences Branch], 1977.

Y

Yard waste

As the field of solid **waste management** becomes more developed and specialized, the categories into which **solid waste** is sorted and managed become more numerous. Yard waste—often called vegetative waste—includes leaves, grass clippings, tree trimmings, and other plant materials that are typically generated in outdoor residential settings. Depending on the season and the neighborhood, leaves may account for 5–30% of the total **municipal solid waste** stream, grass clippings may comprise 10–20%, and wood may form about 5–10%. Yard waste does not include food and animal wastes except possibly as impurities. Yard waste is usually perceived as cleanest type of solid waste, in contrast to household refuse or industrial and commercial trash.

Since yard waste is almost entirely vegetative in nature, it lends itself easily to **composting**. Composting is the natural breakdown of organic matter by **microbes**, in the presence of oxygen, to form a stable end product called compost. In the management of yard waste, this natural process decreases the need for **landfill** space and produces **soil** amendment as an end product.

Beneficial use and composting of yard waste may be practiced in the homeowner's backyard or in the waste management areas set up by townships and municipalities. Not all forms of yard waste lend themselves equally well to composting. In fact, woody materials such as tree trunks may be better managed and used by shredding into wood chips or saw dust, since wood is very resistant to composting. Leaves are most suitable for composting and provide the richest and most stable base of material and nutrients. With sufficient oxygen, water, and turnover, leaves can usually be broken down in a matter of weeks. Grass clippings decompose so quickly that they need to be mixed with slower-degrading materials (such as leaves) in order to slow down and regulate the rate of **decomposition**. Control over the decomposition process is essential because of the potential for problems such as noxious odors. If the decomposition process is too rapid and proceeds without sufficient oxygen, the process becomes **anaerobic** and generates odors that can be highly unpleasant. In seasons other than fall, when the volume of leaves in yard waste is relatively small, many experts recommend that grass clippings be used as **mulch** rather than as compost to avoid the potential for odor.

In urban and mixed **land use** neighborhoods, yard waste may inadvertently contain **plastics**, papers, and other non-degradable materials. In some areas, the bagging of yard waste may involve the use of plastic bags. Depending on the efficiency of separation at the waste management area, these inert materials may find their way into the final compost product and lower the esthetics and usefulness of the compost.

[*Usha Vedagiri*]

Resources
Books

Strom, P. F., and M. S. Finstein. *Leaf Composting and Yard Waste Management for New Jersey Municipalities.* New Jersey Department of Environmental Protection and Energy, 1993.

Yellowstone National Park

Yellowstone National Park has the distinction of being the world's first **national park**. With an area of 3,472 sq mi (8,992 sq km), Yellowstone is the largest national park in the lower 48 states. This is an area larger than Rhode Island and Delaware combined. Although primarily in Wyoming (91%), 7.6% of the park is in Montana and the remaining 1.4% is in Idaho.

John Colter, a member of the Lewis and Clark expedition of 1803–1806, was probably the first white man to visit and report on the Yellowstone area. At that time the only Native Americans living year-round in the area were a mixed group of Bannock and Shoshone known as "sheepeaters." In 1859, the legendary trapper and explorer Jim Bridger, who had been reporting since the 1830s about the wonders of the

A grizzly bear with its cubs in Yellowstone National Park. (National Park Service Harpers Ferry Center. Reproduced by permission.)

region, led the first government expedition into the area. The discovery of gold in the Montana Territory in the 1860s brought more expeditions. In 1870, the Washburn-Langford-Doane expedition came to verify the reports about the wonders of the area. They spent four weeks naming the features, including Old Faithful, the most famous geyser in the world. Legend states that while the 19 members of this expedition sat around a campfire, reflecting on the beauty of the area, they came up with the idea of turning the region into a national park. The truth of the legend is debatable, yet there is no doubt that it was the lecturing and writing of these men that prompted the United States **Geological Survey** to send a follow-up group to the park in 1871. Reports and photographs from the U.S. Geological Survey expedition stimulated the drafting of legislation to create the first national park. Because of the prevalent utilitarian philosophy and the country's poor economic condition, the battle for the park was difficult and hard-fought. Eventually the park proponents were successful, and on March 1, 1872, President Ulysses S. Grant signed the bill establishing the park.

Because Congress did not allocate any money for park maintenance or protection, the early years of the park were marked by vandalism, **poaching**, the deliberate setting of forest fires, and other destructive behaviors. Eventually, in 1886, the U.S. Army took responsibility for the park. They remained in the role of park managers until the **National Park Service** was formed in 1916.

Water covers about 10% of the park. The largest body of water is Yellowstone Lake, with a surface area of 136 sq mi (352 sq km). It is one of the largest, highest, and coldest lakes in North America. The park has one of the highest waterfalls in the United States (Lower Yellowstone Falls, 308 ft; 93.87 m) and the top three trout fishing streams in the world. Approximately 10,000 thermal features can be found in the park. In fact, there are more geysers (200–250) and hot springs in the park than in the rest of the world put together.

The park has a great abundance and diversity of **wildlife**. It has the largest concentration of mammals in the lower 48 states. There are 58 **species** of mammals in the park, including two species of bears and seven species of

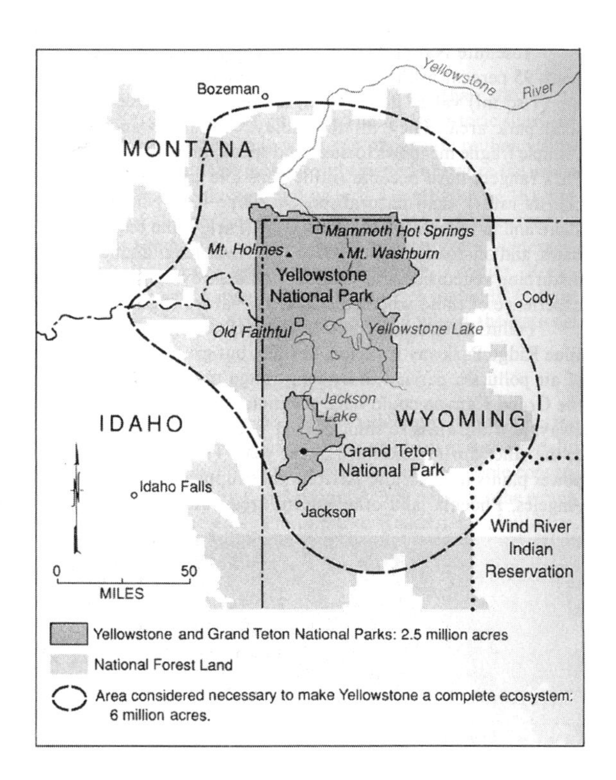

Map of the Yellowstone ecosystem complex or biogeographical region, which extends far beyond the boundaries of the park. (McGraw-Hill Inc. Reproduced by permission.)

ungulates. It is one of the last strongholds of the **grizzly bear** and is the only place where a **bison** herd has survived continuously since primitive times. Yellowstone is noted also for having the largest concentration of elk to be found anywhere in the world. Besides mammals, the park is home for 279 species of birds, 18 species of fish, five species of reptiles, and four species of amphibians.

However, one of the continuing difficulties at Yellowstone and other national parks is that Yellowstone is not a self-contained **ecosystem**. Its boundaries were established through a variety of political compromises, and lands around the park that once provided a buffer against outside events are being developed. Airsheds, watersheds, and animal **migration** routes extend far beyond park boundaries, yet they dramatically affect conditions within the park. Yellowstone is but one example of the need to manage entire biogeographical areas to preserve natural conditions within a national park.

[*Ted T. Cable*]

Resources

Books

Frome, M. *National Park Guide.* 19th ed. Chicago: Rand McNally, 1985.

Other

Yellowstone Fact Sheet. National Park Service, U.S. Department of Interior, 1992.

Organizations

Yellowstone National Park, P.O. Box 168, Yellowstone National Park, WY 82190-0168 (307) 344-7381, Email: yell_visitor_services@nps.gov, <http://www.nps.gov/yell>

Yokkaichi asthma

Nowhere is the connection between industrial development and environmental and human health deterioration more graphically demonstrated than at Yokkaichi, Japan. An international port located on the Ise Bay, Yokkaichi was a major textile center by 1897. The shipping business shifted to nearby Nagoya in 1907, and Yokkaichi filled in its coastal lowlands in a successful bid to attract modern industries, especially chemical processing, steel production, and oil and **gasoline** refining.

Spurred by both the World War II demand and the postwar recovery effort, several more **petrochemical** companies were added through the 1950s, creating an oil refinery complex called the Yokkaichi Kombinato. In 1959 it began 24-hour operations, and the sparkle of hundreds of electric lights became known as the "million-dollar night view." Although citizens took pride in the growing industrial complex, their enthusiasm waned when **air pollution** and **noise pollution** created human health problems. As early as 1953, the central government sent a research group to try to discover the cause, but no action was taken. Instead, the petrochemical complex was expanded.

As citizens began to complain about breathing difficulties, scientists documented a high correlation between airborne **sulfur dioxide** concentrations and bronchial **asthma** in schoolchildren and chronic **bronchitis** in individuals over 40. Despite this knowledge, a second industrial complex was opened in 1963. In the Isozu district of Yokkaichi, the average concentration of sulfur dioxide was eight times that of unaffected districts. Taller smokestacks spread **pollution** over a wider area but did not resolve the problem; increased production also added to the volume discharged. Despite **resistance**, a third industrial complex was added in 1973, one of the largest **petroleum** refining and ethylene producing facilities in Japan.

As the petrochemical industries continued to expand, local citizens' quality of life deteriorated. In the early years, heavy **smoke** was emitted by **coal combustion**, and parents worried about the exposure of schoolchildren whose playground was close to the emissions source. Switching from coal to oil in the 1960s seemed to be an improvement, but the now-invisible stack gases still contained large quantities of sulfur oxides, and more people developed **respiratory**

diseases. By 1960, fish from the local waterways had developed such a bad taste that they were unsalable, and fishermen demanded compensation for their lost livelihood. By 1961, 48% of children under six, 30% of people over 60, and 19% of those in their twenties had respiratory abnormalities. In 1964, a pollution-free room was established in the local hospital where victims could take refuge and breathe freely.

Even so, two desperate people committed suicide in 1966, and 12 Yokkaichi residents who had been trying to resolve the problem by negotiation finally filed a damage suit against the Shiohama Kombinato in 1967. In 1972, the judge awarded the plaintiffs $286,000 in damages to be paid jointly by the six companies. This was the first case in which a group of Japanese companies were forced to pay damages, making other kombinatos vulnerable to similar suits. As a consequence of the successful litigation by the Yokkaichi victims, the Japanese government enacted a basic antipollution law in 1967. Two years later, the Law Concerning Special Measures for the Relief of Pollution-Related Patients was enacted. It applied to chronic bronchitis/bronchial asthma victims not only from Yokkaichi but also Kawasaki and Osaka. In addition, national air-pollution standards were strengthened to require that oil refineries adhere to air pollution abatement policies.

By 1975, the annual mean sulfur dioxide levels had decreased by a factor of three, below the target level of 0.017 **parts per million** (ppm). The harmful effects on residents of Yokkaichi also decreased. In 1973, the Law Concerning Compensation for Pollution-Related Health Damages and Other Measures aided sufferers of chronic bronchitis and bronchial asthma from the other affected areas of Japan, especially Tokyo. By December 1991, 97,276 victims throughout Japan, including 809 from Yokkaichi, were eligible for compensation. *See also* Air-pollutant transport; Environmental law; Industrial waste treatment; Tall stacks

[*Frank M. D'Itri*]

Resources

Books

Huddle, N., and M. Reich. *Island of Dreams: Environmental Crisis in Japan.* Tokyo, Japan: Autumn Press, 1975.

Periodicals

Kitabatake, M., H. Manjurul, P. Feng Yuan, et al. "Trends of Air Pollution Versus Those of Consultation Rate and Mortality Rate for Bronchial Asthma in Individuals Aged 40 Years and Above in the Yokkaichi Region." *Nihon Eiseigaku Zasshi* 50, no. 3 (1995): 737.

Other

"Diseases By Air Pollution." In *Quality of the Environment in Japan.* Tokyo, Japan: Environmental Agency, Government of Japan, 1990.

Yosemite National Park

Yosemite National Park is a 748,542-acre (303,160-ha) park, located on the western slope of the Sierra Nevadas in northern California. The name Yosemite comes from the name of an Indian tribe, the *U-zu-ma-ti*, who were massacred in 1851 by soldiers sent by the governor of California for refusing to attend a reservation agreement meeting.

By the mid-1800s, Yosemite had become a thriving tourist attraction. As word of Yosemite's wonders spread back to the East Coast, public pressure led Congress to declare that the Yosemite Valley and the Mariposa Grove of giant sequoias be held "inalienable for all time." President Abraham Lincoln signed this law on June 30, 1864, turning this land over to the state of California and thereby giving Yosemite the distinction of being the first state park. Concern about sheep **overgrazing** in the high meadows led to the designation of the high country as a **national park** in 1890.

The person most strongly associated with the protection of Yosemite and its designation as a national park is naturalist and philosopher **John Muir**. Muir, sometimes called the "Thoreau of the West," first came to Yosemite in 1868. He devoted the next 40 years of his life to ensuring that the **ecological integrity** of the region was maintained. In 1892, Muir founded the **Sierra Club** to organize efforts to gain national park status for the Yosemite Valley.

Yosemite National Park is famous for its awesome and inspiring scenery. Geologic wonders include the mile-wide Yosemite valley surrounded by granite walls and peaks such as El Capitan (7,569 ft [2,307 m] above sea level; 3,593 ft [1,095 m] from base) and Half Dome (8,842 ft [2,695 m] above sea level; 4,733 ft [1,443 m] from base). Spectacular waterfalls include Yosemite Falls (Upper 1,430 ft [436 m], Middle 675 ft [206 m], Lower 320 ft [98 m]), Ribbon Falls (1,612 ft [491 m]), Bridalveil Falls (620 ft [189 m]), Sentinel Falls (2,001 ft [610 m]), Horsetail Falls (1,001 feet [305 meters]), and Vernal Falls (317 ft [97 m]).

The park supports an abundance of plant and animal life. Eleven **species** of fish, 29 species of reptiles and amphibians, 242 species of birds, and 77 species of mammals can be found in the park. There are approximately 1,400 species of flowering plants, including 37 tree species. The most famous of the tree species found here is the giant sequoia (*Sequoiadendron giganteum*). The park has three groves of giant sequoias, the largest being the Mariposa Grove with 500 mature trees, 200 of which exceed 10 ft (3 m) in diameter.

Because of the high degree of development and the congestion from the large numbers of visitors (over four million per year as of 2002), some have disparagingly referred to the Valley as "Yosemite City." Yosemite has become a

focal point in the controversy of whether we are "loving our parks to death." People and traffic management has become as critical as traditional resource management. At Yosemite, balancing the public's needs and desires with the protection of the **natural resources** is the National Park Service's greatest challenge as they enter the twenty-first century.

[*Ted T. Cable*]

RESOURCES

BOOKS

Frome, M. *National Park Guide.* 19th ed. Chicago: Rand McNally, 1985.

OTHER

Yosemite 1992 Fact Sheet. National Park Service, U. S. Department of Interior, 1992.

ORGANIZATIONS

Yosemite National Park, P.O. Box 577, Yosemite National Park, CA 20240 (209) 372-0200, Email: comments@yosemite.com, <http://www.yosemitepark.com>

National Park Service, 1849 C Street NW, Washington, D.C. 20240 (202) 208-6843, <http://www.nps.gov>

Yucca Mountain

Yucca Mountain is a barren **desert** ridge about 90 mi (161 km) northwest of Las Vegas, Nevada. The **U.S. Department of Energy** plans to build its first high-level nuclear waste storage facility there in an extensive series of tunnels deep beneath the desert surface. Choosing this site has been a long and divisive process. Although over 20 years and more than $7 billion have been spent designing and testing the storage facility, scientists are still not clear whether the facility will remain dry and stable for the thousands of years it will take for natural decay processes to make the wastes less radioactive.

Most of the nuclear waste to be stored at Yucca Mountain will come from nuclear **power plants**. As radioactive **uranium** 235 is consumed in fission reactions, the fuel becomes depleted and accumulating waste products reduce energy production. Every year about one-third of the fuel in a typical reactor must be removed and replaced with a fresh supply. The spent (depleted) fuel must be handled and stored very carefully because it contains high concentrations of dangerous radioactive elements, in addition to natural uranium, thorium, and **radon** isotopes. Some of these materials have very long half-lives, requiring storage for at least 10,000 years before their **radioactivity** is reduced to harmless levels by natural decay processes.

Originally, the United States intended to reprocess spent fuel to extract and purify unused uranium and **plutonium** as fuel for advanced breeder reactors (a type of nuclear reactor). Reprocessing plants released unacceptable amounts of radioactivity, however, and breeder reactors proved to be too expensive and dangerous to serve as civilian power sources. For a time in the 1950s, nuclear wastes were dumped in the ocean, but that practice was stopped when it was discovered that the corroding metal waste containers allow radio-isotopes to dissolve in seawater.

In 1982, the U.S. Congress ordered the Department of Energy (DOE) to build two sites for permanent **radioactive waste** disposal on land by 1998. Understandably, no one wanted this toxic material in their backyard. In 1987, after a long and highly contentious process, the DOE announced that it could only find one acceptable site: Yucca Mountain.

The storage facility would be a honeycomb of tunnels more than 1,000 ft (303.3 m) below ground. When finished, the total length of the tunnels would be more than 112 mi (180 km)—approximately the size of the world's largest subway system in New York City. Altogether, the tunnels would have room for more than 80,000 tons of nuclear waste packed in corrosion-resistant metal canisters. Because Yucca Mountain is located in one of the driest areas in the United States, and the bedrock is composed of highly stable volcanic tuff (compacted ash), the DOE hopes that storage areas will remain intact and free of leaks for the thousands of years it will take for the wastes to decay to a harmless state. Total costs for this facility have been estimated to be as high as $58 billion for construction and the first 50 years operation.

Ever since its designation as the nation's sole repository for high-level nuclear waste, Yucca Mountain has been embroiled in both scientific and political controversy. Citizens of Nevada feel that their state was chosen only because they don't have the political clout to resist. Ironically, Nevada doesn't have a **nuclear power** plant. State officials and the Nevada Congressional delegation have vowed to do everything in their power to try to stop construction and operation of the nuclear waste facility.

Scientists are divided about the safety of the site. Although some studies suggest that Yucca Mountain has been geologically stable for thousands of years, others show a history of fairly recent volcanic activity and large earthquakes in the area. Fractures and faults have been discovered in the bedrock that could allow ground water to seep into storage vaults if the **climate** changes. One hydrologist suggested that **groundwater** levels, while low now, could rise suddenly and fill the tunnels with water. Opponents of the site claim that over 200 technical problems had not been solved by the DOE by 2002.

The storage project has been continually delayed. After over $1 billion had been spent on exploratory studies and trial drilling at Yucca Mountain, the DOE announced in 1986 that it could not meet the 1998 deadline imposed by Congress. The earliest date by which the repository might be operational is now 2010. States and power companies,

having already paid more than $12 billion to the DOE to build the repository, have threatened to sue the government if it does not open as originally promised. Some utilities have begun to negotiate privately with Native American tribes and foreign countries to take their nuclear wastes.

While underground storage of nuclear wastes is fraught with problems and uncertainties, none of the alternatives now available appears better. Currently, about 45,000 tons of spent fuel is sitting at 131 sites, including 103 operating nuclear plants, in 39 states. Another 20,000 tons are expected to be generated before a storage facility can be opened in 2010, or about 2,000 tons of new waste per year. Most of this waste is being stored in deep water-filled pools inside nuclear plants. These pools were never intended for long-term storage and are now over-crowded. Many power plants have no more room for spent fuel storage and will be forced to shut down by 2010 unless some form of long-term storage is approved. In 1997, the DOE asked Congress for $100 million to purchase large steel casks in which power companies can stockpile spent fuel assemblies in outdoor storage yards. Neighbors who live near these storage sites worry about the possibility of leakage and accidents. Nevertheless, several utilities have already begun on-site **dry cask storage**. It is estimated that 165 million Americans live within a two-hour drive of stored high-level nuclear waste.

In February 2002, under the recommendation of Energy Secretary Spencer Abraham, President George W. Bush approved Yucca Mountain as the nation's nuclear waste storage site, to be opened in 2010. It would receive a minimum of 3,000 tons of nuclear waste per year for 23 years, storing up to a legal maximum of 77,000 tons. After 2007, another energy secretary may consider expanding the storage facility.

As expected, and as allowed by federal law, the state of Nevada filed a formal objection to the site approval. In May 2002 the U.S. House of Representatives voted to override Nevada's objections to the site. A major political battle was then taken to the U.S. Senate, which had until July 26, 2002 to decide upon the future of U.S. nuclear waste disposal. In July 2002 the site was approved.

The nuclear power industry is the main supporter of the storage site, in order to continue producing nuclear energy, which supplies about 20% of U.S. electricity. Opponents of the site claim that long-term storage and **transportation** of the waste is far too dangerous, given inconclusive scientific studies. Furthermore, if nuclear power is continued, opponents note that Yucca Mountain will eventually run out of space and the nuclear waste problem will still be unsolved.

Waste storage at Yucca Mountain is a complex, expensive, and potentially high-risk problem that exemplifies the complications of many environmental issues. It also demonstrates the difficulty of making democratic decisions in issues with consequences that affect large geographical areas and **future generations** in ways that are difficult to predict and potentially control.

[*William P. Cunningham Ph.D. and Douglas Dupler*]

Resources

Books

Gerrard, Michael B. *Fairness in Toxic and Nuclear Waste Siting.* Cambridge, MA: MIT Press, 1994.

Rahm, Dianne, ed. *Toxic Waste and Environmental Policy in the Twenty-first Century United States.* Jefferson, NC: McFarland & Co., 2002.

Shrader-Frechette, K. S. *Burying Uncertainty: Risk and the Case Against Geological Disposal of Nuclear Waste.* Berkeley: University of California Press, 1994.

Streissguth, Thomas, ed. *Nuclear and Toxic Waste.* San Diego, CA: Greenhaven Press, 2001.

Periodicals

"Battle of Yucca Mountain." *Washington Post*, April 30, 2002: A18.

"How Safe is Safe?" *Christian Science Monitor* April 18, 2002: 14.

Kerr, R. A. "A New Way to Ask the Experts: Rating Radioactive Waste Risks." *Science* November 8, 1996: 913.

Whipple, C. G. "Can Nuclear Waste be Stored Safely at Yucca Mountain?" *Scientific American* 274, no. 6: 72–77.

Other

Environmental Protection Agency. *Yucca Mountain Standards.* June 19, 2002 [cited July 4, 2002]. <http://www.epa.gov/radiation/yucca/index.html>.

Organizations

U.S. Department of Energy, 1000 Independence Avenue SW, Washington, DC 20585 (202) 586-6151, Fax: (202) 586-0956, <http://www.doe.gov>

Yucca Mountain Project, PO Box 364629, North Las Vegas, NV 89036-8629 (800) 225-6972, Fax: (702) 295-5222, <http://www.ymp.gov>

Z

Zebra mussel

The zebra mussel (*Dreissena polymorpha*) is a small bivalve mollusk native to the freshwater rivers draining the Caspian and Black Seas of western Asia. This **species** of shellfish, which gets its name from the dark brown stripes on its tan shell, was introduced into the **Great Lakes** of the United States and Canada and became established sometime between 1985 and 1988.

The zebra mussel was first discovered in North America in June of 1988 in the waters of Lake St. Clair. The introduction probably took place two or three years prior to the discovery, and it is believed that a freighter dumped its ballast water into Lake St. Clair, flushing the zebra mussel out in the process. By 1989 this mussel had spread west through Lake Huron into Lake Michigan and east into **Lake Erie**. Within three years it had spread to all five of the Great Lakes.

Zebra mussel. (Michigan Sea Grant Society. Reproduced by permission.)

Unlike many of its other freshwater relatives, which burrow into the sand or **silt** substrate of their **habitat**, the zebra mussel attaches itself to any solid surface. This was the initial cause for concern among environmentalists as well as the general public; these mollusks were attaching themselves to boats, docks, and water intake pipes. Zebra mussels also reproduce quickly, and they can form colonies with densities of up to 100,000 individuals per square meter. The city of Monroe, Michigan, lost its water supply for two days because its intake pipes were plugged by a huge colony of zebra mussels, and Detroit Edison spent half a million dollars cleaning them from the cooling system of its Monroe power plant. Ford Motor Company was forced to close its casting plant in Windsor, Ontario, in order to remove a colony from the pipes which send cooling water to their furnaces.

Although the sheer number of these filter feeders has actually improved **water quality** in some areas, they still pose threats to the ecological **stability** of the aquatic **ecosystem** in this region. Zebra mussels are in direct **competition** with native mussel species for both food and oxygen. Several colonies have become established on the spawning grounds of commercially important species of fish, such as the walleye, reducing their reproductive rate. The zebra mussel feeds on algae, and this may also represent direct competition with several species of fishes.

The eradication of the zebra mussel is widely considered an impossible task, and environmentalists maintain that this species will now be a permanent member of the Great Lakes ecosystem. Most officials in the region agree that money would be wasted battling the zebra mussel, and they believe that it is more reasonable to accept their presence and concentrate on keeping intake pipes and other structures clear of the creatures. The zebra mussel has very few natural predators, and since it is not considered an edible species, it has no commercial value to man. One enterprising man from Ohio has turned some zebra mussel shells into jewelry, but currently the supply far exceeds the demand.

[*Eugene C. Beckham*]

Resources
Periodicals

Griffiths, R. W., et al. "Distribution and Dispersal of the Zebra Mussel (*Dreissena polymorpha*) in the Great Lakes Region." *Canadian Journal of Fisheries and Aquatic Science* 48 (1991): 1381–1388.

Holloway, M. "Musseling In." *Scientific American* 267 (October 1992): 30–36.

Schloesser, D. W., and W. P. Kovalak. "Infestation of Unionids by *Dreissena polymorpha* in a Power Plant Canal in Lake Erie." *Journal of Shellfish Research* 10 (1991): 355–359.

Walker, T. "*Dreissena* Disaster." *Science News* 139 (May 4, 1991): 282–284.

Zebras

Zebras are striped members of the horse family (*Equidae*) native to Africa. These grazing animals stand 4–5 ft (1.2–1.5 m) high at the shoulders and are distinctive because of their striking white and black or dark brown stripes running alternately throughout their bodies. These stripes actually have important survival value for zebras, for when they are in a herd, the stripes tend to blend together in the bright African sunlight, making it hard for a lion or other predator to concentrate on a single individual and bring it down.

The zebra's best defense is flight, and it can outrun most of its enemies. It is thus most comfortable grazing and browsing in groups on the flat open plains and **grasslands**. Zebras are often seen standing in circles, their tails swishing away flies, each one facing in a different direction, alert for lions or other predators hiding in the tall grass.

Most zebras live on the grassy plains of East Africa, but some are found in mountainous areas. They live in small groups led by a stallion, and mares normally give birth to a colt every spring. Zebras are fierce fighters, and a kick can kill or cripple predators (lions, leopards, cheetahs, jackals, hyenas, and wild dogs). These predators usually pursue newborn zebras, or those that are weak, sick, injured, crippled, or very old, as those are the easiest and safest to catch and kill. Such predation helps keep the **gene pool** strong and healthy by eliminating the weak and diseased and ensuring the "survival of the fittest."

Zebras are extremely difficult to domesticate, and most attempts to do so have failed. They are often found grazing with wildebeest, hartebeest, and gazelle, since they all have separate nutritional requirements and do not compete with each other. Zebras prefer the coarse, flowering top layer of grasses, which are high in cellulose. Their trampling and grazing are helpful to the smaller wildebeest, who eat the leafy middle layer of grass that is higher in protein. And the small gazelles feed on the protein-rich young grass, shoots, and herb blossoms found closest to the ground.

One of the most famous African **wildlife** events is the seasonal 800-mi (1,287-km) **migration** in May and November of over a million zebras, wildebeest, and gazelles, sweeping across the Serengeti plains, as they have done for centuries, "in tides that flow as far as the eye can see." But it is now questionable how much longer such scenes will continue to take place.

Zebras were once widespread throughout southern and east Africa, from southern Egypt to Capetown. But **hunting** for sport, meat, and hides has greatly reduced the zebra's range and numbers, though it is still relatively numerous in the parks of East Africa. One **species**, once found in South Africa, the quagga (*Equus quagga quagga*) is already extinct, wiped out by colonists in search of hides to make grain sacks.

Several other species of zebra are threatened or endangered, such as Grevy's zebra (*Equus grevyi*) in Kenya, Ethiopia, and Somalia; Hartmann's mountain zebra (*Equus zebra hartmannae*) in Angola and Namibia; and the mountain zebra (*Equus zebra zebra*) in South Africa. The Cape mountain zebra has adapted to life on sheer mountain slopes and ravines where usually only wild goats and sheep can survive. In 1964, only about 25 mountain zebra could be found in the area, but after two decades of protection, the population had increased to several hundred in Cradock **National Park** in the Cape Province's Great Karroo area.

Probably the biggest long-term threat to the survival of the zebra is the exploding human population of Africa, which is growing so rapidly that it is crowding out the wildlife and intruding on the continent's parks and refuges. Political instability and the proliferation of high powered rifles in many countries also represent a threat to the survival of Africa's wildlife.

[*Lewis G. Regenstein*]

Resources
Periodicals

D'Alessio, V. "Born-Again Quagga Defies Extinction." *New Scientist* 132 (November 30, 1991): 14.

Grubb, P. "Equus burchelli." *Mammalian Species* no. 157, 1981.

Miller, J. A. "Telling a Quagga By Its Stripes." *Science News* 128 (August 3, 1985): 70.

Penzhorn, B. L. "*Equus zebra*." *Mammalian Species* no. 314, 1988.

Zero discharge

Zero **discharge** is the goal of eliminating discharges of pollutants by industry, government, and other agencies to air, water, and land with a view to protect both public health and the integrity of the **environment**. Such a goal is difficult to achieve except where there are point sources of **pollution**, and even then the cost of removing the last few percent of

Zebras and wildebeasts at Ngorongoro Crater, Serengeti National Park, Tanzania. (Photograph by Carolina Biological Supplies. Phototake. Reproduced by permission.)

a given pollutant may be prohibitive. However, the goal is attainable in many specific cases (e.g., the electroplating industry), and is particularly urgent in the case of extremely toxic pollutants such as **plutonium** and dioxins. In other cases, it may be desirable as a goal even though it is not likely to be completely attained.

Zero discharge was actually proposed in the early 1970s by the United States Senate as an attainable goal for the Federal **Water Pollution** Control Acts Amendments of 1972. However, industry, the White House, and other branches of government lobbied intensely against it, and the proposal did not survive the legislative process. *See also* Air quality criteria; Pollution control; Water quality standards

Zero population growth

Zero **population growth** (also called the replacement level of fertility) refers to stabilization of a population at its current level. A population growth rate of zero means that people are only replacing themselves, and that the birth and death rates over several generations are in balance. In more developed countries (MDC), where infant **mortality** rates are low, a fertility rate of about 2.2 children per couple results in zero population growth. This rate is slightly more than two because the extra fraction includes infant deaths, infertile couples, and couples who choose not to have children. In **less developed countries** (LDC), the replacement level of fertility is often as high as six children per couple.

Zero population growth, as a term, lends its name to a national, non-profit organization founded in 1968 by Paul R. Ehrlich, which works to achieve a sustainable balance between population, resources, and the **environment** worldwide. *See also* Family planning

Resources

Organizations

Zero Population Growth, 1400 16th Street NW, Suite 320, Washington, D.C. 20036

Zero risk

A concept allied to, but less stringent than, that of **zero discharge**. Zero risk permits the release to air, water, and land of those pollutants that have a **threshold dose** below which public health and the **environment** remain undamaged. Such thresholds are often difficult to determine. They may also involve a time element, as in the case of damage to vegetation by **sulfur dioxide** in which less severe fumigations over a longer period are equivalent to more severe exposures over a shorter period. *See also* Air quality criteria; Pollution control; Water quality standards

Zone of saturation

In discussions of **groundwater**, a zone of saturation is an area where water exists and will flow freely to a well, as it does in an **aquifer**. The thickness of the zone varies from a few feet to several hundred feet, determined by local geology, availability of pores in the formation, and the movement of water from recharge to points of **discharge**.

Saturation can also be a transient condition in the **soil profile** or **vadose zone** during times of high precipitation or **infiltration**. Such saturation can vary in duration from a few days or weeks to several months. In soils, the saturated zone will continue to have unsaturated conditions both above and below it; when a device used to measure saturation, called a piezometer, is placed in this zone, water will enter and rise to the depth of saturation. During these periods, the **soil** pores are filled with water, creating reducing conditions due to lowered oxygen levels. This reduction, as well as the movement of iron and manganese, creates distinctive soil patterns that allow the identification of saturated conditions even when the soil is dry. *See also* Artesian well; Drinking-water supply

Zoo

Zoos are institutions for exhibiting and studying wild animals. Many contemporary zoos have also made **environmental education** and the **conservation** of **biodiversity** part of their mission. "Zoo" is a term derived from the Greek *zoion*, meaning "living being." As a prefix it indicates the topic of animals, such as in zoology (knowledge of animals) or zoogeography (the distribution and evolutionary **ecology** of animals). As a noun it is a popular shorthand for all zoological gardens and parks.

Design and operation

Zoological gardens and parks are complex institutions involving important factors of design, staffing, economics, and politics. Design of a zoo is a compromise between the needs of the animals (e.g., light, temperature, humidity, cover, feeding areas, opportunities for natural behavior), the requirements of the staff (e.g., offices, libraries, research and veterinary laboratories, garages, storage sheds), and amenities for visitors (e.g., information, exhibitions, restaurants, rest areas, theaters, **transportation**).

Zoos range in size from as little as 2 acres (0.8 ha) to as much as 3,000 acres (1,231 ha), and animal enclosures may range from small cages to fenced-in fields. The proper size for each enclosure is determined by many factors, including the kind and number of **species** being exhibited and the extent of the enclosure's natural or naturalistic landscaping. In small zoos, space is at a premium. Cages and paddocks are often arranged into taxonomic or zoogeographic pavilions, with visitors walking among the exhibits. As a collection of animals in a confined and artfully arranged space, these zoos are called zoological gardens. In larger zoos, more space is generally available. Enclosures may be bigger, the landscaping more elaborate, and cages fewer in number. Nonpredatory species such as hoofed mammals may roam freely within the confines of the zoo's outer perimeter. Visitors may walk along raised platforms or ride monorail trains through the zoo. Lacking the confinement and precise arrangement of a garden, these zoos are called zoological parks. Some zoos take go a step farther and model themselves after Africa's national parks and **game preserves**. Visitors drive cars or ride buses through these "safari parks," observing animals from vehicles or blinds.

A zoo's many functions are reflected in its staff, which is responsible for the animals' well-being, exhibitions, environmental education, and conservation programs. There are several job categories: administrators, office personnel, and maintenance staff tend to the management of the institution. Curators are trained **wildlife** biologists responsible for animal acquisition and transfer, as well as maintaining high standards of animal care. Keepers perform the routine care of the animals and the upkeep of their enclosures. Veterinarians focus on preventive and curative medicine, guard against contagious disease, heal stress-induced ailments or accidental injuries, and preside over births and autopsies. Well-trained volunteers serve as guides and guards, answering questions and monitoring visitor conduct. Educators and scientists may also be present for the purpose of environmental education and research.

Zoos are expensive undertakings, and their economy enables or constrains their resources and practices. Few good zoos produce surplus revenue; their business is service, not profit. Most perpetually seek new sources of capital from both public and private sectors, including government subsidies, admission charges, food and merchandise sales, concessionaire rental fees, donations, bequests, and grants. The funds are used for a variety of purposes, such as daily opera-

tions, animal acquisitions, renovations, expansion, public education, and scientific research. The precise mixture of public and private funds is situational. As a general rule, public money provides the large investments needed to start, renovate, or expand a zoo. Private funds are better suited for small projects and exhibit startup costs.

A variety of public and private interests claim a stake in a zoo's mission and management. Those concerned with the politics of zoos include units of government, regulatory agencies, commercial enterprises, zoological societies, non-profit foundations, activist groups, and scholars. They influence the availability and use of zoo resources by holding the purse strings and affecting public opinion. Conflict between the groups can be intense, and ongoing disputes over the use of zoos as entertainment, the acquisition of "charismatic megafauna" (big cute animals), and the humane treatment of wildlife are endemic.

History and purposes

Zoos are complex cultural phenomena. Menageries, the unsystematic collections of animals that were the progenitors of zoos, have a long history. Particularly impressive menageries were created by ancient societies. The Chinese emperor Wen Wang (circa 1000 B.C.) maintained a 1,500 acre (607 ha) "Garden of Intelligence." The Greek philosopher Aristotle (384-322 B.C.) studied animal taxonomy from a menagerie stocked largely through the conquests of Alexander the Great (356-323 B.C.). Over 1,200 years later, the menagerie of the Aztec ruler Montezuma (circa A.D. 1515) rivaled any European collection of the sixteenth century.

In the main, these menageries had religious, recreational, and political purposes. Ptolemy II of Egypt (300-251 B.C.) sponsored great processions of exotic animals for religious festivals and celebrations. The Romans maintained menageries of bears, **crocodiles**, **elephants**, and lions for entertainment. Powerful lords maintained and exchanged wild and exotic animals for diplomatic purposes, as did Charlemagne (A.D. 768-814), the medieval king of the Franks. Dignitaries gawked at tamed cheetahs strolling the botanical gardens of royal palaces in Renaissance Europe. Indeed, political and social prestige accrued to the individual or community capable of acquiring and supporting an elaborate and expensive menagerie.

In Europe, zoos replaced menageries as scholars turned to the scientific study of animals. This shift occurred in the eighteenth century, the result of European voyages of discovery and conquest, the founding of natural history museums, and the donation of private menageries for public display. Geographically representative species were collected and used to study natural history, taxonomy, and physiology. Some of these zoos were directed by the outstanding minds of the day. The French naturalist Georges Leopold Cuvier (1769-1832) was the zoological director of the Jardin de Plants de Paris (founded in 1793), and the German geographer Alexander von Humboldt (1769-1859) was the first director of the Berlin Zoological Garden (founded in 1844). The first zoological garden for expressly scientific purposes was founded by the Zoological Society of London in 1826; its establishment marks the advent of modern zoos.

Despite their scientific rhetoric, zoological gardens and parks retained their recreational and political purposes. Late nineteenth century visitors still baited bears, fed elephants, and marveled at the exhausting diversity of life. In the United States, zoos were regarded as a cultural necessity, for they reacquainted harried urbanites with their **wilderness** frontier heritage and proved the natural and cultural superiority of North America. As if **recreation**, politics, and science were not enough, an additional element was introduced in North American zoos: conservation. North Americans had succeeded in decimating much of the continent's wildlife, driving the **passenger pigeon** (*Ectopistes migratorius*) to **extinction**, the American **bison** (*Bison bison*) to endangerment, and formerly common wildlife into rarity. It was believed that zoos would counteract this wasteful slaughter of animals by promoting their wise use. At present, zoos claim similar purposes to explain and justify their existence, namely recreation, education, and conservation.

Evaluating zoos

Critics of zoos contend they are antiquated, counterproductive, and unethical; that field ecology and **wildlife management** make them unnecessary to the study and conservation of wildlife; and that zoos distort the public's image of **nature** and animal behavior, imperil rare and endangered wildlife for frivolous displays, and divert attention from saving natural **habitat**. Finally, critics hold that zoos violate humans' moral obligations to animals by incarcerating them for a trivial interest in recreation. Advocates counter these claims by insisting that most zoos are modern, necessary, and humane, offering a form of recreation that is benign to animal and human alike and is often the only viable place from which to conduct sustained behavioral, genetic, and veterinary research.

Zoos are an important part of environmental education. Indeed, some advocates have proposed "bioparks" that would integrate aquariums, botanical gardens, natural history museums, and zoos. Finally, advocates claim that animals in zoos are treated humanely. By providing for their nutritional, medical, and security needs, zoo animals live long and dignified lives, free from hunger, disease, fear, and predation.

The arguments of both critics and advocates have merit, and in retrospect, zoos have made substantial progress since the turn of the century. In the early 1900s, many zoos collected animals like postage stamps, placing them in cramped and barren cages with little thought to their comfort. Today, zoos increasingly use large naturalistic enclo-

sures and house social animals in groups. They often specialize in zoogeographic regions, permitting the creation of habitats which plausibly simulate the **climate**, **topography**, **flora**, and **fauna** of a particular **environment**. Additionally, many zoos contribute to the protection of biodiversity by operating captive-breeding programs, propagating **endangered species**, and restoring destroyed species to suitable habitat.

This progress notwithstanding, zoos have limitations. Statements that zoos are "arks" of biodiversity, or that zoos teach people how to manage the "megazoo" called nature, greatly overstate their uses and lessons. Zoos simply cannot save, manage, and reconstruct nature once its genetic, species, and habitat diversity is destroyed. That said, the promotion by zoos of environmental education and biodiversity conservation is an important role in the defense of the natural world.

[*William S. Lynn*]

Resources

Books

Croke, V. *The Modern Ark: The Story of Zoos: Past, Present, and Future.* Collingdale: DIANE Publishing Co., 2000.

McKenna, V., W. Travers, and J. Wray, eds. *Beyond the Bars: The Zoo Dilemma.* Northamptonshire, UK: Thorsons Publishing Group, 1988.

Page, J. *Zoo: The Modern Ark.* New York: Facts On File, 1990.

Periodicals

Foose, T. J. "Erstwild & Megazoo." *Orion Nature Quarterly* 8 (Spring 1989): 60–63.

Robinson, M. H. "Beyond the Zoo: The Biopark." *Defenders* 62 (November-December 1987): 10–17.

Stott, J. R. "The Historical Origins of the Zoological Park in American Thought." *Environmental Review* 5 (Fall 1981): 52–65.

Zooplankton

Aquatic animals and protozoans whose movements are largely dependent upon currents. This diverse assemblage includes organisms that feed on bacteria, **phytoplankton**, and other zooplankton, as well as organisms that may not feed at all. Zooplankton may be divided into *holoplankton*, organisms which spend their entire lives as **plankton**, such as **krill**, and *meroplankton*, organisms that exist in the plankton for only part of their lives, such as crab larvae. Fish may live their entire lives in the water column, but are only classified as zooplankton while in their embryonic and larval stages.

ZPG

see **Environmental Stress Index**

Historical Chronology

1798 *Essay on the Principle of Population* published by Thomas Robert Malthus, in which he warned about the dangers of unchecked population growth.

1830 World population is one billion.

1849 U.S. Department of the Interior established.

1854 Henry David Thoreau publishes *Walden*, a work that inspired many people to live simply and in harmony with nature.

1864 Yosemite in California becomes the first state park in the United States.

1864 George Perkins Marsh publishes *Man and Nature*, described by some environmentalists as the fountainhead of the conservation movement.

1869 Ernst Haeckel coins the term ecology to describe "the body of knowledge concerning the economy of nature."

1872 Yellowstone in Wyoming becomes the first national park.

1875 American Forestry Association founded to encourage wise forest management.

1879 U.S. Geological Survey established.

1890 Yosemite becomes a national park.

1892 John Muir founds the Sierra Club to preserve the Sierra Nevada mountain chain.

1892 Henry S. Salt publishes *Animal Rights Considered in Relation to Social Progress*, a landmark work on animal rights and welfare.

1892 Adirondack Park established by New York State Constitution, which mandated that the region remain forever wild.

1898 Rivers and Harbors Act established in an effort to control pollution of navigable waters.

1900 Lacey Act regulating interstate shipment of wild animals in the United States is passed.

1902 U.S. Bureau of Reclamation established.

1905 National Audubon Society formed.

1908 Chlorination is used extensively in U.S. water treatment plants for the first time.

1913 Construction of Hetch-Hetchy Valley Dam approved to provide water to San Francisco; however, the dam also floods areas of Yosemite National Park.

1914 Martha, the last passenger pigeon, dies in the Cincinnati Zoo.

1916 United States National Park Service established.

1918 Save-the-Redwoods League founded.

1918 U.S. and Canada sign treaty restricting the hunting of migratory birds.

1920 Mineral Leasing Act enacted to regulate mining on federal land.

1922 Izaak Walton League founded.

1924 Gila National Forest in New Mexico is designated the first wilderness area.

1930 Dust Bowl.

1933 Tennessee Valley Authority created to assess impact of hydropower on the environment.

1934 Taylor Grazing Act enacted to regulate grazing on federal land.

1935 U.S. Soil Conservation Service established to study and curb soil erosion.

1935 Wilderness Society founded by Aldo Leopold.

1936 National Wildlife Federation established.

1943 Alaska Highway completed, linking lower United States and Alaska.

1944 Norman Borlaug begins his work on high-yielding cropvarieties.

1946 U.S. Bureau of Land Management created.

1946 Atomic Energy Commission established to study the applications of nuclear power. It was later dissolved in 1975, and its responsibilities were transferred to the Nuclear Regulatory Commission and Energy Research and Development Administration.

1947 Defenders of Wildlife founded, superseding Defenders of Furbearers and the Anti-Steel-Trap League, to protect wild animals and their habitat.

1949 Aldo Leopold publishes *A Sand CountyAlmanac*, in which he sets guidelines for the conservation movement and introduces the concept of a land ethic.

1952 Oregon becomes first state to adopt a significant program to control air pollution.

1954 Humane Society founded in United States.

1956 Construction of Echo Park Dam on the Colorado River is aborted, due in large part to the efforts of environmentalists.

1959 St. Lawrence Seaway is completed, linking the Atlantic Ocean to the Great Lakes.

1961 Agent Orange is sprayed in Southeast Asia, exposing nearly 3 million American servicemen to dioxin, a probable carcinogen.

1962 *Silent Spring* published by Rachel Carson to document the effects of pesticides on the environment.

1963 First Clean Air Act passed in the United States.

1963 Nuclear Test Ban Treaty signed by the United States and the Soviet Union to stop atmospheric testing of nuclear weapons.

1964 Wilderness Act passed, which protects wild areas in theUnited States.

1965 Water Quality Act passed, establishing federal water quality standards.

1966 Eighty people die in New York City due to pollution-related causes.

1967 Supertanker *Torrey Canyon* spills oil off the coast of England.

1967 Environmental Defense Fund established to save the osprey from DDT.

1967 American Cetacean Society founded to protect whales, dolphins, porpoises, and other cetaceans. Considered the oldest whale conservation group in the world.

1968 Wild and Scenic Rivers Act and National Trails System Act passed to protect scenic areas from development.

1969 Greenpeace founded.

1970 First Earth Day celebrated on April 22.

1970 National Environmental Policy Act passed, requiring environmental impact statements for projects funded or regulated by federal government.

1970 Environmental Protection Agency (EPA) created.

1971 Consultative Group on International Agricultural Research (CGIAR) founded to improve food production in developing countries.

1972 U.N. Conference on the Human Environment held in Stockholm to address environmental issues on a global level.

1972 Clean Water Act passed.

1972 Use of DDT is phased out in the United States.

1972 Coastal Zone Management Act and Marine Protection, Research, and Sanctuaries Act passed.

1972 Oregon becomes first state to enact bottle-recycling law.

1972 *Limits to Growth* published by the Club of Rome, calling for population control.

1973 Convention on International Trade in Endangered Species of Wild Fauna and Flora (CITES) signed to prevent the international trade of endangered or threatened animals and plants.

1973 Endangered Species Act passed.

1973 Arab members of the Organization of Petroleum Exporting Countries (OPEC) institute an embargo preventing shipments of oil to the United States.

1973 Cousteau Society founded by Jacques-Yves Cousteau and his son to educate the public and conduct research on marine-related issues.

1973 E. F. Schumacher publishes *Small Is Beautiful*, which advocates simplicity, self-reliance, and living in harmony with nature.

1974 Safe Drinking Water Act passed, requiring EPA to set quality standards for the nation's drinking water.

1975 Atlantic salmon is found in the Connecticut River after a 100-year absence.

1975 *The Monkey Wrench Gang* published by Edward Abbey, who advocates radical and controversial methods for protecting the environment, including "ecotage."

1976 Resource Conservation and Recovery Act passed, giving EPA authority to regulate municipal solid and hazardous waste.

1976 Poisonous gas containing 2,4,5-TCP and dioxin is released from a factory in Seveso, Italy, causing massive animal and plant death. Although no human life was lost, a sharp increase in deformed births was reported.

1976 Land Institute founded by Wes and Dana Jackson to encourage more natural and organic agricultural practices.

1978 Residents of Love Canal, New York, are evacuated after Lois Gibbs discovers that the community was once the site of a chemical waste dump.

1978 Oil tanker *Amoco Cadiz* runs aground, spilling 220,000 tons of oil.

1979 Three Mile Island Nuclear Reactor almost undergoes nuclear melt-down when the cooling water systems fail. Since this accident no new nuclear power plants have been built in the United States.

1980 Mount St. Helens explodes with a force comparable to 500 Hiroshima-sized bombs.

1980 Comprehensive Environmental Response, Compensation, andLiability Act (Superfund) enacted to clean up abandoned toxic waste sites.

1980 *The Global 2000 Report* published, documenting trends in population growth, natural resource depletion, and the environment.

1980 Alaska National Interest Lands Conservation Act enacted, setting aside millions of acres of land as wilderness.

1980 Thomas Lovejoy proposes the idea of debt-for-nature swap that helps developing countries alleviate national debt by implementing policies to protect the environment.

1980 Earth First! founded by Dave Foreman, with the slogan "No compromise in the defense of Mother Earth."

1982 Bioregional Project founded to promote the aims of the bioregional movement in North America.

1984 Emission of poisonous *methylisocyanate* vapor, a chemical by-product of agricultural insecticide production, from the Union Carbide plant kills more than 2,800 people in Bhopal, India.

1985 Rainforest Action Network founded.

1985 Ozone hole observed over Antarctica.

1986 Chernobyl Nuclear Power Station undergoes nuclear coremelt-down, spreading radioactive material over vast parts of the Soviet Union and northern Europe.

1986 Evacuation of Times Beach, Missouri, due to high levels of dioxin.

1987 *Ecodefense: A Field Guide toMonkeywrenching* published by Dave Foreman, in which he describes

spiking trees and other "environmental sabotage" techniques.

1987 *Our Common Future* (The Brundtland Report) is published.

1987 Montreal Protocol on Substances that Deplete the Ozone Layer signed by 24 nations, declaring their promise to decrease production of chlorofluorocarbons(CFCs).

1987 Yucca Mountain designated the first permanent repository for radioactive waste by the U.S. Department of Energy.

1987 World population is five billion.

1988 Ocean Dumping Ban Act established.

1988 Global ReLeaf program inaugurated with the motto "Plant a tree, cool the globe" to address the problem of global warming.

1989 Oil tanker *Exxon Valdez* runsaground in Prince William Sound, Alaska, spilling 11 million gallons of oil.

1990 Oil Pollution Act signed, setting liability and penalty system for oil spills as well as a trust fund for clean up efforts.

1990 Clean Air Act amended to control emissions of sulfur dioxide and nitrogen oxides.

1991 Mount Pinatubo in Philippines erupts, shooting sulfur dioxide 25 miles into the atmosphere.

1991 Persian Gulf War.

1991 Train containing the pesticide *metasodium* falls off the tracks near Dunsmuir, California, releasing chemicals into the Sacramento River. Plant and aquatic life for 43 miles downriver die as a result.

1991 Over 4,000 people die from cholera in Latin Americanepidemic.

1992 U.N. Earth Summit held in Rio de Janeiro, Brazil.

1992 Mexico City, Mexico, shuts down as a result of incapacitating air pollution.

1992 Captive-bred California condors and black-footed ferrets reintroduced into the wild.

1992 United Nations calls for an end to global drift net fishing by the end of 1992.

1993 *Braer* oil tanker runs aground in the Shetland Islands, Scotland, spilling its entire cargo into the sea.

1993 Forest Summit convened in Portland, Oregon, by President Bill Clinton, who met with loggers and environmentalists concerned with the survival of the northern spotted owl.

1993 Norway resumes hunting of minke whales in defiance of a ban on commercial whaling instituted by the International Whaling Commission.

1993 Eight people from *Biosphere 2* emerge after living two years in a self-sustaining, glass dome.

1995 Ken-Sara Wiwa is executed in Nigeria for protesting and speaking out about oil industry practices in the country. Shell Oil Co. takes criticism for its role in the matter.

1997 Forest fires worldwide burn a total of five million hectares of forest

1997 Julia Butterfield Hill climbs a 180 ft (55 m) redwood tree in California to protest the logging of the surrounding forest as well as to protect the tree. The tree was eventually named Luna and Julia Butterfield Hill removed herself from the tree in 1999 after she negotiated a deal to save the tree and an additional three acres of the forest.

1997 Dolly, the world's first cloned sheep, is born.

1997 Kyoto Protocol mandates a reduction of reported 1990 emissions levels by 6–8% by 2008.

1997 Monserrat volcano erupts.

1999 World population reaches six billion.

1999 World Trade Organization (WTO) conference in Seattle, Washington, is marked by heavy protests, highlighting WTO's weak environmental policies.

2000 During his presidency, Bill Clinton appropriated a total of 58 million acres of wilderness as conservation land—the largest amount of land to be set aside for conservation by any other president to date.

2000 Russian nuclear submarine the *Kursk* sinks off the coast of Minsk, Russia.

2000 West Nile virus discovered in the eastern United States.

2001 Draft of human genome sequence published.

2001 The United States does not ratify the Kyoto Protocol.

2001 The World Trade Center towers in New York collapse after being struck by two commercial airplanes commandeered by terrorists. A third airplane is crashed into the Pentagon building just outside Washington, D.C., causing loss of life and major damage to the building.

2001 War on Terrorism begins.

2002 U.N. Earth Summit held in Johannesburg, South Africa.

2002 EPA adopts California emissions standards for off-road recreation vehicles to be implemented by 2004.

2002 President George W. Bush introduces the Clear Sky Initiative that will restrict the amounts of sulfur dioxide (SO2), nitrogen oxides (NOx), and mercury emitted into the atmosphere by industry. If passed, the Clear Sky plan will build upon the Clean Air Act.

2002 EPA announces its Strategic Plan for Homeland Security to support the National Strategy for Homeland Security enacted after the September 11, 2001; terrorist attacks in the United States.

2002 Three Gorges Dam is 70% complete. Filling of the reservoir will begin in 2003 and will demand the relocation of 1.13 million people. Critics worry about the stability of the dam, as well as the ecological effects it will have when fully operational in 2009.

2002 President George W. Bush signs Bill approving the using of Yucca Mountain as a nuclear waste storage site.

2002 Severe Acute Respiratory Syndrome (SARS) virus is found in patients in China, Hong Kong, and other Asian countries. The newly discovered coronavirus is not identified until early 2003. The spread of the virus reaches epidemic proportions in Asia and expands to the rest of the world.

2003 EPA Administer Christine Todd Whitman resigns after only two years in office. The Bush Administration's resistance to environmental progress is cited as one of the primary reasons for the resignation.

Environmental Legislation In The United States

1862 **Homestead Act** makes free homesteads on unappropriated land available on a vast scale, speeding settlement of the Plains states and destruction of the region's prairie ecosystem.

1872 **Mining Law** allows any person finding mineral deposits on public land to file a claim which grants him free access to that site for mining or similar development.

1891 **Forest Reserve Act** authorizes the President to create forest reserves from the public domain.

1899 **Refuse Act** authorizes the Army Corps of Engineers to issue discharge permits.

1900 **Lacey Act** makes interstate shipment of wild animals a federal offense if taken in violation of state laws.

1902 **Reclamation Act** provides funding for the "reclamation" of drylands in the western United States through irrigation and damming of rivers.

1916 **National Park Service Act** establishes the National Park Service, an agency in the U.S. Department of the Interior, the first such agency in the world.

1918 **Migratory Bird Treaty Act** establishes federal authority over migratory game and insectivorous birds.

1920 **Mineral Leasing Act** regulates the exploitation of fuel and fertilizer minerals on public lands.

1934 **Migratory Bird Hunting Stamp Act** provides funding for waterfowl reserves.

1934 **Taylor Grazing Act** establishes federal management policy on public grazing lands, the last major category of public lands to come under the active supervision of the government.

1935 **Soil Conservation Act** establishes the Soil Conservation Service to aid in combating soil erosion following the Dust Bowl.

1938 **Food, Drug and Cosmetics Act** sets extensive standards for the quality and labeling of foods, drugs, and cosmetics.

1938 **Natural Gas Act** gives the Federal Power Commission the right to control prices and limit new pipelines from entering the market.

1948 **Water Pollution Control Act** is the first statute to provide state and local governments with the funding to address water pollution.

1954 **Atomic Energy Act** grants the federal government exclusive regulatory authority over nuclear-power facilities.

1955 **Air Pollution Control Act** grants funds to assist the states in their air pollution control activities.

1956 **Federal Water Pollution Control Act** (amended 1965, 1966, 1970, 1972) increases federal funding to state and local governments to address water quality issues and calls for the development of water quality standards by the newly-created Federal Water Pollution Control Administration.

1957 **Price-Anderson Act** limits the liability of civilian producers of nuclear power in the case of a catastrophic nuclear accident.

1960 **Hazardous Substances Act** (amended 1966) authorizes the Secretary of the Department of Health, Education, and Welfare (HEW) to require warning labels for household substances deemed hazardous.

1960 **Multiple Use-Sustained Yield Act** mandates that management of the national forests be balanced between ecological and economic interests.

1963 **Clean Air Act** (amended 1970, 1977, 1990) serves as the backbone of efforts to control air pollution in the United States.

1964 **Classification and Multiple Use Act** instructs the Bureau of Land Management to inventory its lands and classify them for disposal or retention, the first such inventory in the United States.

1964 **Public Land Law Review CommissionAct** establishes a commission to examine the body of public land laws and make recommendations as to how to proceed in this policy area.

1964 **Wilderness Act** creates the National Wilderness Preservation System to preserve wilderness areas for present and future generations of Americans.

1965 **Shoreline Erosion Protection Act** provides assistance to protect shorelines and stream banks from erosion.

1965 **Solid Waste Disposal Act** (amended 1970, 1976) addresses inadequate solid waste disposal methods. It was amended by the **Resource Recovery Act** and the **Resource Conservation and Recovery Act**.

1966 **Animal Welfare Act** designates U.S. Department of Agriculture as responsible for the humane care and handling of warm-blooded and other animals used for biomedical research and calls for inspection of research facilities to insure that adequate food, housing, and care are provided.

1966 **Laboratory Animal Welfare Act** (amended 1970, 1976, 1985, 1990) regulates the use of animals in medical and commercial research.

1968 **Wild and Scenic Rivers Act** creates three categories of rivers—wild, scenic, and recreational—and provides for safeguards against degradation of wild and scenic rivers.

1969 **Coal Mine Health and Safety Act** addresses safety problems in the mining industry.

1970 **National Environmental Policy Act** (NEPA) ushers in a new era of environmental awareness in the United States, requiring all federal agencies to take into account the environmental consequences of their plans and activities.

1970 **National Mining and Minerals Act** directs the Secretary of the Interior to follow a policy that encourages the private mining sector to develop a financially viable mining industry while conducting research to further "wise and efficient use" of these minerals.

1970 **Occupational Safety and Health Act** (OSHA) requires employers to provide each of their employees with a workplace that is free from recognized hazards, which may cause death or serious physical harm.

1970 **Pollution Prevention Packaging Act**

1970 **Resource Recovery Act**, anamendment to **Solid Waste Disposal Act**, funds recycling programs and mandates an extensive assessment of solid waste disposal practices.

1972 **Clean Water Act** (amended 1977, 1987), which replaced the language of the **Federal Water Pollution Control Act**, is the farthest reaching of all federal water legislation, setting as a national goal the attainment of "fishable and swimmable" quality for all surface waters in the United States.

1972 **Coastal Zone Management Act** (reauthorized 2001) establishes a federal program to help states in planning and managing the development and protection of coastal areas.

1972 **Federal Insecticide, Fungicide, and Rodenticide Act** regulates the registration, marketing and use of pesticides.

1972 **Marine Mammals Protection Act** (amended 1988 and reauthorized 2002) protects, conserves and encourages research on marine animals. It places a moratorium on harassing, hunting, capturing, killing or importing marine mammals with some exceptions (e.g. subsistence hunting by Eskimos).

1972 **Marine Protection, Research and Sanctuaries Act** (amended 1990) (also known as the **Ocean Dumping Act**) regulates ocean dumping, authorizes marine pollution research, and establishes regional marine research programs. It also establishes a process for designating marine sanctuaries of significant ecological, aesthetic, historical, or recreational value. The **Ocean Dumping Ban Act**

amends the law, prohibiting sewage sludge and industrial wastes from being dumped at sea after December 31, 1991.

1972 **Ports and Safe Waterways Act**regulates oil transport and the operation of oil handling facilities.

1973 **Endangered Species Act** empowers the Secretary of the Interior to designate any plant or animal (including subspecies, races and local populations) "endangered" (imminent danger of extinction) or "threatened" (significant decline in numbers and danger of extinction in some regions). The act protect habitats critical to thesurvival of endangered species and prohibits hunting, killing, capturing, selling, importing or exporting products from endangered species.

1974 **Federal Non-Nuclear Research and Development Act** focuses government efforts on non-nuclear research.

1974 **Forest and Rangeland Renewable Resources Planning Act** establishes a process for assessing the nation's forest and range resources every ten years. It also stipulates that every five years the Forest Service provide a plan for the use and development of these resources based on the assessment.

1974 **Safe Drinking Water Act** (amended 1996) requires minimum safety standards for every community water supply, regulating such contaminants as bacteria, nitrates, arsenic, barium, cadmium, chromium, fluoride, lead, mercury, silver, and pesticides.

1974 **Solar Energy Research, Development andDemonstration Act** establishes a federal policy to "pursue a vigorous and viable program of research and resource assessment of solar energy as a major source of energy for our national needs." The act also establishes two programs: the Solar Energy Coordination and Management Project and the Solar Energy Research Institute.

1975 **Eastern Wilderness Act** creates 16 eastern wilderness areas and directs the Forest Service to alter its methods so that it considered areas previously affected by humans.

1975 **Hazardous Materials TransportationAct** establishes minimum standards of regulation for the transport of hazardous materials by air, ship, rail, and motor vehicle.

1976 **Energy Policy and Conservation Act**, an amendment to **The Motor Vehicle Information and Cost Savings Act**, outlines provisions intended to decrease fuel consumption. The most significant provisions are the Corporate Average Fuel Economy (CAFE) standards, which set fuel economy standards for passenger cars and light trucks.

1976 **Federal Land Policy and ManagementAct** gives the Bureau of Land Management the authority and direction for managing the lands under its control. The act also sets policy for the grazing, mining and preservation of public lands.

1976 **Magnuson-Stevens Fishery Conservation and Management Act** manages fishing activities in waters extending from the edge of State waters to the 200-mile limit. Its goal is to phase out foreign fishing within the area adjacent to the United States coastline.

1976 **National Forest Management Act** grants the Forest Service significant administrative discretion in managing the logging of national forests based on the general philosophies of multiple use and sustained yield.

1976 **Resource Conservation and Recovery Act** (RCRA), an amendment to the **Solid Waste DisposalAct**, regulates the storage, shipping, processing, and disposal of hazardous substances and sets limits on the sewering of toxic chemicals.

1976 **Toxic Substances Control Act**(TSCA) categorizes toxic and hazardous substances and regulates the use and disposal of poisonous chemicals.

1977 **Surface Mining Control and Reclamation Act** limits the scarring of the landscape, erosion and water pollution associated with surface mining. It empowers the Department of the Interior to develop regulations that impose nationwide environmental standards for all surface-mining operations.

1978 **Energy Tax Act** creates a "Gas Guzzler" tax on passenger cars whose individual fuel economy value fall below a certain threshold, starting in 1980.

1978 **Port and Tanker Safety Act**empowers the U.S. Coast Guard to supervise vessel and port operations and to set standards for the handling of dangerous substances.

1978 **Public Utilities Regulatory PoliciesAct** promotes the development of renewable energy and requires that utilities purchase power at "just" rates from producers who use such alternative sources as wind and solar.

1978 **Uranium Mill-Tailings Radiation Control Act** orders Department of Energy to stabilize and maintain 24 uranium processing sites and approximately 5,000 vicinity sites that are known sources of wind-blown ore tailings.

1980 **Alaska National Interest Lands Conservation Act** protects 44 million ha (109 million acres) of land in Alaska, establishing 11 new parks, 12 new wildlife refuges and setting aside wilderness areas comprising 22.7 million ha (56 million acres).

1980 **Comprehensive Environmental Response,Compensation, and Liability Act** (Superfund) (amended 1986) permits direct federal response to remedy the improper disposal of hazardous waste. It establishes a multi-billion dollar cash pool (Superfund) to finance government clean-up actions.

1980 **Low-Level Radioactive Waste PolicyAct** (amended 1985) places responsibility for disposal of low-level radioactive waste on generating states and encourages those states to create compacts to develop centralized regional waste sites.

1980 **Nongame Wildlife Act** provides limited federal aid to state wildlife agencies.

1982 **Nuclear Waste Policy Act** (amended 1987) instructs the Department of Energy to develop a permanent repository for high-level nuclear wastes by 1998.

1986 **Asbestos Hazard Emergency Response Act** requires schools to inspect its building for evidence of asbestos and provides guidelines for clean-up if any asbestos is found.

1986 **Emergency Planning and CommunityRight-to-Know Act** requires federal, state and local governments and industry to work together in developing plans to deal with chemical emergencies and community right-to-know reporting on hazardous chemicals.

1987 **Marine Plastic Pollution Research and Control Act** prohibits the dumping of plastics at sea, severely restricts the dumping of other ship-generated garbage in the open ocean or in U.S. waters, and requires all ports to have adequate garbage disposal facilities for incoming vessels.

1987 **National Appliance Energy ConservationAct** sets minimum efficiency standards for heating and cooling systems in new homes as well as for such new appliances as refrigerators and freezers.

1988 **Alternative Motor Fuels Act** encourages automobile manufactures to design and build cars that can burn alternative fuels such as methanol and ethanol.

1988 **Indoor Radon Abatement Act** calls for indoor air to be as free of radon as outdoor ambient air in as much as it can be.

1988 **Lead Contamination Control Act** recalls and bans the continued sales of lead-lined water coolers.

1988 **Medical Waste Tracking Act** creates a "cradle-to-grave" tracking system based on detailed shipping records, similar to the program in place for hazardous waste.

1988 **Ocean Dumping Ban Act** prohibits dumping of any sewage sludge or industrial waste into ocean waters and penalizes those who do so.

1988 **Shoreline Protection Act**

1990 **National Environmental Education Act** establishes a program within the EPA to increase environmental education in the United States.

1990 **Oil Pollution Act**, passed in response to the *Exxon Valdez* disaster, initiates a comprehensive federal liability system for all oil spills. It also establishes a federal trust fund to help pay for cleanups, strengthens civil and criminal penalties against parties involved in spills, and requires companies to have spill contingency and readiness plans.

1990 **Pollution Prevention Act** creates a new EPA office intended to help industry limit pollution through information collection, assistance in technology transfer, and financial assistance to state pollution prevention programs.

1992 **Federal Facility Compliance Act** requires the military to clean up its nuclear waste sites.

1996 **Food Quality Protection Act** amends Insecticide, Fungicide, and Rodenticide Act (FIFRA) and the Federal Food Drug, and Cosmetic Act by requiring

that all pesticides used on food products be of no harm to the population.

1996 **National Invasive Species Act** attempts to prevent the introduction of invasive marine species into the Great Lakes from ballast water.

1998 **Harmful Algal Bloom and Hypoxia Research and Control Act** identifies the need for a comprehensive effort to research and monitor algal blooms in an attempt to control and ultimately prevent them from occurring.

1999 **Chemical Safety Information, Site Security, and Fuels Regulatory Relief Act**

Organizations

The following is an alphabetical compilation of organizations listed in the *Resources* section of the main body entries. Although the list is comprehensive, it is by no means exhaustive. It is a starting point for further information, as well as other online and print sources. Many of the organizations listed have links to additional related websites. E-mail addresses and web addresses listed were provided by the associations; Gale Group is not responsible for the accuracy of the addresses or the contents of the websites.

A

African Wildlife Foundation
1400 16th Street, NW
Washington, DC, USA 20036
Phone: (202) 939-3333
Fax: (202) 939-3332
e-mail: africanwildlife@awf.org
website: <http://www.awf.org>

Agricultural Stabilization and Conservation Service
10500 Buena Vista Court
Urbandale, IA, USA 50322-3782
Phone: (515) 254-1540
Fax: (515) 254-1573

Air & Waste Management Association
420 Fort Duquesne Blvd, One Gateway Center
Pittsburgh, PA, USA 15222
Phone: (412) 232-3444
Fax: (412) 232-3450
e-mail: info@awma.org
website: <http://www.awma.org>

The Alaska Coalition
419 6th St, #328
Juneau, AK, USA 99801
Phone: (907) 586-6667
Fax: (907) 463-3312
e-mail: info@alaskacoalition.org
website: <http://www.alaskacoalition.org>

Alliance for Nuclear Accountability
1914 N. 34th St., Ste. 407
Seattle, WA, USA 98103
Phone: 206-547-3175
Fax: 206-547-7158
e-mail: ananuclear@earthlink.net
website: <http://www.ananuclear.org>

Alliance for the Prudent Use of Antibiotics
75 Kneeland Street
Boston, MA, USA 02111
Phone: (617) 636-0966
Fax: (617) 636-3999
website: <http://www.healthsci.tufts.edu/apua>

America's Clean Water Foundation
750 First Street NE, Suite 1030
Washington, DC, USA 20002
Phone: (202) 898-0908
Fax: (202) 898-0977
website: <http://yearofcleanwater.org/>

American Cetacean Society
P.O. Box 1391
San Pedro, CA, USA 90733-1391
Phone: (310) 548-6279
Fax: (310) 548-6950
e-mail: acs@pobox.com
website: <http://www.acsonline.org>

American Chemical Society
1155 Sixteenth St. NW
Washington, DC, USA 20036
Phone: (202) 872-4600
Fax: (202) 872-4615
Toll Free: (800) 227-5558
e-mail: help@acs.org
website: <http://www.chemistry.org>

American Dental Association
211 E. Chicago Avenue
Chicago, IL, USA 60611
Phone: (312) 440-2500
Fax: (312) 440-2800
e-mail: publicinfo@ada.org
website: <http://www.ada.org>

American Eagle Foundation
P.O. Box 333
Pigeon Forge, TN, USA 37868
Phone: (865) 429-0157
Fax: (865) 429-4743
Toll Free: (800) 2EAGLES
e-mail: EagleMail@Eagles.Org
website: <http://www.eagles.org>

The American Farmland Trust (AFT)
1200 18th Street, NW, Suite 800
Washington, DC, USA 20036
Phone: (202) 331-7300
Fax: (202) 659-8339
e-mail: info@farmland.org
website: <http://www.farmland.org>

American Forests
P.O. Box 2000
Washington, DC, USA 20013
Phone: (202) 955-4500
Fax: (202) 955-4588
e-mail: info@amfor.org
website: <http://www.americanforests.org>

American Indian Environmental Office
1200 Pennsylvania Avenue, NW
Washington, DC, USA 20460
Phone: (202) 564-0303
Fax: (202) 564-0298
website: <http://www.epa.gov/indian>

American Oceans Campaign
600 Pennsylvania Avenue SE, Suite 210
Washington, DC, USA 20003
Phone: (202) 544-3526
Fax: (202)544-5625
e-mail: info@americanoceans.org
website: <http://www.americanoceans.org>

American Petroleum Institute
1220 L Street, NW
Washington, DC, USA 20005-4070
Phone: (202) 682-8000
website: <http://www.api.org>

American Public Works Association
2345 Grand Blvd., Suite 500
Kansas City, MO, USA 64108-2641
Phone: (816) 472-6100
website: <http://www.pubworks.org>

American Rivers
1025 Vermont Ave., NW Suite 720
Washington, DC, USA 20005
Phone: (202) 347-7550
Fax: (202) 347-9240
e-mail: amrivers@amrivers.org
website: <http://www.amrivers.org>

American Wildlands
40 East Main #2
Bozeman, MT, USA 59715
Phone: (406) 586-8175
Fax: (406) 586-8242
e-mail: info@wildlands.org
website: <http://www.wildlands.org>

American Wind Energy Association
122 C Street, NW, Suite 380
Washington, DC, USA 20001
Phone: (202) 383-2500
Fax: (202) 383-2505
e-mail: windmail@awea.org
website: <http://www.awea.org>

Animal Legal Defense Fund
127 Fourth Street
Petaluma, CA, USA 94952
Fax: (707) 769-7771
Toll Free: (707) 769-0785
e-mail: info@aldf.org
website: <http://www.aldf.org>

Animal Welfare Institute
P.O. Box 3650
Washington, DC, USA 20007
Phone: (202) 337-2332
e-mail: awi@awionline.org
website: <http://www.awionline.org>

The Antarctica Project
1630 Connecticut Ave., NW, 3rd Floor
Washington, DC, USA 20009
Phone: (202) 234-2480
e-mail: antarctica@igc.org
website: <http://www.asoc.org>

Arctic Council, Ministry for Foreign Affairs, Unit for Northern Dimension
P.O. Box 176
Helsinki, Finland FIN-00161
Phone: +358 9 1605 5562
Fax: +358 9 1605 6120
e-mail: Petri.Ojanpera@formin.fi
website: <http://www.arctic-council.org>

Argonne National Laboratory
9700 S. Cass Avenue
Argonne, IL, USA 60439
Phone: (630) 252-2000
website: <http://www.anl.gov>

Association of National Estuary Programs
4505 Carrico Drive
Annandale, VA, USA 22003
Phone: (703) 333-6150
Fax: (703) 658-5353
website: <http://www.anep-usa.org>

Atlantic States Marine Fisheries Commission
1444 Eye Street, NW, Sixth Floor
Washington, DC, USA 20005
Phone: (202) 289-6400
Fax: (202) 289-6051
e-mail: comments@asmfc.org
website: <http://www.asmfc.org>

The ATSDR Information Center
Phone: (404) 498-0110
Fax: (404) 498-0057
Toll Free: (888) 422-8737
e-mail: ATSDRIC@cdc.gov
website: <http://www.atsdr.cdc.gov>

B

Beltsville Agricultural Research Center
Rm. 223, Bldg. 003, BARC-West, 10300 Baltimore Ave.
Beltsville, MD, USA 20705
website: <http://www.ars.usda.gov>

Beyond Pesticides
701 E Street, SE, Suite 200
Washington, DC, USA 20003
Phone: (202) 543-5450
Fax: (202) 543-4791
e-mail: info@beyondpesticides.org
website: <http://www.beyondpesticides.org>

BirdLife International
Wellbrook Court, Girton Road
Cambridge, United Kingdom CB3 0NA
Phone: +44 1 223 277 318
Fax: +44 1 223 277200
e-mail: birdlife@birdlife.org.uk
website: <http://www.birdlife.net>

British Government Committee of Inquiry into Hunting with Dogs in England and Wales
England
e-mail: huntingwithdogs@defra.gsi.gov.uk
website: <http://www.huntinginquiry.gov.uk/mainsections/huntingframe.htm>

Bureau of Reclamation
1849 C Street NW
Washington, DC, USA 20240-0001
website: <http://www.usbr.gov>

Bushmeat Crisis Task Force
8403 Colesville Road, Suite 710
Silver Spring, MD, USA 20910-3314
Fax: (301) 562-0888
e-mail: info@bushmeat.org
website: <http://www.bushmeat.org>

The Bushmeat Project, The Biosynergy Institute
P.O. Box 488
Hermosa Beach, CA, USA 90254
e-mail: bushmeat@biosynergy.org
website: <http://www.bushmeat.net>

The Bushmeat Research Programme, Institute of Zoology, Zoological Society of London
Regents Park
London, UK NW1 4RY
Phone: 44-207-449-6601
Fax: 44-207- 586-2870
e-mail: enquiries@ioz.ac.uk
website: <http://www.zoo.cam.ac.uk/ioz/projects/bushmeat.htm>

C

Canadian Forest Service, Natural Resources Canada
580 Booth Street, 8th Floor
Ottawa, Ontario, Canada K1A 0E4
Phone: (613) 947-7341
Fax: (613) 947-7397
e-mail: cfs-scf@nrcan.gc.ca
website: <http://www.nrcan-rncan.gc.ca/cfs-scf>

Canadian Wildlife Service, Environment Canada
Ottawa, Ontario, Canada K1A 0H3
Phone: (819) 997-1095
Fax: (819) 997-2756
e-mail: cws-scf@ec.gc.ca
website: <http://www.cws-scf.ec.gc.ca>

Cape Cod Commercial Hook Fishermen's Association
210 Orleans Road
North Chatham, MA, USA 02650
Phone: (508) 945-2432
Fax: (508) 945-0981
e-mail: enichols@ccchfa.org
website: <http//www.ccchfa.org>

The Carter Center
One Copenhill 453 Freedom Parkway
Atlanta, GA, USA 30307
e-mail: carterweb@emory.edu
website: <http://www.cartercenter.org>

Catskill Watershed Corporation
PO Box 569 Main St.
Margaretville, NY, USA 12455
Phone: (845) 586-1400
Fax: (845)586-1401
Toll Free: (877) 928-7433
e-mail: invest@cwconline.org
website: <http://www.cwconline.org>

Center for Environmental Philosophy, University of North Texas
370 EESAT, P.O. Box 310980
Denton, TX, USA 76203-0980
Phone: (940) 565-2727
Fax: (940) 565-4439
e-mail: cep@unt.edu
website: <http://www.cep.unt.edu>

The Center for Marine Conservation
1725 DeSales Street, NW, Suite 600
Washington, DC, USA 20036
Phone: (202) 429-5609
Fax: (202) 872-0619
e-mail: cmc@dccmc.org
website: <http://www.cmc-ocean.org>

Center for Respect of Life and Environment
2100 L Street, NW
Washington, DC, USA 20037
Phone: (202) 778-6133
Fax: (202) 778-6138
e-mail: info@crle.org
website: <http://www.crle.org>

Center for Rural Affairs
101 S Tallman Street, P.O. Box 406
Walthill, NE, USA 68067
Phone: (402) 846-5428
Fax: (402) 846-5420
e-mail: info@cfra.org
website: <http://www.cfra.org>

Center for Science in the Public Interest
1875 Connecticut Ave., NW, Suite 300
Washington, DC, USA 20009
Phone: (202) 332-9110
Fax: (202) 265-4954
e-mail: cspi@cspinet.org
website: <http://www.cspinet.org>

Center for the Study of Bioterrorism and Emergency Infections—Saint Louis University
3545 Lafayette, Suite 300
St. Louis, MO, USA 63104
website: <http://bioterrorism.slu.edu>

The Center for Urban Ecology
4598 MacArthur Blvd., NW
Washington, DC, USA 20007-4227
Phone: 202-342-1443
Fax: 202-282-1031
website: <http://www.nps.gov/cue/cueintro.html>

Centers for Disease Control and Prevention
1600 Clifton Road
Atlanta, GA, USA 30333
Phone: (404) 639-3534
Toll Free: (800) 311-3435
website: <http://www.cdc.gov>

Centers for Disease Control and Prevention - Bioterrorism Preparedness & Response Program (CDC)
1600 Clifton Road
Atlanta, GA, USA 30333
Phone: (404) 639-3534
Toll Free: (888) 246-2675
e-mail: cdcresponse@ashastd.org
website: <http://www.bt.cdc.gov>

Central Plains Center for BioAssessment
2021 Constant Avenue
Lawrence, KA, USA 66047-3729
Phone: (785)864-7729
e-mail: dbaker@ukans.edu
website: <http://www.cpcb.ukans.edu>

CGIAR Secretariat, The World Bank
MSN G6-601, 1818 H Street NW
Washington, DC, USA 20433
Phone: (202) 473-8951
Fax: (202) 473-8110
e-mail: cgiar@cgiar.org
website: <http://www.cgiar.org>

Chesapeake Bay Foundation, Philip Merrill Environmental Center
6 Herndon Avenue
Annapolis, MD, USA 21403
Phone: 410/268-8816
website: <http://www.cbf.org/about_cbf/contact_us.htm>

Chlorine Chemistry Council (CCC)
1300 Wilson Boulevard
Arlington, VA, USA 22209
Phone: 703) 741-5000
website: <http://c3.org/index.html>

CITES Secretariat, International Environment House
Chemin des Anémones
Châtelaine, Geneva, Switzerland CH-1219
Phone: (+4122) 917-8139/40
Fax: (+4122) 797-3417
e-mail: cites@unep.ch
website: <http://www.cites.org>

Citizens for a Better Environment
1845 N. Farwell Ave., Suite 220
Milwaukee, WI, USA 53202
Phone: (414) 271-7280
Fax: (866) 256-5988
Toll Free: (414) 271-5904
e-mail: cbewi@cbemw.org
website: <http://www.cbemw.org>

Citizens for Health
5 Thomas Circle, NW, Suite 500
Washington, DC, USA 20005
Phone: (202) 483-1652
Fax: (202) 483-7369
e-mail: cfh@citizens.org
website: <http://www.citizens.org>

Clean Water Action Council
1270 Main Street, Suite 120
Green Bay, WI, USA 54302
Phone: (920) 437-7304
Fax: (920) 437-7326
e-mail: CleanWater@cwac.net
website: <http://www.cwac.net/index.html>

The Club of Rome
Rissener Landstr 193
Hamburg, Germany 22559
Phone: +49 40 81960714
Fax: +49 40 81960715
e-mail: mail@clubofrome.org
website: <http://www.clubofrome.org>

The Coastal Society
P.O. Box 25408
Alexandria, VA, USA 22313-5408
Phone: (703) 768-1599
Fax: (703) 768-1598
e-mail: info@thecoastalsociety.org
website: <http://www.thecoastalsociety.org>

Coastal Waters Project/Task Force Atlantis
418 Main Street
Rockland, ME, USA 04841
Phone: (207) 594-5717
e-mail: coastwatch@acadia.net
website: <http://www.atlantisforce.org>

Coastal Zone Management Program, OCRM, NOS, NOAA
Coastal Programs Division, N/ORM3; 1305 East-West Highway, SSMC4
Silver Spring, MD, USA 20910
Phone: (301) 713-3155
Fax: (301) 713-4367
website: <http://www.ocrm.nos.noaa.gov/czm/>

Commission for Environmental Cooperation
393, rue St-Jacques Ouest, Bureau 200
Montréal, Québec, Canada H2Y 1N9
Phone: (514) 350-4300
Fax: (514) 350-4314
e-mail: info@ccemtl.org
website: <http://www.cec.org>

Conservation International
1919 M Street, NW, Suite 600
Washington, DC, USA 20036
Phone: (202) 912-1000
Toll Free: (800) 406-2306
e-mail: inquiry@conservation.org
website: <http://www.conservation.org>

Contraceptive Research and Development Program (CONRAD)
Eastern Virginia Medical School
1611 North Kent Street, Suite 806
Arlington, VA, USA 22209
Phone: (703) 524-4744
Fax: (703) 524-4770
e-mail: info@conrad.org
website: <http://www.conrad.org>

Cornell Lab of Ornithology
P. O. Box 11
Ithaca, NY, USA 14851
Toll Free: (800) 843-2473
website: <http://www.birds.cornell.edu>

Council on Environmental Quality
722 Jackson Place, NW
Washington, DC, USA 20503
Fax: (202) 395-5750
Toll Free: (202) 456-6546
website: <http://www.whitehouse.gov/ceq>

Countryside Alliance
The Old Town Hall, 367 Kensington Road
London SE11 4PT, England
Phone: (011) 44-020-7840-9200
Fax: (011) 44-020-7793-8899
e-mail: info@countryside-alliance.org
website: <http://www.countryside-alliance.org>

The Cousteau Society
870 Greenbrier Circle, Suite 402
Chesapeake, VA, USA 23320
Phone: (800) 441-4395
e-mail: cousteau@cousteausociety.org
website: <http://www.cousteausociety.org>

Dante B. Fascell North–South Center, University of Miami
1500 Monza Avenue
Coral Gables, FL, USA 33146
Phone: (305) 284-6868
Fax: (305) 284-6370
e-mail: nscenter@miami.edu
website: <http://www.miami.edu/nsc>

Defenders of Wildlife
1101 14th Street, NW #1400
Washington, DC, USA 20005
Phone: (202) 682-9400
e-mail: info@defenders.org
website: <http://www.defenders.org>

Department of Energy, Carlsbad Field Office
PO Box 3090
Carlsbad, NM 88221
Phone: (505) 234-7352
website: <http://www.wipp.carlsbad.nm.us>

Division for Sustainable Development/DESA
United Nations Plaza, Room DC2-2220
New York, NY, USA 10017
Phone: (212) 963-3170
Fax: (212) 963-4260
e-mail: dsd@un.org
website: <http://www.un.org/esa/sustdev/dsdgen.htm>

DOE Office of Energy Efficiency and Renewable Energy (EERE)
Department of Energy, Mail Stop EE-1
Washington, DC, USA 20585
Phone: (202) 586-9220
website: <http://www.eren.doe.gov>

Ducks Unlimited, Inc.
One Waterfowl Way
Memphis, TN, USA 38120
Phone: (901) 758-3825
Toll Free: (800) 45DUCKS
website: <http://www.ducks.org>

The Earth Charter Initiative, The Earth Council
P.O. Box 319-6100
San Jose, Costa Rica
Phone: +506-205-1600
Fax: +506-249-3500
e-mail: info@earthcharter.org
website: <http://www.earthcharter.org>

Earth Island Institute
300 Broadway, Suite 28
San Francisco, CA, USA 94133-3312
Phone: (415) 788-3666
Fax: (415) 788-7324
website: <http://www.earthisland.org>

Earth Pledge Foundation
122 East 38th Street
New York, NY, USA 10016
Phone: (212) 725-6611
Fax: (212) 725-6774
website: <http://www.earthpledge.org>

Earth Policy Institute
1350 Connecticut Ave. NW
Washington, DC, USA 20036
Phone: (202) 496-9290
Fax: (202) 496-9325
e-mail: epi@earth-policy.org
website: <http://www.earth-policy.org>

Earth Systems
508 Dale Avenue
Charlottesville, VA, USA 22903
Phone: 434-293-2022
website: <http://www.earthsystems.org>

Earthwatch
3 Clock Tower Place, Suite 100 Box 75
Maynard, MA, USA 01754
Phone: (978) 461-0081
Fax: (978) 461-2332
Toll Free: (800) 776-0188
e-mail: info@earthwatch.org
website: <http://www.earthwatch.org>

Ecological Society of America
1707 H St, NW, Suite 400
Washington, DC, USA 20006
Phone: (202) 833-8773
Fax: (202) 833-8775
e-mail: esahq@esa.org
website: <http://www.esa.org>

Electronic Industries Alliance
2500 Wilson Boulevard
Arlington, VA, USA 22201
Phone: (703) 907-7500
website: <http://www.eia.org>

Energy Information Administration
1000 Independence Avenue, SW
Washington, DC, USA 20585
Phone: (202) 586-8800
e-mail: infoctr@eia.doe.gov
website: <http://www.eia.doe.gov>

Environmental Defense
257 Park Avenue South
New York, NY, USA 10010
Phone: (212) 505-2100
Fax: (212) 505-2375
Toll Free: (800) 684-3322
e-mail: members@environmentaldefense.org
website: <http://www.environmentaldefense.org>

Environmental Defense Society
PO Box 95 152 Swanson
Aukland, New Zealand 1008
Phone: (64-9) 810-9594
Fax: (64-9) 810-9120
e-mail: manager@eds.org.nz
website: <http://www.eds.org.nz>

Environmental Law & Policy Center
35 East Wacker Drive, #1300
Chicago, IL, USA 60601
Phone: (312) 673-6500
Fax: (312) 795-3730
website: <http://www.elpc.org>

Environmental Law Institute
1616 P St, NW, Suite 200
Washington, DC, USA 20036
Phone: (202) 939-3800
Fax: (202) 939-3868
e-mail: law@eli.org
website: <http://www.eli.org>

Environmental Monitoring and Assessment Program
e-mail: emap@epa.gov
website: <http://www.epa.gov/emap>

Environmental Protection Agency (EPA)
1200 Pennsylvania Avenue, NW
Washington, DC, USA 20460
Phone: (202) 260-2090
e-mail: public-access@epa.gov
website: <http://www.epa.gov>

Environmental Protection Agency California Aquatic Bioassessment Workgroup, Department of Fish and Game
1200 Pennsylvania Avenue, NW
Washington, DC, United States 20460
Phone: (202) 260-2090
website: <http://www.epa.gov>
website: <http://www.dfg.ca.gov>

Environmental Protection Agency Office of Air and Radiation
Ariel Rios Building, 1200 Pennsylvania Ave. NW
Washington, DC, USA 20460
Phone: (202)564-7400
website: <http://www.epa.gov/air/oarofcs.html>

Environmental Protection Agency Office of Solid Waste
1200 Pennsylvania Avenue NW
Washington, DC, USA 20460
Phone: (703) 412-9810
Toll Free: (800) 424-9346
website: <http://www.epa.gov>

Environmental Protection Agency Office of Wastewater Management
Mail Code 4201, 401 M Street, SW
Washington, DC, USA 20460
Phone: (202) 260-7786
Fax: (202) 260-6257
website: <http://www.epa.gov/owm/sso>

Environmental Protection Agency Office of Water (4101M)
1200 Pennsylvania Avenue, NW
Washington, DC, USA 20460
Phone: (202) 566-0388
Fax: (202) 566-0409
e-mail: OW-GENERAL@epa.gov
website: <http://www.epa.gov/waterscience/beaches/>

Environmental Protection Agency - Oil Spill Program
UNIDO New York Office
Toll Free: 800-424-9346
e-mail: oilinfo@epa.gov
website: <http://www.epa.gov/oilspill/>

Environmental Protection Agency primary contact: Carolyn Offutt
1200 Pennsylvania Avenue, NW, Mail Code 5202G
Washington, DC, USA 20460
Phone: (703) 603-8797
e-mail: offutt.carolyn@epa.gov
website: <http://www.epa.gov/superfund/contacts/index.htm>

Environmental Working Group
1718 Connecticut Ave., NW, Suite 600
Washington, DC, USA 20009
Phone: (202) 667-6982
Fax: (202) 232-2592
e-mail: info@ewg.org
website: <http://www.ewg.org>

Estuarine Research Foundation
P.O. Box 510
Port Republic, MD, USA 20676
Phone: (410) 586-0997
Fax: (410) 586-9226
website: <http://www.erf.org>

Euro Chlor: Chlorine Online Information Ressource
Avenue E Van Nieuwenhuyse 4, box 2
Brussels, Belgium B-1160
Phone: + 32 2 676 7211
Fax: + 32 2 676 7241
e-mail: eurochlor@cefic.be
website: <http://www.eurochlor.org>

European Wind Energy Association
Rue de Trone 26
B-1000 Brussels, Belgium
Phone: +32 2 546 1940
Fax: +32 2 546 1944
e-mail: ewea@ewea.org
website: <http://www.ewea.org>

The Exxon Valdez Oil Spill Trustee Council
441 West Fifth Avenue, Suite 500
Anchorage, AL, USA 99501
Phone: (907) 278-8012
Fax: (907) 276-7178
Toll Free: (800) 478-7745 (within Alaska)
(800) 283-7745 (outside Alaska)
e-mail: restoration@oilspill.state.ak.us
website: <http://www.oilspill.state.ak.us/>

F

Federal Bureau of Investigation
935 Pennsylvania Ave.
Washington, DC, USA
Phone: (202) 324-3000
website: <http://www.fbi.gov>

Federal Energy Regulatory Commission
888 First Street, NE
Washington, DC, USA 20009
website: <http://www.ferc.gov>

Federation of American Scientists
1717 K Street, NW, Suite 209
Washington, DC, USA 20036
Phone: (202) 546-3300
website: <http://www.fas.org>

Fish and Wildlife Reference Service
54300 Grosvenor Lane, Suite 110
Bethesda, MD, USA 20814
Phone: (301) 492-6403
Toll Free: (800) 582-3421
e-mail: fw9fareferenceservice@fws.gov
website: <http://www.lib.iastate.edu/collections/db/usfwrs.html>

The Food Irradiation Website
website: <http://www.food-irradiation.com>

F.-A. Forel Institute
10 route de Suise
Geneva, Switzerland
Phone: +4122-950-92-10
Fax: +4122-755-13-82
e-mail: wyss@terre.unige.ch
website: <http://www.unige.ch/forel>

Forest Service, U. S. Department of Agriculture
Sidney R. Yates Federal Building 201 14th Street, SW at Independence Ave., SW
Washington, DC, USA 20250
e-mail: wo_fs-contact@fs.fed.us
website: <http://www.fs.fed.us>

Forest Stewardship Council United States
1155 30th Street, NW, Suite 300
Washington, DC, USA 20007
Phone: (202) 342 0413
Fax: (202) 342 6589
e-mail: info@foreststewardship.org
website: <http://www.fscus.org>

Free Trade Area of the Americas "FTAA" Administrative Secretariat
Apartado Postal 89-10044, Zona 9
Ciudad de Panamá, Republica de Panamá
Phone: (507) 270-6900
Fax: (507) 270-6990
website: <http://www.ftaa-alca.org/alca_e.asp>

Friends of the Earth
1025 Vermont Avenue, NW
Washington, DC, USA 20005
Phone: (202) 783-7400
Fax: (202) 783-0444
Toll Free: (877) 843-8687
e-mail: foe@foe.org
website: <http://www.foe.org>

The Fund for Animals
200 West 57th Street
New York, NY, USA 10019
Phone: (212) 246-2096
Fax: (212) 246-2633
e-mail: hdquarters@fund.org
website: <http://fund.org>

Global Forum
East 45th St., 4th Floor
New York, NY, USA 10017

Great Whales Foundation
PO Box 6847
Malibu, CA, USA 90264
Phone: (310) 317-0755
Fax: (310) 317-1455
Toll Free: (800) 421-WAVE
e-mail: whales@elfi.com

Green Seal
1001 Connecticut Avenue, NW, Suite 827
Washington, DC, USA 20036-5525
Phone: (202) 872-6400
Fax: (202) 872-4324
e-mail: greenseal@greenseal.org
website: <http://www.greenseal.org>

Greenpeace
702 H Street NW
Washington, DC, USA 20001
Toll Free: (800) 326-0959
website: <http://www.greenpeaceusa.org>

Halon Alternatives Research Corporation, Halon Recycling Corporation
2111 Wilson Boulevard, Eighth Floor
Arlington, VA, USA 22201
Phone: (703) 524-6636
Fax: (703) 243-2874
Toll Free: (800) 258-1283
e-mail: harc@harc.org
website: <http://www.harc.org>

Harvard Center for the Study of World Religions: Religions of the World and Ecology
42 Francis Avenue
Cambridge, MA, USA 02138
Phone: (617) 495-4495
Fax: (617) 496-5411
e-mail: cswrinfo@hds.harvard.edu
website: <http://www.hds.harvard.edu/cswr/ecology/index.htm>

The Human Cloning Foundation
website: <http://www.humancloning.org>

Human Genome Project Information, Oak Ridge National Laboratory
1060 Commerce Park MS 6480
Oak Ridge, TN, USA 37830
Phone: (865) 576-6669
Fax: (865) 574-9888
e-mail: genome@science.doe.gov
website: <http://www.ornl.gov/hgmis>

The Humane Society of the United States
2100 L Street, NW
Washington, DC, USA 20037
Phone: (202) 452-1100
website: <http://www.hsus.org>

I

INFORM, Inc.
120 Wall Street
New York, NY, USA 10005
Phone: (212) 361-2400
Fax: (212) 361-2412
e-mail: brown@informinc.org
website: <http://www.informinc.org>

Institute for Global Communications
P.O. Box 29904
San Francisco, CA, USA 94129-0904
e-mail: support@igc.apc.org
website: <http://www.igc.org/igc/gateway/enindex.html>

International Atomic Energy Agency
P.O. Box 100, Wagramer Strasse 5
Vienna, Austria A-1400
Phone: (413) 2600-0
Fax: (413) 2600-7
e-mail: official.mail@iaea.org

International Fabricare Institute
12251 Tech Road
Silver Spring, MD, USA 20904
Phone: (301) 622-1900
Fax: (301) 236-9320
Toll Free: (800) 638-2627
e-mail: techline@ifi.org
website: <http://www.ifi.org>

International Fund for Animal Welfare
411 Main Street, P.O. Box 193
Yarmouth Port, MA, USA 02675
Phone: (508) 744-2000
Fax: (508) 744-2009
Toll Free: (800) 932-4329
e-mail: info@iafw.org
website: <http://www.iafw.org>

International Institute for Sustainable Development
161 Portage Avenue East, 6th Floor
Winnipeg, Manitoba, Canada R3B 0Y4
Phone: (204) 958-7700
Fax: (204) 958-7710
e-mail: info@iisd.ca
website: <http://www.iisd.org>

International Joint Commission
1250 23rd Street, NW, Suite 100
Washington, DC, USA 20440
Phone: (202) 736-9000
Fax: (202) 735-9015
website: <http://www.ijc.org>

International Primate Protection League
P.O. Box 766
Summerville, SC, USA 29484
Phone: (843) 871-2280
Fax: (843) 871-7988
e-mail: ippl@awod.com
website: <http://ippl.org>

International Register of Potentially Toxic Chemicals
Chemin des Anémones 15
Genève, Switzerland CH-1219
Phone: +41-22-979 91 11
Fax: +41-22-979 91 70
e-mail: chemicals@unep.ch

International Rivers Network
1847 Berkeley Way
Berkeley, CA, USA 94703
Phone: (510) 848-1155
Fax: (510) 848-1008
e-mail: info@irn.org
website: <http://www.irn.org>

International Society for Environmental Ethics, Environmental Philosophy Inc., Department of Philosophy, University of North Texas
P.O. Box 310980
Denton, TX, USA 76203-0980
website: <http://www.cep.unt.edu/ISEE.html>

International Tanker Owners Pollution Federation Limited
Staple Hall, Stonehouse Court, 87-90 Houndsditch
London, U.K. EC3A 7AX
Phone: +44(0)20-7621-1255
Toll Free: 800-424-9346
e-mail: central@itopf.com
website: <http://www.itopf.com/index2.html>

International Tire & Rubber Association Foundation, Inc.
PO Box 37203
Louisville, KY, USA 40233-7203
Phone: (502) 968-8900
Fax: (502) 964-7859
Toll Free: (800) 426-8835
e-mail: itra@itra.com
website: <http://www.itra.com>

International Wildlife Coalition
70 East Falmouth Highway
East Falmouth, MA, USA 02536
Phone: (508) 548-8328
Fax: (508) 548-8542
e-mail: iwchq@iwc.org
website: <http://www.iwc.org>

International Wolf Center
1396 Highway 169
Ely, MN, USA 55731
Phone: (218) 365-4695
e-mail: wolfinfo@wolf.org
website: <http://www.wolf.org>

IPCC Secretariat, C/O World Meteorological Organization
7bis Avenue de la Paix, C.P. 2300, CH- 1211
Geneva, Switzerland
Phone: 41-22-730-8208
Fax: 41-22-730-8025
e-mail: ipcc.sec@gateway.wmo.ch
website: <http://www.ipcc.ch>

IUCN—The World Conservation Union Headquarters
Rue Mauverney 28
Gland, Switzerland CH-1196
Phone: ++41 (22) 999-0000
Fax: ++41 (22) 999-0002
e-mail: mail@hq.iucn.org
website: <http://www.iucn.org>

K

Kesterson National Wildlife Refuge, c/o San Luis NWR Complex
340 I Street, P.O. Box 2176
Los Banos, CA, USA 93635
Phone: (209) 826-3508

The Land Institute
2440 E. Water Well Road
Salina, KS, USA 67401
Phone: (785) 823-5376
Fax: (785) 823-8728
e-mail: thelandweb@landinstitute.org
website: <http://www.landinstitute.org>

The Land Stewardship Project
2200 4th Street
White Bear Lake, MN, USA 55110
Phone: (651) 653-0618
Fax: (651) 653-0589
e-mail: lspwbl@landstewardshipproject.org
website: <http://www.landstewardshipproject.org>

League of Conservation Voters
1920 L Street, NW, Suite 800
Washington, DC, USA 20036
Phone: (202) 785-8683
Fax: (202) 835-0491
website: <http://www.lcv.org>

M

Manchester University Computing Society
Oxford Road, Room G33
Manchester, Great Britain M13 9PL
Phone: 44(0)161-275-7329
e-mail: admin@compsoc.man.ac.uk
website: <http://www.compsoc.man.ac.uk/~samp/nuclearage/effect.html>

Massachusetts Toxics Use Reduction Institute
University of Massachusetts Lowell, One University Avenue
Lowell, MA, USA 01854-2866
Phone: (978) 934-3275
Fax: (978) 934-3050
website: <http://www.turi.org>

Masters of Foxhounds Association
Morven Park, P.O. Box 2420
Leesburg, VA, USA 20177
e-mail: office@mfha.com
website: <http://www.mfha.com>

Ministry of Water, Land, and Air Protection
PO Box 9360 Stn Prov Govt
Victoria, BC, Canada V8W 9M2
Phone: (250) 387-9422
Fax: (250) 356-6464
website: <http://www.gov.bc.ca/wlap>

Mountain States Legal Foundation
707 17th Street, Suite 3030
Denver, CO, USA 80202
Phone: (303) 292-2021
Fax: (303) 292-1980
website: <http://www.mountainstateslegal.com>

N

NAFTA Information Center, Texas A&M International University, College of Business Administration
5201 University Blvd.
Laredo, TX, USA 78041-1900
Phone: (956)326-2550
Fax: (956)326-2544
e-mail: nafta@tamiu.edu
website: <http://www.tamiu.edu/coba/usmtr/>

National Center for Fluoridation Policy and Research (NCFPR) at the University of Buffalo
315 Squire Hall
Buffalo, NY, USA 14214
Phone: (716) 829-2056
Fax: (716) 833-3517
e-mail: mweasley@buffalo.edu
website: <http://fluoride.oralhealth.org>

The National Academies
2101 Constitution Avenue, NW
Washington, DC, USA 20418
Phone: (202) 334-2000
website: <http://www4.nationalacademies.org>

The National Academies, National Research Council
2101 Constitution Avenue, NW
Washington, DC, USA 20418
Phone: (202) 334-2000
website: <http://www.nas.edu/nrc>

National Academy of Sciences
2001 Wisconsin Avenue, NW
Washington, DC, USA 20007
website: <http://www.nas.edu>

National Association of Counties
440 First Street, NW
Washington, DC, USA 20001
Phone: 202-393-6226
Fax: 202-393-2630
website: <http://www.naco.org>

National Audubon Society
700 Broadway
New York, NY, USA 10003
Phone: (212) 979-3000
Fax: (212) 979-3188
website: <http://www.audubon.org>

National Audubon Society Horseshoe Crab Campaign
1901 Pennsylvania Avenue NW, Suite 1100
Washington, DC 20006
Phone: (202) 861-2242
Fax: (202) 861-4290
e-mail: pplumart@audubon.org
website: <http://www.audubon.org/campaign/horseshoe/contacts.htm>

National Earthquake Information Center
P.O. Box 25046, DFC, MS 967
Denver, CO, USA 80225
Phone: (303) 273-8500
Fax: (303) 273-8450
e-mail: sedas@neis.cr.usgs.gov
website: <http://www.neic.usgs.gov>

National Electronics Product Stewardship Initiative
website: <http://eerc.ra.utk.edu/clean/nepsi/index.htm>

National Food Processors Association
1350 I Street, NW Suite 300
Washington, DC, USA 20005
Phone: (202) 639-5900
Fax: (202) 639-5932
e-mail: nfpa@nfpa-food.org
website: <http://www.nfpa-food.org>

National Institute for Urban Wildlife
10921 Trotting Ridge Way
Columbia, MD, USA 21044

National Institute of Environmental Health Sciences
P.O. Box 12233
Research Triangle Park, NC, USA 27709
Phone: (919) 541-3345
website: <http://www.niehs.nih.gov>

National Institute of Occupational Safety and Health
4676 Columbia Pkwy.
Cincinnati, OH, USA 45226
Fax: 513-533-8573
Toll Free: (800) 35-NIOSH
e-mail: eidtechinfo@cdc.gov
website: <http://www.cdc.gov/niosh>

National Marine Fisheries Service Office of Habitat Conservation
1315 E. West Highway, 15th Floor
Silver Springs, MD 20910
Phone: (301) 713-2325
Fax: (301) 713-1043
e-mail: cyber.fish@noaa.gov
website: <http://www.nmfs.noaa.gov>

National Marine Fisheries Service
1315 East West Highway, SSMC3
Silver Spring, MD 20910
Phone: (301) 713-2334
Fax: (301) 713-0596
e-mail: cyber.fish@noaa.gov
website: <hhttp://www.nmfs.noaa.gov>

National Marine Sanctuary System, National Ocean Service, National Oceanic and Atmospheric Administration
1305 East-West Highway, 11th Floor
Silver Spring, MD, USA 20910
Phone: (301) 713-3125
Fax: (301) 713-0404
e-mail: nmscomments@noaa.gov
website: <http://www.sanctuaries.nos.noaa.gov/>

National Nuclear Security Administration, Sandia National Laboratories, New Mexico
P. O. Box 5800
Albuquerque, NM, USA 87185
Phone: (925) 294-3000
website: <http://www.sandia.gov>

National Oceanic and Atmospheric Administration
14th St and Constitution Ave, NW, Room 6013
Washington, DC, USA 20230
Phone: (202)482-6090
Fax: (202)482-3154
e-mail: answers@noaa.gov
website: <http://www.noaa.gov>

National Office for Marine Biotoxins and Harmful Algal Blooms; Woods Hole Oceanographic Institution
Biology Dept., MS #32
Woods Hole, MA, USA 02543
Phone: (508) 289-2252
Fax: (508) 457-2180
e-mail: jkleindinst@whoi.edu
website: <http://www.redtide.whoi.edu/hab/>

National Park Service
1849 C Street NW
Washington, DC 20240
Phone: (202) 208-6843
website: <http://www.nps.gov.>

National Parks and Conservation Association
1300 19th Street, NW, Suite 300
Washington, DC, USA 20036
Fax: (202) 659-0650
Toll Free: (800) 628-7275
e-mail: npca@npca.org
website: <http://www.eparks.org>

National Pollution Prevention Roundtable
11 Dupont Circle, NW, Suite 201
Washington, DC, USA 20036
Phone: (202) 299-9701
Fax: (202) 299-9704
e-mail: talk@p2.org
website: <http://www.p2.org>

National Recycling Coalition
1325 G Street NW, Suite 1025
Washington, DC, USA 20005-3104
Phone: (202) 347-0450
Fax: (202) 347-0449
e-mail: info@nrc-recycle.org
website: <http://www.nrc-recycle.org>

National Renewable Energy Laboratory
1617 Cole Blvd.
Golden, CO, USA 80401-3393
Phone: (303) 275-3000
e-mail: client_services@nrel.gov
website: <http://www.nrel.gov>

National Weather Service, National Oceanic and Atmospheric Administration, U. S. Department of Commerce
1325 East West Highway
Silver Spring, USA 20910
website: <http://www.nws.noaa.gov>

National Wildlife Federation
111000 Wildlife Center Drive
Reston, VA, USA 20190
Phone: (703)438-6000
Toll Free: (800) 822-9919
e-mail: info@nwf.org
website: <http://www.nwf.org>

The Natural History Museum of the Adirondacks
P. O. Box 897
Tupper Lake, NY, USA 12986
Phone: (518) 359-2533
Fax: (518) 523-9841
website: <http://www.adkscience.org>

Natural Resources Defense Council
40 W 20th Street
New York, NY 10011
Phone: (212) 727-2700
Fax: (212) 727-1773
e-mail: nrdcinfo@nrdc.org
website: <http://www.nrdc.org/>

The Nature Conservancy
1815 N. Lynn Street
Arlington, VA, USA 22209
website: <http://www.tnc.org>

Neighborhood Cleaners Association
252 West 29th Street
New York, NY, USA 10001
Phone: (212) 967-3002
Fax: (212) 967-2240
e-mail: sales@nca-i.com
website: <http://www.nca-i.com>

New Jersey Work Environment Council (NJWEC)
1543 Brunswick Avenue
Lawrenceville, NJ, USA 08648-4627
Phone: (609) 695-7100
Fax: (609) 695-4200
e-mail: rickengler@aol.com
website: <http://www.njwec.org>

North American Association for Environmental Education
410 Tarvin Road
Rock Spring, GA, USA 30739
Phone: (706) 764-2926
Fax: (706) 764-2094
e-mail: email@naaee.org
website: <http://www.naaee.org>

North American Earth Liberation Front Press Office, James Leslie Pickering
P.O. Box 14098
Portland, OR, USA 97293
Phone: (503) 804-4965
e-mail: elfpress@tao.ca
website: <http://www.animalliberation.net/library/facts/elf>

North Carolina State University Center for Applied Aquatic Ecology
620 Hutton Street, Suite 104
Raleigh, NC, USA 27608
Phone: (919) 515-3421
Fax: (919) 513-3194

Northeast Fisheries Science Center
166 Water Street
Woods Hole, MA, USA 02543-1026
Phone: (508) 495-2000
Fax: (508) 495-2258
website: <http://www.nefsc.nmfs.gov>

The Northeast Regional Aquaculture Center, University of Massachusetts Dartmouth
Violette Building, Room 201 285 Old Westport Road
Dartmouth, MA, USA 02747-2300
Phone: (508) 999-8157
Fax: (508) 999-8590
Toll Free: (866) 472-NRAC (6722)
e-mail: nrac@umassd.edu
website: <http://www.umassd.edu/specialprograms/NRAC/>

Northwest Fisheries Science Center, NMFS, NOAA
2725 Montlake Blvd. E
Seattle, WA, USA 98112
Phone: (206) 860-3200
e-mail: NWFSC.Webmaster@noaa.gov
website: <http://www.nwfsc.noaa.gov/>

Nottingham Hunt Saboteurs, The Sumac Centre
245 Gladstone Street
Nottingham NG7 6HX, England

Nuclear Information and Resource Service
1424 16th Street NW, #404
Washington, DC, USA 20036
Phone: (202) 328-0002
Fax: (202)462-2183
e-mail: nirsnet@nirs.org
website: <http://www.nirs.org>

The Ocean Conservancy
1725 DeSales Street, Suite 600
Washington, DC, USA 20036
Phone: (202) 429-5609
e-mail: info@oceanconservancy.org
website: <http://www.oceanconservancy.org>

Office of Civilian Radioactive Waste Management
Toll Free: (800) 225-6972
e-mail: mowebmaster@rw.doe.gov
website: <http://www.rw.doe.gov>

Office of Information and Regulatory Affairs (OIRA), The Office of Management and Budget
725 17th Street, NW
Washington, DC, USA 20503
Phone: (202) 395-4852
Fax: (202) 395-3888
website: <http://www.whitehouse.gov/omb/inforeg>

Office of Management and Budget
725 17th Street, NW
Washington, DC, USA 20503
Phone: (202) 395-3080
website: <http://www.whitehouse.gov/omb>

Office of Surface Mining
1951 Constitution Ave. NW
Washington, DC, USA 20240
Phone: (202) 208-2719
e-mail: getinfo@osmre.gov
website: <http://www.osmre.gov>

Office of Technical Assistance for Toxics Use Reduction, Commonwealth of Massachusetts, Executive Office of Environmental Affairs
251 Causeway Street, Suite 900
Boston, MA, USA 02114-2136
Phone: (617) 626-1060
Fax: (617) 626-1095
website: <http://www.state.ma.us/ota>

Office of Wastewater Management, U. S. Environmental Protection Agency
Mail Code 4201, 401 M Street, SW
Washington, DC, USA 20460
Phone: (202) 260-5922
Fax: (202) 260-6257
website: <http://www.epa.gov/npdes>

Oil Spill Response Ltd (ORSL)
1 Great Cumberland Place
London, U.K. W1H 7AL
Phone: +44 (0)20-7724-0102
Toll Free: 800-424-9346
e-mail: orsl@orsl.co.uk
website: <http://www.oilspillresponse.com/>

The Online Fuel Cell Information Center
website: <http://www.fuelcells.org>

Organization of American States
17th Street and Constitution Ave., NW.
Washington, DC, USA 20006
Phone: (202) 458-3000
website: <http://www.oas.org>

Organization of Petroleum Exporting Countries
Obere Donaustrasse 93
Vienna, Austria A-1020
Phone: + 43-1-21112 279
Fax: + 43-1-2149827
website: <http://www.opec.org>

P

Pesticide Action Network North America
49 Powell St., Suite 500
San Francisco, CA, USA 94102
Phone: (415) 981-1771
Fax: (415) 981-1991
e-mail: panna@panna.org
website: <http://www.panna.org>

PETA
501 Front St.
Norfolk, VA, USA 23510
Phone: (757) 622-PETA
Fax: (757) 622-0457
e-mail: info@peta-online.org
website: <http://www.peta-online.org>

Pew Initiative on Food and Biotechnology
1331 H Street, Suite 900
Washington, DC, USA 20005
Phone: (202) 347-9044
Fax: (202) 347-9047
e-mail: inquiries@pewagbiotech.org
website: <http://pewagbiotech.org>

Physicians for Social Responsibility
1875 Connecticut Avenue, NW, Suite 1012
Washington, DC, USA 20009
Phone: (202) 667-4260
Fax: (202) 667-4201
e-mail: psrnatl@psr.org
website: <http://www.psr.org>

Population Council
1 Dag Hammarsrkjold Plz
New York, NY, USA 10017
Phone: (212) 339-0500
Fax: (212) 755-6052
e-mail: pubinfo@popinfo.org
website: <http://www.popcouncil.org>

Priory of Saint Cosme
Loire, France
Phone: 33-(2) 47 37 32 70
website: <http://www.cg37.fr/>

Project Eco-School
881 Alma Real Drive, Suite 301
Pacific Palisades, CA, USA 90272

Public Citizen, Critical Mass Energy & Environmental Program
1600 20th St. NW
Washington, DC, USA 20009
Phone: (202) 588-1000
e-mail: CMEP@citizen.org
website: <http://www.citizen.org/cmep/foodsafety/food_irrad>

Public Policy Center, National Cattlemen's Beef Association (Public Lands Council)
1301 Pennsylvania Avenue, NW, Suite 300
Washington, DC, USA 20004-1701
Phone: (202) 347-0228
Fax: (202) 638-0607
e-mail: jcampbell@beef.org
website: <htttp:/hill.beef.org/files/fedlnds.htm>

Puget Sound Water Quality Action Team
PO Box 40900
Olympia, WA, USA 98504-0900
Phone: (360) 407-7300
Toll Free: (800) 54-SOUND
website: <http://www.wa.gov/puget_sound>

Puget Sound/Georgia Basin International Task Force
e-mail: jdohrmann@psat.wa.gov
website: <http://www.wa.gov/puget_sound/shared/shared.html>

Rachel Carson Council, Inc.
8940 Jones Mill Road
Chevy Chase, MD, USA 20815
Phone: (301) 652-1877
Fax: (301) 951-7179
e-mail: rccouncil@aol.com
website: <http://members.aol.com/rccouncil/ourpage/index.htm>

Rails-to-Trails Conservancy
1100 17th Street, 10th Floor, NW
Washington, DC, USA 20036
Phone: (202) 331-9696
e-mail: railtrails@transact.org
website: <http://www.railtrails.org>

Rainforest Action Network
221 Pine Street, Suite 500
San Francisco, CA, USA 94104
Phone: (415) 398-4404
Fax: (415) 398-2732
e-mail: rainforest@ran.org
website: <http://www.ran.org>

Rainforest Alliance
65 Bleecker Street
New York, NY, USA 10012
Phone: (212) 677-1900
e-mail: canopy@ra.org
website: <http://www.rainforestalliance.com>

The Ramsar Convention Bureau
Rue Mauverney 28
Gland, Switzerland CH-1196
Phone: +41 22 999 0170
Fax: +41 22 999 0169
e-mail: ramsar@ramsar.org
website: <http://www.ramsar.org/index.html>

Rechargeable Battery Recycling Corporation
1000 Parkwood Circle, Suite 450
Atlanta, GA, USA 30339
Phone: (678) 419-9990
Fax: (678) 419-9986
e-mail: corporate@rbrc.com
website: <http://www.rbrc.org>

Recycled Materials Resource Center
220 Environmental Technology Building
Durham, NH, USA 03824
Phone: (603) 862-3957
e-mail: rmrc@rmrc.unh.edu
website: <http://www.rmrc.unh.edu>

Renew America
1200 18th Street, NW, Suite 1100
Washington, DC, USA 20036
Phone: (202) 721-1545
Fax: (202) 467-5780
e-mail: renewamerica@counterpart.org
website: <http://sol.crest.org/environment/renew_america>

Renewable Natural Resources Foundation
5430 Grosvenor Lane
Bethesda, MD, USA 20814-2193
Phone: (301) 493-9101
Fax: (301) 493-6148
e-mail: info@rnrf.org
website: <http://www.rnrf.org>

Resources for the Future
1616 P Street NW
Washington, USA 20036-1400
Phone: (202) 328-5000
Fax: (202) 939-3460
website: <http://www.rff.org>

Restore America's Estuaries
3801 N. Fairfax Drive, Suite 53
Arlington, VA, USA 22203
Phone: (703) 524-0248
Fax:)703) 524-0287
website: <http://www.estuaries.org>

Riverkeeper, Inc.
25 Wing & Wing
Garrison, NY, USA 10524-0130
Phone: (845) 424-4149
Fax: (845) 424-4150
e-mail: info@riverkeeper.org
website: <http://www.riverkeeper.org>

Rocky Mountain Institute
1739 Snowmass Creek Road
Snowmass, CO, USA 81654-9199
Phone: (970) 927-3851
website: <http://www.rmi.org>

The Rodale Institute
611 Siegfriedale Road
Kutztown, PA, USA 19530-9320
Phone: (610) 683-1400
Fax: (610) 683-8548
e-mail: info@rodaleinst.org

Rubber Manufacturers Association
1400 K Street, NW
Washington, DC, USA 20005
Phone: (202) 682-4800
e-mail: info@rma.org
website: <http://www.rma.org>

Sandia National Laboratories: Energy and Critical Infrastructure
PO Box 5800
Albuquerque, NM, USA 87185
Phone: (505) 284-5200
website: <http://www.sandia.gov>

Santa Barbara Wildlife Care Network
P.O. Box 6594
Santa Barbara, CA, USA 93160
Phone: (805) 966-9005
e-mail: sbwcn@juno.com
website: <http://www.silcom.com/~sbwcn/spill.htm>

Save-the-Redwoods League
114 Sansome Street, Room 1200
San Francisco, CA, USA 94104-3823
Phone: (415) 362-2352
Fax: (415) 362-7017
e-mail: info@savetheredwoods.org
website: <http://www.savetheredwoods.org>

Save the Whales
P.O. Box 2397
Venice, CA, USA 90291
e-mail: maris@savethewhales.org
website: <http://www.savethewhales.org>

Save the Whales
PO Box 3650, Georgetwon Sta.
Washington, DC, USA 20007
Phone: (202) 337-2332
Fax: (202) 338-9478
e-mail: awi@animalwelfare.com
website: <http://www.awionline.org/whales/indexout.html>

Scientific Committee on Problems of the Environment (SCOPE)
%1, blvd. De Montmorency F-75016
Paris, France
Phone: 33-1-45250498
Fax: 33-1-42881466
e-mail: secretariat@icsu-scope-org
website: <http://www.icsu-scope.org>

Sea Shepherd International
22774 Pacific Coast Highway
Malibu, CA, USA 90165
Phone: (310) 456-1141
Fax: (310) 456-2488
e-mail: seashepherd@seashepherd.org
website: <http://www.seashepherd.org>

Secretariat of the United Nations Commission on Sustainable Development
United Nations Plaza, Room DC2-2220
New York, NY, USA 10017
Phone: (212) 963-3170
Fax: (212) 963-4260
e-mail: dsd@un.org
website: <http://www.un.org/esa/sustdev/csd.htm>

Sierra Club
85 Second St., Second Floor
San Francisco, CA, USA 94105-3441
Phone: (415) 977-5500
Fax: (415) 977-5799
e-mail: information@sierraclub.org
website: <http://www.sierraclub.org>

Smithsonian Environmental Research Center
647 Contees Wharf Road
Edgewater, MD, USA 21037
Phone: (443) 482-2200
Fax: (443) 482-2380
website: <http://www.serc.si.edu>

Society for Animal Protective Legislation
P.O. Box 3719
Washington, DC, USA 20007
Phone: (202) 337-2334
Fax: (202) 338-9478
e-mail: sapl@saplonline.org
website: <http://www.saplonline.org>

Society for Conservation Biology
4245 N. Fairfax Drive
Arlington, VA, USA 22203
Phone: (703) 276-2384
Fax: (703) 995-4633
e-mail: information@conservationbiology.org
website: <http://conbio.net/scb>

Society for Developmental Biology
9650 Rockville Pike
Bethesda, MD, USA 20814
Phone: 301–571–0647
Fax: 301–571–5704
e-mail: ichow@faseb.org
website: <http://www.sdb.bio.purdue.edu>

Society of American Foresters
5400 Grosvenor Lane
Bethesda, MD, USA 20814
Phone: (301) 897-8720
Fax: (301) 897-3690
e-mail: safweb@safnet.org
website: <http://www.safnet.org>

Soil and Water Conservation Society
7515 NE Ankeny Road
Ankeny, IA, USA 50021-9764
Phone: (515) 289-2331
Fax: (515) 289-1227
e-mail: webmaster@swcs.org
website: <http://www.swcs.org>

State Hydrological Institute
23 Second Line VO
St. Petersburg, Russia 199053

Stratospheric Ozone Information Hotline, United States Environmental Protection Agency
1200 Pennsylvania Avenue, NW
Washington, DC, USA 20460
Phone: (202) 775-6677
Toll Free: (800) 296-1996
e-mail: public-access@epa.gov
website: <http://www.epa.gov/ozone>

Student Environmental Action Coalition
P.O. Box 31909
Philadelphia, PA, USA 19104-0609
Phone: (215) 222-4711
Fax: (215) 222-2896
e-mail: seac@seac.org
website: <http://www.seac.org>

The Styrene Information and Research Center
1300 Wilson Boulevard, Suite 1200
Arlington, VA, USA 22209
Phone: (703) 741-5010
Fax: (703) 741-6010)
website: <http://www.styrene.org>

T

Tennessee Valley Authority
400 W. Summit Hill Dr.
Knoxville, TN, USA 37902-1499
Phone: (865) 632-2101
e-mail: tvainfo@tva.gov

Texas Department of Health Bioterrorism Preparedness Program (TDH)
1100 West 49th Street
Austin, TX, USA 78756
Phone: (512) 458-7676
Toll Free: (800) 705-8868
website: <http://www.tdh.state.tx.us/bioterrorism/default.htm>

Toxics Release Inventory
Phone: (202) 566-0250
e-mail: tri.us@epa.gov
website: <http://www.epa.gov/tri>

TRAFFIC East/Southern Africa - Kenya
Ngong Race Course, PO Box 68200, Ngong Road
Nairobi, Kenya
Phone: (254) 2 577943
Fax: (254) 2 577943
e-mail: traffic@iconnect.co.ke
website: <http://www.traffic.org>

U.S. Bureau of Land Management, Office of Public Affairs
1849 C Street, Room 406-LS
Washington, DC, USA 20240
Phone: (202) 452-5125
Fax: (202) 452-5124
website: <http://www.blm.gov≷>

U.S. Bureau of Reclamation, Freedom of Information Act (FOIA) Office
PO Box 25007, D-7924
Denver, CO, USA 80225-0007
Phone: (303) 445-2048
Fax: (303) 445-6575
Toll Free: (800) 822-7646
e-mail: borfoia@usbr.gov
website: <http://www.usbr.gov>

U.S. Department of Agriculture
Washington, DC, USA 20250
e-mail: vic.powell@usda.gov
website: <http://www.usda.gov>

U.S. Department of Energy
1000 Independence Avenue SW
Washington, DC, USA 20585
Phone: (202) 586-6151
Fax: (202) 586-0956
website: <http://www.doe.gov>

U.S. Department of Energy, Hanford Site, Richland Operations Office, Freedom of Information Office
825 Jadwin Avenue, P.O. Box 550
Richland, WA, USA 99352
Phone: (509) 376-6288
Fax: (509)376-9704
e-mail: FOIA@rl.gov
website: <http://www.hanford.gov/FOIA/index.cfm>

U.S. Department of Health and Human Services
200 Independence Avenue, SW
Washington, DC, USA 20201
Phone: (202) 619-0257
Toll Free: (877) 696-6775
e-mail: HHS.Mail@hhs.gov
website: <http://www.hhs.gov>

U.S. Department of Labor, Occupational Safety & Health Administration
200 Constitution Avenue
Washington, DC, USA 20210
Toll Free: (800) 321-OSHA
website: <http://www.osha.gov>

U.S. Department of State
2201 C Street NW
Washington, DC, USA 20520
Toll Free: (202) 647-4000
website: <http://www.state.gov/g/oes>

U.S. Department of the Interior
1849 C. Street NW
Washington, DC, USA 20240
Phone: (202) 208-3100
website: <http://www.doi.gov>

U.S. Department of the Interior, United States Geological Survey, Minerals Information
988 National Center
Reston, VA, USA 20192
Toll Free: (888) ASK-USGS (888-275-8747)
e-mail: Contact person: Joseph Gambogi, jgambogi@usgs.gov
website: <http://www.minerals.usgs.gov/minerals/>

U.S. Department of Transportation
400 7th Street NW
Washington, DC, USA 20590
Phone: 202-366-4000
e-mail: dot.comments@ost.dot.gov
website: <http://www.transportation.gov>

U.S. Fish and Wildlife Service
Washington, DC, USA
website: <http://www.fws.gov>

U.S. Food and Drug Administration
5600 Fishers Lane
Rockville, MD, USA 20857-0001
Toll Free: (888) INFO-FDA
website: <http://www.fda.gov>

U.S. Geological Survey (USGS), Biological Resources Division (BRD), Western Regional Office (WRO)
909 First Ave., Suite #800
Seattle, WA, USA 98104
Phone: (206) 220-4600
Fax: (206) 220-4624
e-mail: brd_wro@usgs.gov
website: <http://biology.usgs.gov>

U.S. GLOBEC Georges Bank Program
Woods Hole Oceanographic Institution
Woods Hole, MA, USA 02543-1127
Phone: (508) 289-2409
Fax: (508) 457-2169
e-mail: rgroman@whoi.edu
website: <http://globec.whoi.edu>

U.S. Nuclear Regulatory Commission
Office of Public Affairs (OPA)
Washington, DC, USA 20555
Phone: (301) 415-8200
Toll Free: (800) 368-5642
e-mail: opa@nrc.gov
website: <http://http://www.nrc.gov>

U.S. Nuclear Regulatory Commission, Office of Public Affairs
Washington, DC, USA 20555
Phone: (301) 415-8200
Toll Free: (800) 368-5642
e-mail: opa@nrc.gov
website: <http://www.nrc.gov>

U.S. Public Interest Research Group
215 Pennsylvania Avenue
Washington, DC, USA 20003
e-mail: webmaster@pirg.org
website: <http://pirg.org>

UNEP World Conservation Monitoring Centre
219 Huntingdon Road
Cambridge, UK CB3 0DL
Phone: +44 (0)1223 277314
Fax: +44 (0)1223 277136
e-mail: info@unep-wcmc.org
website: <http://www.unep-wcmc.org>

UNEP-Infoterra/USA
MC 3404 Ariel Rios Building, 1200 Pennsylvania Avenue
Washington, DC, USA 20460
Fax: (202) 260-3923
e-mail: library-infoterra@epa.gov

UNIDO—Climate Change/Kyoto Protocol Activities
UNIDO New York Office
New York, NY, USA 10017
Phone: (212) 963-6890
e-mail: office.newyork@unido.org
website: <http://www.unido.org/doc/310797.htmls>

Union of Concerned Scientists
2 Brattle Square
Cambridge, MA, USA 02238-9105
Phone: (617) 547-5552
Fax: (617) 864-9405
e-mail: ucs@ucsusa.org
website: <http://www.ucsusa.org>

United Nations
New York, NY, USA 10017
website: <http://www.un.org>

United Nations Commission on Sustainable Development
United Nations Plaza, Room DC2-2220
New York, NY, USA 10017
Phone: 212-963-3170
Fax: 212-963-4260
e-mail: dsd@un.org
website: <http://www.un.org/esa/sustdev/>

United Nations Environment Programme
United Nations Avenue, Gigiri, PO Box 30552
Nairobi, KenyaPhone: (254-2) 621234
Fax: (254-2) 624489/90
e-mail: eisinfo@unep.org
website: <http://www.unep.org>

United Nations Environment Programme, Chemicals Division
11-13 chemin des Anemones, 1219 Chatelaine
Geneva, Switzerland
Phone: (41) (22) 917-8191
Fax: (41) (22) 797-3460

United Nations Environment Programme, Division of Technology, Industry and Economics, Energy and OzonAction Programme
Tour Mirabeau, 39-43 quai André Citroën
73759 Paris Cedex 15, France
Phone: (33-1) 44 37 14 50
Fax: (33-1) 44 37 14 74
e-mail: ozonaction@unep.fr
website: <http://www.uneptie.org/ozonaction>

United Nations Environment Programme: Chemicals
11-13, chemin des Anémones, 1219 Châtelaine
Geneva, Switzerland
e-mail: opereira@unep.ch
website: <http://www.chem.unep.ch>

United States Golf Association
P.O. Box 708
Far Hills, NJ, USA 07931-0708
Fax: 908-781-1735
e-mail: usga.org
website: <http://www.usga.org/>

University of Geneva
24 street Rothschild
Geneva, Switzerland
e-mail: Jacques.Grinevald@iued.unige.ch
website: <http://www.unige.ch>

University Leaders for a Sustainable Future
2100 L Street, NW
Washington, DC, USA 20037
Phone: 202-778-6133
Fax: 202-778-6138
website: <http://www.ulsf.org>

Urban Ecology
414 13th Street, Suite 500
Oakland, CA 94612
Phone: 510-251-6330
Fax: 510-251-2117
e-mail: urbanecology@urbanecology.org
website: <http://www.urbanecology.org/>

Izaak Walton League of America
707 Conservation Lane
Gaithersburg, MD, USA 20878
Phone: (301) 548-0150
Fax: (301) 548-0146
Toll Free: (800) IKE-LINE
e-mail: general@iwla.org
website: <http://www.iwla.org>

Washington D.C. Field Office, National Wildlife Federation
1400 16th Street NW
Washington, DC, USA 20036
Phone: (202) 797-6800
Fax: (202) 797-6646
website: <http://www.nwf.org>

Water Environment Federation
601 Wythe Street
Alexandria, VA, USA 22314-1994
Phone: (703) 684-2400
Fax: (703) 684-2492
Toll Free: (800) 666-0206
e-mail: csc@wef.org
website: <http://www.wef.org>

Waterkeeper Alliance
78 North Broadway, Bldg. E
White Plains, NY, USA 10603
Phone: (914)422-4410
Fax: (914)422-4437
e-mail: info@waterkeeper.org
website: <http://www.waterkeeper.org>

Western Watersheds Project (National Public Lands Grazing Campaign)
P.O. Box 1770
Hailey, ID, USA 83333
e-mail: andykerr@andykerr.net
website: <http://www.publiclandsranching.org>

Whale and Dolphin Conservation Society
P.O. Box 232
MelkshamWiltshire, United Kingdom SN12 7SB
Phone: 44 (0) 1225-354333
Fax: (44) (0) 1225-791577
e-mail: webmaster@wdcs.org
website: <http://www.wdcs.org>

The Wilderness Society
1615 M St. NW
Washington, DC, USA 20036
Toll Free: 800–843–9453
e-mail: member@tws.org
website: <http://www.wilderness.org/standbylands/orv>

World Conservation Union Headquarters
Rue Mauverney 28
Gland, Switzerland 1196
Phone: 41 (22) 999-0000
Fax: 41 (22) 999-0002
e-mail: mail@hq.iucn.org
website: <http://www.iucn.org>

World Health Organization Regional Office
525 23rd Street NW
Washington, DC, USA 20037
Phone: (202) 974-3000
Fax: (202) 974-3663
e-mail: postmaster@paho.org
website: <http://www.who.int>

World Resources Institute
10 G Street, NE, Suite 800
Washington, DC, USA 20002
Phone: (202) 729-7600
Fax: (202) 729-7610
e-mail: front@wri.org
website: <http://www.wri.org>

World Wildlife Fund
1250 24th Street, NW, P.O. Box 97180
Washington, DC, USA 20090-7180
Fax: 202-293-9211
Toll Free: (800)CALL-WWF
website: <http://www.panda.org>
website: <http://www.worldwildlife.org>

Worldwatch Institute
1776 Massachusetts Ave., NW
Washington, DC, USA 20036-1904
Phone: (202) 452-1999
Fax: (202) 296-7365
e-mail: worldwatch@worldwatch.org
website: <http://www.worldwatch.org/>

Dr. Charles F. Wurster
Marine Sciences Research Center, State University of New York at Stony Brook
Stony Brook, NY, USA 11794-5000
Phone: (631) 632-8738
e-mail: Charles.F.Wurster@stonybrook.edu

Yellowstone National Park
P.O. Box 168
Yellowstone National Park, WY 82190-0168
Phone: (307) 344-7381
e-mail: yell_visitor_services@nps.gov
website: <http://www.nps.gov/yell>

Yosemite National Park
P.O. Box 577
Yosemite National Park, CA 20240
Phone: (209) 372-0200
e-mail: comments@yosemite.com
website: <http://www.yosemitepark.com>

Yucca Mountain Project
PO Box 364629
North Las Vegas, NV 89036-8629
Phone: (800) 225-6972
Fax: (702) 295-5222
website: <http://www.ymp.gov>

Zero Population Growth
1400 16th Street NW, Suite 320
Washington, DC 20036

General Index

A

C

D

E

F

G

H

I

J

K

L

M

N

O

P

S

T

U

V

X

Y

Z